Design of Coastal Hazard Mitigation Alternatives for Rising Seas

ADVANCED SERIES ON OCEAN ENGINEERING

Editor-in-Chief
Philip L-F Liu (*Cornell University and National University of Singapore*)

Published

Vol. 49 Design of Coastal Hazard Mitigation Alternatives for Rising Seas
by David Basco

Vol. 48 Introduction to Coastal Engineering and Management (3rd Edition)
by J. William Kamphuis

Vol. 47 Coastal Engineering: Theory and Practice
by Vallam Sundar and Sannasi A. Sannasiraj

Vol. 46 Tsunami: To Survive from Tsunami (Second Edition)
by Tomotsuka Takayama, Kazumasa Katoh, Fumihiko Imamura, Yoshiaki Kawata, Susumu Murata and Shigeo Takahashi

Vol. 45 Ocean Surface Waves: Their Physics and Prediction (Third Edition)
by Stanisław Ryszard Massel

Vol. 44 Satellite SAR Detection of Sub-Mesoscale Ocean Dynamic Processes
by Quanan Zheng

Vol. 43 Japan's Beach Erosion: Reality and Future Measures (Second Edition)
by Takaaki Uda

Vol. 42 Theory and Applications of Ocean Surface Waves (Third Edition)
by Chiang C. Mei, Michael Aharon Stiassnie and Dick K.-P. Yue

Vol. 41 Dynamics of Coastal Systems (Second Edition)
by Job Dronkers

Vol. 40 Design and Construction of Berm Breakwaters
by Jentsje Van der Meer and Sigurdur Sigurdarson

Vol. 39 Liquefaction Around Marine Structures
by Mutlu Sumner

Vol. 38 An Introduction to Hydraulics of Fine Sediment Transport
by Ashish J. Mehta

Vol. 37 Computational Wave Dynamics
by Hitoshi Gotoh, Akio Okayasu and Yasunori Watanabe

Vol. 36 Ocean Surface Waves: Their Physics and Prediction (Second Edition)
by Stanislaw R. Massel

Vol. 35 Dynamics of Water Waves: Selected Papers of Michael Longuet-Higgins
edited by S. G. Sajjadi

*For the complete list of titles in this series, go to https://www.worldscientific.com/series/asoe

Advanced Series on Ocean Engineering — Volume 49

Design of Coastal Hazard Mitigation Alternatives for Rising Seas

David Basco
Old Dominion University, USA

World Scientific

NEW JERSEY • LONDON • SINGAPORE • BEIJING • SHANGHAI • HONG KONG • TAIPEI • CHENNAI • TOKYO

Published by

World Scientific Publishing Co. Pte. Ltd.
5 Toh Tuck Link, Singapore 596224
USA office: 27 Warren Street, Suite 401-402, Hackensack, NJ 07601
UK office: 57 Shelton Street, Covent Garden, London WC2H 9HE

Library of Congress Cataloging-in-Publication Data
Names: Basco, David R., author.
Title: Design of coastal hazard mitigation alternatives for rising seas /
David Basco, Old Dominion University, USA.
Description: Singapore : World Scientific Publishing Co. Pte. Ltd., 2019. |
Series: Advanced series on ocean engineering, 1793-074X ; vol 49 |
Includes bibliographical references and index.
Identifiers: LCCN 2019027078 | ISBN 9789811206931 (hardcover) | ISBN 9789811206948 (ebook)
Subjects: LCSH: Shore protection.
Classification: LCC TC330 .B37 2019 | DDC 627/.42--dc23
LC record available at https://lccn.loc.gov/2019027078

British Library Cataloguing-in-Publication Data
A catalogue record for this book is available from the British Library.

For any available supplementary material, please visit
https://www.worldscientific.com/worldscibooks/10.1142/11462#t=suppl

Typeset by Stallion Press
Email: enquiries@stallionpress.com

Printed in Singapore

Preface

The objective of this book is to provide a textbook for graduate students in coastal engineering and to serve as a reference book for professional engineers in government, and the consulting engineering profession. Design is the key word in the title. Most graduate-level courses emphasize the coastal sciences and theory and say little about design. However, climate change and rising seas are quickly changing what material is covered in courses and how we do the design work. In the United States and Canada "*only a licensed professional engineer may prepare, sign and seal, and submit engineering plans and drawings to a public authority for approval, or seal engineering work for public and private clients*" (National Society of Professional Engineers, 2007).

Ideally, the book can be used for several graduate-level courses after a first "Introduction to Coastal Engineering" course is completed by undergraduates in Civil Engineering or first semester masters-level students. The basic, introductory material presented in each chapter of this book can be employed for this first course. The graduate degree program in Coastal Engineering at Old Dominion University (ODU) was established 34 years ago (1986). The online, internet-based "Coastal Engineering Certificate Program" (CECP) was established 16 years ago (2004) as the first in the US (and world). Over the 10-year period (2004–2014) 102 graduates from 22 US states and 6 foreign countries have earned the Certificate. A total of 297 students from 29 of the 30 US coastal states and from 14 foreign countries took online courses over this 10-year period.

At ODU, on-campus graduate students and online students in the Certificate program take four 3-credit courses over a two-year cycle (one each semester) that are taught in the following order to cover the basics, science, engineering practice, and principles of design.

Year 1 Fall	CEE 482/582 Introduction to Coastal Engineering (one text)
Year 1 Spring	CEE 787 Dredging, Beach Engineering, and Tidal Hydraulics (multiple texts)
Year 2 Fall	CEE 788 Coastal Hydrodynamics and Sediment Transport (multiple texts)
Year 2 Spring	CEE 782 Design of Coastal Structures (multiple texts).

The excellent text for the first course (CEE482/582) is *Introduction to Coastal Engineering and Management* (3rd Edition) by William Kamphuis (World Scientific, 2020). The subsequent courses expanded the theory and culminated in the last course devoted to design. Many different textbooks, manuals, and references were used in these courses. The material presented in Chapters 1–10 in this book is essentially that covered in these four courses. Chapter 11 in this book is all about sea level rise (SLR) with Section 11.3 devoted to how SLR impacts each of the preceding chapters. Instructors can jump to Section 11.3 after each chapter and/or wait until Chapter 11 is covered in class. For example, subsection 11.3.2 explains how SLR has impacted Chapter 2 "Design Criteria" with 11.3.2 (b) summarized from Kriebel *et al.* (2015), Kriebel (2018) regarding nuisance flooding events. Microsoft Excel spreadsheets developed and provided to the students in these courses and employed by the author for engineering application are included with the book. They have been checked by graduate students and used for homework problems but may contain errors. They should be first checked using simple examples before reliance on the results for design. Responsibility for their accuracy is solely up to the professional engineer to validate their correctness when applied for design. Most include both the metric system and imperial units (English system) and a read me page with further explanation.

The author began a draft book manuscript in January 2013 at the University of Salento in Lecce, Italy. Much has happened in the past seven years regarding our knowledge of climate change and sea level rise that has been necessary to incorporate in this book. For example, Chapter 8 on economics and risk assessment now includes a subsection 8.7 on quantification of social resilience to flooding events. Quantification of community resilience to flooding using Bayesian networks is mathematically, statistically, and socially the best probabilistic approach. "*We should be thinking probabilistically if we want to make good choices*" (Alan Turing, 1968).

Society now has the moral obligation to include both risk and resilience metrics as the basis for project evaluation of flood mitigation alternatives and making choices under a future of rising seas. The global challenge is how to wisely invest limited resources to reduce risk and increase resilience for all the people living and working along the coast.

As a freshman in the Civil Engineering department at the University of Wisconsin-Madison back in 1956, we all visited Bascom Hall where in the main entrance, a tablet cast in bronze on the wall says:

> "*Whatever may be the limitations which trammel inquiry elsewhere, we believe that the great state University of Wisconsin should ever encourage that continual and fearless sifting and winnowing by which alone the truth can be found.*"

Sifting and winnowing is a metaphor for the academic pursuit of truth affiliated with the University of Wisconsin–Madison (UW). It was coined by UW President Charles Kendall Adams in a 1894 final report from a committee exonerating an economics professor of censurable charges from the state education superintendent. The phrase became a local byword for the tenet of academic freedom. Looking back over my 45-year teaching/research/service career at Texas A&M University and Old Dominion University, and as a registered professional engineer, these words ring with absolute truth about what we strive to do in academia and the civil/coastal engineering profession. The undergraduate and graduate-level courses we teach; the science and engineering literature we read; the conferences we attend; the engineering projects we work on; the daily written and graphic material distributed in the media all provide an overwhelming amount of information that must be sifted and winnowed to separate the "wheat from the shaft". That work of **sifting and winnowing** all the world's information for the design of coastal hazard mitigation alternatives under climate change and unprecedented sea level rise conditions is what has ended up in this book.

The Wheat Sifters, Gustave Courbet, 1854

Winnowing Grain, Eastman Johnson, 1873–1879

Sifting sorts by size, while winnowing sorts by weight.

David Basco

Acknowledgements

When and where to begin? All knowledge is cumulative. I was first introduced to water aspects of Civil Engineering thanks to Professors Jim Woodburn, Jim Villemonte, and Peter Monkmeyer at the University of Wisconsin-Madison. I still remember the sight of Professor Woodburn paddling his canoe across Lake Mendota to the Hydraulics Lab in Madison to teach a class. I wanted to try teaching and learn about dredge pumps as part of Coastal Engineering under the leadership of Professor John Herbich at Lehigh and then at Texas A&M-College Station (TAMU) where Professors Bob Sorensen, Dick Dominguez, and Wes James were inspirational colleagues in the new Ocean Engineering department. My good friends Dave Norton (Aerospace) and Skip Porter (Electrical) enlarged my appreciation of all good things that engineers do. And my friend Bob Reid contributed greatly to all the new research oceanography scientists undertake that civil/coastal engineers must understand to be successful.

But I needed more education in Coastal Engineering and where best to go? — the Netherlands to attend the Dutch "International Courses". Professors Eco Bijker, Kees d'Angremond, Jurjen Battjes, Jan van der Graff, (and many others: de Koning, van der Veen, van Ette, Kalkwijk, Abraham, Kruk, de Vries, de Koning, Breusers, Bakker, etc.) provided more than simply knowledge and facts during their lectures. The teaching methods, styles, and dedication displayed by the instructors have inspired my teaching career. The ideas, topics, and enthusiasm of the instructors have benefited my research career. And the ability of Dutch coastal engineers to cut through all the side issues, focus on the key variables, and come up with *practical* solutions to coastal problems has aided my consulting career. The knowledge gained from my Delft experience was *uitstekend* (outstanding)! And, the additional benefit was provided by Dr. Mike Abbott stemming from his background and experience in mathematics, fluid mechanics, and computational hydraulics at the Danish Hydraulics Institute. Thanks, Mike for all you taught me.

And then it was on to Old Dominion University (ODU) resulting from the foresight of Leon Wang where I benefited greatly from the long-term excellence of the Oceanography program: the research efforts of Jack Ludwick and George Oertel; the weekly seminars organized by Eileen Hoffman and the Center for Coastal Physical Oceanography directed by John Klinck. I want to thank all the MS and PhD degree (and online, Certificate program) students at TAMU and ODU and all the professional engineers/coastal scientists working together on engineering projects from 1969–2019 (50 years). I especially want to thank Tom Langley, George Sadler, John Headland, and Tim Kana for allowing me to participate and hopefully help your consulting engineering efforts. There are far more and far too many to cite here and I trust and hope you understand. As one graduate-student said to me many years later, "*I learned more about Coastal Engineering from that engineering project you paid me to work on than from that research project and my thesis for my degree*" (anonymous). The wide variety of graduate and online, internet students from all over the world required my fresh look at the wide variety of coastal shorelines, hydrodynamics, and sediments that require a wide variety of alternatives to reduce the coastal hazards at those locations.

Over the years, I was fortunate to attend almost all the bi-annual International Conference on Coastal Engineering (ICCE) events since 1984 as held in many countries. Although organized by the Coastal Engineering Research Council of the American Society of Civil Engineers they require countless hours of help from the academics, consulting engineers, and government employees in each country where the meetings are held. Over the years, I have met and enjoyed the friendship of many who did the work with special thanks to Marcel Stive and Dano Roelvink (Holland), William Allsop (UK), Ida Broker, Per Madsen, and Hans Burcharth (Denmark), Hans Hanson and Magnus Larson (Sweden), Felice d'Alessandro, and Roberto Tomasicchio (Italy), Luc Hamm (France), Peter Nielsen (Australia), and Takao Yamashita, Prof Tsuchiya, and Prof Goda (Japan). There are many more but my memory needs refreshing by attendance at the next ICCE.

I would also like to thank all my friends and colleagues in the US Army Corps of Engineers for their help and support over the years with far too many names to mention without forgetting a few. And lastly, I wish to acknowledge the many, many academicians in the US and Canada who have led the way forward in coastal engineering teaching, research, and service. My special thanks to Bob Dean (deceased), Nick Kraus (deceased), and

Tony Dalyrmple (still kicking) in the US and to Bill Kamphuis (alive and well in Canada). Bill said yes to my including his example for Lake Huron and the Peak-Over-Threshold analysis method for extreme wave heights. There are over 750 references cited in this book. My heartfelt thanks to all the authors for their contributions to our knowledge for application in the design practice of Coastal Engineering.

The writing of this book began at the University of Salento in Lecce, Italy while on sabbatical leave from Old Dominion University starting January 2013. Without the initial support of Dr. Roberto Tomasicchio to provide my office there, it would never have happened. Many, many thanks, Roberto! You may ask why southern Italy? My answer is simply a lot of warm sunny weather, many beautiful beaches, much delicious red wine and 'la dolce vita' of the wonderful people we met, still know and love in Italy.

Editorial support was provided by Yu Shan Tay, World Scientific (Singapore). I could not have done it without you, Yu Shan.

About the Author

Dr. David R. Basco, Distinguished Professor Emeritus of Civil and Environmental Engineering at Old Dominion University was born in Chicago, Illinois on January 30, 1939. He earned the B.S. degree in Civil Engineering in 1960 and the M.S. degree in Civil Engineering in 1962 from the University of Wisconsin, Madison. His Ph.D. degree in Civil Engineering was awarded in 1970 from Lehigh University, Bethlehem, PA. Post-graduate education was facilitated by attending the Delft International Hydraulic Engineering courses in the Netherlands, where he earned the IHE-diploma in 1976.

Professor Basco began his academic career as an Instructor at Lehigh University (1965–1967), and then as Assistant Professor (1969–1973), Associate Professor (1973–1983), Full Professor (tenured) (1983–1986) of Civil and Ocean Engineering at Texas A&M University, College Station. He joined Old Dominion University as Full Professor of Civil Engineering and Director, the Coastal Engineering Centre in 1986 where he served until retirement in May 2014. He has organized short courses on coastal engineering and given lectures in the Caribbean (Antigua, Dominica, Grenada, St. Lucia, and Trinidad in 2001), Japan (2002), Turkey (2002), Korea (2005), and Italy (2007, 2013). He is presently Professor Emeritus at Old Dominion University, Norfolk, Virginia, US.

Throughout his career, Dr. Basco has worked for the US federal government (Bureau of Reclamation, Geological Survey, Corps of Engineers), in industry (Allis-Chalmers Mfg. Co.), and as a consulting engineer for many federal agencies, state and city governments, private associations and foundations, legal firms, consulting engineering companies, dredging contractors, and international governments. These efforts were through his own firms: E_2O Consultants in College Station, Texas and Beach Consultants,

Inc. Norfolk, VA. This has been in engineering feasibility studies and final design; laboratory, field, and computer research; project management; as an expert witness in court disputes.

Dr. Basco has profited from three, extremely fruitful sabbatical leaves in Europe (the Netherlands, 1975–1976; Denmark, 1982; and Italy, 2013) which led to significant, long-lasting international collaborations in coastal engineering. His time at the International Institute for Hydraulic Engineering in Delft and at the Danish Hydraulic Institute in Demark led to the publication of his first book titled *Computational Fluid Dynamics: An Introduction for Engineers* (Addison Wesley Longman, United Kingdom, 1989, 4th printing 1997) with Michael Abbott. This book has been well received by the coastal engineering community.

Dr. Basco has authored and co-authored 92 journal and proceeding papers, 119 technical reports and two chapters in handbooks. One handbook chapter is "Storm Hazard Mitigation Structures" (Chapter 30) in the *Springer Handbook of Ocean Engineering* (2016) and the second is "On the Hydrodynamic Strength of a Coastal Storm", (Chapter 50) in *Handbook of Coastal and Ocean Engineering, 2nd Edition* (World Scientific, Volume 2, 2018). Dr. Basco's contributions to coastal engineering practice include two chapters in the US Army Corps of Engineers Coastal Engineering Manual (CEM, 2010) that has been widely used by coastal engineers. One chapter is Part V, Section 3 "Shore Protection Projects" and the second is Part V, Section 7 "Example Problems". Summaries of this material is also included herein.

Examples of his fundamental contributions of research in hydraulic dredging (analytic model), surf-zone hydrodynamics (Boussinesq theory), breaking waves (Relative Trough Froude Number), barrier island breaching (numerical model), the interaction of seawalls and beaches (field studies at Sandbridge), and development of the Coastal Storm Impulse (COSI) parameter (to complement the Saffir-Simpson wind speed scale) are found in this book. Also included are examples of major design works for the CORE-LOC armor units for repair of the Lajes Airfield refueling breakwater in the Azores; and the buried rock revetment beneath the rebuilt dune and beach nourishment at Dam Neck, Virginia.

Dr. Basco began the manuscript for this book in January 2013 at the University of Salento, Lecce Italy as visiting professor in the "Dipartimento Di Ingegneria Della Innovazione" (Innovation Engineering Department) and sponsored by Dr. Roberto Tomasicchio, Professor of Civil Engineering.

Supplementary Materials

Thank you for purchasing Design of Coastal Hazard Mitigation Alternatives for Rising Seas by David Basco.

To complement the contents of this book, we have prepared supplementary materials available for download from the World Scientific website (www.worldscientific.com).

These materials are spreadsheets to calculate key design variables using the equations presented in the book. They are found in Chapter 2 for water wave mechanics, Chapter 5 for functional design, Chapter 6 for structural design, and Chapter 8 for economics.

One key geometrical variable that determines cost is the seaward slope, cot ϕ that dictates the crest elevation, volume of material, and size/strength of the material required for design. Therefore, many of the spreadsheets present the results for a range of typical structure slopes, cot $\phi = 1.5$, 2.0 2.5, 3.0, 4.0 and 5.0. For the given input variables in the equation, the designer can immediately learn the impact of structure slope (i.e. structure costs). Each spreadsheet has separate pages for the metric and English (imperial) unit systems and a read me page for the subsection where the equations are fully discussed. The author created these spreadsheets over many years and most have been well tested by students over many years. But there is always room for error, so please let me know about any problems you have uncovered in their application.

"Only the application makes the rod into a lever."
(Ludwig Wittgenstein (1921)

Downloading the supplementary materials

1. Enter the following URL into your browser (note: this access URL should be activated ONLY ONCE):

https://www.worldscientific.com/r/11462-SUPP

2. When you enter the site, you will be asked to login. If you have not previously registered on the World Scientific site, a simple (free) registration process will be required.
3. Once you have successfully logged in and reached the page for *Design of Coastal Hazard Mitigation Alternatives for Rising Seas*, select the "Supplementary" tab. This tab contains a direct link to the supplementary materials.
4. Clicking on the link will allow you to download the zipped file containing the figures.
5. If your computer automatically places downloaded files in a folder named "Downloads", you will want to relocate the zipped file to a convenient permanent directory before unzipping. The zipped file is named "Spreadsheets.zip" and after unzipping, it will produce an unzipped folder of the same name.

Please contact sales@wspc.com if you have any problems downloading the material.

Contents

Chapter 1

Introduction

1.1 History and Heritage

1.1.1 *Time periods*

Civil engineering is arguably the world's second oldest profession. The earliest records of human existence reveal transportation via ships, needing ports and harbors at the land–sea boundary. We divide the history of the present-day branch of civil engineering called "coastal engineering" into four time periods:

(1) "Naturalistic" antique age (circa 3000 BCE–20 BCE);
(2) "Experience" technology age (20 BCE–1950 CE);
(3) "Engineering/science" computer age (1950 CE–1990 CE);
(4) "Internet" digital information age (circa 1990 CE–present).

In the antique age, natural land–sea harbor sites were employed for the location of ports. Little, if any, 'engineering' was required. This was the situation in general for the Mediterranean Sea for many centuries with years designated as Before Common Era (BCE). The Common Era (CE) is also often labeled as AD and begins as year 1.

The world's first modern harbor constructed at a site lacking a protective cape or bay was in 20 BCE at Caesara Maritima on the coast of Palestine (Israel). Nautical archeologists have explored, uncovered, and presented construction details (Hohlfelder *et al.*, 1987) as graphically depicted in Figure 1.1(a). Technological firsts were said to be the following: the use of small stones beneath the caisson on the ocean floor to prevent settlement: a large, reusable wooden form with double-walled, top channel to pour concrete to sink the caissons; the large-scale use of volcanic ash (from Naples) to make "hydraulic" concrete that hardens under water; a smaller, parallel breakwater seaward of the southern breakwater to break the waves before hitting the main structure; and a rock revetment directly in front of the north seawall. These were engineers sent from Rome by Caesar Augustus

Figure 1.1(a). Caisson breakwater construction (from Hohlfelder *et al.*, 1987).

(Emperor) at the request of Herod, King of Judaea. There were no known precedents for the technical challenges of this harbor site. Figure 1.1(b) displays the harbor entrance towers, seawalls, storage areas, and quays as depicted in Hohlfelder *et al.* (1987).

Many concrete blocks, vertical breakwaters and quay walls from ancient Roman harbors are still visible today. A catalog listing ancient Mediterranean ports and their locations as mentioned in the writing of 65 ancient authors has been published (De Graauw, 2018). At some sites, the ships simply landed on open beaches.

Figure 1.1(b). Harbor entrance, seawalls and quay (from Hohlfelder *et al.*, 1987).

Technical expertise was developed and handed down through the centuries. Storm damage mitigation structures and erosion protection projects were also planned, designed and built during this "technological" age ending circa 1950 CE for reasons detailed in the following.

For example, Figure 1.2 shows cross-sections of the Galveston, Texas (Gulf of Mexico) seawall constructed following the devastating hurricane of September 8, 1900 that killed over 6,000 people. The original design (1902) Figure 1.2(a) was modified and constructed (1904) Figure 1.2(b) and later modifications Figures 1.2(c)–1.2(f) took place as the wall length was extended and experience gained over time. This was the first hurricane protection project designed and built by the US Army Corps of Engineers and is presently 16 km long. It has now saved millions of dollars and an untold number of lives in its 115-year existence to mitigate the damage of hurricane storm events.

The second example in the technical/experience age goes back to the first European settlement at Jamestown, Virginia in 1607. The original fort was assumed lost to erosion. Around 1895, a civil engineer (Yonge, 1926) estimated the original shoreline position and fort location; designed a seawall/revetment to protect the site from further erosion; and convinced the US congress to fund construction in 1904. In the 1990s, archeological investigations (Kelso *et al.*, 1994) proved that the location and orientation of

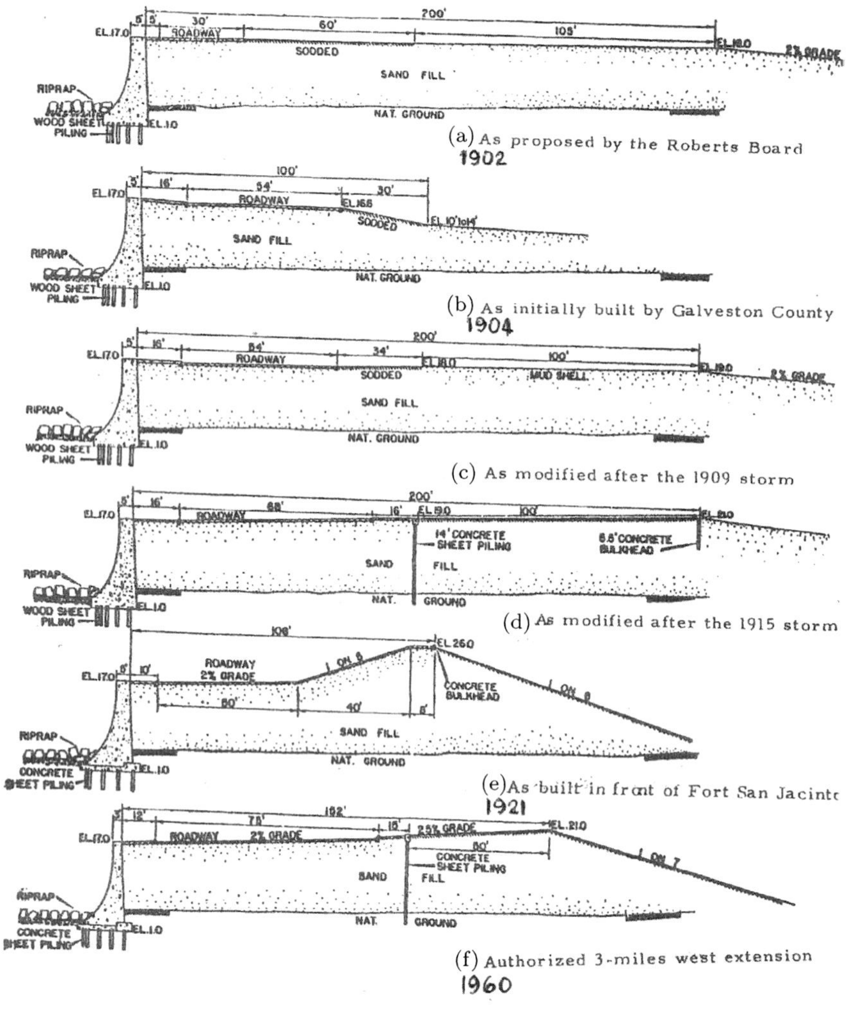

Figure 1.2. Galveston, Texas (US) hurricane protection seawall (1904–1960).

the James Fort on Yonge's map from 1907 is very close to the now proven location as shown in Figure 1.3 (APVA, 2000). This schematic view of Jamestown fort (solid red line) as preserved today by the seawall/revetment displays the original fort area (dotted red line) and 20% lost by the long-term erosion of the James River tidal estuary. Archeological excavation areas are the white boxes (Jamestown Rediscovery, 2019). About 80% of the fort remains. The most historically significant archeological treasures of

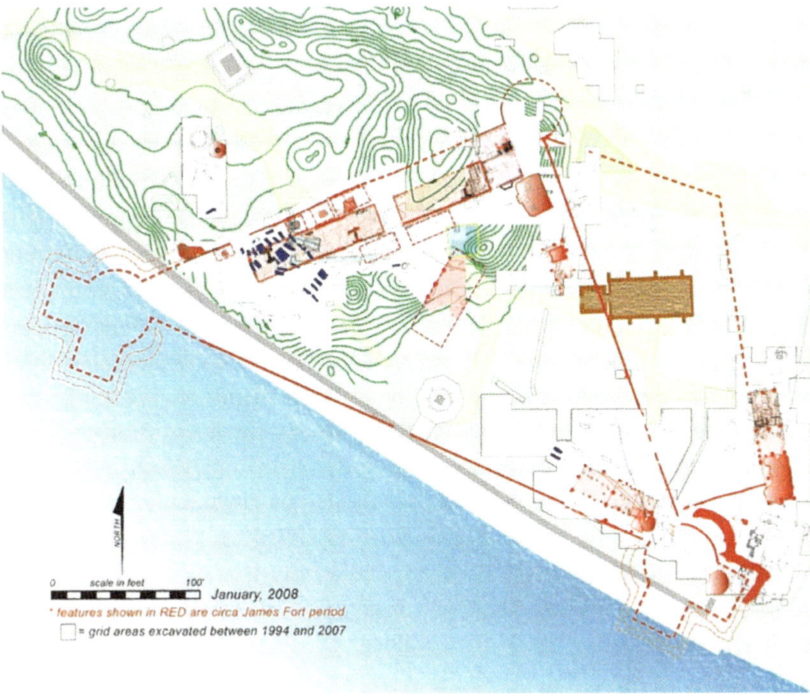

Figure 1.3. Schematic view of Jamestown fort (red line) as preserved today by the seawall/revetment with original fort area (dotted red line) showing 20% lost by long-term erosion of the James River tidal estuary. Archeological excavation areas are the white boxes (courtesy Jamestown Rediscovery, Preservation Virginia).

incalculable value have been preserved and protected by the seawall at the historic Jamestown Virginia, the "cradle of the republic" (Basco, 2014a).

In October 1950, the "engineering/science" age of coastal engineering began. The first International Conference on Coastal Engineering (ICCE) was held at Long Beach, California where the term "coastal engineering" was first used in print. The purpose was

> ... *to help engineers by summarizing the state of the art and science related to the planning and design of coastal works* (Wiegel and Saville, 1996).

The development of coastal "scientific" analysis methods in the early 1900s accelerated in the 1940s during World War II (WW II) for forecasting wave conditions to plan amphibious landing on northern European and African beaches (O'Brien and Johnson, 1947). Coastal engineering knowledge for harbor breakwaters also advanced in the planning, design, and placement of the structures for the logistics supply of materials across the beaches of Normandy, France. The success of these "temporary" harbors is often mentioned as a key to ending WW II with details presented in Chapter 5 (Arromanches, 1997).

A fourth time period, the "internet" digital information age began in the early 1990s. The free exchange of ideas and information over the internet has changed the course of events and ushered in the digital information age of coastal engineering. The University of Delaware began the "Coastal_List" to facilitate worldwide, instant communication opportunities on September 28, 1995. Graduate-level, full-semester courses and continuing education "webinars" are available on the "net". Consulting engineering firms exchange files, drawings, photos, etc. for design projects. Software for research, planning, studying, and designing coastal engineering works is exchanged and some are even available free for downloading over the internet.

Websites have sprouted to provide a wealth of information that has been developed and organized by the sponsoring organizations and agencies. For example, the US Army Corps of Engineers, *Coastal Engineering Manual* (CEM, 2000 and updates) is now freely available over the internet. These websites summarize the latest information and can be readily updated with new research results. The Google Company was founded in 1998 with the goal: "*...to organize the world's information and make it universally accessible and useful.*" Google Earth maps of the worlds' coastlines dating back in time are now used in coastal engineering research, planning, and design.

The Proceedings of the ICCE, dating back to the first conference in 1950, are now all available for free from the internet. The goals of the ICCE can still be well presented by the preface from the first conference in 1950:

> *The Conference on Coastal Engineering was conceived originally as a local meeting of engineers and scientists interested in shoreline problems... The aim was to aid engineers by summarizing the present state of the art and science related to the design and planning of coastal works... Although much remains to be done in the way of developing reliable design methods, the series of papers presented at the conference and published in this volume do represent a rather thorough summary of coastal engineering as now practiced.*
>
> — J. W. Johnson

In some respects, too much information is now available. Fewer coastal engineers are available than needed in the world and faster decisions are being required. The need exists to "sift and winnow" all the available information to aid the design engineer in the practice of his/her profession and to do a better job. This is one aim in the writing of this book. The ultimate use of all this information is for planning, design, construction, and management.

1.1.2 *Problem evolution*

The Silver (25th) anniversary of the ICCE conferences was in 1996, and it was the occasion for documentation of the *History and Heritage of Coastal Engineering* (Kraus, 1996). It is strongly recommended to those wanting to dig much deeper into the history of the coastal engineering profession. The conferences had been held in 15 countries since 1950 and each country contributed a chapter. In all 15 countries, the earliest coastal engineering activity was port and harbor development. The evolution of coastal engineering problems in all countries was said to occur in three cycles and we have here added a fourth as listed in the following:

(1) Exploitation and utilization of the coast,
(2) Development of protection from coastal hazards (flooding, high waves, erosion),
(3) Preserving and creating harmony between nature and coastal uses, and
(4) Planning for climate change and accelerated sea level rise.

Some countries are in stage (3) while others are in stages (1) and (2). (Kraus, 1996).

Concern for the coastal environment (water quality, coastal ecology, etc.) in stage (3) saw the planning, design, and construction of movable gated structures in tidal estuaries. The gates are normally open for the daily tidal exchanges necessary for the salinity and ecology of the tidal bays and lagoons and only closed to prevent upland flooding during storm events. Figure 1.4 schematically depicts the movable gates in the MOSES Barrier project to protect the Venice Lagoon from coastal flooding in the Adriatic Sea. The gates are normally full of water and flush with the seabed inside the caisson structure. Daily tidal cycles flow unimpeded through the three tidal inlets to the Venice Lagoon. Air is pumped into each hollow gate to raise the barrier and protect Venice from flooding. The operation has begun in 2018.

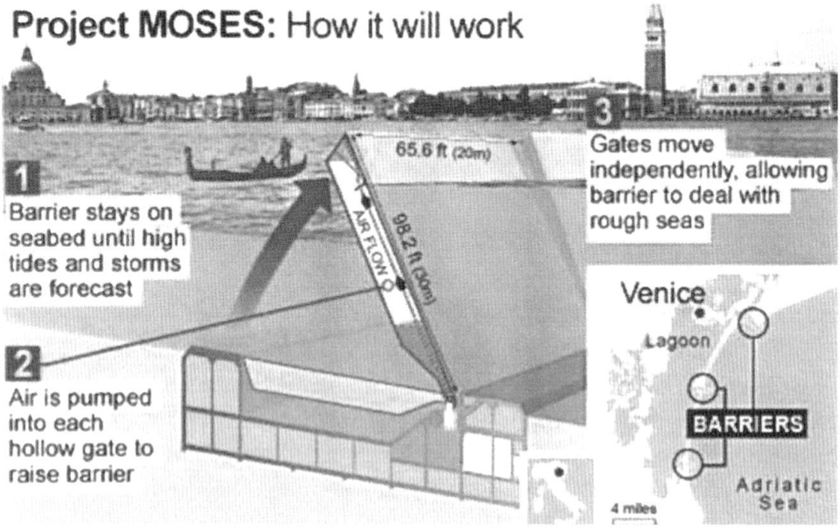

Figure 1.4. Schematic of the MOSES Barrier project to protect the Venice Lagoon (Telegraph Media Group Limited 2008).

All the world's countries are now in stage (4), facing a future of climate change with the potential for accelerated sea level rise at their coast. Flooding events will take place more often. Coastal storms may occur more frequently causing more beach erosion. Dealing with the conflicts between stages (1)–(3) is now known as Coastal Zone Management (CZM) and will grow more intense. Chapter 11 is devoted to stage (4) of the evolution of problems in coastal engineering design.

A second theme of the 25th ICCE in 1996 (Silver Anniversary) was the repeated mention of the same "pivotal workers" in the field with worldwide influence. We owe a tremendous debt of gratitude to these individuals and benefit from this heritage. They have published in journals, books, and recently on websites, and many will be referenced in this book.

1.2 The Feasibility Study

This book is about the design of coastal structures that (1) mitigate storm damage, (2) mitigate erosion, and (3) enhance/create coastal ecosystems. Few books are available that discuss coastal engineering design (see for example, Kamphuis, 2010; Reeve *et al.*, 2012). The operable word in this book is "design". It is written primarily for the professional engineer. In the

Table 1.1. The feasibility study.

1. Defining the problem
2. Gathering site-specific information
3. Developing design criteria
4. Reviewing design constraints
5. Reviewing alternatives for solution
6. Developing consequences of failure
7. Doing a feasibility study
8. Preparing, presenting, recommending results
9. Deciding what comes next?

United States and Canada "*only a licensed professional engineer may prepare, sign, seal, and submit engineering plans and drawings to a public authority for approval, or seal engineering work for public and private clients*" (NSPE, 2007).

From the author's experience as a licensed, professional engineer, the feasibility study process entails the nine steps as illustrated in Table 1.1 in the following. A step 10 could be added to remember to archive the data and results for efficient use the next time they are needed. There usually is a "next time". This is only the initial part of the full design process as discussed in Section 1.3 and shown in Table 1.2.

1.2.1 *Defining the problem*

1.2.1.1 *What is the problem?*

The initial step is to define the problem. Too often, this initial step is ignored or forgotten. We avoid the term "shore protection" because it conveys a false sense of security that can never be completely attained. We can only "mitigate" or reduce the damages caused by storms and coastal erosion. This will be further discussed in Chapter 4.

The site of the problem is located within a physical system. A system can be defined as "*... all the processes, inputs, outputs, and interactions that affect a process or event*" (Kamphuis, 2010). The system size selected depends on many factors and is related to problem definition.

For example, Figure 1.5(a) displays the Southeastern region of Virginia and outer banks of North Carolina at a regional scale that can be divided into smaller subregions for problem definition. The subregion in Figures 1.5(b) and 1.5(c) shows the location of 19 nearshore breakwaters that are providing erosion mitigation directly behind but may be

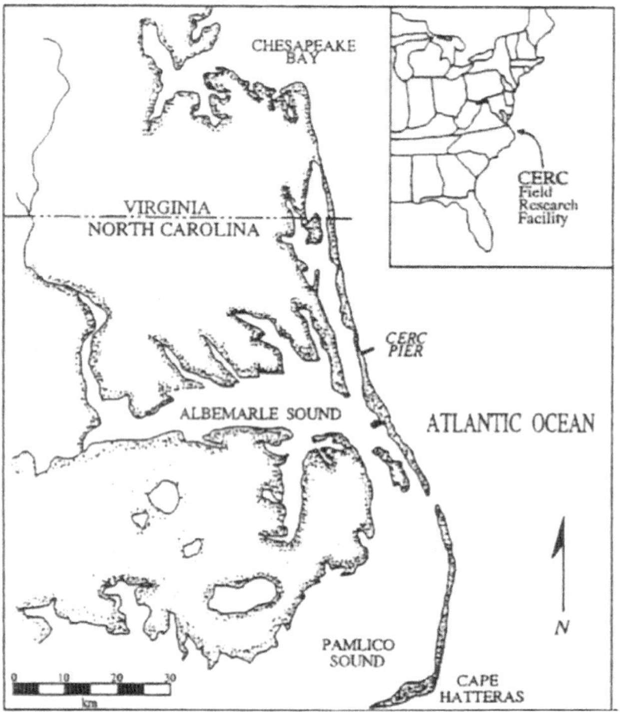

Figure 1.5(a). Regional view of Southeastern Virginia and outer banks of North Carolina.

responsible for down-drift erosion as discussed in Chapter 8. The Atlantic Ocean and entrance to the Chesapeake Bay form the water boundaries. Tidal inlets interact with the bays, drain the interior, and connect with the sea. Sediment transports occur that produce spits, fill the tidal inlet channels, move on- and off-shore, and create sediment cells that typically define the smaller subregions of the physical system. This regional system has political boundaries (private property, cities, states, federal government, etc.) that differ from the physical boundaries and are equally important in problem definition. The regional system has wave climates that come from both the full Atlantic Ocean and the Chesapeake Bay.

The regional system of Southeastern Virginia and outer banks of North Carolina illustrated in Figure 1.5. Subregions and their many problems (elevated water levels and flooding, wave damage, and coastal erosion) will be referenced and used as examples throughout this book.

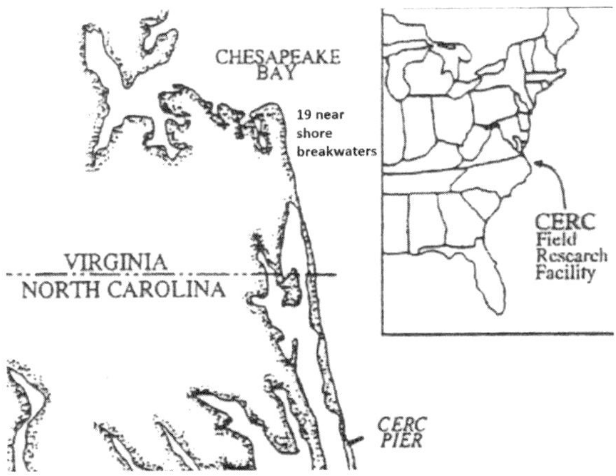

Figure 1.5(b). Subregional view and location of 19 nearshore breakwaters at Fort Story, VA.

Figure 1.5(c). View of 19 nearshore breakwaters at Fort Story, VA (courtesy Norfolk District, Corps of Engineers).

1.2.1.2 *What is causing the problem?*

Determining the primary cause of the problem should be initiated in this first step but this usually requires more time. In many cases, the problem was caused by some "anthropogenic" (man-made) effort to solve another problem. The classic example is the installation of a groin to mitigate erosion by trapping sand on one property that causes "down-drift" erosion on the adjacent property. As discussed in Chapter 10, this "problem" may be due to many other factors, and coastal engineers have long learned how to design the "terminal" groin and combine groin field construction with beach nourishment.

The design team (coastal engineers, oceanographers, coastal geologists, and other civil disciplines (structural, geotechnical, and environmental)) should all be involved in determining the cause of the problem. The team should organize an initial meeting with the client (or general public). It is strongly recommended to first get the problem clearly defined in writing. A review meeting with the client about one-quarter into the preliminary design effort will ensure that what the client expects and what the engineering team will deliver are the same. We will refer to the "client" throughout but also mean the general public when public-funded projects are involved.

The history of the coastal engineering profession is full of examples where the "solution" to the problem made matters worse because the primary "cause" of the problem was never identified at the start. A clear definition of the problem and what is the cause of the problem will aid the remaining tasks explained in the following sections.

1.2.2 *Gathering site-specific information*

A site visit is strongly recommended, which will enable the physical appreciation of the scale of the problem with dated photographs taken for later review. Previous field constructions nearby can also be observed. Viewing the site from water and air is always a bonus if funding permits.

Previous studies of the problem and site as found in reports, conference proceedings, journal papers, newspaper articles, etc. should be gathered and reviewed. They can be obtained from government agencies, consulting firms, universities, and the local news media.

Historical maps and charts dating as far back in time as available should be collected. They provide a record of the shoreline position for analysis of erosion rates. Ground-level and aerial photographs are also invaluable records of how the site has evolved over time. Here, the client may help by providing private records and photos.

Most importantly, field data of water levels (tides, storm surge) waves, geotechnical, bathymetry, etc. should be obtained for use in developing the design criteria and for design as described and discussed throughout this book.

1.2.3 *Developing design criteria*

The "rules" that govern and constrain the design process must be clearly articulated. Design practice has evolved from deterministic to probabilistic. In deterministic design, one "solution" is obtained for a given level of recurrence-interval event. For example, storm-surge "protection" for the 100-year water level. Probabilistic design uses the entire exceedance probability curve of water levels for the analysis of the economics and risk. Chapter 2 presents the full details for developing design criteria and Chapter 4 summarizes the probabilistic design.

1.2.4 *Reviewing design constraints*

Finding solutions within constraints is what engineers do. One good definition of engineering is "design under constraint." Engineering is creating and designing what can be, but it is constrained by our understanding of nature, economics (costs), concerns of environmental impact, institutional, social, and legal issues, aesthetics, and possibly the media. We limit the discussion here to an introduction and refer to the upcoming chapters for details.

1.2.4.1 *Scientific and engineering understanding of nature*

The coastal setting is dynamic and influenced by the land, water, and air interactions and processes. It is a regime of extremes, surprises, and constant motion as the coast and structures respond to changing conditions. A fully dynamic, three-dimensional (3D), numerical model of water levels, waves, and sediment transport to simulate bathymetric and shoreline change is still under development. It remains a long way from routine application for the functional design of coastal structures. This entire book is devoted to presenting the latest scientific and engineering understanding of nature for use in engineering design. But our complete understanding of how nature works is always a design constraint. This is now especially important today because of the accelerated rise in global mean sea levels as fully discussed in Chapter 11.

1.2.4.2 *Economics*

A key constraint for each alternative is life-cycle cost. In general, the level of storm-damage mitigation and, hence, costs (high, medium, low) can only be justified when the corresponding value of property and infrastructure to be protected (benefits) are comparable (high, medium, low). Costs are not subjected to any other constraints. Life-cycle costs combine initial construction costs and maintenance costs over the design life of the project. Benefits, however, can be restricted and limited to only those that are perceived to benefit the funding authority. Chapter 8 gives complete details of how to compute maintenance costs with examples.

1.2.4.3 *Environmental*

A third major constraint is the impact on the environment. For example, the Eastern Scheldt storm surge barrier in the Netherlands was the focus of much discussion in the early 1970s. Environmental scientists favored raising the dikes around the periphery to maintain the saltwater ecology of the tidal estuary. Agricultural and water boards favored a solid dam across the mouth that would create an inland, freshwater lake. A compromise was reached: a storm surge barrier with movable gates which stay open under normal conditions but are closed at very high storm surge events. The final design, construction methods, and equipment required much research and challenged the ingenuity and technical process of the Dutch coastal engineers. In the final analysis, the environmental constraint, to maintain the saltwater ecology, dictated the final design. The additional engineering and construction cost proved to not be the deciding factor. Chapter 10 presents the types of environmental concerns, impacts of structures on natural sediment transport systems, and mitigation measures to avoid, minimize, and/or compensate for negative impacts.

In the US, all proposed alterations of the physical environment at the coast require a permit. In Virginia, the local "Wetlands Board", the state's Virginia Marine Resources Commission (VMRC), and the US Army Corps of Engineers all must approve the project and jointly issue the permit. An example is presented in Chapter 10.

1.2.4.4 *Institutional, political, legal*

A fourth area that has a formidable influence on the design process are the institutional, political (social), and legal requirements for all projects. Institutional policies and guidelines vary around the world. They depend on which stage ((1): exploitation/utilization, (2): development/protection, or (3): preservation/harmony) the site and country are located in. A brief summary of some US state and Federal government institutional policies will be presented in Chapter 3 when design alternatives are discussed. The situation is similar for the political and legal constraints on the coastal structures' design around the world. In some countries, there are no institutional, political, or legal constraints. See the *Coastal Engineering Manual* (2006) Part V, Section 3.1(c) for a detailed discussion of the institutional, political, and legal constraints on coastal designs in the US.

1.2.4.5 *Aesthetics*

A fifth and especially challenging area in design pertains to the sense of beauty and accepted notions of good taste. Many people feel that natural shorelines (e.g. wide, sandy beaches, rocky cliffs, or vegetated marshes and trees) are more aesthetically pleasing than the ones artificially manipulated for shore protection. An uninterrupted, uncluttered, and natural view of the sea is desirable for most people. Therefore, an aesthetically balanced and consistent appearance replicating natural systems is preferred, but not always possible. This has become the "green" versus "gray" argument. See Chapter 10.

1.2.4.6 *The media*

A final constraint may be the media which has an influence on public opinion. Educating the newsprint writers and television producers on the value of both proposed and constructed projects is an important aspect of coastal engineering. The public perception is often that either the projects are too expensive or only the wealthy home owners living at the shore receive the benefits. The media reports these perceived negative aspects. After a storm event, making sure that the monetary value of the damages prevented is well publicized by the media is equally important. Public funding by governments for needed storm-damage and erosion-mitigation projects may never be realized if the complete story of how the benefits exceed the costs is never told by the media.

Engineering is not an applied science. Our limited understanding of nature is one design constraint, but it is far from the only one, seldom the most important one, and almost never the limiting constraint for coastal engineering design.

1.2.5 ***Reviewing alternatives for solution***

The alternatives for a potential solution depend on the definition of the problem, the site conditions, and the constraints. As stated before, this book

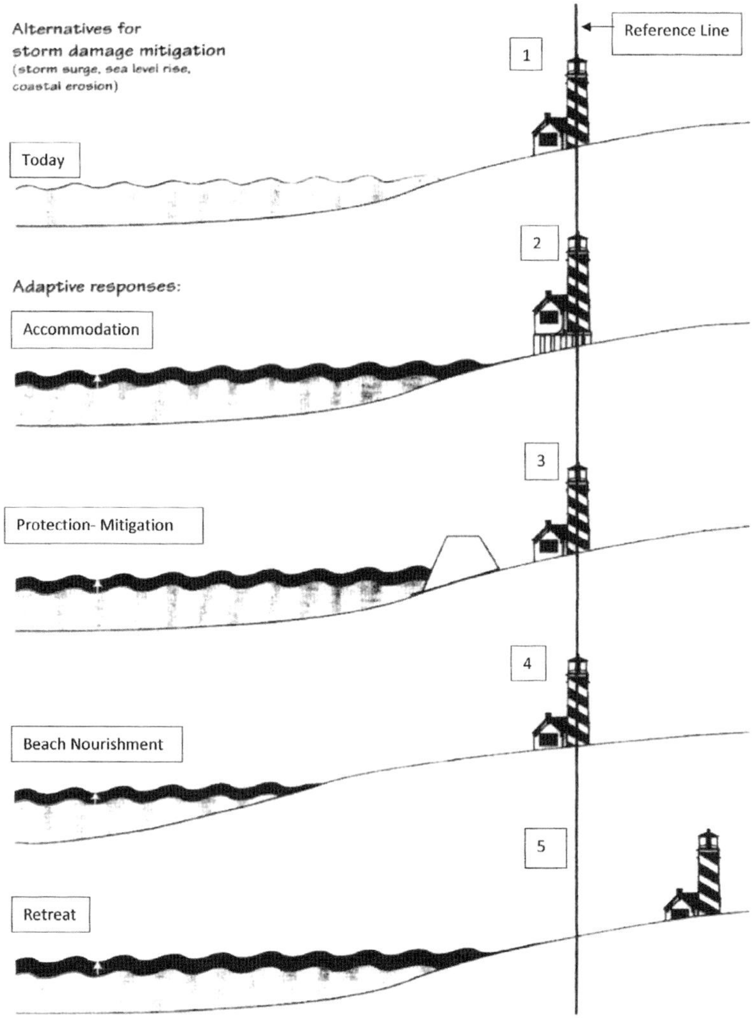

Figure 1.6. Alternatives for mitigation of coastal hazards.

is about the design of coastal structures that (1) mitigate storm damage, (2) mitigate erosion, and (3) enhance/create coastal ecosystems.

Storm damage is due to both high-water events causing flood damage and excessive wave energy causing wave damage. There are many alternatives for damage mitigation as illustrated in Figure 1.6. The top image (panel 1) indicates today's normal sea conditions, with the light house representing the threatened location.

Elevated water levels (black region) are then shown in the lower four panels that illustrate two different types of response. Changes to the natural, physical system are possible using some type of coastal structure (panel 3) or beach nourishment (panel 4). These are often called the "structural" solution. Changes to man's system for living near the sea can be accommodation (panel 2) or retreat (panel 5) and are called the "non-structural" solutions. Often, combinations of structural and non-structural solutions are employed. The final alternative is to do nothing. This book covers the functional (Chapter 5) and structural (Chapter 6) designs of seawalls, bulkheads, flood walls, and revetments that are fixed in location (panel 3). Also discussed are movable gated structures (not shown). Beach Engineering is covered in Chapter 7.

The second major coastal hazard is erosion. Shoreline movements landward and earth embankment/dune volume losses also take place during storm events. The same structural alternatives illustrated in Figure 1.6 can be employed to mitigate coastal erosion. The functional (Chapter 5) and structural (Chapter 6) designs of headland and nearshore breakwaters, groins, and revetments are the alternatives considered.

Chapter 3 presents complete details on design alternatives. The correct choice in many cases depends on the design constraints for the particular site.

1.2.6 *Developing the consequences of failure*

Failure is now an anticipated and expected occurrence in coastal structures design (Kamphuis, 2010). Therefore, the consequences of the functional failure and structure failure of the alternatives considered for solution must be considered from the start. Chapter 4 on Probabilistic Design and Chapter 8 covering Economics, Risk, and Resilience will review all the details on what Kamphuis (2010) calls Contemporary Design.

1.2.7 *Doing a feasibility study*

Using the results of steps 1–5 listed in Sections 1.2.1–1.2.3, the latest design formulations as summarized in Chapters 5 and 6, and the spreadsheets included in this text, the preliminary designs of selected alternatives for problem solution are prepared including cost estimates. These are life-cycle cost estimates including estimates for anticipated, relative sea level rise over the design life (Chapter 11). Environmental impacts (Chapter 10) must be addressed. A design philosophy for failure consequences and resilience of the system must also be considered for each alternative (Chapters 8 and 11). All aspects must be considered including the potential social, political, and institutional design constraints at the site. The level of study detail for each alternative must be sufficient to give confidence that a further study will not produce any surprises and major changes in the results.

1.2.8 *Preparing, presenting, and recommending results*

A detailed, written technical report should be prepared giving all the calculations and information to support the final results. It should summarize the cause of the problem and substantiate the reasons for the final recommended design alternative selected. It should be written so that its contents can be understood and reused in the future. A summary of the study and final recommendations should be prepared and presented orally to the client. A recommendation that none of the alternatives is feasible is possible and just as valuable to the client. All the solutions may be technically, or economically, or environmentally and/or politically infeasible. This final result is the reason for doing a "feasibility" study.

1.2.9 *Deciding what comes next?*

In most cases, the study will prove that one design alternative is feasible in all aspects (technical, economic, environmental, political, social, institutional, and aesthetic). The selected alternative to solve the problem must now be studied in full detail as summarized in Section 1.3. If proven to be infeasible, the reason(s) may be investigated at this stage and a restudy may be conducted, if warranted.

1.2.10 *Archiving the data and results*

The valuable site information and meteorological data collected are needed for the final design. Or, it may be reused for a future study. For example, water-level information from local tidal gauges is used to develop extreme frequency curves for use in storm-surge estimates of rare events. With each new year of data, the revised analysis shifts the trend line accordingly. Spreadsheets can be easily updated and reapplied if initially designed for the yearly addition of new data. Digital data storage files facilitate the saving of the finding and reuse of the previous study results. The main problem is remembering to do it before the next project.

1.2.11 *Other structural design elements beyond the scope of this book*

This book does not consider the plan-form layout of navigation channels and jetties, harbor breakwaters, and quays (harbor design.) These coastal structures mitigate the excessive wave energy environment for safe ship navigation between the jetties, ship and boat moorings behind the breakwaters in the tranquil harbor waters, and alter the longshore currents and sediment transports near the pier.

However, the structural design for stable rock structures and caisson structures used to construct jetties and harbor breakwaters will be presented in Chapter 6. Wave forces on slender piles is also reviewed in Chapter 9 for the stability of piers and monopole foundations of nearshore wind turbines.

1.3 Design of Coastal Hazard Mitigation Alternatives

Coastal design is complex and challenging because each site is unique in terms of the physical, ecological, and social/community aspects of the location. There are no cookie-cutter solutions. The complete process of the design of coastal hazard mitigation alternatives is summarized in Table 1.2. We present only a brief overview of each aspect and add more details throughout this book.

1.3.1 *The feasibility study and new data*

The feasibility study is discussed in detail in Section 1.2, as summarized in Table 1.1. The analysis of alternatives is generally only based on the empirical formulas presented in Chapter 5 (Functional Design) and Chapter 6 (Structural Design) which can be called "analytical" models. The results

Table 1.2. Design of coastal hazard mitigation alternatives.

1. The feasibility study (Table 1.1) and new data
2. Modeling (numerical and/or physical)
3. The final design
4. Plans and specifications
5. Final permit and funding approval
6. Monitoring (construction and impacts)
7. Storm performance (the media)
8. Documentation of experience

are approximate and often produce conservative designs. Physical (scale) models and modern numerical models give much more insight and details about the physics and should always be considered for use in preparing the final design. Both types are discussed in Section 1.3.2.

Field measurements may also be considered with the additional data collected at the start. For example, the feasibility study often reveals the need for geotechnical data to determine the actual soil conditions for a new site. Soil borings can be specified at the exact locations where the coastal structure will be built. Instruments may be installed at the site to measure the daily tidal variations and wave conditions, although the record lengths can never be long enough to capture the extreme storm events.

A bathymetric survey should also be made to learn about the current bottom contours at the site. They may be significantly different than those found on older maps and charts. The latest bathymetric conditions will be needed in the modeling.

Finally, subaerial surveys of the beach and dune system (if present) at the site and for some distance on both sides of the adjacent property should be considered. This "baseline" data are needed to determine if any "down-drift" impacts to adjacent property can be attributed to the coastal structure after construction. Further details are found in Chapter 10.

1.3.2 *Modeling*

Modeling is a tool for design optimization. Modeling provides the designer and the client the opportunity to visualize the water levels and waves and their interactions with the structures. Modeling is an indispensable tool in coastal design (Kamphuis, 2012).

Modeling improves the analysis and final design and generally reduces the life-cycle costs of the project. It almost always pays for itself in the long run. For example, wave runup and wave overtopping formulas permit the determination of the crest elevation of coastal structures as discussed in detail in Chapter 5. Modeling may determine that a lower crest elevation can be employed and hence will reduce the initial costs. Or modeling may determine that higher crest elevations are required and hence will reduce the overtopping rates and lower the costs of interior drainage systems. As a second example, rubble-mound stability formulations as discussed in detail in Chapter 6 will determine the median diameter of the rock for a stable design. Modeling may result in smaller rock size for a stable breakwater to reduce the initial structure cost. Or, modeling may determine that larger rocks are needed for a stable design and hence reduce the amount of damage under higher wave conditions and reduce the maintenance costs. In both examples, modeling reduces the total life-cycle costs and saves money in the long run.

Both physical (scale) and numerical (computer) models used in the final design require additional funding. Choosing which one to utilize depends on what aspect of the feasibility study using the analytical models is the weakest in producing the final design. In many cases, both models are employed in the final design.

1.3.2.1 *Physical (scale) models*

Physical models built to some reduced scale of the full dimensions in nature have long been employed to study the flow of water. The empirical formulas presented in Chapters 5 and 6 have often resulted from generic, basic research done at reduced scales in laboratories around the world. See Hughes (1993) for a recent book on physical (scale) modeling practice.

Wave tanks producing irregular, two-dimensional (2D) breaking waves with varying water levels during storm events are most commonly employed for the study of wave runup elevations and the stability of rock armorlayers. The permeability of the under-layers in the structure is a key variable that is still being researched (see Chapters 5 and 6). Wave basins producing irregular, 3D waves that are transformed by the local bathymetry (shoaling, refraction, diffraction, reflection, breaking; see Chapter 2) are employed for similar studies especially to understand the armor-layer stability on the terminal end of the structure. Wave transformations in the feasibility study are usually based on simplified bathymetric conditions that may produce incorrect wave heights at the structure for design.

Physical (scale) models require special facilities available in government laboratories and at universities. The expertise to design, construct, operate, and obtain the data is generally available for hire, and the designer's job is to supervise the effort and utilize the results in the final structure design. The designer does not need to also become an expert in physical modeling which remains an "art" more than a science. The designer can spend his/her time concentrating on the final design.

Physical (scale) models remain the only way to study the stability of coastal rock structures. See Section 6.7 (Chapter 6) for the role of physical models in structural design.

1.3.2.2 *Numerical (computer) models*

Numerical modeling on high-speed, digital computers is now the traditional practice in coastal structures design. Numerical wave basins producing 3D waves that are transformed by the local bathymetry to produce the needed design wave conditions at the structure are the most common application. Storm surges, tidal inlet flows, and other current conditions at the site can be generated using these "hydrodynamic" models.

Numerical (computer) models require high-speed computers. Software companies specializing in coastal engineering applications market their systems along with the training for understanding and use. Software packages now also include visualization tools. The design organization needs to invest in the computing power required and continual need for upgrades and maintenance. Training for use by the designer is also required. Data entry is labor-intensive and takes the most time for model setup.

The advantage of the numerical model is that the actual structure dimensions can be employed. There are no "scale" effects. However, numerical modeling of the structure of turbulent flows remains a long-term challenge.

Both modeling types have limitations for sediment transport simulations. However, both types produce good results for the 3D wave transformations over the nearshore bathymetry which is assumed constant in time in the modeling process.

The perception exists that physical (scale) modeling is more expensive than numerical (computer) modeling. This may not always be correct when the full costs of numerical modeling (hardware, space, maintenance, software, training, upgrades, application labor) are considered. But it is an indisputable fact that whether the type employed is physical (scale) or numerical (computer) "*... modeling is an indispensable tool in coastal design*" (Kamphuis, 2012). See Section 5.10 (Chapter 5) for the role of numerical models in functional design.

1.3.3 *The final design*

The additional field data and modeling results are now utilized to produce the final design. All the steps in Table 1.1 are repeated. The design criteria may be revised because new field data and the modeling results produce altered wave heights at the structure. Also, when the soil conditions are found to be weak at certain locations, the structure foundation design may need to be altered significantly along the structure. In most cases, the functional and structural elements of the design will not change significantly from that learned in the feasibility study. If radically different, the reason(s) should be determined for future reference.

1.3.3.1 *The permit application*

In the US, all proposed alterations of the physical environment at the coast require a permit. In Virginia, the local "Wetlands Board", the state's VMRC and the US Army Corps of Engineers all must approve the project and jointly issue the permit. The time required to obtain the permit may be lengthy, so the permit application materials should be completed and submitted as early as possible in the final design process.

The following four basic elements are needed in the permit application:

(1) determining the position (location) of the hazard mitigation alternative on the coast;
(2) determining the crest elevation of a structural alternative;
(3) determining the dimensions and materials of construction (to make a cost estimate);
(4) determining how the structural alternative interacts with the adjacent coastal environment.

This information can be summarized on three drawings (small projects) that form the nucleus of the permit application. The horizontal and vertical controls (datum) must be clearly displayed on these drawing.

Latitude and longitude are the universally accepted coordinates for the horizontal position datum as illustrated in Figure 1.7(a) at the Lynnhaven Inlet on the Chesapeake Bay. The horizontal line is Lat 36° and 54 min and the vertical line is Long 76° and 6 min. In the US, each state has its' own "State Plane Coordinate" system and the US Army Corps of Engineers software product CORPSCON permits the ready conversion of most horizontal coordinate systems for all US states. The Site Location (1st drawing) for the Permit for a seawall project on the Eastern edge of the Lynnhaven Inlet at the Lesner Bridge crossing is shown in Figure 1.7(a). The Site Layout (2nd drawing) is shown in Figure 1.7(b). The north direction arrow gives the orientation and all dimensions are in US feet units for this US permit application drawing. Many notes appear to aid what is displayed. The horizontal position datum on this drawing must be referenced to some physical monument found in the field. This monument must be established by surveyors giving both horizontal position and vertical elevation. Modern, differential, geographical positioning systems (DGPS) are now commonly

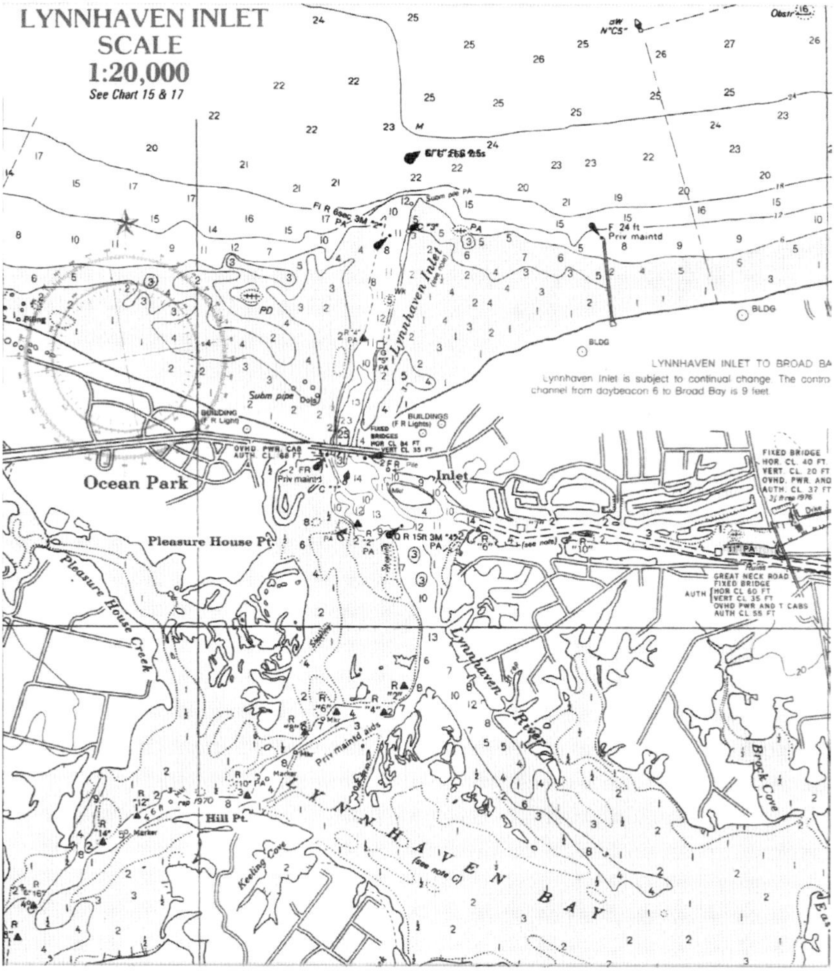

Figure 1.7(a). Example site location for seawall project permit at Lynnhaven Inlet (courtesy ADC Chartbook for Chesapeake Bay).

employed. The horizontal control datum is relatively easy to establish on the drawings and in the field. The author of this book designed the seawall and aided in obtaining the permit for this project for MS&A Consulting engineers (Basco, 2014a).

The vertical, elevation datum requires more thought and care to establish. For coastal engineering projects, the vertical datum may be related to some "mean sea level" but this is no longer useful when we know that the sea levels are changing relative to the local land elevations, which may also be

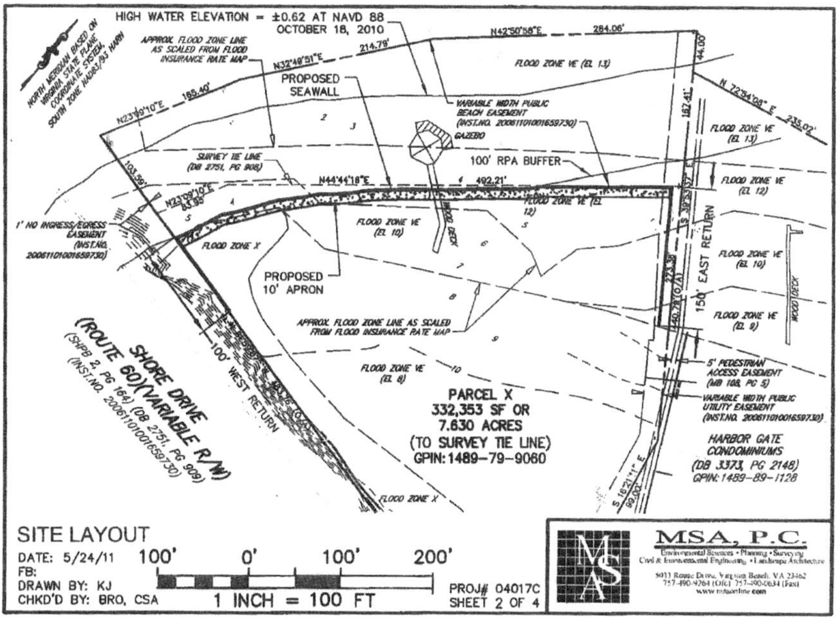

Figure 1.7(b). Example site layout for seawall project permit (dimensions shown in US feet units) (MSA, P.C. Virginia Beach, VA).

changing in time (subsidence). Or, it can come from some land-based, vertical datum. Every country is different. In the US, the Federal Government's National Ocean Service (NOS) maintains the system of tide gauges throughout the country and the mean lower-low water (MLLW) = 0.00 is the water-level based vertical datum. The land-based datum is the North American Vertical Datum (NAVD88) established in 1988 and is commonly related to MLLW = 0.00. Further details are found in Chapter 2.

Figure 1.7(c) displays the cross-sections (3rd drawing) with elevations for the permit application in this example. The vertical datum is NAVD88. Unfortunately, a note specifying this vertical datum is *Not Shown* on this third drawing for the permit application.

1.3.3.2 *Final design report*

The permit application drawings together with other design sketches, spreadsheet information, graphs, etc. form the basis for the final design and are normally summarized in a final report. This will include all the dimensions, notes, and details required to prepare the construction drawing

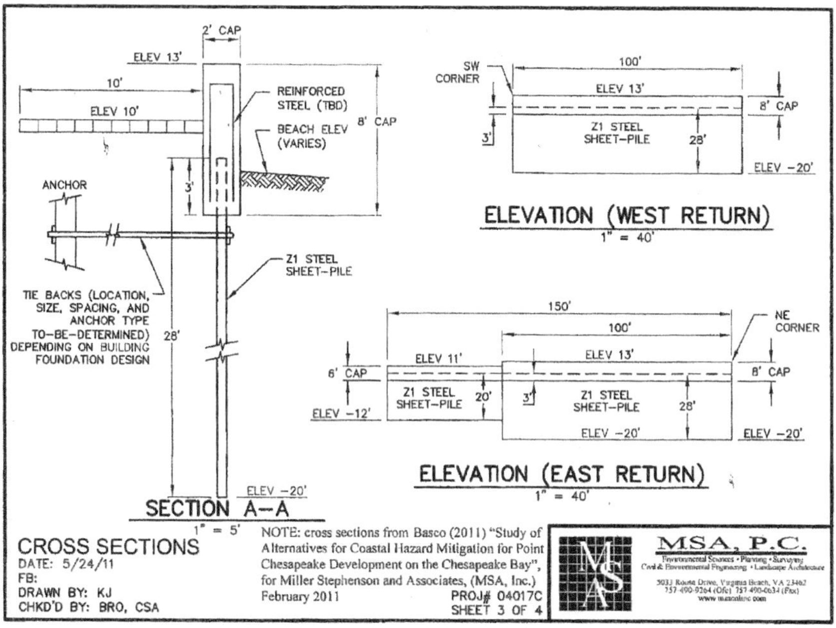

Figure 1.7(c). Example cross-sections for seawall project permit at Lynnhaven Inlet (elevations shown in US feet above the NAVD88 vertical datum) (MSA, P.C. Virginia Beach, VA).

(Plans) and the contract documents (Specifications) used for cost estimates and bid documents to the perspective contractors and for construction (Section 1.3.4).

An internal design review process and approval of the final design by all members of the design team follows to ensure everyone agrees on the final design. An external review may sometimes be needed. The final design is then turned over to the graphics specialist (draftsman), who prepares the final plans for construction using software products (e.g. AutoCAD).

1.3.4 *Plans and specifications*

Preparation of the drawings (Plans) for cost estimates, contractor bids, and construction in the field is a critical aspect of the design process. The final plan views, side views, end views, and cross-section views of the structures involved sometimes reveal aspects that were not previously evident to the designer. For example, when full plan-view layouts are complete, some aspects of construction in the field may be difficult. Modifications are then possible. The designer works closely with the graphics specialist. The coastal engineer also works closely with other civil engineering specialists in structures, geotechnical, etc. in preparing of the final construction drawings.

To avoid possible construction delays, disputes, and potential legal action, the construction specifications (Specs) must be carefully written. Although they commonly consist of many pages of routine (boilerplate) material that is required in a legally binding document, the designer is responsible for those sections that spell out the special requirements for construction. Some of these special requirements also appear as general notes on the drawings.

The construction drawings (Plans) may contain hundreds of sheets of drawings, and the construction specifications (Specs) may run to hundreds of pages for large projects. For example, the Hurricane Protection Project for Virginia Beach, VA as designed by the US Army Corps of Engineers (Corps, 1995). Figure 1.8 shows the sheet 27 of the 34 sheets of the construction drawings for the main seawall from 27th street to 34th street. Note the level of detail required in each cross-sectional view showing all the steel reinforcement sizes and placement locations for construction. Figure 1.8 is complete in all the details and is far more detailed than the Permit application drawing (Figure 1.7(c)).

The construction plans should function as a standalone plan, and are the sole documents that can be referred for obtaining the details necessary for construction of the structure.

After everything is reviewed, checked, and approved, the designer signs and seals the documents as required by all US states. The seal is issued to registered professional engineers (P.E.) by licensing boards in each state.

1.3.5 *Final permit and funding approval*

If proper planning and design was made, the Environmental Permit is issued for construction. If not, further delays may ensue. Delays involved in modifying the design to ensure permit approval can be avoided by consideration of the environmental constraints in the feasibility study. Further details are found in Chapter 10. Delay caused by the permitting agencies cannot be avoided. In no case should construction begin without the written permit approval.

The second written approval is for the funding for construction. This document will include the construction schedule and method for funding

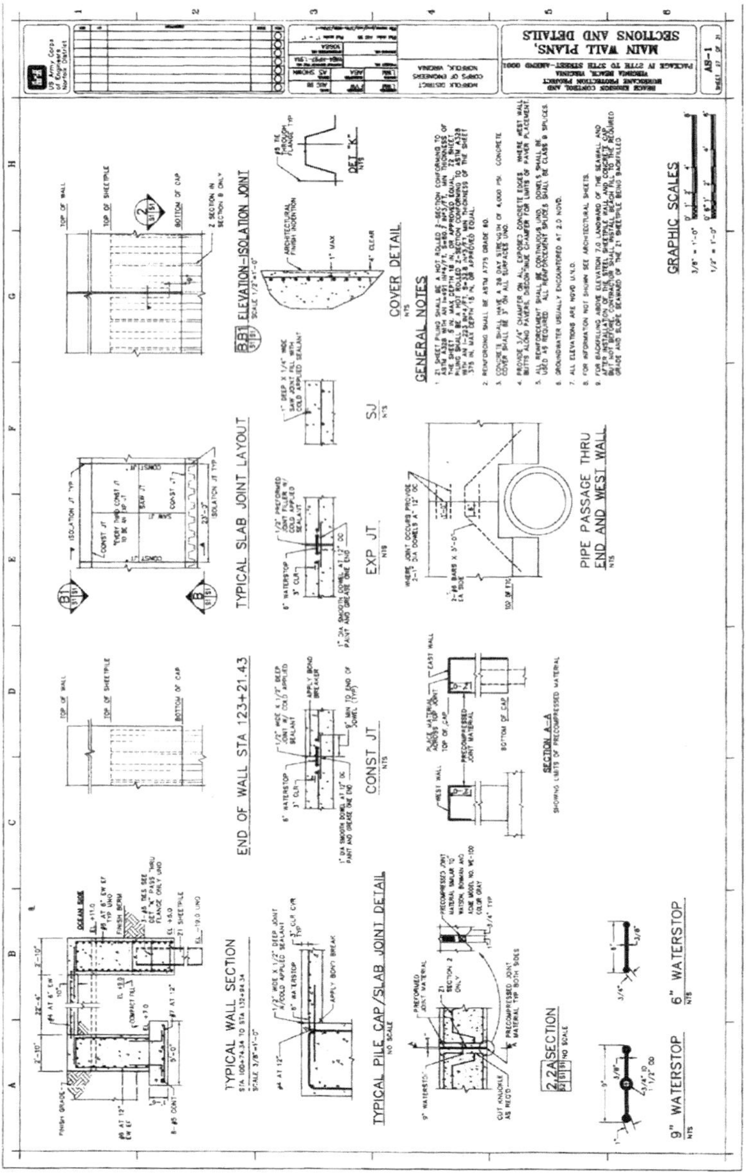

Figure 1.8. Example of construction drawing (sheet 27of 34) for Virginia Beach Seawall as designed by the US Army Corps of Engineers (Corps, 1995).

over the construction period. In no case should construction begin without a clear, written plan for funding of the complete project. If funding for only a part of the project is available, and funding is pending for the rest of the project's completion, then the start of project construction should be delayed until full funding is appropriated for project completion. The fault of a half-finished project is that of the design engineer and not the contractor.

When both the construction permit and construction funding are in place, the work can then be sent out for bids to those contractors with experience in this type of work. The design engineer is responsible for knowing those contractors with experience to complete the work in a competent manner, on time, and within the budget.

1.3.6 *Monitoring (construction and impacts)*

One of the truly gratifying aspects of coastal engineering design is being physically present in the field to monitor the construction. And, if needed, the designer may participate in the monitoring of the beach and dune property adjacent to the construction site. After days, weeks, and months in the design office, the opportunity to spend time in the field is the most gratifying to many design engineers.

1.3.6.1 *Construction monitoring*

The contractor will have professionals responsible for the supervision of the construction. That is not the responsibility of the design engineer. But when questions arise in the field during construction for any reason that is related to what appears on the construction drawings (Plans) or what is said in the construction specifications (Specs), then the designer is consulted. The designer is responsible for reviewing the problem, determining the reason for the problem, and solving the problem to the satisfaction of all parties involved. When the solution requires more funding, written documentation is needed.

The designer will also want to document the construction with photographs for use at later times. Figure 1.9 shows the hurricane protection project seawall during construction at Virginia Beach, VA. The steel sheet piles (Z-section) and steel reinforcement bars are clearly visible and the worker gives a scale to the dimensions. The cross-section shown in Figure 1.9 is that drawn in the upper left-hand corner of the Corps of Engineers drawing (Figure 1.8). Things not shown in Figure 1.9 include the wooden form

Figure 1.9. Hurricane protection project seawall under construction at Virginia Beach VA (author's photo).

box placed over the steel and the completed concrete seawall that will be partially buried when the sand trench is filled at a later time. The designer should visit the construction site for capturing photos on days with good, warm weather.

In some cases, the construction drawings must be modified to show the actual constructed project and these modified drawings must be included in the final reports.

1.3.6.2 *Impact monitoring*

All coastal structures have limited dimensions with project boundaries that terminate somewhere. In some cases, the boundary is water as a tidal inlet and the structure is a "terminal" groin. In most cases, land boundaries are present at both ends with ownership at all levels of governments, or commercial, or private property interests. The Permit application must define the adjacent property owners, and the ensuring application should be reviewed and must be approved by the adjacent property owners for issuing of the Permit. In some cases, the environmental Permit will be issued on the condition that the down-drift impacts of the construction must be monitored. The designer must be aware of this possibility and will be involved to some degree. Further details are presented in Chapter 10.

1.3.7 *Storm performance (the media)*

Coastal storms may occur before, during, or after the construction is complete. The designer is the most qualified person to answer questions regarding the storms' effect on the structure's design and performance. For example, if a storm occurs during the design process, a new bathymetric survey may be needed. Greater water depths may occur at the planned structure location and may require larger stones and more volume of materials to increase the costs of the nearshore breakwaters. The designer is then responsible for the modifications of the Plans and Specs and finding additional funds for covering the construction costs.

Most of the time, the media is interested in the performance of the structure after the construction is complete and the first storm of the season impacts the storm damage mitigation structure.

1.3.7.1 *No one pays attention when everything works*

Everything works as planned, designed, and constructed. No damage is experienced by the structure. And most importantly, all the infrastructure (roads, public facilities, etc.), commercial buildings, and private property that were at risk are saved from considerable storm damage. Little is said about the project to the client (often the general public) and to the media. The thousands and millions of dollars in damages that were saved by the project are almost never mentioned by the media. The coastal engineer responsible for the design is the most knowledgeable person in this regard and should report the dollar savings to the client and the media.

It is unfortunate that this rarely occurs in practice. Consequently, the client (private, public officials, or general public) is not routinely made aware of the value of the project. Because the vast majority of storms fall under this category, the damages in dollars saved add up and are considerable, and the repeated reminder to the public in the media would demonstrate that the money invested was well spent. The benefits of the project are delivered to the client (public) by evidence of the dollar saving after each storm event.

1.3.7.2 *The media pays attention when something appears not to work*

The storm exceeds the level of damage mitigation attainable under the design constraints of the project. Usually, the key constraint is cost. Some aspect of the structure was "damaged" and some level of damage occurred to the infrastructure, commercial, and private property at risk. The design organization and designer are called upon to explain why the project "failed". No matter what is said regarding all the constraints that resulted in the final design (e.g. the key constraint of available funding), the public perception is failure. The public expects "shore protection" and does not yet understand "storm-damage mitigation".

Failure is now an anticipated and expected occurrence in the coastal structure's design. The consequences of the functional and structural failures are considered in the Feasibility Study of Alternatives (see Section 1.2.6). Chapters 4 and 8 review the details on what Kamphuis (2010) calls "Contemporary Design". The key aspects are the needed change of philosophy of the designer and his/her communication of the risks of failure to the client and the media. Chapter 11 is devoted to these challenging new responsibilities of the designer in the face of accelerating sea level rise when more "failures" are anticipated.

A start can be made now by simply reporting the "successes" of the project (dollar damages saved) to the client and the media each time a storm occurs. When the rare, extreme storm event occurs to cause "failure", the client and media will have some perspective on the event. We can only blame ourselves as "coastal designers" for not publicizing our many "successes" and wondering why we are always blamed for the "failures". We need to tell both sides of the story.

1.3.8 ***Documentation of experience***

Preparation of a detailed, technical, final report with construction plans and specifications is required to fully document the work. In some cases, the "as built" drawings will differ from the original construction plan drawing due to the modifications made during construction. The "as built" drawings must be kept on file.

The final report then forms the basis for an oral, summary presentation to the client, the media, and to local engineering club meetings, seminars, and association events. Presentations at civic club meetings are also important to explain the design and its working especially after storm events. Preparation of these summary materials is also needed for presentations to show potential, future clients to demonstrate the experience of the designer and the design organization in solving similar problems in coastal engineering.

Preparation for engineering conferences (abstracts, oral presentations, and proceedings papers) takes additional time that should be considered of value to the designer and the design organization, and this should not be considered as "extra work". The International Conference of Coastal Engineering (ICCE) is recognized by coastal engineers throughout the world and has become a valuable resource for finding the latest information on coastal hazard mitigation design.

Publication of unique, challenging, and innovative solutions to coastal design problems should be considered for engineering journals. The American Society of Civil Engineers (ASCE) Coastal, Ocean, Ports, and Rivers Institute (COPRI) and the international journals should be considered. Both are always interested in publication of more engineering "design" papers because they are mostly filled with "research" results.

As an example, consider the contribution of civil engineer Sam Yonge to fully document his findings at Jamestown, Virginia (Yonge, 1926) as summarized in Section 1.1.1 and as shown in Figure 1.3. Not only did Yonge design a seawall to protect Jamestown from further erosion, but he also published his work. As a result, in 1990, archeological investigations proved his fort location and have since uncovered thousands of historical treasures at the birth place of the US. Some would argue that his publication (Yonge, 1926) is as important as his seawall design.

All the publication and presentation efforts cited in this chapter may be recorded with MS Office (Excel, Word, Powerpoint) software. These "desktop" publishing tools have significantly increased the productivity of office work to document the design. The coastal engineering design profession needs more published examples of projects and these tools greatly facilitate the effort.

Finally, the design engineer should take their partner or spouse and kids to see the finished project. The parents and possibly grandparents should see it too. They have heard about it and lived through it and deserve to see the final results.

1.4 Summary of Project Design

The term "designer" may indicate a single person in a small organization or even a number of individuals working together and sharing the work. But only one person will be the leader and he/she should be the coastal engineer.

See Section 11.3.1 of Chapter 11 for the impacts of sea-level rise on Chapter 1.

References

APVA (2000) Association for the Preservation of Virginia Antiquities (https://preservtionvirginia.org).

Arromanches (1997) *The Story of D-Day and the Mulberry Harbours*, Musee Du Debarquement, Arromanches-les-Bains, France (http://www.arromanches-museum.com).

Basco, D. R. (2014a) "Study of Requirements for Point Chesapeake Property in Flood Zone X: Reanalysis of Wave Overtopping Rates With Re-curved Parapet at Seawall Crest," Final Report for MS&A, Inc, Virginia Beach, VA, October.

Basco, D. R. (2014b) "The Incalculable Benefits of Hardening the Shoreline at Historic Jamestown, Virginia," *Shore & Beach,* Vol. 82, No.3, pp. 3–10.

CEM (2000) *Coastal Engineering Manual*, US Army Corps of Engineers, Washington, DC.

CEM (2006) *Coastal Engineering Manual*, US Army Corps of Engineers, Washington, DC.

Corps (1995) "Hurricane Protection Project for Virginia Beach, Virginia," Construction Drawings, Corps of Engineers, Norfolk District Office, Norfolk, VA.

De Graauw, A. (2018) "A Catalogue of Ancient Ports and Harbours," Four Volumes (http://www.ancientportsantiques.com/docs-pdf/).

Hohlfelder, R. L., W. Curtsinger, and J. R. Teringo (1987) "Herod the Great's City on the Sea: Caesarea Maritima," *National Geographic*, February.

Hughes, S. A. (1993) *Physical Models and Laboratory Techniques in Coastal Engineering*, World Scientific, Singapore.

Jamestown Rediscovery (2019) Archaeological Project, Association for the Preservation of Virginia Antiquities, (https://historicjamestowne.org).

Kamphuis, J. W. (2010) *Introduction to Coastal Engineering and Management*, 2nd edn., World Scientific Press, Singapore.

Kamphuis, J. W. (2012) "Coastal Engineering Education and Coastal Models," *Proceedings, of the 33rd International Conference Coastal Engineering*, ASCE.

Kelso, W., N. M. Luccketti, and B. A. Staube (1994) "Jamestown Resiscovery-1994 Field Report," Association for the Preservation of Virginia Antiquities, APVA Archives, College of William and Mary, Williamsburg, VA.
Kraus, N. C. (Ed.) (1996) *History and Heritage of Coastal Engineering*, American Society of Civil Engineers, New York.
MS&A (2011) Miller-Stephenson & Associates, Civil Engineering/Architecture Consulting firm, Virginia Beach, Virginia, USA (http://www.msaonline.com/).
NSPE (2007) National Society of Professional Engineers, Washington, DC.
O'Brien, M. P., and J. W. Johnson (1947) "Wartime Research on Waves and Surf," *The Military Engineer*, Vol. 39, pp. 239–242.
Reeve, D., A. Chadwick, and C. Fleming (2012) *Coastal Engineering Processes, Theory and Design Practice*, 2nd edn., Spon Press, London.
Wiegel, R. L. and T. Saville (1996) "History of Coastal Engineering in the USA," in N.C. Kraus (Ed.), *History and Heritage of Coastal Engineering*, American Society of Civil Engineers, New York, pp. 513–600.
Yonge, S. (1926) "The Site of Old 'JamesTowne', 1607–1698," Association for the Preservation of Virginia Antiquities, Tercentenary Edition, APVA Archives, College of William and Mary, Williamsburg, VA.

Chapter 2

Design Criteria

2.1 Introduction

2.1.1 *Description of the site*

Coasts may be loosely classified as two types, namely, consolidated (rocky, firm, stable material) or unconsolidated (sandy, erosion/depositional, unstable material). See Chapter 10, Figure 10.16 from Shephard (1976) for a coastal classification system identifying 11 types of shorelines found around the worlds' coasts and in open oceans. Adequate analysis and characterization of the coastal site is important to minimize the potential for failure of the design. The coastal environment (sediment processes, waves, underlying geology, etc.) varies spatially and, most importantly, in time at the site. For example, a site experiencing long-term erosion with a protective revetment will continue to erode and lower the seaward profile such that wave heights for stable rubble-mound design must be based on water depths estimated at the end of the project, and not today's bathymetric conditions.

The project area, boundaries, and adjacent property that are potentially impacted by the project should be identified and delineated on drawings with horizontal control (Lat., Lon.; State Plane Coordinates, etc.). Topographic, bathymetric, and cadastral information (land ownership, value, etc.) from maps can now be combined using computer-aided drafting tools or graphical information systems (GIS). All maps should have compass directions, vertical datum, and dates clearly indicated especially when shoreline positions at the land/water interface are shown. Sources of map data vary widely by country. The online accessible software product Google Earth includes historic photos on a timeline that provides a valuable reference source for the world's shorelines.

All components of the system (physical, biological, cultural, environmental, economic, etc.) are important to characterize the project site and define the design constraints.

2.1.2 *Design criteria*

The design criteria are the physical conditions (water levels, wave characteristics, soil properties, etc.) and the other "rules" stemming from the design constraints that govern the design at the site. In this chapter, we focus on the physical conditions. However, this is not an introductory textbook on the basics of tidal physics or wave mechanics or sediment transport. The primary focus is the introduction (review) of all the physical variables that are employed for the functional design (Chapter 5) and structural design (Chapter 6) of coastal structures for beaches and dunes (Chapter 7), and for pile-supported structures in Chapter 9.

2.2 Water Levels

2.2.1 *Introduction*

Water depth is arguably the most important independent variable in the coastal structures' design. By definition, the structure is at or near the land–water interface (the coast) where shallow water depths limit the wave height due to wave breaking. We begin with the simple fact that "*Coastal design is primarily a function of water levels*" (Kamphuis, 2010, p. 145). It all starts with the daily predicted tidal variation.

2.2.2 *Tides*

A daily (diurnal) or twice daily (semi-diurnal) variation of the water levels (tide) is experienced at most coasts. The predicted levels depend on many astronomical and physical factors (see any physical oceanography text, e.g. Neumann and Pierson, 1966). A tidal range is microtidal (0–2 m) meso-tidal (2–4 m) or macrotidal (>4 m). The tidal currents are also very complex. And, because tidal amplitudes are normally measured with gauges located on piers and in confined places, the tidal currents in the region are reported for different locations than at the tide gauges.

For navigational interests and engineering purposes, the predicted daily tidal variations are available from government agencies. The nearest tide gauge station to the project site is located and its horizontal position is determined relative to the project site. The measured tidal variations at the gauge must then be related to the expected tidal variations at the site. For this purpose, a clear understanding of the vertical datum must be first established.

2.2.2.1 *Vertical datum*

Table 2.1 presents the vertical tidal elevations as published by the US government (National Ocean Service (NOS)) for the Sewells Point, Virginia, Tide Gauge (Station 8638610). The elevations are relative to two vertical datums and can be obtained in both the Imperial (English) and Metric systems. One vertical datum (0.00 elevation) is water-based for navigation interests and is the mean lower, low-water (MLLW) level as averaged over the 18.6-year interval (tidal epoch) for the 1983–2001 time period. The second vertical datum (0.00) is the land-based, North American Vertical Datum of 1988 (NAVD88). Both MLLW and NAVD88 vertical datums are employed by US coastal engineers. Note that the NAVD88 land-based datum lies 0.501 m (1.65 ft) above the MLLW ocean-based datum permitting the ready transfer between both datums.

Figure 2.1 displays in graphical form, the various water-level definitions relative to both datums for the Sewells Pt gauge (Table 2.1). The mean tidal range (MTR) (MHW–MLW) is 0.740 m (2.43 ft). Tidal water levels displayed to scale have been found to aid in the physical appreciation and understanding of the many water-level definitions of the tides. These graphs to scale should appear on the drawings. There is nothing worse than a project drawing with no mention of the vertical datum for elevations shown for the structure. For example, a drawing showing a crest elevation of 5.0 m for a seawall has no meaning without reference to a vertical datum.

The column labeled "Value (STND)" in Table 2.1 are elevations above the station datum (STND). The station datum is unique to a tide station. It is set as a permanent, fixed elevation that does not change over time. However, the station datum must be periodically checked for long-term land movements (subsidence or rebound) through advanced surveying techniques. This is a function of the government agency responsible for tidal measurement.

2.2.2.2 *Tidal variations at the project site*

The information displayed in Table 2.1 is for a location with tide gauge for continuous, recorded measurement of the actual water levels as displayed in Figure 2.2 and discussed in Section 2.2.3. At other locations, tidal "corrections" of amplitude and phase (time variations of MHW and MLW values) are published in the US as related to that predicted for the primary measurement location. For the feasibility study, these tidal "corrections" can

Table 2.1. Elevations on station datum, MLLW datum, and NAVD88 datum for Metric and Imperial units (Sewell Pt., Virginia, Station 8638610, 1983–2001 Epoch, NOS (NOAA).

		Metric			Imperial		
		Value (STND)	Elevation (MLLW)	Elevation (NAVD88)	Value (STND)	Elevation (MLLW)	Elevation (NAVD88)
Description	Datum	meters	meters	meters	feet	feet	feet
Mean higher-high water	MHHW	2.176	0.840	0.339	7.14	2.76	1.11
Mean high water	MHW	2.114	0.778	0.277	6.94	2.56	0.91
North American vertical datum of 1988	NAVD88	1.837	0.501	0.00	6.03	1.65	0.00
Mean diurnal tide level	DTL	1.756	0.420	−0.081	5.76	1.38	−0.27
Mean seal level	MSL	1.748	0.412	−0.089	5.73	1.35	−0.30
Mean tide level	MTL	1.744	0.408	−0.093	5.72	1.34	−0.31
Mean low water	MLW	1.374	0.038	−0.463	4.51	0.13	−1.52
Mean lower-low water	MLLW	1.336	0.00	−0.501	4.38	0.00	−1.65
Station datum	STND	0			0		

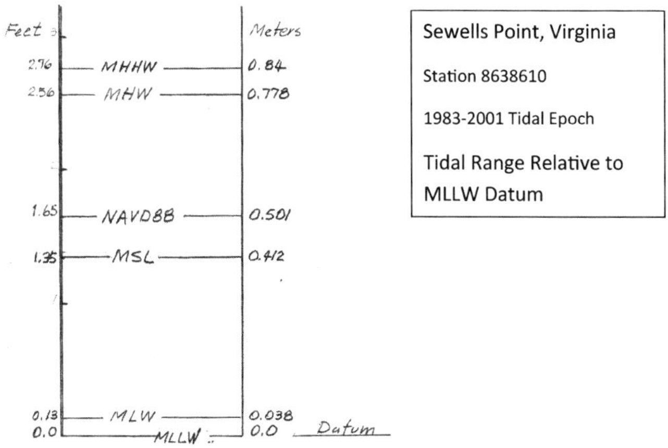

Figure 2.1. Graphic display of Sewells Pt., VA tidal data relative to MLLW datum.

be employed to estimate the predicted tidal conditions at the project site by simple linear interpolation. For the final design, three-dimensional (3D), tidal hydrodynamic, numerical models are often employed to refine the analysis at the site. In this case, the tide-gauge information and the tidal "corrections" are used to "calibrate" the numerical model.

The best way to verify the interpolated or modeled tides at the site is by field measurement. Expensive instruments can be employed. Or, a simple board (staff gauge) with painted water-level intervals (3 cm or 0.1 ft markings) clearly visible for reading can be installed at the site. The elevation of 0.0 on the board after installation must be established by survey. Then, the water-level elevation is read from the board at 1–1/2 hour intervals over 13 hours (semi-diurnal tidal period is about 12.42 hours). For relatively calm days with little winds, the measured tidal amplitude and phase can then be employed to determine the actual tidal "corrections" at the site by relating to the predicted tides at the gauge for the same time interval as the measurements. This procedure can be repeated to ensure accuracy of the "corrections".

The "corrections" for the predicted tides may be different under windy and storm conditions. The only way to truly verify these modified "corrections" is to again compare the tide-gauge records with the actual measurements at the site. This may not always be possible. For design purposes under extreme storm conditions, the use of calibrated and verified,

3D numerical models is recommended. Field data of high-water elevations during the storm event are needed for calibration.

2.2.3 *Storm-surge hydrographs*

Low-pressure front storms and hurricanes (typhoons) with strong winds create storm surges that can significantly increase the local water depths during the storm event. The measured elevations at the gauge will be higher than predicted. The surface wind shear stress (caused by the strong pressure gradients from the high pressure ridge to the low-pressure trough or in the circulating wind field) is the most important factor. We herein define storm surge S, as the vertical elevation of the observed (measured) water level minus the predicted tidal elevation. The storm-surge hydrograph is the storm surge variation over time for the duration of the storm event, $S(t)$. The units of S are simply meters or feet.

Figure 2.2 displays the predicted (blue), measured (red) and difference (measured — predicted, green) tidal record at the Sewells Pt. Virginia tide gauge for Hurricane Isabel, at September 18, 2003. The highest elevation recorded during Hurricane Isabel was 2.38 m (7.82 ft) relative to the MLLW datum and second highest water level recorded since the records began in 1927. The green curve displays the storm surge hydrograph. The rising slope of the storm surge was on the order of 7–8 cm per hour (3 inches per hour)

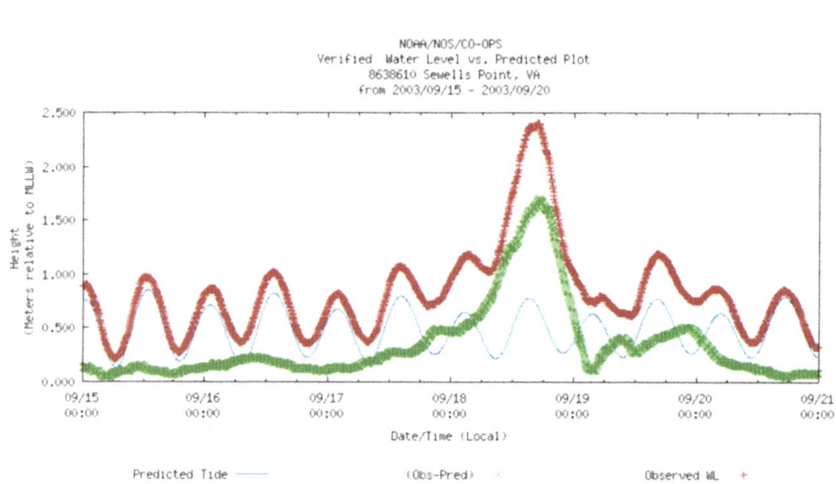

Figure 2.2. Measured tide gauge record at Sewell Pt. VA for Hurricane Isabel (September 18, 2003).

for Hurricane Isabel. Storm surge is not a "wall of water" slamming into the coast. Unfortunately, the media (newsprint and weather channels) continues to mislead the general public in this regard.

When the storm starts (and ends) is subject to debate. The initiation of a coastal storm based on water level is primarily related to the "consequences" of the storm event. In urban coastal areas throughout the world, the threshold, coastal "storm" water levels are often defined by impacts to transportation infrastructure (roads, bridges, railway lines, etc.). When roads are flooded deep enough to impede traffic, a coastal "storm" is said to exist. Low-lying transportation corridors are constructed above the normal, high-tide levels plus some additional elevation at a given frequency of exceedance that is based on local economic and environmental consequences. See Basco and Mahmoudpour (2012) for a detailed literature review on the definition of a coastal storm.

Acqua alta (high water) events in the Piazza San Marco in Venice, Italy, are a good example. Whenever the water levels in the Piazza impede the normal foot traffic of citizens and tourists, a coastal "storm" is felt to be present in the Adriatic Sea. The elevated water levels may also be caused by atmospheric (barometric) pressure gradients and long-period wave oscillations (seiche). They will be part of the elevated water level measured at the tide gauge. De Boni *et al.* (1993) summarized what happened at the start of the International Conference on Coastal Engineering (ICCE) October 4–5, 1992 in Venice. The total storm surge in Piazza San Marco was estimated at 75–78 cm and coincided exactly with two successive high tides about 23 hours apart. The total water levels were measured at 124 cm and 112 cm for these two high-tide events. Kamphuis (2010) showed that the barometric pressure gradients accounted for only about 10% of the total surge with 90% due to the wind stress. And, that 23-hour oscillation in the tidal peaks coincided with the natural period of oscillation (seiche) of the Adriatic Sea. See Section 6.5.3 of Chapter 6 for the complete details of this acqua alta event and its analysis by Kamphuis (2010).

The 10 highest flooding elevations recorded in Venice beginning from 1900 ranged from 140 to 194 cm (which shows that the above-mentioned example of flooding is only at a moderate level). The frequency of flooding events is increasing rapidly due to both sea-level rise and the sinking of Venice (see Chapter 11). The Italian response has been the MOSES project as illustrated in Figure 1.4. A key aspect of the design was the development of the water level versus recurrence interval (probability of exceedance) curve for extreme water-level events.

2.2.4 *Extreme water levels*

There are several methods that have been historically applied for coastal flood frequency analysis. They are as follows:

(1) estimates based on historical water-elevation data recorded at tide gauges;
(2) the empirical simulation technique (EST);
(3) the joint probability method (JPM);
(4) the Monte Carlo simulation method.

For the first method, the length of record (years) for the historical data must be adequate (>75 years). The storm events may be caused by both tropical and extra-tropical storms. Maximum water level each year is found and the data is ordered (ranked) in increasing order from the smallest ($m = 1$) to the largest ($m = N$). The cumulative probability of non-exceedance, P, is then first determined. Secondly, a line is fitted to the ranked values by some extreme value fitting procedure. Finally, the fitted line may be extrapolated to estimate the probability of extreme values.

The formula to calculate the cumulative probability of non-exceedance, P "plotting positions" (Weibull, 1951) is

$$P = \frac{m}{N+1}. \tag{2.1}$$

The recurrence interval (also called the return period) T_r is then found from

$$T_r = \frac{1}{Q} = \frac{1}{1-P}, \tag{2.2}$$

where Q is the probability of exceedance, $1 - P$.

For example, Figure 2.3 displays the water elevations (above the National Geodetic Vertical Datum, 1929[1]) versus the probability of exceedance, Q, for the Sewells Point, VA, tide gauge in Norfolk for the period 1928–1978 (Corps of Engineers, 1988). The plotting positions are the black dots and the solid line is one possible curve-fit formula. But the curved line is not easy to extrapolate for extreme events.

As mentioned before, the elevation of Hurricane Isabel (September 18, 2003) was 2.38 m (7.82 ft) and it is the second highest water level recorded since the records began in 1927.

[1] *Note*: National Geodetic Vertical Datum, NGVD 1929 was based on the mean sea level as determined in 1929. It has been replaced by the MLLW and NAVD88 vertical datums shown in Table 2.1 and Figure 2.1.

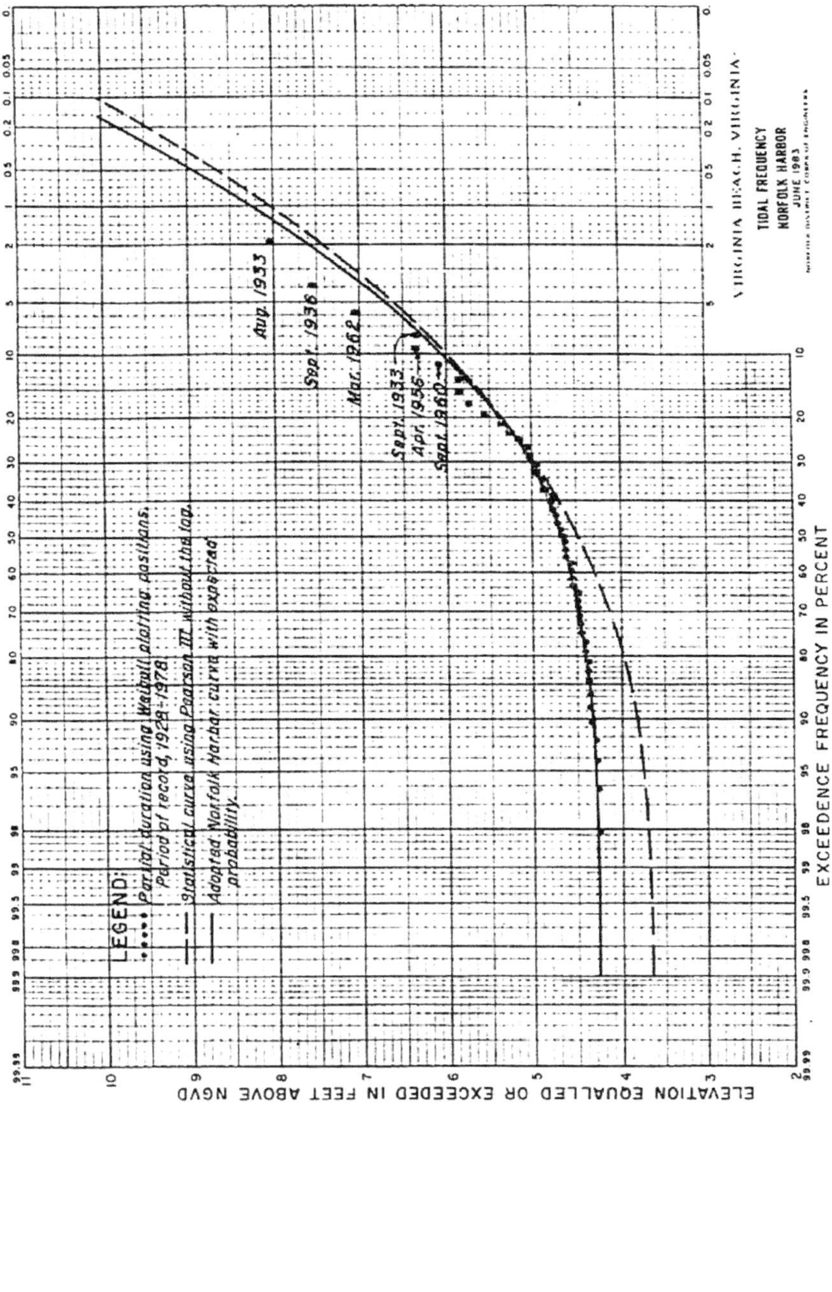

Figure 2.3. Exceedance probability curve for water levels, Norfolk, Virginia, tide gauge at Sewells Pt. for 51 years data set (adapted from Corps of Engineers, 1988).

The main criticism of this "ordered data" method is that only one extreme value for any one year is employed. In many years, the second highest annual value is far greater than some lower extreme annual events. The "ordered data" set can be extended by using all the extreme water levels above some peak level. The peak-over-threshold (POT) method, the extreme value probability distributions, and curve identical to fitting procedures are those employed for the analysis of extreme wave heights and will be discussed in detail in Section 2.3.8.

The second method, the EST, has been developed to generate a synthetic sequence of storms. The historical storm record is extended by assuming that each storm could have occurred at various combinations of tidal phase and tidal range. For example, if Hurricane Isabel (September 18, 2003) had occurred 10 days later during Spring Tide conditions, the water-surface elevation measured at the tide gauge would have been 0.26 m (0.85 ft) higher and the highest ever recorded since 1927. See Scheffner *et al.* (1999) for complete details about the EST method. The EST method can be applied for both tropical and extra-tropical storms.

The results of the extreme analysis by the "ordered data" method and extended by the EST are only applicable at the tide-gauge locations where the data was taken or extended. For design at any coastal location, flood levels must be determined by numerical modeling of the physical processes by simulating numbers of storms over a long period of record and then deriving flood statistics from the results. The EST provides the "training set" of storms to drive the numerical model. The JPM and Monte Carlo Methods also require the use of the numerical model.

The JPM has been primarily applied for tropical storms (hurricanes) since the joint, wind-variable probabilities within the hurricane (central pressure, storm size, storm speed, etc.) that produce the storm surge along the coast are reasonably well known. A modified JPM–OS method has been developed (Toro *et al.*, 2010). The JPM for extra-tropical storms produced by low-pressure meteorological events is related to the response parameters (surge, wave height, etc.) and not the wind field.

The Monte Carlo method is similar to the JPM and relies on the estimates of the probability distributions established for the parameters needed to characterize a storm. But a random-number generator is employed to select the values for each parameter. Each storm selected by this Monte Carlo procedure is equivalent to simulating a long period of history, and many thousands of storms can be generated with high-speed computers to improve the extrapolations to extreme storm events.

In the US, the Federal Emergency Management Agency (FEMA) and their mapping partners (consulting firms) provide flood insurance rate maps (FIRMs) for general use by the cities and general public. The exceedance probability, Q, is at the 1% level (100-year recurrence interval) within the flooding-level boundaries displayed on the maps. The US Federal government provides flood insurance at actuarial rates to property owners within the 1% level. The flood insurance premiums pay for the flooding damages that occur each year. All new constructions within the 1% chance flood zone must be above the 1% chance flood base level plus the additional elevation for waves and wave setup.

See FEMA (2007) for an excellent summary of these four methods and a review of procedures for developing the FIRMs. All the three methods (EST, JPM, and Monte Carlo) can be employed.

For coastal structures designs subjected to tropical storms, Resio (2007), Agbley and Basco (2008), and Toro *et al.* (2010) recommend the JPM method.

2.2.5 *Climate change and sea-level rise*

Long-term water-level changes (decades of years) on the order of 2 m have occurred on the Great Lakes in the US and Canada. Periods of coastal damage to structures are related to higher waves coming closer to shore during higher water levels lasting for many months or even many years. These fresh-water lake experiences provide evidence of what could happen in a future of climate change and sea-level rise in the worlds' oceans.

Tidal amplitudes will not be altered. However, since the plan-form area controls the volume of water in the tidal prism, greater plan-form areas in bays and estuaries will produce larger tidal prisms and stronger currents and tidal inlets.

The position of the probability of exceedance versus water elevation curve (Figure 2.3) will shift laterally, i.e. a given water-level elevation will occur more frequently. For example, the 1% chance water elevation may increase to the 2–3% chance level. The FEMA FIRMs will need to be upgraded more frequently.

For coastal structures' design, the storm frequency increase and deeper water levels will bring larger breaking waves to impact the structures. Chapter 11 will have a full discussion of the impacts on water-level change as the potential for an accelerated sea-level rise due to climate change is experienced in the future.

2.2.6 *Water levels for design*

The complete range of water levels for the full range of probabilities of exceedance (and beyond) as shown in Figure 2.3 are needed and applied in the probabilistic design of coastal-hazard-mitigation alternatives. The risk-based approach employs the entire exceedance probability curve and it is the modern approach as discussed fully in Chapters 3 and 7. Some probability of exceedance will cause failure. To plan for failure is now the correct philosophy of coastal engineering design.

Some level of the recurrence interval (e.g. the 100-year level) is often still selected in the deterministic design method. The deterministic method using the 100-year flood is no longer valid. What are the consequences of flooding when levels exceed the 100-year level? Is the analysis, methods, and record length sufficient to have confidence in the 100-year flood level? Why is the 100-year flood level selected? What is the magic in the 100-year flood level? For a 30-year period (average home mortgage period), the probability that a 1% water level will be experienced is 26%. See Chapter 4 for a full discussion of probabilistic design.

Finally, the wave conditions at the structure can produce wave setup $\bar{\eta}$ which is the time-averaged increase in water level due to the presence of waves at the coast. Local beach slope is important. Dean and Walton (2009) suggests using $\bar{\eta}/Hs = 0.191 \pm 0.100$ as a useful guide for the maximum setup at the coast. For example, if the offshore, significant wave height, Hs is 2 m, the maximum wave setup would be on the order of 0.4 m. This is a significant increase in mean water level that cannot be ignored. See Chapter 5 for a detailed review of wave setup.

At a minimum, the local tidal data can be analyzed and used to estimate the water level, and probability of exceedance distribution curve (e.g. Figure 2.3) for application at the project site. The probability of exceedance curve is the basic water-level information needed as design criteria.

2.3 Water Waves

2.3.1 *Introduction*

The water-surface variation, η, in space (x, y) and time (t) of irregular, ocean waves is complex as seen in Figure 2.4. The surface varies greatly in different directions and at each point (x, y) in time. Herein, we simply summarize the many science and engineering simplifications of the dimensions, definitions, and statistical distributions that have evolved over time to define the wave height, wave period, and wave direction of irregular

Figure 2.4. Photo of irregular water surface variation in ocean waves (courtesy of Natural News, https://www.naturalnews.com).

waves for use in the empirically formulated equations for engineering design.

This is not a text for the introduction of water-wave mechanics, so only the results of the theory are given. See Dean and Dalrymple (1984) for all the derivation details of linear and nonlinear wave theory.

2.3.2 *Regular waves*

2.3.2.1 *Linear wave theory*

The first and most fundamental wave theory is to consider one space dimension, x, and only one wave frequency, f (wave period, $T = 1/f$)), as needed for regular (sinusoidal waves) in linear wave theory. These are for progressive (propagating) waves. Standing waves are discussed in Chapter 6.

Figure 2.5 displays the wave surface variation, η, with amplitude, a, so that $2a$ equals the wave height, H, as the vertical distance from crest to trough. If Figure 2.5 is a snap shot in the x-direction, the variation of $\eta(x)$ repeats over wave length, L, as a sinusoidal, normalized function $2\pi(x/L)$. If the horizontal axis in Figure 2.5 is time, t, the variation of $\eta(t)$ repeats

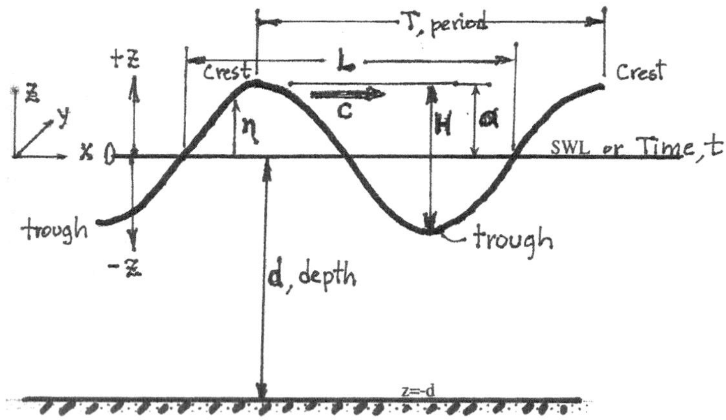

Figure 2.5. Fundamental definitions of progressive, sinusoidal wave.

over time period, T, as a sinusoidal, normalized function $2\pi(t/T)$. Since the wave varies in both the x-distance and in time, t, and using the cosine function (cos),

$$\eta(x,t) = \frac{H}{2} cos(\kappa x - \omega t), \tag{2.3}$$

where $\kappa = 2\pi/L$, the wave number, and $\omega = 2\pi/T$, the radian wave frequency.

The propagation speed of the progressive wave is the wave celerity, $C = L/T$. The vertical coordinate, z is taken from the still water level (SWL) with $+z$ (positive) and $-z$ (negative) from the SWL surface $z = 0$. Water depth, d, is then relative to the vertical datum of the SWL with the horizontal bottom at $-z = d$.

The wave length L depends on both the wave period T and the water depth, d

$$L = \frac{g}{2\pi} T^2 \tanh\left(\frac{2\pi d}{L}\right) \tag{2.4}$$

and is one of the most fundamental of all basic equations in water wave mechanics. In deep water with d/L greater than 0.5, the hyperbolic tangent function, tanh, approaches 1.0. The deep-water wave length, L_o is then simply $(g/2\pi)T^2$. In the metric system, $L_o = 1.56\ T^2$, and in the imperial unit system, $L_o = 5.12\ T^2$. A 10-s wave is $L_o = 156$ m (512 ft) in length in deep water. As the wave moves into shallower coastal waters, the wave length, L, decreases in length for a given wave period, T.

The equation for L has an implicit meaning, and it cannot be directly (explicitly) solved, but only by trial and error, tables or approximation methods must be employed. For engineering purposes and spreadsheet solution, the approximate formula (Hunt, 1979) has been found to be most useful.

$$\frac{C^2}{gd} = [y_o + (1 + 0.6522y_o + 0.4622(y_o)^2 + 0.0864(y_o)^4 + 0.0675(y_o)^5)^{-1}]^{-1}, \tag{2.5}$$

where $y_o = 2\pi d/L_o$, and $L_o = (g/2\pi)T^2$ is the deep-water wave length.

For a given wave period, T and water depth, d, the wave celerity C is found and then the wave length, $L = CT$. For example, for a 10-s wave period in 5 m water depth (16.4 ft), the Hunt formula gives $L = 67.7$ m (220 ft). It is adequate for engineering purposes because our knowledge of the wave period to employ is also approximate. If the wave period was 10.1 s, $L = 68.4$ m (224 ft).

The online supplementary materials accompanying this book contains a subgroup of spreadsheets called Water Wave Mechanics. One is called "Wave Lengths Hunts Formula" as shown in the example Figure 2.6. (not interactive). Simply enter the wave period, T (seconds), and the water depth, d (meters), and find the relative water depth ratio, d/L_o. Then insert this value in Column 1 which contains generic values from 1.0 down to 0.002 for the d/L_o ratio. Insert a blank new row in the interpolated position. The spreadsheet contains a second sheet for the imperial unit system.

The second important formula is for the wave-energy density, E (i.e. the wave energy per unit surface area of the ocean surface),

$$E = \frac{1}{8}\rho g H^2 \tag{2.6}$$

with H, the wave height, and ρ, the mass density of the water. For seawater, ρ equals about 1,030 kg/m^3 (64 lb/ft^3). The energy over one wave length (per unit width) is then simply E times L.

Ocean wave energy propagates over the ocean surface at a speed that is called the group celerity, C_g. This is called the wave energy flux or wave power, P, and is found from the third useful formula

$$P = EC_g. \tag{2.7}$$

The group celerity, C_g, is related to the wave celerity, C, and determined from

$$C_g = nC, \tag{2.8}$$

Water Wave Mechanics

Wave Lengths Versus Water Depth for Wave Periods
Group Parameter and Group Celerity

Small Amplitude (Linear) Wave Theory
Hunt (1979) Formula for C

Metric System	
Wave Period, *T* sec	10.0
Gravity Constant, g,	9.81
Deep Water Wave Length, *Lo*, *m*	156.13
Deep Water Celerity, *Co* m/s	15.61
Wave Number, *ko*,	0.201215
Water Depth of Interest, *d* m	**5.0**
Relative Water Depth, *d/Lo*	**0.0320**

Insert the Relative Water Depth Ratio, *d/Lo* in column 1 by first inserting a blank row at the interpolated position

WAVE KINEMATICS Table 2.2

d/Lo	*d* meters	*yo* (Hunt)	*C* m/sec	*L* m	*d/L*	*kd*	tanh *kd*	*n*	*Cg* m/s	Remarks
1.0000	156.13	6.283	15.61	156.1	1.0001	6.2838	0.999993034	0.500	7.81	
0.7500	117.10	4.712	15.61	156.1	0.7504	4.71473	0.999839367	0.501	7.81	
0.5000	78.07	3.142	15.55	155.5	0.5022	3.15519	0.996371933	0.511	7.95	Deep water
0.4000	62.45	2.513	15.42	154.2	0.4050	2.54477	0.987753821	0.531	8.19	
0.3000	46.84	1.885	15.02	150.2	0.3119	1.9596	0.961059	0.578	8.68	
0.2000	31.23	1.257	13.87	138.7	0.2252	1.41468	0.888484823	0.668	9.26	
0.1000	15.61	0.628	11.08	110.8	0.1410	0.88571	0.709267472	0.810	8.97	
0.0900	14.05	0.565	10.63	106.3	0.1322	0.83038	0.680680254	0.827	8.80	
0.0800	12.49	0.503	10.14	101.4	0.1232	0.7738	0.649136088	0.845	8.57	
0.0700	10.93	0.440	9.60	96.0	0.1139	0.71555	0.614147354	0.863	8.28	
0.0600	9.37	0.377	8.99	89.9	0.1043	0.65503	0.575046584	0.881	7.92	
0.0500	7.81	0.314	8.29	82.9	0.0941	0.59136	0.530875159	0.900	7.46	Shallow water
0.0400	6.25	0.251	7.50	75.0	0.0833	0.5232	0.480168416	0.919	6.89	
0.0320	5.00	0.201	6.77	67.7	0.0738	0.46397	0.433311814	0.935	6.33	
0.0300	4.68	0.188	6.57	65.7	0.0713	0.44828	0.420481796	0.939	6.16	
0.0200	3.12	0.126	5.42	54.2	0.0576	0.36215	0.347109709	0.959	5.19	
0.0100	1.56	0.063	3.87	38.7	0.0403	0.25338	0.248091698	0.979	3.79	
0.0050	0.78	0.031	2.75	27.5	0.0284	0.17821	0.17634587	0.990	2.72	
0.0020	0.31	0.013	1.75	17.5	0.0179	0.11234	0.111873972	0.996	1.74	

Figure 2.6. Wave lengths calculated using Hunts formula for metric units (not interactive).

where n is the group celerity parameter given by

$$n = \frac{1}{2}\left[1 + \frac{2\kappa d}{\sinh 2\kappa d}\right]. \tag{2.9}$$

The "Wave Lengths Hunts Formula" spreadsheet (Figure 2.6) also includes columns displaying n and C_g from deep water to very shallow water ($d/L_o = 0.002$). For the example above with $T = 10\,\text{s}$ and $d = 5\,\text{m}$, $n = 0.935$ and $C_g = 6.33\,\text{m/s}$ (20.77 ft/s). Before the common use of spreadsheets and Hunts (1979) approximation formula, further simplifications of the wave-length formula in "shallow" water were employed starting about $d/L_o = 0.05(1/20)$ to 0.04 (1/25), where the tanh κd is approximately equal to κd. These are no longer needed.

The final formula introduced here is for the wave-pressure distribution, $p(z, t)$ beneath the progressive wave. Because of streamline curvature in the orbital motion of the water particles beneath the surface, the pressure distribution in the vertical, $p(z)$, is not hydrostatic. And, we formulate the pressure-head relationship $p/(\rho g)$ since civil (coastal) engineers trained in hydraulics are most familiar with columns of liquid "head" to express the vertical pressure distribution in an open channel flow.

$$\frac{p}{\rho g} = -z + K_p \eta, \tag{2.10}$$

where K_p is the pressure-response factor to account for the wave-orbital motion and given by

$$K_p = \frac{\cosh \kappa(z + d)}{\cosh \kappa d}. \tag{2.11}$$

The z term is the hydrostatic pressure-head distribution beneath the $z = 0$ water surface and the negative sign is needed since z is negative beneath the water surface. To facilitate the computation, the Wave Mechanics folder in the online supplementary materials contains the spreadsheet "Pressure Calculation Generic" with an example as illustrated in Figure 2.7 for the pressure head and pressure distribution beneath the wave crest. Examples and the application of the "Pressure Calculation" spreadsheet are detailed in Chapter 6, which discusses the wave forces on seawalls.

Many other formulas for the wave-particle velocities, accelerations, and orbital dimensions could be reviewed here as derived from the linear wave theory. These formulas are summarized in Table 2.2. Their application is found in the Wave Mechanics folder, "Particle Velocity Calculations" spreadsheet. We return to this spreadsheet in Chapter 9 to understand about the wave forces on slender piles.

Water Wave Mechanics **Wave Dynamics**

Pressure Versus Water Depth for Wave Height and Period

Wave Energy Density, Wave Power

Small Amplitude (Linear) Wave Theory

Hunt (1979) Formula for *C* and *L*

Metric System	
Wave Period, *T* sec	10
Wave Height, *H* m	6.6
Water Depth, *d* m	20
Relative Crest Height, *z/d*	0.165
Gravity constant, *g* m2/s	9.81
Mass Density, ρ kg/m^3	1035
Deep Water, *Lo* m	156.13
Wave Length, *L* m	121.19
Relative Depth, *d/L*	0.1650265
Wave Number, *kd*	1.0368922
Deep Water (*kd*)o	0.8048607
Celerity, C *m*/s	12.12
Group Celerity parameter, *n*	0.765
Group Celerity, *Cg*	9.27
Wave Energy Density, *E*	55285.0
Wave Power, *P*	512479.5
Miche Breaking Criterion, *H/L*	0.11

Relative crest height, z/d automatically located in column A, row 24

WAVE DYNAMICS BELOW CREST

z/d	1+z/d	z meters	Kp	Kp η meters	p/(ρ*g) z+ Kpη meters	Static Head h meters	Head of Waves meters	p N/m^2	
0.165	1.165				0	0		0.0	**Crest**
0.0	1	0.00	1	3.30000	3.3000	3.30	0.00	33506.1	**SWL**
–0.1	0.9	–2.00	0.924705	3.051527	5.0515	5.30	–0.25	51289.9	
–0.2	0.8	–4.00	0.859361	2.835892	6.8359	7.30	–0.46	69407.2	
–0.3	0.7	–6.00	0.803265	2.650773	8.6508	9.30	–0.65	87834.3	
–0.4	0.6	–8.00	0.755812	2.494181	10.4942	11.30	–0.81	106551.1	
–0.5	0.5	–10.00	0.716493	2.364428	12.3644	13.30	–0.94	125540.4	
–0.6	0.4	–12.00	0.684885	2.260119	14.2601	15.30	–1.04	144788.0	
–0.7	0.3	–14.00	0.660646	2.180131	16.1801	17.30	–1.12	164282.5	
–0.8	0.2	–16.00	0.643516	2.123604	18.1236	19.30	–1.18	184015.3	
–0.9	0.1	–18.00	0.633312	2.08993	20.0899	21.30	–1.21	203980.1	
–1.0	0	–20.00	0.629923	2.078745	22.0787	23.30	–1.22	224173.2	**Bottom**

Figure 2.7. Pressure distribution beneath crest, example using spreadsheet in the online supplementary materials.

Table 2.2. Formulas from linear wave theory for progressive water waves.

Water surface (m)	$\eta = \frac{H}{2}\cos(kx - \omega t)$
Velocity of propagation (m/s) (Dispersion equation)	$C = \frac{L}{T} = \frac{\omega}{k} = \frac{gT}{2\pi}\tanh kd$
	$= \sqrt{\frac{gL}{2\pi}\tanh kd}$
Wave length (m)	$L = CT = \frac{gT^2}{2\pi}\tanh\ kd$
Pressure (m of water)	$\frac{p}{\rho g} = -z + K_p\eta$
Pressure response factor	$K_p = \frac{\cosh k(z+d)}{\cosh kd}$
Energy density (j/m^2)	$E = \frac{1}{8}\rho g H^2; \quad KE = PE = \frac{E}{2}$
Wave power (w/m)	$P = EC_g$
Group velocity (m/s)	$C_g = nC$
Group velocity parameter	$n = \frac{1}{2}\left[1 + \frac{2kd}{\sinh 2kd}\right]$
Horizontal component of orbital velocity (m/s)	$\mathrm{u} = \frac{\pi H}{T}\frac{\cosh k(z+d)}{\sinh kd}\cos(kx - \omega t)$
Vertical component of orbital velocity (m/s)	$\mathrm{w} = \frac{\pi H}{T}\frac{\sinh k(z+d)}{\sinh kd}\sin(kx - \omega t)$
Horizontal orbital semi-axis (m)	$\mathrm{A} = \frac{H}{2}\frac{\cosh k(z+d)}{\sinh kd}$
Vertical orbital semi-axis (m)	$\mathrm{B} = \frac{H}{2}\frac{\sinh k(z+d)}{\sinh kd}$

Table 2.2 summarizes the important formulas from linear wave theory which are cited and applied throughout this book. The equations for η, L, C, E, C_g, and P should become familiar to the reader so that they can be easily remembered and applied. The others can always be referred to make sure no mistakes are made in their use. The spreadsheets are provided to make them immediately useful in coastal engineering design.

2.3.2.2 *Nonlinear wave theories*

As waves move into shallow water, the surface at the crest becomes more peaked and the trough region flattens as schematically depicted in Figure 2.8. Nonlinear water-wave theories provide a better fit to the free

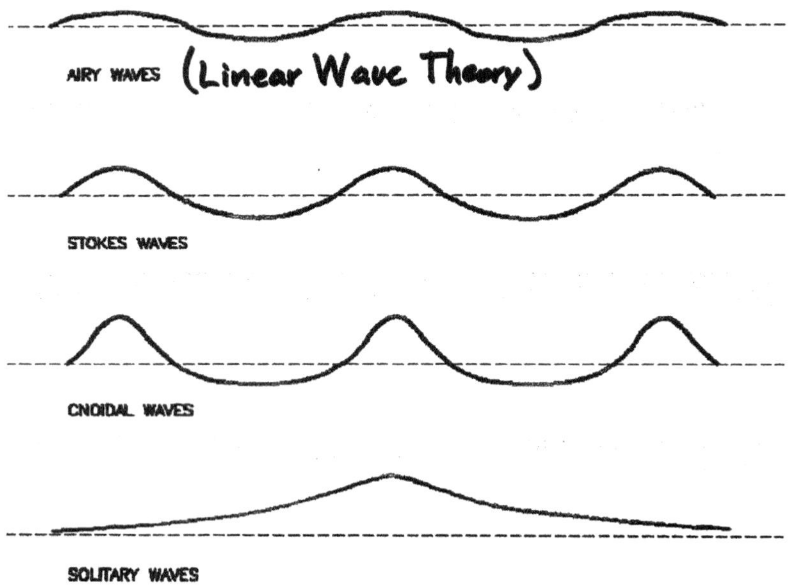

Figure 2.8. Schematic of water surface shape for linear (top) and three nonlinear wave theories.

surface variation in time and space. And, they are far more exact in the velocity, acceleration, and orbital motions that produce wave forces on slender piles. We shall return to the nonlinear wave theory in Chapter 9.

In summary, the basic formulas for regular, linear wave theory have been presented and some spreadsheets have been identified for use in the design. However, real ocean waves are irregular in space and time so that statistical means must be employed to characterize all the same variables that have been introduced for regular waves.

2.3.3 *Irregular waves*

2.3.3.1 *Basic statistics*

A wave gauge located at some point in Figure 2.4 will record the water-level variation, $z(t)$, in time and produce an irregular pattern as displayed in Figure 2.9(a) for a 20-min (1,200 s) record. The values for elevation, z, range from about 1.8 to 3.4 m above the vertical datum. The water-level record can be found for waves coming from different directions. It is simply a time-series record of the irregular water level at the location, with hour and date recorded.

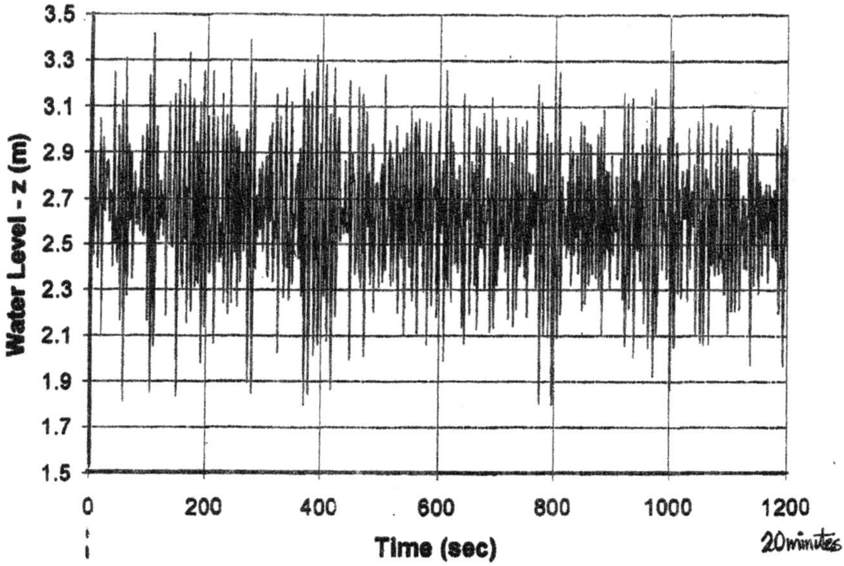

Figure 2.9(a). Water-level variation recorded over 20 min (from Kamphuis, 2010).

Figures 2.9–2.12 are adapted from Kamphuis (2010).

Wave-gauge data are usually recorded (or sampled) at time intervals, Δt of 0.25 or 0.5 s. We shall assume that $\Delta t = 0.5$ s herein [see Section 2.3.3.3]. Taking $j = 1$ at time $t = 0$ gives the number of samples $N = 1,200/0.5 = 2,400$ at times $j\Delta t$ where $1 \leq j \leq N$. They can be readily employed to find the basic water-level statistics, namely, the mean value, $\bar{z}$, and the standard deviation of the mean, σ.

$$\bar{z} = \frac{1}{N}\sum_{j=1}^{N} z_j, \tag{2.12}$$

$$\sigma = \mathrm{SQRT}\left[\frac{1}{N}\sum_{j=1}^{N}(z_j - \bar{z})^2\right]. \tag{2.13}$$

The distribution of z is assumed Gaussian about the mean value, $\bar{z}$, for the 20-min record, since it is commonly assumed that the mean water level and wave conditions are steady for this relatively short time interval. Note that nothing yet can be said about the wave heights or wave periods in this record.

2.3.3.2 *Time domain analysis (the significant wave height)*

The water surface variation, η_j, at time steps $j\Delta t$ can now be determined as follows:

$$\eta_j = z_j - \bar{z} \quad \text{for all } 1 \leq j \leq N \quad \text{with } \bar{\eta} = 0. \tag{2.14}$$

The water surface variation in time $\eta(\Delta t)$ is shown in Figure 2.9(b) for the water levels from Figure 2.9(a). An enlarged view from 40 to 80 s is shown in Figure 2.10. As shown in Figure 2.5, a wave height is defined as the vertical distance from the wave crest and the preceding wave trough. The vertical distance between the maximum and minimum water levels that lie between two locations where η crosses zero on the way down (down-crossing) is the correct way to identify a wave trough preceding the following wave crest. The zero down-crossing intersections (crosses) in Figure 2.10 show that each trough (t_i) must precede the following crest (c_i) in time to give the correct wave height, H_i. The times, between each zero down-crossing are the wave periods, T_i.

The time domain analysis proceeds as follows.

(1) The zero down-crossing method is applied and all the wave heights, H_i, in the 20-min record are determined.

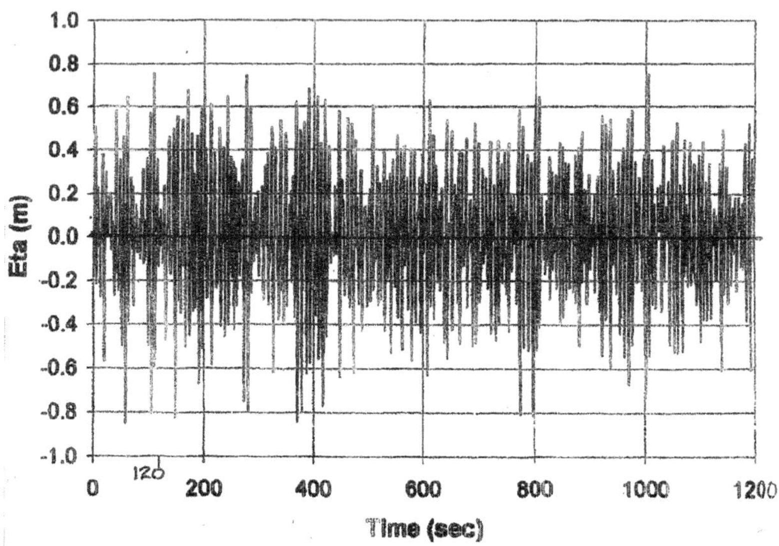

Figure 2.9(b). Water-surface variation for the water levels in Figure 2.9(a).

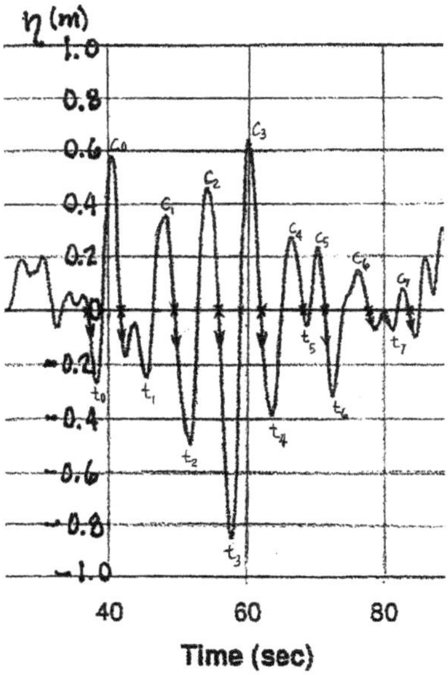

Figure 2.10. Time series of water-surface variation $\eta(t)$ from 40 to 80 s.

(2) The maximum height is found and divided into 10 (or more) equally spaced "bins".
(3) The wave heights are sorted and counted by the number found in each bin to produce a wave height histogram as illustrated in Figure 2.11 for the data of Figure 2.9(b). Here, the bin size was 0.15 m and the number in each bin is plotted at the mid-point of each bin.
(4) A spreadsheet is developed to determine various statistical properties of the wave-height distribution which is always non-Gaussian . Figure 2.12 is an example (see supplementary materials).
(5) The wave periods, T_i, can also be analyzed in this way.

As shown in Figure 2.12, the mean wave height, $H_{\text{mean}} = 0.67\,\text{m}$, the root mean square (rms) wave height, $H_{\text{rms}} = 0.75\,\text{m}$, and the significant wave height, $H_s = H_{1/3} = 1.04\,\text{m}$.

The average of the highest 1/3 (33%) of all the wave heights has come to be called the "significant" wave height, H_s. The original definition resulted from work by the oceanographer Munk (1944) during World War II (WWII). The significant wave height was intended to mathematically express the height estimated by a "trained observer". It is commonly used as a measure of the height of ocean waves. It is arguably the key

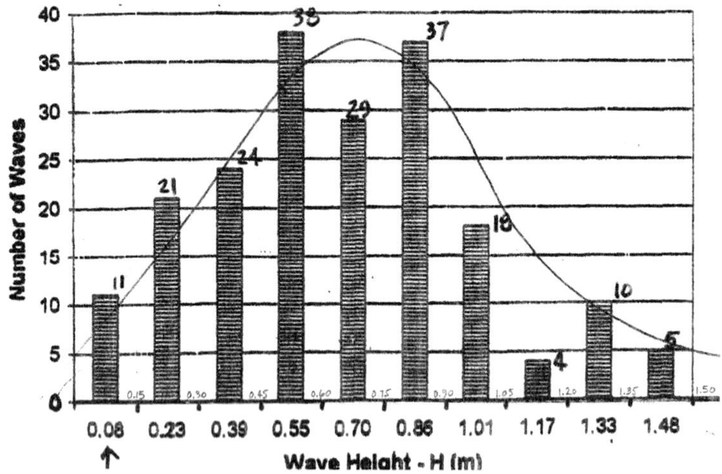

Figure 2.11. Histogram for wave-height distribution of waves shown in Figure 2.10.

statistically developed design wave variable needed in coastal structures design. In Figure 2.10, there were 197 waves so that 66 are in the highest 1/3 (33%) range, and the last three columns in Figure 2.12 demonstrate how the mean was determined for the highest 1/3 giving the significant wave height, $H_s = 1.04\,\text{m}$. We will use the symbols $H_s = H_{1/3}$ interchangeably in this book.

Five kinds of representative waves can be defined, namely, the highest (maximum) wave; the mean of the highest 1/10 (10%) of all the waves; the mean of the highest 1/3 (33%) of all the waves; the rms wave height, H_{rms}; and the mean or average value of all the waves heights in the record (see Goda, 2010 for how the wave periods associated with these representative wave heights are determined).

This time domain analysis procedure works for any time series of water-surface variation, $\eta(t)$. Under many conditions, the wave periods associated with the wave heights found in the above procedure do not vary greatly (i.e. a narrow frequency band). In such cases, a theoretical distribution for the wave heights is possible, and this is called the Rayleigh Distribution.

2.3.3.3 *The Rayleigh Distribution*

The probability that η has a certain value is called the probability density function (PDF). Longuet-Higgins (1952) developed the PDF for the maximum water surface at each peak, η_{max}, and made the assumption that

H range (bin) meters Low	High	Average in bin, m	Number of Waves	Probability p(H)	Cum Dis Fun CDF P(H'<H)	Probability of Exceedence Q(H'>H)	Average Wave Ht, Contribution	Squared Wave Ht. Contribution	Significant Wave Height Contribution, Average Highest 1/3 (33%)		
0.00	0.14	0.08	11	0.056	0.056	0.944	0.004467	0.00036			
0.15	0.29	0.23	21	0.107	0.162	0.838	0.0245178	0.00564			
0.30	0.44	0.39	24	0.122	0.284	0.716	0.0475127	0.01853			
0.45	0.59	0.55	38	0.193	0.477	0.523	0.1060914	0.05835			
0.60	0.74	0.70	29	0.147	0.624	0.376	0.1030457	0.07213	Number	Contribution	
0.75	0.89	0.86	37	0.188	0.812	0.188	0.1615228	0.13891	29	0.4394	0.377879
0.90	1.04	1.01	18	0.091	0.904	0.096	0.0922843	0.09321	18	0.2727	0.275455
1.05	1.19	1.17	4	0.020	0.924	0.076	0.0237563	0.02779	4	0.0606	0.070909
1.20	1.34	1.33	10	0.051	0.975	0.025	0.0675127	0.08979	10	0.1515	0.201515
1.35	1.49	1.48	5	0.025	1.000	0.000	0.0375635	0.05559	5	0.0758	0.112121
		N=	197			Mean Height=	0.668	0.56031	66	1.0000	1.037879
		N/3=	66				Hrms=	0.75		H 1/3=	1.04

Figure 2.12. Spreadsheet to analyze wave height histogram shown in Figure 2.11.

the wave height, H, associated with each peak was equal to twice $\eta_{\max}$. The Rayleigh Distribution, $p(H)$, for a relatively narrow frequency band then becomes

$$p(H) = \frac{1}{4}\frac{H}{\sigma^2} exp\left[\frac{-H^2}{8\sigma^2}\right], \tag{2.15}$$

where σ is the standard deviation of the mean as found in Equation (2.13).

Integrating the $p(H)$ distribution from 0 to any H gives the cumulative distribution, $P(H^* < H)$, for any wave height H^* below H. We normally are interested in the exceedance probability Q $(H^* > H) = 1 - P$ that is found from

$$Q(H^* > H) = exp\left[\frac{-H^2}{8\sigma^2}\right], \tag{2.16a}$$

from which is found both (1) the wave height, H_Q, with a probability of exceedance, Q, and (2) the average wave height with a probability of exceedance of Q, written $\overline{H}_Q$. Table 2.3 presents commonly defined wave heights for both these definitions relative to the standard deviation of the mean, σ (from Kamphuis, 2010). The significant wave height, H_s, is most important, and from its definition as the average of the highest 1/3 (33.3%) or $\overline{H}_{1/3} = 4\sigma$ from Table 2.3 gives $Q_s = 0.135$. This means that 13.5% of all the waves are larger than the significant wave height, H_s, which can be written as $H_{0.135}$.

Table 2.3. Commonly used statistical definitions of wave heights from the Rayleigh Distribution.

Symbol	Description	Value
$\overline{H}_{.01}$	Average of highest 1% of the waves	6.67σ
$H_{.01}$	Height, exceeded by 1% of the waves	6.07σ
$\overline{H}_{.02}$	Average of the highest 2% of the waves	6.23σ
$H_{.02}$	Height, exceeded by 2% of the waves	5.59σ
$\overline{H}_{.01}$	Average of highest 10% of the waves	5.09σ
$H_{.01}$	Height exceeded by 10% of the waves	4.29σ
$H_s = \overline{H}_{1/3}$	Significant wave height (average height of the highest 1/3 of the waves)	4.0σ
$\overline{H}$	Average wave height	$\sqrt{2\pi}\sigma$
$H_{0.5}$	Median wave height	2.35σ
H_{mode}	Most probable wave height	2.0σ
H_{rms}	$\sqrt{\frac{H_1^2+H_2^2+H_3^2+\cdots}{N}}$	$2\sqrt{2}\sigma$

The rms wave height, $H_{\rm rms}$, gives the same wave energy, E, as determined for a regular wave height, H, in Section 2.3.2.1. Using the σ relationships in Table 2.3, we find that $H_s = 1.414 H_{\rm rms}$. Since wave energy, $E \sim H^2$, using the significant wave height in structural stability formulas doubles the energy load on coastal structures. There is always even more energy present since 13.5% of all the waves are greater in height than H_s.

As a check on whether (or not) the measured wave-height distribution fits a Rayleigh Distribution, a "Rayleigh parameter", R can be calculated from Equation (2.16a) giving

$$H = (2\sqrt{2}\sigma)R. \tag{2.16b}$$

This plots as a straight line through zero with slope $(2\sqrt{2}\sigma)$. The wave data in Figure 2.10 gave $H = 0.6872R$ with a correlation coefficient of 0.9875, which means this 20-min recording fits the Rayleigh Distribution.

2.3.3.4 *Frequency domain analysis (the wave spectrum)*

The wave spectrum is a plot of the wave-variance spectral density function, $S(f)$ (m^2/Hz), on the vertical axis versus the wave frequency, f (Hertz, Hz), on the horizontal axis. Recall the wave frequency, $f = 1/T$. An example is shown in Figure 2.13 for the water surface η_j as found in Figure 2.9(b) and after the raw data spectrum is smoothed numerically.

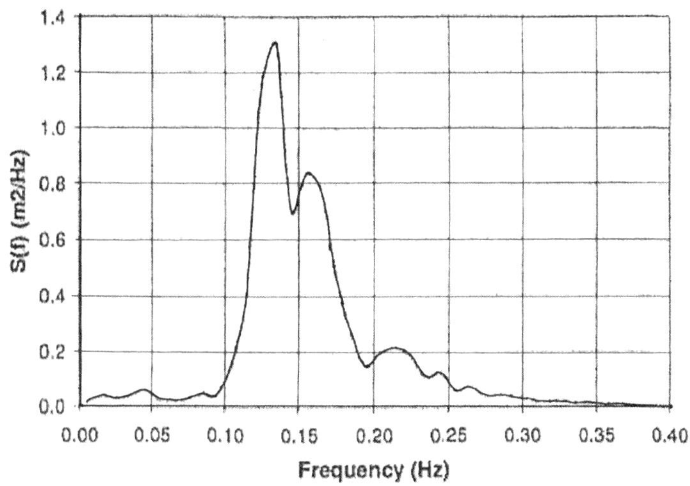

Figure 2.13. Smoothed spectrum for data in Figure 2.9(b).

The variance, $S(f)$, is a statistical term and is simply the physical wave energy $E(f)$ divided by the unit weight of the water (ρg) so that

$$E(f) = \rho g S(f). \tag{2.17}$$

Sometimes it is referred to as the energy-density spectrum. Note that in the example shown in Figure 2.13, there is little energy below $f = 0.1\,\text{Hz}$ (above $T = 10\,\text{s}$) and above $f = 0.3\,\text{Hz}$ (below $T = 3.3\,\text{s}$).

The area under the spectrum is the total variance, σ_f^2 (the square of the standard deviation of the mean). Assuming a Rayleigh Distribution, from Table 2.2, the significant wave height is 4 σ_f and the subscript f means that σ_f was determined in the frequency domain. We now define H_{mo} as the spectral, significant wave height and see that

$$H_{mo} = 4\sqrt{\text{Area under the spectrum}}, \tag{2.18}$$

to distinguish this definition of H_s from that determined in the time domain $\overline{H}_{1/3}$.

One additional result of calculating the spectrum is that there are other moments, m_n, that can be calculated from the $S(f)$ versus f distribution. In general, any moment can be found from the product of the moment arm, f^n times the value of $S(f)$, where $n = 0, 1, 2, 3, 4$, etc. are the exponents of the moment arm, f.

$$m_n = \int_0^{0\alpha} f^n \; S(f)\; df. \tag{2.19}$$

It is now clear H_{mo} is the zeroth moment, f^{o} the area under the spectrum from $f = 0$ to $f =$ infinity.

Three wave periods are important for combining with H_{mo} and defining the irregular wave characteristics, namely, the peak period, T_p, the mean period, T_m, and the spectral wave period, $T_{m-1.0}$ (sometimes called the negative, first-moment wave period). The spectral wave period, $T_{m-1.0}$, is equal to the ratio (m_1/m_o) and has been found useful because it gives more weight to the longer periods in the spectrum and is independent of the type of spectrum. It is commonly employed for wave runup and overtopping calculations on coastal structures and is discussed in full detail in Chapter 5.

The wave spectrum may contain two peaks indicating waves from different wave directions that produce both sea and swell conditions needed in structural design. In Figure 2.13, there are two peaks with the main $f_p = 0.13\,\text{Hz}$ or $7.7\,\text{s}$. A second smaller peak is at a higher frequency, $f = 0.16\,\text{Hz}$, or lower wave period, $T = 6.3\,\text{s}$. Coastal engineers must continually switch between wave frequency and wave period when reviewing wave spectrums.

The calculated $S(f)$ is always an estimate of the true $S(f)$ because only one estimate of η_j is available and for a relatively short time period (20 min), and for finite increments of frequency, Δf defined as 1 over the length of the record (20 min). Therefore, the raw data-wave spectrum is typically smoothed to produce the relatively smooth curves displayed for the example spectrum illustrated in Figure 2.13.

To calculate $S(f)$, the water surface variation, η_j, at time steps $j\Delta t$ is expressed as a Fourier Series with discrete frequency components, f_n. It is known as a discrete fourier transform (DFT) and is for an infinite time series. Because all record lengths of waves are relatively short (17–20 min), the smallest frequency (longest wave period, T) that can be found is on the order of the record length. The highest frequency that is possible to resolve (smallest wave period, T) is simply $2\Delta t$ since it takes at least three time points to define the smallest wave. This is why we chose $\Delta t = 0.5$ s which gives the Nyquist frequency, f_N, of 1.0 s and corresponding wave period $T_N = 1$ s. There is very little wave energy in the ocean of interest at wave periods of only 1 s. The result of a finite record length and finite number of points, η_j, for analysis results in the finite discrete fourier transform (FDFT). Details on the theory of water-wave spectral analysis (e.g. Goda, 2010 or Kamphuis, 2010) is beyond the scope of this book. A spreadsheet that applies the FDFT is available in the online supplementary materials on Wave Mechanics for use in design.

For the water surface variation, $\eta(t)$ shown in Figure 2.9, the results by the three methods are summarized in Table 2.4. For this example, all three methods give similar results but this is not always the case. The main advantages of the wave spectrum are that the peak wave energy can

Table 2.4. Summary of statistical analysis of wave-surface time series in Figure 2.9.

Wave Statistic	Basic Statistics	Time Domain	Frequency Domain
	Assumed Rayleigh	Zero-Down Crossing	Wave Spectrum
σ(m)	$\sigma = 0.28$	$\sigma_z = 0.24$	$\sigma_f = 0.28$
N (number of waves)		197	
T(s)		$\mathrm{T_m} = 6.1$	$T_p = 7.6$
			$T_2 \sim Tm = 6.1$
			$T_{m-1.0} = 6.92$
H_s(m)	$H_{0.135} = 1.12$	$\overline{H}_{1/3} = 1.05$	$H_{mo} = 1.12$
H_{rms}(m)	0.79	0.76	0.79

be identified (possible multiple peaks) and that the spectral wave period, $T_{m-1.0}$, can be found as needed for coastal structures design.

These results are for one 20-min record at the location, date (time, day, month, year), and water depth of the wave gauge (or wave-information study position). The complete, long-term wave-climate data (or wave information) needed for design is discussed in Section 2.3.8.

2.3.3.5 *Theoretical wave spectra*

Development of theoretical equations coupled with empirically found coefficients to calculate wave spectra has produced formulas for "fully developed" sea conditions (Pierson and Moskowitz, 1964) called the PM spectrum, for "developing" seas called the JONSWAP spectrum (Hasselmann *et al.*, 1973), and then when waves become depth-limited in shallow water called the TMA spectrum (Bouws *et al.*, 1985). The formulas require knowledge of the distance over which the wind blows called the fetch, F, and the wind speed, U. Since the area under the spectrum is related to the spectral significant wave height, H_{mo}, the theoretical wave spectra form the basis for wind-wave hindcasting methods that are fully discussed in Section 2.3.7. The concepts of "fully developed" and "developing" seas will also be defined.

Wave direction and storm duration are additional variables required for both functional and structural designs.

2.3.3.6 *Wave direction*

Multiple water-level gauges need to be deployed and their simultaneous records related in order to estimate the mean direction, α, of the waves. See Earle *et al.* (1995) for details regarding the design of directional wave gauges. For coastal structures design, the complete wave climate (height, period, and direction) is needed for the long-term analysis of extreme wave events as discussed in Section 2.3.8. We shall return to wave direction and storm duration in Section 2.3.8.

The theoretical, directional wave spectrum is a function of both wave frequency f, and wave direction α. A complete discussion may be found in Goda (2010) and is beyond the scope of this book.

2.3.4 *Wave breaking*

In general, coastal engineers now use H_{mo}, T_p, and α as found in deep water near the design site. As irregular waves propagate into shallower water, the largest waves break to alter the zero down-crossing histogram and wave spectrum. There are fewer large waves present due to the depth-limited wave breaking in shallow water. The histogram (Figure 2.11) is truncated and the spectrum (Figure 2.13) has less energy (variance).

2.3.4.1 *Types of wave breaking*

Waves' breaking occurs in different forms across beaches and on structures, namely, spilling, plunging, surging, and collapsing as illustrated in Figure 2.14 (from CEM, 2006).

2.3.4.2 *The dimensionless surf-similarity parameter, ξ*

A dimensionless ratio of the beach (or structure) slope, tan α and wave steepness, H/L, has been found useful to quantify the breaker type. It is called the surf-similarity parameter (or Irrabaren number), ξ. Complications arise as to which wave height to use (deep water H_o or at breaking H_b). Battjes (1974) defined two formulations but employed the deepwater

(a) Spilling breaking wave

(b) Plunging breaking wave

(c) Surging breaking wave

(d) Collapsing breaking wave

Figure 2.14. Types of breaking waves (from CEM, 2006).

wave length L_o in both as follows:

$$\xi_o = \tan \propto /\sqrt{(H_o/L_o)} \quad \text{and} \quad \xi_b = \tan \propto /\sqrt{(H_b/L_o)}. \tag{2.20}$$

Using the experimental data from Galvin (1968), it was then learned (Battjes, 1974) that the breaker types could be quantified by the following ranges for the surf-similarity parameter:

$$\begin{array}{llll} \text{Spilling Breaker:} & \xi_o < 0.5 & \text{or} & \xi_b < 0.4, \\ \text{Plunging Breaker:} & 0.5 < \xi_o < 3.3 & \text{or} & 0.4 < \xi_b < 2.0, \\ \text{Surging (collapsing) Breaker:} & 3.3 < \xi_o & \text{or} & 2.0 < \xi_b. \end{array} \tag{2.21}$$

These early developments were for regular waves. Further developments in use of the surf-similarity parameter in irregular wave formulations for runup and wave overtopping are detailed in Chapter 5.

2.3.4.3 *Regular wave breaking*

For regular waves defined by H and T and for all water depths, d, Miche (1944) proposed the following formula for the maximum H/L ratio or when breaking begins (incipient conditions):

$$\left(\frac{H}{L}\right)_b = 0.142 \tanh\left(\frac{2\pi}{L}d\right). \tag{2.22}$$

In deep water, the tanh () approaches 1.0 so the wave height is steepness-limited, namely, $(H/L)_b < 0.14$. In shallow water, tanh () approaches the argument () so that $(H/d)_b \sim 0.8$. Many refinements exist in the published literature to account for other factors such as the slope (rise/run), m, of the bathymetry from deep to shallow water. But all involve regular waves for only one H and T, are not practical and are not discussed herein.

2.3.4.4 *Irregular wave breaking*

For irregular waves defined by the significant wave height, H_s, and peak period, T_p, at water depth d on an average beach slope, m, the following formula is herein proposed for the incipient breaking condition. The wave length corresponding to the peak period at water depth, d, is denoted as L_p

$$\left(\frac{H_s}{L_p}\right)_b = 0.085[\exp(3.75\,\text{m})] \tanh\left(\frac{2\pi d_b}{L_p}\right). \tag{2.23}$$

It is a compromise of the criteria proposed by Kamphuis (1991) and has its roots in the Miche (1944) formulation for all water depths.

Kamphuis (1991) proposed separate criteria for the initiation of wave-steepness breaking and depth-limited breaking based on laboratory-scale model testing. The significant wave height H_s is the defined wave height at breaking and the spectral period T_p is used at water depth d to find the wave length L_p from the linear wave theory at the wave breaking water depth, d. Kamphuis (1991) found that

$$\left(\frac{H_s}{L_p}\right)_b = 0.095[\exp(4.0\,\mathrm{m})]\tanh\left(\frac{2\pi d_b}{L_p}\right)$$

$$\text{for wave-steepness breaking, and} \qquad (2.24)$$

$$\left(\frac{H_s}{d}\right)_b = 0.56[\exp(3.5\,\mathrm{m})] \quad \text{for depth-limited breaking.} \qquad (2.25)$$

The influence of beach slope, m is included and the compromise value (3.75) in Equation (2.23) is simply the average coefficient in the two exponential expressions. In shallow water, the tanh() approaches its argument () so that the L_p cancels leaving

$$\left(\frac{H_s}{d}\right)_b = 0.085[\exp(3.75\,\mathrm{m})]2\pi \qquad (2.26)$$

or

$$\left(\frac{H_s}{d}\right)_b = 0.534\exp(3.75\,\mathrm{m})$$

which again is close to the depth-limited coefficient 0.56 proposed by Kamphuis (1991). Our knowledge of the correct T_p is limited to further justify the compromise reached in Equation (2.23).

References in the literature give additional confidence to the compromise formula. For depth-limited breaking on flat slopes ($m = 1/100$), exp (3.75 m) equals 1.04. Using the Rayleigh distribution (Table 2.2), $H_s = 1.414 H_{\mathrm{rms}}$ to find H_{rms}/d gives

$$\left(\frac{H_{\mathrm{rms}}}{d}\right)_b = (0.534\exp(3.75\,\mathrm{m}))/1.414 = (0.543 \times 1.04)/1.414 = 0.40.$$

Thornton and Guza (1983) used the ratio $(H_{\mathrm{rms}}/d)_b = 0.42$ in their work on breaking waves in the surf zone at the Scripps Institute of Oceanography in La Jolla, CA. Sallenger and Holman (1985) suggest that $(H_{\mathrm{rms}}/d)_b$ is in the 0.3–0.5 range and gave the following formula:

$$\left(\frac{H_{\mathrm{rms}}}{d}\right)_b = 3.2\,\mathrm{m} + 0.32, \quad \text{which for } m = (1/100) \text{ gives } 0.352.$$

Oumeraci *et al.* (2001) do not consider the approach slope and suggest the following for steepness-limited breaking:

$$\left(\frac{H_s}{L_p}\right)_b = 0.093 \tanh\left(\frac{2\pi d_b}{L_p}\right), \tag{2.27}$$

which is further confirmation that the compromise formula (2.23) is reasonably accurate. See further discussion in Chapter 6 for the calculation of breaking wave forces on vertical walls.

Goda (1975, 2010) has developed relatively complex empirical formulas which use H'_o, an equivalent deepwater, unrefracted wave height corresponding to the significant wave height as reference. The compromise formula cited previously gives similar results.

We herein employ the compromise formula (Equation (2.23)) for use in the spreadsheets for breaking waves. The deepwater wave conditions will be transformed by the near-shore bathymetry as the waves move into shallow water.

2.3.5 *Wave transformations in coastal waters*

2.3.5.1 *Basic theory for regular waves*

Figure 2.15 (adapted from Svendsen *et al.*, 1978) schematically displays the many ways that waves approaching the shore at an angle are transformed in near-shore, coastal waters. The solid lines represent the mean wave crests that bend (refraction) on the open coast and spread behind structures (diffraction) and bounce off (reflection) structures. The wave length L and wave direction α change significantly in all cases in the shallower water. One wave ray (orthogonal) is depicted in Figure 2.15. Along each wave ray, the wave height is modified by the decreasing water depths (shoaling). Adjacent wave rays (orthogonals) may converge (focus) or diverge (spread apart) and the wave heights will, respectively, increase (convergence) or decrease (divergence) as depicted in Figure 2.16. Finally, as indicated by the dotted "breaker" line (Figure 2.15), the waves break to produce a surf zone and generate long-shore currents. Wave transformations in coastal

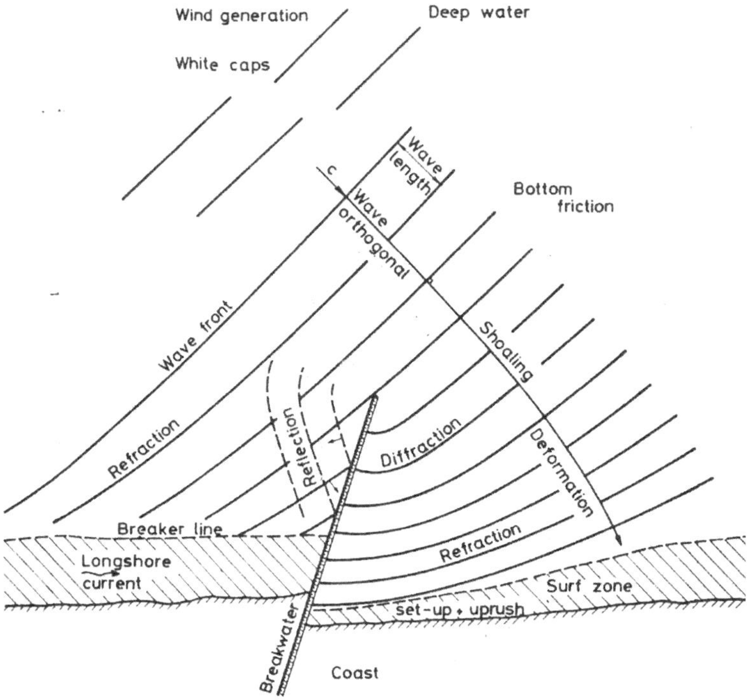

Figure 2.15. Schematic of wave transformations in coastal waters (adapted from Svendsen *et al.*, 1978).

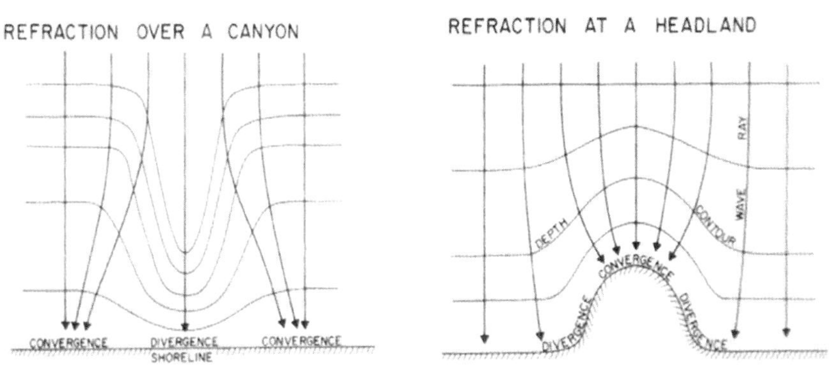

(a) (b)

Figure 2.16. Schematic of wave ray convergence and divergence (a) over submarine canyon and (b) at protruding headland.

waters where the $d/L \leq 0.5$ is a very complicated process. Knowledge of the wave height at the structure is required for design.

(1) Wave shoaling

The discussion of water-wave transformation always begins in deep water ($d/L_o \geq 0.5$). For regular waves and linear wave theory, the deep-water wave length, $L_o = (g/2\pi)T^2$. The wave transformation always begins in deep water because the wave length L_o only depends on the wave period in deep water.

For regular waves approaching the shoreline at right angles ($\alpha = 0$), wave power, $P = nCE =$ constant. Using the equation for the wave energy, E, related to the wave height squared gives for any two water depths $nCH^2 = n_1C_1H_1^2 = n_2C_2H_2^2 = n_0C_0H_0^2 =$ constant since the $1/8\rho g$ terms cancel. Therefore, the wave height at any depth, H, can be related to the deep-water (un-refracted) wave height, H_o', by the relationship

$$\frac{H}{H_o'} = \text{SQRT}\left[\left(\frac{n_o}{n}\right)\left(\frac{C_o}{C}\right)\right].$$

Using the group celerity parameter, n, from Equation (2.9) with deep water, $n_o = 1/2$, and the wave celerity, $C = C_o \tanh kd$ from Equation (2.4) for wave length L and wave celerity, $C = LT$ gives

$$\frac{H}{H_o'} = \text{SQRT}\left[\left(\frac{1}{2}\frac{1}{n}\right)\left(\frac{1}{\tanh kd}\right)\right].$$

The deep-water wave height is given as a prime symbol H_o', called the unrefracted wave height. As shown below, this wave-transformation process is called wave shoaling. A wave-shoaling coefficient K_s can then simply be defined as

$$K_s = \frac{H}{H_o'} = \text{SQRT}\left[\left(\frac{1}{2}\frac{1}{n}\right)\left(\frac{1}{\tanh kd}\right)\right]. \tag{2.28}$$

(2) Wave refraction

If the lateral distance between any two wave rays (orthogonal) in deep water is b_o, and b is the transformed lateral distance at any depth d as depicted in Figure 2.17, then the wave refraction coefficient K_r is found from

$$K_r = \sqrt{\frac{b_o}{b}}. \tag{2.29}$$

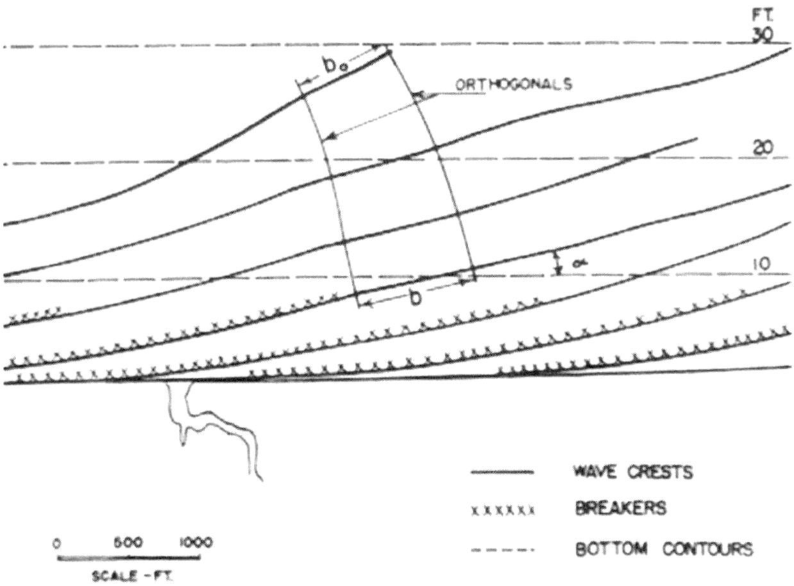

Figure 2.17. Wave refraction on open coast.

(3) Wave shoaling and refraction

The bathymetry significantly alters the width b between each adjacent orthogonal as seen in Figure 2.17. The combined result of shoaling and refraction gives for any depth d

$$H = K_s K_r H_o'. \tag{2.30}$$

2.3.5.2 *Basic theory for irregular waves*

The analysis for irregular wave transformation is made practical by introducing some statistical definition of the wave characteristics. What to use for the deep-water wave height, period, and direction is subject to debate. We herein use the significant wave height, H_s, the mean wave period, T_m, and the mean wave direction, α. To avoid complications with the nomenclature, H_o' will again be the un-refracted, deep water, significant wave height. The deepwater, un-refracted wave height H_o' is called the "effective deepwater wave" in the Japanese literature (Goda, 2010).

Irregular waves in shallow water are far more nonlinear in shape so that wave shoaling is different than with linear wave theory. Goda (1975) presented a design diagram for a nonlinear wave shoaling coefficient, K_s^*, and Kweon and Goda (1996) derived the following closed-form equation using coefficients to approximate the diagram:

$$K_s^* = K_s + 0.0015\left(\frac{d}{L_o}\right)^{-2.87}\left(\frac{H_o'}{L_o}\right)^{1.27}, \tag{2.31}$$

where K_s is the same as above for linear waves. The second term is always positive so that nonlinear wave shoaling increases the shoaled wave height. The deep-water wave length L_o is found using T_m.

The refraction coefficient for irregular waves is far more complex. Real waves in deep water come from different directions with different frequencies in the spectrum. See Goda (2010) for a method to estimate the effective refraction coefficient K_{eff} and/or by use of an energy-balance equation.

For preliminary design, assuming straight, parallel-depth contours relative to the shoreline significantly facilitates the calculation of the refraction coefficient.

2.3.5.3 *Wave transformations over straight, parallel-depth contours (spreadsheet solution)*

We begin with regular waves defined by H, T, and α. On plane, monotonically decreasing beach profiles that have relatively straight and parallel-depth contours, the ratio of the wave celerity, C, and the sine of the wave angle, $\sin\alpha$, is constant and known as Snell's Law of refraction, namely,

$$\frac{C}{\sin\alpha} = \frac{C_o}{\sin\alpha_o} = \text{constant}. \tag{2.32}$$

Because each wave orthogonal refracts the same way, the ratio $b/\cos\alpha$ is also constant and gives the Snell's Law refraction coefficient, K_r, for regular waves

$$K_r = \sqrt{\frac{b_o}{b}} = \sqrt{\frac{\cos\alpha_o}{\cos\alpha}}. \tag{2.33}$$

A spreadsheet which is available in the online supplementary materials, called "ShoalRefractBrk", can be applied for preliminary design. The waves are transformed by shoaling, refraction, and breaking processes. A trial-and-error procedure is required to determine the water depth at the incipient breaking point. The spreadsheet uses H_s, T_m, and α as the characteristic, irregular wave conditions in deep water.

The complete refraction process for irregular waves with different directions and frequencies in deep water is complicated even for straight, parallel-depth contour, bathymetric conditions. However, Goda (1975) have developed design curves using a directional spectrum spreading parameter $S_{\max}$ to facilitate the calculation. Unfortunately, the design curves (Goda, 2010, Fig. 3.10, Fig. 3.11) have yet to be put in analytical form for use in spreadsheet calculations. In general, the refraction coefficients are lower due to directional spreading so that use of the Snell's Law approximation is conservative for design. Use of wave-spectrum transformation numerical models has also become more commonplace for design. Or, a physical, hydraulic model may be employed.

2.3.5.4 *Wave transformations over complex bathymetry (physical and numerical models)*

Complex, irregular bathymetry at the project site may require some type of model solution to determine the transformed wave conditions needed for design. Both physical (scale) and numerical models can be employed. The latest bathymetry at the proposed site is found by field survey. The vertical datum is again important to insure that the bottom contour elevations are correctly constructed in the physical model and are correctly located at the discrete space positions $(\Delta x, \Delta y)$ in the numerical model. Both model types have their advantages and disadvantages.

The physical model does not need to discretize the bathymetry, but must construct it to some physically reduced scale for producing scale effects. The physical model will shoal, refract, diffract, and break the incoming waves to produce the correct (scaled) wave characteristics at the structure. And, it may also be used to study the stability of scale-modeled armor units on the structure. The physical model employed by the Corps of Engineers to correctly reproduce near-shore wave transformations and to study the stability of the breakwater armor units at the Latjes Airforce Base, Terceria Azores, Portugal (from Melby and Myrick, 2004) is shown in Figure 2.18.

The boundary condition waves produced in deep water may be either regular, single frequency waves or irregular waves to reproduce some directional wave spectrum depending on the wave-making capabilities of the laboratory. Figure 2.18 shows regular waves that are shoaling and refracting before striking the breakwater structure.

Numerical models of the bathymetry can be implemented at full scale but still suffer from scale effects due to the need to simulate turbulence

Figure 2.18. The physical model employed by the Corps of Engineers to study the stability of the breakwater armor units at the Latjes Airforce Base, Terceria, Azores, Portugal (see Melby and Myrick, 2004, Section 6.2, Example 6.3).

in wave breaking and the need to use a discrete number of space girds $(\Delta x, \Delta y)$. An example of complex bathymetry is shown as depth contours (dotted lines) in Figure 2.19. The offshore wave conditions are regular waves of uniform wave height propagating in the shore-normal (upward) direction as depicted by the arrows in the top half of the diagram. The complex bathymetry causes the adjacent rows of arrows (wave rays) to converge and diverge in a complicated pattern which results in the wave-height contours (solid lines, bottom) that would be needed for design.

The shoreline location is not shown in Figure 2.19, but the green area of the deep water may be part of a tidal inlet with shallow water shoals in the yellow areas on each side of the inlet. Very shallow water at the tip of the two protruding yellow areas causes wave shoaling and refraction to produce 4-feet waves before breaking in 5-feet water depths. Wave focusing at these locations would be considered in the jetty design.

Both phase-averaged and phase-resolving numerical models have been employed for wave-transformation studies. Phase-averaging models are most common. Early versions only considered regular waves (one wave frequency), but now wave-specturm transport models are routinely utilized.

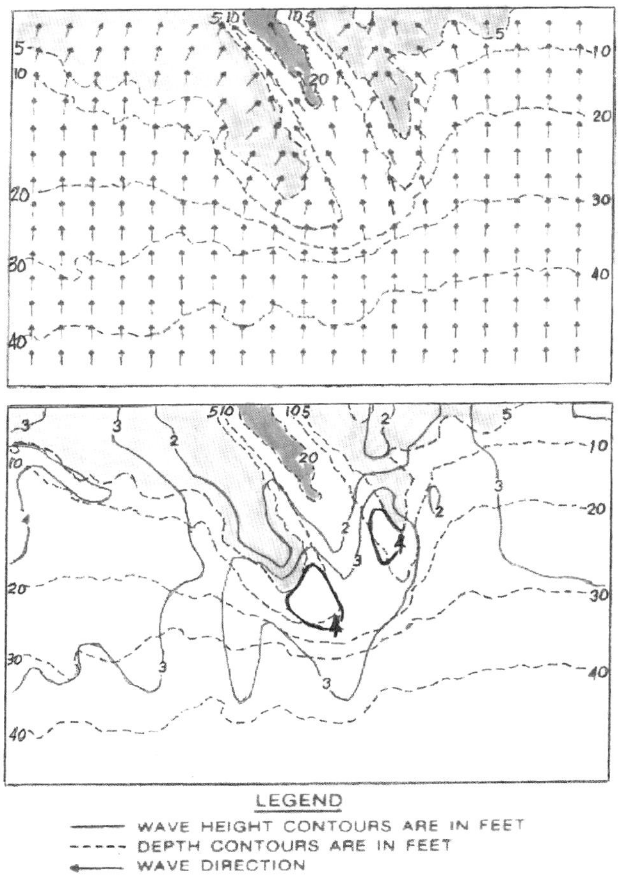

Figure 2.19. Example of wave shoaling and refraction in numerical model.

The phase-averaged, wave characteristics commonly extracted at each grid point in space and time are the spectral, significant wave height, the peak wave period, and the mean wave direction. The grid resolution can be relatively large to aid the computation times and costs.

Phase-resolving models (Boussinesq-type) propagate the water-surface variation, $\eta(x, y, t)$, in space and time so that the full details of the irregular, water surface motions are computed. Finely spaced grids are required to ensure the numerical accuracy of resolving short period waves in the computation. The result is very large computation times and costs even for relatively small regions of the coast.

A full discussion of the theory and application of numerical models for water-wave transformation studies is well beyond the intended scope of this book. See Lin (2008) and Ma (2010), for more details on the numerical modeling of water waves. The numerical models as developed by large software organizations (Deltares, Danish Hydraulics, Corps of Engineers, etc.) continue to evolve over time and some are now available at no cost. Their websites are the best source for the latest information. Unfortunately, none have been developed for spreadsheet applications, so we cannot provide a numerical model for the wave transformations study in this book.

Finally, wave diffraction, reflection, and transmission will be discussed in the upcoming chapters. Figure 2.20 clearly shows wave diffraction behind a near-shore breakwater at a small harbor entrance with jetties.

Figure 2.20. Wave diffraction behind an offshore breakwater at the mouth of a small harbor with jetties (courtesy Corps of Engineers).

2.3.6 *Wave set-down, set-up, and wave-induced currents*

2.3.6.1 *Wave-induced radiation stresses*

The depth-integrated and time-averaged momentum produced by progressive waves minus the hydrostatic pressure force is the horizontal wave thrust called the "radiation stress" (Longuet-Higgins and Stuart, 1964). The radiation stresses produce mean water-level changes called wave setdown and wave setup. They also create wave-induced currents (longshore and undertow) as presented in Chapter 5.

When linear wave theory is employed, the principal stress components along a wave ray S_{XX} and along the wave crest (normal to the ray) S_{YY} are

$$S_{XX} = E\left(2n - \frac{1}{2}\right), \tag{2.34}$$

$$S_{YY} = E\left(n - \frac{1}{2}\right), \tag{2.35}$$

where $E = 1/8\rho g\, H^2$ and $n = C_g/C$.

In general, these principal stress components must be transformed to the onshore direction x and the shore parallel direction y by a coordinate transformation method (formula or Mohr's Circle) to give the onshore directed component, S_{xx}, the shore-parallel direction component, S_{yy}, and a shear stress component, S_{xy}. The wave-crest angle with the shoreline is α, and Figure 2.21 schematically depicts the radiation stress components.

$$S_{xx} = E\left(\frac{3}{2}n - \frac{1}{2}\right) + E\left(\frac{\mathrm{n}}{2}\right)\cos 2\alpha, \tag{2.36}$$

$$S_{yy} = E\left(\frac{3}{2}n - \frac{1}{2}\right) - E\left(\frac{\mathrm{n}}{2}\right)\cos 2\alpha, \tag{2.37}$$

$$S_{xy} = En\sin\alpha\cos\alpha. \tag{2.38}$$

Gradients of the radiation stresses produce the wave setup.

2.3.6.2 *Wave setup*

Mean water levels along beaches are higher (setup) than measured by the nearby tide gauges in sheltered waters. The magnitude of the wave setup, $\bar{\eta}$, at the new shoreline location is on the order of 10–20% of the offshore wave height as schematically shown at the top of Figure 2.21. This additional

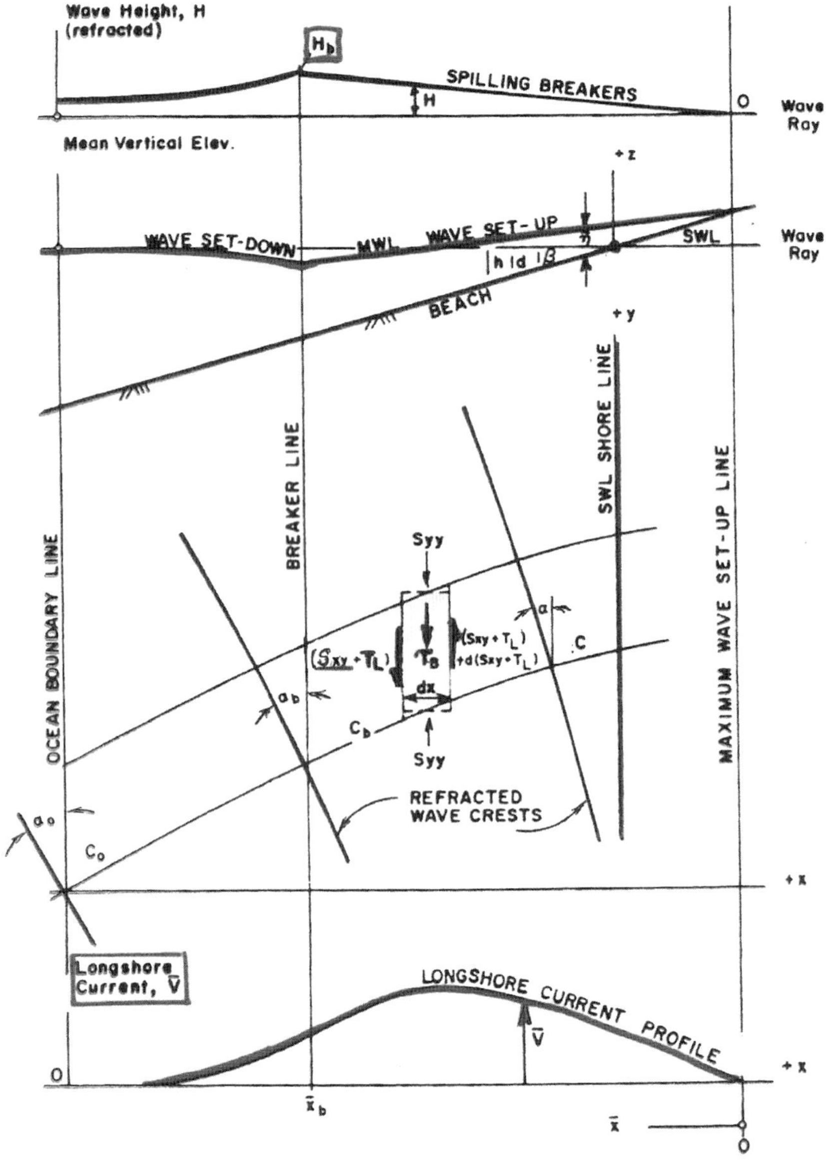

Figure 2.21. Cross-shore gradients in wave height produce wave setdown and wave setup.

water level is on top of that produced by the wind stress and other factors causing storm surge as discussed in Section 2.2.4. Consequently, wave setup produces higher flooding elevations and must be included in coastal structures design.

The simplest case is for shore normal waves where the gradient of S_{xx} is balanced by the water level presssue gradient of still water depth, d, plus the wave setup level, i.e. $(d + \bar{\eta})$. The gradient of S_{xx} is positive offshore of the break point, H_b, due to wave shoaling and negative landward of the break point due to wave breaking. Seaward of the breakpoint, the mean water level is below the SWL and is called the wave setdown.

Longuet-Higgins and Stewart (1964) derived the following equation for wave setup, $\bar{\eta}$,

$$\bar{\eta} = \frac{\kappa H_b}{16} + \frac{3\kappa^2/8}{1 + (3\kappa^2/8)}(d_b - d) \tag{2.39}$$

where $\kappa = H_b/(d_b + \bar{\eta}_b)$ is the depth-limited breaking condition, the bottom shear stresses are neglected, and the first term (negative) on the right-hand side is the starting condition seaward for the wave setup. For κ about 0.8, the maximum setup at the new shoreline ($x = 0$) location is 0.24 H_b. A 1-m breaking wave will produce about 24 cm (0.8 ft) of wave setup for this simple case with all the assumptions.

Bowen *et al.* (1968) conducted laboratory experiments and some results are shown in Figure 2.22 for single period waves ($T = 1.14$ sec) on a plane beach slope, m $= 1/12$. The maximum measured setup is on the order of 0.28 H_b.

Dean and Walton (2009) present an exhaustive review of the literature on wave setup over the past 50 years including published guidelines for engineering applications. For irregular waves using the significant wave height at breaking, H_{sb}, they recommend using $\bar{\eta}/H_{sb} = 0.191 \pm 0.100$ as a useful "guide" for the maximum static wave setup at the shoreline. They also warn that a full quantitative understanding is not yet available.

2.3.6.3 *Wave-induced currents*

The breaking waves in the near-shore region also produce time-averaged, currents in the surf zone (longshore currents) as illustrated in Figure 2.21 (bottom) and time-averaged currents beneath the broken waves in the surf zone (undertow) as illustrated in Figure 2.23. These currents transport sediments in and around coastal structures and will be further discussed in Chapters 5 and 7.

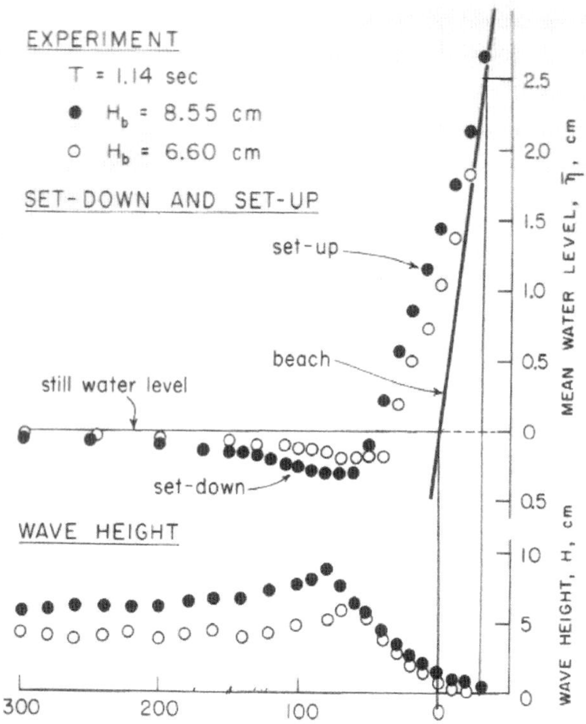

Figure 2.22. Wave height (bottom) and wave setdown and wave setup levels measured by Bowen *et al.* (1968).

2.3.7 *Wave hindcasts from wind statistics*

2.3.7.1 *Wave data or wave information*

In the above-mentioned sections, knowledge of the wave conditions (height, period, direction) is assumed known in deep water. This knowledge may come from field measurements of instantaneous water-elevation variations in time as discussed in Section 2.3.3. These measurements produce wave "data". Wave data is expensive to obtain over extended lengths of time and the number of wave gauges is limited to a few locations along many of the worlds' coasts.

As a result, physical oceanographers and coastal engineers studied the theory, conducted field experiments, and developed empirical equations relating wind conditions (fetch, speed, duration) to the wave conditions produced by the winds. The procedure is called wave "hindcasting" because

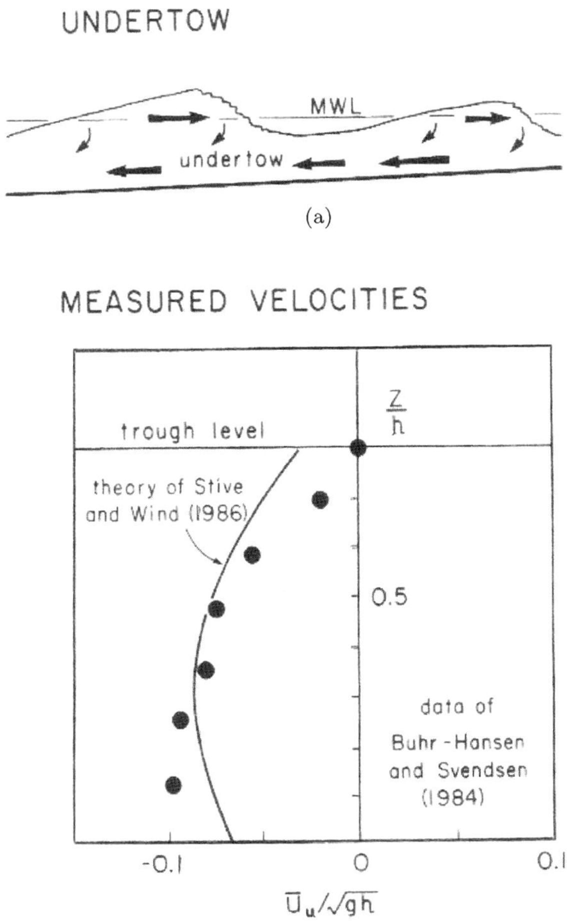

(b)

Figure 2.23. Measured time-averaged current profile (undertow) in the wave setup region (from Buhr-Hansen and Svendsen, 1984).

wind statistics are employed using the historical measurement of wind data. The wave results are called wave "information" to distinguish between the actual field measurement of real wave "data" and the computer-generated wave "information" from wind data. We always try to distinguish between real, measured wave "data" and computer-generated wave "information" in this book.

Wind data from airports located near coastal waters are a common source because of the long history of wind measurements, collection, and storage. There is far more knowledge of wind data near coastal waters than of wave "data". Wave "hindcasting" of wave conditions is often employed to generate the design criteria for waves at the project site.

2.3.7.2 *Wave hindcasting by simple parameter method*

It is somewhat intuitive that the speed of the wind over the water, U, the distance over the water which the wind blows called the fetch distance, F, and the length of time the winds of speed U blow over the fetch distance, F, called the wind duration, t, are the three key variables (parameters) that produce the wave conditions (height, period). Early relationships for defining waves from weather forecasts for planning the invasion of Northern Europe during WW II are summarized in Sverdrup and Munk (1947).

In the early 1970s, a series of field experiments were conducted in the North Sea of Europe that measured both the wave conditions and the wind conditions that produced the waves. The growth of the wave spectrum with fetch distance was quantified as schematically illustrated in Figure 2.24. The size of the spectrum increases with the fetch distance so that growth of the spectral significant wave height, H_{mo}, can be documented. And, the peak of the spectrum decreased in frequency (down-shifts) so the wave-period peak, $T_p(1/f_p)$, increase can also be determined. The project was called (JONSWAP), (*JO*int *N*orth *S*ea *WA*ve *P*roject), with results reported in Hasselmann *et al.* (1973).

The JONSWAP method employs four dimensionless ratios (superscript * means normalized values for F, H_{mo}, T_p, and t variables defined before) as follows:

$$F^* = \frac{gF}{U^2}, \tag{2.40}$$

$$H^*_{mo} = \frac{gH_{mo}}{U^2}, \tag{2.41}$$

$$T^*_p = \frac{gT_p}{U}, \tag{2.42}$$

$$t^* = \frac{gt}{U}, \tag{2.43}$$

and the following three dimensionless formulas (with coefficients for the metric system):

$$H^*_{mo} = 0.0016(F^*)^{1/2}, \tag{2.44}$$

$$T^*_p = 0.286(F^*)^{1/3}, \tag{2.45}$$

$$t^* = 68.8(F^*)^{2/3}, \tag{2.46}$$

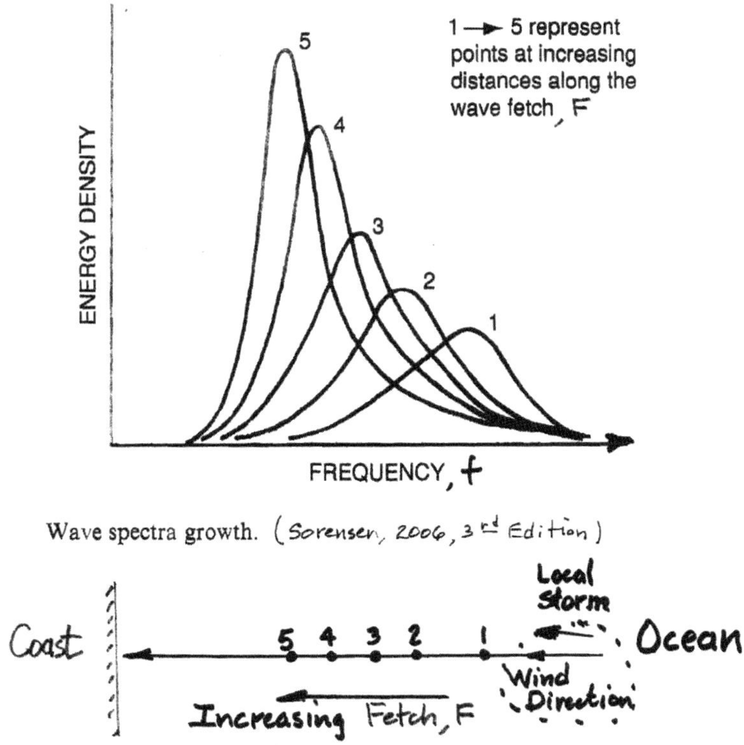

Figure 2.24. Schematic of wave spectral growth with increasing fetch distance (from Sorensen, 2006).

to estimate the spectral, significant wave height, H_{mo}, and spectral peak period, T_p, for a given site defined by the wind speed, fetch distance, and wind duration, i.e. U, F, t. The following example best illustrates the application of the JONSWAP method.

Given a wind speed, U, of 20 m/sec blowing over a 100 km water surface, i.e. the fetch distance, F, with wind conditions remaining relatively the same for a duration, t of 6 hours. Using the four dimensionless ratios gives F^* equal to 2,451.5, H^*_{mo} equal to 0.07922, T^*_p equal to 3.85625, and t^* equal to 12,508.9. Then, using the three formulas with coefficients for the metric system gives $H_{mo} = 3.2\,\text{m}$, $T_p = 7.9\,\text{s}$, and $t = 7.09$ hours. This example illustrates the important fact that the formula for t is for the *minimum* duration the winds must blow to produce these H_{mo} and T_p results. In other words, the given wind duration, t of 6 hours, is not sufficient and is below the minimum of 7.09 hours. This example illustrates how the wind-duration limits wave growth and therefore the wave height and wave period will be *less* for the 6 hours duration.

For a given wind speed, U, the key variable is the fetch distance, F, for a given wind duration, t. In the above example, a shorter fetch distance or "effective" fetch distance, F^*_{eff}, would produce a minimum duration, t of 6 hours. Rewriting the formula for t^* gives the following relation for F^*_{eff}:

$$F^*_{\text{eff}} = \left(\frac{t^*}{68.8}\right)^{3/2}, \tag{2.47}$$

where the actual duration, t is employed to find t^*.

Using the above example with t of 6 hours gives t^* equal to 10,590 (convert t in hours to seconds) and the resulting, dimensionless, effective fetch, F^*_{eff} is equal to 1,909.8 and the actual, effective fetch distance, F_{eff} is equal to 77.9 km. The effective fetch distance is always less than the actual fetch distance if the wind duration, t, is less than the minimum wind duration found from the JONSWAP formula.

To finish this example, using the effective fetch distance, F^*_{eff}, gives for the dimensionless ratios, H^*_{mo} equal to 0.06992 and T^*_p equal to 3.55. Then, using the JONSWAP formulas gives H_{mo} equal to 2.9 m and T_p equal to 7.2 s. The resulting wave conditions are "duration limited" in height and period because they could not develop further along the fetch due to the short duration of the given conditions. In general, always use the smaller value (F^* or F^*_{eff}) in the JONSWAP formulas.

Figure 2.25 is a nomograph as developed by the US Army Corps of Engineers (CEM, 2006) for graphic application of the JONSWAP formulas. The vertical axis is the wind speed, U (m/s). It can be used for a check on the numerical results using the formulas. For long fetch distances and low wind speeds (lower right corner) the results are labeled as "fully arisen sea". This simply means that no additional wave growth is possible no matter how long the wind duration.

When fetch distances are given in statute miles (5,280 ft per mi), the conversion is 2.23 mi per hour equals 1.0 m/s. If the fetch distance is nautical miles (6,073 ft per mi), the conversion is 1.945 knots per hour that equals 1.0 m/s.

2.3.7.3 *Quantification of U, F, t for use in parametric method*

The above is a classic, academic example when the U, F, and t are given information. In practice, determination of the correct wind speed, fetch

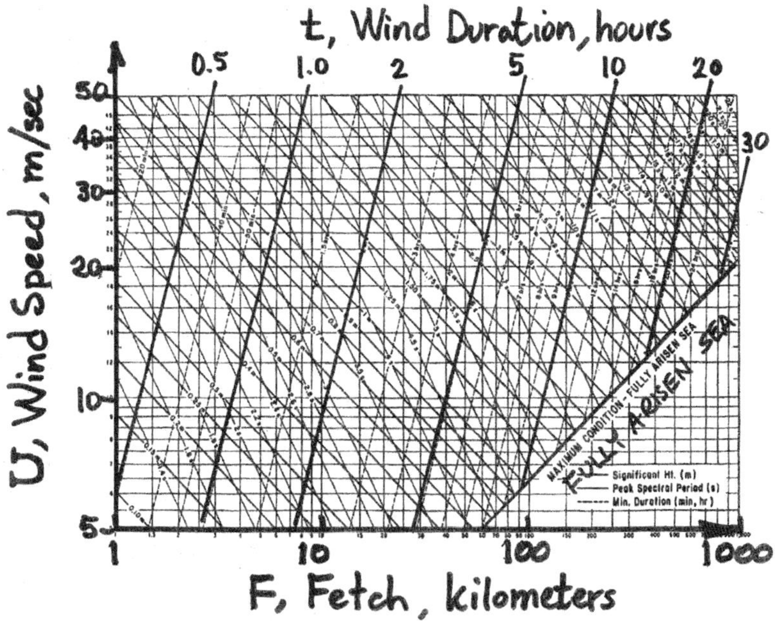

Figure 2.25. Nomograph for graphic application of the JONSWAP formulas (from CEM, 2006).

distance, and wind duration is the hardest part. The observed wind speed must be adjusted to the (1) correct elevation above the ground, (2) proper duration since hourly averages are needed, (3) whether wind data is measured over land (inland airports) or over water (coastal airports), and (4) for air–water temperature difference and stability of the atmosphere. The procedures for adjusting the observed winds are far beyond the intended scope of this book. The reader should consult Part II, Chapter 2, Section II-2-1-f (1) in CEM (2006) for all the details. Corps of Engineers software ACES (Leenknecht *et al.*, 1992) is also available to aid the effort. The ratios and formulas listed above in the JONSWAP method require the use of the adjusted wind speed, U, for proper application. And, a statistical analysis of the adjusted wind-speed data is performed to determine the probability of exceedance curve for the wind-speed data. The methods are similar to the probabilistic analysis of the measured wave data discussed in Section 2.3.8. Finally, the observed wind speed data are directional. Quadrants with 22.5° compass headings are normally applied to sort the wind-speed directions.

Airport wind data is also used to estimate the appropriate wind-speed durations, t, that have been observed, on average, for various levels of wind speed. In general, the higher the level of the adjusted wind speed, the shorter the sustained wind-speed duration. If possible, the joint probability relationship between wind speed and duration of wind speed for each quadrant of wind direction should be established. Even if crudely approximated, it will produce more realistic wave hindcast results.

Finally, the wind fetch distance(s) must be determined. Topographic and bathymetric maps are employed. The fetch distances for various dominant wind directions are simply scaled from the maps as illustrated in Figure 2.26 for a coastal reach on the northern shoreline for winds from the southwesterly, south, and southeasterly directions. The fetch lines 1, 2, 3, 4, and 5 are drawn in the middle of the 22.5° quadrants. The longest fetch direction may not coincide with the dominant wind-speed direction. The width of the fetch area is no longer a factor. CEM (2006) recommends that the straightline fetch should be used to define the fetch length in applications. For open ocean fetches, the wind duration, t, often governs the design. Waves are duration-limited.

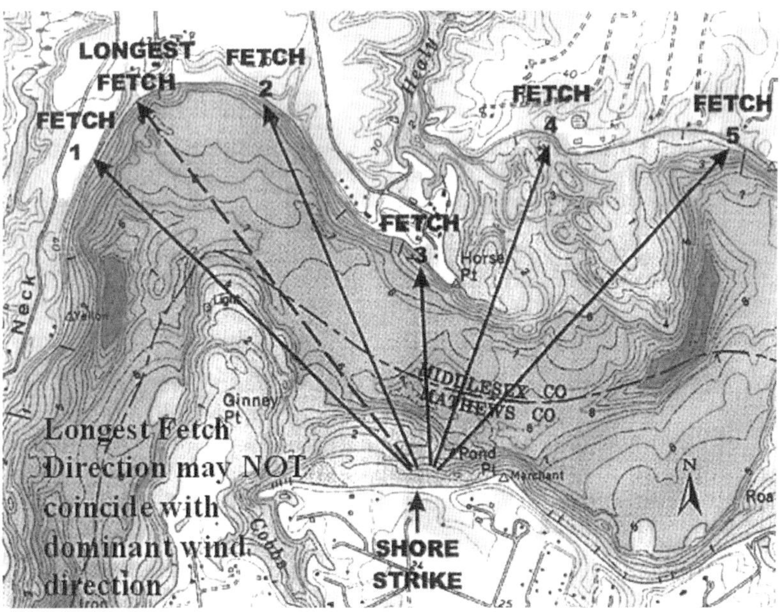

Figure 2.26. Graphic display of fetch distances over shallow water (from https://www.vims.edu/research/departments/physical/programs/ssp/shoreline_management/index.php).

Figure 2.26 also reveals the bathymetry along the various fetch distances. Early studies felt that water depth (shallow water) reduced wave growth by bottom friction. However, Jensen (1983) determined that the fetch-limited wave growth in shallow water was very similar to deepwater for comparable adjusted wind speeds. CEM (2006) recommends using the deep-water formulas (JONSWAP method) for all water depths but with one constraint, namely, that the peak wave period, T_p, cannot exceed that estimated by the following equation (Vincent, 1985):

$$(T_p)_{\max} = 9.78 \left(\frac{d}{g}\right)^{1/2} \quad \text{in shallow water.} \tag{2.48}$$

In the metric system, this equation approximately reduces to

$$d_{\max} = \frac{T_p^2}{10}. \tag{2.49}$$

If the weighted average water depth along the fetch, $\bar{d}$, is possibly less than $d_{\max}$, then the procedure is to first calculate T_p from the JONSWAP formulas; then calculate $d_{\max}$ and if $\bar{d} < d_{\max}$, to apply Equation (2.48) to find the reduced T_p.

2.3.7.4 *Two-dimensional hindcasting models*

The JONSWAP hindcast model is one-dimensional (1D) and parametric, i.e. empirically related to the parameters U, F, and t along the one-fetch direction. Wave-spectrum propagation models in two horizontal space dimensions and based on conservation of wave energy or wave-action density principles are far more complex. They have been developed and are operated by meteorological and oceanographic organizations in many countries of the world to produce wave predictions and hindcasts on global scales for the worlds' oceans. International collaboration in the 1970s–1980s produced the WAM models that spawned the WAVEWATCH III (series) (WAMDI Group, 1988) for the US NOS at ocean scales. The details are far beyond the scope of this book. They have been employed to develop long-term wave information to estimate extreme water wave conditions. See Section 2.3.8.3 for more details.

2.3.8 *Extreme water waves*

Long-term wave data (or wave information) are necessary to determine extreme wave conditions for design. The analysis of one, 20-min record in

Section 2.3.3 produced one realization of H_s ($\overline{H}_{1/3}$ or H_{mo}) and T (T_m or T_p). For ease of notation, we omit the subscripts with the understanding that H, T are for irregular waves. If hourly recording (or hindcasts) were made daily for 35 years, there will be over 300,000 values of H. Calm sea conditions will result in less than that number. During storm conditions lasting many hours, these hourly wave heights are not statistically independent events. They are related to the previous values in the time series, and consequently the time series of $H(t)$ must be separated into individual "storm" events which are independent to permit statistical analysis.

2.3.8.1 *Peak-over-threshold (POT) method*

The POT method is routinely employed to separate the time series of $H(t)$ values into separate storm events. A threshold wave height, H^*, to define the start of the coastal storm must first be established. This can be somewhat arbitrary or statistically related to the mean and standard deviations of the mean wave-height conditions at the site (Basco and Mahmoudpour, 2012). For example, at the Corps of Engineers, Field Research Facility (FRF) at Duck, NC, the threshold wave height, H^*, is the mean wave height. $\overline{H}$, as measured at the near-shore, end-of-pier wave gauge (8-m water depth), since 1985, plus two standard deviations of the mean, σ_H. In 2010, this was $H^* = \overline{H} + 2\sigma_H = 0.90 + 2(0.57) = 2.04\,\text{m}$ and approximately 2 m (Birkemeier, 2010, personal communication). The storm "ends" whenever the significant wave height drops below 2 m. The measured H must be above 2 m for a minimum of 8-hours duration to be considered as one "storm" event. The mean $\overline{H}$ should be based on a minimum of 1 year of data.

See Basco and Mahmoudpour (2012) for additional definitions of the start of a coastal storm. There is no generally accepted definition for the start of a coastal storm event.

The POT method proceeds as outlined in the following steps:

(1) The long-term time series of $H(t)$ values is separated into a number of "bins" of equal magnitude. For example, if $H_{\max}$ is 6 m and the bin size is 0.250 m, there will be 24 bins. Each bin will then contain the number of waves within the bin limits. The smallest bin (0.000–0.250) may not contain the largest number but the largest bin (5.751–6.000) will normally contain the fewest to produce a non-Gaussian wave-height histogram.

(2) The POT wave height, H^*, is invoked to separate all the waves, $H(t) > H^*$ into individual storm wave events. If $H^* = 1.5\,\text{m}$, the first bin range is now (1.501–1.750). The maximum wave height in each individual storm is determined and counted as one storm in the bin size that contains the magnitude of that storm wave height. For example, Figure 2.27 displays one month of hourly wave information on Lake Huron with a threshold wave height, H^*, of 1.5 m (from Kamphuis, 2010, Figure 4.1, p. 88) Consequently, eight "storms" are identified ($H(t) > 1.5\,\text{m}$) and the *maximum* wave height for each storm is determined and counted as one wave height in the bin containing that wave height. For this one month, two storms labeled (2) and (5) were in the (3.251–3.500) bin range. The number of storms above the threshold level is on the order of 1% of the total number of hourly wave heights. Increasing the threshold level significantly decreases the number of storm events. For a threshold wave height, $H^* = 1.5\,\text{m}$, there are now 18 bins remaining for the above case with H_{max} of 6 m and a bin size of 0.250 m. The histogram is now generally always decreasing in the number of storms with every increasing maximum wave height.

(3) The probability that any wave height, H', is equal to or less than a specified wave height, H, is now determined. In the above example,

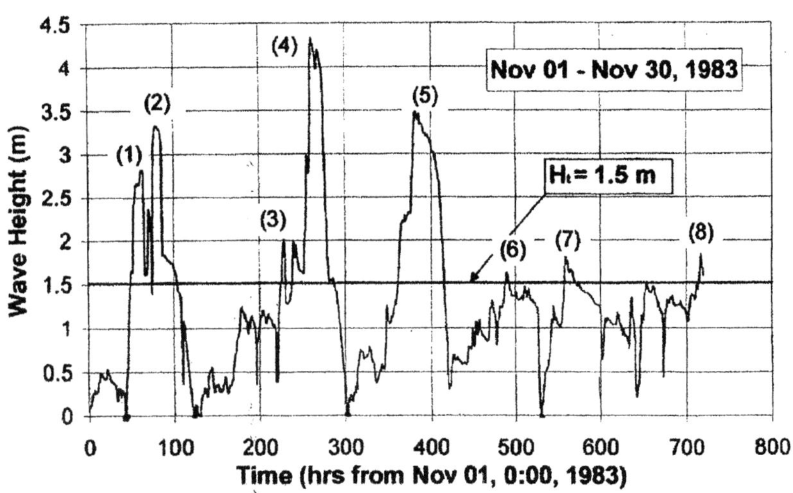

Figure 2.27. Example of determining the number of storms within each bin for a given POT wave height, $H^* = 1.5\,\text{m}$ (from Kamphuis, 2010).

since the threshold wave height $H^* = 1.5\,\text{m}$ and the bin size is $0.250\,\text{m}$, the number of storm events in the first bin (1.501–1.750) are all below 1.750 m.

A spreadsheet can then be developed where the first column is the largest value of the bin and the second column is the number of storms N in the bin. Table 2.5 is an example as modified from Kamphuis (2010) for the Lake Huron wave information cited above. This example contained 2,738 storms over a 34.9-year period using the threshold H^* of 1.5 m. We are thankful to Dr. Kamphuis for permitting the use of this example.

The first bin (1.501–1.750) contained $N = 1{,}019$ storms as shown in Column (2).

Formally,

$$P = P(H' \leq H), \tag{2.50}$$

where P is the probability of non-exceedance, Column (3) in Table 2.5 and is the cumulative number of storms $\sum N$ below that bin level divided by the total number of storms. The probability of non-exceedance below $H = 1.750$

Table 2.5. Weibull distribution analysis of Lake Huron wave information (from Kamphuis, 2010).

(1) H	(2) N	(3) P	(4) Q	(8) W $\alpha = 0.8$	(9) W $\alpha = 1.3$
1.75	1019	0.372	0.628	0.384	0.555
2.00	549	0.573	0.427	0.816	0.883
2.25	382	0.712	0.288	1.316	1.184
2.50	254	0.805	0.195	1.848	1.459
2.75	174	0.869	0.131	2.421	1.723
3.00	113	0.910	0.090	2.996	1.964
3.25	81	0.939	0.061	3.627	2.210
3.50	60	0.961	0.039	4.366	2.477
3.75	40	0.976	0.024	5.176	2.750
4.00	27	0.986	0.014	6.105	3.044
4.25	19	0.993	0.007	7.326	3.406
4.50	10	0.996	0.004	8.638	3.769
4.75	4	0.998	0.002	9.632	4.031
5.00	2	0.99854	0.00146	10.436	4.234
5.25	1	0.99890	0.00110	11.014	4.377
5.50	2	0.99963	0.00037	13.276	4.910
5.75	0	0.99963	0.00037	13.276	4.910
6.00	1	1.00000	0.000		
Total	2738				

is 1,019/2,738 = 0.372, below $H = 2.00$, P is $(1{,}019 + 549)/2{,}738 = 0.573$, etc. All the storm maximum wave heights were below $H = 6.00$ so $P = 1.000$. A plot of probability of non-exceedance (Column 3) as the y-axis versus wave height (Column 1) on the x-axis produces a curved line that is difficult to extrapolate for extreme wave heights.

2.3.8.2 *The Weibull extreme value distribution for POT method*

(4) A coordinate transformation is performed on the probability of non-exceedance P values (Column 3) on the y-axis to produce a straight-line $Y = AX + B$ that can be extended for extreme-wave-height storm events. The Weibull extreme-value distribution employs three parameters (α, β, γ) and is a special case of the generalized extreme value (GEV) distribution. In probability theory and statistics, the GEV distribution is a family of continuous probability distributions developed within extreme-value theory to combine the Gumbel, Fréchet, and Weibull families also known as Type I, II, and III extreme-value distributions.

The extreme-value non-exceedance probability P for the Weibull distribution is

$$P(H' < H) = 1 - \exp\left(-\left\{\frac{H-\gamma}{\beta}\right\}^{\alpha}\right), \tag{2.51}$$

where γ is the position parameter, β is the scale parameter, and α is the shape parameter. For design, we are interested in the exceedance probability, Q which is

$$Q = Q(H' > H) = 1 - P \tag{2.52}$$

or

$$Q = \exp\left(-\left\{\frac{H-\gamma}{\beta}\right\}^{\alpha}\right) \tag{2.53}$$

as shown in Column (4) in Table 2.5 for the Lake Huron example. Taking the natural log (ln) of both sides and transforming the exponent, α gives

$$\left(\ln\frac{1}{Q}\right)^{1/\alpha} = \frac{H-\gamma}{\beta}. \tag{2.54}$$

For a straight line given by

$$Y = AX + B, \tag{2.55}$$

the transformed Y-axis (called the Weibull reduced variate, W) becomes

$$Y = W = \left(\ln \frac{1}{Q}\right)^{1/\alpha} \tag{2.56}$$

with

$$A = \frac{1}{\beta}, \quad X = H, \quad \text{and} \quad B = -\left(\frac{\gamma}{\beta}\right). \tag{2.57}$$

(5) The Weibull reduced variate, W, becomes the transformed y-axis and is calculated in Columns (8) and (9) in Table 2.5. Plotting W (y-axis) versus H (x-axis) generally gives a straight line and a least-squares curve fit permits determination of the coefficients A and B from which β and γ can be determined by the relations in Equation (2.57). But the only difficulty is what to take for the third free parameter α needed to calculate W? There is no explicit solution. Repeated, trial and error with different values of α and determination of the straight line of best fit (i.e. the largest correlation coefficient, R) is the only way to apply the Weibull extreme-value distribution.

In the example (from Kamphuis, 2010), using $\alpha = 0.8$ gave the results in Column (8) which was not a straight line. The results for a best-fit value of $\alpha = 1.3$ in Column (9) for reduced variate, W, is shown in Figure 2.28 as the black data points. A least-squares line of best fit gave $y = 1.128x - 1.403$ with $R^2 = 0.997$ so that $A = 1.128$ and $B = -1.403$. Using the relationships in Equation (2.57) gives

$$\beta = \frac{1}{A} = \frac{1}{1.128} = 0.886,$$
$$\gamma = -\beta \times B = -0.886\,(-1.403) = 1.243,$$

for $\alpha = 1.3$.

(6) Knowledge of the three parameters (α, β, and γ) for the Weibull distribution that best fits the data can now be employed to give the central estimate of the significant wave height with a recurrence interval T_r years ($T_r = 1/Q$), herein noted as $\hat{H}_s^T$. The caret symbol (ˆ) over the top means the central tendency estimate. We present a full discussion of the meaning of $\hat{H}_s^T$ in Section 4.3. The POT method ranks all the wave heights above the threshold value, H^*, so that consequently each one represents one storm. For the example, there were 2,738 storms over the 34.9 years giving the number of storm events per year, λ, equal to

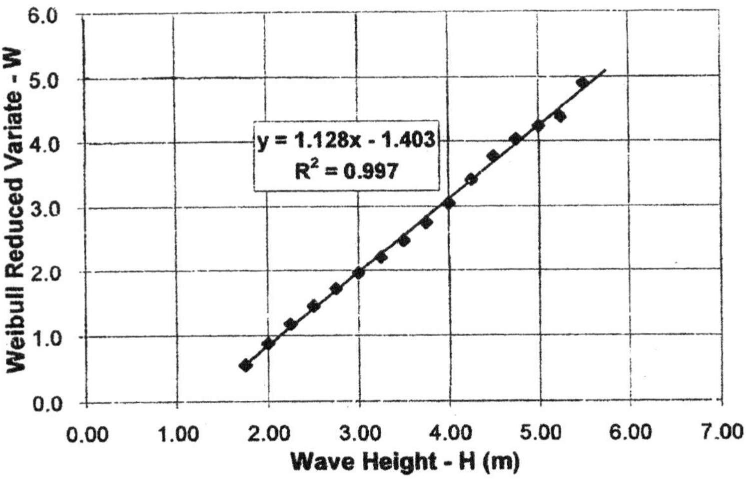

Figure 2.28. From Kamphuis (2010) for Lake Huron example for Weibull Distribution ($\alpha = 1.3$).

78.45. Consequently, the exceedance probability, Q, for one event with a recurrence interval of T_r years becomes

$$Q = \frac{1}{\lambda T_r}. \tag{2.58}$$

Rearranging the straightline equation and solving for $\hat{H}_s^T$ gives in general

$$\hat{H}_s^T = \gamma + \beta\left(\ln\frac{1}{Q}\right)^{1/\alpha}, \tag{2.59}$$

$$\text{or} \qquad \hat{H}_s^T = \gamma + \beta\,(\ln(\lambda T_r))^{1/\alpha}. \tag{2.60}$$

For example, Lake Huron data with $\alpha = 1.3$,

$$\hat{H}_s^T = 1.243 + 0.866\,(\ln(78.45T_r))^{1/1.3},$$

and now central tendency estimates of $\hat{H}_s^T$ can be found for a range of Q and T_r values as illustrated in Table 2.6. The central tendency is the mean line of best fit for the data.

A spreadsheet has been provided in the online supplementary materials with this example. The spreadsheet can be modified for general use with the Weibull method with any long-term wave data or wave information generated from wind data.

Table 2.6. Estimates of $\hat{H}_s^T$ for various recurrence intervals, T_r, for Lake Huron example.

Recurrence Interval, T_r, years	Probability of Exceedence, Q	Wave Height, H_s, m
20	0.05	5.26
50	0.02	5.64
100	0.01	5.92
200	0.005	6.20
500	0.002	6.56

In effect, the POT method uses wave heights for all the individual storm events each year in the analysis. In contrast, using only the maximum storm event each year restricts the analysis to only a few years of available data or information. For this reason, the POT method is preferred. See Chapter 4, Section 4.3 for how to estimate the confidence interval (C.I.) associated with the central tendency (mean value) of $\hat{H}_s^T$.

2.3.8.3 *Wave data or wave information*

In all the preceding sections, long-term knowledge of wave heights is assumed known in deep water at the site. This knowledge may come from field measurements of waves "wave data" or from computer-generated "wave information" using long-term wind data. We always try to distinguish between the real, measured wave "data" in the field and the computer-generated wave "information" in this book.

Because there are very few wave gauges operating in coastal waters, the US Army Corps of Engineers has developed the Wave Information System (WIS) (Jensen, 1983). The WIS project produces a high-quality online database of hind-cast, wave conditions in 15–20 m water depths covering US coasts. The hind-cast information provides a 20-year long wave record needed in coastal engineering design. It has relatively dense spatial coverage, and a level of temporal continuity not available from measurements. Updated hind casts are available for download for the Atlantic and Gulf coasts (1980–1999), Pacific (1981–2011), the five Great Lakes (record length varies), Alaska (1985–2011), and in some locations in the Caribbean. The website for the Wave Information Study is http://wis.usace.army.mil.

For example, one Atlantic Ocean Station 63213 is located at latitude 36.420°, longitude 75.420° in 26-m water depths offshore the Corps' FRF in Duck, NC. The products available for this station location (typical) are shown in Figure 2.29 as downloaded from the USACE website. One of the

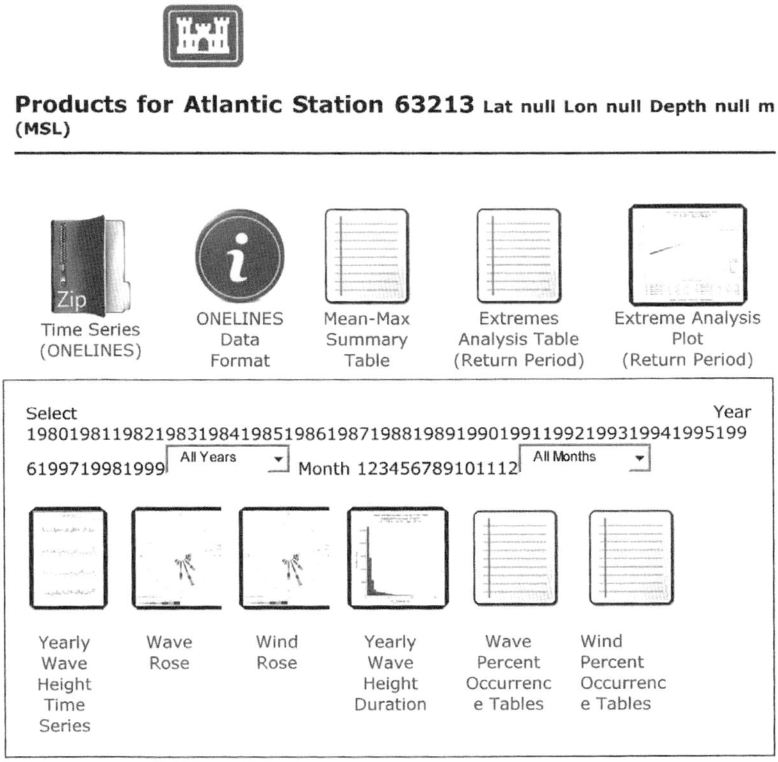

Figure 2.29. Products available for downloading for Station 63213 (typical) from WIS online (courtesy Corps of Engineers).

key products is the "Wave Percent Occurrence Tables" which summarize the number of 3-hour hindcasts in joint wave height, H_{mo}, bins (0.5 m increments) versus wave period, T_p, bins (1–2 s intervals) for each of the 16 wave direction bands for 22.5° quadrants on the 360° compass. This information is summarized by calendar years and for all 20 years (1980–1999) as shown in Figure 2.30.

Figure 2.31 gives the wave-direction information for 1998 as a wave "rose" plot for all 16 quadrant directions. Note the different colors indicate various wave-height bins and the length of the colors illustrate the frequency of occurrence to readily determine the dominant wave directions.

A second important product is the extreme event analysis table which ranks all the storms with wave heights greater than a specified threshold wave height, H^*. For this Station 63213, the threshold wave height was

ATLANTIC WAVE HINDCAST : ST63213_v01
ALL MONTHS FOR YEARS PROCESSED : 1980 - 1999
STATION LOCATION : (-75.42 W / 36.42 N)
DEPTH : 26.0 m

PERCENT OCCURRENCE (X1000) OF HEIGHT AND PERIOD
FOR ALL DIRECTIONS

NO. CASES : 58437
NO. CALMS : 41

HEIGHT IN METERS	PARABOLIC FIT OF PEAK SPECTRAL WAVE PERIOD (IN SECONDS) <5.0	5.0-5.9	6.0-6.9	7.0-7.9	8.0-8.9	9.0-9.9	10.0-11.9	12.0-13.9	14.0-15.9	16.0-LONGER	TOTAL
0.00-0.10	.	.	.	.	.	.	.	.	.	.	70
0.10-0.49	3001	910	2830	2623	1209	930	973	516	138	29	13159
0.50-0.99	17779	3607	4546	3157	2209	1728	2116	1168	231	88	36629
1.00-1.49	10515	5404	1860	1312	1064	1132	1406	1102	231	77	24103
1.50-1.99	1659	5339	2570	566	513	484	600	474	154	27	12386
2.00-2.49	78	1033	3725	900	330	159	335	242	65	13	6880
2.50-2.99	.	23	1798	1194	232	104	155	142	53	8	3709
3.00-3.49	.	.	61	1023	330	65	119	75	30	6	1709
3.50-3.99	.	.	.	169	297	123	66	53	29	6	743
4.00-4.49	.	.	.	17	112	61	63	30	10	13	306
4.50-4.99	.	.	.	.	17	37	27	35	10	.	126
5.00-5.99	.	.	.	.	.	30	22	44	5	.	101
6.00-6.99	.	.	.	.	.	.	13	17	3	.	33
7.00-7.99	.	.	.	.	.	.	.	1	1	.	2
7.00+	.	.	.	.	.	.	.	1	1	.	2
TOTAL	33032	16316	17390	10961	6313	4853	5895	3899	960	267	

MEAN Hmo(M) = 1.2 LARGEST Hmo(M) = 7.4 MEAN TPP(SEC) = 6.5 FINITE

Figure 2.30. WIS joint wave height versus wave period information for Sta 63213 for all directions.

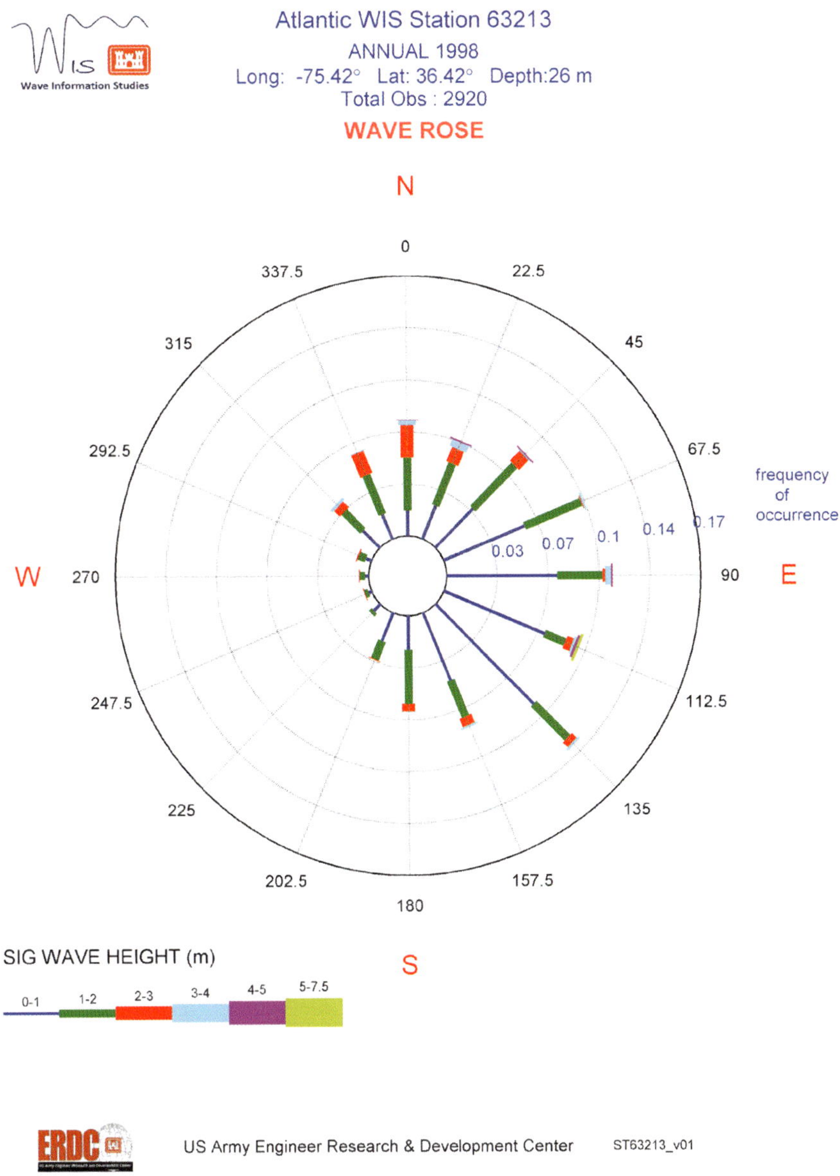

Figure 2.31. Wave rose for Station 63213 for year 1998 from WIS information at 26-m water depth (courtesy Corps of Engineers).

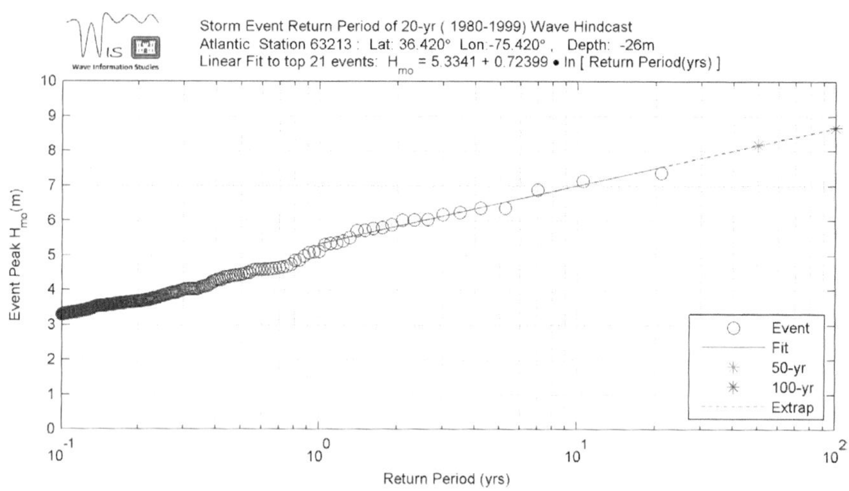

Top 10 events based on Peak H_{mo}

Event	Date/Time(UTC)	H_{mo}	T_p	θ_{mean}	Event	Date/Time(UTC)	H_{mo}	T_p	θ_{mean}
1	1998/08/28 00:00	7.37	13.38	121.0	6	1990/10/26 12:00	6.23	10.88	14.0
2	1999/08/31 18:00	7.13	14.05	56.0	7	1992/09/25 12:00	6.17	11.06	69.0
3	1993/09/01 03:00	6.88	12.29	78.0	8	1986/08/18 00:00	6.03	10.31	160.0
4	1994/11/18 15:00	6.36	14.01	99.0	9	1999/09/16 15:00	6.02	10.19	151.0
5	1993/03/14 00:00	6.35	11.12	173.0	10	1982/10/25 00:00	6.01	12.26	57.0

An event is defined as any period when H_{mo} > 3.00m θ_{mean} is direction that waves are arriving from

ERDC US Army Engineer Research & Development Center ST63213_v01

Figure 2.32. Extreme event analysis product for Station 63213 for the 20-year period (1980–1999) (courtesy Corps of Engineers).

3.0 m and 350 storm events were identified and can be downloaded from the extreme analysis table link given in Figure 2.29. The extreme analysis plot link in Figure 2.29 is displayed in Figure 2.32 with wave height H_{mo} on the vertical axis (linear scale) and recurrence interval T_r on the horizontal axis (logarithmic scale). The Corps of Engineers simply fit a linear curve to the top 21 wave heights to produce the equation shown. The 100-year ($Q = 0.01$) recurrence interval wave height is 8.67 m using this linear fit.

Using all 350 storm events in the POT method and making a Weibull, extreme-value analysis for $\alpha = 1.15, \beta = 1.063943$, and $\gamma = 2.228$, the 100-year recurrence interval wave height is 8.34 m. We feel this result is more accurate since it uses all the 350 storm wave heights above the threshold level of 3.0 m and is the product of applying the well-known Weibull, three-parameter analysis theory.

A complete review of all the products available for the WIS stations as illustrated in Figure 2.32 is beyond the scope of this book. The WIS information base presents spectral-significant wave heights H_{mo}, peak wave period T_p, and wave direction θ (azimuth from true north) at three-hour intervals for long time periods (20 years). It is an extremely valuable source of wave information for coastal structures design in US waters.

2.3.9 *Climate change and water waves*

Chapter 11, Section 11.3.2 includes a review of climate change, sea-level rise, and current research to understand their impact on water wave statistics.

2.3.10 *Design wave conditions*

Some level of the recurrence interval (e.g. $T_r = 100$ years) is often still selected in the deterministic design method. What are the consequences of damage and/or functional non-performance when the levels exceed the 100-year level? Is the analysis, methods, and record length sufficient to have confidence in the 100-year wave-height level? Why is the 100-year wave-height level selected? What is the magic in the 100-year wave height level?

The risk-based design approach employs the entire exceedance probability curve and is the modern approach as discussed fully in Chapters 4 and 8. The complete range of water-wave conditions for the full range of probabilities of exceedance as shown in Figure 2.33 (extrapolated region in red from Weibull analysis for above example) are needed and applied in the probabilistic design of coastal structures. Some probability of exceedance will cause structural or functional failure. To plan for failure is now the correct philosophy of coastal structures design.

Wave conditions at the structure produce wave setup, $\bar{\eta}$, as discussed in Section 2.3.6. This is a significant increase in mean water level that cannot be ignored.

2.4 Geotechnical (Soil) Conditions

Coastal structures sit upon, are acted upon, or push on the sediment beneath and adjacent to the structure. Consequently, knowledge of the local sediments is important to properly design the structure. We believe the following geotechnical conditions are important for coastal structures design. And, only the basic characteristics (size, density, porosity, and bulk density) are reviewed. See CEM (2006, Part III, Chapter 1) for a thorough discussion of coastal sediment properties.

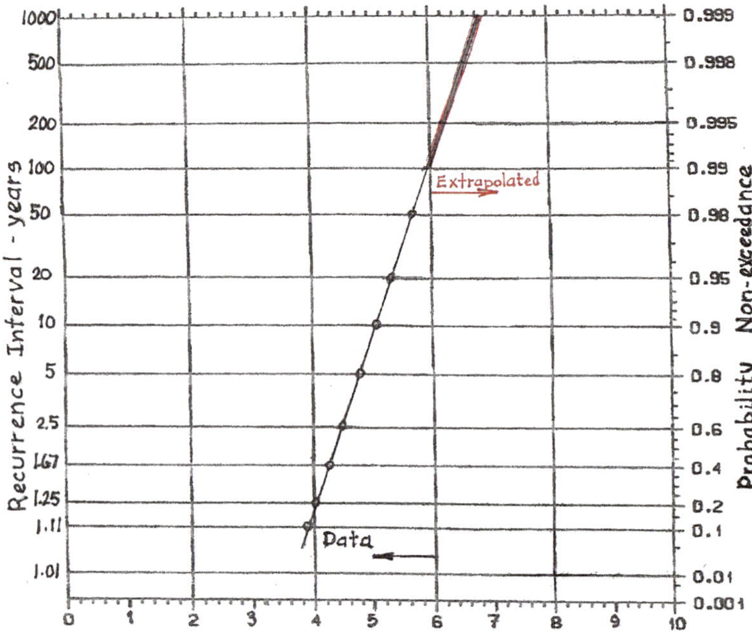

Figure 2.33. Spectral significant wave height, H_{mo}, versus recurrence interval, T_r, for example from Lake Huron using 34.9 years of wave information. Extrapolated information in red.

2.4.1 *Sediment characteristics*

2.4.1.1 *On the surface*

The most important sediment parameter is size as determined for a sample taken at the surface. The diameter, D, passing through a certain standard size sieve opening is the most common. (*Some references use the symbol, d, for grain size*). Figure 2.34 displays the sieve size in mm for US Standard Sieve (ASTM and Unified Soil Classification System).

For example, a Number-10 sieve will pass all sizes smaller than 2.00 mm and retain all sizes greater than 2.00 mm. Since all samples will contain a range of sizes, sorting by passing through a range of sieves will produce the grain-size (diameter) distribution with example shown in Figure 2.35. The percent finer is shown on the vertical axis for the range of "sieve" diameters (mm) on the horizontal axis. In Figure 2.35, about 90% of the sample is finer than 2 mm (passes through the Number 10 sieve). Other countries have different sizes for the mesh in the sieves and other methods (e.g. the fall

Table III-1-2
Sediment Particle Sizes

ASTM (Unified) Classification[1]	U.S. Std. Sieve[2]	Size in mm	Phi Size	Wentworth Classification[3]
Boulder		4096.	-12.0	
		1024.	-10.0	Boulder
	12 in. (300 mm)	256.	-8.0	
Cobble		128.	-7.0	Large Cobble
		107.64	-6.75	
		90.51	-6.5	Small Cobble
	3 in. (75 mm)	76.11	-6.25	
		64.00	-6.0	
		53.82	-5.75	
		45.26	-5.5	Very Large Pebble
Coarse Gravel		38.05	-5.25	
		32.00	-5.0	
		26.91	-4.75	
		22.63	-4.5	Large Pebble
	3/4 in. (19 mm)	19.03	-4.25	
		16.00	-4.0	
		13.45	-3.75	
		11.31	-3.5	Medium Pebble
		9.51	-3.25	
Fine Gravel	2.5	8.00	-3.0	
	3	6.73	-2.75	
	3.5	5.66	-2.5	Small Pebble
	4 (4.75 mm)	4.76	-2.25	
	5	4.00	-2.0	
Coarse Sand	6	3.36	-1.75	
	7	2.83	-1.5	Granule
	8	2.38	-1.25	
	10 (2.0 mm)	2.00	-1.0	
	12	1.68	-0.75	
	14	1.41	-0.5	Very Coarse Sand
	16	1.19	-0.25	
	18	1.00	0.0	
Medium Sand	20	0.84	0.25	
	25	0.71	0.5	Coarse Sand
	30	0.59	0.75	
	35	0.50	1.0	
	40 (0.425 mm)	0.420	1.25	
	45	0.354	1.5	Medium Sand
	50	0.297	1.75	
	60	0.250	2.0	
	70	0.210	2.25	
Fine Sand	80	0.177	2.5	Fine Sand
	100	0.149	2.75	
	120	0.125	3.0	
	140	0.105	3.25	
	170	0.088	3.5	Very Fine Sand
	200 (0.075 mm)	0.074	3.75	
	230	0.0625	4.0	
Fine-grained Soil:	270	0.0526	4.25	
	325	0.0442	4.5	Coarse Silt
Clay if PI ≥ 4 and plot of PI vs. LL is on or above "A" line and the presence of organic matter does not influence LL.	400	0.0372	4.75	
		0.0312	5.0	
				Medium Silt
		0.0156	6.0	
				Fine Silt
		0.0078	7.0	
				Very Fine Silt
		0.0039	8.0	
Silt if PI < 4 and plot of PI vs. LL is below "A" line and the presence of organic matter does not influence LL.				Coarse Clay
		0.00195	9.0	
				Medium Clay
		0.00098	10.0	
				Fine Clay
		0.00049	11.0	
(PI = plasticity limit; LL = liquid limit)		0.00024	12.0	
		0.00012	13.0	Colloids
		0.000061	14.0	

[1] ASTM Standard D 2487-92. This is the ASTM version of the Unified Soil Classification System. Both systems are similar (from ASTM (1994)).
[2] Note that British Standard, French, and German DIN mesh sizes and classifications are different.
[3] Wentworth sizes (in mm) cited in Krumbein and Sloss (1963).

Figure 2.34. Sediment particle sizes (from CEM, 2006, Corps of Engineers).

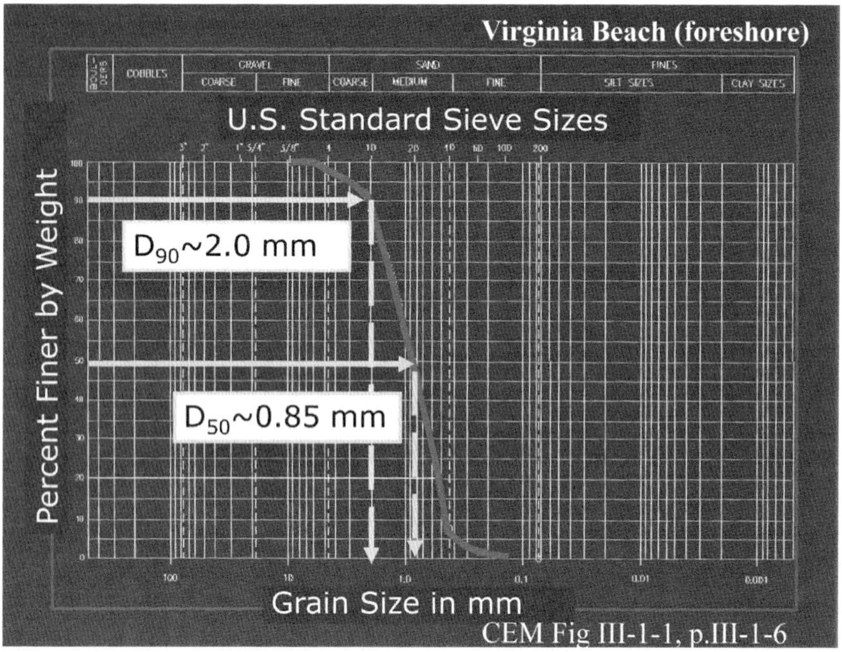

Figure 2.35. Example grain-size distribution curve for percent finer at Virginia Beach foreshore (CEM, 2006).

velocity) can be employed to determine the equivalent fall velocity "size" of the sediment.

For most coastal engineering applications, the median, D_{50}, size in mm is taken as representative of the sediment size distribution. Fifty percent is smaller and 50% is larger. And, the ASTM Classification is employed to separate the sizes into fine-grain soil (clay, silt) below 0.074 mm, fine sand $0.074 < D < 0.420$ mm, medium sand $0.420 < D < 2.00$ mm, etc. (see Figure 2.34). For the example in Figure 2.35, D_{50} is about 0.85 mm and a medium sand is taken from the foreshore at Virginia Beach, VA.

Coastal geologists employ the Wentworth classification system (Figure 2.34) which has more subnames for size ranges and uses the Phi Size, φ, defined as

$$\phi = -\log_2 D, \tag{2.61}$$

to define the sediment size in φ units with D the grain diameter in mm. For $D = 2$ mm in physical size, $\varphi = -1.0$ in φ units. Note the negative sign in front of the number of φ units and note that there is no length scale (mm, ft) associated with the φ-unit scale. As the sediment sizes increase in the physical diameter, D, the φ units get smaller (larger in negative numbers). And, as the physical size of the sediments D get smaller, the φ-scale numbers increase and become positive numbers. The φ-unit scale has no physical meaning, it has no dimensions to be used in dimensional analysis, and is very ambiguous and counterintuitive for coastal engineers. Unfortunately, it remains a part of the science and therefore is referenced here.

A more useful form for Equation (2.61) is

$$\phi = \frac{-\ln(D)}{\ln 2}. \tag{2.62}$$

As an example, when $D = 0.5\,\text{mm}$, then

$$\phi = \frac{-\ln(D)}{\ln 2} = \frac{-\ln(0.5)}{\ln(2)} = \frac{0.693}{0.693} = 1,$$

as shown in Figure 2.34 and the division between medium and coarse sand in the Wentworth classification.

The conversion from φ units to sediment size D in mm is convenient to use in the following form:

$$D = 2^{-\varphi}. \tag{2.63}$$

The following list displays four additional statistics from the grain-size distribution curve (Figure 2.35) namely, the mean, $M_{D\varphi}$, the standard deviation of the mean (sorting), σ_φ, the skewness (symmetry), α_φ, and the kurtosis (peakedness), β_φ, with formulas using φ values from the cumulative distribution curve.

$$\text{Median} = \varphi_{50} \text{ (or } D_{50})$$

$$\text{Mean, } M_{\text{D}\varphi} \equiv \frac{\phi_{34} + \varphi_{16}}{2} \quad \text{or} \quad \frac{\phi_{34} + \varphi_{50} + \varphi_{16}}{3}$$

$$\text{Std Dev. of Mean, } \sigma_\varphi \equiv \frac{\phi_{34} - \varphi_{16}}{2} \quad \text{(Sorting)}$$

$$\text{Skewness, } \alpha_\varphi \equiv \frac{M_{\text{D}\varphi} - \varphi_{50}}{\sigma_\varphi} \quad \text{(Symmetry)}$$

$$\text{Kurtosis, } \beta_\varphi \equiv \frac{(\varphi_{16} - \varphi_5) + (\varphi_{95} - \varphi_{84})}{2\sigma_\phi} \quad \text{(Peakedness)}$$

These formulas can be found in CEM (2006, Part III, Section 1) with examples. Unfortunately, it is not made clear which cumulative distribution curve (percent finer or percent coarser) should be employed to obtain the needed φ values. Engineers use the sediment size D in the mm scale and plot the percent finer (passing) to obtain the cumulative, percent finer distribution curve shown in Figure 2.35. Geologists use the sediment size as the φ-scale and plot the percent coarser (retained) to obtain the cumulative, percent coarser distribution curve shown in Figure 2.36. A review of the original paper by Professor Inman (1952) facilitated the learning of the answer.

> "In all formulas, use the percent coarser cumulative distribution curve for φ values." (Inman, 1952).

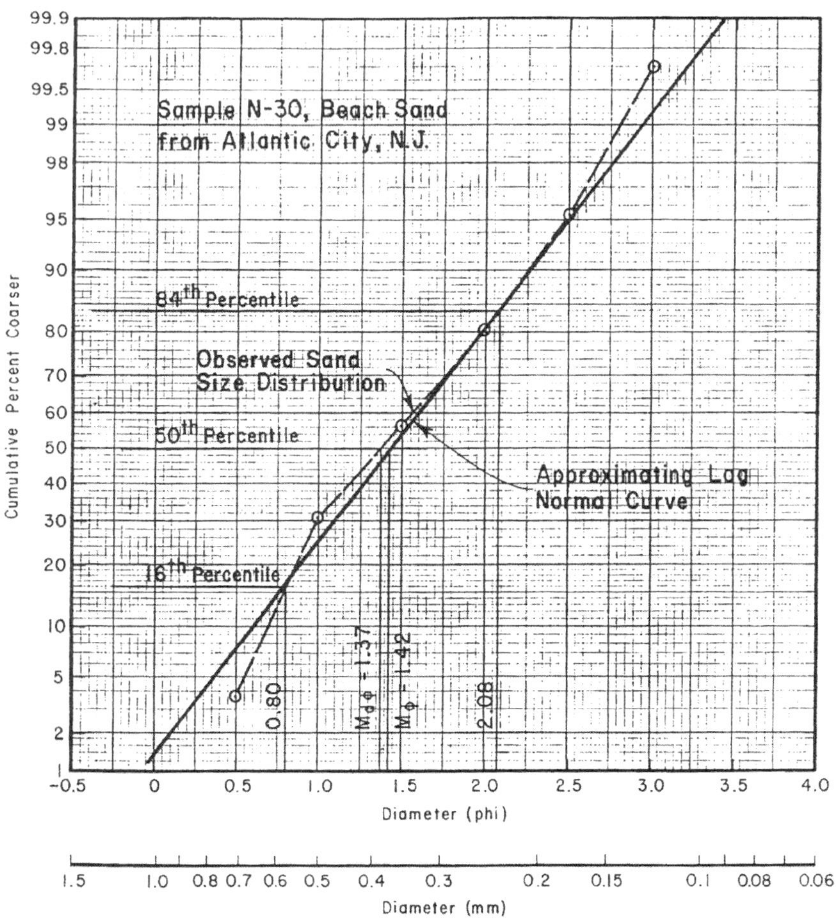

Figure 2.36. Cumulative percent coarser grain size distribution curve for Atlantic City, NJ (from CEM, 2006).

Quartz and feldspar are the minerals that make up most of the sediments. The specific gravity s is on the order of 2.65, so that the mass density ρ_s is about 2,650 kg/m^3 (5.14 slugs/ft^3). Since in the metric system, the gravitational constant g is 9.807 m/s^2, then the unit weight γ_s of the sediment is ρ_s times g or about 25,995 N/m^3 or 165.4 lbs/ft^3. The proper unit of force (weight) in the metric system is Newtons, N and we shall never use kg for force units in this book.

All sediments have pore space (voids). The porosity p is defined as the ratio of the pore space to the total volume. For coastal sediments composed of sand, p varies from 0.35–0.40. When the sediment volume is submerged and the voids are full of water of mass density, ρ, the bulk mass density of the submerged unit volume can be determined.

$$\text{Submerged bulk mass density} = (1 - p)\rho_s + p\rho. \tag{2.64}$$

Using common values for $\rho_s = 2{,}650$ (kg/m^3), $\rho = 1{,}000$ (kg/m^3), and $p = 0.35$ gives submerged bulk mass density $= (1 - 0.35) \times 2{,}650 + 0.35 \times 1{,}000 = 2{,}075$ kg/m^3 (4.02 slugs/ft^3) and the submerged bulk unit weight $=$ 20,350 (N/m^3) or 129 (lbs/ft^3).

CEM (2006, Part III, Chapter 1) gives tables for the mass density of common coastal sediments (Table III-1-4) and bulk mass densities (Table III-1-6) along with a thorough discussion of additional soil bulk properties including permeability, angle of repose, and other sediments (clay, silt, and mud). These results are for a grab sample taken at the surface location of the coastal structure.

2.4.1.2 *Beneath the surface (soil borings)*

For the feasibility study, it is commonly assumed that the surface sediments are representative of the soil conditions with the depth beneath the structure. Or available information from the previous projects in the area can be applied. For the final design, new soil-boring data will quantify subsurface conditions to prove that the structure weight will not cause undue settlement to lower the crest elevation and/or cause structure instability and failure. The soil-boring investigation identifies the elevation, thickness, and physical properties of the general soil group in each layer (sand, clay, etc.) down to bed rock (if present) and groundwater level. Figure 2.37 is an example where the panel (a) extends from the ground surface to

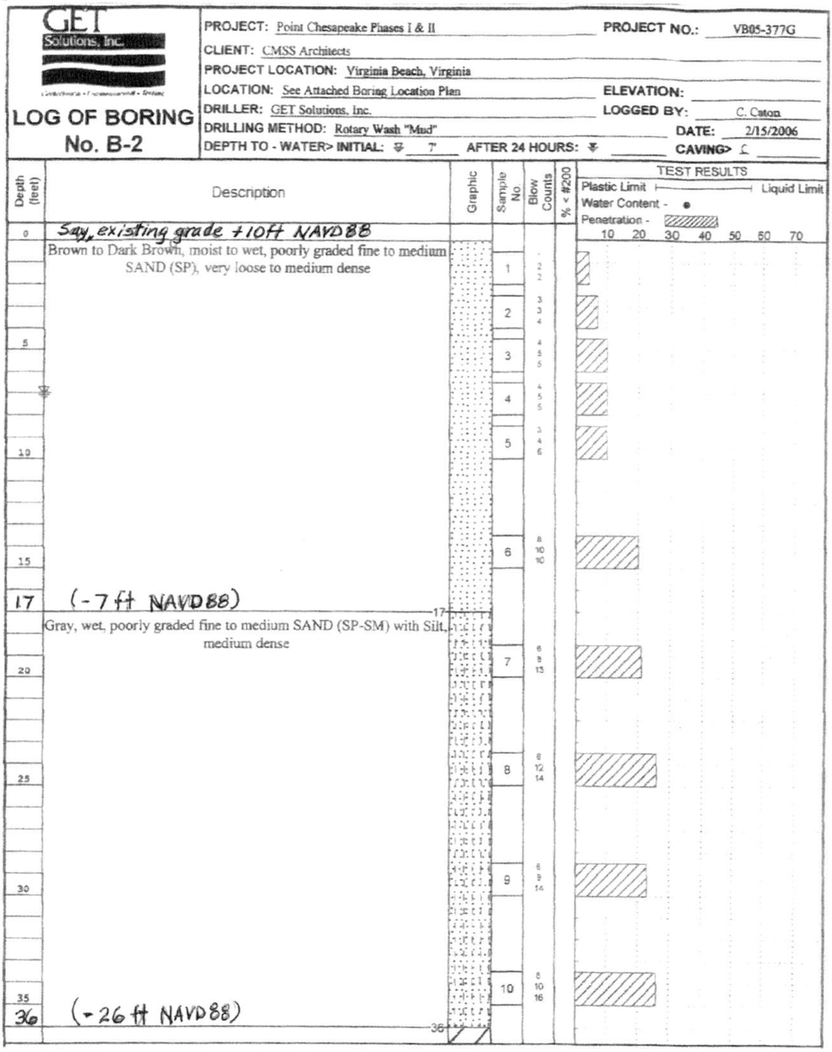

Figure 2.37(a). Soil-boring log for location B2 at Point Chesapeake, Virginia (surface to 36 ft depth) (courtesy GET Solutions, Virginia Beach, VA).

36 ft (11 m) beneath the surface, and the panel (b) continues from 36 ft (11 m) to 57 ft (17.4 m) beneath the surface as shown along the left side of each boring log. The second (wide) column describes the soil type (Unified Soil Classification System). The remaining columns present the

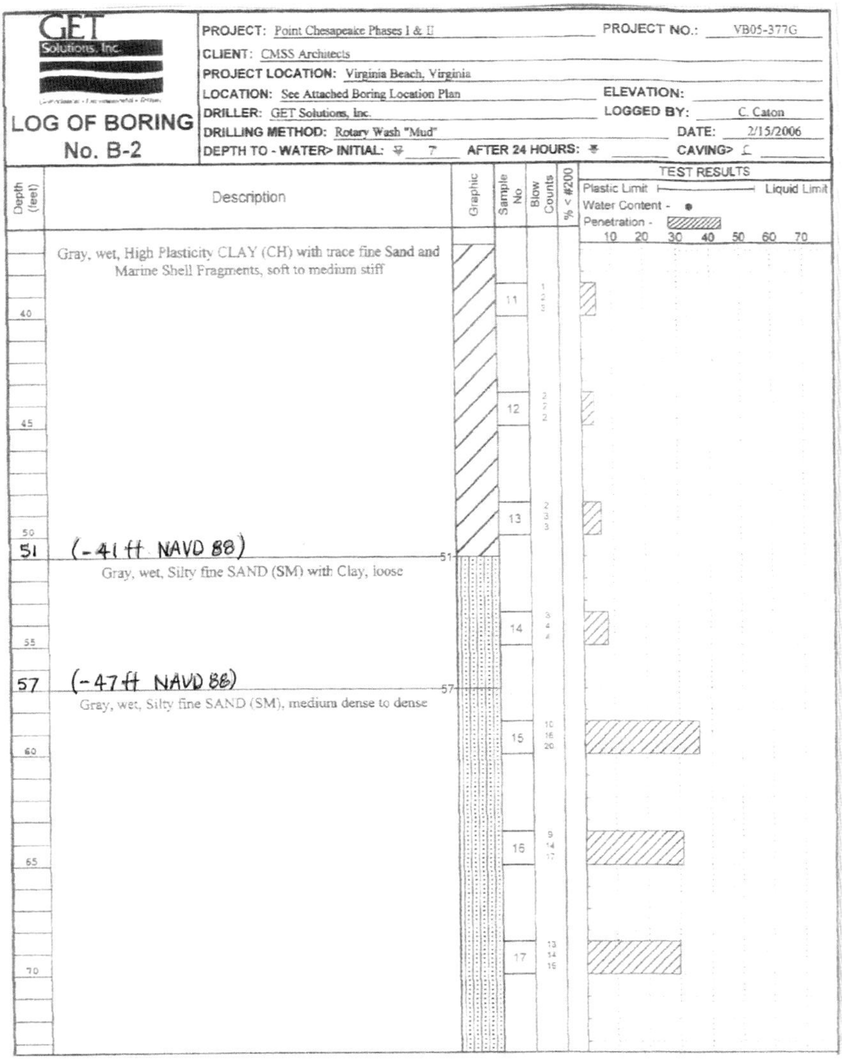

Figure 2.37(b). Soil-boring log for location B2 at Point Chesapeake, Virginia (36 ft depth to 57 ft depth) (courtesy GET Solutions, Virginia Beach, VA).

results of representative samples at these depths for the geotechnical tests conducted.

Standard practice of the soil-boring company is to simply reference the depth log as measured from the "surface" at each drilling location. The ground-or near-shore elevation of the "surface" must be established relative to the vertical datum. The location of this soil-boring log B2 on the seaward side of a proposed seawall is shown in Figure 2.38 along with other logs made at the site. The soil-boring locations should also be georeferenced to the horizontal datum employed on each plan-view drawing.

2.4.2 *Foundations*

The above information about the sediments is needed to develop the criteria for the design of the structure foundation. The experience and expertise of a geotechnical engineer should be consulted to ensure the success of the foundation design. The report by Eckert and Callender (1987) presents an excellent summary review of the geotechnical requirements for the foundation design in the coastal environment.

2.4.3 *Bulkheads and earth dikes*

Bulkheads and earth dikes (sometimes called revetments) also require the expertise of the geotechnical engineer. For example, the primary loads on the bulkhead (aka retaining wall) are the active and passive soil pressure distributions that produce the forces causing overturning of the vertical bulkhead wall. One of the results of the analysis of the soil-boring log is the determination of the active soil-pressure coefficient k_a and passive soil pressure coefficient k_p for each soil layer. An example of a bulkhead design will be given in Chapter 6 (Section 6.3.6) to demonstrate the method.

The criteria required for the design of earth dikes is far beyond the scope of this book. Again, Eckert and Callender (1987) present an introduction. Coastal engineers are called upon to study and present recommendations for the crest elevation and for the prevention of erosion of the earth dike, but are not generally responsible for stability design.

2.5 Coastal Geomorphology

Coastal structures harden the shoreline or their location in near-shore waters and consequently interfere with water levels, wave patterns, and the natural movements of sediments. The science of coastal hydrodynamics and the resulting coastal sediment transport is discussed in Chapter 7, Section 7.2 for the design of beach and dunes. For the functional and structural design of coastal structures and resulting sediment transport, see Chapters 5 and 6. The book *Coastal Processes with Engineering Applications* (Dean

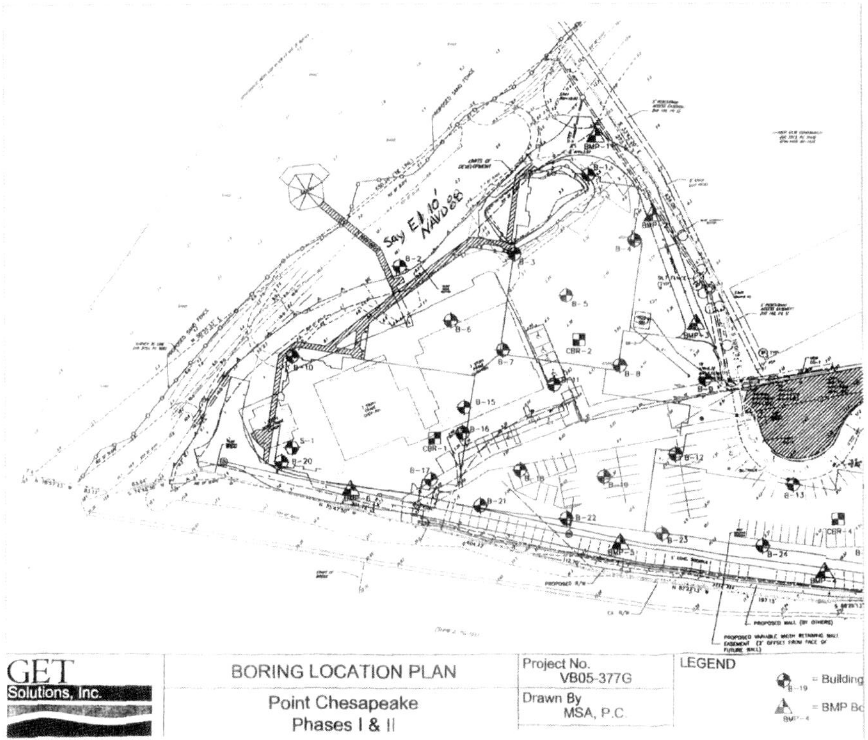

Figure 2.38. Soil-boring locations on the Point Chesapeake Virginia site (from MSA, 2014) (from MSA, P.C. Virginia Beach, VA).

and Dalrymple, 2002) is an excellent summary of the state-of-the-art for coastal engineers.

2.5.1 *Littoral drift and sediment transport*

Breaking waves in the near-shore region produce time-averaged currents in the surf zone (longshore currents) as illustrated in Figure 2.21 (bottom) and time-averaged currents beneath the broken waves in the surf zone (undertow) as illustrated in Figure 2.23. These currents may transport sediments in and around coastal structures if the currents are sufficient to cause bed load and suspended load sediment transport. The longshore currents transport sediments in the longshore direction called littoral drift and the undertow currents transport sediments in the offshore direction called cross-shore transport. Other cross-shore transport mechanisms (e.g. wave asymmetry) significantly complicate the cross-shore transport physics. And, differences in the mean water levels caused by differences in wave setup behind structures create additional currents that transport sediments behind and past coastal structures.

2.5.2 *Sources and sinks*

Knowledge of the source of sediment at the study site is one key criterion for design. In the feasibility study (Section 1.2.1), this knowledge is necessary to properly identify the cause of the problem especially if the problem is erosion. Upland rivers bringing sediments to the coast is the classic source, but offshore sediments transported landward by the waves and shoreline erosion of the adjacent beaches, dunes, and cliffs may be the primary source of sediments at the site. Aeolian transport is also a source. The most common source of sediment to a site is the unbalance of littoral drift rates produced by near-shore wave conditions over time. When more sediment enters than leaves the study-site region, the net balance is positive and considered a sediment source. The time-scale which is important for storms, seasons, and long-term trends may be different. Sediment brought to the site to renourish the beach and rebuild the dunes is also an artificial source.

Clearly, when the opposite is true, and more littoral drift sediment leaves than enters the site, the net balance is negative and considered a “sink” or loss of sediment to the site. Classic sediment sinks are tidal inlets which move sediments through the inlet during each tidal exchange, but retain more sediment in the bay or lagoon during the flooding stage than released during the ebbing stage. Storms may drag sediments far offshore in deep water where it may never be mobilized again for waves to return it to the shallower, near-shore waters. The subterranean canyons off the coast of California are the well-known-sinks for sediments as illustrated in the classic movie (now video) "The Beach: A River of Sand" (Inman and Shelton, 1967). Aeolian transport of sediments is also a common sink term.

Storm surges creating wave overtopping of dunes may transport sediments far inland (over-wash). The return of the over-wash sands to the beach/dune system will not occur naturally and is considered a sediment sink.

2.5.3 *Shoreline change trends*

All the above-mentioned, listed sources, and sinks of sediment and sediment transports produce shoreline changes at the site that may be positive (accretion), negative (erosion), or no change (equilibrium) for a representative "shoreline" at the land–water interface for the time interval of interest. And, these "shoreline" changes are relative to a fixed reference or baseline.

For estimates of historic long-term trends in shoreline change, topographic and bathymetric maps are commonly employed. The meaning of the line on the map delineating the land boundary from the water area, i.e. the "shoreline," must be clearly stated. In the US, most maps produced by Federal government agencies display the mean high-water (MHW) shoreline as determined from nearby tide gauge measurements for the tidal epoch when the map was produced. Relative to a fixed baseline position, the difference in MHW shoreline position can be determined. The shoreline change rate is then simply the shoreline change distance (m or ft) divided by the time interval for when the maps were produced. Seasonal variability should be considered. Aerial photographs can also be employed.

The shoreline change trends may be due to natural causes due to offshore bathymetry-producing regions of wave focusing adjacent to regions of wave spreading, or due to the relative sea-level changes at the site. Or, the shoreline change trends may be due to the anthropogenic (man-induced) effects. Because of the widespread interest in coastal erosion, many local, state and federal government agencies, consulting firms, universities, institutes, etc. have made shoreline change studies of the coastal regions.

For example, Figure 2.39 (left panel) displays the shoreline change rates on the horizontal axis (m/year) for the southeast region of Virginia's coast along the Atlantic Ocean shown as the vertical axis (Everts *et al.*, 1983).

The solid line is for 1858–1925 (68 years) and the dotted line for 1925–1980 (55 years) and both time periods gave a similar, maximum negative (erosion) rate of about 3 m/year (about 10 ft/year) at the south end of Sandbridge, Virginia. Wright *et al.* (1987) employed a 2D numerical model for wave transformation over the near-shore bathymetry displayed in the middle panel to calculate the average, breaking wave height variation along the coast for the year 1982, which is displayed in the third, right side panel. The yearly mean breaking wave height for the entire region was 0.58 m and the yearly average was greater than the mean in the Sandbridge region where the 10-m contour was the closest to shore and the erosion rate was the greatest. Note also that the shoreline change was positive at the north end (near the entrance to the Chesapeake Bay) and even more positive at the southern end called False Cape. Even over this 123-year period of relative sea-level rise, some reaches experienced shoreline accretion. We will return to this example at Sandbridge Virginia in Chapter 7 for a sediment budget (Figure 7.19) and in Chapter 10 when discussing the impacts of seawalls on adjacent beaches.

The historic shoreline change rate at the site and development of a sediment budget are the key design criteria.

2.5.4 *Sediment budget*

A sediment budget is a coastal engineering design and management tool to aid in defining the problem, and shows the present trends and future sediment movements before and after the project. Sources and sinks of sediment are identified (Section 2.5.2) and the shoreline change rate is quantified (Section 2.5.3). The coastal region is divided into littoral cells or compartments (Bowen and Inman, 1966) usually defined between two rocky barriers which mark the ends of a beach and have a fixed sediment budget. Littoral cells can either be free or fixed and can occupy a hierarchy of scales, from individual rip cells to entire beaches. Tidal inlets are also commonly employed as the boundary of a sediment cell. The region of maximum sediment-transport rate shown in Figure 2.39 is considered a null "point" for sediment transport in both the south and north directions, and can also be considered a boundary for sediment budget cells to the north and south regions of the coast.

The conservation of the sediment mass for a coastal reach (cell) with known sediment-transport conditions at the cell boundaries is the cornerstone equation for determination of a sediment budget. Figure 2.40 schematically displays the sources (credits, black) and sinks (debits, white) for

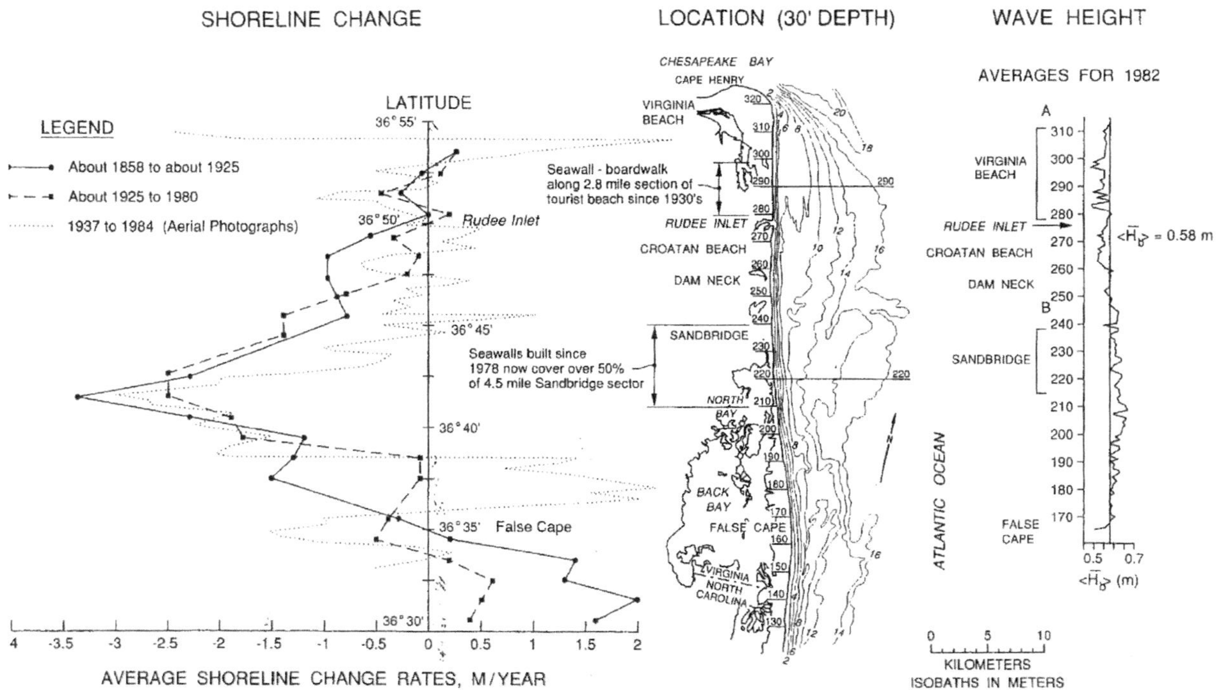

Figure 2.39. Composite of location and depth contours (middle) and breaking wave heights (right) as found in Wright *et al.* (1987). The shoreline change information (left) is added by the author and found in Everts *et al.* (1983) (from Basco *et al.*, 1997).

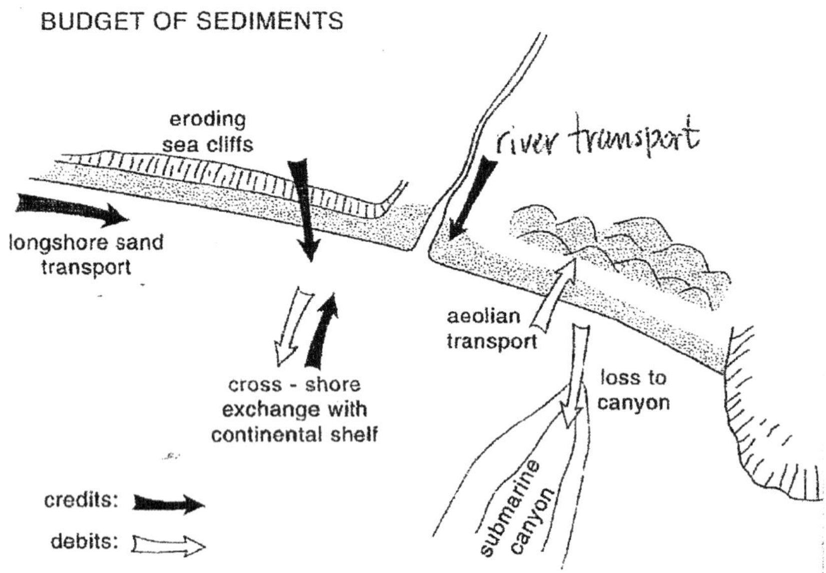

Figure 2.40. Schematic of sediment transports and directions (black toward shoreline called credits and white away from shoreline called debits) from Komar (1976).

quantification of a sediment budget (from Komar, 1976). The difficult part is assigning volume transport rates (m^3/yr) for each of the transports and sources/sinks to develop a sediment balance.

2.5.4.1 *Near-shore bathymetry*

The latest bathymetric charts and topographic maps available at the site must be employed for the design. In most cases, a new bathymetric survey should be made to learn the current bottom contours at the site. They may be significantly different from those found on older maps and charts. The latest bathymetric conditions will be needed in the modeling efforts.

Finally, subaerial surveys of the beach and dune system (if present) at the site and for some distance on both sides of adjacent property should be considered. This "baseline" data is needed to determine if any "down-drift" impacts to adjacent property can be attributed to the coastal structure after construction. Further details are found in Chapter 7.

The coastal setting is dynamic and influenced by land, water, and air interactions and processes. It is a regime of extremes, surprises, and constant motion as the coast and structures respond to changing conditions.

A fully dynamic, three-dimensional (3D), numerical model of water levels, waves, and sediment transport to simulate the bathymetric and shoreline change is still under development. It remains a long way from routine application.

2.6 Other Design Factors/Considerations

2.6.1 *Available materials for construction*

Results from the feasibility study (see Chapter 1) may demonstrate that natural rock structures with armor layer of a certain size (weight) are required for a stable design. The availability of this material from nearby or distant quarries must be determined. The final design may require concrete armor units if the natural rock size is not available. Or a modified rock design may be necessary (see Chapter 6, berm breakwater design).

2.6.2 *Construction aspects and accessibility of site*

The feasibility study may demonstrate the need to construct breakwaters in near-shore waters. Water-based construction equipment or the need for a temporary land fill for land-based construction must be determined. The final design may be dictated by the availability of water-based construction equipment of sufficient size.

When land-based construction is considered, the equipment must have transport accessibility to reach the site. This may require the acquisition of temporary easements for use of adjacent property for the transport path. If not feasible, this may force the use of water-based construction. The construction cost could be significantly different for land-based and water-based constructions.

2.6.3 *Environmental considerations*

The need for an Environmental Permit from local, state, and Federal government agencies for the construction of any type of coastal structure is required in the United States. Some US states prohibit hard, structural alternatives (seawalls, revetments, breakwaters, etc.) on open coastal shorelines. For example, the State of North Carolina in the US for the Atlantic Ocean coast does not permit coastal structures of any type.

The designer must determine the specific, local laws, and requirements for obtaining the Environmental Permit at the start of the design process. The choice of alternatives to solve the problem may be dictated by what is permitted (and what is not permitted) at the site. The environmental constraints become a design criteria that governs the final design.

Chapter 10 presents full details and examples.

2.7 Summary

See Section 11.3.2 for understanding the impacts of sea-level rise which have been discussed in this chapter.

References

Agbley, S. K., and D. R. Basco (2008) "An Evaluation of Storm Surge Frequency of Occurrence Estimators," *Proceedings of the Solutions to Coastal Disasters*, ASCE, pp. 185–197.

ASTM Standards (D 2487-92) "Unified Soil Classification System," American Society Testing Materials.

Basco, D. R., and N. Mahmoudpour (2012) "The Modified Coastal Storm Impulse (COSI) Parameter and Quantification of Fragility Curves for Coastal Design," *Proceedings of the International Conference Coastal Engineering*, ASCE, New York.

Basco, D. R., D. A. Bellomo, J. M. Hazelton, and B. N. Jones (1997) "Influence of Seawalls on Subaerial Beach Volumes with Receding Shorelines," Contract Report, CHL-97-2, Waterways Experiment Station, Corps of Engineers, Vicksburg, MS.

Battjes, J. A. (1974) "Surf Similarity," *Proceedings of the 14th International Conference Coastal Engineering*, American Society of Civil Engineers, pp. 466–480.

Birkemeier, W. (2010) Personal Communication.

Bouws, E., H. Gunther, and C. L. Vincent (1985) "Similarity of Wind Wave Spectrum in Finite-Depth Water, Part I: Spectral Form," *Journal Geophysical Research*, Vol. 85, No. C3, pp. 1524–1530.

Bowen, A. J., and D. L. Inman (1966) "Budget of Littoral Sand in the Vicinity of Point Arguello, California," Technical Memorandum No. 19, U.S. Army Coastal Engineering Research Center, Fort Belvoir, VA.

Bowen, A. J., D. L. Inman, and V. P. Simmons (1968) "Wave Set Down and Set-Up," *Journal Geophysical Research*, Vol. 73, pp. 2569–2577.

Buhr-Hansen, J. and I. A. Svendsen (1984) "A Theoretical and Experimental Study of Undertow," *Proceedings of the 19th International Conference Coastal Engineering*, American Society of Civil Engineers, pp. 2246–2262.

CEM (2006) *Coastal Engineering Manual*, US Army Corps of Engineers, Washington, D.C. (online).

Corps of Engineers (1988) "Design of Hurricane Protection Project for Virginia Beach, Virginia," Norfolk District Office, Corps of Engineers, Norfolk, VA.

De Boni, Cecconi, Nuova, Canestrelli, Franco, and Egori (1993) "A Description of the 'Acqua Alta Event on 5th October, 1992," *Proceedings of the 23rd*

International Conference Coastal Engineering, the Conference Organizing Committee.

Dean, R. G., and T. L. Walton (2009) "Wave Setup," in Y. C. Kim (ed.), *Handbook of Coastal and Ocean Engineering*, World Scientific, Singapore, pp. 1–23.

Dean, R. G., and R. A. Dalrymple (1984) *Water Wave Mechanics for Engineers and Scientists*, World Scientific, Teaneck, NJ.

Earle, M. D., D. McGehee, and M. Tubman (1995) "Field Wave Gaging Program, Wave Data Analysis Standard," Instruction Report CERC-95-1, Corps of Engineers, Waterways Experiment Station, Vicksburg, MS.

Eckert, J., and G. Callender (1987) "Geotechnical Engineering in the Coastal Zone," Instruction Report CERC-87-1, Coastal Engineering Research Center, Waterways Experiment Station, Corps of Engineers, Vicksburg, MS.

Everts, C. H., J. P. Battley Jr., and P. N. Gibson (1983) "Shoreline Movements: Cape Henry Virginia to Cape Hatteras, NC, 1849–1980," Report No. 1, Technical Report CERC 83-1, Waterways Experiment Station, Vicksburg, MS, July, 123p.

FEMA (2007) "Guidelines and Standards for Flood Risk Analysis and Mapping," Federal Emergency Mapping Agency, Washington, D.C. (and revisions in later years).

Galvin, C. J. (1968) "Breaker Type Classification on Three Laboratory Beaches," *Journal of Geophysical Research*, Vol. 73, No. 12, pp. 3651–3659.

Goda, Y. (1975) "Computation of Refraction and Diffraction of Sea Waves with Mitsuyasu's Directional Spectrum," Technical Note of Port and Harbour Research Institute, Vol. 230, pp. 1–45 (in Japanese).

Goda, Y. (1975) "Irregular Wave Deformation in the Surf Zone," *Coastal Engineering in Japan*, Vol. 18, pp. 13–26.

Goda, Y. (2010) *Random Seas and Design of Maritime Structures*, 3rd edn. World Scientific, Singapore.

Hasselmann, K., T. P. Barnett, E. Bouws, H. Carlson, D. E. Cartwright, K. Enke, J. A. Ewing, H. Gienapp, D. E. Hasselmann, P. Kruseman, A. Meerburg, P. Mller, D. J. Olbers, K. Richter, W. Sell, and H. Walden (1973) "Measurements of Wind-Wave Growth and Swell Decay During the Joint North Sea Wave Project (JONSWAP)," *Deutsche Hydrograph Zeitung, Erganzungsheft Reihe A* (80), No. 12.

Hunt, J. N. (1979) "Direct Solution of Wave Dispersion Equation," *Journal Waterway, Port, Coast, and Ocean. Engineering*, 105, 457–459.

Inman, D., and J. S. Shelton (1967) "The Beach: A River of Sand," Movie/Video, *Encyclopedia Britannica Films, Inc. and American Geological Institute* (WorldCat.org).

Inman, D. L. (1952) "Measures for Describing the Size Distribution of Sediments," *Journal of Sedimentary Petrology,* Vol. 22, No. 3, pp. 125–145.

Jensen, R. E. (1983) "Atlantic Coast Hindcast, Shallow-Water, Significant Wave Information," WIS Report 9, Waterways Experiment Station, Corps of Engineers, Vicksburg, MS.

Kamphuis, J. W. (1991) "Incipient Wave Breaking," *Coastal Engineering*, Vol. 15, pp. 185–203.

Kamphuis, J. W. (2010) *Introduction to Coastal Engineering and Management*, World Scientific, Singapore.

Komar, P. D. (1976) *Beach Processes and Sedimentation* (2nd edn.), Prentice-Hall, Upper Saddle River, NJ.

Kweon, H.-M., and Y. Goda (1996) "A Parametric Model for Random Wave Deformation by Breaking on Arbitrary Peach profiles," *Proceeding of the 25th International Conference on Coastal Engineering*, American Society of Civil Engineers, pp. 261–274.

Leenknecht, D. A., A. Szuwalski, and A. R. Sherlock (1992) "Automated Coastal Engineering System, User Guide and Technical Reference, Version 1.07," Coastal Engineering Research Center, U.S. Army Engineer Waterways Experiment Station, Vicksburg, MS.

Lin, P. (2008) *Numerical Modeling of Water Waves*, Taylor & Francis, London.

Longuet-Higgins, M. S., and R. W. Steward (1963) "A Note on Wave Set-up," *Journal Marine Research,* Vol. 21, No. 1, pp. 4–10.

Longuet-Higgins, M. S., and R. W. Stewart (1964) "Radiation Stresses in Water Waves: A Physical Discussion with Applications," *Deep Sea Research*, Vol. 11, pp. 529–562.

Longuet-Higgins, M. S. (1952) "On the Statistical Distribution of the Wave Heights of Sea Waves," *Journal Marine Research*, Vol. 11, pp. 245–266.

Ma, Q. (2010) *Advances in Numerical Simulation of Nonlinear Water Waves*, World Scientific, Singapore.

Melby, J. A., and G. Myrick (2004) "Lajes Field, Azores Breakwater Stability Model Study," Corps of Engineers ERDC/CHL TR-04-xx, Vicksburg, MS.

MSA (2014) "Boring Location Plan: Point Chesapeake Virginia," from GET Solutions, Inc.

Miche, R. (1944) "Mouvements Ondulatoires des Mers en Profondeur Constante on Decroisante," *Annales des Ponts et Chaussees*, pp. 25–78, 131–164, 270–292, 369–406.

Munk, W. H. (1944) "Proposed Uniform Procedure for Observing Waves and Interpreting Instrument Records," Scripps Institute of Oceanography, Wave Project, La Jolla, CA.

Munk, W. H. (1949) "The Solitary Wave Theory and Its Applications to Surf Problems," *Annals of the New York Academy of Sciences*, Vol. 51, pp. 376–462.

Neumann, G., and W. J. Pierson (1966) *Principles of Physical Oceanography*, Prentice-Hall, Englewood Cliffs, NJ.

Oumeraci, H., W. Allsop, M. de Groot, R. Crouch, H. Vrijling, and H. Voortman (2001) *Probabilistic Design Tools for Vertical Breakwaters*, Swets & Zeitlinger B. V. Lisse, Germany.

Pierson, W. J., and L. Moskowitz (1964) "A Proposed Spectral Form for Fully-Developed Wind Sea Based on the Similarity Law of S. A. Kitaigorodoskii," *Journal of Geophysical Research*, Vol. 69, pp. 5181–5203.

Resio, D. T. (2007) "White Paper on Estimating Hurricane Inundation Probabilities," US Army Corps of Engineers, Engineer Research and Development Center, Vicksburg, MS.

Sallenger, A. H., and R. A. Holman (1985) "Wave Energy Saturation on a Natural Beach of Variable Slope," *Journal Geophysical Research*, Vol. 90, Issue C6, pp. 11939–11944.

Scheffner, N. W., J. E. Clausner, A. Militello, L. E. Borgman, B. L. Edge, and P. E. Grace (1999) "Use and Application of the Empirical Simulation Technique: Users Guide," Technical Report CHL-99-21, U.S. Army Engineer Waterways Experiment Station, Vicksburg, MS.

Shephard, F. P. (1976) "Coastal Classification and Changing Coastlines," *Geoscience and Man*, Vol. 14, pp. 53–64.

Sorensen, R. M. (2006) *Basic Coastal Engineering*, 3rd edn., Springer Publishing Company, New York City.

Svendsen, I. A., P. A. Madsen, and J. B. Hansen (1978) "Wave Characteristics in the Surf Zone," *Proceedings of the* 16th *Coastal Engineering Conference*, American Society of Civil Engineers, pp. 520–539.

Sverdrup, H. U., and W. H. Munk (1947) "Wind, Sea, and Swell: Theory of Relations for Forecasting," Pub. No. 601, U.S. Navy Hydrographic Office, Washington, DC.

Thornton, E. B., and Guza, R. T. (1983) "Transformation of Wave Height Distribution," *Journal of Geophysical Research*, Vol. 88, No. C10, pp. 5925–5938.

Toro, G., D. T. Resio, A. W. Niedoroda, D. Divoky, and C. W. Reed (2010) "Efficient Joint Probability Methods for Hurricane Surge Frequency Analysis," *Ocean Engineering*, Vol. 37, No. 1, pp. 125–134.

Vincent, C. L. (1985) "Depth-Controlled Wave Height," *Journal, Waterway, Port, Coastal and Ocean Engineering*, Vol. 111, No. 3, pp. 459–475.

WAMDI Group (1988) "The WAM Model — A Third Generation Wave Prediction Model," *Journal of Physical Oceanography,* Vol. 18, pp. 1775–1810.

Weibull, W. (1951) "A Statistical Distribution Function of Wide Applicability," *Journal of Applied Mechanics, Transactions*, Vol. 18, No. 3, pp. 293–297.

Wright, L. D., C. S. Kim, C. S. Hardaway, S. M. Kimball, and M. O. Green (1987) "Shoreline and Beach Dynamics of the Coastal Region from Cape Henry to False Cape Virginia," Technical Report, Virginia Institute of Marine Science, Gloucester Point, VA.

Chapter 3

Alternatives for Hazard Mitigation

3.1 Introduction

3.1.1 *Major concerns for "shore protection"*

Flooding of low-lying regions is the major concern of coastal storms and the primary hazard that can be mitigated by coastal structures. Property (private, commercial, public) and infrastructure (water supply, waste-water treatment facilities, utilities, etc.) are damaged or destroyed; transportation is disrupted or stopped; and agricultural lands made unproductive by salt water. High waves on top of the elevated water levels also produce damage to property and infrastructure and are a hazard of coastal storms. Strong currents generated by flooding and high waves may also cause damage and erode sediments at the coast.

Erosion of shorelines is a major concern all over the world. Erosion may occur quickly during a storm or be due to long-term natural and/or anthropogenic causes. Mitigation of coastal erosion reduces the potential damage during storms by providing increased distance and elevation of the land between the ocean and the property and infrastructure to be protected. The same coastal structures can be employed to mitigate coastal erosion, but they alter the natural hydrodynamic and sediment transport processes at the site.

A third need in developed countries is to restore, enhance, or create coastal ecosystems. Again, the same coastal structures as discussed in this book may play a role in their design. For example, low rock sills reduce wave energy to permit marsh grasses to thrive in their wake and the combination mitigates coastal erosion. The marsh grasses play an important rolc in the total ecosystem. The loss of coastal ecosystems can be considered a coastal hazard for the biological resources at the coast.

We herein consider flooding, high waves, strong currents, erosion, and ecosystem loss *together* as the hazard to be mitigated by coastal structures.

And, as mentioned previously, we avoid the term "shore protection" and always refer to "hazard mitigation" in this book. We can never completely "protect" shorelines and man's artifacts from these hazards as the result of living near coasts. But we can reduce or "mitigate" the damages and slow down erosion rates and ecosystem loss. That is the goal for coastal hazard mitigation design as found in this book.

3.1.2 *Alternatives for coastal hazard mitigation*

There are many alternatives for the mitigation of these coastal hazards that impact human resources (property, infrastructure, transportation systems, and agriculture), physical resources (shoreline erosion), and biological resources (ecosystems) that make up the coastal zone. These alternatives can be categorized as (A) structural changes to the natural, physical environment as detailed in Figure 3.1 (CEM, 2006), or (B) non-structural changes as required by man's social system, or (C) combinations, or (D) the do-nothing alternative shown in Figure 3.2 (CEM, 2006).

Structural changes to the natural, physical environments are often further categorized as armoring (hardening) the shoreline (Figure 3.1(a)); or beach stabilization (moderating) of the shoreline (Figure 3.1(b)), or beach renourishment (softening) of the shoreline (Figure 3.1(c)). These three alternative types are the primary focus of this book. Note that the engineering design of a beach nourishment project or the "soft" alternative is a structural change to the natural, physical environment so that the design of a beach nourishment project is considered as a coastal structures design even though it is considered the "soft" alternative.

3.1.3 *Movable gate structures*

Concern for the coastal environment (water quality, coastal ecology, etc.) led to the planning, design, and construction of movable gated structures in tidal estuaries. The gates are normally open for the daily tidal exchanges necessary for the salinity and ecology of the tidal bays and lagoons and only closed to prevent upland flooding during storm events. Movable gate structures in the Netherlands (Ooesterschelde, Maas), the UK (Thames), and Italy (Venice Lagoon) will be summarized in Chapter 6. Hurricane SANDY (July 2012) impacting New York City has increased interest in movable gate structures in the US.

APPROACH	(A) CHANGES TO THE NATURAL, PHYSICAL SYSTEM		
Class	1. Armoring Structures		
Type	Seawall	Bulkhead	Dike/Revetment
Geometry (Configuration) or Location	Vertical • Curved • Gravity • Composite	• Crib • Stepped/Terraced • Cantilevered • Tie-Backed	• Sloped
Construction Material	• Concrete • Rock	Sheet-pile (steel, timber, concrete, aluminum)	• Earth • Rock • Geotextiles (bags) • Precast concrete blocks • Gabions

(a) Armoring Structures

APPROACH	(A) CHANGES TO THE NATURAL, PHYSICAL SYSTEM			
Class	2. Beach Stabilization Structures & Facilities			
Type	Breakwaters	Groins	Sills& (Vegetation)	Groundwater Drainage
Geometry (Configuration) or Location	• Headland • Detached • Single • System • Tuned • Shoreline	• Normal • Angled • Single • System • Notched • Permeable • Adjustable Shaped (T or L)	• Shoreline • Submerged • Perched beach (Submerged Aquatic Vegetation)	• Beach drain • Bluff dewatering • Interior drainage
Construction Material	• Rock • Precast concrete units • Sheet-pile (steel, timber, concrete, timber) • Geotextiles bags			• System of pipes and pumps with sumps

(b) Beach-Stabilization Structures

APPROACH	(A) CHANGES TO THE NATURAL, PHYSICAL SYSTEM	
Class	3. Beach Restoration	
Type	Beach Nourishment	Sand Passing
Geometry (Configuration) or Location	• Subaerial • Dune • Feeder • Profile • Underwater Berms	• Bypassing • Backpassing
Construction Material	• Borrow Sites * offshore * land • Dredged Material • Artificially Mode (crushed rock)	• Littoral Traps • Smooth Out "hot-spots" • Downdrift material returned updrift

(c) Beach Restoration

Figure 3.1. Changes (structural) to the natural physical system.

(B) CHANGES TO MAN'S SYSTEM			(C) CHANGES IN BOTH		(D) NO CHANGE
4. Adaptation & Accomodation			5. Combinations		6. Do Nothing
Flood Proofing	Zoning	Retreat	Structural and Restoration	Structural and Restoration and Adaptation	
• Elevated Structures • Raise grade • Sandbags • Flow diversion	• Setbacks • Land use restrictions • Public lands (Institutional)	• Individuals • Communities • Infrastructure • Move structures	Any Combination of 1, 2, or 3 alternatives	Any Combination of all alternatives except retreat.	Let Nature take its course.
• Single-family homes on timber piles					
V-3-4 V-3-4b. V-3-4b. V-3-4c.			V-3-5		V-3-6

Figure 3.2. Changes to (B) man's system, (C) combinations, and (D) the do-nothing alternative.

3.1.4 *The role of the design constraints*

The design constraints reviewed in Section 1.2.4 play a significant role in what emerges as the alternative selected for the final design. In the review of the following alternatives, some common constraints on the use of the alternative will be discussed when applicable.

3.2 Structural Alternative — Armoring

3.2.1 *Seawalls*

Armoring the shoreline with seawalls, bulkheads, revetments, or dikes (with protective revetments) has been the historic choice for flooding and wave-damage mitigation for centuries. Figure 3.3 illustrates different types of cross-sections of vertical seawalls and sloping revetments or dikes (from Pilarczyk, 1990). Seawalls are monolithic (normally concrete) vertical walls

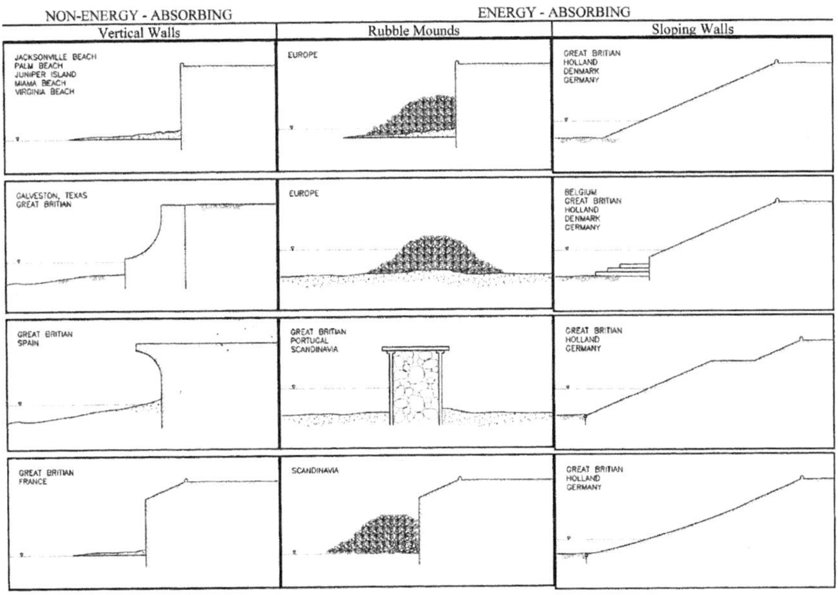

Figure 3.3. Examples of vertical and sloping seawalls (from Pilarczyk, 1990).

with curved or sloping sections and do not absorb but reflect wave energy. When rubble mounds are placed in front or the entire front face is sloping, wave energy is adsorbed and only partial wave reflection takes place. The length of seawalls must be sufficient to prevent flanking by extreme water levels to mitigate coastal flooding.

Seawalls are designed to withstand the full forces of the sea during extreme high-water events that are accompanied by high waves. They are normally backed by natural or fill sediments that help to absorb the horizontal forces of the water levels and the waves. Consequently, their structural design is quite complicated. Chapter 6 will give further details. Figure 1.2 displayed the Galveston, Texas, curved seawall that has existed for over 110 years.

3.2.2 *Bulkheads*

At the opposite extreme are relatively weak bulkheads with the primary purpose to prevent the sediments behind the structure from sliding into the sea. Figure 3.4 displays some typical bulkhead types. Now, the design loads come from the active and passive soil pressures on the vertical wall which are influenced by the groundwater levels behind the wall. Anchor rods and

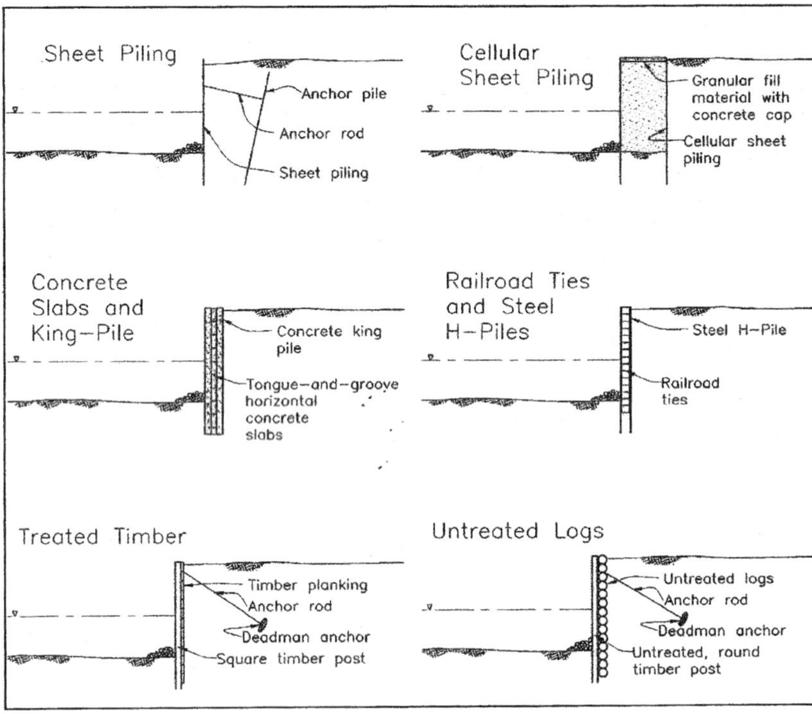

Figure 3.4. Examples of typical types of bulkheads (courtesy Corps of Engineers).

anchor piles (dead man) help to hold back the wall when a cantilever design is not practical. Chapter 6 presents an example of an anchored, sheet-pile, bulk head design. Geotechnical engineers should be consulted for bulkhead design.

3.2.3 *Revetments*

Natural coastal embankments of loose sediments can be armored with protective materials (revetments) to prevent bank erosion during high-water and-wave events. Figure 3.5 shows 11 different types of surface materials ranging from rough stones and concrete blocks to smooth revetment surfaces constructed from many materials. Wave runup levels and overtopping rates vary with surface roughness as quantified in Chapter 5. Structural stability for the material size (weight) and type of surface is found in Chapter 6. One type of revetment shown in Figure 3.5 is simply a tidal wetland at the toe with grasses on the bank for use at relatively protected, low-wave energy sites.

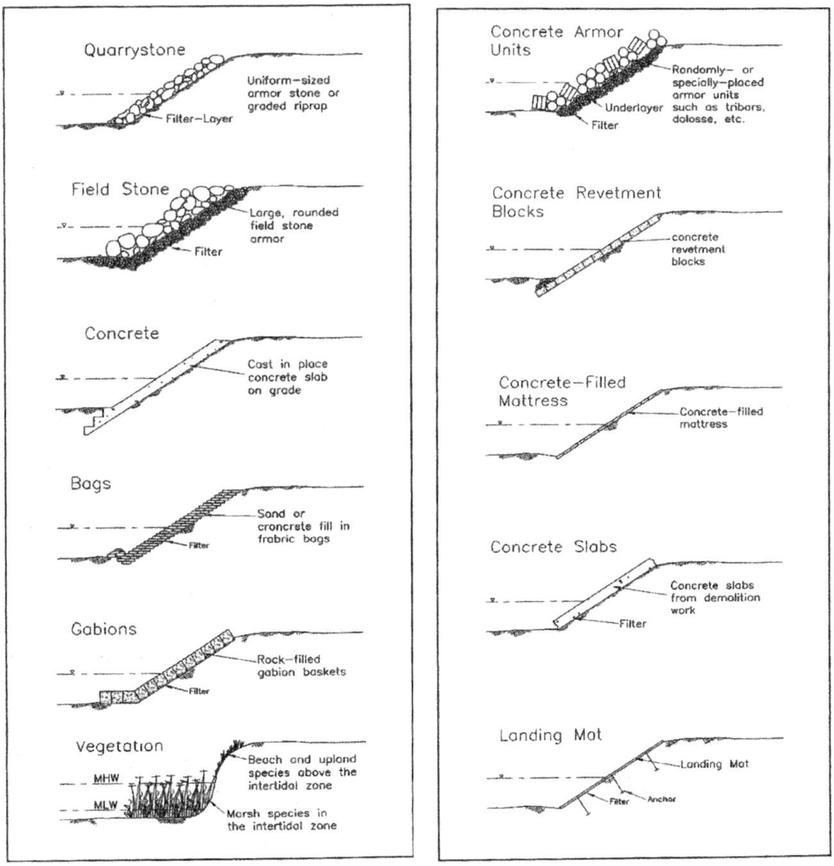

Figure 3.5. Different types of revetments (courtesy Corps of Engineers).

3.2.4 *Dikes and flood walls*

Earth dikes or levees are long, naturally occurring or artificially constructed embankments to control water levels and keep low-lying areas behind from flooding during storm events. They are normally covered with native grasses but may need revetment protection on the seaward side from wave attack and on the leeward side from scour during wave overtopping. The crest elevation and side slopes require space on both sides for the relative large cross-sectional volume compared to flood walls.

Flood walls are typically narrow, vertical, concrete barriers to contain elevated water levels from flooding the property behind the wall. They are often used in cities where space is limited (nearby buildings, commercial property, bridges, etc.) and require movable flood gates to provide passage except when closed during flood events.

New Orleans, Louisiana, experienced catastrophic flooding from Hurricane Katrina in September 2005 when its system of dikes (levees) and flood walls failed because of inadequate designs for the existing soil conditions. Dikes and floodwalls are discussed in Chapter 6.

3.2.5 *Hardening the shoreline*

Armoring the coast fixes (hardens) the shoreline position in space relative to the flexible sediments that are in front and behind. Each type prevents the sediments behind the wall or beneath the revetment surface from being released into the sea during coastal storm events. The complete story of how armored structures interact with adjacent beaches is found in Chapter 10. Some US states (e.g. North Carolina) do not permit any type of permanent hardening of the shoreline with an armoring structure on the Atlantic Ocean coast of the state. This is a design constraint for the type of alternative allowed for all coastal hazard mitigation projects on the ocean shorelines in that state.

3.3 Structural Alternative — Beach Stabilization

3.3.1 *Naturally stable shorelines*

Coasts with marine deposits (barrier islands and sandy beaches) are one of the most widely found geomorphic forms in the world. Figure 3.6 displays (a) an example of a dynamically stable shoreline shape between two rocky headlands (from Silvester and Hsu, 1993) and (b) the concept of a minimum dry beach width, $Y_{\min}$ in the profile cross-section. The dry beach width is much narrower behind the headland than that receiving direct wave attack during storms. This volume of sediment protects the uplands (cliffs, dunes, vegetation, infrastructure, or structures) from damage under normal storm conditions. In 1960, Silvester wrote as follows:

> *"... to allow for storm-cycles and the short-term reversals of drift, a sufficient width of beach should be allowed as working capital on which the sea can operate. Once the coast has been stabilized by preventing the net movement of sediment, no long-term erosion need be anticipated and the 'active' beach width can be minimized"* (Silvester, 1960, p. 469).

However, many shorelines are eroding due to both natural causes (e.g. sea-level rise) and for anthropogenic reasons (e.g. down drift of jetties that

(a)

(b)

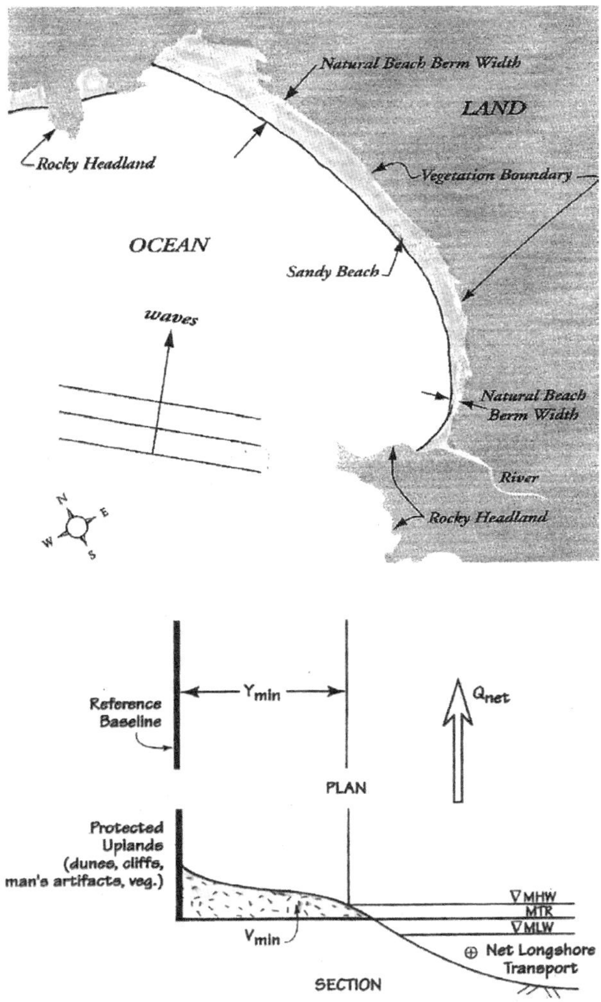

Figure 3.6. (a) Plan form of naturally stable shoreline (from Silvester and Hsu, 1993) with the "Natural Beach Width" label from MSL to upland vegetation and (b) concept of minimum dry beach width, $Y_{\min}$ (from CEM, 2006, V-3-3).

block the natural sand movements past tidal inlets). Shoreline stabilization structures can provide upland protection beyond the reference baseline for storm events and hence can mitigate the impacts of coastal erosion. At a minimum, these structures should be designed to provide the minimum, dry beach width, $Y_{\min}$ for shore protection for the storm conditions at the site.

Figure 3.7(a) shows Y_{min} for a sandy beach and (b) for a rocky headland-bay beach.

3.3.2 *Shoreline-stabilization structures*

Figure 3.7 also displays the three most common types of beach erosion mitigation structures, namely, (c) headland breakwaters, (d) nearshore breakwaters, and (e) a groin field. Their primary purpose is to retain sand on the beach during storm events. Present design practice is to always combine their construction with beach renourishment to minimize their down-drift impact. Clearly, headland and nearshore breakwaters work best when cross-shore sediment transport processes dominate, and a groin field works best where longshore sediment transport (LST) processes are most important.

As illustrated in Figure 3.7, the response of the shoreline behind the structure due to waves, currents, and resulting sediment transport can be either a tombolo ((c), shoreline attached to structure) or a salient ((d), no shoreline attachment). Field data published by the Corps of Engineers (Pope and Dean, 1986) reveal that the key variables are:

L_s: the length of the structure, L_g: the length of the gap between the structures, Y: the structure distance offshore, from the design, renourished beach shoreline, and d_s: the mean water depth offshore at the toe of the structure.

When the L_s/L_g ratio is large (long structure/small gap) so that less wave energy reaches the shoreline and/or when the Y/d_s ratio is small (structure close to shore in shallow water) tombolos form behind the structures which are called "headland" breakwaters. When the opposite is true with these key ratios, more wave energy reaches the shoreline to maintain LST processes and only salients are present behind what are called "near shore" breakwaters. Clearly, as illustrated in Figure 3.7 when short structures with large gaps are located far offshore in relatively deep water, they may have little or no influence on the shoreline (no sinuosity) and offer little shore damage mitigation.

The functional design of shoreline stabilization structures is covered in complete detail in Chapter 5.

3.3.3 *Reefs, sills, and wetlands*

Many coastlines of the world are found along relatively smaller water bodies with limited fetch distances that restrict wave heights. An example is the

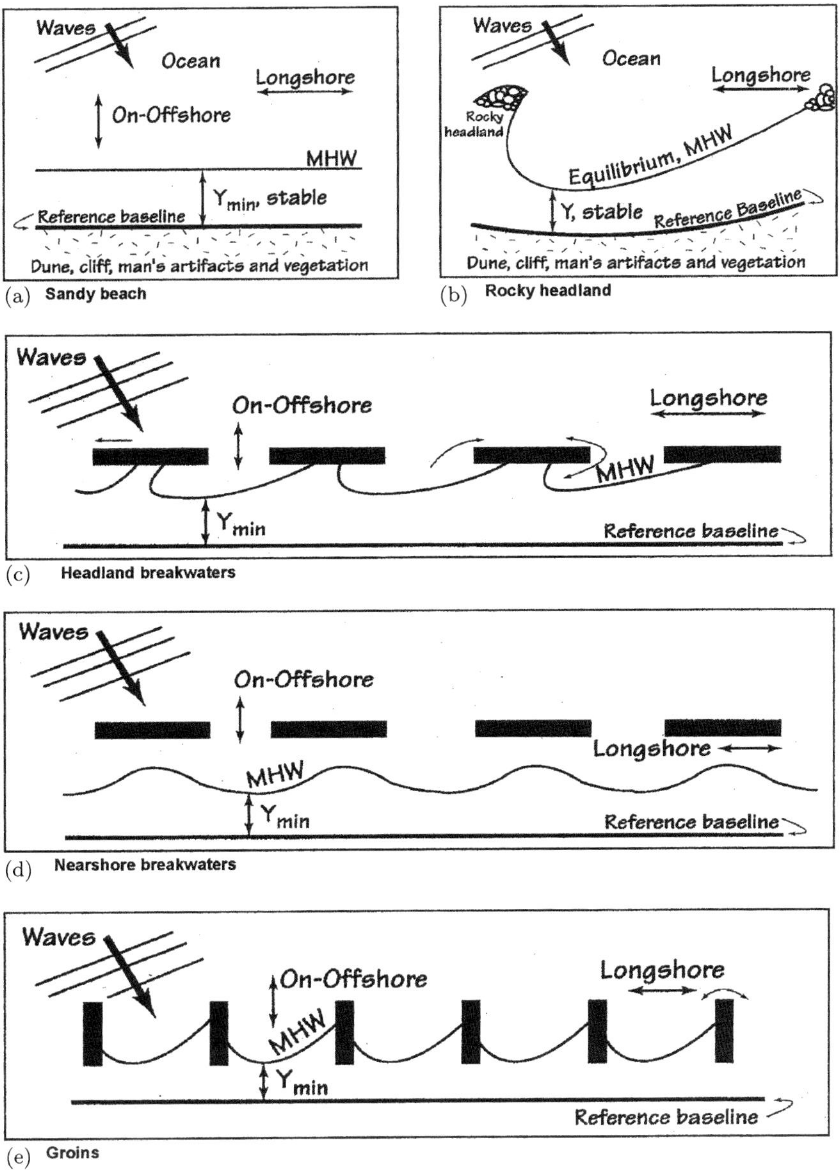

Figure 3.7. Natural (a, b) and artificial stable shorelines with minimum dry beach width, Y_{min} for (c) headland breakwaters, (d) near shore breakwaters, and (e) groin field.

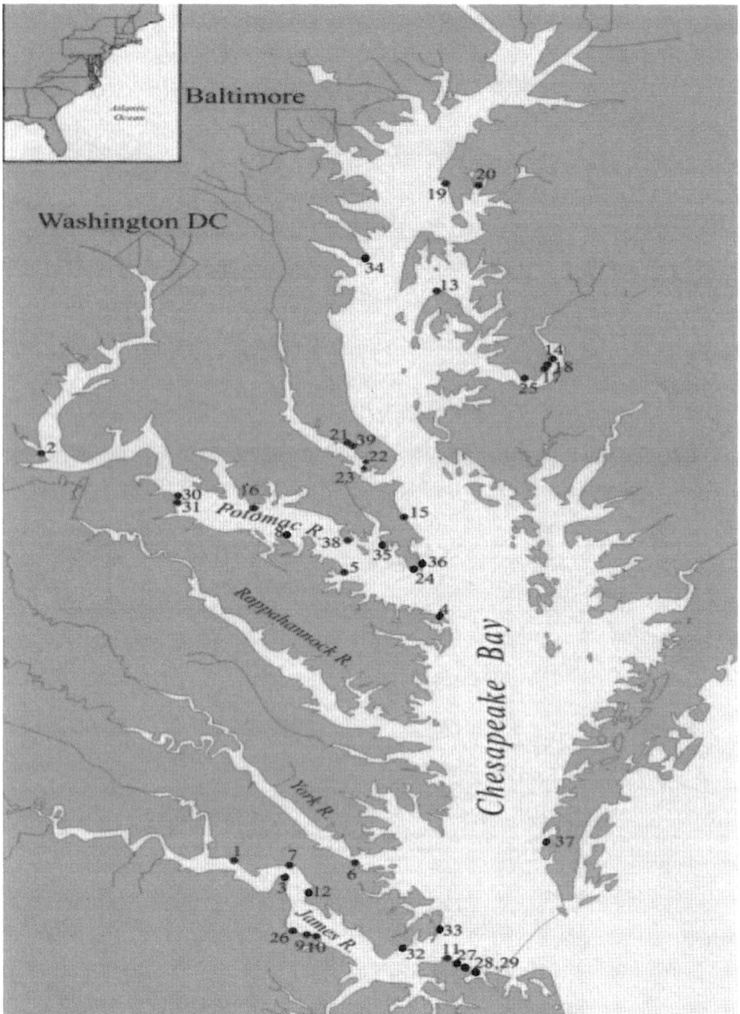

Figure 3.8. Chesapeake Bay and locations (black dots) of nearshore breakwater projects (from Hardaway and Byrne, 1999).

Chesapeake Bay as seen in Figure 3.8 with entrance in Virginia on the Atlantic Ocean and many of the large tidal estuaries and small tributaries to the Bay. These shorelines are considered to occur along "sheltered coasts" (National Research Council, 2007). Alternatives in these low-wave energy environments to mitigate damage and erosion may be simple rock reefs and sills or tidal wetlands composed of marsh grasses. Combinations are common.

Natural reefs are found in shallow, nearshore waters and are biotic platforms at a fixed level below the low tide. Rock reefs submerged below the surface are continuous dikes to hold sand (also called a perched beach) and are designed to mimic nature to reduce wave energy before striking the coast.

Sills are also low-profile, continuous dikes to hold a marsh but located at the shoreline. Figure 3.9(a) shows a nearshore breakwater and rock sill with marsh grass planted behind the sill as constructed on the Choptank river, Talbot County, MD, on the Chesapeake Bay (from Hardaway and Byrne, 1999). The marsh grass has grown to cover the rock sill after five years as shown in Figure 3.9(b).

Coastal salt marshes are meadows of herbaceous plants in low-lying areas that require periodic water-level fluctuations between the mean water level and mean high-water level of the daily tides. They survive on very low-wave-energy shorelines and have become a priority of many governments for restoration and new development. Natural and nature-based features (NNBF) "Living Shorelines" are discussed in Section 10.4.

Figure 3.9(a). Rock sill with marsh grass planted behind sill. Headland breakwaters with beach nourishment as constructed on the Choptank River, Talbot County, MD on the Chesapeake Bay (from Hardaway and Byrne, 1999).

Figure 3.9(b). Marsh grass has grown to cover rock sill after five years.

3.4 Structural Alternative — Beach Nourishment

The "soft" structural alternative for coastal hazard mitigation is beach nourishment. A simple definition is the placement of large quantities of good quality sand on the beach to advance it seaward as schematically illustrated in Figure 1.6 (alternative No. 4). The distance from the sea to the reference baseline is increased and the water depth is decreased to reduce the wave energy striking the dune, cliff, or man's artifacts that may be damaged in a storm. The nourishment material may also be placed as an underwater berm in nearshore waters. For flood hazard mitigation under storm surge conditions, the beach nourishment must be combined with dune construction. The sediment volume in the renourished beach and rebuilt dune serves as a reservoir of sand that moves offshore during the storm event to create an offshore bar for wave-energy dissipation. This is the natural defense mechanism of most of the world's coastlines with sediments of all sizes, colors, and shapes of natural materials that have found their way to the beach. Swell waves move the offshore bar materials landward to rebuild the beach. The beach is said to be in "dynamic equilibrium" when no net loss of sediment occurs in time and the shoreline position remains constant in time. However, most beach renourishment projects are constructed on eroding shorelines that will continue to erode after the initial nourishment. As a result, the long-term maintenance of all beach nourishment projects is an important economic aspect for design.

3.4.1 *History of beach nourishment in the US*

The first large-scale, public beach nourishment project in the United States was on Coney Island, near New York City in 1923. Over 1.7 million cubic yards (cy) of sand was placed along 1.5 miles of shoreline. Figure 3.10 shows some details of what some call urban parks for the masses (Dornhelm, 1995).

The Army Corps of Engineers is the Federal Government agency in the US with overall responsibility for coastal hazard mitigation projects. Many projects cross different political boundaries (cities, counties, states) so that only a project at the federal government level can coordinate the many layers of bureaucracy involved. Sand movement along coasts knows no political boundaries.

A cost study of the Corps' "Shore Protection Projects" since 1950 was conducted in the 1990s and Figure 3.11 presents the results (Hillyer, 1996). The trend toward the "soft" beach nourishment alternative since the 1970s continues today. More details on why this reversal from the previous use of

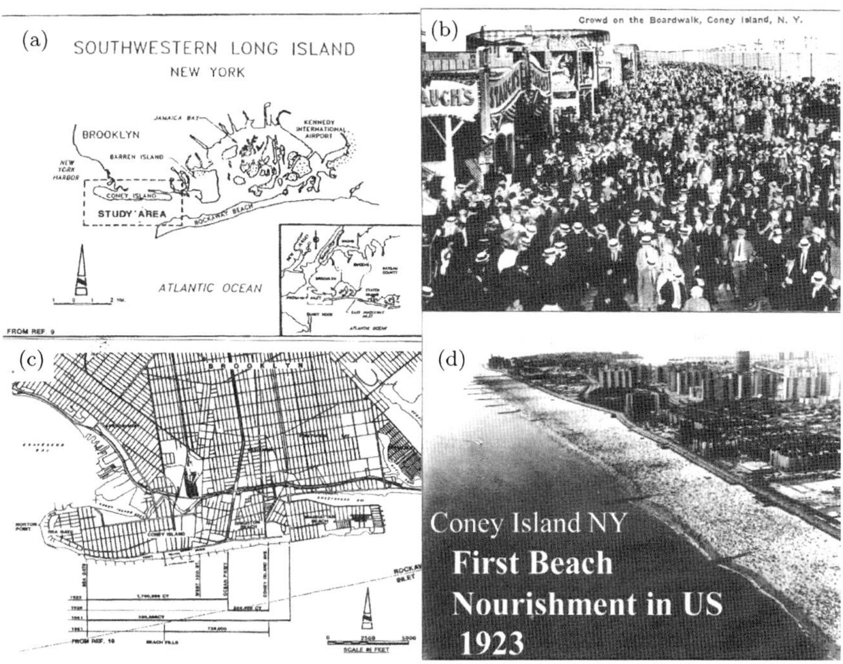

Figure 3.10. (a) Location, (b) crowd on the Boardwalk, (c) site plan, and (d) aerial view on the lower right shows the thousands of people at the beach in this July 4, 1966 photo (from Dornhelm, 1995).

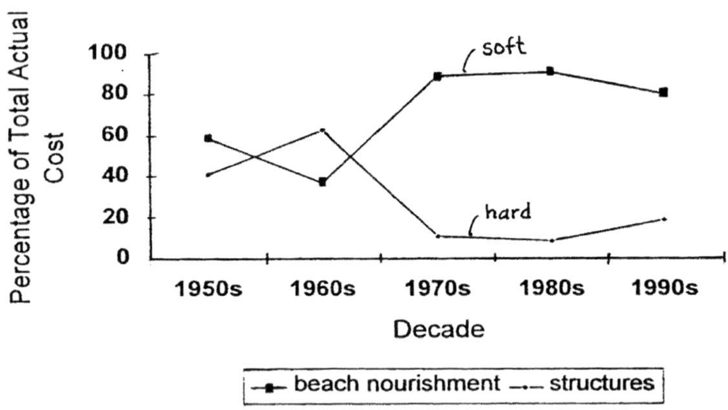

Figure 3.11. History of US Army Corps of Engineers shore protection projects since 1950s. Trend toward "soft" beach nourishment projects (Hillyer, 1996).

hard structures in the 1950s and 1960s occurred in the 1970s can be found in Chapters 7 and 8.

3.4.2 *Evolution of beach nourishment shoreline protrusion*

A beach nourishment project sticks out and is a protrusion of the existing, natural shoreline into the sea. Waves from all directions impact the protrusion causing sand transport in both the longshore and offshore directions away from the project boundaries as schematically illustrated in Figure 3.12 (Dean, 2002).

3.4.2.1 *Planform spreading*

In the longshore direction parallel to the shoreline, the beach sand spreads laterally in both directions and is labeled "spreading out" losses from the original volume placed on the beach (Figure 3.12(a)). Combining the conservation of mass and a LST equation results in an equation for the time and space variation of the shoreline position known as the "diffusion" equation (Pelnard-Considere, 1956) because it contains one coefficient of diffusivity similar to all mass, momentum, and energy (heat) diffusion equations of mathematical physics (Bird *et al.*, 1960).

The diffusion equation has many analytical solutions for simple initial and boundary conditions (see Larson *et al.*, 1997). For example, a rectangular beach fill of length L will have spreading losses of 50% (half-life) in a time $t_{50\%}$ that varies directly with the square of the project length L^2.

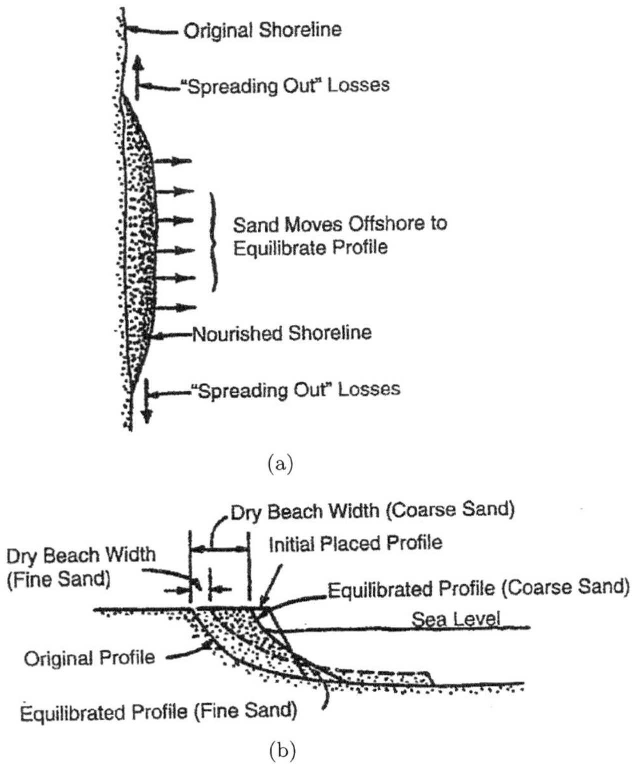

Figure 3.12. (a) Plan view for spreading losses and (b) elevation view for equilibration of a beach nourishment project (from Dean and Rosati, 2010).

Doubling the length of a project will increase its half-life by 400%. A second key variable is the breaking wave height, H_b. The half-life varies inversely as $H_b^{5/2}$ so that doubling H_b produces a five-fold decrease in half-life. These spreading losses do not include the local, historic erosion-rate losses. Chapter 7 includes the equations to compute the spreading losses of a beach nourishment project.

3.4.2.2 *Cross-shore equilibration*

In the cross-shore direction, (Figure 3.12(b)), the steeply constructed beach profile spreads seaward in the offshore direction to "equilibrate" the beach profile; in other words, to return the constructed profile to one that is like the original, natural beach profile. The sand volume is conserved in the offshore direction out to a depth where repeated beach profiles show little if any variation in time, even after storm events. However, severe storms may drag sand past this survey "closure depth" so that some volume loss from the project occurs. The beach equilibration process takes a minimum of one full year of wave conditions with both storm and swell waves present in the normal wave climate. Chapter 7 includes equations to calculate the "closure depth" and the time to equilibration.

The design, dry beach width is always less than the constructed beach width for the initially placed steep profile as schematically illustrated in Figure 3.12(b). This is because construction of the beach fill template (berm elevation, construction width, and fill volume per unit length) is cheaper for the steep nearshore profile at a slope of 1-vertical to 20-horizontal. The construction width is reduced after the first storm and continues to adjust in time (1–2 year) until equilibration is reached. Unfortunately, this fact of beach design and construction is rarely explained and understood by the media so that the public is often told and believes that "*...half the new beach is lost after the first storm*". More discussion of the need for this to be explained to the public and the media is found in Chapter 7.

The simplest design case is when the grain size of the beach fill material from the "borrow" site is the same as the native beach. The beach profile will return to the prefill beach profile in the equilibration process. When the beach fill material is coarser than the native beach material, the profile with coarse sand will be steeper after equilibration. And the opposite is true when the beach is renourished with finer material. The profile with fine sand will be flatter after equilibration. These profile adjustments for beaches renourished with non-compatible sand sizes follow from the equilibrium beach profile theory of Dean (1991) to provide formulas for the design width of the dry beach and the resulting volume of material per unit length of beach as fully presented and discussed with examples in Section 7.3 of Chapter 7.

3.4.3 *Design beach width and volume for variable sediment sizes*

The equilibrium beach profile is a simplification of the complicated bar-trough beach profiles found on the world's sandy beaches. The equation was first proposed by Bruun (1954) and developed further by Dean (1977, 1991) and Moore (1982). When y is the distance from the shoreline, then the water depth, $h(y)$ at any distance y is given by

$$h(y) = Ay^{2/3} \tag{3.1}$$

with A as the "scale parameter" of the profile and dependent on the median grain size, D of the beach sand. Table 7.1 is A versus D in mm units from 0.100 to 1.09 mm and A has units of $m^{1/3}$(Dean, 2002). Values of A increase as D increases so that the well-known fact that beaches with coarse sands are steeper than beaches with fine sands (Bascom, 1964) is preserved by this simple, equilibrium beach profile equation.

Pioneering relationships for the volume of sand required for a given design dry beach-width of a beach nourishment project when the borrow-site (fill) material with scale parameter A_F was *greater* (coarser) than the native beach material, A_N (i.e. $A_F/A_N > 1$), were developed by Dean (1991). An example is displayed in Figure 3.13 for a berm height $B = 1.5$ m and a closure depth $h* = 6$ m when ratio $A_F/A_N = 1.2$. The fill size and native beach sizes are shown in the figure. For a given design dry beach width, using coarser fill material results in a large saving in fill volume relative to when the same size material is used ($A_F/A_N = 1.0$) to nourish the beach. Conducting a search for borrow site material with $A_F/A_N > 1$ is recommended.

The relationships for the opposite case when the borrow-site (fill) material with scale parameter A_F was *less* (finer) than the native beach material, A_N (i.e. $A_F/A_N < 1$), were also developed by Dean (1991). The example in Figure 3.13 shows that when $A_F/A_N = 0.8$, 3–4 times more fill volume

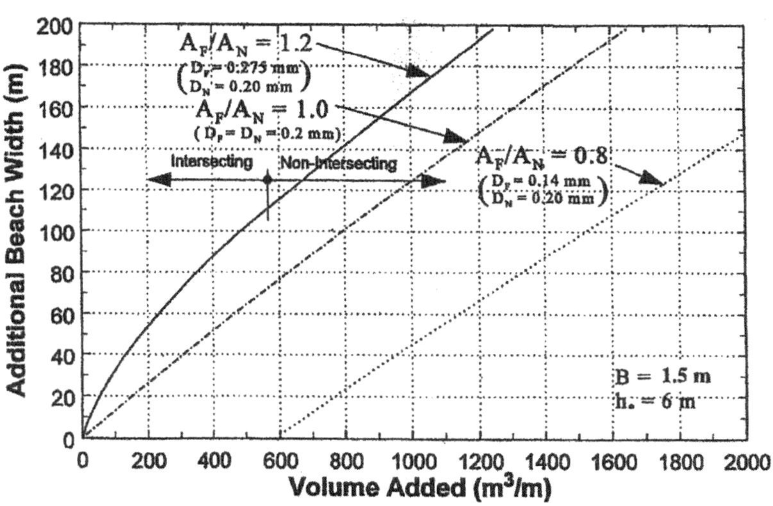

Figure 3.13. Volumes of beach fill and relative increase in fill width as example (from Dean, 1991).

is needed. Using borrow-site (fill) material with $A_F/A_N < 1$ is not recommended but sometimes unavoidable in practice.

Complete details including all design formulations for conditions when the borrow material is coarser or finer than the native beach material is presented in Chapter 7, Section 7.3.3 for the equilibrium beach profile.

3.4.4 *Summary — beach nourishment*

Environmental concerns, monitoring, and economics are found in Chapter 7 including comments by critics (Pilkey *et al.*, 1982; Pilkey and Wright, 1988; and Dean, 2001). Chapter 7 presents a full summary of beach nourishment design. The book *Beach Nourishment: Theory and Practice* is highly recommended (Dean, 2002). Chapter 30 "Beach Nourishment" (Dean and Rosati, 2010) is an update as found in Kim (2010). We summarize with a quote from Professor Robert Wiegel (1994), as follows:

> *Our beaches are the playground of the nation and represent the greatest single public park system we have, stretching as they do along most of its coastline. Many miles of beaches are located in our towns and cities, or nearby, readily available to local residents with short or no traveling distance. They serve as the vital function of urban parks, as well as recreational centers for visitors from inland regions of the USA and from abroad.* (Wiegel, 1994, 2002)

The three main structural alternatives for storm damage mitigation as listed in Figure 3.1 that interact and change the physical environment are armoring structures, shoreline-stabilization structures, and beach nourishments. Changes to man's "system" through adaptation and accommodation (Figure 3.2) are called non-structural alternatives and include flood proofing, zoning, and retreat.

3.5 Non-Structural Alternatives

Society has developed ways to mitigate damages by setting requirements for the elevation of buildings, providing insurance and planning for continual erosion with setback limits for new construction. The final non-structural alternative is to retreat by relocation, abandonment, or demolition.

3.5.1 *Building elevations and insurance*

Building codes that require structural changes in the design, construction, or alteration of a building to reduce damage caused by flooding or related factors (waves, currents, and erosion) are considered a flood proofing alternative by the US Federal Emergency Management Administration (FEMA). The most common is to elevate the first-floor living area above the "100 year" flood level. As discussed in Chapter 2, this is the 1% chance in any one-year water level. The FEMA in the US also produces Flood Insurance Rate Maps (FIRM) for this purpose to alert the general public of the areas where water levels at this 100-year recurrence elevation will occur. Figure 3.14 is an example of a FIRM for the east side of the Lynnhaven Inlet in Virginia Beach, VA (FEMA, 2004). It is panel 37 of the 300 needed at this scale to provide adequate visual coverage of the large city of Virginia Beach. The dark area is the location of the elevated water levels, and the elevations shown beneath each "ZONE VE" include wave setup and wave heights to the crest elevation of waves estimated to occur under the storm conditions that produce the 1% chance event. The region is on the Chesapeake Bay (Figure 1.7(a)) with the black line showing the submerged shoreline and the white area regions above the 1% chance

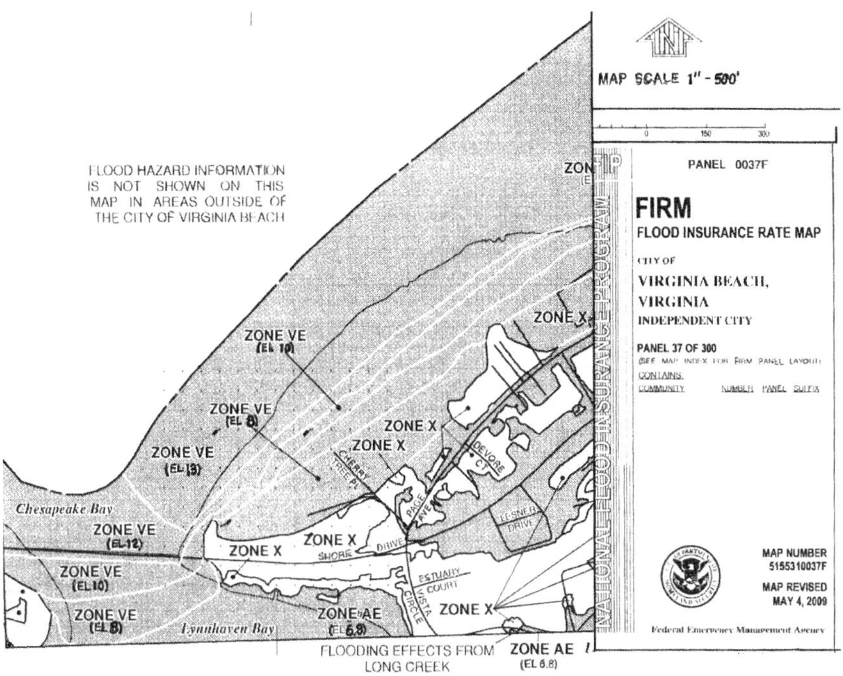

Figure 3.14. FIRM for the Lynnhaven Inlet of Virginia Beach, VA. See Figure 1.7(a) for location on the Chesapeake Bay.

event. A storm-surge elevation at the 1% exceedance level (100-year recurrence interval) plus waves are employed to determine risk and insurance rates for individual properties located on the flood maps. Insurance rates are much lower for structures elevated above the 100-year flood level and are a requirement for all new construction in the coastal, high-hazard VE Zone. In effect, these regulations become the floodplain zoning laws applicable to individual property owners.

The FEMA National Flood Insurance program in the US is actuarial but currently in debt because of recent major flooding events such as Hurricanes Katrina, Irene, and Superstorm Sandy. Currently, the FEMA and US Congress are studying ways to pay off these debts and restore the FEMA national flood insurance program to its actuarial basis as required by law for this program.

3.5.2 *Setback limits*

A second way to mitigate damage is to limit the construction close to the shoreline and establish setback limits. The National Oceanic and Atmospheric Administration (NOAA) of the US National Ocean Service has identified land-use planning and the location of construction as one effective means to reduce coastal storm hazards, particularly on eroding coasts. Here, the mechanism to require change in old construction practices in the US is the Coastal Zone Management Act (CZMA) of 1972. Through the CZMA, the NOAA provides funds to individual states to help solve their own coastal hazard problems. As a result, many states have developed coastal construction setback lines and zones that include historic erosion rates at each site. The methods, definitions, widths, etc., vary from state to state. A good example is Florida (Coastal Construction Control Line, CCCL, Florida Statutes, Section 161.052, 1970). A key element is the historic, average erosion rate at each site. Presently, FEMA does not include delineation of erosion zones and erosion hazards on its flood maps. Methods to incorporate both coastal erosion and beach nourishment (National Research Council, 1995) in the national, flood insurance program have been proposed but have yet to be formally adopted. Clearly, coastal erosion increases the risk and beach nourishment reduces the risk of coastal flood and wave damage.

3.5.3 *Retreat*

Retreat is the final hazard mitigation option. Relocation here also considers abandonment and demolition. To some (Pilkey and Wright, 1988) retreat is the only option. But practically, all constraints (economic, environmental, social, legal, etc.) must also be evaluated for the retreat alternative as well as for all others as previously discussed.

The US Army Corps of Engineers considers non-structural alternatives for hazard mitigation. CEM (2006) provides two examples. In one case at Cape Shoalwater, Washington, the average annual erosion rate (30–40 m/year) was so high and the relative rural area benefits so low, that both structural (jetty) and non-structural (land purchase) alternatives prohibited a Federal government project. The "do-nothing" alternative was the choice.

At Baytown, Texas, storm surge flooding on Galveston Bay and excessive land subsidence (oil and gas extraction and groundwater withdrawal) resulted in an analysis of both structural (earth levee, concrete floodwall) and non-structural (permanent relocation and evacuation) alternatives to mitigate the flooding hazard. The Galveston District Office of the Corps (Galveston, 1976) recommended the retreat alternative and Congress authorized funding the retreat alternative (WRDA, 1996). Unfortunately, the project floundered on local disagreement over the value of the land to be purchased and was never funded. In 1980, Congress was ready to authorize funding, but the local, 20% cost share could not be provided. The local residents decided to stay, and the Corps placed the project in its inactive category. Because the benefit/cost (B/C) ratio was 1.3, the annual benefits to the Federal government exceeded the 20%, local cost share amount. In other words, the Federal government could have paid for the entire project and still realized net economic benefits to the Federal government. In summary, a non-structural solution for part of the community of Baytown, Texas, proved impossible to implement because the entire community would not share in the local cost requirement. The lesson learned is that in some cases with adequate B/C ratio, the Federal government should waive the cost-share requirement and implement the retreat alternative. See Pendergrass and Pendergrass (1990) for full details. This would be an "adaptation" to the present institutional constraint of the cost-sharing method now in place.

A managed retreat alternative for South Louisiana was discussed by an expert panel of the National Academy of Sciences in 2004 (NAS, 2004). Williamson (2012) studied a managed retreat alternative for five southeast Louisiana communities (Delacroix, Reggio, Hopedale, Ycloskey, and Shell Beach). It was learned that the most recent comprehensive master plan for a sustainable coast (2012) in Louisiana did not mention a managed retreat as a last resort option. Williamson (2012) believes that the time has come for preliminary discussions with major stakeholders (at a minimum) in these communities regarding the managed retreat alternative. It could also be called the Socially Acceptable Retreat Alternative (SARA).

3.6 Combinations and New Technologies

3.6.1 *Structural combinations*

Beach-stabilization structures together with beach nourishment are discussed in Section 3.3. The combination mitigates down-drift impacts and/or increases the fill life of the renourished beach. Together, their life-cycle costs and environmental impact are usually less than if selectively implemented. Construction of the beach stabilization structures without fill is likely to damage adjacent beaches.

The original design of the new seawall for hurricane protection at Virginia Beach was to be a curved, concrete-type structure as at Galveston, Texas (Figure 1.2). To accommodate a lowered seawall crest for aesthetic reasons (see Figures 1.8 and 1.9), a wide beach nourishment project was added to the design to reduce flooding and wave-damage (Corps of Engineers, Norfolk, 1994). Together with improved interior drainage and pumping equipment, this combined design provides the same hurricane flooding and wave-damage protection as the original seawall design. Figure 3.15(a) provides an artists' sketch of the steel, sheet-pile bulkhead over the old boardwalk and Figure 3.15(b) an aerial photograph of the completed project including the seawall combined with the beach nourishment for storm damage mitigation at Virginia Beach, Virginia (courtesy Norfolk District Office, Corps of Engineers).

The soft structural alternative (beach and dune) can be strengthened by adding a buried rock seawall/revetment within the dune as shown in Figure 3.16. The combination was determined to be both environmentally and economically advantageous when compared against an armored revetment for storm protection against the 1% chance water-level event at Dam Neck, Virginia, on the Atlantic Ocean (Basco and Shin, 1996). The final design cross-section is shown in Figure 3.16(a) and includes a buried rock seawall/revetment beneath and set well back within the dune. In the event of a major storm causing severe dune erosion, the buried seawall will prevent storm damage if a second major storm occurs in the same season. Figure 3.16(b) shows a photograph of dune construction with the buried, rock seawall (Basco, 2000). Hurricane Isabel (September 18, 2003) with elevated water levels (Figure 2.2) eroded one-half of the dune and exposed the toe of the rock structure for the first time. The dune was rebuilt to original

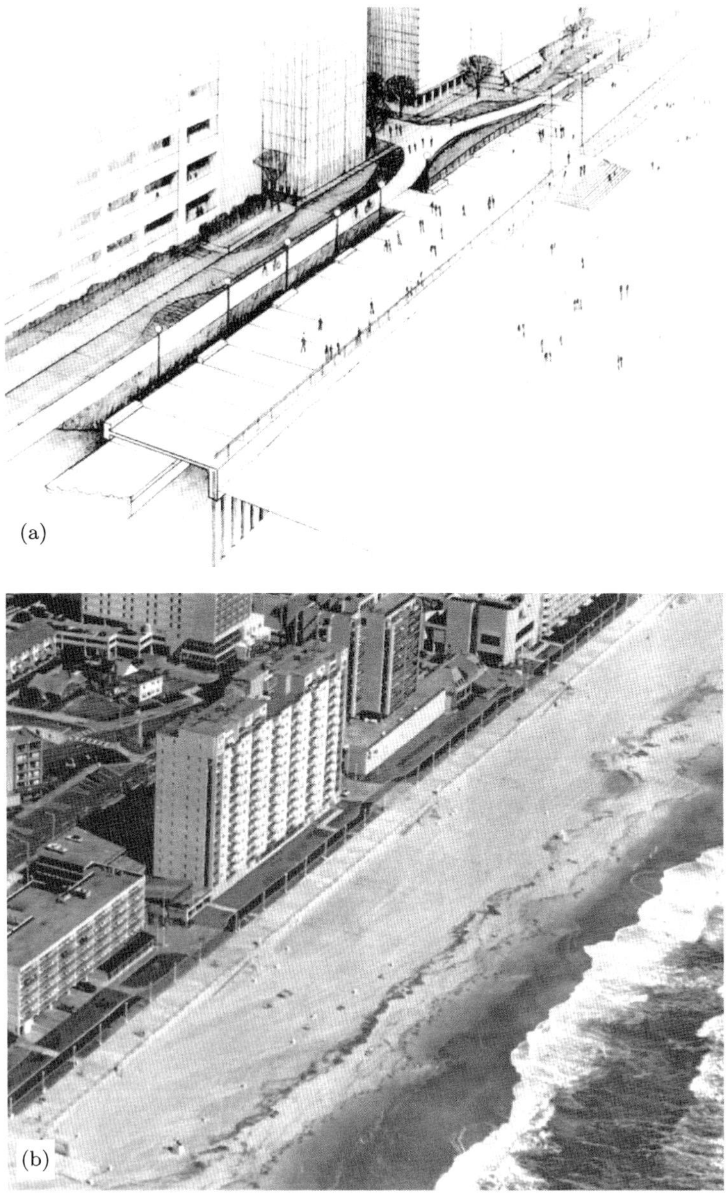

Figure 3.15. Structural combination. (a) Steel sheet-pile seawall with concrete cap and (b) beach nourishment project for storm damage mitigation at Virginia Beach, VA (courtesy City of Virginia Beach).

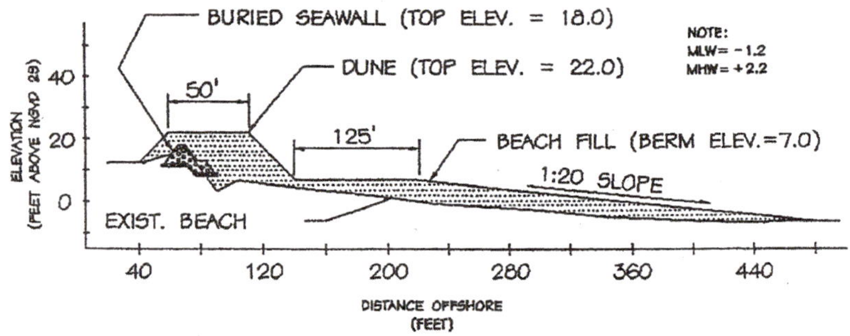

(a)

(b)

Figure 3.16. Beach nourishment, rebuilt dune, and buried rock seawall at Dam Neck, VA.

design dimensions when renourishment of the beach took place in 2004. See Section 7.4 in Chapter 7 for further discussion of this example.

3.6.2 *Non-structural and structural combinations*

At many locations, elevated structures combined with some type of armoring, shoreline stabilization structure, and/or beach nourishment are employed in combination. Presently, 32 of 35 US coastal states and territories have some type of setback requirements for new construction and existing structures found uninhabitable after a storm (Heinz Center, 2000). These non-structural, adaptive measures and structural alternatives are often combined to address the wide range of coastal problems previously cited. An example is the barrier island of Grand Isle, Louisiana, where beach nourishment, rebuilt dunes, a groin field, and near-shore breakwaters are used for a community where the first floor of most residences is constructed above the 1% chance flood level (with waves) and the public lands on the rapidly eroding east end of Grand Isle are restricted from any development.

3.6.3 *New technologies*

Many non-traditional ways to armor, stabilize, or restore the beach including the use of patented, precast concrete units, geotextile-filled bags, and beach dewatering systems have been tried in the field. Their success depends on their stability during storm events and durability over the economic, design life. Their initial cost and cost for removal if environmental impacts warrant can be less than traditional methods, at some sites. But the long-term maintenance costs and life-cycle costs are unknown. These new technologies often involve non-traditional materials or shapes but are employed in a traditional manner, e.g. nearshore breakwaters.

3.6.3.1 *Precast, concrete units*

Patented, modular, precast concrete units that can be interconnected to form one or more nearshore breakwaters have been tested in Florida, Georgia, New Jersey and other sites around the world. One unit called the "Beachsaver" system (Breakwaters International, Inc., Flemington, NJ) was field tested as two 50-m long breakwater sections with a 34-m gap placed in a 0.9-m water depth about 37 m offshore in 1989. Independent monitoring revealed significant settlement (over 1.2 m) and scant sand trapping (Sorensen and Weggel, 1992).

Another design labeled the prefabricated erosion protection (P.E.P) reef system by developer American Coastal Engineering, Inc. West Palm Beach, FL was field tested in the early 1990s in West Palm Beach and then again in late 1990s in Vero Beach, Florida. The first effort constructed a 1,270-m long reef (called a reef system because the crest was below mean sea level) and placed 76-m offshore in 2.9-m water depth. It consisted of 330 interlocking units each 3.7 m long weighing 22 metric tons. Independent monitoring revealed settlements of 17% of the units (0.5–0.65 m) and that wave overtopping, wave setup, pounding, increased longshore currents and sediment transport produced an *increase* in the normal erosion trend at the West Palm Beach site (Dean and Chen, 1996). As a result, the reef was removed.

The second P.E.P installation at Vero Beach, FL tried a staggered and gapped configuration. Analysis of nearshore profiles over 25 months (August 1996–June 1999) when compared with historic volume change (1972–1986) revealed that erosion has increased landward of the reef. And, background erosion decreased north but remained the same in control areas outside the reef area (Stauble and Smith, 1999).

Both these concrete unit designs are pyramid-shaped with narrow crests and flatter slopes on the seaward side. In contrast, a vertical, cylindrical shape forms the Hollow Core Reef System (HCR) as patented by Hollow Core Reef Enterprises, Inc. Newport, News, VA. Another shape is the Reef Ball that is a spherical-shaped unit with many sized holes to provide habitat for marine life (Harris, 2003). See Stauble and Tabar (2003) for analysis of the submerged, narrow-crested type for erosion control.

Chapter 6 and Section 6.6.2 "Artificial Reef Breakwaters" includes sketches and photographs of the "Beachsaver", P.E.P. reef system, and "Reef Ball" (Figure 6.107) and design of "Low-Crested Structures" (LCS) in Europe.

Articulated concrete blocks and mats used in revetment design are discussed in detail in Section 6.4.3.2 as shown in Figure 6.88.

3.6.3.2 *Geotextile containers filled with sand*

Geotextile materials have a long history for foundation mats beneath rubble-mound structures (see Chapter 6). And they have been used as silt curtains to contain dredged materials in the water column. They have also been formed into small bags, long, sausage-shaped cylinders (called Longard tubes) and flat mattress shapes. When the various shapes are filled with sand, groups of the containers have been deployed in many ways as illustrated in Figure 3.17 (Oumeraci and Recio, 2009). In the 1980s and 1990s, there has been significant improvement in the quality and durability of geotextile fabrics making them suitable for a variety of coastal applications. But their long-term strength decreases under ultraviolet rays and storm waves weaken the fabric ultimately leading to their deterioration and destruction. As a result, their structural integrity is a function of time. They have been called "soft" rock structures.

The design life of a geotextile container filled with sand depends on many factors. It is generally always less than properly designed rock structures serving the same function. However, if found to cause negative impacts to adjacent shorelines, the containers can be cut open and removed with the filled sand remaining on the beach. It is for this reason

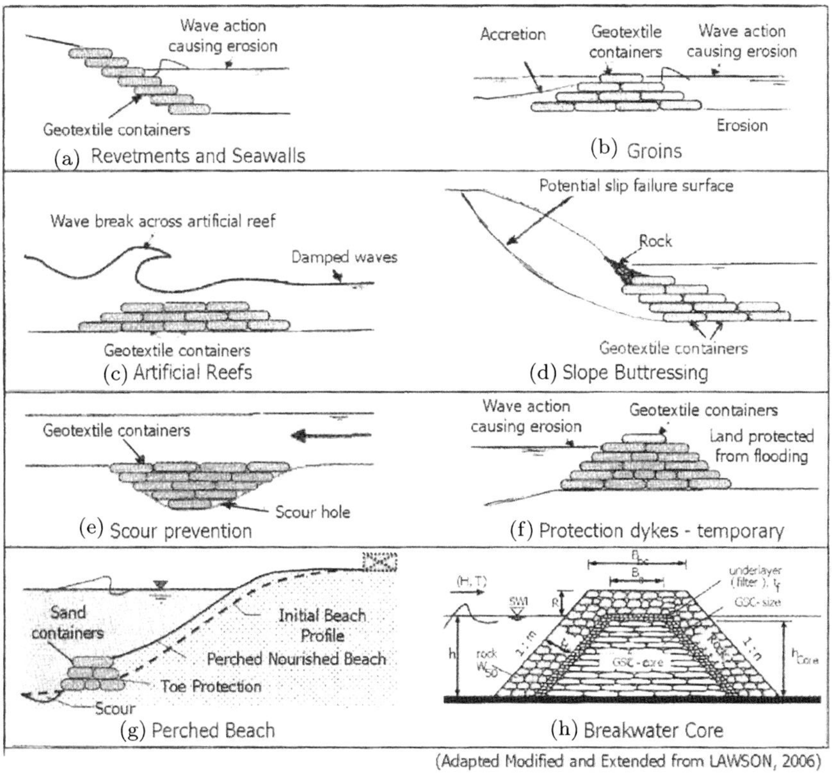

Figure 3.17. Possible coastal engineering applications of geotextile containers filled with sand (from Oumeraci and Recio, 2009).

that a soft groin field was permitted by the North Carolina Coastal Resource Commission (CRC) together with a beach nourishment project for South Beach on the western end of Bald Head Island, North Carolina (Denison, 1998). The North Carolina CRC prohibits hardened structures on its ocean coast. The long-term survivability of this system has yet to be determined.

Pioneering, comprehensive, laboratory testing of geotextile sand containers (GSCs) has been performed in the early 2000s in Germany. Both small-scale model studies (1-l bags, 20 cm waves) and large-scale experiments (150-l bags, 1.6 m significant wave heights) have been employed to determine the stability of containers on slopes and at the crests of structures. All the details and formulas as found in Oumeraci and Recio

(2009) will be presented in Chapter 6, Section 6.4.3 including sketches and photographs shown in Figures 6.91 and 6.92.

3.6.3.3 *Beach dewatering systems*

The location of the water table beneath the beach surface influences the sediment transport dynamics. A permeable beach with low water table is more stable and has led to the development of beach dewatering (drainage) systems to artificially lower the groundwater table. Buried pipes with spaced holes and pumps to remove the groundwater and lower the water table beneath the beach surface have been studied in the laboratory, and the first field application in Australia produced evidence that the beach width increased (accreted) by lowering the water table elevation at that location (Chappel *et al.*, 1979). Since then, beach dewatering systems have been extensively studied in field tests in Denmark (Vesterby, 1994) and many other sites. A history of technology development from 1940–1995 can be found in Turner and Leatherman (1997) which also included independent evaluations of the effectiveness of some applications at some sites.

3.6.3.4 *Others*

In Japan, a concrete mat with a cross-sectional shape to mimic sand ripples of a large size (distorted ripples) has been tested in the laboratory and field and found to aid the landward movement of sand beneath waves in the nearshore region (Ono *et al.*, 2004). Woodruff (2006) gave a historical overview of the many innovative erosion control ideas that have been field tested in Florida. Those not mentioned before include groins made of netting material and artificial seaweed (grasses) planted in the nearshore region. Both did not survive the first storm event.

Since the *Coastal Engineering Manual* (CEM, 2006) was made available, there have been many additional papers highlighting or summarizing other new technologies for coastal hazard mitigation. The proceedings of the International Conference on Coastal Engineering (ICCE) should be consulted for other possible new technologies.

3.6.3.5 *Summary — New technologies*

The coastal engineer would be wise to bring an open mind, but healthy degree of skepticism to all new technologies with broad claims to stop erosion and save the beach. In some cases (not all), proponents can be likened to "snake oil" salesmen of the distant past who promised a cure for all bodily ills of the person buying the product. The key fact is that sand cannot be created by these devices. The volume of sand in the system remains constant and if retained at one location by the device, it must be lost at another nearby region at the site.

In addition, the initial cost of the material and placement (construction) in the field is only that, the *initial cost*. Many new technologies are advertised as the "low cost" solution to erosion. The relationship between excess storm wave energy and damage of the device, i.e. the "damage" curve is unknown because the device has never been tested in the laboratory or the field under wave conditions that cause damage. Consequently, the annual maintenance costs cannot be determined so that the long-term life-cycle costs are unknown. One favorite question is to request the damage curve from laboratory or field tests under excessive wave conditions from proponents of their device. This author has yet to obtain such information from any of the manufactures of the new pre-cast concrete units on the market. The one exception known to the writer is for geotextile containers filled with sand as summarized by Oumeraci and Recio (2009) and discussed in detail in Chapter 6.

See Chapter 11 for some possible "adaptation" strategies for existing designs when facing the unknown levels of future sea-level rise.

3.7 Do-Nothing

One final alternative that must always be evaluated is the do-nothing or no-project case. The risk of flooding and wave damage continues or increases if historic erosion is also present at the site. When this response is appropriate, what happens to the area, and what government programs are available to help?

It is almost certain that the area impacted will decline economically. Social and economic stresses will continue. Examples include social stresses from apprehension and helplessness, economic stresses of depressed property values, personal property losses, cost and inconvenience of restoration after repetitive flooding, reduced recreational opportunities for citizens, reduced tourism benefits, reduced employment opportunities for tourism, property values decrease, and related property taxes diminish. The no-action alternative may perpetuate a more costly, federal commitment than would be realized otherwise, because other federal assistance programs exist.

When the no-action plan remains, the US Federal government relies on three methods to mitigate coastal damages. As previously discussed, the National Flood Insurance Program (NFIP) provides compensation for flooding and wave damages, but it does not protect property from flooding. The NFIP also only encourages adaptation measures; but they are not mandatory, except for new construction where local authorities have adopted flood zoning ordinances. The US Federal government also participates in measures for emergency, evacuation route planning. Because many coastal communities lack sufficient bridge and highway capacities, emergency evacuation is not a dependable means of hazard mitigation. The third US program is Emergency Assistance from FEMA when the President declares the region eligible for this financial package after a major, coastal storm event.

The United States does not have a national plan of coastal hazard mitigation.

3.8 Summary — Alternatives for Coastal Hazard Mitigation

The alternatives for coastal hazard (flooding, wave damage, erosion) mitigation discussed in this chapter can be summarized in the four major categories listed in Table 3.1. Full details for the structural alternatives and combinations are found in the remainder of this book.

Needs for coastal structures for navigation jetties, ecosystem enhancement, and pile-supported structures are also benefited for their design by

Table 3.1. Alternatives for coastal hazard mitigation.

STRUCTURAL
Armoring (seawalls, bulkheads, revetments, dikes and floodwalls) and movable gates
Beach stabilization (headland breakwaters, nearshore breakwaters, groins, reefs, and wetlands)
Beach nourishment (beaches, dunes), sand bypassing of inlets
NON-STRUCTURAL
Zoning and insurance (building elevations, flood insurance)
Setback limits
Retreat
COMBINATIONS AND NEW TECHNOLOGIES
Structural combinations
Non-structural and structural combinations
New technologies
DO NOTHING

the remaining material as covered in this book. See Section 11.3.3 for appreciating the effect of sea-level rise impacts that are discussed in Chapter 3.

References

Basco, D. R. (2000) "Beach Monitoring Results and Management Plan, FCT-CLANT, Dam Neck, Virginia Beach, VA," Final Report for Department of the Navy, NAVFAC and Glenn and Sadler, Norfolk, VA

Basco, D. R., and C. S. Shin (1996) "Dune Damage Curves and Their Use to Estimate Dune Maintenance *Costs," Proceedings of the International Conference Coastal Engineering*, ASCE, Orlando, Florida, Vol 3.

Bascom, W. (1964) *Wave and Beaches*, Doubleday Anchor Press, New York.

Bird, R. B., W. S. Stewart, and E. N. Lightfoot (1960) *Transport Phenomena*, John Wiley & Sons, New York.

Bruun, P. (1954) "Coast Erosion and the Development of Beach Profiles," Technical Memorandum No. 44, Corps of Engineers, Coastal Engineering Research Center, Fort Belvoir, VA, 82 p.

CEM (2006) *The Coastal Engineering Manual*, US Army Corps of Engineers, Washington, DC.

Chappel, J., I. G. Eliot, M. P. Bradshaw, and E. Lonsdale E. (1979) "Experimental Control of Beach Face Dynamics by Water-table Pumping," *Engineering Geology*, Vol. 14, pp. 29–41.

Corps of Engineers, Norfolk. (1994) "Hurricane Protection Project for Virginia Beach, VA," General Design Memorandum-Revised, Norfolk District Office, VA.

Dean, R. G. (1977) "Equilibrium Beach Profiles: US Atlantic and Gulf Coasts," Technical Report 12, Department of Civil Engineering, University of Delaware, Newark.

Dean, R. G. (1991) "Equilibrium Beach Profiles: Characteristics and Applications," *Journal of Coastal Research*, Vol. 7, pp. 53–84.

Dean, C. (2001) *Against the Tide: The Battle for America's Beaches*, Columbia University Press, New York.

Dean, R. G. (2002) *Beach Nourishment: Theory and Practice*, World Scientific, Singapore.

Dean R. G., and R. Chen (1996) "Performance of the Midtown Palm Beach PEP Reef Installation", Final Report, Coastal and Ocean Engineering Laboratory, University of Florida, Gainesville FL.

Dean, R. G., and J. Rosati (2010) "Beach Nourishment," Chapter 30, in Y. C. Kim (ed.), *Handbook of Coastal and Ocean Engineering*, World Scientific, Singapore.

Denison, P. S. (1998) "Beach Nourishment/Groin Field Construction Project: Bald Head Island, North Carolina," Shore & Beach, American Shore and Beach Preservation Association.

Dornhelm, R. B., (1995) "The Coney Island Public Beach and Boardwalk Improvement of 1923," *Shore and Beach*, Vol. 63 No.1, pp. 7–11.

FEMA (2004) "Flood Insurance Rate Map, Panel 37, City of Virginia Beach, VA," Federal Emergency Management Agency, Washington, DC, May 4, 2000 (now available online).

Galveston (1976) "Baytown, Texas," Phase I, Phase II, Design Memorandum, Galveston District Office, Corps of Engineers, TX.

Hardaway, C. S., and R. J. Byrne (1999) "Shoreline Management in Chesapeake Bay," Special Report in Applied Marine Science and Ocean Engineering, No. 356, College of William and Mary, Virginia Institute of Marine Science, Gloucester Point, VA.

Harris, L. E. (2003) "Submerged Reef Structures for Beach Erosion Control," *Proceedings of the Coastal Structures Conference,* ASCE.

Heinz Center (2000) "Evaluation of Erosion Hazards," The H. John Heinz III Center for Science, Economics and the Environment, Washington, DC, p. 203.

Hillyer, T. M. (1996) "An Analysis of the US Army Corps of Engineers Shore Protection Program," Final Report, IWR 96-PS-1, Institute of Water Resources, Corps of Engineers, Fort Belvoir, VA.

Kim (2010, ed.) *Handbook of Coastal and Ocean Engineering,* World Scientific, Singapore.

Larson, M., H. Hanson, and N. C. Kraus (1997) "Analytical Solutions of One-line Model for Shoreline Change Near Coastal Structures," *Journal Waterway, Port, Coastal, Ocean Engineering,* Vol. 123, pp. 180–191,

Moore, B. (1982) "Beach Profile Evolution in Response to Changes in Water Level and Wave Heights," MS Thesis, University of Delaware, Newark, DE.

NAS (2004). "Managed Retreat from the South Louisiana Shore," Marine Board, National Research Council, National Academy Press, Washington, DC.

National Research Council (1995) *Beach Nourishment and Protection*, Marine Board, National Research Council, National Academy Press, Washington, DC.

National Research Council (2007) *Mitigating Shore Erosion Along Sheltered Coasts*, Ocean Studies Board, The National Academies Press, Washington, DC.

Ono, N., I. Irie, and H. Yamaguchi (2004) Preserving System of Nourished Beach By a Distorted Ripple Mat, *Proceedings of the 29th International Conference on Coastal Engineering*, ASCE.

Oumeraci, H., and J. Recio (2009) "Geotextile Sand Containers for Shore Protection," in *Handbook of Coastal and Ocean Engineering*, pp. 553–600.

Pelnard-Considére, R. (1956) "Essai de Theorie de l'Evolution des Formes de Rivage en Plages de Sable et de Galets," *4th Journeess de l'Hydraulique, Les Energies de la Mer*, Question III, Rapport No.1.

Pendergrass, B. B., and L. F. Pendergrass (1990) "In the Era of Limits: A Galveston District History Update 1976–1986," U.S. Army Engineer District, Galveston, Galveston, TX.

Pilarczyk, K. W. (ed.) (1990) *Coastal Protection*, A.A. Balkema, Rotterdam.

Pilkey, O. H., and H. L. Wright, (1998). "Seawalls Versus Beaches," *Journal of Coastal Research*, Special Issue No. 4, pp. 41–46.

Pilkey, O. H., J. D. Howard, B. Brenninkmeyer, R. W. Frey, A. C. Hine, J. C. Kraft, R. Morton, D. Nummedal, and H. Wanless (1982) "I- Saving the American Beach: A Position Paper by Concerned Coastal Geologists," *Shore & Beach*, American Shore and Beach Preservation Association, April, pp. 3–4.
Pope, J., and J. L. Dean (1986) "Development of Design Criteria for Segmented Breakwaters," *Proceedings of the 20th International Conference on Coastal Engineering*, ASCE, NY, pp. 2144–2158.
Silvester, R. (1960) Stabilization of Sedimentary Coastlines, *Nature*, Vol. 188, No. 4749, pp. 467–469.
Silvester, R., and J. R. C. Hsu (1993) *Coastal Stabilization: Innovative Concepts*, Prentice-Hall, Englewood Cliffs, NJ.
Sorensen, R. M., and Weggel, J. R. (1992) "Field Monitoring of a Modular Detached Breakwater System," *Proceedings of the Coastal Engineering Practice*, ASCE, New York, pp. 189–203.
Stauble, D., and J. Tabar, (2003). "The Use of Submerged Narrow-Crested Breakwaters for Shoreline Erosion," *Journal of Coastal Research*, Vol. 19, pp. 684–722.
Stauble, D. K., and J. B. Smith (1999) "Performance of the PEP Reef Submerged Breakwater Project, Vero Beach, Florida," 25 Months Report, Coastal Engineering Research Center, Corps of Engineering, Vicksburg, MS.
Turner, I. L., and S. P. Leatherman (1997) "Beach Dewatering as a 'Soft' Engineering Solution to Coastal Erosion-A History and Critical Review," *Journal of Coastal Research*, Vol. 13 No. 4, pp. 1050–10.
Vesterby, H. (1994) "Beach Face Dewatering-The European Experience," *Proceedings of the 7th National Conference on Beach Preservation Technology*, Florida Shore and Beach Preservation Association., Tallahassee, FL.
Wiegel, R. L. (1994) "Ocean Beach Nourishment on the USA Pacific Coast," *Shore & Beach*, Vol. 62, No. 1, pp. 11–36.
Wiegel, R. L. (2002) "Seawalls, Seacliffs, Beachrock: What Beach Effects?," *Shore & Beach*, Vol. 70, Nos. 1, 2, and 3.
Williamson, S. (2012) "Applicability of a Managed Retreat Alternative for the Southeast Louisiana Communities of Delacroix, Reggio, Hopedale, Ycloskey, and Shell Beach, Louisians," MS Thesis, Department of Civil and Environmental Engineering, Old Dominion University, Norfolk, VA.
Woodruff, P. E. (2006) "Historical Overview of Innovative Erosion Control Technology in Florida," Bureau of Beaches and Coastal Systems, Florida Department of Environmental Protection, Gainesville, FL.
WRDA (1996) "National Shoreline Erosion Control Development and Demonstration Program," Section 227, Water Resources Development Act (WRDA) of 1996, Congressional Record, Washington, DC.

Chapter 4

Probabilistic Design

4.1 Introduction

Many of the alternatives for the mitigation of flooding, high waves, and erosion as discussed in Chapter 3 were designed in the recent past by *deterministic* methods. The modern approach is to follow *probabilistic* design methods. A brief summary of each method helps to understand the differences and why the probabilistic method is required to quantify risk and resilience in coastal structures design.

4.1.1 *Deterministic method*

Deterministic systems do not permit any randomness in the future development of states of the system. For example, all the formulas in Chapter 5 for Functional Design (wave runup, wave overtopping, etc.) and in Chapter 6 for Structural Design (i.e. armor layer stability) present the relationships between water level, wave characteristics, and the structural response. These formulas are deterministic models that always produce the same output from a given set of initial conditions. They are the basis for homework and examination questions so that the correct answer is "determined" by the correct use of the formula and input conditions. They are deterministic.

Some believe that if using a probabilistic level for the design conditions such as the water level with a recurrence interval of 100 years (probability of exceedance in any one year of 1%), then this is a probabilistic design. This is not correct. Use of this water level in the wave runup and overtopping formulas will again only produce one deterministic result. Adding in some safety factor again only produces one correct answer and tells nothing about the failure probability and the reliability of the selected design.

4.1.2 *Probabilistic method*

A stochastic (random) process in probability theory employs variables that are random and represent some random value or system.

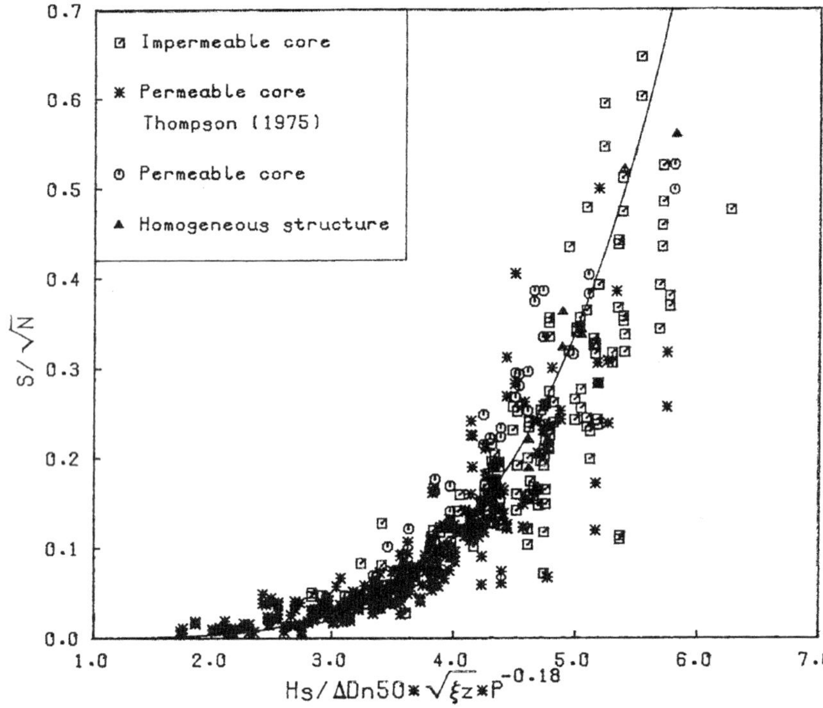

Figure 4.1. Plot of test data employed to develop stability formula for plunging waves on rubble-mound structures (from van der Meer, 1987).

The formulas for design in Chapters 5 and 6 were developed from physical insight and theory, and require laboratory experiments and field measurements to obtain the data needed to quantify the model. The empirical coefficients were determined by fitting the resulting formulas to the data. Considerable scatter is not uncommon. For example, Figure 4.1 (van der Meer, 1987) displays the original data as measured and calculated to determine the stability formula for plunging waves on rubble-mound structures. The original measured data from model scale experiments can be found in the thesis by van der Meer (1988). The solid curve is the mean, central-tendency fit to the data to produce the empirical, constant coefficient (6.2) needed to quantify the formula.

All the variables to determine the horizontal and vertical coordinates in Figure 4.1 will be defined and discussed further in the following. They all have probability distributions about their own mean values. They represent the strengths of the structure and the loads on the structure, and all are statistical quantities that taken together determine the probability of failure of the structure to perform its function or be stable.

4.1.3 *Probabilistic design practice (Level I design)*

The emphasis in this Chapter is to provide some understanding of the tools for use in today's practice of probabilistic design. It is called Level I design (Section 4.2) and employs partial safety factors that have been derived from higher levels of theory (Levels II and III) to take into account probability distributions of the load and resistance variables. The key result is that a specified level for the probability of failure can be established at the outset depending on the goals and importance of the structure.

The material summarized in this chapter is primarily taken from CEM (2006, Chapter VI-6) as written by Dr. Hans Burchardt of Alborg University, Denmark, and from Kamphuis (2010).

4.2 Probability of Failure — Level I Design Practice

4.2.1 *The limit state equation*

Let the linear failure function, G, be defined as

$$\mathrm{G} = \mathrm{R} - \mathrm{S}, \tag{4.1}$$

where R is the resistance to failure or strength of the structure and S is the load or forces that could produce failure. Equation (4.1) is called the limit state equation and is a linear equation. When the resistance is less than the load ($\mathrm{G} < 0$), there is a failure in the design condition.

4.2.2 *Probability of failure of the design condition, P_f*

The probability of failure, P_f, of the design condition is defined by

$$P_f = \text{the probability}\,(\mathrm{G} < 0). \tag{4.2}$$

In the deterministic method, there is no way to quantify the probability of failure, P_f. It means all or nothing, either $\mathrm{R} > \mathrm{S}(P_f = 0)$ or $\mathrm{R} < \mathrm{S}$ and $(P_f = 1)$.

In the probabilistic method, the resistance R and load S in the limit state equation represent all the many variables that are important for the *mode of failure* of the structure under the design conditions. And each variable in R or S has its own statistical probability distribution. As an example, consider a rubble-mound structure with the mode of failure being the armor layer stability on the seaward side of the structure.

4.2.2.1 *An example of deterministic design (stability formula for plunging waves on rubble-mound structures)*

Arguably, one of the most important equations in this book is the one developed over 30 years ago for the stability of rock armor layers on sloping surfaces in deep water exposed to plunging-type waves by Dr. Jentse van der Meer (van der Meer, 1987, 1988). It has stood the test of time. The structure could be a breakwater, groin, jetty, or protective revetment on an earth slope. Rocks to stabilize the sloped surface (armor layer) against wave forces continue to be one of the most commonly used materials to construct coastal structures as discussed in detail in Chapter 6.

The original formulation of the equation for plunging breakers was

$$(H_s/\Delta D_{n50})\,\sqrt{\xi_{mo}} = 6.2\;P^{0.18}(S/\sqrt{N})^{0.2}, \tag{4.3}$$

where H_s is the significant wave height in deep water at some level of probability of exceedance, Δ is the relative submerged rock density $=$ $(\rho_s - \rho_w)/\rho_w$, ρ_s is the mass density of armor stone, ρ_w is the mass density of water, D_{n50} is the nominal (cubical) diameter of stone, ξ_{mo} is the dimensionless surf-similarity parameter in deep water $=$ $\tan\alpha/\mathrm{SQRT}(2\pi H_s)/gT_m^2$, T_m is the mean wave period, α is the structure slope, P is the permeability (nominal) of the core, S is the dimensionless damage level (here italics to distinguish from load, S) is the A/D_{n50}^2, A is the eroded, cross-sectional area of the profile, N_z is the average number of waves in the coastal storm, and 6.2 is the calibration coefficient in deep water.

The symbols in Figure 4.1 are now defined except ξ_z, which is the same as ξ_{mo} since we have used the symbol T_m throughout this book. Also, the values for the power-law exponent for P (0.18) and curve-fit constant (6.2) are now evident. A full discussion of the effects of shallow water on the coefficient "6.2" and on the use of other wave periods is found in Chapter 6.

The original formulation can be rearranged as a limit state equation

$$\mathrm{G} = \mathrm{R} - \mathrm{S}, \quad \text{where} \quad \begin{cases} < 0 & \text{(failure)}, \\ = 0 & \text{limit state (failure)}, \\ > 0 & \text{no failure (safe region)}, \end{cases} \tag{4.4}$$

by grouping all the terms that provide resistance, R to failure, and all the terms, S, that produce the load and cause failure. We then obtain

$$G = [6.2\, S^{0.2} P^{0.18} \Delta D_{n50}\, (\xi_{mo})^{-0.5} N_z^{-0.1}] - [H_s] = 0. \qquad (4.5)$$

The variable within the first bracket that provides the greatest resistance to failure is the rock diameter, D_{n50}. All the other variables have probability distributions about their mean values, but their mean values form the structural design (slope, permeability, rock density, and allowable damage level). The number of waves is important but appears to the negative power in the resistance term so that the longer the storm, the more this term decreases, meaning a larger rock diameter is needed to resist the load. The surf-similarity parameter also appears to the negative power in the resistance term for plunging waves. The full meaning and separate formulation for surging waves can be found in Chapter 6.

The only load term, S, is the significant wave height and the second bracketed term. Clearly, as the significant wave height increases, larger armor stone is required for a stable design.

The above formulation is deterministic, in that, for a given structure design and wave conditions, only one rock diameter will be calculated for G = 0. For example, given a rubble-mound breakwater of the following design: $\cot\alpha = 2$, $\rho_s = 2{,}650\,\text{kg/m}^3$, $\rho_w = 1{,}025\,\text{kg/m}^3$, Δ is 1.585, P = 0.1, and S = 2 (dimensionless damage level), with storm and wave conditions $H_s = 3.0\,\text{m}$ at an annual probability of exceedance level of 10%, $T_n = 6.0\,\text{s}$, $N_z = 2{,}000$ waves, a rock diameter, $D_{n50} = 1.265\,\text{m}$ (4.15 ft) will be required. If the wave height is larger, the breakwater will fail the design condition, meaning some rocks will move down the slope. If the stone size is smaller, the breakwater will also fail the design conditions. There is no probability of failure possibilities in the design.

4.2.2.2 *Partial safety factors for probabilistic design*

In the early 1990s, the Permanent International Association of Navigation Congresses (PIANC) formed Working Group 12 to develop a system of partial safety factors that are related to the probability of failure and hence to determine the reliability of the design (PIANC, 1992). The results were published by Burcharth (1991) and Burcharth and Sørensen (1999).

Consider again the limit state equation, G, but now with a partial safety factor for the load, γ_H, and a partial safety factor for the resistance, γ_z

$$G = \frac{1}{\gamma_z}R - \gamma_H S = 0. \qquad (4.6)$$

Partial safety factors are equal to or greater than 1.0, so the design resistance will be less (R/γ_Z), requiring a larger rock diameter, and the design load (γ_H S) will be more, also requiring a larger rock diameter. The larger rock diameter will provide the margin for safety.

Full details of how the partial safety factors were determined are presented in Chapter 6. The main result is that for five levels of the probability of failure, P_f (0.01, 0.05, 0.10, 0.20, and 0.40) and two levels of confidence in knowledge of the wave height conditions (high and low), tables of the resistance and load partial safety factors have been developed and published for many of the most important functional and structural design formulas. The best way to illustrate this is by example.

4.2.2.3 *An example of probabilistic design (stability formula for plunging waves on rubble-mound structures)*

The limit state equation for the stability formula with plunging waves on rubble-mound structures now becomes, with partial safety factors γ_H for load and γ_Z for resistance,

$$G = \frac{1}{\gamma_Z}[6.2\, S^{0.2} P^{0.18} \Delta D_{n50} (\xi_{mo})^{-0.5} N^{-0.1}] - \gamma_H [H_s] = 0. \qquad (4.7)$$

Partial safety factor tables have been published by the Corps of Engineers in CEM (2006, Table VI-6-5) for the stability failure of rock armor under plunging waves for the formula given above (van der Meer, 1987). It is for designs when laboratory-scale model tests are not performed. Model tests could result in lowering the safety factors. Table 4.1 reproduces CEM (2006, Table VI-6-5).

Table 4.1 shows two major column headings for the coefficient of variation of the quality of the significant wave heights, σ'_{FH_S}. The high-quality wave data corresponding to measured wave *data* is $\sigma'_{FH_S} = 0.05$. Recall the coefficient of variation is the ratio of the standard deviation of the mean, σ, divided by the mean value, μ, so that σ' is this ratio. The meaning of the subscript F will be explained in Chapter 6. The second major column is for low-quality wave *information* with $\sigma'_{FH_S} = 0.2$ as would be determined from wind-wave formulas for a given wind speed, fetch distance, and wind duration.

Table 4.1. Partial safety factors for stability failure of rock armor, plunging waves, van der Meer (1987) formula, design without model tests (from CEM, 2006, Table VI-6-5).

	$\sigma'_{FH_S} = 0.05$		$\sigma'_{FH_S} = 0.2$	
P_f	γ_H	γ_Z	γ_H	γ_Z
0.01	1.6	1.04	1.9	1.00
0.05	1.4	1.02	1.5	1.06
0.10	1.3	1.00	1.3	1.10
0.20	1.2	1.00	1.2	1.06
0.40	1.0	1.08	1.0	1.10

Table 4.2. Damage level, S for two-layer, rock armor (van der Meer, 1988; CEM, 2006, Table VI-5-21).

Unit	Slope	Initial Damage, S	Intermediate Damage, S	Failure
Rock	1:1.5	2	3–5	8
Rock	1:2	2	4–6	8
Rock	1:3	2	6–9	12
Rock	1:4–1:6	3	8–12	17

The rows in Table 4.1 are the five levels of the probability of failure that were chosen by the PIANC (1992) Working Group 12. To understand their meaning, we must first define what is meant by "failure". The definitions of "failure" in the van der Meer formulations for rock-armor stability are given in Table 4.2 and related to the level of damage, S, as defined above, the structure slope, and for two layers of armor stone. In general, "failure" of a rock armor structure is when the under-layer or filter layer is exposed to direct wave attack. The armor layer has "failed" in its job to protect the smaller sized rocks or earth material beneath it from direct wave attack.

Consider again the same values for the variables in the deterministic design example which gave $D_{n50} = 1.265\,\text{m}$. This is essentially with both partial safety factors equal to 1.00. Assume that the target safety level in terms of probability of failure is at the 5% level so that $P_f = 0.05$. The design wave height, $H_s = 3\,\text{m}$, is based on estimates from wind data, and the quality is low. In Table 4.1, using the second column for $\sigma'_{FH_S} = 0.2$ gives $\gamma_H = 1.5$ and $\gamma_Z = 1.06$. Applying the van der Meer (1987) formulation for rock stability under plunging waves (Equation (4.7)) with $P_f = 0.05$

and $S = 2$ with partial safer factors for $\sigma'_{FH_S} = 0.2$ gives $D_{n50} = 2.01\,\text{m}$. Selecting a low probability of failure, $P_f = 0.05$, has resulted in a larger rock size than the deterministic design to be stable. If the purpose of the breakwater is for the safe unloading of ships in a harbor, then the target safety level for damage probability may be justified. The larger rock size is needed to ensure that the function of the breakwater (reduced waves for ship unloading) will not be impeded during the design wave conditions.

The rubble-mound breakwater in this example could be one of a system of nearshore breakwaters that together with beach nourishment mitigate shoreline erosion. The target safety level in terms of the probability of damage could be set at 40% so that $P_f = 0.40$. In this case, $\gamma_H = 1.0$ and $\gamma_Z = 1.10$, giving $D_{n50} = 1.39\,\text{m}$ and only slightly more than the deterministic design.

Note that the Damage Level given in Table 4.2 for the value of S when failure occurs could have been employed *together* with the Partial Safety Factor given in Table 4.1 when applying the partial safety factors in the van der Meer formulations. Using the value of $S = 8$ (failure condition) together with the partial safety factors in Table 4.1 will result in smaller rock diameters. A complete discussion of when to use the S values at failure will be found in Chapter 8.

Clearly, the partial safety factors for various levels of safety in terms of the probability of failure, P_f, give the design engineer the ability to match the function and importance of the project to the structural stability, size of the armor layer, and ultimately, the cost of the design.

The example above and values employed for the variables are similar to those found in Reeve *et al.* (2012, p. 301). Spreadsheets are provided and discussed in detail in Chapter 6 for the van der Meer (1988) formulations for both plunging and surging waves in deep and shallow water environments.

4.2.3 *Format for partial safety factors*

In general, there are multiple random variables, X_i that contribute to the resistance, R_i $(R_1, R_2, R_3 \ldots R_n)$ and to the load, S_i $(S_1, S_2, S_3 \ldots S_n)$ in the limit state equation (4.1). A partial safety factor, γ_i, can be applied to each resistance variable, $R_i/\gamma_i^{\text{res}}$, and each load variable, $\gamma_i^{\text{load}} S_i$. Or, the partial safety factors, γ_i, could be applied to combinations of resistance and combinations of load variables. For example, if applied to each resistance variable in the van der Meer (1988) formula for plunging waves (Equation (4.5)), a total of nine partial safety factors would be needed for the resistance variables in the formula. The water level would also need to be considered as discussed in Chapter 6. This large number would greatly complicate the analysis.

In addition, the partial safety factors, γ_i, are related to a characteristic value of each random variable, $X_{i,\mathrm{ch}}$. If a variable, X_i, has a normal (Gaussian) bell-shaped probability distribution as illustrated in Figure 4.4, the mean value, μ_i, and standard deviation of the mean, σ_i, determine the random variable, $X_i = \mu_i \pm \mathrm{n}\sigma_i$. The *characteristic* value $X_{i,\mathrm{ch}}$ choosen by PIANC is the mean value, i.e. $X_{i,\mathrm{ch}} = \mu_i$. The common assumption is that the resistance variables, R_i are normally distributed. For ease of notation, the caret symbol ($\widehat{\ }$) over the variable indicates the characteristic, mean value, $X_{i,\mathrm{ch}} = \mu_{i^\cdot} = \widehat{X}_i$. See CEM (2006, Part VI, Section 6) for a treatment if R_i are non-normal, random variables.

For design, the load variable, S_i, is commonly related to an extreme-value, non-normal probability distribution. In the van der Meer (1988) formulation, the single load variable is the characteristic, significant wave height, $H_{s,\mathrm{ch}}$ and is related to some level of exceedance of the non-normal probability distribution as discussed in Section 4.3.

The goal of the PIANC Working Group 12 was to have a simple system with few partial safety factors but not to compromise the accuracy of the design equations. The system selected was for two partial safety factors; one for the resistance γ_z, that combined all the resistance variables and one for the load γ_H, that was related to the characteristic wave height, $H_{s,\mathrm{ch}}$. The result for the van der Meer (1988) formulation (Equation (4.7)) is repeated here with the caret symbol ($\widehat{\ }$) over the variables to emphasize their characteristic mean values.

$$\mathrm{G} = \frac{1}{\gamma_Z}[6.2\,\widehat{S}^{0.2}\widehat{P}^{0.18}\widehat{\Delta}\widehat{D}_{n50}\widehat{\xi}_{om}^{-0.5}\widehat{N_z}^{-0.1}] - \gamma_H[H_{s,\mathrm{ch}}] = 0. \tag{4.8}$$

This equation could also be written as

$$\widehat{D}_{n50} = \gamma_Z\gamma_H\frac{H_{s,\mathrm{ch}}}{6.2\,\widehat{S}^{0.2}\widehat{P}^{0.18}\widehat{\Delta}(\widehat{\xi_{om}^{-0.5}})\widehat{N_z}^{-0.1}}. \tag{4.9}$$

The product of the partial safety coefficients $\gamma_Z\gamma_H$ in Equation (4.9) will produce the same rock size, $\widehat{D}_{n50}$, as Equation (4.8) only if all the other parameters are equal. As seen in the example given in Section 4.2.2.3, this would not be the case when using Table 4.2 for the damage parameter, S, under failure conditions for a given probability of failure, P_f.

What to use for the design value of the characteristic, significant wave height, $H_{s,\mathrm{ch}}$ is a fundamental question in all of coastal engineering design.

4.3 Statistical Analysis of Extreme Wave Heights

For a complete summary of all the many distribution functions and academic concerns in the statistical analysis of extreme waves, see Goda (2010, 3rd edn.). We herein focus solely on the Weibull distribution function and how to estimate the confidence interval wave heights at a given recurrence interval. The range of possible wave heights for various recurrence intervals and within the bounds of the confidence interval far exceeds the relatively minor differences between the various distribution functions. Consequently, we summarize the practical aspects using only the Weibull distribution in the hope of providing some guidance to answer the fundamental question: what to use for $H_{s,\mathrm{ch}}$?

4.3.1 *The Weibull extreme-value distribution*

The Weibull extreme-value distribution was introduced in Chapter 2 (Section 2.3.8.2). The cumulative, non-exceedance probability, P (Equation (2.51)) is repeated here as Equation (4.10).

$$\mathrm{P}(H < H') = 1 - exp\left(-\left\{\frac{H-\gamma}{\beta}\right\}^{\alpha}\right). \tag{4.10}$$

For the cumulative, exceedance probability, $Q = 1-P$, the Weibull, reduced variate, W, is

$$W = \left(\ln\frac{1}{\mathrm{Q}}\right)^{1/\alpha} = \frac{H-\gamma}{\beta}, \tag{4.11}$$

with the three Weibull coefficients defined as β the scale parameter as related to the linear scale for H, γ is the location parameter to fix the horizontal axis for H, and α is the shape parameter to define the functional shape of the Weibull distribution.

In these formulas, $H = H_s$, the significant wave height from measured wave data or computed wave information.

The probability density of the Weibull distribution, p is given as

$$p(H) = \frac{\alpha}{\beta}\left\{\frac{H-\gamma}{\beta}\right\}^{\alpha-1} \exp\left(-\left\{\frac{H-\gamma}{\beta}\right\}^{\alpha}\right), \tag{4.12}$$

and Figure 4.2 displays the shape of the distribution for four values of $\alpha = 0.75, 1.0, 1.4$, and 2.0, when the value for the scale parameter, $\beta = 1$ (from Goda, 2010).

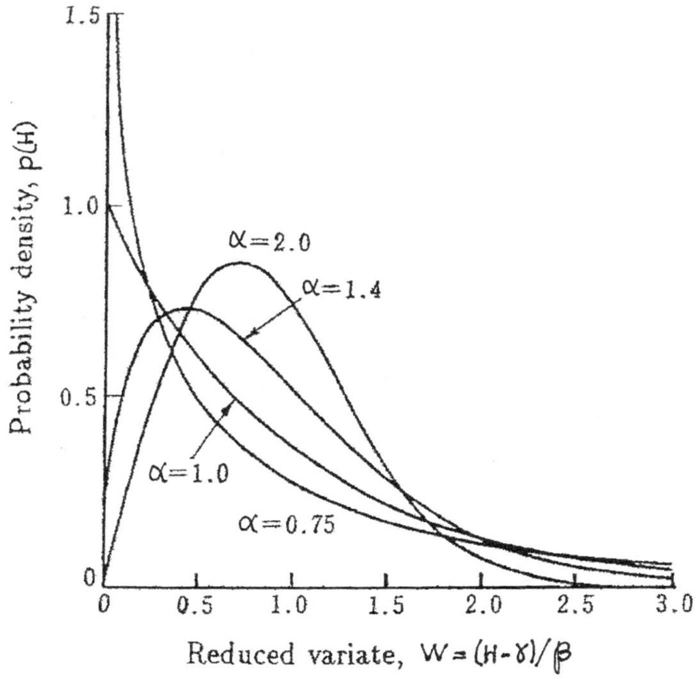

Figure 4.2. Weibull distribution for four values of α when the scale parameter, $\beta = 1$ (from Goda, 2010).

$\alpha = 1$ is the exponential, decay function and the distribution flatten as α decreases. A lognormal distribution is very similar to taking $\alpha = 2$ (Goda, 2010).

As demonstrated in Chapter 2, plotting the Weibull reduced variate, W, versus the wave height requires selecting a value for the shape parameter, α. Fitting a straight line to the plotted data permits determination of the two base parameters, β and γ, and the highest correlation coefficient (line-of-best-fit) by trial and error determines the shape parameter α.

We now formally define the following variables for use in the subsequent discussion. Let N_R equal the number of years of wave data or wave information (partial years are acceptable), N_S be the total number of storms with peak wave heights above the threshold wave height, H_s^*, N be the number of waves or wave height bins used in the statistical analysis, and N^* be the censored number of waves or wave height bins used in the statistical analysis.

The mean, annual rate of the number of storms, λ is

$$\lambda = N_S/N_R, \tag{4.13}$$

and a censoring parameter is introduced (Goda, 2010)

$$\nu = N/N^* \tag{4.14}$$

when some of the wave data are omitted in the statistical analysis as discussed in the following.

The relationship between the cumulative probability of exceedance, Q in any one year and the annual recurrence interval, T_r, then becomes

$$Q = \frac{1}{\lambda T_r}. \tag{4.15}$$

When the straight line of best fit is determined and rearranged, the equation for the significant wave height, $\widehat{H}_s^T$, becomes

$$\widehat{H}_s^T = \gamma + \beta\left(\ln\frac{1}{Q}\right)^{1/\alpha}, \text{or} \qquad (2.59)/(4.16)$$

$$\widehat{H}_s^T = \gamma + \beta(\ln(\lambda T)^{1/\alpha}. \qquad (2.60)/(4.17)$$

The caret symbol ($\widehat{\ }$) means the straight line gives only the central tendency or mean value estimate for $\widehat{H}_S$ at a given recurrence interval, T_r. Use of the peak-over-threshold (POT) method for the sampled years of storm records and the number of storms in that period produces only one "central tendency" estimate of the statistical relationship between the significant wave height, $\widehat{H}_S$, and the recurrence interval, T_r.

4.3.2 *Statistical variability of central tendency estimate*

In general, the statistical variation of the extreme wave height distribution *increases* as the return period increases (low probability) for storm events. Figure 4.3 displays the results of 100, numerically generated, storm wave records with 100 storm wave heights in each record (Goda *et al.*, 1993). The horizontal axis is the Weibull reduced variate ($\alpha = 1.4$), which can be related to the recurrence interval, T_r and the vertical axis is H_s (meters). The numerical model used the central tendency values of $H_s = 8$ m for $T_r = 1$ year and $H_s = 13$ m for $T_r = 100$ years for a mean annual storm rate $\lambda = 5$, which is equivalent to $N_R = 20$ years of wave data or information.

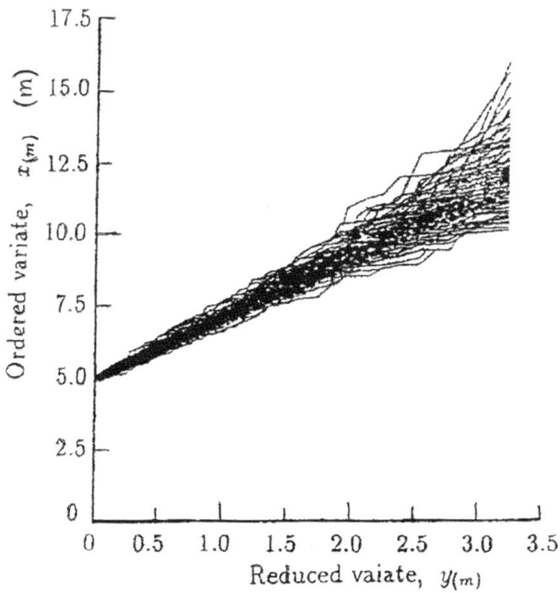

Figure 4.3. Plot of 100 numerically simulated, extreme wave height distributions (from Goda *et al.*, 1993).

Each line in Figure 4.3 represents one 20-year wave record with 100 storms as analyzed with the Weibull distribution theory. Clearly, the statistical variation increases as the Weibull reduced variate (i.e. the return period) T_r increases. The standard deviation of the mean or central tendency values must also then be a function of the reduced variate and recurrence interval.

4.3.3 *Standard deviation estimates — Goda method*

Calculation of the standard deviation σ_H of the mean of a sample, $\widehat{H}$, is the way to determine the statistical variability of the sample. The unbiased variance (standard deviation squared) of a sample is defined as

$$\sigma_H^2 = \sum_{i=1}^{N_s} (H_i - \widehat{H})^2/(N_s - 1). \tag{4.18}$$

But this is the overall population variance and not the total variance of the extreme wave height distributions, which increases as the recurrence interval increases as illustrated in Figure 4.3.

There is no known analytical theory for the estimate of the standard deviation of the Weibull distribution about the central tendency estimate over the range of recurrence intervals. Goda (1988) developed an empirical formula for the estimate of the standard deviation of the reduced variate, σ_W, as detailed in the following. Then, the product of this empirical formula for σ_W times the unbiased standard deviation of the sample, σ_H is defined as the standard deviation, $\sigma(\widehat{H}_s^T)$, that is,

$$\sigma(\widehat{H}_s^T) = \sigma_W \cdot \sigma_H. \tag{4.19}$$

To do this, Goda (1988, 2004) analyzed the information from Monte Carlo simulations of 20,000 samples in which normal (Gaussian) distributions were assumed at each reduced variate (i.e. recurrence interval). The technique is also found in Mathiesen *et al.* (1994), and this was the result of a Working Group of the IAHR (International Association for Hydro-Environment Engineering and Research) Section of Maritime Hydraulics of which Professor Goda was a member.

The empirical formula for σ_W involves many coefficients that vary with the shape parameter, α, as shown in Table 4.3 (modified from Goda, 2010).

The formula for the standard deviation of the reduced variate, σ_W, is

$$\sigma_W = \{[1.0 + a(W - c + (alpha)(\ln \nu))^2]^{1/2}\}/\sqrt{N}, \tag{4.20}$$

where the coefficient a is given by

$$a = a_1 \exp[a_2 N^{-1.3} + \kappa(-\ln \nu)^2]. \tag{4.21}$$

Recall that N is the number of data points used in the Weibull analysis and is the number of wave height bins when the storm wave heights are grouped into increments (bins) covering the full range of wave height values above the threshold value. We have used "α" for the shape parameter in this book, and so we have used word "alpha" for the last coefficient in Table 4.1 and in Equation (4.20) because Goda used this symbol for this coefficient in his work.

Table 4.3. Constants for the standard deviation of the formula for σ_W (from Goda, 2010).

Distribution	a_1	a_2	κ	c	(alpha)
Weibull ($\alpha = 0.75$)	1.65	11.4	−0.63	0.0	1.15
Weibull ($\alpha = 1.0$)	1.92	11.4	0.0	0.3	0.90
Weibull ($\alpha = 1.4$)	2.05	11.4	0.69	0.4	0.72
Weibull ($\alpha = 2.0$)	2.24	11.4	1.34	0.5	0.54

To facilitate a spreadsheet solution, second-order polynomials have been fit to the four values in Table 4.3 to permit the calculation of the three coefficients for any shape parameter, α. The value for a_2 is constant. The equation for c was split into two parts for better fit. The equations are as follows:

$$a_1 = -0.3061\alpha^2 + 1.2861\alpha + 0.8848, \tag{4.22}$$

$$\kappa = -0.7918\alpha^2 + 3.739\alpha - 2.9755, \tag{4.23}$$

$$\text{alpha} = 0.3297\alpha^2 - 1.3781\alpha + 1.9817, \tag{4.24}$$

$$c = 1.2\alpha - 0.9 \text{ when } (0.75 < \alpha < 1.0), \text{ and}$$

$$c = -0.0833\alpha^2 + 0.450\alpha - 0.0667 \text{ when } (\alpha \geq 1.0). \tag{4.25}$$

Illustration of the Goda method to make estimates of the standard deviation of the central tendency estimate, $\sigma(\widehat{H}_s^T)$, is best demonstrated by an example.

In Chapter 2, the 34.9 years of wave information available for Lake Huron (Table 2.5) from Kamphuis (2010) was analyzed by the Weibull method and produced the following results for $\alpha = 1.3$ when the threshold wave height H^* was 1.5 m giving 2,738 storm events.

$$\beta = 0.886,$$

$$\gamma = 1.243,$$

$$\lambda = 78.45,$$

which gives for the central tendency estimate

$$\widehat{H}_s^T = \gamma + \beta(\ln(\lambda T_r)^{1/\alpha} \tag{2.60}$$

or

$$\widehat{H}_s^T = 1.243 + 0.886(\ln(78.45\text{T})^{1/1.3}). \tag{4.26}$$

And the formula for the Weibull reduced variate, W becomes

$$W = 1.128\widehat{H}_s^T - 1.403 \text{ as shown in Figure 2.28.} \tag{4.27}$$

The wave information in Table 2.5 can also be used to determine the mean wave height of 2.103 m and the unbiased, standard deviation of the mean, $\sigma_H = 0.3167$ m.

Now, using Equations (4.22)–(4.25) for $\alpha = 1.3$ gives $a_1 = 2.0394$, $\kappa = 0.5477$, $c = 0.3775$, and alpha $= 0.74736$, with $a_2 = 11.4$, we obtain from Equation (4.21) with $N = 18$ and $\nu = 1.0$,

$$\begin{aligned} a &= a_1 \exp[a_2 N^{-1.3} + \kappa(-\ln\nu)^2] \\ &= 2.0394 \exp[11.4 \times 18\hat{}(-1.3) + 0.5477(-\ln 1.0)\hat{}2], \\ a &= 2.661. \end{aligned}$$

Now, take the first case for the 100-year recurrence interval event ($T_r = 100$ years) which gives for the central tendency estimate

$$\begin{aligned} \widehat{H}_s^T &= 1.243 + 0.886(\ln(78.45T_r)^{1/1.3} \\ &= 1.243 + 0.886(\ln(78.45 \times 100)\hat{}(1/1.3)), \\ \widehat{H}_s^T &= 5.9241\,\text{m}. \end{aligned}$$

And the associated Weibull reduced variate becomes

$$\begin{aligned} W &= 1.128\widehat{H}_s^T - 1.403, \\ W &= 1.128 \times 5.9241 - 1.403 = 6.6824 - 1.403, \\ W &= 5.2796. \end{aligned}$$

The standard deviation of the Weibull reduced variate is then

$$\begin{aligned} \sigma_W &= \{[1.0 + a(W - c + (alpha)(\ln\nu))^2]^{1/2}\}/\sqrt{N} \\ &= \{[1.0 + 2.661(5.2796 - 0.3775 + (0.74736)(\ln 1.0)\hat{}2]\hat{}(1.2)\}/18\hat{}(1/2), \\ \sigma_W &= 1.8995. \end{aligned}$$

Finally, the standard deviation of the central tendency of the significant wave height at recurrence interval $T_r = 100$ years is

$$\begin{aligned} \sigma(\widehat{H}_s^T) &= \sigma_W \cdot \sigma_H \\ &= 1.8995 \times 0.3167, \\ \sigma(\widehat{H}_s^T) &= 0.6016\,\text{m}. \end{aligned}$$

A spreadsheet greatly facilitates the calculation for a range of recurrence intervals and wave heights. As expected, $\sigma(\widehat{H}_s^T)$ increases as the recurrence interval increases. This means the spread in our estimate of the central tendency value increases and can be quantified by a formal definition of statistical measures such as a "confidence interval".

CEM 2006 (Part VI-6-3) gives an analytical expression for $\sigma(\widehat{H}_s^T)$ from the Gumbel extreme-value probability distribution as found from maximum likelihood estimates.

4.3.4 *Confidence intervals*

Figure 4.4 displays the normal, Gaussian (bell-shaped) probability distribution for the standard deviation of the mean with areas beneath the curve for one (68.27%), two (95.45%), and three (99.73%) standard deviations (plus and minus) about the mean value. The "confidence interval" is a statistical measure to indicate the reliability of the mean value estimate (central tendency value) and the probability that the true mean lies within the range around the sample mean.

The 90% confidence interval is commonly used in structural analysis and is in the range of ±1.645 times the standard deviation $\sigma(\widehat{H}_s^T)$ for the given recurrence interval, T_r. Figure 4.5 displays the 34.9 years of wave information available for Lake Huron (Table 2.5) from Kamphuis (2010) as analyzed by the Weibull POT (Peak Over Threshold) method for the

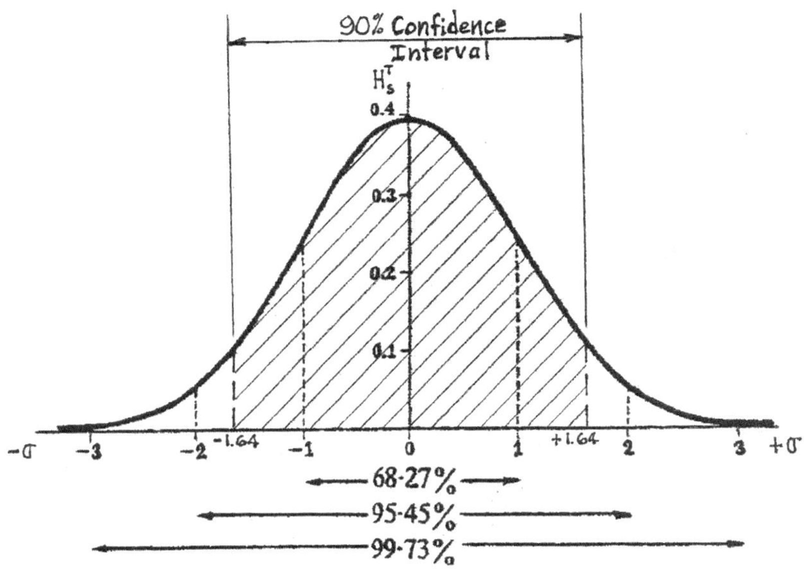

Figure 4.4. Normal (bell-shaped) probability distribution showing areas beneath the 1, 2, and 3 standard deviations and the 90% confidence interval for $\sigma = 1.64$.

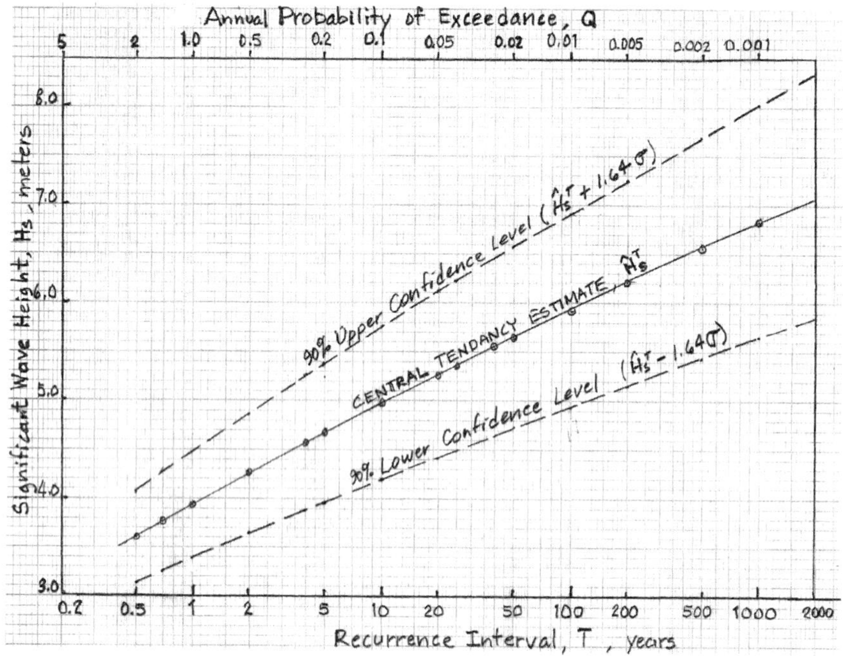

Figure 4.5. Central tendency estimate, and 90% confidence intervals for range of recurrence intervals for the Lake Huron example in Chapter 2 for significant wave heights.

central tendency estimate of $\widehat{H}_s^T$ and the 90% confidence interval levels, $\widehat{H}_s^T \pm 1.64\sigma(\widehat{H}_s^T)$. For the T_r = 100-year event, the 90% confidence interval is 4.9–6.9 m for the central tendency (mean) $\widehat{H}_s^T = 5.9\,\mathrm{m}$.

The spread of the wave heights at the 90% confidence interval is not insignificant for this example and when using the Goda (1988, 2004) method to estimate σ. In fact, the spread at the 90% confidence interval level is far greater than that found using other extreme-value distribution methods to estimate the central tendency distribution. This is the reason we have focused solely on the Weibull method in this book.

A word of caution, however. The Goda method was derived over 30 years ago. Storm wave height measurements in the oceans began in the early 1960s and wave numerical model simulations using wind statistics in the 1980s. We do not understand what wave climate changes took place in the past and are just now beginning to study what will transpire in the future, as will be discussed in Chapter 11. Extreme wave height statistics and the use of the Goda method to estimate the standard deviation of the central tendency estimate must now be placed in this context of deep uncertainty regarding future climate change.

It is important to estimate the spread of the extreme-value distribution about the Weibull method than use other extreme-value distribution methods.

4.3.5 *Design wave heights*

There is no universally accepted recurrence interval, T_r for selection of a "design wave height" in coastal engineering design. In other words, using the $T_r = 100$-year recurrence interval (probability of exceedance of 1% each year) for design is not an acceptable design practice in coastal engineering. As shown in Figure 4.5 for the 90% confidence level, H_s^{100} equals 6.9 m, which would be at the $T_r = 100$-year level for the central tendency estimate, $\widehat{H}_s^T = 5.9\,\text{m}$.

For the economic evaluation of the life-cycle costs that include both initial construction costs and design-life maintenance costs, a range of wave heights is employed to determine the optimum, balanced design. The probabilistic approach is far superior to the deterministic approach. Complete details are found in Chapter 8. All methods to estimate the central tendency of the extreme, significant wave height assume that the future wave climate will remain the same as in the past. This assumption is incorrect, and therefore the use of the empirical method for estimating the standard deviation of the extreme wave heights (Goda 1988, 2004) is questionable for the future.

4.4 Tables of Partial Safety Factors

The Coastal Engineering Manual (CEM, 2006) includes Table VI-6-3 which lists the partial safety factor tables for: (a) rubble-mound structures (17 tables) covering armor stability, toe berm stability, breakage, runup, and scour and (b) vertical-wall caisson structures (six tables) covering sand soil foundations, clay soil foundations, sliding failure, overturning failure, scour, and toe berm failure. Table 4.4 displays this table (CEM 2006, table VI-6-3) and many of the tables listed are discussed in further detail in Chapter 5 (Functional Design) and Chapter 6 (Structural Design) in this book.

Table 4.4. List of partial safety factor tables for probabilistic design of rubble-mound and vertical-wall caisson structures from CEM (2006, Table VI-6-3).

Structure	Failure	Armor	Table(s)
Rubber-mound structures	Armor stability	Rocks	VI-6-4–VI-6-6
		Cubes	VI-6-7
		Tetrapods	VI-6-8
		Dolosse	VI-6-9 & VI-6-10
		Hollowed cubes	VI-6-11 & VI-6-12
	Toe berm		VI-6-13
	Breakage	Dolosse	VI-6-14 & VI-6-15
		Tetrapods	VI-6-16
	Runup	Rock	VI-6-17
		Hollowed cubes	VI-6-18
		Dolosse	VI-6-19
	Scour		VI-6-20 & VI-6-21
Vertical-wall caisson structures	Foundation: Sand subsoil		VI-6-22
	Foundation: Clay subsoil		VI-6-23
	Sliding failure		VI-6-24
	Overturning failure		VI-6-25
	Scour		VI-6-26
	Toe berm		VI-6-27

4.5 Parameter Uncertainty

Uncertainty is a general term that includes randomness as quantified by statistics, measurement errors, and our fundamental lack of knowledge of the underlying physics of the problem. Uncertainty can be best described by a probability distribution. The true distribution is rarely known, so a normal (bell-shaped curve, Figure 4.4) distribution is often assumed. Therefore, a "measure" of uncertainty is the coefficient of variation, σ', given as

$$\sigma' = \frac{\rho}{\mu} = \frac{\text{standard deviation of mean}}{\text{mean value}}. \tag{4.28}$$

Uncertainty can be categorized by three types: the equation developed to describe the failure mode, the many environmental variables (waves, water levels, etc.) in the failure formula, and the parameters that describe the structure. The environmental parameters associated with the short-term and long-term sea-state statistics are the most important.

4.5.1 *Short-term sea-state uncertainty*

Environmental conditions (waves, water levels, etc.) over a short term (normally 20 min) are assumed constant to permit the short-term statistics to be estimated (e.g. the significant wave height, H_s) as discussed in Chapter 2. For time-series measurements, instrument response errors and the influence of measurement location over complex seabed bathymetry produce significant uncertainty in the H_s estimates. Variability also comes from different spectral analysis and smoothing techniques or when using the zero-crossing approach and if the Rayleigh Distribution is assumed to represent the water surface variation statistics in the short term. Wind-wave hindcast (deep and shallow-water numerical) models all rely on the quality of the wind field input (forcing) data.

Burcharth (1992) published estimates for short-term sea-state uncertainties for wave heights, periods, and directions and for water levels as shown in Table 4.5 (adopted from CEM, 2006, Table VI-6-1). The top section is for significant wave heights with the most accurate (least uncertainty) for offshore wave rider buoys ($\sigma' = 0.05$–0.1), and the least accurate (most uncertainty) for hindcast numerical models ($\sigma' = 0.1$–0.2). Visual ship board operations are also listed with a coefficient of variability of 0.2. Nearshore significant wave height estimates from numerical models have similar uncertainty ($\sigma' = 0.1$–0.2) as in deep water but can be much larger in cases with varying shallow-water bathymetry. (*The "Bias" column is simply an estimate of the difference in calculated standard deviation of the mean from a data set that is not infinite in length, Benjamin and Cornell, 1970.*)

Mean wave period uncertainty (second from top) is least in deep water for wave rider buoys ($\sigma' = 0.02$–0.05). Hindcast numerical models have a coefficient of variability for the spectral peak frequency in the 0.1–0.2 range.

Mean wave direction uncertainty is quantified in degrees in Table 4.5 (third from top). Again, wave-rider buoys are the most accurate (5 degrees), while hindcast numerical models based on the wind fields can be 15–30 degrees different from the mean direction. Linear arrays of wave gauges in nearshore waters can give much better accuracy in mean wave direction than a single instrument at one location.

See Kamphuis (2010, Section 3.7) for a more recent summary of wave measurement uncertainty.

As shown in the bottom rows for water levels, predicted astronomical tides can be very accurate ($\sigma' = 0.001$) and the upper boundary (0.07) is usually when tidal elevations must be estimated in areas some distance from tidal measurement locations. The last line shows the range of the coefficient of variation for storm surge numerical models as $\sigma' = 0.1$–0.25. Since Burcharth (1992) published this estimate, there has been significant

Table 4.5. Representative coefficients of variation, σ' for sea-state parameters (from Burcharth, 1992) and CEM (2006, Table, VI-6-1).

Parameter	Methods of Determination	Estimated Typical Values		Comments
		σ'	Bias	
Significant wave height, OFFSHORE	Accelerometer buoy, pressure cell, vertical radar	0.05–0.1	~0	
	Horizontal radar	0.15	~0	
	Hindcast numerical models	0.1–0.2	0–0.1	Very dependent on quality of weather maps
	Hindcast, SMB method	0.15–0.2	?	Valid only for storm conditions in restricted sea basins
	Visual observations from ships	0.2	0.05	
Significant wave height NEARSHORE determined from offshore significant wave height accounting for shallow-water effects	Numerical models	0.1–0.20	0.1	σ' can be much larger in some cases
	Manual calculations	0.15–0.35		

Wave Height

(*Continued*)

Table 4.5. (*Continued*)

Parameter	Methods of Determination	Estimated Typical Values		Comments
		σ'	Bias	
Mean wave period offshore on condition of fixed significant wave height	Accelerometer buoy	0.02–0.05	∼0	Wave Period
	Estimates from ampl. Spectra	0.15	∼0	Wave Period
	Hindcast, numerical models	0.1–0.2	∼0	Wave Period
Duration of sea state with significant wave height exceeding a specific level	Direct measurements	0.02	∼0	Wave Period
	Hindcast, numerical models	0.05–0.1	∼0	Wave Period
Spectral peak frequency offshore	Measurements	0.05–0.15	∼0	Wave Period
	Hindcast, numerical models	0.1–0.2	∼0	Wave Period
Spectral peakedness offshore	Measurements and hindcast numerical models	0.4	∼0	Wave Period
Mean direction of wave propagation offshore	Pitch-roll buoy	5 degrees		Wave Direction
	Measurements η, u, v or ρ, u, v'	10 degrees		Wave Direction
	Hindcast, numerical models	15–30 degrees		Wave Direction
Astronomical tides	Prediction from constants	0.001–0.07	∼0	Water Level
Storm surge	Numerical models	0.1–0.25	±0.1	Water Level

Note: 1: Two horizontal velocity components and water-level elevation or pressure.

development and improvement in storm surge numerical models. Our knowledge of the wind fields to drive the models is better and the computer resolution is vastly improved, permitting variable grid sizes with much smaller grids in the nearshore regions. An example is the finite element model ADCIRC (1992) (Luettich *et al.*, 1992). No known recent reference has been found for today's estimate of σ' for storm surge numerical models, but it is now probably in the 0.05–0.1 range.

4.5.2 *Long-term sea-state uncertainty*

Our knowledge of the uncertainty of long-term sea states is far less understood, and no published estimates are known to the author. The statistical analysis of extreme wave heights was discussed in detail in Chapter 2 (Section 2.3.8) and in Section 4.3. Long-term wave measurements (or wave information from wind data) are needed. Unfortunately, the number of wave gauges for measurement is small and, where available, the length of wave measurement records is relatively short so that uncertainty in the estimates for extreme wave heights at low probabilities of exceedance is large. There is no agreement on what is meant by "long term". However, by counting the more numerous numbers of "storm" events each year of wave data (or information) and considering the extreme wave height statistics on these "storm" events, the number of events can be increased to greatly improve the statistical estimates of the extreme wave heights. The key uncertainty is the assumption that the values in the data set are uncorrelated values of H_s, so that the methods for fitting the data are statistically valid.

In this chapter, we have only discussed the fitting of the data using the theoretical, Weibull distribution method and the POT analysis method to define "independent" storm events that form the data set. Uncertainty exists in the choice of the threshold value, the minimum length of a storm event, and whether the Weibull distribution is the proper choice. But as shown in Section 4.3.3 and Figure 4.5, estimates of the standard deviation of the mean and associated confidence intervals for the central tendency estimate is the area of largest uncertainty in the long term. We have shown only the Goda (1988) method to estimate $\sigma(\widehat{H}_s^T)$. Much more research is needed to understand variance about the central tendency (mean) estimates than methods such as maximum likelihood method, the method of moments, the least square method, visual graphical fit methods, etc. to estimate the central tendency curve.

The possible impacts of climate change and potential, accelerated sea-level rise on the long-term sea state are discussed in Chapter 11.

4.6 Probability of Failure — Level II and Level III Theories

4.6.1 *Level II theory (linear failure functions of normally distributed random variables)*

Consider again the linear limit state equation (4.6) but now for a characteristic load, S_{ch}, with partial safety factor, γ_s, and for a characteristic resistance, R_{ch}, with partial safety factor, γ_r,

$$\mathrm{G} = \frac{1}{\gamma_r} R_{\mathrm{ch}} - \gamma_s\, S_{\mathrm{ch}} = 0. \tag{4.29}$$

Assume both the characteristic resistance R_{ch} and the characteristic load S_{ch} have normal (Gaussian) probability distributions. This assumption of normal, bell-shaped distributions is central to the Level II theory applied for estimates of the partial safety factors in Level I practice as discussed in the following. Therefore,

$$R_{\mathrm{ch}} = \mu_r + z_r \sigma_r, \tag{4.30}$$

$$S_{\mathrm{ch}} = \mu_s + z_s \sigma_s, \tag{4.31}$$

with the mean values, μ, and the standard deviation of the mean, σ. The symbol z stands for the number of multiples (1, 2, 3, etc.) of standard deviations from the mean of interest to give the probability (area under the bell-shaped normal as shown in Figure 4.4. For example, at the 90% probability level or confidence interval, $z = \pm 1.64$. This means that 10% lies outside or $\pm 5\%$ lies outside each tail of the normal distribution. For conservative design, taking $z_r = -1.64$ (negative) means that the resistance, R_{ch} is exceeded 95% of the time in Equation (4.29). Also, for a conservative design, taking the positive value $z_s = +1.64$ gives the characteristic load, S_{ch} larger than the mean load in Equation (4.29). Solving Equation (4.29) for R_{ch} is design. One must calculate the needed structural strength (resistance) to withstand the loads imposed on the structure.

$$R_{\mathrm{ch}} = (\gamma_r \gamma_s) S_{\mathrm{ch}} = \Gamma S_{\mathrm{ch}} \tag{4.32}$$

and inserting the statistical distributions for R_{ch} and S_{ch} (Equations (4.30) and (4.31) into (4.32)) gives

$$\mu_r + z_r\sigma_r = (\gamma_r\gamma_s)(\mu_s + z_s\sigma_s), \tag{4.33}$$

or

$$\mu_r = -z_r\sigma_r + (\gamma_r\gamma_s)(\mu_s + z_s\sigma_s). \tag{4.34}$$

This is the standard design equation for a single, linear failure mode in coastal structures design.

4.6.1.1 *Two, independent, normal probability distributions and failure probability*

The resistance R_{ch} and load S_{ch} probability distributions are independent of each other so that the failure probability, P_f, can be calculated as the area beneath the probability density curves for the resistance, $p(r)$, and load, $p(s)$, respectively, in the region where they overlap as illustrated in Figure 4.6 (adapted from Kamphuis, 2010).

$$P_f = \int_0^{\infty} p(r)dr \times \int_0^{\infty} p(s)ds. \tag{4.35}$$

Clearly, the larger the mean resistance, μ_r, the further apart will be the two probability distribution curves producing a smaller overlapping area and lower probability of failure, P_f. Therefore, the magnitude of failure probability is determined by the product of the two partial safety factors $(\gamma_r\gamma_s)$; the magnitudes of the standard deviations of the means, $\sigma_r\sigma_s$; and the levels chosen for the z_r and z_s factors in the design equation (4.34). All these six variables and μ_s are needed to evaluate Equation (4.34) and then the integrals and their product in Equation (4.35), which is difficult. Fortunately, the two independent probability distributions (if normal, bell-shaped distributions) can be combined into one normal-shaped distribution by the work of Ang and Tang (1984).

4.6.1.2 *Single, normal probability distribution and failure probability*

Figure 4.7 displays a single, normal, probability distribution, $p(g)$ for the limit state variable, G in Equation (4.29) as adapted from Kamphuis (2010).

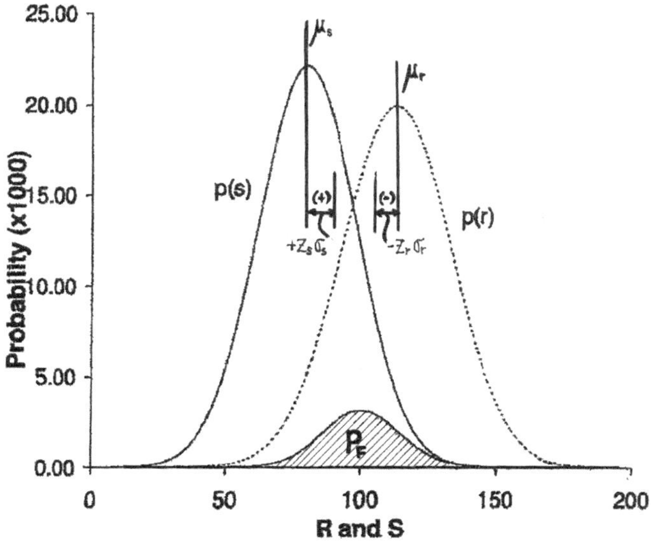

Figure 4.6. Schematic of two probability distributions for resistance, $p(r)$, and load, $p(s)$, and overlapping region where the integrated area is the failure probability, P_f (from Kamphuis, 2010).

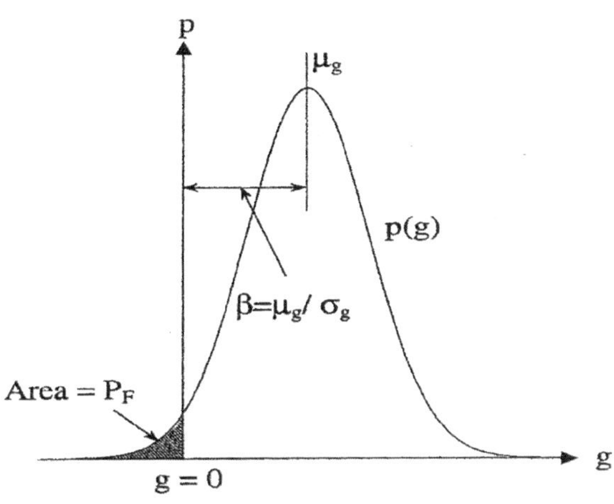

Figure 4.7. Single probability distribution for the combined resistance and load normal distribution curves and the failure probability as the area beneath the curve between $g = 0$ and $g = -\infty$ (from Kamphuis, 2010).

The probability of failure, P_f, is found from the equation

$$P_f = 1 - \Phi(\beta), \tag{4.36}$$

where Φ = the cumulative standard, normal distribution function (see Abramowitz and Stegun, 1965)

$$\beta = \frac{\mu_g}{\sigma_g}, \tag{4.37}$$

$$\mu_g = \mu_r - \mu_s, \tag{4.38}$$

and

$$\sigma_g = \sqrt{(\sigma_r^2 + \sigma_s^2)}. \tag{4.39}$$

Application of the single probability distribution method of Ang and Tang (1984) to determine the failure probability, P_f, is best illustrated by an example.

4.6.1.3 *Example of single distribution method — Rubble mound structure*

The linear, limit state equation (4.8) for the stability formula with plunging waves on rubble-mound structures with partial safety factors γ_H for load and γ_Z for resistance as employed in Section 4.2.3 is repeated here. Recall that the caret symbol ($\widehat{\ }$) over the variables signifies characteristic, mean values.

$$G = \frac{1}{\gamma_Z}[6.2\,\widehat{S}^{0.2}\widehat{P}^{0.18}\widehat{\Delta}\widehat{D}_{n50}(\widehat{\xi}_{om}^{-0.5})\widehat{N}^{-0.1}] - \gamma_H[H_{s,\mathrm{ch}}] = 0. \tag{4.40}$$

Comparing (4.29) and (4.40), we see that

$$R_{\mathrm{ch}} = \mu_r + z_r\sigma_r = [6.2\,\widehat{S}^{0.2}\widehat{P}^{0.18}\widehat{\Delta}\widehat{D}_{n50}(\widehat{\xi}_{om}^{-0.5})\widehat{N}^{-0.1}], \tag{4.41a}$$

$$S_{\mathrm{ch}} = \mu_s + z_s\sigma_s = [H_{s,\mathrm{ch}}]. \tag{4.41b}$$

For the example in Section 4.2.2.1 with a rubble-mound structure of the following design: $\cot\alpha$ is 2 (structure slope), ρ_s is 2,650 kg/m^3 (rock), ρ_w is 1,025 kg/m^3 (sea water), Δ is 1.585, P is 0.1 (impermeable core) and S is 2 (no damage), and with storm and wave conditions, H_s is the 3.0 m at the annual, probability of exceedance level of 10%, T_m is the 6.0 s, and N_z is the 2,000 waves in the coastal storm.

4.6.1.4 *Example deterministic design — probability of failure*

With the partial safety factors = 1.0 and $z_r = z_s = 0$, the deterministic, limit state equation becomes

$$G = \mu_r - \mu_s = [6.2\,\widehat{S}^{0.2}\widehat{P}^{0.18}\widehat{\Delta}\widehat{D}_{n50}(\widehat{\xi_{om}^{-0.5}})\widehat{N_z}^{-0.1}] - H_{s,\mathrm{ch}} = 0. \tag{4.42}$$

Consequently, in the single-distribution method of Ang and Tang (1984),

$$\mu_g = \mu_r - \mu_s = 0$$

so that $\beta = 0$ and $\Phi(\beta) = \Phi(0) = 0.500$ (see the MS Excel software function NORMSDIST makes the computation of Φ straightforward). Therefore $P_f = 1 - \Phi(0) = 0.5$, which means that for the deterministic design, the probability of failure is 50%. The deterministic design rock diameter, D_{n50}, was 1.265 m for the above variables in Equation (4.42). If the wave height is larger, G < 0, the breakwater will fail the design condition meaning some rocks will move down the slope. If the stone size is smaller, again G < 0 and the breakwater will also fail the design conditions. There are no failure probabilities associated with the deterministic design.

4.6.1.5 *Example probabilistic design — probability of failure*

Clearly the goal is to make G > 0 in the limit state equation (4.4) using partial safety factors greater than 1.0 along with appropriate values of the standard deviations, σ about the mean, characteristic values, μ. We will apply the single distribution method of Ang and Tang (1984) and Equations (4.36–4.39) to demonstrate how the five levels of probability of failure in Table 4.1 are determined.

In Table 4.1, the two major column headings are for relatively good quality of wave data ($\sigma'_{FH_S} = 0.05$) and for relatively poor quality of wave data ($\sigma'_{FH_S} = 0.2$) for the coefficients of variation to estimate the standard deviation of the load, σ_s in Equation (4.39). The coefficient of variation of the resistance, σ'_r, is related to the calibration coefficient, "6.2", in Equation (4.42) since all the other variables are mean values (caret symbol). The CEM for the plunging waves formulation, (2006, Table VI-5-23) recommends the coefficient of variation for the resistance, $\sigma'_r = 0.065$. No reference is provided.

For the variables of this example with $H_s = 3.0$ m, Table 4.6 presents the results for the six probabilities of failure (target) levels in Column (1) with the specified partial safety factors in Columns (2) and (3) for relatively

Table 4.6. Example of probability of failure calculation for plunging waves on rock slopes using van der Meer (1988) ($H_s = \mu_s = 3.0$ m and given data).

Target $\sigma' F H_s = 0.2$ (Table 4.1)

Probability of Failure P_F (1)	Partial Safety Factors Load $\gamma_S = \gamma_H$ (2)	Partial Safety Factors Resistance $\gamma_r = \gamma_z$ (3)	van der Meer Formula μr (4)	Load Uncertainty $\sigma' s$ (5)	Load Std Devia σs (6)	Resistance Uncertainity $\sigma' r$ (7)	Resistance Std Devia σr (8)	One Eqn Std Devia σg (9)	One Eqn ($\mu r - \mu s = \mu g$) (10)	Reliability Index β (11)	Cumulative Std Normal Dist Funct $\phi(\beta)$ (12)	Formula Probability of Failure $P_F = 1 - \phi(\beta)$ (13)
0.01	1.9	1.00	5.69992	0.2	0.600	0.140	0.86645	1.05391	2.69992	2.56180194	0.9948	0.01
0.05	1.5	1.06	4.77006	0.2	0.600	0.140	0.86645	1.05391	1.77006	1.67951018	0.9535	0.05
0.10	1.3	1.10	4.29006	0.2	0.600	0.140	0.86645	1.05391	1.29006	1.22406523	0.8895	0.11
0.20	1.2	1.06	3.81600	0.2	0.600	0.140	0.86645	1.05391	0.81600	0.77425642	0.7806	0.22
0.40	1.0	1.10	3.29996	0.2	0.600	0.140	0.86645	1.05391	0.29996	0.28461514	0.6120	0.39
0.50	1	1	3.00000	0.2	0.600	0.140	0.86645	1.05391	0	0	0.500	0.50

Notes:
Column 1: Target value for probability of failure;
Columns 2,3: Partial safety factors from Table 4.1 for weak knowledge of wave conditions;
Column 4: Calculation using van der Meer (1988) formula for plunging waves;
Column 5: Knowledge of wave conditions is poor (see Table 4.1);
Column 6: Standard deviation of the load ($H_s \times$ https://humanityplus.org/philosophy/ Column (5));
Column 7: The value given in Table VI-5-23 (CEM, 2006) is 0.065. Here we used $2.15 \times 0.065 = 0.140$;
Column 8: Standard deviation of the resistance ($6.2\times$ Column (7));
Column 9: Standard deviation of the single distribution model (Equation 4.39);
Column 10: From the definition of $\mu g = (\mu r - \mu s)$ (Equation 4.37);
Column 11: From the definition of $\beta = \mu g / \sigma g$;
Column 12: The cumulative standard, normal distribution function (in MS.EXCEL the statistics function is NORMSDIST());
Column 13: The probability of failure using the van der Meer formula.

Remarks: All probability of failure estimates near target values in Column 1 Deterministic.

low-quality wave data ($\sigma'_{FH_S} = 0.2$). An explanation for each column is provided, and Column (13) shows that all probability of failure estimates are close to the target levels in Column (1). The greatest uncertainty is in the value for the coefficient of uncertainty for the resistance (σ'_r) and/or to the calibration coefficient "6.2" in the plunging formulation of van der Meer (1988). To obtain the probability of failure estimates shown in Column (13), we needed to roughly double the uncertainty of the resistance from the recommended value of 0.065 to a value of 0.140 in Column (7). Similar results were found for the relatively high-quality wave data ($\sigma'_{FH_S} = 0.05$).

The key to understanding these results in Table 4.6 is graphically illustrated in Figure 4.6. As the distance between the mean values increases, the overlapping area of the two probability distributions decreases so that the probability of failure also decreases. Note the change in μ_g in Column (10). Increasing the uncertainty of the resistance coefficient from 0.065 to 0.140 increased the standard deviation of the resistance, σ_r, and hence the *spread* of the probability distribution increased to cover more overlap and increase the probability of failure. The results using 0.065 were too low. The mean value and standard deviation of the mean value for the load are relatively certain compared with the uncertainty surrounding the resistance, which must be found from the formula that contains many variables and a calibration coefficient. A full discussion of the uncertainty of the calibration coefficient "6.2" is presented in Chapter 6. In addition, the partial safety factors could also be modified in Table 4.6 to achieve reasonable probabilities of failure for this example. A complete review of how the partial safety factors in Table 4.1 were determined can be found in Chapter 6.

4.6.2 *Other Level II theories*

The above discussion for Level II theories only presented the linear failure function (Equation (4.1) or (4.6)) and the normally distributed probability distributions for two characteristic load and resistance variables. Nonlinear limit state equations for the failure function and non-normally distributed probability distribution models are also possible. See CEM (2006, Part IV) for a full discussion and also Reeve *et al.* (2012, pp. 298–316), including worked examples. An excellent summary of Level II and Level III methods is also found in Voortman *et al.* (2001)

4.6.3 *Level III theory*

The highest and most comprehensive theory for the probability of failure is Level III. In general, there are multiple random variables, X_i that contribute to the resistance, R_i $(R_1, R_2, R_3 \ldots R_n)$ and to the load S_i $(S_1, S_2, S_3 \ldots S_n)$ in a limit state equation that is nonlinear. Numerical simulation (Monte Carlo) methods have been employed (Everts, 2016) for prediction of damage to a rock-armored structure. Four failure mechanisms were considered. The probability of failure levels obtained is compared with the partial safety factor method (Level I) and design rules given in the Rock Manual (2007), which is discussed in Chapter 6. In general, the Level III design using fully probabilistic calculations with the Monte Carlo simulations produced higher levels of failure probability than that given in the Rock Manual (2007), which gave more conservative results. See Everts (2016) for complete details.

4.7 Probability Analysis of Single and System Modes of Failure

4.7.1 *Single mode of failure*

The earlier example (Section 4.2.2.3) of the probability of failure analysis focused on only one mode of failure, namely, the stability of the rock-armor layer. As illustrated in Figure 4.8, this erosion and possible breakage of the armor layer (on both sides) is only one of many potential failure modes of a rubble structure together with a concrete cap to reduce wave runup and overtopping on the leeward side.

The structural stability of the toe berm supporting the armor layer, geotechnical concerns (scour, slip failure, settlement, etc.) and concrete cap

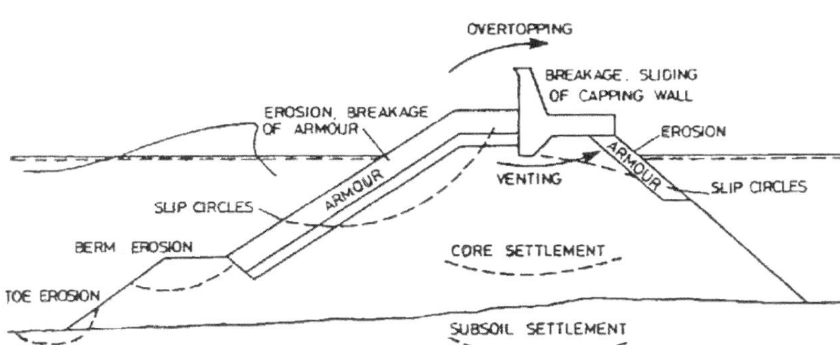

Figure 4.8. Schematic of failure modes for rubble-mound breakwater with concrete cap and road way (adapted from Burcharth and Sørensen, 1999).

failure (sliding, tilting, breakage, etc.), must be considered. The structure may also "fail" to perform its design function as a storm damage mitigation structure or "breakwater" to reduce wave energy in the lee of the structure.

4.7.2 *Systems of failure modes*

Coastal structures are complex systems of components that interact so that failure of one may cause failure of another and lead to failure of the entire structure. Figure 4.9 shows an example of a "fault tree" for a rubble-mound breakwater that relates excessive wave transmission (top of tree) over the breakwater to events leading to the failure of the system (all the OR/AND branches below). The more complete "cause–effect consequences" diagram for the rubble-mound breakwater is shown in Figure 4.10, and the simplified fault tree in Figure 4.9 is only one part of this. A full discussion of the complete failure diagram, risk, and resilience is found in Chapters 8 and 10 regarding the economic, environmental, and social consequences of system failure. There are two fundamental types of systems, series and parallel, as schematically illustrated in Figure 4.11.

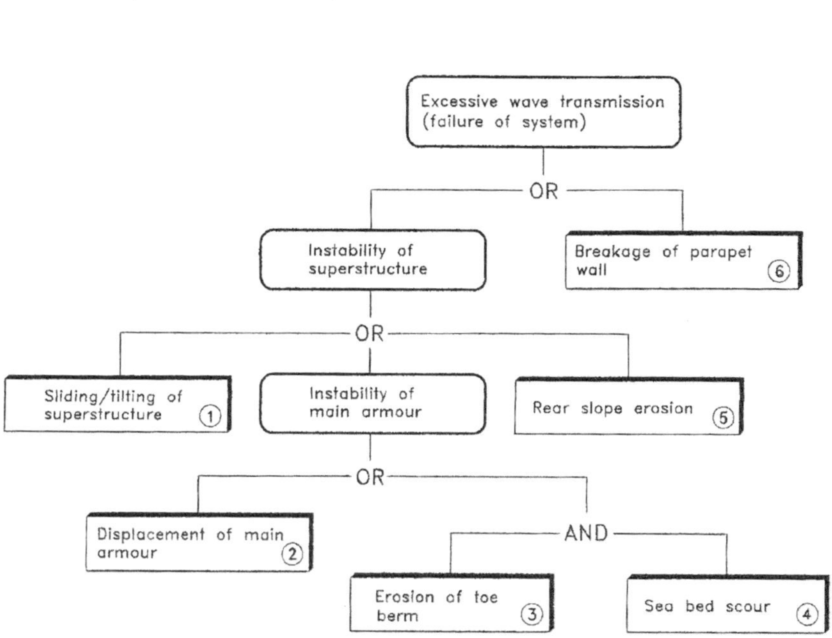

Figure 4.9. Simplified fault tree for rubble-mound breakwater (from CEM, 2006 Part VI, Section 6-4, Figure VI-6-7).

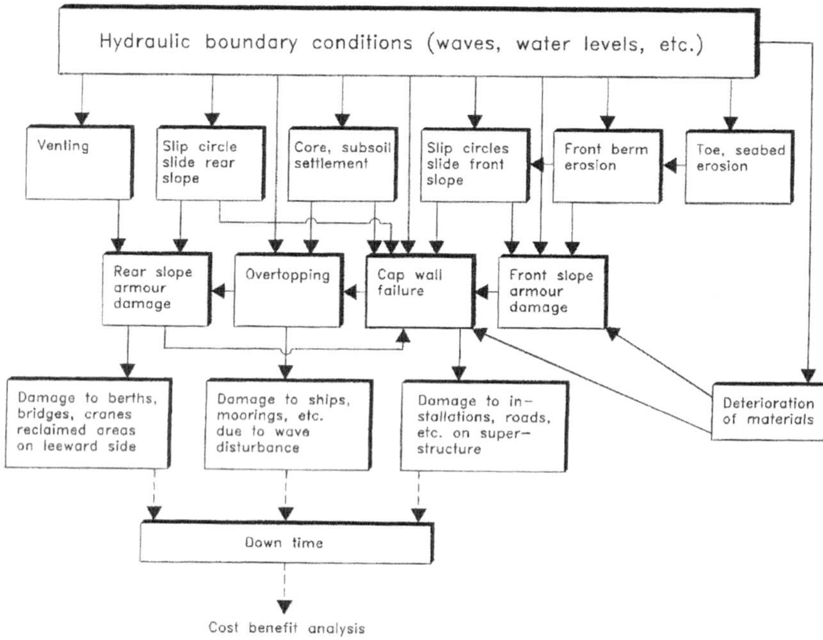

Figure 4.10. An example "cause-effect-consequences" diagram with causes of partial failures and interactions between failure modes (from CEM, 2006, Figure VI-6-8).

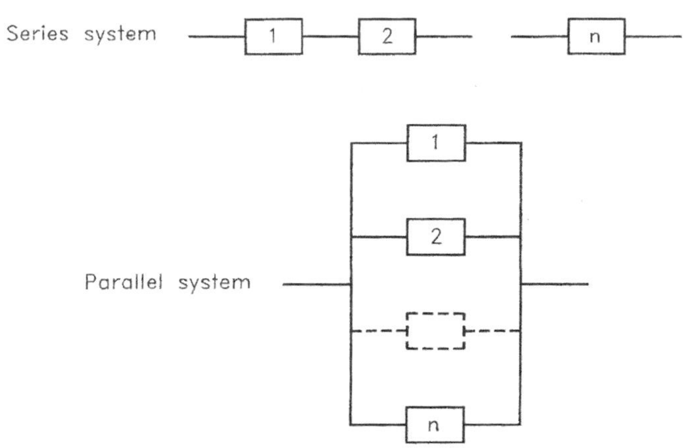

Figure 4.11. Schematic of a series system (top) and parallel system (bottom) (from CEM, 2006, Figure VI-6-9).

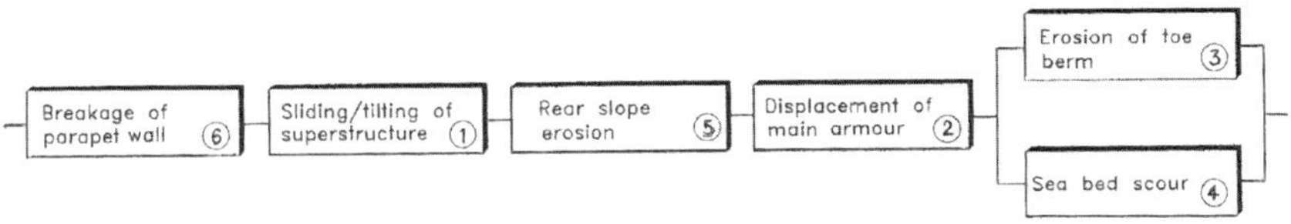

Figure 4.12. Rearrangement of the fault tree in Figure 4.9 into series and parallel components (from CEM, 2006, Figure VI-6-10).

4.7.2.1 *Series systems*

In series systems, *if any one* of the many elements linked together in series *fails*, the entire system fails. For example, the fault tree in Figure 4.9 has been rearranged into series and parallel components as shown in Figure 4.12. If any one of the in-line (horizontal) elements fails, e.g. parapet wall breakage, superstructure sliding or tilting, rear slope erosion, or main armor instability (displacement), the system will fail in its function to prevent excessive wave transmission over the breakwater to protect the leeside harbor. Series components are dominant in breakwater failure analysis. The upper bound of the failure probability of a series system of n components is simply one minus the products of $(1 - P_{fn})$ where the probability of failure, P_{fn} of $1 \ldots n$ number of components is known. The lower bound is the maximum for P_{fn} The upper bound is for no correlation between failure modes, and the lower bound is for full correlation.

4.7.2.2 *Parallel systems*

For parallel systems, *all* the elements in line (vertical) *must fail* to lead to failure of the entire system. In Figure 4.12, seabed scour could cause toe-berm erosion, and consequently armor-layer instability, and all taken together thus can cause structure failure. The upper bound of the failure probability of a parallel system is simply the minimum value estimated for P_{fn}. The lower bound is the product of all the component probabilities of failure. The upper bound is for full correlation between failure modes, the lower bound is for no correlation.

The material presented in this section has been extracted from CEM (2006, Part VI-6-4). Two example problems are included in CEM (2006) using assigned annual failure probabilities for each failure component. The first example uses the fault tree in Figure 4.9 and its rearrangement into series and parallel components shown in Figure 4.12. However, no discussion of the assigned annual failure probabilities for each component is presented, so the numerical example results demonstrate the theory with numbers but little else. Only the annual failure probability for the main armor-layer stability was discussed in detail earlier in Section 4.2.2.3 (and methods to estimate the annual failure probabilities of the erosion of the toe berm and the rear-slope erosion as found in Chapter 6). Methods to estimate the parapet wall breakage and concrete seawall sliding and tilting probabilities are unknown to the writer.

The second example in CEM (2006, Part VI-6-4) includes other possible ways to construct fault trees and their decomposition into series and parallel systems than illustrated in Figures 4.9 and 4.12. Using the same assigned annual probabilities of failure results in significantly different system failure probabilities. This means that starting with the correct fault tree and system decomposition is important in system analysis of failure probabilities. The reader is encouraged to review these examples in CEM (2006) for further understanding of systems of failure modes as introduced herein. But our full knowledge and abilities to compute the annual failure probabilities of many modes of rock-structure failure (Figure 4.9) are completely unknown as of the writing of this chapter.

The probability of failure, P_f, of single components (e.g. rock armor layers) or series or parallel systems, P_{fs}, only considers the probability of failure on an annual (yearly) basis. We must also consider the lifetime of the project and probability of failure over the design life of the project.

4.8 *N*-year Probability and Project Lifetime Failure Probability

4.8.1 *N-year probability, P_N*

All the probabilities discussed in this chapter have been for *annual* chance of exceedance in any one year. For example, the significant wave height, $H_s = 3.0\,\text{m}$, used in the rock, armor-layer stability calculations in Sections 4.2.2.1, 4.2.2.2 and 4.6.1.3 was for an *annual* probability of exceedance, $P_> = 10\%$ (*note: the symbol Q was previously used for the cumulative probability of exceedance*). And, in Table 4.1 for the five target levels of probability of failure in Level I analysis, each level of P_f is for an *annual* chance of failure in any one year. If each years' annual probability of exceedance for H_s and annual probability of failure (for armor layer) are independent, what are the probabilities for any extended number of years, N?

We herein define the N-year probability, P_N, as the probability over a period of N years of the annual, probability of exceedance, $P_>$, for the failure level, P_f, of interest. The annual probability of exceedance, $P_>$ and its related recurrence interval, $T_r = 1/P_>$ are variations in the formula that is widely used in the statistical description of floods, waves, water levels, etc. We use the symbol T_r here with subscript for the recurrence interval to distinguish from wave period, T.

The N-year probability, P_N, is

$$P_N = 1 - (1 - P_>)^N = 1 - \left(1 - \frac{1}{T_r}\right)^N. \tag{4.43}$$

For example, for an $N = 30$-year period, the $P_> = 10\%$ annual chance wave height $H_s = 3\,\text{m}$ will have an N-year probability of 96% [(1 − (1 − 0.10)^30 = 0.958]. The so-called 100-year recurrence interval event (1% chance in any one year) will have a 26% chance of occurrence in 30 years [1 − (1 − 0.01)^30 = 0.260]. Equation (4.43) is a product of the binomial distribution and each annual probability must be independent each year. The number of encounter years, N, could be related to the design life of the project. The number of years N has no relationship to the recurrence interval T_r. As defined in Section 4.3.1, the superscript "T" in the significant wave height $\widehat{H}_s^T$ stands for the recurrence interval and will be used throughout this book when it is clearly not the symbol for wave period.

Kamphuis (2010) defines P_N as an encounter probability and gave it the symbol P_E.

4.8.2 *Lifetime project probability of failure, P_L*

If only one failure mode was dominant to cause failure of a structure (for example, the rock-armor layer of a rubble-mound structure) and only one load variable was dominant in the analysis of the limit state equation (for example, the significant wave height), then a simple project lifetime probability of failure, P_L could be calculated as

$$P_L = P_N \times P_f, \tag{4.44}$$

where P_N is the N-year probability for the significant wave height, $\widehat{H}_s^T$, for the project design life, N years, T_r is the selected recurrence interval for the significant wave height (not necessarily N years), and P_f is the selected probability of failure, using the Level I, partial safety factors (e.g. in Table 4.1).

For the above-mentioned example, taking a design life of $N = 30$ years, and using $H_s = 3.0\,\text{m}$ as the annual probability of exceedance level of 10% gave $P_N = 0.96$ for the N-year probability. Over a 30-year period, there is a 96% chance of experiencing the $H_s = 3.0$-m significant wave height. Using partial safety factors at the $P_f = 0.05$ level and low-quality wave data, and armor-layer rock diameter $D_{n50} = 2.01$, was needed for stability using the plunging-wave formulation of van der Meer (1988). Therefore, the project lifetime probability of failure, $P_L = P_N \times P_f = 0.96 \times 0.05 = 0.048$ or about 5% during the lifetime of the project for this one failure mode and one loading variable. Note that if the $H_s = 3.0$-m wave height was at the exceedance probability level of 1% annual chance ($T_r = 100$ years), then $P_N = 0.26$ and the project lifetime probability of failure $P_L = P_N \times P_f = 0.26 \times 0.05 = 0.013$ and was over 3.5 times less. Clearly, many combinations of P_N and P_f are possible to achieve a goal for P_L. We shall return to this in Chapters 6 and 8 when discussing and seeking a "balanced design" for the total lifecycle costs that include both initial construction costs and life-cycle maintenance costs over the project design life.

But as discussed in Section 4.7, coastal structures have many possible failure modes and two ways (series and parallel) to decompose their failure trees for determination of project failure of the system, $P_{fsystem}$. And, this analysis would only be for the *annual* probability of failure of the system. To apply Equation (4.43) for the N-year probability requires that each years' events are independent. For some failure modes of the "resistance," e.g. the concrete strength of the seawall, this requirement cannot be met. Therefore, use of the N-year probability equation together with the $P_{fsystem}$ will produce totally incorrect numbers for the lifetime project probability of failure, P_L, and is thus not recommended (see CEM 2006, Part VI-6-4 and the two example problems of system probability of failure analysis for further discussion).

4.9 Summary

Probabilistic design of coastal structures is now the accepted standard method in the practice of coastal engineering. The key result is that a specified level for the probability of failure can be established at the outset depending on the goals and importance of the structure. As an example, a rubble-mound structure with the mode of failure being the armor layer stability on the seaward side of the structure is considered. The van der Meer (1988) formula for plunging waves in deep water is employed in the Limit State equation using Level I, and partial safety factors derived from Level II theory for probabilistic design. Level I is the most practical design method currently employed by coastal engineers. Tables of partial safety factors are presented for later study in Chapter 5 (Functional Design) and Chapter 6 (Structural Design).

This chapter also reviewed the Weibull extreme-value distribution for estimating the probabilities of rare wave heights using wave data (or wave information) and the POT analysis method. A key aspect is ways to determine the standard deviation of the data as a function of the recurrence interval and hence the confidence interval of the method.

Finally, all coastal structures have a number of different failure modes that can be organized as a fault tree and decomposed into a system of failure modes that form series and/or parallel failure systems. It would be possible to estimate the annual probability of failure of the system if the physical interactions between various failure modes were known and could all be described by a formula. However, correlations between the involved parameters are not well known, and these correlations not fully developed for practical application.

The probabilistic method introduced here is required to quantify economics, risk, and resilience in coastal structures design. Chapter 8 will present the full details.

See Section 11.3.4 for the impacts of sea-level rise on probabilistic design in this chapter.

References

Abramowitz, M., and I. A. Stegun (1972). *Handbook of Mathematical Functions*, Dover Publications, New York.

ADCIRC (1992) "ADvanced CIRCulaton Model," Corps of Engineers (https://www.erdc.usace.army.mil/) (See Luettich, Westerink and Scheffner, 1992 for initial published reference).

Ang, A. H.-S., and W. H. Tang (1984) *Probability Concepts in Engineering Planning and Design*, Vol. II, John Wiley & Sons, New York.

Benjamin, J. R., and C. A. Cornell (1970) *Probability, Statistics, and Decision for Civil Engineers*, McGraw-Hill Book Company, New York.

Burcharth, H. F. (1991) "Introduction of Partial Coefficient in the Design of Rubble Mound Breakwaters," *Proceedings of the Conference on Coastal Structures and Breakwaters*, Institution of Civil Engineers, London, Thomas Telford Publishing, London, UK, pp. 543–565.

Burcharth, H. F. (1992) "Uncertainty Related to Environmental Data and Estimated Extreme Events," Final Report of PIANC Working Group 12, Group B, June.

Burcharth, H. F., and J. D. Sørensen (1999) "The PIANC Safety Factor System for Breakwaters," *Proceedings of Coastal Structures' 99,* Vol. 2, Santander, Spain, pp. 1125–1144.

CEM (2006) "Reliability Based Design of Coastal Structures," Chapter VI-6, in *Coastal Engineering Manual*, US Army Corps of Engineers, Washington, DC.

Everts, P. S. (2016) "Probabilistic Design of a Rubble Mound Breakwater," Master Thesis, Department of Civil and Geotechnical Engineering, Delft University of Technology, Delft, The Netherlands.

Goda, Y. (1988) "On the Methodology of Selecting Design Wave Height," *Proceedings of the 21st International Conference Coastal Engineering, ASCE*, Delft, pp. 899–913.

Goda, Y. (2004) "Spread Parameter of Extreme Wave Height Distribution for Performance-based Design of Maritime Structures," *Journal Waterway, Port, Coastal, and Ocean Engineering*, Vol. 130, No. (1), pp. 29–38.

Goda, Y. (2010) *Random Seas and Design of Maritime Structures*, 3rd edn., World Scientific, Singapore.

Goda, Y., P. Hawkes, E. Mansard, M. H. Martin, M. Mathiesen, E. Peltier, E. Thompson, and G. van Vledder (1993) "Interecomparison of Extreme Wave Analysis Method Using Numerically Simulated Wave Data," *Proceedings, International Symposium Ocean Wave Measurement and Analysis, ASCE*, New Orleans, USA, pp. 963–977.

Kamphuis, J. W. (2010) *Introduction to Coastal Engineering and Management*, World Scientific, Singapore.

Luettich, R. A., Jr., J. J. Westerink, and N. W. Scheffner (1992), "ADCIRC: An Advanced Three-Dimensional Circulation Model for Shelves, Coasts, and Estuaries'. Report 1: Theory and Methodology of ADCIRC-2DDI and ADCIRC-3DL," Dredging Research Program, Technical Report DRP-92-6, U.S. Army Engineers Waterways Experiment Station, Vicksburg, MS, 137 p.

Mathiesen, M., Y. Goda, P. Hawkes, E. Mansard, M. J. Martin, E. Peltier, E. Thompson, G. van Vledder (1994) "Recommended Practice for Extreme Wave Analysis," *Journal of Hydraulic Research*, Vol. 32, No. 6, pp. 803–814.

PIANC (1992) "Analysis of Rubble Mound Breakwaters (1992–1994)," Marcom Working Group (WG 12), Permanent International Association of Navigation Congress, Paris (https://www.pianc.org/).

Reeve, D., A. Chadwick, and C. Fleming (2012) *Coastal Engineering: Processes, Theory, and Design Practice*, Spon Press, London and New York.

Rock Manual (2007) "The Rock Manual: The Use of Rock in Hydraulic Engineering," 2nd edn., CIRIA C683, London, UK.

van der Meer, J. W. (1987) "Rock Slopes and Gravel Beaches Under Wave Attack," Ph.D. Dissertation., Delft University of Technology, The Netherlands.

van der Meer, J. W. (1988) "Rock Slopes and Gravel Beaches Under Wave Attack," Publication No. 396, Delft Hydraulics, Delft, the Netherlands.

Voortman, H. G, Gelder, P. H. A. J. M. van, J. K. Vrijling (2001) "Probabilistic Description of Hydraulic Loads on Sea Defences," *Proceedings of the European Safety and Reliability Conference*, Vol. 1, pp. 545–552.

Chapter 5

Functional Design

5.1 Introduction

5.1.1 *Functional design of coastal structures*

This book is about the design of coastal structures whose purposes or functions are to lessen storm damage from flooding and high waves; reduce wave heights for ship/boat navigation and harbor mooring; mitigate coastal erosion; and enhance/create coastal ecosystems. All of these functions require the determination of the elevation of the crest of the structure. In general, the higher the crest elevation, the less flooding and wave energy is transmitted in the lee of the structure to reduce storm damage, permit safer ship/boat mooring, and reduce erosion on beaches and within estuarine marshes. Functional design of coastal structures also determines other geometrical features including structure slopes, locations in the nearshore region or on the shoreline, structure lengths, and structure spacings. In addition, construction and maintenance constraints may influence the geometrical and hence functional design.

Many of the worlds' harbor breakwaters are described and discussed in the summaries of the history and heritage of coastal engineering for the 15 countries listed in Kraus (1996). But possibly no breakwater has more worldwide importance than that constructed for the "Mulberry Harbor" at Normandy, France.

5.1.2 *Harbor breakwaters at Normandy, France (WWII)*

On June 6, 1944, Allied forces began the D-Day invasion on the beaches in Normandy, France. To ease and speed up the unloading process so that the Allied troops were supplied promptly as they advanced across France after breaking out from Normandy, two artificial harbors were constructed. Concrete caissons manufactured in England were towed across the English Channel to form breakwaters at Arromanches (Gold) and on Omaha beach 15 km (9 mi) west as illustrated in Figure 5.1.

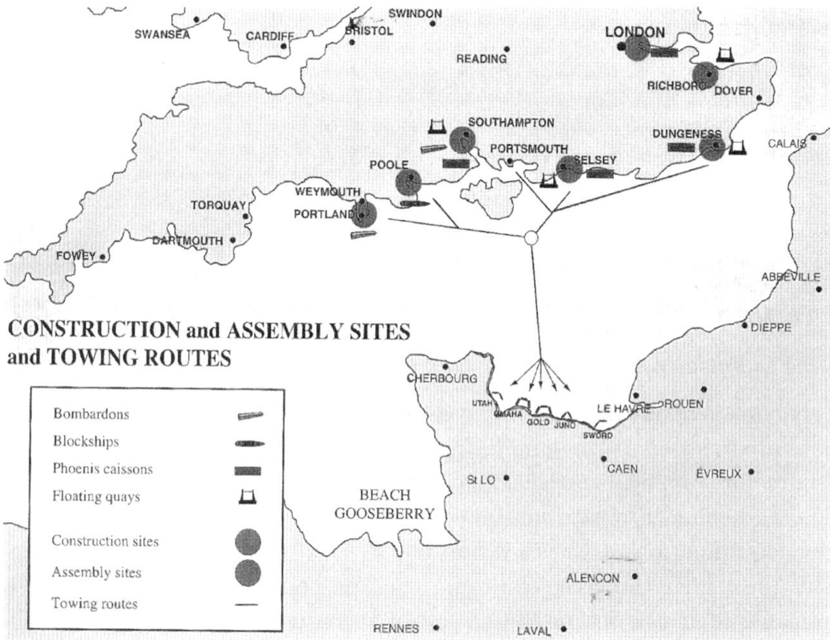

Figure 5.1. Construction and assembly sites, towing routes and Normandy, France beaches. The two "Mulberry" harbors were constructed at Omaha beach (American) and Gold beach (British) at Arromanches, France (from Arromanches, 1997 courtesy OREP Editions).

Figure 5.2 displays the harbor configuration at Arromanches showing the outer line of both (1) scuttled vessels and (2) concrete "Phoenix" caissons forming a harbor with reduced wave energy to unload ships at (3) movable quays on jack-up piles with (5) two floating roadways to the beach. The tidal range of 5 m drove many decisions regarding the harbor design. An outer ring of floating breakwaters (Barbardons, not shown) was initially deployed and the movable quay could be elevated above the tide to permit 24-hour unloading operations.

Over 115 reinforced concrete caissons (the largest was 60 m long, 17 m wide, and 18 m high as illustrated in Figure 5.3) were placed in 9 m mean water depth forming a 9.5 km long breakwater (with scuttled vessels) and enclosing a harbor area of 500 hectares. By June 18th, both harbors were in full use, as shown in an aerial photograph (Figure 5.4). They were meant to stay in use until the capture of Chebourg Harbor in the north of France. However, a violent storm began on June 19. By June 22, the harbor

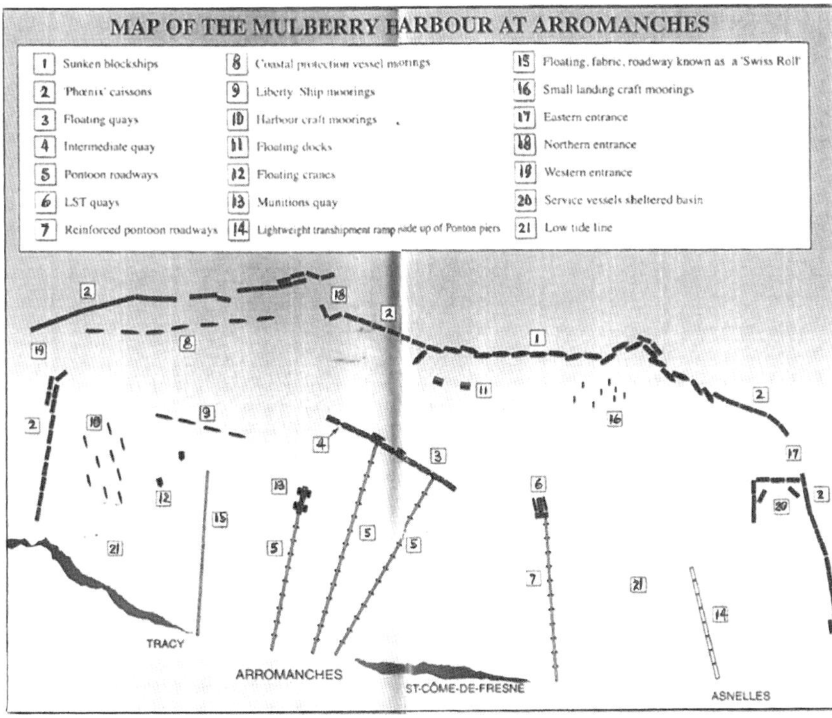

Figure 5.2. Plan-view schematic diagram of the layout of the Mulberry (British) harbor at Arromanches. The North direction is up. The inset table gives details of each structural element to build the harbor and other details (from Arromanches, 1997 courtesy OREP Editions, France).

serving the Americans at Omaha beach was heavily damaged. Parts of it were salvaged to repair the British harbor at Arromanches (Gold beach) which worked for 10 months. In that time this harbor landed 2.5 million men, 500,000 vehicles, and 4 million tons of goods (Arromanches, 1997).

Both temporary harbors used the same combination of structures and design configuration as shown in Figure 5.2 and both experienced the same storm energy. However, the American harbor at Omaha beach was extremely vulnerable to storm damage because of the way it was constructed. Large gaps in the breakwater existed because the scuttled ships were not close together; the Phoenix caissons were sunk in water depths that were too deep and were quickly overtopped by the high storm waves; the outer line of floating breakwaters (Bombardons, Figure 5.3) did little to lower the storm-swell wave heights; and they were not securely moored

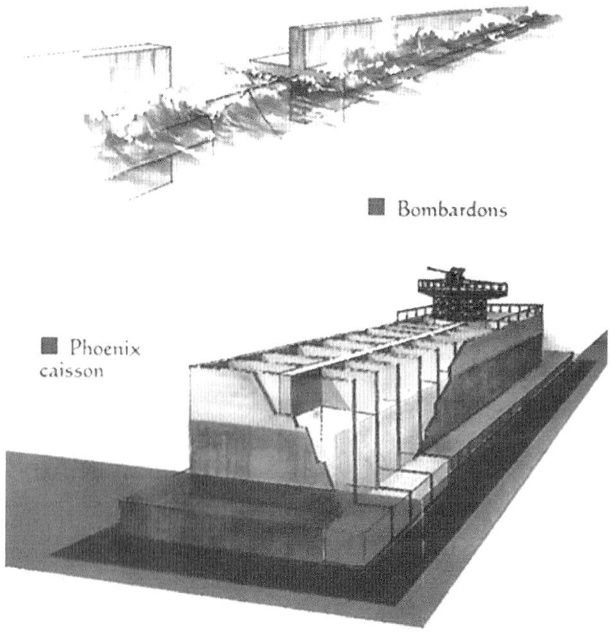

Figure 5.3. The top shows the floating structures (Bombardons) that were 60 m long and 8 m high and anchored to the floor with chains that also tied the units together. A 2-m wave was thought to reduce the wave height to 1 m in the lee of the structure. Unfortunately, the floating units rode the storm-swell waves causing little wave-energy reduction and broke loose from the anchor chains during the storm. The bottom displays the reinforced, concrete caisson which was sunk by filling with water. Some were equipped with anti-aircrafts as shown (from Arromanches, 1997 courtesy OREP Editions).

by the bottom anchors. The British harbor at Arromanches did not suffer from these construction deficiencies and withstood the force of the storm. This example illustrates the fact that the crest elevation, planform layout, and all the other "functional design" variables discussed in Chapter 5 are equally as important as the "structural design" considerations in Chapter 6.

By the end of December 1944, the harbor at Arromanches was no longer needed and the dismantling work had begun. But the remains of a few caissons are still there today after 70 years. The success of this "temporary" harbor is often mentioned as a key to ending World War II. It was probably the greatest military engineering achievement since the Persian armies built a pontoon bridge across the Bosphorous in 460 BC (Arromanches, 1997).

Figure 5.4. Aerial view of Arromanches harbor taken some time between June 23 and December 1944 when the Allied use of the harbor was in full operation for materials, supplies, and troops being transported for WWII (from Arromanches, 1997 courtesy OREP Editions, France).

Coastal engineering knowledge for harbor breakwaters was improved in the planning, design, and placement of the structures across the beaches of Normandy, France.

5.1.3 *Overview of Chapter 5*

Functional design can be divided into two parts that stem from a *sectional* view of the coast and structures, namely **Part I: Wave Runup, Overtopping, Reflection and Transmission** and a *plan* view of the coast and structures, namely **Part II: Planform Layout, Location, and Spacing**. The organization of Chapter 5 follows these two views of the coast and structures.

PART I Wave Runup, Overtopping, Reflection, Transmission

5.2 Wave Runup

All the functions of coastal structures require the determination of the elevation of the crest of the structure. In general, the higher the crest elevation, the less the overtopping (flooding), and wave energy is transmitted in the lee of the structure to reduce storm damage, permit safer ship/boat mooring, and reduce erosion on beaches and within estuarine marshes. Establishment of the crest elevation requires computation of the wave runup together with consideration of other constraints such as economics and the environment.

5.2.1 *Definition and common formula*

The wave runup, R_u, is defined as the *vertical distance* above the still-water level (SWL) that can be reached by the waves as illustrated in Figure 5.5. Common practice for design is to consider the irregular wave heights following the Rayleigh Distribution and define a runup level exceeded by $i_{\%}$ of the waves as $R_{ui\%}$. One dimensionless formulation (Battjes, 1974a, 1974b) is

$$\frac{R_{ui\%}}{H_s} = (A\xi + C)\gamma_r\,\gamma_b\,\gamma_h\,\gamma_\beta, \tag{5.1}$$

where $R_{ui\%}$ is runup level exceeded by $i_{\%}$ of the waves, H_s is the significant wave height (both deep water and toe of structure locations have been employed in the literature), T is the irregular wave period (both mean, T_m, peak T_p, and spectral period, $T_{m-1.0}$ have been employed in the literature), ξ is the surf-similarity parameter (various definitions for wave period and wave height) $= (\tan\alpha)/\sqrt{s}$, s is the wave steepness (H_s/L_o), L_o is the wave length in deep water from linear wave theory depending on wave period, A, C are the empirical coefficients from laboratory and field data depending on the value of $i_{\%}$, and γ are reduction factors (subscripts) for roughness (r), fronting berm (b), shallow water depths (h), and wave-approach angle (β).

One key independent variable is the surf-similarity parameter, ξ. Three definitions have been employed in the literature depending on which

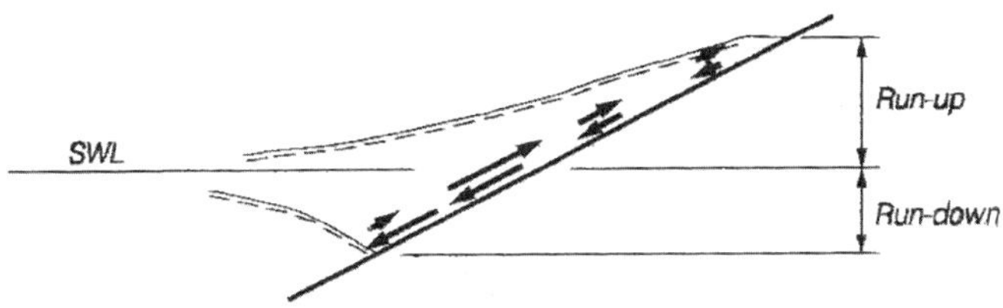

Figure 5.5. Wave runup and rundown.

irregular wave period, T, is specified to define the deep-water wave length, L_0. The mean wave period, T_m; the spectral peak wave period, T_p, and the spectral wave period, $T_{m-1.0}$, have all been employed by various researchers. The spectral period $T_{m-1.0}$ is defined in Section 5.3.4.

The location to define the significant wave height, H_s, is also important since the wave steepness s (H_s/L) appears in the definition of ξ. Wave breaking due to the shallow water-depth limitation in front of the structure may cause lower H_s values at the structure's toe than in relatively deep water where a Rayleigh distribution is commonly assumed.

Since the structure slope and wave steepness trigger different types of breaking waves (spilling, plunging, collapsing, and surging), the coefficients A and C vary with ξ for a given structure. Hunt (1959) was the first to introduce Equation (5.1) but only for regular waves.

For design of crest elevations with essentially non-overtopping wave conditions, the $i = 2\%$ value for runup has long been the choice. See the *EurO-top Manual* (Die Küste, 2007, p. 77) for the history of this choice for design dating back to 1932 in the Netherlands. All the results in Sections 5.2.2 and 5.2.3 are for waves striking the structure at right angles except where noted.

5.2.2 *Impermeable revetments, dikes, and levees*

Impermeable embankments (dikes and levees) with smooth, rough, and rock cover were studied by de Waal and van der Meer (1992). Using the peak wave period T_p to define ξ_{op}, the runup at the $i = 2\%$ exceedance level $R_{u2\%}$ depended on the breaker type. These researchers also employed the significant wave height H_s in relatively deep water with a wave-height distribution close to a Rayleigh distribution.

5.2.2.1 *Smooth impermeable slopes*

Figure 5.6 displays the results for long-crested waves on smooth slopes. The data were taken for two sizes of wave tanks in the laboratory and for slopes ranging from 1:3 to 1:8.

The solid line represents the mean of the data and in Equation (5.1) with all reduction factors, $\gamma = 1.0$,

$$\frac{R_{ui\%}}{H_s} = \begin{cases} 1.5\,\xi_{\text{op}} & \text{for} \quad 0 < \xi_{\text{op}} \le 2 \ (C = 0), \\ 3.0 & \text{for} \quad 2 < \xi_{\text{op}} < 3 - 4 \ (A = 0). \end{cases} \tag{5.2}$$

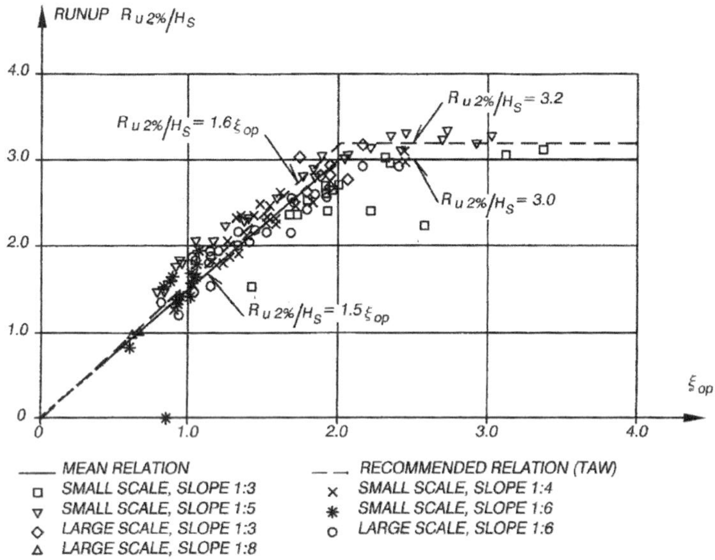

Figure 5.6. Wave runup, R_u for $i = 2\%$ on smooth slopes (from CEM, 2006).

The coefficient of variation $\sigma' = 0.085$ is on the factor 1.5 which was considered a stochastic variable with a normal distribution by de Waal and van der Meer (1992). The dotted line is a safety factor as applied in the Netherlands. At one standard deviation of the mean, $+\sigma = 0.085 \times 1.5 = 0.1275$ so that the slope factor would be $1.5 + 0.1275 = 1.6275$ rounded to 1.6 for $\xi_{\text{op}} < 2$.

5.2.2.2 *Roughened impermeable slopes*

When the impermeable surface is roughened, the roughness reduction factor γ_r in Equation (5.1) has been updated in CEM (2006) including large-scale tests with random waves. The results are presented here as Table 5.1 and valid for $\xi_{\text{op}} < 3.4$. For example, for two layers of rock revetment on an impermeable slope, $\gamma_r = 0.5 - 0.55$ so that the runup height is reduced by about one-half by the rock roughness.

5.2.2.3 *Other reduction factors on impermeable slopes*

Equation (5.1) also shows reduction factors for γ_h (shallow water), γ_β (wave approach direction), and γ_b (berm) fronting the runup slope.

Table 5.1. Surface roughness reduction factor, γ_r, in Equation (5.1).

Type of Slope Surface	γ_r
Smooth, concrete, asphalt	1.0
Smooth block revetment	1.0
Grass (3-cm length)	0.90–1.0
One layer of rock, diameter D ($H_s/D = 1.5$–3.0)	0.55–0.6
Two or more layers of rock ($H_s/D = 1.5$–6.0)	0.50–0.55
Roughness elements on smooth surface	
(length parallel to waterline $= P$, width $= b$, height $= h$)	
Quadratic blocks, $P = b$	
h/b b/H_s area coverage	
0.88 0.12–0.19 1/9	0.70–0.75
0.88 0.12–0.24 1/25	0.75–0.85
0.44 0.12–0.24 1/25	0.85–0.95
0.88 0.12–0.18 1/25 (above SWL)	0.85–0.95
0.18 0.55–1.10 1/4	0.75–0.85
Ribs	
1.00 0.12–0.19 1/7.5	0.60–0.70

For shallow water conditions at the toe of the structure, some waves will break so that the wave height distribution will no longer follow the Rayleigh distribution. Therefore, de Waal and van der Meer (1992) recommend using the representative significant wave height at the toe of the structure $H_{(s)\text{toe}}$ and

$$\gamma_h = \frac{H_{2\%}}{1.4H_{(s)\text{toe}}}, \tag{5.3}$$

where $H_{2\%} = 1.4H_s$ for the deep-water Rayleigh distribution and deep-water significant wave height, H_s. See Chapter 2 and Section 2.3.4 for how to estimate $H_{(s)\text{toe}}$ when breaking waves occur at the toe of the structure.

As expected, for wave angles other than at right angles to the structure ($\beta = 0$), the wave angle direction factor, γ_β is less than 1.0. For short-crested waves, de Waal and van der Meer (1992) showed a linear relationship for γ_β versus β with the minimum value of $\gamma_\beta = 0.8$ at $\beta = 90°$. For long-crested waves, γ_β dropped nonlinearly to 0.6 for $\beta = 60°$ and then was constant at 0.6 for 60°–90°. For conservative design practice, it is recommended to take $\gamma_\beta = 1.0$, since storm waves from offshore directions refract as they approach coastal structures and strike the structure at close to right angles.

Coastal structures with fronting berms can take many geometrical forms. The method suggested by de Waal and van der Meer (1992) is to replace ξ_{op} in Equation (5.2) by a breaking wave, surf-similarity parameter

Table 5.2. Coefficients in Equation (5.1) for runup of long-crested, irregular waves on smooth impermeable slopes (from the CEM, 2006, Table VI-5-2).

ξ	R_u	ξ−Limits	A	C	σ_{Ru}/R_u
ξ_{op}	$R_{u2\%}$	$\xi_p \leq 2.5$	1.6	0	
		$2.5 < \xi_p < 9$	−0.2	4.5	≈ 0.15
ξ_{op}	R_{us}	$\xi_p \leq 2.0$	1.35	0	
		$2.0 < \xi_p < 9$	−0.15	3.0	≈ 0.10

based on an equivalent slope, ξ_{eq}. Further details can be found in CEM (2006, Part VI, Section 5).

5.2.2.4 *Other formulations*

Studies by Van Oorschot and d'Angremond (1968) and Ahrens (1981) for impermeable, smooth slopes are discussed in CEM (2006). Table 5.2 presents the coefficients A and C for both the $R_{u2\%}$ and R_{us} levels of runup used in Equation (5.1). Note that the runup level reaches a maximum at the middle value of the ξ-limits and then decreases for higher values of ξ_p

A spreadsheet has been developed using the formulas developed by these researchers and will be further discussed in Section 5.2.5 for some example problems.

5.2.3 *Permeable, rock-armored structures*

Permeable structures experience less wave runup. The volume of pore space in the voids simply fills with the water moving up the slope so that less water volume is available to contribute to the elevation of the runup. The larger the permeability of the structure, the less the wave runup.

Rock-armored slopes with permeable (and impermeable) cores, e.g. rubble-mound breakwaters, were studied in great detail by van der Meer (1988) for Delft Hydraulics (1989). The permeability resulted in a more complicated form for the dimensionless, wave runup formulation

$$\frac{R_{ui\%}}{H_s} = A\,\xi_{om} + B(\xi_{om})^C + D. \tag{5.4}$$

The significant wave height H_s is at the front (toe) of the structure.

5.2.3.1 *Surf-similarity parameter,* ξ_{om}

The mean wave period, T_m was employed to calculate the surf-similarity parameter, ξ_{om}, for use in correlating the experimental data for permeable structures. Structure slopes (cot α) tested were 1.5, 2, 3, 4, and 6 with a foreshore slope of 1:30. Irregular waves modeled after the Pierson and Moskowitz (1964) spectrum were employed with full details of the test program summarized in CEM (2006, Table VI-5-4). Note that the ratio $T_m/T_p = 0.71$–0.82 for the Pierson–Moskowitz spectrum.

5.2.3.2 *Notational permeability*

The core of the rubble-mound structure was considered "impermeable" for half of the tests conducted and "permeable" for the other half except one test with a "homogeneous" core construction. For development of the wave runup formulas (and rubble-mound stability formulas discussed in Chapter 6), van der Meer (1988) devised a notational permeability parameter, P. It is not the true "permeability" of the structure (i.e. the ratio of the volume of the voids to the total volume of the structure), but simply a notational parameter that gives a number to the relative composition of the breakwater core for the sizes of the materials in the sublayers and a filter layer (if present).

The "impermeable" core was assigned a notational permeability parameter, $P = 0.1$, and is shown in the upper-left corner of Figure 5.7 as taken from van der Meer (1988). A relatively "impermeable" filter layer is present beneath a narrow sublayer of stones beneath the armor layers. At the other extreme, a notational permeability parameter $P = 0.5$ is shown in the lower-left corner of Figure 5.7 for a "permeable" core of the breakwater structure with larger stones in the sublayer and no filter layer. If there was no filter layer and no core of smaller stones, the structure was homogeneous for all similar stone sizes and $P = 0.6$ was assigned as illustrated in the lower-right corner of Figure 5.7.

The core construction shown in the upper-right corner of Figure 5.7 with $P = 0.4$ was estimated by van der Meer (1988) but not tested. Recent research to model a core layer with $P = 0.4$ (Kik *et al.*, 2012) and to develop a method to quantify the notational permeability using numerical models (Verhagen *et al.*, 2012) for structural stability formulas is discussed in Chapter 6.

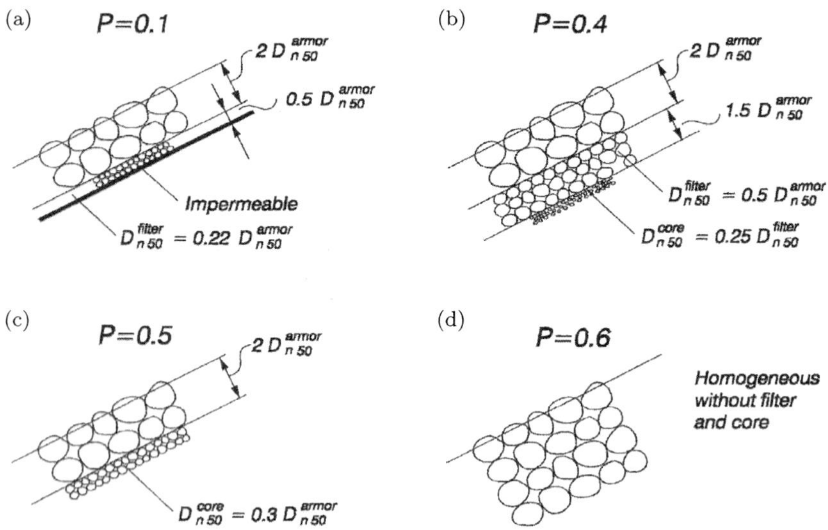

Figure 5.7. Rubble-mound breakwater core construction details for notational permeability parameters as devised by van der Meer (1988).

5.2.3.3 *Test results and formulas*

The results for the wave runup tests by Delft Hydraulics (1989) are plotted in Figure 5.8 for both the impermeable core ($P = 0.1$) and permeable core ($P = 0.5$). The maximum wave runup measured was on the order of $3H_s$ at the 2% level, and was only $2H_s$ at the significant runup level for the highest mean surf similarity value tested, ξ_{om}. For the long-crested, irregular waves tested, this corresponds to the longest wave periods tested. Note that the scatter is much larger for the 2% level test results and for the permeable core ($P = 0.5$) measurements. The solid lines shown in Figure 5.8 are for a central fit to the data and the empirical formulas and coefficients A, B, C, and D in Equation (5.4) have been summarized for various exceedance probability levels, namely, the 0.1, 2.0. 5, 10, 13.5 (significant), and 50% (mean) levels as presented in Table 5.3 (from the CEM, 2006).

For impermeable cores ($P = 0.1$),

$$\frac{R_{ui\%}}{H_s} = \begin{cases} A\xi_{om} & \text{for} \quad 1.0 < \xi_{om} \leq 1.5, \\ B(\xi_{om})^C & \text{for} \quad \xi_{om} > 1.5. \end{cases} \tag{5.5}$$

The coefficient of variation for the A, B, and C coefficients in Table 5.3 is 7%. The data presented by Ahrens and Heinbaugh (1988) for rough (riprap)

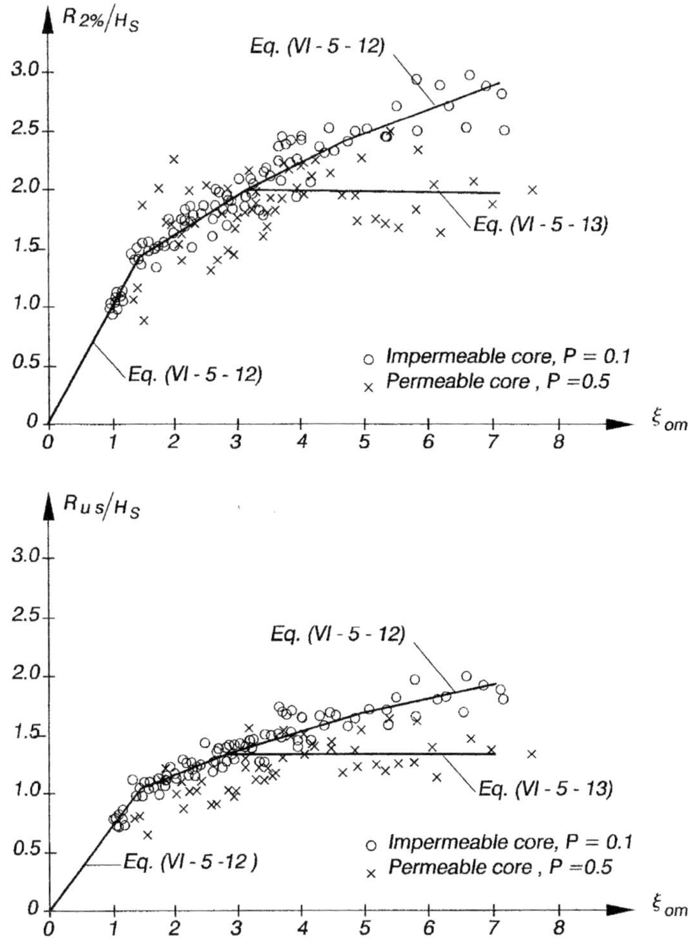

Figure 5.8. (Top) $R_{u2\%}/H_s$ versus ξ_{om} at the 2% runup level and (bottom) R_{us}/H_s versus ξ_{om} for the significant runup level for both impermeable ($P = 0.1$) and permeable ($P = 0.5$) cores of a rock, breakwater (from CEM, 2006).

on impermeable slopes are in agreement with these results for the maximum ($i = 0.1\%$) runup.

For permeable cores ($P = 0.5$)

$$\frac{R_{ui\%}}{H_s} = \begin{cases} A\xi_{om} & \text{for } 1.0 < \xi_{om} \leq 1.5, \\ B(\xi_{om})^C & \text{for } 1.5 < \xi_{om} \leq \left(\frac{D}{B}\right)^{1/C}, \\ D & \text{for } \left(\frac{D}{B}\right)^{1/C} \leq \xi_{om} < 7.5. \end{cases} \tag{5.6}$$

Table 5.3. Coefficients, A, B, C, and D in Equations (5.5) and (5.6) for impermeable and permeable cores for runup on rock-armored slopes at various exceedance levels, i percent[1] (from CEM, 2006).

Percent	A	B	C	D^2
0.1	1.12	1.34	0.55	2.58
2.0	0.96	1.17	0.46	1.97
5	0.86	1.05	0.44	1.68
10	0.77	0.94	0.42	1.45
(Significant)	0.72	0.88	0.41	1.35
50 (mean)	0.47	0.60	0.34	0.82

Notes: [1]Exceedence level related to the number of waves.
[2]Only relevant for permeable slopes.

Here, the coefficient of variation has increased to 12% for the A, B, C, and D coefficients on permeable slopes. Tests with homogenous rock (no core) and $P = 0.6$ gave similar results as shown in Figure 5.8 for $P = 0.5$.

The values of A, B, C, and D for the 2% level of wave runup are normally used for design.

5.2.3.4 *Partial safety factors for runup on rock-armored slopes*

The above-mentioned factors are estimated for some level of the significant wave height, H_s, at the toe of the structure. A full, probabilistic treatment for wave runup on rock-armored slopes with impermeable and permeable cores using $\hat{H}_S^T$, the mean significant wave height at some recurrence interval, has been determined for a range of partial safety factors and levels of probability of failure, P_f. The following is from CEM (2006, Table VI-6-17).

For impermeable cores ($P = 0.1$), the design equations and applicable partial safety factors γ_H (load) and γ_Z (resistance) are separated and different for the *two surf-similarity regions* that quantify the results.

For $(0 < \xi_{om} \leq 1.5)$ region,

$$\frac{R_{ui\%}}{H_s} = A\,\xi_{om} \tag{5.7}$$

so that the failure function, G becomes

$$G = \frac{1}{\gamma_z} R_{ui\%}\, A^{-1}(\xi_{om})^{-1} - \gamma_H \hat{H}_S^T \tag{5.8}$$

Table 5.4(a). Partial safety factors for runup on rock slopes ($0 < \xi_{om} \leq 1.5$).

	$\sigma'_{FHS} = 0.05$		$\sigma'_{FHS} = 0.2$	
P_f	γ_H	γ_Z	γ_H	γ_Z
0.01	1.7	1.04	2.0	1.00
0.05	1.4	1.06	1.6	1.02
0.10	1.3	1.04	1.4	1.06
0.20	1.2	1.02	1.3	1.00
0.40	1.0	1.08	1.1	1.00

Table 5.4(b). Partial safety factors for runup on rock slopes ($\xi_{om} > 1.5$).

	$\sigma'_{FHS} = 0.05$		$\sigma'_{FHS} = 0.2$	
P_f	γ_H	γ_Z	γ_H	γ_Z
0.01	1.5	1.08	1.8	1.02
0.05	1.3	1.06	1.4	1.10
0.10	1.2	1.06	1.3	1.08
0.20	1.1	1.08	1.2	1.06
0.40	1.0	1.06	1.0	1.10

and the partial safety factors are found in Table 5.4(a) for this lower range of surf-similarity region.

For ($\xi_{om} > 1.5$) region,

$$\frac{R_{ui\%}}{H_s} = B(\xi_{om})^C \tag{5.9}$$

so that the failure function, G becomes

$$G = \frac{1}{\gamma_z} R_{ui\%}\ B^{-1}(\xi_{om})^{-C} - \gamma_H \hat{H}_S^T \tag{5.10}$$

and the partial safety factors are found in Table 5.4(b) for the upper range of surf-similarity region.

For permeable cores ($P > 0.4$) as seen in Equation (5.6), an additional failure equation is needed.

For the region, $\left(\frac{D}{B}\right)^{1/c} \leq \xi_{om} < 7.5$,

$$G = \frac{1}{\gamma_z} R_{ui\%}\ D^{-1} - \gamma_H \hat{H}_S^T. \tag{5.11}$$

The lower limit $\left(\frac{D}{B}\right)^{1/c}$ can be found from Table 5.3 for the seven levels of exceedance and ranges from $\xi_{om} = 3.3$ (0.1) to $\xi_{om} = 2.5$ (50%) and in all cases greater than 1.5. CEM (2006) does not give a separate table of partial safety factors for this region for permeable cores. All that is stated is

> *For permeable structures, $P > 0.4$, the upper limit of $R_{ui\%}$ is given by $R_{ui\%}/\hat{H}_S^T = D$.*

We herein recommend using the partial safety factors found in Table 5.4(b) for use in the failure Equation (5.11) for permeable core structures since in all cases for this region, $\xi_{om} > 1.5$.

5.2.4 *Other research works*

Many other formulations are available. See Goda (2010), Mase and van Gent (2001) and Allsop *et al.* (1985) for additional empirical formulations from physical model tests. See also Die Küste (2007) for calculation of wave runup heights under the range of experimental conditions employed to develop the *EurOtop Manual* for wave-overtopping prediction. These authors have employed the spectral period $T_{m-1.0}$ to define $\xi_{m-1.0}$ in all their work to better incorporate the longer period waves in the empirical, wave runup formulations and coefficients. Further discussion is presented in Section 5.3.

The US Army Corps of Engineers has developed a hydrodynamic, numerical model to predict wave runup for coastal structures and beaches called CSHORE. The Federal Emergency Management Agency (FEMA) in the US sponsored a study (Melby, 2012) which concluded that EurOtop "*...is likely to have better skill than CSHORE...*" for structures that were within the range of experimental conditions (bathymetry and wave-water level). However, for complex nearshore bathymetry and complex structures with waves and water levels outside the range of EurOtop, the "*... CSHORE is likely to show better skill.*" CSHORE also predicts cross-shore beach morphology change during storm events that can significantly impact the bathymetry and hence the resulting wave heights and wave runup.

It is highly recommended that hydraulic model studies be considered to determine $R_{u2\%}$ for use in the design determinations of the crest elevation of coastal structures whenever possible. This is especially important when artificial, concrete armor units are employed on breakwaters. An example for the new "Ecopode" armor units is found in Calabrese *et al.* (2010). No partial safety factors are presently available for smooth surfaces on impermeable slopes to facilitate a probabilistic design. This is not the case for impermeable and permeable, rubble-mound structures as discussed before.

The above-mentioned details are best illustrated by an example.

5.2.5 *Example calculation of wave runup*

5.2.5.1 *Deterministic design — given data*

Given a relative mild foreshore up to a sloping (1.0 vertical: 2.5 horizontal; 1 on 2.5) structure as shown in Figure 5.9. At the design SWL, the water depth at the toe of the structure is 3.0 m (9.8 ft). The design significant wave height $H_s = 2.0$ m (6.6 ft) and peak-wave period $T_p = 8$ s were measured at a wave gauge located in $d = 4.5$ m (14.8 ft). Assume the design waves are at right angles to the structure. All the formulas for determination of the wave runup level need calculation of the surf-similarity parameter ξ_p (or ξ_m) using the significant wave height H_s at the toe of the structure.

Of interest is the wave runup at the $R_{u2\%}$ level for impermeable (smooth and rough) structures and permeable cores in rubble-mound structures.

5.2.5.2 *Surf-similarity parameter, ξ*

Assume irregular wave shoaling of the significant wave height according to the linear, regular wave theory (ideally a spectral wave-transformation model should be employed). The unrefracted, deep water wave height H_o'

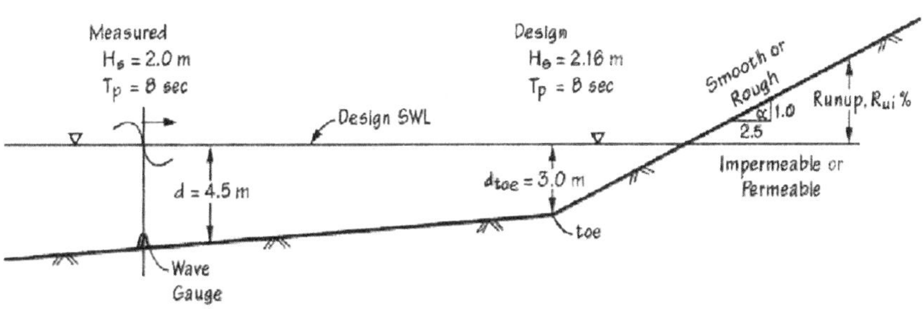

Figure 5.9. Example for calculation of wave runup on smooth and rough, and impermeable or permeable slopes.

can be determined using an approximation method (Nielsen, 1984), where

$$k_o = \frac{2\pi}{L_o} = \frac{4\pi^2}{gT^2} = \frac{4\pi^2}{g8^2} = 0.062879, \quad k_o d = (0.062879 \times 4.5) = 0.28296,$$

$$\frac{C_g}{C_o} = \sqrt{k_o d}\left[1 - \frac{1}{2}k_o d + \frac{7}{72}(k_o d)^2\right]$$

$$= \sqrt{0.28296}\,[1 - 0.5 \times 0.28296 + 0.09722(0.28296)^2] = 0.46082,$$

$$K_s = \text{shoaling coefficient} = \sqrt{0.5\,C_o/C_g} = \sqrt{0.5}/0.46082 = 1.0416,$$

$$K_s = H/H_o',$$

$$H_o' = H/K_s = 2.0/1.0416 = 1.920\,\text{m}\ (6.3\,\text{ft}).$$

In turn, this deep-water wave can then be shoaled to the toe of the structure, where $d = 3.0$ m using a similar procedure, and the approximation method of Nielsen (1984) gives,

$$k_o d = (0.062879 \times 3.0) = 0.188637,$$

$$\frac{C_g}{C_o} = 0.39486,$$

$$K_s = 1.1253 = H/H_o' = H_{\text{toe}}/1.920, \text{so that}$$

$$H_{\text{toe}} = 1.1253 \times 1.920 = 2.161\,\text{m}\ (7.09\,\text{ft}).$$

Other linear wave shoaling methods such as that in SPM (1984, Table C-1) or computer software (e.g. ACES, Leenknecht *et al.*, 1992) could be employed to determine the shoaling coefficients.

When the spectral peak wave period in deep water T_p is employed, the surf-similarity parameter can be written as follows:

$$\xi_{op} = \frac{\tan\alpha}{\sqrt{S_{op}}} = \frac{\tan\alpha}{\sqrt{\frac{H_{\text{stoe}}}{L_{op}}}},$$

with S_{op} as the wave steepness and as a hybrid using the deep-water spectral peak wave length L_{op} from linear wave theory, and with the significant wave height H_{stoe} at the structure toe. The additional subscript zero in the notation is to emphasize that the deep-water wave length, i.e. L_o, is employed in the definition.

Therefore,

$$L_{\text{op}} = \frac{g}{2\pi}T_p^2 = \frac{9.81}{2\pi}8^2 = 99.92\,\text{m},$$

$$S_{\text{op}} = \frac{H_{\text{stoe}}}{L_{op}} = \frac{2.161}{99.92} = 0.02162,$$

$$\text{and} \quad \xi_{\text{op}} = \frac{1/2.5}{\sqrt{0.02162}} = 2.72.$$

This value of ξ_{op} will be employed in the following formulas for wave runup on impermeable, smooth, and rough slopes. Note that as the wave period increases, the wave length increases, the wave steepness decreases, and therefore the surf-similarity parameter increases, generally resulting in higher wave runup.

In some formulations, the mean spectral wave period T_m is employed. Development of theoretical equations coupled with empirically found coefficients to calculate wave spectra has produced formulas for "fully developed" sea conditions (Pierson and Moskowitz, 1964) called the PM spectrum, for "developing" seas called the JONSWAP spectrum (Hasselmann *et al.*, 1973), and then when the waves become depth-limited in shallow water called the TMA spectrum (Hughes, 1984; Bouws *et al.*, 1985). The commonly accepted relationships for the ratio T_m/T_p are the following: Pierson–Moskowitz spectrum, $T_m/T_p = 0.71$–0.82; JONSWAP spectrum, $T_m/T_p = 0.79$–0.87; but the relationship for T_m/T_p for the TMA spectrum is unknown to the author.

To continue, assume $T_m/T_p = 0.76$ so that $T_m = 0.76\ T_p = 0.76\ (8) =$ 6.1 s so that

$$L_{om} = \frac{g}{2\pi}T_m^2 = \frac{9.81}{2\pi}6.1^2 = 58.05\,\text{m},$$

$$S_{om} = \frac{H_{\text{stoe}}}{L_{om}} = \frac{2.161}{58.052} = 0.03718,$$

$$\text{and} \quad \xi_{om} = \frac{1/2.5}{\sqrt{0.03718}} = 2.07.$$

This value of ξ_{om} will be employed in the following formula for wave runup on a permeable, rubble-mound structure.

5.2.5.3 *Impermeable structures*

For smooth, impermeable slopes, when $\xi_{\text{op}} > 2$, de Waal and van der Meer (1992) in Equation (5.2) state

$$\frac{R_{u2\%}}{H_s} = 3.0$$

so that $R_{u2\%} = 3.0 \times 2.16 = 6.48$ m (21.3 ft) at the 2% runup level to prevent wave overtopping.

An alternate formulation by van Oorschot and d'Angremond (1968) together with Ahrens (1981) gave for $2.5 < \xi_{\text{op}} < 9$, in Equation (5.1) using Table 5.2,

$$\frac{R_{u2\%}}{H_s} = (-0.2\,\xi_{op} + 4.5)$$

so that $R_{u2\%} = 2.16[(-0.2 \times 2.72) + 4.5] = 2.16(-0.544 + 4.5) = 2.16\ (3.956) = 8.55$ m (28.0 ft). There is considerable scatter in the relatively small-scale, laboratory tests of van Oorschot and d'Angremond (1968) combined with data from Ahrens (1981). Also, data presented by Allsop *et al.* (1985) showed runup levels considerably lower than these alternate formulation results.

For rough, impermeable slopes, the surface roughness reduction factor γ_r is found in Table 5.1. Consider two or more layers of rock on an impermeable rubble-mound structure so that $\gamma_r = 0.55$. Using de Waal and van der Meer for $R_{u2\%} = 6.48\ \gamma_r = 6.48 \times 0.55 = 3.56$ m (11.7 ft). For comparison, consider the Delft Hydraulics (1989) test results for a rubble-mound structure with impermeable core ($P = 0.1$) and Equation (5.5). Since $\xi_{om} = 2.07 > 1.5$ and using Table 5.2 with $A = 0$, $B = 1.17$, $C = 0.46$, and $D = 0$ gives

$$\frac{R_{ui\%}}{H_s} = A\,\xi_{om} + B(\xi_{om})^{\mathbf{C}} + D = 1.7(2.07^{0.46}) = 1.635,$$

$$R_{u2\%} = 2.16\ (1.636) = 3.35\,\text{m}(11.6\,\text{ft}),$$

this result is very close to that given by the de Waal and van der Meer (1992) formulation.

Therefore, it is recommended that the formulations by de Waal and van der Meer (1992) be used for wave runup on impermeable, smooth, or rough slopes.

5.2.5.4 *Permeable rock-armored structures*

For rock-armored (aka rubble-mound) structures with permeable cores ($P = 0.5$), the correct formulation in Equation (5.6) for $R_{u2\%}$ depends on the values $B = 1.17, C = 0.46$, and $D = 1.97$ in Table 5.2 so that

$$\left(\frac{D}{B}\right)^{1/C} = \left(\frac{1.97}{1.17}\right)^{0.46} = 3.10.$$

Recall that for this example, $\xi_{om} = 2.07$ and therefore $\left(1.5 < \xi_{om} \leq \left(\frac{D}{B}\right)^{\frac{1}{C}}\right)$ so that

$$\frac{R_{u2\%}}{H_s} = B(\xi_{om})^C = 1.17\ (2.07^{0.46}) = 1.635$$

$$\text{and} \quad R_{u2\%} = 2.16\ (1.635) = 3.53\,\text{m}\ (11.6\,\text{ft}).$$

This result is similar to that for impermeable cores ($P = 0.1$) as seen in Figure 5.8 (top).

For $\xi_{om} > 3.10$, the runup is constant so that

$$\frac{R_{u2\%}}{H_s} = D = 1.97$$

and $R_{u2\%} = 2.16\ (1.97) = 4.25$ m (13.9 ft).

Working back to find the wave period T_m for $\xi_{om} \geq 3.10$,

$$\xi_{om} = \frac{\tan\alpha}{\sqrt{S_{om}}} = \frac{\tan\alpha}{\sqrt{\frac{H_{stoe}}{L_{om}}}} = 3.10,$$

$$L_{om} = 129.2\,\text{m}, \text{ and}$$

$$T_{om} \geq 9.1\,\text{s so the } T_{\text{op}} \approx 12\,\text{s}.$$

It is clear that the longest period wave expected at the structure should be considered along with the largest wave height to produce $\xi_{om} \geq 3.10$ which will give the highest wave runup. Knowledge of the uncertainty of the wave height and target level of failure probability P_f (runup levels higher that predicted) requires a probabilistic design.

5.2.5.5 *Probabilistic design — partial safety factors*

For this example, assume a probability of failure $P_f = 0.10$ so that the runup elevation calculated has a 90% chance of success and assume a relatively high level of wave-height accuracy since waves were "measured" at a gauge in 4.5-m water depth so that the coefficient of uncertainty $\sigma'_{FHS} = 0.05$.

Partial safety factors are available for rock-armored slopes with impermeable core ($P = 0.1$) as found in Table 5.4. Since $\xi_{om} = 2.07$, the partial safety factors in Table 5.4(b) apply with $\gamma_H = 1.2$ and $\gamma_Z = 1.06$.

The failure function G in Equation (5.10) with coefficients $B = 1.17$ and $C = 0.46$ from Table 5.2 for $i = 2.0\%$.

$$G = \frac{1}{\gamma_z} R_{ui\%}\, B^{-1}(\xi_{om})^{-1C} - \gamma_H \hat{H}_S^T$$

$$= \frac{1}{1.06} R_{u2\%}\, (1.17^{-1})(2.07)^{-0.46} - 1.2 \times 2.16.$$

We have used $\hat{H}_S^T = 2.16$. At $G = 0$ and solving for $R_{u2\%}$ gives

$$R_{u2\%} = 4.49\,\mathrm{m}(14.7\,\mathrm{ft}).$$

If the core is permeable ($P = 0.4$–0.5), the result would be the same since the partial safety factors are also found in Table 5.4(b).

A full probabilistic design would consider many design water levels and wave characteristics (H_s, T_p) so that the above-mentioned example calculations would need to be repeated many times for the structure slope, tan α. Other structure slopes would need to be evaluated for economic reasons. Spreadsheet solutions make these calculations manageable.

5.2.5.6 *Spreadsheets for functional design*

In general, wave runup increases as ξ_{op} (or ξ_{om}) increases. Consider an altered form for the surf-similarity parameter

$$\xi_o = \frac{\tan\alpha}{\sqrt{S_o}} = \frac{1}{\cot\alpha\sqrt{S_o}}. \tag{5.12}$$

The slope, tan α is the sole structural variable that governs the functional design. As the slope, tan α decreases (cot α increases), the structure flattens, ξ_o decreases and the wave runup is decreased. But the flatter structure requires more volume for construction. The key structural design variable in the spreadsheet is the structure slope, tan α (or cot α).

Spreadsheets have been developed for wave runup calculations for the formulas above and can be downloaded as supplementary material from the website for this book. All employ the software product, Microsoft Excel. See

(1) **Metric System**

(2) **VAN DER MEER (1988) and Delft Hyd (1989)** P = 0.5
, Equations 5.6 and Table 5-4
Partial Safety Factors for Run-up, Rock armored slopes

Permeable core structure.
Slope, Cot alpha = 1.5 to 5.0

(5) **Calculated Results for Wave Runup**

Slope	Surf Similarity p.	RUN UP		
Cot alpha	Em	Rus(m)	Ru2%(m)	Ru0.1%(m)
1.5	3.44	2.92	4.26	5.57
2.0	2.58	2.81	3.91	4.88
2.5	2.07	2.56	3.53	4.32
3.0	1.72	2.38	3.25	3.90
4.0	1.29	2.01	2.68	3.13
5.0	1.03	1.61	2.14	2.50

(6) Ru=A*Hs*Em [Em≤1.5]
G=1/gamaZ*Ru*A^ -1*{(Cot alpha)*(Som^0.5)} - gamaH*Hs
Ru=B*Hs*(Em)^C [1.5<Em < ((D/B)^(1/C)]
G=1/gamaZ*Ru*B^ -1*{(Cot alpha)*Som^0.5}^C - gamaH*Hs
Ru = Hs*D [(D/B)^(1/C)≤ Em < 7.5]
G=1/gamaZ*Ru*D^ -1 - gamaH*Hs

(8)

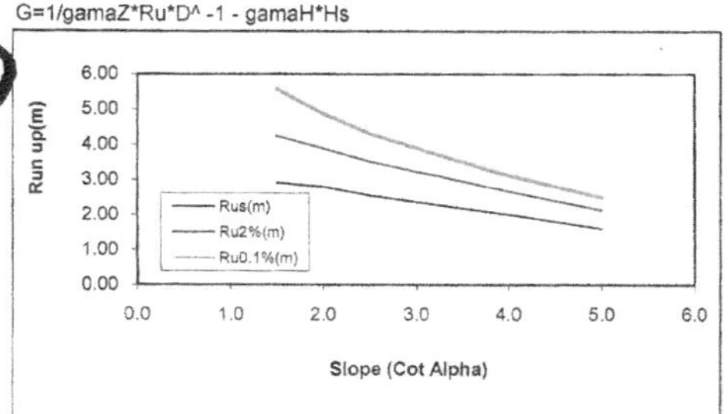

WAVE RUNUP
(Metric System)

(3) **Enter Data**

Enter Data		Unit
Wave Height,Hs=	2.0	m
Peak Wave Period=	8.0	sec
Depth at Wave Gauge=	4.5	m
Depth at Toe of Strucure=	3.0	m
*Pf(Probability of Failure)	0.01	-
SigmaFHS(Coefficient of Variation)	0.05	-
GamaH(load p. s.factor for Em<=1.5)	1	-
GamaZ(resistance p. s. factor -Em<=1.5)	1	-
P(notational permeability,cf. FigureVI-5-11	0.5	-
GamaH(load p. s.factor for Em>1.5)	1	-
GamaZ(resistance p. s. factor -Em>1.5)	1	-

(4) **Calculated Information**

Calculated Information		
Mean Wave Period=	6.1	sec
Wave Length,Lom=	57.668	m
Wave Length,Lop=	99.840	m
Ko(wave number at deep water) =	0.063	-
Cg(group vel.at gauge)/Co(deep water vel.)	0.461	-
Ks = (Hs/Ho) =	1.042	-
Ho (deep water)=	1.920	m
Cg(group vel.at toe)/Co(deep water vel.) =	0.395	-
Ks= (Htoe/Ho) =	1.125	-
Htoe=	2.161	m
Steepness, Som=	0.037	-

FOR PARTIAL SAFETY FACTORS
SEE TABEL 5-4 on READ ME PAGE
SEE DVD FUNCTIONAL DESIGN

(7)

Exceedence Probability (%)	A	B	C	D**	(D/B)^(1/C)
0.10	1.12	1.34	0.55	2.58	3.291
2.00	0.96	1.17	0.46	1.97	3.104
Significant	0.72	0.88	0.41	1.35	2.840

* different probability of failures can be chosen and the corresponding values of GamaH and GamaZ can be taken from Table VI-6-18.
**For Permeable Structures, P>0.4, the upper limit of Ru is given by Ru/Hs=D

Figure 5.10. Example spreadsheet for calculation of wave runup for range of structure slopes, cot α. For van der Meer (1988) and Delft Hydraulics (1989), metric, rock-armor, permeable ($P = 0.5$) core structures.

the folder for Functional Design/Wave Runup. Figure 5.10 is an example for the van der Meer (1988) and Delft Hydraulics (1989) formulas in Equation (5.5) for permeable cores ($P = 0.5$) for rock-armor slopes and in metric units.

There is a considerable amount of information on this spreadsheet. However, the pattern layout for all the wave runup spreadsheets is the same and generally is as follows:

(1) The top of the sheet indicates the units are in the *Metric System.* A second sheet in the MS EXCEL file gives an identical spreadsheet for the Imperial (English) system of units.
(2) The upper-left corner gives the *name and dates of the author(s) for the equations employed* and reference to the text tables and equations that were the source. The notational permeability (P) is given in red.
(3) The upper-right side is the given *Enter Data* of which the significant wave height H_s is the most important and highlighted in blue. The location of this wave height is very important, for the water depth must also be specified at this location. Then the peak wave period, water level (water depths), failure probability, partial safety factors, etc. are specified.
(4) On the right-side middle is the *Calculated Information.* The wave conditions at the specified location are shoaled to deep water and then back to the toe of the structure. The peak and mean wave heights can be calculated. The bottom two lines are for the significant wave height at the toe of the structure and the wave steepness in deep water.
(5) The *Calculated Results for Wave Runup* are found on the left-side middle and highlighted in yellow. These results are for the full range of possible structure slopes ($\cot\alpha = 1.5, 2.0, 2.5, 3.0, 4.0,$ and 5.0) which appears in the far left-side column. The second column is the surf-similarity parameter, ξ. Three wave runup heights above the SWL are then shown as calculated using the given data. The example problem mentioned before was for $\cot\alpha = 2.5$ and the 2% runup value, $R_{u2\%} = 3.53$ m is shown. Runup values for other levels of runup ($R_{us}, R_{uo.1\%}$) and other structure slopes are also shown.
(6) The formulas employed are listed on the left-side middle and also included on a Read Me sheet that is included with each spreadsheet.
(7) The coefficients needed in the formulas for the three levels of wave runup are shown in the lower right side along with the value for the formula boundary at $\left(\frac{D}{B}\right)^{1/C}$.

(8) Finally, the lower left side shows a plot of the results for the three levels of runup over the possible ranges of structure slope. Clearly, runup decreases as the cot α increases.

Spreadsheets for wave runup calculations for use in design are included as supplementary material (Functional Design, Wave Runup) for the following:

Impermeable Slopes (Smooth and Rough)

- de Waal and van der Meer (1992) Equation (5.2) (recommended)
- van Oorschot and d'Angremond (1968) and Ahrens (1981) Table 5.2 (not recommended)

Permeable and Impermeable, Rock-armor Structures

- van der Meer (1988)/Delft Hydraulics (1989) Impermeable Equation (5.5) (recommended)
- van der Meer (1988)/Delft Hydraulics (1989) Permeable Equation (5.6) (recommended)

All spreadsheets have three pages (sheets) with one for the Metric system, one for the English unit's system and one for a Read Me page that includes all the equations, coefficients, and partial safety factors. They should be helpful for design.

5.2.6 *Summary for wave runup*

CEM (2011) presents a method for handling wave runup on composite slopes (varying slopes with berms) and additional example calculations. Wave rundown (see Figure 5.5) estimate methods are also discussed but not generally of interest for design.

For many reasons (mainly economic), the crest elevations of coastal structures must be set below the $R_{u2\%}$ level so that some wave overtopping results.

5.3 Wave Overtopping

The most important structural variable is the crest elevation. It determines the distance R_c from the design SWL to the structure crest and is called the "freeboard" as shown in Figure 5.11.

5.3.1 *Definitions and common formulations*

Clearly, whenever the wave runup $R_{ui\%}$ is greater than the freeboard R_c, some wave overtopping of the structure must occur. The wave overtopping

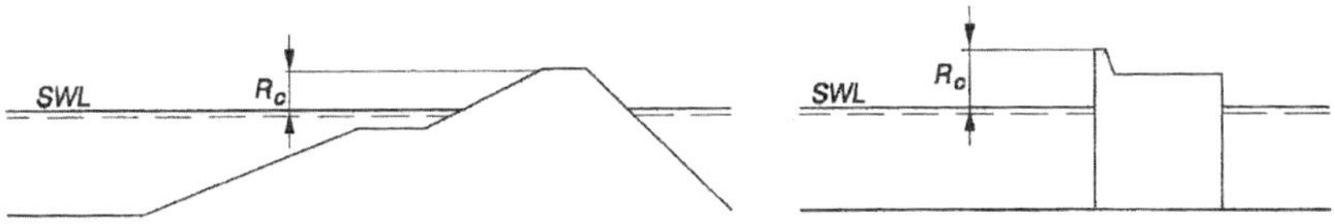

Figure 5.11. Definition of crest freeboard of structure (from CEM, 2011, Fig. VI-5-13).

volume varies in time and space because the quantity of overtopping water varies from wave to wave for the irregular wave conditions at the crest. The overtopping volume per unit time per unit length of structure for each wave is *time-averaged* to define the time average, volume discharge rate q with units of m^3/s (or ft^3/s) per structure length.

Two dimensionless variables Q and R appear in the many empirical, wave overtopping formulas resulting from curve fitting the measured data to the hydraulic model test results. The dimensionless, average volume discharge rate is Q and R is a dimensionless freeboard. These dimensionless formulations have two basic forms as follows:

$$Q = a \exp(-b \times R), \tag{5.13}$$

and

$$Q = aR^{-b}, \tag{5.14}$$

where a and b are dummy variables standing for various combinations of the wave conditions (H_s, T_o), wave angle of incidence β, water depth h_{toe} at the toe of the structure, the gravity constant, and many other structure geometry variables. CEM (2011) lists eight formulations for research results from 1982 to 1999 and all are for the five structure profile geometries shown in Figure 5.12.

5.3.2 *Impermeable revetments and breakwaters*

5.3.2.1 *Smooth, rough, straight, and berm slopes*

The earliest formulation for cases (a) and (b) in Figure 5.12 are those of Owen (1980, 1982). The exponential equation form (Equation (5.13)) was employed to model the measured, irregular wave data quantified by the significant wave height H_s at the toe of the structure and the mean, deep wave period, T_{om}. The formulation became

$$\frac{q}{gH_sT_{om}} = a \exp\left(-b\frac{R_c}{H_s}\sqrt{\frac{s_{om}}{2\pi}}\frac{1}{\gamma_r}\right) \tag{5.15}$$

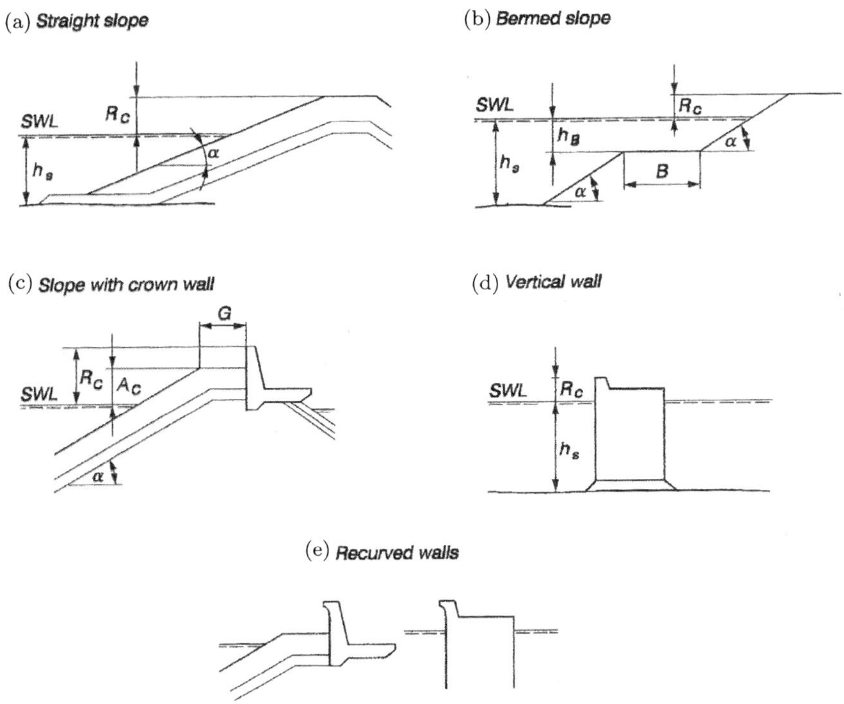

Figure 5.12. Five basic geometrical shapes employed in wave-overtopping formulations (from CEM, 2011).

so that in Equation (5.13), the dimensionless variables are

$$Q = \frac{q}{gH_sT_{om}} \tag{5.15a}$$

$$\text{and } R = \frac{R_c}{H_s}\sqrt{\frac{s_{om}}{2\pi}}\frac{1}{\gamma_r} \tag{5.15b}$$

with γ_r = roughness coefficients as found in Table 5.1, and s_{om} = the wave steepness, H_s/L_{om} using the deepwater, linear theory wave length and T_{om}.

For non-depth limited waves (no breaking waves), the coefficients a and b in Equation (5.15) for straight, smooth slopes originally given by Owen (1982) are given in Table 5.5.

The values presented in Table 5.5 have been slightly revised following tests by Besley (1999) for slopes 1:1 to 1:5 and have been extended to slopes 1:6 to 1:15 by Le Fur *et al.* (2005). See Rock Manual (2007, Table 5.5) for a complete summary of these revised coefficients for straight, impermeable

Table 5.5. Coefficients a and b in Equation (5.15) as originally presented by Owen (1982).

Slope	a	b
1:1	0.008	20
1:1.5	0.010	20
1:2	0.013	22
1.3	0.016	32
1.4	0.019	47

slopes. A full table of coefficients a and b for smooth, berm slopes can be found in CEM (2011, Table VI-5-8).

Note there are no restrictions in Equation (5.15) (as found for wave runup in Figure 5.2 and Equation (5.2)) related to the surf-similarity parameter, ξ_{om}. Hawkes *et al.* (2002) suggested that the Owen's formulation is only applicable for $\xi_{om} < 2.5$. These authors proposed an overtopping reduction factor F, for the following ranges of ξ_{om} with the reduced value for the overtopping rate, simply $q^* = F \times q$.

$$2.5 < \xi_{om} < 3.0, \quad F = 0.3,$$

$$3.0 < \xi_{om} \leq 4.3, \quad F = 0.2,$$

$$\xi_{om} > 4.3, \quad F = 0.1.$$

These overtopping reduction factors will produce a considerable reduction in the overtopping rates. The lower rates of overtopping have been confirmed by van der Meer and Janssen (1995) as discussed in the following.

The dimensionless, wave overtopping rate Q versus dimensionless freeboard R data are shown in Figure 5.13 as presented by van der Meer and Janssen (1995) for a wide range of conditions. The data are divided into two groups by the surf-similarity parameter, ξ_{op} defined by the *peak*, deep-water wave period, T_{op}.

These researchers also employed the exponential form (Equation (5.13)) of the basic equation and now separated into the following two regions:

When $\xi_{\mathrm{op}} < 2$ (Figure 5.13, top),

$$\frac{q}{\sqrt{gH_s^3}}\sqrt{\frac{s_{\mathrm{op}}}{\tan\alpha}} = 0.06\exp\left(-5.2\frac{R_c}{H_s}\frac{\sqrt{s_{\mathrm{op}}}}{\tan\alpha}\frac{1}{\gamma_r\gamma_b\gamma_h\gamma_\beta}\right) \tag{5.16}$$

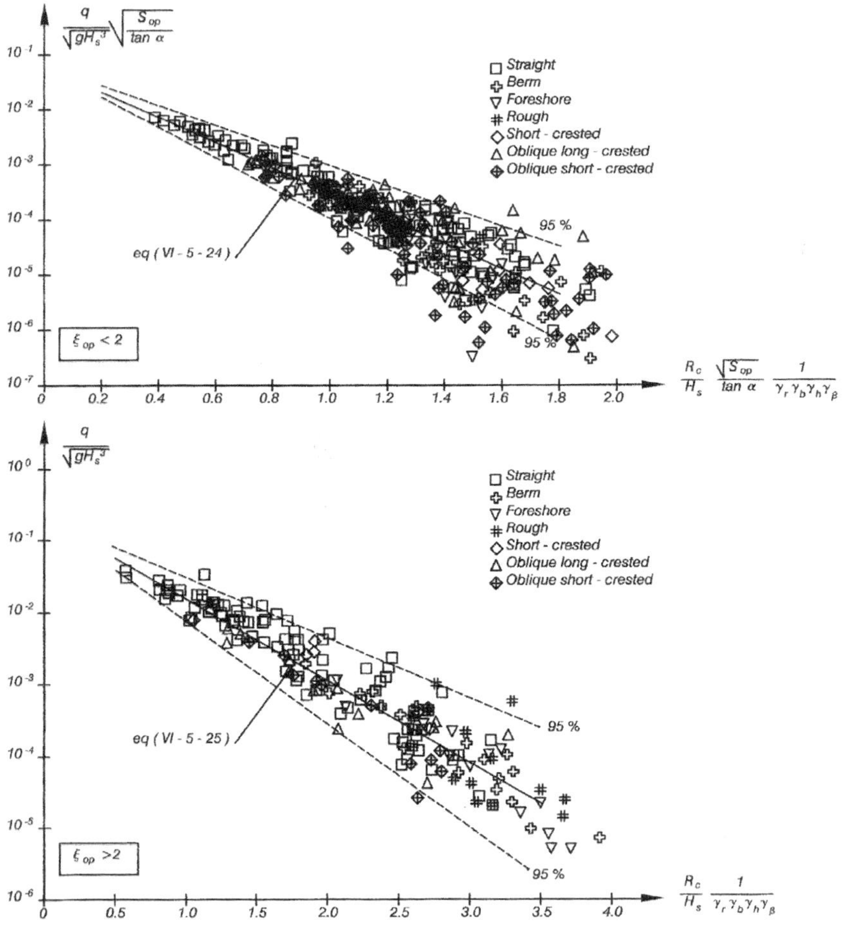

Figure 5.13. Wave-overtopping data (van der Meer and Janssen, 1995) from CEM (2011).

so that

$$Q = \frac{q}{\sqrt{gH_s^3}}\sqrt{\frac{s_{\text{op}}}{\tan\alpha}} \tag{5.16a}$$

and

$$R = \frac{R_c}{H_s}\frac{\sqrt{s_{\text{op}}}}{\tan\alpha}\frac{1}{\gamma_r\gamma_b\gamma_h\gamma_\beta} \tag{5.16b}$$

with the condition that $0.3 < R < 2$. The standard deviation is 0.55 on the coefficient 5.2.

When $\xi_{\text{op}} \geq 2$ (Figure 5.13, bottom),

$$\frac{q}{\sqrt{gH_s^3}} = 0.2 \exp\left(-2.6 \frac{R_c}{H_s} \frac{1}{\gamma_f \gamma_b \gamma_h \gamma_\beta}\right) \tag{5.17}$$

so that

$$Q = \frac{q}{\sqrt{gH_s^3}} \tag{5.17a}$$

and

$$R = \frac{R_c}{H_s} \frac{1}{\gamma_f \gamma_b \gamma_h \gamma_\beta}. \tag{5.17b}$$

As seen in Figure 5.13 bottom, the data scatter is less so that the uncertainty coefficient is 0.35 on coefficient "2.6."

The reduction factors for $\gamma_r, \gamma_h, \gamma_b$ are the same as for wave runup in Section 5.2.2.3. However, the wave direction coefficient, γ_β, now depends on the crestedness of the irregular wave train. For short-crested waves (storms), $\gamma_\beta = 1 - 0.0033\beta$, and for long-crested waves (swell),

$$\begin{aligned} 0^\circ \leq \beta \leq 10^\circ, &\quad \gamma_\beta = 1.0, \\ 10^\circ < \beta \leq 50^\circ, &\quad \gamma_\beta = \cos^2(\beta - 10^\circ), \\ \beta > 50^\circ, &\quad \gamma_\beta = 0.6. \end{aligned} \tag{5.18}$$

The minimum total reduction when combining all γ-factors should be 0.5. In the example shown in Section 5.3.6, the difference between the wave overtopping results of Owen (1980, 1982) and van der Meer and Janssen (1995) will be discussed in detail. Most importantly, the work of van der Meer and Janssen (1995) has been incorporated into the *EurOtop Manual* (Die Küste, 2007) as discussed in detail in Section 5.3.4.

5.3.2.2 *Armored slopes fronting vertical walls*

Two references, Bradbury and Allsop (1988) and Pedersen and Burcharth (1992) studied case (c) in Figure 5.12 for rock-armored, relatively impermeable slopes with a flat berm fronting a vertical seawall (crown wall). Aminti and Franco (1988) included artificial armor units (cube, Tetrapod) in similar efforts. Pedersen (1996) extended this effort to rock-armored, permeable slopes fronting crown walls. Ahrens and Heimbaugh (1988) investigated seven different designs for the sloping revetment and seawall combination.

Formulas for all these efforts can be found in CEM (2011, Table VI-5-7) and are not included here. Further discussion is found in Section 5.3.4 where the results of the *EurOtop Manual* (Die Küste, 2007) are summarized in complete detail.

5.3.3 *Vertical walls*

Wave striking a vertical wall are deflected vertically upward but a significant water volume may overtop the wall as illustrated in Figure 5.14.

Pioneering research efforts in Japan by Goda *et al.* (1975) produced design diagrams for vertical seawalls and rock-armor slopes fronting crown walls which Goda called "block-mound seawalls". However, only the diagrams were published (no empirically derived equations) and only the deep water, unrefracted wave height H'_{os} was employed in the diagrams. Goda (2009) reanalyzed the Japanese laboratory data and developed a design formula using the wave height at the toe of the structure but did not include any reference to the wave periods that produced the results. He included the work of Tamada *et al.* (2002) for smooth, impermeable slopes in his universal formula (Goda, 2010). See Section 5.3.4 for further details.

Figure 5.14. Waves overtopping a vertical seawall.

The Coastal Engineering Manual (CEM, 2011) presents only the research by Franco and Franco (1999) for estimation purposes on vertical seawalls subjected to non-breaking and oblique waves. The equation given is said to be applicable for long-crested and short-crested waves, but the tested data range is only for one value of wave steepness, $s_{\mathrm{op}} = 0.04$ using the peak spectral wave period. The tests however did include a recurved seawall at the crest as depicted in Figure 5.12(e) to reflect the waves seaward and reduce the overtopping rate. This important feature is described in Section 5.3.4.

5.3.4 *The wave EurOtop manual*

Our understanding of wave overtopping has improved significantly as a result of the European Communities' (EC 5th Framework) *Crest Level Assessment of Coastal Structures by Full Scale Monitoring, Neural Network Prediction and Hazard Analysis on Permissible Wave Overtopping* (CLASH, 2004) project, that was summarized in the *EurOtop Manual* (Die Küste, 2007). Extensive research studies involved 12 partners at government laboratories (Germany, Netherlands, and UK) and universities across Europe, and took over five years to finish, and has replaced three previous manuals that were employed for wave overtopping guidance in Germany, the Netherlands, and the UK. Wave overtopping measurements were made at full scale at three coastal sites in Europe with coastal processes simulated in the laboratory wave tanks to understand possible scale effects. It was determined that small-scale laboratory tests with rough, permeable structures underpredicted wave overtopping rates. The *EurOtop Manual* may be downloaded (free) from http://www.overtopping-manual.com/manual.html (January, 2015).

One key result of the CLASH project was that all the worlds' wave overtopping data were reviewed and put on a common baseline in terms of the wave spectral period, $T_{m-1.0}$.

5.3.4.1 *The wave spectral period and surf-similarity parameter*

As stated in the *EurOtop Manual* (Die Küste, 2007),

> *The wave period used for wave runup and overtopping formulae is the spectral period, $T_{m-1.0}$. This period gives more weight to the longer periods in the spectrum and is independent of the spectrum type. In this way, wave overtopping can be easily determined for double-peaked and flattened spectra without the need for other difficult procedures.*

Spectral moments of a wave-energy density spectrum are defined as

$$m_n = \int_{f1}^{fz} f^n S(f)df = nth \text{ moment of the spectral density, and} \quad (5.19)$$

the zeroth moment, m_o becomes with $n = 0$

$$m_0 = \int_{f1}^{fz} f^0 S(f)df = \int_{f1}^{fz} 1 \times S(f)df = \text{area under the spectrum, and}$$

the negative first moment becomes with $n = -1$

$$m_{-1} = \int_{f1}^{fz} f^{-1} S(f)df = \int_{f1}^{fz} \frac{1}{f} S(f)df,$$

so that finally we obtain the definition of the spectral wave period,

$$T_{m-1.0} = \frac{m_{-1}}{m_0}. \quad (5.20)$$

In many cases, the actual wave spectrum is not available, and a single peaked spectrum must be assumed for design with the peak period, T_p, known or specified. In this case, the relationship between $T_{m-1.0}$ and T_p is fixed as $T_{m-1.0} = 0.909\ T_p$. Recall that the ratio of the mean spectral period T_m and the peak spectral period T_p depended on the shape of the single peaked spectrum. For example, for the Pierson–Moskowitz spectrum, $T_m = 0.71$–$0.82\ T_p$. The use of the wave spectral period $T_{m-1.0}$ was then carried forward to define the surf-similarity parameter, $\xi_{m-1.0}$.

$$\xi_{m-1.0} = \frac{\tan\alpha}{\sqrt{S_{mo}}}, \quad (5.21)$$

where $S_{mo} = \frac{H_{mo}}{L_{m-1.0}}$ and $L_{m-1.0} = \frac{g}{2\Pi} T^2_{m-1.0}$.

Working back through the above relationships, we see that the ratio of the peak and spectral period surf-similarity parameters is

$$\frac{\xi_{\text{op}}}{\xi_{m-1.0}} = 1.1 \quad (5.22)$$

and the same as the wave period ratio, $T_p = 1.1\ T_{m-1.0}$.

The three major types of wave breaking on a slope can then be found with the following results:

$$\begin{aligned} &\text{Spilling Breakers:} \quad 0.2 < \xi_{m-1.0} \\ &\text{Plunging Breakers:} \quad 0.2 < \xi_{m-1.0} < 2.3 \text{ and} \\ &\text{Surging Breakers} \quad \xi_{m-1.0} > 2.3. \end{aligned} \quad (5.23)$$

The *EurOtop Manual* divided the analysis of the data and results into three major types of structures: (1) impermeable, sloping revetments and dikes; (2) permeable, armored slopes and mounds; and (3) vertical and steep seawalls. We follow this same order in the following and present the key formulas for probabilistic design.

5.3.4.2 *Impermeable, sloping revetments, and dikes*

The results presented in the *EurOtop Manual* are basically the same as found in the Netherlands' TAW (2002) manual as edited by van der Meer. They are identical to that in Equation (5.16) (van der Meer and Jannsen, 1995) except for the stated range of applicability for $\xi_{m-1.0}$. In addition, the formulas now employ the spectral, significant wave height H_{mo}, and are in terms of $\xi_{m-1.0}$. The results in Equation (5.24) are identical with Equation (5.16) when Equation (5.22) is used to adjust the coefficients. Note also that Equation (5.24) now also includes a berm coefficient, γ_b, outside the exponential function.

For $\xi_{m-1.0} < 5$ (non-breaking and breaking waves),

$$\frac{q}{\sqrt{gH_{mo}^3}} = \frac{0.067\gamma_b}{\sqrt{tan\alpha}}\xi_{m-1.0}\exp\left(-4.75\frac{R_c}{H_{mo}}\frac{1}{\xi_{m-1.0}}\frac{1}{\gamma_r\gamma_b\gamma_h\gamma_\beta}\right). \tag{5.24}$$

The *maximum* value for the wave overtopping rate, q, in this $\xi_{m-1.0} < 5$ range is found from

$$\frac{q}{\sqrt{gH_{mo}^3}} = 0.2\exp\left(-2.6\frac{R_c}{H_{mo}}\frac{1}{\gamma_r\gamma_\beta}\right). \tag{5.25}$$

Note this was Equation (5.17) in van der Meer and Jannsen (1995) for the range $\xi_{op} \geq 2$ and is no longer recommended. This is a broad range for $\xi_{m-1.0} < 5$ that includes non-breaking and breaking waves of the types in Equation (5.23). The standard deviation of the mean $\sigma = 0.5$ is on coefficient 4.75 and 0.35 on coefficient 2.6, respectively. This will be discussed further in Section 5.3.5.

In the *EurOtop Manual*, we now have an entirely new formula for the range $\xi_{m-1.0} > 7$. On shallow, very mild sloping foreshores with heavy wave breaking, the spectrum is very flat with no major peak. Long waves are present that significantly increase the surf-similarity parameter.

For $\xi_{m-1.0} > 7$ (heavy breaking on shallow foreshores),

$$\frac{q}{\sqrt{gH_{mo}^3}} = 0.12 \exp\left(-1.0 \frac{R_c}{H_{mo}} \frac{1}{\gamma_\beta(0.33 + 0.022\xi_{m-1.0})\gamma_r}\right). \tag{5.26}$$

In Equation (5.26), the coefficient 0.12 was determined from a mean value $c = -0.92$ for the exponent in the equation $10^c = 10^{-0.92} = 0.12$ and the standard deviation for c was 0.24. We shall return to this in Section 5.3.5.

For the region $5.0 < \xi_{m-1.0} < 7$, linear interpolation of the results of the above-mentioned regions is recommended. The application of Equations (5.24)–(5.26) will be demonstrated in the examples in Section 5.3.8.

The *EurOtop Manual* also presents many details on how to treat and compute the four additional variables $\gamma_r, \gamma_b, \gamma_h, \gamma_\beta$ in Equation (5.24) which are not covered in this book. Note that only the coefficients γ_r and γ_β appear in Equation (5.26). Zero and negative freeboard conditions are also covered following the efforts of Schüttrumpf (2001). Please refer to the *EurOtop Manual* (Die Küste, 2007) for the special, mean wave overtopping rate formulas for impermeable, sloping revetments and dikes.

5.3.4.3 *Permeable armor layers and mounds*

As expected, the porosity (and roughness) of the permeable armor layers significantly reduces the overtopping rate q, as illustrated in Figure 5.15 (from Die Küste, 2007). The upper solid line (triangles) is for an impermeable, smooth $\cot\alpha = 1.5$ slope using Equation (5.24) and $\gamma_r = 1.0$ and the lower solid line is for the mean of the 10 armor-layer conditions tested, using $\gamma_r = 0.45$ and the following equation for probabilistic design:

$$\frac{q}{\sqrt{gH_{mo}^3}} = 0.2 \exp\left(-2.6 \frac{R_c}{H_{mo}} \frac{1}{\gamma_r \gamma_\beta}\right). \tag{5.27}$$

This is identical to Equation (5.25) and the maximum for impermeable slopes where the relative crest height R_c/H_{mo} approaches zero.

Armor-layered structures have steep slopes of $\cot\alpha = 1.5$ or 2.0 so that the slope is not a factor in Equation (5.27). But the roughness coefficient γ_r is certainly important, and Table 5.6 presents the results from the CLASH project. Two-layer systems (open symbols in Figure 5.15) have more pore space and less wave overtopping.

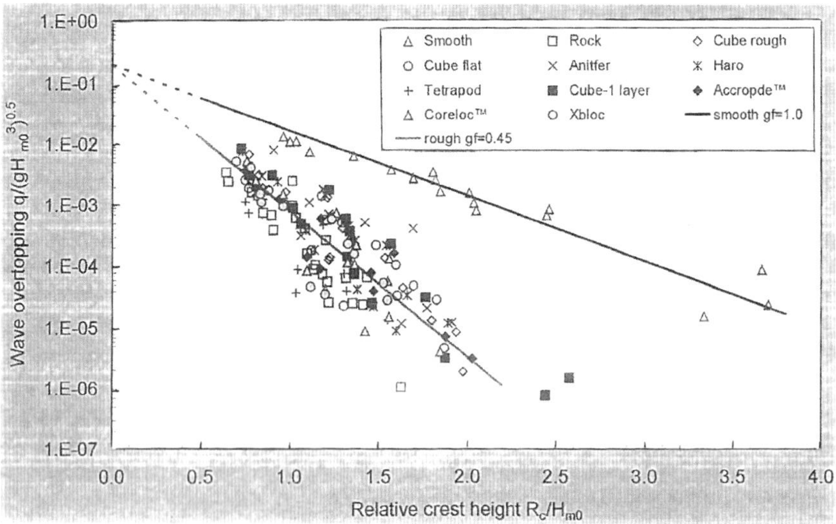

Figure 5.15. Comparison of mean wave overtopping rates for impermeable, smooth (top, triangles,) and permeable, rough armor layer (rubble-mound) slopes, cot $\alpha = 15$ (from Die Küste, 2007).

The *EurOtop Manual* gives further details for the effects of armor-crest width, oblique wave attack, composite slopes, reshaping or berm-type breakwaters, all of which are not treated herein.

But many armored structures also have a wall on the crest (aka, crown wall, capping wall, wave wall, crest unit) as illustrated in Figure 5.12. The purpose is to end the armor layer or to provide an access road on the breakwater or to reduce the wave overtopping rate. The available test data for this case were found to fall within the scatter of the data for armored structures as displayed in Figure 5.15. Consequently, it was concluded that when the wall height is greater than the crest elevation of the structure, simply use the wall height to determine R_c and apply Equation (5.27) to determine the mean wave overtopping rate.

5.3.4.4 *Vertical and steep seawalls*

This is the most complicated case. The wave (and water-level) conditions in front of the wall produce three major differences in the wave processes striking the wall that in turn create different wave overtopping regimes. Criteria have been established to discriminate these different wave overtopping regimes.

Table 5.6. Roughness coefficient γ_r values for rock-armored structures with cot $\alpha = 1.5$ slopes (from Die Küste, 2007, Table 6.2). Note that if the core of the structure is impermeable, the roughness coefficient is higher giving more overtopping.

Type of Armor Layer	γ_r
Smooth impermeable surface	1.00
Rocks (One layer, impermeable core)	0.60
Rocks (One layer, permeable core)	0.45
Rocks (Two layers, impermeable core)	0.55
Rocks (Two layers, permeable core)	0.40
Cubes (One layer, random positioning)	0.50
Cubes (Two layers, random positioning)	0.47
Antifers	0.47
HARO's	0.47
Accropode™	0.46
Xbloc®	0.45
CORE-LOC®	0.44
Tetrapods	0.38
Dolosse	0.43

Non-impulsive (non-breaking, pulsating) wave conditions occur in relatively deep water for low steepness wave conditions. The wave overtopping the structure is "green water" overtopping. Conversely, "impulsive" wave conditions are when a wave breaks directly (violently) onto the wall and is the result of larger waves in shallow, local water depths at the toe of the structure. White water conditions characterize the overtopping. A third scenario is when wave breaking occurs seaward of the wall and broken waves impact the structure. The horizontal wave forces on the wall vary considerably for these three cases and will be further discussed in Chapter 6.

Criteria are needed to distinguish between the "non-impulsive" and "impulsive" regimes which are the dominant scenarios for a given structure and design sea state. For plain vertical walls, the dimensionless criterion h_* is

$$h_* = 1.35 \frac{d_{\text{toe}}}{H_{mo}} \frac{2\pi h_s}{gT_{m-1.0}^2}, \tag{5.28}$$

where d_{toe} is the water depth at the toe of the vertical wall, h_s is the water depth seaward of a possible rubble-mound in front of the structure, and all other terms as previously defined. Note that if there is a plain vertical wall with no rubble-mound in front of the wall, then $d_{\text{toe}} = h_s$. If a rubble-mound is present in front of the wall, then d_* replaces h_* in Equation (5.28). This configuration is termed a "composite vertical wall" in the *EurOtop manual.*

(1) Plain vertical walls ($d_{toe} = h_s$)

"Non-impulsive" (pulsating) regime, $h_* > 0.3$:

For probabilistic design, the formula to estimate the mean wave overtopping rate for "non-impulsive" (pulsating) conditions becomes

$$\frac{q}{\sqrt{gH_{mo}^3}} = 0.04 \exp\left(-2.6\frac{R_c}{H_{mo}}\right), \tag{5.29}$$

when $h_* > 0.3$ and is applicable for the region $0.1 < \frac{R_c}{H_{mo}} < 3.5$. With zero freeboard, $\frac{q}{\sqrt{gH_{mo}^3}} = 0.062$. The standard deviation of the mean $\sigma = 0.8$ on coefficient 2.6.

"Impulsive" (breaking) regime, $h_* \leq 0.2$:

For probabilistic design, the formula to estimate the mean wave overtopping rate for "impulsive" (breaking) conditions becomes

$$\frac{q}{h_*^2\sqrt{gh_s^3}} = 1.5 \times 10^{-4}\left(h_*\frac{R_c}{H_{mo}}\right)^{-3.1}, \tag{5.30}$$

when $h_* \leq 0.2$ and is applicable for the region $0.03 < h_*\frac{R_c}{H_{mo}} < 1.0$. No data are available for the zero-freeboard condition.

The standard deviation of the mean σ = unknown on the coefficient 1.5×10^{-4}.

For conditions when $0.2 \leq h_* < 0.3$, then both Equations (5.29) and (5.30) should be employed and the larger value is used in design.

For broken waves reaching the wall or when the toe of the wall is visible above the SWL, see the *EurOtop Manual* (Die Küste, 2007) for additional formulas. In both cases, less overtopping results.

Plain but non-vertical (battered) seawalls have higher overtopping rates. For impulsive conditions, it is recommended to increase the results by 30% for near-vertical walls with a 10:1 slope. For lower slopes, the equations for impermeable, smooth, sloping revetments should be employed as discussed in Section 5.3.2.

(2) Vertical walls with rubble-mound structures in front ($d_{toe} < h_s$)

Rock mounds fronting vertical walls are common to protect from toe scour and they trip the waves to alter the wave overtopping rates. Now, the water depth at the toe of the vertical wall d_{toe} is less than the seaward water depth h_s, and the dimensionless criterion becomes d_* in Equation (5.28).

"Non-impulsive" (pulsating) regime, $d_* > 0.3$: For this "non-impulsive" regime and $d_* > 0.3$, use the same Equation (5.29) as for the plain, vertical wall case.

"Impulsive" (breaking) regime, $d_* \leq 0.2$: For the "impulsive" (breaking) regime when $d_* \leq 0.2$, use

$$\frac{q}{d_*^2\sqrt{gh_s^3}} = 4.1 \times 10^{-4}\left(d_*\frac{R_c}{H_{mo}}\right)^{-2.9} \tag{5.31}$$

and valid for the region, $0.05 < d_*\frac{R_c}{H_{mo}} < 1.0$ and $h_* < 0.3$.

The standard deviation of the mean σ = unknown on the coefficient 4.1×10^{-4}.

Again, for the transition region $(0.2 < d_* < 0.3)$ use the larger result for design.

(3) Parapet, wave-return crests (bullnose) on vertical seawalls

The presence of an overhanging parapet (aka, wave return, bullnose, recurve, see Figure 5.12(e)) at the crest of a vertical seawall may significantly reduce the wave overtopping rate. Waves striking the wall and water rushing upward is deflected back seaward. Research by Kortenhaus *et al.* (2002) and Pearson *et al.* (2004) using physical model studies forms the basis for the formulas discussed in the following. For a future facing the increase of sea levels and water depths with less freeboard at existing seawalls, retrofitting the crest with an overhanging parapet may be a relatively economic alternative for managing the overtopping rate.

Defining k as the ratio of the overtopping rate with the parapet to the overtopping rate without the parapet, all other things being equal,

$$k = \frac{q_{\text{with parapet}}}{q_{\text{without parapet}}}. \tag{5.32}$$

Kortenhaus *et al.* (2002) studied plain and composite (with rubble-mound) walls, non-impulsive and impulsive wave conditions at two different model scales. An example for a given overhanging parapet geometry for k as a function of the relative freeboard R_c/H_s is shown in Figure 5.16. The denominator is the significant wave height H_s, but could be the spectral value H_{mo}. In some cases, a three-fold reduction in the overtopping rate was measured relative to that for plain walls. In some tests, k was below 0.05. Kortenhaus *et al.* (2002) divided the results into three regions

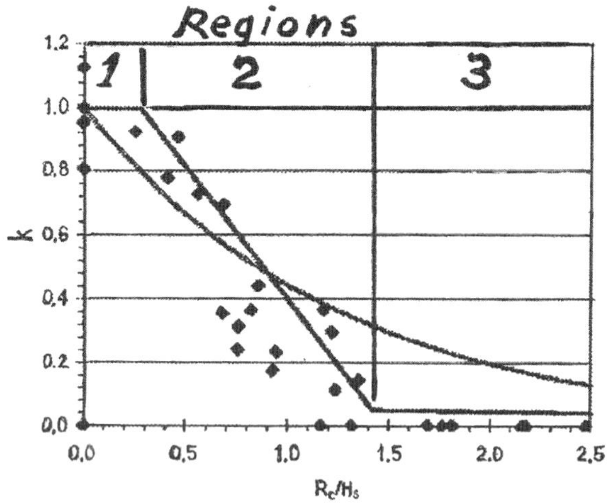

Figure 5.16. The parapet factor, k, as determined for one parapet geometry and one wall configuration as a function of the relative freeboard, R_c/H_s, showing three analysis regions (modified from Pearson *et al.*, 2004).

of the relative freeboard and employed the definitions of all the variables as illustrated in Figure 5.17. The key parapet variables are P_c, the vertical distance from the SWL to the base of the parapet; h_r, the height of the parapet; and B_r, the width of the parapet. For overhanging parapets, the angle $\alpha < 90°$.

Pearson *et al.* (2004) summarized the k-reduction factors for the three regions in a "decision chart" for what was labeled "tentative guidance" and Figure 5.18 is a modified version. Row 6 at the bottom of the chart shows the dimensionless variables R_o^*, m, and m' needed in the k-factor formulas given earlier. Note the recommended value for the variable k_{23} is 0.2.

Row 2 shows the relative freeboards R_c/H_{mo} that divide the data into the three regions, 1, 2, and 3 for calculation of the k-factor. In region $1, k = 1.0$ (Row 3) where the overtopping rate with the parapet was little or no different than with no parapet. In region 2 (transition, outlined in red), the formula for the k-factor is given as listed in Row 3 and depends on R_o^* and m.

Region 3 (outlined in black) is broken into three subregions (Row 4) that depend on another relative freeboard defined as R_c/h_s. This region is when the freeboard is large and the k-factors are very small. The k-factor

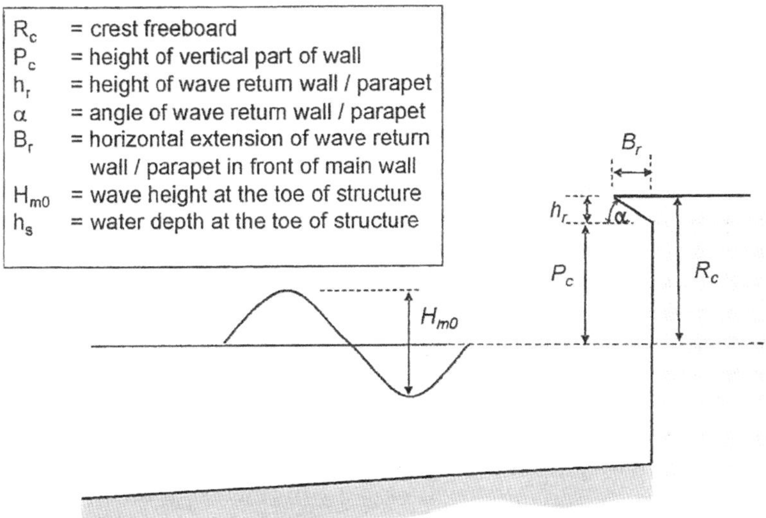

Figure 5.17. Definition sketch of variables in parapet formulas for calculation of k-factor (from Die Küste, 2007, *EurOtop Manual)*. The water depth h_s at the toe of the structure is not shown.

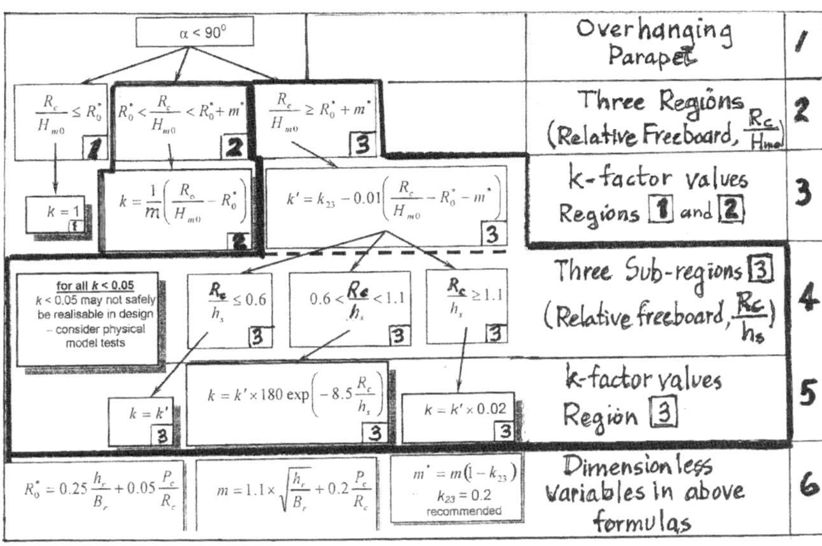

Figure 5.18. A "decision chart" to estimate the k-factors for overhanging parapets at the crest of vertical walls (adapted from Pearson *et al.*, 2004). The k-factor equations are assigned as Equation (5.33).

formulas for these three subregions of region 3 are shown in Row 5. When these formulas determine that $k < 0.05$, the chart cautions these results based on the physical model tests may not be safe. The *EurOtop Manual* (Die Küste, 2007) recommends that detailed, physical model studies should be conducted if values of k below 0.05 are necessary for the design. See Section 5.3.7 for guidance on tolerable wave overtopping limits. Example calculations are found in Section 5.3.8.

Cornett *et al.* (1999) presents results for when the wall is chamfered backwards at the crest ($\alpha > 90°$) which are not included herein. Greater wave overtopping occurs.

5.3.5 *Probabilistic design*

There are no partial safety factors (known to the author) for the loads and the resistance variables to be incorporated in the above formulas for calculation of wave overtopping rates at some specified level of probability of failure, P_f. The work to accomplish this has yet to be attempted. Consequently, all that is available for some level of safety are the values of the standard deviations σ, for the mean values of the key constants in the formulas presented.

For example, consider Equations (5.24) and (5.25) for impermeable, sloping revetments when $\xi_{m-1.0} < 5$ and repeated here as follows:

$$\frac{q}{\sqrt{gH_{mo}^3}} = \frac{0.067\gamma_b}{\sqrt{\tan\alpha}}\xi_{m-1.0}\exp\left(-4.75\frac{R_c}{H_{mo}}\frac{1}{\xi_{m-1.0}}\frac{1}{\gamma_r\gamma_b\gamma_h\gamma_\beta}\right) \quad (5.24)$$

and for the maximum value,

$$\frac{q}{\sqrt{gH_{mo}^3}} = 0.2\exp\left(-2.6\frac{R_c}{H_{mo}}\frac{1}{\gamma_r\gamma_\beta}\right). \quad (5.25)$$

The standard deviation σ is 0.5 for coefficient -4.75 in Equation (5.24) and 0.35 on coefficient -2.6 in Equation (5.25). These are the coefficients for the mean, central tendency curve shown as the solid line in Figure 5.19 (from Die Küste, 2007). At the 90% confidence level, in other words, ±5% about the mean, central tendency curve, the dashed +5% and −5% dotted lines in Figure 5.19 display how many data points fell outside the 90% confidence interval (C.I.) assuming a normal, Gaussian distribution for the data. For the normal distribution, 90% of the area beneath the curve is for

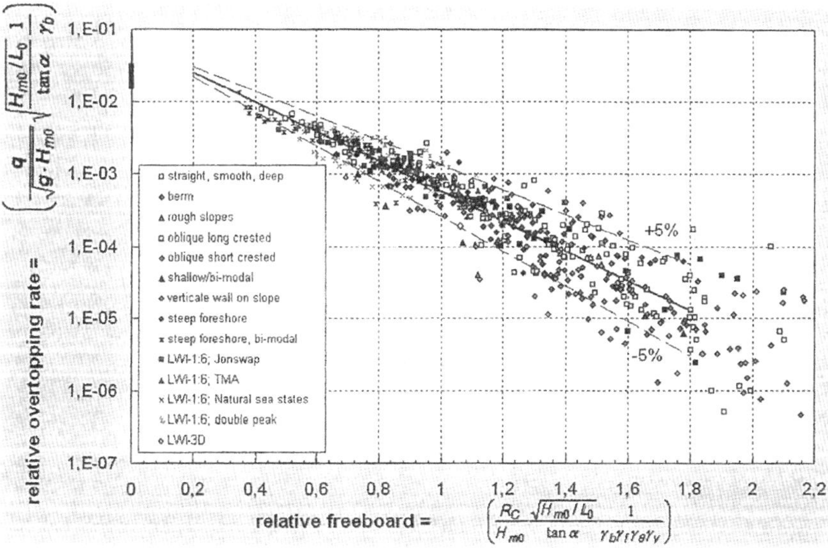

Figure 5.19. Wave overtopping data showing central tendency (solid) curve for Equation (5.24) and ±5% exceedance limit curve (dotted) at the 90% C.I. (from CEM, 2006).

the mean value, $\mu \pm z\sigma$ ($z = 1.645$) applied to the mean coefficients, -4.75 and -2.6 in Equations (5.24) and (5.25), respectively. Therefore,

$$-4.75 + 1.645\ (0.5) = -3.9275 = -3.9\ \text{(rounded), and}$$

$$-2.6 + 1.645\ (0.35) = -2.02425 = -2.0\ \text{(rounded)}.$$

At the upper (conservative) limit for the 90% C.I., Equations (5.24) and (5.25) become

$$\frac{q}{\sqrt{gH_{mo}^3}} = \frac{0.067\gamma_b}{\sqrt{\tan\alpha}}\xi_{m-1.0}\exp\left(-3.9\frac{R_c}{H_{mo}}\frac{1}{\xi_{m-1.0}}\frac{1}{\gamma_r\gamma_b\gamma_h\gamma_\beta}\right) \tag{5.34}$$

and for the maximum value,

$$\frac{q}{\sqrt{gH_{mo}^3}} = 0.2\exp\left(-2.0\frac{R_c}{H_{mo}}\frac{1}{\gamma_r\gamma_\beta}\right). \tag{5.35}$$

These equations should be employed for safety considerations in design.

Other values of the multiplier z are possible in $\mu \pm z\sigma$ to determine the area beneath the normal distribution curve. For example, taking $z = 1.0$ gives $\mu \pm \sigma$ or 68.3% of all the data. This level of safety was recommended

Table 5.7. Adjusted coefficients in the wave overtopping equations at the 90% confidence level.

		Mean Coefficient		Std Dev.	90%	1.645σ
Structure Type	Eqn. No.	a	b	1.0σ	a	b
Sloping revetment, $\xi_{m-1.0} < 5$	5.24	0.067	−4.75	0.5	0.067	−3.9
Maximum, $\xi_{m-1.0} < 5$	5.25	0.2	−2.6	0.35	0.2	−2.0
Sloping revetment, $\xi_{m-1.0} > 7$	5.26	0.12	−1.0	0.24 on c	0.30	−1.0
Permeable armor layer	5.27	0.2	−2.6	0.35	0.2	−2.0
Vert. wall, plain, non-impulsive	5.29	0.04	−2.6	0.8	0.04	−1.3
Vert. wall, plain, impulsive	5.30	1.5×10^{-4}	—	UNK	UNK	UNK
Vert. wall, comb, non-impulsive	5.29	0.04	−2.6	0.8	0.04	−1.3
Vert. wall, comb, impulsive	5.31	4.1×10^{-4}	—	UNK	UNK	UNK
Vertical wall, parapet	Figure 5.18	—	—	UNK	UNK	UNK

Note: UNK — unknown.

in (Die Küste, 2007, *EurOtop Manual*) but is less conservative for safety assessment. All the equations presented under the heading "deterministic design or safety assessment" in the *EurOtop Manual* are actually for the coefficients that have been modified by $+1.0\ \sigma$. In addition, the words "probabilistic" and "deterministic" are employed with a different meaning in the *EurOtop Manual* than used throughout this book.

The structure types, equation numbers, mean coefficients, standard deviations, and adjusted coefficients at the 90% C.I. for all the equations sited above are summarized in Table 5.7. Determination for the adjusted coefficients follows the above example in all cases except for the sloping revetment when $\xi_{m-1.0} > 7$ in Equation (5.26). The mean for coefficient "a" was obtained from $10^c = 10^{-0.92} = 0.12$. The standard deviation for the exponent coefficient "c" is 0.24. Therefore, $10^c = 10^{-0.92+(1.645\times0.24)} = 10^{-0.525} = 0.293 = 0.3$ (rounded) as shown in Table 5.7. Unfortunately, the standard deviations of the mean σ are unknown for Equations (5.30) and (5.31) and for the k-factor Equation (5.33) in Figure 5.18.

Finally, all the above is directed toward estimation of the mean, wave overtopping discharge rate q. The significance of overtopping *volumes* of individual waves is also of interest.

5.3.6 *Wave-overtopping volumes*

Damaging wave impacts on coastal structures, people, and vehicles are mostly caused by individual wave volumes V and not by the time-averaged, wave overtopping rate, q. As should be expected, the formulations for the individual wave volumes are different for sloping and vertical structures.

5.3.6.1 *Sloping revetments and dikes*

The probability per wave for overtopping volume P_{ov} is assumed to follow a Rayleigh distribution (Chapter 2, Section 2.3.3) for wave runup. Using the runup $R_{u2\%}$ as the scaling variable, the *EurOtop Manual* gives

$$P_{\text{ov}} = \exp\left[-\left(\sqrt{-\ln 0.02}\frac{R_c}{R_{u2\%}}\right)^2\right], \tag{5.36}$$

to estimate $P_{\text{ov}} = N_{ow}/N_w$, where N_w is the number of incoming waves and N_{ow} is the number of overtopping waves.

The overtopping volume of individual waves V can then be described by a two-parameter Weibull distribution as introduced in Chapter 2, Section 2.3.8. The coefficients a and b are "shape" and "scale" factors, respectively. For overtopping volumes, they were first presented by Franco *et al.* (1994) and van der Meer and Janssen (1995).

The exceedance probability $P_v = P(v > V)$ of wave overtopping volume is

$$P_v = 1 - \exp\left[-\left(\frac{V}{a}\right)^b\right], \tag{5.37}$$

where V is the specified overtopping volume (per wave) per unit width (m^3/m).

$$a = 0.84\frac{T_m q}{P_{\text{ov}}} = 0.84\ T_m\ q\frac{N_w}{N_{ow}} = C\ T_m\ q\ \frac{N_w}{N_{ow}},$$
$$b = 0.75 \quad \text{and} \quad C = 0.84, \tag{5.38}$$

where T_m is the mean wave period (seconds), q is the average overtopping discharge rate per unit width (m^3/sec per meter), and P_{ov} is the probability of overtopping per individual wave (Equation (5.36)).

Solving Equation (5.37) for the overtopping volume per wave for a specified probability of exceedance, P_v,

$$V = a\left[-\ln(1 - P_v)\right]^{1/b}. \tag{5.39}$$

To determine the overtopping volume per wave requires knowledge of $R_c, R_{u2\%}, q, T_m$, the probability of exceedance level P_v of interest for design, and the duration of the storm to find N_w. For the maximum overtopping volume in a storm of specified duration, the number of overtopping waves, N_{ow} can be used in Equation (5.39) to estimate the maximum overtopping volume, $V_{\max}$ as

$$V_{\max} = a[\ln(N_{ow})]^{1/b}. \tag{5.40}$$

5.3.6.2 *Plain vertical walls*

As expected, the formulations depend on whether the waves are of the non-impulsive (pulsating) or impulsive (breaking) type so that Equation (5.28) must first be determined to find, h_*. Then, the major difference with that mentioned above for sloping structures is for calculation of the probability of overtopping volume per wave, P_{ov}.

Non-impulsive (pulsating) waves ($h_* > 0.3$):

Franco *et al.* (1994) used the same linkage above (runup related to the Rayleigh Distribution) for sloping structures and determined that for $h_* > 0.3$, the probability of the overtopping volume per wave is

$$P_{\text{ov}} = \exp\left[-1.21\left(\frac{R_c}{H_{mo}}\right)^2\right]. \tag{5.41}$$

Then Franco *et al.* (1996) determined a weak dependency on the wave steepness for the shape parameter a and the scale parameter b. Using the spectral period $T_{m-1.0}$ to replace T_m and to evaluate the coefficient C in Equation (5.38),

for $s_{m-1.0} = 0.02$, $C = 0.74$ in Equation (5.38) to calculate coefficient a, and $b = 0.66$.

for $s_{m-1.0} = 0.04$, $C = 0.90$ in Equation (5.38) to calculate coefficient a, and $b = 0.0.82$.

Use linear interpolation for other $s_{m-1.0}$ values.

Impulsive (breaking) waves ($h_* < 0.3$):

Besley (1999) and Pearson *et al.* (2002) found that for impulsive wave breaking, a better formula for P_{ov} is

$$P_{\text{ov}} = 0.031\frac{H_{mo}}{R_c}\frac{1}{h_*} \tag{5.42}$$

with the coefficients $C = 0.92$ in Equation (5.38) to calculate coefficient a and $b = 0.85$.

The formulas to calculate P_v (Equation (5.37)), a (Equation (5.38), using correct C coefficient and $T_{m-1.0}$), V (Equation (5.39)) and $V_{\max}$ (Equation (5.40)) remain unchanged for both non-impulsive and impulsive conditions. An example is found in Section 5.3.8.

See (Die Küste, 2007, *EurOtop Manual*) for further details for composite walls and oblique waves. Information on overtopping currents and erosion of landward slopes is found in Chapter 6.

5.3.7 *Guidance on tolerable wave-overtopping limits*

Relatively crude guidance for the tolerable limits of wave overtopping mean discharges and maximum volumes is available for humans, vehicles, property, and coastal structures. A wide variety of situations are possible, and the guidelines are for locations at or behind the crest. Table 5.8 gives some limits and ranges for safety (or damage) for the specific conditions cited and for the "notes" that accompany these conditions. Slight changes in the specified conditions may significantly modify the risks. The unit for mean discharge q is liters per second per unit meter of the structure width (l/s/m) and that for maximum volume, $V_{\max}$ is liters per meter (l/m).

The researchers involved in the CLASH project employed previous guidance and their own observations and measurements to develop the tolerable overtopping limits summarized in Table 5.8 as adapted from Die Küste (2007, *EurOtop Manual*) for both mean discharge q and maximum volume $V_{\max}$.

Please consult the *EurOtop Manual* for more details about Table 5.8. For example, for (a) pedestrians: with no clear view of incoming waves, or if not properly dressed and can get wet, or if easily frightened and upset, or on a narrow walkway to trip and fall, a limit of $q \leq 0.03$ (l/s/m) may be more appropriate. For (b) if vehicles: the upper limit in the listed range applies to non-impulsive (pulsating) flows across the roadway, whereas the lower limit is for pulsating (breaking) flows at some elevation above the roadway

Table 5.8. Tolerable mean discharges and maximum volumes (from *EurOtop Manual*).

a. Limits for Overtopping for Pedestrians	Mean Discharge q (l/s/m)	Max Volume[1] $V_{\max}$(l/m)
Trained staff well-shod and protected, expecting to get wet, overtopping flows at lower levels only, no falling jet, low danger of fall from walkway	1–10	500 at low level
Aware pedestrian, clear view of the sea, not easily upset or frightened, able to tolerate getting wet, wider walkway[2]	0.1	20–50 at high level or velocity

Notes: [1]These limits relate to overtopping velocities well below $v_c \approx 10\,m/s$. Lower volumes may be required if the overtopping process is violent and/or overtopping velocities are higher.
[2]Not all of these conditions are required, nor should failure of one condition on its own require the use of a more severe limit.

b. Limits for Overtopping for Vehicles	Mean Discharge q (l/s/m)	Max Volume $V_{\max}$ (l/m)
Driving at low speed, overtopping by pulsating flows at low flow depths, no falling jets, vehicle not immersed	10–50[1]	100–1,000
Driving at moderate or high speed, impulsive overtopping giving falling or high velocity jets	0.01–0.0.5[2]	5–50[2] at high level or velocity

Notes: [1]These limits probably relate to overtopping defined at highway.
[2]These limits relate to overtopping defined at the defence, but assumes the highway to be immediately behind the defence.

c. Limits for Overtopping for Property Behind the Defence	Mean Discharge q (l/s/m)	Max Volume $V_{\max}$ (l/m)
Significant damage or sinking of larger yachts	50	5,000–50,000
Sinking small boats set 5–10 m from wall. Damage to larger yachts	10[1]	1,000–10,000
Building structure elements	1[2]	—
Damage to equipment set back 5–10 m	0.4[1]	—

Notes: [1]These limits relate to overtopping defined at the defence.
[2]This limit relates to the effective overtopping defined at the building.

(*Continued*)

Table 5.8. (*Continued*)

d. Limits for Overtopping for Damage to the Defence Crest or Rear Slope	Mean Discharge q (l/s/m)
Embankment seawalls/sea dikes	
No damage if crest and rear slope are well protected	50–200
No damage to crest and rear face of grass covered embankment of clay	1–10
No damage to crest and rear face of embankment if not protected	0.1
Promenade or revetment seawalls	
Damage to paved or armored promenade behind seawall	200
Damage to grassed or lightly protected promenade or reclamation cover	50

with unexpected suddenness of the event. Little data were available for the limits in (c) property behind the coastal defense structure. The limits in (d) coastal structures and rear slopes is mainly from early work in Japan (Goda *et al.*, 1975; Fukuda *et al.*, 1973). The results of damage to the rear slopes of coastal dikes using a wave overtopping simulator in the field as developed in the Netherlands (van der Meer *et al.*, 2006) will be discussed in detail in Chapter 6.

The US Army Corps of Engineers has summarized the published literature (1968–1994) as shown in Table 5.9 (see CEM, 2011, Part VI, Section 5-5, p. VI-5-23). For pedestrians and vehicles, the unsafe and dangerous regions are shaded but little detail is presented. Property (buildings) is listed under structural safety and the potential damage region is shaded. Three types of coastal structures are identified and again, the potential damage region is shaded and tolerable limits for mean discharge rate q are much higher than for people, vehicles, and buildings. The limits on overtopping volumes are not given.

It is not possible to combine Tables 5.8 and 5.9 because most of the detailed subcases mentioned in each table are not the same. Because it is based on recent studies in the CLASH project, the tolerable limits for wave overtopping in Table 5.8 are recommended.

The US FEMA has published *Guidelines and Specifications for Flood Hazard Mapping Partners* (FEMA, 2007) that includes suggestions for interpretation of mean wave overtopping rates to establish flood insurance risk zones (see FEMA, 2007, Table D.2.8-6). Table 5.10 summarizes these "suggestions for interpretation" of the mean overtopping rate q in both the English and Metric systems.

Table 5.9. Tolerable mean discharges (from CEM, 2011).

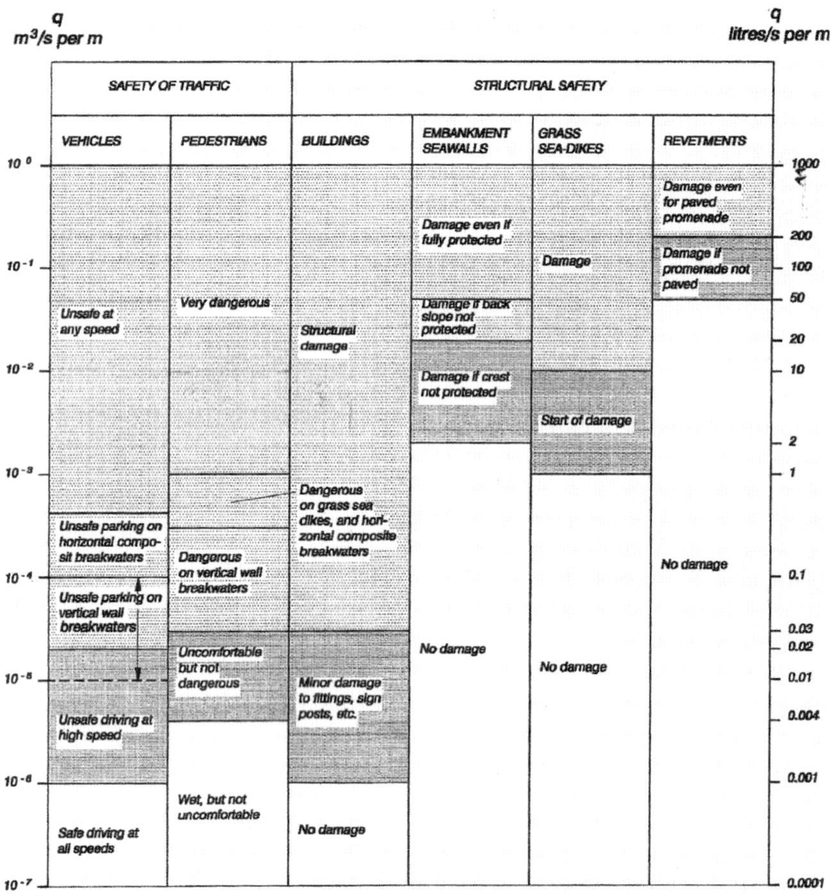

The flood-risk zones extend from (1) the area outside the 1% and 0.2% annual-chance floodplains (Zone X, minimal risk, no flood insurance required), to (2) seaward through areas subject to inundation by the 1% annual-chance shallow flooding (Zone AO (1–3 ft, (30–94 cm)), high risk, flood insurance required), to (3) coastal waters subject to the 1% annual-chance flood elevations plus wave setup and high waves and currents (Zone VE, high risk, federal floodplain management regulations and mandatory flood insurance purchase requirements apply). Flood insurance rates vary across the flood zones. See FEMA (2007) for complete details.

Table 5.10. Suggestions for interpretation of mean overtopping rates (modified from FEMA, 2007).

Flood Insurance Risk Zone Behind Barrier	Order of Magnitude for q	Order of Magnitude for q
	English Units, cfs/ft	Metric Units. l/s/m
Zone X	<0.0001	<0.009
Zone AO — 1 foot (30 cm) depth	0.0001–0.01	0.009–0.9
Zone AO — 2 feet (62 cm) depth	0.01–0.1	0.9–9.3
Zone AO — 3 feet (94 cm) depth	0.1–1.0	9.3–93
Zone VE, 30 ft (9 m) width (elevation 3 ft (1 m) above barrier crest)	>1.0	>93

The limits given in Table 5.10 are much lower than shown in Tables 5.8 and 5.9. For example, for pedestrians at the structure crest, both Tables 5.8 and 5.9 give a limit of 0.03 l/s/m as perhaps uncomfortable but not dangerous, whereas the FEMA criteria in Table 5.10 says 0.009 l/s/m for Zone X. No reference supporting the numbers in Table 5.10 is given in FEMA (2007).

5.3.8 *Example calculations*

5.3.8.1 *Deterministic design: Given data*

Given the same data as for the wave runup example in Section 5.2.5 as shown in Figure 5.9. A relatively mild foreshore approaches a plain, sloping (1.0 vertical: 2.5 horizontal; (1 on 2.5)) structure. At the design SWL, the water depth at the toe of the structure is 3.0 m (9.8 ft). The design significant wave height $H_s = 2.0$ m (6.6 ft) and peak wave period $T_p = 8$ s were measured at a wave gauge located in $d = 4.5$ m (14.8 ft). Assume the design waves are at right angles to the structure. The formulas for determination of wave overtopping need calculation of the surf-similarity parameters, ξ_m, $\xi_{m-1.0}$, and ξ_p using the significant wave height H_s at the toe of the structure. Reverse shoaling the measured wave to deep water gave the unrefracted, deep-water wave height, $H'_{so} = 1.920$ m (6.3 ft) and shoaling the measured wave to the toe of the structure gave, $H_{\text{stoe}} = 2.161$ m (7.09 ft.)

Of interest are the wave overtopping rate q and maximum overtopping volume $V_{\max}$ when the crest distance $R_c = 2.0$ m (6.56 ft) above the SWL for both impermeable (smooth and rough) structures and for permeable, rubble-mound structures. Take $\gamma_b = \gamma_h = \gamma_\beta = 1.0$.

5.3.8.2 *Surf-similarity parameters*

For the spectral peak wave period, $T_p = 8$ s, $\xi_{\text{op}} = 2.72$. And, for the mean wave period $T_m = 6.1$ s, it was found that $\xi_{om} = 2.07$. From Equation (5.22), it was learned that $\xi_{m-1.0} = \xi_{\text{op}}/1.1 = 2.72/1.1 = 2.47$.

5.3.8.3 *Impermeable, plain slope structure*

For impermeable, sloping revetments and dikes when $\xi_{m-1.0} < 5$, the *EurOtop Manual* (Equation (5.24)) gives for a smooth slope ($\gamma_r = 1.0$).

$$\frac{q}{\sqrt{gH_{mo}^3}} = \frac{0.067\gamma_b}{\sqrt{\tan\alpha}}\xi_{m-1.0}\exp\left(-4.75\frac{R_c}{H_{mo}}\frac{1}{\xi_{m-1.0}}\frac{1}{\gamma_r\gamma_b\gamma_h\gamma_\beta}\right),$$

$$q = \sqrt{gH_{mo}^3}\frac{0.067\gamma_b}{\sqrt{\tan\alpha}}\xi_{m-1.0}\exp\left(-4.75\frac{R_c}{H_{mo}}\frac{1}{\xi_{m-1.0}}\frac{1}{\gamma_r\gamma_b\gamma_h\gamma_\beta}\right)$$

$$= \sqrt{9.81(2.16)^3}\frac{0.067\times 1.0}{\sqrt{1/2.5}}2.47\exp\left(-4.75\frac{2.0}{2.16}\frac{1}{2.47}\frac{1}{1\times 1\times 1\times 1}\right),$$

where $q = 0.4403\,\text{m}^3/\text{s/m}$. (5.24)

For a rough slope taking $\gamma_r = 0.55$ gives $q = 0.1028\,\text{m}^3/\text{s/m}$.

These results are identical to that given by van der Meer and Janssen (1995) using Equation (5.16) for $\xi_{\text{op}} < 2$ even though outside this range ($\xi_{\text{op}} = 2.72$). Using the formulation by Owen (1980,1982) in Equation (5.15) gave $q = 0.3142\,\text{m}^3/\text{s/m}$ for the smooth slope. Use of Equation (5.24) from the *EurOtop Manual* is recommended to estimate wave overtopping rates on impermeable slopes.

5.3.8.4 *Permeable, rubble-mound structures*

For permeable armor layers and mounds, the *EurOtop Manual* gave Equation (5.27) for $\cot\alpha = 1.5$.

$$\frac{q}{\sqrt{gH_{mo}^3}} = 0.2\exp\left(-2.6\frac{R_c}{H_{mo}}\frac{1}{\gamma_r\gamma_\beta}\right). \tag{5.27}$$

As shown, the structure slope is not a factor since this equation was derived from 10 test conditions all with the same slope. Assume two layers of rock with an impermeable core ($\gamma_r = 0.55$ in Table 5.5). Applying Equation (5.27) for the example data gives

$$q = \sqrt{gH_{mo}^3}\, 0.2 \exp\left(-2.6\frac{R_c}{H_{mo}}\frac{1}{\gamma_r \gamma_\beta}\right),$$

$$q = \sqrt{9.81(2.16^3}\, 0.2 \exp\left(-2.6\frac{2.0}{2.16}\frac{1}{0.55 \times 1}\right),$$

$$q = 9.94 \times 0.2 \exp\left(-4.377\right) = 0.02495\,\mathrm{m^3/s/m},$$

which is five times less than the impermeable, rough slope above. For two layers of rock on a permeable core ($\gamma_r = 0.40$ in Table 5.5), the mean, wave overtopping rate, $q = 0.00484\,\mathrm{m^3/s/m}$. Clearly, the structure permeability significantly lowers the overtopping rate. The actual numbers would be even lower for a structure slope, cot $\alpha = 2.5$.

5.3.8.5 *Plain vertical walls, no berm ($d_{\text{toe}} = h_s$)*

The dimensionless criterion, h_* to distinguish "non-impulsive" and "impulsive" wave regimes (Equation (5.28)) is

$$h_* = 1.35\frac{d_{\text{toe}}}{H_{mo}}\frac{2\Pi h_s}{gT_{m-1.0}^2}$$

$$= 1.35\frac{3.0}{2.16}\frac{2 \times 3.1416 \times 3.0}{9.81 \times 7.27^2}, \tag{5.28}$$

$$h_* = 0.0682.$$

Therefore, for the impulsive (breaking) regime ($h_* \leq 0.2$), use Equation (5.30),

$$\frac{q}{h_*^2\sqrt{gh_s^3}} = 1.5 \times 10^{-4}\left(h_*\frac{R_c}{H_{mo}}\right)^{-3.1},$$

$$q = h_*^2\sqrt{gh_s^3}\, 1.5 \times 10^{-4}\left(h_*\frac{R_c}{H_{mo}}\right)^{-3.1}$$

$$= 0.0682^2\sqrt{9.81(3)^3}\, 1.5 \times 10^{-4}\left(0.0682\frac{2.0}{2.16}\right)^{-3.1},$$

$$q = 0.07569 \times 1.5 \times 10^{-4} \times 5234.58,$$

$$q = 0.0594\,\mathrm{m^3/s/m}. \tag{5.30}$$

This level of overtopping rate is similar to the rock-armor layers with permeable core. See Figure 5.18 for the impact of a parapet at the crest of the vertical seawall.

5.3.8.6 *Probabilistic design — 90%-confidence level*

As discussed above, there are no partial safety factors for the loading (waves) and structure resistance at some level of the probability of failure P_f for wave overtopping. However, as presented in Table 5.7, for an analysis at the 90% C.I. level, use of the standard deviations for the key coefficients in some of the equations is possible. At this level, we have employed the mean coefficient plus 1.645 times the standard deviation ($\mu + 1.645\,\sigma$) to determine the coefficient at the 90% C.I. level. For example, let us consider the following cases.

5.3.8.6.1 Impermeable, plain slope structure

As shown in Table 5.7, the coefficient 0.067 is unchanged but the coefficient -4.75 becomes $-\mathbf{3.9}$ (in bold).

$$q = \sqrt{9.81(2.16)^3}\,\frac{0.067 \times 1.0}{\sqrt{1/2.5}} 2.47 \exp\left(-\mathbf{3.9}\frac{2.0}{2.16}\,\frac{1}{2.47}\,\frac{1}{1 \times 1 \times 1 \times 1}\right),$$

giving $q = 0.603\,\mathrm{m}^3/\mathrm{s}/\mathrm{m}$ at the 90%-confidence level.

Recall at the deterministic level ($\mu = -4.75$) giving $q = 0.440\,\mathrm{m}^3/\mathrm{s}/\mathrm{m}$.

Note that the *EurOtop Manual* takes only one standard deviation ($\mu + 1.0\,\sigma$), so that in effect 68.3% of all the data are accounted for in their approach and the coefficient becomes -4.3 ($-4.75 + 0.3$) in their Equation (5.9) on (Die Küste, 2007, p. 80).

5.3.8.6.2 Permeable, rubble-mound structures

As shown in Table 5.7, the coefficient 0.2 is unchanged but the coefficient -2.6 becomes $-\mathbf{2.0}$ (in bold).

$$q = \sqrt{9.81(2.16)^3}\;0.2 \exp\left(-\mathbf{2.0}\frac{2.0}{2.16}\frac{1}{0.55 \times 1}\right),$$

giving $q = 0.06859\,\mathrm{m}^3/\mathrm{s}/\mathrm{m}$ at the 90% confidence level for two rock layers with an impermeable core. The deterministic level $q = 0.02495\,\mathrm{m}^3/\mathrm{s}/\mathrm{m}$.

5.3.8.6.3 Plain vertical walls, no berm

As shown in Table 5.7, for plain vertical walls and impulsive waves, Equation (5.30) is applicable and the coefficient 1.5×10^{-4} becomes

$\mathbf{3.64 \times 10^{-4}}$ (in bold).

$$q = 0.0682^2 \sqrt{9.81(3)^3}\ \mathbf{3.64 \times 10^{-4}} \left(0.0682 \frac{2.0}{2.16}\right)^{-3.1},$$

giving $q = 0.1441\,\text{m}^3/\text{s/m}$.

Multiply $q(\text{m}^3/\text{s/m}) \times 1000$ to obtain q (l/s/m) as found in Tables 5.8 and 5.9.

5.3.8.7 *Maximum overtopping volumes*

To determine the overtopping volume per wave requires knowledge of $R_c, R_{u2\%}, q, T_m$, the probability of exceedance level P_v of interest for design, and the duration of the storm to find N_w. We again consider the three example cases of interest raised earlier. Assume the probability of exceedance level $P_v = 10\%$ for a storm duration of 6 hours.

5.3.8.7.1 Impermeable, plain slope, smooth structure

For the maximum overtopping volume, the number of overtopping waves N_{ow} can be used in Equation (5.39) to estimate the maximum overtopping volume $V_{\max}$ repeated here.

$$V_{\max} = a\,[\ln(N_{\text{ow}})]^{1/b}. \tag{5.40}$$

Summarizing the above-mentioned example information, $R_c = 2.0$ m and $T_m = 6.1$ s and for smooth slopes, $R_{u2\%} = 6.48$ m (de Waal and van der Meer, 1992) and $q = 0.603\,\text{m}^3/\text{s/m}$ at the 90% C.I. level.

Therefore,

$$N_w = \frac{\text{Storm Duration in Seconds}}{\text{Mean Wave Period}, T_m}$$

$$= \frac{6 \times 60 \times 60}{6.1} = \frac{21{,}600}{6.1} = 3{,}540 \text{ incoming waves.}$$

$$P_{\text{ov}} = \exp\left[-\left(\sqrt{-\ln 0.02}\frac{R_c}{R_{u2\%}}\right)^2\right] \tag{5.36}$$

$$= \exp\left[-\left(\sqrt{-\ln 0.02}\,\frac{2}{6.48}\right)^2\right]$$

$$= \exp(-0.3726),$$

where $P_{\text{ov}} = 0.6889$ is the probability per wave for overtopping volume, $N_{\text{ow}} = P_{\text{ow}} \times N_w = 0.6889 \times 3{,}540 = 2{,}439$ is the number of overtopping waves, b is 0.75, and $C = 0.84$.

$$a = CT_m\, q\frac{N_w}{N_{\text{ow}}} = 0.84\ (6.1)\ (0.603)\ (1/0.6889),$$

$$a = 4.485, \tag{5.38}$$

and therefore,

$$V_{\max} = a[\ln(N_{\text{ow}})]^{1/b} = 4.485\ [\ln(2{,}439]^{1/0.75},$$

$$V_{\max} = 4.485\ (15.46) = 69.3\,\text{m}^3/\text{m}.$$

Clearly, the wave runup and wave overtopping discharge rate and maximum volume for a smooth slope with a crest elevation $R_c = 2.0$ m are exceedingly dangerous for both people and property as shown in Tables 5.8 and 5.9.

5.3.8.7.2 Impermeable, plain slope, rough structure

Roughening the slope ($\gamma_r = 0.55$) gave $R_{u2\%} = 3.56$ m and $q = 0.1823\,\text{m}^3/\text{s/m}$ at the 90% C.I. level. This greatly reduces the probability per wave overtopping $P_{\text{ov}} = 0.291$ and the number of overtopping waves $N_{\text{ow}} = 1{,}030$. Now, $a = 3.21$ and $V_{\max} = 42.4\,\text{m}^3/\text{m}$, but this is still a large value.

5.3.8.7.3 Permeable, rubble-mound structures

The same equations (as for impermeable structures) are applicable for rubble-mound structures with the appropriate values for $R_{u2\%}$ and q. Recall the value for $R_{u2\%} = 3.53$ m was similar to that for rough, impermeable structures. At the 90% C.I. level, $q = 0.0686\,\text{m}^3/\text{s/m}$. The probability per wave overtopping $P_{\text{ov}} = 0.291$ and the number of overtopping waves $N_{\text{ow}} = 1{,}030$ remain the same. But now, $a = 1.21$ and $V_{\max} = 15.91\,\text{m}^3/\text{m}$. These levels are still high so that a vertical seawall should be considered. Or, if possible, the crest elevation should be increased.

5.3.8.7.4 Plain vertical wall, no berm

Since $h_* = 0.0682 < 0.3$, then using Equation (5.42) for the probability of overtopping volume per wave gives

$$P_{\text{ov}} = 0.031\frac{H_{mo}}{R_c}\frac{1}{h_*}$$

$$= 0.031\frac{2.16}{2.0}\frac{1}{0.0682} = 0.4909, \tag{5.42}$$

so that $N_{\text{ow}} = 0.4909 \times 3540 = 1738$, and with $C = 0.92$ and $b = 0.85$, with $q = 0.0393$ at the 90% confidence level, so that $a = CT_m\, q\frac{N_w}{N_{\text{ow}}} = 0.92$ (6.1) (0.03933) (1/0.0.4909), $a = 0.4496$, and finally, $V_{\max} = a[\ln(N_{\text{ow}})]^{1/b} = 0.4496\,[\ln(1738)]^{1/0.85}$, $V_{\max} = 4.78\,\text{m}^3/\text{m}$, and the lowest value relative to all above. Adding an overhanging parapet will further reduce these results but raising the crest elevation may be the best solution.

For English units, 1.0 (l/s/m) = (0.010753) cubic feet per second per foot (cfs/ft).

5.3.9 *Spreadsheets and the EurOtop calculation tool*

5.3.9.1 *Spreadsheets for this book*

The three key geometric variables in all the above-mentioned equations to calculate q and $V_{\max}$ are the structure slope, cot α (or vertical wall), the water depth h_s at the toe of the structure, and the crest elevation, R_c. Many possible combinations of these geometric variables with the hydrodynamic variables (wave heights and wave periods) make the repeated application of these formulas tedious and prone to errors in computation.

Spreadsheets have been developed to facilitate the efficient study of these geometric variables on wave overtopping. They are found in the online supplementary materials which accompanies this book. All employ the software product, Microsoft Excel. See the folder for Functional Design/Wave Overtopping to locate the files. Figure 5.20 is an example for impermeable, variable sloping structures and follows the same layout as in Figure 5.10 for wave runup. The results for cot $\alpha = 2.5$ are shown along with those for other possible structure slopes permitting immediate comparison of the impact of structure slope. Of course, the true benefit of the spreadsheet is the ability to change any input, hydrodynamic variable (e.g. wave height in deep water, H_s) and obtain the results. A similar spreadsheet is for variable freeboard as the key design condition for a constant slope.

Spreadsheets for wave overtopping for use in design are included in the online supplementary materials (Functional Design, Wave Overtopping) for the following:

Impermeable Slopes (Smooth and Rough)

- *EurOtop Manual* (Die Küster, 2007) Variable Slope Equations (5.24)–(5.26) Recommended
- *EurOtop Manual* (Die Küster, 2007) Variable Freeboard Equations (5.24)–(5.26) Recommended

Metric System

EurOtop Manual Impermeable, Plain Slopes

Enter initial conditions
Impermeable structure, smooth slope,rough slope
Slope, Cot α = 1.5 to 5.0
(γ_r = 0.55 for two or more layers of rock)
Valid for [$\xi_{om-1.0}$ < 6] and for ξom-1.0 >6

Slope	Surf Sim Par.	OVERTOPPING q (m^3/sec/m)	OVERTOPPING q (m^3/sec/m)
COT alpha	$\xi_{m-1.0}$	Smooth	Rough
1.5	4.122	**1.1576**	**0.4836**
2.0	3.091	**0.7025**	**0.2194**
2.5	2.473	**0.4403**	**0.1028**
3.0	2.061	**0.2817**	**0.0492**
4.0	1.546	**0.1198**	**0.0117**
5.0	1.236	**0.0526**	**0.0029**

WAVE Overtopping
(Metric System)

Enter Data		**Units**
Sign. Wave Height in deep water,H_s=	1.92	m
Peak Wave Period, T_p=	8.0	sec
Depth at Toe of Structure, h_s =	3.0	m
Crest Freeboard, R_c =	2.0	m
Roughness coefficient, γ_r =	0.55	-
Berm coefficient, γ_b =	1	-
Shallow depth coefficient, γ_h =	1	-
Wave angle coefficient, γ_β =	1	-
Combined coefficients, berm,depth,angle	1	
Calculated Information		
Wave Spectral Period, $T_{m-1.0}$ =	7.3	sec
Wave Length, $L_{m-1.0}$ =	82.582	m
Steepness, $S_{om-1.0}$ =	0.0262	
Wave Length deep water,L_{op}=	99.873	m
K_o(wave number at deep water)=	0.0629	-
C_g(local group vel.)/C_o(deep water vel.) =	0.395	-
Shoaling Coef., K_s = (Htoe/Ho) =	1.125	-
Wave Height at toe, H_{toe}=	**2.16**	m
Deep Water Steepness, S_{op}=	0.021633	-
R_c/H_{toe} =	0.9257	

Figure 5.20. Spreadsheet for impermeable structures with calculated overtopping rates for smooth and rough slopes from cot $\alpha = 1.5$ to 5.0.

- van der Meer and Janssen (1995) Equations (5.16) and (5.17)
- Owen (1980, 1982) Equation (5.15)

Armor-Layer Structures and Mounds

- *EurOtop Manual* (Die Küster, 2007) Equation (5.27) (Table 5.5)

Vertical Seawalls

- *EurOtop Manual* (Die Küster, 2007) Equations (5.29) and (5.30)

The above-listed spreadsheets incorporate the ability to alter the key coefficients (Table 5.7) for determinations of results at the 90% confidence level. In some cases, maximum overtopping volumes are also calculated but require additional computation of wave runup as input data.

All spreadsheets have three pages (sheets), with one for the Metric system, one for the English units' system, and one for a Read Me page that includes all the equations and coefficients. They should be helpful for design.

5.3.9.2 *EurOtop calculation tool*

The CLASH Project that developed the *EurOtop Manual* (Die Küste, 2007) also produced an online calculation tool to calculate overtopping rates as summarized above. The calculation tool website is http://www.overtopp

ing-manual.com/calculation-tool (January, 2015). Twenty-seven structure types are delineated and grouped into five categories (with number of sub-types in parentheses) as follows:

- Vertical and Steep Seawalls (Five);
- Stepped Structures (Six);
- Coastal Dikes and Embankment Seawalls (Seven);
- Armored Rubble Slopes and Mounds (Six);
- Stepped and Composite Structures (Three).

Each of the 27 structure types is graphically represented, and Figure 5.21 shows three categories with only the first type in each category studied in this book. By clicking on the graphic of the structure type of interest, the link will display this graphic with dimensions and hydrodynamic variables needed for the calculation of the overtopping rate. For example, Figure 5.22 displays the plain, simple slope graphic needing a wave period T, wave height at toe H_{mo}, freeboard R_c, water depth at toe h_s, and structure slope to make a calculation. All dimensions are in the Metric system and the wave period entered must be defined as one of three types. Once these five variables are placed in the boxes provided, hitting the "Calculate Overtopping Rate" button will generate the results for q as the mean wave-overtopping rate q in units of l/s/m.

A word of caution. The "Method Selection" buttons near the top of Figure 5.22 have the reverse meaning than as defined in this book. The "Probabilistic" button will calculate the mean tendency results and the "Deterministic" button will calculate the mean value μ plus one standard deviation σ and, not the 90% C.I. value as found using the formulas given before.

The online calculation tool also presents the option to use a neural network, generic overtopping prediction tool that was also a product of the CLASH project. A neural network consists of a family of statistical learning algorithms that depends on a large number of inputs presented as a system of interconnected "neurons" to compute values from the inputs. A neural network is capable of pattern recognition based on their adaptive nature.

The neural-network prediction tool has been established for 18 of the 27 structure types that employ empirical calculation methods. It is not recommended for use in design but only in the concept phase of the study. It should only be used for estimates of the mean overtopping discharge rates.

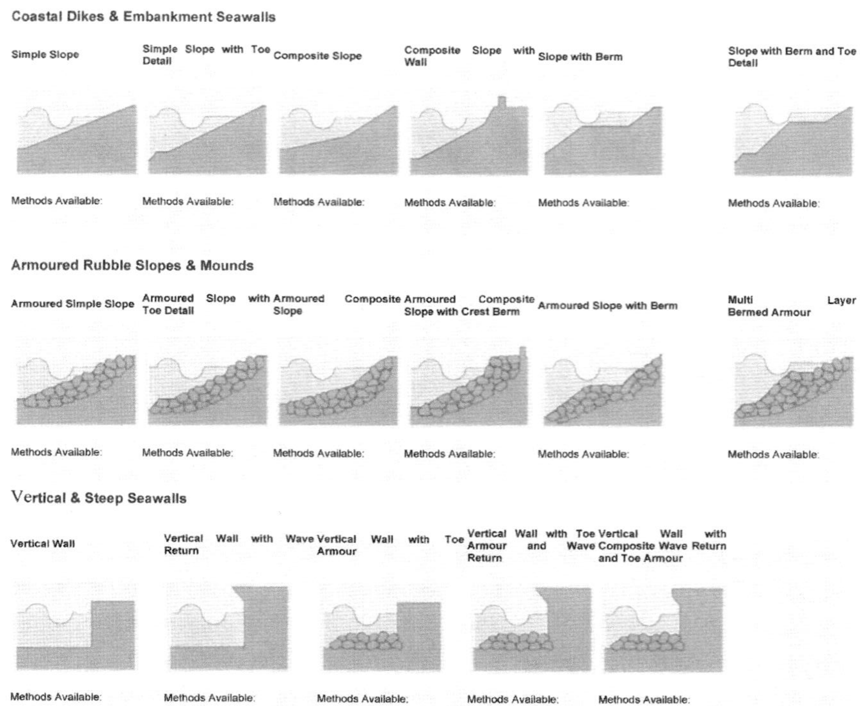

Figure 5.21. Coastal dikes and embankment seawalls (top), armor rubble slopes and mounds (middle), and vertical and steep seawalls (bottom). Three of the five structure categories as defined in the *EurOtop* calculation tool (courtesy Taylor&Francis).

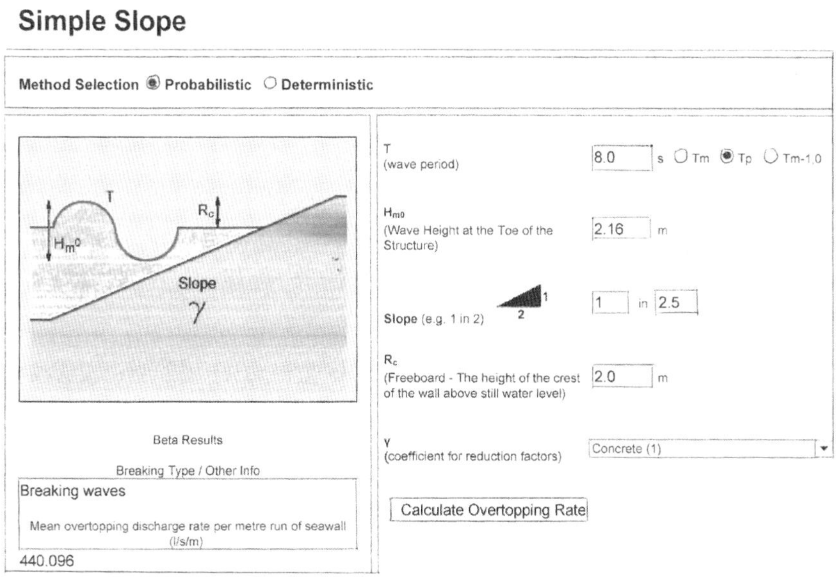

Figure 5.22. *EurOtop* graphic, data input for example and calculation sheet for wave overtopping rate on simple slope. The calculation tool result $q = 0.440\,\mathrm{m}^3/\mathrm{s}/\mathrm{m} = 440\,\mathrm{l}/\mathrm{s}/\mathrm{m}$ is shown in the lower left corner (courtesy Taylor&Francis).

In general, the advantage of the spreadsheets provided herein is that they give the user an immediate, quantitative understanding of the influence of the variable structure slope and structure freeboard on the overtopping rates. The advantage of the EurOtop, online Calculation Tool is the wide range of possible structure configurations available for calculation results. Use of both is recommended for design.

5.4 Wave Reflection

Nearshore waves are also reflected seaward by coastal structures. Figure 5.14 illustrates the extremely confused sea conditions in front of a vertical seawall that were created by the irregular waves approaching the wall and their interaction with the reflected waves from the wall. The wall curvature adds to the complexity of the intersecting wave trains. Reflected waves increase the horizontal velocities and shear stresses at the bottom and propagate wave energy away from the structure that may reach a normally calm area. We only introduce some basic features of wave reflection for sloping structures in this section. Wave reflection on vertical structures is treated in detail in Chapter 6 where it is very important to estimate the horizontal forces for structural design.

5.4.1 *Wave reflection coefficient*

We define a coefficient of wave reflection K_R as simply

$$K_R = \frac{\text{Significant wave height of reflected waves}}{\text{Significant wave height of incoming waves}} = \frac{H_{sR}}{H_{si}}, \quad (5.43)$$

where the subscript s indicated the significant wave height, the subscript i is for the incident (incoming) waves, and subscript R for reflection. (Recall from Chapter 2, the symbol K_r was for the refraction coefficient.) The value of K_R approaches 1.0 for vertical walls and can be as low as 0.05 (or lower) for beaches. The value of K_R is a measure of the amount of wave-energy dissipation since $E \approx H^2$.

5.4.2 *Wave reflection at sloping structures*

As seen before for wave runup and wave overtopping, the surf-similarity parameter ξ is an important variable for its correlation with the type of wave breaking on the slope, $\tan\alpha$. Energy dissipation by wave breaking is greater than dissipation by roughness and porosity of the structure. Therefore, early efforts correlated the physical model test data for measured values of K_R and ξ. See the US Army Corps of Engineers, *Coastal Engineering Manual* (CEM 2006, Part VI, Chapter 5, Table VI-5-14) for a basic equation and results for impermeable, smooth, rough, rock armor, artificial armor, and other geometry formulations from both laboratory and field experiments conducted from the 1980s to the mid-1990s.

More recently, much of this work along with new data (over 6,000 data) was summarized by Zanuttigh and van der Meer (2007, 2008) for the European CLASH program. They proposed the following new formula for the reflection coefficient of sloping structures:

$$K_R = \tanh[a(\xi_{m-1.0})^b], \quad (5.44)$$

where the coefficients a and b are defined for plain slopes as follows:

$$a = 0.167[1 - \exp(-3.2\gamma_f)], \quad (5.44a)$$

$$b = 1.49(\gamma_f - 0.38)^2 + 0.86, \quad (5.44b)$$

with the usual definitions for $\xi_{m-1.0}$ which permits application to bimodal and flat spectra.

$$\xi_{m-1.0} = \frac{\tan\alpha}{\sqrt{S_{mo}}},$$

$$S_{mo} = \frac{H_{mo}}{L_{m-1.0}},$$

$$L_{m-1.0} = \frac{g}{2\pi}T^2_{m-1.0}.$$

A table for coefficients a and b and roughness coefficient γ_f for smooth (impermeable), rock armor (impermeable and permeable cores), and artificial armor units (see Chapter 6) is reproduced in Table 5.11 (from Verhagen *et al.*, 2012).

For relatively smooth ($\gamma_f = 1.0$) impermeable slopes, $a = 0.16$ and $b = 1.43$. For impermeable slopes covered with rocks, $a = 0.14, b = 0.90$, and $\gamma_f = 0.55$. For the given data in Section 5.3.8, $\xi_{m-1.0} = 2.47$ so that on a smooth slope, the wave reflection coefficient, $K_R = 0.66$. For the rough, rock surface, $K_R = 0.30$ and less than half. In the above, the significant wave height H_s and the spectral, significant height H_{mo} are interchangeable. See Zanuttigh and van der Meer (2008) for other structure conditions including how to incorporate composite slopes with berms.

General tables of reflection coefficients are also available (Goda, 2010). Non-overtopping wave conditions are assumed in all cases for reflected waves.

Table 5.11. Coefficients a and b and roughness coefficients γ_f for estimation of wave reflection coefficient K_R using Equation (5.44).

Structure	a	b	γ_f
Rock,permeable	0.12	0.87	0.40
Rock, impermeable	0.14	0.90	0.55
Smooth slopes	0.16	1.43	1.0
Tetrapods, two layers	0.102	0.87	0.38
Core-Loc, one layer	0.113	0.87	0.44
Xbloc, one layer	0.112	0.87	0.45
Accropode, one layer	0.115	0.87	0.46
Antifer, two layers	0.115	0.87	0.47
Cubes, two layers	0.108	0.87	0.47
Cubes, one layer	0.120	0.87	0.50

5.4.3 *Wave reflection — Vertical walls*

This topic is covered in detail in Chapter 6, because wave reflection at vertical walls significantly influences the water levels and hence horizontal water forces on the vertical plane for structural design.

5.5 Wave Transmission

Nearshore waves are also transmitted over, around, and through coastal structures. Figure 5.23(a) is an aerial photo of the breakwater at the Lajes, US Air Force Base on Terceira Island, in the Azores, and the pier for unloading fuel tankers at this "gas station for jet planes in the Atlantic". In December 2001, a severe Atlantic Ocean storm with wave heights over 6 m overtopped the breakwater and the transmitted waves prevented tanker unloading (i.e. "functional damage") from the pier (Figure 5.23(b)). The storm damaged the rock breakwater and full details are presented in Section 6.2.10.

Shore-parallel, nearshore, and low-crested rubble-mound breakwater systems have emerged as an effective alternative to increase the fill life of renourished beaches. We only introduce the basic formulations for simple structure configurations and only for the two-dimensional (2D), sectional case for transmission over and through the structure. Wave transmission around the ends for the three-dimensional (3D) case is covered in Part II of this chapter.

5.5.1 *Wave transmission coefficient*

We define a coefficient of wave transmission K_t as simply

$$K_t = \frac{\text{Significant wave height of transmitted waves}}{\text{Significant wave height of incoming waves}} = \frac{H_{st}}{H_{si}}, \qquad (5.45)$$

where the subscript s indicates the significant wave height, the subscript i is for incident (incoming) waves, and subscript t for transmitted waves past the structure. The value of K_t is a measure of the amount of wave energy dissipation since $E \approx H^2$. For permeable, rubble-mound structures as schematized in Figure 5.24, the waves transmitted, H_{st}, are due to both waves overtopping the crest and waves passing through the structure. The variables shown in Figure 5.24 are the same as for wave runup and wave overtopping and will be again defined in the following equations.

(a)

(b)

Figure 5.23. (a) Aerial photo of Lajes, AFB breakwater and jet fuel unloading pier on Terceria Island in the Atlantic Ocean (the Azores). (b) Wave overtopping and transmission during Atlantic Ocean storm in December 2001 (courtesy TranSystems, Norfolk, VA).

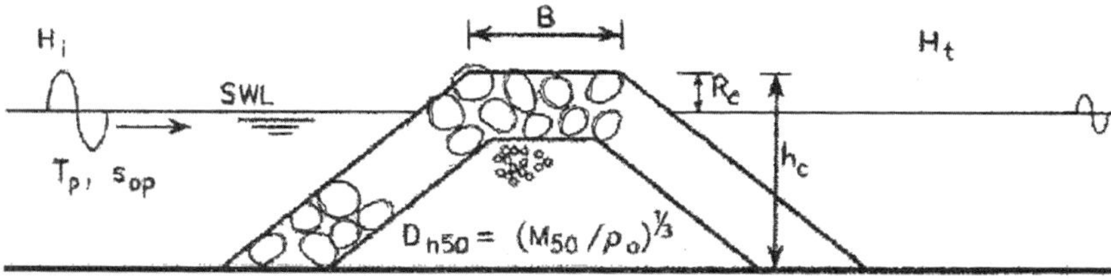

Figure 5.24. Definition sketch for low-crested breakwater structures (from van der Meer *et al.*, 2005).

5.5.2 *Low-crested structures (LCS)*

As illustrated in Figure 5.24, a key variable is R_c, the crest elevation above the SWL (freeboard) and taken as positive upward and zero at the SWL. Negative values mean the crest is submerged. A second key structure variable is the crest width B and clearly important when $R_c = 0$. Many laboratory experiments using irregular waves have been carried out over the years, and regression analysis is employed to develop empirical formulations depending on how a non-dimensional crest freeboard R_c^* is defined. Many formulations exist, see e.g. Tomasicchio and D'Alessandro (2013).

5.5.2.1 $R_c^* = R_c/H_{si}$ *(d'Angremond et al., 1996)*

As noted for wave runup in Section 5.2 (Equation (5.1)) and for wave overtopping in Section 5.3 (Equation (5.15)), the significant wave height H_s is the key variable relative to the crest freeboard R_c. The empirical formulation developed should also hold for relatively impermeable and smooth structures and not just permeable, rubble-mound structures. The d'Angremond *et al.* (1996) formula is

$$K_t = -0.4\frac{R_c}{H_{si}} + a\left(\frac{B}{H_{si}}\right)^{-0.31}(1 - e^{-0.5\xi_{op}}), \tag{5.46}$$

where a is 0.64 for permeable structures, and 0.80 for impermeable structures, and ξ_{op} is the surf-similarity parameter using the peak, incident wave period, i.e. Equation (5.1).

Figure 5.25 (from d'Angremond *et al.*, 1996) demonstrates how wave transmission decreases as the relative freeboard, R_c/H_{si} increases. Even when the water level is at the crest elevation ($R_c = 0$), roughly 50% of the wave energy is removed from the incoming waves by the LCS. The

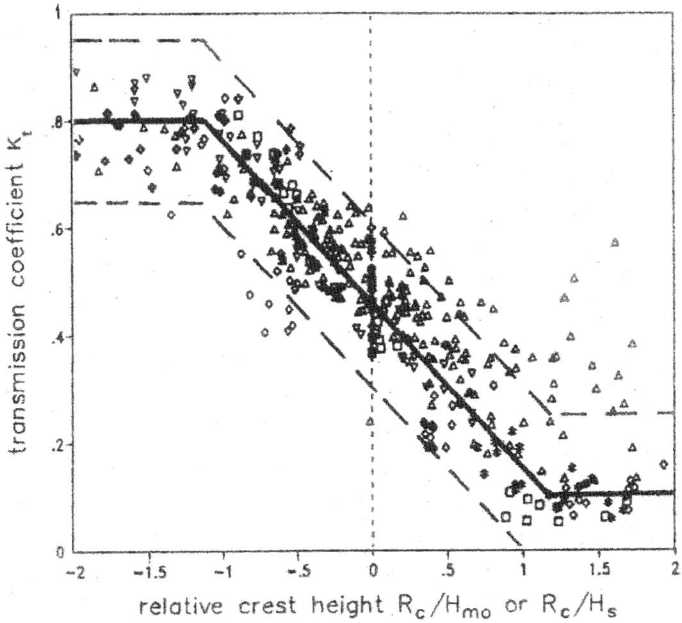

Figure 5.25. Importance of relative freeboard R_c/H_{si} in reducing the wave-transmission coefficient for LCS (from d'Angremond *et al.*, 1996).

upper boundary is about $K_t = 0.8$ and the lower boundary is $K_t = 0.08$. Equation (5.46) is applicable for relatively narrow crested breakwaters when $B/H_{si} < 8$.

In the early 2000s, the European Union funded the DELOS (2005) project that created a very large database (over 2,300) of wave transmission coefficients and some with large crest widths. An analysis by van der Meer *et al.* (2005) showed that for relatively wide-crested structures ($B/H_{si} < 12$) the formula for rough, permeable structures should be

$$K_t = -0.35\frac{R_c}{H_{si}} + a\left(\frac{B}{H_{si}}\right)^{-0.65}(1 - e^{-0.41\xi_{op}}), \tag{5.47}$$

where a is 0.51 for permeable structures, and ξ_{op} is the surf-similarity parameter using the peak, incident wave period, i.e. Equation (5.1).

Now, the upper boundary is about 0.93 and the lower boundary is 0.05. Interpolation is recommended for the intermediate range $8 < B/H_{si} < 12$.

The DELOS project also found that for smooth sloping, relatively narrow ($B/H_{si} < 4$) impermeable structures, the following prediction formula for K_t should be applied:

$$K_t = -0.30\frac{R_c}{H_{si}} + 0.75(1 - e^{-0.5\xi_{,\mathrm{op}}}). \tag{5.48}$$

5.5.2.2 $R_c^* = R_c/D_{n50}$ *(van der Meer and Daemen, 1994)*

Van der Meer and Daemen (1994) proposed the formula

$$K_t = a\frac{R_c}{D_{n50}} + b, \tag{5.49}$$

where a was a function of $\frac{H_{si}}{D_{n50}}$ and b was a function of $\frac{H_{si}}{D_{n50}}, \frac{B}{D_{n50}}$, and the wave steepness, s_{op}. The DELOS data set analyzed by van der Meer *et al.* (2005) concluded that the design formula of d'Angremond *et al.* (1996) given here as Equation (5.46) for narrow crests was best. Equation (5.47) for wide crests and Equation (5.48) for smooth slopes should also be used for design.

5.5.2.3 *Goda and Ahrens (2008)*

These authors separated the wave transmission coefficient into two parts, one for the contribution of the wave overtopping and one for the contribution of the waves passing through the LCS. The theory was that for submerged structures, the contribution of the overtopping transmission is much larger than the contribution passing through. The original paper should be reviewed for model formulation and empirical equations developed using 14 data sets from around the world and 774 data points.

Tomasicchio and D'Alessandro (2013) added 19 more data sets that expanded the number of data points to over 3,300 and recalibrated the original formulation based on the full database. The new formulation gave slightly larger values of the transmission coefficient for LCS than that of Goda and Ahrens (2008).

5.5.2.4 *Other recent efforts*

The DELOS project also employed the 2,300 data points to develop a separate transmission model based on neural networks. Details are found in Van Oosten *et al.* (2006) and nine input parameters are required, which is more than the empirical formulas given earlier.

Numerical analysis tools for wave transmission over and through low-crested breakwaters have also been employed in recent years. For example,

Peng *et al.*, (2008) employed the full Navier–Stokes equations and volume of fluid method (RANS-VOF) to study wave transformation over an impermeable LCS. Many other studies also employ the physical model data sets to test and calibrate the numerical models. For design, Equations (5.46)–(5.48) are more than adequate given the wide range of input wave and water-level conditions present at the site.

5.5.3 *Wave transmission past vertical walls*

Vertical walled breakwaters have a long history of development in Japan as jetty structures at the entrance to harbors and marinas (Goda, 2010). Relatively deep water makes this caisson-type breakwater more economic than the rubble-mound type. However, the caisson structure was always constructed on a rubble-mound base that varied in thickness so that the ratio of the water depth at the vertical wall d to the still-water depth h_s behind the structure (i.e. d/h_s) was important. As a result of physical model tests with regular waves, Goda (1969) published the results shown in Figure 5.26. Goda *et al.* (1974) (in Japanese) showed that these results were also good for irregular waves using H_{si} as the incident wave height. The wave transmission formula given by Goda (1969) is

$$K_t = \{0.25[1 - \sin(\pi/2\alpha)(R_c/H_{si} + \beta)]^2 + 0.01(1 - d/h_s)^2\}^{0.5}, \quad (5.50)$$

where α is 2.2 and β ranges nonlinearly from -0.2 for $d/h_s = 0$ up to 0.4 at $d/h_s = 1.0$. A simple second-order power equation gives

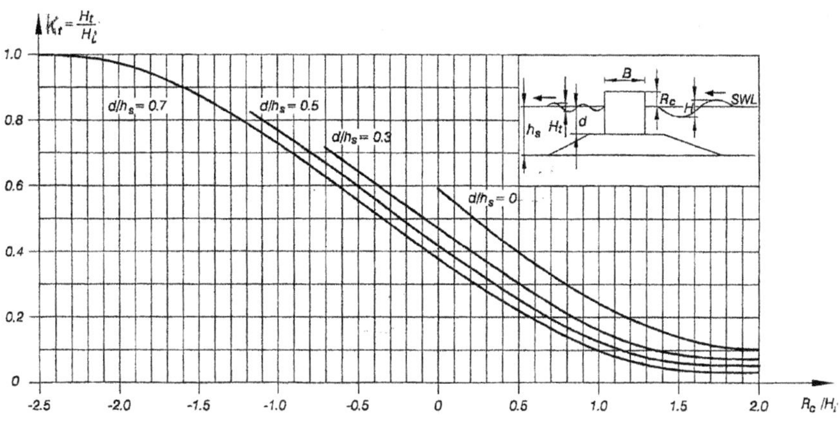

Figure 5.26. Wave transmission for vertical wall structure on rubble-mound base (from Goda, 1969).

$\beta = 2.1548(d/h_s)^2 + 1.1012(d/h_s) + 0.1621$ to order $R^2 = 0.9835$ to readily permit a spreadsheet solution for Equation (5.50).

In most cases, the d/h_s ratio is equal to or greater than 0.7. Alberti *et al.* (2001) proposed the following formula resulting from irregular wave transmission tests over vertical breakwaters for $R_c/H_{si} > -0.2$:

$$K_t = \exp[-(1.14 + 1.16(R_c/H_{si}))]. \tag{5.51}$$

This formula gives slightly lower results than in Figure 5.26 for $d/h_s = 0.7$.

Note that the caisson width B does not appear in Equation (5.50). Cuomo *et al.* (2011) studied four different caisson widths ($B = 0.05, 0.25$, 0.44, and 0.64 m) for irregular wave transmission tests of a low-crested caisson breakwater model in the laboratory and found that K_t decreased with increasing crest width. These authors modified the original Equation (5.50) of Goda (1969) to become

$$K_t = \{c[1 - \sin(\pi/2a)(R_c/H_{si} + b)]^2 + 0.01(1 - d/h_s)^2\}^{0.5}, \tag{5.52}$$

where the coefficients, a, b, and c are best-fit empirical parameters as summarized in Table 5.12.

See Goda (2010) and CEM (2011) for many additional wave-transmission results for other conditions of the vertical wall and subbase.

5.5.4 *Wave period changes*

The full spectrum of wave energy at different amplitudes and frequencies are each transmitted over and through coastal structures with varying degrees of energy reduction that together produce a modification of the transmitted energy spectrum. As a result, the peak period of the transmitted spectrum will be different than the incoming spectral peak period. And clearly, the change will be different for permeable and non-permeable LCS and for vertical-walled structures.

Table 5.12. Best-fit parameters for Equation (5.52) from Cuomo *et al.* (2011).

Relative Caisson Width, B/H_{si}	a	b	c	R^2
0.125	2.2	0.0034	0.118	0.73
0.620	2.2	0.31	0.093	0.84
1.11	2.2	0.49	0.081	0.87
1.60	2.2	0.61	0.075	0.87

Van der Meer *et al.* (2005) discuss spectral change using data from the DELOS project compared with the method proposed in van der Meer *et al.* (2000). Zanuttigh and Martinelli (2008) present a conceptual model for prediction of the transmitted wave energy spectrum past permeable LCS. In general, the extremely limited results from these research efforts preclude their detailed review herein. But the interested reader should consult them for insight.

PART II Planform Layout, Location, and Spacing

5.6 Planform Analysis of Natural Shorelines and Shoreline Stabilization Structures

5.6.1 *Naturally stable shorelines*

The concept of a naturally stable shoreline and a minimum dry beach width was discussed in Chapter 3, Section 3.3.1 and was graphically displayed in Figure 3.6 (from Silvester and Hsu, 1993), and is reproduced here as Figure 5.27.

The shoreline configuration between the natural, rocky outcroppings called "headlands" along the coast is illustrated schematically in Figure 5.28 (Silvester and Hsu, 1993) for wave rays (solid lines, 1, 2, 3, etc.), which shoal, refract, and diffract in the shallow-water contours (dashed lines) and strike the shoreline. If the wave rays break at right angles at the shoreline, no longshore sediment transport (LST) can take place. This is the "equilibrium" shoreline position for the given offshore wave direction. The wave crest approach angle β is 50° in Figure 5.28.

When the shoreline is in non-equilibrium as shown in Figure 5.29 (dotted lines), it will continue to evolve (erode or fill) until again reaching equilibrium. The furthest non-equilibrium position is after a coastal storm when the minimum dry beach width, $Y_{\min}$ is reached.

Major storm events may produce breaching of narrow barrier beaches between headlands. Figure 5.33 shows an artificial, headland system to illustrate the natural, rocky headland case from which the empirical models for the shoreline shape have been developed.

An empirical approach developed from the shoreline data taken from bays known to be in static planform equilibrium and physical hydraulic models has been developed by Hsu *et al.* (1989) and is called the "parabolic" model. Figure 5.30 illustrates the two, key independent variables R_o, the

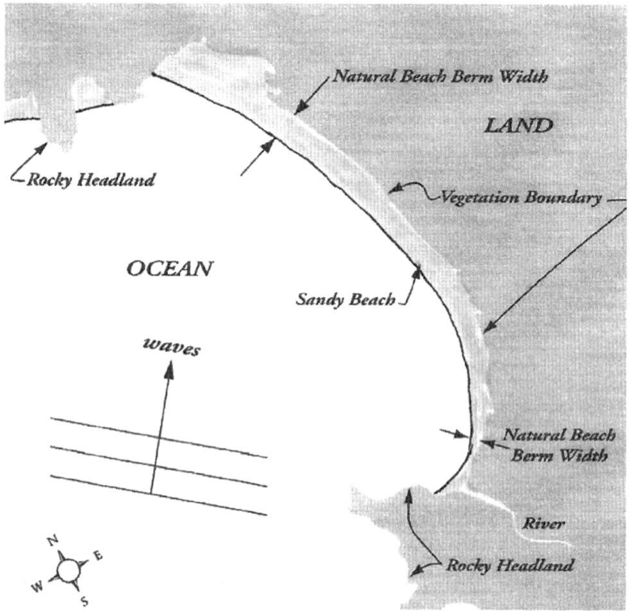

(a) Naturally stable shoreline between rocky headlands

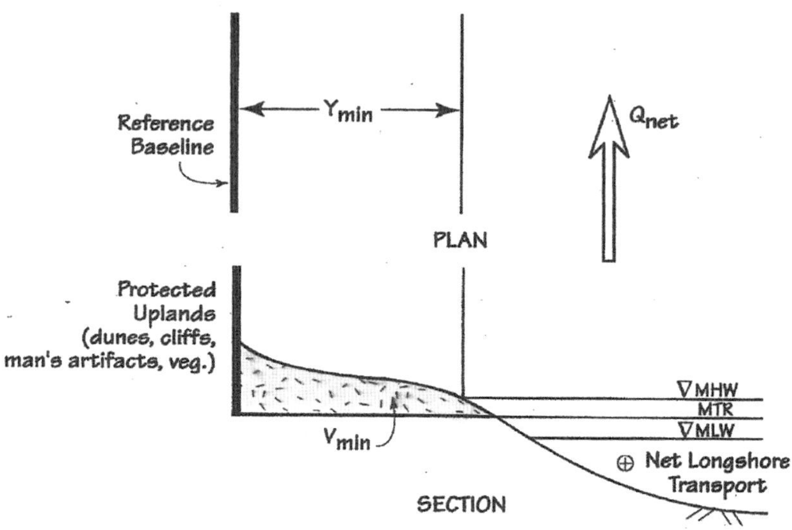

(b) Minimum dry beach width, Y_{min} relative to reference baseline

Figure 5.27. (a) Planform of naturally stable shoreline (from Silvester and Hsu, 1993) with the "Natural Beach Width" label from MSL to upland vegetation and (b) concept of minimum dry beach width, $Y_{\min}$ (from CEM, 2006, V-3-3).

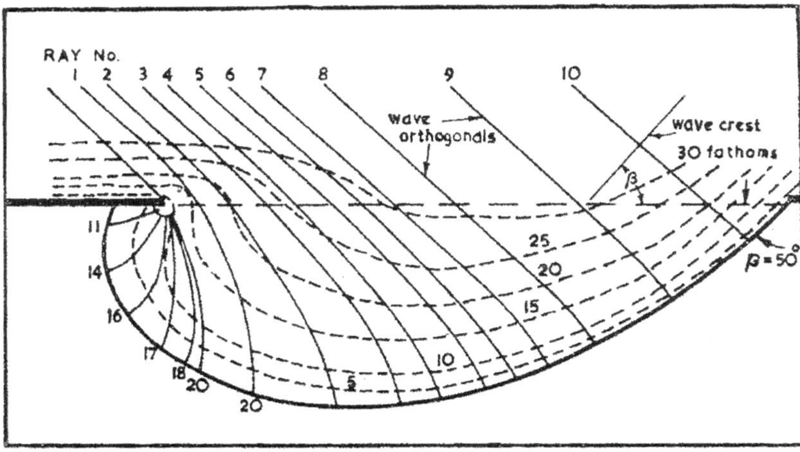

Figure 5.28. Wave rays shoaling, refracting, and diffracting to break at right angles so that no LST results on the equilibrium planform shape (Silvester and Hsu, 1993).

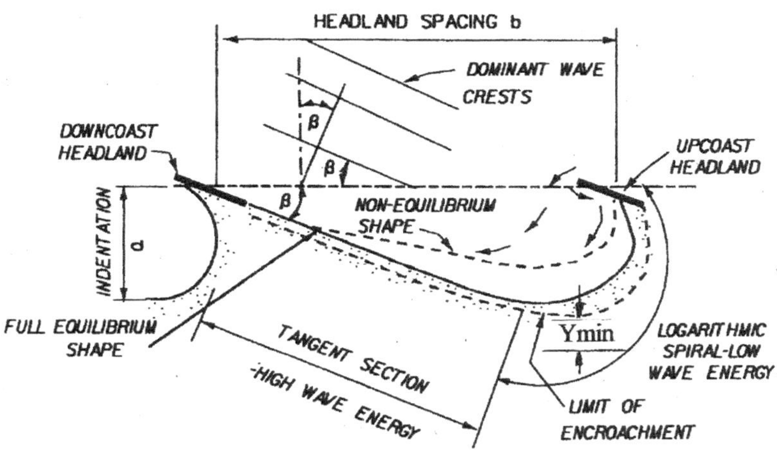

Figure 5.29. Schematic diagram of headland breakwaters and beach planforms that are in equilibrium shape (solid line) and non-equilibrium position (dotted lines) (from CEM, 2006).

distance of a control line from the upland coast headland (tip) to the downcoast headland (tip) and the approach wave angle β between the wave crest and the control line, $R_o(\beta = 24°$ in Figure 5.30). These variables are site–specific. The two dependent variables in the model are R and θ where R

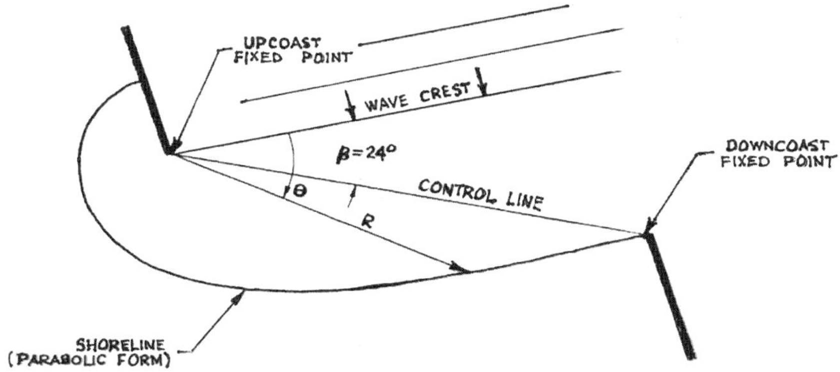

Figure 5.30. Schematic diagram of parabolic shoreline-form model defined by four variables, R_o, β, R, and θ (from Hsu *et al.* 1989).

is the radius and θ is the angle from the wave crest to radius R as defined in Figure 5.30. Using 10 to 12 R, θ pairs defines a parabola given by the formula (Hsu *et al.* 1987),

$$\frac{R}{R_o} = C_o + C_1\left(\frac{\beta}{\theta}\right) + C_2\left(\frac{\beta}{\theta}\right)^2. \tag{5.53}$$

The coefficients C_o, C_1, and C_2 in the parabolic model are shown in Figure 5.31 as a function of the approach wave angle β developed empirically by Hsu *et al.* (1989). Simple, second-order polynomials have been applied to curve-fit these coefficients as a function of the approach wave angle β and give

$$C_o = -0.0002\beta^2 + 0.0079\beta - 0.0209, \tag{5.54a}$$

$$C_1 = 0.0002\beta^2 + 0.0004\beta + 0.9437, \tag{5.54b}$$

$$C_2 = (-5E - 05)\beta^2 - 0.0093\beta + 0.0879. \tag{5.54c}$$

The accuracy of the C_2 expression may be improved by using a higher order polynomial. To facilitate the efficiency of application in the design process, a software package MEPBAY has been developed from computation to graphically display the results (Klein *et al.*, 2003). The accuracy of Equation (5.53) to model equilibrium-shaped bays of natural and man-made beaches in Spain has been confirmed by Gonzalez and Medina (2001). See also Silvester and Hsu (1997).

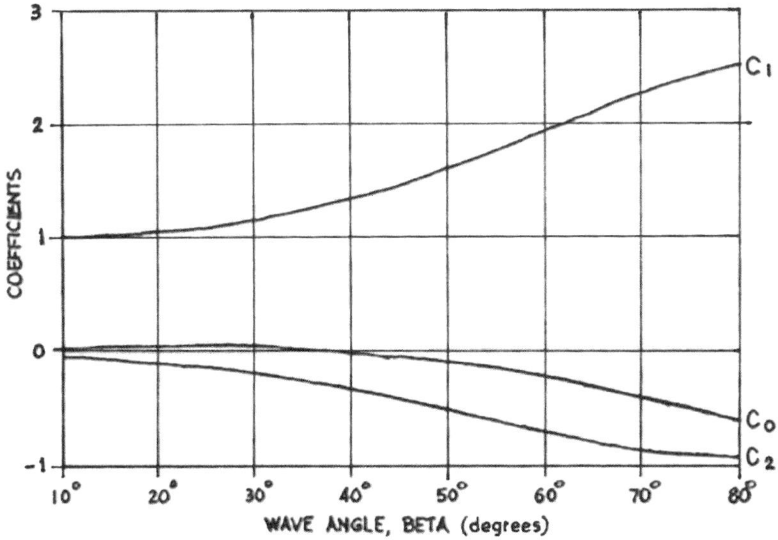

Figure 5.31. Coefficients, C_o, C_1, C_2 in the parabolic model of Hsu *et al.* (1989). (From CEM, 2006, Fig. 111-2-27 b)

5.6.2 *Shoreline stabilization structures*

Many shorelines are eroding due to both natural causes (e.g. bathymetric transformations of waves causing unbalance of LST) or by coastal structures that interrupt longshore or cross-shore transport that create the unbalance. Shoreline stabilization structures can aid the retention of sand on the beach to dissipate wave energy during storm events and hence provide upland damage mitigation beyond the reference baseline.

5.6.2.1 *Salients or tombolos*

As illustrated schematically in Figure 5.32, the response of the shoreline behind the structure due to waves, currents, and resulting sediment transport can be either a tombolo (shoreline attached to the structure) or a salient (no shoreline attachment with variable shoreline shape). The line representing the "shoreline" is dependent on the tidal range and definition of the shoreline, e.g. the mean tidal elevation. Therefore, in some cases, a tombolo will be present at low-tide conditions and only the salient form would be visible under high-tide conditions. The tidal range is therefore an important factor.

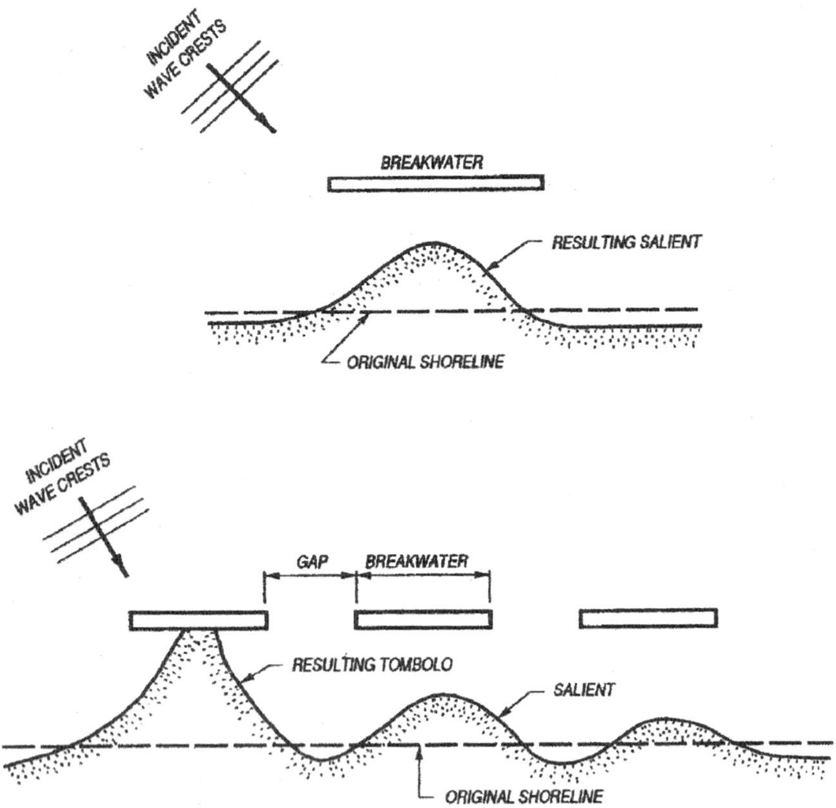

Figure 5.32. Schematic illustration of tombolo attached to near-shore breakwater (bottom, left) and a salient shaped shoreline (top, and bottom-middle, and right) (from the CEM, 2006).

Figure 5.33(c)–5.33(e) displays the three most common types of beach erosion mitigation structures, namely, (c) headland breakwaters, (d) nearshore, detached, shore-parallel breakwaters, and (e) a groin field. Note that the (c) headland breakwater type mimics the (b) natural, rocky headland and the shoreline is attached by a tombolo in the lee of each structure. The (d) nearshore breakwaters are detached (salients form behind) and are shore-parallel. Clearly, the only difference in the (c) headland and (d) nearshore types may be the tidal range for definition of the "shoreline" position.

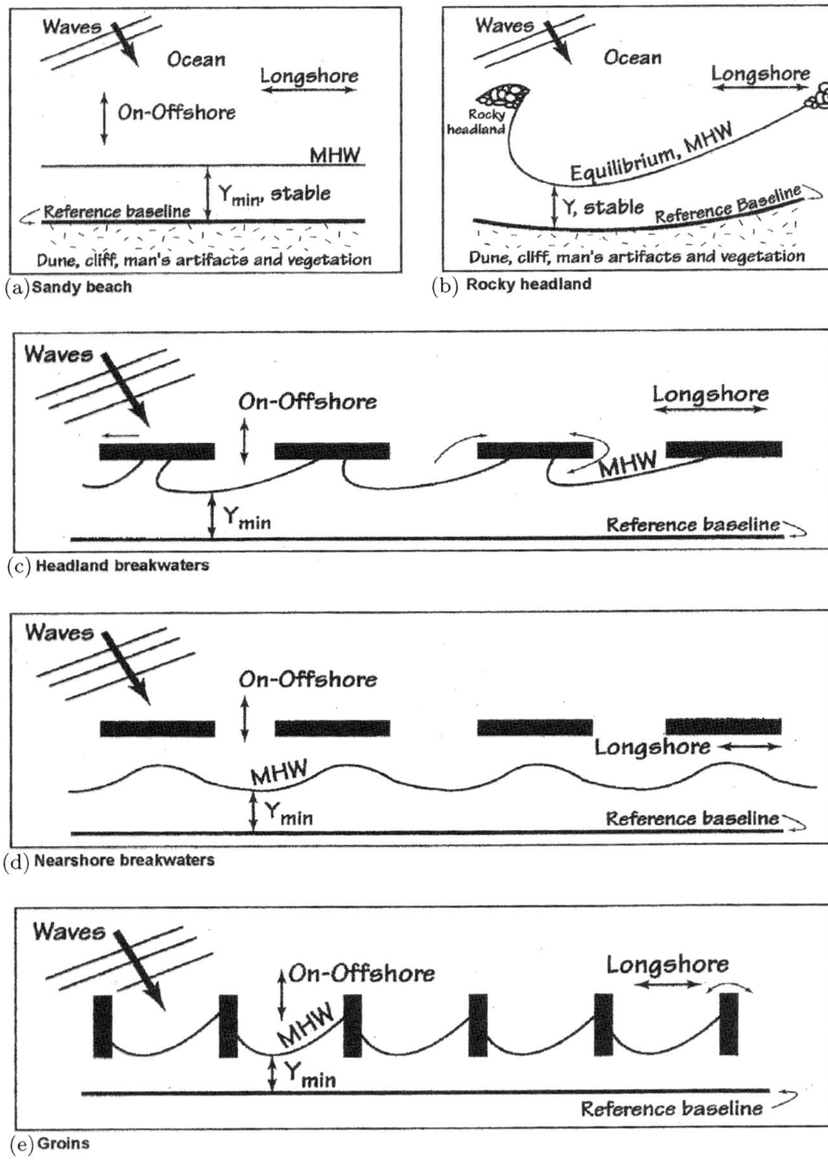

Figure 5.33. Natural (a,b) and artificial stable shorelines with minimum dry beach width, $Y_{\min}$ for (c) headland breakwaters, (d) nearshore breakwaters, and (e) groin field (from CEM, 2006).

5.6.2.2 *Microtidal beaches*

The vast majority of prototype, laboratory, and numerical model studies of salient or tombolo formations behind structures on beaches have concentrated on microtidal beaches (tide range less than 2 m). Twenty references are cited from 1966 to 1990 in Chasten *et al.* (1993) with only the key ones discussed in the following. Interestingly, none mention the importance of tidal range in their work. All relate morphological response to structural geometry and structural position relative to the original shoreline position and beach profile. Few include the wave and sediment characteristics.

An analysis of these references by Chasten *et al.* (1993) is presented in Figure 5.34 for conditions when tombolos formed (Table 4), salients formed (Table 5), and when there was minimal shoreline response (Table 6). Much of their analysis is credited to the work of Rosati (1990). The dimensionless "condition" was the L_s/Y ratio where L_s is the length of the structure, and Y is the structure distance offshore from the average beach shoreline or the renourished beach shoreline.

Clearly, the larger this dimensionless ratio, L_s/Y the longer the structure or the closer to the shoreline so that in Figure 5.34 (Table 4 Tombolos) (tombolos) almost all the references show the ratio, L_s/Y much greater than 1.0. At the opposite extreme is Table 6 (minimum or no shoreline response) where the condition ratio, L_s/Y, is far less than unity meaning the short breakwater lengths far offshore have minimal impact on the morphological response of the shoreline. Then, in the middle, Table 5 for salient formation, the ratio, L_s/Y is near unity. But there is much variation in these general results and possibly due to tidal variations.

Single Breakwater: Suh and Dalrymple (1987) using movable-bed, laboratory-scale experiments, and zero tidal change conditions found that tombolos usually formed behind a single breakwater when $L_s/Y \geq 1.0$. Dally and Pope (1986) recommend that $L_s/Y = 1.5$ to 2.0 for design of a tombolo behind a single breakwater, from examination of prototype results in the field of microtidal conditions, which is more conservative.

Multiple Breakwater Structures: The gap distance, L_g, between successive, shore-parallel breakwater structures is now important. The large gap distance will let more wave energy through the gap to keep the morphological response salient for a given breakwater length and offshore distance. Small breakwater gaps will allow less wave energy to penetrate behind the structures and form tombolos. The ratio of breakwater length to breakwater gap, L_s/L_g, will govern the beach morphological response. Large L_s/L_g

Table 4
Conditions for the Formation of Tombolos

Condition	Comments	Reference
$L_s/Y > 2.0$		SPM (1984)
$L_s/Y > 2.0$	Double tombolo	Gourlay (1981)
$L_s/Y > 0.67$ to 1.0	Tombolo (shallow water)	Gourlay (1981)
$L_s/Y > 2.5$	Periodic tombolo	Ahrens and Cox (1990)
$L_s/Y > 1.5$ to 2.0	Tombolo	Dally and Pope (1986)
$L_s/Y > 1.5$	Tombolo (multiple breakwaters)	Dally and Pope (1986)
$L_s/Y > 1.0$	Tombolo (single breakwater)	Suh and Dalrymple (1987)
$L_s/Y > 2\,b/L_s$	Tombolo (multiple breakwaters)	Suh and Dalrymple (1987)

Table 5
Conditions for the Formation of Salients

Condition	Comments	Reference
$L_s/Y < 1.0$	No tombolo	SPM (1984)
$L_s/Y < 0.4$ to 0.5	Salient	Gourlay (1981)
$L_s/Y = 0.5$ to 0.67	Salient	Dally and Pope (1986)
$L_s/Y < 1.0$	No tombolo (single breakwater)	Suh and Dalrymple (1987)
$L_s/Y < 2\,b/L_s$	No tombolo (multiple breakwaters)	Suh and Dalrymple (1987)
$L_s/Y < 1.5$	Well-developed salient	Ahrens and Cox (1990)
$L_s/Y < 0.8$ to 1.5	Subdued salient	Ahrens and Cox (1990)

Table 6
Conditions for Minimal Shoreline Response

Condition	Comments	Reference
$L_s/Y \leq 0.17$ to 0.33	No response	Inman and Frautschy (1966)
$L_s/Y \leq 0.27$	No sinuosity	Ahrens and Cox (1990)
$L_s/Y \leq 0.5$	No deposition	Nir (1982)
$L_s/Y \leq 0.125$	Uniform protection	Dally and Pope (1986)
$L_s/Y \leq 0.17$	Minimal impact	Noble (1978)

Figure 5.34. Table 4 (Tombolo), Table 5 (Salient), and Table 6 (Minimal Response) from Chasten *et al.* (1993) for L_S/Y ratio.

ratios should produce tombolos, and conversely, small L_s/L_g ratios will develop salients or no sinuosity of the shoreline.

As for a single breakwater, the offshore distance of the structure, Y, is also a key factor. To normalize this factor, Pope and Dean (1986) choose

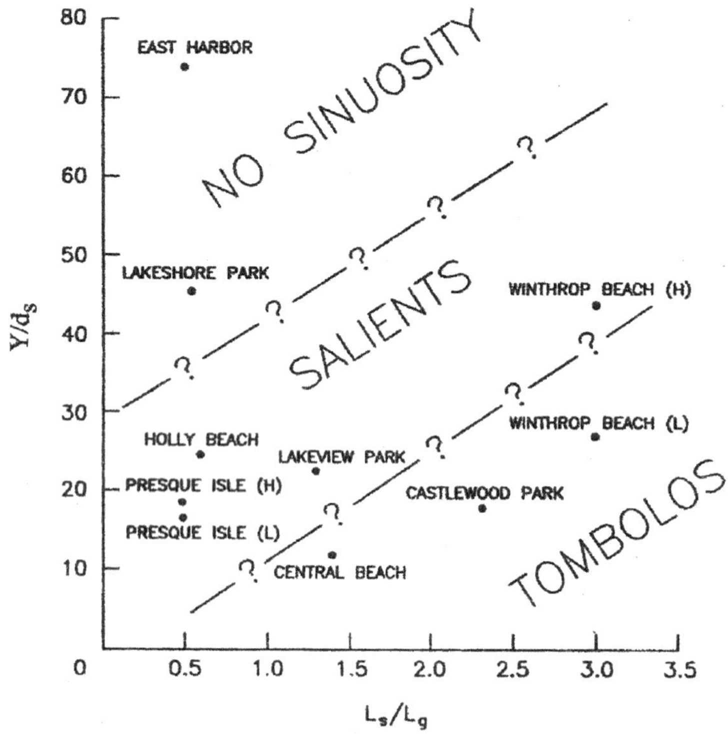

Figure 5.35. Field data from the Corps of Engineers studies of breakwaters constructed around the United States (from Pope and Dean, 1986).

the average water depth at the structure, d_s, to give the offshore distance ratio, Y/d_s, and used data from the field to prepare the dimensionless plot of L_s/L_g versus Y/d_s as shown in Figure 5.35.

When the L_s/L_g ratio is large (long structure, small gap) so that less wave energy reaches the shoreline and/or when the Y/d_s ratio is small (structure close to shore in shallow water), tombolos form behind multiple structures which are called "headland" breakwaters. When the opposite is true with these key ratios, more wave energy reaches the shoreline to maintain LST processes and only salients are present behind what are called "nearshore" breakwaters. Clearly, as illustrated in Figure 5.35, when short structures with large gaps are located far offshore in relatively deep water, they may have little or no influence on the shoreline (no sinuosity) and offer little shore damage mitigation.

Table 5.13. Beach response index I_s for five types of beach response (Pope and Dean, 1986; Ahrens and Cox, 1990).

Type of Beach Response	Beach Response Index, I_s
Permanent tombolo formulation	1
Periodic tombolos	2
Well-developed salients	3
Subdued salient	4
No sinuosity	5

The sloping lines (with "?" symbols) are somewhat arbitrary. To further define a shoreline-classification scheme, Pope and Dean (1986) defined five types of shoreline response to a system of detached, shore-parallel, nearshore breakwaters, and Ahrens and Cox (1990) quantified the five types by a beach response index, I_s, as formulated in Equation (5.55) with index levels shown in Table 5.13.

$$I_s = \exp[1.72 - 0.41 \times (L_s/Y)]. \tag{5.55}$$

For example, assume a breakwater length $L_s = 30$ m ($\approx$100 ft) and located at $Y = 20$ m ($\approx$67 ft) offshore so that $L_s/\mathrm{Y} = 1.5$ gives a beach response index, $I_s = 3.0$, so that a well-developed salient should occur. There are many more possible combinations of L_s and Y that give the same results.

5.6.2.3 *Meso- and Macrotidal Beaches*

When the tidal range is 2–4 m, the range is mesotidal and if greater than 4 m, the range is called macrotidal. By contrast with the many studies discussed before, very few field, laboratory, and numerical model studies have been made of the functional performance of shoreline stabilization structures under these meso- and macrotidal range conditions. The differences are explained in Section 5.8, where the new design guidance found in Environmental Agency (2010) is summarized in detail.

5.7 Planform Analysis of Headland Breakwater

Figures 5.28 and 5.29 schematically illustrate the theory for one wave direction which in nature is rarely the case. Waves approach the shoreline from a wide variety of directions as illustrated by a wave rose in Chapter 2, Section

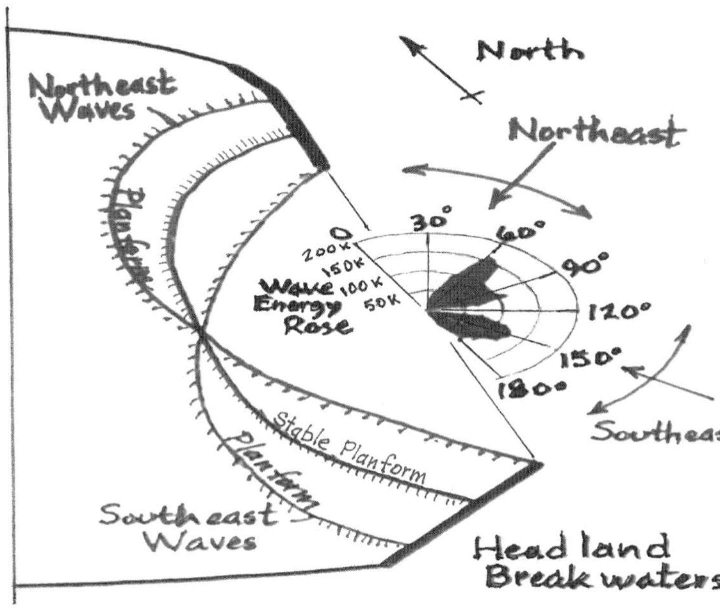

Figure 5.36. Stable planform (red) estimated from two dominant wave directions using the parabolic model.

2.3.8.3, Figure 2.31. Many sites have waves approaching from two dominant directions as schematically illustrated in Figure 5.36 showing wave roses from the Northeast (blue) and the Southwest (green) directions. The parabolic model can then be applied twice for the two dominant wave directions and the resultant, stable planform shape (red) be estimated (weighted average position) as the result.

For design, the ratio of $L_s/\mathrm{Y} = 1.8$ is recommended for a headland breakwater system (Dally and Pope, 1986).

Over 40 headland breakwater systems have been built around the Chesapeake Bay (Hardaway *et al.*, 1991) in Virginia, USA. They recommend using $L_s/L_g = 1$ and setting $Y_g/L_g = 0.6$, where Y_g is the gap distance from a line connecting the structures and the renourished beach shoreline. These shore-parallel structures have been combined with a beach fill at the time of construction and the berm elevation set so that the subaerial beach is visible (tombolo) at high tide.

At some shoreline locations with large LST rates in one dominant direction, an initially constructed, detached, nearshore breakwater system will

Figure 5.37. The system of 19 nearshore breakwaters at Fort Story, Virginia that have been transformed into headland breakwaters by longshore sediment from bottom-left to upper-right corner (courtesy City of Virginia Beach, VA).

become filled with sediment and be transformed to a headland system with tombolos. This is what took place at Fort Story, Virginia on the Atlantic Ocean near Virginia Beach, Virginia as shown in Figure 5.37. Sediments moving to the north (bottom-left to upper-right) have filled in behind the breakwaters and trapped a large volume of sediment that was not transported to down-drift beaches and is now producing down-drift impacts. The designer must always be aware of the potential for sand retention behind a system of detached, nearshore breakwaters that could cause down-drift erosion.

In Italy, D'Alessandro and Frega (2007) proved that Equation (5.53) modeled the shoreline behind a series of headland breakwaters along the Tyrrhenian sea coast. And, they also showed that the formula to calculate the surface area in the tombolo created by the breakwater could be modeled by the method given by Ming and Chiew (2000).

5.8 Planform Analysis of Nearshore Breakwaters

5.8.1 *Micro-tidal conditions*

Nearshore breakwaters (Figure 5.33(d)) are found mainly in the US, Japan, and in the Mediterranean Sea. These locations are primarily along coasts with small (microtidal, 0–2 m) daily variations in tidal range. The empirically derived relationships for the primary variables shown in Figure 5.33 and listed in the following, are for microtidal coasts and from the US experience as reported in CEM (2006).

Nearshore breakwater systems are generally shore-parallel and detached structures that are located and spaced such that only salients form behind the structure. Figure 5.38 presents the key variables in the planform and profile views. Their purpose is the same as that for headland breakwater systems, namely, to (1) increase the fill life (longevity) of the beach nourishment project behind the breakwaters; (2) provide upland protection from storm damage; (3) provide a wide recreational beach; and in some cases (4) create or stabilize an upland wetland area.

The key geometric variables are the structure length, L_g and the distance offshore, Y, from the renourished beach at the design beach width, W. Dally and Pope (1986) recommend $L_g/Y = 0.5$ to 0.67 for shore system and even further offshore (i.e. $L_g/Y = 0.125$) for very long systems of nearshore breakwaters as, for example, constructed at Presque Isle, Pennsylvania, US.

A detached, nearshore breakwater system with salient formation is the preferred breakwater shoreline stabilization system recommended by the US Army Corps of Engineers as stated in CEM (2006). This design permits LST to continue to move through and past the project area to down-drift shorelines and minimize down-drift impacts. Wave action and longshore currents are strong enough to keep the salient from connecting to the structure. An extensive literature review of both nearshore and headland breakwater systems can be found in Chasten *et al.* (1993) as applied on microtidal beaches.

Note in Figure 5.38 that the salient distances from the design beach width, W, shoreline positions are not symmetrical ($Y_s \neq Y_g$) and that the storm erosion distance, e, in the gap must be included to predict $Y_{\min}$ for the design storm. The planform configuration of the salient-shaped beach, and the storm erosion distance, e, have been studied extensively over the years, but no general analytical procedure exists. See Rosati (1990) for a plot of the gap erosion for a number of breakwater cases in the US. The breakwater crest elevation and crest width, the structure permeability,

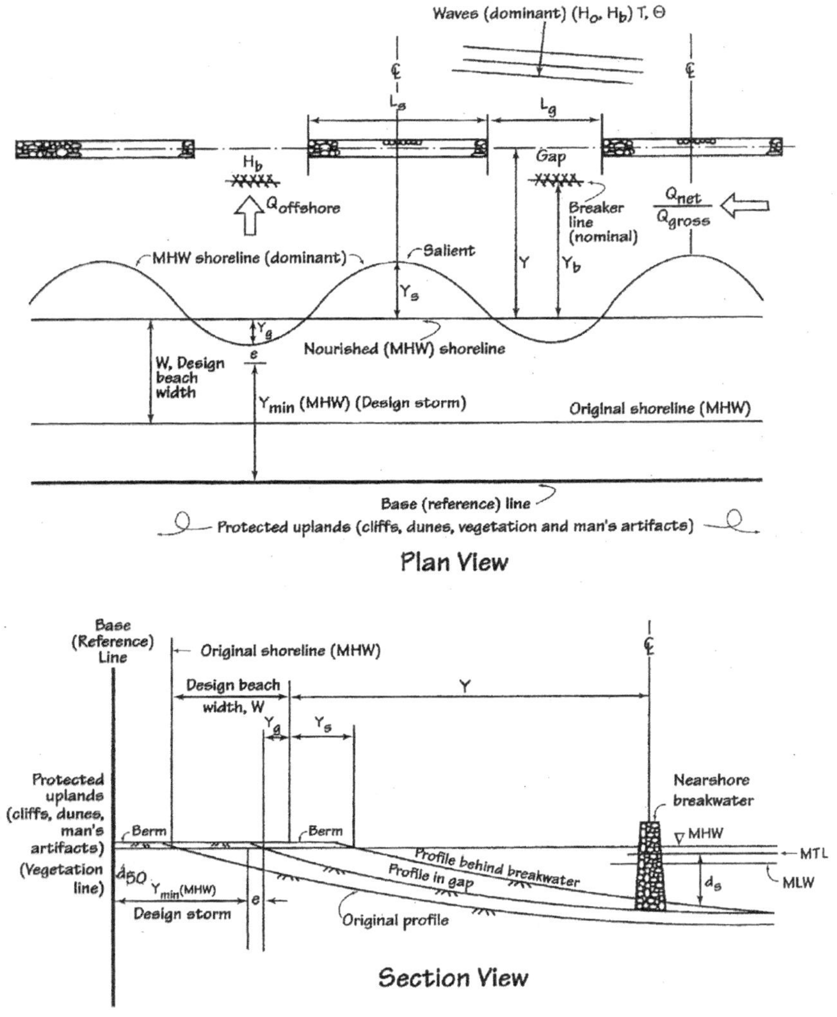

Figure 5.38. Nearshore breakwater system in plan and profile views (from CEM, 2006).

the structure slope, and other factors all influence the functional performance of nearshore breakwater systems. For these reasons, numerical models of waves, wave-induced currents, sediment transport, and bathymetric change have been employed to study the shoreline modifications created by nearshore breakwater systems as discussed further in Section 5.10.

5.8.2 *Macro-tidal conditions*

Because over 75% of the UK coastline is classified as meso- or macrotidal, Rodgers *et al.* (2006) concluded that UK needed to study and develop guidelines for breakwater performance on sandy, macrotidal coasts. Two morphological, numerical models were employed: PISCES (developed by HR Wallingford) and MIKE 21 CAMS (developed by Danish Hydraulic Institute (DHI) Water and Environment, Denmark) as reported in Environmental Agency (2009) and gave conflicting results. This is not surprising since the numerical models require four coupled, submodules (waves, currents, sediment transport, and bathymetric change) and use different empirical formulations for sediment transport. In fact, the design guidelines cited before for microtidal coasts have been used to confirm some numerical model results (see e.g. Karambas, 2012). However, results of the numerical modeling study did reveal some important differences compared with microtidal results that have been summarized in Environmental Agency (2010).

In general, the length of the salient Y_s decreases with increasing tidal range. However, for large values of $L_s/Y(> 1.3)$ where the breakwater is relatively close to the shoreline, the influence of tidal range is smaller if the breakwater crest is above the mean high water (MHW) tide level. The numerical models also showed that the salient length is less as the breakwater crest is reduced. And, as expected, the beach elevations are lower as the breakwater crest level is reduced. Finally, both models produced less reduction in the nearshore profile in the gap region with increasing tidal range. Equation (5.53) as developed by Hsu *et al.* (1989) was still found to be applicable on macrotidal coasts. See Environmental Agency (2009) for examples.

As a result, the existing design guidance (Fleming and Hamer, 2000) was replaced with a new design procedure for macrotidal coasts as found in the Environmental Agency (2009) report.

5.9 Planform Analysis of Groin Systems

5.9.1 *Negative perceptions of groins*

Groins are constructed to retain sand on the subaerial beach. Groins are the oldest, most-commonly applied, and arguably the most misused, non-engineered, and improperly constructed of all coastal structures. This is because in the distant past they were either constructed in the wrong sequence or formed by themselves without the simultaneous placement of new sand on the beach between the groins as a beach nourishment project.

Modern coastal engineering practice combines beach nourishment with a groin design suitable to permit immediate sand bypassing of the groin field with minimal loss to the sand-sharing system. Groins are normally shore-perpendicular structures of relative short lengths (when compared with navigational jetties at tidal inlets) as illustrated in Figure 5.33(e). If LST is present, the groin interrupts the longshore transport such that sand is deposited on the up-drift side and consequently absent on the down-drift side as illustrated in Figure 5.39(a) for a single groin. The net longshore transport direction is revealed (sand accumulates on the side of the groin

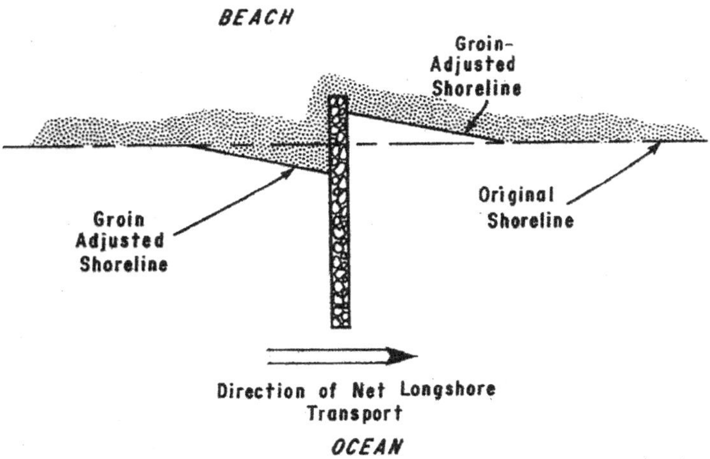

(a) Single groin

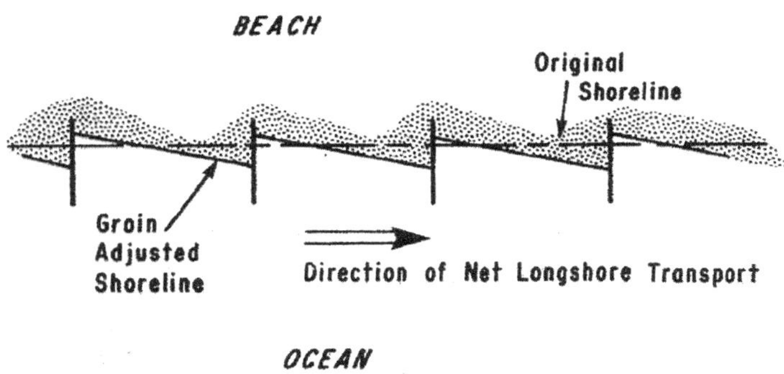

(b) Groin field

Figure 5.39. Simple schematic diagram of (a) single groin and (b) groin field where longshore transport of sand exists, no background erosion is present, and no sediment gets transported past the groin (from CEM, 2006).

from the dominant, up-drift wave direction) for locations where the annual wave climate produces LST in both directions but volumes that are unbalanced.

Figure 5.39(b) schematically depicts the traditional, academic view of groin fields that has led to their widespread, negative view by most oceanographic scientists and coastal zone management specialists. In practice, at many locations in the world, the down-drift impacts may be due to the combined influence of several factors, namely,

(1) the level of historic, background, shoreline retreat (erosion) at the site that does not change;
(2) the possible presence of up-coast revetments (and/or seawalls) that have reduced sediment volumes in the groin construction area;
(3) the amount of longshore sediment that gets past the groin (*bypassing* around the end, *over-passing* across the groin profile, *shore-passing* on the beach berm, and *through-passing* via internal voids).

It is extremely difficult to separate which of these factors is the dominant factor so that the simplistic assumption that the negative volume on the down-drift side is solely caused by the groin is not correct.

5.9.2 *Literature review*

Special Issue No.33 of the *Journal of Coastal Research* (Kraus and Rankin, 2004) focused on the functioning and design of coastal groins. It included papers on the background and literature review, management, design, case studies, and numerical and physical modeling of groin fields. It contains a wealth of information and should be consulted for both functional and structural designs.

In the Corps of Engineers, *Coastal Engineering Manual*, CEM (2000), a fundamental discussion of groins is found in Part V, Section 3 "Shore Protection Projects" as written by the author of this book. A summary has also been published by Basco and Pope (2004) in Special Issue No. 33 of the *Journal of Coastal Research.*

Kraus *et al.* (1994) listed 13 functional properties attributed to groins and present a critical evaluation of each as reproduced in Table 5.14. The first five are well-accepted properties that have led to the general rule of thumb (ROT) to make the groin spacing to length ratio about 2 to 4. The length controls water depth at the end and hence the amount of sediment *bypassing* around the tip.

Table 5.14. Functional properties of groins (from Kraus *et al.* 1994)

Property	Comment
1. Wave angle and wave height are leading parameters (longshore transport).	Accepted. For fixed groin length, these parameters determine bypassing and the net and gross longshore transport rates.
2. Groin length is a leading parameter for single groins (length controls depth at tip of groin.)	Accepted, with groin length defined relative to surfzone width.
3. Groin length to spacing ratio is a leading parameter for groin fields.	Accepted. See previous item.
4. Groins should be permeable.	Accepted. Permeable groins allow water and sand to move alongshore, and reduce rip current formation and cell circulation.
5. Groins function best on beaches with a predominant longshore transport direction.	Accepted. Groins act as rectifiers of transport. As the ratio of gross to net transport increases, the retention functioning decreases.
6. The updrift shoreline at a groin seldom reaches the seaward end of the groin. (This observation was not found in the literature review and appears to be original to the present paper.)	Accepted. Because of sand bypassing, groin permeability, and reversals in transport, the updrift shoreline cannot reach the end of a groin by longshore transport processes alone. Onshore transport is required for the shoreline to reach a groin tip, for a groin to be buried, or for a groin compartment to fill naturally.
7. Groin fields should be filled (and/or feeder beaches emplaced on the downdrift side).	Accepted. Filling promotes bypassing and mitigates downdrift erosion.
8. Groin fields should be tapered if located adjacent to an unprotected beach.	Accepted. Tapering decreases the impoundment and acts as a transition from regions of erosion to regions of stability.
9. Groin fields should be built from the down-drift to up-drift directions.	Accepted, but with the caution that the construction schedule should be coordinated with expected changes in seasonal drift direction.
10. Groins cause impoundment to the farthest point of the updrift beach and erosion to the farthest point of the downdrift beach.	Accepted. Filling a groin field does not guarantee 100% sand bypassing. Sand will be impounded along the entire updrift reach, causing erosion downdrift of the groin(s).
11. Groins erode the offshore profile.	Questionable and doubtful. No clear physical mechanism has been proposed.
12. Groins erode the beach by rip current jetting of sand far offshore.	Questionable. Short groins cannot jet material far offshore, and permeable groins reduce the rip current effect. However, long impermeable jetties might produce large rips and jet material beyond the average sutizone width.
13. For beaches with a large predominant wave direction, groins should be oriented perpendicular to the breaking wave crests.	Tentatively accepted. Oblique orientation may reduce rip current generation.

The cross-sectional elevation in the swash zone controls *over-passing*, the length and elevation on the beach berm control *shore-passing*, and the structure materials control *through-passing* that occur within rubble-mound and permeable groins. In general, the groin crest line should be designed to vary in the cross-shore direction to mimic a natural cross-shore beach profile. Philosophically, groins should be designed to function as a cross-shore beach template (or brace), and not as a block (or "dam") to longshore transport. The permeability of a groin will allow hydrodynamic exchange across and through the groin which is important in reducing rip currents and offshore losses.

Kraus *et al.* (1994) concluded that the up-drift shoreline will rarely reach the seaward end of the groin (No. 6, Table 5.14). Properties No. 7–9 of the table have long been accepted standard conditions for groin-field design and construction. But as noted in No. 10, filling a groin field does not guarantee 100% sand bypassing.

The entire, filled groin field system may function as a headland, impounding sand on the up-drift side and creating a down-drift impact. Properties No. 11 and 12 are often cited by opponents of groins but are labeled "questionable" in Table 5.14 because they cannot be supported by physical mechanisms nor sand conservation principles. No. 13 is tentatively accepted.

Tidal range, predominant wave characteristics, net and gross sediment transports, grain size, and profile slope are all key hydrodynamic and sediment parameters. These factors will lead to the optimum spacing and planform configuration of the shoreline within each compartment for average climate conditions.

5.9.3 *Functional design*

5.9.3.1 *Planform layout*

Boundary conditions for groins in numerical models of shoreline change give insight into how they must function (e.g. for the GENESIS model of Hanson and Kraus (1989). Three key factors are (1) bypassing, (2) permeability, and (3) longshore transport. These process-based factors controlling groin function are summarized in Table 5.15 with the parameters and in Figure 5.40.

Bypassing around the seaward end depends on the water depth and breaking wave height at the tip. The depth of active LST can be taken as some multiple of the significant breaking wave height (e.g. Hallermeier,

Table 5.15. Three main process factors controlling groins (from Kraus *et al.*, 1994).

Process	Parameter	Description
1. Bypassing	D_g/H_b	Depth at groin tip/breaking wave height
2. Permeability		
• Over-passing	$Z_g(y)$	Groin elevation across profile, tidal range
• Through-passing	$P(y)$	Grain permeability across shore
• Shore-passing	Z_b/R	Berm elevation/runup elevation
3. Longshore transport	Q_n/Q_g	Net rate/gross rate

Notes: $D_g = d_g$ = mean water depth at groin tip; H_b = breaking wave height (significant wave height at breaking); $Z_g(y)$ = groin elevation across the profile and a function of the tidal range; $p(y)$ = permeability of the groin structure design (armor layer, sublayers, core); Z_b = elevation of the beach berm; R = wave runup elevation; Q_n = net LST rate; Q_g = gross LST rate.

1983 who suggested 1.6). The key ratio is water depth at the tip (end) of the groin, Dg or (dg) to the breaking wave height, H_b. Groin length relative to the surf zone width could also be employed to estimate the bypassing factor.

The permeability factor combines groin elevation (over-passing), groin porosity (through-passing), and tidal range/crest elevation/land location (shore-passing) as sketched in Figure 5.40. A 100% permeability factor means transport rates and shoreline change patterns identical to that with no groins present. The third key factor is the ratio of the net transport rate Q_n to the gross rate Q_g (Bodge, 1992). When the Q_n/Q_g ratio is zero, a perfectly balanced transport (no net) produces symmetrical fillets on both sides of a single groin. The opposite extreme is Q_n/Q_g equals unity, meaning unidirectional transport and a single fillet.

Kraus *et al.* (1994) exercised the GENESIS model for a single groin and studied the bypassing formulation. They found that

- rapid filling was followed by gradual build-up over time,
- filling to capacity was only possible for $Q_n/Q_g = 1$, the unidirectional case,
- as $Q_g > Q_n$, seaward growth of the shoreline decreased,
- an increased amount of material bypassing occurs as the shoreline grows out toward the tip.

These tests mean groins need cross-shore sediment transport processes to obtain beach elevations and widths that are beyond that reproduced in one-line, shoreline change models. Realistic distributions of longshore

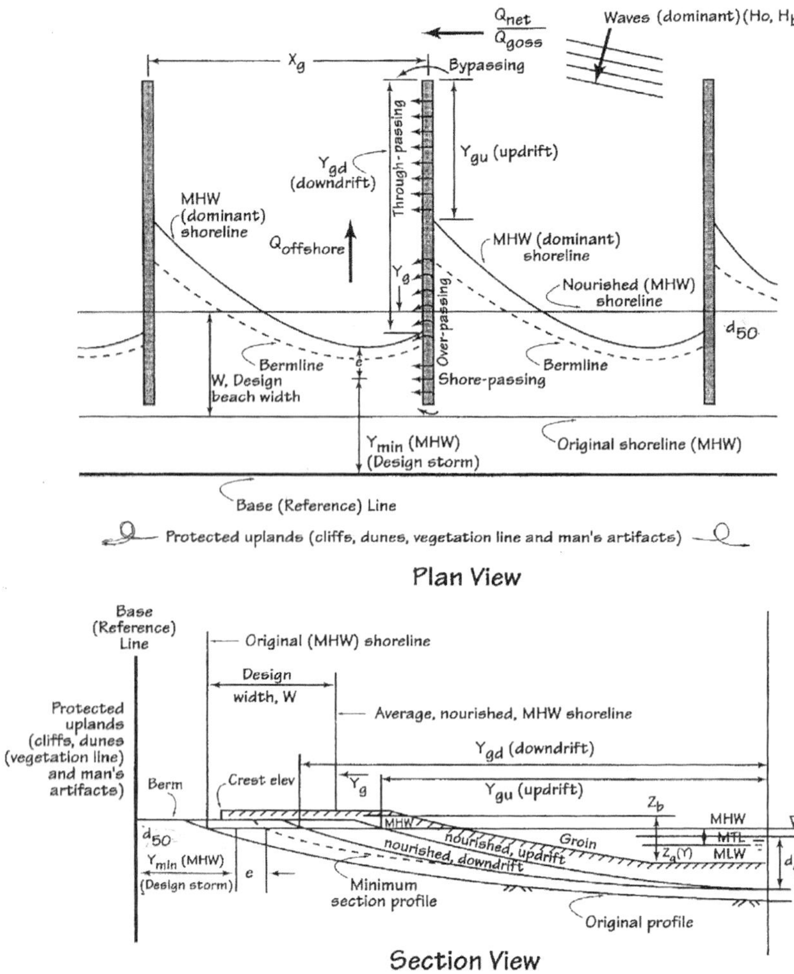

Figure 5.40. Definition sketch of key variables in functioning of groins (top plan and bottom section) (from CEM, 2006).

current and sediment transport across the surf zone, beach profile shapes with bars and troughs, and other sediment transport mechanisms (wind, tide) are further complicating factors that have yet to be fully addressed in numerical models.

These three process factors incorporate most of the 20 or more fundamental variables (Kraus *et al.*, 1994). The geometry ratio of spacing X_g to length Y_g is also the controlling factor for groin systems, as found in the literature. In Figure 5.40, Y_g represents a *mean* groin length measured from the average, nourished beach, MHW shoreline. Using Y_{gu} (up-drift) gives a larger ratio or using Y_{gd} (down-drift) produces a smaller ratio that may account for some variability. The *Shore Protection Manual* (SPM, 1984) suggested $X_g/Y_g = 2$–3 for the proper functioning of shore-normal groins and this ratio still is valid today.

5.9.3.2 *Groin profile*

The cross-shore profile of the top of the groin should mimic the local, normal beach profile. A typical groin profile with inshore (berm) section, sloping middle section, and horizontal outer section is shown in Figure 5.41. In general, the landward end is set at the elevation of the natural, existing beach berm, the sloping section matches the slope of the beach in the swash zone and the outer, seaward section as set at mean low water or lower.

5.9.3.3 *Groin field transition (taper)*

Generally, groin lengths are decreased along a line converging to the natural shoreline from the last full-length groin, making an angle of about 6° with

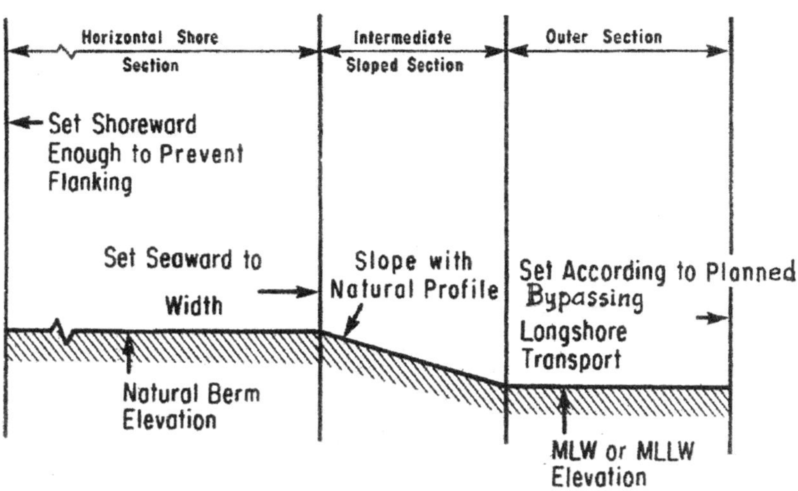

Figure 5.41. Recommended groin profile with sloping section to permit sand bypassing both over the groin and around the outer end to mitigate down-drift impacts.

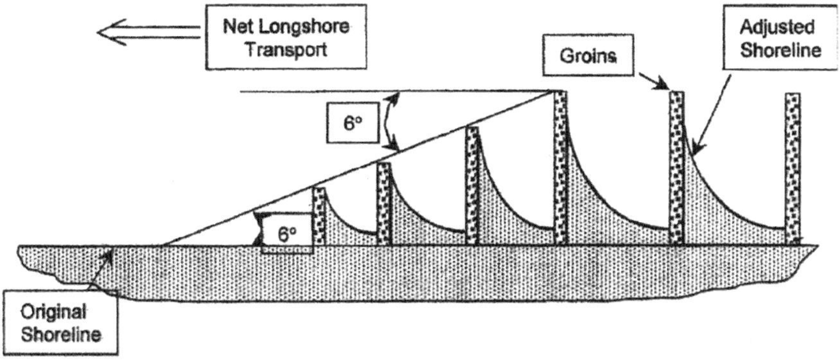

Figure 5.42. Transition from groin field to natural beach. The number of shorter groins in the taper is not limited to that shown in this schematic drawing. One groin in the taper distance may be sufficient (from CEM, 2006).

the original shoreline as depicted in Figure 5.42. The spacing is also reduced to maintain the same X_g/Y_g ratio (2–3) used in the design.

Further details regarding permeability, terminal groins, order of construction, non-traditional configurations, and spur groins can be found in the *Coastal Engineering Manual* (CEM, 2006, Part V, Section 3). See also Chapter 10 for an example design of a terminal groin with beach nourishment and dune construction.

5.9.3.4 *Basic rules for the functional design of groins*

There are 10 basic "rules" recommended for the functional design of groins (Basco and Pope, 2004).

Rule 1. Conservation of mass for transport of sediment alongshore and cross-shore means groins neither create nor destroy sediment.

Rule 2. To mitigate erosion of adjacent beaches, always include a beach fill in the design.

Rule 3. Agree on the minimum, dry beach width, $Y_{\min}$ for upland protection during storm events as a measure to judge success.

Rule 4. Begin with $X_g/Y_g = 2$–3 at time of construction. The groin length Y_g is from the renourished beach.

Rule 5. Use a numerical, one-line model to estimate shoreline change around single groins and groin fields (e.g. Hanson and Kraus, 1989, the GENESIS model of the Corps of Engineers).

Rule 6. Use a cross-shore, profile-change model to estimate the minimum, dry beach width $Y_{\min}$ during storm events (e.g. Larson and Kraus, 1989, the SBEACH model of the Corps of Engineers).

In the near future, Rules 5 and 6 may be combined by using a numerical, beach evolution model that combines both longshore and cross-shore transport processes in one model. See Section 5.10.

Rule 7. Bypassing, structure permeability, and the balance between net and gross longshore transport rates are the three key factors in the functional design. Use model simulation to iterate a final design to meet the $Y_{\min}$ criterion.

Rule 8. Consider tapered ends, alternate planforms, and cross-sections to minimize the impacts on adjacent beaches. For example, consider T-groins, permeable groins, notched groins, submerged groins, and other shapes (see Basco and Pope, 2004). See Chapter 10 for an example with a T-head groin.

Rule 9. Establish a field-monitoring effort to determine if the project is successful and to identify adjacent beach impacts.

Rule 10. Establish a "trigger" mechanism for decisions to provide modification (or removal) if adjacent beach impacts are found not acceptable.

A "Rule 0" could be added to emphasize the fact that groins are only useful where LST processes are dominant. If cross-shore transport processes control, then nearshore or headland breakwaters should be considered first.

A statement made in CEM (2006) has been misapplied and taken out of context by opponents of groins. The statement says "... probably the most misused and improperly designed of all coastal structures" (page V-3-59) when reviewing the history of groins. These opponents then fail to also state what follows (page V-3-60):

> *Modern coastal engineering practice is to combine beach nourishment with groin construction to permit sand to immediately begin to bypass the groin field.*

This modern practice to always combine groin construction with beach nourishment fills the compartment on the up-drift side so that sand bypassing begins immediately and mitigates the down-drift impacts. But this statement in CEM (2006) is often not mentioned by opponents of groins.

See CEM (2006) for example problems in the functional design of groins.

5.10 The Role of Model Tests in Functional Design

5.10.1 *Process-based numerical models*

Physical scale models and process-based, numerical models can be employed to study and understand the functional design of shoreline stabilization structures. However, large-scale physical models are expensive and suffer from the scale effects for sediment transport especially when waves are important as a forcing function. Consequently, the scientific and coastal engineering communities are employing dynamic, morphological models, or "morpho-dynamic models" that attempt to solve the mathematical equations of the underlying physics written in discrete, numerical form for solution on computers.

Process-based, morpho-dynamic numerical models calculate the 3D change in the sandy beach surface elevation (both subaerial and subaqueous) near the coastal structures illustrated in Figure 5.33. Four numerical models (modules) that describe the physics of waves, water levels (flow) sediment transports, and beach surface (bed) changes must be coupled together in x–y–z space and time, t. Each numerical model is a mathematical representation in a discrete number of space elements and time steps of the physical processes that includes empirical formulations and simplifications for solution on computers. These four coupled models can be called "coastal morpho-dynamic numerical models".

Figure 5.43(a) is a graphic illustration for the bed changes surrounding a tidal inlet with a normal jetty on the upper side and a slanted, weir jetty on the lower side of the inlet. The bottom, right-side depicts two main sediment deposition areas (red) with one at the end of the upper-side jetty and the second on the inside of the lower, uniform weir section of the lower jetty. The sediment deposits (red) would be as expected for longshore sediment transport from the top to bottom due to the wave climate and some sediment transport due to tidal and wave-induced currents from bottom to top. The weir jetty system at Rudee Inlet, Virginia is shown in Figure 5.43(b) and is similar to this graphic as found in the Corps of Engineers, Coastal Modeling System (CMS) website, http://www.erdc.usace.army.mil/Media/Fact-Sheets.

Note the red arrows in the middle of Figure 5.43(a). Starting with the initial bathymetric conditions, the flow field, waves, and resulting sediment transport are computed in the three respective modules and then the bathymetric (bed) change is estimated in the fourth module. Consequently, this bottom change must then be sent back (red arrows) to calculate the flow,

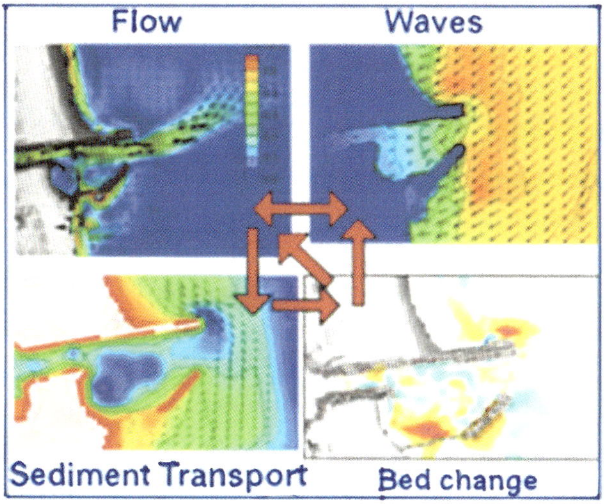

Figure 5.43(a). Schematic diagram of flow, waves, sediment transport and bottom (bed) change (from Corps of Engineers, CMS website http://www.erdc.usace.army.mil/Media/Fact-Sheets%20for%20a%20%20tidal%20inlet).

Figure 5.43(b). Photo of Rudee Inlet, Virginia Beach looking East with normal jetty on left side and slanted weir jetty on right side. The schematic diagram in Figure 5.43(a) is similar when rotated 90° (courtesy Norfolk District, Corps of Engineers).

waves, and sediment transport that results. In other words, the changing bottom (morphological) conditions dynamically influence the results within or in the subsequent time steps in the numerical calculation. Although the illustration is for conditions surrounding a tidal inlet, the application of a coastal morpho-dynamic numerical model for the shoreline stabilization structures in Figure 5.33 is the same.

All four modules need boundary conditions specified for all four sides of each area covered by the model and over the length of time required for the simulation. Because the morphological changes of the bed are much slower than the hydrodynamic forcing conditions of the waves and currents, different timescales are often employed for the sediment module than the flow and wave modules. Wetting–drying subroutines at the land–water interface permit calculation of shoreline change.

The gird system in Figure 5.43(a) is uniform as evident by the regular spacing of the arrows shown to delineate the flow, wave, and sediment transport directions. In many cases, finer grid spacing is needed to capture the physical details more accurately. Nested grids are possible with finer scales, but boundary-fitted flexible mesh (FM) grids are becoming more popular for ease in implementing the smaller scales and accurately representing the curvilinear shorelines and structural elements in the model. See Roelvink *et al.* (2016) for an excellent summary of the principles of process-based morpho-dynamic models and an extensive list of references that apply the software products of the three major developers of these numerical models.

5.10.2 *Software development organizations*

The development, maintenance, and continual improvement of a coastal morpho-dynamic numerical model is complex, time-consuming, and expensive. There are only three software development organizations that have done this in the world at the time of writing this book that are known to the author. A brief introduction to each is presented in the following along with the current contact information through a published website. No attempt will be made to rate or rank them. All are currently being applied by coastal engineers and scientists for research and to study the existing coastal problems and in engineering design. They each require substantial time and effort to learn how to use them properly.

5.10.2.1 *Deltares (The Netherlands)*

Delft3D is the brand name for the Dutch-based, coastal morpho-dynamic numerical model (see Delft 3D, 2019). As of January 1, 2011, the hydrodynamic module Delft3D (FLOW), the waves module Delft3D (WAVE), and the morphology module Delft3D (MOD) were available free of charge as open-source software models. The resulting bathymetric changes are handled in the morphology module (MOD). The hydrodynamic (FLOW) module can be both 2D (horizontal, depth-averaged) or fully 3D on a curvilinear, boundary-fitted grid or with spherical coordinates. For 3D, the vertical grid is a sigma coordinate or z-layer grid. The wave module (Delft3D, WAVE) is the robust, short-crested SWAN model (see e.g. Holthuijsen *et al.*, 1993; Booij *et al.*, 1999; Ris *et al.*, 1999). Sediment transport (suspended, bed, and total load) and bottom changes for the cohesive and non-cohesive fractions of sediment grain sizes are estimated in the Delft3D (MOD) module.

In 2015, the Delft3D Flexible Mesh Suite 2016 was made public (Delft3D FM). The new generation of such models are called *FM Models* in which the local model resolution can be adjusted as required. This technique provides the possibility of modeling large areas in one single model without shifting to several layers of finer grid models. The FM models significantly improve the time required to set up and operate the numerical model.

The Delft3D morpho-dynamic numerical model is available at the independent, Dutch-based research institute Deltares website, http://oss.deltares.nl/web/delft3d/about. On this home page, links are provided to: Get Started, Download, Get Help, Forum, Research, and About Delft3D. The long-term need for further research with coastal morpho-dynamic numerical models is enhanced by making Delft3D open source so that a large number of worldwide users can freely apply, learn, and improve the modules.

One example application of Delft3D in the author's experience was for the study of barrier island breaching on Long Island, New York as reported in Roelvink (2006). An empirical, breaching algorithm was required to simulate the widening of the breach width in time. Basco (2004) related this empirical breaching algorithm to the conservation of the sediment momentum equation. This application is fully presented in Section 7.4.3 in Chapter 7. See Roelvink *et al.* (2016) for many more recent references that applied the Delft3D model.

5.10.2.2 *Danish Hydraulic Institute (Denmark)*

Up until about 2010, the brand name for the Danish-based morpho-dynamic model was MIKE21.CAMS standing for the *C*oastal *A*rea *M*odeling *S*ystem (Johnson *et al.*, 1994) as marketed by the Danish Hydraulic Institute (DHI).

The discretization in the two horizontal dimensions was by the finite-difference method, and the processes in the vertical direction were depth-averaged as originally developed by Mike Abbott giving rise to the long-standing brand name MIKE 21. Water flows, waves, and sediment transport modules were employed together with a bathymetric change module as illustrated in Figure 5.43(a). A deterministic, intrawave period, sediment transport model calculated total transport for combined wave motions and water current flow. Pioneering work was made in treating the time-scale differences in the hydrodynamic (current and wave) processes and the morphological bed-change processes (see Niemann *et al.*, 2006).

Upgrade to FM models began in 2005. In addition, considerable research efforts focused on "model reduction" techniques, i.e. methods to take advantage of the difference in timescales between hydrodynamic and morphological changes. Jones *et al.* (2007) discuss two main ways to achieve process-based model reduction. The classic way is to increase the morphological time step and they focus on this method in their paper. The other way is "formal model reduction" involving a fundamental analysis of the underlying equations and assumptions in the process models (Hudson *et al.*, 2005).

Long-term predictions for shoreline change behind and adjacent to coastal structures is driving the need to improve "model reduction" techniques in process-based models. Application of morpho-dynamic models to simulate long-term predictions become unreliable perhaps due to the complex, interactions of the time series for the flow, waves, and sediment transports at the boundaries needed for forcing the bathymetric and resulting shoreline changes. Hybrid models that combine the morpho-dynamic model and one-line (shoreline) models have been developed (Kaergaard and Droenen, 2015).

At this date, the DHI morpho-dynamic model that has replaced MIKE21.CAMS is called MIKE21 Coupled FM. Further details can be found at the website http://www.mikepoweredbydhi.com.

5.10.2.3 *Army Corps of Engineers (US)*

The brand name is the CMS for numerical modeling of the coastal physical processes as developed by the Engineer Research and Development Center (ERDC) for the US Army Corps of Engineers (formerly the Waterways Experiment Station) in Vicksburg, MS. Only Cartesian-type grids are possible (uniform, non-uniform, and telescoping) so that grid nesting is necessary for large domains and solely for two horizontal dimensions (2DH). All flow and wave processes are depth-averaged.

There are two models, namely CMS-Flow and CMS-Wave in the CMS. The CMS-Flow model handles hydrodynamics (water currents and water levels), sediment transport, and morphology change. These models can be run independently or coupled together in a process called "steering". The CMS-Wave (spectral, wave transformation) model is run first over a constant time interval called the "steering interval" and the steering process is controlled by an internal steering module (versions 4.0 and newer). Wave information (significant wave height, period, direction, radiation stress gradients including surface roller stresses in breaking waves, if present) and wave dissipation are passed to the CMS-Flow model and the flow model grid. The CMS-Flow model is then run until the next steering interval is reached. The wave variables are linearly interpolated over the specified time in the steering interval. Water depths and current velocities, are employed to update the wave length and bottom orbital velocities at each time step in the CMS-Flow model calculations. Finally, water levels, current velocities, and bed elevations are estimated for the next wave time step. Interpolation is used to move from the CMS-Flow grid to the CMS-Wave grid, if different. The CMS-Wave is run again for the next time step and the process is repeated.

The "steering interval" is related to the "model reduction" technique of increasing the morphological time step relative to the hydrodynamic time step. The interested reader is referred to the CMS Website, http://www.erdc.usace.army.mil/Media/Fact-Sheets for further details. See especially the Documentation Portal page http://cirpwiki.info/wiki/CMS#Documentation_Portal.

The cost for the CMS is free. It is interfaced and distributed through the Surface-Water Modeling System (SMS) as marketed by Aquaveo Water Modeling Solution, LLC but the SMS is not free. Their website is http://www.aquaveo.com. The SMS significantly aids the application of the CMS to real-world problems with the aid of new graphical user interfaces. See the Aquaveo website for costs of the SMS.

5.11 Summary

The key questions for planform design are the following: (1) what to use for the minimum, dry beach width, $Y_{\min}$, and (2) how to minimize down-drift impacts on adjacent shorelines? Present design practice is to always combine construction of shoreline stabilization structures with beach nourishment so that sand bypassing begins immediately during and continues after the construction is complete. Clearly, headland and nearshore breakwaters work best when the cross-shore sediment transport processes are dominant and a groin field where LST processes are most important. Process-based numerical models hold the key for our future understanding and prediction of coastal morpho-dynamic impacts of shoreline stabilization structures for their functional design, but are still in the development stage for everyday use.

See Section 11.3.5 of Chapter 11 for understanding the impacts of sea-level rise discussed in Chapter 5.

References

Ahrens, J. P. (1981) "Irregular Wave Runup on Smooth Slopes," Technical Aid No. 81-17, U.S. Army Engineer Waterways Experiment Station, Coastal Engineering Research Center, Vicksburg, MS.

Ahrens, J. P., and Cox, J. J. (1990) "Design and Performance of Reef Breakwaters," *Journal of Coastal Research,"* Vol. 6, No.1 pp. 61–75.

Ahrens, J. P., and M. S. Heimbaugh (1988) "Approximate Upper Limit of Irregular Wave Runup on Riprap," Technical Report CERC-88-5, U.S. Army Engineer Waterways Experiment Station, Coastal Engineering Research Center, Vicksburg, MS.

Ahrens, J. P., and M. S. Heinbaugh (1988). "Seawall Overtopping Model," *Proceedings of the 21st International Conference Coastal Engineering*, American Society of Civil Engineers, Vol. 1, pp. 795–806.

Alberti, P., T. Bruce, J. Pearson, L. Franco, J. Burgon, and C. Ecob (2001). "Wave Transmission Behind Vertical Walls Due to Overtopping," in *Proceedings of the Conference on Shorelines, Structures & Breakwaters*, ICE, London. Paper No. 21.

Allsop, N. W., P. I. Hawkes, F. A. Jackson, and L. Franco (1985). "Wave Run-Up on Steep Slopes — Model Tests Under Random Waves," Report No. SR2, Hydraulics Research Station, Wallingford, England.

Aminti, P., and L. Franco, (1988). "Wave Overtopping on Rubble Mound Breakwaters," *Proceedings of the 21st International Conference Coastal Engineering*, American Society of Civil Engineers, Vol. 1, pp 770–781.

Arromanches (1997) *The Story of D-Day and the Mulberry Harbours*, Musee Du Debarquement, Arromanches-les-Bains, France (http://www.arromanches-museum.com).

Basco, D. R. (2004) "Calibration of a Process Model for Barrier Island Breaching," *Proceedings of the 29th International Conference Coastal Engineering,* ASCE, NY, Vol. 2, pp. 2108–2119.

Basco, D. R., and J. Pope (2004) "Groin Functional Design Guidance from the Coastal Engineering Manual," *Special Issue, Journal of Coastal Research,* Vol. 33, No. 33, pp. 121–130.

Battjes, J. A. (1971) "Run-up Distributions of Waves Breaking on Slopes," *Journal, Waterways, Harbors and Coastal Engineering Division*, Vol. 97, pp. 91–114.

Battjes, J. A. (1974a) "Computation of Set-Up, Longshore Currents, Run-Up, and Overtopping Due to Wind Generated Waves," Report 74-2, Committee on Hydraulics, Department of Civil Engineering, Delft University of Technology, The Netherlands.

Battjes, J. A. (1974b) "Surf Similarity," *Proceedings of the 14th International Coastal Engineering Conference*, American Society of Civil Engineers, Vol. 1, pp. 466–479.

Besley, P. (1999) "Overtopping of Seawalls Design and Assessment Manual," Technical Report W 178, Environmental Agency, Bristol, UK.

Bodge, K. R. (1992) "Gross Transport Effects at Inlets," *Proceedings, of the 6th National Conference on Beach Preservation Technology*, Florida Shore and Beach Preservation Assoc., Tallahassee, FL, pp. 112–127.

Booij, N., R. C. Ris, and L. H. Holthuijsen (1999) "A Third-generation Wave Model for Coastal Regions, Part I, Model Description and Validation," *Journal Geophysical Research*, C4, Vol. 104, pp. 7649–7666.

Bradbury, A. P., and N. W. Allsop (1988) "Hydraulic Effects of Breakwater Crown Walls," *Proceedings of the Breakwaters '88 Conference*, Institution of Civil Engineers, Thomas Telford Publishing, London, UK, pp. 385–396.

Broker, I., J. A. Zyserman, E. Ostergaard, and K. Mangor (2007) "Morphological Modelling: A Tool for Optimization of Coastal Structures," *Journal of Coastal Research,* Vol. 23 No.5, pp. 1148–1158 (MIKE 21 CAMS, Coastal Area Modelling System).

Calabrese, M., M. Buccino, F. Ciardulli, and P. Di Pace (2010) "Wave Run-up and Reflection at Rubble Mound Breakwaters with Ecopode Armor Layer," *Proceedings of the 32nd International Conference Coastal Engineering*, ASCE, Washington, DC.

CEM (2006) *The Coastal Engineering Manual*, US Army Corps of Engineers, Washington, DC.

CEM (2011) *The Coastal Engineering Manual*, US Army Corps of Engineers, Washington, DC (Upgrade of CEM 2006).

Chasten, M. A., J. D. Rosati, J. W. McCormick, and R. E. Randall (1993) "Engineering Design Guidance for Detached Breakwaters as Shoreline Stabilization Structure," Technical Report CERC-93-19, U.S. Army Engineer Waterways Experiment Station, Vicksburg, MS.

CLASH (2004) "Crest Level Assessment of Coastal Structures by Full Scale Monitoring, Neural Network Prediction and Hazard Analysis on Permissible Wave Overtopping," Wave overtopping database, European Union Project, EVK3-CT-2001-00058. Online (http://www.clash-eu.org.)

Cornett, A., Y. Li, and A. Budvietas (1999). "Wave Overtopping at Chamfered and Overhanging Vertical Structures," in *Proceedings of the International Workshop on Natural Disasters by Storm Waves and Their Reproduction in Experimental Basins*, Kyoto, Japan.

Cuomo, G., A. Polidoro, E. Pasquini, J. Alderson, D. Robinson, W. Allsop, and S. Dunn (2011) "Wave Loading, Overtopping and Transmissioin of Low Crest Caisson Breakwaters," *Proceedings, of the 6th Coastal Structures Conference*, HR Wallingford Report HRPP484.

D'Alessandro, F. and F. Frega (2007) "A Verification of Static-Equilibrium Parabolic Formulation at the Protected Shoreline of Pizzo Calabro (Italy)," *Coastal Structures 2007*, World Scientific, Vol. II, pp. 1217–1228.

d'Angremond, K., J. W. Van der Meer, and R. J. De Jong (1996) "Wave Transmission at Low-crested Structures," *Proceedings of the 25th International Conference Coastal Engineering*, ASCE, pp. 2418–2427.

Dally, W. R., and Pope, J. (1986) "Detached Breakwaters for Shore Protection," Technical Report CERC-861, U.S. Army Engineer Waterways Experiment Station, Vicksburg, MS.

de Waal, J. P., and van der Meer, J. W. (1992) "Wave Run-Up and Overtopping on Coastal Structures," *Proceedings of the 23rd International Conference Coastal Engineering,* American Society of Civil Engineers, Vol. 2, pp. 1758–1771.

Delft Hydraulics (1989) "Rock Slopes and Gravel Beaches Under Wave Attack," *Publication No. 396,* Delft, The Netherlands.

Delft3D (2019) Open Source Platform for Water, Sediment Transport and Bathymetric Change in Coastal Sites, see http://oss.deltares.nl/ for home page Deltares Consultants, Inc. Delft, the Netherlands.

DELOS (2005) Extensive Database for Low-crested Rubble Mound Structures for Wave Transmission, European Union funded project. Over 2300 data points.

Die Küste (2007) *EurOtop, Wave Overtopping of Sea Defenses and Related Structures*, Assessment Manual, Heft 73, UK, NL, and DE team (http://www.overtopping-manual.com).

Dietrich, J. C., S. Tanaka, J. J. Westerink, C. N. Dawson, R. A. Luettich, M. Zijlema, L. H. Holthuijsen, J. M. Smith, L. G. Westerink, and H. J. Westerink (2012) "Performance of the Unstructured-mesh, SWAN+ADCIRC Model in Computing Hurricane Waves and Surge," *Journal Scientific Computing*, Vol. 52, pp. 468–497.

Drønen, N., P. Sørensen, R. Deigaard, S. Kristensen, and O. Ries (2017) "A Model Based Study of Sand Nourishment Decay," *Coastal Dynamics*, Paper No. 220.

Ecopode (2009) "The Coastal Protection Solution that Blends Most Efficiently into the Surrounding Landscape," (http://www.concretelayer.com/).

Environmental Agency (2009) "Guidance for Outline Design of Nearshore Detached Breakwaters on Sandy Macro-Tidal Coasts," Project SC 060026/R1, Environmental Agency, London, UK.

Environmental Agency (2010) "Development of Estuary Morphological Models," Department for Environment, Food and Rural Affairs (Defra), Environmental Agency, UK (http://www.gov.uk/).

FEMA (2007) "Guidelines and Specifications for Flood Hazard Mapping Partners," Final Report, US Federal Emergency Management Agency, Washington, DC.

Fleming, C., and B. Hamer (2000) "Successful Implementation of an Offshore Reef Scheme," in *Proceedings of the 27th International Conference on Coastal Engineering* (Sydney), ASCE, NY.

Formentin, S. M., and B. Zanuttigh (2013) "Prediction of Wave Transmission Through a New Artificial Neural Network Developed for Wave Reflection." *Proceedings of 7th International Conference on Coastal Dynamics* (http://www.coastaldynamics2013.fr/).

Franco, C., J. W. van der Meer, and I. Franco (1996) "Multi-directional Wave Loads on Vertical Breakwaters," in *Proceedings of the 25th International Conference on Coastal Engineering* (Orlando), ASCE, NY.

Franco, C., and L. Franco (1999) "Overtopping Formulas for Caisson Breakwaters with Nonbreaking 3D Waves," *Journal of Waterway, Port, Coastal, and Ocean Engineering*, American Society of Civil Engineers, Vol. 125, No. 2, pp. 98–108.

Franco, L., M. de Gerloni, and J. W. van der Meer (1994) "Wave Overtopping on Vertical and Composite Breakwaters," *Proceedings of the 24th International Conference on Coastal Engineering*, American Society of Civil Engineers, Vol. 1, pp. 1030–1045.

Fukuda, N., T. Uno, and I. Irie (1973) "Field Measurements of Wave Overtopping of Seawalls," in *Proceedings of the 20th Japanese Conference on Coastal Engineering* (in Japanese).

Goda, Y. (1969) "Reanalysis of Laboratory Data on Wave Transmission Over Breakwaters," Report of Port and Harbour Research Institute, Japan, Vol. 8, No. 3, pp. 3–18.

Goda, Y. (2009) "Derivation of Unified Wave Overtopping Formulas for Seawalls With Smooth, Impermeable Surfaces Based on selected CLASH datasets," *Coastal Engineering*, Vol. 56, pp. 385–399.

Goda, Y. (2010) *Random Seas and Design of Maritime Structures*, 3rd edn., World Scientific, Singapore.

Goda, Y., and J. P. Ahrens (2008) "New Formulation of Wave Transmission Over and Through Low-crested Structures," *Proceedings of the 31st International Conference Coastal Engineering*, ASCE, pp. 3530–3541.

Goda Y., Y. Kishira and Y. Kamiyama (1975) "Laboratory Investigation on the Overtopping Rate of Seawalls with Irregular Waves," Report for Port and Harbor Research Institute, No. 14(4), pp. 3–44 (In Japanese).

Gonzalez, M., and R. Medina (2001) "On the Application of Static Equilibrium Bay Formulations to Natural and Man-made Beaches," *Coastal Engineering*, Vol. 43, No. 3, pp. 209–225.

Gosse J. S., J. van der Meer, A. van Hoven, and A. Labrujere (2016) "Wave Run-up Simulations on Real Dikes", *Proceedings, International Conference Coastal Engineering*, ASCE.

Hallermeier, R. J. (1983) "Sand Transport Limits in Coastal Structure Design," *Proceedings of the Coastal Structures '83*, 9–11 March, Arlington, VA, American Society of Civil Engineers, pp. 703–716.

Hanson, H., and N. C. Kraus (1989) "GENESIS: Generalized Model for Stimulating Shoreline Change," Technical Report, CERC-89-19, U.S. Army Engineer Waterways Experiment Station, Vicksburg, MS.

Hardaway, C. S., G. R. Thomas, and J.-H. Li (1991). "Chesapeake Bay Shoreline Study: Headland Breakwaters and Pocket Beaches for Shoreline Erosion Control," Final Report No.313, Virginia Institute of Marine Science, Gloucester Point, VA.

Hasselmann, K., T. P. Barnett, E. Bouws, H. Carlson, D. E. Cartwright, K. Enke, J. A. Ewing, H. Gienapp, D. E. Hasselmann, P. Kruseman, A. Meerburg, P. Müller, D. J. Olbers, K. Richter, W. Sell, and H. Walden (1973) "Measurements of Wind-Wave Growth and Swell Decay During the Joint North Sea Wave Project (JONSWAP)," *Deutsche Hydrograph Zeitung, Erganzungsheft Reihe A* (80), No. 12.

Hawkes P. J., B. P. Gouldby, J. A. Tawn, and M. W. Owen (2002) "The Joint Probability of Waves and Water Levels in Coastal Engineering Design," *Journal of Hydraulic Research*, Vol. 40, No. 3, pp. 241–251.

Holthuijsen, L. H., N. Booij, and R. C. Ris, (1993) "A Spectral Wave Model for the Coastal Zone," *Proceedings of the 2nd International Symposium on Ocean Wave Measurement and Analysis*, New Orleans, Louisiana, pp. 630–641. Wave model SWAN.

Hsu, J. R. C., R. Silvester, and Y. M. Xia (1989) "Static Equilibrium Bays: New Relationships," *Journal Waterways, Port, Coastal and Ocean Engineering*, ASCE, Vol. 115, No. 3, pp. 285–298.

Hudson, J., J. Damgaard, N. Dodd, T. Chesher, and A. Cooper (2005) "Numerical Approaches for 1D Morphodynamic Modelling," *Coastal Engineering*, Vol. 52, pp. 691–707.

Hughes, S. A. (1984) "The TMA Shallow-Water Spectrum Description and Applications," Techical Report CERC-84-7, Corps of Engineers, Waterways Expt. Station, Vicksburg, MS.

Hunt, I. A. (1959) "Design of Seawalls and Breakwaters," *Journal Waterways and Harbors Division*, Vol. 85, No. WW3.

Johnson, H. K., I. Brøker, and J. A. Zyserman (1994) "Identification of Some Relevant Processes in Coastal Morphological Modelling," *Proceedings of the 24th International Conference on Coastal Engineering*, ASCE, pp. 2871–2885.

Jones, O. P., O. S. Petersen, and H. Kofoed-Hansen (2007) "Modelling of Complex Coastal Environments: Some Considerations for Best Practice," *Coastal Engineering*, Vol. 54, pp. 717–733.

Karambas, T. V. (2012) "Design of Detached Breakwaters for Coastal Protection: Development and Application of An Advanced Numerical Model," in *Proceedings of the 33rd International Conference on Coastal Engineering* (Santander), World Scientific, Singapore.

Kærgaard, K., and N. Drønen (2015) "A Hybrid Shoreline Model for the Sand Engine: Comparison with Observations and Long Term Predictions," *Proceedings of the Coastal Sediments* 2015, ASCE, 14 pp.

Kitaigorodoskii, S. A. *et al.* (1975) "On Phillips' Theory of Equilibrium Range in the Spectra of Wind Generated Gravity Waves," *Journal Physical Oceanography*, Vol. 5, No. 3, pp. 410–420.

Klein, A. H. R., A. Vargas, A. Raabe, L. A. André, and J. R. C. Hsu (2003) "Visual Assessment of Bayed Beach Stability With Computer Software," *Computers & Geosciences*, Vol. 29, pp. 1249–1257.

Kortenhaus, A., H. Oumeraci, J. Geeraerts, J. De Rouck, J. R. Medina, and J. A. Gonzalez-Escriva (2002) "Laboratory Effects and Further Uncertainties Associated With Wave Overtopping Measurements," *Proceedings of the 29th International Conference Coastal Engineering*, ASCE, pp. 4456–4468.

Kraus, N. C. (ed.) (1996) *History and Heritage of Coastal Engineering*, American Society of Civil Engineers, New York.

Kraus, N. C., H. Hanson, and S. H. Blomgren (1994) "Modern Functional Design of Groin System," *Proceedings of the 24th International Conference on Coastal Engineering*, ASCE, NY, pp. 1327–1342.

Kraus, N. C., and R. Rankin (eds.) (2004) "The Functioning and Design of Coastal Groins," Special Issue No. 33, *Journal of Coastal Research.*

Larson, M., and N. C. Kraus (1989) "SBEACH: Numerical Model for Simulations Storm-Induced Beach Change", Report 1. Empirical Foundation and Model Development, Technical Report CERC-89-9, US Army Engineer Waterways Experiment Station, Vicksburg, MS.

Le Fur, A., T. Pullen, and N. W. H. Allsop (2005) "Prediction of Low Overtopping Rates on Embankment Structures," pp. 4391–4403. doi:10.1142/979898127.1916_0354.

Leenknecht, D. A., A. Szuwalski, and A. R. Sherlock (1992) "Automated Coastal Engineering System (ACES): User's Guide," Corps of Engineers, ERDC, Washington, DC.

Lesnick, J. R. (1979) "An Annotated Bibliography on Detached Breakwaters and Artificial Headlands," Miscellaneous Report 79-1, Coastal Engineering Research Center, U.S. Army Engineer Waterways Experiment Station, Vicksburg, MS.

Mase, H. (1989) "Random Wave Runup Height on Gentle Slope," *Journal, Waterway, Port, Coastal, and Ocean Engineering*, ASCE, Vol. 115, No. 5, pp. 649–661.

Melby, J. A. (2012) "Wave Runup Prediction for Flood Hazard Assessment," ERDC/CHL TR-12-24, Corps of Engineers, Engineering Research and Development Center, Vicksburg, MS.

Ming, D., and Y.-M. Chiew (2000) "Shoreline Changes Behind Detached Breakwater," *Journal of Waterway, Port, Coastal, and Ocean Engineering*, ASCE, Vol. 126, No. 2, pp. 63–70.

Netherlands, TAW (2002) "Wave Run-up and Wave Overtopping at Dikes," Technical Advisory Committee for Flood Defense in the Netherlands (TAW), Delft.

Nielsen, P. (1984) "Explicit Solutions to Practical Wave Problems," in *Proceedings of the 19th International Conference on Coastal Engineering* (Houston), ASCE, NY.

Nielsen, P., and D. J. Hanslow (1991) "Wave Runup Distributions on Natural Beaches," *Journal of Coastal Research*, Vol. 7, No. 4, pp. 1139–1152.

Niemann, S. L., J. H. Jensen, J. A. Zyserman, I. Brøker, and S. Baek (2006) "Morphological Modelling of a Danish Tidal Inlet," *Proceedings of International Conference Coastal Engineering*, ASCE.

Owen, M. W. (1980) "Design of Seawalls Allowing for Wave Overtopping," Report No. 924, Hydraulics Research Station, Wallingford, UK.

Owen, M. W. (1982) "The Hydraulic Design of Seawall Profiles," *Proceedings of the Coastal Protection Conference*, Institution of Civil Engineers, Thomas Telford Publishing, London, UK, pp. 185–192.

Panizzo, A., and R. Briganti (2007) "Analysis of Wave Transmission Behind Low-crested Breakwaters Using Neural Networks," *Coastal Engineering*, Vol. 54, pp. 643–656.

Pearson, J., T. Bruce, W. Allsop, and X. Gironella (2002) "Violent Wave Overtopping-Measurements at Large and Small Scale," *Proceedings of the 28th International Conference Coastal Engineering*, ASCE.

Pearson, J., T. Bruce, W. Allsop, A. Kortenhaus, and J. van der Meer (2004) "Effectiveness of Recurve Walls in Reducing Wave Overtopping on Seawalls and Breakwaters," *Proceeding, International Conference Coastal Engineering*, ASCE.

Pedersen, J. (1996). "Experimental Study of Wave Forces and Wave Overtopping on Breakwater Crown Walls," Series paper 12, Hydraulics & Coastal Engineering Laboratory, Department of Civil Engineering, Aalborg University, Denmark.

Pedersen, J., and H. F. Burcharth (1992) "Wave Forces on Crown Walls," *Proceedings of the 23rd International Conference Coastal Engineering*, American Society of Civil Engineers, Vol. 2, pp. 1489–1502.

Peng, Z., Q. Zou, B. Wang, and D. Reeve (2008) "Transformation of Wave Skewness and Asymmetry Over Smooth Low-crested Breakwaters," School of Engineering, University of Plymouth, UK.

Pierson, W. J., and L. Moskowitz (1964) "A Proposed Spectral Form for Fully-Developed Wind Sea Based on the Similarity Law of S. A. Kitaigorodoskii," *Journal of Geophysical Research*, Vol. 69, pp. 5181–5203.

Pope, J., and J. L. Dean (1986) "Development of Design Criteria for Segmented Breakwaters," *Proceedings of the 20th International Conference Coastal Engineering*, ASCE, NY, pp. 2144–2158.

Price, D. M., T. J. Chesher, and H. N. Southgate (1995) "PISCES: A Morphodynamic Coastal Area Model," Final Report SR 411, HR Wallingford, UK.

Ris, R. C., N. Booij, and L. H. Holthuijsen (1999) "A Third-generation Wave Model for Coastal Regions, Part II, Verification," *Journal Geophysical Research*, C4, Vol. 104, pp. 7667–7681.

Rock Manual (2007) *The Use of Rock in Hydraulic Engineering* 2nd edn., CIRIA, C683, London, UK.

Rodgers, J., T. Chester, and B. Hamer (2006) "LEACOAST2: Practical Guidance Scoping Study," Techical Report, Department for Environment, Food, and Rural Affairs, London, UK.

Roelvink, J. A. (2006) "Coastal Morphodynamic Evolution Techniques," *Coastal Engineering*, Vol. 53, Nos. 2–3, pp. 277–287.

Roelvink, J. A. D., D.-J. R. Walstra, M. van Der Wegen, and R. Ramasinghe (2016) "Modeling of Coastal Morphological Processes," Chapter 28, In *Handbook of Ocean Engineering*, Springer, New York.

Roelvink, J. A., and A. J. H. M. Reniers (2011) "A Guide to Modelling Coastal Morphology," *Advances in Coastal and Ocean Engineering*, World Scientific, Singapore.

Rosati, J. D. (1990) "Functional Design of Breakwaters for Shore Protection: Empirical Methods," Technical Report, CERC-90-15, U.S. Army Engineer Waterways Experiment Station, Vicksburg, MS.

SPM (1984) "Shore Protection Manual," US Army Corps of Engineers, Waterways Experiment Station, Vicksburg, MS.

Schuttrumpf, H. (2001) "Wellenuberlaufstromung Bei Seedeichen", Experimentelle und Theoretische Untersuchungen, Hannover University, PhD dissertation (in German).

Silvester, R., and J. R. C. Hsu (1993) *Coastal Stabilization: Innovative Concepts*, Prentice-Hall, Englewood Cliffs, NJ.

Silvester, R., and J. R. C. Hsu (1997) *Coastal Stabilization*, World Scientific, Singapore (Reprint of Silvester R, and J. R. C. Hsu (1993)).

Suh, K., and R. A. Dalrymple (1987) "Offshore Breakwaters in Laboratory and Field," *Journal, Waterway, Port, Coastal, and Ocean Engineering*, ASCE, Vol. 113, No. 2, pp. 105–121.

Tamasicchio, G. R., and F. D'Alessandro (2013) "Wave Energy Transmission Through and Over Low-crested Breakwaters," *Journal of Coastal Research*, Vol. 65, No. 1, doi: 10.2112/s165-068.1.

Tamada, T., M. Inoue, and T. Tezuke (2002) "Experimental Studies on Diagrams of Wave Overtopping Rate on Gentle Slope-type Seawalls and Their Reduction on Wave Overtopping," *Proceedings Coastal Engineering, JSCE,* Vol. 49, pp. 641–645.

TMA (1975) Shallow water spectrum from data collected at Texel, Marsen and Arsloe tidal estuaries and following Kitaigorodoskii, S. A. *et al.* (1975) "On Phillips' Theory of Equilibrium Range in the Spectra of Wind Generated Gravity Waves," *Journal Physical Oceanography*, Vol. 5, No. 3, pp. 410–420.

van der Meer, J. W. (1988) "Rock Slopes and Gravel Beaches Under Wave Attack," Ph.D. Dissertation, Delft University of Technology, The Netherlands.

van der Meer, J. W., and I. F. R. Daemen (1994) "Stability and Wave Transmission at Low-crested Rubble Mound Structures," *Journal of Waterways, Ports, Coastal and Ocean Engineering*, ASCE, Vol. 120, No. 1, pp. 1–19.

van der Meer, J. W., and W. Janssen (1995) "Wave Run-up and Wave Overtopping at Dikes," In *Wave Forces on Inclined and Vertical Wall Structures*, Kobayashi and Demirbilek (eds.), American Society of Civil Engineers, pp. 1–27.

van der Meer, J., R. Schrijver, B. Hardeman, A. van Hover, H. Verheij, and G. J. Steendam (2010) "Guidance on Erosion Resistance of Inner Slopes of Dikes from Three Years, of Testing With the Wave Overtopping Simulator," Report, Rijkswaterstraat, Rotterdam, The Netherlands.

van der Meer, J. W., H. J. Regeling, and J. P. de Waal (2000) "Wave Transmission: Spectral Changes and its Effects on Run-up and Overtopping," *Proceedings of the 27th International Conference Coastal Engineering*, pp. 2156–2168.

van der Meer, J. W., R. Briganti, B. Wang, and B. Zanuttigh (2004) "Wave Transmission at Low-crested Structures Including Oblique Wave Attack," *Proceedings of the 29th International Conference Coastal Engineering,* ASCE.

van der Meer, J. W., W. Snijders, and E. Regeling (2006) "The Wave Overtopping Simulator," *Proceedings of the International Conference Coastal Engineering*, ASCE, pp. 4654–4666.

van der Meer, J. W., R. Briganti, B. Zanuttigh, and B. Wang (2005) "Wave Transmission and Reflection at Low-crested Structures: Design Formulae, Oblique Wave Attack and Spectral Change," *Coastal Engineering*, Vol. 52, Nos. 10–11, pp. 915–929.

van Gent, M. (2001) "Wave Runup on Dikes with Shallow Foreshores," *Journal, Waterway, Port, Coastal, and Ocean Engineering,* ASCE, Vol. 127, No. 5, pp. 254–262.

van Gent, M., M. R. A. van den Boogaard, B. Pozueta, and J. R. Medina (2007) "Neural Network Modelling of Wave Overtopping at Coastal Structures," *Coastal Engineering*, Vol. 54, No. 8, pp. 586–593.

van Oorschot, J. H., and K. d'Angremond (1968) "The Effect of Wave Energy Spectra on Wave Run-Up," *Proceedings of the 11th International Conference Coastal Engineering,* American Society of Civil Engineers, New York, Vol. 2, pp. 886–900.

van Oosten, R. P., J. P. Marco, J. W. van der Meer, M. R. A. van Gent, and H. J. Verhagen (2006) "Wave Transmission at Low-crested Structures Using Neural Networks," *Proceedings of the International Conference Coastal Engineering*, ASCE.

Verhagen, H. J., K. d'Angrremond, and F. van Roode (2012) *Breakwaters and Closure Dams*, 2nd edn., VSSD, Delft, The Netherlands.

Zanuttigh, B., and J. W. Van der Meer (2008) "Wave Reflection from Coastal Structures in Design Conditions," *Coastal Engineering*, Vol. 55, pp. 771–779.

Zanuttigh, B., and L. Martinelli (2008) "Transmission of Wave Energy at Permeable Low-crested Structures," *Coastal Engineering*, Vol. 55, No. 12, pp. 1135–1147.

Zanuttigh, B., S. M. Formentin, and R. Briganti (2013) "A Neural Network for the Prediction of Wave Reflection from Coastal and Harbor Structures," *Coastal Engineering*, Vol. 80, pp. 49–67.

Zanuttigh, B., S. Formentin, and J. W. Van der Meer (2014) "Advances in Modelling Wave-Structure Interaction through Artificial Neural Networks," *Proceedings of the 34th International Conference Coastal Engineering*, Vol. 1.

Chapter 6

Structural Design

6.1 Introduction

Calculation of the ability of coastal structures to resist structural damage for a range of probabilistic design levels of hydrodynamic forces is the subject of this chapter. Empirical formulas have been developed over many years of physical model studies in laboratories, field measurements, and numerical models on computers. In recent years, large-scale physical model investigations with wave heights in the 2-m range (6.5 ft) over a controlled range of water depths have greatly improved these formulas. There are no national, structural, steel, and concrete design codes (as those for the design of buildings, bridges, etc.) that deal with the stresses and strains that the loads produce. All we can say is how the many material, shape, and hydrodynamic variables combine so that no (or very little) structural change (movement) occurs or that some (percentage) change or "damage" takes place.

Four types of coastal structures are considered in this chapter. One type (rubble-mound) is the most common and complete details are presented in Section 6.2. This type is more than just a "... pile of rocks in the water" (d'Angremond, 1975). A second type are vertical walls and caisson-type structures with structural design formulas and details given in Section 6.3. When space permits, earthen dikes can be constructed as reviewed in Section 6.4. In Section 6.5, we review some of the structural design considerations for movable gates in tidal estuaries. Design formulas for piles that support coastal piers and wharfs are found in Chapter 9.

Breakwaters provide shelter from wave action on the leeward side and both the rubble-mound type and the vertical wall, caisson-type structures can be employed. The choice is relatively easy when the conditions at the site are similar to those found in Table 6.1. When the conditions make no clear choice, both types should be considered and a cost analysis must be performed to determine the least, total, life-cycle costs.

Table 6.1. Conditions to employ for clear choice in breakwater type.

Rubble-Mound Breakwater	Vertical Wall, Caisson-Type Breakwater
Relatively shallow water depth	Water depths are medium or large
Stone materials available	Large quantities of stone materials unavailable
Mooring not required on leeward side	Mooring on leeward side is required
Soils weak to support caisson-type	Soils strong to support caisson-type
Reflections on seaward side are important	No restrictions on seaward-side reflections
No time limit on construction time	Minimum time needed for construction
Breaking waves impact structure	Location seaward of breaker zone

Alternative designs for coastal hazard mitigation are discussed in detail in Chapter 3. We begin with the rubble-mound type of design using natural rock for the outer, armor layer.

6.2 Rubble-Mound Structures (Breakwaters, Groins, Jetties, and Revetments)

6.2.1 *Introduction*

Natural stones (rocks) mined from quarries are the materials employed to form the cross-sections needed for stable designs of headland and nearshore breakwaters, groin fields, and jetties at tidal entrances. These rock materials are also placed on natural, earth slopes, and dikes to prevent erosion and are called revetments. Together, these structures have come to be called "rubble-mound" structures. The outer layers of stones (armor layer) are most critical for stability. The following sections of this chapter consider the best formulations for stable designs of natural rock and artificial concrete armor layers, low-crested designs, cross-section designs for both conventional and berm-type structure shapes, toe protection, construction details, and design examples from field applications including spreadsheets for the design equations.

6.2.2 *Hydraulic stability — Hudson formula*

Waves rushing up and down the sloping surface of the top layer of stones can cause hydraulic instability as illustrated in Figure 6.1. One or more stones may (a) rock back and forth, (b) rotate upward and slide down the slope during wave down-rush, (c) rotate upward and be pushed up the slope during wave up-rush, and (d) several stones may slide down the slope during the down-rush of the wave. Armor layer stability involves rocking,

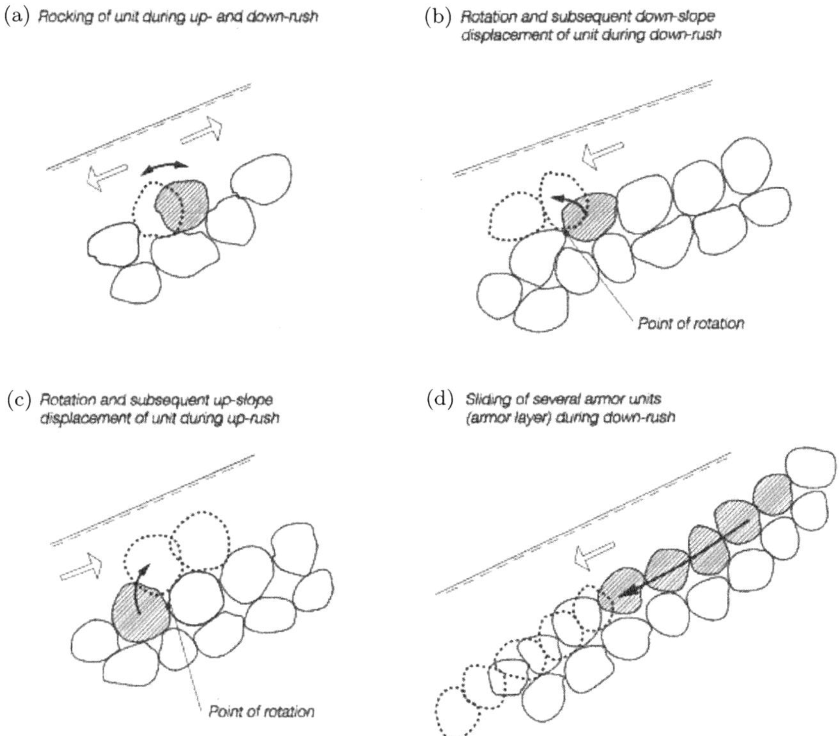

Figure 6.1. Possible hydraulic instabilities of the armor layer of a rubble-mound structure under non-overtopping conditions (from CEM, 2006, Part VI; Burcharth, 1993).

displacement from a layer, sliding, and possibly even settlement of the layer. Note that hydraulic instability is not the breakage of individual stones in the layer.

A metaphysical experiment can be helpful to appreciate the relative importance of the water wave, fluid/stone physical properties, and the structure slope that interacts to maintain the hydraulic stability of the armor layer. Clearly, these hydrodynamic, fluid, and physical properties are as follows: H is some measure of the wave height, T is some definition of a wave period, W is the weight of the armor layer stone in air, D is some nominal dimension of the size of the rock, ρ_s is the mass density of rock (stone) material, ρ_w is the mass density of water, g is the gravitational constant, and α is the structure slope angle with the horizontal.

If D is the diameter of a cube, then W equals $\rho_s g D^3$. Intuitively, the larger the wave height, the heavier the stone weight is required for hydraulic stability. Rock weight, W, must be directly proportional to some power of the wave height, H. But the rock weight W is proportional to the rock size cubed D^3. Therefore, rock size cubed, D^3, is directly proportional to the wave height cubed, H^3. But the rock weight must be that weight in water of mass density ρ_w since if completely submersed, a buoyant force is present and is equal to the weight of the water displaced, $\rho_w g D^3$. This buoyant force increases the required weight of rocks (in air) for hydraulic stability by a dimensionless factor, Δ defined as

$$\Delta = \rho_s/\rho_w - 1 = s_r - 1, \tag{6.1}$$

where s_r is the specific gravity of the rock.

Therefore, a simple, dimensionless, numerical stability parameter, N_s, can be written as

$$N_s = \frac{H}{\Delta D} = \frac{\text{The nominal wave height}}{\text{The nominal rock dimension immersed in water}}. \tag{6.2}$$

The second important relationship for hydraulic stability is the weight (or rock dimension) relative to the structure slope angle α and may not be intuitively obvious. The steeper the structure slope, the heavier the stone weight is required for stability. As seen in Figure 6.1, a stone that moves upward can more readily roll down a steep slope than a flatter slope. For a given rock size, flatter slopes are more stable than steeper slopes. Therefore, the stone weight is inversely proportional to the structure slope. The inverse of the structure slope $(1/\tan\alpha)$ is the $\cot\alpha$. The stone size cubed is inversely proportional to the structure slope, $\tan\alpha$ and directly proportional to the $\cot\alpha$.

A stability equation can then be formulated by introducing a stability coefficient, K, to equate the stability parameter (Equation (6.2)) and its relationship with the structure slope given by $\cot\alpha$ to give

$$\frac{H}{\Delta D} = (K \cot \propto)^{1/3}. \tag{6.3}$$

The stability equation (6.3) was first proposed by Hudson (1958, 1959) and is the basis for the Hudson stability formula (Hudson, 1974) as introduced in the *Shore Protection Manual* (SPM, 1977) and rearranged into the following form for the stable rock weight, W:

$$W = \frac{(\rho_s g)H^3}{K_D(s_r - 1)^3 \cot\alpha}, \tag{6.4}$$

where K_D is the Hudson stability coefficient.

The Hudson stability coefficients for rubble-mound breakwaters were developed using single-frequency (monochromatic) regular waves with two armor layers and for non-overtopping conditions with slope angles, $\cot\alpha =$ 1.5–3.0 in the laboratory. SPM (1977) recommended K_D values for breaking and non-breaking waves using the significant wave height, H_s, in Equation (6.4). Depth-limited wave breaking in front of the armor slope was the "breaking-wave" condition with lower K_D values than the non-breaking case (no breaking in front of the armor slope). Then in SPM (1984), the value of the recommended wave height in Equation (6.4) was increased to $H_{1/10}$ and with completely different K_D values. For a Rayleigh distribution for the wave heights, $H_{1/10} = 1.27\,H_s$, so that $(1.27\,H_s)^3$ is approximately double and partly explains the completely different K_D values in SPM (1984). But confusion continues today regarding what K_D values to apply and when to use breaking or non-breaking conditions.

In addition, the Hudson stability formula does not consider many additional hydrodynamic and structure variables that are important in the rubble-mound structure design. Ahrens (1975) discussed the importance of the wave period that was missing in the Hudson formula. The permeability of the rock structure, the level of damage (rock movement or displacement), and the number of waves impacting the rock structure are all important but not considered in the Hudson formulation. For all these reasons, the use of the Hudson formulation for calculation of the stable weight of rock armor layers for rubble-mound structures is not recommended. Much could be written here regarding the history, limitations, and current application of the Hudson formula. The interested reader is referred to the *Coastal Engineering Manual* (CEM, 2006) and Table VI-5-22 for further details. We shall return to the Hudson formula for calculation of the stability of concrete armor units in Section 6.2.4. See also Burcharth and Thompson (1983).

6.2.3 *Rock armor layer stability — Modern formulations*

6.2.3.1 *van der Meer (1987) for non-overtopping waves*

In the early 1980s, the Delft Hydraulics Laboratory in the Netherlands conducted an extensive series of laboratory experiments using irregular wave spectrums with variable (1) spectral shapes, (2) core permeability, (3) structure slopes, (4) rock grading, and (5) relative mass density of the rock structures at the test section. The results were published as *Design Formulae (for) Stability of Breakwater Armour Layers* by van der Meer

(1987) that has become a standard reference work (e.g. Rock Manual, 2007). It was learned that the type of wave breaking as quantified by the surf-similarity parameter ξ_m (using T_m the average wave period) was critical to *separate* the data set into that for plunging-type wave breaking and that for surging-type waves on the structure. In other words, for irregular, head-on waves impacting the rock test section, two separate dimensionless formulations were developed (van der Meer, 1987, 1988), namely,

plunging waves: $\xi_m < \xi_{mc}$

$$\frac{H_s}{\Delta D_{n50}} = 6.2 \cdot S^{0.2} P^{0.18} N_z^{-0.1} \xi_m^{-0.5} \tag{6.5}$$

and

surging waves: $\xi_m > \xi_{mc}$

$$\frac{H_s}{\Delta D_{n50}} = 1.0 \cdot S^{0.2} P^{-0.13} N_z^{-0.1} (\cot\alpha)^{0.5} \xi_m^P, \tag{6.6}$$

where the critical, dividing surf-similarity parameter, ξ_{mc} is defined as

$$\xi_{mc} = [6.2\, P^{0.31} (\tan\alpha)^{0.5}]^{1/(P+0.5)}, \tag{6.7}$$

where H_s is significant wave height in front of breakwater, D_{n50} is equivalent cube length of median rock, ρ_s is mass density of rock, ρ_w is mass density of water, Δ is relative mass density $= [(\rho_s/\rho_w) - 1]$, S is relative eroded area of the armor layer, P is notation permeability parameter, N_z is number of waves, α is structure slope angle, s_m is wave steepness $= H_s/L_{om}$, and L_{om} is deepwater wavelength corresponding to mean wave period, T_m,

so that

$$\xi_m = \text{surf-similarity parameter} = \tan\alpha/\mathrm{s}_m^{-0.5}. \tag{6.8}$$

We have introduced a number of these same variables in Chapter 4 for probabilistic design and in Chapter 5 for wave runup and wave overtopping formulations. Four additional, key variables are now required as inputs to solve the van der Meer (1987) formulations for rock armor layer stability. They are mean wave period, T_m, notational permeability, P, the relative eroded area, S, and the number of waves, N_z, in the coastal storm.

The mean wave period T_m can be obtained from the wave spectrum. Recall that $T_m/T_p = 0.71 - 0.82$ for the Pierson–Moskowitz spectrum.

The notational permeability parameter P was shown in Figure 5.7 and has been reproduced here as Figure 6.2. See Sections 5.2.3 and 5.2.3.2 for a discussion of the notational permeability. It is not a true permeability of the

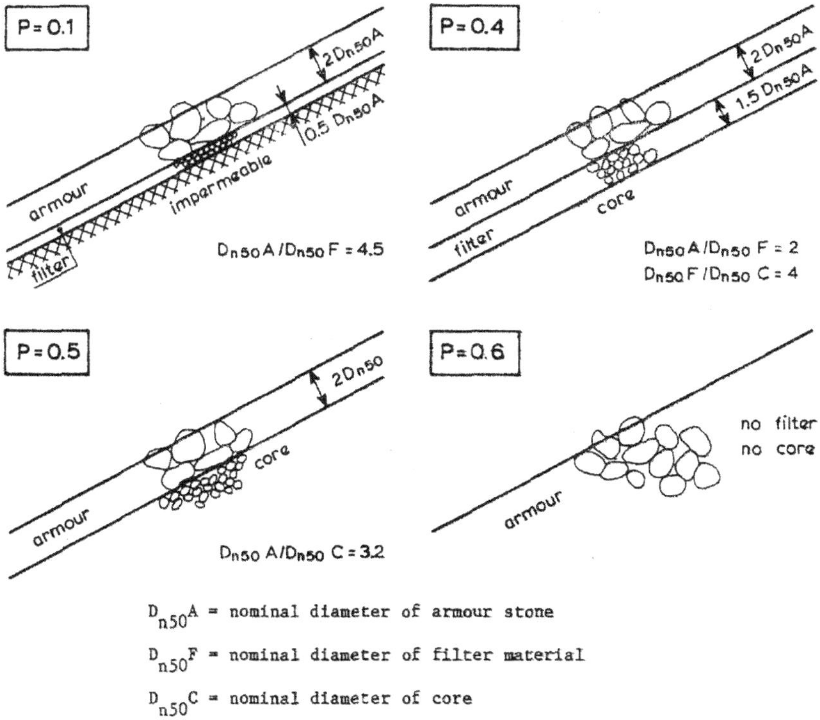

Figure 6.2. Notational permeability parameter P as defined by van der Meer (1987) with details of relative sizes of various layers beneath the armor layers (from CEM, 2006).

structure but a surrogate value and hence called the "notational" permeability parameter. Figure 6.2 displays typical revetment structure design with under-layers of smaller sizes stones. The construction shown in the upper-right corner of Figure 6.2 with $P = 0.4$ was estimated by van der Meer (1988) but not tested. Recent physical model experiments (Kik *et al.*, 2012) with stone size ratios for the $P = 0.4$ structure gave a notational permeability of $P = 0.37$ so that a value of $P = 0.35$ was recommended for practical use. A method to correlate the notational permeability to both the actual structural permeability as a porous media and the internal hydrodynamics using numerical, volume of fluid (VOF) models has been presented in a M.Sc. Thesis at the Delft Technical University. This research is ongoing (Verhagen *et al.*, 2009). This P parameter appears as a coefficient with various powers and as an exponent in the structural stability formulas of van der Meer (1987) and consequently is a very important variable that was never considered in the Hudson formulation. Complete details of the cross-sectional design of breakwaters, groins, and jetties is found in Section 6.2.4 for the size of smaller stones in the under-layers and the core.

The outer layers of stones (armor layer) are most critical for stability. If the armor layer is eroded so that the smaller stones in the under-layers are exposed, then the entire rubble-mound structure is vulnerable to structural failure. The relative level of the armor layer damage can be defined several ways. See CEM (2006, Table VI-5-18) for five different definitions of the damage parameter. We only consider one definition herein. Figure 6.3 displays the eroded area A_e as defined in the relative eroded area, damage formulation S (Broderick, 1983) as employed by van der Meer (1987), namely,

$$S = A_e/D_{n50}^2. \tag{6.9}$$

Typically, the armor layer consists of two layers as schematically shown in Figure 6.2. For natural rock, the various levels of damage S for a two-layer armor as determined by van der Meer (1988) is summarized in Table 6.2. Flatter slopes require a larger S value at the failure level.

The initial damage levels for S (2–3) are a practical lower limit recognizing that a few units may move and that taking $S = 0$ would cause

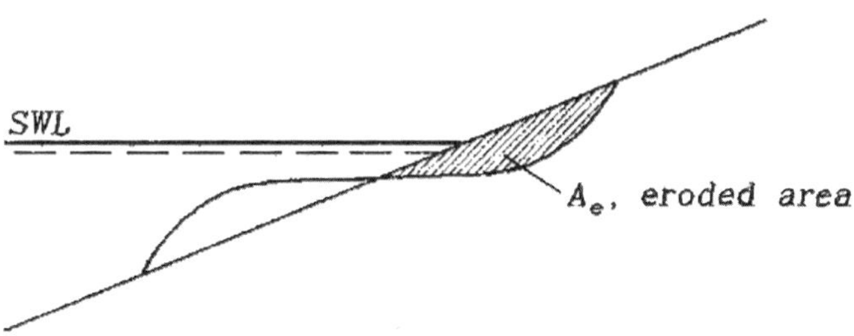

Figure 6.3. Definition of eroded area A_e in relative damage formulation, S.

Table 6.2. Damage levels, S for two-layer, rock armor (van der Meer, 1988).

Unit	Slope, $\cot\alpha$	Initial Damage, S	Intermediate Damage, S	Failure Damage Level, S
Rock	1.5	2	3–5	8
Rock	2	2	4–6	8
Rock	3	2	6–9	12
Rock	4–6	3	8–12	17

Equations (6.5) and (6.6) to be zero. The intermediate damage levels recognize that some rocks are displaced but without exposure of the under-layer (or filter layer) by direct wave attack. The rubble-mound structural damage can be repaired, and the structure still performs its design function. The damage levels of S for failure (8–17) means that the under-layer (or filter layer) is exposed to wave attack and the structure may require redesign and may not perform its design functions as a breakwater, groin, jetty, or revetment. Table 6.2 is also presented in Chapter 4 as Table 4.2.

6.2.3.2 *van der Meer design equations — Deep water (non-overtopping waves)*

These two formulae of van der Meer (1988), Equations (6.5) and (6.6) can be converted into design equations with partial safety factors γ_H and γ_z, where the caret (hat symbol) indicates mean values of the variable, and each equation has its own table of partial safety factors for the full range of failure probability P_f and wave height uncertainty σ'_{FHs}. The partial safety factors are from PIANC (1992), Working Group 12, and Burcharth and Sorensen (2000). See Chapter 4 and Sections 4.2.2 and 4.2.3 for a complete discussion of probabilistic design using partial safety factors. The probabilistic mean value of the significant wave height at some design value for the recurrence interval, T_r, is given the symbol, $\hat{H}_S^T$. The plunging wave coefficient C_{pl} and surging wave coefficient C_s are explained further in the following.

Plunging waves ($\widehat{\xi_m} < \widehat{\xi_{mc}}$),

$$G = \frac{1}{\gamma_z} C_{\mathrm{pl}} \hat{S}^{0.2} \hat{P}^{0.18} \hat{\Delta} \hat{D}_{n50} \hat{f} (\hat{\cot} \propto)^{0.5} (\hat{s}_{om})^{0.25} \hat{N}_Z^{-0.1} - \gamma_H \hat{H}_S^T, \quad (6.10)$$

with partial safety factors as given in Table 6.3.

Surging waves ($\hat{\xi_m} > \hat{\xi_{mc}}$),

$$G = C_s \frac{1}{\gamma_z} \hat{S}^{0.2} \hat{P}^{0.13} \hat{\Delta} \hat{D}_{n50} \hat{f} (\hat{\cot} \propto)^{0.5-p} (\hat{s}_{om})^{-0.5P} \hat{N}_Z^{-0.1} - \gamma_H \hat{H}_S^T, \quad (6.11)$$

with partial safety factors in Table 6.4.

Critical surf-similarity parameter,

$$\widehat{\xi_{mc}} = \left[\frac{C_{\mathrm{pl}}}{C_s} \hat{P}^{0.31} (\tan \hat{\alpha})^{0.5} \right]^{1/(\hat{P}+0.5)}. \quad (6.12)$$

Table 6.3. Partial safety factors for plunging waves (from CEM, 2011, Table VI-6-5).

	$\sigma'_{FHs} = 0.05$		$\sigma'_{FHs} = 0.2$	
P_f	γ_H	γ_Z	γ_H	γ_Z
0.01	1.6	1.04	1.9	1.00
0.05	1.4	1.02	1.5	1.06
0.10	1.3	1.00	1.3	1.10
0.20	1.2	1.00	1.2	1.06
0.40	1.0	1.08	1.0	1.10

Table 6.4. Partial safety factors for surging waves (from CEM, 2011, Table VI-6-6).

	$\sigma'_{FHs} = 0.05$		$\sigma'_{FHs} = 0.2$	
P_f	γ_H	γ_Z	γ_H	γ_Z
0.01	1.7	1.00	1.9	1.02
0.05	1.3	1.10	1.6	1.00
0.10	1.3	1.02	1.4	1.04
0.20	1.1	1.10	1.2	1.08
0.40	1.0	1.08	1.1	1.00

All the above is valid for non-overtopping wave conditions. Wave overtopping was discussed in Chapter 5, Section 5.3, and Figure 5.11 presented the definition of the crest freeboard, R_c. Equations (6.10) and (6.11) include a reduction factor $\hat{f}$ which models the effect of low-crested, overtopped breakwaters as discussed in detail in Section 6.2.5.

The publication by van der Meer (1988) was his Ph.D. dissertation at the Technical University of Delft, and acknowledges the earlier work of Thompson and Shuttler (1975) as the starting point for this basic research using irregular, random waves in the laboratory. He recognized that the coefficient 6.2 (defined as coefficient C_{pl} for the plunging breaker in Equation (6.10)) and coefficient 1.0 (defined as coefficient C_s for the surging breaker in Equation (6.11)) were stochastic, mean values with standard deviations of mean of 0.4 and 0.08, respectively. See Figure 4.1 in Chapter 4 for the plot of the original test data for plunging waves to visualize the scatter of the data giving a standard deviation of the mean coefficient, C_{pl} of 0.4. Hence, the coefficient in the transition Equation (6.12) separating both formulas was $C_{\mathrm{pl}}/C_s = 6.2$. Van der Meer (1988) also recognized that waves moving into shallow water would be depth-limited in height and suggested using the Rayleigh distribution ratio of $H_{2\%}/H_s = 1.4$ to adjust the C_{pl} and C_s coefficient and to use $H_{2\%}$ in the formulas. Finally, van der Meer (1988) called for more research to understand the permeability parameter P and stone shape (roundness) on static stability.

The permeability parameter P as defined by van der Meer (1987) is shown in Figure 6.2. At one extreme are impermeable structures ($P = 0.1$) such as revetments on dikes and natural earth slopes. Rubble-mound, breakwater structures as depicted in Section 6.2.5 with a core use $P = 0.4$ or 0.5. The most permeable structure is made up of all one size rocks with no core and has a permeability parameter $P = 0.6$ in the van der Meer (1987) formulation.

The van der Meer (1987) equations can be used to develop damage (fragility) curves for a given design, stable rock size D_n by calculating S as the wave heights increase above the design wave height. We shall return to calculation of damage (fragility) curves in Chapter 8, Section 8.3.

The surf-similarity parameter ξ_m brings the mean wave period T_m into the formulation. Finally, the number of waves N_z is a variable in the van der Meer (1987) formulation. For short storms lasting 4–7 hours, $N_z = 2{,}500$ is recommended for mean wave periods from 5 to 10 s. The maximum N_z is 7,500 giving (15–30) hour storms for waves in the (5–15) s, mean period range.

6.2.3.3 *The van der Meer equations in shallow water (non-overtopping waves)*

There are many variables, coefficients, and exponents in the original van der Meer (1987, 1988) formulations as seen in Equations (6.10)–(6.12). Graduate students at the Delft Technical University and engineers at Delft Hydraulics (now Deltares) have been conducting research investigation over the past 15–20 years to further understand and improve the original vdM formulas. The following summarizes some past and recent developments as found in the literature with more probably to follow.

It has been recognized that most of the over 300 tests conducted by van der Meer (1988) were with relatively deep water at the toe of the structure. Additional tests (over 200) of rock stability with shallow foreshores have been conducted at Delft Hydraulics (van Gent, 2000) as reported by van Gent *et al.* (2004). These additional tests demonstrated that (1) less scatter was evident if the mean wave period T_m was replaced by $T_{m-1.0}$, the spectral wave period to find the surf-similarity parameter $\xi_{m-1.0}$; (2) less scatter was evident if the ratio of $H_{2\%}/H_s$ is employed because of depth-limited wave breaking; and (3) the additional data reduced the standard deviations of the mean values of the modified C_{pl} coefficient in the original formulations of van der Meer (1988). For "shallow water" wave conditions, van Gent *et al.* (2004) recommend using $C_{\mathrm{pl}} = 8.4(H_s/H_{2\%})$ and $C_s = 1.3(H_s/H_{2\%})$. These values are also given in Rock Manual (2007)* and require/allow the user to specify the appropriate $(H_s/H_{2\%})$ ratio. These modified values for C_{pl} and C_s for shallow water may result in lighter (or heavier) rock for stability than the original, deep water, van der Meer (1988) values. This is in keeping with shallow-water wave breaking and truncation of the wave spectrum from that of the deep water, Rayleigh distribution. Rock Manual (2007) defines "shallow water" when the water depth, h is less than $3H_s$ at the toe of the structure. Van Gent (2004) suggested using $H_{s(\mathrm{toe})} = 0.7H_{so}$(deep water) as the "shallow water" condition. No partial safety factors have yet been published for the shallow-water, van der Meer formulations. *This reference also CIRCA, CUR, CETMEF (2007).

6.2.3.4 *The Corps of Engineers*

The Corps of Engineers has a long history of research on the static stability of rubble mound breakwaters. The Hudson (1959) formulation was the standard for many years and is still applied for artificial, concrete armor layer units as discussed in the following.

Melby and Kobayashi (2011) propose new, static-stability formulations for plunging and surging waves on sloping, armor-stone structures that depart from the traditional approach where the primary load variable is the significant wave height H_s at the toe of the structure. In their approach, the maximum, nonlinear wave momentum flux (MF_{max}) as defined by Hughes (2004) is employed as the key load variable in the static stability formulations. The results in Melby and Kobayashi (2011) expand and improve the earlier work in Melby and Hughes (2003); utilize the small-scale physical data from van der Meer (1988); and employ the same definitions for S, N_z, and P as originally proposed by van der Meer (1988). Two new coefficients are needed and defined. One coefficient a_m appears in the final equation, but has two separate formulations, one for plunging and one for surging waves. The second new coefficient K_s is added for an analysis of damage progression and will be further discussed in Chapter 8. The new, dividing equation separating the plunging and surging formulations was formulated to only depend on the structure slope and does not include the permeability parameter P as found in van der Meer (1988) and Equation (6.12). The analysis and formulas presented by Melby (1999), Melby and Kobayashi (2011) are relatively new and are not included herein.

6.2.3.5 *Summary of van der Meer equations for design (non-overtopping waves)*

The key mean values and standard deviations of the mean values of the coefficients, C_{pl} and C_s for both deep-water and shallow-water condition at the toe of structure are summarized in Table 6.5. The water depth h at the toe of the structure is a key variable. When $h > 3H_s$, then the original van der Meer coefficients as shown in Equations (6.4)–(6.6) should be employed. And, the mean wave period T_m is used to calculate the surf-similarity parameter, ξ_m. For normal, single-peaked spectrums, $T_m \sim 0.82T_p$, where T_p is the spectral peak wave period.

When $h < 3H_s$ (shallow water), then a value of the ratio $H_{2\%}/H_s$ must be selected. Using the Rayleigh distribution, $H_{2\%}/H_s = 1.4$. And, $T_{m-1.0}$, the spectral wave period must be employed to find the surf-similarity parameter, $\xi_{m-1.0}$. As in Chapter 5 for wave overtopping and the *EuroTop manual*, the wave period ratio, $T_p = 1.1T_{m-1.0}$, so that $T_{m-1.0} \sim 0.91T_p$. Clearly, if the wave spectral shape is known and different than these shapes, different values for T_m and $T_{m-1.0}$ will result.

The mean values in Equations (6.10)–(6.12) are assumed and the hat symbol is omitted in what follows. The equations for calculating the nominal diameter, D_n, for design of a non-overtopped structure become

plunging waves ($\xi_m < \xi_{mc}$),

$$D_{n50} = \gamma_z \gamma_H \frac{H_s^T}{C_{\text{pl}} S^{0.2} P^{0.18} \Delta f(\xi_m^{-0.5}) N_z^{-0.1}}, \tag{6.13}$$

Table 6.5. Key coefficients in the van der Meer formulas.

Breaker Type Wave Period and Surf Parameter	Deep Water $h > 3H_s$ Mean Value, μ	Deep Water $h > 3H_s$ Std Deviation, σ	Shallow Water $h < 3H_s$ Mean Value, μ	Shallow Water $h < 3H_s$ Std Deviation, σ	Equation
Plunging, C_{pt}	6.2	0.4	8.4 $(H_s/H_{2\%})$	0.70	6.13
Surging, C_s	1.0	0.08	1.3 $(H_s/H_{2\%})$	0.15	6.14
Surf Parameter, ξ	ξ_m		$\xi_m - 1.0$		
Wave Period, T	T_m		$T_m - 1.0$		
Critical, C_{pt}/C_s	$6.2/1.0 = 6.2$		$8.4/1.3 = 6.46$		6.15

surging waves ($\xi_m > \xi_{mc}$),

$$D_{n50} = \gamma_z \gamma_H \frac{H_s^T}{C_s S^{0.2} P^{-0.13} \Delta f(\xi_m^P) N_z^{-0.1} (\cot\alpha)^{0.5}}, \tag{6.14}$$

critical surf-similarity parameter,

$$\xi_{mc} = \left(\frac{C_{\text{pl}}}{C_s} P^{0.31} (\tan\alpha)^{0.5} \right)^{1/(P+0.5)}. \tag{6.15}$$

In these equations, it is understood that if shallow-water conditions prevail at the toe of the structure ($h < 3H_s$) then the spectral wave period, $T_{m-1.0}$, must be employed to find the surf-similarity parameter, $\xi_{m-1.0}$.

Once the nominal rock diameter D_{n50} is computed, then the rock armor-layer weight W (or mass, M) can be readily determined with the stone mass density, ρ_s.

Weight of rock armor,

$$W = (\rho_s g)(D_{n50})^3. \tag{6.16}$$

Mass of rock armor,

$$M = (\rho_s)(D_{n50})^3. \tag{6.17}$$

Rock Manual (2007) uses the standard deviations σ of the C_{pl} and C_s coefficients in Table 6.5 to present a sensitivity analysis for deterministic design. For example, at the 5% exceedance probability limit, the normal, bell-shaped (Gaussian) curve factor is 1.64 so that the C_{pl} coefficient for plunging waves is reduced and becomes $(6.2 - 1.64 \times 0.4) = 5.5$. This lower value will produce a larger armor rock diameter since the C_{pl} coefficient is in the denominator of Equation (6.13). But this remains a deterministic design as discussed in Chapter 4 since nothing can be said regarding the probability of failure, P_f, for this design. Rock Manual (2007, p. 573) mentions a Level I probabilistic design using partial safety factors (PIANC, 1992) but does not present the details.

We strongly recommend retaining the mean values for the C_{pl} and C_s coefficients in Table 6.5 and employing the probabilistic design methods as discussed in detail in Chapter 4 that use the partial safety factors $\gamma_z \gamma_H$ that appear in Tables 6.3 and 6.4. In this way, knowledge of both the level of confidence in the significant wave heights σ'_{FHs} and the probability of failure P_f become explicit design criteria for the design.

We now have over 25 years of experience in applying the "deep-water", van der Meer (1988) formulations for static stability of rock structures.

Adjustments have been added for the "shallow-water", van der Meer versions as summarized in Table 6.5. Equation (6.15) for finding the critical, dividing line surf-similarity parameter ξ_{mc} is still valid but uses the C_{pl}/C_s values in Table 6.5 for shallow-water applications.

6.2.3.6 *Spreadsheets for armor layer, probabilistic design applications*

Development of spreadsheets with the structure slope $\cot\alpha$ as the key design variable greatly facilitates the application of these equations and tables for design of the stable size, or weight (mass) of the armor layer. They are found in the online supplementary materials which accompanies this book. All employ the software product Microsoft Excel and are located in the folder Structural Design/ Rubble Mound. Figure 6.4(a) is an example for plunging waves under deep-water conditions and Figure 6.4(b) is also for plunging waves but under shallow-water conditions, when applicable as explained in the following. Similar spreadsheets are included for surging wave conditions and separate spreadsheets are provided for both English and Metric unit systems with a Read Me page giving the formulas and additional information.

There is a considerable amount of information on these spreadsheets. But the pattern layout is the same for all the rock-armor stability spreadsheets and follows the pattern (numbered) as explained in the following.

(1) The top-middle gives the *Title* and that this sheet is for the *Metric System.* It also shows that this sheet is for either *Deep Water Conditions* (Figure 6.4(a)) or *Shallow Water Conditions* (Figure 6.4(b)).
(2) The top-left corner gives the *name, dates* and possible reference to the equations employed. In deep water (Figure 6.4(a)) the reference is CEM (2006).
(3) The middle is for the given *Enter Data* of which the significant wave height H_s is most important and highlighted in red. The rock and fluid density values are entered, delta is calculated, and the partial safety factors are entered (see point number 8). On Figure 6.4(b) for shallow-water conditions, the water depth, h, at the structure toe is entered here but not required in Figure 6.4(a) for deep-water conditions.
(4) The middle-right side shows *Additional Data.* The four key additional variables in the van der Meer (1988) formulations are the nominal damage level, the notational permeability, the number of waves in the storm, and the wave period. Because the peak wave period T_p is often known with the significant wave height, we have chosen to enter the peak wave period and calculate other wave periods (see point number 5).

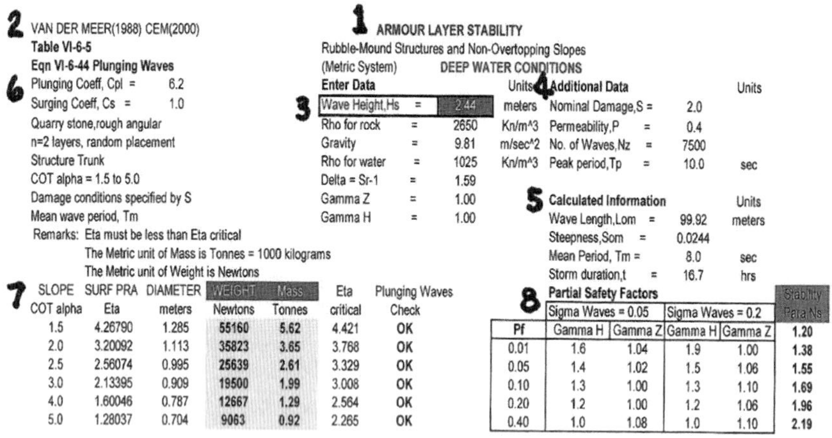

2 VAN DER MEER(1988) CEM(2000)
Table VI-6-5
Eqn VI-6-44 Plunging Waves
6 Plunging Coeff, Cpl = 6.2
Surging Coeff, Cs = 1.0
Quarry stone,rough angular
n=2 layers, random placement
Structure Trunk
COT alpha = 1.5 to 5.0
Damage conditions specified by S
Mean wave period, Tm
Remarks: Eta must be less than Eta critical
The Metric unit of Mass is Tonnes = 1000 kilograms
The Metric unit of Weight is Newtons

1 ARMOUR LAYER STABILITY
Rubble-Mound Structures and Non-Overtopping Slopes
(Metric System) DEEP WATER CONDITIONS

3 Enter Data			Units
Wave Height,Hs	=	2.44	meters
Rho for rock	=	2650	Kn/m^3
Gravity	=	9.81	m/sec^2
Rho for water	=	1025	Kn/m^3
Delta = Sr-1	=	1.59	
Gamma Z	=	1.00	
Gamma H	=	1.00	

4 Additional Data		Units
Nominal Damage,S =	2.0	
Permeability,P =	0.4	
No. of Waves,Nz =	7500	
Peak period,Tp =	10.0	sec

5 Calculated Information		Units
Wave Length,Lom =	99.92	meters
Steepness,Som =	0.0244	
Mean Period, Tm =	8.0	sec
Storm duration,t =	16.7	hrs

7 SLOPE COT alpha	SURF PRA Eta	DIAMETER meters	WEIGHT Newtons	Mass Tonnes	Eta critical	Plunging Waves Check
1.5	4.26790	1.285	55160	5.62	4.421	OK
2.0	3.20092	1.113	35823	3.65	3.768	OK
2.5	2.56074	0.995	25639	2.61	3.329	OK
3.0	2.13395	0.909	19500	1.99	3.008	OK
4.0	1.60046	0.787	12667	1.29	2.564	OK
5.0	1.28037	0.704	9063	0.92	2.265	OK

8 Partial Safety Factors

	Sigma Waves = 0.05		Sigma Waves = 0.2		Stability Para Ns
Pf	Gamma H	Gamma Z	Gamma H	Gamma Z	1.20
0.01	1.6	1.04	1.9	1.00	1.38
0.05	1.4	1.02	1.5	1.06	1.55
0.10	1.3	1.00	1.3	1.10	1.69
0.20	1.2	1.00	1.2	1.06	1.96
0.40	1.0	1.08	1.0	1.10	2.19

Figure 6.4(a). Example spreadsheet (van der Meer, 1988) for calculation of rock diameter and weight (or mass) for a range of structure slopes, $\cot\alpha$. The waves are plunging, and deep-water conditions prevail.

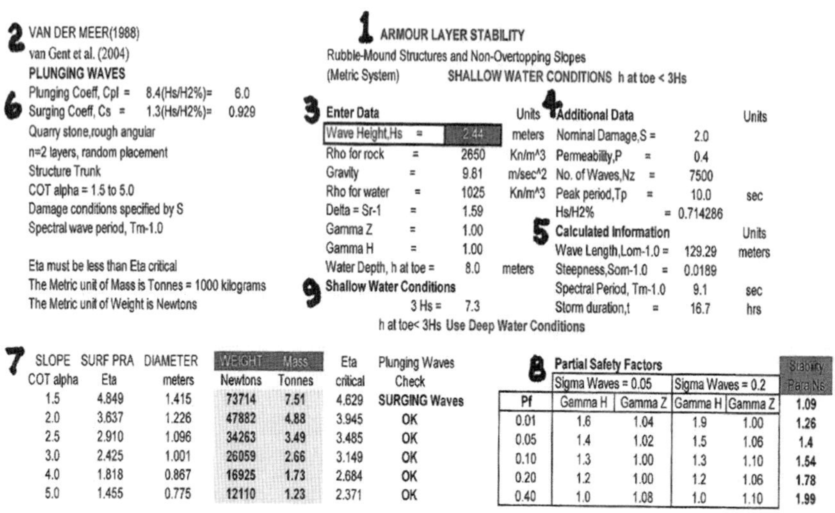

2 VAN DER MEER(1988)
van Gent et al. (2004)
PLUNGING WAVES
6 Plunging Coeff, Cpl = 8.4(Hs/H2%)= 6.0
Surging Coeff, Cs = 1.3(Hs/H2%)= 0.929
Quarry stone,rough angular
n=2 layers, random placement
Structure Trunk
COT alpha = 1.5 to 5.0
Damage conditions specified by S
Spectral wave period, Tm-1.0

Eta must be less than Eta critical
The Metric unit of Mass is Tonnes = 1000 kilograms
The Metric unit of Weight is Newtons

1 ARMOUR LAYER STABILITY
Rubble-Mound Structures and Non-Overtopping Slopes
(Metric System) SHALLOW WATER CONDITIONS h at toe < 3Hs

3 Enter Data			Units
Wave Height,Hs	=	2.44	meters
Rho for rock	=	2650	Kn/m^3
Gravity	=	9.81	m/sec^2
Rho for water	=	1025	Kn/m^3
Delta = Sr-1	=	1.59	
Gamma Z	=	1.00	
Gamma H	=	1.00	
Water Depth, h at toe =		8.0	meters

9 Shallow Water Conditions
3 Hs = 7.3
h at toe< 3Hs Use Deep Water Conditions

4 Additional Data		Units
Nominal Damage,S =	2.0	
Permeability,P =	0.4	
No. of Waves,Nz =	7500	
Peak period,Tp =	10.0	sec
Hs/H2% =	0.714286	

5 Calculated Information		Units
Wave Length,Lom-1.0 =	129.29	meters
Steepness,Som-1.0 =	0.0189	
Spectral Period, Tm-1.0	9.1	sec
Storm duration,t =	16.7	hrs

7 SLOPE COT alpha	SURF PRA Eta	DIAMETER meters	WEIGHT Newtons	Mass Tonnes	Eta critical	Plunging Waves Check
1.5	4.849	1.415	73714	7.51	4.629	SURGING Waves
2.0	3.637	1.226	47882	4.88	3.945	OK
2.5	2.910	1.096	34263	3.49	3.485	OK
3.0	2.425	1.001	26059	2.66	3.149	OK
4.0	1.818	0.867	16925	1.73	2.684	OK
5.0	1.455	0.775	12110	1.23	2.371	OK

8 Partial Safety Factors

	Sigma Waves = 0.05		Sigma Waves = 0.2		Stability Para Ns
Pf	Gamma H	Gamma Z	Gamma H	Gamma Z	1.09
0.01	1.6	1.04	1.9	1.00	1.26
0.05	1.4	1.02	1.5	1.06	1.4
0.10	1.3	1.00	1.3	1.10	1.54
0.20	1.2	1.00	1.2	1.06	1.78
0.40	1.0	1.08	1.0	1.10	1.99

Figure 6.4(b). Example spreadsheet (van der Meer, 1988) for calculation of rock diameter and weight (or mass) for a range of structure slopes, $\cot\alpha$. The waves are plunging, and shallow-water condition may prevail.

For shallow-water conditions, the $H_s/H_{2\%}$ ratio must also be specified as shown in Figure 6.4(b).

(5) The *Calculated Information* is found on the middle-right side. For deep water, the mean wave period, T_m, is calculated by some given relationship with the peak period, e.g. $T_m/T_p = 0.71$–0.82 for the Pierson–Moskowitz spectrum and then used to calculate the deep-water wave length and wave steepness. The storm duration is then calculated knowing the number of waves. For shallow water, the spectral period $T_{m-1.0}$ is calculated by some given relationship with the peak period, e.g. $T_{m-1.0} \sim 0.91T_p$ and then the deep-water wave length and wave steepness follow. However, the storm duration is the same since the mean wave period is employed.

(6) The middle-left side presents the *Plunging Wave and Surging Wave Coefficients* in the van der Meer (1988) formulas. They are constants in deep water but dependent on the $H_s/H_{2\%}$ ratio for shallow-water conditions. The following list presents further details of the limitations of the van der Meer (1988) equations that are applicable to these spreadsheets.

(7) The *Calculated Results* for rock *Diameter, Weight*, and *Mass* (highlighted in yellow) are found on the lower-left side of the spreadsheet, for the range of structure slopes, $\cot\alpha$ from 1.5 to 5.0. The unit for mass is the metric ton. For this example of input data, a structure slope $\cot\alpha = 2.0$ requires a 3.65 metric ton mass for the armor layer. The surf-similarity parameters are also calculated along with the critical values to determine whether plunging or surging wave conditions are applicable. As shown in Figure 6.4(a) for all structure slopes, the calculated surf-similarity parameters are all less than the critical values (Equation (6.15)) so that the check column reads "OK" in all cases. This confirms that for this example input data, the plunging wave formulations of van der Meer (1988) are the correct ones to apply.

(8) The bottom-right side displays the *Partial Safety Factors* as found in Table 6.3 for plunging waves. Since the input data was set for both γ_Z and $\gamma_H = 1.0$, the results shown do not include any level of probabilistic design. This is a good way to begin. Now, for a specified level of probability of failure P_f and confidence level of the wave data or wave information, the partial safety factors applicable can be inserted in the appropriate boxes (see note 3) for entered data. The resulting rock diameter, weight, and mass will increase accordingly.

(9) Finally, a check is made on *Shallow Water Conditions* as found in the middle of Figure 6.4(b). In this case, the criteria for h at the toe being

less than $3H_S$ is not satisfied with note "Use Deep Water Conditions" added. This simply confirms that the results presented in Figure 6.4(a) for deep water are the correct answers.

We hope this detailed explanation helps to understand the considerable amount of information on these spreadsheets and that they are found useful in your design of rubble-mound armor layers. See the Read Me sheets on each spreadsheet for more details. Additional spreadsheets found in the online supplementary materials are for overtopped, submerged, and reef-type breakwaters and are discussed in the following.

Further design details surrounding the van der Meer static stability formulas include the influence of rock gradations, the influence of rock shapes, the influences of packing and rock placement, and the influence of steep approach slopes and can be found in Rock Manual (2007).

When the rock quarry cannot produce the rock size (weight/mass) required or is not located at an economical distance from the construction site for transport, then artificial armor units manufactured from concrete must be employed.

6.2.4 *Artificial concrete armor units*

Armor layer units manufactured from concrete can be designed in many shapes. Some shapes have greater interlocking ability so that their dynamic stability is increased for the same weight (mass) under equal wave conditions. This means they can withstand larger wave heights and remain stable or lighter weight (mass) units can be employed for the given design wave conditions. However, gravel, sand, cement, and water are needed at the manufacture site along with the steel forms to shape the unit. Their manufacture is labor-intensive for concrete mixing, pouring, and steel-form assembly and disassembly after curing. In general, artificial concrete units are more costly than natural rock for the armor layer(s) of coastal structures when two layers are employed.

Figure 6.5 displays a sample of some shapes that have been employed over many years (from Kamphuis, 2010).

The US Army Corps of Engineers has developed and patented the CORE-LOC® unit (Melby and Turk, 1994) as shown in Figure 6.6 which is similar in shape to the Accropod® as shown in Figure 6.7. The Xbloc unit was patented by Delta Marine Consultants, in the Netherlands in 2001.

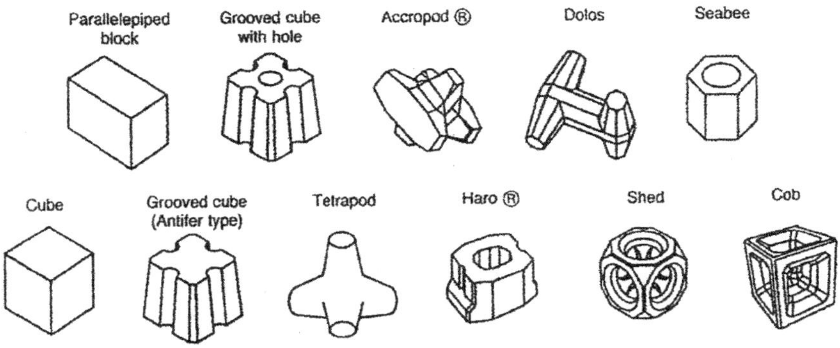

Figure 6.5. Sample of some concrete armor units (from Kamphuis, 2010). Some have registered trademarks shown by the symbol®.

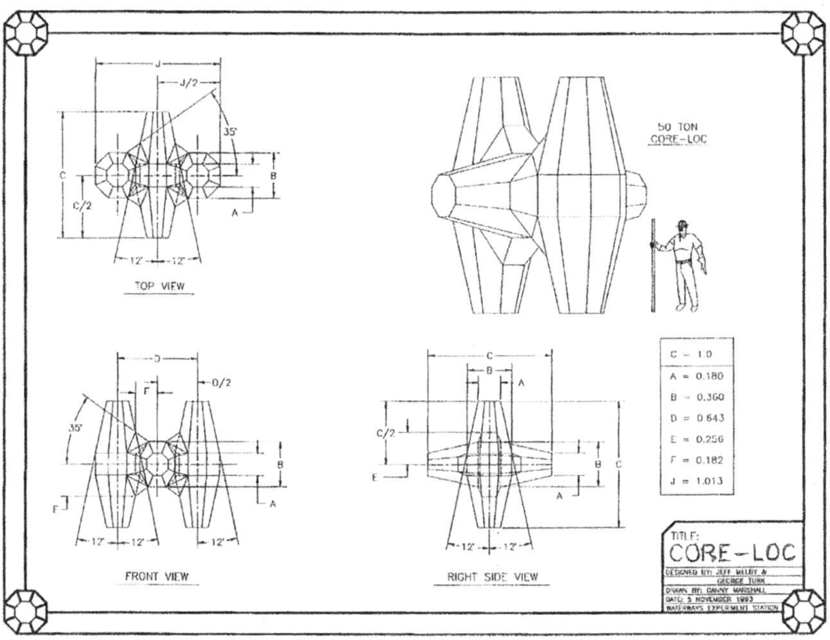

Figure 6.6. The US Army Corps of Engineers, CORE-LOC® unit with dimensions scaled by length "C" = 1.0 in length (from Reeve *et al.*, 2012).

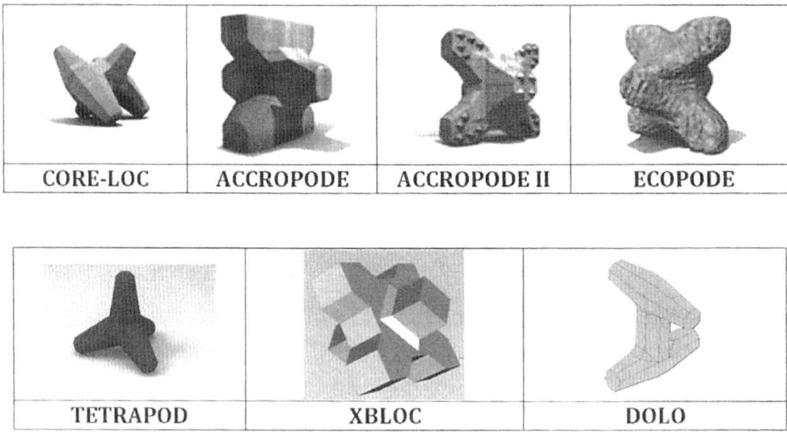

Figure 6.7. The most popular concrete armor layer shapes used all over the world (courtesy Concrete Layer Innovations, Inc.).

The Tetrapod unit (France, SOGREAH, 1950) is no longer protected by a patent. The Accropod® (France, 1981), CORE-LOC® (US, 1994; CERCular, 1995), and Xbloc® (Netherlands, 2001) units have registered trademarks and a license must be obtained for their manufacture. All four have similar interlocking ability and high structural integrity. The Dolo (US, circa 1970s) also has a high interlocking ability but has a relatively poor structural integrity (the units are weak at the slender connections) and break apart to act as smaller, individual pieces and are then far less stable under wave attack. Their use is no longer recommended. The great advantage of the four types with high structural strength and interlocking ability is that they can be placed in a single-layer system as the armor layer and hence are cost-effective. A one-layer system uses far less concrete in the armor layer and should always be compared with a two-layer rock system. Photographs of the (a) scale for 35 metric ton units, (b) the cannon fashion placement method, and (c) manufacture technique for CORE-LOC units are shown in Figure 6.8 as found at the Concrete Layer Innovations (CLI) website www.concretelayer.com.

Simple plain cube or parallel blocks and those with grooves are also popular for relative ease of manufacture and placement.

The Hudson stability formula (Hudson, 1974) was introduced in Section 6.2.2 and repeated here as Equation (6.4) for the stable weight, W.

$$W = \frac{(\rho_s g) H^3}{K_D (S_r - 1)^3 \cot \alpha}. \tag{6.4}$$

(a) The scale of 31.6-metric ton CORE-LOC® units

(b) CORE-LOC® units placed in cannon fashion

(c) CORE-LOC® units stacked after manufacture and the steel forms ready for concrete poring

Figure 6.8. Photographs giving the (a) scale, (b) cannon placement method, and (c) manufacture of CORE-LOC units (from CLI website: www.concretelayer.com) (courtesy Concrete Layer Innovations, Inc.).

The only new variable is the Hudson stability coefficient, K_D. The Hudson formula is popular for design of concrete armor units because the K_D values are available for many of the common shapes as illustrated in Table 6.6. The Hudson formula does not include the influence of the four additional

Table 6.6. Hudson stability coefficients for non-breaking waves on concrete armor units with zero damage, non-overtopping conditions, structure trunk location, and using $H_{1/10}$ as the design wave height.

Armor Unit	Hudson Stability Coefficient, K_D
Natural rock	4 (non-breaking waves)
Natural rock	2 (breaking waves)
Tetrapod	6–8
Dolos	32 (8)
Accropod®	9.5–15
CORE-LOC®	13–16
Antifer (grooved) Cube	6.0
Cube, modified	7.5
Cube	6.5
Tribar	10
XBloc®	13–16

Note: All the concrete units have K_D values greater than natural rock.

variables found to be important by van der Meer (1988) for rock armor. Unfortunately, it is not clear which wave height to apply for design (H_s or $H_{1/10}$) and different K_D values appear in the literature for breaking and non-breaking waves. The K_D values in Table 6.6 come from many sources as summarized by Reeve *et al.* (2012, Table 9.12) and are coupled with using $H_{1/10}$ as the recommended design wave height.

The CLI company (Echirolles, France) is the licensing agent for both Accropod® and CORE-LOC® units. Their website www.concretelayer.com includes a computing tool calculator for preliminary design, but requires the input of the significant wave height, H_s. The Xbloc® unit website is www.xbloc.com for Delta Marine Consultants (Gouda, the Netherlands). Detailed documentation is available from these organizations and should be consulted when considering concrete armor units for design. However, these organizations have advocated the use of steep, $\cot\alpha = 1.33$ for these units. This practice is not recommended, and the steepest slope employed should be $\cot\alpha = 1.5$ for design. Note that all three of these popular units needing a license for manufacture have similar Hudson stability coefficients, K_D, so that the ultimate choice will depend on many other factors.

The above-mentioned discussion for both rubble-mound (rocks) and concrete armor layer stability is for non-overtopping wave conditions where the crest elevation is elevated sufficiently to prevent waves from running up and over the top of the structure.

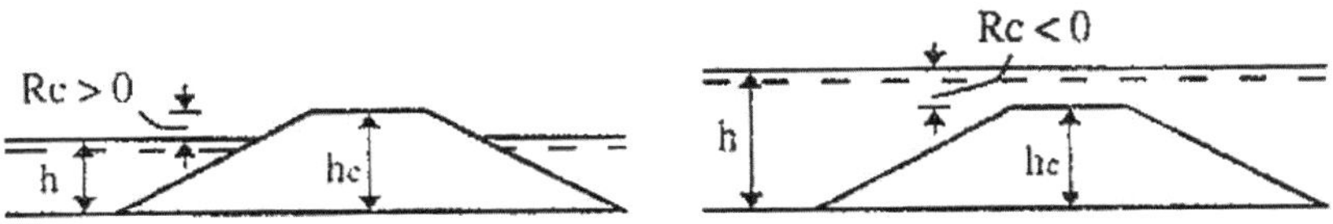

Figure 6.9. Definition of freeboard, $R_C = h_C - h$.

6.2.5 *Overtopped and submerged breakwater stability*

6.2.5.1 *Moderate wave overtopping — No submergence*

Wave overtopping was discussed in Chapter 5, Section 5.3. Figure 6.9 shows the definition of the crest freeboard R_c as the vertical distance between the crest elevation h_C and the water depth h at the toe of the structure. The freeboard R_C is positive when the water depth is below the crest and negative when the structure crest is completely submerged.

First, consider a crest elevation moderately overtopped by waves but the crest is not completely submerged. Equations (6.10) and (6.11) for rubble-mound structures include a reduction factor $\hat{f}$ which models the effect of overtopped breakwaters, and $\hat{f}$ is given by the following expression (van der Meer, 1991):

$$\hat{f} = \left(1.25 - 4.8\frac{R_c}{H_s}\sqrt{\frac{s_{op}}{2\pi}}\right)^{-1}, \quad 0.8 \leq \hat{f} \leq 1.0, \tag{6.18}$$

$$0 < \frac{R_c}{H_s}\sqrt{\frac{s_{op}}{2\pi}} < 0.052. \tag{6.19}$$

When the freeboard R_c is zero and the still-water level (SWC) is at the crest of the structure $\hat{f}$ reduces to a minimum of 0.8 and when the argument $\frac{R_c}{H_s}\sqrt{\frac{s_{op}}{2\pi}}$ reaches 0.052, there is no reduction in the stone size since $(4.8 \times 0.052) = 0.25$ so that $\hat{f} = 1.0$. Note that in Equations (6.18) and (6.19), the peak wave period T_P must be used to find the deep-water wave length L_{op} and then the wave steepness, $s_{op} = H_s/L_{op}$. The spreadsheets found in the online supplementary materials that accompanies this book include the crest elevation so that the freeboard R_c and reduction factor $\hat{f}$ can be calculated for design with rubble-mound (rock) armor layers. Rock movement (damage) is again mainly located on the seaward-side, front slope of the structure.

6.2.5.2 *Low-crested structures (LCS)*

As the water level increases, it may become higher in elevation than the structure crest to completely submerge the structure. Rock movement and damage may now also be on the crest and on the shoreward, back slope of the structure. Low-crested structures (LCS) are by definition, designed to be significantly overtopped by waves and completely submerged under some water-level conditions. Typically, LCS are located in shallow water as shore-parallel, detached, and segmented breakwater systems to mitigate beach erosion. Consequently, rock movement and damage may also occur on the more vulnerable end sections since each stone is less supported by neighboring stones and subject to stronger currents in the gaps between the structures. Their planform layout was discussed in Section 5.6.2 in Chapter 5.

In the early 2000s, the EU Commission funded the "Environmental Design of Low-Crested Coastal Defense Structures" (DELOS) project with the results summarized in Special Issue 52 of the *Coastal Engineering Journal* (Lamberti *et al.*, 2005). The purpose of the DELOS project was to "*... promote effective and environmentally compatible design of low crested structures (LCS) to defend European shores against erosion and to preserve the littoral environment and the coast economic development.*" The project involved 18 partners from seven European countries and end users. The environmental impacts of LCS's (biological, ecological, etc.) are discussed in Chapter 10.

The structural stability of detached, low-crested breakwater structures was summarized by Burcharth *et al.* (2006). Papers by Allsop (1983), Givler and Sorensen (1986), van der Meer and Pilarczyk (1990), Vidal *et al.* (1992), Burger (1995a, 1995b), Kramer and Burcharth (2003), and Kramer *et al.* (2005) were all reviewed and employed. These papers presented results from two-dimensional (2D) and three-dimensional (3D) laboratory-scale model tests in which initial rock movement (damage) occurred at some location on the front slope, crest, or back slope of the trunk section, and/or on the roundhead section at the end of the structure for both head-on and directional wave attack. The following stability formula represents the start of damage at some location on the structure:

$$\frac{H_s}{\Delta D_{n50}} = 0.06\left(\frac{R_c}{D_{n50}}\right)^2 - 0.23\frac{R_c}{D_{n50}} + 1.36 \qquad (6.20)$$

and is for the same armor size for the entire structure for the normalized freeboard, range $-3 \leq R_c/D_{n50} < 2$ when exposed to non-depth-limited, short-crested waves on a slope, $\cot\alpha = 1.5$. Recall that R_c is positive when the water depth is below the crest and negative where the structure is submerged. Equation (6.20) has been verified against 23 existing prototype structures inventoried in the DELOS project (Denmark (2), Greece (3), United Kingdom (2), Spain (4), and Italy (12)) with good agreement between predictions and the observed structure response. See Burcharth *et al.* (2006, Table 4, p. 389) for two cases in Italy where Equation (6.20) was not okay.

When waves break due to depth-limited conditions, Burcharth *et al.* (2006) present further analysis which suggests that for breakwater crest elevations in the 2–4 m range, the most critical conditions are when slightly submerged with $R_c = -0.36\,h_c$. As shown in Chapter 2, the depth-limited breaking wave formulation (Kamphuis, 1991) was given as Equation (2.25) repeated here (note, the water depth, d, is now h in Chapter 6).

$$\left(\frac{H_s}{h}\right)_b = 0.56[\exp(3.5\,m)] = \gamma \tag{2.25}$$

and often with the breaking ratio symbol γ employed. For very gently foreshore slopes ($m = 1/100$) and depth-limited breaking conditions, Burcharth *et al.* (2006) provide a rule of thumb (RoT) for Equation (6.20), namely $D_{n50} \geq 0.29\,h_c$. They also suggest using a flatter structure slope, $\cot\alpha = 2$ when applying Equation (6.20) for stability under depth-limited, breaking wave conditions. Using the RoT value for D_{n50} as the first iteration for the RHS of Equation (6.20) in a trial-and-error solution procedure greatly facilitates its solution.

6.2.5.3 *An example*

The following example is for the same wave and structure conditions as discussed in Section 4.2.2 except we now take $P = 0.4$, namely, $\cot\alpha = 2$, $\rho_s = 2{,}650\,\text{kg/m}^3$, $\rho_w = 1{,}025\,\text{kg/m}^3$, $\Delta = 1.585$, $P = 0.4$, $S = 2$, $\gamma_Z = \gamma_H = 1.0$, with storm and wave conditions, $H_s = 3.0\,\text{m}$, $T_p = 7.5\,\text{s}$, $N_z = 2{,}000$ waves, and we now also need the structure crest elevation and water depth, $h_c = 5\,\text{m}$, $h =$ variable.

Table 6.7 summarizes the deterministic results for a stable armor rock diameter D_{n50} for a wide range of water depth conditions that result in non-overtopping and overtopping wave conditions, submerged structure conditions, and deep- and shallow-water depth conditions at the toe of the structure. The structure may be a low-crested breakwater structure (LCS).

Based on the laboratory tests of Givler and Sorensen (1986) with regular waves and those of van der Meer (1988) with irregular, non-breaking waves, van der Meer and Pilarczyk (1990) proposed the following formula

Table 6.7. Summary of example results for the armor layer, rock diameter, D_{n50}.

Reference	Text Equation	Conditions	Freeboard, R_c	D_{n50}, m
van der Meer (1988)	(6.13)($P = 0.4$)	Deep water, plunging	Non-overtopping	0.986
van der Meer/van Gent	(6.13, Table 6.5)	Shallow water, plunging	Non-overtopping	1.115
van der Meer (1991)	(6.18)($P = 0.4$)	Overtopping, plunging	None, $R_c = 0$	0.789
van der Meer/Pilarczyk	(6.21)	Submerged	None, $R_c = 0$	0.809
Burcharth *et al.*(2006)	(6.20)	Low-crested structures	None, $R_c = 0$	1.391

for completely submerged, rock breakwaters in irregular, head-on waves:

$$\frac{h_c}{h} = (2.1 + 0.1S)\exp(-0.14N_S^*) \leq 1.0, \tag{6.21}$$

where

$$N_S^* = \frac{H_s}{\Delta D_{n50}} S_p^{-1/3}, \tag{6.22}$$

and $S_P = \frac{H_s}{L_P}$ is the wave steepness at the toe of the structure, with L_P is the wave length at the toe of the structure using the peak wave period, T_P, and the other variables are the same as defined before.

Clearly, the wave length at the toe of the structure, L_P depends on the water depth, h, and must be calculated using some method such as Hunt's formula (Equation (2.5)) as given in Section 2.3.2. This is different than all the equations found before which employed the deep-water wave length, L_o. However, no mention is made of the notational permeability in Equation (6.21). A separate spreadsheet has been provided for the solution of Equations (6.20) and (6.21) for submerged and LCS breakwaters.

The first four formulations in Table 6.7 involving van der Meer are all for the trunk section of the structure. For the non-overtopping cases, shallow-water conditions produce a slightly larger rock diameter for stability. In general, as water levels increase to produce wave overtopping, a smaller armor layer rock diameter is required but these results clearly also depend on the internal structure as found in the notational permeability, P. When the water depth coincides with the crest elevation (Freeboard, $R_c = 0$), the lower limit of Equation (6.18) is reached ($f = 0.8$) and taking $P = 0.4$ for a relatively porous structure produces a rock diameter that is similar to Equation (6.21) for completely submerged breakwaters. In all four of

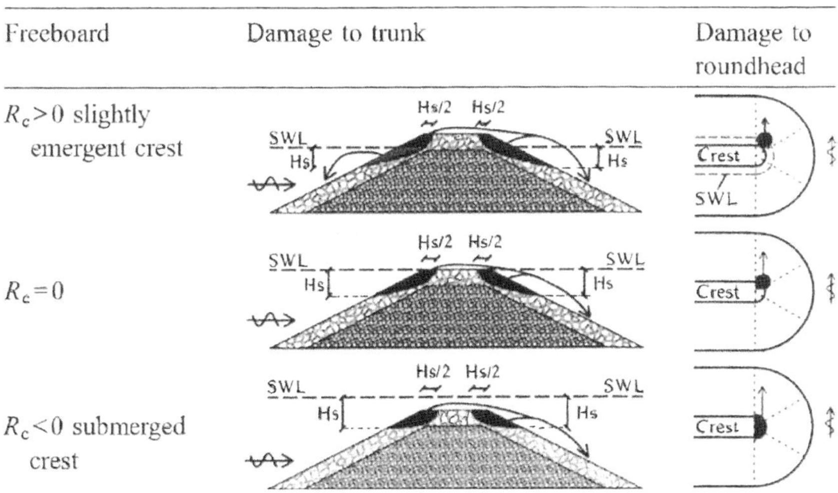

Figure 6.10. Black areas showing locations of initial armor layer movements (damage) to LCS subject to head-on wave attack (from Kramer and Burcharth, 2003).

these formulations involving van der Meer, rock movements (initial damage) occurred on the seaward, front slope or crest of the structure.

The formulation of Burcharth *et al.* (2006) is for the armor layer stability of the *entire structure*. Figure 6.10 illustrates the observed displacement (black areas) for damage initiation of head on the waves from Kramer and Burcharth (2003). Damages to both the trunk front slope, crest, and rear-side slope and damage to the roundhead crest and rear-side slope were noted for all three freeboard conditions shown. Locations for damage to the roundhead were slightly different for other wave angles of approach.

Clearly, Equation (6.20) which gives the largest armor layer rock diameter for this example (Table 6.7) reflects this design need for stability of the armor layer for the entire low-crested breakwater structure.

Additional information on damage to the trunk and head regions of low-crested breakwaters is found in Garcia and Kobayashi (2014) who used a numerical model. Agreement between predicted and observed damage for 188 tests was only within a factor of 2.

6.2.5.4 *Reef breakwaters*

In very shallow water, the RoT $D_{n50} \geq 0.29h_c$ may mean that only one homogenous class of stone is possible for the entire structure (no core stones) and this has become known as a reef breakwater. Formulas for the trunk section using data from Ahrens *et al.* (1982) and Ahrens (1984) as developed by Powell and Allsop (1985) and developed by van der Meer (1990) are found in the *Coastal Engineering Manual* (CEM, 2006, Table V1-6-27). However, it is recommended that formulation (Equation (6.20)) as developed by Burcharth *et al.* (2006) for LCS be employed since it is applicable for the entire structure including the roundhead ends.

6.2.6 *Cross-sectional design of breakwaters*

6.2.6.1 *Conventional shape — Trunk section*

The conventional shape for a breakwater trunk section as recommended by CEM (2006) is shown in Figure 6.11. The cross-sectional geometry of the breakwater remains constant unless damaged in a coastal storm of major intensity that exceeds the initial design criteria for the structure. The top figure has multiple under-layers over a core and the bottom has only one primary under-layer and the core.

A top-down approach is employed to determine the dimensions. The armor layer (sometimes referred to as the cover layer) is calculated first to determine the stable weight W illustrated in Figure 6.11. We have presented formulas and spreadsheets to calculate the nominal stable stone size D_{n50} from which the stable weight W (or mass, M) is found from Equation (6.16) (or Equation (6.17)). Many of the sublayer rock sizes and dimensions are then found relative to the armor layer weight, W. We herein present 10 "rules" for the cross-sectional design of the trunk section of rubble-mound breakwaters.

(1) Crest elevation, E_c

The crest elevation defines the crest height, h_c, at the toe of the structure and depends on the purpose of the structure. Wave runup and overtopping formulations were presented in Chapter 5, Part I, to estimate crest elevations for non-overtopping conditions required for ship unloading or to mitigate erosion on the lee side of the structure. The design crest elevation should be the lowest required for the desired purpose of the structure. Knowledge of the maximum design SWL is required and will be influenced by the potential for accelerated sea-level rise in the future as discussed in Chapter 11.

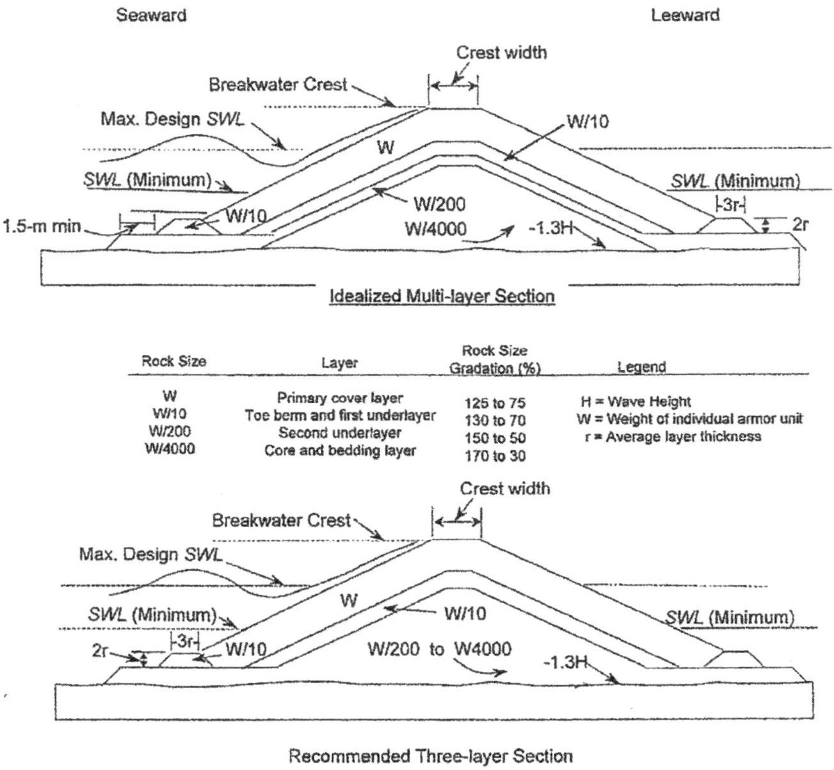

Figure 6.11. Conventional shape for trunk section as recommended by CEM (2006) for both an idealized, multi-layer section (top) and a three-layer section (bottom).

(2) Crest width, B

The number of stones, n, determines the crest width B and a minimum of three stones is recommended. However, the nominal stone size D_{n50} must be modified by a "layer" coefficient, k_Δ to account for the actual, average size of the quarry stone or artificial concrete units as found in Table 6.8. The crest width formula then becomes

$$B = n(k_\Delta D_{n50}) \quad (n = 3\,\text{minimum}). \tag{6.23}$$

(3) Thickness of rock armor layer, r

A similar formulation for the average layer thickness, r, of the armor layer is employed but here the minimum number of stones is two so that if the cover

Table 6.8. Layer coefficients and porosity for various armor units (from CEM, 2006).

Armor Unit	n	Placement	Layer Coefficient k_Δ	Porosity P (percent)
Quarrystone (smooth)[a]	2	Random	1.02	38
Quarrystone (rough)[b]	2	Random	1.00	37
Quarrystone (rough)[b]	∃3	Random	1.00	40
Quarrystone (parallelepiped)[c]	2	Special	—	27
Quarrystone[d]	Graded	Random	—	37
Cube (modified)[a]	2	Random	1.10	47
Tetrapod[a]	2	Random	1.04	50
Tribar[a]	2	Random	1.02	54
Tribar[a]	1	Uniform	1.13	47
dolos[e]	2	Random	0.94	56
Core-Loc[f]: Vol. < 5 m[c]	1	Random	1.51	60
5 < Vol. < 12 m[c]				63
12 < Vol. < 22 m[c]				64
Vol. < 5 m[c]				57
Accropod[g] < Vol. < 12 m[c]	1	Random	1.51	59
12 < Vol. < 22 m[c]				62

Notes: [a]Hudson (1974).
[b]Carver and Davidson (1983).
[c]Layer thickness is twice the average long dimension of the parallelepiped stones. Porosity is estimated from tests on one layer of uniformly placed modified cubes (Hudson, 1974).
[d]The minimum layer thickness should be twice the cubic dimension of the W_{50} riprap. Check to determine that the graded layer thickness is ∃1.25 the cubic dimension of the $W_{\max}$ riprap (see Equations VI-5-119 and VI-5-120).
[e]Carver and Davidson (1977).
[f]Turk and Melby (1997).
[g]Accropod informational brochure.

layer is moved, a second layer is still present to protect the smaller stones in the under-layers. The layer coefficient k_Δ is again found in Table 6.8.

$$r = n(k_\Delta D_{n50}) \quad (n = 2\,\text{minimum}). \tag{6.24}$$

This same formula is applied for the thickness of all under-layers that may be used in the cross-section using the appropriate D_{n50} for that weight of the under-layer, and again the minimum number of stones should be two ($n = 2$). The number of concrete armor units per unit surface area N_a is discussed in the following (point 9) and the weight and thickness of a layer of a well-graded stone (riprap) is covered in (point 10).

(4) Weight of primary under-layer, $W/10$

Both cross-sectional designs shown in Figure 6.11 have a primary under-layer with weight that is about one-tenth of the weight of the overlying armor layer. This ratio will provide stone sizes in the primary under-layer from being pulled upward and through the pore spaces in rock armor layers. As a formula,

$$W/10 = \frac{\text{Weight of stone armor layer}, W}{10}. \tag{6.25}$$

However if concrete armor units are CORE-LOC®, then the primary under-layer should be heavier in weight and larger in size and only one-fifth of the weight of the overlying armor layer ($W/5$) to promote interlocking in the pore spaces of the concrete units (CEM, 2011). Recall that the four basic designs for concrete armor units (Figure 6.7) can be placed in a single layer, and thus, the larger and heavier primary under-layer is important to ensure a stable design.

(5) Weight of additional under-layers and core, $W/200$ to $W/4000$

The gradation (range) of material sizes available from the quarry will generally determine whether (or not) a multi-layer or three-layer section can be designed. The multi-layer design makes fuller use of all the quarry materials but is more difficult and expensive to construct. In general, a three-layer section design is recommended for ease of construction and lower cost.

If a multi-layer design (Figure 6.11, top), the second under layer should weigh 20 times less than the primary under-layer, i.e.

$$W/200 = \frac{\text{Weight of stone primary underlayer}, W/10}{20} = \frac{W}{200}. \tag{6.26}$$

It therefore follows that the core in the multi-layer design, there should be stones that weigh 20 times less than the second under-layer, i.e.,

$$W/4000 = \frac{W/200}{20} = \frac{W}{4000} = \text{weight of core stones}. \tag{6.27}$$

The three-layer design simply combines the $W/200$ and $W/4000$ stones in the core, i.e. it consists of all the remaining stones mined from the quarry and does not have a second under-layer (Figure 6.11, bottom).

Figure 6.11 (middle) also displays these median rock sizes for these "top-down" layers and the rock size gradations (ranges in % of mean rock size) to prevent smaller rocks in the under-layers from being pulled through the upper layer by wave action. For example, consider an armor (cover) layer weight, $W = 454$ kg weight (1,000 lbs) for quarry stone with a unit weight, $\rho_a g = 2{,}650 \times 9.81 = 2.59 \times 10^4$ Newton/m^3 (165 lbs/ft^3). The nominal, D_{n50} size becomes

$$D_{n50} = \left(\frac{W}{\rho_a\, g}\right)^{1/3} = \left(\frac{1000}{165}\right)^{1/3} = 1.82\,\text{ft} = 0.56\,\text{m}.$$

But the nominal stone size D_{n50} is not the same as the size measured by a sieve, or by visual inspection. Thomsen *et al.* (1972) found that the sieve size of angular stone is approximately

$$D_{\text{sieve}} \approx 1.15\, D_{n50}, \tag{6.28}$$

so that for this example, the sieve stone size, $D_{\text{sieve}} = 0.64$ m (2.10 ft). Therefore, the range of armor layer stones sizes (125% to 75%) becomes 0.80 m (2.63 ft) to 0.48 m (1.58 ft). Weights and size dimensions using Equation (6.28) for quarry stone ranging from 20 US tons (18.144 metric tons) down to 0.025 lbs (0.01 metric ton) are given in CEM (2006, Table VI-5-50).

As a final check on the ranges given in Figure 6.11 (middle) for the various layers to prevent small rocks being pulled up through the overlayers, let us define

D_{15} (upper layer) as the diameter exceeded by the coarsest 85% of the diameters of the upper layer immediately above the under layer, and

D_{85} (under layer) as the diameter exceeded by the coarsest 15% of the diameters in the under layer.

Therefore, to prevent pulling of rocks upward through the voids in the upper layer (Ahrens, 1981),

$$D_{15}(\text{upper layer}) \leq 5\, D_{85}\ (\text{under layer}). \tag{6.29}$$

The locations (elevations) of the ends of these layers depend on the design wave height.

(6) Bottom elevation of armor (cover) layer, E_{ba}

The orbital motions of waves decrease with depth below the surface, and down to the bottom elevation E_b over the water depth h at the toe of the structure. Therefore, the location of the bottom elevation of the armor layer E_{ba} is related to the design wave height H_s employed to determine the stable design weight, W and to the water depth, h.

(a) Water depth $h > 1.5\, H_s$. Recall that waves break in shallow water when $H_s \approx 0.56\, h$ (Equation 2.25) so that this is a relatively deep water non-breaking condition. The recommended (CEM, 2006) location of the bottom elevation of the armor layer, E_{ba} is one H_s below the design SWL:

$$E_{ba} = \text{SWL} - 1.0\,H_s. \qquad (6.30\text{a})$$

This has been labeled as the "primary" armor layer.

Below this elevation, this is labeled the "secondary" armor layer. The weight of this second armor layer should be at least one-half of the primary armor layer, $W/2$, and extend down to the toe protection for the structure. The armor layer thickness must match that for the upper section of the armor so that $n \geq 3$ in Equation (6.24) for the appropriate, D_{n50}.

(b) Water depth $h \leq 1.5\,H_s$. These are conditions for wave breaking so that the recommended location of the bottom elevation of the armor layer E_{ba} is at the toe of the structure where the toe protection is constructed as discussed in Section 6.2.8. Hence, it is all built using the stone size in the primary armor layer.

$$E_{ba} = \text{Elevation of the toe protection} = E_b. \qquad (6.30\text{b})$$

(c) Bottom elevations of under-layers. As shown in Figure 6.11, the under-layers should extend completely beneath any upper layer of a given stone size.

(7) Toe protection for armor layer stability

In many studies of excessive damage and failure of rubble-mound structures, the primary cause was the inadequate design of the toe protection at the base of the structure. Therefore, we have provided an entire Section 6.2.7 to adequately address this subject.

(8) Granulated material and geotextiles as bedding filters

The sublayers and core material must be placed on some type of bedding filter. The most common method is to first lay down a geotextile filter cloth on the sandy bottom and then a layer of small stones 15–30 cm (6–12 inches) thick over the top to hold the geotextile in place during construction. The bedding stones prevent the larger, heavier rocks in the sublayers and core from tearing the filter cloth and also provide a relatively uniform surface to support the layers above. The purpose of the bedding filter is to prevent the entire structure from sinking into the sandy bottom.

Theoretically, the same material retention criterion as described above (Equation (6.29)) is applicable for this combination of geotextile filter fabric and overlaying granulated (stone) material,

$$D_{15}\,(\text{upper filter material}) \leq (4\text{–}5)\,D_{85}\,(\text{underlying material}). \qquad (6.31)$$

In theory this same formulation can be applied for the relation between the geotextile porosity and the sandy bottom grain-size distribution.

In practice, the geotextile fabric material is woven, nylon, 1–2 mm thick with variable porosity (see CEM, 2006, Part VI, Chapter 4 for more details) and the granulated material consists of small pebbles and stones.

(9) Concrete armor layers

The great advantage of the four types (Tetrapod, Accropod®, CORE-LOC®, and Xbloc®) with high structural strength and interlocking ability is that they can be placed in a single-layer system as the armor layer and hence are cost-effective. A one-layer system uses far less concrete in the armor layer and should always be compared with a two-layer rock system.

However, many of the above "rules" for cross-sectional design with rock armor layers are different. For example, the placing density is defined as the number of concrete armor units N_a per unit surface area, A, and is estimated from the following equation:

$$\frac{N_a}{A} = n\,k_\Delta \left(1 - \frac{P}{100}\right)\left(\frac{\rho_a\, g}{W}\right)^{2/3}. \qquad (6.32)$$

Values for porosity P and k_Δ are found in Table 6.8.

Special rules are numerous for the weights (sizes) and bottom elevations of the under-layers with many discussed in the CEM (2006, Part VI, Section 5) and depend on the value of the Hudson stability coefficient, K_D as found in Table 6.6. The licensee should always be consulted for details.

(10) Graded riprap armor layers

Again, the "rules" are different if graded rocks (riprap) are employed as the cover, armor layer.

Riprap is used more for shore- and bank-protection structures (revetments) that are not exposed to high waves or strong currents. The wider size distribution of riprap provides a less uniform armor layer that is more susceptible to damage by strong waves and currents. Riprap is less expensive than uniform stone, and placement on the slope is usually less precise (e.g. dumping from trucks). Many states have graded riprap "standards" for specification as employed mainly by their respective highway departments for erosion protection on land slopes. These are relatively inexpensive materials for emergency repairs of coastal structures. There exists no universal "standard" for the size variation ranges in graded riprap. CEM (2006) recommends that the design wave height should be less than 1.5 m for employment of graded riprap used as armor layer for coastal structures.

Let W_{50} be the weight of the 50% size rock in the riprap gradation. The Hudson (1974) stability formula (Equation (6.4)) is the only one known to this writer to estimate the median stable weight, W_{50}, for a graded riprap armor layer. The riprap stability coefficient is now defined by the symbol K_{RR}, and recommended values are $K_{RR} = 2.2$ for breaking waves and $K_{RR} = 2.5$ for non-breaking waves (Ahrens, 1981). For a conservative design, it is recommended to use $H_{1/10}$ as the design wave height.

Rule 3, to estimate the thickness of the rock armor layer, r, needs special treatment if graded riprap is employed.

(a) Using the W_{50} riprap size,

$$r = 2.0\left(\frac{W_{50}}{\rho_a\, g}\right)^{1/3} \quad \text{or} \tag{6.33a}$$

(b) using the heaviest stone in the gradation, $W_{\max}$,

$$r = 1.25\left(\frac{W_{\max}}{\rho_a\, g}\right)^{1/3}, \quad \text{or} \tag{6.33b}$$

(c) using 0.3-m (1 ft) thickness,

$$r = 0.3\,\text{m}\,(1\,\text{ft}). \tag{6.33c}$$

Simply use the largest value r, estimated using three ways as in Equation (6.33).

The minimum requirements for the under-layer beneath the cover, armor layer again follows the same material retention criteria given above as Equation (6.29). However, rules 4 and 5 for the under-layers may also need special treatment. The primary under-layer (rule 4) requires larger stones to prevent material from moving up through the larger voids in the riprap armor layer resulting from stone movements during storms. Therefore, use a stone size with a 50% weight, $W_{50}/5$, for the primary under-layer and make it three times the thickness of the $W_{50}/5$ stone rather than the coefficient 2 in Equation (6.33a). This larger, under-layer stone is more permeable and will reduce wave runup and will also increase the interlocking of stones between the cover and primary under-layer. The core material (rule 5) will consist of all the remaining riprap material that is readily available and only a three-layer section should be considered.

The above 10 "rules" should permit the calculation of a draft cross-section for the trunk of a rubble-mound breakwater, groin, jetty, or revetment. The CLI website www.concretelayer.com includes online tools for making AutoCAD drawings of the truck cross-sections.

6.2.6.2 *Conventional shape — Roundhead ends and leeside armor layer*

All coastal structures have end sections that require special treatment for the armor layer. Wave focusing, strong overflow currents, and less lateral support from adjacent armor layer units can produce damage in the critical area (shaded) as illustrated in Figure 6.12 (from Burcharth, 1993). One common method to increase the stability is to simply increase the structure slope on the ends to make what has become known as "roundhead" ends.

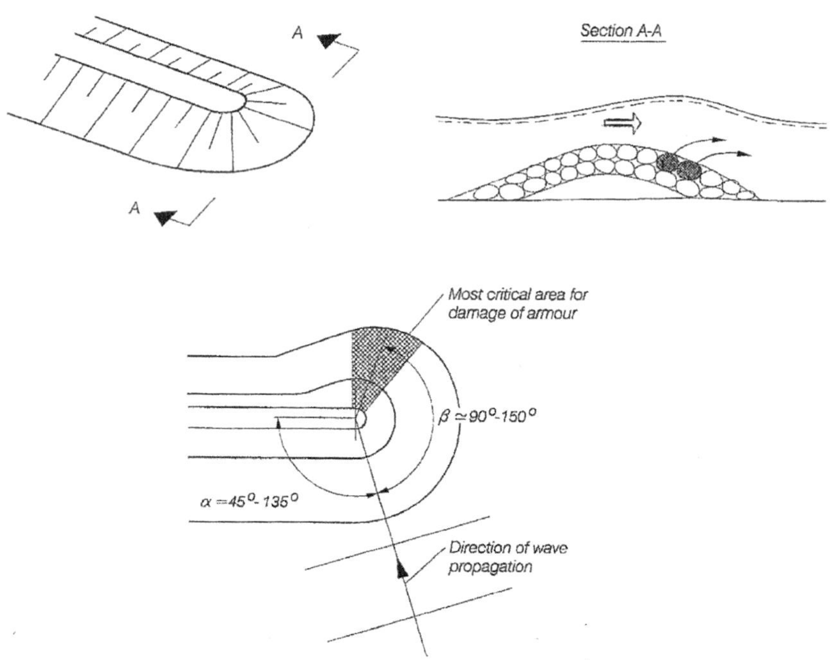

Figure 6.12. Critical damage region for end sections of breakwaters (from Burcharth, 1993).

As noted in the spreadsheets (Figure 6.4) for calculation of the stable rock diameter for the trunk section, as the slope $\cot\alpha$ increases, the stable rock size decreases. This method is far better for the contractor during construction since there is no need to increase the rock size on the ends. Note that for LCS as discussed in Section 6.2.5, the formulation developed by Burcharth *et al.* (2006) and given here as Equation (6.20) considers the entire structural stability and especially damage to the roundhead (Figure 6.10) so that there is no need to modify the slopes on the ends.

The leeside of nearshore breakwater structures is exposed to far lower wave height conditions. The leeside, armor layers could be smaller, lighter rocks if that were the only concern. But the leeside also experiences strong forces from waves that overtop the crest and from currents when water levels overtop the crest. Consequently, the seaward side armor is also employed on the leeside of the structure as shown in Figure 6.11 whenever moderate wave overtopping and/or submergence conditions are present. Because the potential exists for accelerated, sea-level rise in the future (Chapter 11), it is recommended to extend the seaside armor layer to the bottom on the leeside even for non-overtopping designs.

6.2.6.3 *Conventional shape — Concrete caps*

Figure 6.13 displays a small harbor breakwater in France with a concrete section (cap) on the crest (from CLI website). Accropod® armor units are employed on the seaward side and on the roundhead end and the leeside uses small, concrete blocks as the cover layer. Concrete may be placed to fill the pore spaces of rock armor layers on the crest and along the seaward side slope or used to construct a vertical wall with level road or walkway section as shown in Figure 6.13. Concrete caps on breakwaters are built to (1) increase the crest height to limit wave overtopping on the leeward side, (2) provide an access roadway for construction and maintenance, and (3) provide a walkway for recreational use. In some instances, concrete has been used to fill void spaces on jetties to reduce sand movement through the structure that decreases water depths in navigation channels.

For all these reasons, the design of concrete caps on coastal structures is a very specialized topic and beyond the scope of this book. CEM (2006, Part VI, Chapter 5) includes a brief introduction to the wideranging literature on this subject. Rock Manual (2007) also includes many recent references in Chapter 6.

Figure 6.13. Bandeau 235 Cerbere, France. Marne Harbor breakwater (Accropod® armor units) with roundhead end section and concrete cap for recreational use (from CLI website) (courtesy Concrete Layer Innovations, Inc.).

6.2.7 *Toe protection and scour prevention*

6.2.7.1 *Toe berms for armor layer stability*

The conventional shaped breakwaters shown in Figure 6.11 include a small berm at the toe (base) of the armor layer. The primary purpose of the toe berm is to support the armor layer and prevent it from sliding down during wave attack. Figure 6.14 shows toe berm designs in three water depths where the wave heights H_s total water depth h, and water depth at top of berm h_b have been found to influence the required stone diameter, D_{n50} for a stable toe-berm design. In very shallow water where waves break, the stone size of the berm should remain the same as the stone size in the primary armor layer (Figure 6.14(a)). At some shallow to intermediate water depths (Figure 6.14(b)), the stone size in the berm can be reduced at the base of the structure. In deep water (Figure 6.14(c)), the location of the toe berm can be moved upward and the stone size in the toe berm reduced. Some guidelines for these three water depth cases as provided in CEM (2006) and Rock Manual (2007) are as follows.

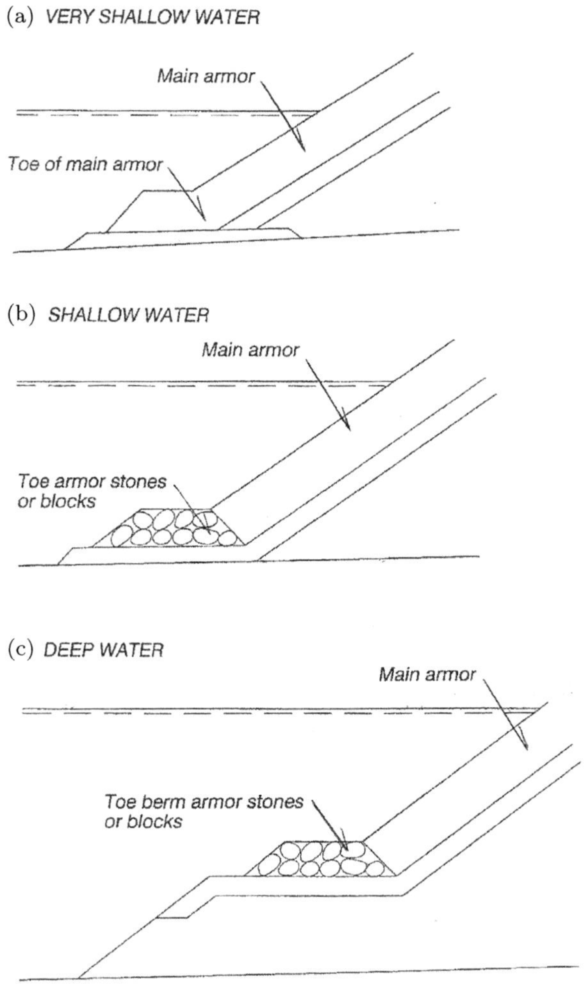

Figure 6.14. Schematic diagram for toe berm designs for three water depths (from CEM, 2006).

In relative small-scale physical model experiments, van der Meer *et al.* (1994) studied the stability of a two-layer stone berm located at the base (Figure 6.14(b)) of a rubble-mound breakwater with steep slope of $\cot\alpha = 1.5$ and attacked by irregular, head-on waves. The toe berm size was kept 3–5 stones wide and 2–3 stones high with a front slope also at $\cot\alpha = 1.5$. Besides the three key independent variables, H_s h, and h_b, the number of armor layer units, N_{od}, moved out of a strip width, D_{n50} was also noted and the following damage levels were defined:

$$N_{od} = \begin{Bmatrix} 0.5 \text{ no damage} \\ 2 \text{ acceptable damage} \\ 4 \text{ severe damage} \end{Bmatrix}. \tag{6.34}$$

The following formulation was developed for the toe berm stability as a function of h_b/h:

$$\frac{H_s}{\Delta D_{n50}} = \left(2.0 + 6.2\left(\frac{h_b}{h}\right)^{2.7}\right) N_{od}^{-0.15} \tag{6.35}$$

for the range of $0.4 < h_b/h < 0.9$. Using the no damage level 0.5 for N_{od} in Equation (6.35) gives the following results for this range of depth ratios as shown in Table 6.9. As the dimensionless stability parameter, $H_s/\Delta D_{n50}$ increases, the nominal stone size, D_{n50} decreases for a give wave height, H_s.

Our recommended design for toe berm stability is also shown in Table 6.9 for the three water depth conditions in Figure 6.14.

The Corps of Engineers, CEM (2006) conventional shape design in Figure 6.11 shows using a toe berm size of armor layer weight, W, divided by 10, i.e. $W/10$ but with no mention of the water depths. Rule 3 for the thickness of the rock armor layer r is also shown and recommended for the toe berm elevation, $2r$ (a minimum of two rock layers) and a toe berm crest width of $3r$. The nominal rock diameter D_{n50} as found using Table 6.9 is then applied in rule 3 to calculate r for the toe berm. Burcharth *et al.* (1995) slightly modified Equation (6.35) for using rectangular-shaped, concrete blocks for the toe berm.

Table 6.9. Toe stability as a function of relative water depth over the toe berm.

Depth Ratio (h_b/h)	$H_s/\Delta D_{n50}$	Recommended Toe Stability Design
0.4	2.88	Use armor layer stone for toe design
0.5	3.28	Use armor layer stone for toe design
0.6	3.95	Use Equation (6.35) for toe stability design
0.7	4.85	Use Equation (6.35) for toe stability design
0.8	5.99	Use Equation (6.35) for deep water, Figure 6.14(c)
0.9	7.40	Use Equation (6.35) for deep water, Figure 6.14(c)

6.2.7.2 *Scour prevention*

Scour can be simply defined as the removal of granular material near the bottom (bed region) of coastal structures. Scour may undermine the toe berm producing local stone movement, damage, and potentially complete failure of the structure. Figure 6.15 (from CEM, 2006) illustrates potential scour problems for detached breakwaters, seawalls (and revetments), bulkheads, and tidal inlet jetties that are the focus of this chapter. The other scour problems associated with vertical piles (piers) and pipelines are discussed in Chapter 9.

Excessive shear stresses created by waves, currents, and breaking wave turbulence produce scour. The presence of the coastal structure may increase these hydrodynamic forces due to their location, geometry, roughness, and permeability to cause sediment transport and scour. Combining knowledge of the hydrodynamics and resulting sediment movement is extremely difficult and has resulted in some simple RoTs for scour depth,

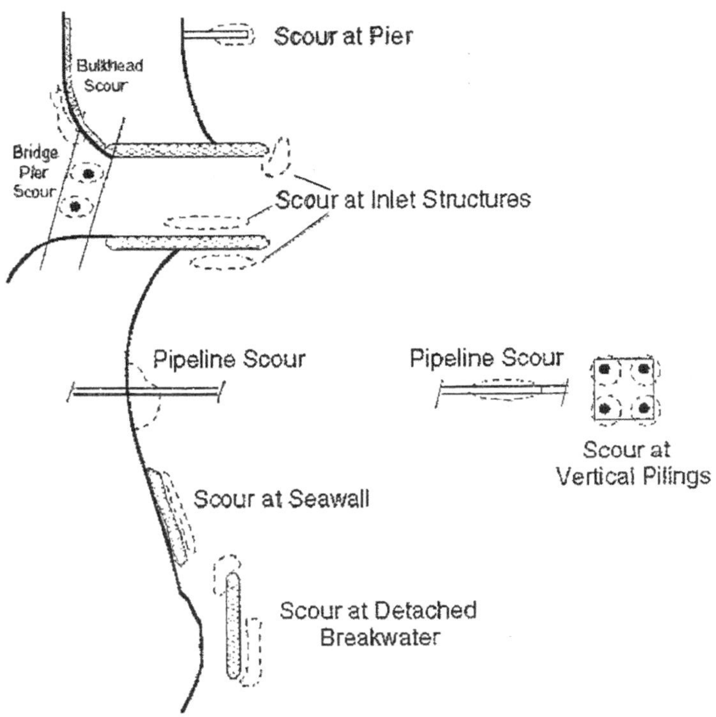

Figure 6.15. Typical examples of scour at coastal structures (from CEM, 2006).

d_s, prediction. The most information available is for scour depths in front of vertical walls so that the prediction and prevention of scour shall be addressed fully in Section 6.3.

6.2.8 *Berm breakwaters*

The design of the conventional, fixed-shaped breakwaters shown in Figure 6.11 is driven by the design wave climate, the stones available in the selected quarry, and the construction cost. When the design wave height is exceeded, some stones (or concrete armor units) will move so that the trunk shape is damaged, and becomes curved (S-shaped) as shown in Figure 6.16 (top). The relatively flattened section along the S-shaped section resembles a berm in front of the original, fixed-shape design. This berm section breaks the waves and reduces the wave heights that strike the remaining armor layers so that a statically stable, reshaped structure is the result. This is the basic principle behind what has become known as a berm breakwater as illustrated in Figure 6.16 (bottom). The design is also known as the S-slope or S-shaped breakwater.

The conventional, fixed slope design with little or no armor layer movement for the design wave height was driven by the weight, W, requirements of the armor layer and all sublayer weights are submultiples ($W/10, W/200, W/4000$) as illustrated in Figure 6.11. The designer has little flexibility in this "demand-based" design driven by W.

In contrast, the berm breakwater design is based on the yield of stones from the quarry, i.e. is "supply-based" that gives many more options for the designer. Clearly, the potential, gradation yield from the quarry must be investigated at an early stage in the design process. Ways to predict the quarry yield are beyond the limits of this book. See Rock Manual (2007, Section 3.9.5) for more details.

The design of berm breakwaters has evolved since the early 1980s. The National Research Council of Canada organized a "Berm Breakwaters: Unconventional Rubble-Mound Breakwaters" Workshop in 1987 with the proceedings published by the American Society of Civil Engineers (ASCE) (Willis *et al.*, 1988). The stability of the seaward slope of berm breakwaters was investigated by Delft Hydraulics and produced a computer program "BREAKWAT" (see van der Meer, 1992). Many of the first berm breakwaters were constructed in Iceland as summarized in Sigurdarson *et al.* (2000). A "State of the Art of Design and Construction of Berm Breakwaters" report was developed and published by (PIANC/MarCom 40, 2003). As

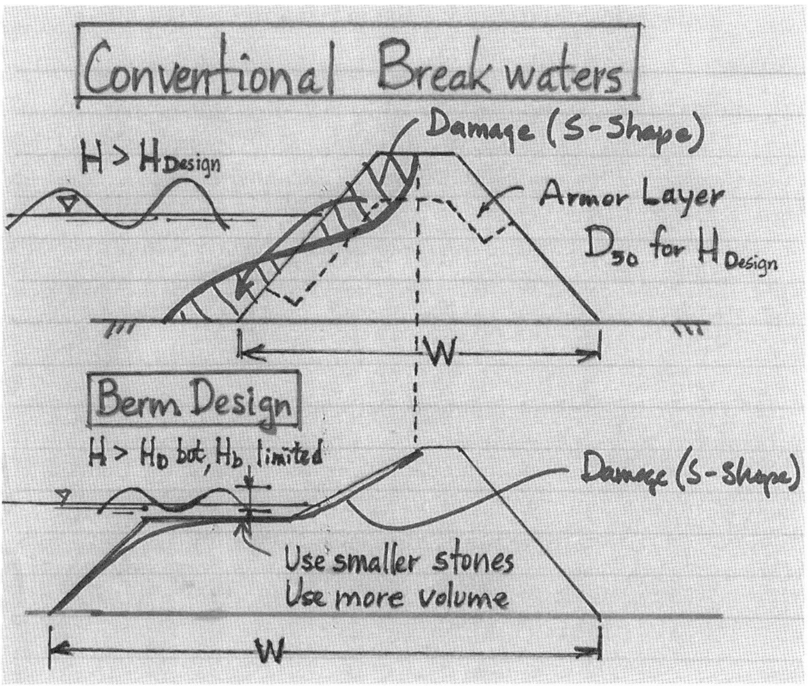

Figure 6.16. Conventional breakwater section (top) and berm breakwater shape (bottom).

deduced from the many fine references summarized in these works, three main types of berm breakwater designs have been developed. In all three types, the cross-sectional stone volume is greater than the conventional design. Artificial, concrete armor units are never employed in a berm breakwater design. See also Sigurdarson *et al.* (2005).

(1) Statically stable, non-reshaping, berm design

The armor stone in the crest area and the berm is the largest stone class available in the quarry. Little to zero stone movement is permitted similar to the conventional design. However, the presence of the berm causes wave breaking so that the stone size required is less than that required for the conventional design. Many sublayers can then be employed to make full use of the size gradations coming from the quarry. The goal is to employ 100% of the quarry yield. This type is called the Icelandic-type breakwater for the many designs constructed in Iceland.

(2) Statically stable, reshaping berm design

When the size and volume of large stones from the quarry are not sufficient, the flat berm section will be constructed of stones that will move seaward during the design storm conditions. The constructed, flat berm section is expected to recede and form an S-shaped berm profile as illustrated in Figure 6.17(a). This type has a wider cross-section with more volume than the non-reshaping berm design. Once the new S-shaped profile is formed, the profile and individual stones along the profile are stable. Figure 6.17(b) is an example of a statically stable, reshaping berm breakwater design.

(3) Dynamically stable, reshaping berm design

This type is only dynamically stable during the reshaping process in that the surface stones continue to roll up and down along the S-slope after the berm profile has been reshaped. Unavoidable stone breakage may result to further weaken the armor layer. See Rock Manual (2007, Section 3.8.5) and

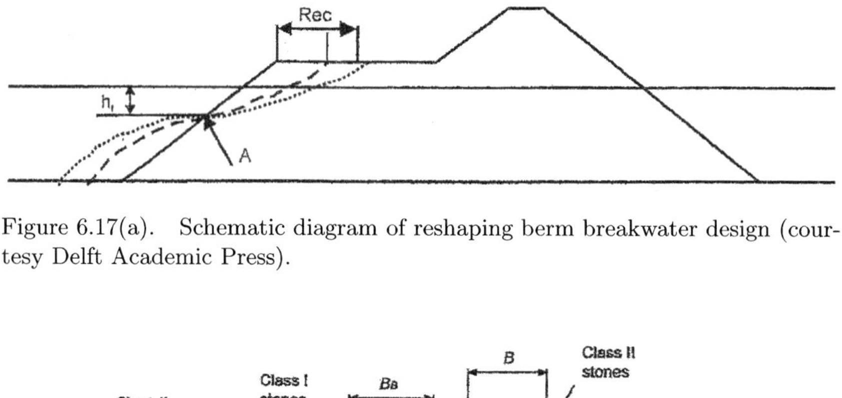

Figure 6.17(a). Schematic diagram of reshaping berm breakwater design (courtesy Delft Academic Press).

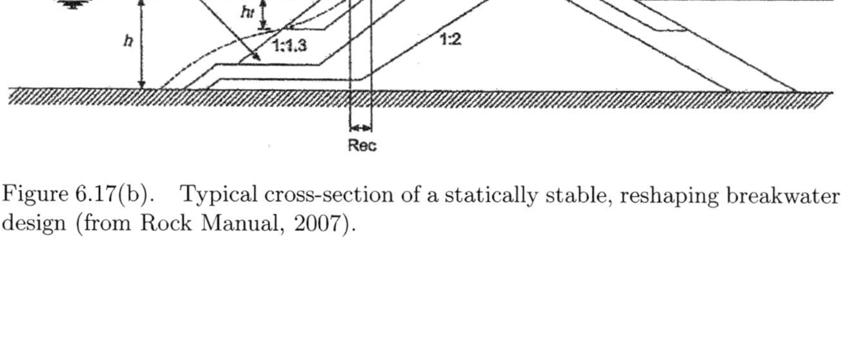

Figure 6.17(b). Typical cross-section of a statically stable, reshaping breakwater design (from Rock Manual, 2007).

PIANC (2003) report for methods to estimate the strength of stones from a quarry. This type of design should be avoided.

A comparison between dynamically stable and statically stable, non-reshaping breakwaters of the Icelandic-type can be found in Verhagen *et al.* (2009) along with a procedure to select "... the most appropriate type" as described by Siguardarson *et al.* (2004).

6.2.8.1 *Berm recession, Rec*

The horizontal distance that the flat berm crest elevation moves inland is called the berm recession, Rec, as illustrated in Figures 6.17(a) and 6.17(b). The constructed berm width B_B and elevation must be sufficient to prevent movement of the stones (Class II) in the upper crest of the breakwater and exposure of the sublayer of stones (Class II) below the statically stable, S-shaped profile in the reshaping profile design. The numerical model called "BREAKWAT" as developed by Delft Hydraulics (van der Meer, 1990, Report H986) predicts the recession distance Rec and statistically stable profile. Analytical expressions for the S-shaped profile can be found in van der Meer (1992). The primary under-layer and core must always be covered by at least two or three thicknesses of the armor layer of Class I stones. Trial and error is required.

To aid the design process, the analytical formula for the recession distance, Rec, can be employed as developed by Torum *et al.* (2003). When normalized by the nominal D_{n50} size of the Class I armor layer stones, this becomes

$$\frac{\text{Rec}}{D_{n50}} = f(H^*, T^*), \tag{6.36}$$

where

$$H^* = \frac{H_s}{\Delta D_{n50}} \quad \text{and} \quad T^* = T_m\sqrt{\frac{D_{n50}}{g}}, \tag{6.37a,b}$$

so that

$$\frac{\text{Rec}}{D_{n50}} = A\,(H^*T^*)^3 + B\,(H^*T^*)^2 + C(H^*T^*) - f(f_g) - f(f_h), \tag{6.38}$$

with

$$A = 2.7 \times 10^{-6}, \quad B = 9.0 \times 10^{-6}, \quad C = 0.11, \tag{6.39a,b,c}$$

and

$$f(f_g) = -9.9f_g^2 + 23.9f_g - 10.5, \tag{6.40}$$

$$f_g = \frac{D_{n85}}{D_{n15}} \quad (1.3 < f_g < 1.8), \tag{6.41}$$

and

$$f(f_h) = -0.16f_h + 4.0, \tag{6.42}$$

$$f_h = \frac{h}{D_{n50}} \quad (12.5 < f_h < 25). \tag{6.43}$$

Although somewhat involved, the key variables for design are the design wave conditions (H_s, T_m), the quarry yield (D_{n50} and D_{n85}/D_{n15}) and the water depth, h, at the toe of the structure. Clearly, the smaller the Class I armor stones D_{n50} available from the quarry, the greater the berm recession Rec so that a wider reshaping berm width B_B must be placed requiring a larger volume of stones from the quarry.

6.2.8.2 *Additional recommendations for statically stable, reshaping berm design*

The crest elevation of the berm breakwater is again governed by the purpose of the structure and the acceptable wave runup and overtopping rate. Part I of Chapter 5 presents design formulas for the conventional, fixed-shape breakwater profiles. The constructed berm, reshaping profile, and highly porous nature of the berm breakwater all make these formulations inaccurate. Rock Manual (2007) recommends setting the crest height, $R_c = (1\text{–}1.4)H_s$, but for harbor facilities for unloading materials from ships moored behind the breakwater, clearly, physical hydraulic model studies should be investigated when non-overtopping conditions are required. CEM (2006) and Rock Manual (2007) present wave runup and overtopping formulas for breakwaters with fixed berms in front of the conventional shaped design. Pillai *et al.* (2017) present a review and sensitivity analysis of prediction models for wave overtopping of reshaping berm breakwaters. Models differ significantly in accounting for variations in berm width and berm elevation.

The recommended berm height h_B should be in the range $h_B \approx (0.5\text{–}0.9)H_s$ above the design water level. Torum (1999) found little change in recession for berm heights in the range $h_B/D_{n50} = 2\text{–}4$. The berm crest elevation is generally set to permit safe passage of trucks during construction.

The berm width B_B must be some safe, minimum width greater than the recession distance, Rec, estimated from the maximum level of design wave heights, H_s. No acceptable ratio of $B_B/$ Rec is known to this author. Rock Manual (2007) suggests making the berm as wide as possible and at least the expected recession from, for example, the 1:1,000-year, extreme wave-height storm event. The berm width is then governed by balancing breakwater costs against the probability of storm damage to the breakwater.

The Class I stones in the reshaped breakwater profile should extend a distance h_f below the design water level. The design for the Class II under-layer and toe berm in Figure 6.17(b) then follows the same as for a conventional breakwater as discussed in Section 6.2.7.1 using Table 6.9 and Equation (6.25) as illustrated in Figure 6.14.

The above-mentioned discussion is for the trunk section of the berm breakwater. The head (ends) of the berm breakwater requires special attention and Juhl *et al.* (1996) should be consulted for details.

6.2.8.3 *Formal design methodology for berm breakwaters*

In September 2016, the book *Design and Construction of Berm Breakwaters* as co-authored by van der Meer and Sigurdarson (2017) was published by World Scientific. The book includes chapters on *geometrical designs of the berm breakwater cross-section, including berm reshaping and wave overtopping, quarry and project management, blasting and sorting techniques, designs for various wave conditions and available rock classes, and case studies of already constructed berm breakwaters.*

The book is recommended for researchers and practicing design engineers for the latest information and a formal design methodology for berm breakwaters.

6.2.9 *Rubble-mound construction details*

Construction of rubble-mound structures (breakwaters, groins, jetties, revetments) requires the location and purchase, transport to the site, and then placement of extremely large amounts of loose materials. This is a problem in logistics. Construction also entails obtaining access to the site, moving and maintaining proper construction equipment at the site, and providing office space for personnel on the site. These many details are best left for the construction contractor awarded the contract for the project.

But the method of construction is the job of the designer. The method of construction may influence the initial design details that produced the crest elevation, cross-section of the trunk, size of the armor layer materials, sublayer and core elevations and volumes, toe protection design, and interface between the bottom and the structure. The method of construction is equally as important as the analysis and design drawings of the structure and may provide construction cost savings that are many times more than slight changes in the preliminary design. The overriding concern is that construction always occurs under the potential for storm conditions that could cause damage to exposed core and under layers. The protective over layers and armor layers should be placed as soon as possible.

Construction can be either from the land (land-based) or from the water (marine-based) equipment or with combinations. When the material is transported to the site from the quarry by land, land-based construction is usually more economic. The opposite is equally true for material transported by water barges from the quarry, especially for nearshore, detached breakwater construction.

6.2.9.1 *Land-based construction*

Figure 6.18 (from Rock Manual, 2007) displays typical plan view and trunk cross-section for a land-based breakwater construction. The numbers (1)–(6) indicate the sequence of material placement by the equipment listed in chronological order as follows:

(1) quarry run core material placed by dump trucks,
(2) core material (or larger) placed by crawler-type crane (or excavator),
(3) scour protection material placed by crawler crane,
(4) under-layer material placed by crawler crane or excavator,
(5) toe material on seaward side place by crane of excavator,
(6) armor layer material placed on seaward slope by crane or excavator.

For this example of land-based breakwater construction, the core elevation and width must be sufficient for the dump trucks, crane, and/or excavator to operate safely. Core material placed at or near the water line is very vulnerable to movement by waves so that the under-layer material protecting the core must be placed along with the core material as the construction proceeds in the seaward direction from land. See Rock Manual (2007) for further details regarding space requirements and land-based construction of the toe, berm, scour protection, and armor layer.

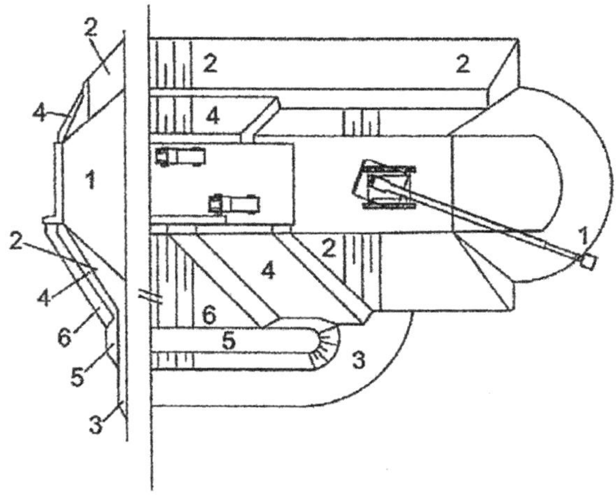

Figure 6.18. Schematic diagram of plan view and trunk cross-section for land-based breakwater construction. Numbers indicate chronological sequence of construction as listed in Section 6.2.9.1 (from Rock Manual, 2007).

Groins and nearshore, detached breakwaters have relatively narrow crest widths and may be submerged during tidal conditions to prohibit construction from the surface of the core. Land-based construction is limited by depths and exposed conditions. However, a temporary sand bridge from the shore may provide access and permit land-based equipment to reach the structure. When water depths and wave conditions permit, floating equipment may be used to construct the offshore sections of groins and the entire detached breakwater structure.

6.2.9.2 *Marine-based construction*

A typical plan view, side view, and cross-section view (trunk) for construction of a nearshore breakwater by floating, water-based equipment is shown in Figure 6.19. The numbers again refer to the construction sequence using typical equipment as follows:

(1) small stones for filter cloth cover and scour protection placed by a stone-dumping vessels,
(2) quarry run core stones placed by a split-hopper barge (depth-limited) then dumped from flat-top barge,
(3) under-layers placed and slopes trimmed using a floating hydraulic excavator or a floating crane,

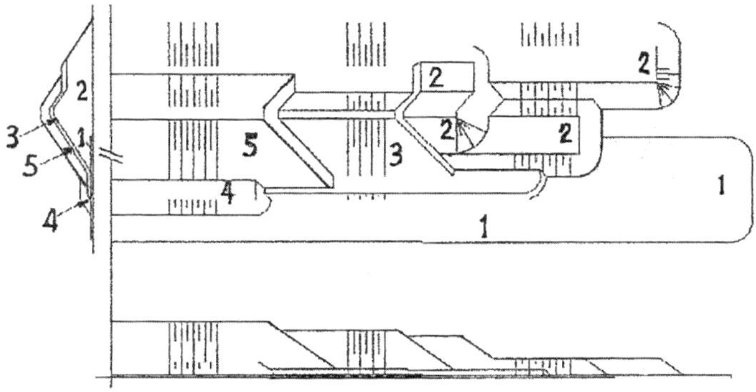

Figure 6.19. Typical marine-based breakwater construction sequence (plan view, side view, cross-section view) from Rock Manual (2007).

(4) seaward toe placed by a side dumping vessel or a floating crane,
(5) armor layer placed by a floating crane.

The main reasons for water-based construction using floating equipment are the following: (1) limiting elevation and width of the crest, (2) the quarry may be located near water making stone transport by barge more economical than land transport, and (3) the location of the structure making land-based cranes unable to reach far enough for stone placement. However, wave conditions (heights, periods) may restrict times for safe operation of floating equipment and cause more downtime. See Rock Manual (2007) for additional information on core, toe, under-layer and armor layer marine-based construction.

Table 6.10 is taken directly from Rock Manual (2007) and is a summary of nine issues for consideration in using land-based or marine-based equipment for construction. All nine areas should be carefully reviewed by the design engineer. Many additional topics must be considered including both positive and negative reasons why one method should be used and not the other. In many cases, a combination is best in that both methods can be taking place at the same time to overcome wave-erosion issues and speed up the construction time for the entire structure, if required.

6.2.9.3 *Concrete armor layer placement*

The placement of concrete armor units is highly specialized and requires a significant percentage of the total time for rubble-mound structure construction. As mentioned before, the Tetrapod unit (France, SOGREAH,

Table 6.10. Comparison of land-based and marine-based construction methods (from Rock Manual, 2007).

Land-Based Operation	Water-Based Operation
Cross-section	
• Crest elevation and crest width are determined by the dimensions of the cranes and trucks and acceptable workability due to overtopping by waves and spray • Flat slopes and wide berms may exceed maximum reach of cranes	• The crest level is determined by hydraulic stability and overtopping requirements • The core level is preferably 3 m below LW to allow free dumping of material from barges
Length section	
• The workfront is necessarily short. The various construction phases follow each other closely since works are concentrated around the position of the crane(s)	• The workfront is extended over a large area to allow for sufficient maneuvring and anchor spaces
Logistics	
• All supplies are over the breakwater; the narrow access limits the daily supply • Short breakwaters are most suitable for land-based operation • Existing infrastructure can normally be used for transport between the quarry and the site • Work is generally restricted to one front	• A loading terminal has to be provided for the barges before actual construction can start • The length of the breakwater forms no logistic limitation • Work may be initiated and proceed at different locations
Morphology	
• Current concentrations may create scour holes at the temporary head, as it is difficult to construct extensive scour protection	• A scour protection can be made in advance to avoid the development of scouring holes

(Continued)

Table 6.10. (*Continued*)

Land-Based Operation	Water-Based Operation
Limiting factors	
• Reach and capacity of the cranes effectively limit the cross-section and progress of the work	• A water depth of 2–4 m is required. Sometimes work is only possible in part of the tidal cycle
• Breakwaters in deep water can become troublesome	• Barges and other waterborne equipment are not as widely available as land-based equipment and their use requires specialised personnel
Environmental constraints	
• Available construction time is determined by the freeboard between working platform and water level	• Available construction time is determined by the allowable motions of barges and floating equipment. Downtime is longer than for land-based operations
	• Prane pontoons are particularly sensitive. Excessive impact between barges and pontoons, forces on the crane, and risk of collision with the breakwater precludes operations in waves with $H_s > 0.5$ m to 0.75 m
Damage during execution	
• There is a high risk of damage because the core and small-size under-layer extend above the water level	• The risk of damage to core and under-layer can be limited by keeping the top of this material at a low level
• The length of the unprotected structure is limited	• If damage occurs, it is generally over a considerable length
• In the case of an extended construction period, the risk of damage increases proportionately	

(*Continued*)

Table 6.10. (*Continued*)

Land-Based Operation	Water-Based Operation
Planning	
• Lead time for construction is determined by quarry preparation • Mobilization time for land-based equipment can be short • Progress of construction depends on one or two cranes that are critical for a number of construction phases • Equipment is readily available	• The lead time for construction might be long if special terminal facilities have to be constructed • The mobilization time for marine equipment is longer • At critical phases, production can be increased by introducing more equipment and working in parallel • Large stockpiles or high production rates are required for the efficient operation of marine equipment • Availability is limited because of the specialized nature of equipment
Maintenance and repair	
• If the breakwater is designed with a sufficiently wide crest element, maintenance from the superstructure can be done with standard land-based equipment. Frequently, this is not the case and floating equipment is required	• Maintenance and repair can only be done with floating equipment

1950) is no longer protected by a patent, but the Accropod® (France, 1981), CORE-LOC® (US, 1994), and Xbloc® (Netherlands, 2001) units have registered trademarks. The great advantage of the four types with high structural strength and interlocking ability is that they can be placed in a single-layer system as the armor layer and hence are cost-effective. But, how they should be placed varies somewhat so that the licensee should be consulted for construction placement details.

6.2.9.4 *Summary — Construction details*

It is recommended that potential construction contractors be invited to review preliminary designs and consulted to ensure that the design and the method chosen for its construction (land-based or marine-based) are practical. Design modifications can then be made (if needed) so that the final design when submitted for bids will result in the lowest bid cost for the project.

6.2.10 *Examples*

6.2.10.1 *Liquid natural gas (LNG) from Algeria — Port of Arzew breakwater*

Algeria has over 10% of the world's reserves of natural gas. Liquefaction reduces natural gas to 1/600th of its volume and makes transport via ship tankers profitable for export to countries all over the world. In the mid-1970s, a new harbor was designed and 2 km (6500 ft) long, outer breakwater was constructed in 1978–1979 at the Port of Arzew, Algeria to provide wave protection for three new berths for LNG tankers. A mock-up is shown in Figure 6.20 (from Rossard, 1977). The existing structure may be viewed via the website for Google Earth, with the coordinates (Lat. N35°49′30″, Lon. W0°15′18″) for Berth No. 1 (top, left in Figure 6.20).

Figure 6.20. Mock-up of the Port of Arzew, Algeria showing three LNG berths at the new, outer breakwater (from Rossard, 1977). Berth No. 1 is in the top, left end with ship moored at the dock.

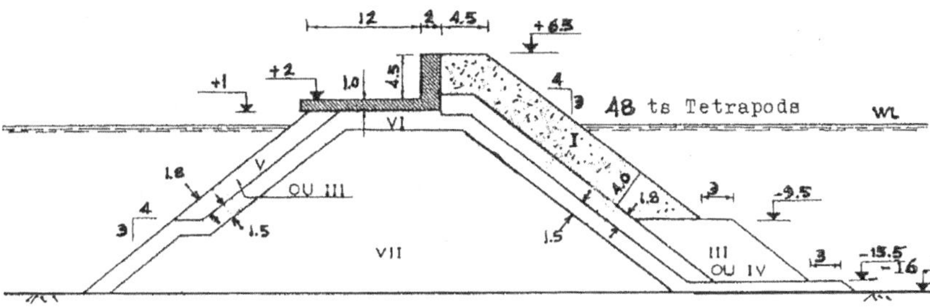

Figure 6.21. Trunk cross-section for Arzew, Algeria outer breakwater (from Bruun, 1985).

(a) Original breakwater design and damage

The trunk cross-section of the new, outer breakwater as originally designed and constructed in water depth of 16.5 m (54 ft) is shown in Figure 6.21 (from Bruun, 1985). The armor layer was a 48-metric ton tetrapod in two layers (layer thickness of 4.0 m, 13 ft) and built on a steep slope of $\cot \alpha =$ 1.33. The crest elevation of the armor layer was at 6.5 m (21 ft) above the design water level (datum unknown) and extended −9.5 m (−31 ft) below this design water level where a toe berm of material labeled III was placed down to 1 m above the bottom. The concrete cap is 12 m (39 ft) wide to provide room for a road, pipelines, and other docking features.

Soon after construction was completed, a storm on December 28–29, 1980 produced extensive damage to a section of the tetrapod armor layer and concrete cap. The reasons for the damages were analyzed by Bruun (1981) based on the available data, a site visit, and a report by divers. The following summarizes comments made by Per Bruun as found in Bruun (1985, pp. 671–675).

Physical model studies using irregular waves were conducted for significant wave heights, H_s = 7–8 m (23–26 ft) and wave periods of 12, 14, and 16 s (whether mean or peak wave period is not specified). Some wave data as measured elsewhere along the 1,650 km (1,025 mi) Algerian coast and wave hindcast information were employed, but the model tests did not include waves felt to be "*... of particular dangerous character*" and "*... proved to be a severe shortcoming*". In conclusion,

> *If the advice on heavier armor given in the preliminary stages of the project had been followed in design and construction, damage would undoubtedly have been less or perhaps non-existing. It is nature which determines wave action — not the designer.* (Bruun, 1985, p. 675).

No comment was made regarding the extremely steep slope $\cot \alpha = 1.33$ employed in the design and construction.

(b) Preliminary design — International Courses in Hydraulic Engineering (Delft)

In 1975–1976, the author was a postgraduate student attending the International Courses in Hydraulic Engineering (IHE) in Delft, the Netherlands. The specialty areas "Computational Hydraulics" and "Coastal Engineering" were followed and after one year, the "Diploma Hydraulic Engineering" was awarded (1976). The first specialty area resulted in the book *Computational Hydraulics* (Abbott and Basco, 1989). The second specialty area has been the subject of study for over 40 years and is the foundation material in this book.

One course assignment was the design of an outer harbor breakwater for the Port of Arzew. This preliminary design was a many-student, team effort for a wide range of design water height conditions to produce total (initial plus maintenance) cost estimates for the design. See Chapter 8 for further details of cost estimates using this example. Our team (Basco and Stadler, 1976) developed the design for a non-overtopping breakwater to withstand a significant wave height with a 100-year return period (1% chance event every year). The design criteria employed is summarized in the following.

Wind data for the Port of Arzew region were provided for three wind speeds (26, 40, and 60 knots) over a range of 10 directions (22.5° intervals from west–southwest through north to east) as exceedance frequencies in percentage of time. The wind data were taken from the derived wind roses from ship records, but no references were provided. Because of the relatively short fetch in this western end of the Mediterranean Sea, the wave heights were said to be duration-limited. The duration values given for the three levels of wind speed were 16, 12, and 6 hours, respectively. No references were provided.

A wave-hindcasting nomograph was employed to estimate the significant wave heights H_s (similar to Figure 2.25 in Chapter 2) for these three wind speeds and wind durations. The north–northwest, north, north–northeast, and northeast wave directions impacting the breakwater from across the Mediterranean Sea were combined to estimate the probabilities of exceedance $p_>$ (number of storms per year). The results for the three wind speeds were ($H_s = 3.05\,\text{m}$, $p \geq 13.7$), ($H_s = 5.27\,\text{m}$, $p \geq 1.67$) and

($H_s = 7.01$, $p \geq 0.283$) respectively. Plotting these results and using linear extrapolation gave for the design wave heights, H_s,

(1) Wave runup and overtopping ($p_> = 1.0$, 1-year return interval): $H_s = 6.0\,\text{m}$, $T_s = 10\,\text{s}$
(2) Breakwater armor layer stability ($p_> = 0.01$, 100-year return interval): $H_s = 10.6\,\text{m}$

Tides in this region are a maximum of 0.5 m. Storm-surge calculations were not performed. The breakwater depth for this preliminary design was said to be 24–25 m. The soil was sandy overlying bedrock with little to no settling.

Costs were provided based on local experience for the Port of Arzew for quarry stone in four size categories and for concrete armor units (cube and tetrapod). Three seaside breakwater slopes were considered, $\cot\alpha = 1.5$, 2.0, and 2.5. The most economical design was for the slope $\cot\alpha = 1.5$, and used tetrapod as the armor layer since the available quarry stone was far too small for stability.

The Hudson formula (Equation (6.4)) was used to calculate the armor layer weight of tetrapod,

$$W = \frac{(\rho_s g) H_s^3}{K_D (S_r - 1)^3 \cot\alpha}, \tag{6.4}$$

taking, $H_s = 10.6\,\text{m}$ (34.8 ft), $\rho_s = 2{,}400\,\text{kg/m}^3$ (4.65 slugs/ft^3) for concrete, $S_r = 2.33$ for concrete in sea water, $\cot\alpha = 1.5$, steepest seaward slope recommended in SPM (1973), and $K_D = 8.3$, Hudson stability coefficient for tetrapod recommended in SPM (1973), to give the result $W = 97.5$ metric tons (97.5 long tons).

The latest Corps of Engineers, SPM (1973) was employed for the Hudson stability coefficient, $K_D = 8.3$, for non-breaking waves and the required tetrapod weight, W rounded to 100 metric tons as shown in Figure 6.22 (from Basco and Stadler, 1976).

Wave runup R_u was estimated using a plot of the surf-similarity parameter ξ (using the $p_> =$ 1-year wave values given above) versus the ratio, R_u/H for permeable slopes (no reference) to give $R_u = 7.2\,\text{m}$ (23.6 ft). Adding 0.5-m tide, the crest elevation was set at 7.7 m (25.3 ft) above the design water level, mean sea level (MSL).

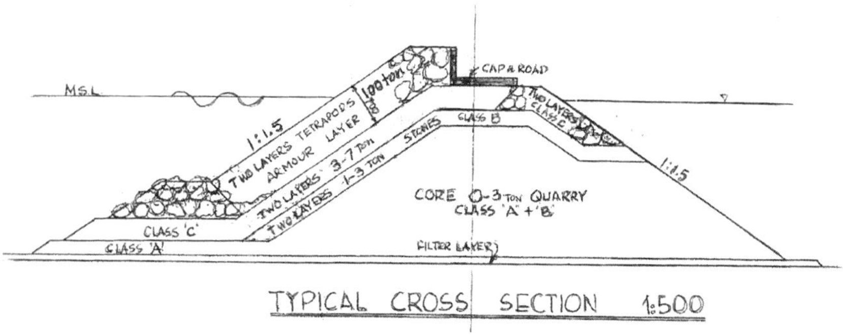

Figure 6.22. Cross-sectional design for non-wave overtopping and 100-year wave height conditions for outer breakwater, Port of Arzew, Algeria (from Basco and Stadler, 1976 for Delft, International Courses).

(c) Discussion

The preliminary design performed in 1976 for the Delft International Courses gave the required stable weight $W = 100$ metric tons for the tetrapod at the 100-year recurrence interval design level. This is about double the construction weight of 48 tons. This preliminary design slope for $\cot \alpha = 1.5$ is flatter than the construction slope, $\cot \alpha = 1.33$. Using a heavier tetrapod (on a flatter slope) would have prevented the damage that took place in December 1980 soon after construction. Information regarding how the outer breakwater for the Port of Arzew was repaired was not available for this book.

6.2.10.2 *Chesapeake Bay bridge tunnel — Storm of the century*

The Chesapeake Bay Watershed is the largest tidal estuary in the United States and third largest in the world. The Chesapeake Bay water body is about 326 km (200 mi) long from Havre de Grace, Maryland in the north down to Norfolk, Virginia at the southern end. The bay mouth is roughly 27 km (17 mi) wide between the City of Virginia Beach, Virginia and the Eastern Shore of Virginia where the bay entrance connects to the Atlantic Ocean. Vehicle ferry services which operated since the 1930s connected the southern side (South Hampton Roads) to the northern side on the Virginia Peninsula and further north to Delaware, Maryland, i.e. the Delmarva Peninsula. A fixed crossing study was authorized by the Virginia General Assembly and conducted in 1956. The conclusion was that a fixed crossing was feasible.

Figure 6.23. (Author's photo) The relative scale for about a 5-ton tetrapod installed on a nearshore breakwater off the Sea of Japan coast near Murakami.

The Chesapeake Bay Bridge Tunnel (CBBT) was designed by the Sverdrup & Parcel consulting engineering firm out of St. Louis, Missouri in the early 1960s. An information sign available on the CBBT website (http://www.cbbt.com) has been reproduced here as Figure 6.24 (north direction up). The CBBT was first opened on April 15, 1964 with two-way traffic along one fixed, causeway-bridge transport structure (southside structure). A second structure was built and opened on April 19, 1999 so that today each fixed causeway bridge structure carries two lanes of traffic in one direction. Commercial and navy ships have access to interior terminal and pier facilities by crossing two tunnels that are key elements in the fixed-crossing design. The Thimble Shoal Ship Channel crosses above the south side tunnel between Islands 1 and 2 and the Chesapeake Ship Channel crosses above

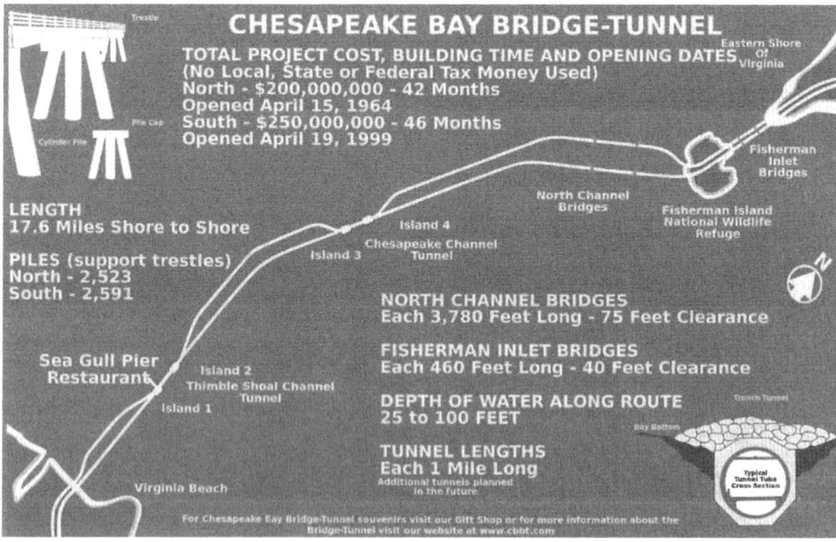

Figure 6.24. By Sarregouset — CBBT_Info_Sign.jpg. This vector image was created with Inkscape, CC BY-SA 3.0, http://www.cbbt.com/wp-content/uploads/2015/11/Project-Information-Board.jpg.

the north side tunnel between Islands 3 and 4. However, as shown in Figure 6.24, the tunnels continue to carry only two-way traffic and second, parallel tunnels have yet to be built. The CBBT was renamed the Lucius J. Kellam Jr. Bridge Tunnel in 1987 in honor of the first chairman of the CBBT Commission.

(a) Original island revetment design and damage

The two tunnels are bordered by four man made islands that were constructed to permit tunnel access and egress from the elevated causeways. Figure 6.25 is an aerial view of Island 1 looking Northeast at the time when only the southside causeway structure existed. The islands were built up from bottom sands, quarry rock materials, and large armor stones as a protective revetment as depicted in Figure 6.26.

The CBBT islands were constructed with 10-ton (US units) armor stones as a single layer revetment on a $\cot \alpha = 2$ slope. These revetments have been relatively stable and survived many tropical and extra-tropical storms since 1964. However, one location was damaged and in what now is being called as "The Storm of the Century".

Figure 6.25. Aerial view looking northeast at causeway and Island 1 and location for revetment damage (courtesy Chesapeake Bay Bridge Transportation Commission).

On October 31, 1991, the famous Halloween storm caused severe damage to the Island 1 revetment at the location shown in Figure 6.25. This storm was the subject of the best-selling novel "*The Perfect Storm*" (Junger, 1997) and a Hollywood movie *The Storm of the Century* (1999). Hydrographic surveys determined the extent of the damage and will be discussed in detail in Chapter 8. Original design details were not available from the consulting firm so that it is not known how the armor stone weight of 10 tons (US) was determined in the early 1960s. Therefore, an indirect method was employed by working backward and taking the stone weight, $W_{50} = 10\,\text{tons} = 20{,}000$ lbs US (89,000N) as given and determining the stable, design wave height for a single layer of stone on an impermeable revetment slope with non-overtopping wave conditions. The equivalent cube dimension, $D_{n50} = 4.94\,\text{ft}$ (1.51 m) for the 10-ton stone weight. This permitted a study of the evolution in armor stability design since the 1960s including such factors as alteration

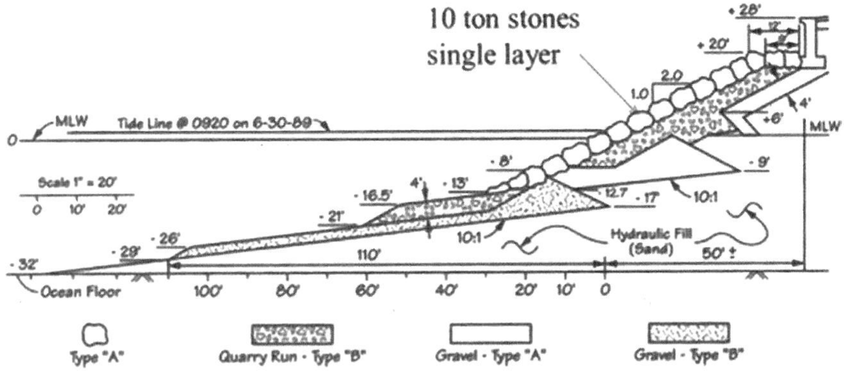

Figure 6.26. Typical cross-section of island materials and single layer of 10-ton (US units) stones as revetment armor (from the CBBT Commission, circa 1995).

of coefficients, effect of wave period, and incorporation of partial safety factors for design. Complete details are found in the CEM (2006, Part VI, Example Problem, VI-7-9). A summary of these results is presented in the following.

Wave conditions employed were those measured for the October 31, 1991 storm event at the US Army Corps of Engineers, Field Research Facility (FRF) located about 105 km (65 mi) south in 8 m (26.2 ft) water depth and at the Virginia Beach wave gauge (VA001) also located in 8 m water depth. These measured results were for peak wave conditions. These are extremely rare, very long period swell waves for the Atlantic Ocean.

(1) At the Corps' FRF, $H_s = 4.6\,\text{m}$ (15.1 ft), $T_p = 22\,\text{s}$
(2) At the VB gauge VA001, $H_s = 2.6\,\text{m}$ (8.53 ft), $T_p = 23\,\text{s}$

Other assumptions were as follows:

(3) Sea water, $\rho_w = 1{,}030\,\text{kg/m}^3$ ($64\,\text{lbs/ft}^3$)
(4) Rock, $\rho_s = 2{,}650\,\text{kg/m}^3$ ($5.14\,\text{slugs/ft}^3$)
(5) One layer ($n = 1$), rough angular stones, random placement
(6) No damage criteria

The results are best summarized in a Table 6.11. The column for the significant wave height displays the values of H_s that each formulation for armor layer stability would employ to calculate a stone weight, $W_{50} = 10$ tons

Table 6.11. Wave heights of various formulations to require stone armor weight of 10 tons.

Armor Layer Stability Formula	Hudson Coefficient, K_D	Wave Period T_p s	Breaking or Surging	Safety Factors	Stable Significant Wave Height, H_s	Remarks
Hudson (1974) SPM (1977)	Non-breaking 4.0					Non-breaking waves
Hudson (1974) SPM (1977)	Breaking 3.5				14.8 ft (4.5 m)	Breaking waves
Hudson (1974) SPM (1984)	Non-breaking $H_{0.1} = 1.27\,H_s$ 4.0					Non-breaking waves
Hudson (1974) SPM (1984)	Breaking $H_{0.1} = 1.27\,H_s$ 2.0				9.7 ft (2.32 m)	$H_{0.1} = 1.27\,H_s =$ 12.3 ft (3.75 m)
van der Meer (1988)		22.0	Surging	No	9.5 ft (2.90 m)	$P = 0.1$ $N_Z = 2360$ $S = 2$
van der Meer (1988)		22.0	Surging	Yes Partial safety factors	5–9 ft (1.5–2.7 m)	$P = 0.1$ $N_Z = 2360$ $S = 2$ $\sigma'_{FHs} = 0.2$

(US). For example, assuming the Hudson (1974) formula and K_D coefficient (3.5) for breaking waves as found in the earlier version of the *Shore Protection Manual* (SPM, 1977), a $H_s = 4.5\,\text{m}$ (14.8 ft) would produce the need for a rough, angular armor stone weighing 10 tons. The next edition of the *Shore Protection Manual* (SPM, 1984) was more conservative. The suggested design wave height was increased to the $H_{0.1} = 1.27H_s$ value, but to compensate the K_D coefficient was reduced to 2.0. This combination of changes resulted in a wave height 3.8 m (12.3 ft) needed to produce the armor stone weight of 10 tons. Using the Hudson (1974) formulation for non-breaking waves with larger values for the K_D coefficient would have lowered the design wave heights for a stone weight of 10 tons. The Hudson (1974) formulation is no longer recommended for calculation of stone weights for armor layers. The importance of the wave period is missing and especially in this example when very long period swell waves were observed during the October 31, 1991 storm event.

With the van der Meer (1988) formulation, there is no guessing whether (or not) the waves are breaking on the structure and the design wave height is fixed as the significant wave height, H_s. The notational permeability for a revetment is $P = 0.1$, the length of the storm gave the number of waves, $N_z = 2{,}360$, and the no-damage parameter was set as $S = 2$.

The critical surf-similarity parameter, $\xi_{mc} = 3.57$. For the measured very long peak wave period $T_p = 22\,\text{s}$ ($T_m = 18.3\,\text{s}$), even relatively low wave heights produce surging wave conditions ($\xi_m > \xi_{mc}$) for use in the van der Meer (1988) formulation. The resulting $H_s = 2.9\,\text{m}$ (9.5 ft) produced the need for a 10-ton armor stone. Lowering the peak wave period for this same design wave height ($H_s = 2.9\,\text{m}$) increased the required armor stone weight for surging wave conditions. However, for $T_p = 10\,\text{s}$ (and lower), plunging waves are present that results in lower stone weight.

Finally, Table 6.11 displays the range of H_s results when partial safety factors are applied in the design. Assuming relatively poor knowledge of wave conditions at the site ($\sigma'_{FHs} = 0.2$), the range for relatively low probability of failure ($p_f = 0.01$) to high probability of failure ($p_f = 0.4$) gives $H_s = 1.5$ to $2.7\,\text{m}$ (5–9 ft), respectively. In other words, to ensure a low probability of failure, on using 10-ton armor stone on a $\cot\alpha = 2$ slope with long period waves, the wave heights should only be in the 1.5 m (5 ft) range.

The wave heights and wave periods experienced in the October 31, 1991 "Halloween" storm were clearly strong enough to damage the rock revetment at Island 1 at the CBBT.

Figure 6.27. Island 1 on Thimble Shoal (South) Tunnel and seaward side abutment where damage occurred on rounded corner (Map data: Google, Maxar Technologies).

(b) Repair of the damaged revetment on the CBBT island

The CBBT island revetment was repaired in August 1994. The W_{50} for repair was 13.5 tons (US, 27,000 lbs) with allowable range of 12–15 tons. Figure 6.27 displays the damaged location on the seaward corner of Island 1 (from Google Earth, Lat. N 36° 57′ 46″, Lon. W 76° 06′ 50″). The increase in the size of the repair stones is clearly apparent in this picture. It is unknown how this size and stone weight were determined.

(c) Repair recommended using van der Meer (1988)

It is instructive to investigate what size and weight of stone for repair would be recommended using the van der Meer (1988) formulations. First, the design significant wave height H_s and peak wave period T_p must be selected. As indicated in Table 6.11, with no partial safety factors, $H_s = 2.90$ m (9.5 ft) for $T_p = 22$ s that produced the damaged revetment. The ratio of wave heights with damage ($S > 2$) to the design wave heights with no damage ($S = 2$), i.e. $H_s(S > 2)/H_s(S = 2)$ and this ratio equals 1.35 at the damage level of $S = 10$ and when about 10% of the armor layer is damaged. This level of damage is considered as failure of the armor layer because the under-layers are exposed to wave attack. This level of damage was estimated from the damage profile surveys taken by the engineering staff, CBBT District from the October 31, 1991 Halloween storm. Chapter 8 presents all the details, and this is discussed fully in CEM (2006, Example Problem, VI-7-10). Therefore, for design use $H_s = 1.35\ (2.90\,\text{m}) \approx 4.0\,\text{m}$ rounded (13.0 ft) and variable T_p. This is about the 50-year recurrence interval level for wave heights.

With no partial safety factors applied and for a storm duration $N_z = 2360$, Table 6.12 presents the influence of wave period starting with peak $T_p = 22\,\text{s}$. The recommended repair weight is roughly twice the $W_{50} = 13.5$ tons (US) that was actually constructed. The recommended repair weight increases slightly as the wave period decreases to more common values in the 14–16 s range. At about 12 s, wave conditions change to plunging-type waves and the required stable weight decreased significantly. As expected, increasing the number of waves in the storm (duration) also results in heavier and larger stone sizes for repair at the no-damage ($S = 2$) level.

Finally, consider if an "intermediate" level of damage in the $S = 4$–6 range (say $S = 5$) is permitted as found by van der Meer (1988) and presented in Table 6.2 for two-layer armor of rocks on a $\cot\alpha = 2$ slope. Table 6.13 presents the armor stone weight for three wave periods and two storm durations when $S = 5$ in the van der Meer (1988) formulations.

Permitting some level of damage significantly reduces the armor stone weight required for repair. This means that the current repair weight of 13.5 tons will probably again suffer future damage. See Chapter 8 for a full discussion on the use of damage curves to estimate maintenance (repair)

Table 6.12. Influence of wave period for repair when $H_s = 4.0\,\text{m}$ (13.0 ft).

Peak Wave Period, T_p s	Armor Layer Weight, tons US	Armor Layer Size, D_{n50} m (ft)	Plunging or Surging	Storm Duration, hours
22	27.2	2.1 (6.9)	Surging	12.0
20	27.9	2.12 (6.96)	Surging	10.4
18	28.9	2.15 (7.04)	Surging	9.8
16	29.0	2.17 (7.12)	Surging	8.7
14	31.1	2.20 (7.22)	Surging	7.6
12	25.8	2.07 (6.78)	Plunging	6.6
10	19.5	1.88 (6.18)	Plunging	5.4

Table 6.13. Influence of intermediate level of damage ($S = 5$) on the recommended repair weight.

Wave Period T_p s	Armor Stone Weight, tons (US) $N_Z = 2360$	Armor Stone Weight, tons (US) $N_Z = 3500$
22	15.7	17.6
18	16.7	18.8
14	18.0	20.2

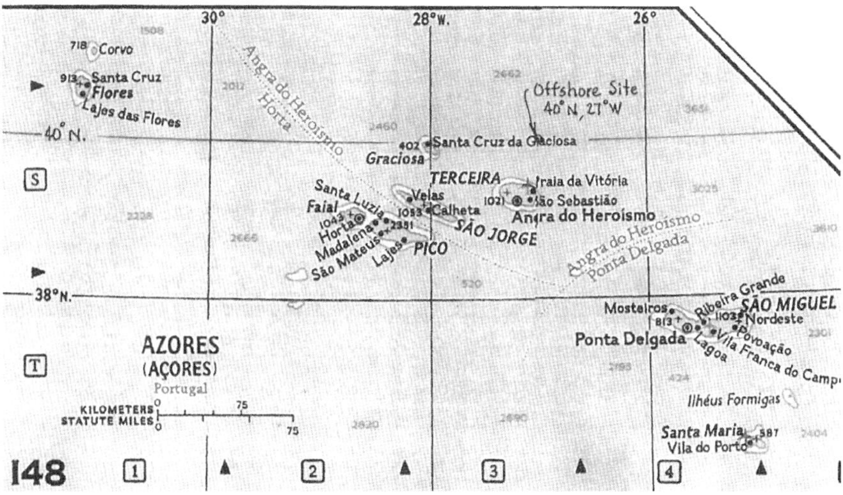

Figure 6.28. Azores (Portugal), Island of Terceria, Praia da Vitoria and Lajes Air Field Base (airport symbol on eastern side of island), Atlantic Ocean. Lisbon, Portugal is approximately 1,565 km (970 mi) due East at Lat. N38.7°.

costs for rubble-mound structures and the final "balanced design" for the lowest total construction plus maintenance cost.

An additional repair alternative would be the use of artificial, concrete armor units.

6.2.10.3 *Azores, Terceira island, Lajes field air base — Gas station in the Atlantic*

US Department of Defense (Air Force, Navy, Army) aircraft flying across the Atlantic Ocean to Europe stop to refuel at the Lajes Air Field Base on Terceira Island in the Azores (Portugal), see Figure 6.28. A 600-m (1970-ft) long harbor breakwater was constructed at Praia da Vitoria on the Eastern (lee) side of the island in 1962–1963 to protect the pier where ocean tankers unload jet fuel as shown in Figure 6.29. This military fueling facility has come to be called the "gas station in the Atlantic". See Google Earth (Lat. N38° 43′36″, Lon. W27° 03′ 03″) for the present view of the end of the breakwater.

(a) Original breakwater design and damages

Design of the breakwater took place in 1960 by TAMS Engineering Co. (NYC). The original breakwater was 585 m (1920 ft) long with crest elevation at 6.7 m (22 ft) above MSL. Design wave height was 5.2 m (17.0 ft) for 12-s period waves as hindcast by the SMB method recommended in the original Corp of Engineers, Technical Report No. 4, TR4 (1954) report. Using the Hudson formula with coefficient $K_D = 3$ for quarry stone on a $\cot\alpha = 2$ slope gave (13–15)-ton armor rock (two layers). The under-layer was 4–6 ton quarry stone over a core of graded material from the local quarry (Carvalho, 1964). The breakwater has a north–south orientation and is exposed to the Atlantic Ocean waves arriving from the east.

Damage occurred on December 25, 1962 before the construction was complete. Wave overtopping took place for over 30 hours by wave heights in the 8 m (26 ft) range with periods 13–14 s. Special 16-ton concrete blocks were added before construction was completed in 1963. Breakwater damage due to wave heights exceeding design levels for constructed stone weights was confirmed by model tests (Carvalho, 1965). Breakwater repair took place in 1963 when concrete units (Tetrapod) weighing 16 tons were used in the damaged sections.

Damage to the 16-ton tetrapod again took place in October–November 1966 and March 1970. Repair occurred in the fall of 1970 with 1,205, 16-ton tetrapods placed on the seaward face. A severe storm in December 1973 destroyed the navigation light on the head with repair taking place in 1973–1974 using about 1,600, 16-ton tetrapods. The storm of March 1984 exposed the stone under-layers at the outer end with major slumping of the seaward slope. A Corps of Engineers report (Davidson, 1984) presents a good summary of all the above-mentioned storm events. Repair took place in 1984 by again using the 16-ton (number unknown) tetrapod.

The Lajes Air Base wind speeds recorded in the tower and the recorded wave height estimates (Burns, 2001) provided valuable information for the major storm of December 21–24, 2001. At the peak on Sunday, December 23, the wave heights were estimated at over 9 m (30 ft) with long

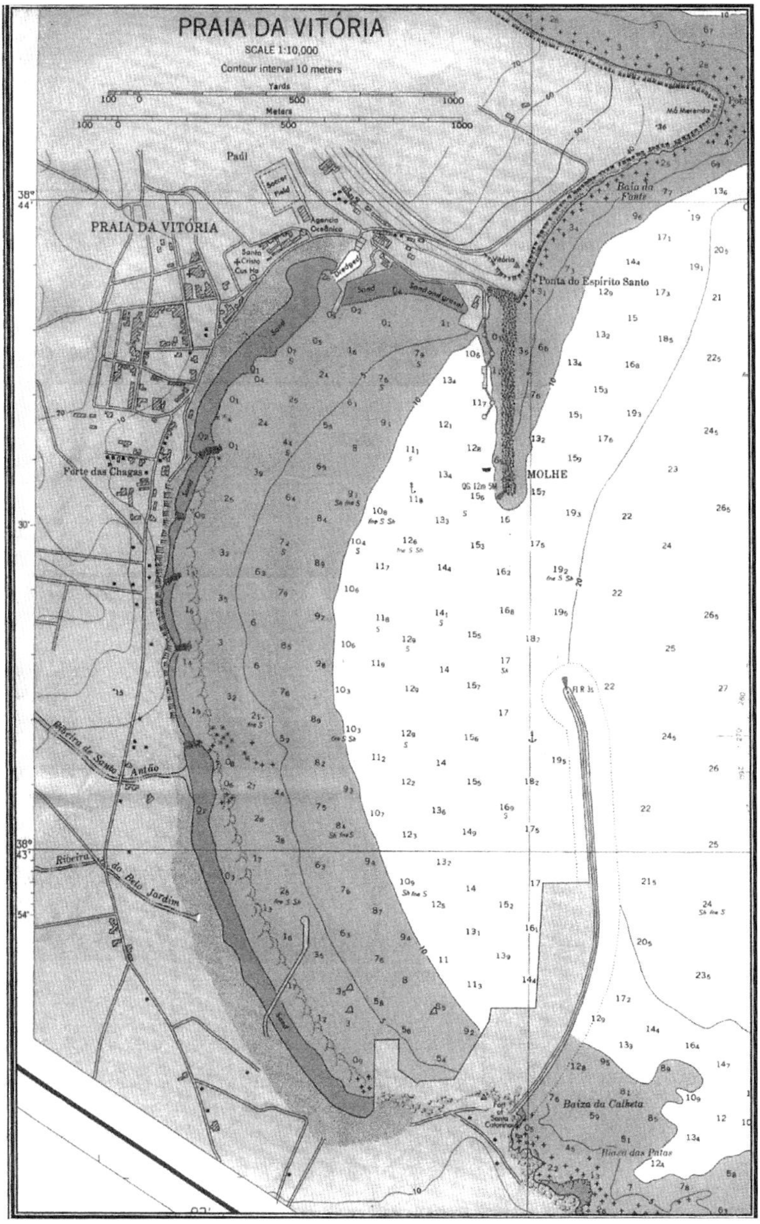

Figure 6.29. Praia Harbor Chart, showing Lajes Air Field breakwater (top) and jet fuel pier on landward side for unloading tankers. Harbor breakwater (bottom) for City Marina with harbor light at end (from US Defense Mapping Agency, Chart No. 51064, April 13, 1996). Soundings (m) 1 m below MSL.

Figure 6.30. Photos taken during (upper-left) and after the storm on December 21–24, 2001 (courtesy TranSystems Inc., Norfolk).

periods and the breakwater collapsed at the end destroying the outer 100 m (330 ft) as shown in Figure 6.30. The upper-left picture taken during the storm shows the tremendous wave overtopping that took place. The right-side photo shows the outer-end damage that moved tetrapod and under-layer stones on both the seaward side below the water level and curving landward at the end. The bottom-left photo displays how the breakwater crest was lowered down to the under-layer material and the core. The latter two photos were taken from a helicopter the next day after the storm.

Since construction in 1962, six storms were strong enough to cause damage to the breakwater (every 6–7 years) with the 16-ton tetrapod as armor layer. This last storm in December 2001 completely destroyed the outer end and warranted a new analysis and design.

(b) Major repair and redesign

Due to annual budgetary constraints, the repair effort was spread out in three phases: (1) emergency repair, (2) intermediate repair, and (3) final design and repair that took place from 2002 to 2006.

(1) Emergency repair (20-ton Antifer blocks)

For the next winter season (2002), a single layer of 20-ton Antifer cubes were placed over the new underlayer material and new core at the breached location. It was recognized that a more substantial "interim" repair would be needed. Figure 6.5 shows a picture of a cube with groves (Antifer block). The commercial harbor breakwater for the City of Praia da Vitoria was rebuilt using 25–40-ton Antifer blocks.

(2) Interim repair (31.6 ton, CORE-LOC® units)

Wave conditions in deep water (Figure 6.28, Lat. N40° W27°) were obtained from satellite altimeter data (Topex/Poseidon and Geosat Mission) by OCEANOR, Inc., Trondheim, Norway (Barstow, 2002). Wave-transformation analysis to the site near the end of the breakwater in 20 m (66 ft) water depths was performed using the EuroWaves software package (Cavaleri and Malanotte-Rizzoli, 1981) for each 6-hour directional wave spectra reading of H_{mo}, T_m, T_p, and the mean wave direction. Then, wave statistical analysis was performed and produced the following results for the annual, exceedance probability levels of wave height as summarized in Table 6.14 (Cardone, 2002). These results compared favorably with wave hind casts by the European Centre for Medium Range Weather Forecasts. Complete details are found in Basco (2002a, 2002b). Additional analysis (Baird, 2003) gave similar but slightly lower wave height results.

The concrete armor units called CORE-LOC™ were first introduced by Melby and Turk (1993) and via the Corps of Engineers' newsletter "The CERCular" in March, 1995 (Vol. CERC-95-1). For design, the K_D values in the Hudson stability formulation were set by Turk and Melby (1997). Wave runup and overtopping studies are summarized in Melito and Melby (2002). Lajes Air Base and tanker unloading dock is a US Department

Table 6.14. Extreme probabilities for transformed, satellite, wave-height data at breakwater (from Barstow, 2002; Basco, 2002a, 2002b).

Annual Exceedance Probability, $P_>$	Recurrence Interval, T_R years	Significant Wave Height, H_{mo} m	Std Dev, σ m	Two Std Dev, 2σ m	Design Wave Height 95% Confidence Interval (C.I.), m (ft)
0.10	10	6.40	0.45	0.90	7.30 (24.0)
0.04	25	7.10	0.55	1.10	8.20 (26.9)
0.02	50	7.65	0.65	1.30	8.95 (29.4)
0.01	100	8.20	0.80		9.00 (29.5) *$H_{mo} + 1\,\sigma$

of Defense facility; therefore, it was logical to employ the CORE-LOC™ units in the design for interim repair since the US Army Corps of Engineers, Research and Development Program at WES in Vicksburg, MS developed the design under basic research. The (™) was the trademark symbol when first developed and has been replaced by the ® symbol for the patent issued on January 28, 2008.

The Hudson formulation was employed in the CORE-LOC®, armor unit design using $K_D = 16$ as found in Table 6.6 for the trunk section and the design wave heights listed in Table 6.14. A slope $\cot\alpha = 1.5$ produced the results summarized in Table 6.15. The recommended recurrence interval for construction was $T_r = 50$ years, since the wave height at the 95% C.I. was about the same as for the 100-year level but only plus one standard deviation, σ. This recommendation was after a full analysis of the total, life-cycle costs (initial construction plus maintenance) as discussed in Chapter 8. Full design details are presented in Basco (2002b) including crest elevation, under-layer, toe weights and depth, and the end (head) region. Full details are also presented in Eversen and Basco (2003).

Figure 6.31 presents a cross-sectional drawing of the recommended, interim repair design and Figure 6.32 displays a photo during placement of one 31.6-ton CORE-LOC® unit during construction. Note the primary size dimension, C is 3.91 m (12.8 ft).

Melby and Hughes (2003) reviewed this breakwater design and concluded that it followed a "... *reasonable standard of care*", but concluded that excavation at the toe may be impractical and that physical model studies need to be conducted.

The interim repair was carried out in the summer of 2003 using the 31.6-ton CORE-LOC® units except that toe area excavation was not carried out. Instead the 20–ton Antifer blocks were moved to the toe region and placed on top of the debris field and used to protect the toe of the CORE-LOC®s.

Table 6.15. Summary of CORE-LOC® weights and sizes for three recurrence intervals at trunk section.

Annual Exceedance Probability, $P_>$	Recurrence Interval, T_R years	Design Wave Height, H_{mo}	Crest Elevation, m	CORE-LOC, Weight, W tons	CORE-LOC, Volume, V m^3	CORE-LOC, Size C m
0.04	25	8.2	10.6	24.3	10.2	3.574
0.02	50	8.95	11.2	31.6	13.25	3.901
0.01	100	9.0	11.75	31.6	13.25	3.901

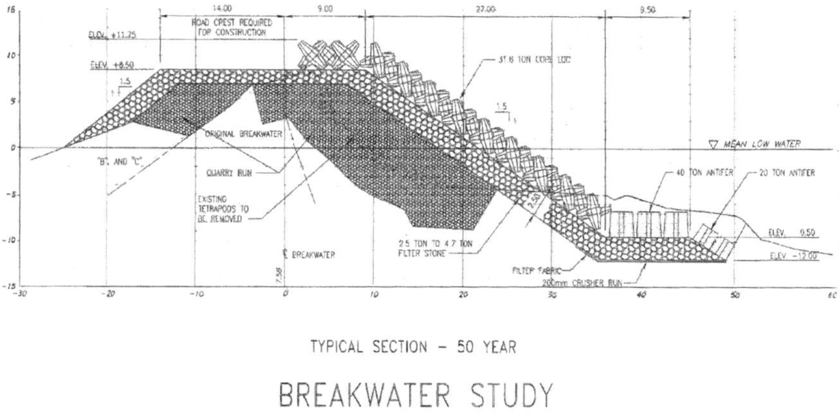

Figure 6.31. Trunk cross-section for recommended interim repair of Lajes Field breakwater (from TranSystems, 2003).

Figure 6.32. Photo of one CORE-LOC® unit for placement during interim repair of the breakwater during the summer of 2003 (courtesy TranSystems Inc., Norfolk).

This interim repair using the CORE-LOC® was for the outer section of breakwater shown in Figure 6.30.

Storms occurred in October 2003 and March 2004 that continued to cause damage to the interior reaches still protected by tetrapod and the CORE-LOC® region because the toe cubes moved and some downslope movement of the CORE-LOC® occurred (Scott *et al.*, 2006).

(3) Final design and repair (hybrid CORE-LOC and berm breakwater)

The final design solution for repair was undertaken by W. F. Baird & Assoc. Ltd., Madison, WI (Werren, 2004) with physical model testing performed by the US Army Corps of Engineers, Engineering and Research and Development Center, Vicksburg, MS (Melby and Myrick, 2004). The major challenge for the final repair/design was the stabilization of the toe region. This was because a very large debris field (stones, broken tetrapod, broken concrete blocks, etc.) was present from over 40 years of damage. The debris field is mobile during storms and expensive to remove.

Both 2D and 3D model studies were conducted to study various design solutions. The length scale ratio L_R for the 2D model was 55.76. This translates to a primary size dimension for the model CORE-LOC unit, $C_m = 7\,\text{cm}$ (0.23 ft, 2.75 inches). Model specific gravity was 2.294 and less than the prototype unit (2.38) so the mass of the model CORE-LOC unit was 187 g which is 0.41 pounds in weight units for a prototype CORE-LOC® unit weighing 31.6 metric tons (31.6 US long tons). The prototype water depth was 11.5 m (0.2 m, 0.67 ft in model). For a range of significant wave heights $H_{mo} = 4.6$–$7.55\,\text{m}$ (0.08–0.135 m, 0.26–0.44 ft in model) and peak wave periods $T_p = 13$–$16\,\text{s}$ (1.74–2.14 s in model), it was found that two basic alternatives emerged for consideration in the final design.

One was the interim design with an armor layer of 31.6-ton CORE-LOC units but with an excavated toe region similar to Figure 6.30. The second was a "hybrid" solution with CORE-LOC units extending down to MSL and then an extensive berm width of armor stone that would reshape during storm events as depicted in Figure 6.33 (from Werren, 2004). The rock berm could be constructed over the existing material in the debris field which would save a large volume of rock to build the berm.

An undistorted scale, 3D model was also employed using the same $L_R = 55.76$. Figure 6.34 is an end view photo of the 3D model in the Coastal Hydraulic Laboratory (CHL) in Vicksburg, MS (from Melby and Myrick, 2004). The 600 m (1970 ft) long breakwater is 10.75 m (35 ft) in the model.

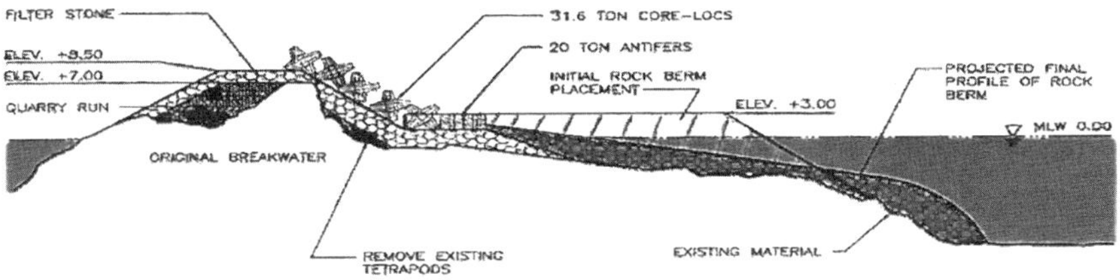

Figure 6.33. Schematic diagram of "hybrid" solution combing CORE-LOC armor units and rock berm (from Werren, 2004).

Figure 6.34. Photograph of 3D model of breakwater in wave basin (from Melby and Myrick, 2004).

For the final repair, the final design selected is shown in Figure 6.35. It consists of a 20–30 m wide berm (65–100 ft) at elevation +3.0 m (10 ft) above MSL and extending out from the toe of the CORE-LOC armor layer. Along the trunk, rock weight is 5–18 ton in the berm. The head end is more vulnerable and 10–18 ton rock is required for the berm. Unfortunately, berm recession, Rec (Equation (6.36)) was not measured and recorded in both the 2D and 3D model tests.

Construction began in 2004 with berm size stones imported from Norway because of the size limitations of the local quarry. A fast-moving storm occurred in February, 2005 with wave heights measured in front of

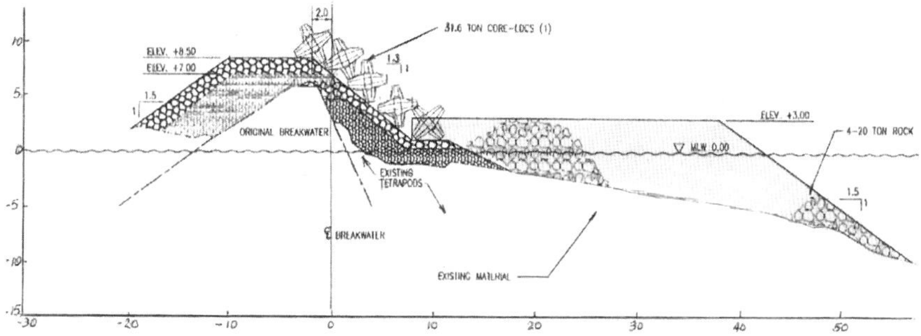

Figure 6.35. Final cross-section of the "hybrid" design selected for final repair of the Lajes breakwater (from Werren *et al.*, 2006).

the breakwater at 7.5 m (24.5 ft) with 14 s peak period but only impacting the site for a few hours. The "hybrid" sections of the repaired breakwater in place (Phase I, 150 m of central potion) experienced no loss of material or fracture of the CORE-LOC armor units. However, heavy damage occurred to those sections not protected by the berm and especially at the head end of the breakwater. See also Werren *et al.* (2002, 2006).

Construction was finally completed for the entire length in 2006. See Google Earth (Lat. N38° 43′36″, Lon. W27° 03′ 03″) for the present view of the Lajes Air Field breakwater.

In summary of all three examples, the age-old adage applies, "*we only learn something when something goes wrong*". Inadequate knowledge of the wave climate (wave height, period, direction) and its application to existing, inadequate, armor layer stability formulas at the time of design and construction in the 1960s and 1970s are the first things to consider. But these scientific and engineering constraints are not the only ones involved. We need to also consider the economic (budget) constraints of the funding organizations as discussed in detail in Chapter 8.

6.3 Vertical-Walled Seawalls, Caissons, Bulkheads, and Flood Walls

6.3.1 *Introduction*

Vertical-walled coastal structures are monolithic in cross-section consisting of one solid surface separating the water side from the interior. The water side is subject to hydrostatic water pressures plus dynamic wave pressures which integrated together over the water depth produce a horizontal force on the wall. Design of the structural integrity and stability of this vertical wall requires the ability to calculate this horizontal force on the wall. It is only one of other forces at work on the four main types of vertical-walled structures. The four types are mainly distinguished by their purpose (function) in coastal engineering design.

Seawalls mitigate the water level and wave damage to what is on the interior side. Caissons are breakwaters that reduce wave heights on the leeward side for safe navigation and shoreline erosion mitigation. Bulkheads prevent loose soil materials in the interior side from eroding into the water on the seaside. Flood walls mitigate water-level damage to the interior side (like seawalls) but may have little or no wave forces to consider in their design. The design of all four types will be reviewed in this section, but first we review the basic equations to calculate the wave forces on vertical walls.

6.3.2 *Oscillating wave forces on vertical walls*

6.3.2.1 *Linear wave theory — Progressive, non-breaking waves*

As introduced in Chapter 2, the wave pressure distribution, $p(z,t)$ beneath a progressive, regular wave is

$$\frac{p}{\rho g} = -z + K_p \eta, \tag{2.10}$$

where K_p is the pressure response factor to account for the wave orbital motion and is given by

$$K_p = \frac{\cosh\, \kappa\,(z+d)}{\cosh\, \kappa d}. \tag{2.11}$$

Because of streamline curvature in the orbital motion of the water particles beneath the surface, the pressure distribution in the vertical, $p(z)$ is not hydrostatic and $K_p\eta$ accounts for the waves. We have formulated the pressure head relationship $p/(\rho g)$, since civil (coastal) engineers trained in hydraulics are most familiar with columns of liquid "head" to express the vertical pressure distribution in statics and open channel flows with no streamline curvature. The z term is the hydrostatic pressure head distribution beneath the $z = 0$ water surface and the negative sign is needed since z is negative beneath the water surface. To facilitate the computation, the Water Waves folder in the online supplementary materials contains the spreadsheet "Pressure Calculation Generic" with example in Figure 2.7 for the pressure head and pressure distribution beneath the wave crest for a progressive wave in linear wave theory.

Recall that the $\kappa = \frac{2\pi}{L}$, the wave number so that the expression for K_p can be written as

$$K_p = \frac{\cosh\left(2\frac{\pi(Z+d)}{L}\right)}{\cosh\left(\frac{2\pi d}{L}\right)},$$

so that the wave pressure response factor K_p depends upon both the water depth d and the wave length L, and hence clearly also on the wave period, T.

For design, we are primarily interested in the maximum horizontal force, $f_{\max}$ when the wave crest is at the wall, hence, for $\eta_{\max}$.

Integrating Equation (2.10) from $z = 0$ to $z = -d$ at the bottom gives for the combined wave and hydrostatic force, f,

$$f = \frac{\rho g \eta_{\max}}{2\pi/L}\tanh(2\pi d/L) + \frac{1}{2}\rho g d^2,$$

where the first term is the wave force component and the second is the hydrostatic force over water depth, d. Linear wave theory only applies below the free surface, $z = 0$. The standard assumption is that above the $z = 0$ surface, the pressure distribution is hydrostatic so that the maximum horizontal force, $f_{\max}$ beneath the wave crest becomes

$$\begin{aligned} f_{\max} &= \frac{\rho g \eta_{\max}}{2\pi/L}\tanh(2\pi d/L) + \frac{1}{2}\rho g d^2 + \frac{1}{2}\rho g \eta_{\max}^2 \\ &= \text{term}(1) + \text{term}(2) + \text{term}(3). \end{aligned} \tag{6.44}$$

The Water Waves/Wave Dynamics folder in the online supplementary materials also includes a “Pressure and Force Calculation” spreadsheet to facilitate calculation of Equation (6.44).

For example, consider a wave height $H = 1.52\,\text{m}$ (5 ft) in $d = 6.1\,\text{m}$ (20 ft) water depth with wave periods, $T = 15\,\text{s}$, $10\,\text{s}$, and a deep-water wave period. The water is sea water, $\rho = 1035\,\text{kg/m\^{}3}$ (64 lbs/ft^3). Find the pressure distribution $p(z)$ beneath the wave crest, the horizontal force, $f_{\max}$, the relative contribution of the three force components in Equation (6.44), and the location of the centroid of the horizontal force. In addition, calculate the maximum hydrostatic force $f^*_{\max}$ assuming shallow water, very long period wave theory where the wave streamlines are almost horizontal so

Table 6.16. Wave forces on a vertical plane for linear wave theory and progressive waves.

Wave Period T, s	Total Force f_{max}, N/m	% Wave Force Term (1)	% Water Depth Force Term (2)	% Surface Force Term (3)	Hydrostatic Crest Force, f^*_{max}
15	237, 065	19.2%	79.6%	1.2%	238, 767
10	234, 979	18.5%	80.3%	1.3%	238, 767
2.8	206, 554	7.2%	91.3%	1.4%	238, 767

that the total water depth at the crest $(d + \eta_{\text{max}})$ is employed

$$f^*_{\text{max}} = \frac{1}{2}\rho g \left(d + \eta_{\text{max}}\right)^2. \tag{6.45}$$

In "deep water", the relative water depth, $d/L_o \geq 0.5$ so that the wave length, $L_o = 12.2\,\text{m}$ (40 ft) giving, $T = 2.8\,\text{s}$ for a deep-water wave period. The spreadsheet results are summarized in Table 6.16.

It is seen that the higher the wave period, the larger the horizontal force and that the hydrostatic force from the crest using Equation (6.45) produces a similar yet conservative result. Note also that the water wave term (1) is roughly 20% of the total and not to be neglected for design (1 lb/ft $=$ 14.59 N/m force).

To determine the location of this total force, the centroid of the pressure prism can be found by taking moments about the crest surface and dividing by the total force. The resulting location is essentially two-thirds the distance below the crest and for this example $= 0.667\ (d + \eta_{\text{max}}) = 0.667(6.1 + 0.76) = 4.57\,\text{m}$ (15 ft).

6.3.2.2 *Linear wave theory — Wave reflection and standing waves*

Water waves are reflected from vertical walls and reflection increases the water surface elevation at the wall above that for the incident wave. When the progressive (traveling) incident wave reflects from the vertical wall, the water surface variation at the wall η_w is the average of the sum of the incident H_i and reflected wave height H_r. Defining a wave-reflection coefficient $\chi = H_r/H_i$ gives a simple expression for the water surface variation at the wall

$$\eta_w = (1 + \chi)\frac{H_i}{2}. \tag{6.46}$$

Perfect reflection means $\chi = 1.0$, i.e. $\eta_w = H_i$. Then Equation (6.44) becomes

$$f_{\max} = \frac{\rho g H_i}{2\pi/L} \tanh(2\pi d/L) + \frac{1}{2}\rho g d^2 + \frac{1}{2}\rho g H_i^2. \tag{6.47}$$

The Water Waves/Wave Dynamics folder in the online supplementary materials with the "Pressure and Force Calculation" spreadsheet can again be applied using 2 H in the Wave Height, H box (J4) to calculate $f_{\max}$ in Equation (6.47). For example, for the 10 s wave period in the above given example, $f_{\max} = 287,197\,\text{N/m}$ (19,491 lbs/ft). Note that this is a 22.2% increase and not double the horizontal force because the pressure distribution on the wall is a triangular prism. For practical purposes, some wave energy is absorbed or transmitted past the wall, so use $\chi \leq 0.9$ for design.

A standing wave (*clapotis*, French) pattern is when the water surface does not travel (progress) horizontally, but moves in a fixed vertical pattern at the antinodes and horizontal pattern at the nodes where the water surface elevation remains constant at $z = 0$. Standing wave patterns can readily be found in laboratory wave tanks with horizontal bottoms and a fixed wave paddle at one end but are seldom found in nature with sloping beaches that permit reflected waves from vertical walls to radiate back out to sea.

6.3.2.3 *Finite-amplitude wave theory*

As schematically illustrated in Figure 2.8, finite-amplitude (nonlinear) waves have asymmetrical shapes in space with higher peaks and broad, flatter troughs. As a result, when the higher crest elevation is at the vertical wall, the horizontal force against the vertical plane is larger. In addition, the mean water level (MWL) is above the SWL by some distance, so that the water depth term (2) is also greater. A complete review of finite-amplitude wave theories is beyond the scope and needs of this book.

For nonlinear, shallow-water waves in "coastal" waters, the wave profile, $\eta(x,t)$ given by the Jacobian elliptical cosine function, usually designated "cn" for the cnoidal functions is often applied in coastal engineering. See Abramowitz and Stegun (1972) for the mathematical handbook of cnoidal functions. For the 10-s wave period and the example given above ($H = 1.52\,\text{m}$, $d = 6\,\text{m}$), cnoidal wave theory gives $\eta_{\max} = 1.00\,\text{m}$ (3.29 ft), and $f_{\max} = 243,982\,\text{N/m}$ (16,558 lbs/ft) or about a 4% increase. See Wiegel (1964), SPM (1984), or ISVA (1974) from the Technical University of Denmark for cnoidal wave formulas and tables. Clearly, selection of an appropriate wave reflection coefficient χ outweighs the importance of these nonlinear wave theory results.

A long history of research efforts throughout the world has produced many formulas over the years for calculation of wave pressures and forces on vertical walls. Pioneering efforts of Japanese coastal engineers have led the way to develop practical formulas for the wave pressures and forces on caisson-type structures for irregular waves under both non-breaking and breaking wave conditions. We focus on these herein.

6.3.2.4 *Goda formula*

Figure 6.36 presents the pressure distribution and key variables for irregular waves as formulated by Goda (1974) and Tanimoto *et al.* (1976) and herein called the "Goda" formula for the wave force F_H on a vertical wall. The static water-depth pressure distribution is also shown to give the static force F_s so that the total force on the wall is $F_T = F_H + F_s$. To begin, we assume the wave direction is at right angles to the wall, the bottom is horizontal, the wall extends vertically to the bottom and the water depth, d is for the design, SWL. The wall crest height is h_w and the crest is the

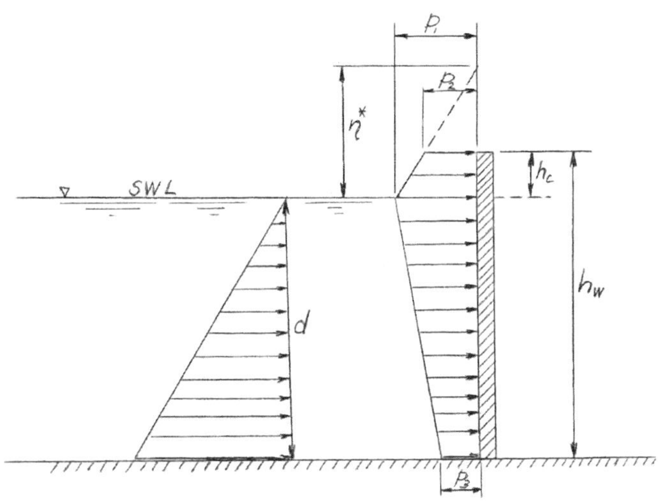

Figure 6.36. Wave pressure distribution and key variables in Goda formula for wave force F_H on a vertical wall. The static water depth pressure distribution is also shown to give the static force F_s so that the total force on the wall, $F_T = F_H + F_s$.

vertical distance h_c above the SWL. The Goda formula for the horizontal wave force F_H on the vertical wall is then simply the sum of two trapezoidal, pressure prisms,

$$F_H = U_{FH}\left[\frac{1}{2}(p_1 + p_2)h_c + \frac{1}{2}(p_1 + p_3)(h_w - h_c)\right], \tag{6.48}$$

with U_{FH} as a statistical variable to account for the uncertainty and bias in the horizontal wave force. Goda's formulas to calculate the four key variables (η^*, p_1, p_2, p_3) are as follows:

$$\eta^* = 1.5H_{\text{design}}, \tag{6.49}$$

$$p_1 = (\alpha_1)\rho g\, H_{\text{design}}, \tag{6.50}$$

$$p_2 = \left(1 - \frac{h_c}{\eta^*}\right) p_1, \tag{6.51}$$

$$p_3 = \alpha_3 p_1, \tag{6.52}$$

$$\alpha_1 = 0.6 + 0.5\left[\frac{4\pi d/L}{\sinh(4\pi d/L)}\right]^2, \tag{6.53}$$

$$\alpha_3 = \frac{1}{\cosh\,(2\pi d/L)}, \tag{6.54}$$

where ρ is the mass density of the water.

The wave conditions needed are the design wave height H_{design} and a representative wave period to calculate the wave length L in water depth d. For wave period, the Goda formula employs a significant wave period T_s and suggests using $T_s \approx 1.1\,T_m$ (the mean wave period).

Recommendations for the design wave height H_{design} are more complicated. They depend on where the wall is located and include a factor of safety that is related to the coefficient of uncertainty, U_{FH}.

(1) Wall within surf zone for breaking waves

Estimate the highest of the irregular, breaking, significant wave heights, H_{sbmax} at a distance approximately $5H_s$ in front of the wall and use this as H_{design}. Recall from Chapter 2 for depth-limited, irregular wave breaking:

$$\left(\frac{H_s}{d}\right)_b = 0.56\exp(3.5\,m). \tag{2.25}$$

For a relatively flat bottom $(H_s)_b \approx 0.56d$ so that in many cases, the design SWL water depth determines the design wave height, H_{design} within the breaking wave surf zone.

(2) Wall outside (seaward) of the surf zone

Goda (1985) recommends using

$$H_{\text{design}} = 1.8H_s. \tag{6.55}$$

For an irregular wave spectrum that fits a Rayleigh Distribution, the significant wave height H_s is exceeded by 13.5% of the wave heights ($H_{0.135}$) and is the average of the highest 1/3 of the wave heights ($\bar{H}_{0.333}$). This $H_{\text{design}} = 1.8H_s$ wave height is the average of the highest 1/250 of the wave heights ($\bar{H}_{0.004}$) in one 20-min, irregular wave record (see Section 2.3.3). But Goda (2010) in his discussion never addresses the long-term, probability distribution $P(H_S)$ of wave heights for design (see Section 2.3.8). In Section 6.3.5, we define H_s^T as the significant wave height at recurrence interval T_r for caisson design and a complete discussion is found in Section 4.3.

(3) Uncertainty and bias in the formula for F_H

A reanalysis of the Goda formula has been carried out by Juhl and van der Meer (1992), Bruining (1994), and van der Meer *et al.* (1994) using model scale data carried out by Delft Hydraulics and the Danish Hydraulic Institute. A mean value was found for $U_{FH} = 0.90$ with the standard deviation from the mean, σ, equal to 0.25 for desktop calculations. If model tests are performed, use $\sigma = 0.05$. Therefore, this result modifies the safety factor of 1.8 as recommended by Goda (1985) for H_{design}.

(4) An example

As an example, again employ a significant wave height $H_s = 1.52\,\text{m}$ (5 ft) in $d = 6.1\,\text{m}$ (20 ft) water depth with significant wave period, $T_s = 10\,\text{s}$. The water is sea water, $\rho = 1{,}035\,\text{kg/m}^3$ ($64\,\text{lbs/ft}^3$). The wall height h_w will be calculated so that it extends a distance η^* above the SWL so that the pressure at the top of the wall is zero. Assume the wall is both seaward and within the surf zone. The results for total force are summarized in Table 6.17 and compared with both linear and cnoidal wave theories with full reflection at the wall.

For this simple example with $T = 10\,\text{s}$, the Goda formula using $H_{\text{design}} = H_s$ inside the surf zone gives very similar results to linear and cnoidal wave theory when full wave reflection is used at the wall. When seaward of the surf zone, using $H_{\text{design}} = 1.8\,H_s$ gives far larger results. Applying the uncertainty coefficient, $U_{FH} = 0.90$ reduces this magnitude.

Table 6.17. Comparison of linear wave theory, cnoidal wave theory (with full reflection at wall), and Goda theory both outside and inside the surf zone for $T = 10$-s wave period.

	H_{design} ($T = 10\,\text{s}$)	Goda, F_T	Linear Theory (full reflection)	Cnoidal Theory (full reflection)
Wall location	m (ft)	Newton/m (lbs/ft)	Newton/m (lbs/ft)	Newton/m (lbs/ft)
Inside surf zone	1.52 (5.0)	288,706 (19,693)	287,197 (19,491)	297,795 (20,210)
Outside surf zone	2.74 (9.0)	393,720 (26,855)	287,197 (19,491)	297,795 (20,210)

Increasing the wave period will increase the total force in the Goda formula, as expected. Further discussion of the Goda formula, $F_{h\,\text{Goda}}$ is given in Section 6.3.3.

Using the Goda formula, there is no need to apply a relatively complicated, nonlinear wave theory such as cnoidal theory and there is no need to estimate a reflection coefficient at the wall. We strongly recommend using the Goda formula to calculate the oscillating wave forces against vertical walls. Further discussion of the Goda formula, its modification (Takahashi *et al.*, 1994), and the value to use for H_{design} is found in Section 6.3.3, and in the section on Caisson design.

6.3.3 *Impact wave forces on vertical walls — PROVERBS*

The original Goda method is used normally for design of relatively long structures where the structure length in effect spreads the wave loads along the length of the structure. This horizontal spreading of the wave loads calculated per unit length of structure is an added factor of safety in applying the original Goda formulations for the pressure p_1 at the water surface as shown in Figure 6.36 for both non-breaking and breaking waves.

However, under some wave-breaking conditions, very high wave impact (aka, impulsive, slamming, hammering) loads of short duration occur that produce greater wave loads than estimated by the original Goda (1974) formula. Takahashi *et al.* (1994) have modified the pressure p_1 at the SWL to take this slamming force into account. CEM (2011) advises that these relatively small-scale model results do not correctly include air pockets trapped between the wall and the slamming wave.

The need for improved knowledge for the wave impact loads, geotechnical aspects, and structural analysis aspects of vertical breakwater design resulted in the *PRO*babilistic Design Tools for *VE*rtical *B*reakwater*S* (*PROVERBS*) project in the European Union (1994–1998). The results of the PROVERBS project are summarized by Oumeraci *et al.* (2001). Only the wave impacts (Hydraulic Aspects: Allsop and Kortenhaus, Chapter 2) are presented here. The interested reader should consult this reference for Geotechnical Aspects (de Groot, Chapter 3), Structural Aspects (Crouch, Chapter 4), and Probabilistic Design Tools (Vrijling and Voortman, Chapter 5) that are applicable for a vertical, monolithic structure design.

Three major advances have resulted from the PROVERBS project, namely,

(1) A parametric decision map has been developed to provide a calculation procedure to identify the breaker-type and time variation of the horizontal force as illustrated in Figure 6.37.
(2) The breaker-type depends on the structure as well as the wave parameters approaching the structure. The four structure types identified and studied in PROVERBS are shown in Figure 6.38.
(3) A 12-step design procedure has been developed that combines the four wave-type conditions and the four types of vertical walls in the structures to produce a parametric map as illustrated in Figure 6.39 (from Kortenhaus and Oumeraci, 1998).

The major advance is improvement in the wave impact load formulation. As schematically depicted in panel (c) in Figure 6.37, the wave impact force, $F^*_{h\,\max}$ can be 2.5 times (or many more) larger than the quasi-standing wave force, F^*_{hq}. Note the ratios of $F^*_{h\,\max}/F^*_{hq}$ for three wave-type conditions in Figure 6.37. Broken waves and quasi-standing waves have $F^*_{h\,\max}/F^*_{hq} \approx 1.0$.

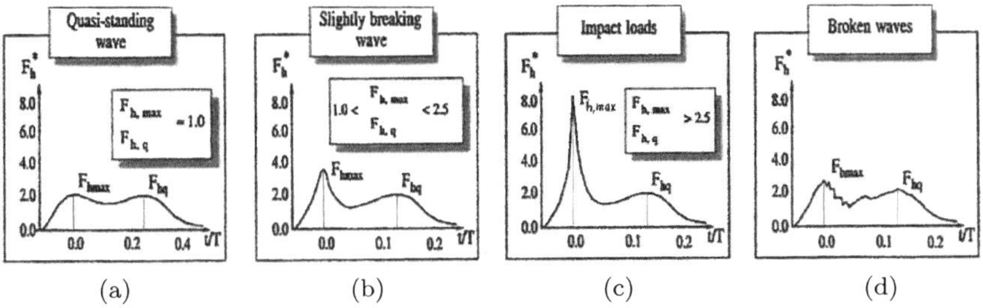

Figure 6.37. Time variation (t/T) of relative horizontal wave force F^*_h on vertical wall structures exposed to head-on, long-crested, irregular waves of (a) oscillating, quasi-standing wave, (b) slightly breaking wave, (c) directly impacting, breaking waves, and (d) broken waves (modified from Kortenhaus and Oumeraci, 1998).

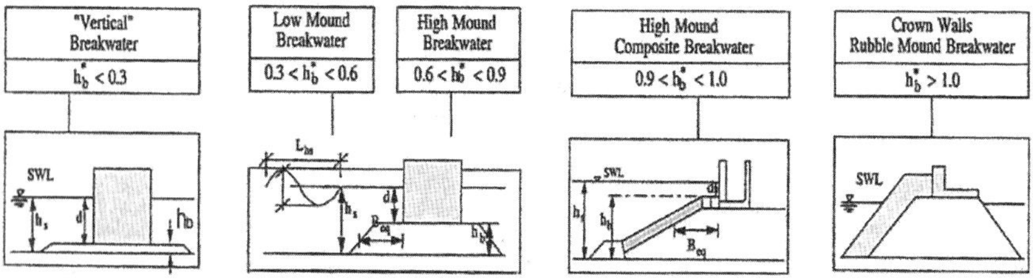

Figure 6.38. Four structure types as identified and studied in PROVERBS (modified from Kortenhaus and Oumeraci, 1998).

As seen in Figure 6.38, the four structure types are: a "vertical" breakwater, a low- or high-mound breakwater, a composite breakwater, and a rubble-mound breakwater with crown wall. They are identified by the ratio of the berm height h_b on which the vertical, monolithic structure wall sits and the water depth h_s in front of the structure. This is defined as $h_b^* = h_b/h_s$. Note the ranges for h_b^* for the four types in Figure 6.38. Note that the depth d is the water depth immediately in front of the vertical wall. A final key structure variable is the berm "equivalent" width, B_{eq}, defined as the berm width, $B_b + (mh_b/2)$ where m is the berm slope in front of the wall. When normalized by a wave length, L_{pi}, the value for the relative berm width, $B^* = B_{\text{eq}}/L_{pi}$ may trigger various kinds of wave breaking on the vertical wall. The peak wave period T_p in the incident wave spectrum is used in linear wave theory to calculate the peak period wave length L_{pi} in water depth, h_s. As shown in Figure 6.38, this wave length is also labeled L_{hs}.

The key wave parameter is $H_s^* = H_{si}/h_s$, where H_{si} is the incident, significant wave height in water depth, h_s in front of the structure. After wave breaking, the breaking wave height, H_b, is estimated using a modification of Miche's formula (Miche, 1944, see Equation (2.22)) for both irregular waves, wave reflection, and the relative berm width, B_{eq}/d. This wave breaking formula is

$$\frac{H_b}{L_{pi}} = (0.1025 + 0.0217C^*) \tanh(2\pi\ k_b h_s/L_{pi}), \tag{6.56}$$

where

$$C^* = \frac{(1-\chi)}{(1+\chi)}, \text{ a new coefficient related to the overall wave reflection coefficient, } \chi, \tag{6.57}$$

$$k_b = 0.0076(B_{eq}/d)^2 - 0.1402(B_{\rm eq}/d) + 1,$$

$$\text{an empirical factor,} \qquad (6.58)$$

with the following suggested values for χ:

$\chi = 1.0$ for highly reflective vertical walls, i.e. perfect reflection so $C^* = 0$,
$\chi = 0.95$ simple vertical walls and small mounds below high crest,
$\chi = 0.8 + 0.1\ R_c/H_{si}$, for low-crest walls ($0.5 < R_c/H_{si} < 1.0$), where R_c is crest freeboard above water level, and
$\chi = 0.5 - 0.7$ for composite wall, large mounds, and significant wave breaking.

From Equation (6.57), C^* varies almost linearly from 0 ($\chi = 1.0$) to 0.333 ($\chi = 0.5$) with very little effect on Equation (6.56). The empirical coefficient k_b varies nonlinearly from 1.00 where $B_{\rm eq} = 0.0$ (no berm) to about 0.4 when ($B_{\rm eq}/d$) equals 6–10. From Figure 6.39, only moderate berm widths ($0.12 < B^* < 0.4$) produce breaking wave impact loads. Small, large, and very large waves can produce all four types of wave-type conditions shown in Figure 6.37.

6.3.3.1 *Twelve steps in recommended procedure for design*

The 12 steps in the recommended design procedure are as follows.

Step 1. Identify the main geometric and wave parameters. All the geometric variables are introduced above except α, the bed slope in front of the entire structure. The angle of the structure to the direction of the design waves is β. Various possible water levels should be considered since they produce different values of the relative berm height h_b^* and relative wave height, H_s^*.

It is recommended to use Goda's design wave height, $H_{\rm design} = 1.8\,H_s = H_{\max b}$, the maximum breaking wave height at a distance, $5H_s$ seaward of the structure. Recall that this $H_{\rm design} = 1.8\,H_s$ wave height is the average of the highest 1/250 of the wave heights ($\bar{H}_{0.004}$) if fitting the Rayleigh distribution.

The authors of PROVERBS also present Shuto's method (Shuto, 1974) for calculating the maximum, wave breaking height $H_{\max b}$ using an irregular shoaling coefficient when the deep water, significant wave height is given.

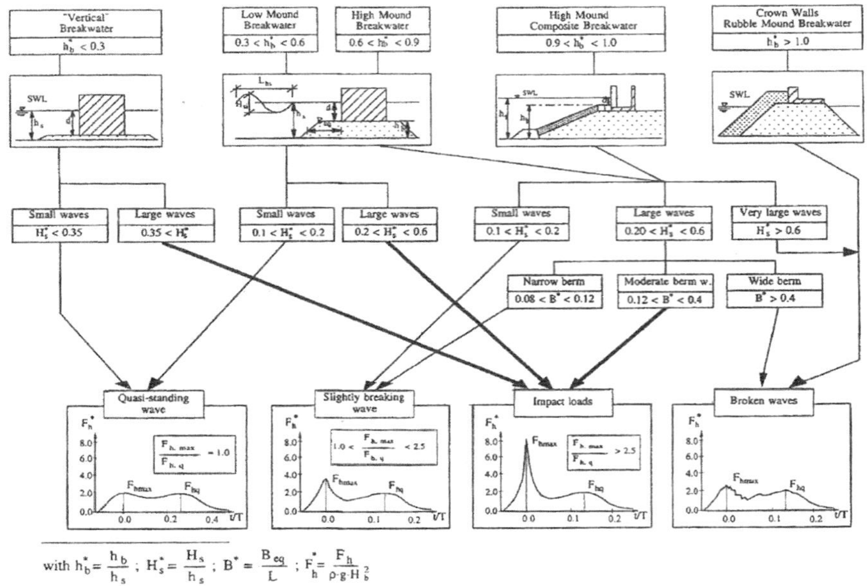

Figure 6.39. The parametric decision map for four structure types and four classes of irregular wave types that result in different wave loadings on the structures (from Kortenhaus and Oumeraci, 1998).

Step 2. Estimate the mean pressure and mean wave force over the wall height. The average wall pressure, $p_{\text{av}} = 1.5\rho g H_s$ is assumed uniform over a vertical wall height h_f extending up to $1.25H_s$ above the SWL ($h_f = d + 1.25H_s$), so that the mean wave force (Hiroi, 1919), $F_{\text{Hiori}} = 1.5\rho g H_s h_f$. This initial estimate of horizontal wave force is needed for caisson design in Section 6.3.5. It does not include the hydrostatic water level force.

Step 3. Use Goda formula to calculate the oscillating wave force. The full method is given in Section 6.3.2.4. Use $H_{\text{design}} = 1.8H_s$ for the design.

Step 4. Revise estimates of caisson size. See Section 6.3.2.

Step 5. Identify loading case using the parameter decision map (Figure 6.39).

The key decision parameters are h_b^*, H_s^*, and B^*. For example, if there is a "vertical" breakwater ($h_b^* < 0.3$) with large waves ($0.35 < H_s^*$), then the impact load formulas must be applied to check the design dimensions.

Step 6. Initial calculation of impact force. If the results from Step 5 indicate a "slightly breaking" or "impact load", then estimate the horizontal force, $F_{hAV} = 15\rho_w g d^2 (H_{si}/d)^{3.134}$ (Allsop and Vicinanza, 1996). This can be used as a reference value. However, the following information is needed first in Step 7.

The wave breaking formula, Equation (6.56), for calculating the breaking wave height H_b requires knowledge of the overall wave reflection coefficient χ, empirical factor k_b, and the wave length L_{pi} based on the incident peak wave period, T_p.

Step 7. Estimate percentage of breaking waves ($P_{b\%}$) that produce impacts. Use results to decide and confirm the loading case type from the following levels of $P_{b\%}$:

$$P_{b\%} < 2\% \left(\frac{H_{si}}{H_b} \leq 0.6 \right) \tag{6.59}$$

Little breaking. Wave loads are primarily non-breaking.

$$2\% < P_{b\%} < 10\% \left(0.6 < \frac{H_{si}}{H_b} < 1.2 \right) \tag{6.60}$$

Breaking waves give impacts.

$$P_{b\%} > 10\% \left(\frac{H_{si}}{H_b} \geq 1.2 \right) \tag{6.61}$$

Heavy breaking may give impacts or broken wave forces.

First calculate H_b, the breaking wave height using Equations (6.56)–(6.58). In PROVERBS, Oumeraci *et al.* (2001) use the subscript H_{bc} and note that this is a fictional breaking wave height which may differ "significantly" from other methods. Then calculate $P_{b\%}$ the estimated proportion of breaking waves using the following formula:

$$P_{b\%} = \exp[-2(H_b/H_{si})^2] \tag{6.62}$$

and is considered an upper limit for the percentage of impact loadings. Note that in Oumeraci *et al.* (2001), the symbol $P_{i\%}$ is employed for the percent of breaking waves.

Step 8. Estimate maximum impact force. This is a key result within the PROVERBS project that gives a new procedure based on the statistical distribution of forces (Oumeraci and Kortenhaus, 1997).

Recall from Chapter 2, Sections 2.3.8 that three parameters were found by fitting a straight line to the wave-height data. But first, the Weibull reduced variate, W, defined as

$$W = \left(\ln \frac{1}{Q}\right)^{1/\alpha} \tag{2.56}$$

was determined for plotting with $Q(H) = 1 - P(H) =$ the exceedance probability of the data and $\alpha =$ the free shape parameter selected to find the least squares, straight line of best fit of the wave height, H versus W plot. The Weibull distribution is a special case of the generalized extreme value (GEV) distribution (Frechet, 1927; Wikipedia, 2005).

To avoid confusion, we herein employ the following three parameter definitions and symbols to define the GEV distribution as found in Oumeraci *et al.* (2001) for PROVERBS, namely: β, the position parameter, α, the scale parameter, and γ, the shape parameter.

The comparable GEV-reduced variate plotted on the y-axis is defined as

$$\text{GEV} = \frac{1}{\gamma}\{1 - [-\ln(P(F_h^*))]^\gamma\}, \tag{6.63}$$

where $P(F_h^*)$ is the probability of non-exceedance of the relative wave impact force.

The equation of the straight line with F_h^* plotted on the x-axis is

$$Y = AX + B = \frac{F_h^*}{\alpha} - \frac{\beta}{\alpha}, \tag{6.64}$$

so that as before, $A = 1/\alpha$, $B = -\beta/\alpha$, and $X = (F_h^*)$.

Recall, by definition

$$F_h^* = \frac{F_h}{\rho g H_b^2}. \tag{6.65}$$

The values for β, α, and γ depend on the bed slope, the number of tests, and scale of the tests conducted to measure the impact wave force, F_h. The summary PROVERBS report (Oumeraci *et al.*, 2001) recommends using $\beta = 7.86$, $\alpha = 3.97$, and $\gamma = -0.32$ as determined from large-scale model tests closest to prototype conditions.

Finally, combining Equations (6.63) and (6.64) we find,

$$F_h^* = \frac{\alpha}{\gamma}\{1 - [-\ln(P(F_h^*))]^\gamma\} + \beta. \tag{6.66}$$

Oumeraci *et al.* (2001) in PROVERBS also recommend taking $P(F_h^*)$ larger than 90% in Equation (6.66) for a reasonable estimate of $F_{h\,\max}^*$ during the design storm. Using the three recommended parameter values and taking 0.9 as the probability of non-exceedance gives $F_{h\,\max}^* = 20.9$. The value of $F_{h\,\max}^*$ in Figure 6.37 is about 8.5 which would have an $P(F_h^*)$ value of non-exceedance of less than 50%.

Step 9. Impact rise time and duration.

An example of the actual measured time variation of the horizontal force $F_h(t)$ is shown in Figure 6.40 (left side) and the schematized, triangular impact force illustrated on the right. The rise time measured, t_{rFh} to $F_{h\,\max}$ is equated to the triangular rise time t_r by the conservation of mass and gives

$$t_r = 0.086\frac{\sqrt{d_{\text{eff}}/g}}{F_{h\,\max}^*}(\text{std deviation}, \sigma = \pm 0.084 \text{ for coefficient } 0.086), \tag{6.67}$$

with d_{eff} as a measure of the effective water depth in front of the vertical wall and normally identical to the water depth in which the wave breaks. See Oumeraci *et al.* (2001) for the method for calculating d_{eff} for effects of berm width and berm slope. For simple vertical walls, $d_{\text{eff}} = d$.

The measured time duration, t_{dFh} was related to the schematized, triangular duration, t_d by a statistically derived relation with the schematized rise time, t_r. The relationship was

$$t_d = -\frac{2.17}{\ln t_r}(\text{std deviation}, \sigma = \pm 1.08 \text{ on the coefficient } 2.17). \tag{6.68}$$

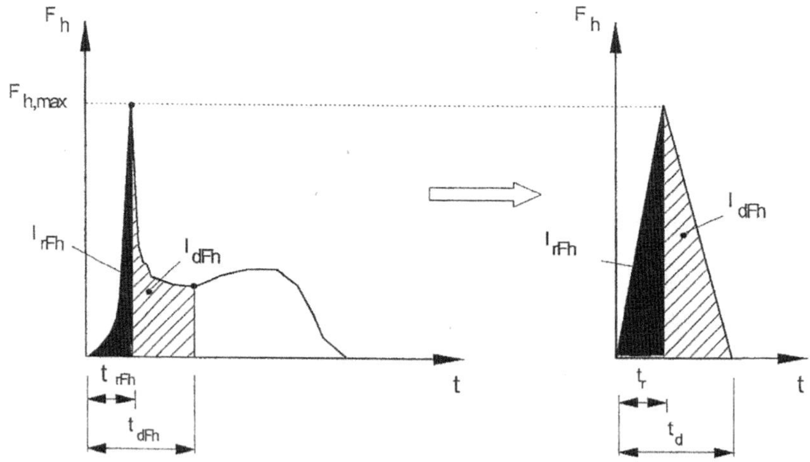

Figure 6.40. Measured impact force variation in time (left) and schematized, triangular shape (right) (from Oumeraci *et al.*, 2001).

Step 10. This step only applies to the design of caisson-type structures where uplift forces are calculated under horizontal force impacts. Further details are found in Section 6.3.5.

Step 11. Scale corrections. When conditions in Steps 6 and 7 warrant impact loads, PROVERBS includes some scaling corrections in the force calculation. Major factors are the aeration present in salt water and the amount of entrapped air at full, prototype scale relative to the fresh water used in the small to mid-scale laboratory force measurements. The compressibility of the impacting fluid/air mixture is questionable in the laboratory results. Another factor is the response of the vertical structure to the impact load. The laboratory-scale vertical structure may be more (or less) flexible to absorb the impact force to distort the time history of $F_h(t)$.

The percentage of aeration in the breaking waves, P_a (given in percent), has been empirically determined from field measurements (Allsop and Calabrese, 1998) and found to depend on the following three factors:

(1) $P_{b\%}$, the estimated proportion of breaking waves using

$$P_{b\%} = \exp[-2(H_b/H_{si})^2],$$

(2) t_{tot} = the duration of the design storm, and
(3) N_w = the number of waves in the storm.

The percentage of aeration at impact of waves at full scale in seawater is then found from

$$P_a = 2.0 + 5.3\left\{P_{b\%}\frac{N_w}{t_{tot}}\right\} \; (P_{b\%} \text{ must be given in percent}) \qquad (6.69a)$$

or since the mean wave period in the storm, $T_M = t_{\text{tot}}\,(s)/N_w$, this simply becomes

$$P_a = 2.0 + 5.3(P_{b\%}T_M). \qquad (6.69b)$$

This estimate assumes that the minimum value of P_a is about 2%.

A force reduction factor, k_{fa} has been found (Hewson *et al.*, 1998) to be

$$k_{fa} = \frac{4.0}{\{(2.5 + P_a)(97.5 - P_a)\}^{0.252}}. \qquad (6.70)$$

This empirical estimate for the force reduction factor, k_{fa} varies from about 0.87 when the percentage of aeration is the minimum ($P_a = 2\%$) to about 0.6 when the percentage of aeration is large ($P_a > 50\%$).

The reduced wave impact force due to aeration and scale correction $F_{h\,\text{impact}}$ then becomes

$$F_{h\,\text{impact}} = k_{fa}F_{h\,\text{max}}. \qquad (6.71)$$

Step 12. Pressure distributions. When conditions in Steps 6 and/or 7 indicate that impact forces are important, then the following four variables determine the vertical pressure distribution illustrated in Figure 6.41. Note that some of the symbols employed are the same as by Goda (Figure 6.36) but the formulas to calculate the values are entirely different.

(1) Elevation of pressure distribution η^* above the design SWL. This is the same symbol as employed by the Goda formula (Figure 6.36) but the PROVERBS formula says,

$$\eta^* = 0.8\,H_b. \qquad (6.72)$$

(2) Maximum pressure at the SWL, $p_1(t)$. The maximum pressure p_1 at the SWL varies in time with the force history, $F_h(t)$. The area of the pressure prism varies in time and represents the force history, $F_h(t)$.

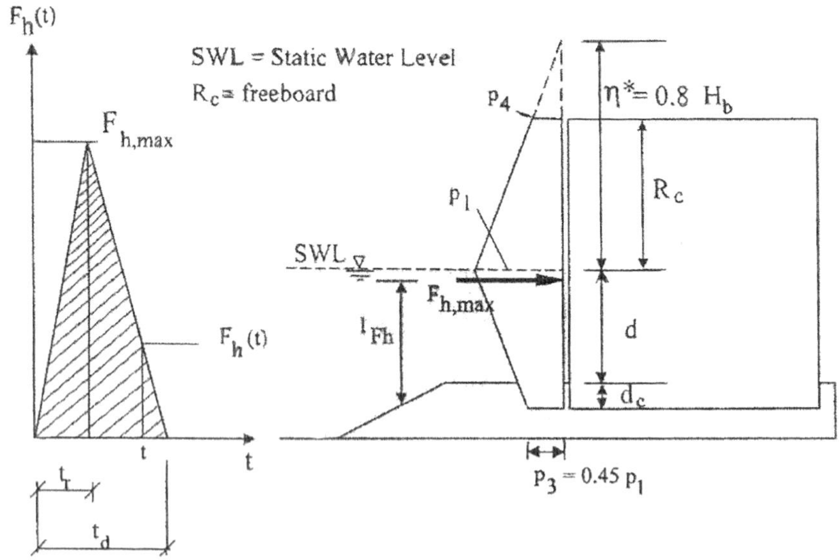

Figure 6.41. Schematized impact load variation in time and resulting pressure distributions on a vertical wall (from Kortenhaus and Oumeraci, 1998).

Calculating the area of the pressure prism requires knowledge of p_1, p_3, and p_4 as given in the following. Using these relationships gives

$$p_1(t) = \frac{F_h(t)}{0.4H_b + 0.7(d + d_c)}, \tag{6.73}$$

where d_c is the vertical distance to the berm elevation from the structure bottom.

The main difference between the Goda method and PROVERBS is that in the Goda method p_1 is a function of H_{design} (Equation (6.50)) that dictates the size of the pressure prism that determines the wave force, F_H. Whereas in PROVERBS, the impact wave force F_h is a function of the probability of non-exceedance distribution as specified by the designer (Equation (6.66)) and the breaking wave height, H_b (Equation (6.56)) and both calculated first to then determine p_1 using Equation (6.73).

(3) The pressure at the bottom of the pressure prism, p_3. The authors (Oumeraci *et al.*, 2001) simply took

$$p_3 = 0.45\, p_1. \tag{6.74}$$

(4) The pressure at the crest of the structure, p_4. Using linear interpolation (as in Goda, Figure 6.36) gives

$$p_4 = \left(1 - \frac{R_c}{\eta^*}\right) p_1 \quad (p_4 = 0 \text{ for } \eta^* < R_c). \tag{6.75}$$

Finally, location of the horizontal impact force F_h must be determined. Taking moments of the centroids of the two pressure prism forces about the bottom of the vertical wall and equating to the moment of the horizontal impact force, i.e. $F_h(t)$ times $l_{Fh}(t)$ will determine the location l_{Fh} above the wall bottom. In other words,

$$l_{Fh}(t) = \frac{\begin{array}{c}\text{Upper Prism force} \times \text{centroid location}\\ +\text{Lower Prism force} \times \text{centroid location}\end{array}}{F_h}. \tag{6.76}$$

The upper pressure prism centroid location will depend on which distance (R_c or η^*) is minimum and the lower prism centroid location will depend on the distance, $d + d_c$. The spreadsheet employed to calculate impact forces and location on caissons is found in Section 6.3.5.

Step 12 Continued. When overtopping occurs.
When wave overtopping occurs, the total wave impact force, $F_{h\,\max ov}$ is less than $F_{h\,\max}$. This case is illustrated in Figure 6.41 and Equation (6.75) is employed to calculate p_4 at the top of the wall. The triangular pressure prism above the wall crest is missing and is simply the freeboard, vertical distance $(\eta^* - R_c)$ times $0.5p_4$. Therefore,

$$F_{h\,\max ov} = F_{h\,\max} - 0.5p_4\,(\eta^* - R_c). \tag{6.77}$$

6.3.3.2 *Simple example*

An example calculation will help to understand Steps 1–12 to calculate wave impact forces on vertical walls. Again, employ a significant wave height, $H_s = 1.52\,\text{m}$ (5 ft) in $d = 6.1\,\text{m}$ (20-ft) water depth with significant wave period, $T_s = 10\,\text{s}$. The water is sea water, $\rho = 1035\,\text{kg/m}^3$ ($64\,\text{lbs/ft}^3$). The wall height h_w will be calculated so that it extends a distance η^* above the SWL so that the pressure at the top of the wall is zero.

Steps 1–4. The results are the same as above for the Goda method using $H_{\text{design}} = 1.8H_s$ for the design, which gave $H_{\text{design}} = 2.74\,\text{m}$ (9.0 ft) and $F_{h\,\text{Goda}} = 2{,}06{,}038\,\text{N/m}$ (14,051 lbs/ft). However, check the result in

Step 12 for the value of η^* and hence the total length of the vertical wall which may be slightly different.

Step 5. The loading case is identified using the parameter decision map (Figure 6.39). The berm height, h_b is zero so $h_b^* = 0$ and far less than 0.3 for a "vertical" breakwater. The key wave parameter, $H_s^* = H_{si}/h_s = 2.74/6.095(9/20) = 0.450$ and clearly "large waves" since greater than 0.35. Wave impact loads must be considered. Note that $h_s = d$ since no berm is present in front of the vertical wall.

Step 6. Estimate of average impact force. Using the formula proposed by Allsop and Vicinanza (1996) gives $F_{hAV} = 461{,}250\,\mathrm{N/m}$ (31,450 lbs/ft). This is roughly double the Goda force.

Step 7. Estimate the breaking wave height H_b and then the percentage of breaking waves, $P_{b\%}$. Using Equation (6.58), the empirical factor $k_b = 1$ since the equivalent berm width, B_{eq} is zero. The new reflection coefficient C^* is also zero since we assume perfect reflection $\chi = 1$ from the wall. Therefore, using Equation (6.56) gives the breaking wave height $H_b = 3.61\,\mathrm{m}$ (12.3 ft) where it must be reemphasized that this is a nominal breaking wave height determined using the PROVERBS Equation (6.56).

Equation (6.62) then estimates the proportion of breaking waves and gives $P_{b\%} = 3.1\%$. As shown in Equation (6.60), this amount of wave breaking gives wave impacts.

Step 8. Estimate the maximum impact force, $F_{h\,\max}$. Using the GEV distribution with non-exceedance probability $P(F_h^*)$ of 90%, and recommended position $\beta = 7.86$, scale $\alpha = 3.97$, and shape $\gamma = -0.32$ parameters gives (Equation (6.66)) the relative impact force, $F_h^* = 20.945$. Then, using Equation (6.65) we find $F_{h\,\max} = 2{,}758{,}113\,\mathrm{N/m}$ (188,138 lbs/ft). This impact force is over 13 times larger than the Goda force calculated in Step 1 ($F_{h\,\mathrm{Goda}} = 206{,}038\,\mathrm{N/m}$ (14,051 lbs/ft).

In the original Goda formula, the key pressure at wave impact is p_1 as shown in Figure 6.36 and calculated by Equation (6.50). Takahashi *et al.* (1994) modified the Goda formula for pressure p_1 for impact forces due to head-on wave breaking and this has come to be called the modified Goda formula. See CEM (2006, Table VI-5-54) for the full modification formulas. $F_{h\,\mathrm{GodaMod}} = 303{,}181\,\mathrm{N/m}$ (20,676 lbs/ft). The impact force from PROVERBS is still over nine times larger than the modified Goda formula.

Step 9. Calculate the impact rise time and duration. Using Equation (6.67), the rise time to the peak impact force, $t_r = 0.0032\,\mathrm{s}$ and the impact duration

$t_d \approx 0.4\,\mathrm{s}$ from Equation (6.68). For this case of a simple vertical wall, $d_{\mathrm{eff}} = d$. Note, however, the large standard deviations of the mean values of the coefficients in these equations. Using $2\sigma = 2\,(1.08)$ in Equation (6.68) gives $t_d \approx 0.8\,\mathrm{s}$ or twice the duration of the impact event.

Step 10. This step only applies to caisson-type structures as discussed in Section 6.3.5.

Step 11. Scale and aeration effect corrections. As shown in Step 7, $P_{b\%} = 3.1\%$ so that from Equation (6.69b) with $T_m = 10\,\mathrm{s}$ gives the percentage of aeration in the breaking waves, $P_a = 3.65\%$. Then the force reduction factor, $k_{fa} = 0.806$ from Equation (6.70). As a result of aeration and model scale effects, the reduced wave impact force becomes

$$F_{h\,\mathrm{impact}} = k_{fa} F_{h\,\max} = 0.806(2{,}758{,}113) = 2{,}223{,}039\,\mathrm{N/m}(151{,}639\,\mathrm{lbs/ft}).$$

This reduced wave impact force in PROVERBS is about seven times larger than the modified Goda formula.

Step 12. The pressure distribution as shown in Figure 6.41 is necessary to calculate the trapezoidal loads on the wall for both the pressure prism above the SWL and below the SWL to the bottom and to locate the centroid of these two pressure prisms to calculate moments about the base. Finally, the location of the horizontal impact force l_{Fh} can be determined by Equation (6.73).

The PROVERBS formula (Equation (6.69)) says, $\eta^* = 0.8\,H_b = 0.8\,(3.61) = 2.89\,\mathrm{m}$ (9.48 ft). Note this is different than the Goda formula which gave 4.11 m (13.5 ft). Since this example assumed the vertical wall extends above the SWL to the distance to where the pressure is zero, some of the differences in the impact loads are a result of the vertical wall heights being different. The total wall height in the Goda method was 10.2 m (33.5 ft) and in the PROVERBS was 8.985 m (29.5 ft). Since the pressure goes to zero at the top of the wall, this difference in pressure prisms is relatively minor when integrated to calculate the wave force.

The major difference is in the calculation of pressure p_1 at the SWL. For this example, $p_{1\mathrm{GodaMod}} = 38{,}935\,\mathrm{N/m^2}$ ($809\,\mathrm{lbs/ft^2}$) and from PROVERBS, $p_{1\mathrm{PROVERBS}} = 482{,}987\,\mathrm{N/m^2}$ ($10{,}040\,\mathrm{lbs/ft^2}$). The difference is about a factor of 12.5 and similar to that for the difference in maximum impact wave force calculated in Step 8.

Finally, the location, l_{Fh} of $F_{h\,\max}$ is found to be 4.44 m (14.6 ft) above the bottom using Equation (6.73). Note that this is only for the wave force and does not include the hydrostatic force on the vertical wall and the location of the hydrostatic force.

Step 13. Discussion. The example mentioned above illustrates the importance of selection of the probability of non-exceedance, $P(F_h^*)$. Using $P(F_h^*) = 0.7$ would reduce the $F_{h\,\mathrm{PROVERBS}}/F_{h\,\mathrm{GodaMod}}$ ratio to about 5.5. Also, the significant wave height H_s selected will impact this result. Doubling H_s to 3 m and reducing the $P(F_h^*) = 0.7$ will again lower the $F_{h\,\mathrm{PROVERBS}}/F_{h\,\mathrm{GodaMod}}$ ratio to about 1.7. This is because in PROVERBS, the breaking wave height H_b does not change but in Goda, $H_{\mathrm{design}} = 1.8\,H_s$ to significantly impact pressure, p_1.

Finally, the maximum wave impact force $F_{h\,\max}$ strikes the vertical wall over a time duration t_d of less than 1 s (Equation (6.68), Step 9) in this example. The full meaning of the wave impact force and its importance will depend on the structure type (seawall, caisson, or bulkhead) under design. When the vertical breakwater is made up of relatively short, individual sections, the sudden, direct impact of a plunging, breaking wave may be along the length of one, short, walled section. Then clearly, the Impact Load formulas from the PROVERBS project should be used to check the design.

The *Probabilistic Design Tools for Vertical Breakwaters* book (Oumeraci *et al.*, 2001) also includes sections on broken waves hitting the vertical wall and what are labeled "alternative, low-reflection structures", namely perforated vertical walls. This book is also a valuable reference for material presented in Chapter 5 on probabilistic design tools for vertical wall design.

An Excel spreadsheet entitled "Goda and PROVERBS Method — Wave Force on Vertical Walls" is found in the Seawalls and Caissons folder in the online supplementary materials that accompanies this book.

In summary, for many designs, using Figure 6.39 to determine the loading case will reveal that the Goda method is adequate. When it is not, following Steps 6–12 for the impact loading case will provide a safe design.

6.3.4 *Seawalls*

Seawalls mitigate water-level inundation damage and wave damage to what is on the interior side of this water–land interface. There are many different types of seawalls as shown in Figure 6.42(a) as adapted from Pilarczyk (1990) and in Figure 6.42(b) from Reeve *et al.* (2012). Seawalls harden

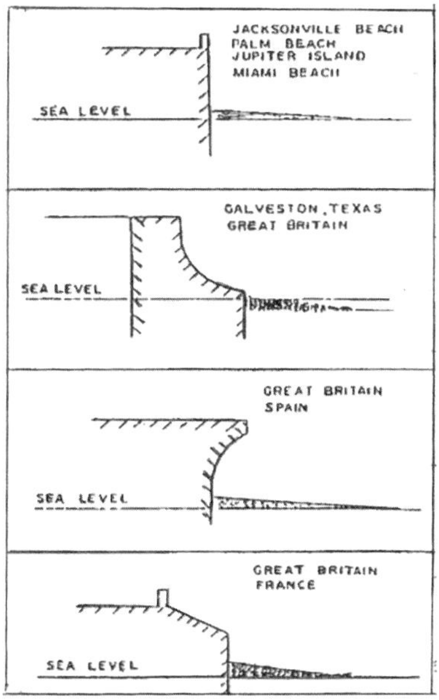

Figure 6.42. Typical seawall types from Reeve *et al.* (2012).

(armor) the shoreline to fix its location, and they have some configuration and crest elevation that determines the height of the wave runup and the amount of wave overtopping for the range of water levels and wave conditions at their location on the coast. Functional failure mechanisms and crest design to reflect waves in the seaward direction to minimize wave overtopping were discussed in Chapter 5.

Reports of structural damage and failure mechanisms of seawalls are relatively few. In many cases, the surface elevation of the material at the seaward toe of the vertical wall (sand, shingle, or rubble) is lowered during storm events or from long-term erosion at the wall location. The original design did not take this factor into account when considering the water depth d at the toe of the structure for design. Other mechanisms for structural damage are washout of fine-grained sediment (fill) behind the wall and concrete spalling, cracking, and disintegration producing partial crest failure. Concrete seawalls can also end up buried beneath dunes and sandy beaches (e.g. the Embarcadero seawall in San Francisco, Wiegel, 2001).

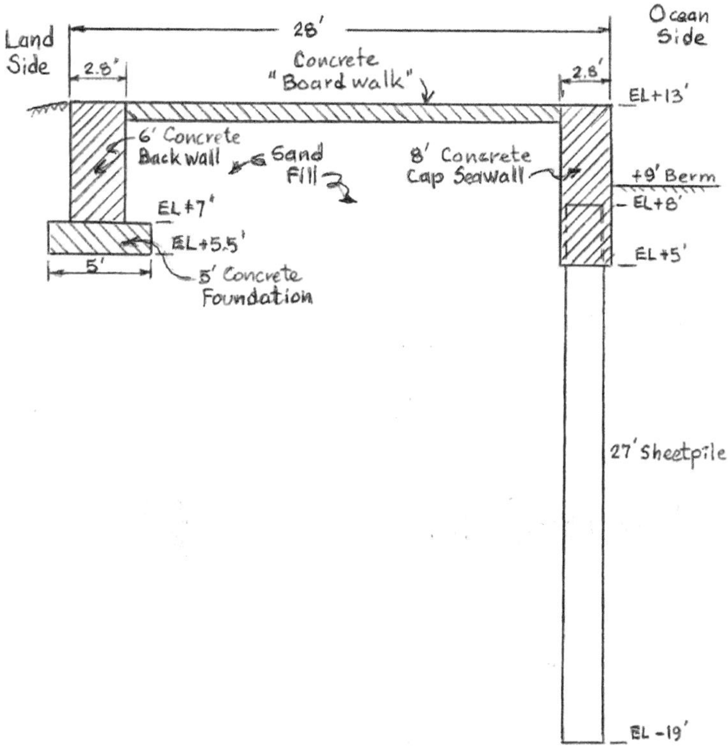

Figure 6.43. Virginia Beach seawall as designed by the Norfolk, District Office of the US Army Corps of Engineers. All dimensions in imperial (ft) length and elevation units.

As illustrated in Figure 6.42, seawalls are backed by soils which provide substantial resistance to water level and wave forces. The active and passive geotechnical loadings of the soils on the wall may govern the design and is certainly the case for bulkheads as discussed in Section 6.3.6. In some designs, the seawall dimensions provide sufficient gravitational weight to prevent horizontal movement (sliding) of the wall as for caissons discussed in Section 6.3.5. And, as illustrated in Figure 6.43, the vertical sea wall (right, seaward side) may be a concrete cap above the relatively long, steel sheet piles driven down into the soil. In addition, the concrete "boardwalk" abuts a back wall sitting on a concrete footing as foundation. In between, a sandy fill region supports the concrete slab and ties the entire structure together. This is the design for the Virginia Beach seawall by the Norfolk District Office of the US Army Corps of Engineers as constructed

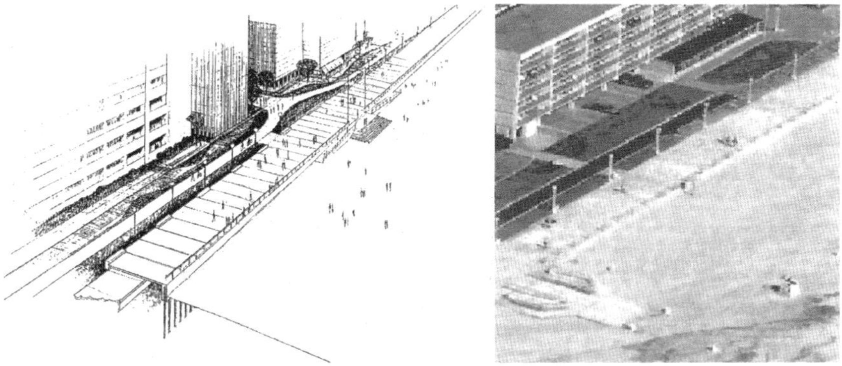

Figure 6.44. Artists schematic drawing of the Virginia Beach seawall constructed over the old boardwalk (left) and aerial photo of seawall/boardwalk structure with sandy beach fill (right) (courtesy, US Army Corps of Engineers, Norfolk District Office).

in 2001–2003. Figure 1.8 in Chapter 1 presented an example design drawing including reinforcing steel and Figure 6.43 shows the true dimensions of a typical cross-section. Figure 1.9 showed a photo of the steel reinforcing being installed in the concrete cap during construction.

Proper seawall design requires the combined efforts of coastal engineers (wave- and water-level forces on the vertical wall; toe scour at the base of the wall to locate the lowest beach elevation; potential down-drift impacts; etc.), geotechnical engineers (soil properties behind the wall; active and passive soil pressures on the wall, active and passive soil forces and moments on the wall, etc.), and structural engineers (loads, stresses, strength of materials, etc.) for concrete, steel, and possible wood structures. The design of seawalls is any extremely complicated process and we shall return to this subject in Chapter 11. Figure 6.44 shows an artist's schematic drawing of the Virginia Beach seawall constructed over the old boardwalk (left) and aerial photo of seawall/boardwalk structure with sandy beach fill (right).

6.3.5 *Caisson-type Structures*

6.3.5.1 *Introduction*

The choice between the use of a vertical-walled, caisson-type structure, or a rubble-mound breakwater depends on many factors with water depth being one of the most important. In deep water, the cross-sectional area (hence initial cost) of the caisson-type is lowest, whereas in shallow water

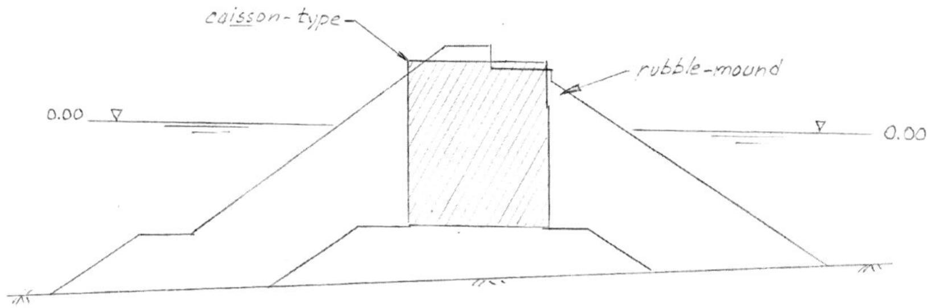

Figure 6.45. Schematic diagram of rubble-mound and caisson-type breakwaters.

the material cost often favors the rubble-mound type. No general design rules apply so that both types should be considered in the feasibility study. Figure 6.45 schematically illustrates a relatively deep-water example where the cross-sectional area of the rubble-mound structure is far greater than the caisson-type.

A caisson is typically a concrete box with vertical, concrete walls for strength and filled with sand as illustrated in Figure 6.46(a). The (b) composite-type sits on a rock base, the (c) perforated-type has the front section open to the sea for wave absorption, and the (d) armored-type has a rubble-mound structure in front for wave force reduction on the vertical wall.

In contrast to seawalls, reports of structural damage and failure mechanisms of caissons are many, thanks to their extensive use for breakwater jetties in Japan for more than 75 years (Goda, 2010). Figure 6.47 illustrates 12 different failure mechanisms. The most dominant are: (3) Translation (sliding), (5) Rotation (overturning), and (9, 11, 12) Foundation failures. A special issue on failure mechanisms for vertical breakwaters was published in the *Coastal Engineering Journal* (1994, Vol. 22).

The need for improved knowledge for the wave impact loads, geotechnical aspects, and structural analysis aspects for caisson-type breakwater design resulted in the **PRO**babilistic Design Tools for **VE**rtical Breakwater**S** (**PROVERBS**) project in the European Union (1994–1998). Only the wave impacts (Hydraulic Aspects: Allsop and Kortenhaus, Chapter 2) were presented in Section 6.3.3. Oumeraci *et al.* (2001) cite the following reasons for research on monolithic, caisson-type structures: catastrophic failures in Europe, need for breakwaters at greater depths, need for environmentally friendly structures, need for multi-purpose structures, and the potential for application in seawall designs. The interested reader should

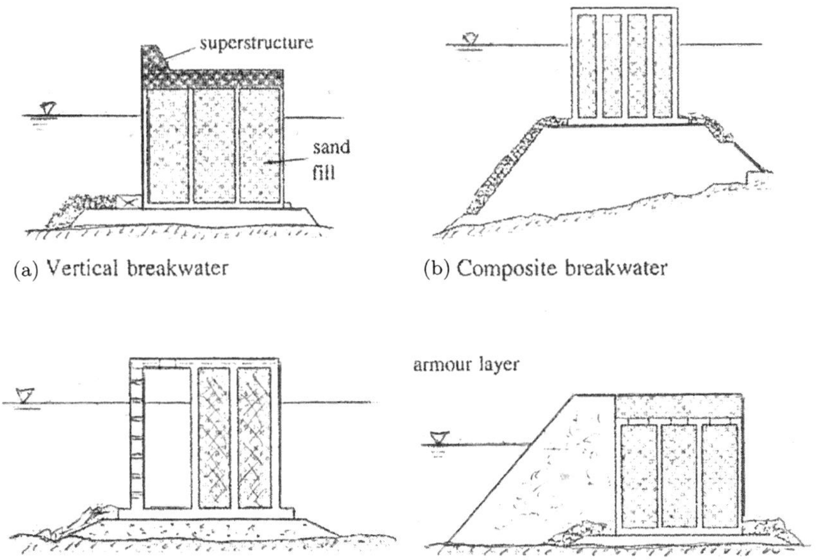

(a) Vertical breakwater (b) Composite breakwater

(c) Perforated breakwater (d) Armoured caisson breakwater

Figure 6.46. Four vertical breakwater types using a caisson as the basic wave blocking structure (from Oumeraci *et al.*, 2001).

consult this reference for Geotechnical Aspects (de Groot, Chapter 3), Structural Aspects (Crouch, Chapter 4), and Probabilistic Design Tools (Vrijling and Voortman, Chapter 5) that are applicable for caisson-type breakwater design.

6.3.5.2 *Complete Goda formulas for caissons of various types*

Figure 6.36 and Equations (6.49)–(6.54) presented the *basic* Goda formulas for a simple vertical wall on a horizontal bottom with head-on wave conditions (β, angle of incidence of waves $= 0°$). The complete Goda formulas for pressures and coefficients relative to the various types of caisson structures on a rubble-mound foundation (Figure 6.48) are as follows:

$$\eta^* = 0.75(1 + \cos\beta)\lambda_1 H_{\text{design}}, \tag{6.78}$$

$$p_1 = 0.5(1 + \cos\beta)(\lambda_1\alpha_1 + \lambda_2\alpha_* \cos^2\beta)\rho g H_{\text{design}}, \tag{6.79}$$

$$p_2 = \begin{cases} \left(1 - \frac{h_c}{\eta^*}\right) p_1 & \text{for} \quad \eta^* > h_c, \\ 0 & \text{for} \quad \eta^* \leq h_c, \end{cases} \tag{6.80}$$

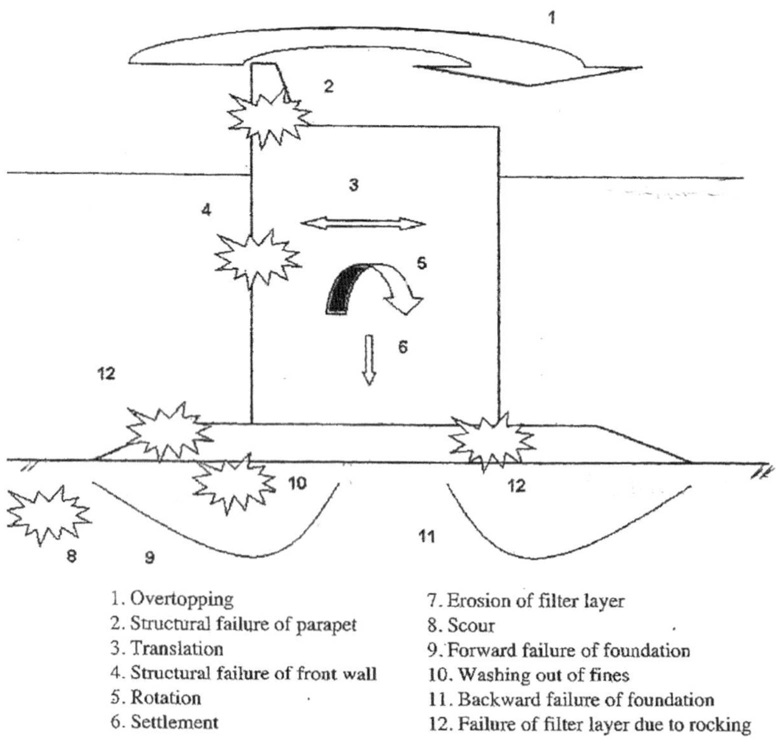

Figure 6.47. Twelve potential caisson structure damage and failure modes (from Oumeraci *et al.*, 2001).

$$p_3 = \alpha_3 p_1, \tag{6.52}$$

$$p_u = 0.5(1 + \cos\beta)\lambda_3 \alpha_1 \alpha_3 \rho g H_{\text{design}}, \tag{6.81}$$

where β is the angle of incidence of waves, and

$$\alpha_* = \alpha_2, \tag{6.82}$$

$$\alpha_1 = 0.6 + 0.5\left[\frac{4\pi h_s/L}{\sinh(4\pi h_s/L)}\right]^2, \tag{6.53}$$

$$\alpha_2 = \text{the smallest of } \frac{h_b - d}{3h_b}\left(\frac{H_{\text{design}}}{d}\right)^2 \text{ and } \frac{2d}{H_{\text{design}}}, \tag{6.83}$$

$$\alpha_3 = 1 - \frac{h_w - h_c}{h_s}\left[1 - \frac{1}{\cosh(2\pi h_s/L)}\right], \tag{6.84}$$

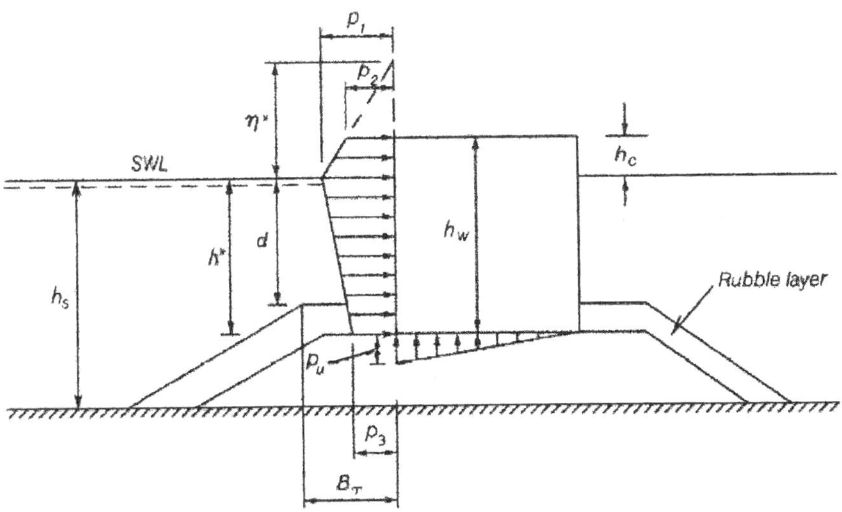

Figure 6.48. Definitions of terms in *complete* Goda formula for caissons with β, the angle of wave incidence, and λ_1, λ_2, and λ_3 modification factors depending on the structure type (from CEM, 2006).

where L is the wavelength at water depth h_b corresponding to that of the significant wave $T_s \simeq 1.1\,T_m$, where T_m is the average period, and h_b is the water depth at a distance of $5H_s$ seaward of the breakwater front wall. λ_1, λ_2, and λ_3 are modification factors depending on the structure type. For conventional vertical wall structures, $\lambda_1 = \lambda_2 = \lambda_3 = 1$.

For example, $\eta^* = 0.75(1 + \cos\beta)\lambda_1 H_{\text{design}}$ is the complete formula (Equation (6.74)). Taking $\beta = 0°$ gives $\cos 0° = 1$ and with $\lambda_1 = 1$ for a vertical wall, then $\eta^* = 0.75(1 + 1)H_{\text{design}} = 1.5H_{\text{design}}$ which is the basic formula (Equation (6.49)). Similar reductions for $\beta = 0°$ and $\lambda_1 = \lambda_2 = \lambda_3 = 1$ were applied to Equations (6.74)–(6.82) to give the basic Goda formulas for pressures, p_1, p_2, and p_3 in Equations (6.50)–(6.54).

These should be considered as a "lower bound" from the oscillation wave loads on the vertical wall.

6.3.5.3 *Uplift pressures*

6.3.5.3.1 Oscillating waves (Goda)

For a caisson, we now have an additional, wave-induced uplift pressure p_u on the bottom at the seaward-side corner of the caisson given by Equation (6.78) with a linear pressure distribution down to zero at the landward side (heel) as shown in Figure 6.48. This is again for the oscillation wave loads on the structure and should be considered as a "lower bound" for uplift pressure, p_u. Note that the hydrostatic pressure is not considered since the water level is the same on both sides of the caisson. Using the same given information for the example above for a caisson on a flat bottom ($H_{\text{design}} = 2.74\,\text{m}$, $T_m = 10\,\text{s}$, $d = 6.096\,\text{m}$) gives

$$\begin{aligned} p_u &= 0.5(1 + \cos\beta)\lambda_3\alpha_1\alpha_3\rho g H_{\text{design}} \\ &= 0.5(2)1(0.9546)(0.88005)(1030)(9.81)(2.74) \\ &= 23{,}258\,\text{N/m}^2\,(483.5\,\text{lbs./ft}^2). \end{aligned} \tag{6.81}$$

When the parametric decision map (Figure 6.39) for different wave loadings indicates that wave impacting conditions are present, then the PROVERBS method to calculate the uplift pressures should be employed.

6.3.5.3.2 Impacting waves (PROVERBS)

During the large-scale, wave tank experiments on wave impacts, it was discovered that the uplift pressure at impact on the landward bottom edge (heel) of the caisson p_{ru} was greater than hydrostatic pressure (Oumeraci and Kortenhaus, (1997)). The measured test data were employed to determine a simple equation to calculate p_{ru}. Equation (6.56) is again used to calculate H_b.

$$p_{ru} = \rho g H_b \left\{ \frac{H_b}{h_s} - 0.1 \right\}. \tag{6.85}$$

This produced an uplift force F_u and scaled the same way as the horizontal impact force giving the relative uplift force,

$$F_u^* = \frac{F_u}{\rho g H_b^2}. \tag{6.65}$$

The relative uplift force F_u^* was again based on the hydraulic model test data using "upper-bound" limits producing conservative estimates. The GEV distribution was again employed using the following three parameter definitions and symbols to define the GEV probability of non-exceedance distribution, $P(F_u^*)$: β is the position parameter, α is the scale parameter, and γ is the shape parameter, but with different statistical parameter values to fit the distribution. For the relative uplift force, F_u^* calculated from Equation (6.65) use $\alpha = 2.17$, $\beta = 4.384$, and $\gamma = -0.11$

(Oumeraci *et al.*, 2001, p. 104).

$$F_u^* = \frac{\alpha}{\gamma}\{1 - [-\ln(P(F_u^*))]^\gamma\} + \beta. \tag{6.66}$$

For a given probability of non-exceedance level, e.g. $P(F_u^*) = 90\%$, a near-maximum uplift force, $F_{u\max}$ can be estimated under wave impacting conditions. This near-maximum uplift force can then be used together with the estimate for the heel-uplift pressure p_{ru} and a straightline pressure distribution to find the near-maximum uplift pressure $p_{u\max}$ on the seaward side of the caisson, namely,

$$p_{u\max} = \frac{2\,F_{u\max}}{B} - p_{ru}, \tag{6.86}$$

where B is the width of the caisson.

The uplift pressure, $p_{u\max}$, calculated from Equation (6.86) should be considered an upper bound.

Finally, the location of the uplift force l_{Fu} relative to the heel position can be found from

$$l_{Fu} = \frac{B^2(p_{ru} + 2p_u)}{6\,F_{u\max}}. \tag{6.87}$$

6.3.5.4 *Goda formulas for design*

6.3.5.4.1 Forces and moments (bias and uncertainty)

Figure 6.49 illustrates the forces that dominate the stability of a caisson design. The horizontal wave force, F_H causes sliding and overturning. The size and hence weight W of the caisson resists sliding and overturning but since submerged in water, the caisson is buoyed up by the buoyant force B that contributes to overturning. The wave uplift force F_U also contributes to overturning, and the sliding is resisted by a sliding force F_S. The overturning bearing pressures of the caisson on the soil at the heel of the caisson (location "e") are also of major concern for foundation damage and failure.

Three stability tests are normally conducted (sliding, overturning, and heel-bearing pressure) to determine the design caisson width B as illustrated in Figure 6.49. The gravity force F_G is simply the difference between the caisson weight in air W and the buoyant force B, and when multiplied by a sliding friction coefficient μ produces the sliding friction resistance force F_S that opposes the horizontal wave force F_H. The wider the caisson width B, the greater the sliding friction resistance force to produce a stable structure against sliding failure.

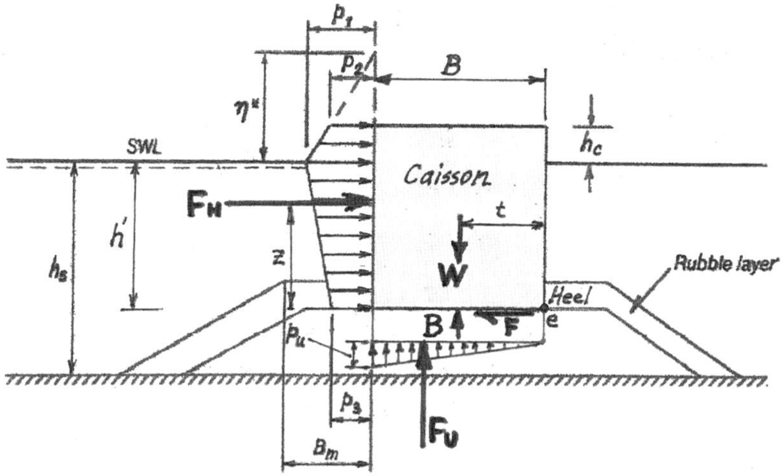

Figure 6.49. Forces on the caisson structure that cause sliding, overturning, and bearing pressures.

The uplift force F_U is the result of an assumed triangular uplift pressure distribution with p_u on the front edge. Note that p_u is not equivalent to p_3. The uplift moment arm is $2/3\,B$ and the horizontal wave force moment arm is the distance z for clockwise moments about the heel of the caisson at location, e. The gravity force, F_G (W–B) times its moment arm t produces the counterclockwise moment about the heel location e to prevent overturning of the caisson. Here again, the wider the caisson width B, the greater the resistance to the overturning moments to produce a stable structure against the overturning failure.

The third "failure" test normally considered is the soil bearing pressure at the heel of the caisson structure. Further details are found in the following.

The wave pressure and uplift pressure distributions produce forces and moments per unit length of structure that are readily calculated by the following formulas. The gravity force F_G is simply the difference between the caisson weight in air W and the buoyant force B. The forces are given as follows:

$$F_H = U_{FH}\left[\frac{1}{2}(p_1 + p_2)h_c + \frac{1}{2}(p_1 + p_3)h'\right], \tag{6.88}$$

$$F_U = U_{FU}\frac{1}{2}p_u B, \tag{6.89}$$

$$F_G = \rho_c g\, Bh_w - \rho g\, Bh', \tag{6.90}$$

where ρ_c is the mass density of the structure (complex mixture, concrete, sand, etc.), ρ is the mass density of the water, U_{FH} is the stochastic variable signifying the bias and the uncertainty related to the horizontal *force* in Goda formula, U_{FU} is the stochastic variable signifying the bias and the uncertainty related to the uplift *force* in Goda formula, h' is the submerged height of the wall from the toe to the SWL, h_w is the height of the caisson (not shown), and B structure width.

The corresponding moments taken about the heel of the structure are as follows:

$$M_H = U_{MH}\left[\frac{1}{6}(2p_1 + p_3)h'^2 + \frac{1}{2}(p_1 + p_2)h'h_c + \frac{1}{6}(p_1 + 2p_2)h_c^2\right], \tag{6.91}$$

$$M_U = U_{MU}\frac{1}{3}p_u B^2, \tag{6.92}$$

$$M_G = \frac{1}{2}B^2 g(\rho_c h_w - \rho h'), \tag{6.93}$$

where U_{MH} is the stochastic variable signifying the bias and the uncertainty related to the horizontal *moment* in Goda formula, and U_{MU} is the stochastic variable signifying the bias and the uncertainty related to the uplift *moment* in Goda formula.

A reanalysis of the Goda formula has been carried out by Juhl and van der Meer (1992), Bruining (1994), and van der Meer *et al.* (1994) using the model scale data carried out by Delft Hydraulics and the Danish Hydraulic Institute. The results for the stochastic variables signifying the bias and the uncertainty are summarized in Table 6.18. For example, the mean value for U_{FH} was 0.90 with the standard deviation from the mean σ equaling 0.05 when the scale model tests are conducted. If only desktop calculations are performed, use $\sigma = 0.25$. Therefore, this result modifies the safety factor of 1.8 as recommended by Goda (1985) for H_{design}.

Global safety factors can be applied for the sliding, overturning, and bearing pressure formulations but are not recommended because they are not related to the probability of failure P_f of the structure. Modern, probabilistic design methods now employ partial safety factors.

Partial safety factors have been developed for sliding, overturning, and foundation failure. For calculations of F_H and F_u the Corps of Engineers,

Table 6.18. Stochastic variables for bias and uncertainty for horizontal wave-induced forces, uplift force, horizontal moment, and uplift moment in Goda formulas, Equations (6.88)–(6.93).

Stochastic Variable, X_i	Mean Value, μ_{X_i}	Standard Deviation of Mean, σ_{X_i} (model)	Standard Deviation of Mean, σ_{X_i} (Desktop)
U_{FH}	0.90	0.05	0.25
U_{FU}	0.77	0.05	0.25
U_{MH}	0.81	0.10	0.40
U_{MU}	0.72	0.10	0.37

Coastal Engineering Manual (CEM, 2011) recommends using wave height $\gamma_H H_s^T$, where γ_H is the load partial safety factor for wave height and H_s^T is the significant wave height at recurrence interval, T.

Two levels of knowledge of the wave height coefficient of variability $\sigma_{FH_s} = 0.05$ for good knowledge and $\sigma_{FH_s} = 0.2$ for relatively poor knowledge are presented and five levels (0.01, 0.05, 0.10, 0.20, 0.40) of probability of failure, P_f, for the probabilistic design of caisson breakwaters.

6.3.5.4.2 Partial safety factors

(1) Sliding failure

The friction coefficient f for sliding depends on the composition of the bed material beneath the bottom of the concrete caisson. Typical values are $f = 0.4$ for sandy bottom and $f = 0.6$ for rock, gravel bottom (CEM, 2011). A serrated base slab (Hutchinson *et al.*, 2010) or protruding bottom rails will increase the coefficient. Friction coefficients for caisson sliding on hard clay bottoms have not been found in the literature. This is perhaps because a sand or gravel layer would be first placed as a sublayer. See "Geotechnical Aspects" in Oumeraci *et al.*, (2001, Chapter 3) for a full review of how friction coefficients are related to the geotechnical properties of the foundation material.

For sliding failure, the partial safety factors for deep and shallow water and for design with and without model tests are presented in Table 6.19 for both the load partial safety factor γ_H for wave height and for the resistance partial safety factor γ_Z for the friction coefficient, f. For the friction coefficient, use $(1/\gamma_Z)f$. For calculations of F_H and F_u, the Corps of Engineers, *Coastal Engineering Manual* (CEM, 2011) recommends using

Table 6.19. Partial safety factors for sliding failure of vertical wall caissons (CEM 2011, Table VI-6-24).

(a) Deep-water conditions

	Design Without Model Tests				Design Load Determined With Model Tests			
	$\sigma_{FH_s} = 0.05$		$\sigma_{FH_s} = 0.20$		$\sigma_{FH_s} = 0.05$		$\sigma_{FH_s} = 0.20$	
P_f	γ_H	γ_Z	γ_H	γ_Z	γ_H	γ_Z	γ_H	γ_Z
0.01	1.4	1.7	1.5	1.7	1.3	1.5	1.4	1.5
0.05	1.3	1.4	1.4	1.4	1.2	1.4	1.3	1.4
0.10	1.3	1.2	1.4	1.3	1.2	1.2	1.3	1.2
0.20	1.2	1.2	1.3	1.2	1.1	1.2	1.2	1.2
0.40	1.1	1.0	1.1	1.1	1.0	1.2	1.1	1.0

(b) Shallow-water conditions

	Design Without Model Tests				Design Load Determined With Model Tests			
	$\sigma_{FH_s} = 0.05$		$\sigma_{FH_s} = 0.20$		$\sigma_{FH_s} = 0.05$		$\sigma_{FH_s} = 0.20$	
P_f	γ_H	γ_Z	γ_H	γ_Z	γ_H	γ_Z	γ_H	γ_Z
0.01	1.3	1.9	1.4	1.9	1.2	1.6	1.3	1.6
0.05	1.2	1.6	1.3	1.6	1.1	1.5	1.2	1.5
0.10	1.2	1.4	1.3	1.4	1.1	1.3	1.2	1.3
0.20	1.1	1.3	1.2	1.3	1.1	1.2	1.1	1.2
0.40	1.0	1.2	1.0	1.2	1.0	1.1	1.0	1.1

wave height $\gamma_H H_s^T$ where H_s^T is the significant wave height at recurrence interval, T.

(2) Overturning failure

For overturning failure, partial safety factors for design with and without model tests are presented in Table 6.20 for the load partial safety factor γ_H for wave height. It is not possible to specify partial safety factors when the design is made without model tests for $P_f = 0.01$ and with poor knowledge of the wave conditions for $P_f = 0.05$. Failure of a caisson by "overturning" should be called "rotation" failure. The foundation beneath the structure will fail and the allowable bearing pressures will be exceeded long before the caisson is overturned by wave and uplift loads. There is no resistance partial safety factor, γ_Z for overturning.

Table 6.20. Partial safety factors for overturning failure of vertical wall caissons (CEM 2011, Table VI-6-25).

	Design Without Model Tests		Design Load Determined With Model Tests	
	$\sigma_{FH_s} = 0.05$	$\sigma_{FH_s} = 0.20$	$\sigma_{FH_s} = 0.05$	$\sigma_{FH_s} = 0.20$
P_f	γ_H	γ_H	γ_H	γ_H
0.01	—	—	2.1	2.3
0.05	2.7	—	1.7	1.9
0.10	2.0	2.5	1.4	1.6
0.20	1.6	1.7	1.3	1.4
0.40	1.2	1.2	1.1	1.2

(3) Foundation failure

For foundation failure, partial safety factors for deep and shallow water and for design with and without model tests are presented in Table 6.21. For the partial safety factor γ_H for wave height, the values are used for both sand and rubble-mound subsoil conditions. The values in parentheses are for clay subsoil.

The resistance partial safety factor γ_Z is for the effective friction angle of the material, and values shown are again for both sand and rubble-mound subsoil conditions. The values in parentheses are when clay is the subsoil beneath the structure. The clay partial safety factor γ_c is for the undrained shear strength of clay. A full discussion of foundation failure under clay subsoil conditions is beyond the scope of this book. See the CEM (2012, Section VI-5-5) for details.

The preferred foundation failure method to consider for design is the largest soil-bearing pressures p_e beneath the heel of the caisson.

6.3.5.4.3 Bearing capacity of sandy soil foundations

For locations where the seabed is a dense sand layer with good bearing capacity, Goda (2010) presents the following method to avoid the foundation failure of the caisson.

Define the vertical summation of forces W_e and combine the resisting and load moments M_e as in the following:

$$W_e = (W - B) - F_u, \tag{6.94}$$

Table 6.21. Partial safety factors for foundation failure of vertical wall caissons (CEM 2011, Tables VI-6-22 and VI-6-23). Values shown are for both sand and rubble-mound subsoil conditions. The values in parentheses are for clay subsoil.

(a) Deep-water conditions

	Design Without Model Tests						Design Load Determined With Model Tests					
	$\sigma_{FH_s} = 0.05$			$\sigma_{FH_s} = 0.20$			$\sigma_{FH_s} = 0.05$			$\sigma_{FH_s} = 0.20$		
P_f	γ_H	γ_Z	γ_C	γ_H	γ_Z	γ_C	γ_H	γ_Z	γ_C	γ_H	γ_Z	γ_C
0.01	1.4 (1.3)	1.3 (1.5)	(1.6)	1.4 (1.4)	1.3 (1.5)	(1.6)	1.3 (1.2)	1.2 (1.5)	(1.6)	1.4 (1.3)	1.2 (1.5)	(1.6)
0.05	1.3 (1.2)	1.2 (1.4)	(1.5)	1.3 (1.3)	1.2 (1.4)	(1.5)	1.3 (1.1)	1.1 (1.3)	(1.5)	1.4 (1.2)	1.1 (1.3)	(1.5)
0.10	1.2 (1.1)	1.2 (1.3)	(1.5)	1.2 (1.2)	1.2 (1.3)	(1.5)	1.2 (1.0)	1.1 (1.3)	(1.5)	1.3 (1.1)	1.1 (1.3)	(1.4)
0.20	1.1 (1.0)	1.1 (1.3)	(1.4)	1.1 (1.0)	1.2 (1.3)	(1.5)	1.1 (1.0)	1.1 (1.2)	(1.3)	1.1 (1.0)	1.1 (1.3)	(1.3)
0.40	1.1 (1.0)	1.0 (1.1)	(1.1)	1.1 (1.0)	1.0 (1.1)	(1.2)	1.1 (1.0)	1.0 (1.1)	(1.1)	1.1 (1.0)	1.0 (1.0)	(1.1)

(b) Shallow-water conditions

	Design Without Model Tests						Design Load Determined With Model Tests					
	$\sigma_{FH_s} = 0.05$			$\sigma_{FH_s} = 0.20$			$\sigma_{FH_s} = 0.05$			$\sigma_{FH_s} = 0.20$		
P_f	γ_H	γ_Z	γ_C	γ_H	γ_Z	γ_C	γ_H	γ_Z	γ_C	γ_H	γ_Z	γ_C
0.01	1.3 (1.2)	1.4 (1.5)	(1.6)	1.3 (1.3)	1.4 (1.5)	(1.6)	1.3 (1.2)	1.2 (1.3)	(1.4)	1.4 (1.3)	1.2 (1.3)	(1.4)
0.05	1.2 (1.1)	1.3 (1.4)	(1.5)	1.3 (1.2)	1.3 (1.4)	(1.5)	1.3 (1.1)	1.1 (1.2)	(1.4)	1.4 (1.2)	1.1 (1.2)	(1.4)
0.10	1.2 (1.1)	1.2 (1.3)	(1.3)	1.2 (1.2)	1.2 (1.3)	(1.3)	1.2 (1.1)	1.1 (1.2)	(1.3)	1.3 (1.1)	1.1 (1.2)	(1.3)
0.20	1.1 (1.0)	1.1 (1.3)	(1.3)	1.1 (1.1)	1.2 (1.2)	(1.3)	1.1 (1.0)	1.1 (1.1)	(1.3)	1.1 (1.1)	1.1 (1.1)	(1.2)
0.40	1.1 (1.0)	1.0 (1.1)	(1.1)	1.1 (1.1)	1.0 (1.1)	(1.1)	1.1 (1.0)	1.0 (1.0)	(1.0)	1.1 (1.1)	1.0 (1.0)	(1.0)

$$M_e = \{(W - B)t\} - \{F_u \left(\frac{2}{3}B\right) - F_H z\}$$

$$= \text{(resisting moment} - \text{load moment) about the heel,} \tag{6.95}$$

using the bold terms defined in Figure 6.49.

The location of the "average" momentum arm t_e is found from

$$t_e = \frac{M_e}{W_e}. \tag{6.96}$$

Assuming that a triangular or trapezoidal bearing pressure distribution exists beneath the caisson at the start of overturning, the largest soil-bearing pressure p_e at the heel depends on the location on the average moment arm, t_e. Recall that B is the width of the caisson.

$$\text{For } t_e \leq \frac{1}{3}B \quad p_e = \frac{2W_e}{3t_e}. \tag{6.97}$$

$$\text{For } t_e > \frac{1}{3}B \quad p_e = \frac{2W_e}{B}\left(2 - 3\frac{t_e}{B}\right). \tag{6.98}$$

To avoid the foundation failure, the largest soil-bearing pressures at the heel p_e should be kept below a critical level, $p_b = 400$–$500\,\text{kN/m}^2$ (8,000–10,000 lb/ft^2) (Goda, 1985). However, Goda (2010) states that recent deeper water constructions sites have permitted p_b as high as $600\,\text{kN/m}^2$ (12,500 lbs/ft^2).

6.3.5.4.4 Spreadsheets for design

Consider again the limit state equation G (Equation (4.6)) for a characteristic load S_{ch} with a partial safety factor for the load γ_H, and for the characteristic resistance R_{ch} with a partial safety factor for the resistance, γ_z.

$$G = \frac{1}{\gamma_Z}R_{ch} - \gamma_H S_{ch} = 0. \tag{4.6}$$

Stability against sliding requires

$$(F_G - F_u)f \geq F_H. \tag{6.99}$$

Stability against overturning requires

$$M_G \geq (M_H + M_u). \tag{6.100}$$

Foundation bearing pressure stability requires

$$p_e < p_b. \tag{6.101}$$

The characteristic loads (forces and moments), characteristic resistances (forces and moments), and characteristic bearing pressure all employ partial safety factors in their calculation. Therefore, it is not necessary to also employ overall safety factors against sliding, overturning, and bearing pressure as suggested by Goda (2010, p. 188). However, if the partial safety factors are taken as 1.0, then we again have a deterministic design.

What is the most important geometric design variable for a vertical caisson? Should we assume the crest elevation has been established to limit wave overtopping? (recall Sections 5.3 and 5.3.3). The simple answer is caisson width, B. The size and hence net weight in water F_G resists sliding and overturning and spreads the bearing pressure over the base. Therefore, a generic, spreadsheet for the preliminary design of caisson breakwaters has been developed that begins with the caisson width B as column A and covers 10 possible rows starting at width $B = 3\,\text{m}$ (10 ft) and ending at width $B = 30\,\text{m}$ (100 ft). Any intermediate design width may then be inserted or added as required.

An extremely wide spreadsheet has been developed that spans 41 columns and far too large to be displayed here. This spreadsheet for the preliminary design of caisson breakwaters has been provided in the online supplementary materials which accompanies this book, and found in the Seawall and Caisson folder. Figure 6.50 describes what is found in each column and what text equation was utilized in the spreadsheet solution. Application of this spreadsheet titled "Goda&PROVERBS Caisson Design" is best illustrated by an example.

6.3.5.4.5 Examples

(1) Deterministic design

Verification of the calculations performed in the spreadsheet were made by comparing the results with the Example Problem VI-7-16 as found in the Corps of Engineers, Coastal Engineering Manuel (CEM, 2006). The problem was to determine the width of a vertical walled caisson-type breakwater/jetty structure for the north jetty site at Barnegat Inlet, New Jersey, US. Table 6.22 presents the given information and additional assumptions made regarding a concrete/sand-filled caisson structure. This example used Imperial Units. The spreadsheet provided in the online supplementary materials includes both Imperial and Metric System units.

Figure 6.50 displays this example problem to scale with all the dimensions defined in Table 6.22. Note the relatively small freeboard distance h_c for the given storm surge, S. The 50-year recurrence interval is associated with this design storm surge as specified in (CEM, 2006, Example Problem VI-7-15). No information was presented on the recurrence interval for the

Table 6.22. Example Problem VI-7-16 as found in CEM (2006). See Figure 6.50 for variable symbols.

Variable	Value	Imperial Units	Description
H_{mo}	13.8	ft	Spectral significant wave height for design
T_p	14	s	Peak spectral period
β	0	$^\circ$	Wave direction for design
h	24	ft	Water depth near structure head (MLLW)
S	8.1	ft	Storm surge for design (MLLW) $T_R = 50$ years
T_r	50	years	Recurrence interval for design (water level)
h_{crest}	9.0	ft	Structure crest elevation (MLLW)
γ_w	64	lb/ft^3	Unit weight of seawater
γ_c	140	lb/ft^3	Unit weight of concrete
γ_s	125	lb/ft^3	Unit weight of wet sand
Assumptions			
h^*	20	ft	Caisson bottom depth (MLLW) ≈ 0.85 hour (rounded)
h_w	29	ft	Total caisson structure height $(20 + 9)$
t	2.0	ft	Caisson wall thickness
nt	4.0	ft	Caisson cap thickness $(n = 2)$
r	3.0	ft	Armor layer thickness $= 0.75$ $(h - h^*)$
B_m	6.0	ft	Armor layer berm width = 2r
Calculated Dimensions			
h_s	32.1	ft	Total water depth $= h + S$
h'	28.1	ft	Submerged structure depth $= h^* + S$
d	25.1	ft	Armor layer depth $= h' - r$
h_c	0.9	ft	Freeboard $= h_w - h'$
$h_f = h - h^*$	4.0	ft	Foundation thickness
Wave Height For Design			
H_{design}	24.8	ft	Goda design wave height $= 1.8H_s (H_{mo} = H_s)$

design significant wave height H_{mo} and it was assumed that $H_s = H_{mo}$. The design wave height $H_{\text{design}} = 1.8H_s$ as recommended by Goda. The peak wave period, T_p was 14 s for head-on waves.

Sliding failure controls the design width. Figure 6.51 describes the columns in the spreadsheet and presents the results of (CEM, 2006,

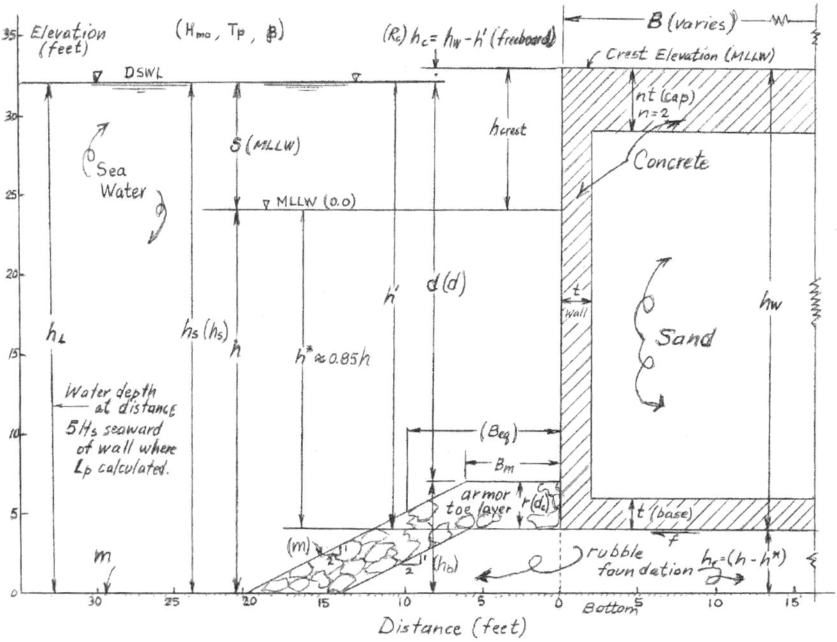

Figure 6.50. Schematic diagram of proposed concrete/sand-filled caisson structure with design conditions. Variables in parentheses are employed in the PROVERBS method for wave impact force calculations.

Example Problem VI-7-16) for $B = 40$ ft (12.2 m) and $B = 60$ ft (18.3 m). Using a caisson width $B = 40$ ft was not safe from sliding since the sliding ratio, SR was less than 1 (Column U, red). It was determined that a caisson width $B = 60$ ft was safe from both sliding and overturning (Columns U and V, yellow) with heel-bearing pressures well below the critical levels (Column AB, yellow). This agreed with the conclusions found in (CEM, 2006, Example Problem VI-7-16).

The above results were for a deterministic design in which no failure probability P_f was specified. It is instructive to repeat this example under probabilistic design conditions.

(2) Probabilistic design

The design method of Goda using $H_{\text{design}} = 1.8H_s$ does not permit any understanding of the probability of failure of the design results. We can employ the same spreadsheet by making the following changes for probabilistic design. This is found as a new tab (Stability Cassions — Goda

Stability of Vertical Walled Caissons- Goda Method
Spreadsheet Column Explanation and Example Solutions

			EXAMPLE PROBLEM VI-7-16 (CEM 2006)			SPREADSHEET SOLUTION		
Column	**Description**	**Text Equation/Remarks**	**Value**	**Imperial Units**	**Remarks**	**Value**	**Inperial Units**	**Remarks**
A	Caisson Width, B	Design Variable	40	ft	assumed	60	ft	
B	Weight in Air, **W**	Concrete box plus sand fill	149980	lbs/ft		224280	lbs/ft	
C	Buoyant Force, **B**	Submerged depth	71936	lbs/ft		107904	lbs/ft	
D	Net Wt in Water	**W-B**	78044	lbs/ft		116376	lbs/ft	
E	η*	6.74	37.3	ft		37.3	ft	
F	coefficient, α1	6.80	0.98			0.98		
G	coefficient, α2a	6.81	0.0764			0.0764		
H	coefficient, α2b	6.81	2.021			2.021		
I	coefficient, α2 (min)	6.81 Smallest of Column G or H	0.0764			0.0764		
J	coefficient, α3	6.82	0.913			0.913		
K	pressure, p1	6.75	1680	lbs/ft^2		1680	lbs/ft^2	
L	pressure, p2	6.76	1639	lbs/ft^2		1639	lbs/ft^2	
M	pressure, p3	6.77	1538	lbs/ft^2		1538	lbs/ft^2	
N	Uplift pressure, pu	6.78	1426	lbs/ft^2		1426	lbs/ft^2	
O	Horiz Force, FH	6.88	42031	lbs/ft		42031	lbs/ft	
P	Uplift Force, FU	6.89	21962	lbs/ft		32944	lbs/ft	
Q	Net Wt, FG	6.90	78044	lbs/ft		116376	lbs/ft	
R	Horiz Moment, MH	6.91	556625	ft-lb/ft		556625	ft-lb/ft	
S	Uplift Moment, MU	6.92	547634	ft-lb/ft		1232176	ft-lb/ft	
T	Buoyant Moment,MG	6.93	1560880	ft-lb/ft		3491280	ft-lb/ft	
U	Sliding Ratio, SR	6.99 (Load - Resistance)	0.801		Too narrow	**1.191**		okay
V	Overturnng Ratio, OR	6.100(Positive Mom - Negative Mom)	1.414		okay	**1.952**		okay
W	Heel Moment, Me	6.94	456621	ft-lb/ft		1702479	ft-lb/ft	
X	Net Vert Force, We	6.95	56082	ft-lb/ft		83432	ft-lb/ft	
Y	Heel Arm, te	6.96	8.14	ft		20.403	ft	
Z	Is te < (B/3)	yes or no	yes			No		
AA	Case A (Pe)A	6.101	4592		okay	NA		
AB	Case B (Pe)B	6.101	NA			**2725**		okay
AC	Width, B	repeat column A	40	ft		60	ft	
AF-AP	Partial Safety Factors	as needed	1		N.A.	1		

Figure 6.51. Description of columns in spreadsheet "Stability of Vertical Walled Caissons — Goda Method" and results of (CEM, 2006, Example Problem VI-7-16) for $B = 40$ ft and $B = 60$ ft.

Method) in the online supplementary materials and applies when sliding failure controls the design.

(1) $H_{\text{design}} = H_s$.
(2) In column K, in equation for p_1, use $\gamma_H H_{\text{design}}$ and the partial safety factors γ_H for sliding failure as found in Table 6.19.
(3) In column U, for sliding resistance, use f/ γ_Z and the partial safety factors γ_Z for sliding failure as found in Table 6.19.

For the example problem described above,

Case A (Conservative). Take $P_f = 0.01$, $\sigma'_{FHs} = 0.2$, $\gamma_H = 1.4$, and $\gamma_Z = 1.9$ for conditions in shallow water without model test (i.e. a desk top, preliminary design study using hind-cast wave information). The result is again that $B = 60$ ft. (18.3 m) and the same as the deterministic result above. We can therefore conclude that Goda's method using $H_{\text{design}} = 1.8H_s$ generally gives conservative results.

Case B (Less Conservative). Take $P_f = 0.05$, $\sigma'_{FHs} = 0.0.05$, $\gamma_H = 1.2$, and $\gamma_Z = 1.6$ for conditions in shallow water without model test using locally measured wave data and taking some additional risk (95% chance of no sliding failure). The result is now $B = 50$ ft (15.2 m) and a far less expensive design.

All the above must now also be checked for possible wave impact pressures and wave forces using the methods outlined above (Oumeraci *et al.*, 2001) in Figure 6.39 for PROVERBS.

6.3.5.5 *PROVERBS formulas for design*

6.3.5.5.1 Effects of caisson length, wave angle and short-crested waves

The results presented in Section 6.3.3 for wave impact loadings are only for one unit (m or ft) of structure length. Oumeraci *et al.* (2001) in PROVERBS discuss the 3D effects of wave attack on impact loadings on caissons. In general, wave impact loads are lower under oblique wave angles of attack, are less when averaged over typical caisson lengths of 10–20 m (33–65 ft), and are less for short-crested waves than for long-crested waves.

Figure 6.52 presents the results of Schüttrumpf *et al.* (2001) who derived a simple, regression line formula for the reduction factor of horizontal impact forces as follows:

$$C_{Fh} = 1 - k_b(L_c/L_{\text{op}}), \tag{6.102}$$

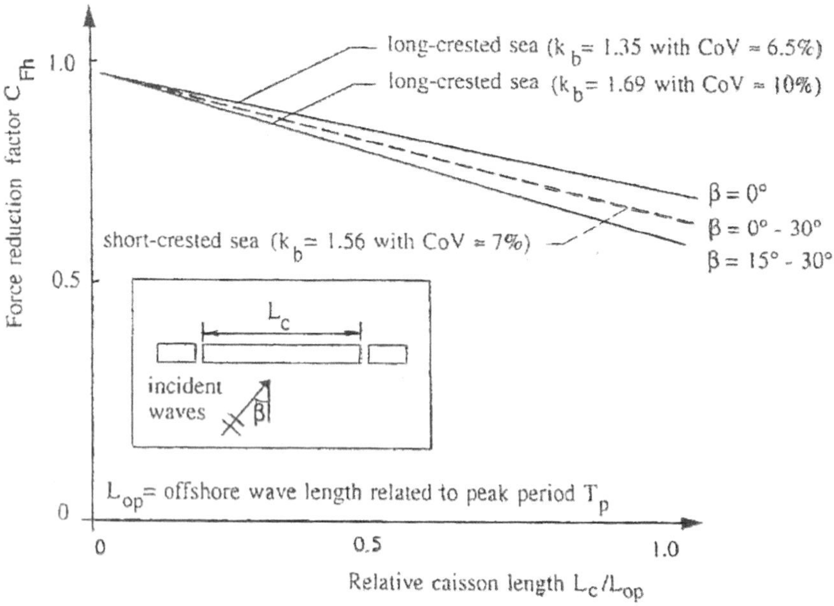

Figure 6.52. Effect of caisson length on impact forces for normal ($\beta = 0°$) and oblique, short-crested, and long-crested waves (from Schüttrumpf *et al.*, 2001).

Table 6.23. Impact force reduction coefficients k_b for various wave conditions (adapted from Oumeraci *et al.*, 2001).

Wave Condition	Coefficient, k_b	Coef. Variation, %	Correlation, r^2
Long-crested, $\beta = 0°$	1.35	6.6	0.82
Long-crested, β = 15–30°	1.69	9.2–10.4	0.77–0.79
Short-crested	1.56	7.0	0.77

where k_b is a reduction coefficient dependent on the wave conditions in Table 6.23, L_c is the length of the caisson, and L_{op} is the deep-water wave length from linear wave theory for peak wave period.

Therefore, the maximum wave impact force $F_{h\,\max}$ is further reduced by the factor C_{Fh} using Equation (6.102). Considerable further research can be found in the literature on the relationship between $F_{h\,\text{impact}}$ as determined per unit length of caisson and the length of the caisson structure L_c which is called the horizontal "coherence". See for example, Martinelli and Lamberti (2011) and Kim *et al.* (2004).

6.3.5.5.2 Horizontal impact force $F_{h\,\text{impact}}$

To apply the parametric decision map (Figure 6.39) to determine if wave impact loads are important, the following three key dimensionless variables are first determined as Step 5 in the method of analysis. Note that many variables are defined the same in the Goda method (Figure 6.49) and in PROVERBS (Figure 6.50) and where different, they are enclosed in parentheses in Figure 6.50 for PROVERBS.

The ratio berm height/water depth $h_b^* = 0.218$ and hence this Example Problem VI-7-16 is a "vertical breakwater". The wave parameter, $H_{si}/h_s = H_s^* = 0.430$ and clearly a large wave impacts the caisson wall. The given significant wave height, $H_s = 13.8\,\text{ft}$ (4.21 m) is employed in PROVERBS. Finally, the relative berm width $B^* = 0.018$ and is a very "narrow berm". As illustrated in the parametric decision map (Figure 6.39) wave impact loads are important and must be considered for design.

In Step 6, for this very low-crested wall ($R_c/H_{si} = 0.065$), use the lowest value for $\chi = 0.5$ giving little or no reflection so that $C^* = 0.33$ from Equation (6.57). The equivalent berm width $B_{\text{eq}} = 7.8\,\text{ft}$ so that using Equation (6.58) gives the empirical factor, $k_b = 0.96$. Finally, applying Equation (6.56) repeated here gives

$$\frac{H_b}{L_{pi}} = (0.1025 + 0.0217C^*)\tanh(2\pi k_b h_s/L_{pi}) = 0.0453 \tag{6.56}$$

using $L_{pi} = 439\,\text{ft}$ as found from linear wave theory for $T_p = 14\,\text{s}$.

The key result is that the breaking wave height, $H_b = 19.92\,\text{ft}$.

In Step 7, an estimate is made of the percentage of breaking waves using Equation (6.62) and gives $P_{b\%} = 1.5\%$ for the relative breaking height ratio, $H_b/H_{si} = 1.44$ (breaking waves give impacts, Equation (6.60)). Note that the estimated breaking wave height is larger than the incident wave height resulting from some wave reflection and from the fact that Equation (6.56) is from a "modified Miche theory" that has resulted from the research conducted in PROVERBS.

Estimates for the dimensionless, maximum wave force at impact F_h^* are made in Step 8 using the GEV distribution for the probability distribution, $P(F_h^*)$. Using the recommended values for $\beta = 7.86$, $\alpha = 3.97$, and $\gamma = -0.32$ and taking $P(F_h^*) = 0.9$ in Equation (6.66) gives, $F_h^* = 20.9$.

Using Equation (6.65) repeated here gives for an estimated, maximum impact force, $F_{h\,\max}$,

$$F_h^* = \frac{F_h}{\rho g H_b^2} = 20.9$$

so that $F_{h\,\max} = 532,016\,\text{lbs/ft}$. Note that this is for a full vertical wall crest height above the elevation of the pressure distribution η^* and must be reduced to account for the wave overtopping.

To calculate the triangular pressure prism missing above the crest elevation of the wall as shown in Figure 6.49, estimates of the trapezoidal pressure prism must be made first. This gave $\eta^* = 15.94\,\text{ft}$, $p_1 = 19{,}249\,\text{lbs/ft}$, and $p_4 = 18{,}162\,\text{lbs/ft}$ using Equations (6.72), (6.73), and (6.75), respectively. Then using Equation (6.77) repeated here gave

$$F_{h\,\mathrm{maxov}} = F_{h\,\max} - 0.5p_4(\eta^* - R_c) = 395{,}460\,\text{lbs/ft},$$

which is a 25.7% reduction resulting from the relatively low freeboard in this example.

Finally, in Step 11, aeration and scale corrections are applied. Recall the estimate for percentage breaking is $P_{b\%} = 1.5\%$ so that using Equation (6.69b) gives for the percentage of aeration, $P_a = 3.10\%$. The force reduction factor, $k_{fs} = 0.822$ from Equation (6.70). The reduced wave impact force due to aeration and scale effects, $F_{h\,\mathrm{impact}}$ is then from Equation (6.71) repeated here,

$$F_{h\,\mathrm{impact}} = k_{fa} F_{h\,\mathrm{maxov}} = 0.822(395{,}460) = 325{,}114\,\text{lbs/ft}, \qquad (6.71)$$

which results in a 17.8% reduction from the overtopping result. However, this lowered wave impact force found from PROVERBS is still about 7.7 times larger than that calculated by the Goda method.

The trapezoidal pressure prism pressures now become $p_1 = 11{,}763\,\text{lbs/ft}$, $p_3 = 5{,}293\,\text{lbs/ft}$, and $p_4 = 11,099\,\text{lbs/ft}$ and should be employed to calculate the location of $F_{h\,\mathrm{impact}}$ using Equation (6.76). These pressures and horizontal impact force will be used in point(3) to calculate sliding and overturning stability and heel-bearing pressures.

Breaking wave impact force rise time using Equation (6.67) is extremely short ($t_r = 0.004\,\text{s}$) and the impact duration t_d was only 0.77 s.

6.3.5.5.3 Uplift pressures and uplift force

As discussed above, the uplift pressure on impact at the landward bottom edge (heel) of the caisson p_{ru} was not found to be hydrostatic in

PROVERBS. Using Equation (6.83) gave $p_{ru} = 664\,\text{lbs/ft}^2$. The GVE was again employed to find the three parameters for the probability distribution $P(F_u^*)$ using the measured uplift force data. The recommended values for $\beta = 4.384$, $\alpha = 2.17$, and $\gamma = -0.11$ with $P(F_u^*) = 0.9$ gave $F_u^* = 9.92$. This resulted in $F_{u\,\max} = 252{,}102\,\text{lbs/ft}$. Equation (6.86) was then employed to find the maximum uplift force, $p_{u\,\max}$ at the seaward bottom end of the caisson and depended on the caisson width, B. Equation (6.87) determined the location of $F_{u\,\max}$.

6.3.5.5.4 Caisson sliding, overturning, and bearing pressure

The spreadsheet labeled "Goda & PROVERBS Caisson Design" as found in the supplementary materials accompanying this book includes separate sheets for Goda and PROVERBS with and without partial safety factors and for Imperial and Metric systems of units. It was employed for this Example Problem VI-7-16 and it was determined that the caisson width $B = 60\,\text{ft}$ was insufficient to prevent sliding failure. In fact, only when the width was set very wide ($B = 300\,\text{ft}$) was sliding failure prevented, which is not practical for design. Overturning stability was obtained when $B = 65\,\text{ft}$ but this produced excessively high bearing pressures under the heel.

These results using $F_{h\,\text{impact}} = 325{,}114\,\text{lbs/ft}$ should be further reduced using the force reduction factor C_{Fh} for the caisson length, possible wave approach angle, and wave short-crestedness using Equation (6.102) and Table 6.23. But this Example Problem VI-7-16 does not give any further information in this regard. This would further reduce the ratio $F_{h\,\text{impact}}/F_{\text{Goda}}$. Oumeraci *et al.*, (2001) in PROVERBS give the following suggestions regarding the force reduction factor C_{Fh} that is often related to the "horizontal coherence length", defined as $L_{\text{op}}C_{Fh}$, for

(a) Light Wave Force Impacts ($F_{h\,\text{impact}}/F_{\text{Goda}} < 2$) take $C_{Fh} \leq 0.25$;
(b) Heavy Wave Force Impacts ($F_{himpact}/F_{Goda} \gg 2.5$) take $C_{Fh} \leq 0.0625$.

Typical coherence lengths of caissons are $L_{\text{op}}C_{Fh} \approx 0.125$ which for a peak wave period $T_p = 10\,\text{s}$ give a wave length $L_{\text{op}} = 512\,\text{ft}$ $(156\,\text{m})$ and hence caisson length $L_c = 64\,\text{ft}$ $(19.5\,\text{m})$.

Considerable further research can be found in the literature on the relationship between $F_{h\,\text{impact}}$ as determined per unit length of caisson, the length of the caisson structure relative to the deep-water wave length, and the wave angle/wave crestedness since the published work of Oumeraci *et al.* (2001) for PROVERBS.

Most importantly, these deterministic results in Example Problem VI-7-16 depend on the chosen significant wave height, $H_s = 13.8\,\text{ft}$ (4.21 m). For calculations of F_H and F_u, the Corps of Engineers, *Coastal Engineering Manual* (CEM, 2011) recommends using wave height $\gamma_H H_s^T$ where H_s^T is the significant wave height at recurrence interval T. No information regarding the recurrence interval T_r for this chosen wave height was initially given in Example Problem VI-7-16.

6.3.5.5.5 Partial safety factors — probabilistic design

Partial safety factors for vertical breakwaters have been calibrated on a probabilistic basis in PROVERBS (Oumeraci *et al.*, 2001, Chapter 5) and are summarized in Table 6.24 for partial safety factors γ_H for wave height H_s. However, they are presented in terms of the design life recurrence interval T_r (years) and not in terms of the desired failure probability, P_f. The values shown for the uncertainty related to wave height, σ_{FH_s} are the estimates of the author of this book.

Detailed discussion of the partial safety factors γ_m for geotechnical parameters is also found in Chapter 5 and is summarized in Table 6.25. But here, the values are for four levels of the Safety Class (very high, high, normal, and low) for P_f, respectively. See Oumeraci *et al.* (2001, Section 3.7.2) for the limit state equations that utilize the angle of internal friction of the foundation rubble material, φ_r, and the angle of internal friction of the subsoil material, φ_s. These limit state equations are for bearing capacity

Table 6.24. Partial safety factors γ_H for wave height H_s in calculations of wave-impact forces on vertical wall caissons using PROVERBS (modified from Oumeraci *et al.*, 2001, Chapter 5).

	Design Without Model Tests		Design With Model Tests	
	$\sigma_{FH_s} = 0.05$	$\sigma_{FH_s} = 0.20$	$\sigma_{FH_s} = 0.05$	$\sigma_{FH_s} = 0.20$
T_R, years	γ_H	γ_H	γ_H	γ_H
100	1.2	1.2	1.0	1.0
50	1.1	1.1	0.9	0.9
20	1.1	1.0	0.9	0.9

Table 6.25. Partial safety factors γ_m for geotechnical parameters in calculations of wave impact forces on vertical wall caissons using PROVERBS (modified from Oumeraci *et al.*, 2001, Chapter 5).

	Sliding Failure, Friction Factor f		Limit State Equations	
	Drained Subsoil	Undrained Subsoil	Drained Subsoil	Undrained Subsoil
P_f	γ_m	γ_m	γ_m	γ_m
0.01	1.5	1.6	1.7	1.8
0.05	1.3	1.4	1.4	1.5
0.1	1.2	1.3	1.3	1.4
0.2	1.1	1.2	1.2	1.3

failure due to rotation of the caisson and replace the general bearing pressure limits as suggested by Goda (2010, p. 188).

No information regarding the geotechnical parameters were given in Example Problem VI-7-16 so the effects of wave impacts using these partial safety factors could not be made.

In general, many water levels, wave conditions, and potential caisson locations must be considered. The breakwater structure will encounter both pulsating and impact wave loads during its design life. The theoretical, non-exceedance, probability distribution function, $P(F < f)$ for all the wave forces is

$$P(F < f) = P_{b\%}P(F < f_{\text{impact}}) + (1 - P_{b\%})P(F < f_{\text{no impact}}), \qquad (6.72)$$

where $P_{b\%}$ is the probability of breaking wave impacts, $P(F < f_{\text{impact}})$ is the probability of wave impact forces, and $P(F < f_{\text{no impact}})$ is the probability of no wave impact forces (pulsating wave conditions).

For PROVERBS, the models used for wave forces are Goda (1985) for pulsating (quasi-static) waves and the impact wave force models described above. The complexity of both the water level and wave conditions, the two types of wave loading model, and the need for the parametric decision map (Figure 6.39) have led to the use of Monte Carlo simulation methods for reliability analysis of caissons where wave impact loads may be present. Many Excel spreadsheet algorithms are now available for this purpose but are not included in this book.

6.3.5.6 *Methods to reduce wave impact forces*

Application of impact loading leads to an increase in the probability of failure. Various methods to reduce wave impact forces have been suggested in the literature and also proven in the field.

The parametric decision map (Figure 6.39) should be reviewed to learn what could be changed to reduce the wave impact forces. For example, the "High-Mound"-type places the concrete caisson on a rock base to increase the berm height h_b and a wide equivalent berm width, B_{eq} (Figure 6.46) both help to decrease the breaking wave heights H_b and the amount of wave breaking that produce impact forces. The seaward face of the caisson may include small holes, and the first inner compartment on the left filled only with seawater to absorb wave energy.

The base of the caisson may be cast on a corrugated formwork to give a rough underside to increase the resistance to sliding. See Hutchinson *et al.* (2010) for a good example of caisson design for sliding. Their field investigation with an effective serrated depth of 90 mm ($\approx$0.3 ft) produced a sliding friction coefficient, $f = 0.78$. Working Group 28 of the World Association for Waterborne Transport Infrastructure (PIANC, PTC II, 2003) recommends making the depth of the underside serrations roughly the same as the average stone size used for the rubble bed.

Shear keys along the bottom (aka "skirts") to reduce sliding and connecting caissons to each other to spread wave impact loads horizontally are also ways to mitigate possible failure from impact forces.

6.3.5.7 *Summary recommendations for caisson design*

(1) Probabilistic design for both pulsating and impact wave loads

Use partial safety factors at the five levels of probability of failure P_f for a probabilistic design to prevent sliding, rotation, and bearing pressure failure. For the design wave height H_{design}, use the significant wave height at designated levels of recurrence interval H_s^T for pulsating (oscillating) waves and the formulas of Goda (1985) for wave and uplift forces. Use the parametric decision map (Figure 6.39) and the methods of PROVERBS (Oumeraci *et al.*, 2001) for wave impact and uplift forces (when warranted) that also requires use of H_s^T. Acquire geotechnical data for subsoil and foundation rubble material to obtain partial safety factors γ_m for geotechnical parameters needed for design. Consider alterations in the design geometry to reduce wave impact loads and reduce the probability of failure.

(2) Forces in seaward direction

Remember that the above wave and uplift force formulations all assume fully hydrostatic water-level pressure distributions (Figure 6.36) on *both* sides of the caisson structure. When freeboards are small, wave overtopping may produce water levels on the landward side of the caisson that are different than on the seaward side. Forces in the seaward direction on vertical-walled caissons are discussed in detail in Oumeraci *et al.* (2001, Section 2.5.2).

(3) Fault tree for vertical breakwaters

The main purpose of a vertical breakwater is to provide relatively calm wave conditions for entrance, maneuvering, and berthing of ships in harbors. Figure 6.53 presents a fault tree where both the functional aspects as discussed in Chapter 5 (wave overtopping and transmission) and structural design are shown. Structural failure is also considered in Oumeraci *et al.* (2001) and Chapter 4 which present many details. Chapter 5 includes partial safety factors for concrete strength and steel reinforcement for concrete caisson design.

6.3.6 *Bulkheads*

6.3.6.1 *Introduction*

A coastal bulkhead is a vertical walled structure that divides the land from the sea. Its main function is to prevent the soil behind the wall from sliding into the water in front of the structure. Figure 6.54 schematically shows the two most common types of sheet-pile bulkhead (b) cantilevered and (c) anchored. It also shows a less common type (a) rigid retaining wall that could also be labeled a seawall.

All three types are retaining wall structures that keep the earth material from moving laterally. The soil behind the vertical wall exerts an "active" lateral soil pressure (force) on the wall. The "cantilevered" bulkhead needs an imbedded length of wall (sheets) beneath the ground level, aka "mud line" or "dredge line" in front of the wall to resist the soil pressure force on the wall. The "anchored" bulkhead also employs tie-back rods to help restrain the wall from the soil pressure force on the wall. Both the

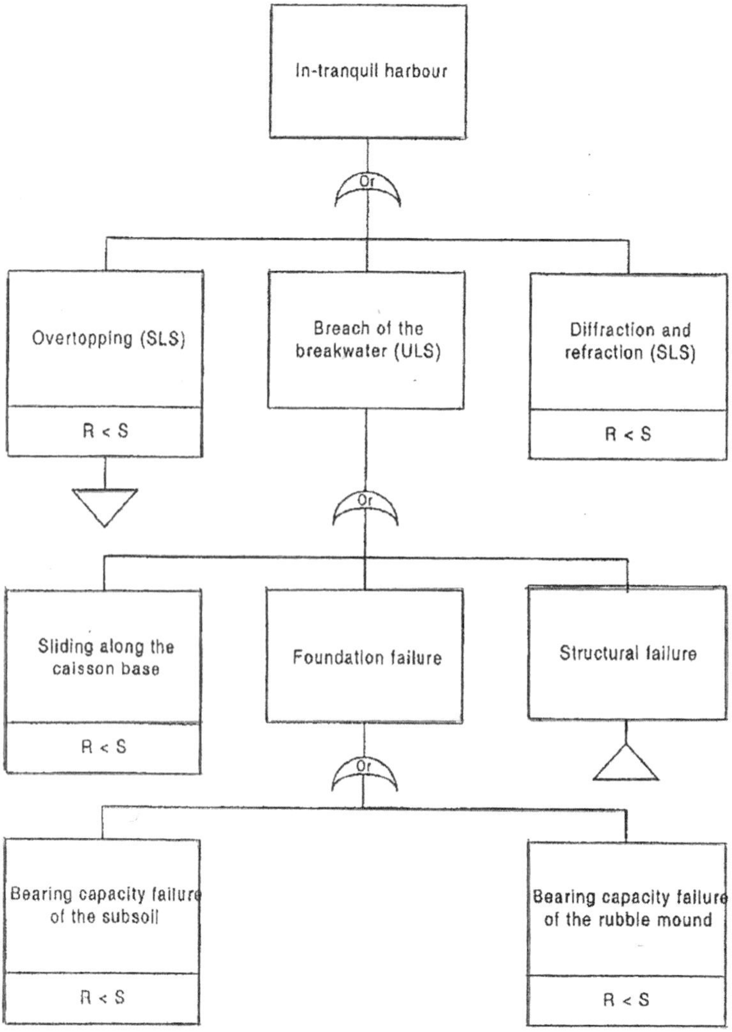

Figure 6.53. Fault tree for vertical, caisson-type breakwater structures (from Oumeraci *et al.*, 2001).

cantilevered and anchored type bulkheads are the least expensive vertical wall structures to construct. In contrast, the rigid retaining wall requires its size and weight to restrain the lateral soil pressure force and is generally far more expensive to construct. Bulkhead structures are called embedded retaining walls in the United Kingdom.

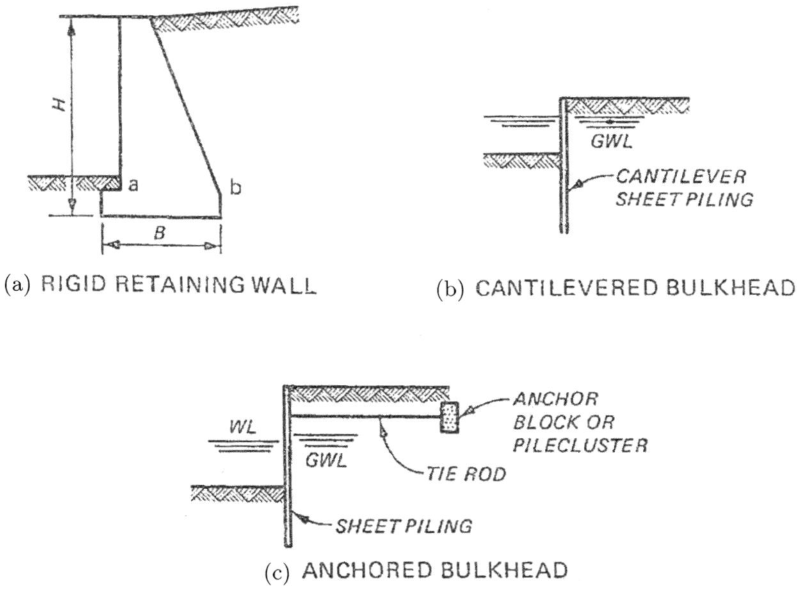

Figure 6.54. Lateral retaining wall structures (from Eckert and Callender, 1987).

The vertical-walled bulkhead can be constructed of many different types of materials as illustrated in Figure 6.55. Steel sheets are the most common because those can be readily driven vertically downward by standard pile-driving equipment through both sandy and clay soils. Concrete and steel piles can be driven at various spacings with both concrete slabs and wood planks (railroad ties, logs, etc.) laid horizontally between the piles to provide the bulkhead wall.

Bulkhead design is primarily the responsibility of geotechnical engineers. The soil mechanics parameters for both the original soil conditions and possible backfilled soil behind the wall are needed to determine the lateral soil pressures and forces against the wall. The pressure diagrams are then used to determine the shear force and bending moment diagrams using structural engineering principles. Then materials for construction with strength properties for a stable structural design are selected. An example for the proper engineering design of an anchored bulkhead is presented in Figure 6.55.

Unfortunately, too often geotechnical and structural engineers are not consulted for design. General contractors with experience in bulkhead

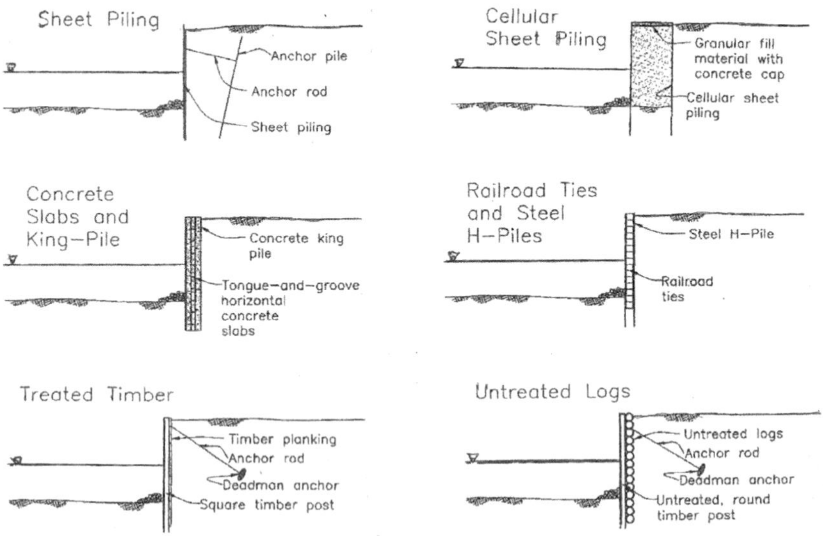

Figure 6.55. Common materials used to construct bulkheads of various types (from CEM, 2006).

construction are hired by home owners with waterfront property that is being eroded. Figure 6.56 humorously depicts disagreements among the adjacent property owners that typically results (figure courtesy of Kevin Bodge, coastal engineer, Olsen Associates, Jacksonville, FL). The primary concern of the waterfront property owner is the initial construction expense. Annual maintenance costs and costs to repair damage from coastal storms bringing high water levels and wave conditions on the seaward side are rarely considered.

Coastal engineers should also be involved as consultants in bulkhead design. This section of this book is intended to simply introduce bulkhead design as practiced by geotechnical and structural engineers to the coastal engineering community. Civil engineers should have sufficient undergraduate background in geotechnics and structures for this introduction. Those areas of coastal bulkhead design that are most susceptible to failure because of lack of attention to potential, damaging high-water events bringing wave overtopping and wave impact loads to the bulkhead are then addressed. Unfortunately, most failures of coastal bulkheads are due to

Figure 6.56. Property owners disagree on contractor "design" of a bulkhead (courtesy, Kevin Bodge).

coastal hydrodynamic forces on the seaward side and not to geotechnical forces on the landward side.

6.3.6.2 *Active and passive soil pressures — Geotechnical*

(1) Soil properties

We limit the following introduction to non-cohesive soil conditions for both the existing soil sublayer and the sand backfill. When the existing soil sub-layer(s) include cohesive (clay) soils, the analysis is more complex. We define the following variables for the non-cohesive soil conditions as:

Φ is the angle of internal friction (shear strength) in degrees (25°–45°), δ is the wall friction angle in degrees, γ_d is the unit weight of dry soil (lbs/ft^3) or (N/m^3), γ_{SAT} is the unit weight of wet (saturated) soil (pore spaces filled with water), and γ_{SUB} is the unit weight of submerged soil (below water table, buoyance effect) $= \gamma_{SAT} - \gamma_W$ (γ_W = unit weight of seawater or fresh water).

(2) Active and passive soil pressures

For the mass of soil adjacent to the wall, several theories are available for estimating the minimum (active) soil pressures exerted on the wall and the maximum (passive) pressures exerted by the wall on the soil. The most common are the Coulomb Theory and the Rankine Theory (see geotechnical engineering text). For a horizontal ground level for the backfill and no wall friction ($\delta = 0°$), both give the same results for the active pressure coefficient K_A which simply depends on the angle of internal friction, ϕ.

$$K_A = \tan^2\left(45° - \frac{\phi}{2}\right) \tag{6.103}$$

and used to find the active soil pressure P_A from the simple relation

$$P_A = K_A(\gamma_d \text{ or } \gamma_{SUB})Z, \tag{6.104}$$

where Z = the vertical distance below the surface, and use γ_d above the water table and γ_{SUB} below the water-table elevation.

For increasing vertical distance Z below the surface, the active soil pressures become additive and dependent on the active soil pressures above the level Z of interest.

The active soil pressure is zero at the ground-level surface where $Z = 0$. However, some surface surcharge q is usually present (vehicles, people, structures, etc.) on the ground at the surface to increase the active soil pressures beneath the surface and must be accounted for in the calculation.

Similarly, both theories give a similar result for the passive pressure coefficient K_P on a horizontal ground level surface, namely,

$$K_P = \tan^2\left(45° + \frac{\phi}{2}\right) \tag{6.105}$$

and used to find the passive soil (resistance) pressure

$$P_P = K_P(\gamma_{SUB})Z \tag{6.106}$$

and usually below the water table.

All the above are clearly explained in the following example.

(3) Active and passive forces

The active pressure coefficient K_A and passive pressure coefficient K_P depend on the rotation of the wall (assumed rigid) and are related to more

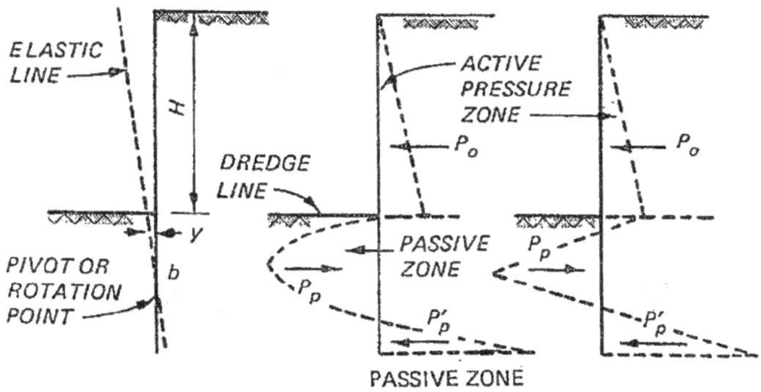

(a) Assumed elastic line of the sheetpiling (b) Probable soil-pressure distribution obtained in finite element solution (c) Simplified pressure diagram for computational purposes granular

(*after Bowles 1977*)

Figure 6.57. Simplified, straightline pressure prisms of active and passive pressures resulting from a cantilevered bulkhead wall rotation (after Bowles, 1977).

fundamental soil properties and effective stresses within the soil and is beyond the scope of this book. Figure 6.57 illustrates how (a) a rigid wall rotates about a pivot point, (b) the active pressure prism (Equation (6.74)) producing the rotation above the pivot point, and the (c) simplified passive pressure prism (Equation (6.76)) for computations of a cantilevered bulkhead.

The pressure prisms for computation are more complex for an anchored bulkhead when surcharge at the surface, the water table, and anchor rod force T at location A are included as schematically illustrated in Figure 6.58. But the pressure prisms are all triangular or trapezoidal in shape. Once the pressure prisms have been established using Equations (6.103)–(6.106), then the active pressure forces and their locations on the wall are readily obtained by simple arithmetic.

Simple engineering statics analysis methods are then applied to determine the penetration depth D for static equilibrium of the anchored bulkhead. The moments for the pressure prism forces are summed about the tie-back rod location since the tie-back rod force is unknown. The only unknown force and its location is the passive (resistance) force and its magnitude and hence moment about the tie-back location depend on its penetration distance D. A trial-and-error solution will determine the penetration distance D so that the summation of the positive (active) moments equals

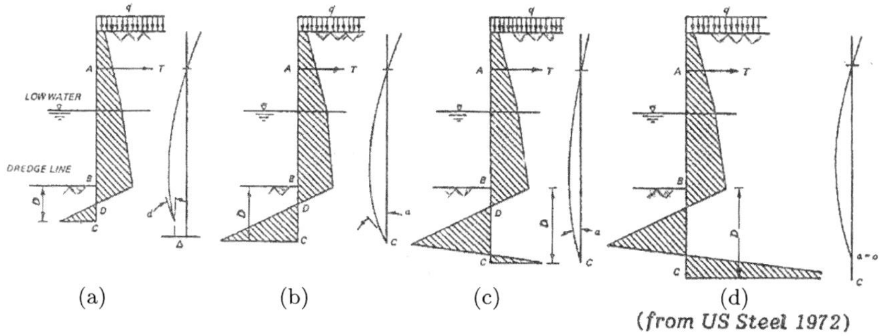

Figure 6.58. Active and passive soil pressure prisms for an anchored bulkhead with surface surcharge, water table effects, and tie-back rod included. Diagrams and curves representing bending moment diagrams for increasing penetration depths D of the wall beneath the dredge line from left to right (from US Steel Design Manual, 1972).

the negative (passive) moments about the tie-back location as illustrated in Figure 6.58(b). Extending the penetration depth D beyond this length again adds more active pressure to the wall as illustrated in Figures 6.58(c) and 6.58(d) and reduces the maximum bending moment. Making the sheet pile longer adds a safety factor to the design. This design method is called the free-earth support method.

6.3.6.3 *Shear and bending moments — Structural design*

All the active and passive pressure prisms are then used to calculate the force, shear force, and bending moment diagrams that they produce on the wall. The pressure distributions of the soil on the wall (active) and wall on the soil (passive) produce force distributions that are similar to a non-uniform distributed load applied to a beam fixed at one end and free to move at the other end as illustrated in Figure 6.59.

The slope of the shear force $V(x)$ at any location x is equal to the negative value of the load $w(x)$ at location x, as follows:

$$\frac{dV(x)}{dx} = -w(x) \tag{6.107}$$

so that the shear force is

$$V(x) = \int -w(x)dx \tag{6.108}$$

to give the shear force diagram.

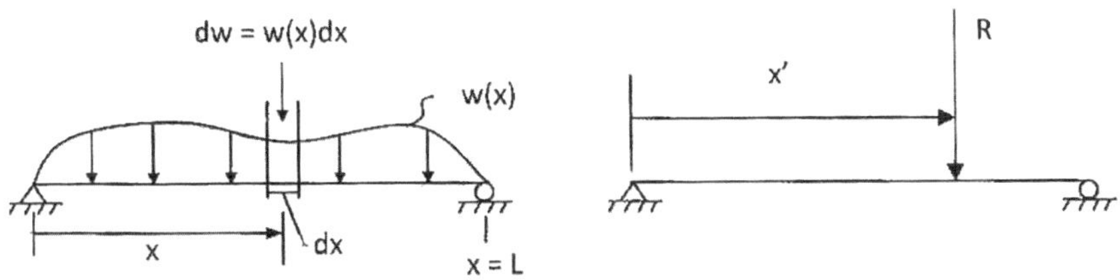

$$R = \int_0^L dw = \int_0^L w(x)dx \qquad x'R = \int_0^L x dw = \int_0^L x w(x)dx$$

Figure 6.59. Simple beam with non-uniform distributed load.

The slope of the bending moment $M(x)$ is equal to the value of the shear force $V(x)$ at location x,

$$\frac{dM(x)}{dx} = V(x) \tag{6.109}$$

so that the bending moment is

$$M(x) = \int V(x)dx = \int -w(x)xdx \tag{6.110}$$

to give the bending moment diagram.

The maximum bending moment is used to specify the type, size, and strength of the material used to construct the sheets of the bulkhead wall.

Once the penetration depth D is determined, all the forces (active and passive) are known and the summation of the horizontal forces is used to determine the force in the tie-back rod, T. The material, size, and strength of the tie-back rod can then be specified for construction.

6.3.6.4 *Example calculation*

(1) Given information

An eroding coastal site with "silty-sand" soil conditions below the "Dredge Line" has the need for a bulkhead to increase the surface ground elevation 32 ft (9.8 m) for industrial development. Normal water levels are 22 ft (6.7 m) in the region above the bottom. Once the bulkhead is in place, it will be backfilled with "coarse sand" to the top of the bulkhead. Figure 6.60 presents these dimensions including the need for a surcharge, q of 400 lbs/ft^3 (62,835 N/m^3) at the surface for expected equipment operating near the bulkhead. An anchored, sheet-pile bulkhead is proposed with tie-back anchor located 8 ft (2.4 m) below the surface. Once the sand backfill

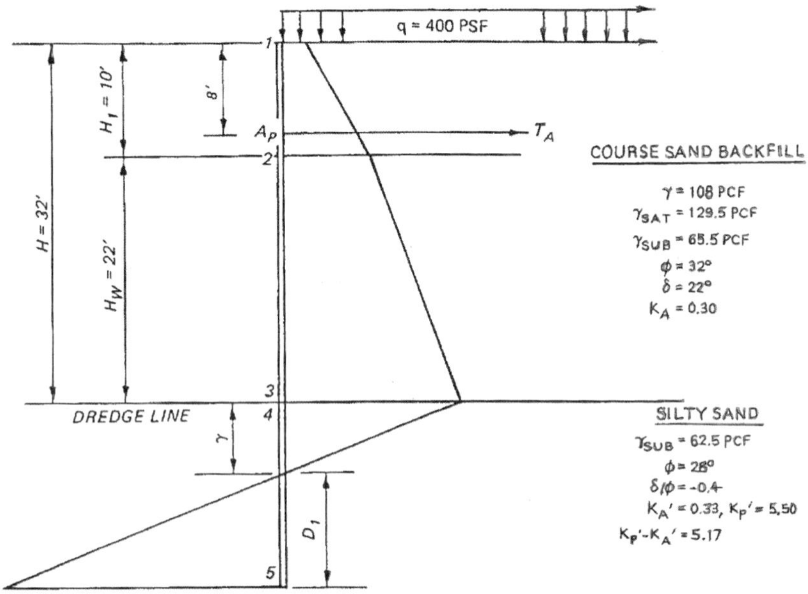

Figure 6.60. Anchored bulkhead dimensions and soil variables used for example calculation to determine the unknown distance, D_1, and force in the tie-back rod T_A for a stable, sheet-pile design.

is in place, the water table elevation beneath the surface is assumed to be the same as the present water level at the site.

Figure 6.60 also gives the soil conditions for this example. The value shown for $K_A = 0.30$ in the coarse sand backfill follows from Equation (6.103) for $\phi = 32°$. But the values for the natural, silty sand at the site for K'_A and K'_P do not follow from these equations and are the result of the mixture of sand and silt in the soil. This illustrates the need to always collect core samples of the soil at the site and have them analyzed by geotechnical engineering firms.

An anchored, sheet-pile design to determine the length L of the sheets and force in the tie-back rod T_A is required for a stable, safe design. The length of a cantilevered (no anchor) bulkhead sheet is also needed for comparison.

(2) Calculation of pressures, forces, and locations

All the results will be in Imperial units. All results are per unit length (ft) of the wall. For the metric system, $1\,\text{lb/ft} = 14.59\,\text{N/m}$, $1\,\text{lb/ft}^2 = 47.88\,\text{N/m}^2$, and $1\,\text{lb/ft}^3 = 157.09\,\text{N/m}^3$.

Soil pressures are needed at five locations along the wall to define the pressure prisms in Figure 6.60. At the surface, the active soil pressure p_1 is due to the surcharge load q.

$$p_1 = K_A \times q = 0.30(400) = 120\,\text{lbs/ft}^2.$$

At $H_1 = 10$ ft below the ground surface at the water table,

$$p_2 = p_1 + K_A \gamma_d H_1 = 120 + (0.30 \times 108 \times 10) = 120 + 324 = 444\,\text{lbs/ft}^2.$$

At $H_2 = 22$ ft below the water table and at the bottom of the coarse sand backfill,

$$p_3 = p_2 + K_A \gamma_{\text{SUB}} H_2 = 444 + (0.30 \times 65.5 \times 22) = 444 + 432 = 876\,\text{lbs/ft}^2.$$

At $H_2 = 22$ ft and at the top of the silty-sand, native material, and still submerged,

$$p_4 = K'_A q + K'_A \gamma_d H_1 + K'_A \gamma_{\text{SUB}} H_2 = 0.33(400) + (0.33 \times 108 \times 10) + (0.33 \times 65.5 \times 22) = 965\,\text{lbs/ft}^2.$$

At distance "y" where $p_y = 0$ ($p_{\text{active}} = p_{\text{passive}}$),

$p_y = 0 = 965 + K'_A \gamma'_{\text{SUB}} y - K'_p \gamma'_{\text{SUB}} y = 965 + (0.33 \times 62.5 \times y) - (5.50 \times 62.5 \times y)$, so that

$y = 3.0$ ft below the Dredge Line where $p = 0$.

Finally, at distance D_1 to the bottom of the sheet pile,

$$p_5 = (p_{\text{active}} - p_{\text{passive}}) = (0 + K'_A \gamma'_{\text{SUB}} D_1) - (0 + K'_p \gamma'_{\text{SUB}} D_1) = (0.33 \times 62.5 \times D_1) - (5.50 \times 62.5 \times D_1),$$

$$p_5 = 323 D_1.$$

Using these pressures at the five locations and at distance "y" where the pressure is zero, the resulting pressure prisms were employed to calculate the resulting active soil forces on the wall (forces, f_1, f_2, and f_3) and the passive wall force on the soil f_4 as shown in Figure 6.61 and summarized in Table 6.26. The locations of these forces within their pressure prisms y_n and then relative to the tie-back location l_n are also summarized in Table 6.26. We only show one example here and the rest are similar to calculate. For example,

$$\begin{aligned} f_1 &= \text{trapezoidal prism} \\ &= (\text{rectangle} + \text{triangle}) = (10 \times 120) + (0.5 \times 10 \times 324) \\ &= 1{,}200 + 1{,}620 = 2{,}820\,\text{lbs}. \end{aligned}$$

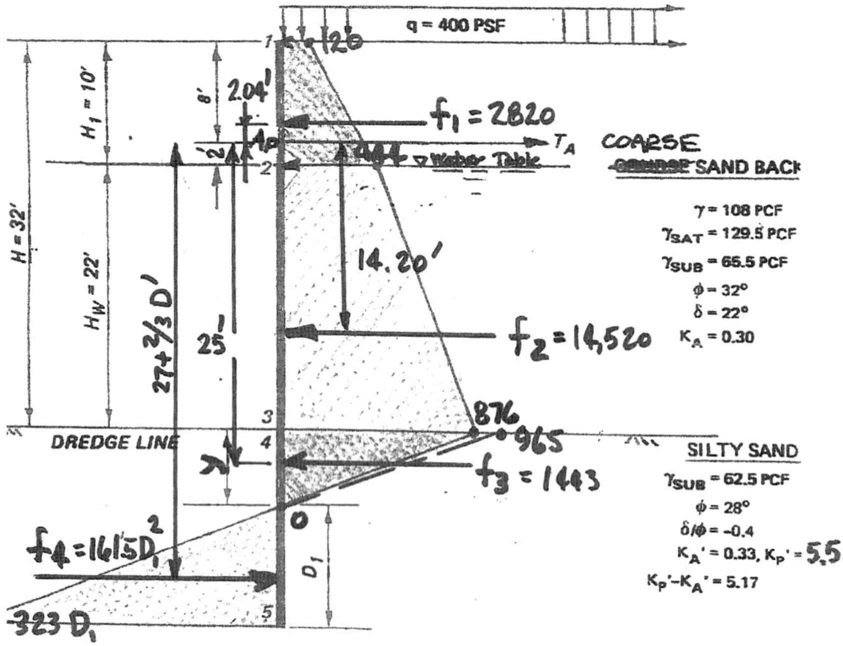

Figure 6.61. Forces and locations for four pressure prisms in the example.

Table 6.26. Forces and locations of example.

Prism, n	Force, f_n lbs	Prism Location, y_n ft	Moment Arm, l_n ft	Moment, M_n ft-lbs CW(+)
1	2,820	4.04	2.04	−5,753
2	14,520	9.80	14.20	+206,184
3	1,433	1.0	25.0	+36,075
4	161.5 D_1^2	$(2/3)D_1$	$27 + (2/3)D_1$	$-161.5D_1^2 \times (27 + (2/3)D_1)$

$$y_1 = (\text{sum of moments about point 2})/f_1 = [(1200 \times 5)/2820] + [(1620 \times 3.33)/2820] = 4.04\,\text{ft and therefore location of } f_1 \text{ is above the tie-back location so that}$$

$$l_1 = 4.04 - 2.0 = 2.04\,\text{ft}.$$

The values for y_n and l_n in Table 6.26 are found in a similar fashion.

(3) Calculation of moments and distance D_1 and tie-back force, T_A

The force f_n times its moment arm l_n gives the moments M_n about the tie-back position located 8 ft below the surface at A_p. The results are also given in Table 6.26. Clockwise direction moments (CW+) are taken as positive for forces, f_2 and f_3. The summation of the moments about A_p must equal zero for static equilibrium. The moment for prism 4 produces a cubic equation for the unknown distance D_1 of the form

$$0 = -5{,}753 + 206{,}184 + 36{,}075 - 107.67D_1^3 - 4{,}360.5D_1^2.$$

By trial and error, $D_1 = 6.812$ ft. say 6.8 ft.

We now see the $p_5 = 323D_1 = 2{,}200\,\text{lbs/ft}^2$ and that $f_4 = 161.5\,D_1^2 = 161.5 \times (6.8)^2 = 7{,}468\,\text{lbs}$.

Summing the horizontal forces acting on the wall (active force, positive) gives

$$f_1 + f_2 + f_3 - f_4 - T_A = 2{,}820 + 14{,}520 + 1{,}443 - 7{,}468 - T_A, \quad \text{so that}$$

$$T_A = 11{,}315\,\text{lbs/ft}.$$

Spreading this load T_A by spacing the tie-back rods, S permits reasonable tie-rod forces $f_A = T_A/S$ for selection of tie-rod materials and dimensions. The above is all best conducted by geotechnical engineers with knowledge and experience and with possible variability of the soil parameters, codes, and safety factors employed for the length of the sheet piles and spacing of the tie-rods for a safe design.

(4) Shear force and bending moment diagrams

Equations (6.108) and (6.110) are then used to construct the (a) shear force and (b) bending moment diagrams shown in Figure 6.62. The maximum bending moment, $M_{\text{MAX}} = 76{,}050$ ft-lbs where the shear force equals zero at a depth, $z = 23.35$ ft (7.1 m) below the surface.

The above analysis was by the "Free-Earth" method assuming a rigid, anchored sheet-pile wall. Rowe (1952) from field observations of actual deflections of flexible, sheet-pile walls developed a maximum bending moment reduction method for anchored bulkheads. See Eckert and Callender (1987) for Rowe's moment reduction diagrams. In the above example, for a cantilever design with no anchor, the pressure prisms are the same and taking moments about the "y" $= 0$ location 35 ft below the surface gives a penetration depth $D_1 = 13.6$ ft and twice the length of the anchored design.

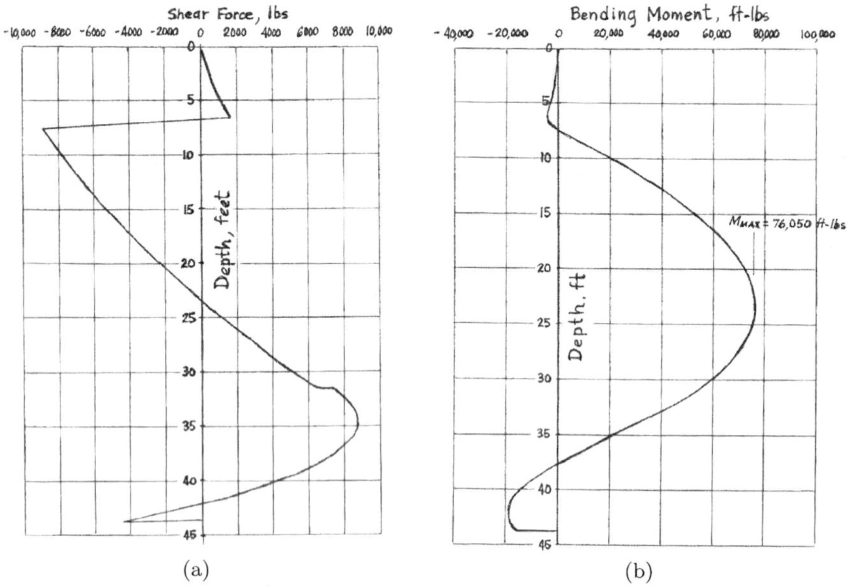

Figure 6.62. (a) Shear force and (b) bending moment diagrams for the example.

Steel, sheet-piles shaped like a horizontally stretched letter "Z" are the most common type specified for bulkheads in North America. They have many applications. Figure 6.63 shows the heavy-duty, PZ design with "ball and socket" connections that allow interlocking of adjacent sheets to create a rigid wall for both earth and water. The interlocked locations are as far away as possible from the neutral axis to ensure high transmission of shear forces and increase the strength-to-weight ratio. Determination of the bending moments and their use to select the correct material, dimensions, and type of sheet piles for construction is best performed by structural engineers with knowledge and experience with codes and safety factors for a reliable design.

Geotechnical and structural engineers have extensive experience with the design of bulkheads for retaining walls, cofferdams, parking garages, Navy piers, etc. Figure 6.64 is a photo of a Z-pile bulkhead constructed on an industrial, deep-water pier site in Bridgeport, CT with the anchor, tie-back system at an angle and spaced along an external, lateral beam.

For these applications, the design criteria remain relatively constant at the site during the design life. As a result, many design charts and nomographs have been developed for both cantilevered and anchored sheet

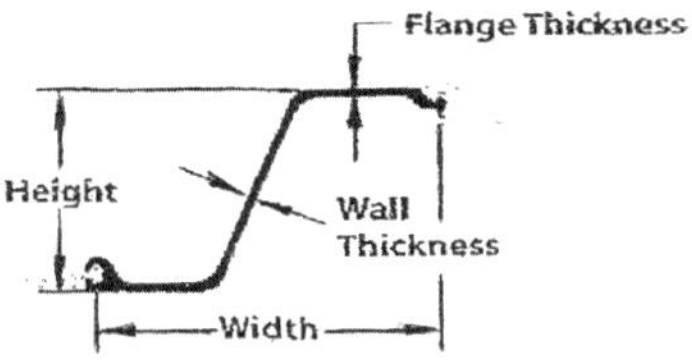

(a) Z-pile

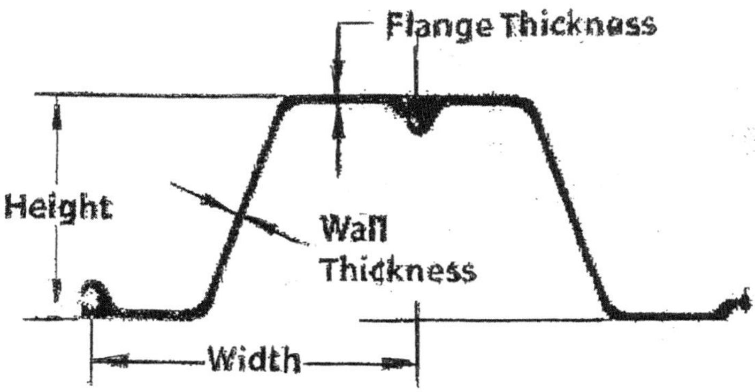

(b) Two Z-piles driven with interlocks connected to form sheet-pile wall

Section	Width	Height	Thickness		Area	Weight		Section Modulus	Moment of Inertia	Coating Area	
			Flange	Web		Pile	Wall			Both Sides	Wall Surface
	in.	in.	in.	in.	in.sq / ft	lb. / ft	lb. / ft. sq.	in. cu. / ft.	in. qu / ft.	ft. sq. / foot of single	ft. sq. / ft. sq.
PZ 22	22	9	0.375	0.375	6.47	40.3	22	18.1	84.38	4.48	1.22
PZ 27	18	12	0.375	0.375	7.94	40.5	27	30.2	184.2	4.48	1.49
PZ 35	22.64	16.1	0.6	0.5	10.29	66	35	48.5	361.2	5.37	1.42
PZ 40	19.69	14.9	0.6	0.5	11.76	65.6	40	60.7	490.8	5.37	1.64
	All Dimensions Are Nominal										

(c) PZ-pile specifications

Figure 6.63. (a) Schematic diagram of a Z-pile. (b) Two sheet-piles interconnected to form sheet-pile wall. (c) Specifications of dimensions, weights, section modulus, and moment of inertia for structural design. The PZ number refers to the weight in lbs/ft^2.

pile design. See US Steel Design Manual (1984), Corps of Engineers (1994) and US Navy, NAVFAC (1982) for example. Updated versions of these references are now available online. See Gaba *et al.* (2017) for UK design guidance.

Figure 6.64. Industrial site in Bridgeport, CT for Z-pile bulkhead with lateral beam to strength anchor system (courtesy, Ocean and Coastal Consultants, Inc., Trumbull, CT) (courtesy Ocean and Coastal Consultants, Inc.).

For applications of bulkheads at open-coast locations exposed to changing design conditions, elevated water levels, waves, and coastal erosion during storm events, the help and advice of coastal engineers is required, but seldom utilized.

6.3.6.5 *Coastal engineering design guidance*

Figure 6.65 is a photo taken on the beach at Sandbridge, a suburb of Virginia Beach, Virginia on the open Atlantic Ocean coast the day after a coastal storm. The author is holding a 25-ft (7.6-m) surveying rod at a severely eroded location in front of the steel, anchored sheet-pile bulkhead, where the depth of local erosion at the base (mud line) of the bulkhead was about 8–9 ft (2.5–2.8 m). Rusted sections in orange show where the sandy beach had been before the storm forming a relatively constant beach berm elevation in front of the bulkhead. Storm surge elevations and high waves overtopped the bulkhead. Water ponded behind the structure to significantly raise the water table level before draining back to the sea beneath the bottom of the sheet piles at this location. Sand eroded is due to both

Figure 6.65. A photo taken on the beach at Sandbridge, a suburb of Virginia Beach, Virginia on the open Atlantic Ocean coast the day after a coastal storm. Author holding a 25-ft survey rod (Author's photo).

interior drainage and wave-toe scour at the former beach elevation at the base of the structure.

At another location in Sandbridge shown in Figure 6.66, the anchored, steel sheet-pile bulkhead failed in this coastal storm. The tie-back rods broke, and the active soil pressure force pushed the bulkhead seaward. This anchored design included an interior, lateral running beam that strengthened the anchor system. The brown color of this lateral beam and the steel sheet piles below the beam indicate the depth of the original backfill at this location. The timber in the foreground is from damaged porches of homes

Figure 6.66. Failed anchored, steel sheet pile bulkhead at Sandbridge, Virginia (Author's photo).

that also received some damage from the failed bulkhead. These structures were not designed by professional geotechnical and structural engineers with aid from coastal engineers.

These two photographs illustrate the major problems with bulkheads designed for open coastal conditions, namely: (1) variable, high-water events, (2) high waves at these deeper water levels, and (3) erosion of the soil materials adjacent to the bulkhead on the water side of the structure. In effect, these open coastal bulkheads should be designed as seawalls and not as bulkheads. The report by Eckert and Callender (1987) titled "Geotechnical Engineering in the Coastal Zone" presents a valuable summary of bulkhead design under these open coast conditions. The material presented so far will be valuable when coastal engineers are requested for design guidance. Chapter 10 presents further information for toe erosion at the base of walls due to wave action. Of major concern is establishing the crest elevation for coastal bulkheads. Coastal engineers must be involved to properly evaluate all the design constraints for establishing the final elevation of bulkheads constructed at coastal sites.

6.3.7 *Flood walls*

6.3.7.1 *Introduction*

Flood walls are thin, concrete, vertical structures to separate high-water elevations on the water side from possible damage to man's property and infrastructure on the land side (L.S.). Figure 6.67 shows the "inverted T-wall" (top) and "I-wall" (bottom) and both types are cantilevered walls

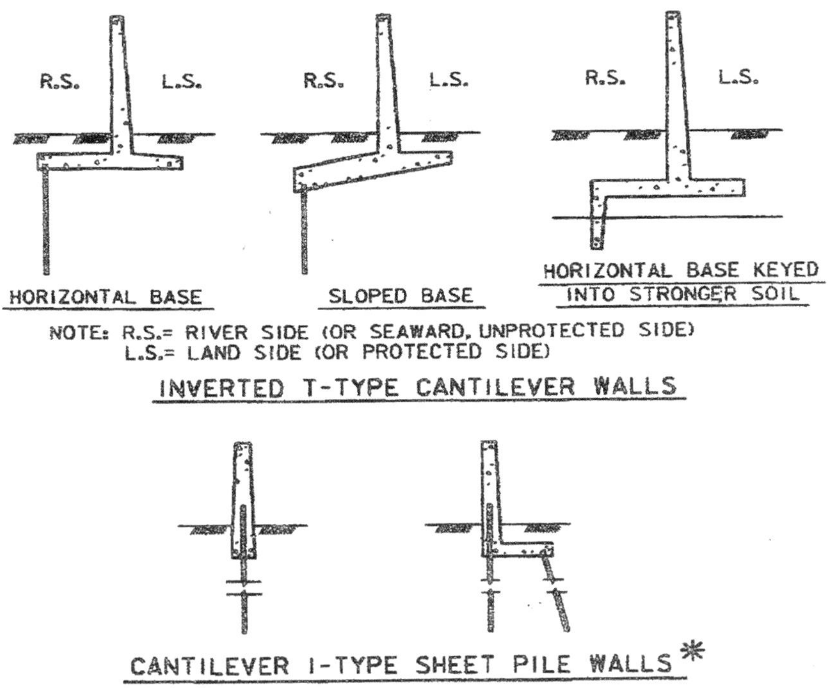

Figure 6.67. Two main types of flood wall design by the US Army Corps of Engineers (from Corps of Engineers, 1989).

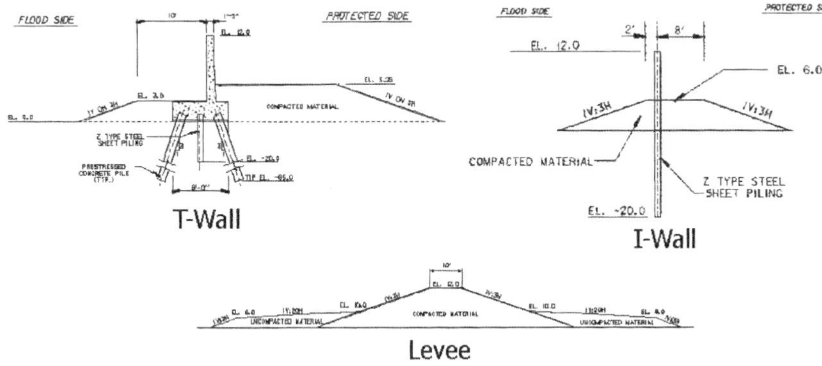

Figure 6.68. Two main types of concrete flood walls and an example earth levee structure as designed by the US Army Corps of Engineers (from Corps of Engineers, 1989).

that hold back the water forces on the seaward (river, R.S.) side from the L.S. They have a long history of applications to mitigate damage from flooding rivers passing thorough the rural and urban areas. For example, see Corps of Engineers (1989) "Engineering and Design: Retaining and Floodwalls" with some details shown in Figure 6.68 including the design of concrete piles, sheet piles, and compacted material to aid the structural design. Figure 6.68 also shows an earthen levee structure for flood damage mitigation where sufficient lateral space is available in relatively rural areas. The "footprint" of the concrete, flood wall structure is far less than the lateral space required for the earthen levees. In some cases, the flood wall may be constructed on top of the levee. Levees, dikes, and revetments are discussed in Section 6.4.

6.3.7.2 *Flood wall design*

Geotechnical and structural engineering are the major components of flood wall design. As for bulkheads, coastal engineering guidance is also required when located at coastal sites. The crest elevation depends on both high water elevations and wave heights from storms causing wave overtopping events. The hydrostatic (water elevation) and hydrodynamic (oscillating and impact wave) pressure distribution for the flood wall structural design were shown in Figure 6.36 and thoroughly reviewed in Section 6.3. But now, these pressure distributions must be added together to give the full water force on the wall since the L.S. is dry.

6.3.7.3 *Case studies*

We shall briefly introduce and discuss three important coastal locations where flood walls are presently or planned to be used to mitigate interior damage. In Chapter 11, these same three sites will be reviewed again relative to the potential for accelerated sea-level rise in the future and the consequences of higher water levels and more flooding events.

(1) City of Norfolk, Virginia

Figure 6.69 is a map of the Chesapeake Bay and Atlantic Ocean and shows the location of Norfolk, Virginia (US) on the Elizabeth River tidal estuary that connects to the Bay. In 1965, a 2,674-ft (815-m) long flood wall (Figure 6.70, yellow markers) with five closure gates and pump station for interior drainage was constructed by the US Army Corps of Engineers in downtown Norfolk. Figure 6.71 shows some views of the closure gates and flood wall dimensions relative to ground-level sidewalk. The I-Wall design

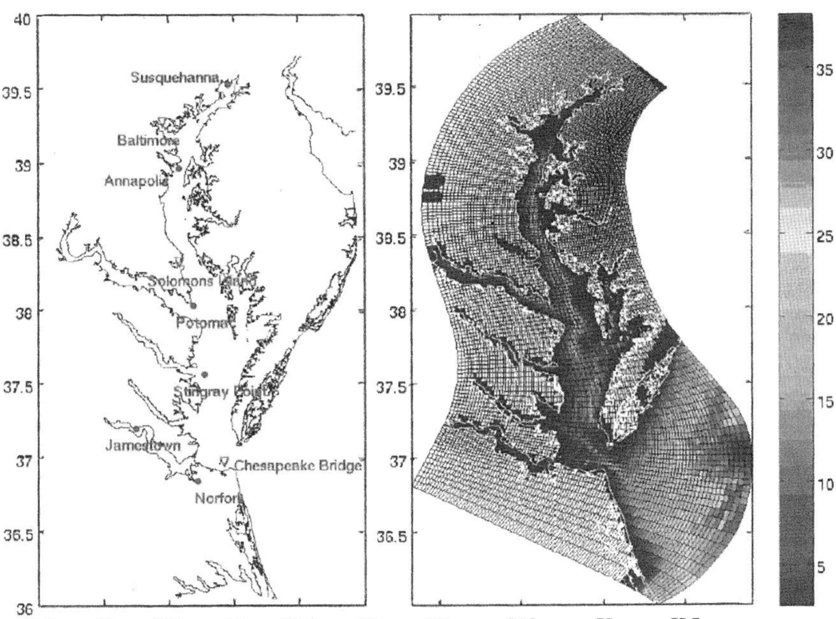

Figure 6.69. The Chesapeake Bay on the Atlantic Ocean and location of Norfolk Virginia at Lat. 38°51′ and Lon. 76°17′ on the Elizabeth River tidal estuary. The bathymetry employed in a numerical model of tidal and storm-surge elevations and currents is on the right (meters water depth) (courtesy Norfolk District, Corps of Engineers).

Figure 6.70. Flood wall in Downtown Norfolk (yellow markers). Wall extends 2,674 linear feet (815 m) from South End (1) and around the curve (Boush Street) to North End (7) with intermediate locations of closure gates (2), (3), (4), and (6) with flood wall pump station at (5). Courtesy, Google Earth (2016). The ship is the Battleship Wisconsin permanently moored in Norfolk (courtesy Map data: Google, Maxar Technologies).

Figure 6.71. Top-left is the right end of Nauticus closure gate, location 3; top-right is the flood wall elevation relative to the ground level; bottom-left is the interior view of rock revetment on the inside of flood wall and heading toward Battleship Wisconsin closure gate entrance shown on bottom right at location 4 (Author's photos).

(Figure 6.68) with concrete cap was constructed. Hurricane Isabel (September 18, 2003) was the second highest storm surge event recorded in Norfolk since 1927 as shown in Figure 2.2 in Section 2.2.3. In over 50 years of operation, the flood wall has never been overtopped and the pump station has always worked to keep the streets open during rainfall events.

The constructed flood wall crest elevation averaged 11 ft (3.4 m) above the MSL datum in 1965 which is roughly the NGVD datum today. A study showed that this crest elevation does not meet the current US Federal Emergency Management Agency (FEMA) standard for the 1% annual storm base flood (+9 ft, NGVD) plus wave run-up (+6.1 ft) plus freeboard (1.0 ft) totaling +16.1 ft (4.9 m) (Martin *et al.*, 2008). How the City of Norfolk and the Norfolk District Office of the Corps of Engineers have addressed this problem along with the future sea-level rise will be summarized in Chapter 11.

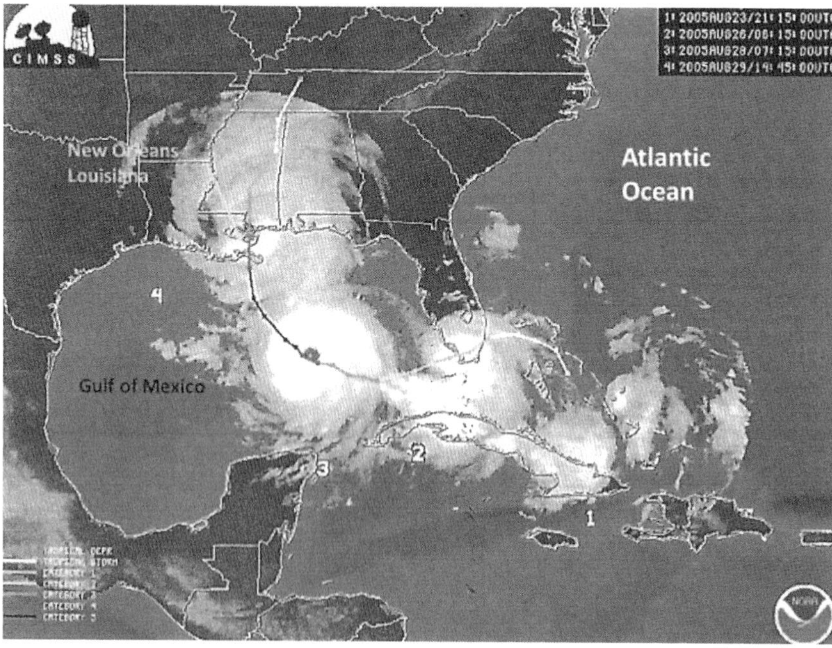

Figure 6.72. New Orleans, Louisiana on the Gulf of Mexico and track of Hurricane Katrina (August, 2005) over four days: (1) August 21, (2) August 26, (3) August 28, and (4) August 29 landfall (courtesy New Orleans District, Corps of Engineers).

(2) City of New Orleans, Hurricane Katrina

The city of New Orleans, Louisiana is located on the Gulf of Mexico (Figure 6.72). Over 50 breaches occurred during Hurricane Katrina (August 29, 2005) in the New Orleans' and adjacent regions' storm surge mitigation (levee and flood walls) system. Over 80% of the region was flooded causing $108 billion in property damage (dollar value in 2005) with loss of life totaling 1,245 people. Flooding impacted over 100,000 homes and businesses (see Wikipedia for many summaries with references).

As illustrated in the land elevation profile (Figure 6.73), parts of the city of New Orleans from Gentilly Ridge to University of New Orleans (UNO) lie below NGVD (0.00 ft datum) which is near MSL. It is the only coastal city in the United States below sea level and similar to the low lands of the Netherlands (Holland) in this regard. As a result, it remained flooded for many days following the storm. Forty-nine percent of the New Orleans and surrounding subdivisions lie below the sea level (Camapanella, 2017).

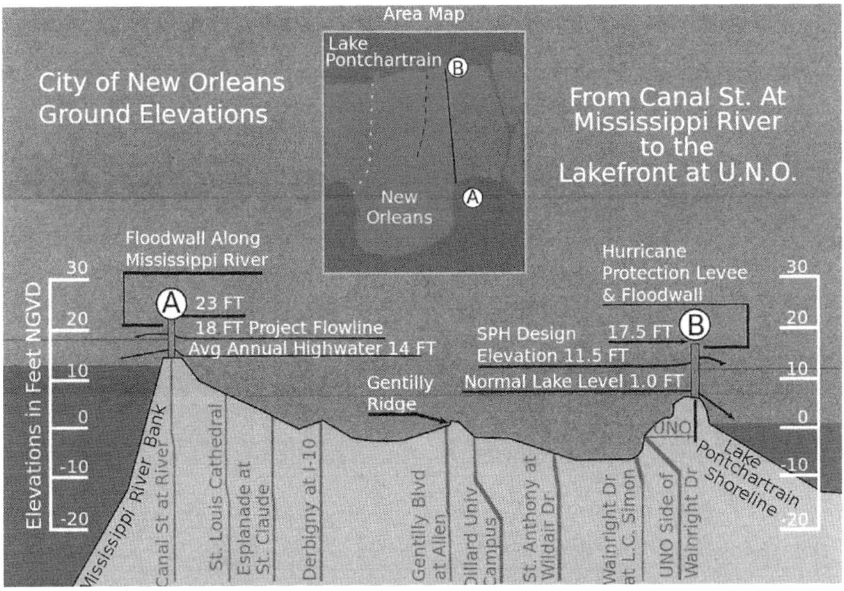

Figure 6.73. Land-elevation profile from (A) Mississippi River Bank (left side) to (B) Lake Pontchartrain Shoreline (right side) where flood wall failures filled basin below NGVD (0.00 ft datum) from Gentilly Ridge to the University of New Orleans (UNO) (courtesy New Orleans District, Corps of Engineers).

Flood wall and levee failures filled this basin below NGVD (0.00 ft datum). The famous French Quarter of New Orleans is on relatively higher ground on the Mississippi River side and only experienced ponding from rainfall.

Three major canals (Industrial, 17th Street, and London Ave) are key components of the regions' storm surge mitigation system as designed by the US Army Corps of Engineers. Six major breaches of the levees and flood walls on these canals occurred during Hurricane Katrina. The 17th Street canal flood wall failed on the west side near the Old Hammond Highway Bridge as seen in Figure 6.74. Many subsequent studies for the reasons for this flood wall failure have been updated in the published literature. We herein summarize the findings and lessons learned for the flood wall failure from Hurricane Katrina (2005).

The flood wall was designed as an "I-Wall" embedded in a levee as shown in Figure 6.68. A report by the American Society of Civil Engineers (ASCE, 2007) cited two engineering oversights for failure of the 17th flood wall. The strength of soil (marsh peat from remains of a swamp) was

Figure 6.74. The 17th Street canal flood wall failure on the west side near the Old Hammond Highway Bridge (foreground) (courtesy New Orleans District, Corps of Engineers).

very low and its strength was overestimated in design calculations (IPET, 2006). Figure 6.75 schematically depicts the weak landward soil, slip-stream failure (red). The soil moved laterally allowing the elevated water levels to push the flood wall laterally and fall over causing a breach in the 17th street canal.

The second reason was the failure to recognize and account for a water-filled gap at the base of the flood wall and the levee soil on the canal side of the wall. It was learned that this gap was known to exist and should have led to design modifications and increased safety factors in the design (Rodgers *et al.*, 2015). For example, underseepage can be mitigated by relief wells, cutoff walls, berms, and longer piles.

The Corps of Engineers, Engineering Research and Design Center (ERDC) performed model studies as depicted and shown in Figure 6.76 (from Pezza, 2008). The top schematic diagram shows I-Wall and levee flood wall deflection under high water; the middle depicts subsequent

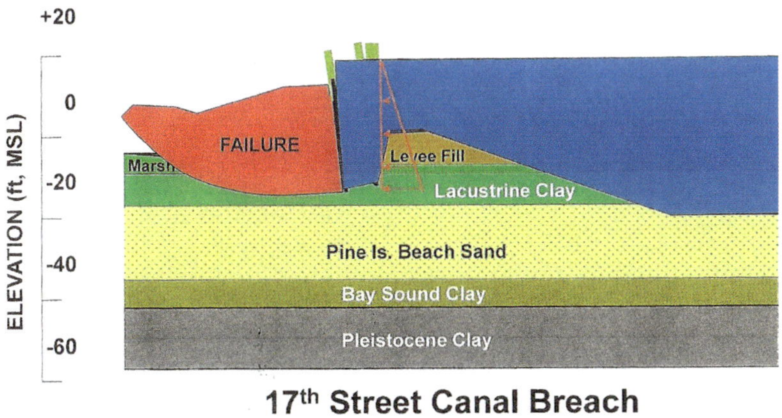

Figure 6.75. Weak marsh soil on L.S. caused slip-stream soil movement and failure of flood wall under high-water levels (from Rodgers *et al.*, 2015).

slip-stream (curved soil boundary) failure of the L.S. levee and soil (gray) with movement of the flood wall; and the (bottom) model shows the test photograph for confirmation of this result. The Chief of the Corps of Engineers (Strock, 2005) in his testimony before the US Senate stated: "*We have now concluded we had problems with the design of the structure*".

Many studies were made of the failure of flood walls and levees in the New Orleans region from Hurricane Katrina (IPET, 2006; ASCE, 2007; NAS, 2006) but they all essentially reach the same conclusions (see Wikipedia, *2005 "Levee Failures in Greater New Orleans"*, for summaries with references for these studies). This failure of the flood wall and breach on the 17th street canal was not due to high water levels overtopping the crest elevation of the flood wall. Many other breaches at levees were caused by water levels overtopping the crest. Levee design is introduced in Section 6.4.

It has been over 14 years since Hurricane Katrina (2005) devastated New Orleans and the surrounding parishes. Much was learned from these flood wall and levee failures that has been incorporated in the redesign and rebuilding of the storm surge damage mitigation system that now exists. For example, Figure 6.77 depicts the scale of a new flood wall crest relative to a construction crew on the landward side. The repaired flood wall sections employed the "T-Wall" type design. Figure 6.78 is an aerial photo of the recently completed (2015) Lake Borgne Surge Barrier that is 1.8 mi (2.9 km) long to greatly reduce future flooding levels in New Orleans. In

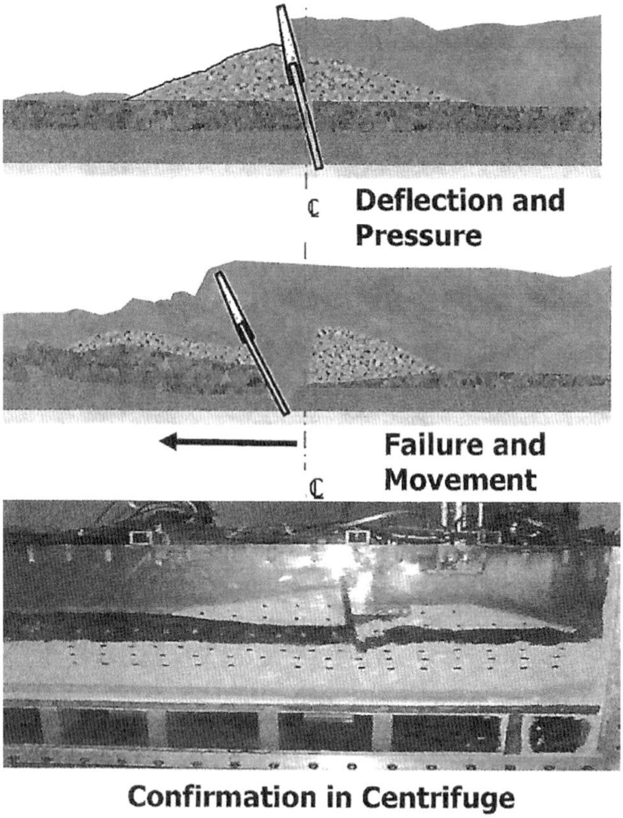

Figure 6.76. Corps of Engineers, Engineering Research and Design Center (ERDC) model studies as depicted and shown (bottom) to confirm failure mechanisms for flood wall and levee system constructed in New Orleans (from Pezza, 2008).

the foreground is the movable tide gate closure structure that will be fully discussed in Section 6.5.

How the City of New Orleans and the Corps of Engineers have improved storm damage mitigation while addressing future sea-level rise will be summarized in Chapter 11.

Hurricane Katrina (2005) was ranked one of the worst major disasters in US history. Storm surge damage impacted four coastal states: Louisiana, Mississippi, Alabama, and Florida. Then came Superstorm Sandy (2012) that caused major coastal flooding problems along the Eastern, Atlantic

Figure 6.77. Scale of a new flood wall crest relative to the construction crew on the landward side (courtesy Corps of Engineers, New Orleans District).

Figure 6.78. Lake Borgne, Louisiana storm surge barrier with movable, tide gate closure structure (courtesy Map data: Google, Maxar Technologies).

Ocean seaboard in 12 states from North Carolina to Maine and especially the subways and tunnels in New York City.

(3) City of New York, Superstorm Sandy

When Superstorm Sandy moved ashore northeast of Atlantic City, New Jersey on October 29, 2012, it was a tropical storm (sustained wind speeds <74 mi per hour) and not classified as a Category 1 hurricane as defined by the Saffir–Simpson scale. It has since then become known as Superstorm Sandy. The City of New York consists of five boroughs: the Bronx, Brooklyn, Manhattan, Queens, and Staten Island as shown in different colors in Figure 6.79. All received flood damage when the storm surge from the Atlantic Ocean funneled into Lower New York Bay, then through the narrow opening between Staten Island and Brooklyn (the Verrazano Bridge), and into the Upper New York Bay to create record water levels in lower

Figure 6.79. The five boroughs: the Bronx, Brooklyn, Manhattan, Queens, and Staten Island that comprise the City of New York on the Atlantic Ocean (from NACCS, 2015).

Manhattan. The NOAA gauge at the NOAA tide gauge station (The Battery, ID 8518750) was 9.40 ft (2.87 m) above the astronomical tide and 11.5 ft (3.50 m) above MSL (NAVD88-MSL = 0.222 ft, 0.0676 m). Superstorm Sandy caused $62 billion property damage in the US, the most expensive since Hurricane Katrina. It caused 125 deaths and over 75 million residences lost power.

Prior to Superstorm Sandy, the US Army Corps of Engineers did not have any coastal flood damage mitigation projects in New York City that employed vertical-walled or dike structures. However, many beach restoration projects that existed were damaged, and this will be discussed in Chapter 7.

Since Superstorm Sandy, the Corps of Engineers now has two projects underway in the New York City region. As seen in Figure 6.80, the Southeastern shoreline of Staten Island is directly exposed to storms in the Atlantic Ocean. A Corps project to identify potential alternatives to mitigate storm damages was authorized in 1993 and as a result of Superstorm Sandy, funds are now available for construction. This project will construct flood walls and dikes (levees) seaward of the existing boardwalk from Fort Wadsworth to Great Kills. Early final design is under way at the time of this writing. The second project began on July 15, 2016 and was one of nine feasibility studies recommended by the Corps' "North Atlantic Coast

Figure 6.80. Artists' sketch of "Big U" project as proposed by the Bjarke Ingels Group (courtesy, Henk Ovink).

Comprehensive Study" (NACCS, 2015) report along the entire Eastern seacoast of the US. It was titled "New York/New Jersey Harbor and Tributaries Focus Area Feasibility Study" so that all the storm tidal waters in the region (see Figure 6.79) were included. The full spectrum of structural and non-structural flood damage mitigation alternatives is being considered including movable tidal gates to be discussed in Section 6.5. This project is the direct result of Superstorm Sandy because New York City was the most severely impacted due to flooding damage to subways and roadway tunnels.

The US Department of Housing and Urban Development (HUD) began the "Hurricane Sandy Rebuilding Task Force" (Donovan, 2013). (HUD is also a Federal Government agency but in the Department of Commerce). A "Rebuild by Design" contest for architects, designers, scientists, engineers, etc. was held and 148 proposals were submitted for a "smarter and more resilient infrastructure along the region's waterfront". In July 2014, six winners were announced and one was called "Big U" (Bjarke Ingels Group, 2014). As graphically illustrated in Figure 6.80 for the Battery region and financial district of lower Manhattan, a 10-mi (16-km) long, U-shaped barrier "berm" would be constructed. The yellow line gives the location of the boundary along the East River (right side) and Hudson River (left side) between flooded and dry areas with Battery Park in the foreground. The Bjarke Ingels Group is a Danish architectural firm and included additional design concepts along the East River, adjacent and below the FDR Drive as shown in Figure 6.81. Top-left is the Brooklyn Bridge and the proposed design concept along the East River. Top-right is the proposed typical "berm" which is a dike (levee) with steps for recreation and as a protective revetment when under storm conditions. Bottom-left is green space park area adjacent to FDR Drive. At some locations (bottom-right) beneath the elevated roadway, shops and seasonal markets would have overhead panels beneath FDR Drive that could be rotated downward to form a flood wall during storm conditions. As of this writing, the "Big U" project has been awarded $335 million by HUD for project implementation but construction has not begun.

The "Big U" project design used the term "berm" for the elevated dike (levee) with protective revetment and for a flood wall to separate the water side from upland property and infrastructure. Dikes (levees) and revetments are covered in Section 6.4. How the City of New York, the Corps of Engineers, and the HUD have improved storm damage mitigation conditions for the present while addressing future sea-level rise concerns will be summarized in Chapter 11.

Figure 6.81. Artists' sketch of details along the East River for the "Big U" project as proposed by the Bjarke Ingels Group (courtesy, Henk Ovink).

On March 30 and 31, 2009 a conference "Storm Surge Barriers to Protect New York City" was held by the ASCE (Hill *et al.*, 2012) in an effort to encourage action on the city's efforts to avoid a repeat of what happened in New Orleans with Hurricane Katrinia. See Section 6.5 for more details regarding the movable tidal gate barriers that were proposed.

6.4 Dike (Levee) and Revetments

6.4.1 *Introduction*

Dike or levee? Both terms are employed for the elevated embankments along the land/water boundary that prevent flooding on the dry side when the water level is increased on the water side of the boundary. Along rivers flowing downstream in one direction, the embankment is usually termed a *levee* to prevent inundation of the adjacent floodplain. Along coastal shorelines with tidal flows in both directions and waves striking the land/water boundary, the embankment is usually termed a *dike* to prevent flooding of the inland area. Both terms (unfortunately) are found in the literature for both riverine and coastal engineering. In this book, we will only use the term dike. In the Dutch language, the word *dijk* is pronounced the same as in English.

Figure 6.82. Photo of sheep grazing on dike in the Netherlands.

Some coastal locations have natural ridges near the water that serve as a dike. Artificially designed and constructed dikes are usually made from earth materials (clay, silt, sand, gravel, rock) with sloping sides as illustrated in Figure 6.68. They are the professional design responsibility of geotechnical engineers for proper specification of the earth materials for construction and for the analysis of possible failure modes of the dike itself and the soils beneath the dike.

The sloping sides constructed from earth materials are subject to erosion by waves and currents. The sloping sides must then be strengthened by a revetment that can be of many materials and types. Or as done in Holland, at some locations, the Dutch graze sheep on the *dijk* to compact the soil as illustrated in Figure 6.82 (see also Rosenberg, 2019).

6.4.2 *Dike design*

6.4.2.1 *Geotechnical engineering*

The design of dikes in coastal regions is mainly the responsibility of geotechnical engineers. Coastal site selection requires subsurface soil exploration in the field, soil testing, soil classification, and determination of engineering properties by soil analysis in the laboratory. An excellent introduction is found in "Geotechnical Engineering in the Coastal Zone" (Eckert and Callender, 1987). See Section 2.4 for a summary of geomechanics in this book.

6.4.2.2 *Failure modes of dikes*

The stability of the dike slope is of major concern. Slope failure for slopes of limited heights such as on dikes are termed finite slopes and generally classed in two groups as shown in Figure 6.83.

Soils with cohesive materials (silt, clay) do not have long, straight failure surfaces that are parallel to the sloping surface of the dike. They have complex (a) curved failure surfaces or (b) combinations of curves and straightline failure surfaces for one group shown in Figure 6.83. The second group is analyzed as (c) sliding block of wedge failure surfaces. Stability analysis methods for both groups are presented in Eckert and Callender (1987) and are well known by geotechnical engineers.

As discussed in Section 6.3.7 regarding flood walls and their failure during Hurricane Katrina that struck New Orleans there were over 50 failures (breaches) of the dikes. Because the Mississippi River flows through New

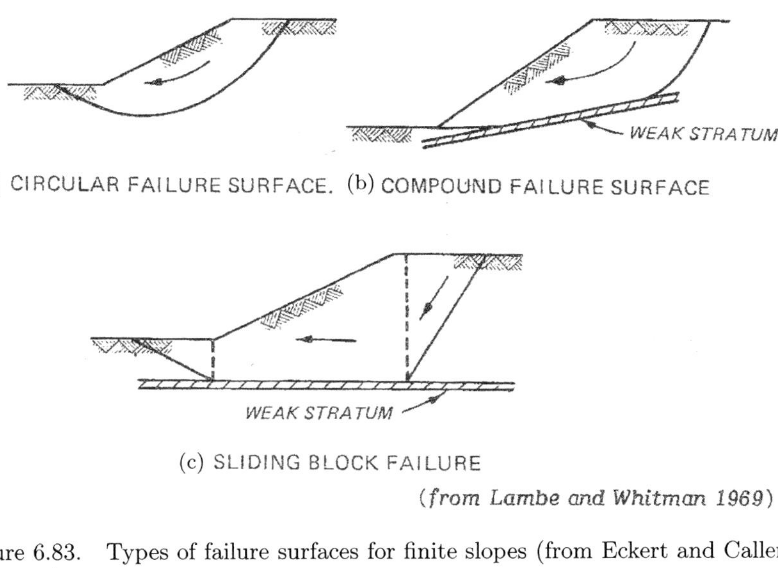

Figure 6.83. Types of failure surfaces for finite slopes (from Eckert and Callender, 1987).

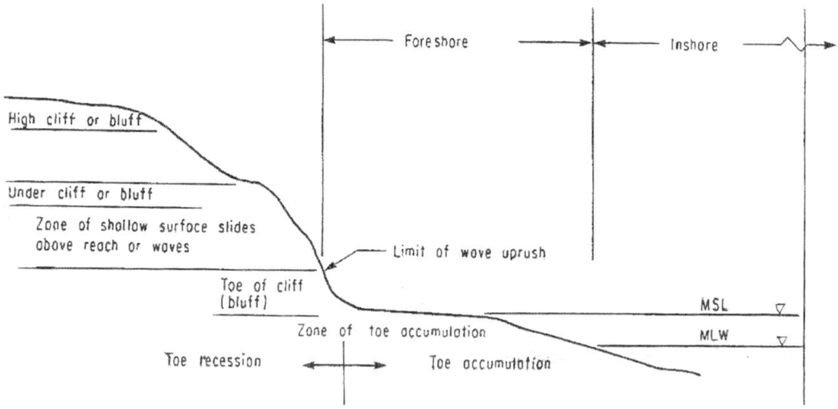

Figure 6.84. Definition of natural slopes in the coastal zone (from Eckert and Callendar, 1987).

Orleans, they are called levees in the storm surge damage mitigation system for the region. The primary mechanism for failure of levees in eastern New Orleans was reported to be the presence of sand instead of the thick Louisiana clay in 10% of the region. All the major breaches were caused by overtopping and erosion. Reduced levee crest elevations increased the amount of overtopping, erosion, and subsequent flooding (IPET, 2006).

Natural coastal slopes fronting high cliffs, bluffs, or banks as illustrated in Figure 6.84 are also reviewed and their stability analysis discussed in Eckert and Callender, 1987 but beyond the scope of this book. Analyzing the erosion and designing a revetment to mitigate the toe recession is the responsibility of coastal engineers.

6.4.2.3 *Herbert Hoover dike*

The dike surrounding Lake Okeechobee in Florida (US) is approximately 143 mi (230 km) in length and averages 30 ft (9 m) in height (NGVD). The first embankments were constructed circa 1915, and in 1960 a new dike system was completed as designed by the Corps of Engineers. Crest elevations were raised to +41 ft (12.5 m) NGVD for storms with a recurrence interval, $T_r \approx 935$ years. However, seepage and embankment stability near culvert locations and foundation problems were found from field investigations in the 1990s. During high-water events, piping through the dike occurs. A plan approved in 2000 for dike rehabilitation has begun dividing the dike into eight reaches (Wikipedia, 2005). It is now called the Herbert Hoover Dike named after the US President who visited Lake Okeechobee in 1928 after a major hurricane and who authorized the US Army Corps of Engineers to undertake flood damage mitigation improvements (Corps of Engineers, 2008).

6.4.3 *Revetment design*

Recall that the purpose of the revetment is the mitigation of erosion of the slope of the natural bank or man-made dike. Many materials have been employed to construct revetments.

6.4.3.1 *Materials for a revetment*

The most common revetment material is natural rock from a nearby quarry. Figure 6.85 displays a classic coastal engineering design on a 2:1 slope for elevated water levels and waves using uniform-sized or graded rip-rap as armor stone of weight, W, an under layer, geotextile filter cloth, toe protection, and a splash apron. The stable weight of the armor layer for rubble-mound structures including revetments with rock armor was covered in detail in Section 6.2.

Many other types of materials have been used to construct revetments as displayed in Figure 6.86. Concrete material can be used to make armor layer units (e.g. CORE-LOC) or cast in place as a slab, or made into small blocks of many shapes fitted together to form the revetment layer. Too often, when concrete roadways are broken up and replaced, the used

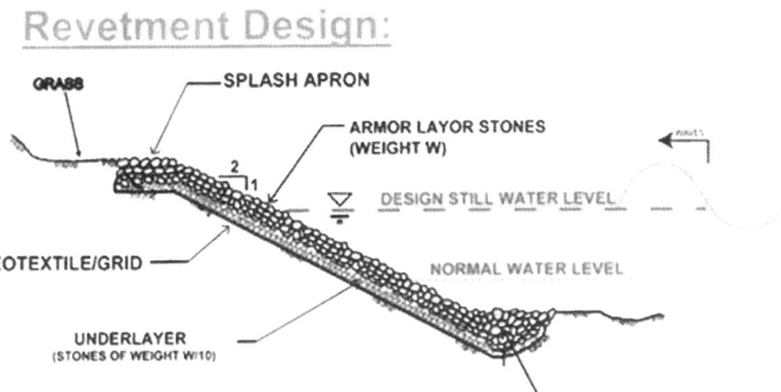

Figure 6.85. Schematic diagram of classic revetment using natural rock as the material for the armor layer (from CEM, 2006).

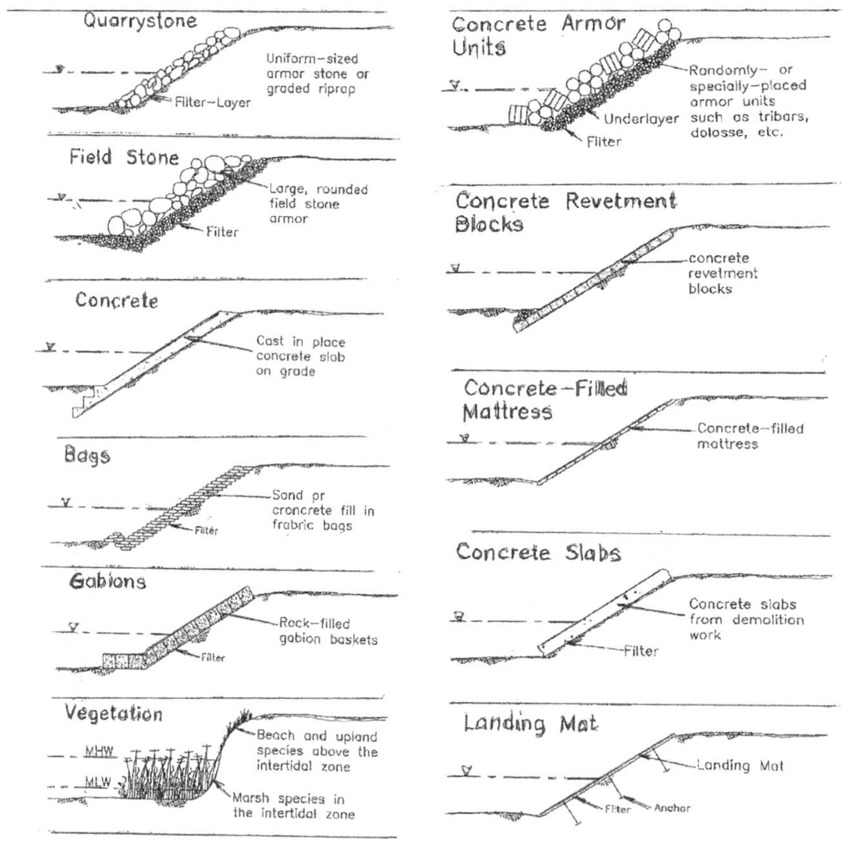

Figure 6.86. Many types of materials used for revetments placed on natural or graded slopes (from Corps of Engineers, *Coastal Engineering Manual*, CEM, 2006).

concrete-rubble material is dumped to make a very unsightly revetment. Gabions are rock-filled wire baskets that can be placed on the slope and/or stacked step-wise to form a revetment. Sand- or concrete-filled geotextile material bags are another option (see in the following). Note that in all the types shown in Figure 6.86, the slope of the revetment is about $\cot \alpha = 2$ (or less) except the steep, natural bank with marsh grass in the intertidal zone. Therefore, when designing a revetment for a steep, natural slope, the bank must be graded as depicted in Figure 6.87 and a splash apron provided.

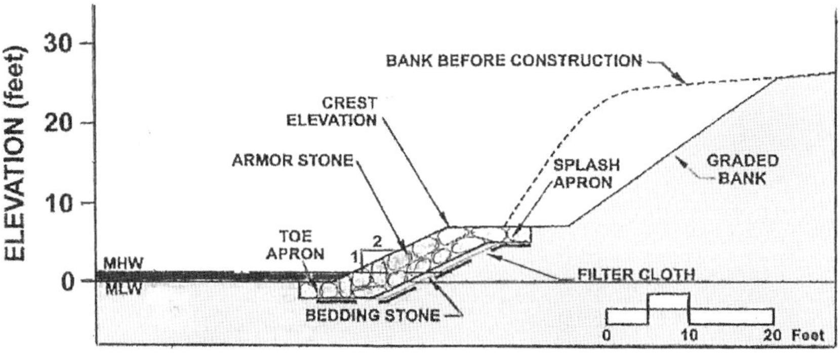

(a) Schematic diagram of a graded bank with wave splash apron

(b) Photograph of a graded bank, rock revetment and wave splash apron on the Chesapeake Bay

Figure 6.87. Revetment design for steep, natural slopes. (a) Schematic diagram. (b) Photograph (courtesy, VIMS, 2006).

6.4.3.2 *Articulated concrete blocks and mats*

Revetments constructed to retain soil materials on earth slopes from erosion are designed by civil engineers for many applications. Highway bridge abutments, retention ponds, drainage ditches, pipelines, etc. are a few examples.

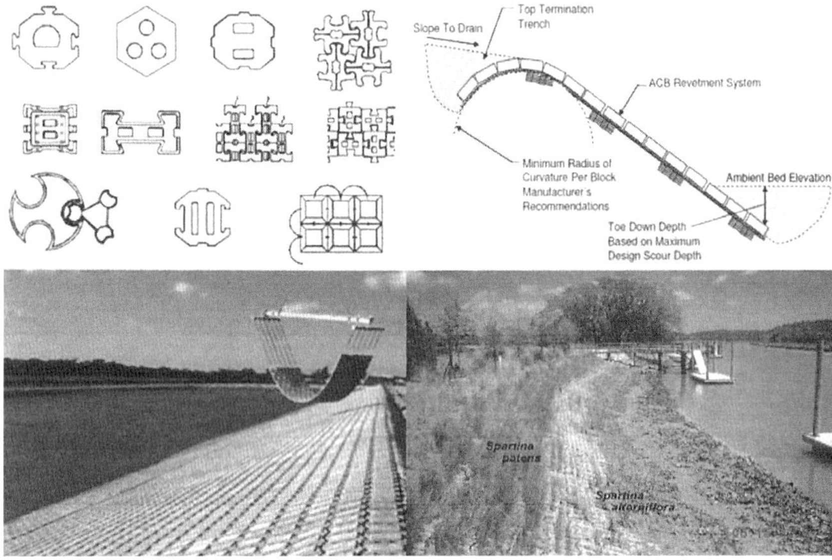

Figure 6.88. Some details and photographs of ACB and ACM systems (courtesy National Concrete Masonry Association).

Articulated concrete block (ACB) and mat (ACM) systems are commonly used in these and in the riverine and coastal revetment applications. The "articulated" means the system of blocks or mat remains interlocked while conforming to changes in the ground profile beneath the revetment. A block system is usually placed by hand or installed as a preassembled mat. Expansion and contraction of the ACB or ACM system takes place naturally. Figure 6.88 shows some details and photographs of ACB and ACM systems.

The concrete blocks come in many shapes as shown in Figure 6.88 (top, left), and when fitted together conform to the dike or bank profile extended below the water level, Figure 6.88 (top, right). Concrete mats (Figure 6.88, bottom-left) are more quickly and economically placed with construction equipment. Many more examples can be found online from the National Concrete Masonry Association (https://ncma.org).

Concrete blocks permit marsh grasses (*Spartina alterniflora* and *patens*) to grow through the block openings on tidal marshes as shown in the photograph, Figure 6.88 (bottom, right). More details on "living shorelines" are presented in Chapter 10.

6.4.3.3 *Geotextile sand containers*

Geotextile materials cut and sewn to form bags and tubes of various lengths and filled with sand have come to be called "geotextile sand containers" (GSC). They have been around for over 60 years. As shown in Figure 6.89 they have been used as a coastal revetment when stacked together to form a sloping surface from the beach to the crest of the dike or bank. Figure 6.89 (top, left) shows three rows of relatively long tubes under construction and (top, right) shows a slope of 7–8 rows of sand-filled bags and the eroded bank before construction. Unfortunately, little research has been conducted over the years to document the relative strength of the GSC system to resist damage from elevated water levels and high-wave conditions. Too often, storms cause damage as illustrated in Figure 6.89 (bottom, left). Some bags have fallen down the slope which has slumped to form a bulge in the revetment.

Since 1999, the Technische Universitat Braunschweig, Leichweiss Institute (LWI) in Germany has been conducting basic research to help understand GSC employed as geotextile sand container structures (GSCS) in coastal engineering (see Oumeraci and Recio (2009) and GeoBag @ TU

Figure 6.89. GSCs and possible mechanisms for failure (from https://www.tu-braunschweig.de/lwi and from Leichtweiss-Institute, Germany).

Figure 6.90. GSCs used as a groin (from https://www.tu-braunschweig.de/lwi).

Braunschweig website for details). Figure 6.89 (bottom, right) displays potential failure modes of a GSC revetment (from Dassanayake and Oumeraci, 2009). Recent publications discussing key engineering properties of GSC and their effects on the hydrodynamic stability of GSCs in coastal engineering are by Dassanayake and Oumeraci (2012a, 2012b). Effects of the geotextile material, sand-fill ratio, and crest freeboard are discussed based on pullout and wave impact tests in the large LWI wave tank. Additional work can be found in van Steeg and Vastenburg (2010) and van Steeg and Breteler (2008).

Figure 6.90 shows a groin constructed from geotextile sand bags. They have also been used as seawall/dike structures buried beneath dunes. They are becoming popular for environmental concerns as alternatives to conventional rock structures, but their use is still an emerging technology. Design guidelines for GSCs have been summarized by Bezuinen and Vastenburg (2013). Chapter 10 presents more details.

6.4.4 *Coastal engineering aspects of dike and revetment design*

Design guidance for estimating wave run-up and wave overtopping of dikes with smooth or roughened slopes was presented in Chapter 5. Determination of the appropriate crest elevation is then a compromise considering all five design constraints as discussed in Chapter 1 (Section 1.2.4). The revetment structural stability when the material is natural rock or concrete armor units is presented in detail in Chapter 6 (Section 6.2). Dikes are often overtopped by waves or completely submerged below elevated water levels so that damage mitigation revetments are often needed on both sides of the dike.

In low-lying coastal areas with large urban populations and infrastructure, combinations of flood walls, dikes with protective revetments, and movable gate structures for access/egress are all the design responsibility of Coastal Engineers. This combination of coastal structures will be more and more important in the coming years of accelerated sea-level rise as fully discussed in Chapter 11.

6.5 Storm Surge Barriers with Movable Tidal Flood Gates

6.5.1 *Introduction*

Daily tidal cycles sustain the ecology of the coastal zones of the world's oceans. Storms cause elevated water levels (surge) above the daily levels for brief time periods. The problem is how to mitigate damage to property and infrastructure (and save lives) from infrequent storm surge events with minimal impact on the daily ecological, environmental, and navigational importance of the coastal zone.

Barriers in coastal waters change the salinity distributions (lateral, vertical, and temporal), alter fish and shellfish habitat (especially juveniles), disrupt the normal sediment transport distribution patterns, alter marshes and tidal wetlands, and impact the overall quality of tidal waters.

The daily variations of the tides may cover many hectares (acres) of surface area and reach many km (mi) inland from the ocean (Section 2.2.2). In many locations, high and low tide levels occur twice each day (semi-diurnal) bringing nutrients with the tidal currents to distant shorelines from the ocean. Numerical models of the daily tidal amplitudes, phases, and currents in three space dimensions and time are now routinely applied to simulate local tidal conditions anywhere in the tidal estuary (e.g. ADCIRC, Luettich *et al.*, 1992). When directly coupled to transport models of advection and diffusion of the discrete water volumes in the numerical model, they are now also routinely applied to study the salinity, sediment transport, and other water quality issues in the estuary. Model calibration employs the daily tidal variations (amplitudes and velocities) measured in the field.

In contrast to these daily spatial and temporal excursions of the tides, storm surge hydrograph durations (Section 2.2.3, Figure 2.2) last for hours or extend only for a few days before the average daily tidal conditions are restored. And, the spatial area is increased as the flooded areas are included in the numerical model simulations of storm surge. In highly populated urban areas, coupling the storm surge model with a flood inundation model at high resolution to estimate flood damage is an active research topic (e.g. Yin *et al.*, 2016).

A potential solution to the problem to mitigate storm damage with minimal ecological impacts are storm surge barriers with "movable" tidal flood gates that are only closed for short durations during the storm. (*In the legal profession, it is called "an open and shut case".*) The tidal flood gates are "movable" and remain open throughout the year so that they have the least possible negative impact on the existing ecology, environment, and possible navigational uses. When a "significant" coastal storm occurs, the "movable" tidal flood gates are closed to prevent the storm surge from upland inundation and flood damage. The gates are reopened when storm surge levels return to near-normal tidal conditions.

This section summarizes the possible functions of storm-surge barriers and focuses on the types of "movable" tidal flood gates that have been constructed throughout the world. The "movable" in quotes notation is employed to emphasize that the storm surge barrier considered here is not a fixed dam across the tidal estuary that would forever change the upland water body to a fresh-water lake. In the early stages of planning (1960s) of the Delta Project in the Netherlands following the 1953 floods when over 1,800 Dutch citizens lost their lives, a fixed protective barrier (Oosterscheldekering) across the entrance to the Eastern Scheldt tidal estuary was under consideration (Figure 6.91). The ecological and environmental significance of the estuary won out and even though more expensive, the now world famous, Eastern Scheldt (Oosterschelde 1978–1986) storm surge barrier to the North Sea was conceived and constructed. It has 62 movable, vertical lift gates each weighing 480 metric tons. In Figure 6.92, the top photograph shows closed gates and the bottom aerial view when the gates are open between the 65 concrete piers that range from 30 m to 39 m height (99–127 ft) and weigh 18,000 metric tons. Note that the Haringvlietdam and Browwersdam were earlier constructions of fixed barriers (closure dams) and produced freshwater lakes (light blue color) in Figure 6.91. The Dutch word "*kering*" comes from the Indonesian language and roughly translates to "dry" or almost free from liquid or moisture.

The Maeslantkering project (1991–1997) is also shown in Figure 6.91 on the Rotterdam waterway and will be discussed further in the following. The Dutch Delta Project was honored by the American Society of Civil Engineers (ASCE) in 2016 as one of the "Seven Wonders of the Modern World".

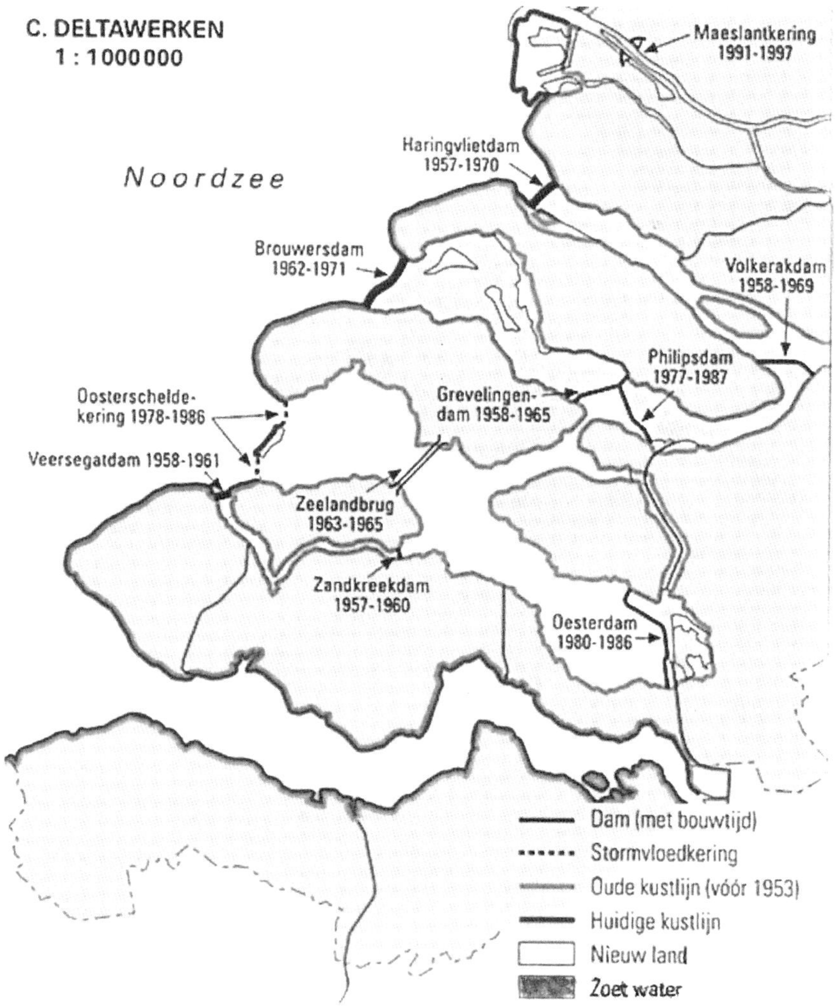

Figure 6.91. The Delta Works, The Netherlands (courtesy, Delta Works online, www.deltawerken.com). The Eastern Scheldt (Oosterschelde-kering).

A third type of coastal "barrier" has been constructed across locations with large tidal range to produce tidal power. An example is the Rance Tidal Power Station located on the estuary of the Rance River in Brittany, France. It opened in 1966 as the world's first tidal power station. This coastal barrier also serves to provide upland damage mitigation from storm

Figure 6.92. Photographs of the Eastern Scheldt storm surge barrier (top — gates closed) and (bottom — gates open) (courtesy, Delta Works online, www.deltawerken.com).

surge in the Atlantic Ocean. This third type of barrier is called a "tidal barrage" and the fixed barrier a "closure dam". Storm surge barriers come with some type of movable tidal flood gate and with locks for ship/boat navigation and are the focus of Section 6.5.

6.5.2 *Constructed storm-surge barriers*

6.5.2.1 *Definition of terms*

At the time of writing, there have been 20 storm-surge barriers constructed in Western Europe and the United States that fit the definition for three sections as schematically illustrated in Figure 6.93 (from Mooyaart *et al.*, 2014), namely, dam section, the movable, gated section, and a lock for ship and boat navigation past the barrier. The total barrier length L is the sum of these three sections but in some cases the gated section is not present, and the barrier only has locks for navigation. The gated section has a number of gates n with span (openings) of widths $a, b, \ldots$ so that the total width of the openings in the gated section is simply the sum of all the span widths. When the gate-span widths (a, b, etc.) are constant,

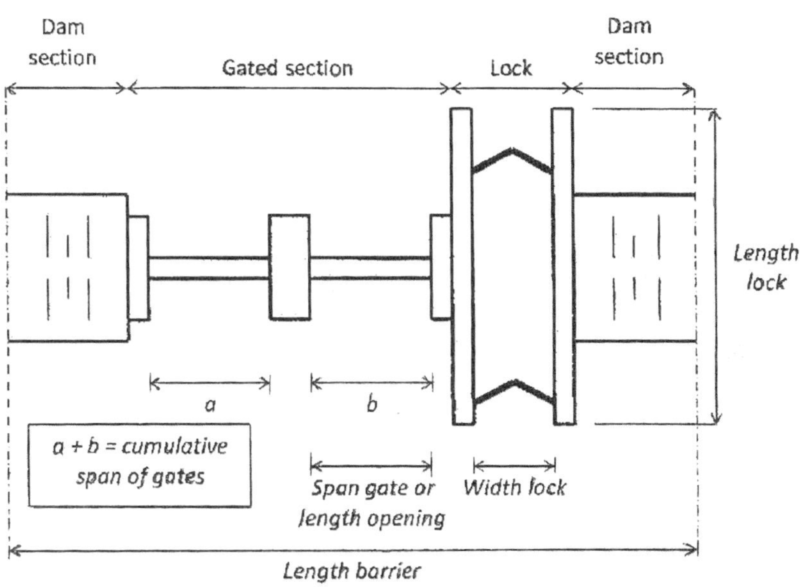

Figure 6.93. Schematic plan view of the three elements of a storm surge barrier (from Mooyaart *et al.*, 2014).

w_g, the total gated section opening, $W_g = nw_g$. Note the length of the gated section L_g is usually greater that the span openings W_g because of the machinery required to open and close the gates.

The function of the gated section of a storm surge barrier is to maintain the natural ecology of the tidal estuary by keeping the gates open except during storm events. The function of the lock section of the barrier is to maintain navigation. The locks of width(s) w_l are also normally kept open for the daily flood and ebb tidal flows past the barrier. In some cases, multiple locks are present so that the total lock width opening is W_l. Therefore, the total combined storm surge barrier opening $W_t = W_g + W_l =$ the gated section (ecology) plus the lock section (navigation) opening. For a total storm surge barrier length L_B, the ratio W_t/L_B represents the percentage of the tidal estuary that is impacted by the storm surge barrier.

6.5.2.2 *Types of movable tidal flood gates*

Movement to close and open the gates is possible in the vertical and/or horizontal directions. There are six types of tidal flood gates that move in both the gated section and the lock section as schematically illustrated in Figure 6.94 (from Dirke *et al.*, 2012). Photographs of examples of each type are shown in Figure 6.95.

(1) Mitre gates (top, left side)

Mitre gates have a long and successful history of use in locks on rivers and canals for ships moving from one water level to another level. When open and housed within the side wall panels, they have minimum interference with the daily tidal flows but small span widths ($<$30 m) restrict their use. The locks on the Panama Canal are shown in Figure 6.95 (top, left).

(2) Vertical lifting gates (top, middle)

Vertical lift gates are employed in the gated section for the storm surge barriers in the Eastern Scheldt (the Netherlands), the Ems (Germany), and at Seabrook, New Orleans (US). The gates are lifted vertically by hoisting cables or hydraulic pressure cylinders in adjacent towers. Large spans, ease of inspection and maintenance, and proven field experience are advantages of this type.

Figures 6.91 and 6.92 are photos of the Eastern Scheldt barrier that has become a popular tourist destination as shown in Figure 6.95 (top, middle) that helps pay for the annual operation and maintenance expenses. Visit www.deltawerken.com for complete technical details of the Delta Project in

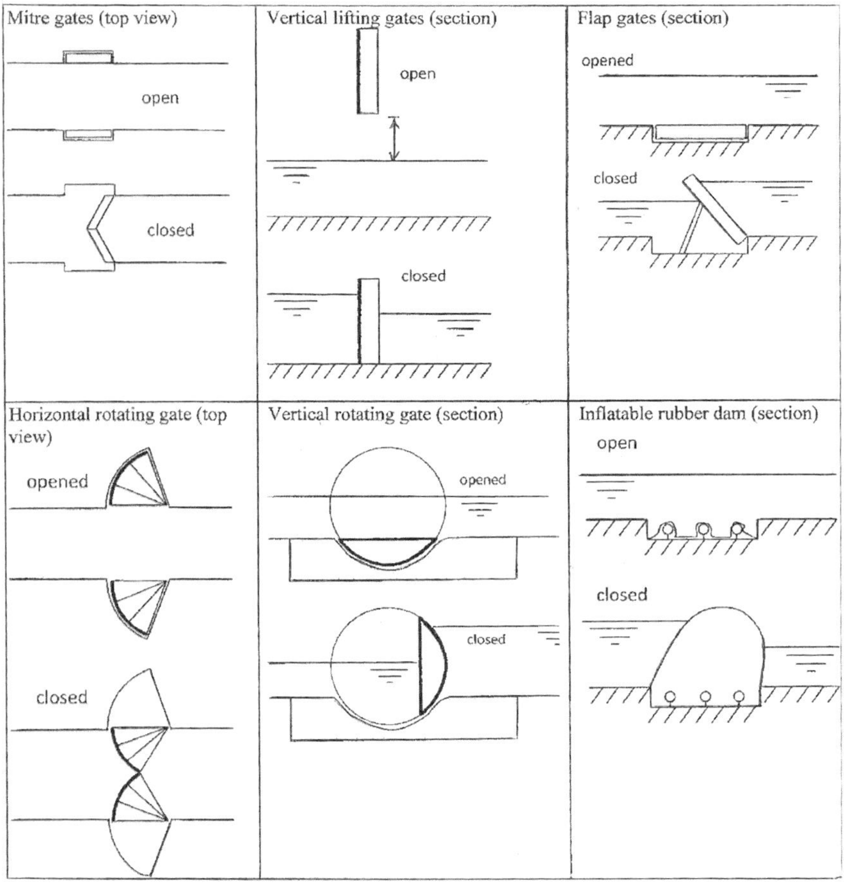

Figure 6.94. Six types of movable, tidal flood gates (from Dircke *et al.*, 2012).

the Netherlands. The gated section of the barrier can be closed in one hour. The Eastern Scheldt storm surge barrier is highly visible and provides a 9-km (5.6-mi) long roadway across the entrance to the Eastern Scheldt Estuary which is now a national park.

(3) Flap gates (top, right side)

Flap gates are not visible when open and lying on the bottom and not in use. The flap can be lifted by either hydraulic cylinders (Lock section, Stamford, Connecticut, US) or pneumatic (air) pressure in the hollow gate for buoyancy (gated section, MOSE, Venice, Italy).

Figure 6.95. Photographic examples of the six types of movable, tidal flood gates. Images courtesy of Rijkswaterstaat.nl (top middle, bottom left and right); MOSE project, Italy (top right); and Environmental Agency, UK (bottom middle).

Individual flaps of width w_g can be added together (n) for practically no limit on the total gated section opening, $W_g = nw_g$. Each individual flap can be housed in its own caisson and floated to the location site for deployment on a prepared bedding layer or piles. Sediment deposition on the flap and within the gate recess in the caisson may be a maintenance issue.

The three tidal inlets (Lido, Malamocco, and Chioggia) that funnel storm surges from the Adriatic Sea into the Venice Lagoon and flood the City of Venice are currently (at the time of writing this book) in the final stages of installation of air-lift flap gates for the storm surge barrier. Figure 6.95 (top-right) shows testing at the Malamocco inlet. Each flap gate is 20 m (65.6 ft) in length. See also Figure 1.4 in Chapter 1 for how it will work.

Figure 6.96 displays an aerial view of the Venice Lagoon and the three tidal inlets (top, left). A schematic diagram of the concrete caisson with tunnels for power, air-pipe equipment, inspection, and maintenance displaying the flap gates is on top, right. Dimensions of the caisson and flap gate raised in the closed position are found on bottom, left. An inspection trip of construction photo shows the scale of concrete caissons in the perspective to the tour group (bottom, right, Author's photo, May 2013). Each individual flap gate section can be removed for inspection, maintenance, and repair with spare sections available for immediate replacement.

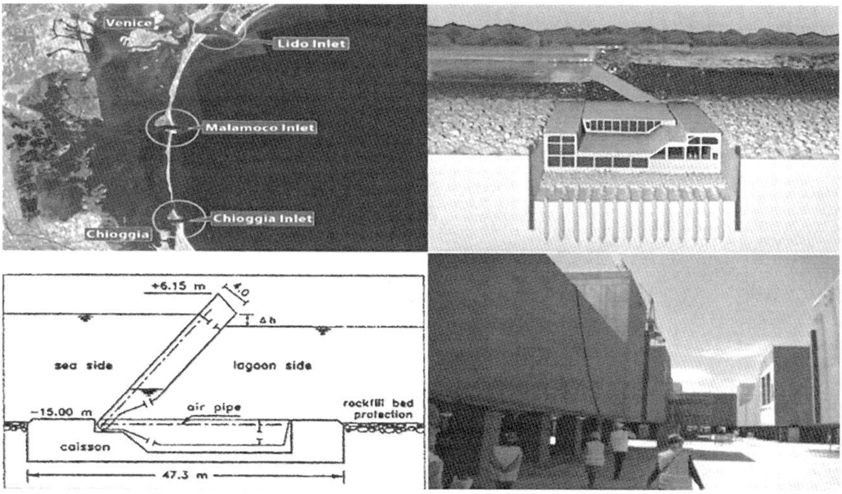

Figure 6.96. Locations of the tidal inlets to the Venice Lagoon; schematic diagram of the caisson on the bottom with flap gates flat (open) and raised (closed); dimensions of flap and supporting caisson structure; and photograph during construction tour (courtesy MOSE project, Italy (top, bottom left); and Author's photo, May 2013 (bottom right)).

The *MO*dulo *S*perimentale *E*lettromeccanico, *Experimental Electromechanical Module* (MOSE) storm surge barrier project for Venice, Italy was scheduled to complete construction and installation of the flap gates by June 2019 (water-technology.net).

(4) Horizontal moving or rotating gates (bottom, left side)

As illustrated in Figure 6.94 (bottom, left) two, quarter-circle structures form a lock section gate when rotating on a horizontal plane about the vertical axis to the fully closed position. When fully open, they rotate horizontally back inside the recessed area within the walls of the lock. The width required is roughly twice the width of the navigation lock. They are also referred to as "sector" gates in the literature (PIANC, 2006). A horizontal rotating gate structure with steel wheels on a concrete sill was designed and installed at New Bedford, MA (US) 1962–1966, by the US Army Corps of Engineers. The lock width was 46 m (150 ft).

Thirty years later, a 360-m (1180-ft) wide, horizontal rotating gate structure that floats was designed and constructed (1989–1997) at Maeslant on the Rotterdam waterway (Figure 6.91). Storm surge damage mitigation for over 1 million people living in the south of Holland and for Europe's busiest seaport that supports the Dutch economy is ensured. Figure 6.95 is an aerial photograph (bottom, left). The buoyancy reduces the hinge forces and permits the sill to be less uniform for seating the gate at closure. When the waterway is open, the gate storage area may be a dry-dock for easier visual inspection, maintenance, and repair. A similar floating, horizontal rotating gate was constructed at St. Petersburg (Russia) in 2011.

Horizontal moving gates that slide (not shown) are not practical for storm surge barriers. Daily tidal flows would be significantly impacted since the water cross-sectional area would be reduced by over 50% because the gate would require the same width (and depth) for storage when the gate section is open.

(5) Vertical rotating gates (bottom, middle)

Rotation now takes place in the vertical plane about a horizontal axis. The gate is a curved "segment" of the circle of rotation and when open lies flush with the bottom and within a curved recess area of the same shape as shown in Figure 6.94 (bottom, middle). In this position, navigation is possible past the gate. To close the opening, the gate is simply rotated 90° so that it stands upright against the elevated water levels on the seaward side. Inspection, maintenance, and repair are all then readily facilitated by further rotation of 90° so that the gate is completely out of the water. It is also called a rotating segment gate (PIANC, 2006).

The storm surge barrier gate section on the Thames tidal estuary near London (United Kingdom) is a world-famous example of this successful design as shown in Figure 6.95 (bottom, middle). It requires relatively complicated structures and machinery on both ends of the gate, but gate closures require only minutes to complete.

A second type of vertical rotating gate is simply a radial segment of a circle with the bottom edge on the bottom when closed against the storm surge. As the gate rotates upward to open water flows beneath the gate until fully open. This second gate type is also called a radial or Tainter gate since it has a long history of successful use on spillways to control the flows past dams on water resource engineering projects.

(6) Inflatable rubber gates (bottom, right side)

The last type of gate is relatively new. It is made of flexible material to form a sealed tube that when inflated with air and/or water inflates to a height and weight to resist elevated water events in its "closed" position (Figure 6.94, bottom, right). The material can be rubber, a synthetic fiber, or laminated plastic. There are no limits to the span width of the gated section, and it is invisible when not in use and uninflated lying on the bottom. Inspection for maintenance and repair is difficult. Sediments and other objects deposited over the gate lying on the bottom may cause problems.

The only known application is the inflatable rubber gates at Ramspol on the Ens tidal estuary in the Netherlands. In operation since 2002, it is installed for both the gate section and the lock as shown in Figure 6.95 (bottom, right). Rubber dams are used all over the world in river engineering but in relatively shallow waters. Closure of the barrier takes about one hour.

All the above six types of movable tidal flood gates have been constructed and are currently functioning in storm surge barriers in Western Europe and the United States. There is no one "best" type. Site-specific conditions and economics will govern the selection. A summary of the existing 20 storm surge barriers may help with a preliminary design.

6.5.2.3 *Summary of constructed storm-surge barriers (1958–2017)*

A summary of key physical dimensions of the 20 storm surge barriers that presently exist is presented in Table 6.27. In most cases they were studied, designed, and constructed following major hurricanes and storms that produced large storm surges and extensive physical damage and loss of lives in the region. For example, Hurricane Carol (1954) in the northeast US resulted in the Fox Point, New Bedford, Stamford, and Charles River projects by the Corps of Engineers in the 1960s and 1970s. The major North Sea storm in 1953 (again in 1962) resulted in the Thames and Eastern Scheldt projects in the UK and the Netherlands in the 1980s. More recently, Hurricane Katrina (2005) resulted in the Inner Harbor Navigation Canal (IHNC) project on Lake Borgne, and the Seabrook, Harvey Canal, and West Closure-GIWW projects near New Orleans on the Gulf of Mexico in 2011–2012. The dates in the fourth column indicate the start and end of construction.

The environment constraint on coastal engineering design began in the 1960s and was not considered by the US Army Corps of Engineers so that none of the northeastern US projects built had gated sections for ecology and environmental concerns. Both projects built in Germany at that same time did include movable tidal gates to aid the openings for the water flows in the daily tides.

Table 6.27. Constructed storm surge barriers in Western Europe and US, 1958–2017 (adapted from Mooyaart *et al.*, 2014).

Storm Surge Barriers		Constructed (1958 to Present date)			Gated Section-Ecology				Lock-Section-Navigation				
Number	Chronological Order Name	Location (Country)	Dates	Total Barrier Length, L_B (m)	Type Movable Tide Gate	Number Gates, n	Gate Width w_g (m)	Total Gated Width W_g (m)	Lock Type	Number Locks	Total Lock Width W_t (m)	Combined Total Opening W_t (m)	Percent Opening W_t/L_B
1	Hollandsche Ijssel	Krimpen (Netherlands)	1954–1958	200	—	—	—	—	Vertical Lift	1	24	80	0.40
2	Fox Point	Providence, RI (US)	1961–1966	215	—	—	—	—	Vertical Rotating (Tainter)	3	12.2	36.6	0.17
3	New Bedford	New Bedford, MA (US)	1962–1966	1,370	—	—	—	—	Horiz. Rotating (rollers)	1	46	46	0.03
4	Billwerder Bucht	Hamburg (Germany)	1964–1966	150	Flap (hydraulic)	2	30	60	Flap (hydraulic)	2	$(2 * 34 = 68)$	128	0.85
5	Stamford	Stamford, CT (US)	1965–1969	870	—	—	—	—	Flap	1	27	27	0.03
6	Eider	Tonning (Germany)	1967–1973	4,900	Vertical Rotating (Tainter)	5	40	200	Mitre	1	14	200	0.04
7	Charles River	Boston, MA (US)	1974–1978	120	—	—	—	—	Mitre	3	$(2 * 6.7)+(12.2) = 26$	26	0.22

(Continued)

Table 6.27. (*Continued*)

Storm Surge Barriers		Constructed (1958 to Present date)			Gated Section-Ecology				Lock-Section-Navigation				
Number	Chronological Order Name	Location (Country)	Dates	Total Barrier Length, L_B (m)	Type Movable Tide Gate	Number Gates, n	Gate Width w_g (m)	Total Gated Width W_g (m)	Lock Type	Number Locks	Total Lock Width W_t (m)	Combined Total Opening W_t (m)	Percent Opening W_t/L_B
8	Hull	Hull (United Kingdom)	1977–1980	40	—	—	—	—	Vertical Lift	1	30	30	0.75
9	Thames	London (UK)	1974–1982	530	Vertical Rotating (Tainter)	4	31	124	Vertical Rotating	6	$(2*31)+(4*61)=30$	430	0.81
10	Eastern Scheldt	Scheldt (Netherlands)	1973–1986	9,000	Vertical Lift	62	42	2604	Mitre	1	16	2,620	0.29
11	Maeslant	Rotterdam (Netherlands)	1989–1997	610	—	—	—	—	Horiz. Rotatings (floating)	1	360	360	0.59
12	Hartel	Spijkenisse (Netherlands)	1993–1997	250	—	—	—	—	Vertical Lift	2	$98+49$	147	0.59
13	Ramspol	Ens (Netherlands)	1996–2002	450	Inflatable Rubber	2	75	150	Inflatable Rubber	1	52	202	0.45
14	Ems	Gamdersum (Germany)	1998–2002	476	Vertical Lift	5	50–63.5	264	Vertical Rotating	2	$(60)+(50)=110$	374	0.79

(*Continued*)

Table 6.27. (*Continued*)

Storm Surge Barriers		Constructed (1958 to Present date)			Gated Section-Ecology				Lock-Section-Navigation				
Number	Chronological Order Name	Location (Country)	Dates	Total Barrier Length, L_B (m)	Type Movable Tide Gate	Number Gates, n	Gate Width w_g (m)	Total Gated Width W_g (m)	Lock Type	Number Locks	Total Lock Width W_t (m)	Combined Total Opening W_t (m)	Percent Opening W_t/L_B
15	St. Petersburg	S. Petersburg (Russia)	1955–2011	25,400	Vertical Rotating (Tainter)	64	24	1,536	Horiz. Rotatings (floating)	2	$200 + 110 = 310$	1,846	0.07
16	IHNC	New Orleans (US)	2008–2011	2,300	—	—	—	—	Vert Rotating+Vert Lift	3	$(2 * 45) + 17 = 107$	107	0.05
17	Seabrook	New Orleans (US)	2008–2011	130	Vertical Lift	2	15	30	Horiz. Rotating (rollers)	1	29	59	0.45
18	Harvey Canal	New Orleans (US)	2008–2011	120	—	—	—	—	Vertical Rotating (Tainter)	1	38	38	0.32
19	West Closure-GIWW	Southeast LA (US)	2008–2012	525	—	—	—	—	Vertical Rotating (Tainter)	1	69	69	0.13
20	MOSE	Venice (Italy)	2003–2018(?)	1,600	Flap (pneumatic)	1	360	360	Flap (pneumatic)	3	$3 * 400 = 1200$	1,560	0.98

Navigation is most important on the Thames tidal estuary for London. But this project also included a gated section that added almost 30% (124 m) to the total percent opening (last column) which was over 81% for this riverine location on the Thames. The Maeslant project is also mainly for navigation on the Rotterdam waterway but did not include a gated section. The side space required to store the floating, horizontal rotating gates reduced the total percent opening (last column) to about 60% on this riverine location.

As mentioned above, the gated section (ecology/environmental benefits) provides almost the total, opening width (2,620 m, last column) for the Eastern Scheldt project. The total barrier length L_B is 9,000 m (5th column) and includes over 2,400 m as the length of the Dam section (Figure 6.93) and called Neeltje Jans Island. This island was artificially built to construct the Eastern Scheldt project and resulted in a total percent opening (last column) of 29%. Today it is the location for the tourist office. Projects finished in the late 1990s and early 2000s (Ramspol and Ems) included gated sections that significantly increased the total percent opening. The vertical rotating gate section (64 Tainter gates each 24 m in width) also greatly aided the ecology of the Gulf of Finland at the Eastern end of the Baltic Sea for the St. Petersburg project. The total barrier length L_B was by far the longest in the list (25,400 m, 15.75 mi) so that the total percent opening was only 7%.

It is of interest to note that the IHNC project barrier on Lake Borgne was 2,300 m (1.43 mi) in length but did not include a gated section for ecological and environmental concerns. As a result, the total percent opening was only 5%. This was true also for the Harvey Canal and West Closure — Gulf Inter-Coastal Water Way (GIWW) projects. Perhaps, recent memory of the Hurricane Katrina (2005) disaster due to storm surge was too fresh during the design phase to warrant the potential delay needed to secure approval of the project when adding a gated section for ecology/ environmental reasons. Or, economics was a constraint.

The impacts of numerous storm surge events on the tourism industry is the primary reason for a storm surge barrier now being constructed for Venice in Italy. Section 2.2.3 in Chapter 2 reviewed the *acqua alta* (high-water) events that are increasing in frequency and magnitude which impedes tourists in San Marco Piazza. Planning actually began in the early 1970s following a major storm in the Adriatic Sea in November 1966. The MOSE project for Venice provides pneumatic flap gates for the three inlet openings to the Venice Lagoon for navigation but also includes a gated section that adds 23% to the total opening. The result is that the total percent opening is 98% of the barrier length and by far the most efficient from an ecological and environmental perspective. In addition, the flap gates are not a visible distraction to tourists visiting the City of Venice or the barrier island beaches fronting the Venice Lagoon. Only time will tell regarding the functionality of this flap gate design to mitigate storm surges and the economics of maintenance and repairs over the design life.

The data found in Table 6.27 has been compiled and checked from the following references: Morang (2007), Hillen *et al.* (2010), Hill *et al.* (2011), Dircke *et al.* (2012), Mooyaart *et al.* (2014), Mooyaart and Jonkman (2017), and PIANC (2006).

6.5.3 *Engineering design*

6.5.3.1 *Wind storm surge*

Wind blowing across the water surface creates waves (Section 2.3.7), produces wave setup (Section 2.3.6), and also creates horizontal momentum forces that push the water mass up against the shoreline boundary to cause storm surge, S. On tidal coasts, storm surge is the difference between the measured and the predicted tidal amplitude as illustrated in Figure 2.2 and is called the storm surge hydrograph. The complete, 3D, differential equations of motion (momentum conservation) and the continuity equation (mass conservation) are beyond the needs of this book. In principle, both local and advection of momentum, Coriolis forces, bottom-surface slope, surface atmospheric pressure, astronomical tides, wind stress, bottom boundary stress, and rainfall on the surface must all be taken into account. Computer models using numerical methods (see e.g. Luettich *et al.*, 1992) to discretize these differential equations are now employed, but many still need to make many further assumptions such as neglecting the vertical accelerations of the flow so that only the depth-averaged, two horizontal-direction momentum equations need to be employed as illustrated in Figure 6.97(a). Further simplification is possible considering only one horizontal direction as illustrated in Figure 6.97(b).

In the shoreward x-direction, the surface wind speed is $U \cos\varphi$ and, neglecting all other terms except the surface wind stress τ_w and bottom shear stress τ_b, a simple, one-dimensional (1D) model of storm surge s can be written where $D = d + s =$ the total water depth and ρg is the unit

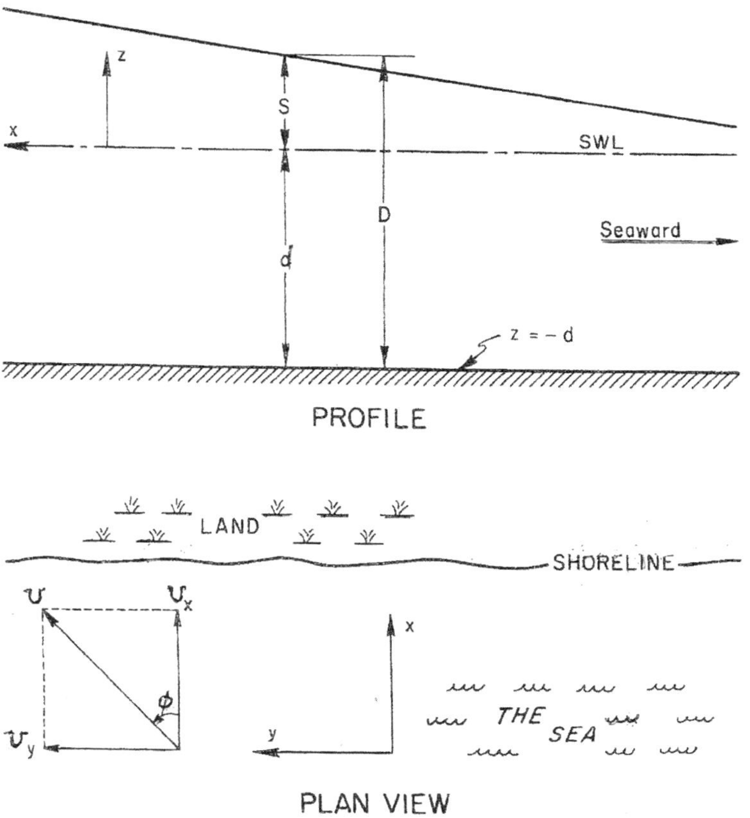

Figure 6.97(a). Definition sketch for depth-averaged, 2D horizontal storm surge calculations (from SPM, 1984).

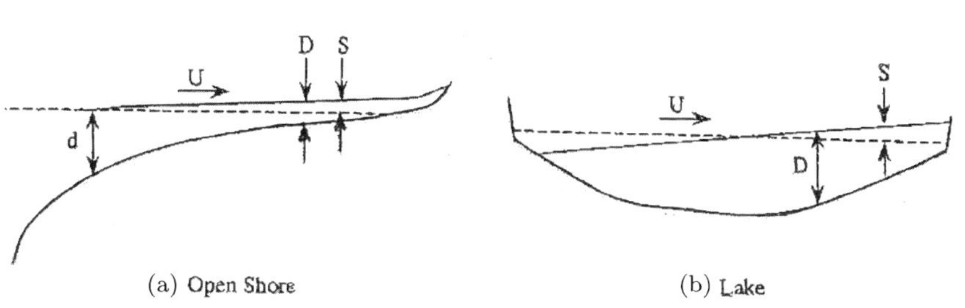

Figure 6.97(b). Definition sketch for depth-averaged, 1D (x-direction) horizontal storm-surge calculations on open shoreline and within enclosed basin (Lake) (from Kamphuis, 2010).

weight of water.

$$\frac{ds}{dx} = \frac{(\tau_w + \tau_b)}{\rho g\, D}. \tag{6.111}$$

The wind stress at the surface produces an extra hydrostatic pressure head ds over a distance dx that is just balanced by gravity and the hydrostatic pressure head for water depth, d.

The surface wind stress is commonly related to a near surface wind speed U and a surface drag coefficient C_D of the form

$$\tau_w = \rho_{\text{air}} C_D U^2. \tag{6.112}$$

A solution for Equation (6.111) still requires a formulation for C_D and for the bottom shear stress, τ_b. Further simplification is possible by relating the surface wind stress and bottom shear stress to one coefficient, λ.

Kamphuis (2010) suggests taking $\lambda = 3.2 \times 10^{-6}$ (a constant, metric units) so that Equation (6.111) becomes

$$\frac{ds}{dx} = \frac{\lambda (U\cos\varphi)^2}{g\, D}. \tag{6.113}$$

Clearly, as the water depth d decreases ($D = d + s$), the storm surge s increases. Solution of Equation (6.113) depends on the bottom profile dd/dx. A spreadsheet solution using a discrete, finite-difference form of Equation (6.113) (i.e. $\Delta s/\Delta x$) is possible if the wind speed U is assumed constant. An example explained in the following applies the spreadsheet that can be found in the online supplementary materials.

6.5.3.2 *Atmospheric pressure differential*

The HIGH pressure (H) and LOW pressure (L) symbols are common graphics shown on weather maps by the local TV weather man/woman on the evening news. They highlight the isobars of atmospheric pressures and explain why wind fields may be strong when the HIGH and LOW pressure differences are large and the distances (i.e. the pressure gradients) are small. When this pressure difference, Δp occurs over the water, an additional water-level change results. Since atmospheric pressures are usually measured by mercury barometers, some have called this the "barometric surge" (Kamphuis, 2010).

Civil/coastal engineers are comfortable when working with pressure "heads" so that the additional water-level rise, Δh is simply

$$\Delta h = \frac{\Delta p}{\rho g}, \tag{6.114}$$

where ρ is the mass density of water. For sea water, $\rho = 1025\,\text{kg/m}^3$ ($1.99\,\text{slugs/ft}^3$). Recall that pressure in the metric system using SI units is Pascal (Pa) or Newton/m^2 (force/unit area) and as pounds/ft^2 (lbs/ft^2) in Imperial units ($1.0\,\text{N/m}^2 = 0.0208854\,\text{lbs/ft}^2 = 0.00014504\,\text{lbs/in}^2$). Since pascals is a very small pressure unit, Δp is usually expressed as kPa (kilopascal). One kilopascal = 0.2953 inches of mercury = 10 millibars at 32°F.

One standard of atmospheric pressure (sea level at 32°F) is 101.325 kPa or 1.01325 "bars" and again a reference to measurements with mercury barometers. One standard atmosphere equals 29.92 inches of mercury or 14.696 lbs/in^2 (psi) or 1,013.25 millibars. When atmospheric conditions produce tropical storms (hurricanes) and extra-tropical depressions (Northeasters' in the US), the center of low pressure is referenced to these standard atmospheric pressures. Pressure drops of about 0.3 inches of mercury (1.0 kPa) produce 0.1 m (0.1 ft) of water-level rise. When the pressure drops by about 3 inches of mercury (10.0 kPa) and the barometer reads near 27 inches of mercury (913 millibars), major hurricanes are possible. Hurricane Katrina in the Gulf of Mexico recorded 26.64 inches of mercury (902 millibars) to theoretically create a pressure head differential, $\Delta h = 1.1\,\text{m}$ (3.6 ft) above the storm surge when applying Equation (6.114). The record low pressure in the US Atlantic Ocean is 26.04 inches of mercury (882 millibars) to give $\Delta h = 1.3\,\text{m}$ (4.3 ft) in Hurricane Wilma (2005). The lowest recorded low pressure was 880 millibars (25.99 in Hg) during Hurricane Patricia (2015) in the eastern Pacific Ocean.

6.5.3.3 *Fundamental oscillation modes (seiche)*

A third mechanism for elevated water levels along the coast is the potential for the oscillation (periodic swinging) of the entire water body at fundamental modes and called a *seiche* (French). A standing wave is the first (fundamental) mode. Recall that in relatively shallow water of depth d, the celerity c (L/T) $= \sqrt{gd}$ for a wave moving across the surface (Chapter 2). For a 1D basin with open end as illustrated in Figure 6.97(b, left) at the start with a flat surface, a wave disturbance starting at the open end would have to travel the basin length L four times to complete a full oscillation at the fundamental mode. The total time for one complete oscillation is called the fundamental period, $T_m(L/c)$. Hence, we get the following.

(1) Open-ended basin

$$T_m = \frac{4\,L}{\sqrt{gd_{\text{mean}}}}, \tag{6.115}$$

where d_{mean} is the average or constant depth of the basin.

When the basin is closed at both ends (Figure 6.97 (b, right)), the wave disturbance would only need to travel one-half the basin length ($L/2$) four times to complete a full oscillation at the fundamental mode. Therefore, we obtain the following.

(2) Closed basin

$$T_m = \frac{2\,L}{\sqrt{gd_{\text{mean}}}}. \tag{6.116}$$

Both the winds blowing along the water surface to create a storm surge and the atmospheric pressure differential are potential ways this fundamental mode of oscillation (*seiche*) can be produced. Higher harmonic modes of oscillation are possible and may be estimated by using multiples of the celerity c in the denominator of the above equations but are rarely found in practice.

6.5.3.4 *An example*

Participants who attended the 23rd International Conference on Coastal Engineering (ICCE) on the opening day (Monday October 5, 1992) were treated to a spectacular sight. Many of the world's most famous coastal engineers came barefoot with pants rolled up and briefcase, socks and shoes held high above the flood waters in Venice, Italy, site of the conference. The Italians could not have planned for a more fitting introduction to many of the papers to be presented that week at the conference. The final paper in the proceeding (ICCE, 1993) presented a summary of this *acqua alta* event and was written by the conference organizing committee (De Boni *et al.*, 1993). Our thanks to these gentlemen for descriptions of the weather conditions that produced the measured tidal conditions and atmospheric pressures in piazza San Marco (St. Marks' square) reproduced here as Figure 6.98. The vertical datum (0.0 approximately MLW) is for the Punta Della Salute tide station maintained by the Permanent Service for Mean Sea Level (PSMSL) for the Italian government. During the peak,

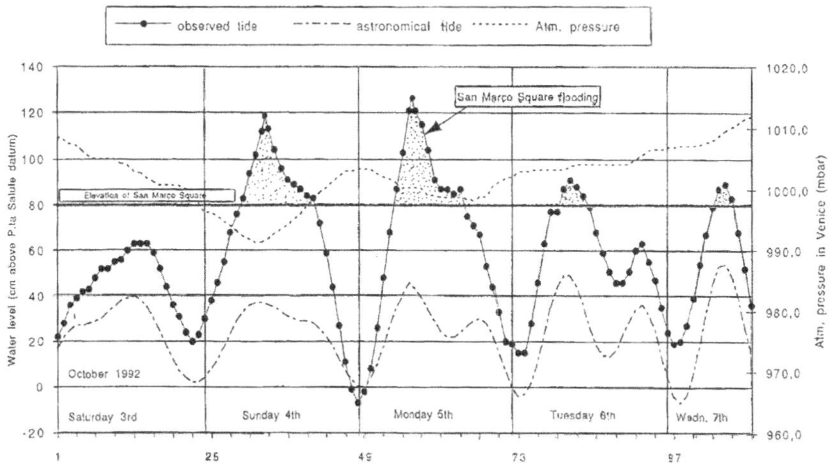

Figure 6.98. Total water levels (cm) and atmospheric pressures (millibar) measured in St. Mark's square from Saturday, October 3 to Wednesday, October 7, 1992 relative to the Venice city datum (from De Boni *et al.*, 1993).

about 46 cm (1.5 ft) of water covered piazza San Marco on that Monday, October 5 morning at 9:00 am, at the time of the opening presentations for the Conference which is 126 cm above the vertical datum. The highest water level ever recorded (November 4, 1966) was 114 cm (3.74 ft) above piazza San Marco or 194 cm (6.37 ft) above vertical datum. It produced extreme flooding damage throughout all of Venice. This record elevation reached 2.5 times higher above piazza San Marco than shown in Figure 6.98.

A low-pressure center located over the north end of Italy created steady winds of about 25 knots (13 m/s) in the north-westerly direction along the Adriatic Sea (Figure 6.99). The measured atmospheric pressure drop, Δp was $(1005-991) = 14$ mbar from early Saturday October 3 to the first peak tide on Sunday, October 4 when participants were arriving in Venice for the conference. The Adriatic Sea is approximately 800 km (500 mi) in length, L, from the open end at the heel of the boot of Italy up to Venice. The depth decreases from the entrance with d_{mean} about 150 m (500 ft) and is sea water ($\rho = 1{,}026\,\text{kg/m}^3$).

The above data presented in Figure 6.98 and obtained via the internet provided sufficient detail to apply the theory of Equations (6.113)–(6.114) to check the measurements which produced a total storm surge, $S =$ (measured − predicted) $=$ (126 cm − 46 cm) $=$ 80 cm (2.6 ft) on Monday, October 5, 1992.

Figure 6.99. Map of Italy and the Adriatic Sea (courtesy World Maps, free Adriatic Sea on Europe Map).

(1) Wind-surge component, s

Writing Equation (6.113) in finite-difference form gives Equation (6.117) which is Column [5] in Figure 6.100 for this example using the spreadsheet developed for this book.

$$\Delta s = [(\lambda U^2)/(g(d_{\text{mean}} + s)]\Delta x. \qquad (6.87)$$

Only six sections were utilized (column [1]) to demonstrate how Equation (6.117) is solved by the finite-difference method. The first few

Storm Surge Calculator in 1-D
Example -Adriatic Sea (Oct 3-4, 1992)

Length, L (m)	800000	Wind speed, U (m/sec)	13	Gravity, g (m/sec^2)	9.81	Sea water, ρ (kg/m^3)	1023
Average depth, d (m)	150	Constant, λ	0.0000032	Initial s (m)	0	pressure diff, Δp (mbar)	14

[1] Section	[2] Δx (m)	[3] Water depth, d (m)	[4] Total depth, D (m)	[5] Δs (m)	[6] Storm Surge, s (m)	[7] Weighted Depth (m)
0	0	250	250		0	
1	240000	250	250.000	0.05292	0.05292	60000000
2	160000	200	200.053	0.04409	0.09701	32000000
3	160000	150	150.097	0.05876	0.15578	24000000
4	80000	50	50.156	0.08793	0.24371	4000000
5	80000	35	35.244	0.12513	0.36884	2800000
6	80000	15	15.369	0.28696	0.65580	1200000
	800000					124000000
Seiche					Mean depth, d (m)	155
Open-ended basin	Seconds	Hours				
Period, Tm (sec)	82063.42251	22.8				
Tm= 4 L/ Sqrt (gd)			Δs =[(λ*U^2)/(g*(dmean+ s)]*Δx			

Figure 6.100. Calculation of wind-storm surge s, for example Adriatic Sea storm and Venice.

sections (Δx in column [2]) are spaced further apart since Δs is small to begin. Column [3] gives the estimated water depths d in each section. Equation (6.117) is employed for Section 1 using $\Delta x = 240{,}000\,\text{m}$ and taking the initial wind surge $s = 0$ to give $\Delta s = 0.05292$ and hence $s = 0.05292$ (column [6]) for Section 2). Adding $d + s = 200 + 0.053 = 200.053$ for D in column [4] for Section 2. Applying Equation (6.87) for Section 2 ($\Delta x = 160{,}000$) gives $\Delta s = 0.04409$ and hence $s = 0.05292 + 0.04409 = 0.09701$ and the calculation continues. The total wind surge, $s = 0.656\,\text{m}$ at the end of Section 6 (Venice) in this relatively crude example as found in column [6].

(2) Atmospheric pressure head component, Δh

Applying Equation (6.114) gives

$$\Delta h = \frac{\Delta p}{\rho g} = \frac{14 \times 100}{1023 \times 9.81} = 0.1395\,\text{m}.$$

Combined, the total storm surge $S = s + \Delta h = 0.796\,\text{m}$ and remarkably close to the 0.8 m measured that morning.

These results are mainly to illustrate the method to calculate s. For example, using $d = 20\,\text{m}$ in Section 6 gives $s = 0.59\,\text{m}$ for the total wind surge. Column [7] is a weighted depth for each section and a check that the mean depth is near 150 m for the entire open-ended basin.

(3) Fundamental oscillation mode, T_m

As reported by the organizing committee (De Boni *et al.*, 1993) a seiche component with period $T_m \approx 21$ hours was also evident in the water-level recordings (Figure 6.98). Scaling the time between peaks in Figure 6.98

gives values in the 22–23 hour range because of differences in the daily peaks of the semi-diurnal tides. For an open-ended basin (Equation (6.115)),

$$T_m = \frac{4\,L}{\sqrt{gd_{\text{mean}}}} = \frac{4 \times 800000}{\sqrt{9.81 \times 155}} = 83699\,\text{s} = 23.3\,\text{hours}$$

which is reasonable. Note that the phase of this fundamental seiche mode coincided with the daily, semi-diurnal tidal periods.

In summary, this example demonstrates the usefulness of relatively simple theoretical relationships for estimating storm surge levels in the early stages of decisions for selection of the type of movable storm-surge barrier (Figure 6.94). For detailed design, fully 3D numerical models are the best.

One final note for this example. The mean predicted tide range was about 40 cm. Should this storm event have occurred during spring tide conditions when the predicted high water level is 75 cm, the total water flood level would have been 35 cm higher and over 80% of Venice would have been flooded. The 23rd ICCE would then probably have been cancelled on that Monday, October 5, 1992. See Kamphuis (2010) for further discussion.

6.5.3.5 *Water level and wave loads on curve walls*

An additional factor in the design of movable, storm surge barriers are water and wave loads on curved surfaces.

(1) Hydrostatic forces

In still water, the pressure p at any depth h is equal in all directions, $p = \rho gh$, and therefore is always normal to the curved surface. Integration of this pressure distribution over the curve is required to determine the resultant hydrostatic force f on the surface. It is often easier to compute the horizontal force on the vertical projection of the curved surface f_h and its location on this vertical plane, $2/3\,h$ below the water surface. To compute the weight of water above the curved surface f_v and the centroid of the area (volume per unit length) to determine the location of f_v. Then, the resultant hydrostatic force f is simply the vector sum of these horizontal and vertical force components. For vertical, rotating gates of the sector type (Tainter gates), this resultant force must pass through the center of rotation. For vertical, rotating gates of the Thames barrier-type, when in the fully closed, upright position (Figure 6.94, bottom-middle), the vertical force is the vector sum of the water volume forces both above and below the curved surface. The upward vertical force is the weight of the total volume of water above the curved surface.

For horizontal, rotating gates, two horizontal force components exist, $f_h(x)$ and $f_h(y)$. Both can be calculated as components on vertical planes in the x- and y-directions and when combined vectorially give the resultant horizontal radial force F_r acting through the center of rotation of the horizontal gate.

(2) Wave forces

Vertical, rotating gates are the only type with a vertical curved surface of concern for wave forces. There exists very little experimental data and references regarding both irregular and breaking wave forces on curved walls. In practice, calculating the horizontal component of the wave force on a vertical projection of the curved surface should produce conservative and satisfactory results.

6.5.3.6 *Costs of constructed projects*

In the early planning stages of alternatives for storm damage mitigation, it is helpful to have cost prediction methods to estimate the costs of storm surge barriers. As shown in Figure 6.93, the three elements of a storm-surge barrier are the dam section, the gated section, and a lock section that together form the total barrier length L_B between the banks of the tidal estuary. Jonkman *et al.* (2013) determined that costs for the gated width W_g and navigation lock width W_l when combined to give the total opening width W_t are 10 to 100 times more expensive than the costs for the dam section.

Mooyaart *et al.* (2014) published total cost data for 15 of the constructed projects listed in Table 6.27 which is adapted in Table 6.28 showing both the total barrier length L_B and the total opening width W_t in meters. Number 16 in the table for the IHNC project in New Orleans is the only one where the construction costs are only for the gated section. All the 15 constructed project costs were converted to costs in year 2014, C_{2014} in millions of Euros, M (€) [second last column] by applying the present worth formula

$$C_{2014} = C_n(1 + r)^{2014-n}, \qquad (6.118)$$

where C_n is the initial construction cost in year, n, and r is the construction-cost index rate (%).

The initial construction costs were in different currencies (f = Dutch Florins, £ = British pounds, €= European Euros, and \$ = US dollars) so all were converted to Euros (€) using currency exchange rates at the time

Table 6.28. Costs of 15 constructed storm-surge barriers in Western Europe and US, 1958–2017 (adapted from Mooyaart *et al.*, 2014).

Storm Surge Barriers Number	Chronological Order Name	Constructed (1958-to date) Location (Country)	Dates	Total Barrier Length, L_B (m)	Combined Total Opening W_t (m)	Cost Year, n	Initial Const. Cost in Year, n Millions, M(*)	Currency Exchange Rate	Construction Cost Index Rate, r(%)	Construction Cost in 2014 M(€)	Cost per Meter in 2014 M(€)/m
1	Hollandsche Ijssel	Krimpen (Netherlands)	1954–1958	200	80	1956	40 M(f)	0.45	4.0	173	2.16
2	Fox Point	Providence, RI (US)	1961–1966	215	36.6						
3	New Bedford	New Bedford, MA (US)	1962–1966	1,370	46	1966	19 M($)	0.74	4.8	122	2.65
4	Billwerder Bucht	Hamburg (Germany)	1964–1966	150	128						
5	Stamford	Stamford, CT (US)	1965–1969	870	27	1969	15 M($)	0.74	4.8	83	3.07
6	Eider	Tonning (Germany)	1967–1973	4,900	200	1973	87 M(€)	1.00	2.9	275	1.38
7	Charles River	Boston, MA (US)	1974–1978	120	26						

(*Continued*)

Table 6.28. (*Continued*)

Storm Surge Barriers Number	Chronological Order Name	Constructed (1958-to date) Location (Country)	Dates	Total Barrier Length, L_B (m)	Combined Total Opening W_t (m)	Cost Year, n	Initial Const. Cost in Year, n Millions, M(*)	Currency Exchange Rate	Construction Cost Index Rate, r(%)	Construction Cost in 2014 M(€)	Cost per Meter in 2014 M(€)/m
8	Hull	Hull (United Kingdom)	1977–1980	40	30	1980	2.7 M (£)	1.20	4.3	19	0.63
9	Thames	London (UK)	1974–1982	530	430	1984	467 M (£)	1.20	3.8	1,667	3.88
10	Eastern Scheldt	Scheldt (Netherlands)	1973–1986	9,000	2,620	1986	2360 M(€)	1.00	2.5	4,602	1.76
11	Maeslant	Rotterdam (Netherlands)	1989–1997	610	360	1997	450 M(€)	1.00	2.5	668	1.86
12	Hartel	Spijkenisse (Netherlands)	1993–1997	250	147	1997	98 M(€)	1.00	2.5	145	0.99
13	Ramspol	Ens (Netherlands)	1996–2002	450	202	2002	100 M(€)	1.00	2.8	136	0.67
14	Ems	Gamdersum (Germany)	1998–2002	476	374	2002	290 M(€)	1.00	2.7	387	1.03

(*Continued*)

Table 6.28. (*Continued*)

Storm Surge Barriers Number	Chronological Order Name	Constructed (1958-to date) Location (Country)	Dates	Total Barrier Length, L_B (m)	Combined Total Opening W_t (m)	Cost Year, n	Initial Const. Cost in Year, n Millions, M(*)	Currency Exchange Rate	Construction Cost Index Rate, r(%)	Construction Cost in 2014 M(€)	Cost per Meter in 2014 M(€)/m
15	St. Peters-burg	St. Petersburg (Russia)	1955–2011	25,400	1,846	2010	4,500 M (£)	1.20	6.8	6,578	3.56
16	IHNC	New Orleans (US)	2008–2011	2,300	107	2011	550 M($)**	0.74	2.5	425	3.97
17	Seabrook	New Orleans (US)	2008–2011	130	59	2011	165 M($)	0.74	2.5	127	2.15
18	Harvey Canal	New Orleans (US)	2008–2011	120	38						
19	West Closure-GIWW	Southeast LA (US)	2008–2012	525	69						
20	MOSE	Venice (Italy)	2003–2018(?)	1,600	1,560	2011	4,700 M(€)	1.00	3.0	4,986	3.20

Notes:
f = Dutch Florins
£ = British pounds
€ = Euros
$ = US dollars
** Only gated section costs

of construction. The values employed for both the construction cost index rate r and the currency exchange rates are subject to discussion, but do not change the final results.

Mooyaart *et al.* (2014) determined that total costs were more accurately predicted when the total opening width W_t was employed to determine cost per unit length. The results for the 15 constructed projects are listed in the last column in Table 6.28 in millions of Euros per meter, M (€/m). The mean of these 15 projects was 2.2 M (€/m) with standard deviation of the mean $\sigma = 1.2\,\text{M}$ (€/m) with a correlation coefficient, $R^2 = 0.84$ for a straight-line, linear fit of the data.

Hence, for early planning stages of a study of alternatives for construction costs, in millions of Euros per meter of length M (€/m) for storm-surge barriers use

$$C_{2014} = 2.2W_t \quad \text{with } \sigma = 1.2, \tag{6.119}$$

where W_t is the total opening width.

Additional cost-prediction methods that separated relatively simple and more complex structures with larger spans and unlimited clearance for navigation, and Dutch versus German barriers gave more accurate prediction formulas. Consult Mooyaart *et al.* (2014) for details but the limited data base may be a factor in these results.

Data in Table 6.28 and Equations (6.118) and (6.119) can be employed to estimate the initial construction costs in the current years and other currencies. New projects have been completed and/or are currently proposed that will add to the data base. Aerts *et al.* (2013) estimated that annual maintenance costs are 0.5–2% of the initial construction costs. Life-cycle costs must always be estimated for each alternative considered to mitigate damages from coastal flooding.

The bottom line is that the storm-surge barrier alternative with wide openings ($W_t/L_B \approx 1$) for minimal interference with daily tidal flows is expensive.

6.5.3.7 *Ecological and environmental concerns*

The gated sections remain open for minimal interference with the normal tidal-flow conditions. Mooyaart *et al.* (2014) also present some flow properties for the 15 competed projects. They plotted the peak volumetric flow rate $Q_p(\text{m}^3/\text{s})$ during average tidal conditions versus the open cross-sectional area $A_w(\text{m}^3)$ at the barrier location. The open cross-sectional area

$A_w = (W_t * d_s)$ where d_s is the average depth of the sill below MSL. As seen in Table 6.27, some projects only had navigation, lock openings, some with only ecology, gated section openings, and some have both lock and gated sections' combined openings. All 15 projects had peak tidal velocities V_p between 0.5 and 5.0 m/s. See Mooyaart *et al.* (2014) for a detailed analysis of all three cases and whether the size of the opening area was driven by navigation (lock) or ecology (gated) reasons or both.

Water quality in the tidal estuary is related to the average residence time, $\tau_{\text{days}} \approx d/(4R)$ for semi-diurnal tides where d is the average water depth (m) and R is the average tidal range (m). It was found that for barriers with high average residence times ($t_{\text{days}} > 10$), the peak tidal velocities were low ($V_p < 0.5$ m/s) and for those with low average residence times ($t_{\text{days}} < 1$) the peak tidal velocities were higher than 0.5 m/s. They concluded that barriers with relatively low average residence times need larger openings to preserve sufficient water quality.

For final design considerations, coupled hydrodynamic and water quality models should be employed to determine an acceptable number of span widths for the total gated section width W_g that satisfies both economic and environmental constraints.

For an update, see Mooyaart and Jonkman (2017) that includes an appendix with photographs and many more details for most of the storm surge barriers listed in Table 6.28.

6.5.3.8 *Multi-disciplinary design team*

For design of storm surge barriers, coastal engineers are part of a multi-disciplinary team that includes civil (structural, geotechnical, environmental), mechanical, and electrical engineers. The gate and lock sections have many moving steel parts and are controlled by electrical signals for their closing and reopening operations. Gauges to measure water levels sent to a control room are critical for "trigger" decisions to be made regarding when to close the gate and lock sections.

Physical oceanography and ecological specialists must also be a part of the multi-disciplinary operation team.

6.5.3.9 *Summary*

The risk of flood damage in coastal areas of the world is increasing due to ever larger populations living in coastal regions and rising sea levels. Storm-surge barriers with movable, gate sections will be increasingly viable as an alternative to mitigate flood damage because they are designed to have minimal impact on the environment. New projects (at the time of this writing) are being considered for New York (US), Houston (US), Norfolk (US), Niewpoort (Belgium), Hamburg (Germany), Gotenborg (Sweden), Tokyo (Japan), Shanghai (China), and many others. The overview and design considerations of the 20 constructed projects till date as summarized herein should provide some help in the future.

6.5.4 *Urban rainfall drain pipe systems*

One additional coastal structure that serves (indirectly) as a storm surge barrier in urban areas is the rainfall/runoff drainage pipe system that starts at a curb/gutter inlet on the edge of the road and ends in the pipe outfall to the sea. The pipe system is designed for some recurrence interval for rainfall intensity to safely drain the roadways for vehicle traffic. Pipe diameter, length, and head difference between the curb/gutter inlet and the invert of the discharge pipe at the seaward end are key variables in the design as the free surface, open channel flow in the pipe. Normally, no valves (gates) are installed in the pipe and that is the problem.

Elevated water levels at the seaward end of drainage pipe systems fill the pipe and discharge tidal water through the curb/gutter (outlet) and flood the roadway. Closures of the roadway may occur that significantly disrupt normal traffic flows, public services (fire, police, ambulance), and evacuation routes during storm events. These tidal flooding events have come to be labeled as "nuisance" or "sunny day" tidal flooding because they often take place when no rainfall occurs. The elevated water levels are caused by wind shear stress, atmospheric pressure differential, and/or oscillation modes (*Seiche*) as discussed earlier and often coincide with the high levels of the predicted, gravitational tide.

The National Weather Service (NWS) within the US National Oceanic and Atmospheric Administration (NOAA) identifies weather threats from tidal flooding and issues NWS Coastal Flood Warnings at the following four levels:

Minor (Nuisance flooding, non-structural, roads) and labeled Coastal Flood Advisory

Moderate (Starts to inundate lowest homes/businesses and roads)

Major (Significant flooding of structures, roads, life, and property)

Record (Higher than ever recorded by NOAA tide gauges)

The local NWS, Weather Forecasting Office (WFO) is responsible for establishing the water level above normal, mean higher high-water (MHHW) conditions that is labeled as "Minor" nuisance flooding for the local region. Recent studies by NOAA (Sweet and Marra, 2014) of 27 long-term tide gauges around the United States have found that the number of days of "nuisance" tidal flooding is increasing every year and is caused by local, relative sea-level rise. For example, in Norfolk, VA, the "nuisance" level is set at 0.53 m (1.7 ft) above MHHW by the local WFO. For the period 1950–2013, the linear trend line gave one day in 1960 and eight days in 2014 with the outlook for 18 days of tidal "nuisance" flooding each year in the near future.

One alternative to mitigate "nuisance" tidal flooding is to raise the roadway and the curb/gutter inlet elevation. A second, more economically viable alternative is to implement a pipe valve (gate) closure system with routine maintenance and trigger levels for closure (and reopening) of the valves. The basic gate type is a flap gate that is self-closing when the water pressure (water depth) forces the gate to seal itself in the closed position. Figure 6.101 displays steel-type (top, middle, circular) and (top, right, square) flap gates. The gate material may also be stiff rubber and curled on the end and called the "Duck-bill" type for closure. A traditional gate value requires manual or automatic means to turn the handle to seat

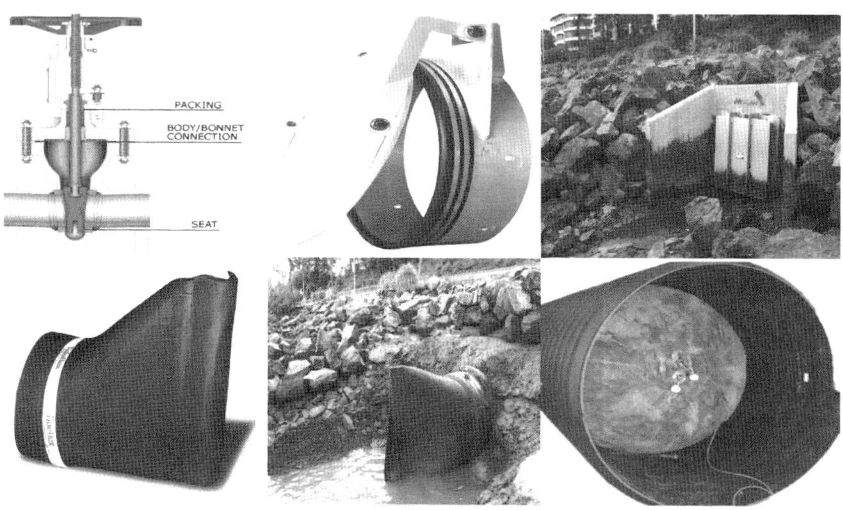

Figure 6.101. Examples of tidal gate values (courtesy Red Valve, Pittsburg).

the gate for closure (top, left). A relatively new type is an inflatable rubber tube that requires manual or automatic operation of a pressure system (air, water, etc.) for closure (bottom, right) in Figure 6.101.

All types require maintenance to keep the closure seals clear of debris and clean for closure. Unfortunately, in most applications they have been installed in the field and then forgotten and expected to last without any periodic maintenance and testing to ensure they function properly. This was the case in Norfolk, VA when the city, Public Works Department studied this alternative and concluded that they:

> "*Often malfunction and are maintenance intensive by clogging with debris, and experience salt water corrosion of the metal and deterioration of the rubber seals*" (Lentz, 2010) as illustrated in Figure 6.102.

This tidal flood gate alternative to mitigate "nuisance" tidal flooding needs a systems' approach to ensure its functional reliability. An example is illustrated in Figure 6.103. The top, left photo is Hampton Blvd in Norfolk, VA (US) which is a major transportation corridor for the Norfolk Naval Base, Old Dominion University, and the Eastern Shore Medical Center in Norfolk and shows the curb/gutter inlet. The sign on the gutter shows a blue crab and reads "DO NOT DUMP — DRAINS TO WATERWAY" (top, middle). The top, right-side photo is the drainage outlet pipe with an open end between some stones and the tidal marsh at low-tide level. A car in the background on the road gives some scale to the elevation drop. During "nuisance" tidal and storm surge tidal conditions, the elevated water levels fill the drainage pipe, backs water up to the curb/gutter opening where it discharges and fills the roadway with water to interrupt normal traffic conditions on Hampton Blvd.

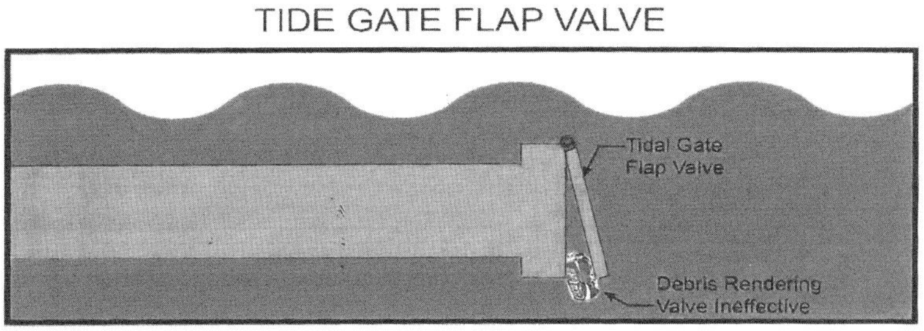

Figure 6.102. Schematic diagram of flap valve-type tidal flood gate (from Lentz, 2010).

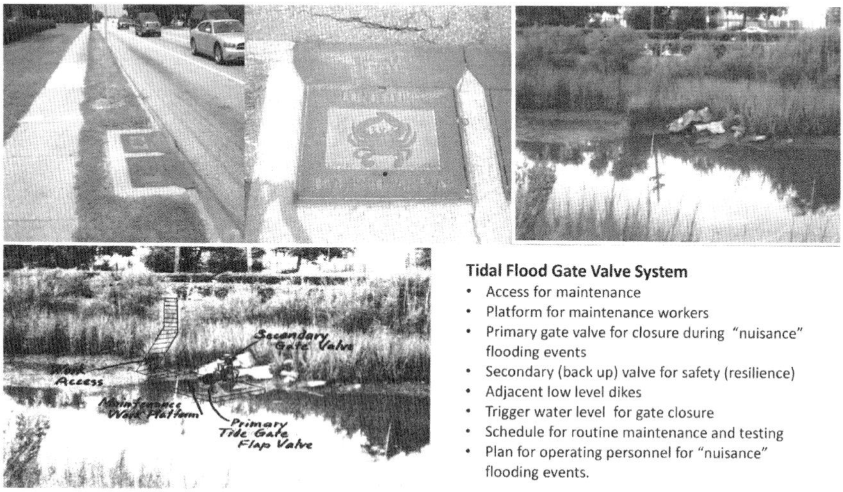

Figure 6.103. Example of tidal flood gate value system for mitigation of "nuisance" tidal flooding events (Author's photos).

A tidal flood gate valve "system" is illustrated in Figure 6.103 (bottom, left) with eight elements required for functional reliability (bottom, right). The first is accessibility for maintenance and a work platform is needed in front of the gate(s) to stand on when performing maintenance and testing. The primary gate could be the flap-valve type and a secondary gate could be a traditional gate value for redundancy. When "nuisance" and/or storm tidal conditions are predicted, the gates should be tested and the secondary gate closed (by hand if required) whenever the primary gate is not working properly. The fifth item listed is "... Adjacent, low level dikes" so that mitigation of roadway flooding from seaside water levels is ensured up to the design elevation of the crest of the dikes.

The water level to "trigger" closing the tidal flood gates in urban rainfall, drainage pipe systems depends on many factors. One is the importance of the roadway as determined by the daily and rush-hour traffic counts and the destinations of the users of the roadway. A second is the availability (or not) of alternate traffic routing pathways. Schedules for routine maintenance, testing, and operating personnel round out the list shown in Figure 6.103. At the time of writing this book, methods to quantify the economic benefits of mitigating interruption to transportation by flooding using tidal flood gate valve systems within the drain pipes was an active research area.

To mitigate flooding problems on low-lying urban streets caused by tidal "nuisance" flooding events, should you raise the road or develop a tidal flood gate valve system? Both alternatives should be investigated and the least life-cycle cost alternative should be selected since they both (theoretically) give the same benefits.

6.6 Special Structures

All the coastal structures discussed above are designed with a foundation that rests on the soil conditions at the site and with a crest elevation that is slightly above (emergent) with the design water level (see Section 5.5.2). Wave transmission past the structure was defined by a wave transmission coefficient K_t in Section 5.5.1 and Equation (5.45), and is repeated here as

$$K_t = \frac{\text{Significant wave height of transmitted waves}}{\text{Significant wave height of incident waves}} = \frac{H_{st}}{H_{si}}. \tag{5.45}$$

In this section, we briefly introduce special breakwater structures that float on the water surface (Section 6.6.1), or with crest elevation below the design water level (Section 6.6.2), or as combined, breakwater and energy conversion structures (Section 6.6.3). All three special breakwater structures require knowledge of the wave transmission coefficient K_t as the key variable in their design.

6.6.1 *Floating breakwaters*

Floating breakwaters reduce the wind- and ship-generated wave heights as protection for moored vessels in small craft harbors, fish farms, swimming areas, dredging, and temporary construction sites and other uses. They are less expensive for deep-water applications, adaptable to water-level fluctuations, and can be relocated and cause less environmental disturbance to water circulation and fish movements. However, their functional performance as measured by the wave transmission coefficient K_t is critically dependent on the period T_p of the incident waves. Figure 6.104 schematically depicts a floating breakwater of the box (prism)-type with incident and transmitted wave conditions.

As shown in Section 2.3.7 and Figure 2.25 for wind-wave hindcasting, at a given wind speed, as the fetch distance increases, the significant wave height H_{si} and the peak wave period T_p increase. The wave length L_p depends on the water depth d and the wave period T_p as found from linear wave theory for the spectral peak of the incident wave spectrum. As perhaps

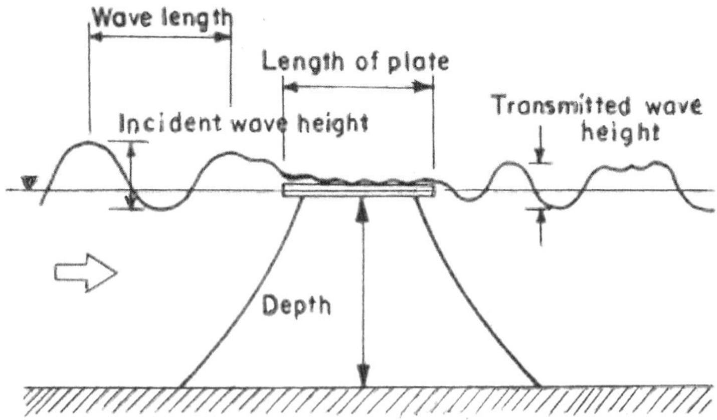

Figure 6.104. Schematic diagram of floating breakwater with incident and transmitted wave heights.

intuitively expected, under long period (swell) wave conditions, when the width W of the floating breakwater (length of plate in Figure 6.104 in the direction of wave propagation) relative to the wave length L_p, i.e. the ratio $W/L_p \approx 0$, then the entire floating structure follows the crest and trough motions of the wave. Little or no wave attenuation occurs so that the wave transmission coefficient, $K_t \approx 1.0$. Clearly, there are very strict limits on where the floating breakwaters are effective. The major factor is the fetch length over which winds blow and the resulting wave periods that can be generated over the fetch distance that produce waves of lengths L_p in the design water depth.

Figure 6.105(a) illustrates three different types of floating breakwaters with their wave transmission coefficients K_t as a function of the W/L ratio shown in Figure 6.105(b) (after Sorensen, 1980). The wave tank experiments employed regular, monochromatic waves (H, T).

Roul (2013) defines these three types as Box-, Pontoon-, and Mat-types of floating breakwater, and one additional type called the Tethered-type which is seldom employed. Incident wave energy is reflected off the front face of the structure, dissipated by wave breaking over the structure top, and dissipated by turbulence generated internally within the wave particle motion beneath the structure. As seen if Figure 6.105(b) for a given width W, as the wave length increases for long period waves, the W/L ratio decreases and the floating breakwater is far less effective since the wave transmission coefficient K_t approaches 1.

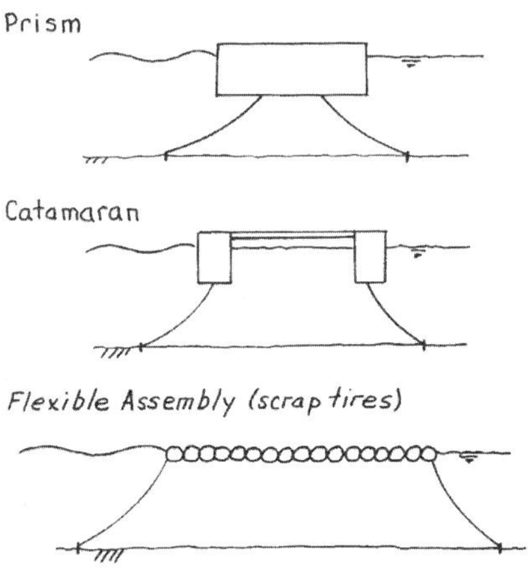

(a) Three major types of floating breakwaters

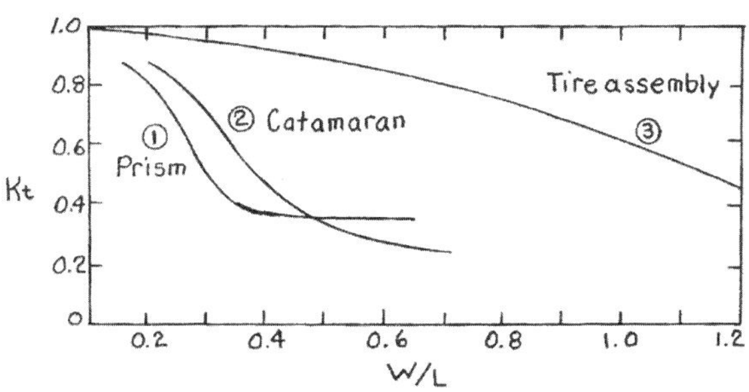

(b) Wave transmission coefficients as function of *W*/*L* ratio

Figure 6.105. (a) types and (b) wave transmission coefficients of floating breakwaters (after Sorensen, 1980) from wave tank experiments using monochromatic, regular waves (H, T).

In addition to wave transmission, fundamental aspects of floating breakwater design also depend on buoyancy and floating stability, mooring forces, and the structural design of each breakwater unit. It is a complicated, iterative process and beyond the scope of this book. Consult McCartney (1985), PIANC (1994), and Biesheuvel (2013) for a historical perspective, a practical guide for design and construction, and the most recent references.

Design of very large floating breakwaters has now moved beyond concept and research stage. Figure 6.106 shows the floating breakwater constructed to enlarge and improve the harbor in Monaco (top, left). It is 352 m (1,155 ft) long, 30 m (98 ft) high, and 30 m (98 ft) wide and weighs 175,000

Figure 6.106. Largest floating breakwater constructed for Monaco Harbor (courtesy Doris Engineering and Monaco).

tons as shown under construction in Spain (top, right). It was towed to the harbor site and requires 12 m (40 ft) draft. Total cost was 150 million Euros in 2013. Remarkably, a shopping mall with automobile access for parking is inside as illustrated in Figure 6.106 (bottom, left). It is the largest concrete honeycomb structure of its type in the world (photos courtesy of FDN Amsterdam, www.fdn-engineering.nl).

6.6.2 *Artificial reef breakwaters*

Artificial nearshore breakwaters that are built completely below MSL are often called "reef" breakwaters because they try to mimic natural, underwater coral reefs that occur at many coastal locations. The coral barrier reef of Australia is one famous example. The effectiveness of coral reefs to dissipate wave energy has been documented (Ferrario *et al.*, 2014) from field measurements over the very wide expanse of natural reefs in the nearshore coast. Shoreline erosion is less where natural reefs exist.

Artificial reefs were previously introduced in Section 6.2.5.4 as shown in Figure 6.9 for the definition of freeboard, $R_c = h_c - h < 0$ when completely submerged. When constructed from a single-size stone, early design and structural stability guidance is found in Ahrens (1984, 1987) and

Ahrens and Cox (1990). Recent research on LCS in Europe as discussed in Section 6.2.5.2 produced Equations (6.20) and (6.21) which should be used for stability calculations of the stone size for reef breakwaters. In these equations, it is assumed that the crest is at least three stone diameters in width, with far narrower, submerged width than natural barrier reefs.

Patented, modular, precast concrete units that are interconnected to form a shore-parallel reef breakwater have been employed since the late 1980s in the US. Figure 6.107 shows some examples. The left side (Beachsaver Reef^{TM}) and middle (Prefabricated erosion protection (P.E.P.) Reef^{TM}) show similar shapes with very narrow crests and were patented. Side pads or groves permit each unit to be joined together in single or multiple rows for wave energy dissipation. The right side (top) of Figure 6.107 displays the Reef Ball design which is also marketed for enhancement of fish and shellfish habitat. New designs are routinely patented (right side, bottom) that are yet to be named (all photos and graphics are taken from the internet).

The arrows in Figure 6.107 (left side, top) show the direction of wave attack from the right on the sloped side and the theoretically upward movement of sand to collect behind the structure. (The small arrows point to

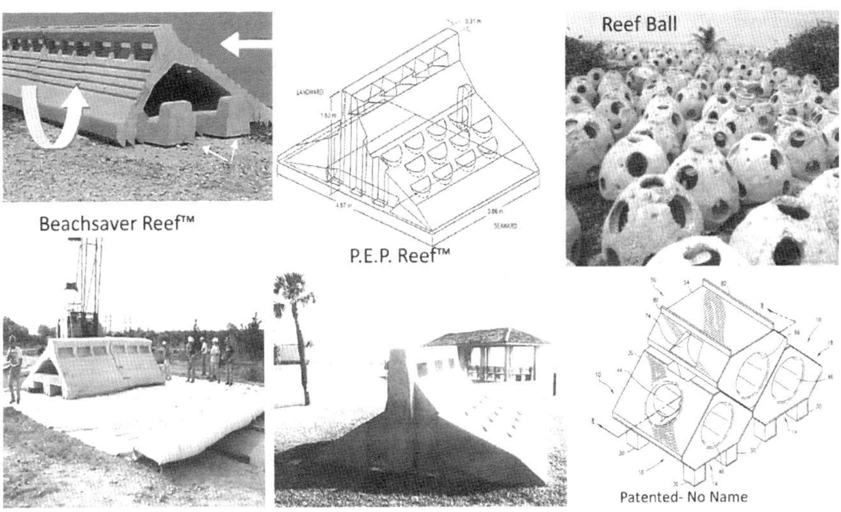

Figure 6.107. Modular, precast concrete units employed as reef breakwaters in the US. Images courtesy Beachsaver Reef (left); P.E.P. Reef (middle); and Reef Ball Foundation (top right).

the pads that hold the units together to form rows). Installation and field testing have been conducted to study wave energy dissipation, structure stability, and sediment volume change in the lee of these pyramid-shaped, reef breakwater structures.

Field tests of the Beachsaver Reef™ in the late 1980s (Sea Isle City, NJ) revealed significant structure settlement and large scour holes on the leeward side when constructed as shore-parallel breakwaters (Sorensen and Weggel, 1992). Additional field experiments in the early 2000s took place at Cape May, NJ within the Corps' National Demonstration Program for "non-traditional" methods when constructed between the ends of two adjacent groins. Again, structure settlement was a problem ($> 1\,\mathrm{m}$) but sand retention between the groins was greater than at groin compartments with no reef breakwater (Stauble *et al.*, 2005). No additional applications of the Beachsaver Reef™ are known to exist. See Chapter 7 for additional information about artificial underwater breakwaters and beach nourishments called a "perched beach".

Companies marketing these modular, precast concrete units point out that they can be easily removed if they do not work to reduce the existing erosion rate. That is exactly what happened at the Town of Palm Beach, Florida in the early 1990s when the P.E.P. Reef™ system was field tested. A detailed analysis of the leeward project area (Dean and Chen, 1996) revealed that wave overtopping, wave setup, pounding, excess longshore currents, and sediment transport produced more erosion than before reef construction. The P.E.P. Reef™ system was removed, salvaged, and used to construct a groin field with beach nourishment (Erickson, 1998). Similar results took place at Vero Beach, Florida in the late 1990s. Erosion landward of the reef increased after two years of monitoring compared to the historic volume change over 15 years (Stauble and Smith, 1999; Wooduff and Dean, 2000). No additional applications of the P.E.P. Reef™ are known to exist.

Comprehensive field testing of the wave dissipation, structural stability, and sediment area and volume changes for the Reef Ball system are unknown to this author. See the Reef Ball Foundation, Inc. website (www.reefball.org) for before and after photographs of beaches and underwater photographs of the many types of flora and fauna attached to the reef balls.[1]

[1]Contact Dr. Lee Harris, Florida Institute of Technology (Lharris@fit.edu) who has extensive practical experience with their design and application.

6.6.3 *Breakwaters and wave energy conversion*

Waves travel hundreds and thousands of kilometers over the worlds' oceans to dissipate almost all their energy at the worlds' coasts. Wave breakwater structures reduce this wave energy, and their design has been the subject of Chapter 5 (functional, overtopping, and planform) and Chapter 6 (structural). The wave energy arriving at the coast can also be utilized in a wave energy convertor (WEC). Combining breakwater and wave energy converter functions in one structure has the potential to bring the added value of energy production to lower breakwater costs over the design life.

Research began in the Hydraulics and Coastal Engineering Laboratory, Department of Civil Engineering, Aalborg University, Denmark in the late 1990s within the Danish Wave Energy program. Early on it was recognized that wave runup and the wave-overtopping water volume can be collected in one or more reservoirs placed above the MWL. This potential energy obtained from wave overtopping can then be converted to electrical energy by directing the water from the reservoir through a low-head turbine connected to a generator and then returning the water to the sea. Masters' and Ph.D. degrees were earned by Jens Peter Kofoed at Aalborg University (Kofoed, 1997; Kofoed, 2002), and his research efforts continue today regarding wave energy conversion within the Danish Wave Energy Program.

WEC devices utilizing wave overtopping were first conceived and tested for offshore, floating structures (e.g. the Wave Dragon, Kofoed *et al.*, 2006b). Kofoed and Burcharth (2002) were the first to propose combining wave overtopping of coastal breakwaters with a collection reservoir for conversion of potential energy in the stored water to electrical power as illustrated in Figure 6.108.

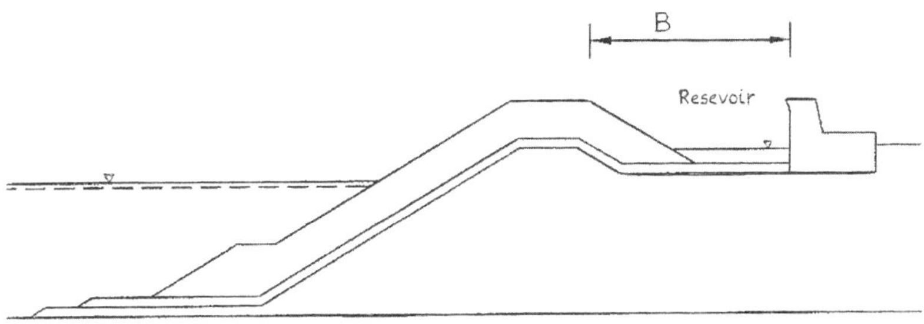

Figure 6.108. Wave-overtopping reservoir behind coastal breakwater (adapted from Burcharth and Andersen, 2006).

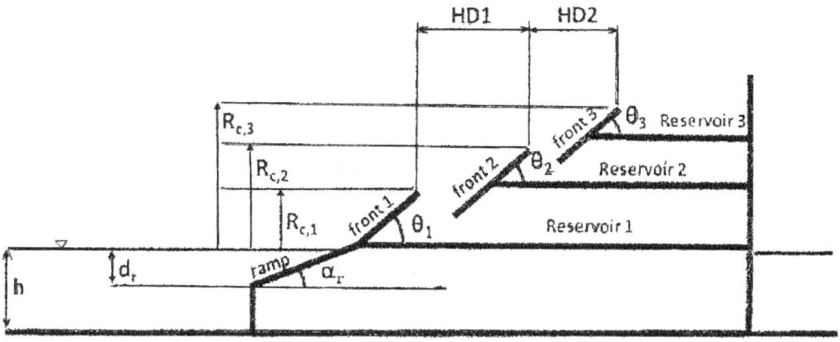

Figure 6.109. Multi-level openings and reservoirs for wave-overtopping energy conversion in an SSG device (from Vicinanza *et al.*, 2012).

It also became obvious (Kofoed, 2002) that multi-levels of overtopping openings and reservoirs would be needed and more efficient since wave runup was highly dependent on the time-varying, statistical distribution of wave heights striking the structure as schematically illustrated in Figure 6.109 (from Vicinanza *et al.*, 2012). This multi-level overtopping device has come to be called Seawave, Slot-cone Generator (SSG) in the WEC literature for floating structures and named the Overtopping Breakwater for Energy Conversion (OBREC) for coastal structures. Sometimes, both abbreviations are found in the wide variety of published literature on this topic.

The European Union (EU) efforts to bring university science and engineering faculties together from many countries in Europe for collaboration in research is the envy of the world. One example was the EU Sixth Framework program 6.1 (Sustainable Energy System) that brought together Dr. Diego Vicinanza of the Civil Engineering Department, Second University of Naples, Aversa, Italy with the researchers at Aalborg University (Vicinanza *et al.*, 2007; Vicinanza and Frigaard, 2009) as mentioned above. Early efforts were for a pilot plant SSG wave energy converter in the North Sea (Margheritini *et al.*, 2009) as schematically illustrated in Figure 6.110, but it was never funded. See also Vicinanza *et al.* (2015).

Small-scale tests of a modified OBREC system were conducted (Vicinanza *et al.*, 2013, 2014) at Aalborg University with the full-scale device installed at the Port of Naples, Italy breakwater as shown in

Figure 6.110. Schematic section view of three-level SSG pilot plant (from Margheritini *et al.*, 2009).

Figure 6.111(a). Two reservoirs at different levels Figure 6.111(c) are filled from two levels of wave overtopping from a 22° sloping ramp Figure 6.111(b) and the test facility is instrumented to measure sea conditions. The modified OBREC design was developed at the University of Campania, Caserta, Italy (Contestabile *et al.*, 2016). This device is the first overtopping, wave energy converter at full scale and installed in an existing rubble-mound breakwater. See Contestabile *et al.* (2016) for complete details.

A second type of WEC is the oscillating water column (OWC) as illustrated in Figure 6.112. The mass of water in the gap on the seaward side rises and falls with the waves to alternately compress and depressurize the air column giving an oscillating flow of air to drive a turbine and extract electrical energy. There are two OWC wave energy converter prototype installations currently operating, and both are installed in caisson-type breakwaters. One is at the Port of Mutriku in Spain as shown in Figure 6.113. See Torre-Enciso *et al.* (2009, 2010) for details. See also Medina-Lopez *et al.* (2015) for a discussion of shortcomings in the Mutriku

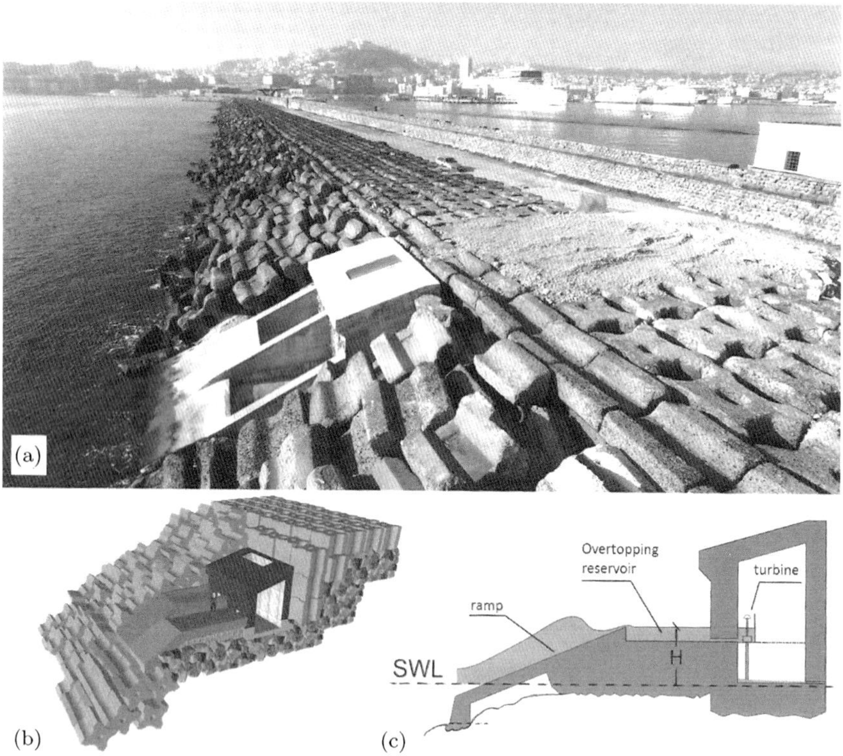

Figure 6.111. Modified OBREC system for field test in Naples, Italy (from Contestabile *et al.*, 2016).

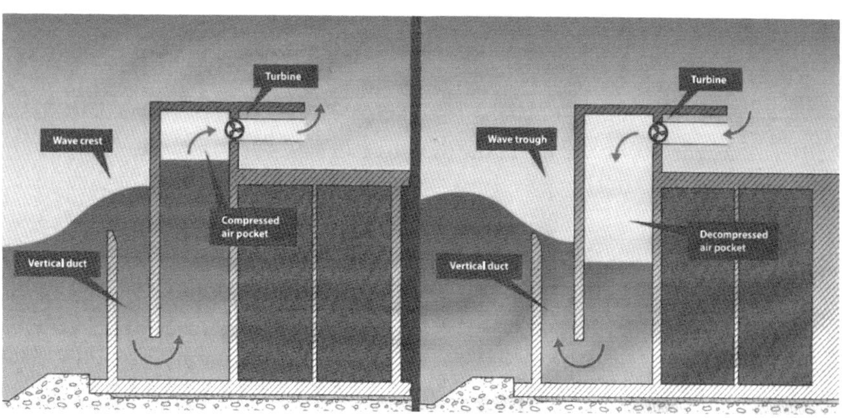

Figure 6.112. Schematic diagram of OWC for wave energy conversion (from Wave Energy, Feb. 9, 2016).

Figure 6.113. Aerial photograph of the breakwater and OWC structure installed at the Port of Mutriku, Spain in 2008 (from Kofoed, 2016).

OWC design. The second installation is for the Port of Civitavecchia near Rome in Italy (see Arena *et al.*, 2013 for details).

As stated above, combining breakwater and wave energy converter functions in one structure has the potential to bring the added value of energy production to lower breakwater costs over the design life. Contestabile *et al.* (2016) present an economic assessment study of nine locations along the southern coast of Western Australia. A thorough review of wave energy conversion devices combined with breakwaters is presented in Mustapa *et al.* (2017). Yet to be demonstrated is the level of economic value achieved over time for the OBREC system.

6.7 The Role of Physical Model Tests in Structural Design

6.7.1 *Introduction*

One role of physical model testing is to implement reliability to further strengthen the partial safety factor system in design. In Section 6.2, the tables presented for partial safety factors (e.g. Table 6.3 Partial Safety

Factors for Plunging Waves) were taken from (CEM, 2011, VI-6-6). For this example, the title of the table was "*Table VI-6-5 Partial Safety Factors for Stability Failure of Rock Armor, Plunging Waves, van der Meer Formula,* ***Design Without Model Tests***." (bold added for emphasis). Clearly, the designer must be aware under what conditions these partial safety factors for rubble-mound structures are to be employed.

In Section 6.3, the tables presented for partial safety factors (e.g. Table 6.21 with title Partial Safety Factors for Foundation Failure of Caissons) were also taken from (CEM, 2011) but included partial safety factors for both "***Design Load Determined With Model Tests***" and "***Design Without Model Tests***". When model tests have been conducted, the partial safety factors are lower. The partial safety factor system is to implement reliability in design and physical model testing increases our confidence in the design.

A second role of physical model testing is the potential for cost saving in coastal engineering design. Many examples exist where the cross-section, stone size, and planform layout of breakwaters can be modified and still function properly, are stable under design conditions, and are found to be less expensive to construct. Too often, the expense of the physical model tests is viewed as adding to the design costs and not considered. However, in many cases the total design and construction costs are less as a result of conducting the physical model tests and proving that the original design can be modified to lower the construction cost.

A third role of physical model testing is the ability of the client to see the design in operation. There is nothing better than being present in the laboratory and experiencing the sights, sounds, and smells of water waves breaking on the structure. Figure 6.114 displays a 3D wave basin with an armor-stone breakwater. Movement of the colored layers of stone under water at the breakwater end are of concern in the design. The walkways permit measurements, photos, and client observations of the model tests.

2D wave flumes and 3D wave basins are the primary laboratory-scale tools for conducting physical model tests in coastal engineering. A laboratory team is responsible for the (1) preparation of the model specifications (model scale) and test methods (instruments, data acquisition), and (2) management and supervision of the personnel running the tests, and (3)

Figure 6.114. Model tests of armor layer stability under breaking waves in a 3D wave basin (Corps of Engineers, ERDC, Vicksburg MS, https://www.erdc.usace.army.mil/Locations/CHL/).

preparation of the test results and reports. The laboratory team is normally responsible for establishing the model scale for the tests to ensure their laboratory flume and/or basin is adequate for the effort. The designer should be familiar with the fundamentals of dynamic similitude as taught in the introductory courses in fluid mechanics.

6.7.2 *Model scales*

A water particle in motion is subject to inertia (mass), gravity, and viscous forces. The ratio of the inertia and gravity forces is the Froude number F_r. For free surface water motions (waves) and flows (velocity), the Froude number is

$$F_r = \frac{V}{\sqrt{gd}}, \tag{6.120}$$

where V is a reference velocity, $\sqrt{gd} = c$ is the celerity of a shallow-water wave, d is a reference water depth, and g is the gravity constant.

The ratio of the inertia and viscous forces in the Reynolds number, R_e gives

$$R_e = \frac{V\,L}{\mu/\rho}, \tag{6.121}$$

where V is a reference velocity, L is a reference length, μ is the absolute viscosity of water, and ρ is the mass density of water.

From fundamental principles of dynamic similitude of all water motions in the prototype (subscript, p) and in the model (subscript, m), both the Froude and Reynolds numbers must be identical as follows:

$$(F_r)_p = (F_r)_m, \tag{6.122}$$

$$(R_e)_p = (R_e)_m. \tag{6.123}$$

The model scale L_r is the ratio of a length (x, y, and z dimensions) in the prototype L_p to the lengths (x, y, and z dimensions) in the model, L_m. This is for undistorted-scale, 2D models.

$$L_r = L_p/L_m. \tag{6.124}$$

All lengths (water depth, pressure head, etc.) are similarly scaled by the model scale ratio, L_r.

In 3D models of tidal flows, the horizontal dimensions (x, y) must cover relatively large surface areas in relation to the water depths so that the horizontal scale ratio $(L_r)_H$ is different than the vertical-scale ratio, $(L_r)_v$. Consequently, 3D models are often constructed with distorted scales.

6.7.3 *Froude number similitude — Undistorted scale models*

Using the definitions in Equations (6.120) and (6.122), for Froude number dynamic similitude gives the following scale ratios for all the commonly employed variables measured and calculated from the results of 2D, undistorted-scale model tests. See any fluid mechanics textbook or Hughes (1993) for derivations. Whenever the gravitational constant g is present in the equations, the location of the model tests is assumed to be near the sea level so that $g_r = 1$. All variable ratios are $X_{\text{prototype}}/X_{\text{model}}$.

(a) Area A_r and volume $\forall_r$ ratios are given as

$$A_r = L_r^2, \quad \forall_r = L_r^3. \tag{6.125}$$

(b) Significant wave height ratio is

$$(H_s)_r = L_r. \tag{6.126}$$

(c) Wave period ratio, T_r

For Froude number similarity using the reference velocity V = length/time = L/T gives for the wave period ratio, $T_r = T_{\text{prototype}}/T_{\text{model}}$

$$T_r = T_p/T_m = \sqrt{L_r}. \tag{6.127}$$

The subscripts for the wave period are for prototype (p) and model (m) and not for peak or mean wave periods as found throughout this book. The wave periods can be for any definition of an irregular wave.

(d) Water depth ratio is

$$d_r = L_r. \tag{6.128}$$

(e) Water mass density ratio, ρ_r

Fresh water is commonly used in wave tanks for model tests. When the prototype is seawater, the mass density ratio is needed in the scale ratios.

$$\rho_r = \frac{\rho_{\text{seawater}}}{\rho_{\text{freshwater}}}. \tag{6.129}$$

(f) Pressure ratio is

$$p_r = \rho_r g_r L_r = \gamma_r L_r, \tag{6.130}$$

where the specific gravity ratio γ_r again depends on whether or not sea water is present at the prototype.

(g) Force ratio is

$$F_r = \gamma_r L_r^3. \tag{6.131}$$

(h) Velocity ratio is

$$V_r = \sqrt{L_r}. \tag{6.132}$$

(i) Volumetric flowrate ratio is

$$Q_r = L_r^{5/2}. \tag{6.133}$$

(j) Energy density ratio is

$$e_r = \gamma_r L_r^2. \tag{6.134}$$

6.7.4 *Reynolds number similitude*

The equality of Reynolds numbers and equality of Froude numbers are *both* necessary for exact dynamic similitude in free-surface flows. Solving for the velocity ratios V_r in both the Froude and Reynolds equations (6.120) and (6.121) equating and solving for the length ratio gives ($g_r = 1.0$),

$$L_r = \left(\frac{\mu_r}{\rho_r}\right)^{2/3} = \upsilon_r^{2/3}, \tag{6.135}$$

where υ_r is the kinematic viscosity ratio.

Whether freshwater or seawater, the kinematic viscosity ratio υ_r is close to unity. Therefore, for both Reynold's and Froude similitude, the scale ratio must also be close to unity, i.e. at full prototype scale. There is no liquid with super-low viscosity that exists so that in all free-surface flow models, the water simply has far too much molecular viscosity. This is one factor that produces "scale effects" in physical models. Fortunately, the eddy viscosity produced by turbulence and boundary shear stresses (frictional boundary roughness) are far more important than molecular viscosity, but still produce "scale effects" in physical models. In large area, 3D models with distorted horizontal and vertical length scales, boundary roughness elements must be used to simulate flows and energy losses in the submerged overbank areas during floods. A considerable amount of judgment and experience is necessary to design and interpret the results.

Exact dynamic similitude is not possible in scale models of free-surface water systems. Therefore, they are designed, operated, and the results determined based only on the Froude number similarity.

6.7.5 *An example*

In Chapter 4, Figure 4.1 displayed the test data for plunging waves on rubble-mound structures (from van der Meer, 1987) that is reproduced here as Figure 6.115. Eleven of the 385 data points were obtained from tests in "Delta Flume (1987)" in which all the physical dimensions of the structures, the water depth, and the wave conditions tested at small-scale in the Delft Technical University laboratory were all increased by using a

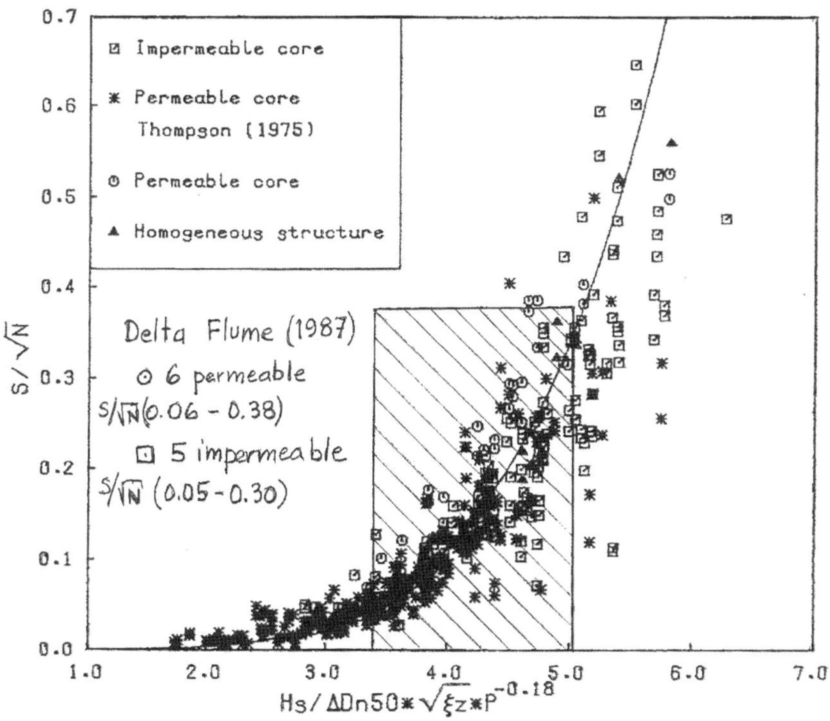

Figure 6.115. Plot of test data and stability formula for plunging waves on rubble-mound structures (from van der Meer, 1987) with added zone where 11 data points are from Delta Flume (1987) tests.

scale ratio, $L_r = 6.25$. The location (zone) where the 11 data points fell in these larger scale tests is shown in Figure 6.115. See van der Meer (1987, Appendix I, Tests 930–940) for the test results.

The small-scale (lab) wave flume was 50 m (164 ft) long, 1.0 m (3.3 ft) wide with water depth, $d = 0.80$ m (2.62 ft). Irregular waves using the Pierson–Moskowitz spectrum produced significant wave heights H_s in the 0.05–0.26 m (0.16–0.85 ft) range with mean wave periods, $T_m = 1.3$–3.2 s. Nominal, median rock size was $D_{n50} = 0.036$ m (0.118 ft, 1.4 inches), with mass $M_{50} = 0.123$ kg (0.625 slugs) and mass density $\rho_a = 2620\,\text{kg/m}^3$ (5.08 slugs/ft^3) in the armor layer. Two gradings were used for the armor layer: for riprap ($D_{85}/D_{15} = 2.25$) and for uniform stones ($D_{85}/D_{15} = 1.25$).

The Delta Flume (DF) dimensions in 1987 were 230 m (755 ft) long, 5.0 m (16.4 ft) wide, and 7.0 m (23 ft) deep. It is instructive to check the Froude scaling laws for these 11 larger scale tests in Delta Flume (1987). All tests were with $\cot \alpha = 3$.

(1) Water depth, d_{DF}

As shown in Equation (6.128), the water depth ratio d_r is simply equal to the length scale ratio, L_r.

$$d_r = \frac{d_{DF}}{d_{\text{lab}}} = 6.25 \text{ so that } d_{DF} = 6.25 \times 0.8 = 5.0\,\text{m}(16.4\,\text{ft}).$$

The Delta Flume is 7.0 m deep, so this water depth $d_{DF} = 5.0$ m allows waves over 1.0 m (3.3 ft) to be produced for the tests. The scale ratio in physical models is often determined by the water depths required to simulate the d/L ratios found in the prototype and therefore from the wave periods for structure design.

(2) Wave periods, T_{DF}

As shown in Equation (6.127), the wave period ratio T_r is equal to the square root of the scale ratio.

$$T_r = \frac{T_{DF}}{T_{\text{lab}}} = \sqrt{L_r} = \sqrt{6.25} = 2.5 \text{ so that}$$

$$T_{DF} = 2.5 \times (1.3 - 3.2)\,\text{s} = (3.25 - 8.0)\,\text{s}.$$

A constant, mean wave period, $T_{DF} = 4.4$ s was employed for all 11 tests in the Delta Flume which scales to a mean wave period, $T_{\text{lab}} = 1.76$ s in the laboratory scale tests.

(3) Significant wave height, $(H_s)_{DF}$

The significant wave height ratio $(H_s)_r$ is equal to the scale ratio, L_r.

$$(H_s)_r = \frac{(H_s)_{DF}}{(H_s)_{\text{lab}}} = 6.25 \text{ so that}$$

$$(H_s)_{DF} = 6.25 \times (0.08 - 0.26)\,\text{m} = (0.5 - 1.625)\,\text{m}.$$

The wave heights varied from 0.5 m to 1.18 m for the Delta Flume tests.

(4) Armor layer, nominal, median stone size, $(D_{n50})_{DF}$

The $(D_{n50})_r$ ratio is equal to the scale ratio, L_r

$$(D_{n50})_r = \frac{(D_{n50})_{DF}}{(D_{n50})_{\text{lab}}} = 6.25 \text{ so that}$$

$$(D_{n50})_{DF} = 6.25 \times 0.036\,\text{m} = 0.225\,\text{m}(0.74\,\text{ft}, 8.9\,\text{inches}).$$

The nominal, median stone size employed was $(D_{n50})_{DF} = 0.210\,\text{m}$ for the gradation of rocks employed in the Delta Flume tests. The small difference illustrates that it is not always possible (practicable) to exactly match the model and prototype rock dimensions. The grading of the stones was $D_{85}/D_{15} = 1.38$ and close to the uniform grading in the laboratory scale tests.

(5) Armor layer median mass, $(M_{50})_{DF}$

The mass ratio M_r equals the weight ratio (gravitational force) divided by the gravity constant, g. From Equation (6.131), for the force ratio, F_r

$$M_r = F_r/g = (\gamma_r L_r^3)/g = \rho_r L_r^3. \tag{6.136}$$

Assuming the same temperature, fresh water was used in both lab and Delta Flume tests, then $\rho_r = 1.0$.

$$M_r = \frac{M_{DF}}{M_{\text{lab}}} = (6.25)^3 = 244.14 \text{ so that}$$

$$(M_{50})_{DF} = 244.14 \times 0.123\,\text{kg} = 30.03\,\text{kg}.$$

The median mass employed was $(M_{50})_{DF} = 26.5\,\text{kg}$ and the difference due to the smaller median rock size since the mass density, $\rho_{DF} = 2700\,\text{kg/m}^3$.

(6) Reynolds numbers $(R_e)_{DF}$

As discussed above, it is impossible to have Reynolds number similitude in free-surface, physical models, $(R_e)_{DF} \neq (R_e)_{\text{lab}}$. Burcharth and Frigaard (1987) and others defined a Reynolds number for armor layer stability tests as

$$R_e = \frac{(\sqrt{gH_s}) \times D_{n50}}{\upsilon}. \tag{6.137}$$

Comparison of terms with Equation (6.121) reveals the terms for the reference velocity which is the celerity of shallow water waves and for the reference length which is the nominal, median armor layer size. The kinematic viscosity, υ of water at 50–60°F (10–15°C) is $(1.4 - 1.2) \times 10^4\,\text{ft}^2/\text{s}$.

Using the range of test conditions in the Delta Flume gave

$$(R_e)_{DF} = (5 \text{ to } 7) \times 10^5.$$

For all the other tests at small-scale, van der Meer (1987) found

$$(R_e)_{\text{lab}} = (4 \text{ to } 8) \times 10^4,$$

and concluded that the scale effects on the armor layer stability due to molecular viscosity in the laboratory scale wave flume were not important since $(R_e)_{\text{lab}} > (1 - 4) \times 10^4$ which is a threshold level (Burcharth and Frigaard, 1987).

(7) van der Meer (1987) in shallow water

Recall Section 6.2.3.3 and Table 6.5 for the modifications suggested by van Gent *et al.* (2004) when the original, van der Meer (1987) equations are applied in shallow water. Note that $d/L_o = 0.1654$ and that $d/L = 0.1962$ in both the lab and Delta Flume tests which is in the wave transition region between deep-water and shallow-water waves. The scatter in the test data shown in Figure 6.115 was reduced. However, all of the Delta Flume tests were found to be under plunging-wave conditions with the surf-similarity parameter, ξ_m less than the critical value, ξ_{mc}. And as seen in Figure 6.115, considerable scatter occurred when the x-axis variable $((H_s/\Delta D_{n50}) \times \sqrt{\xi_{mo}}) \times P^{-0.18})$ exceeded 5.0. It is instructive to consider what scale ratio could be employed if these same tests of van der Meer (1987) were repeated in the new Delta Flume. Perhaps the scatter could be further reduced when both plunging- and surging-wave conditions are tested at nearly prototype scales.

6.7.6 *The new Delta Flume (2015)*

The Dutch have built a new Delta Flume in 2015 that is the largest 2D wave flume in the world. It can produce prototype-scale waves as found on the North Sea coast of Holland. Figure 6.116 displays some photos of the facility located at the Deltares Research Institute near Delft, the Netherlands (internet photos). The flume is 300 m (985 ft) long (top, left), 10 m (33 ft) deep, and 5 m (16.4 ft) wide (bottom, left). Inauguration took place on October 5, 2015 (top, right) showing a large wave breaking at the viewing stand. Irregular wave spectrums with $(H_{mo})_{\max} = 2.2$ m (7.2 ft) and $T_p = 5.7$–13.4 s can be generated with $H_{\max} = 4.5$ m (14.8 ft) in the spectrum. Regular, monochromatic waves with $H_{\max} = 3.3$ m (10.8 ft) and $T = 1$–20 s are also available for tests.

Figure 6.116. Photos of the new Delta Flume (2015) in the Netherlands and dimensions (courtesy Arcadis, the Netherlands).

The first experiments were to test various revetment materials used for Dutch dikes. Figure 6.116 (bottom, middle) displays workman installing a "Basalton" block revetment at scale $L_r = 1.6$ which is almost prototype scale.

At the time of this writing, the author knows of no discussion to repeat the tests of van der Meer (1987) nor to initiate a whole news series of rubble-mound breakwater tests in the new Delta Flume. See the Deltares website https://www.deltares.nl/en/ for more information.

6.7.7 *Large 2D wave flumes*

The length and depth of 2D wave flumes are the important dimensions to consider for model tests. Large flume depths permit a range of water depth-to-wave length ratios (d/L) and larger wave heights in the model to reduce the scale ratio, L_r. The flume length permits mild beach slopes and longer period waves to be tested. However, active wave adsorption technology at the wave paddle end of the flume now permits reflected wave energy in the flume to be radiated through the paddle (i.e. back out to sea) so that the length of the flume can be reduced. Some examples of large 2D wave flumes in the world are listed in the following.

(1) Europe

In Hannover, Germany the Technical University Braunschweig (Fransizus Institute for Water Studies) maintains the Grosser Wellenkanal (GWK) wave flume since 1983. It is 330 m (1,083 ft) long, 7 m (23 ft) deep, and 5 m (16.4 ft) wide (see https://www.lufi.uni-hannover.de/) for details.

The Universitat Politecnica de Catalunya in Barcelona, Spain operates a 100 m (330 ft) long, 4.5 m (15 ft) deep, and 3 m (10 ft) wide wave flume with wedge-shaped paddle to generate deep and intermediate-depth wave conditions (see https://hydralab.eu/facilities--instruments/facilities-in-hydralab/).

The semi-private/government facility, HR Wallingford in Wallingford, UK maintains a large, relatively shallow-depth wave flume 100 m (330 ft) long, 1.8 m (6 ft) deep, and 1.8 m (6 ft) wide. It also has two shorter wave flumes, 45 m (150 ft) long, 1.5 m (5 ft) deep, and 1.2 m (4 ft) wide (see http://www.hrwallingford.com/) for details.

Other shorter, shallower depth wave flumes exist in Belgium, Denmark (DHI), France, Italy, Portugal, Turkey, and other countries in Europe.

(2) United States

In the United States, three organizations maintain large, 2D wave flumes for education and research.

The largest 2D wave flume in North America is located within the O.H. Hinsdale Wave Research Lab at Oregon State University in Corvallis, Oregon. It is 104 m (340 ft) in length, 4.6 m (15 ft) deep, and 3.7 m (12 ft) wide (see http://wave.oregonstate.edu/large-wave-flume).

At the former Waterways Experiment Station (now Engineering Research Development Center) of the US Army Corps of Engineers in Vicksburg, Mississippi, two wave flumes are operated for research. Both are 63 m (205 ft) long, and 1.5 m (5 ft) deep, with one 3.0 m (10 ft) and the second 1.5 m (5 ft) wide. Stability tests for the CORE-LOC armor units were developed here (see http://www.erdc.usace.army.mil).

The third is a ship towing-tank facility that includes a wave paddle system at the US Naval Academy, Annapolis, Maryland. The tank is 116 m (380 ft) in length, 4.9 m (16 ft) deep, and 7.9 m (26 ft) wide. It is located within the Naval Architecture, Ocean and Marine Engineering Dept. in Rickover Hall (see http://www.usna.edu for details).

Many universities in the United States have shorter and shallower depth, 2D wave flumes for education and research. An internet search of graduate education programs in coastal engineering will reveal those facilities currently available for model studies of coastal structures.

(3) Canada

At the Canada Hydraulic Center, National Research Council in Ottawa, a 97 m (320 ft) long, 2.7 m (9 ft) deep and 2 m (6.6 ft) wide is available. A second is 63 m (210 ft) long, 1.22 m (4 ft) deep, and 1.22 m (4 ft) wide (see https://www.canada.ca/en/environment-climate-change/ for details).

The Queens University in Kingston, Ontario has three wave flumes that are each 45 m (150 ft) in length (see www.engineering.queensu.ca for details).

(4) Australia

A new large wave flume facility has opened at the University of New South Wales, Water Research Laboratory in Manly Vale, north of Sydney. It is 44 m (144 ft) long, 1.4 m (4.6 ft) deep, and 0.9 m (3 ft) wide and said to be the largest in Australia (see https://unsw.edu.au for details).

(5) Japan

The Port and Harbor Research Institute in Kanagawa has a 180 m (590 ft) in length, 12 m (39 ft) deep, and 3.5 m (11.5 ft) wide wave flume. The Central Research Institute, Chiba operates a 205 m (670 ft) long, 6 m (20 ft) deep, and 3.4 m (11 ft) wide facility. The Kajima Research Institute, Tokyo has two flumes: one is 62 m (200 ft) long, 2 m (6.5 ft) deep, and 2 m (6.5 ft) wide and another is 60 m (198 ft) long, 1.5 m (5 ft) deep, and 0.7 m (2.3 ft) wide. The websites for these Institutes are in the Japanese language.

Many Japanese companies operate and maintain 2D wave flumes for commercial testing and research. The University of Delaware, Center for Applied Coastal Research has compiled an inventory under the heading "Investigation Card: Large Scale Wave Tank" for Japanese Basins (pdf file).

(6) Other countries

Water wave flumes exist in Russia, China, South Korea, India, Brazil, Mexico, and many other countries throughout the world. They will find more use as coastal regions face rising sea levels in the future.

(7) 3D wave basins

Many of the government laboratories and universities listed above also have 3D wave basins for model tests and research. Their websites listed should be contacted for details which are beyond the needs of this book. Because of space limitations and costs, 3D wave basins must be built at larger scale ratios and therefore experience greater scale effects.

6.7.8 *Summary — Limitations of models*

Scale effects exist in large 2D wave flumes and 3D wave basins employed for physical model tests of coastal structures. Scale effects will always be a limit of physical models.

The designer should ask the following questions regarding the use of physical models. What is the purpose of the physical model tests? Will the results increase our confidence in the desktop design? Will the tests with reduced dimensions and sizes of materials remain structurally (and functionally) sufficient for redesign and hence reduce construction costs? Will the client benefit from seeing the design results in model form and therefore have increased confidence that the money spent is well worth the final cost?

Computers with numerical algorithms for the mass and momentum conservation equations written in discrete space form are applied ever more frequently in design. Some believe they are more accurate because the structure dimensions and hydrodynamics can be implemented at prototype scale. But the eddy viscosity defined as "turbulence" must also be implemented in the numerical model. This is at grid scales below the size of the discrete space form of the conservation equations that must be solved algebraically on the computer. As an example, consider a wave breaking on a rubble-mound armor layer of a breakwater. The turbulence in the breaking wave and the turbulence within the void spaces between the stones must be simulated numerically. This has yet to be done accurately on the world's largest and fastest computers. Therefore, turbulence simulation is a limit of numerical models coded in discrete form.

Smooth particle hydrodynamics (SPH) is a developing alternative to the discrete space form to solving the conservation equations, and future developments may prove useful for the design of coastal structures. See Dalrymple *et al.* (2010) for details regarding the SPH method applied to water waves and free surface flows. We are many years from using the SPH method to replace physical models in coastal structures design.

Physical models will find more use for coastal structures design in coastal regions as a consequence of rising sea levels on the worlds' oceans and more frequent and stronger storms.

6.8 Summary

In some respects, Chapter 6 is the most important chapter in this book.

It has been over 30 years since Jentsje van der Meer did his pioneering work on the stability of rubble-mound structures. By including the effects of wave period, structure permeability, and length of the storm (number of waves), the formulas of van der Meer (1987) have replaced the Hudson formula* for rubble-mound structures and stood the test of time. Storm surges create wave overtopping and submerged structures, and we now benefit from the EU Commission's DELOS project for a stability formula when this occurs (Burcharth *et al.*, 2006) as it will be more often under future rising sea levels. Berm breakwaters also came along over 30 years ago (Willis *et al.*, 1988) to add greater flexibility to breakwater design. Partial safety factors, spreadsheets, and examples add some practical aspects to the material presented. (** The Hudson formula is still applied for concrete armor units*).

It has now been over 20 years since another EU Commission project (PROVERBS) changed the way we think about and design vertical-walled, monolithic structures (Oumeraci *et al.*, 2001). We now have a parametric decision map to identify the breaker-type which depends on the type of structure and the wave parameters approaching the structure. A 12-step procedure greatly aids the designer and includes the pioneering work of Professor Yoshimi Goda and his formula (Goda, 1974) for the wave force on a vertical wall. Caisson design is significantly improved since both breaking and non-breaking waves are included. Partial safety factors, spreadsheets, and examples add some practical aspects to the material presented for caisson-type, vertical-walled structures in Chapter 6.

This chapter also discusses the geotechnical (soil) forces and design of sheet-pile bulkheads, types, and design of urban flood walls, dikes and revetment materials, and storm surge barriers with movable gates. A concluding section on special structures (floating breakwaters, artificial reef breakwaters, and breakwaters with wave energy conversion systems) rounds out the chapter.

Coastal engineers are responsible for the structural design of many types of alternatives for mitigating the damaging impacts of coastal floods, reducing shoreline erosion, and providing safe moorings in the marinas. The use of physical models is often the best way to increase confidence in the structural design and may result in a modified design that lowers the cost, which is always welcomed by the client. Numerical models are helpful in design but must be calibrated and verified for confidence in their results.

References

Abbott, M. B., and D. R. Basco (1989) *Computational Fluid Dynamics: An Introduction for Engineers*, Longman Scientific & Technical and John Wiley & Sons, New York.

Abramowitz, M., and I. A. Stegun (1972, eds.) *Handbook of Mathematical Functions*, Dover Publications.

Aerts, J. C. J. H., W. J. Botzen, and H. De Moel (2013) "Cost Estimates for Flood Resilience and Protection Strategies in New York City," *Annual of the New York Academy Science*, Vol. 1294, pp. 1–104.

Ahrens, J. P. (1975) Large wave tank tests of rip rap stability. TM-57, US Army Corps of Eng., CERC, Fort Belvoir, 1975.

Ahrens, J. P. (1984) "Reef Type Breakwaters," *Proceedings of the 19th International Coastal Engineering Conference*, American Society of Civil Engineers, Vol. 3, pp. 2648–2662.

Ahrens, J. P. (1981) "Design of Riprap Revetments for Protection Against Wave Attack," Technical Paper TP-81-5, U.S. Army Engineer Waterways Experiment Station, Coastal Engineering Research Center, Vicksburg, MS.

Ahrens J. P. (1987) "Characteristics of Reef Breakwaters," Technical Report 87-17, Corps of Engineers, Waterways Experiment Station, Vicksburg, MS.

Ahrens J. P., and J. Cox (1990) "Design and Performance of Reef Breakwaters," *Journal of Coastal Research*, Vol. 7, pp. 61–75.

Ahrens, J. P., G. Viggosson, and K. P. Zirkle, (1982) "Stability and Wave Transmission Characteristics of Reef Breakwaters," Interim Report, U.S. Army Engineer Waterways Experiment Station, Coastal Engineering Research Center, Vicksburg, MS.

Allsop, N. W. (1983) "Low-Crest Breakwaters, Studies in Random Waves," *Proceedings of Coastal Structures '83,* American Society of Civil Engineers, Arlington, pp. 94–107.

Allsop, N. W. H., and D. Vicinanza (1996) "Wave Impact Loadings on Vertical Breakwaters." in *Proceedings of the International Harbour Congress* (Antwerp, Belgium), Vol. II, pp. 275–284.

Allsop, W. and M. Calabrese (1998) "Impact Loadings on Vertical Walls in Directional Seas", *Proceedings of the 26th International Conference Coastal Engineering*, American Society of Civil Engineers, Vol. II, pp. 2056–2068.

Allsop, N. W. H., and A. Kortenhaus (2001) "Hydraulic Aspects," Chapter 2 in PROVERBS (Oumeraci *et al.*, 2001) A.A. Balkema, Lisse, the Netherlands.

Arena, F., A. Romolo, G. Malara, and A. Ascanelli, (2013) "On Design and Building of a U-OWC Wave Energy Converter in the Mediterranean Sea: A Case Study," *Proceedings of the 32nd International Conference on Ocean, Offshore and Arctic Engineering*, Nantes, France.

ASCE (2007) "What Went Wrong and Why," Hurricane Katrina External Review Panel, American Society Civil Engineers, Reston, VA.

Basco, D. R., and L. Stadler (1976) "Design of Rubble-Mound Breakwater for Port Arzew, Algeria," Special Report, Delft International Hydraulic Engineering (IHE) Courses, Delft, the Netherlands.

Basco, D. R. (2002a) "Breakwater Repairs and Wave Study, Lajes Field, Azores," Final Report for TranSystems, Inc., Beach Consultants, Inc. Norfolk, VA.

Basco, D. R. (2002b) "Breakwater Repair," Final Report, TranSystems, Inc., Beach Consultants, Inc., Norfolk, VA.

Baird W. F. (2003) "Lajes Field, Azores Breakwater Repair: Internal Wave Data Report," W. F. Baird & Associates, Coastal Engineers Ltd. Canada.

Barstow, S. F. (2002) "Wave Analysis for Praia da Vitoria, Azores," Final Report, OCN R-22028, Oceanor, Inc. Trondheim, Norway, September 6, 2002.

Bezuijen, A., and E. W. Vastenburg (2013) *Geosystems: Design Rules and Applications*, CRC Press, Boca Raton, FL.

Biesheuvel, A. C. (2013) "Effectiveness of Floating Breakwaters: Wave Attenuating Floating Structures," Civil Engineering and Geosciences, M.S. Thesis, Technical University Delft, The Netherlands.

Bjarke Ingels Group (2014) Architecture, Urban Planning, Consulting (https://www.designboom.com).

Briaud, J., H. Chen, A. Govindasamy, and R. Storesund (2008) "Levee Eosion by Overtopping in New Orleans during the Katrina Hurricane," *Journal of Geotechnical and Geoenvironmental Engineering*, Vol. 134, No. 5, pp. 618–632.

Broderick, L. L. (1983). "Riprap Stability: A Progress Report," *Proceedings of Coastal Structures '83*, American Society of Civil Engineers, pp. 320–330.

Burcharth, H. F. (1993) "The Design of Breakwaters," Department of Civil Engineering, Aalborg University, Denmark.

Burcharth, H. M., and T. L. Andersen (2006) "Overtopping of Rubble Mound Breakwaters with Front Reservoir," in *Proceedings of the 30th International Conference on Coastal Engineering* (San Diego), American Society of Civil Engineers.

Burcharth, H. F., and Z. Liu (1992) "Design of Dolos Armour Units," *Proceedings of the 23rd International Coastal Engineering Conference*, American Society of Civil Engineers, Vol. 1, pp. 1053–1066.

Burcharth, H. F., and Z. Liu (1998) "Force Reduction of Short-Crested Non-Breaking Waves on Caissons," Section 4.3, Part 4, Class II Report of MAST II Project: PROVERBS, Technical University of Braunschweig, Germany.

Burcharth, H. F., and J. D. Sorensen (2000) "Design of Vertical Wall Caisson Breakwaters Using Partial Safety Factors," in *Proceedings of the 26th International Conference on Coastal Engineering* (Copenhagen), ASCE.

Burcharth, H. F., and A. C. Thompson (1983) "Stability of Armour Units in Oscillatory Flow," *Proceedings of Coastal Structures '83*, American Society of Civil Engineers, Washington D.C., USA, pp. 71–82.

Burcharth, H., M. Kramer, A. Lamberti, and B. Zanuttigh (2006) "Structural Stability of Detached Low crested Breakwaters," *Coastal Engineering*, Vol. 53, No. 4, pp. 381–394.

Burger, G. (1995a) "Stability of Low-Crested Breakwaters," M.S. Thesis, Delft Technical University, Delft, The Netherlands.

Burger, G. (1995b) "Stability of Low-crested Breakwaters," *Final Proceedings*, EU Research Project Rubble mound Breakwater Failure Modes, MAST

2 contract MAS2-CT92-0042. Also published in Delft Hydraulics report H1878/H2415.

Cardone, V. (2002) "Praia de Vitoria Azores High Resolution Wave Hindcast and Extreme Statistics," Oceanweather, Inc. Proposal.

Carvalho, J. J. (1964) "Praia de Vitoria Harbor (Azores), Damages in the Breakwater Due to the Storm of the 26th–27th December, 1962," *Proceedings, ICCE (Lisbon)*, Chapter 36, pp. 560–579.

Carvalho, J. J. (1965) "Praia de Vitoria Harbor (Azores), Damages in the Breakwater Due to the Storm of the 26th–27th December, 1962," Memoria No. 271, Laboratorio Nacional De Engenharia Civil, Lisbo, Portugal.

Cavaleri, L., and P. Malanotte-Rizzoli (1981) "Wind Wave Prediction in Shallow Water-Theory and Application," *Journal Geophysical Research*, Vol. 86, pp. 10961.

Campanella, R. (2017) *Cityscapes of New Orleans*, LSU Press, Baton Rouge, LA.

CBBT Commission (circa 1995) "Field Data for Rock Slopes After Oct 31, 1991 Halloween Storm," Chesapeake Bay Bridge Tunnel, Field Office, Cape Charles, Virginia.

CEM (2006) *Coastal Engineering Manual*, US Army Corps of Engineers, Engineering Research and Development Center, Vicksburg, MS.

CEM (2011) *Coastal Engineering Manual*, US Army Corps of Engineers, Engineering Research and Development Center, Vicksburg, MS.

CERCular (1995) "CORE-LOC^{TM} Concrete Armor Units," Vol. CERC-95-1, WES, Vicksburg, MS.

CIRIA,CUR,CETMET (2007) The Rock Manual. *The Use of Rock in Hydraulic Engineering* (2nd Edition) C683, CIRIA, London (also referenced as "The Rock Manual (2007)".

Contestabile, P., F. Vincenzo, E. Di Lauro, and D. Vicinanza (2016) "Full-scale Prototype of an Overtotting Breakwater for Wave Energy Conversion," *Proceedings of the 35th International Conference Coastal Engineering*, ASCE, Washington DC.

Corps of Engineers (1989) "Engineering Design: Retaining and Floodwalls," EM 1110-2-2502, Wash. DC.

Corps of Engineers (1994) "Design of Sheet Pile Walls," US Army Corps of Engineers, EM 1110-2-2504, Washington D.C.

Corps of Engineers (2008) "Lake Okeechobee and the Herbert Hoover Dike," Summary, Engineering Evaluation Seepage and Stability Problems, Jacksonville District Office.

d'Angremond, K. (1975) "Rubble-Mound Breakwaters," Lecture Notes, International Hydraulic Engineering (IHE) Courses, Delft, The Netherlands.

Dalrymple, R. A., A. Herault, G. Bilotta, and R. J. Farahani (2010) "GPU-Accelerated SPH Model for Water Waves and Other Free Surface Flows," *Proceedings of the 32nd, International Conference Coastal Engineering*, ASCE, Shangai, China.

Dassanayake, D. and H. Oumeraci (2008) "Planned Research on the Hydraulic Stability of Geotextile Sand Containers," University of Hannover, Germany.

Dassanayake, D. T., and H. Oumeraci (2012a) "Engineering Properties of Geotextile Sand Containers and Their Effect on Hydraulic Stability and Damage

Development of Low-Crested/Submerged Structures," *International Journal of Ocean and Climate Systems*, Vol. 3, No. 3.

Dassanayake, D. T., and H. Oumeraci (2012b) "Hydraulic Stability of Coastal Structures Made of Geotextile Sand Containers (GSCS): Effect of Engineering Properties of GSCS," *Proceedings of the ICCE 33*, Santander, Spain.

Davidson, D. D. (1984) "Evaluation of Damage and Recommendations for Repair — Lajes Breakwater, Praia de Victoria Harbor, Azores," *Corps of Engineering Report*, WESCW-R, Vicksburg, MS.

De Boni, Cecconi, Nuova, Canestrelli, Franco, and Egori (1993) "A Description of the 'Acqua Alta Event on 5th October, 1992," *Proceedings of the 23rd International Conference Coastal Engineering*, The conference Organizing Committee.

Dean R. G., and R. Chen (1996) "Performance of the Midtown Palm Beach PEP Reef Installation," Final Report, Coastal and Ocean Engineering Laboratory, University of Florida, Gainesville FL.

Dircke, P. T. M., T. H. G. Jongeling, and P. L. M. Jansen (2012) "An Overview and Comparison of Navigable Storm Surge Barriers," *Proceedings, Innovative Dam and Levee Design and Construction, USSD Conference* , New Orleans, LA, April.

Donovan, S. (2013) "Hurricane Sandy Rebuilding Strategy," Final Report, Building Task Force, Dept Housing and Urban Development, Washington DC.

Eckert, J., and G. Callender (1987) "Geotechnical Engineering in the Coastal Zone," Instruction Report, CERC-87-1, Corps of Engineers, Coastal Engineering Research Center, Vicksburg, MS.

Erickson, K. (1998) "Conversion of the P.E.P. Reef Breakwater to a Groin Field for West Palm Beach, FL," in *Proceedings of the Eighth Beach Technology Conference* (Tallahassee, Florida), American Shore and Beach Preservation Association.

Eversen, J., and D. R. Basco (2003) "Study for Repairs to Breakwater at Lajes Field Air Base, Azores," Final Report, TranSystems Corporation, Norfolk, VA.

Ferrario, F., M. W. Beck, C. D. Storiazzi, F. Micheli, C. C. Shepard, and L. Airoldi (2014) "The Effectiveness of Coral Reefs for Coastal Hazard Risk Reduction and Adaptation," *Nature Communications*, Vol. 5, No. 3794.

Fréchet, M. (1927) "Sur la loi de probabilité de l'écart maximum," *Annales de la Societe Polonaise de Mathematique*, 6, 93.

Gaba, A., S. Hardy, L. Doughty, W. Powrie, and D. Selemetas (2017) "Guidance on Embedded Retaining Wall Design," CIRIA- C760, UK,

Garcia, R., and Kobayashi, N. (2014) "Trunk and Head Damage on a Low-crested Breakwater," *Journal of Waterway, Port, Coastal, and Ocean Engineering*, (http://doi.org/10.1061/CASCE)ww.1943-5460.0000276.

Givler, L. D., and R. M. Sorensen (1986). "An Investigation of the Stability of Submerged Homogeneous Rubble Mound Structures Under Wave Attack," H.R. IMBT Hydraulics Report IHL-110-86, Lehigh University, Bethlehem, PA.

Goda, Y. (1974) "A New Method of Wave Pressure Calculation for Design of Composite Breakwater," *Proceedings of the 14th International Conference Coastal Engineering*, ASCE, Vol. 3, pp. 1702–1720.

Hewson, P. J., A. R. Crawford, and M. J. A. Walkden (1998) "Effect of Aeration on Wave Impact Forces," in *Proceedings of the 2nd Project Workshop*, MAST III, PROVERBS, Naples, Chapter 1.5a.

Hill, D., M. J. Bowman, and J. S. Khinda (eds.) (2011) "Storm Surge Barriers to Protect New York City: Against the Deluge," *Conference Proceedings*, Polytechnical Institute of New York City, Brooklyn, NY, March, 30–31 2009 (16 papers and Review Panel).

Hill, D., M. J. Bowman, and J. S. Khinda (eds.) (2012) "Storm Surge Barriers to Protect New York City: Against the Deluge," *ASCE, Conference Proceedings*, Reston, VA., 268 pp.

Hillen, M. M., S. N. Jonkman, W. Kanning, M. Kok, M. A. Geldenhuys, and M. J. F. Stive (2010) "Coastal Defence Cost Estimates," Delft Technical University, Comm. On Hydraulic and Geotechnical Engineering.

Hudson, R. Y. (1958) "Design of Quarry-Stone Cover Layers for Rubble-Mound Breakwaters; Hydraulic Laboratory Investigation," Research Report No. 2-2, U.S. Army Engineer Waterways Experiment Station, Vicksburg, MS.

Hudson, R. Y. (1959) "Laboratory Investigation of Rubble-Mound Breakwaters," *Journal of the Waterways and Harbors Division,* American Society of Civil Engineers, Vol. 85, No. WW3, pp. 93–121.

Hudson, R. Y. (ed.) (1974) "Concrete Armor Units for Protection Against Wave Attack," Miscellaneous Paper H-74-2, U.S. Army Engineer Waterways Experiment Station, Vicksburg, MS.

Hughes, S. A. (2004) "Wave Momentum Flux Parameter: A Descriptor for Nearshore Waves," *Coastal Engineering*, Vol. 51, pp. 1067–1084.

Hutchinson, S., M. Young, and A. Macleod (2010) "Caisson Breakwater Design for Sliding," *Proceedings of the International Conference Coastal Engineering*, ASCE.

IPET (2006) "Performance Evaluation of the New Orleans and Southeast Louisiana Hurricane Protection System", Interagency Performance Evaluation Taskforce, Corps of Engineers, 8 Volumes plus Appendices Government Printing Office, Washington, DC.

ISVA (1974) "Cnoidal Wave Formulas and Tables", Technical University of Denmark, Lyngby, Denmark.

Jonkman, S. N., M. M. Hillen, R. J. Nicholls, W. Kanning, and M. van Ledden, M. (2013) "Costs of Adapting Coastal Defences to Sea-level Rise — New Estimates and Their Implications," *Journal of Coastal Research*, Vol. 29, No. 5, pp. 1212–1226.

Juhl, J. and J. W. van der Meer (1992) "Quasi-static Wave Forces on Vertical Structures, Reanalysis of Data at DHI and Delft Hydraulics," Final Report, MAST G6-S Coastal Structures.

Juhl, J., A. Alikhani, P. Sloth, and R. Archetti (1996) "Roundhead Stability of Berm Breakwaters," *Proceeding of the 25th International Conference on Coastal Engineering*, ASCE.

Junger, S. (1997) *The Perfect Storm: Story of Men Against the Sea*, W.W. Norton & Company, New York.

Kamphuis, J. W. (1991) "Incipient Wave Breaking," *Coastal Engineering*, Vol. 15, pp. 185–203.

Kamphuis, J. W. (2010) *Introduction to Coastal Engineering and Management* (2nd Edition), World Scientific, Singapore.

Kik, R., J. P. van den Bos, J. Maertens, H. J. Verhagen, and J. W. Van Der Meer (2012) "Notational Permeability," *Proceedings of the 33rd International Conference on Coastal Engineering*, ACSE, Reston, No. 84.

Kim, T. M., T. Takayama, and Y. Miyawaki (2004) "Laboratory Experiments on the Sliding Distance and Tilting Angle of a Caisson Breakwater Subject to Wave Impact," *Proceeding of the 29th International Conference on Coastal Engineering*, ASCE, Lisbon, pp. 3762–3773.

Kofoed, J. P. (1997) "Wave Overtopping of Marine Structures for Energy Generation," Master's thesis, Department of Civil Engineering, Aalborg Univresity, Denmark.

Kofoed, J. P. (2002) "Wave Overtopping of Marine Structures: Utilization of Wave Energy," Ph.D. Thesis, Hydraulics & Coastal Engineering Laboratory, Department of Civil Engineering, Aalborg University, Denmark, 171 pp.

Kofoed, J. P. (2006) "Vertical Distribution of Wave Overtopping for Design of Multi-Level Overtopping Based Wave Energy Converters," *Proceedings of the 30th International Conference Coastal Engineering*, ASCE, pp. 4714–4726.

Kofoed, J. P., and H. Burcharth (2002) "Estimation of Overtopping Rates on Slopes in Wave Power Devices and Other Low Crested Structures," *Proceedings of the 28th International Conference Coastal Engineering* ASCE, doi:10.1142/97898/2791306-0184.

Kofoed, J. P., P. B. Frigaard, and M. B. Kramer (2006a) "Recent Developments of Wave Energy Utilization in Denmark," *Proceedings of the Workshop on Renewable Ocean Energy Utilization, 20th Conference*, Korean Society of Ocean Engineers, Seoul.

Kofoed, J. P., P. Frigaard, E. Friis-Madsen, and H. C. Sorensen (2006b) "Prototype Testing of the Wave Energy Converter, Wave Dragon," *Renewable Energy*, Vol. 181, No. 9, pp. 181–189.

Kortenhaus, A. and H. Oumeraci (1998) "Classification of Wave Loading on Monolithic Coastal Structures", *Proceedings of the 26th International Conference on Coastal Engineering*, American Society of Civil Engineering, Vol. I, pp. 867–880.

Kramer, M., and H. Burcharth (2003) "Wave Basin Experiment Final Form, 3-D Stability Tests," Alborg University, Report of DELOS EVK-CT-2000-0004.

Kramer, M., B. Zanuttigh, J. W. Van der Meer, C. Vidal, and F. X. Gironella (2005) "Laboratory Experiments on Low-crested Breakwaters," *Coastal Engineering,* Vol. 52, No. 10–11, pp. 867–885.

Lamberti, A., R. Archetti, M. Kramer, D. Paphitis, C. Mosso, and M. Di Risio, M. (2005) "Prototype Experience Regarding Low-crested Structures," *Coastal Engineering*, Vol. 52, No. 10–11, pp. 841–866.

Lentz, K. M. (2010) "Flooding: Norfolk and the Region," Public Works Department, City Council Briefing, Slide Presentation, Aug 25, 2010.

Luettich, R. A., J. J. Westerink, and N. W. Scheffner (1992) "ADCIRC: An Advanced Three-dimensional Circulation Model for Shelves, Coasts, and Estuaries," Report 1: Theory and methodology of ADCIRC-2DDI and ADCIRC-3DL, Dredging Research Program Technical Report, DRP-92-6, Coastal Engineering Research Center, Vicksburg, Miss.

Margheritini, L., Vicinanza, D., and Frigaard, P. (2009) "SSG Wave Energy Converter: Design, Reliability and Hydraulic Performance of an Innovative Overtopping Device," *Journal Renewable Energy*, Vol. 34, pp. 1371–1380.

Martin, J., T. Mahoney, and B. Burkhart (2008) "City of Norfolk, Downtown Floodwall Evaluation," Certification Report, Moffatt & Nichol, Norfolk, Dec 2.

Martinelli L., and A. Lamberti (2011) "Dynamic Response of Caisson Breakwaters: Suggestions for the Equivalent Static Analysis of a Single Caisson in the Array," *Coastal Engineering*, Vol. 53, No. 1, pp. 1–20.

McCartney, B. L. (1985) "Floating Breakwater Design," *Journal of Waterway, Port, Coastal, and Ocean Engineering*, ASCE, Vol. 111, No. 2.

Melby, J. A. (1999) "Damage Progression on Rubble-Mound Breakwaters," Technical Report CHL-99-17, U.S. Army Engineer Waterways Experiment Station, Coastal and Hydraulics Laboratory, Vicksburg, MS.

Melby, J. A., and N. Kobayashi (2011) "Stone Armor Damage Initiation and Progression Based on Maximum Wave Momentum Flux," *Journal of Coastal Research*, Vol. 27, No 1, pp. 110–119.

Melby, J. A., and G. F. Turk (1993) "CORE-LOC Concrete Armor Units Initial Stability Results," Progress Report Coastal R&D Program, WES, Vicksburg, MS.

Melby, J. A., and G. F. Turk, (1994) "The CORE-LOC: Optimized Concrete Armor," *Proceedings of the 24th International Coastal Engineering Conference*, American Society of Civil Engineers, Vol. 2, pp. 1426–1438.

Melby, J. and G. Myrick (2004) "Lajes Field, Azores Breakwater Stability Model Study," Report prepared for NAVFAC, Atlantic Division, ERDC/CHL TR-04, July.

Melby, J. A., and S. A. Hughes (2003) "Armor Stability Based on Wave Momentum Flux," *Proceedings of Coastal Structures '03 Conference*, American Society of Civil Engineers, New York.

Melito, I., and J. A. Melby (2002) "Wave Runup, Transmission, and Reflection for Structures Armored with CORE-LOC," *Coastal Engineering*, Vol. 45, pp. 33–52.

Mooyaart, L. F., S. N. Jonkman, P. A. L. de Vries, A. van der Toorn, and M. van Ledden (2014) "Storm Surge Barrier: Overview and Design Considerations," *Proceedings of the 34th International Conference Coastal Engineering*, ASCE.

Mooyaart, L. F., and S. N. Jonkman (2017) "Overview and Design Considerations of Storm Surge Barriers," *Journal of the Waterway, Port, Coastal, and Ocean Engineering*, Vol. 143, No. 4.

Morang, A. (2007) "Hurricane Barriers in New England and New Jersey — History and Status After Four Decades," Corps of Engineers, ERDC/CHL TR-07-11.

Mustapa, M. A., O. Yaakob, Y. M. Ahmed, and C.-K. Rheem (2017) "Wave Energy Device and Breakwater Integration: A Review," *Renewable and Sustainable Energy Reviews*, Vol. 77, pp. 433–458.

NACCS (2015) "North Atlantic Coast Comprehensive Study: Resilient Adaptation to Increasing Risk," Final Report, US Army Corps of Engineers, Wash DC, January.

NAS (2006) "Report on Hurricane Katrina," National Academy Sciences, Washington, DC.

NAVFAC (1982) "Foundations and Earth Structures," Naval Facilities Engineering Command, DM-7.2, Alexandria, VA.

Oumeraci, H., and A. Kortenhaus (1997) "Wave Impact Loading-Tentative Formulae and Suggestions for Development of Final Formulae," in *Proceedings of the 2nd Task I Workshop*, MAST III, PROVERBS, Edinburgh, UK.

Oumeraci, H., A. Kortenhaus, W. Allsop, M. deGroot, R. Crouch, H. Vrijling, and H. Voortman (2001) "Probabilistic Design Tools for Vertical Breakwaters (PROVERBS)," Swets & Zeitlinger B.V. Lisse, the Netherlands.

Oumeraci, H., and J. Recio (2009) "Geotextile Sand Containers for Shore Protection," in Y. C. Kim (ed.), *Handbook of Coastal and Ocean Engineering*, World Scientific Publishing, Singapore, pp. 553–600.

Oumeraci, H., A. Kortenhaus, W. Allsop, M. deGroot, R. Crouch, H. Vrijling, and H. Voortman (2001) "Probabilistic Design Tools for Vertical Breakwaters" (PROVERBS), Swets & Zeitlinger B.V. Lisse, the Netherlands.

Pezza, D. (2008) "Overview of Hurricane KATRINA and Corps of Engineers Design," CEE 782 Classroom Presentation, Old Dominion University, Norfolk, VA.

PIANC (1992) "Analysis of Rubble Mound Breakwaters — Formulae for Failure Modes (1992–1994)," Working Group 12 (MarCom), Permanent International Association Navigation Congress, Paris.

PIANC (1994) "Floating Breakwaters: A Practical Guide for Design and Construction," Permanent International Association Navigation Congress, Paris.

PIANC (2006) "Design of Movable Weirs and Storm Surge Barriers," Working Group 26 Permanent International Association Navigation Congress, Paris.

Pilarczyk, K. W. (ed.) (1990) *Coastal Protection*, A.A. Balkema, Rotterdam.

Pillai, K., A. Etemad-Shahidi, and C. Lemckert (2017) "Wave Overtopping at Berm Breakwaters: Review and Sensitivity Analysis of Prediction Models," *Coastal Engineering*, Vol. 120, pp. 1–21.

Powell, K. A., and N. W. Allsop (1985) "Low-Crest Breakwaters, Hydraulic Performance and Stability," Report No. SR 57, Hydraulics Research Station, Wallingford, England.

Reeve, D., A. Chadwick, and C. Fleming (2012) *Coastal Engineering: Processes, Theory, and Design Practice,* Spon Press, London and New York.

Rock Manual (2007) "The Rock Manual: The Use of Rock in Hydraulic Engineering," 2nd edn.), CIRIA C683, London, UK (also referenced as CIRIA,CUR,CETMET (2007)).

Rodgers, J. D., G. P. Kemp, H. J. Bosworth, and R. B. Seed (2015) "Interaction between the US Army Corps of Engineers and the Orleans Levee Board Preceeding the Drainage Canal Wall Failures and Catastropic Flooding of New Orleans in 2005," *Water Policy*, Vol. 17, No. 4, 707–723.

Rosenberg, M. (2019) "Dikes in the Netherlands-Geography," online (https://www.thoughtco.com/polders-and-dikes-of-the-netherlands-1435535).

Rowe, P.W. (1952) "Anchored Sheet-Pile Walls," in *Proceedings of the Institution of Civil Engineers*, London, Part 1, Vol 1., pp. 27–70.

Roul, P. (2013) "Floating Breakwaters," online (http://www.coastalwiki.org).

Schüttrumpf, H., J. Möller, H. Oumeraci, J. Grüne, and R. Weissmann (2001) "Effects of Natural Sea States on Wave Overtopping of Seadikes," *Proceedings of the 4th International Symposium Waves 2001, Ocean Wave Measurement and Analysis*, American Society of Civil Engineers, Vol. 2, pp. 1565–1574.

Scott, D., D. Werren, T. Kenny, and C. Petykowski (2006) "Breakwater Damage and Rehabilitation — Terceira Island, Azores," *Proceedings of the ICCE*, San Diego, pp. 5195–5207.

Shuto, N. (1974) "Nonlinear Long Waves in a Channel of Variable Section," *Coastal Engineerng in Japan*, Vol. 17, pp. 1–12 (in Japanese).

Sigurdarson, S., G. Viggosson, and O. B. Smarason (2005) "Berm Breakwaters," Appendix 2 to Advances in the Design and Construction of Coastal Structures by I. J. Losada, in P. Bruun, (ed.), Port and Coastal Engineering — Developments in Science and Technology, *Journal of Coastal Research*, Special Issue No. 46.

Sorensen, R., and R. Weggel (1992) "Field Studies of Beachsaver Reef at Sea Isle, NJ," Final Report, Civil Engineering Department, Lehigh University, Bethlehem, PA.

Sorensen, R. (1980) "Coastal Engineering Lecture Notes," Texas A&M University, Personal Communication.

SPM (1977) *Shore Protection Manual*, 3rd edn., US Army Corps of Engineers, Engineering Research and Development Center, Washington, DC.

SPM (1984) *Shore Protection Manual*, 4th edn., US Army Corps of Engineers, Engineering Research and Development Center, Washington, DC.

Stauble, D. K., and Smith, J. B. (1999) "Performance of the PEP Reef Submerged Breakwater Project, Vero Beach, Florida," Twenty-Five Months Report, Coastal Engineering Research Center, U.S. Army Engineer and Research Development Center, Vicksburg, MS.

Stauble, D., and J. Tabar (2003) "The Use of Submerged Narrow-Crested Breakwaters for Shoreline Erosion," *Journal of Coastal Research*, Vol. 19, pp. 684–722.

Stauble, D. K., J. B. Smith, and R. W. Wise (2005) "National Shoreline Erosion Control Demonstration and Development Project," Corps of Engineers, Final Report.

Strock, C. A. (2005) "Full Committee Hearing Actions: EPA, Corps of Engineers, FHA Related to Hurricane Katrina," US Senate Committee on Environment and Public Works, Oct 6.

Subramanian, S. (2017) *Up and Out*, Article in *New York Times Magazine*, April 23.

Sweet, W. V., and J. J. Marra (2014) "2014 State of Nuisance Tidal Flooding," White Paper, NOAA, Center for Operational Oceanographic Products and Services, Washington DC.

Sweet, W. V., G. Dusek, J. Obeysekera, and J. J. Marra (2018) "Patterns and Projections of High Tide Flooding Along the US Coastline Using a Common Impact Threshold," NOAA Technical Report, NOS CO-OPS 086, National Oceanic and Atmospheric Administration, Washington, DC.

Tanimoto, K. K. Moto, S. Ishizuka, and Y. Goda (1976) "An Investigation on Design Wave Force Formulae of Composite-type Breakwaters," *Proceedings of the 23rd Japanese Conference Coastal Engineering*, pp. 11–16 (Japanese).

Takahashi, S., K. Tanimoto, and K. Shimosako (1994) "A Proposal of Impulsive Pressure Coefficient for the Design of Composiste Breakwaters," in *Proceedings of the International Conference on Hydro-Technical Engineering for Port and Harbor Construction* (Yokosuka, Japan), Vol. 1, Part 1, pp. 489–504.

Thompson, D. M., and R. M. Shuttler (1975) "Riprap Design for Wind Wave Attack: A Laboratory Study on Random Waves," HRS Wallingford, Report 61, Wallingford, UK.

Thomsen, A. L., P. E. Wohlt, and S. A. Harrison (1972) "Riprap Stability on Earth Embankments Tested in Large- and Small-Scale Wave Tanks," TM-37, U.S. Army, Corps of Engineers, Coastal Engineering Research Center, Fort Belvoir, Va.

Torre-Enciso, Y., J., Marques, and L.I. Lopez de Aguileta (2010) "Mutriku: Lessons Learnt," *Proceedings of the 3th International Conference on Ocean Energy*, Bilbao.

Torre-Enciso, Y., I. Ortubia, L. I. Lopez de Aguileta, and J. Marques (2009) "Mutriku Wave Power Plant: From the thinking Out to the Reality," *Proceedings of the 8th European Wave Tidal Energy Conference*, Uppsala, Sweden, pp. 319–329.

Torum, A., F. Kuhnen, and A. Menze (2003) "On Berm Breakwaters: Stability, Scour, Overtopping," *Coastal Engineering*, Vol. 49, No. 3, pp. 209–238.

TranSystems (2003) "Study for Repairs to Breakwater at Lajes Field Air Base, Azores," Final Report, TranSystems Corp, Norfolk, VA.

Turk, G. F., and J. A. Melby (1997) "CORE-LOC Concrete Armor Units: Technical Guidelines," Misc. Paper CHL-97-6, USAE, Waterways Experiment Station, Vicksburg, MS.

US Steel Design Manual (1972, 1984) "Steel Sheet Piling Design Manual," United States Steel Corporation, Pittsburg, PA.

van der Meer, J. W. (1987) "Rock Slopes and Gravel Beaches Under Wave Attack," Ph.D. Dissertation, Delft University of Technology, The Netherlands (Delft Hydraulics Publication No. 396).

van der Meer, J. W. (1988) "Rock Slopes and Gravel Beaches Under Wave Attack," Publication No. 396, Delft Hydraulics, Delft, The Netherlands.

van der Meer, J. W. (1992) "Estimation of Reshaped Profile in Berm Breakwaters Using Experimental Data," Delft Hydraulics, Delft, The Netherlands.

van der Meer, J. W., and K. Pilarczyk (1990) "Low-Crested and Reef Breakwaters," Report H198/Q638, Delft Hydraulics Laboratory, The Netherlands.

van der Meer, J. W., K. d'Angremond, and J. Juhl (1994) "Probabilistic Calculations of Wave Forces on Vertical Structures," in *Proceedings of the 24th International Conference on Coastal Engineering* (Kobe), ASCE.

van der Meer, J. W., and S. Sigurdarson (2017) *Design and Construction of Berm Breakwaters*, World Scientific, Singapore.

van der Meer, J. W., K. d'Angremond, and E. Gerding (1995) "Toe Structure Stability of Rubble-Mound Breakwaters," *Proceedings of the Coastal Structures and Breakwaters Conference, ICE*, London, UK.

van Gent, M. R. A. (2000) "Wave Run-up On Dikes With Shallow Foreshores," in *Proceedings of the International Conference on Coastal Engineering* (Sydney), ASCE, Vol. 3, pp. 2030–2043.

van Gent, M. R. A. (2004) "On the Stability of Rock Slopes," Springer, NATO Science Series, IV/53, Varna, Italy (keynote address).

van Gent, M. R. A., A. J. Smale, and C. Kuiper (2004) "Stability of Rock Slopes With Shallow Foreshores," *Proceedings of the 4th International Coastal Structures Conference*, ASCE, Reston, VA.

Van Steeg, P. M., and K. Breteler (2008) "Large Scale Physical Model Tests on the Stability of Geotextile Tubes," Deltares report, H4595, Delft, The Netherlands.

Van Steeg, P. M., and E. Vastenburg (2010) "Large Scale Physical Model Tests on the Stability of Geotextile Tubes," Deltares Report, 1200162, Delft, The Netherlands.

Verhagen, H. J., K. d'Angremond, and F. van Roode (2009) *Breakwaters and Closure Dams*, 2nd edn., VSSD Publishing Co. Delft, The Netherlands.

Vicinanza, D., and P. Frigaard (2009) "Wave Pressure Acting on a Seawave Slot-cone Generator," *Coastal Engineering*, Vol. 55, pp. 553–568.

Vicinanza, D., L. Margheritini, J. P. Kofoed, and M. Buccino (2012) "The SSG Wave Energy Converter: Performance, Status and Recent Developments," *Energies*, Vol. 5, No. 2, pp. 193–226.

Vicinanza, D., F. Denale, and M. Buccino (2015) "Structural Response of Seawave Slot-cone Generator (SSG) from Random Wave CFD Simulations," *Proceedings of the Twenty-fifth (2015) International Ocean and Polar Engineering Conference*, Kona, Big Island, Hawaii, USA.

Vicinanza, D., P. Contestabile, J. Q. H. Nørgaard, and T. L. Andersen (2014) "Innovative Rubble Mound Breakwaters for Overtopping Wave Energy Conversion," *Coastal Engineering*, Vol. 88, pp. 154–170.

Vidal, C., Losada, M. A., Medina, R., Mansard, E. P. G., and Gomez-Pina (1992) "A Universal Analysis for the Stability of Both Low-Crested and Submerged Breakwaters," *Proceedings of the 23rd International Coastal Engineering Conference*, American Society of Civil Engineers, Vol. 2, pp. 1679–1692.

VIMS (2006) Virginia Institute of Marine Science, Gloucester Pt. VA (online photo).

Werren, D. (2004) "Lajes Field, Breakwater Repair Design Report," W.F. Baird & Assoc., Ltd., Madison, WI, for NAVFAC, Atlantic Division, December.

Werren, D. J., W. F. Baird, M. Denechere, and M. Fons (2002) "The CORE-LOC^{TM} Armoring Technique: Experience on Recent Projects," *Proceedings of the 28th International Conference Coastal Engineering*, ASCE, NY.

Werren, D. J., C. J. Petykowski, R. D. Scott, and T. F. Kenny (2006) "Breakwater Damage and Rehabilitation," Terceira Island, Azores, Abstract, Paper 114, *30th of the International Conference Coastal Engineering*, ASCE, NY.

Wiegel, R. L. (1964) *Oceanographical Engineering*, Prentice Hall, Englewood Cliffs, NJ.

Wiegel, R. L. (2001) "Seawalls, Seacliffs, Beachrock: What Beach Effects?" *Shore and Beach*, Vol. 70, No. 2, American Shore and Beach Preservation Association.

Wikipedia (2005) "Levee Failures in Greater New Orleans", online search engine.

Willis, D. H., W. F. Baird, and O. Magoon (eds.) (1988) *Berm Breakwaters: Unconventional Rubble-mound Breakwaters,* American Society of Civil Engineers, New York.

Woodruff, P. E., and Dean, R. G. (2000) "Innovative Erosion Control Technology in Florida," in *Book of Abstracts of the 27th International Conference on Coastal Engineering* (New York), Vol. 2, American Society of Civil Engineers.

Yin, J., N. Lin, and D. Yu (2016) "Coupled Modeling of Storm Surge and Coastal Inundation: A Case Study in New York City During Hurricane Sandy," *Water Resource Research*, Vol. 52, pp. 8685–8699.

Chapter 7

Beach Engineering

7.1 Introduction

7.1.1 *The cultural benefits of beaches*

A review of the history of beaches is dramatically told in Lenček and Bosker (1998), as shown in Figure 7.1. Also shown, a list of 10 topics are discussed that reveal the depth of importance of beaches to people living in the Western Cultures of the world. The cultural benefits of beaches are found in the art, literature, music, health, and fashion industries. Figure 7.2 presents some examples for art, dance, movies, and travel.

Lenček and Bosker (1998) claim their book provides "... the History of Paradise on Earth", and if interested, you should read the book to determine this for yourself. "*In every outthrust headland, in every curving beach, in every grain of sand there is the story of the earth*".

7.1.2 *The economic value of beaches*

Beaches and dunes are nature's interface between the land and the sea for much of the world's coast. The loose, non-cohesive materials that collectively shape the dune/beach profile come in many sizes and colors. These materials can be sands, pebbles, and shingles (cobble) and come from local cliffs or distant river sources to the coast where they continue to move along the shore or in the on–offshore directions. The onshore dynamic nature of the world's sandy beaches ($D_{\text{sand}} = 0.0625$–2.00 mm, Figure 2.34, Wentworth scale) is why beach nourishment has become the "soft" alternative for coastal storm damage mitigation. The beach (and dunes if present) are a reservoir of sand volume that is available to move seaward. Storms bring higher, shorter period waves that drag sand offshore to build under-water sand bars causing wave breaking and far less energy reaching the shore face. Wide sandy beaches are considered coastal "infrastructure" that must be maintained to reduce storm damage to upland property, roads, etc., landward of the beach/dune system.

Figure 7.1. Cover of *The Beach: The History of Paradise on Earth* (Lenček and Bosker, 1998) and 10 topics discussed to support the claim in the title (Penguin Random House, LLC).

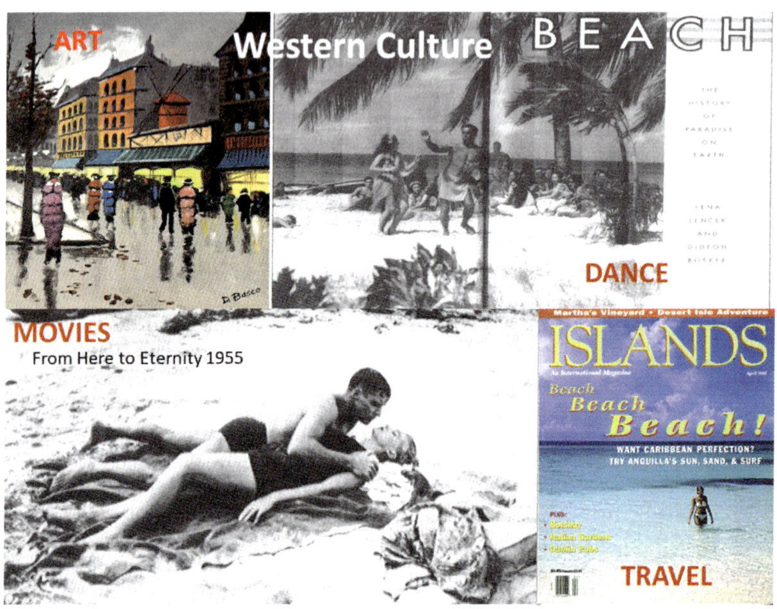

Figure 7.2. Western art (Monet, 1867 "The Beach at Sainte-Adresse", Claude Monet, Art Institute of Chicago), dance (Hawaiian hula), movies (From Here to Eternity, 1965), and travel (Brochure, Islands).

7.1.2.1 *Storm damage reduction benefits*

In the US, all Federal Government projects for shore "protection" as designed by the Corps of Engineers must be justified by the benefits exceeding the costs. However, the only benefits that can be considered are storm damage mitigation and reduction of erosion. Tourism benefits are not allowed for the beach nourishment alternative since the US citizens can selectively choose which states and beaches to visit so that there is no net economic development (NED) benefit to the federal government. Figure 7.3 dramatically shows the shift in Corps of Engineers shore "protection" projects from "hard", fixed structures to "soft" beach nourishment projects beginning in the late 1960s (from Hillyer, 1996). The trend continues today.

The percentage of total costs of all the Corps of Engineers projects for shore "protection" that involves beach nourishment was 83% in 2002 prices (Hillyer, 2003).

As one example, after Hurricane Eloise (1975) a study of 540 structures in Bay County, Florida related their distance from the ocean (vertical axis) to the resulting structural damage (horizontal axis, US$1,000) as shown in Figure 7.4 (adapted from Dean, 1987). Where a wider beach was present, the structural damages were reduced. The zero-distance location

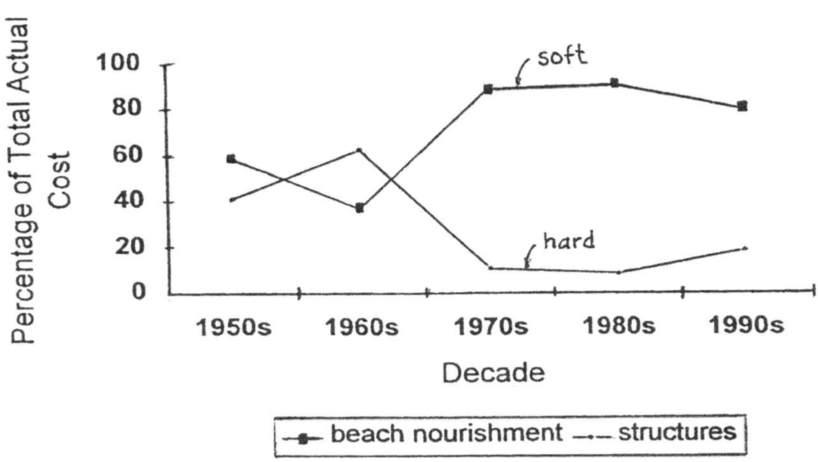

Figure 7.3. The shift from fixed, "hard" structures to "soft" beach nourishment projects as a percentage of the total annual cost of Corps of Engineers projects in the US from 1950 to 1995 (from Hillyer, 1996).

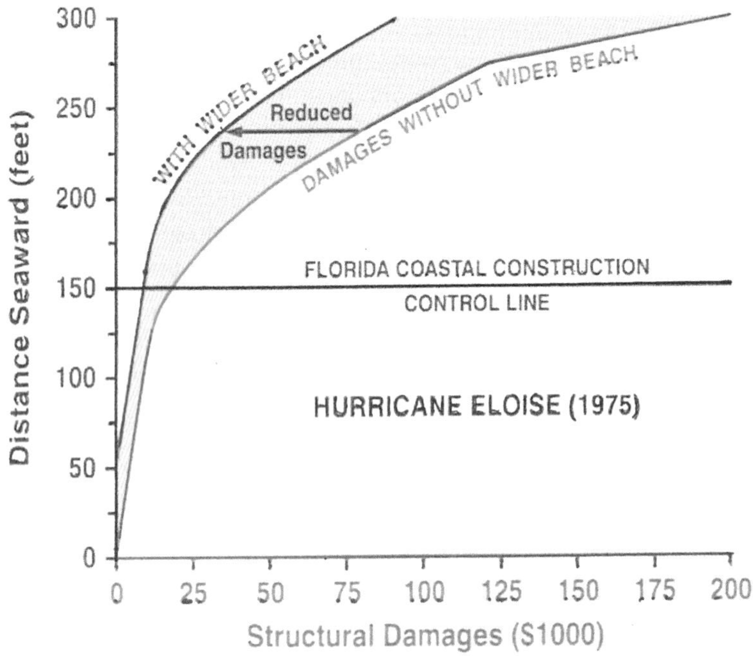

Figure 7.4. Damage to structures from Hurricane Eloise (1975) in Bay County, Florida relative to seaward location and beach width (adapted from Dean, 1987).

is the baseline for Florida's coastal construction control line (CCL) which is discussed further in Chapter 11.

Beach restoration projects reduce storm damage, which in turn, reduces disaster recovery costs which is a NED benefit for the US Federal government.

7.1.2.2 *Travel and tourism benefits*

In a series of articles (Houston, 1995, 1996, 2002, 2008, and 2013) regarding the economic value of beaches to the US travel and tourism (T&T) industry, Houston cites a number of creditable sources (US Bureau of Labor Statics, US Census Bureau, US Department of Commerce, US Department of Interior, US Travel Association, US National Park Service, NY Times, Washington Post, World Travel and Tourism Council, and many others) to reach the following consistent conclusions after almost 20 years of research:

- T&T is the world's largest industry.
- T&T is the world's and America's largest employer.
- T&T has the largest trade surplus of all foreign trade in the US.
- Beaches are the leading US tourist destination. Beaches are also by far the favorite destination for the majority of the world's travelers.
- Beaches receive more day visits than the combined total visits to the Federal and State parks, recreational areas, and public lands in the US.
- The US Federal Government receives the majority share of T&T tax money from beach tourist spending.
- The tax money received by the Federal government is far more than Federal expenditures on beach nourishment projects.
- The US lags behind the world's growth in tourism infrastructure investment including restoration of beaches. Future growth in tourism is projected to fall further behind the rest of the world.

Unfortunately, at the time of this writing, tourism benefits have not been included, though the Corps of Engineers is calculating the benefit/cost ratio for the proposed beach nourishment projects. Fortunately, some states in the US (Florida, New Jersey) provide total funding for beach nourishment projects where the tourism benefits to the state's economy are well documented (e.g. Stronge, 2005; FSBPA, 2017). The current formula for Federal/Local Sponsor shared funding is 65/35% for beach nourishment projects designed by the US Corps of Engineers.

7.1.2.3 *Recreation and sport benefits*

Beaches located in the urban areas of the world's coastal cities provide recreational opportunities for many of the least fortunate citizens living in the area. Families can take public transportation for after-work or day trips so the kids can enjoy the sand and water while the parents relax. Beach nourishment projects are not just for vacationing families driving from the inland areas of the country who can afford hotel, restaurant, amusement park, and automobile expenses as tourists. Beach nourishment projects are not just for wealthy home owners living on the coast as often reported in the media. Urban beaches of the world are the "natural parks" for many people by providing recreational benefits in their lives at little or no expense.

Figure 7.5. Ocean View beach in Norfolk, VA attracting many local citizens in the early evening to view the Chesapeake Bay and enjoy the beach (Author's photo).

Figure 7.5 shows many people enjoying Ocean View beach in Norfolk, Virginia near sunset and lists interesting items visible in the photo. A beach nourishment project jointly funded by the City of Norfolk and the Federal government (Corps of Engineers) was completed at this site in the summer, 2017. The beach width at high tide is now eight times wider than that shown in Figure 7.5.

In recent years, beaches have become sites for individual and team sporting events. Surfing contests, beach volleyball, and sand soccer tournaments attract hundreds of participants who travel from distant locations. These events also attract many spectators and bring beach musical concerts in the evenings for added entertainment.

Beach volleyball was introduced as an official Olympic summer sport in 1996 for both men and women competitors from many countries. In central

Figure 7.6. Beach-volleyball court and spectator stands erected in central London, UK for the 2012 Summer Olympics (photo by Dom Crossley, https://commons.wikimedia.org/).

London, UK, for the 2012 Summer Olympics, a beach volleyball court and spectator stands were erected at the House Gardens Parade location. Figure 7.6 is a photo by Dom Crossley (https://commons.wikimedia.org/). Millions of people watched the sand volleyball matches on television around the world. Most were interested in the countries that won the Olympic medals, but some were more interested in the swimsuit attire worn by the participants.

Lenček and Bosker (1998) claim their book provides "... the History of Paradise on Earth". Television and the internet have now spread Western culture around the world.

7.1.2.4 *Ecology and environmental benefits*

People are not the only beneficiaries of wide sandy beaches. The microorganisms within the air and water void spaces, the crawling worms and crabs, the birds that feed there, the turtles that lay eggs there, and the fish and shellfish living in the water above the subaqueous beach surface are all part of the ecology of beach environments as shown and listed in Figure 7.7.

The economic benefits of beaches include storm damage mitigation, tourism, recreation and sports, and the environment/ecology as quantified by coastal scientists, economists, and engineers. We shall again present more details about the environmental benefits of beaches in Section 7.6.

Figure 7.7. Graphic view of non-cohesive beach materials of many sizes to illustrate the void spaces where the smallest microorganisms live to start the food chain in the ecology of beaches (courtesy Sandgrains, Paia, Hawaii).

7.1.3 *Chapter objectives, topics, and limitations*

The main objective of this chapter is to introduce what is now called "Beach Engineering". In Section 7.2, fundamentals of coastal hydrodynamics resulting in sediment transport are considered. Then, the steps in beach design are reviewed in Section 7.3, with dune design detailed in Section 7.4. Hydraulic dredging machines which are the primary method for beach construction and project monitoring are the topics discussed in Section 7.5. Beach management requires other skills than science and engineering as introduced in Section 7.6. Some real design examples are discussed in Section 7.7 to give an idea about practical application to the theory.

In graduate-level courses taught at the Old Dominion University, one full semester is devoted to the study of *coastal hydrodynamics and sediment transport* using the text "Coastal Processes" by Dean and Dalrymple (2002). One full semester course covers *dredging and beach engineering* with the textbook *Beach Nourishment: Theory and Practice* by Dean (2002). Much of the theory presented in the following is from the innovative research efforts of Professor Bob Dean.

The American Shore and Beach Preservation Association (ASBPA) is the non-profit organization with worldwide membership of coastal scientists, engineers, economists, managers, and the general public, and it is the best source for the latest information on beaches. Their website is http://www.asbpa.org. The ASBPA publishes a journal *Shore & Beach* four times a year with refereed articles that present the latest research developments and design practices in beach engineering.

Beach engineering is the science, engineering, and technology necessary to design a beach nourishment project (including dunes), determine the costs and benefits, supervise construction and monitoring, and provide the guidance needed to successfully manage the beach site.

7.2 Fundamentals of Coastal Hydrodynamics and Sediment Transport

7.2.1 *Wave-induced longshore currents and sediment transport*

7.2.1.1 *Wave-induced radiation stresses*

The depth-integrated and time-averaged momentum produced by progressive waves minus the still-water, hydrostatic pressure force is the horizontal wave thrust and called the "radiation stress" (Longuet-Higgins and Stuart, 1964). Figure 7.8 (bottom) defines wave variables and a wave crest (top) propagating with celerity C with radiation stress S_{XX} in the $+X$-direction and radiation stress S_{YY} at right angles in the $+Y$-direction on a control volume. These are called the principle radiation stress components. For linear wave theory, these principle stress components along a wave ray were derived by Longuet-Higgins and Stuart (1964) and were given as Equations (2.34) and (2.35) for S_{XX} and S_{YY}, respectively, in Section 2.3.6.1.

The direction of wave celerity C is at angle α with the normal, $+x$-axis direction and at right angles to the shoreline which is defined as the y-axis. The $+X$-direction (capital letter X) is in the wave ray direction and the $+x$-direction (lower case letter x) is perpendicular to the shoreline. Using coordinate transformation formulas derived from Mohr's Circle, the onshore directed component S_{xx}, the shore-parallel component S_{yy}, and the shear stress component S_{xy} can be found. Formulas for S_{xx}, S_{yy}, and S_{xy} were given as Equations (2.36)–(2.38), respectively, in Section 2.3.6.1 and are schematically depicted in Figure 7.8 (top). See Basco (1982) *Surf Zone Currents* for detailed derivations of these equations.

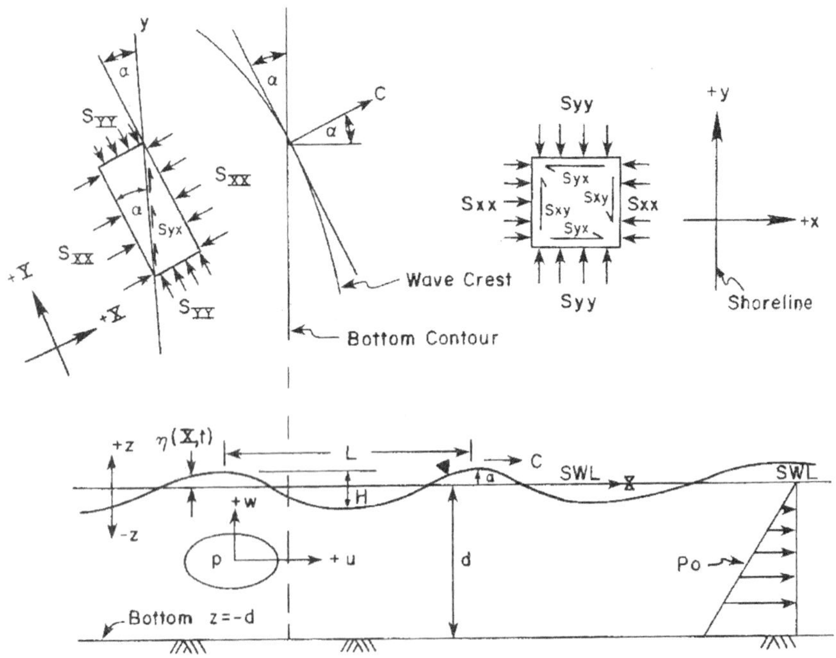

Figure 7.8. Wave-variable definitions (bottom) and radiation stresses (top) depicted for principle component directions (X, Y) and transformed coordinate directions (x, y).

Gradients in the wave heights (Figure 7.9, top) across the surf zone produce gradients in the radiation stress component, S_{xx}. Gradients in S_{xx} in the shore-normal direction produce mean water-level changes called wave setdown and wave setup as shown in Figure 7.9 (top) and are fully discussed in Section 2.3.6.2. The momentum balance equation in the x-direction is employed and decoupled from the y-direction momentum equation.

7.2.1.2 *Wave-induced longshore current, $\bar{v}$*

Gradients in the radiation shear stress component S_{xy} in the horizontal, $+x$-direction across the surf zone drive the longshore current $\bar{v}$ (overbar means time average) and produce the longshore current profile $\bar{v}$(x) as illustrated in Figure 7.9 (bottom). The force components on the control volume width dx between adjacent wave rays are as found in Equation (7.1):

$$S_{xy} = \text{radiation shear stress component} = E\, n\, \sin\alpha\, \cos\alpha, \qquad (2.38)$$

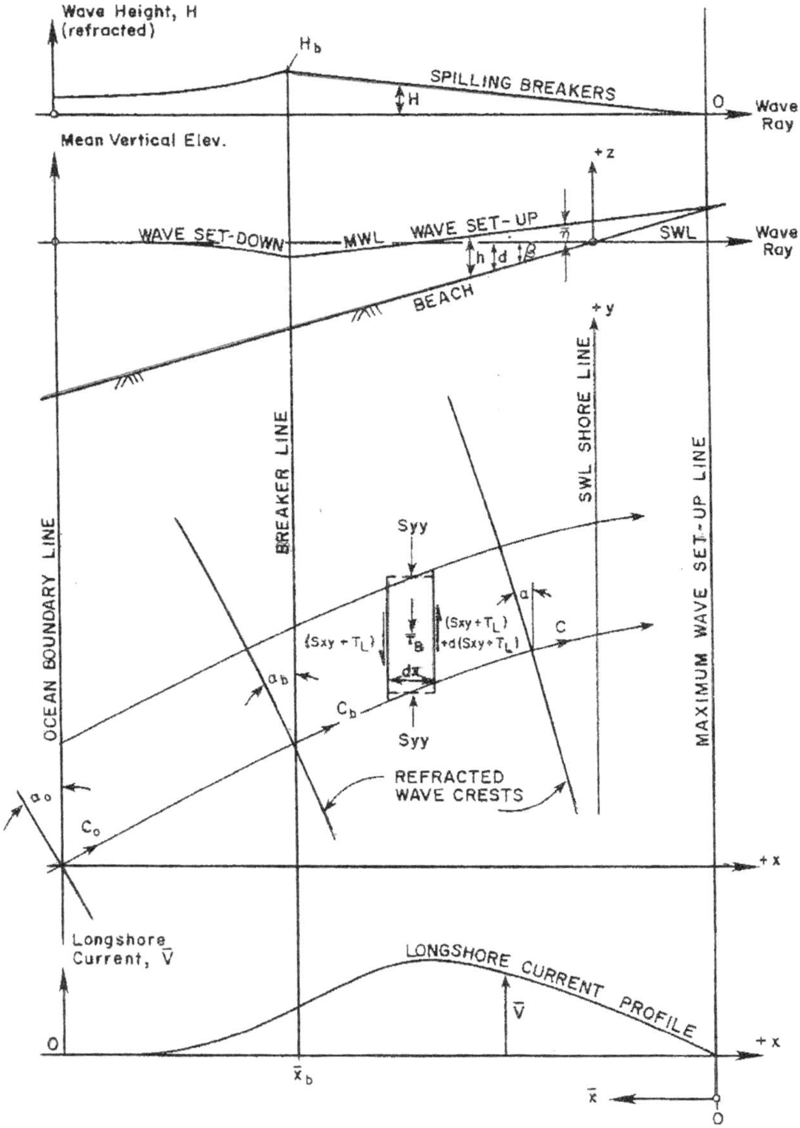

Figure 7.9. Schematic of the key parameters to calculate the longshore current profile as shown in the bottom picture.

where τ_B = the resisting shear stress at the bed ($z = -d$), and T_L = lateral, turbulent mixing stress.

In the y-direction, the momentum balance equation becomes

$$\frac{dS_{xy}}{dx} - \tau_B + \frac{dT_L}{dx} = 0. \tag{7.1}$$

The longshore current velocity $\bar{v}$ appears in the bed shear stress τ_B and turbulent mixing stress T_L terms. With appropriate formulations for these resisting and mixing stress terms and for the radiation shear stress, it is possible to integrate Equation (7.1) to analytically find an expression for $\bar{v}$ (x) across the surf zone.

It is convenient to take a new $+\bar{x}$-direction coordinate system facing seaward so that the maximum wave setup location is $\bar{x} = 0$, and the longshore current, $\bar{v}(0) = 0$. The wave breaker location is $\bar{x}_b$ at the location of the wave height where breaking begins, H_b, as illustrated in Figure 7.9.

The driving force for the longshore current is the gradient of the shear stress component. Outside the surf zone (no wave breaking), the energy dissipation is zero, therefore the gradient of the radiation shear stress is zero (Equation (7.2)). Inside the surf zone, on a plane sloping beach of angle β, Longuet-Higgins (1970) derived the following expression (Equation (7.3)) for shallow-water surf zones ($n \approx 1$), neglecting wave setup ($d \approx h$), for small wave angles ($\cos\alpha \approx 1$), when Snell's law of refraction is applicable [$(\sin\alpha/c)$ = constant] and where the wave-breaking index $\gamma = H/h$ is constant. Therefore we obtain the following:

Outside the surf zone

$$\frac{dS_{xy}}{d\bar{x}} = 0. \tag{7.2}$$

Inside the surf zone

$$\frac{dS_{xy}}{d\bar{x}} \approx \left[\frac{5}{16}\rho g^{3/2}\gamma^2(\tan\beta)^{5/2}\left(\frac{\sin\alpha}{c}\right)\right]\bar{x}^{3/2} = [r]\bar{x}^{3/2}. \tag{7.3}$$

Bottom friction resists the time-averaged, longshore current similar to that for open-channel flow. A time-averaged, bed shear stress $\bar{\tau}_B$ formulation as derived by Longuet-Higgins (1970) is

$$\bar{\tau}_B = \left[\frac{1}{\pi}\rho C_f g^{1/2}\gamma(\tan\beta)^{1/2}\right]\bar{x}^{1/2}\bar{v} = [q]\bar{x}^{1/2}\bar{v}, \tag{7.4}$$

where C_f is the time-averaged, boundary resistance coefficient for combined waves and longshore current.

In this formulation, Longuet-Higgins (1970) linearized the quadratic law for bottom friction in real time for water waves; assumed the longshore current velocity is small compared with the wave orbital velocity; assumed the wave angle is small in the surf zone so the longshore current velocity vector and wave orbits are at right angles; used linear wave theory in shallow water to estimate the wave orbital velocity; and neglected wave setup. Equation (7.4) clearly introduces the longshore current $\bar{v}$ and shows how the bottom shear stress varies as $\bar{x}^{1/2}$.

Finally, the turbulence due to wave breaking spreads laterally in both horizontal directions from the breakpoint so that a longshore current exists outside the breakpoint as illustrated in Figure 7.9 for the longshore current profile, $\bar{v}(\bar{x})$. For the time-averaged, lateral mixing force over the water depth T_L, Longuet-Higgins (1970) derived the following formula for the gradient:

$$\frac{dT_L}{d\bar{x}} = [N\rho g^{1/2}(\tan\beta)^{3/2}]\frac{d}{d\bar{x}}\left\{\bar{x}^{5/2}\frac{d\bar{v}}{d\bar{x}}\right\} = [p]\frac{d}{d\bar{x}}\left\{\bar{x}^{5/2}\frac{d\bar{v}}{d\bar{x}}\right\}, \tag{7.5}$$

where N is a dimensionless, turbulent closure coefficient for lateral wave mixing.

Wave setup is again neglected, and N is related to the lateral, turbulent eddy viscosity due to wave breaking. See Basco (1982) for complete details in derivations of Equations (7.3)–(7.5).

Plugging Equations (7.3)–(7.5) into Equation (7.1) gives a second-order, differential equation for $\bar{v}$ in terms of the bracketed terms, $[r]$, $[q]$, and $[p]$. Since from Equation (7.2) the radiation shear stress vanishes outside the surf zone ($[r]\bar{x}^{3/2} = 0$), the $\bar{v}$ ($\bar{x}$) profile solution is for two ordinary, second-order differential equations. They must match in magnitude and gradient of $\bar{v}$ ($\bar{x}$) at the breaker line and vanish at the shoreline and ocean boundary.

Define a reference longshore current velocity with no turbulence mixing ($N = 0$) as $\bar{v}^*$; hence, $[p] = 0$. Taking the bracketed term $[p] = 0$ in Equation (7.5) greatly simplifies the differential equation inside the breaker zone. The resulting momentum balance gives at the breaker line

$$\bar{v}_b^* = \frac{5}{16}\frac{\gamma}{C_f}(gd_b)\left(\frac{\sin\alpha}{C}\right)_b \tan\beta. \tag{7.6}$$

We introduce a dimensionless distance variable X and a dimensionless longshore current variable V (overbar removed) as follows:

$$X = \frac{\bar{x}}{\bar{x}_b} \quad \text{and} \quad V = \frac{\bar{v}}{\bar{v}_b^*}. \tag{7.7}$$

Solution of the second-order, differential equations is greatly simplified using these dimensionless coordinates.

Inside the breaker zone $(0 \leq X \leq 1)$

The second-order, non-dimensional differential equation becomes

$$PX^{5/2}\frac{d^2V}{dX^2} + \frac{5}{2}PX^{3/2}\frac{dV}{dX} - X^{1/2}V = -X^{3/2}, \tag{7.8}$$

where P is the relative importance of lateral turbulent mixing N compared with bottom friction, C_f

$$P = \frac{N\pi \tan\beta}{C_f \gamma}. \tag{7.9}$$

Theoretically, with no lateral turbulent mixing $(N = 0)$ so that $P = 0$ and Equation (7.8) simply gives

$$V = X$$

for a triangular velocity profile with $V = 1$ where $X = 1$ at the breaker line. Outside the breaker line with no lateral mixing, $V = 0$.

Longuet-Higgins (1970) solved Equation (7.8) with the following results:

Inside the breaker zone $(0 \leq X \leq 1)$

$$V = AX + B_1 X^{P_1}. \tag{7.10}$$

Outside the breaker zone $(X \geq 1)$

$$V = B_2 X^{P_2}, \tag{7.11}$$

where all the coefficients A, B_1, B_2, P_1, and P_2 depend on P and are defined as

$$A = \frac{1}{\left[1 - \left(\frac{5}{2}\right)P\right]}, \tag{7.12}$$

$$P_1 = -\frac{3}{4} + \sqrt{\frac{9}{16} + \frac{1}{P}}, \tag{7.13}$$

$$P_2 = -\frac{3}{4} - \sqrt{\frac{9}{16} + \frac{1}{P}}, \tag{7.14}$$

$$B_1 = \frac{P_2 - 1}{P_1 - P_2}A, \tag{7.15}$$

$$B_2 = \frac{P_1 - 1}{P_1 - P_2}A. \tag{7.16}$$

Note that the coefficient A is undefined ($A = \infty$) when $P = 0.4$. At this singularity, when $P = 0.4$ the following equations must be employed:

Inside the breaker zone $(0 \leq X \leq 1)$

$$V = \frac{10}{49}X - \frac{5}{7}X \ln X. \tag{7.17}$$

Outside the breaker zone $(X \geq 1)$

$$V = \frac{10}{49}X^{-5/2}. \tag{7.18}$$

For a range of P values from 0 to 10, the family of solutions for the dimensionless longshore current (V = 0–1.0) on the vertical axis and the dimensionless cross-shore distance (X = 0–2.0) is shown in Figure 7.10. The unlabeled profile is $P = 0.4$. Other theories for development of longshore current profiles were made by Bowen (1969) and Thornton (1969) starting from the momentum balance Equation (7.1), but Longuet-Higgins (1970) was the first to put his results in dimensionless form.

The longshore currents are depth-averaged. Including the wave setup $\bar{\eta}$ increases the average depth ($h = d + \bar{\eta}$) and decreases the magnitude of the

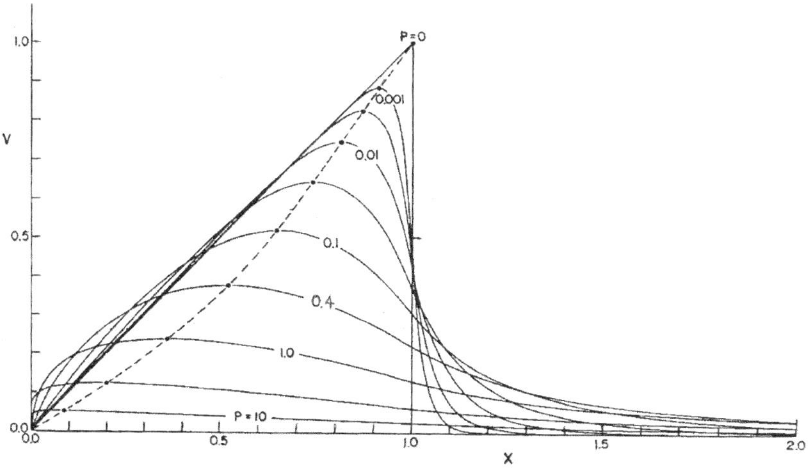

Figure 7.10. The original theory of longshore current profile variation and the influence of the P-parameter, resulting in the profile shape. When $P = 0$, there is no lateral turbulent mixing and the profile is unrealistically triangular. The larger the P value, the more the mixing and the flatter the profile.

longshore current velocity. The original radiation stress theory of longshore currents (Longuet-Higgins, 1970) was improved by including wave setup which increased the cross-sectional area and decreased the current velocity. Bowen *et al.* (1968) used the radiation stress theory for a plane beach slope $\tan\beta$ and showed that for the $\bar{x}$-direction coordinate system (Bowen, 1969) the wave setup slope is

$$\frac{d\bar{\eta}}{d\bar{x}} = -\left[\frac{1}{1+8/(3\gamma^2)}\right]\tan\beta. \tag{7.19}$$

Including the wave setup modifies the gradient of the radiation shear stress term (Equation (7.3)) and modifies the reference velocity with no turbulence at the breaker point (Equation (7.6)). Using $c_b = \sqrt{gh_b}$ for shallow linear waves gives

$$\bar{v}_b^*(\text{modified}) = \frac{5}{16}\frac{\gamma}{C_f}(gh_b)^{1/2}\frac{\tan\beta}{[1+3\gamma^2/8]}\sin\alpha_b\cos\propto_b \tag{7.20}$$

and modifies the P-parameter to

$$P(\text{modified}) = \frac{N\pi}{C_f\gamma}\frac{\tan\beta}{[1+3\gamma^2/8]}. \tag{7.21}$$

For values of the breaking coefficient, $\gamma = H/h = H/(d+\bar{\eta})$ ranging from 0.5 to 1.2, the term $1/[1+3\gamma^2/8]$ in Equations (7.20) and (7.21) varies from 0.91 to 0.65, so that the wave setup modification is not insignificant.

All the dimensionless results of Longuet-Higgins (1970) using Equations (7.10) to (7.18) are still applicable if the modified values of $\bar{v}_b^*$ and P using Equations (7.20) and (7.21) are employed.

Many laboratory-scale and field measurement campaigns have confirmed the longshore current theory of Longuet-Higgins (1970) and the modified theory including wave setup. Many additional modifications to the original theory are found in the literature including wave refraction in the breaking zone; modified bottom shear stress (combinations of current strengths at small and large wave angles); and modified lateral turbulent mixing stresses. All assume a plane beach profile.

The most comprehensive, modified analytic theory for plane beaches is found in Kraus and Sasaki (1979). Figure 7.11 shows field data for a step-type beach profile where for this one data set agreement is good seaward of $\bar{v}_{\max}$ outside the breaker line but the stepped beach profile produced secondary wave breaking, a nonlinear γ ratio, and poor results shoreward of $\bar{v}_{\max}$.

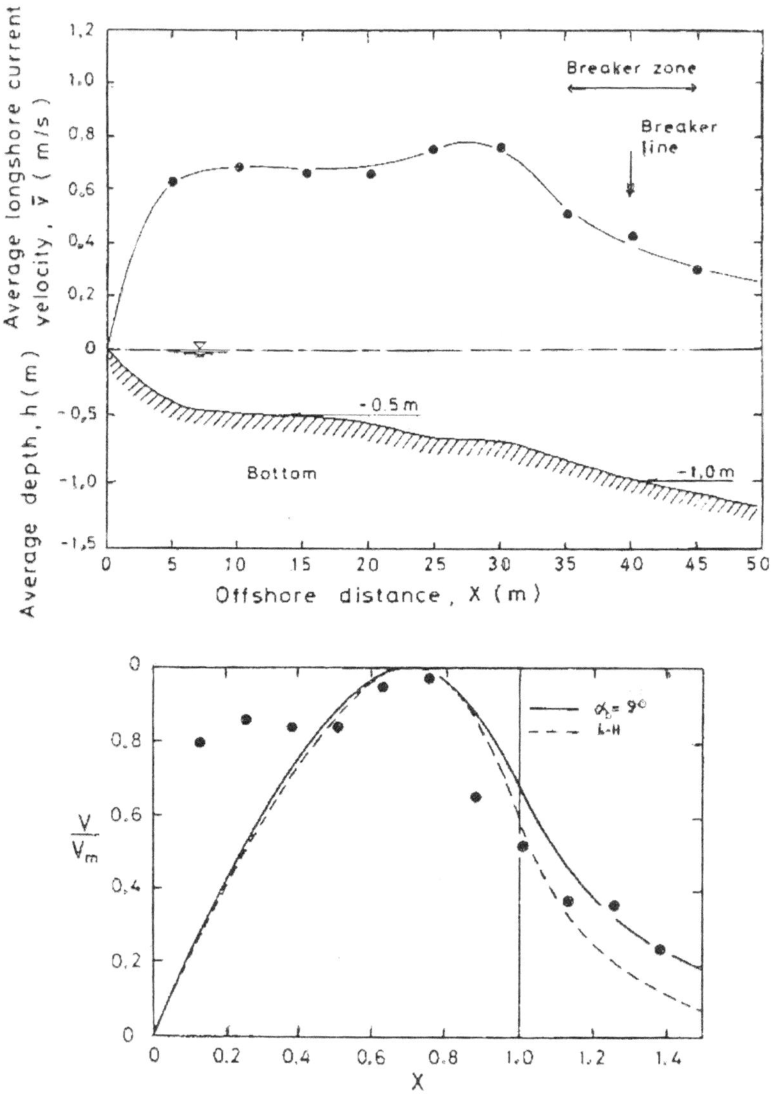

Figure 7.11. Kraus and Sasaki theory compared with field data on step-type beach (after Kraus and Sasaki, 1979).

Many additional refinements include nonlinear and irregular waves, bar profiles, and surf zone, wave height decay empirical models for obtaining numerical solutions on computers. The time-averaged and depth-integrated momentum equations in two horizontal dimensions have also

Figure 7.12. The longshore current is in which direction? (from the Family Circus cartoon by Bill Keane).

been extensively studied for the theoretical models of nearshore circulations and rip currents and employed in 2DH numerical models. The addition of the wave momentum using radiation stress fundamentals first conceived by Longuet-Higgins and Stewart (1964) has paved the way for our physical understanding and numerical modeling efforts today.

Note that $\bar{v}_{\max} = 0.8\,\mathrm{m/s}$ (2.6 ft/s) in Figure 7.11 where the water depth is about 0.6 m (2 ft) and the wading depth is normal on beaches. The time and depth-averaged longshore currents can be strong across the beach profile for transporting sediments and kids along the coast as illustrated in Figure 7.12 (The Family Circus cartoon by Bill Keane, internet). Is the longshore current from right to left or left to right in Figure 7.12?

See Basco (1982, 1983) "Surf Zone Currents" for literature review and summary of all aspects of coastal hydrodynamics prior to 1967 and up about 1980. The definitive publication on this topic is the book by Svendsen (2006).

7.2.1.3 *Wave-induced sediment transport*

In 1967, over 50 years ago, Professor Douglas Inman of the Scripps Institute of Oceanography and John Shelton of the American Geological Institute produced the movie "The Beach: A River of Sand" that has stood the test of time to demonstrate how waves cause longshore currents and longshore sediment transport (LST).

The distribution of the local, LST rate $i_l(y)$ across the surf zone is related to the longshore current profile. (*The nomenclature now follows the standards for x, y space in the literature and that are used in Chapter 5, Part II for the planform layout of coastal structures. The y-coordinate is in*

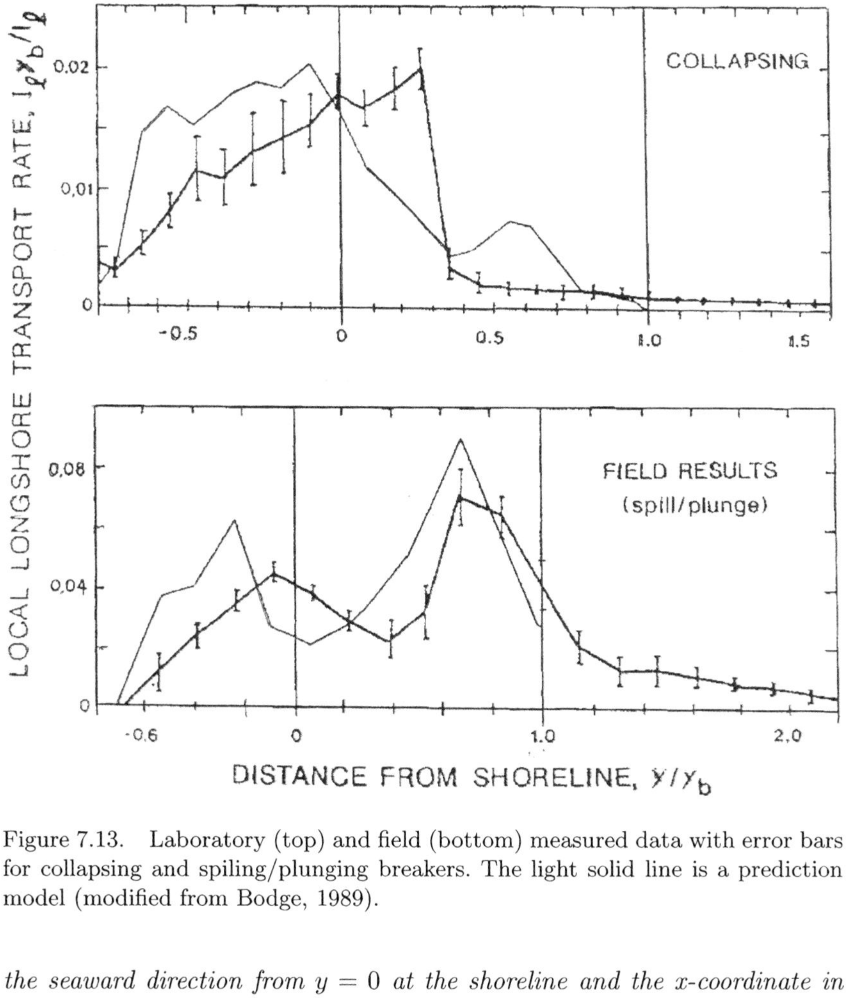

Figure 7.13. Laboratory (top) and field (bottom) measured data with error bars for collapsing and spiling/plunging breakers. The light solid line is a prediction model (modified from Bodge, 1989).

the seaward direction from $y = 0$ at the shoreline and the x-coordinate in the shoreline direction). Figure 7.13 is an example showing the normalized distribution of $i_l(y)$ versus y/y_b across the surf zone and in the swash zone (negatives) where y_b is the distance to the breaker line (from Bodge, 1989). The dark lines with error bars are measured values, with top for laboratory data with collapsing breakers and the bottom from field measurements at the Corps, Field Research Facility (FRF) with spilling/plunging breakers. In both cases, a significant amount of LST occurs in the swash zones. These results are far different than models applied to plane beach profiles using the Longuet-Higgins theory for longshore currents where the maximum sediment transport rate, $i_{l\,\max}(y)$ is predicted between the shoreline and breaker line and there is no swash zone. The light solid line is a prediction model (Bodge, 1989).

Integrating the distribution $i_l(y)$ across the entire width ($-\infty < y < +\infty$) gives the total, bulk sediment transport rate I_l to include the swash zone and well past the breaker line. Formally,

$$I_l = \int_{-\infty}^{+\infty} i_l(y)dy. \tag{7.22}$$

The local and total bulk transport rates based on the *immersed weight* of the sediments is the prefered form for coastal oceanographers and scientists (e.g. Bagnold, 1963; Inman and Bagnold, 1963; Komar, 1971). Formally, $i_l(y)$ is the local *immersed weight* transport rate (Newtons/m^2/s, lbs/ft^2/s), and I_l is the total, bulk *immersed weight* transport rate (Newtons/s, (N/s), lbs/s).

Coastal engineers have employed (Watts, 1953; Caldwell, 1956) the total, bulk *volumetric* transport rate Q, since they have a tradition of open-channel flows with Q as the symbol for volumetric flowrate of water. The relation between Q and I_l is

$$Q = \frac{I_l}{(\rho_s - \rho)g(1-p)} \tag{7.23}$$

with the usual definitions for the ρ_s, ρ, and g symbols and p is the porosity of the sediment. Formally, $q(y)$ is the local *volumetric* transport rate (m^3/m/s, ft^3/ft/s), and Q is the total, bulk *volumetric* transport rate (m^3/s, ft^3/s).

Note that for total LST rate definitions (I_l or Q), the unit of time is seconds. In nature, breaking-wave conditions that produce the longshore currents and resulting sediment transport vary in time so that some level of time averaging is required. In addition, the direction of the waves reaching the coast varies in time so that when facing in the seaward direction, the time-averaged LST rate may be from the right-to-left (−) direction or from the left-to-right (+) direction. When time-averaged over some time interval (usually 1 year), the LST rate may be the "net" rate which is the difference [(+) direction *minus* (−) direction)] or the "gross" rate which is [(+) direction *plus* (−) direction].

There are four ways to estimate the total LST rate. One option is to employ results from nearby sites with well-known LST rates that may be modified by local conditions. These rates are generally net rates in m^3/year. For example, Sandy Hook, NJ (US) is 380,000 m^3/year (net) (CEM, 2006, Vol III, Section 2); recall, 1 cubic yard (cy) $\approx 0.75\text{m}^3$.

A second method is the use of local, historical impoundments, shoreline and/or bathymetric change or dredging volumes. This method is practically the best because it automatically incorporates the time-varying wave conditions at the site. The only assumptions are that both seasonal and long-term wave conditions remain relatively constant. Shoreline changes (impoundments) adjacent to groins, jetties, and behind nearshore breakwaters are good data sources. Rates of shoreline growth at spits and rates of bathymetric change within weir jetty depositon basins are other data sources. Dredged volumes from navigation channels are also good data sources. See CEM (2006, Vol III, Section 2) for a discussion of whether the data are for net or gross transport rates.

A third method is to make measurements in the field. This is the least practical method because it is expensive and the results are only applicable for the time period involved. Experimental measurement methods include instrumentation, sediment traps, tagging sand with tracers, and construction of temporary groins. See CEM (2006, Vol III, Section 2) for details. The method is mainly for research purposes to gather data together with local wave condition measurements to verify the fourth method.

The fourth method is to employ empirically derived, LST rate formulas using wave and beach data. This method is often the only one considered, but the second method cited above is more practical because of the changing wave conditions. The LST rate formula described in the following has come to be called the "CERC" formula since it was first presented by the Coastal Engineering Research Center (CERC) of the US Army Corps of Engineers in their first coastal design manual (1966), in previous editions of SPM (1984), and most recently in the *Coastal Engineering Manual* (CEM, 2006). The proper dimensionless foundation of the CERC formula is based on the transport relationship published by Komar and Inman (1970), where the empirical coefficient K requires that the immersed weight transport rate be employed.

$$I_l = KP_l, \tag{7.24}$$

where I_l is the *immersed weight* transport rate (N/s, lbs/s), P_l is the longshore component of wave energy flux evaluated at the breaker zone (N/s, lbs/s), and K is an empirical, dimensionless coefficient evaluated from field data.

The longshore component of wave energy flux (EC_g) evaluated at the breaker zone is defined as P_l and calculated from the following equation:

$$P_l = (EC_g)_b \sin \propto_{\mathrm{b}} \cos \propto_b, \tag{7.25}$$

where both the wave energy and the wave group celerity are evaluated at the beaker line.

$$E_b = (1/8)\rho g H_b^2 \tag{7.26}$$

and in shallow water depths ($C = \sqrt{gd}$) with a wave-breaking index, $\gamma = (H/d)_b$.

$$C_{gb} = \sqrt{gd_b} = \sqrt{g(H_b/\gamma)}. \tag{7.27}$$

The cos $\propto_b$ term comes from the wave-breaker angle $\propto_b$ and sin $\propto_b$ gives the longshore component. Note that the product of the radiation shear stress component S_{xy} (Equation (2.38)) times the group celerity C_g in shallow water ($n = 1$) gives the same result as Equation (7.25). The longshore current is driven by the gradient in the radiation shear stress and produces the LST.

The correct value for the empirical, dimensionless coefficient K in Equation (7.24) depends on what definition of breaking wave height H_b is used in Equations (7.26) and (7.27). Two different irregular wave height definitions are commonly found in the literature and require the use of two different dimensionless, K coefficients. We herein define K_{rms} when H_{rms} is used and K_s when the significant wave height H_s is used in Equations (7.26) and (7.27).

(1) The root mean square (rms) breaker height, $(H_{\text{rms}})_b$

When $(H_{\text{rms}})_b$, the rms breaker height, is employed in Equations (7.26) and (7.27), then

$$K_{\text{rms}} = 0.92, \quad H_b = (H_{\text{rms}})_b \tag{7.28}$$

as determined from the straightline fit of the data of 14 field research experiments as shown in Figure 7.14 (from CEM, 2006). The data are from California, Florida, Virginia, Alabama, North Carolina, and Wisconsin in the US and from Mexico, Canada, and Japan for sandy beaches with 0.15 mm $< D_{50} <$ 1.0 mm and collected from 1953 to 1991. Using values of K_{rms} that are +50% or −50% different gives the dashed lines in Figure 7.14 and some indication of the relative scatter of the data on this log–log plot.

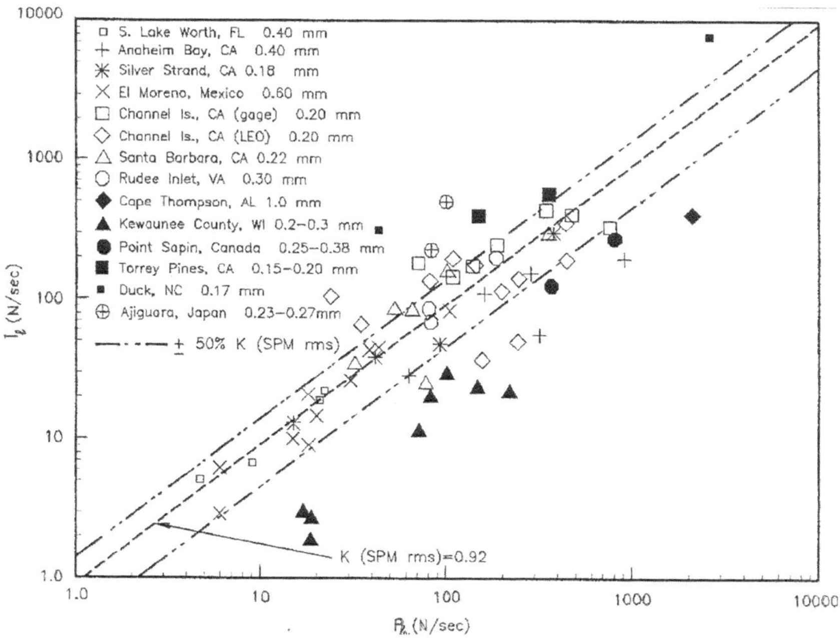

Figure 7.14. Plot of field data from 14 sites for *immersed weight* transport rate, I_l (N/s) as a function of longshore component of wave energy flux at the breaker zone, P_l (N/s) (from CEM, 2006).

(2) The significant wave breaker height, $(\boldsymbol{H}_s)_b$

Inserting Equations (7.26) and (7.27) into Equation (7.25) for P_l gives

$$P_l = (EC_g)_b \sin \propto_b \cos \propto_b = [(1/8)\rho g H_b^2][\sqrt{g(H_b/\gamma)}] \sin \propto_b \cos \propto_b,$$

$$P_l = \left(\frac{\rho g^{3/2}}{8\gamma^{1/2}}\right) H_b^{5/2} \sin \propto_b \cos \propto_b\ . \tag{7.29}$$

Recall from the Rayleigh distribution for irregular waves that the significant wave height H_s is

$$H_s = \sqrt{2} H_{\text{rms}}.$$

$$\text{Therefore, } H_s^{5/2} = (\sqrt{2} H_{\text{rms}})^{5/2} = 2.38 H_{\text{rms}}^{5/2} \tag{7.30}$$

so that $K_s = K_{\text{rms}}/2.38 = 0.92/2.38 = 0.39$.

When $(H_s)_b$, the significant breaker height is employed in Equations (7.26) and (7.27), then

$$K_s = 0.39, \quad H_b = (H_s)_b. \tag{7.31}$$

Since the significant wave height is always bigger than the rms wave height, the reverse must be true for the dimensionless sediment transport coefficients.

A value of $K = 0.77$ as found in the Corps' Shore Protection Manual (SPM, 1984) is no longer valid since it was empirically obtained using both laboratory-scale and field data in Equation (7.24).

7.2.1.4 *An example using the CERC formulation*

The above is all best illustrated by an example. Consider a quartz sand beach with mass density $\rho_s = 2{,}650\,\text{kg/m}^3$ ($5.14\,\text{slugs/ft}^3$) at a sea water location of mass density $\rho = 1{,}025\,\text{kg/m}^3$ ($1.99\,\text{slugs/ft}^3$). The natural porosity p of sandy beaches varies from 0.3 to 0.4, and here $p = 0.4$. The median grain size, $D_{n50} = 1.0\,\text{mm}$. Assume steady, nearshore wave conditions that produce $(H_{\text{rms}})_b = 2.0\,\text{m}$ at $\propto_b = 4.5^o$ on a plane beach with breaking index $\gamma = 0.8$.

From Equation (7.29) and using $H_b = (H_{\text{rms}})_b$ gives for the longshore component of the wave energy flux,

$$P_l = \left(\frac{1{,}025(9.81)^{3/2}}{8(0.8)^{1/2}}\right)(2.0)^{5/2}\sin(4.5)\cos(4.5),$$

$$P_l = 1{,}947.4\,\text{N/s}.$$

From Equation (7.28) for $K_{\text{rms}} = 0.92$ gives for the immersed LST rate,

$$I_l = KP_l,$$

$$I_l = 0.92(1{,}947.4),$$

$$I_l = 1{,}792\,\text{N/s}.$$

In terms of the volume LST rate, Q from Equation (7.23) is

$$Q = \frac{I_l}{(\rho_s - \rho)g(1-p)},$$

$$Q = \frac{I_l}{(2{,}650 - 1{,}025)\,9.81(1 - 0.4)} = \frac{I_l}{9{,}564.75},$$

$$Q = 0.187\,\text{m}^3/\text{s}.$$

Since 1.0 cubic yard (1 cy = 27 cubic feet) is approximately 0.75 m^3, Q in Imperial units,

$$Q = 0.25\,\text{cy/s}.$$

If these wave conditions were persistent for 24 hours, then they could be converted to N/day or m^3/day. But they should never be estimated for longer time periods since the wave conditions are never the same.

7.2.1.5 *Influence of grain size, wave period, and beach profile on LST*

Combining Equations (7.23), (7.24), and (7.29) with the double-angle formula ($1/2 \sin 2x = \sin x \cos x$) gives what many call the "CERC" formula for LST as the volume transport rate,

$$Q = K\left\{\frac{\rho\sqrt{g}}{16(\rho_s - \rho)(1-p)\sqrt{\gamma}}\right\} H_b^{5/2}\sin(2\propto_b). \tag{7.32}$$

Clearly, Q depends on what value is used for H_b (both magnitude and definition) $\propto_b$, and the empirically derived value of K depending on the definition of H_b as discussed above. Continued research has focused on the influence of grain size on the K coefficient, beach cross-section profiles on the γ coefficient, and the missing wave period to describe the wave characteristics.

An extensive review of the literature on the influence of grain size on LST is found in a Corps of Engineers Technical Note (King, 2005, CHETN-II-48). All the data show a clear trend of the decreasing sediment transport with increasing grain size as intuitively expected. The recommended relationship for K_s is simply

$$K_s = \frac{0.1\,\text{mm}}{D_{50}} \tag{7.33}$$

so that the calibrated value of $K_s = 0.39$ (Figure 7.14) gives a nominal $D_{50} = 0.26\,\text{mm}$. For the above example, using $K_s = (0.1\,\text{mm}/1.0\,\text{mm}) = 0.10$ would significantly reduce the LST rate.

Kamphuis (1991) included grain size, wave period, and beach slope effects in his empirical formula for LST. His formulation was for underwater (immersed) units of kilograms mass/sec (kg/s). To be consistent with the nomenclature herein, we have multiplied his formula by the gravity constant g to get the Kamphuis immersed weight transport rate, I_K (N/s). The formula becomes

$$I_K = 2.27g[T_{\text{op}}^{1.5} m_b^{0.75} D_{50}^{-0.25}](H_s)_b^2 \sin^{0.6}(2 \propto_b). \tag{7.34}$$

Kamphuis (2010) compares these results with the CERC formula for a range of wave heights (1, 2, and 3 m) with 8-s period waves and 0.2-mm grain size on a relatively flat slope $m = 0.02$ to show that the Kamphuis (1991) expression gives lower values than the CERC formula. The most important variable remains the breaking wave height. A relatively new total LST rate formula has been put forward by Bayram *et al.* (2007) that is not discussed herein.

The CERC (CEM, 2006), Kamphuis (1991), and Bayram *et al.* (2007) formulas rely on wave characteristics at the edge of the surf zone, some basic sediment properties, and average beach slope. They use empirical coefficients from calibration. They are not process-based models that combine wave transformations in the surf-zone, wave-generated longshore currents, and sediment bed-load plus suspended load in the water column that are all impacted by beach profiles with variable slopes and bars. One example of a process-based model for LST is the UNIBEST-LT model (internet website, Deltares).

The influence of beach profiles on the LST has been studied for the Dutch coast (Mil-homens *et al.*, 2013) using the UNIBEST-LT model. Cross-shore surf-zone features (beach slope and presence of number of bars) were tested using 2,888 bathymetry profiles collected since 1965 as part of the JARKUS data set (Rijkswaterstaat, internet, online) for yearly coastal measurements. It was learned that the calculated LST rates showed significant variability with average profile slopes but no correlation with the number of bars and other bar-related parameters. The research continues.

It is recommended that the CERC formula (Equation (7.32)) be used for preliminary design to estimate the LST rates and that local, historical impoundments, shoreline and/or bathymetric change, or dredging volumes be employed to verify these calculations. This method is practically the best because it automatically incorporates the seasonal and annual time-varying wave conditions at the site. An example is presented in Section 7.2.4 on sediment budgets.

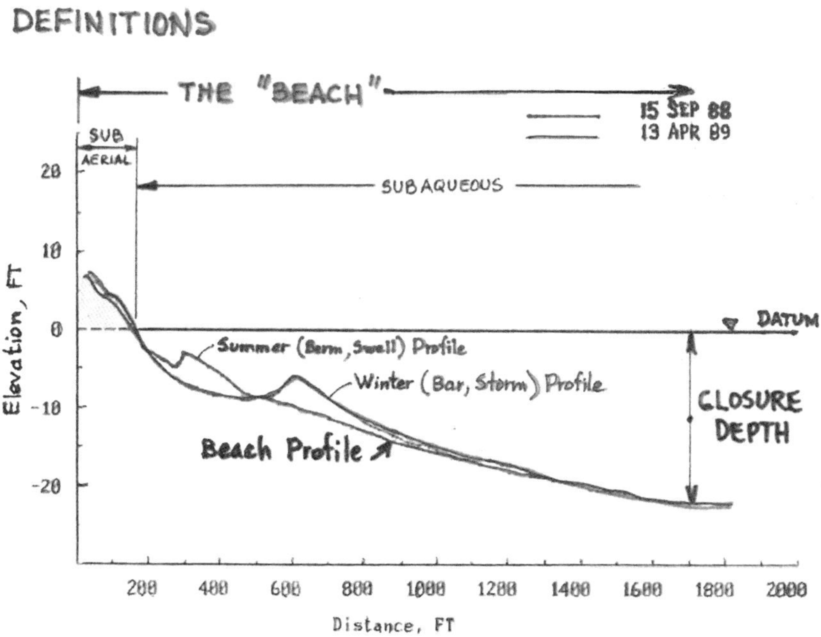

Figure 7.15. Definition of the beach and the two dominant beach profiles at many coastal locations.

7.2.2 *Dynamics of beaches*

Sediments that are non-cohesive materials located at many land–water interfaces of the world's coasts are very dynamic for transport in both the longshore and cross-shore directions. The "beach" cross-sectional profile as defined in coastal science and engineering is depicted in Figure 7.15. The beach profile is the plot of horizontal distance offshore from some land reference "baseline" (usually behind the frontal dune) versus the vertical elevation of the beach surface above some designated vertical datum. The profile location (x-coordinate) and horizontal, vertical (y-, z-coordinate pairs) dimensions define the profile. The beach width includes both a visual, subaerial part and the hidden, subaqueous region out to what is termed the "closure depth". The two beach profiles in Figure 7.15 (topography above vertical datum and bathymetry below datum) are from measurements at profile P168 at the Corps of Engineers, FRF at Duck, NC. The green profile is labeled the "summer profile" and was measured on September 15, 1988 after lower height, longer period summer waves constructed the large

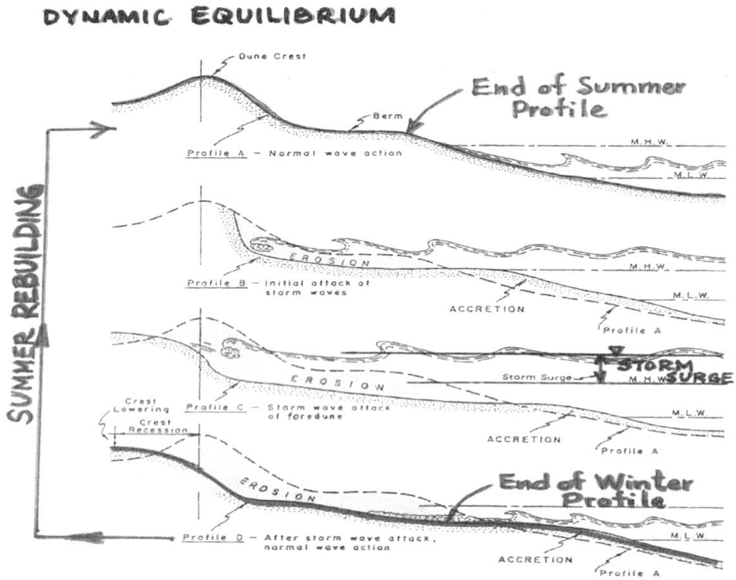

Figure 7.16. Schematic of the dynamic equilibrium of sandy beaches (from CEM, 2006 revised).

nearshore bar and increased elevations on the subaerial part. The red profile is the "winter profile" taken after a full winter season of higher and shorter period waves (April 13, 1989) that dragged the bar further offshore and lowered the elevations on the subaerial beach. A coastal storm in March 1989 did most of the cross-shore transport to build the bar further offshore. The two dominant seasonal profiles are also known as the berm or swell type for the "summer" profile and as the bar or storm type for the "winter" profile. In theory, as depicted in Figure 7.16, the cycle repeats each wave year and is called "dynamic equilibrium". The wave year usually does not coincide with the calendar year but depends on the annual wave climate at the site of interest.

Repeated surveys at the same location reveal a wide variation of the vertical elevation of the beach profile especially in the shallow water, nearshore surf zone as shown in Figure 7.17. The locus of the highest and lowest elevations (NGVD29 vertical datum, m) at each seaward location form an envelope of profile locations and reveal the dynamic nature of the beach for this four-year period (January 1981–December 1984) when 127 profiles were taken on a two-week cycle at location P188 at the Corps' FRF in

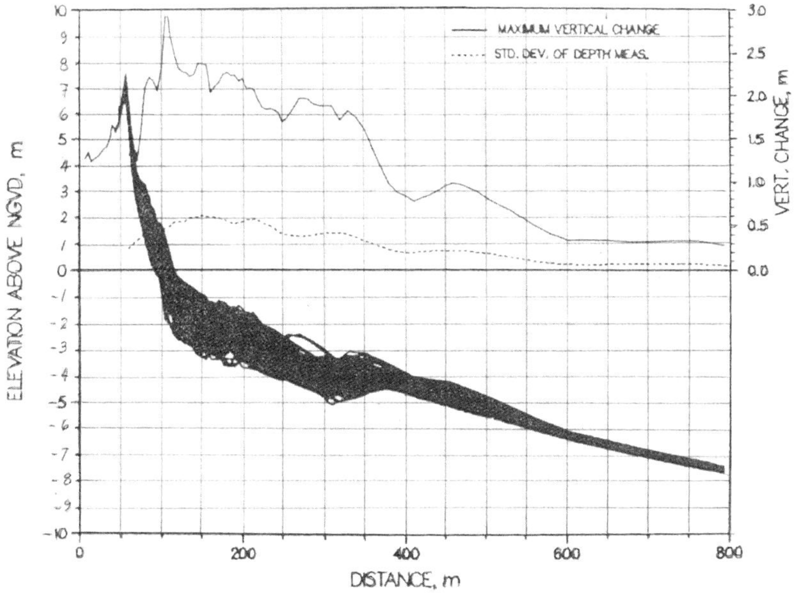

Figure 7.17. Beach profile envelop of 127 measurements (January 1981–December 1984) at location P188 at the Corps' FRF, with standard deviation of the mean (dotted) and maximum vertical change (solid) across the profile.

Duck, NC. The standard deviation of the mean profile location is shown as the dotted line and approaches 2 m (6.6 ft) where the maximum vertical change (solid line) of about 10 m (33 ft) took place. Beaches are truly dynamic environments with the sediments continually moving in both the longshore and cross-shore directions.

7.2.3 *Shoreline change, volume change, and beach erosion*

The position of the land–water interface, i.e. the "shoreline" depends on many factors at any coastal location. One common definition is the mean high-water (MHW) shoreline in locations where daily tidal variations of water levels are present. For example, assume that the MHW level at Duck, NC was 2.0 m above the NGVD29 vertical datum shown in Figure 7.17. The MHW shoreline location shifts back and forth horizontally about 25 m (80 ft) for these 127 measurements. This shifting of the MHW shoreline is the natural, seasonal variabiliy of the shoreline location. The beach is said to be in "dynamic equilibrium" since the mean location of the MHW-level shoreline does not change in time at the Corps' FRF at Duck, NC.

At many beaches in the world, this dynamic environment exists, but the annual cycle is not in equilibrium. As a result, the shoreline position changes with time when determined over many years of the annual cycle. In locations where the annual, mean shoreline movement is landward over time, coastal recession is taking place and relative to a fixed reference monument, erosion rates (m/year, ft/year) are determined. Movement of the mean, annual shore "line" location (erosion rate) is employed as a substitute measure of the volume of sediment lost annually at the site since "lines" have no volume. Estimation of the dynamic variabiity of shoreline erosion rates can be eliminated by using profile volume change in time in place of shoreline erosion rates (Kana, 1993, 2011). Profile volumes are defined as the quantity of sand in the cross-sectional profile per unit width (m^3/m, cy/ft) between the toe of the seaward dune (foredune, if present) and some outer depth contour of interest. When the outer depth contour is the limit of observed or estimated bottom change in time (i.e. the "closure depth", see Section 7.2.4), the full profile volume change is determined. Volume change can be determined for shallower offshore depths such as the normal wading depths of recreational beaches (−1.0 m, −3.0 ft below mean water level (MWL)) or out to the mean low-water (MLW) elevation and hence the MLW tide line. The relationship between shoreline erosion and profile volume change requires detailed surveys of beach profiles over time. When beach profiles are not available, estimates of the relationship are employed as discussed in the following.

At some locations, shorelines are accreting in time as the annual, mean MHW shoreline moves in the seaward direction. The beach is still a dynamic environment but sand volume in the profile is gained each year in the beach cycle. The conservation of sand volume in a fixed space along the coast is the principle leading to development of a sediment budget.

7.2.4 *Closure depths and sediment budgets*

The depth of closure (DOC) is a fundamental concept in coastal engineering. When repeated beach profile surveys are taken as shown in Figure 7.17, the water depth offshore at which there is no significant change in the measured bottom elevation (to accuracy of survey equipment) is defined as the DOC. For this example, for the 127 profiles taken from January 1981 to December 1984 (three years), the maximum vertical elevation change was constant at about 0.3 m (1.0 ft) located 600 m (1970 ft) from the baseline giving a DOC $\approx$ 6.4 m (21.0 ft). At this location, sediment transport is very small or not present in both longshore and onshore–offshore directions. However, subsequent years may produce wave heights and wave periods that increase and/or decrease the closure depth. The accuracy limits of profile survey depths at this offshore distance are about 0.15 m (0.5 ft).

Hallermeier (1978) employed the Corps, FRF field data (Figure 7.17) and wave tank data to develop the first theoretical relationship to estimate the DOC. Birkemeier (1985) revised these equations and obtained the following equations for the DOC, h_*. Only extreme wave conditions (wave heights and wave periods) feel the bottom at deeper water depths. Define an effective wave height H_e and associated, effective wave period T_e that are rare each year and depend on the local wave climate of the coastal location of interest. For the Corps of Engineers, FRF at Duck, NC, Hallermeier (1981) chose

$$\begin{aligned} H_e &= H_{0.00137} \\ &= \text{the wave height exceeded by 0.137\% of the waves each year, i.e.} \\ &= \text{the wave height exceeded only 12 hours each year} \\ &\quad [12/(365 * 24)] = 0.00137, \\ \text{and } T_e &= \text{the wave period most closely associated with } H_e. \end{aligned}$$

The equations defined by Birkemeier (1985) and put in dimensionless form are

$$\frac{h_*}{H_e} = 1.75 - 57.9\left(\frac{H_e}{gT_e^2}\right) \tag{7.35}$$

and a simplified version (CEM, 2006) of this, since the wave period T_e is difficult to determine is given as follows:

$$\frac{h_*}{H_e} = 1.57. \tag{7.36}$$

Equating Equations (7.35) and (7.36) and solving for a range of effective wave periods, T_e, it is clear that the closure depth strongly depends on T_e. The dimensionless relationship is

$$\frac{h_*}{gT_e^2} = 0.00488. \tag{7.37}$$

The correlation coefficient in Equation (7.35) is for beach profile data at the Corps of Engineers' FRF in Duck, North Carolina, US. But they have been shown to be relatively accurate for all US coastlines and other locations (Nicholls *et al.*, 1998). As a result, the Corps of Engineers has developed a web-based tool to download DOC values using WIS hindcast information (Section 2.3.8.3) for all coastal waters in the United States (Alaska, Pacific, Atlantic, Gulf of Mexico, and Great Lakes). Visit the website for all the details (http://cirp.usace.army.mil/products/depth-of-closure.php) or refer Brutsché *et al.* (2016).

It is important to remember that the DOC is **not**

- the depth offshore at which no sediment can move further offshore,
- the depth offshore at which no sediment can move landward due to wave action,
- the outer limit of cross-shore sediment transport,
- a fixed depth that never varies from year to year.

Storms can transport sand beyond the closure depth in thin layers as found within sediment cores taken far offshore.

As shown in Figure 7.18, the closure depth h_* plus the berm elevation B determine the relationship between the shoreline change distance Δy_o (erosion or accretion) and the profile volume change $\forall$ per unit width of shoreline. Formally,

$$\Delta y_o = \frac{\forall}{(h_* + B)}. \tag{7.38}$$

This relationship does not depend on the shape of the profile and the presence of bars but assumes the eroded or accreted profile shape is the same.

For example, assume $B = 6$ ft (1.83 m) and $h_* = 21.0$ ft (6.40 m) so that when the shoreline change $\Delta y_o = 1.0$ ft (0.305 m), the volume change $\forall = (21.0+6.0)\times 1.0 = 27.0$ ft^3 per foot of shoreline (3.56 m^3/m). Since 27.0 ft^3 = 1.0 cubic yard (cy), a rule of thumb often cited is *1.0 ft of shoreline change = 1.0 cy/ft of volume change.* But this special case should not be the general rule and Equation (7.38) should always be employed. In the metric system, when the shoreline change $\Delta y_o = 1.0$ m (3.281 ft), a volume change $\forall = 10.0$ m^3/m requires that $(h_* + B) = 10$ m. For $B = 1.8$ m (6 ft) this means $h_* = 8.2$ m (26.9 ft).

Equation (7.38) is often employed in the determination of a sediment budget. As introduced in Section 2.5 and shown in Figure 7.19(a) for a simple control volume, shoreline erosion is a source of sediment that is

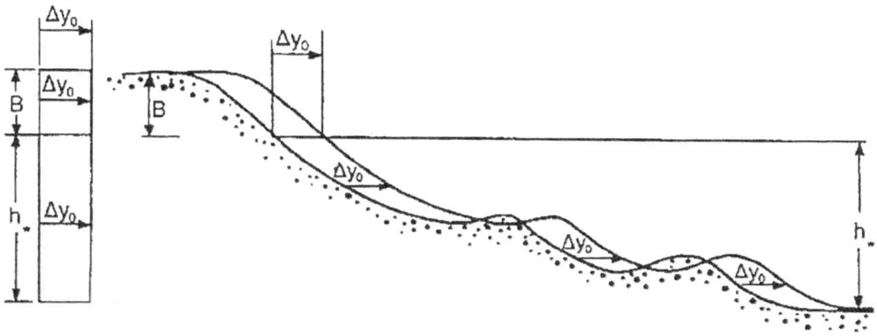

Figure 7.18. Schematic diagram of relation between shoreline change (Δy_o) and ($B + h_*$) to produce volume change $\forall$ per unit width of coast (from Dean, 2002).

available for LST within a coastal cell or compartment with boundaries as delineated in Figure 7.19(b). A sediment budget is an accounting of all possible ways non-cohesive materials (sand, pebbles, cobble) can enter (positive, credit) and be removed (negative, debit) from the coastal cell as illustrated in Figure 2.39. Application of Equation (7.38) in determination of a sediment budget is best illustrated by an example for Sandbridge, Virginia (US).

As shown in Figure 7.20 (repeat of Figure 2.39) for the 50-km (30-mi) coastline of southern Virginia on the Atlantic Ocean below the entrance to the Chesapeake Bay, the historic average shoreline change rates (m/year) vary considerably along this region (left side) between North latitudes 36°55′ and 36°30′. The solid line (1858–1925, 68 years) and dashed line (1925–1980, 55 years) are from MHW shorelines on historic maps as published by Everts *et al.* (1983). The dotted line is from aerial photographs (1937–1964) and gives added confidence to the location of the maximum erosion rate of about −3 m/year (−9.5 ft/year) at the nodal point and southern boundary of a sediment cell from this location north to Rudee Inlet. Sandbridge, Va. is the yellow zone, a 8-km (4.5-mi) subdivision of the City of Virginia Beach, Va.

The nodal point is the location where the net annual longshore sediment tranport divides in opposite directions. The local wave climate and bathymetry alter the wave directions throughout the year so that the nodal point is at about Lat 36°42.5′. Net sediment transport is in the north direction (upward) from the nodal point which becomes the southern boundary of a coastal cell. The northern boundary of the coastal cell is Rudee Inlet. Figure 7.21 presents the details as taken from a spreadsheet.

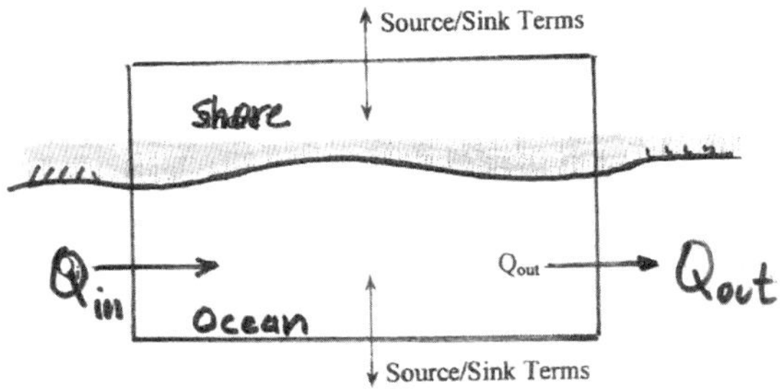

(a) Control Volume

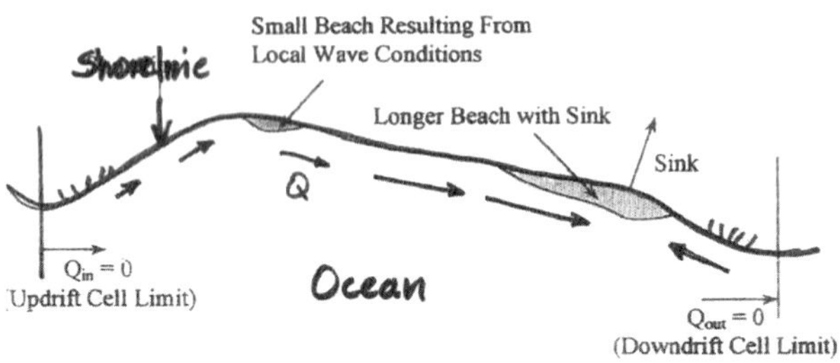

(b) Coastal Cell

Figure 7.19. (a) Control volume within a (b) coastal reach or cell for determination of a sediment budget.

The first column contains subreach numbers for the eroded section that is 43,800 ft (13,350 m) long (numbers 1–10) with number 10 starting at the nodal point. The second column shows the lengths (feet) of the eroded subreaches with average erosion rates (ft/year) for each subreach in the third column as estimated from Figure 7.20. For a berm elevation, $B = 6$ ft from the field data and for a closure depth $h_* = 21$ ft, the conversion factor is simply 1.0 ft/year erosion = 1.0 cy/year per unit length of beach. Eroded volumes for each subreach section are found in Column 4 and are simply the product of Columns 2 and 3. Summing up the total for the average, net erosion gave about $-205,000$ cy/year.

The south side of Rudee Inlet has a weir jetty that blocks some sand forming a short accretion section along Croatan Beach. Columns 1–4 in Figure 7.21 are similar to those above with the total average, net accretion about +10,200 cy/year. Combining for the coastal cell from the Sandbridge nodal point to Rudee Inlet gives about 195,000 cy/year as the net sediment volume reaching the inlet from the eroding beach. For confirmation of this number, to maintain navigation in Rudee Inlet, about 200,000 cy/year are

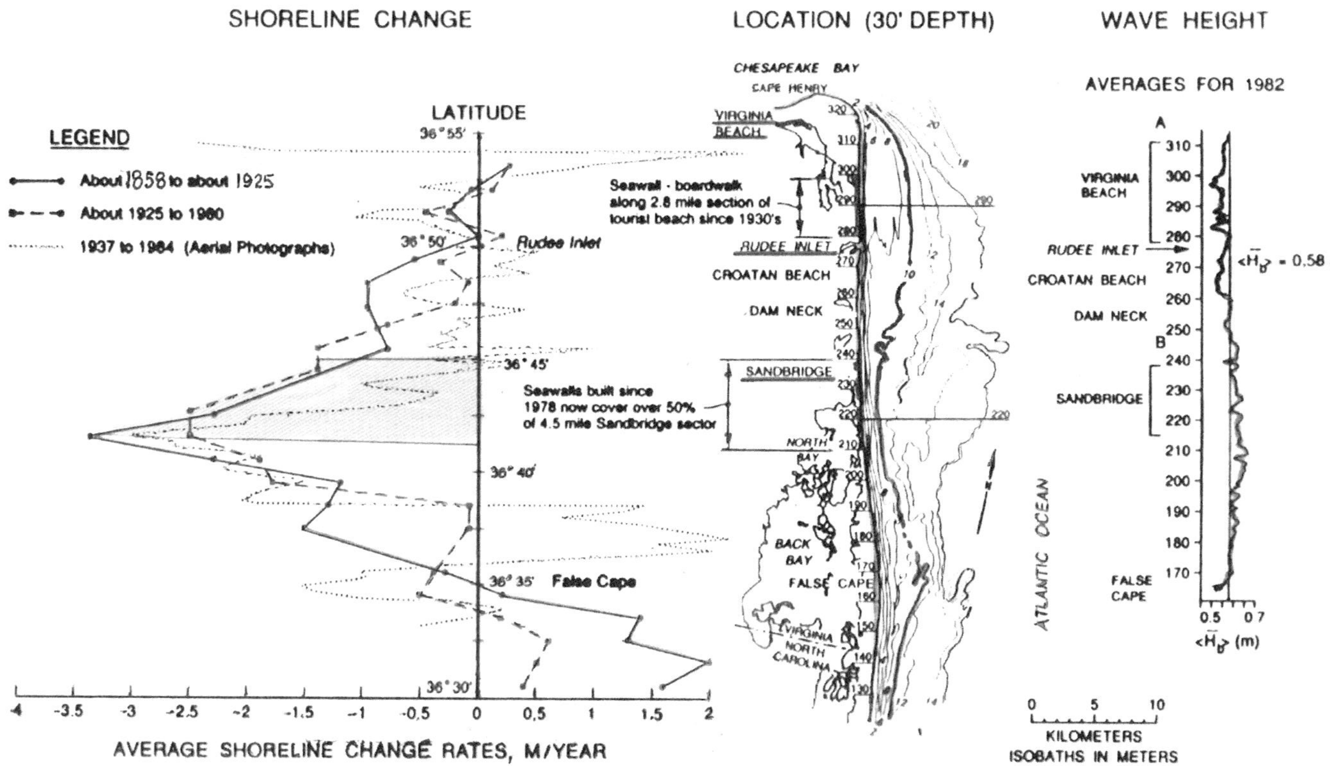

Figure 7.20. Composite of location and depth contours (middle) and breaking wave heights (right) as found in Wright *et al.* (1987). The shoreline change information (left) is added by the author (from Evarts *et al.*, 1983).

Net Sediment Budget
Sandbridge Nodal Point North to Rudee Inlet
Everts, et al (1983) Historic Shoreline Position Change Rates

Given Data

Berm Elevation, B (ft)	6.0	Field Measurements
Closure Depth, h* (ft)	21.0	Hallermeier (1981)
Conversion Factor		
Feet/year to Cy/year	1.00	

ERODED SECTION

Sub Reach Number	Length feet	Ave.Erosion Rate ft/yr	Eroded Volume cy/yr	Location
1	6000	-1.0	-6000	Dam Neck North End
2	5280	-2.3	-12144	
3	6125	-2.5	-15313	
4	3771	-3.5	-13199	
5	3771	-4.5	-16970	
6	3771	-5.5	-20741	
7	3771	-6.5	-24512	
8	3771	-7.5	-28283	
9	3771	-8.5	-32054	
10	3771	-9.5	-35825	Nodal Point
Total,L ft	43802		**-205037**	

ACCRETION SECTION

Sub Reach Number	Length feet	Ave.Accret. Rate ft/yr	Accreted Volume cy/yr	Location
1	4235	1	4235	
2	4000	1.5	6000	Croatan Beach
			10235	

NET SEDIMENT BUDGET AT RUDEE INLET
Cy/Year= **194802**

Average Dredging Volume at Rudee Inlet about **200,000** cy/yr

Figure 7.21. Sediment budget for Sandbridge beach to Rudee Inlet, Virginia.

annually dredged by the City of Virginia Beach (Roehrs, Personal Communication).

Tidal inlets and rocky headlands are commonly used as boundaries for coastal cells in sediment budgets because they provide confidence in all the estimated numbers. Volumes dredged annually from the inlet are known (if dredging takes place) and rocky headlands block sediments from leaving (or entering) the coastal cell. As this example demonstrates, estimates of the yearly average erosion rates (ft/year) and their conversion to volume rates (cy/year) may vary somewhat, but the overall sediment budget numbers should be accurate enough to significally aid in design of beach nourishment projects as explained in the following.

The US Army Corps of Engineers has developed a Sediment Budget Analysis System (SBAS) to formulate, document, and calculate sediment budgets. It includes uncertainty estimation methods to understand the reliability of the sediment budget. See http://cirp.usace.army.mil/products/sbtools.php for the SBAS documents including a web-based tool (Sediment Budget Calculator).

7.3 Beach Design

7.3.1 *Introduction*

The placement of large quantities of "good quality" sand on the beach to advance it seaward is a simple definition of beach nourishment. The renourished beach is a sand reservoir available to move offshore and form subaqueous bars to cause wave breaking and reduce the wave energy striking the coast during storm events. Renourished beaches are designed as coastal "infrastructure" and are the "soft" alternatives for storm damage mitigation. They must be maintained over a design life cycle. Natural erosion, man's interference, or extreme storm events diminish the reservoir volume.

Beach engineering design in the United States has mainly evolved along the sandy coasts of the Atlantic Ocean and the state of Florida as advanced by Dean (2002). The dunes, subaerial, and subaqueous regions out to the closure depth all consist of non-cohesive sediments with grain sizes $D_{50} > 0.0625$ mm (find sand) on the Wentworth scale (Figure 2.34). The Gulf of Mexico coast of Louisiana contains mixed sediments across the dune/beach profile that is a natural combination of sand, silt, and clay materials. See Campbell *et al.* (2005) for a conceptual "morphosedimentary" model of mixed sediment coasts. The main difference is that the total volume of material moved during storm events is not conserved. The silty/clay colloidal sediments suspended within the water column are transported with the longshore and cross-shore currents and may move far offshore or stay in suspension. The introductory materials presented in this book all apply only to sandy coasts as developed by Dean (2002).

In the US, typical beach nourishment (fill) values are design widths (20–30 m, 50–100 ft), design volumes (125–250 m^3/m, 50–100 ft^3/ft), design grain size (0.2–0.5 mm), construction costs (US$ 300–1,000) per linear foot, and unit sand costs (US$ 10–20) per cubic yard (recall that 1 cubic yard (cy) = 0.76 m^3, and 1 cy/ft = 2.5 m^3/m).

Ideal conditions for a successful beach nourishment project over the design life (20–30 years, including periodic maintenance) are the following: (1) a long-term source of beach quality sand at an economic transport distance from the borrow site, (2) a low to moderate background erosional shoreline (≈3–5 years renourishment cycle for maintenance), and (3) substantial upland benefits for storm damage mitigation of private and commercial property, infrastructure, and for recreation and tourism benefits. The benefits to the environment must also be considered. See Basco (1998) for an analysis of the beach nourishment (soft) versus rubble-mound revetment (hard) alternative.

Beach nourishment projects are expensive. The designer must first determine the purpose and objectives of the client (as always) so that the benefit-to-cost ratio is clearly understood. CEM (2006, Volume V, Chapter 4 "Beach Fill Design") should be consulted for many more engineering aspects of Beach Design than can be summarized in this book.

7.3.2 *Parameters for design*

7.3.2.1 *Field work*

A baseline is first established that lies roughly parallel to the shoreline and is located behind the dunes. It becomes the starting location for a system of profiles taken for design and then subsequent monitoring after construction. The baseline location should be a safe distance landward of the shoreline so as not to be lost over the design life of the project. The location should be established in state plane coordinates for baseline location in the field, on hardcopy, and in electronic files.

A system of profiles across the dune(s) and beach out to closure depth is then taken by field survey. The spacing of the profiles varies with the project length. Typical spacing is 300 m (1,000 ft) giving sufficient detail for the calculation of sand volumes that exist on the present beach and to establish the design volume needed for the beach nourishment. Spacing near the tapered ends can be less and regular spacing outside the project boundaries may be needed (five or so profiles) for monitoring after construction. All surveys should be performed by registered land surveyors with experience in both topographic and bathymetric survey work. The surf zone is always difficult, and care must be taken to ensure that the bathymetry matches at the boundaries of land-based measurements and those taken from boats in the water where wave breaking makes both survey methods very difficult. Surveys using towed sleds eliminate this problem. Survey technology continues to advance and LIDAR surveys are now becoming available (See NOAA, NCEI, Coastal Lidar Archive) (https://coast.noaa.gov/digitalcoast/data/coastallidar.html).

Sediment samples are then taken at the surface at beach profile locations. Dean (2002) recommends the following locations of samples to be taken along every other profile: the toe of dune, beach berm, beach face, at 1-m (3-ft) depth intervals out to closure depth h_* and beyond the closure depth ($h \approx 1.25\, h_*$). The Corps of Engineers (CEM, 2006) recommends that samples acquired on the profile between the berm crest (or MHW line) and a water depth corresponding to the position of the typical storm bar should be used to characterize the native beach sand to access the compatibility of sand from potential borrow sources. Each sediment sample has a grain size distribution. How they are combined to determine the mean grain size for the entire "native" beach D_N is explained in the following.

The onshore or offshore site (area) from which sand is removed to be placed on the beach is the "borrow area" for the beach-fill project. The mean grain size of "fill" material D_F must also be determined from sediment samples taken in the field at the borrow area. Four borrow area locations that are land-based (e.g. riverine deposits), back bays (marshes, creeks, etc.), offshore (shoals), and navigation channels (tidal inlets) are the usual sites where beach nourishment projects have found sand that is compatible to put on the native beach. Surface samples may suffice, but sediment cores are needed to locate suitable sand deposits that may lie several meters beneath the surface materials composed of a mixture of fine sands, silts, and clay. Here, geophysical, seismic surveys are needed to help determine the locations for cores to be drilled.

The compatibility of grain size of the native area D_N and borrow area D_F sand is the main variables in beach-fill design. Color is also considered.

When dune rebuilding is involved, sediment samples within the dune field may also be taken in the field. Typically, the same borrow area material is used to fill the dune template so that dune samples are not needed.

7.3.2.2 *Office work*

The system of beach profiles obtained from the field survey work cited above is only a snapshot for the date when the survey was taken. The beach berm elevation, B and closure depth h_* need a time history of similar beach profiles and/or office work to determine suitable values for beach design. Beach berm elevations are relatively stable in time and similar at nearby locations. It is common to use the same "natural" berm elevation B as found at the site. Closure depth calculations using the Equations (7.35) and (7.36) of Birkemeier (1985) give h_*.

Historic map data, studies by the Corps of Engineers and local consulting firms, state agency reports, and university research projects all should be obtained and employed to estimate the background erosion rate, $e(x,t)$. Then using Equation (7.38), the closure depth h_* plus the berm elevation B give the relationship between the shoreline change distance Δy_o (erosion, e) and the profile volume change $\forall$ per unit width of shoreline. Finally, a sediment budget can be prepared and confirmed by again reviewing the work of others for the coastal location of interest.

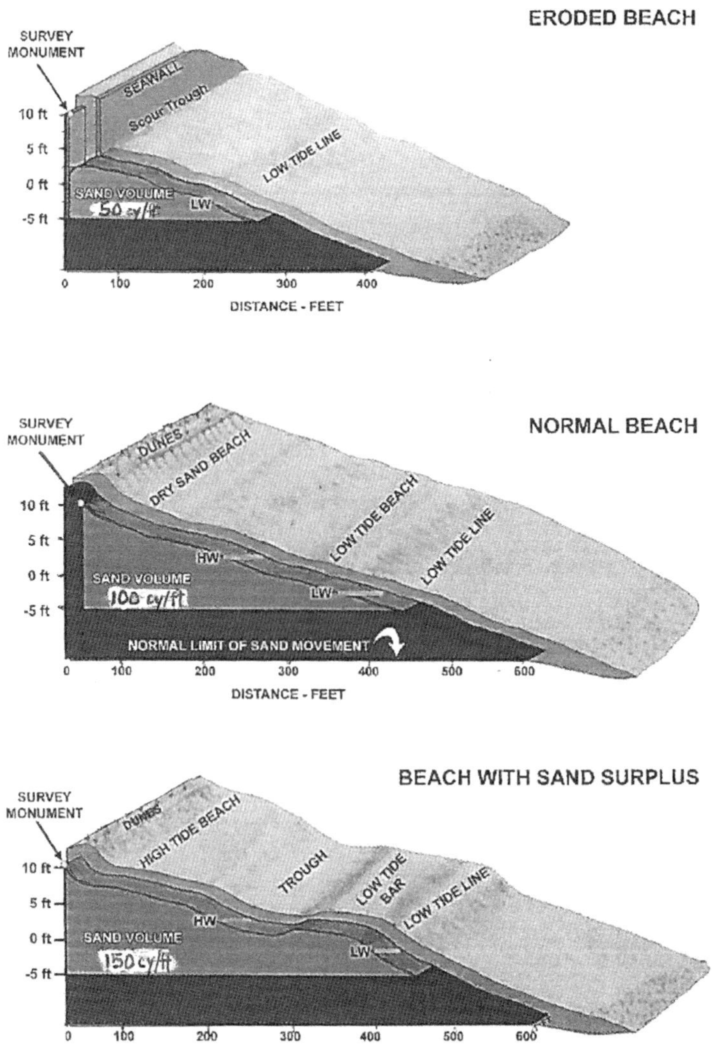

Figure 7.22. Schematic diagram of volumes in three profiles to "normal limits of sand movement" as measures of the health of a sandy beach (from Kana, 1993).

7.3.2.3 *Profile sand volumes of beaches*

Figure 7.22 depicts profile sand volumes within the distance from the survey monument (baseline) out to the "normal limit of sand movement" to provide some measure of the health of a beach (from Kana, 1993, 2011). The profile volumes shown in Figure 7.22 for the three sections depend on the subaqueous contour chosen as the seaward limit (not the closure depth). The value of using profile volumes is that they are independent of the seasonal cycles of shoreline position change (e.g. the MHW shoreline) and they permit comparisons with other locations with known "healthy" profile beaches. A normal, healthy beach (middle, Figure 7.22) is in dynamic equilibrium as illustrated in Figure 7.16. The assumption is that the "normal" profile volume permits the normal range of seasonal (summer (swell) or winter (storm)) beach change with no negative impact on the upland environment (dunes, property, infrastructrue, etc.). This is the minimum profile volume of a healthy beach.

An eroded beach (upper, Figure 7.22) has less sand volume than a normal "healthy" beach even with similar tidal and wave climates. The eroded beach is not in equilibrium. The sand volume deficit is usually caused by a fixed, natural feature (e.g. rocky headland) or a fixed, anthropogenic feature (road, seawall, rock revetment, buildings, etc.) that prevents landward recession and lowers the beach profile in front of the fixed feature. If the feature were not present, shoreline recession would continue permitting the sand volume in a normal beach. Profiles with offshore bars (lower, Figure 7.22) have a sand surplus volume and beach width may grow seaward if the bars attach to the subaerial beach.

In summary, coastal erosion is best quantified by the change in the volume of the sand profile and not by the change in position of a single line such as the MHW shoreline. When profile volumes decrease over time, erosion occurs. The design volume for beach nourishment must include (1) a volume to replace the eroded beach volume (deficit) back to healthy beach conditions and (2) a volume to meet the full objectives of the beach nourishment project.

7.3.2.4 *Design volume for beach nourishment*

Typical objectives for beach nourishment projects are storm damage mitigation, recreation and tourism enhancement, and ecosystem restoration. Benefits for each objective and combinations should exceed the cost which is a primary constraint in determination of the design volume. The client is responsible for the cost over a design life that includes periodic beach renourishments. Different clients have different objectives. No standards exist for the design volume for beach nourishment.

In the United States, when the Federal Government is the client and design is by the Corps of Engineers, the only allowable benefit is storm damage mitigation to justify the cost within a benefit/cost ratio analysis.

Further discussion is given in Section 7.3.7 for the Corps' Beach-*fx* software. When individual states and/or local government is the sole client, recreation and tourism enhancement are important benefits besides damage mitigation. In some cases, the design volume is an initial volume plus an advanced volume for anticipated average annual losses over the design life (Verhagen, 1992; Kana, 1993; Dean, 2002). Or, it can simply be an initial volume plus a safety factor. Increased area for turtle nesting is very rarely the sole benefit for determining a profile volume for beach nourishment.

The US federal government is also responsible for maintaining navigation depths through tidal inlets. When inlet dredging occurs and the material is beach-quality sand, the Corps of Engineers' policy is to place the dredged sand on the downdrift beach. The additional placement costs are greater than costs for offshore disposal of the dredged material. But total benefits to the federal government are more since storm damage mitigation benefits are present on the downdrift beach.

In the Netherlands, government policy has been to "... *maintain the coastline at its 1990 location*". To do this, the Dutch use shoreline change measurements and establish a local erosion rate, estimate a volume loss of sand rate (m^3/year), add an additional 40%, multiply the total by five years, and construct a renourished beach with this volume above the MLW elevation bounded by the dunes on the landward side and the −1.0 m contour on the seaward side.

Beach nourishment is expensive. Assuming \$10/cy and a design volume of 100 cy/ft gives \$1,000/ft for initial construction costs. A project for 1.0 miles of shoreline (5,280 linear feet, 1.61 km) would cost \$5.3 million. Assuming that compatible sand ($D_F = D_N$) was used for construction so that Equation (7.38) was applicable and that the rule of thumb (*1.0 ft of shoreline change =1.0 cy/ft of volume change*) could be applied would give a design beach width, $\Delta y_o = 100$ ft. How these results are far different when the borrow area sand is not the same as the native beach sand ($D_F \neq D_N$) and how the construction width is different than the design width are the subjects of beach cross-section design in Section 7.3.3.

7.3.3 *Beach cross-section design*

7.3.3.1 *Equilibrium beach profiles*

Curve-fitting of over 500 profiles of sandy beaches from the US East Coast, Gulf of Mexico, and other locations (Poland, Black Sea, New Zealand) has resulted in the empirical equation for the relationship between the distance

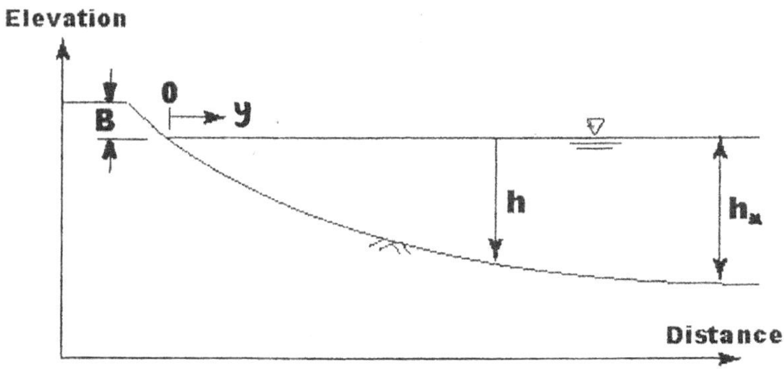

Figure 7.23. Schematic diagram of equilibrium beach profile.

offshore y and the mean water depth h^n (to power n) out to the closure depth h_* as shown in Figure 7.23 and given in Equation (3.1) (Bruun, 1954; Dean, 1977). The correlation coefficient A was found to depend on the the median, grain-size diameter D_{n50} in metric system units of millimeters (mm) and the exponent $n = 2/3$ was found as the best fit for all the data. The beach profiles when surveyed were thought to be in dynamic equilibrium and therefore Equation (3.1) has become known as the theory for "equilibrium beach profiles":

$$h(y) = Ay^{2/3}. \tag{3.1}$$

Formulas exist for the relationship between A and sediment diameter or A and the fall velocity of sediments in water, but the best relationship is Table 7.1 (from CEM, 2006, Table III-3-3) for A versus D_{n50} values, where units of A are $\text{m}^{1/3}$. Unfortunately, the A coefficient in Equation (7.38) is not dimensionless. When imperial units of feet are employed for y and h, then the A coefficients in Table 7.1 must be multipled by 1.5 to get $\text{ft}^{1/3}$.

As demonstrated in Figure 7.24, as the sand grain size D_{n50} (mm) increases, using Table 7.1, the A values also increase producing equilibrium beach profiles that steepen. For example in Figure 7.24, for the same distance $y = 200\,\text{m}$ offshore, $h = -3.5\,\text{m}$ when $D_{n50} = 0.22\,\text{mm}$, and $h = -4.1\,\text{m}$ when $D_{n50} = 0.28\,\text{mm}$. Small differences in the median grain size produce a significant change in the equilibrium beach profile.

A theroetical foundation for the equilibrium beach profile has been put forward as a "dynamic approach" when the constructive (sand pushed onshore) and destructive (sand dragged offshore) forces acting on the bottom sediments are in balance (Dean and Dalrymple, 2002). Forces that

Table 7.1. Summary of recommended A values in Equation (3.1) when units of A are $m^{1/3}$ (from CEM, 2006, Table III-3-3; Dean, 1987).

D (mm)	0.00	0.01	0.02	0.03	0.04	0.05	0.06	0.07	0.08	0.09
0.1	0.063	0.0672	0.0714	0.0756	0.0798	0.084	0.0872	0.0904	0.0936	0.0968
0.2	0.100	0.103	0.106	0.109	0.112	0.115	0.117	0.119	0.121	0.123
0.3	0.125	0.127	0.129	0.131	0.133	0.135	0.137	0.139	0.141	0.143
0.4	0.145	0.1466	0.1482	0.1498	0.1514	0.153	0.1546	0.1562	0.1578	0.1594
0.5	0.161	0.1622	0.1634	0.1646	0.1658	0.167	0.1682	0.1694	0.1706	0.1718
0.6	0.173	0.1742	0.1754	0.1766	0.1778	0.179	0.1802	0.1814	0.1826	0.1838
0.7	0.185	0.1859	0.1868	0.1877	0.1886	0.1895	0.1904	0.1913	0.1922	0.1931
0.8	0.194	0.1948	0.1956	0.1964	0.1972	0.198	0.1988	0.1996	0.2004	0.2012
0.9	0.202	0.2028	0.2036	0.2044	0.2052	0.206	0.2068	0.2076	0.2084	0.2092
1.0	0.210	0.2108	0.2116	0.2124	0.2132	0.2140	0.2148	0.2156	0.2164	0.2172

Notes:
(1) The A values above, some to four places, are not intended to suggest that they are known to that accuracy, but rather are presented for consistency and sensitivity tests of the effects of variation in grain size.
(2) As an example of use of the values in the table, the A value for a median sand size of 0.24 mm is: $A = 0.112\ m^{1/3}$. To convert A values to $ft^{1/3}$, multiply by 1.5.

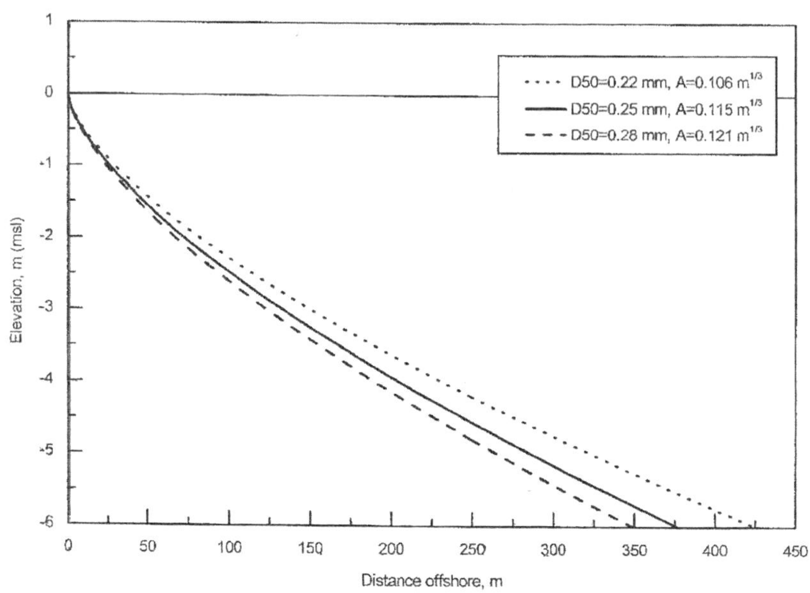

Figure 7.24. Example of equilibrium profiles for three grain sizes and the corresponding A values.

carry sand in the seaward direction are gravity, turbulence, beach slope, and the current undertow. Forces that move sediment toward the shore are the net, onshore shear stresses of the waves, the bottom streaming velocity, and turbulence in the suspended sediment in the trough of the waves. The wave energy dissipation per unit volume is assumed uniform across the surf zone. See Dean and Dalrymple (2002) for the complete derivation of a theoretical expression for A as a function of the sediment grain size and for other models of profile shape. This reference also gives modifications of Equation (3.1) when wave setup and storm surge increase the water levels in front of the beach.

7.3.3.2 *Beach cross-section design for compatibile materials ($D_F = D_N$)*

When the median grain size in the borrow area D_F is identical to that found on the native beach D_N, then the materials are compatible ($D_F = D_N$) and the beach profile is simply translated seaward as illustrated in Figure 7.25. The added beach width W is simply the design beach width, Δy_o as found in Equation (7.38) and repeated here for a design volume, $\forall$ with berm height, B and DOC, h_*

$$\Delta y_o = \frac{\forall}{(h_* + B)}. \tag{7.38}$$

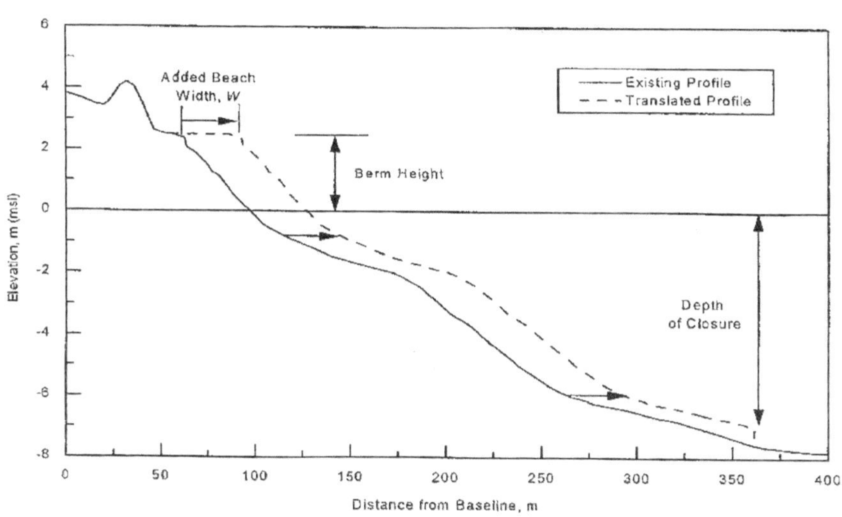

Figure 7.25. Beach cross-section design for compatible materials ($D_F = D_N$).

As an example, for $B = 2.5\,\text{m}$ (8.2 ft), $h_* = 6\,\text{m}$ (19.7 ft), $W = \Delta y_o = 30\,\text{m}$ ($\approx$100 ft), and $D_F = 0.27\,\text{mm} = D_N$ gives $\forall = \Delta y_o(h_* + B) = 255\,m^3/\text{m}$ (102 cy/ft). We shall repeat this example in the following when the borrow area material is not compatible ($D_F \neq D_N$) which is the more common case (*recall that* 1 *cubic yard(cy)*= $0.76\,m^3$, *and 1 cy/ft*= *2.5* m^3/m).

7.3.3.3 *Beach cross-section design when coarser sand is obtained from the borrow area ($D_F > D_N$)*

Professor Bob Dean made a noteworthy contribution in coastal engineering when he published his work (Dean, 1991) for beach cross-section design when the borrow area sand was not compatible with the sand found on the native beach ($D_F \neq D_N$). Three basic types of nourishment profiles were identified as illustrated in Figure 7.26 (from Dean, 1991). In Figure 7.26(a), the added (fill) sand Δy_o from the borrow area D_F is coarser than the native sand ($A_F > A_N$) so that the steeper profile intersects the native profile before reaching the DOC h_* at a distance W_* from the original shoreline. However, if a large volume of coarse sand is added for a wide design beach width, the new profile may extend beyond the DOC h_* at a distance beyond W_* as shown in Figure 7.26(b). Therefore, sand coarser than native sand ($A_F > A_N$) can produce either (a) intersecting or (b) non-intersecting profiles with the native sand beach profile.

(1) Coarser sand, Intersecting profiles

The volume of coarse sand placed per unit length of shoreline for Intersecting profiles, $\forall_I$ is (Dean, 1991)

$$\forall_I = B\Delta y_o + \frac{3}{5}A_N \frac{\Delta y_o^{5/3}}{\left[1 - \left(\frac{A_N}{A_F}\right)^{3/2}\right]^{2/3}}. \tag{7.39}$$

For the same example as given above, now assume coarser sand with $D_F = 0.36$ mm so that from Table 7.1, $A_F = 0.137\,\text{m}^{1/3}$. For the native beach, $D_N = 0.27$ mm and $A_N = 0.119\,\text{m}^{1/3}$ and the beach design was for $B = 2.5$ m and $\Delta y_o = 30$ m ($\approx$100 ft). Putting these values into Equation (7.39) gives $\forall_I = 137\ \text{m}^3/\text{m}$ (55 cy/ft). For the same design beach width, the coarser sand-fill material requires about 54% of the volume needed if compatibile sand was used. This would be a large reduction in the initial cost of the renourished beach.

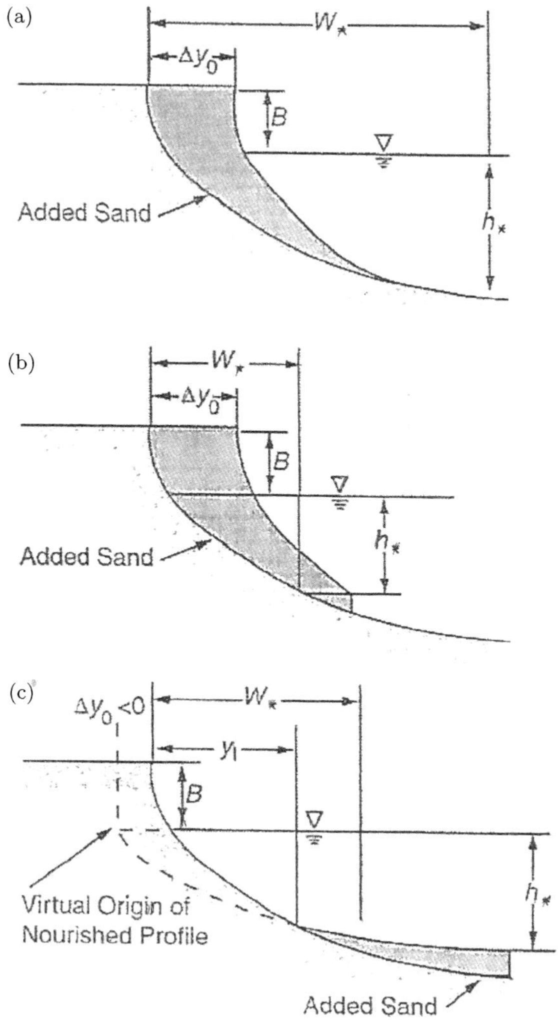

Figure 7.26. Three types of beach nourishment profiles (from Dean, 1991).

(2) Intersecting or non-intersecting profiles

Dean (1991) gave the following formulas to detect intersecting or non-intersecting profiles and this should be checked first to make sure that Equation (7.39) is the correct one to calculate the coarse volume.

$$\Delta y_o \left(\frac{A_N}{h_*}\right)^{3/2} + \left(\frac{A_N}{A_F}\right)^{3/2} < 1, \quad \text{Intersecting,} \qquad (7.40a)$$

$$\text{and} \qquad \Delta y_o \left(\frac{A_N}{h_*}\right)^{3/2} + \left(\frac{A_N}{A_F}\right)^{3/2} > 1, \quad \text{Non-intersecting.} \tag{7.40b}$$

For this example,

$$30\left(\frac{0.119}{6.0}\right)^{3/2} + \left(\frac{0.119}{0.137}\right)^{3/2} = 0.0838 + 0.8095$$

$$= 0.893 < 1, \quad \text{Intersecting, use Equation (7.40a).}$$

Use of Equation (7.39) to calculate $\forall_I$ was correct. Note that if $\Delta y_o = 70$ m (230 ft), then Equation (7.40b) gives $1.005 > 1$ for a non-intersecting profile. This would be a very wide beach nourishment and very expensive.

(3) Coarser sand, Non-intersecting profiles

The volume of coarse sand placed per unit length of shoreline for non-intersecting profiles, $\forall_N$ is

$$\forall_N = B\Delta y_o + \frac{3}{5}\left(\frac{h_*}{A_F}\right)^{5/2}\left\{A_N\left[1 + \Delta y_o\left(\frac{A_F}{h_*}\right)^{3/2}\right]^{5/3} - A_F\right\}. \tag{7.41}$$

Now, the DOC h_* is involved since the sand volume extends past this location as illustrated in Figure 7.26(b). The distance W_* is the offshore distance associated with the DOC h_* or a breaking depth h'_* for the native beach profile and from the equilibrium beach profile equation (Equation (7.39)) using A_N gives

$$W_* = \left(\frac{h'_*}{A_N}\right)^{3/2}, \tag{7.42}$$

$$\text{and} \qquad h'_* = \frac{H_b}{\kappa}, \quad \text{a breaking depth,} \tag{7.43}$$

where H_b is the breaking wave height, and κ is the spilling breaker index (≈ 0.78).

It is interesting to note that in Professor Dean's book *Beach Nourishment:Theory and Practice* (Dean, 2002), the reference breaking depth h'_* can be either Equation (7.43) or the closure depth h_*. Figure 3.8 in his book shows the estimates of the closure depths h_* around the state of Florida coast. The same symbol h_* is employed for both the "reference breaking depth" or the closure depth. However, in the book *Coastal Processes: With Engineering Applications* (Dean and Dalrymple, 2002), the symbol h_c is used for the DOC and h_* for the breaking depth. Therefore, either the DOC h_* or the breaking depth h'_* could be used to find $\forall_N$ in Equation

(7.41). We recommend using the closure depth h_* since Equations (7.35) and (7.36) are available for estimation in the US using WIS hindcast wave data and a web-based tool.

For the same example above, using $\Delta y_o = 70$ m (230 ft) in Equation (7.41) gives $\forall_N = 430\,\text{m}^3/\text{m}$ (170 cy/ft) which is a very wide and expensive beach nourishment. Since $\Delta y_o = 70$ m (230 ft) is exactly when the formula detecting when the coarse profile transitions from intersecting to non-intersecting (≈1.0), using Equation (7.39) also gives $\forall_N = 430\,\text{m}^3/\text{m}$ (170 cy/ft).

7.3.3.4 *Beach cross-section design when finer sand is obtained from the borrow area ($D_F < D_N$)*

When considering equilibrium beach profiles as shown in Figure 7.24, when finer sand from the borrow area ($D_F < D_N$) is employed to nourish the beach, the renourished profile is always the non-intersecting type. Equation (7.41) is applicable to calculate the volume of fine sand per unit length of beach, $\forall_F$ and repeated here as Equation (7.44) to emphasize renourishment with finer sand than the native beach.

$$\forall_F = B\Delta y_o + \frac{3}{5}\left(\frac{h_*}{A_F}\right)^{5/2}\left\{A_N\left[1 + \Delta y_o\left(\frac{A_F}{h_*}\right)^{3/2}\right]^{5/3} - A_F\right\}. \tag{7.44}$$

For the above example, assume that the median sand grain size $D_F = 0.19$ mm in the borrow area ($A_F = 0.0968$) and $\Delta y_o = 30$ m (100 ft) gives $\forall_F = 700\,\text{m}^3/\text{m}$ (280 cy/ft). That is a very expensive project!

As a general rule, beach nourishment using sand finer than the native beach should only be considered when the native material or coarser sand is either not available or is very expensive per unit volume.

The third possible type of beach profile is shown in Figure 7.26(c) as a submerged profile and is non-intersecting. The maximum submerged volume $\forall_{SM}$ is when no dry beach exists ($\Delta y_o = 0$) so that Equation (7.44) reduces to Equation (7.45) which only holds for finer sand from the borrow area ($A_N > A_F$).

$$\forall_{SM} = \frac{3}{5}\left(\frac{h_*}{A_F}\right)^{5/2}\{A_N - A_F\}. \tag{7.45}$$

For the example above, $\forall_{SM} = 400\,\text{m}^3/\text{m}$ (160 cy/ft) for a submerged profile with no dry beach ($\Delta y_o = 0$).

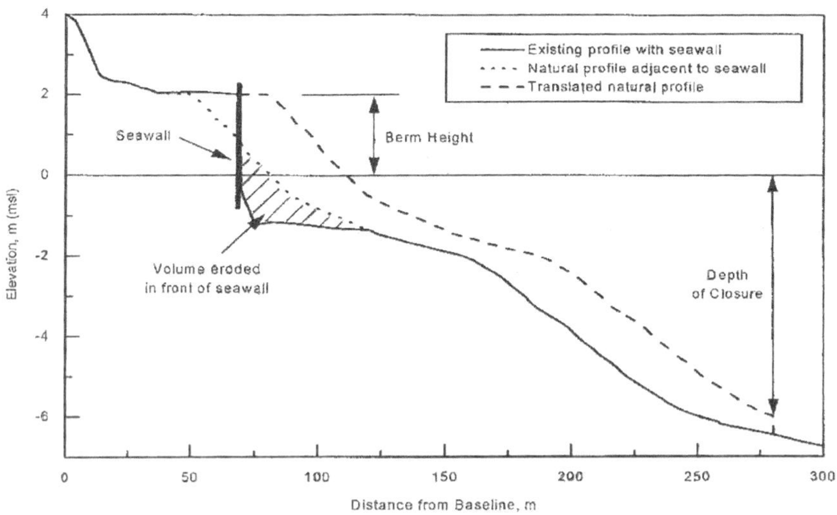

Figure 7.27. Beach profiles needed to estimate the local sand volume deficit $v_d(x)$ (from CEM, 2006).

7.3.3.5 *Beach cross-section design with sand volume deficit*

An eroded beach (upper, Figure 7.22) has less sand volume than a normal "healthy" beach. The sand volume deficit must be initially determined by field survey. Figure 7.27 schematically depicts a beach profile in front of a seawall, the sand volume deficit relative to profiles at adjacent locations, and the nourished beach profile. The sand volume deficit $v_d(x)$ is variable with distance x along the beach and must be integrated along the project length L to give the total, sand volume deficit, $\forall_d$ to be added to that required for the beach nourishment project.

The sand volume deficit when caused by a fixed anthroprogenic feature (road, seawall, rock revetment, buildings, etc.) on naturally eroding shores prevents landward recession and lowers the beach profile in front of the fixed feature. Whether (or not) the fixed feature *increases* the eroded volume is subject to debate and discussed in detail in Chapter 10.

7.3.3.6 *Beach cross-section design using James curves*

Prior to Dean (1991) when the borrow area sand was not compatible with the sand found on the native beach ($D_F \neq D_N$), the "overfill ratio" method of James (1975) was employed using the "James Curves". The grain sizes are determined in Phi units, Φ as related to the grain size D (mm) by Equation

(2.61) (Section 2.4.1 in Chapter 2) for the Wentworth Classification system (Figure 2.34) used by coastal geologists. Mean sediment sizes and standard deviations of sediment sizes as defined in Phi units for the borrow area and native beach sands are employed. The abcissa of the James Curves is the difference of the means normalized by the standard deviation of the borrow area sands (e.g. the relative size) and the ordinate is the ratio of the standard deviations (e.g. the relative degree of sorting). Isolines on the James Curves are the adjusted overfill ratio R_A defined as

$$R_A = \frac{\text{Volume of Borrow Site Material}}{\text{Volume of Native Beach Material}}. \tag{7.46}$$

In other words, R_A is the estimated number of cubic meters of fill material required to produce 1 m^3 of beach material. In James (1975), $R_A \geq 1.0$ for beach nourishment design.

The abcissa scales from $-4.0 \geq 0 \leq +4.0$ (i.e. coarser to finer relative sizes) in cartesian coordinates and the ordinate scales from $0.2 \leq 1.0 \leq 5.0$ (i.e. good to poor relative sorting) in logarithmic coordinates. The isolines of R_A values range from 1.0 to 10.0. Four quadrants result that separate finer and coarser borrow area sand from the native sand beach and whether the borrow area material is more poorly sorted or better sorted than the native sand material. The beach nourishment volume required is not determined.

Use of the James Curves (James, 1975) is not recommended. The Phi unit system has no physical meaning. When $\Phi = 0$, the grain size, $D = 1.0$ mm. As Φ increases positively, the actual sand size decreases and as Φ values get more negative, D increases. This is very confusing for coastal engineers. More importantly, the James Curves are all for $R_A \geq 1.0$ so that for coarse material from the borrow area, the miminum volumes must always be equal to or more than that required for compatible materials ($D_F = D_N$). The beach nourishment volume can never be less by using coarser materials. For this reason, the James Curves are not reproduced in this book. See CEM (2006, Part V, Chapter 4) for full details.

7.3.4 *Planform evolution*

A beach nourishment project is a large perturbation of the original shoreline. It "sticks out" from the natural coast and is immediately impacted by the waves and daily variations of the tides. At the end of the project, even when the constructed width is tapered back to zero, sand spreads laterally to the adjacent beach that has not been nourished. Dean (2002) terms this "spreading out" losses as shown in his Figure 7.28 where the planform of

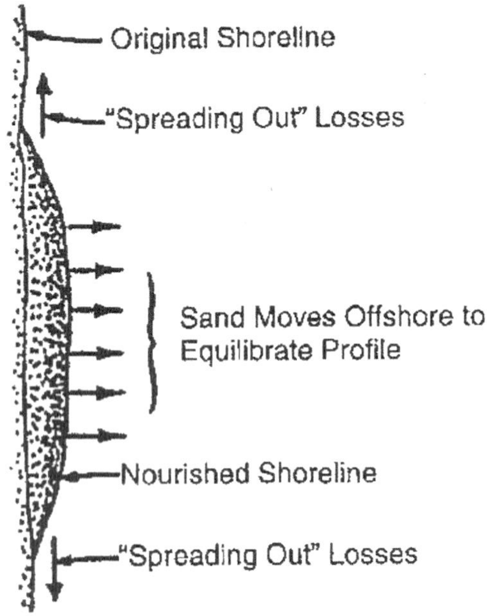

Figure 7.28. Sand spreading laterally and moving offshore in a planview of the constructed beach (from Dean, 2002).

the nourished shoreline evolves in time from the constructed dimensions. In this section, we focus on the "spreading out" losses as modeled by the "diffusion" equation of transport phenomena. In Section 7.3.5, the movement of sand in the offshore direction and the equilibration of the constructed profile will be addressed.

7.3.4.1 *The diffusion equation*

The lateral spreading of sand within and outside the boundaries of a beach nourishment project has been quantified by the diffusion equation of mathematical physics as derived by Pelnard-Considére (1956), namely,

$$\frac{\partial y}{\partial t} = G\frac{\partial^2 y}{\partial x^2}, \tag{7.47}$$

where G is the "longshore diffusivity" coefficient and $y(x,t)$ is the location in time t and space x of the original design width $Y = \Delta y_o(t = 0)$ as schematically illustrated in Figure 7.29 by the dashed line for a rectangular beach nourishment project of length, l. Note that positive distance $+x$ is measured from the center where $y(x,t)$ is symmetrical and that sand moves laterally (diffuses) beyond the project boundaries $\pm a = l/2$. This

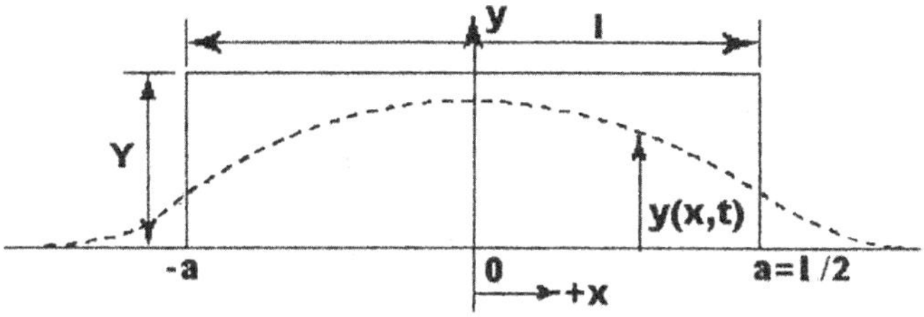

Figure 7.29. Schematic diagram of lateral spreading of sand from rectangular beach nourishment project (from Dean, 2002).

"spreading out" loss of sand is only a "loss" of sand within the project boundaries (Y, l). All the original sand mass is theoretically still present between infinite boundaries $\pm\infty$ along the $\pm x$-axis.

The longshore diffusivity coefficient G has units of m^2/s (ft^2/s) and depends on the breaking wave height, H_b, the scale of the sand volume $(h_* + B)$, and all the same variables as found in the CERC formula for LST rate, Q in Equation (7.32). The formal derivation of the equation for G depends on the conservation of sediment mass, the gradient of the sediment transport rate Q, and simplification of the wave angles and shoreline orientations to the approaching wave conditions. See Dean and Dalrymple (2002) for the full derivation of G which is not given here. The equation to calculate G is as follows:

$$G = K\left[\frac{\sqrt{g/\kappa}}{8(s-1)(1-p)(h_* + B)}\right] H_b^{5/2}, \tag{7.48}$$

so that the same rules apply for the value of the transport coefficient K depending on the definition of H_b. Recall that when the significant wave height at breaking is employed, H_{sb}, then, $K_s = 0.39$.

As an example, for $H_{sb} = 1.0\,\text{m}$ (3.3 ft), $B = 1.83\,\text{m}$ (6 ft), $h_* = 6.4\,\text{m}$ (21 ft), $s = 2.65$ sand with porosity, $p = 0.35$, and a breaking wave height coefficient $\kappa = 0.8$ gives $G = 0.01934\,\text{m}^2/\text{s}$ ($0.208\,\text{ft}^2/\text{s}$). As seen in Equation (7.48), G is extremely sensitive to the breaking wave height to the 2.5 power. Doubling H_b means a 5.6-times increase in G.

7.3.4.2 *Analytical solution for rectangular nourishment project on infinitely long shoreline*

The diffusion Equation (7.47) is a second-order, partial differential equation with analytical, closed-form solutions available for many cases of different initial conditions and boundary conditions (Larson *et al.*, 1987, 1997) for beach nourishment projects. One useful case is shown in Figure 7.29 for initial conditions $[y(x,0)]$ as a rectangle of width Y and length l and boundary conditions $[y(x,0)] = 0$ when $x > \pm l/2$. The solution is given as follows:

$$y(x,t) = \frac{Y}{2}\left\{\text{erf}\left[\frac{l}{4\sqrt{Gt}}\left(\frac{2x}{l}+1\right)\right] - \text{erf}\left[\frac{l}{4\sqrt{Gt}}\left(\frac{2x}{l}-1\right)\right]\right\}, \quad (7.49)$$

where erf is the error function (e.g. Abramowitz and Stegun, 1965).

Fortunately, the error function erf is available in software products for spreadsheets, (e.g. Microsoft Excel) so that tabular solutions are not required. In non-dimensional terms y/Y and defining

$$t' = \frac{16Gt}{l^2} \quad (7.50)$$

gives

$$\frac{y(x,t)}{Y} = \frac{1}{2}\left\{\text{erf}\left[\frac{1}{\sqrt{t'}}\left(\frac{2x}{l}+1\right)\right] - \text{erf}\left[\frac{1}{\sqrt{t'}}\left(\frac{2x}{l}-1\right)\right]\right\}, \quad (7.51)$$

so that a dimensionless plot of y/Y versus $2x/l$ for various values of the dimensionless time value t' reveals how the sand spreads laterally away from the nourished beach in time. Since the lateral spreading is symmetric about $x = 0$, only one half is plotted as shown in Figure 7.30 (from Dean, 2002).

The online supplementary materials which accompanies this book contains a spreadsheet to calculate y/Y versus $2x/l$ values from Equation (7.51) for a given dimensionless time t' as shown in Figure 7.30 from which it is relatively straightforward to convert to dimensional values $y(x,t)$ for use in design in either metric or Imperial units.

As seen in Figure 7.30, when $t' = 4.0$, roughly one-half the sand remains within the project boundary and the other one-half has moved to the adjacent beach. We can be more precise by integrating Equation (7.51) within the project boundaries and dividing by the initial project volume $(Y * l)$ to give that percentage of the sand remaining, $M(t)$. The result is

$$M(t) = \frac{2\sqrt{Gt}}{l\sqrt{\pi}}\left(\exp^{-(l/2\sqrt{Gt})^2} - 1\right) + \text{erf}\left(\frac{l}{2\sqrt{Gt}}\right) \quad (7.52)$$

and $M(t)$ versus $\sqrt{Gt}/l$ is the solid curve plotted in Figure 7.31 (from Dean, 2002). Both scales are plotted with linear coordinates and we see that the percentage of the volume of sand remaining drops rapidly (exponentially) in

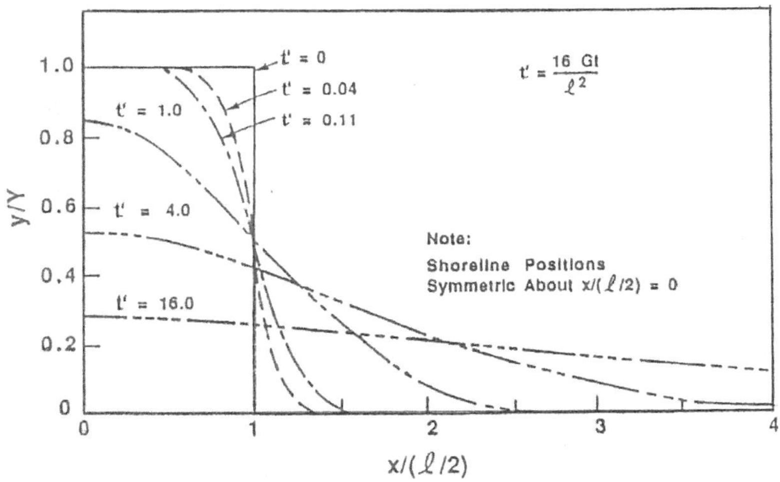

Figure 7.30. Dimensionless shoreline locations y/Y in dimensionless time t' spreading within the positive side boundary, $2x/l \leq 1.0$ and beyond the boundary ($2x/l > 1.0$) for a rectangular beach nourishment project on an infinitely long shore (from Dean, 2002).

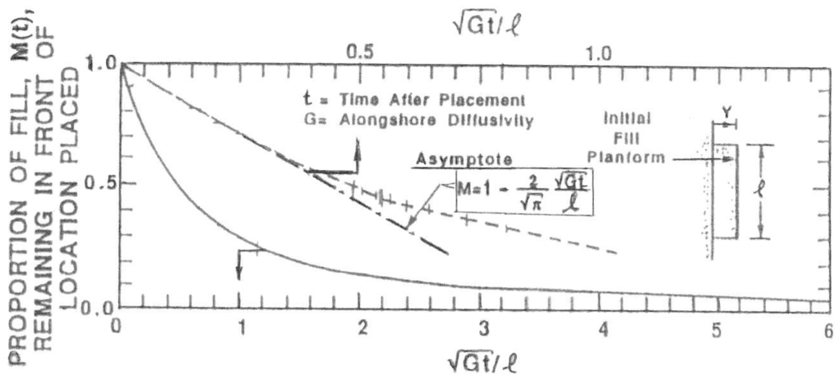

Figure 7.31. The percentage of sand remaining in a rectangular beach fill as a function of $\sqrt{Gt}/l$ (from Dean, 2002).

time before leveling off. The term "half-life" of a beach nourishment project is the time when 50% of the original sand volume remains, $M(t) = 0.5$. Inserting $M(t) = 0.50$ in Equation (7.52) and solving gives

$$\sqrt{Gt}/l = 0.481 \tag{7.53}$$

and solving for the half-life time, t_{50} gives

$$t_{50} = (0.481)^2 \frac{l^2}{G} = 0.231 \frac{l^2}{G}. \tag{7.54}$$

Dean (2002) enployed the asymptote approximation to Equation (7.52) as shown in Figure 7.31 and got slightly different values for the constants in Equations (7.53) and (7.54). But the important results are that the half-life time t_{50} varies as the square of the project length l and is inversely proportional to the longshore diffusivity coefficient G. Doubling the project length l produces a four-fold increase in the half-life of the project. Doubling the breaking wave height, H_b in Equation (7.48) for $G[(H_b^{5/2})^{1/2} = H_b^{5/4}]$ results in over a 50% reduction in the half-life of the project. Since the value employed for the breaking wave height, H_b in Equation (7.48) is a yearly "average" value and G is difficult to estimate for design, making the project length l as long as possible is always good practice in beach nourishment design.

As an example, consider a rectangular, beach nourishment project with design width $Y = 100\,\text{ft}$ (30 m) over a project length $l = 4\,\text{mi}$ (6.4 km) as outlined in yellow in Figure 7.32. The effective breaking wave height for design $H_b = 2.7\,\text{ft}$ (0.82 m), giving $G = 0.12748\,\text{ft}^2/\text{s}$ ($0.01185\,\text{m}^2/\text{s}$) from Equation (7.48) when $B = 1.83\,\text{m}$ (6 ft), $h_* = 6.4\,\text{m}$ (21 ft), $s = 2.65$ sand

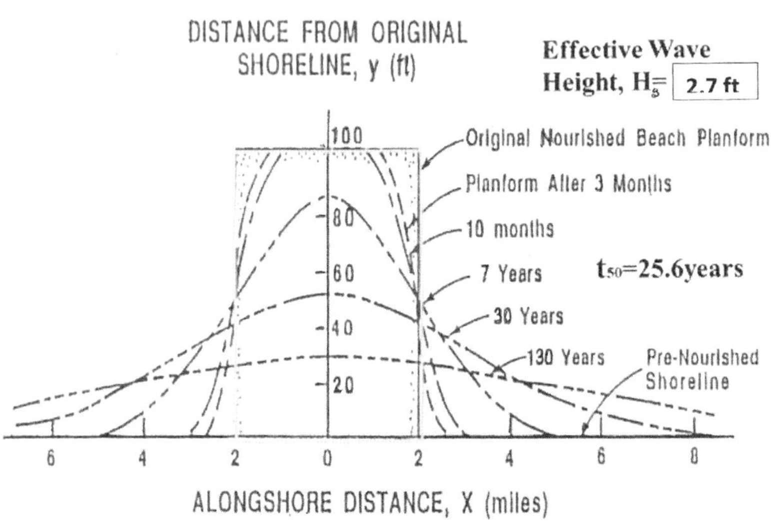

Figure 7.32. Example of rectangular beach nourishment lateral spreading over time using the diffusion equation of Pelnard-Considére (1956).

with porosity, $p = 0.35$, and a breaking wave height coefficient, $\kappa = 0.8$. The half-life, t_{50} of the project is 25.6 years. If the length of fill l was only 2 mi (3.2 km), the half-life t_{50} would be reduced to 6.4 years. Increasing the design wave height, H_b to 4.7 ft (1.4 m) for a 4-mi long project would also reduce the half-life to 6.4 years. Use of Equation (7.54) is important in design.

7.3.4.3 *Analytical solution for shoreline change updrift of a groin or jetty*

The Pelnard-Considére (1956) diffusion Equation (7.47) is also applied to study shoreline change, $y(x, t)$ near coastal structures. One important application is on the side of a groin or jetty that impounds LST moving toward the structure on the "updrift" side. As depicted in Figure 7.33, wave crests making an angle $\propto_o$ with the original shoreline, $y(x, 0)$ produce LST in the negative x-direction (right to left). At the base of the groin $x = 0$, the sand begins to immediately accumulate ($t > 0$). The boundary condition for the solution of the diffusion equation at the groin ($x = 0$) is

$$\frac{dy}{dx} = \tan \propto_o . \tag{7.55}$$

At the instance when impoundment begins, the shoreline orientation must be parallel to the wave crest. At the far updrift boundary ($+x \approx \infty$), the shoreline slope, $dy/dx = 0$. The shoreline slope decays exponentially

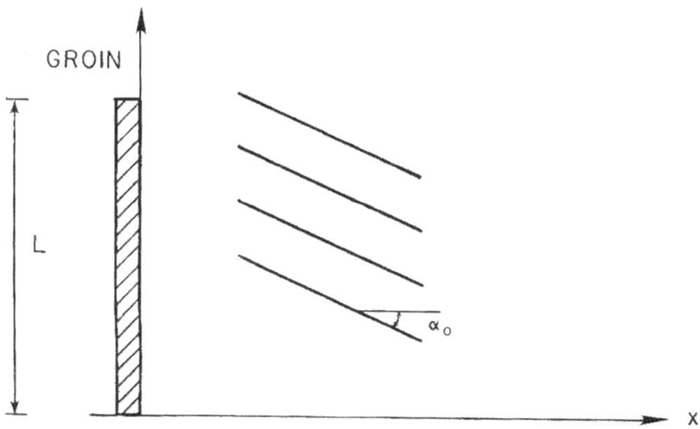

Figure 7.33. Schematic diagram of key variables for shoreline change, $y(x, t)$ updrift of a coastal groin.

updrift of the groin as modified by the complimentary error function, erfc $= (1 - \text{erf})$. A solution for $y(x,t)$ as presented in Larson *et al.* (1987) is

$$y(x,t) = 2\tan \propto_o \left\{ \left(\sqrt{\frac{Gt}{\pi}} \right) \exp\left[-x^2/4Gt\right] - \left[\left(\frac{x}{2}\right) \operatorname{erfc}\left(\frac{x}{2\sqrt{Gt}}\right) \right] \right\} \tag{7.56}$$

for all time, $t < t^*$ as discussed in the following for Equation (7.59).

A similar expression is presented in Dean (2002), except the wave crest angle $\propto_o$ is reduced by the wave-breaking angle $\propto_b$ so that the term tan $(\propto_o - \propto_b)$ is used in Equation (7.56). We recommend using only $\propto_o$ in Equation (7.56) for the simple reason that the diffusion coefficient G is the key independent variable in the equation.

In non-dimensional form, Equation (7.56) can be written in terms of the ordinate $(y/(\tan \propto_o L)$, the abcissa (x/L), and a dimensionless time $t' = Gt/L^2$. Note this definition of t' is different than Equation (7.50) for a rectangular beach nourishment planform spreading in time.

$$\frac{y}{\tan \propto_o L} = 2\left\{ \left(\sqrt{t'/\pi}\right) \exp\left[-\left(\frac{x}{L}\right)^2/4t'\right] - \left[\frac{1}{2}\left(\frac{x}{L}\right) \operatorname{erfc}\left(\frac{x/L}{2\sqrt{t'}}\right)\right]\right\} \tag{7.57}$$

with the result plotted in Figure 7.34.

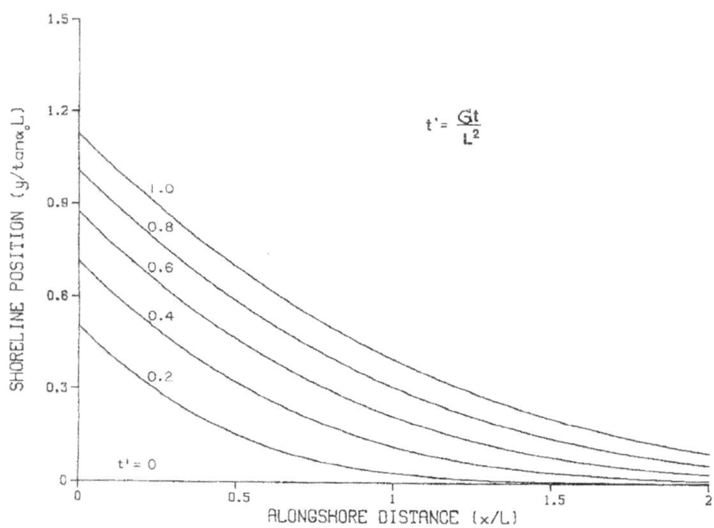

Figure 7.34. Non-dimensional plot of Equation (7.57) for shoreline change updrift of a coastal groin.

At the groin where $x = 0$, Equation (7.57) reduces to

$$\frac{y(0,t)}{\tan \propto_o L} = 2\left(\sqrt{t'/\pi}\right), \tag{7.58}$$

so that the increase in the shoreline position at the groin is equal to the square root of time. Eventually, if wave conditions remain constant, the shoreline position will reach the end of the groin, $y(0,t^*) = L$ at some time, t^*. Setting this condition in Equation (7.58) gives

$$t^* = \frac{\pi L^2}{4\,G\left(\tan^2 \propto_o\right)}. \tag{7.59}$$

Once the shoreline position reaches the end of the groin at time t^*, the boundary condition changes at the groin to $y(0,t^*) = L$. For all time $t > t^*$, a new shoreline position formulation was found by Pelnard-Considére (1956) to be

$$y(x,t) = L\ \mathrm{erfc}\left(\frac{x}{4\,G\,t_{\mathrm{new}}}\right). \tag{7.60}$$

Unfortunately, the shoreline positions $y(x,t)$ are different for these two different solutions Equations (7.56) and (7.60) of the diffusion equation. Matching the sand volumes impounded by the two solutions, it has been found that the time t_{new} given by

$$t_{\mathrm{new}} = t - \left(1 - \frac{\pi^2}{16}\right)t^* \tag{7.61}$$

and employed in Equation (7.60) will produce consistent results (Dean, 2002) for all time $t > t^*$. See Dean (2002) and Dean and Dalyrmple (2002) for additional details regarding impounded areas, volumes, and estimates of the sand-bypassing rates when $t > t^*$.

All that is presented above regarding impoundment for the updrift side of the groin in the (+) x-direction is *theoretically* possible in the downdrift (−) x-direction. The formulas now have a negative sign $-y(x,t)$ so that the total sand volume is conserved. The volume impounded on the updrift side is balanced by the volume missing on the downdrift side of the groin. The classical example often cited in the literature is in Dean (2002, Figure 3.19, p. 54), showing both the direction of LST with sand impounded on the updrift side and sand missing for equal shoreline positions in time on the downdrift side. The groin length is 160 ft (50 m), the ± x-distances cover 4.0 mi (6.5 km), and shoreline positions are 0.5, 1.0, 2.0, and 5.0 years. The longshore transport rate used in this example is 180,000 cubic yards per year (135,000 m^3 per year). This graphic is not reproduced in this book. The calculated results are *theoretically* but not physically possible. Wave diffraction on the downdrift side, daily changes in wave conditions (magnitude and direction) including complete reversal of the LST direction changing the updrift and downdrift sides, and others are neglected. Wave conditions are never constant for five years. We will return to this example in Chapter 10.

7.3.4.4 *Limitations of the diffusion equation*

The Pelnard-Considére (1956) diffusion equation has become known as the "one-line" model of shoreline change. Analytical, closed-form solutions are available for more than 15 cases of different initial conditions and boundary conditions for beach nourishment projects and coastal structures on beaches (Larson *et al.*, 1987). The major assumption is that the entire beach profile stays constant for all $y(x,t)$ shoreline positions in space and time. A generalized, one-line numerical model of shoreline change called GENESIS (Hanson, 1989) is available for application to a wide range of coastal structues interacting with the adjacent beaches.

The one-line model is extremely useful for feasibility-level studies, but has limitations whenever gradients in shoreline change and beach profile change are strong resulting from wave refraction and wave diffraction. Three-dimensional (3D) numerical models must then be employed.

7.3.5 *Profile equilibration*

7.3.5.1 *Construction and design width*

In Figure 7.28, sand also moves in the offshore direction to "equilibrate" the profile. Beach nourishment projects are all constructed with initial berm widths Δy_I greater than the design width Δy_o and initial construction slopes m_I (nominally about 20H:1V) greater than the normal, equilibrium beach profile slope. Beach nourishment construction is less by allowing the daily tidal and wave conditions to reshape the constructed profile over time to natural, equilibrium profile shapes. This process is called profile equilibration. Figure 7.35 schematically depicts (a) a constructed beach-fill width and slope and (b) the equilibrated, reduced width, and adjusted profile shape.

Sand volume is assumed to be conserved so that the constructed fill volume V_I is equal to the design volume V per unit length of shoreline. Dean (2002) presents graphical solutions for the ratio, $\Delta y_o/\Delta y_I$ versus

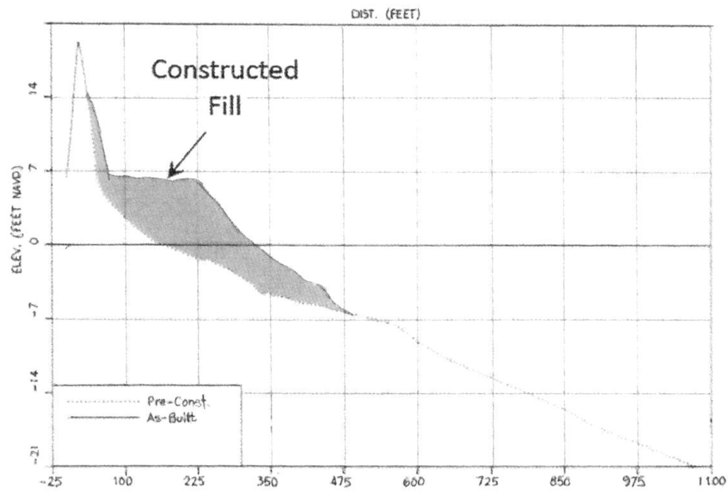

(a)

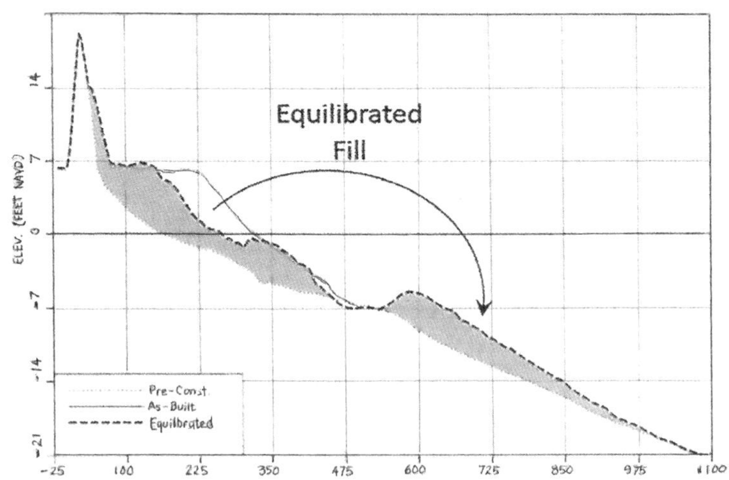

(b)

Figure 7.35. Construction and design width of nourished beaches after profile equilibration (from ASBPA http://www.asbpa.org).

ratios of A_F/A_N, corresponding to various values of a non-dimensional volume (V/BW_*), a non-dimensional closure depth (h_*/B), and a non-dimensional slope parameter $(h_*/(m_i W_*))$. Recall that W_* is the distance offshore to the closure depth h_* and can be found from Equation (7.39).

For compatible beach fill material ($A_F = A_N$), Δy_I can be determined by trial and error for $m_I = 20{:}1$.

7.3.5.2 *Time for equilibration*

A rule of thumb is that it normally takes one full year of "typical" wave climate conditions at the site before the design beach width and profile equilibration have been reached. The seasonal winter storms and summer swell waves all must be realized before the constructed profile is returned to equilibrium. In some cases, more than one full wave year is needed. In other instances, storm waves may occur soon after construction is complete. In this case, an "advance fill" volume as shown in Figure 7.36 may be added to the original design fill volume as a factor of safety in anticipation of a storm event soon after construction (ASBPA, 2017 online graphic).

Two-dimensional (2D), cross-shore numerical models (e.g. SBEACH (Larson and Kraus, 1989), UNIBEST (2011)) can be applied to study the time for profile equilibration. Once the natural, equilibrium beach profile is reestablished, further time-varying shoreline retreat during a coastal storm

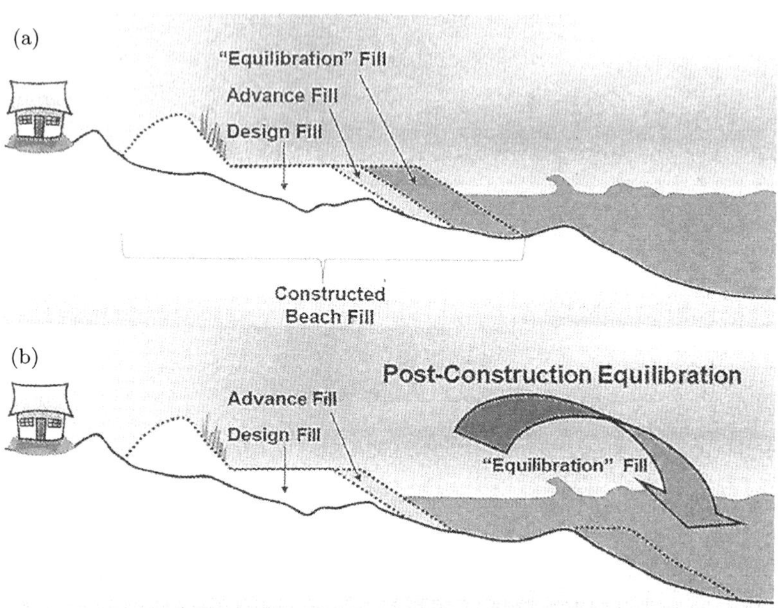

Figure 7.36. Advance fill in anticipation of coastal storm soon after construction (from ASBPA http://www.asbpa.org).

surge event can be studied using the analytical model of Kriebel and Dean (1993). Over extended time periods (years), the historic, shoreline erosion rate, $e(x,t)$ at the site continues at the same rate after beach nourishment. See Thomas and Frey (2013) for general model comparison and literature review.

7.3.5.3 *Background erosion rate combined with planform diffusion rate*

The simplest case is a uniform, background erosion rate $e(x,t)$ over the entire length l of a rectangular beach nourishment project. The eroded distance $y_e(t)$ is then

$$y_e(t) = -et, \tag{7.62}$$

and the proportion of beach fill remaining within the project boundaries $M'(t)$ including the background erosion is

$$M'(t) = \frac{2\sqrt{Gt}}{l\sqrt{\pi}}\left(\exp^{-\left(l/2\sqrt{Gt}\right)^2} - 1\right) + \operatorname{erf}\left(\frac{l}{2\sqrt{Gt}}\right) - \frac{et}{\Delta y_0}. \tag{7.63}$$

Dean (2002) defined a dimensionless "timeline" variable F as follows:

$$F = \frac{el^2}{(\Delta y_o)\, G} \tag{7.64}$$

and isolines of $M'(t)$ as a function of $et/\Delta y_o$ (ordinate) versus $\sqrt{Gt}/l$ (abcissa) for a graphical solution of Equation (7.63). Isolines (dotted) for the level of background erosion were included to give the user some idea when excessively high erosion rates make beach nourishment uneconomic. For other values of time, the dimensionless timeline variable F remains the same so that moving along the timeline gives the new time values of $M'(t)$. The graphs for solution are not reproduced herein.

In many other cases, the background erosion rate, $e(x,t)$ varies with distance x along the project length. See Dean (2002) for how to treat these "hot spot" locations.

7.3.5.4 *Bruun rule*

Over long time periods as the mean of the ocean's sea level rises, the local shoreline position moves landward (retreats). The seasonal beach profiles may be in dynamic equilibrium so that "erosion" is not occuring but shoreline retreat is still occuring. The horizontal recession R of the beach profile (of any shape) response to the sea-level rise S, and the relationship between R and S as put forward by Bruun (1962) has come to be called the Bruun rule, as given in the following:

$$R = S\frac{W_*}{(B + h_*)}. \tag{7.65}$$

The width W_* of the beach profile with active sand movement is usually taken as the horizontal distance to the closure depth h_* for the equilibrium beach profile with berm height, B.

As an example, assume an equilibrium profile with sand of median grain size $D_N = 0.25\,\text{mm}$ so that from Table 7.1, $A = 0.115\,\text{m}^{1/3}$. For a closure depth $h_* = 7\,\text{m}$ (23 ft) and applying Equation (7.39) and solving for W_* gives,

$$W_* = (h_*/A)^{3/2} = (7/0.115)^{3/2} = 475\,\text{m}.$$

For a berm height $B = 2\,\text{m}$ (6.5 ft), the average slope of the active beach profile is

$$\tan\beta = \frac{(B + h_*)}{W_*} = \frac{(2+7)}{475} = 0.01895.$$

Therefore, the Bruun rule could be written as,

$$R = \frac{S}{\tan\beta}. \tag{7.66}$$

Finally, for this example assume a sea-level rise $S = 1\,\text{m}$ (3.3 ft). Applying Equation (7.65) or (7.66) gives

$$R = \frac{1.0}{0.01895} = 53\,\text{m}(173\,\text{ft}).$$

For many locations in the world, the average slope of the active beach profile is small and of the order of $\tan\beta = 0.01 - 0.02$ so that the retreat rates are of the order of 50–100 times the sea-level rise rate as estimated by the Bruun rule. Local sea-level changes may be relative to local land movements (subsidence or lifting) and not the global average. We devote Chapter 11 to the impacts of sea-level rise on coastal engineering design.

7.3.6 *Public perceptions*

The general public has heard about and read the refrain many times that "*... you can't fight mother nature*" as promoted by the radio, TV, newspapers, and now the social media. This statement is then often followed by the words "*...beach erosion is a losing battle*" when related to a beach nourishment project. And then "... *beach nourishment projects don't last, half the beach is lost after the first storm*", when commenting on the visual results as shown in Figure 7.35 regarding a recently constructed project. Over time and all the above repeated many times, the general public has come to believe that "... *beach nourishment is a waste of taxpayer money and beach nourishment is an endless expense*". How has this come to happen? Who is to blame for this misinformation, misunderstanding, and these misconceptions of the general public? Unfortunately, the coastal engineering profession is partly to blame. We simply have not done enough to inform the general public about how the beach nourishment projects work.

Regarding the first refrain "... *you can't fight mother nature, beach erosion is a losing battle*", we must inform the public that beaches naturally recover after storms and that to work with nature we must first understand how nature works. The concept of dynamic equilibrium of beaches is not understood. Late-evening weather forecasts by the TV weathermen often predict beach "erosion" resulting from the next day's coastal storm waves but never mention the beach "recovery" following a few subsequent days of swell-wave conditions. The general public believes beaches always "erode" and never hear about their "recovery".

Regarding the second notion "... *beach nourishment projects don't last, half the beach is lost after the first storm*", we must educate the general public via the media that the construction width is always greater than the design width. It is less costly to pile sand on the beach and let nature work to distribute the sand beneath the waves. We need full wave years' of data to monitor the beach as it reduces in width to what was designed. Beach nourishment projects are "infrastructure" that requires annual maintenance similar to roadways that are maintained to drive to the beach.

And third, "... *beach nourishment is a waste of taxpayer money and beach nourishment is an endless expense*". Beach nourishment is the "soft" alternative for coastal storm damage mitigation benefits. Wide sandy beaches save the taxpayer's money by reducing the federal governments' bailout with disaster relief funds after major hurricane and tropical storm events. Recall Superstorm Sandy (2012) on the East US coast (Seabrook, 2013). Beaches also generate recreation and tourism benefits for the local region and for the ecosystem. Total benefits far exceed costs. Renourished beaches are an endless source of income for governments (federal, state, and local) and they are not an *endless expense.*

In short, "what you (the general public) see is not what you (the general public) get". You, as the designer of a beach nourishment project are the best qualified source for informing the general public via the media as to how beach nourishment projects work. As the author of this book, my advice is for you to take the time needed to get to know the local, radio, TV, and newspaper journalists and let them help you to get this message across. One important first step is to be ready with the facts of the amount of the mitigated damages after the first storm hits your beach nourishment project. One example is Virginia Beach, VA, after Hurricane Isabel, September 18, 2003. The estimated damage mitigation benefits from this one storm alone were $82 million (Roehrs, 2003) resulting from this $125 million project completed only months before in April 2003.

The general public needs to routinely hear about "*when things go right*" with their taxes invested in Corps of Engineers (federal government) beach nourishment projects. Each Corps of Engineers' district office around the country has trained personnel who are responsible for public relations and media education regarding beach nourishment projects. After every coastal storm, the Corps should routinely release a bulletin for the media regarding the damages prevented (dollar amounts saved) by having the beach nourishment project in place. The smaller storms have less damage prevention (mitigation) benefits but occur more frequently so that the dollar amounts of benefits add up over time. If the media (radio, TV, newspapers, and social) were routinely informed of the dollar benefits after each storm (no matter how small) and were to spread this information, then the general public would begin to appreciate and understand the true value of their taxes spent on beach nourishment projects. The media always tell "*when things go wrong*" but seldom mention "*when things go right*".

7.3.7 *Beach-fx software*

In the 1990s, the US Army Corps of Engineers developed a numerical model for desktop computers to evaluate how beaches respond to storm sequences and then to evaluate property damages resulting from water-level inundation, waves, and erosion. The product is called Beach-*fx* and is freely available from the Corps' Engineering Research and Development Center (ERDC) and the Coastal Hydraulics Laboratory (CHL). Both the beach nourishment design (berm width and berm height) as connected to an equilibrium beach profile, and dune design (width, height, slope) in front of a landward boundary are combined for design as illustrated in Figure 7.37 (from Larson and Kraus, 1989). The 2D profile shape is defined by the $z(y, t)$ elevation relative to the vertical datum.

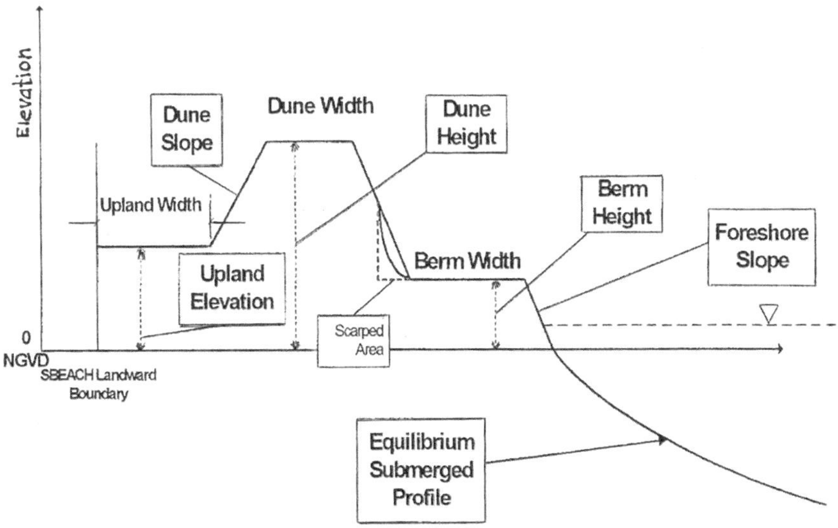

Figure 7.37. Simplified cross-shore profile employed in process model SBEACH (Larson and Kraus, 1989).

A range of plausible storms (Monte Carlo simulations) are applied to design beach and dune profiles using the process model SBEACH (Larson and Kraus, 1989) with the results stored in a database. This stored information is then applied to driving parameters (inundation, wave, erosion) versus percent damage curves to calculate the damage costs to individual property in the Geographic Information System (GIS) framework. Benefits of damages prevented versus costs for each beach/dune geometry considered are the basic final products of the Beach-*fx* tool that allow the designer to select the most appropriate alternative for construction. But this same process could also be employed after construction to determine the damage prevention (mitigation) benefits for media distribution.

The US Army Corps of Engineers has done a good job of developing software planning tools for beach nourishment project design. It could also be used to inform the public of the success of the project in saving the tax-payers' money each time a coastal storm impacts the beach.

7.3.8 *Summary*

The placement of large quantities of "good-quality" sand on the beach to advance it seaward is a simple definition of beach nourishment. The renourished beach is a sand reservoir available to reduce the wave energy striking the coast during storm events. Renourished beaches are designed as coastal "infrastructure" and are the "soft" alternatives for storm damage mitigation. Natural erosion, man's interference, or extreme storm events diminish the reservoir volume.

Dunes adjacent to the beach are a large reservoir of sand and also coastal infrastructure for storm damage mitigation from flooding. Dune design is considered in Section 7.4.

7.4 Dune Design

7.4.1 *Introduction*

If beaches are a "river" of sand (Inman and Shelton, 1967), then dunes are the "reservoir" of sand that feed the river. Where natural dunes are present landward of the beach, they serve as a major reservoir of sand that is available when elevated water levels in storm surge events bring waves to attack the dune face. The elevation and thickness of the dune cross-section may (or may not) be sufficient to prevent flooding behind the dune.

Figure 7.38 displays what happened to a natural dune at Nags Head, NC (US) during Hurricane Isabel (September 18, 2003). The dune cross-over structure and people give some idea of the scale for the large volume lost in the dune between the entrance section, the missing center section, and the steps down to the beach. The dune front face is almost vertical, and the crest elevation is over 6 m (20 ft) at this location. The beach berm elevation is also lower after the storm. The storm surge during Hurricane Isabel at the Corps of Engineers, FRF at Duck, NC (28 km from Nags Head) peaked at 1.26 m (4.1 ft) above MHHW and was the highest ever recorded at Duck since 1980 when the FRF was first built.

This section reviews the structural design of dunes, the breaching of barrier islands, and dune environmental/aesthetic issues.

7.4.2 *Structural design of dunes*

7.4.2.1 *Beach-dune erosion models*

Because much of the country lies below sea level, the Netherlands has a long history of reliance on coastal dunes to prevent inundation and flooding. The probabilistic design of dunes for dune erosion was evaluated by van de Graff (1989). Seven parameters (surge level, wave height, particle size, initial profile, surge duration, surge spikes, and computation accuracy) that contributed to the variance (square of standard deviation of mean) for a

Figure 7.38. Major dune volume lost in Hurricane Isabel (September 2003) at Nags Head, NC (US) (photo credit, USGS Coastal and Marine Science Center, St Petersburg).

data reliability function were analyzed. It was learned that storm surge elevation hydrograph $S(t)$ accounted for over 83% of the total variance associated with dune erosion. All the other variables were far less important so that the change in dune cross-sectional area (per unit width of shoreline) was primarily related to elevated water levels during storm surge events.

An analytical model for the time-varying response of a beach-dune profile has been developed by Kriebel and Dean (1993) and has become known as the "convolution method". The general solution of the time-dependent shoreline response $R(t)$ can be expressed as a convolution integral. The method is based on field and laboratory observations that beaches have a "memory" so that the profile at any time is related to the profile at the preceding time; the new profile in time "lags" behind the conditions causing the profile change (response) and the profile response varies "exponentially" so that the profile response is "damped". In other words, the maximum potential equilibrium profile response R_∞ is less than the full erosion potential of the forcing water level and wave conditions.

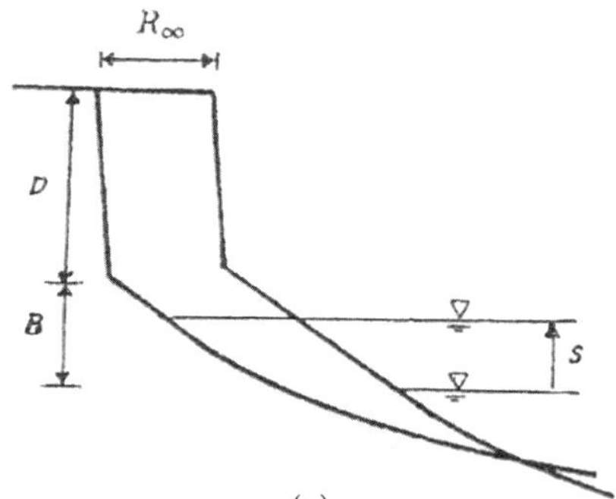

(a)

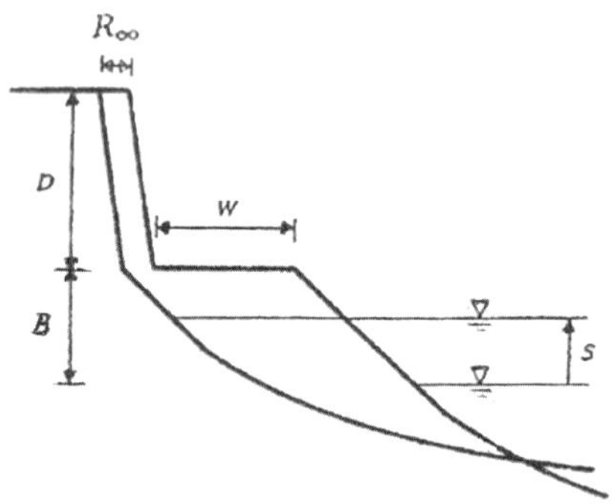

(b)

Figure 7.39. Schematic diagram of two cases for analytical model of beach-dune erosion profile response from Kriebel and Dean (1993). Case (a) is dune with no beach berm and case (b) is for dune with wide beach berm.

Solutions to the convolution integral presented in Kriebel and Dean (1993) were for an equilibrium beach profile with an infinite slope, a square berm, and a linear foreshore slope at the shoreline and for the response of an equilibrium profile *with dunes* as illustrated in Figure 7.39. Case (a) is with no beach berm width in front of the dune and case (b) has a beach width W in front of the dune.

Solutions for the dimensionless shoreline response $R(t)/R_\infty$ with examples can be found in Kriebel and Dean (1993) and in CEM (2006, Part III, Section 3, p. 54). The solutions are for an idealized storm surge hydrograph $S(t)$ of the form

$$S(t) = S\sin^2(\sigma\, t), \tag{7.67}$$

where σ is π/T_D and T_D is the total storm surge duration.

However, solutions are only possible for $S_{\max}$ less than the beach berm height B which makes the analytical solutions less useful, since dunes erode most when the storm surge water levels are above the beach berm heights at the toe of the dune. Numerical models do not have this limitation.

As discussed in Section 7.3.7, the Corps of Engineers numerical, 2D, empirically based model for beach-dune erosion studies is called SBEACH (Larson and Kraus, 1989) that employs the simplified, cross-shore profile shown in Figure 7.37. The 2D profile shape is defined by the $z(y,t)$ elevation relative to the vertical datum. Both the beach nourishment design (berm width and berm height) as connected to an equilibrium beach profile and dune design (width, height, slope) in front of a landward boundary are combined for design. A 2D, process-based numerical model called XBeach has been developed by the consulting firm Deltares (formerly WL/Delft Hydraulics). It is a public-domain model with open-source community website (https://oss.deltares.nl/web/xbeach/home). Both model developers and users find this website helpful for latest improvements and events surrounding the model. Both SBEACH and XBeach are popular numerical models for dune design.

Elevated water levels and storm wave conditions attack the dune front and may cause wave overtopping to erode the rear dune slope. The dune cross-section is flattened in both landward and seaward directions to possibly permit complete dune crest submergence below the maximum storm surge elevation. As a result, inundation of the land area behind the dune and coastal flood damage may result. The US Federal Emergency Management Administration (FEMA) has developed a relatively simple criteria for the minimum dune cross-sectional area to prevent coastal storm surge events from breaking through the dune and flooding the land area behind the dune.

7.4.2.2 *FEMA 540 rule*

The FEMA is responsible for preparing flood elevation maps for flood risk management in the US. At the time of this writing, the elevation of the 1% annual chance water level (100-year recurrence interval) is the standard for flood insurance rate maps (FIRMs) for property insurance within the floodplain boundaries. Where coastal dunes are present, Hallermeier and Rhodes (1988) studied the hurricane and extra-tropical storm events along the US Atlantic coast, the Gulf of Mexico coast, and the Netherlands and plotted the unit width dune erosion volume (cy/ft) above the maximum storm, still-water elevation (SWEL, no wave setup) versus the estimated recurrence interval (years) that produced that dune volume change as shown in Figure 7.40. It was found that a simple, straightline fit to the data gave an erosion volume of 20 cy/ft (540 ft^3/ft) for a 100-year recurrence interval as illustrated in Figure 7.40 (in the metric system, 20 cy/ft equals 50 m^3/m).

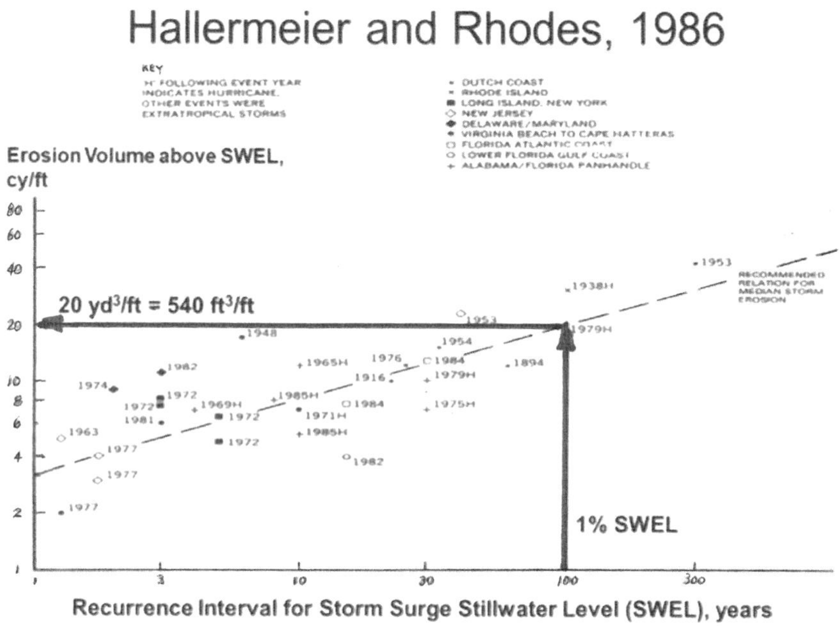

Figure 7.40. Data for development of the FEMA 540 Rule from Hallermeier and Rhodes (1986).

FEMA (1988) defined the "primary frontal dune" as illustrated in Figure 7.41 and established the 540 Rule in its "*Guidelines and Specifications for Flood Hazard Mapping Partners: Appendix D*" (FEMA, 2007). If the primary frontal dune cross-sectional area is equal to, or less than 540 ft^2 above the 1% annual chance storm surge, the presence of the dune is removed in numerical models for flooding studies. The continuous or nearly continuous mound or ridge of sand adjacent to the beach was considered to not be an effective barrier for preventing flooding to upland areas behind the dune for the 100-year recurrence interval storm that is the basis for FIRMs developed by FEMA. This greatly simplified the questions surrounding the effectiveness (or not) of sand dunes. Simply stated,

> *To prevent dune removal during the 1-percent-annual-chance storm, the frontal dune reservoir must typically have a cross-sectional area of at least 540 square feet (or 20 cubic yards volume per foot along the shore) above the 1-percent-annual-chance SWEL (Storm Surge Stillwater Elevation) without wave setup* (FEMA, 1988).

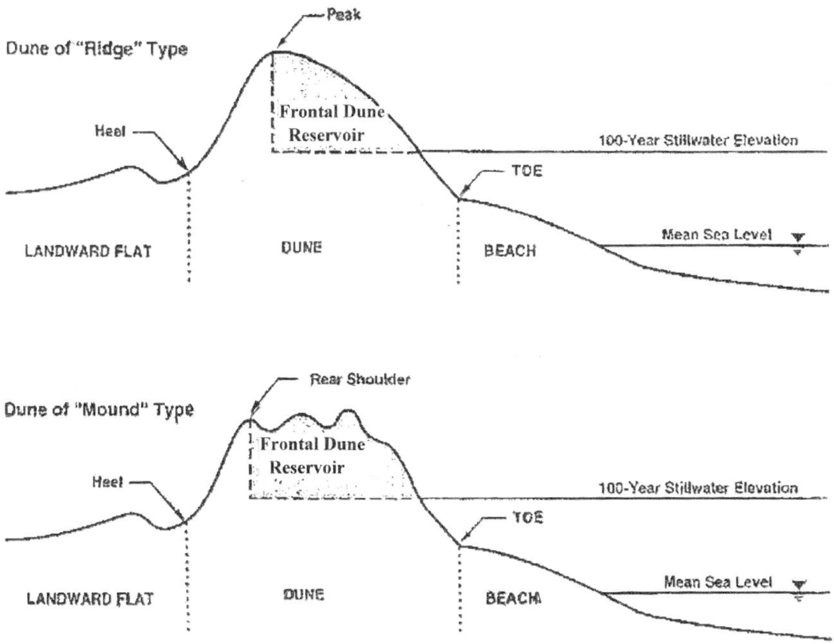

Figure 7.41. Schematic diagram of the "primary frontal dune" for application of the FEMA 540 Rule (from FEMA, 2003, Appendix D).

The 540 Rule continues to be applied for mapping studies on the Atlantic Ocean and along the Gulf of Mexico coasts of the United States.

To test the FEMA 540 Rule, Basco and Payne (2010) employed two real dune/beach profiles located about 13 km (8 mi) south of the Corps' FRF at Duck, NC as shown in Figure 7.42. The survey transects were obtained from FEMA and are about 0.8 km (0.5 mi) apart with cross-sectional areas as shown in Figure 7.43(a) Transect 159 and Figure 7.43(b) Transect 162 with frontal areas 643 ft^2 and 446 ft^2, respectively, above the 1% annual chance, SWEL with no wave setup. These two profile locations are close to where wave and water levels are routinely monitored by the Corps of Engineers; they are near each other so that they both experience similar storm conditions, and they have frontal dune areas that are above (Transect 159: $643 > 540$) and below (Transect 162: $446 < 540$) the FEMA rule. The 1% annual chance SWEL is 6.8 ft (NAVD88 vertical datum) without wave setup (NFIP, 2006).

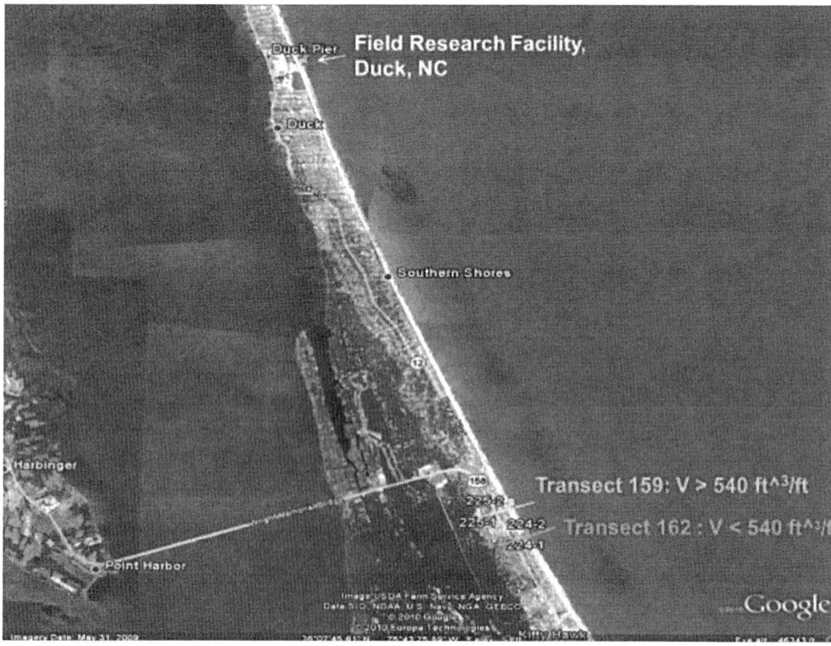

Figure 7.42. Location of Transects 159 and 162 for study of the FEMA 540 Rule (from Basco and Payne, 2010).

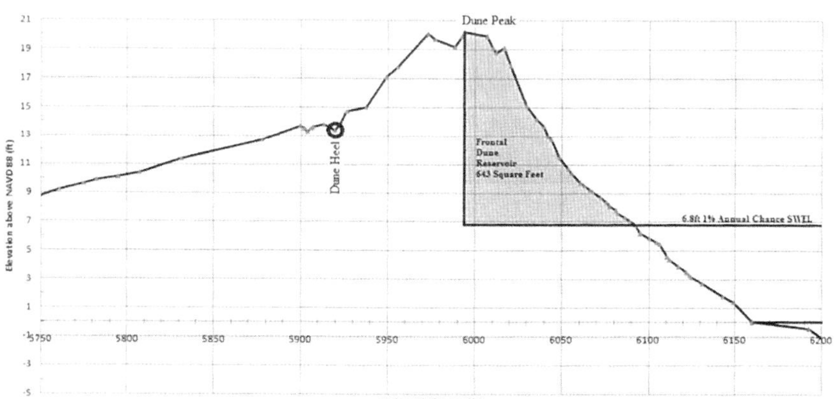

Figure 7.43(a). Transect 159 with frontal dune area = 643 ft^2 above the 1% annual chance SWEL.

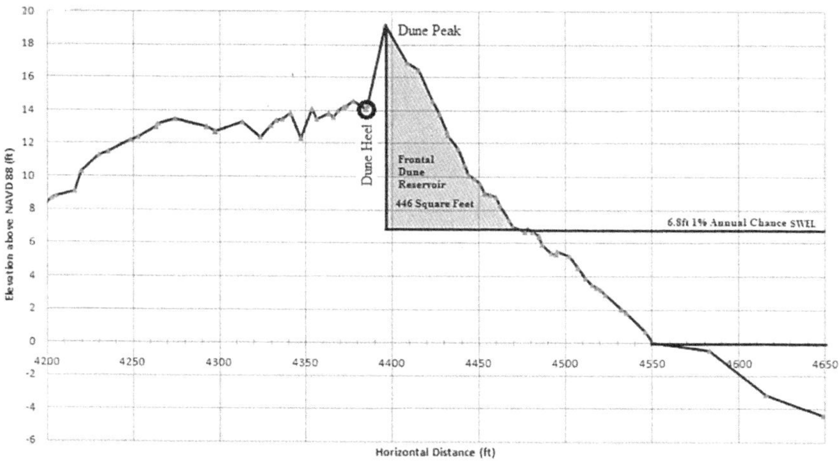

Figure 7.43(b). Transect 162 with frontal dune area = 446 ft^2 above the 1% annual chance SWEL.

The numerical SBEACH model (Larson and Krauss, 1989) was employed in the analysis. Both storm surge hydrograph $S(t)$ and storm wave conditions (significant wave height $H_s(t)$, peak period $T_p(t)$, and storm duration t_d) over the storm time history t are needed in the SBEACH model. Six coastal storms for both hurricanes (Bonnie 1998, Floyd 1999, Isabel 2003) and extra-tropical storms (Twin Northeasters 1998, NorIDA 2009, and Twin A 2008) were modeled. All peak storm water elevations were adjusted to match the 6.8 ft, 1% annual chance SWEL but the wave data were those actually measured at the Corps' FRF at Duck for that particular storm event.

For Hurricane Isabel (September 18, 2003), the profile at Transect 159 (dune area > 540 ft^2) experienced an erosion of 468 ft^2 and proved the FEMA 540 Rule to be correct since the dune cross-section remained intact so that no flooding occurred landward of the dune as shown in Figure 7.44(a). Similarly, for Hurricane Isabel, the profile at Transect 162 (dune area < 540 ft^2) experienced an erosion of 497 ft^2 and proved the FEMA 540 Rule to again be correct since the dune cross-section was completely eroded away as shown in Figure 7.44(b) so that flooding occurred landward of the dune. Similar results for the other five storms also proved the FEMA 540 Rule is correct except at Transect 162 for Hurricane Floyd which had the lowest storm duration of only 7.1 hours. Basco and Payne (2010) concluded that since the FEMA 540 Rule does not consider storm

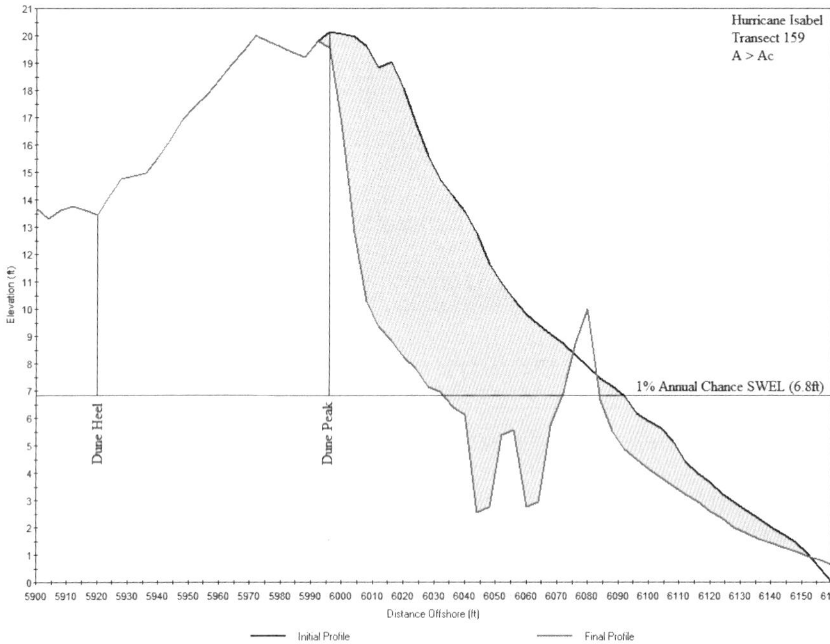

Figure 7.44(a). Results from SBEACH model for Hurricane Isabel at Transect 159 showing the dune erosion area and the cross-section remaining intact to validate the FEMA 540 rule.

duration for fast-moving hurricanes with short storm durations, the FEMA 540 Rule may be conservative. For this one example, the timescale for full dune erosion may not be long enough when the dune area $<540\,\text{ft}^2$.

The FEMA 540 Rule is the *minimum* standard since $540\,\text{ft}^2$ per linear foot of dune volume (ft^3/ft) above the 100-year SWEL is a *median* number. Other investigations have learned that the maximum erosion volume can be 1.5 to over 6 times larger than the median volume (Birkemeier *et al.*, 1988). As a result, FEMA (2000), *Coastal Construction Manual* recommends increasing the critical area to $1{,}100\,\text{ft}^2$ (40.7 cy/ft) which is over double the median number. The reasons are that (1) the median number reflects the dune size at the time of mapping and not future conditions; (2) the median number does not reflect cumulative effects of multiple storms; and (3) using a median volume means that 50% of the time the value is underestimated in a statistical sense. In addition, shoreline change rates (erosion) and the dune condition should also be considered. The state of New Jersey in the US requires more frontal dune volume than the minimum

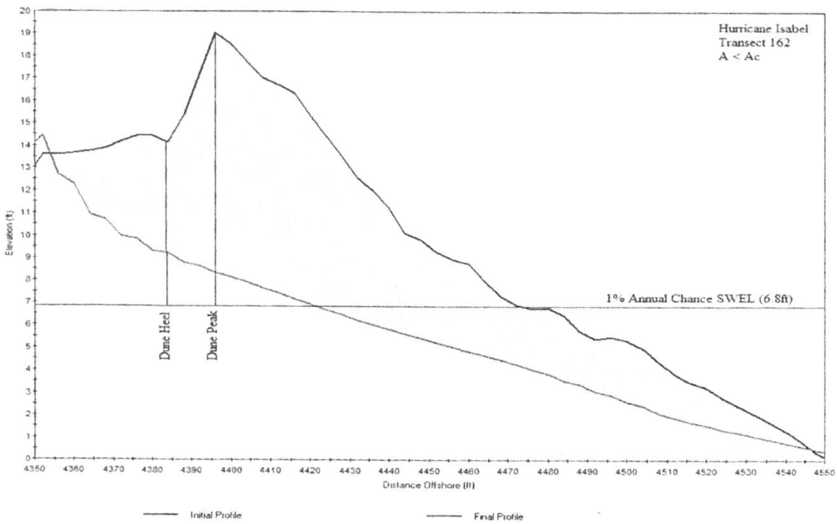

Figure 7.44(b). Results from SBEACH model for Hurricane Isabel at Transect 162 showing the dune completely eroded away to validate the FEMA 540 rule.

standard. We herein recommend using 30 cy/ft ($\approx$800 ft^2) for the critical value.

The probability distribution for water levels at the site is nonlinear with far more times each year that low levels reach the toe of the dune with wave energy to cause dune erosion. Dunes with damage mitigation structures at the toe or within the dune can be effective to strengthen the design of dunes.

7.4.2.3 *Enhanced dune design*

Sand fences, geotextile sand containers (GSCs), rip-rap (rock), and concrete materials have been placed on the beach in front of dunes (toe) to mitigate scour and erosion during minor storm events. Sand fences interrupt the blowing sand across the beach berm causing sand to fall to the surface. The sand deposits build up the frontal area of the dune adding to the reservoir of sand volume available for seaward movement in storms. Multiple rows of sand fencing at different elevations on sloping dunes have also been effective.

Revetment design using GSCs was reviewed in Section 6.4.3, Chapter 6. The same principles apply in that the sand-filled containers resist wave loads that cause toe scour of the dune. When the erosion is excessive, it can produce a complete collapse of the dune face. Similarly, natural rock and concrete materials (concrete segments of roads) offer increased resistance to bigger storm events and can be constructed. However, the elevation of these hardened materials placed in front of the dune face is only a fraction of the total crest elevation of the dune. They are not full slope revetments and their volume (cross-section and crest elevation) is usually an economic constraint.

Major storm events can overtop these low-elevation, hardened structures and cause significant erosion of the dune cross-section. To increase the resilience of dunes, hardened structures (GSCs, rock, and sheet-pile/bulkheads) can be constructed *within* the cross-section of the dune. When a major storm occurs, these interior, hardened structures serve as a second line of defense to prevent the complete washout of the dune as seen in Figure 7.44(b). The location and crest elevation of these interior structures must be carefully considered.

As an example, the US Navy, Fleet Combat Training Center on the Atlantic Ocean at Dam Neck, Virginia (near Virginia Beach, VA) contains the last 20-inch gunnery range in the US and nearshore structures that are continually threatened by historic erosion (0.7 m/year) and serve storm surge events. The Navy requested that a hard, concrete revetment be designed but was overruled when the consulting engineers proved that a beach nourishment, rebuilt dune with buried seawall design was the least life-cycle cost alternative over the 25-year design life (Basco, 2000). The design dune cross-section was that originally designed by Headland (1991, 1992) and is shown in Figure 7.45 (top, left). It was 30 cy/ft ($75\,m^3/m$) above the 1% annual storm surge, i.e. 50% more than the FEMA 540 Rule. The original idea for an enhanced dune design with buried, rock structure within the dune is credited to Headland (1992). The location of the buried seawall is well back beneath the dune in the rear quarter section with crest elevation 4 ft (1.2 m) below the dune crest. The location was to allow over 50% of the dune section in front of the buried seawall to remain as an active reservoir of sand during storm events. The crest of the rock structure was buried so that it would not become an environmental detraction and constraint to the aesthetic aspects of the design.

A key component in the cost analysis was the determination of dune damage curves to estimate dune maintenance costs (Basco and Shin, 1999). The SBEACH model was employed and it was learned that a buried seawall within the dune section significantly prevented dune erosion for storm surge

Figure 7.45. The Dam Neck beach-dune-buried seawall cross-section (top, left), the rock seawall under construction (top, right), the dune with buried seawall under construction (bottom, left), and the finished dune with grass plantings (bottom, right) (Author's photos, 1996).

events that were 80% and higher than the design storm surge level. Even less erosion was found with a wide beach nourishment in front of which was an added component that was not in the original design of Headland (1992). Figure 7.45 (top, right) shows the rock seawall under construction in front of the Enlisted Men's Club, and Figure 7.45 (bottom, left) is a photograph of the buried seawall beneath the dune that is still under construction. The finished dune was planted with dune grass as seen in Figure 7.45 (bottom, right) in front of the BOQ. Everyone seeing the dune is unaware of the buried rock structure beneath the dune.

Hurricane Isabel (September 18, 2003) provided a true test for the combined beach nourishment, rebuilt dune with buried seawall system. Roughly 50% of the front dune section was eroded and the stones at the front (toe) of the rock structure were exposed since the beach berm elevation was also lowered by storm waves dragging sand offshore. The beach was renourished, and the dune was rebuilt to design template dimensions in

2004 and continues to mitigate Atlantic Ocean storm damage today for the over $100 million value of the buildings and gunnery range located behind the dune.

7.4.3 *Barrier island breaching*

Major hurricanes in recent years along the Atlantic Ocean and Gulf of Mexico coastal barrier islands in the United States have produced numerous breaches of the relatively narrow landmass separating the ocean and the back bay. The frequency of these breaching events may increase with accelerated sea-level rise in the future. Roads along these barriers provide the only link for lifeline evacuation, emergency vehicles, daily transportation (schools, commerce, etc.). When breached, serious consequences result.

When all profile elevations $z(y,t)$ along the beach-dune cross-sectional profile fall below MLW as a result of storm erosion at the site, a "breach" in the profile has occurred. Subsequent normal tidal conditions following the storm keep the breach open and a new tidal inlet may form in time if longshore and cross-shore sediment transports are not sufficient to naturally close the breach. As illustrated in Figure 7.46, at the end of a coastal storm,

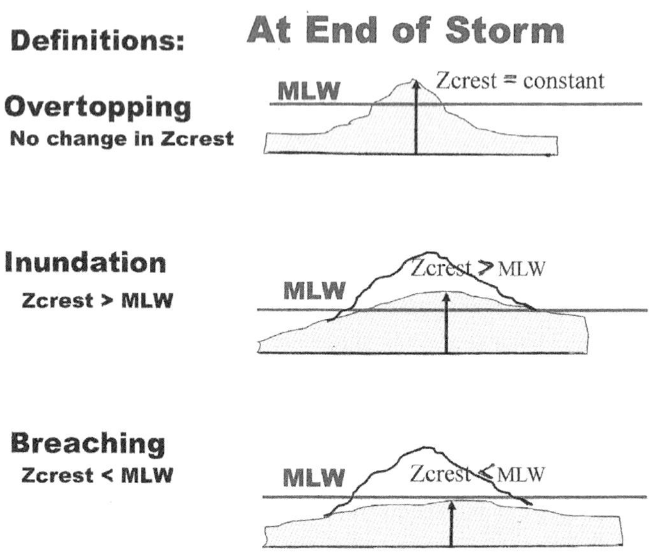

Figure 7.46. Illustration of three conditions of dune/beach profiles at the end of coastal storms regarding the elevation of the highest position z_{crest} relative to the MLW elevation. Breaching is when $z_{\text{crest}} <$ MLW at the end of the storm.

the dune crest may be overtopped by waves but the crest elevation z_{crest} remains unchanged (top). The storm surge may inundate the dune crest but the reduced crest elevation $z_{crest}(t)$ is still above the MLW tidal level (middle). A barrier island breaching event is when the final, lowered elevations $z(y,t)$ along the profile are all below MLW as illustrated in Figure 7.46 (bottom).

Figure 7.47 illustrates a barrier island inundation event (not breaching) after Hurricane Alex (August 5, 2005) on Cape Hatteras Island, NC in the US. The clear distinction between inundation and breaching is important for research of the physical processes (water levels, waves, sediment transports, and profile changes) that must be calculated in numerical models.

Figure 7.48(a) schematically depicts the plan views of a barrier island–bay system with the existing tidal inlet and location of a potential new inlet. In Figure 7.48(b) when the barrier (black) remains above the storm surge elevation, the return flow from the bay must flow back out the existing inlet. However, as illustrated in Figure 7.48(c) when the entire barrier is inundated below the storm surge flood elevation, then the bay water return flow to the ocean may also take place across the top of the barrier island. The lowest barrier island elevation and narrowest width locations are then susceptible to breaching.

Figure 7.47. Inundation of the barrier island at Cape Hatteras Island, NC (US) from Hurricane Alex, August 5, 2005 (from the Virginian Pilot, Drew C. Wilson Photos).

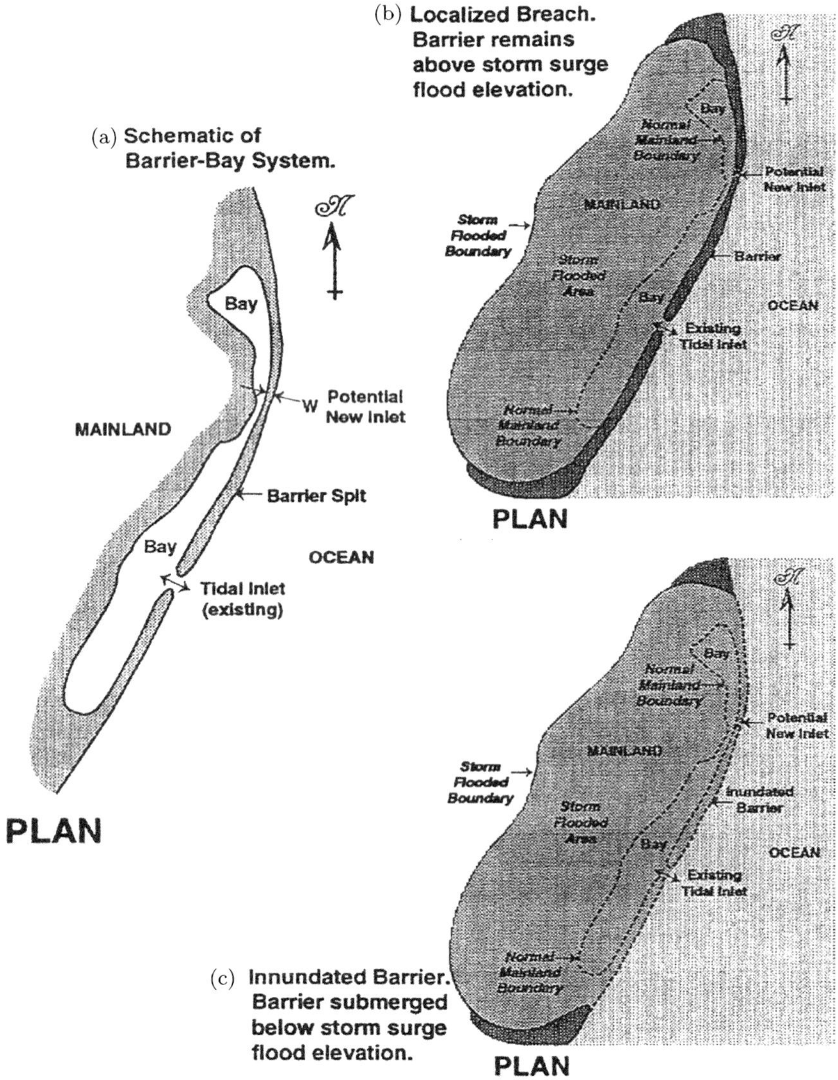

Figure 7.48. Schematic plan views of (a) barrier island–bay system, (b) coastal storm flooding bay but not overtopping barrier island, and (c) coastal storm surge flooding bay and inundating barrier with potential for breaching and new tidal inlet.

7.4.3.1 *A one-dimensional numerical model*

One of the earliest process-based numerical models for studying barrier island breaching was developed and summarized by Basco and Shin (1999). A unit width of the Sandbridge Beach, VA coast with profile shown in Figure 7.49 was modeled from the ocean side (left) to the bay side (right). The top (a) profile is shown with distorted vertical scale to illustrate the key physical and water elevation variables and the bottom (b) profile is undistorted to graphically demonstrate the true, fragile nature of barrier islands. At this location the barrier width $W \approx 500\,\text{m}$ (1,650 ft) with dune crest elevation $D_c \approx 3\,\text{m}$ (10 ft) above the mean lower low-water (MLLW) elevation. The sand grain size $D_{n50} \approx 0.23\,\text{mm}$.

Four stages of the wave, water flow, and sediment transport processes were addressed, namely: (1) dune/beach erosion, (2) overwash/overland flow, (3) storm tide flooding, ocean to bay and (4) storm-tide ebbing,

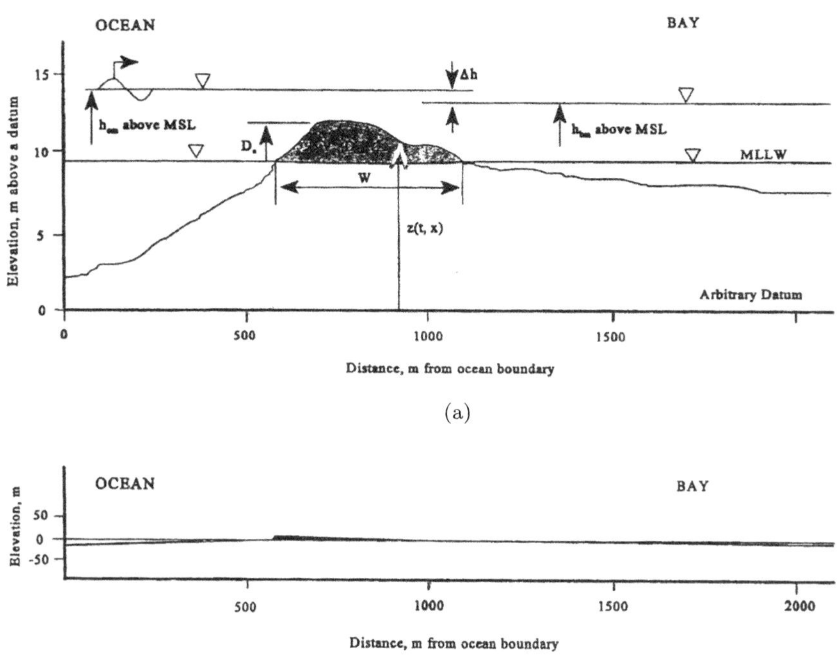

Figure 7.49. Cross-section at Sandbridge Beach, VA employed by Shin (1999). (a) Distorted scales and (b) Undistorted scales. The (a) top section with distorted vertical scale and (b) the bottom section with undistorted vertical scale.

bay to ocean. The SBEACH numerical model was used for stage (1), the open-channel flow equations in conservation form using a Lax–Wendroff scheme (supercritical flows) for stage (2) and the finite difference (Priessmann scheme) for the Eulerian form equations of open-channel flow for stages (3) and (4). See Basco and Shin (1999) for complete details including the sediment transport equations and the formulation of the bottom profile change equation. Dune erosion in stage (1) lowered the dune crest until storm surge inundated the dune and supercritical, open-channel flow quickly reached the bay in stage (2). These first two stages were not critical to produce breaching. Subcritical flow from ocean to bay in stage (3) and then the return flow in stage (4) from bay to ocean both lowered the dune profile.

Figure 7.50 shows one example of storm surge hydrographs on the ocean side (solid, $h_{om} = 5.0\,\text{m}$) and bay side (dotted, $h_{bm} = 3.0\,\text{m}$) giving $\Delta h = 2.0\,\text{m}$ with a time lag $t_{\text{lag}} = 3\,\text{hours}$. When applied to the profile in Figure 7.49, the model produced the resulting profile change in time t as illustrated in Figure 7.51. Note that the center of mass of the barrier island (centroid) first moves landward and is lowered in time for stage (3) when ocean water fills the bay. Breaching occurs during stage (3) at $t = 6{,}000\,\text{s}$ (1.67 hours) for this example. Then the profile continues to

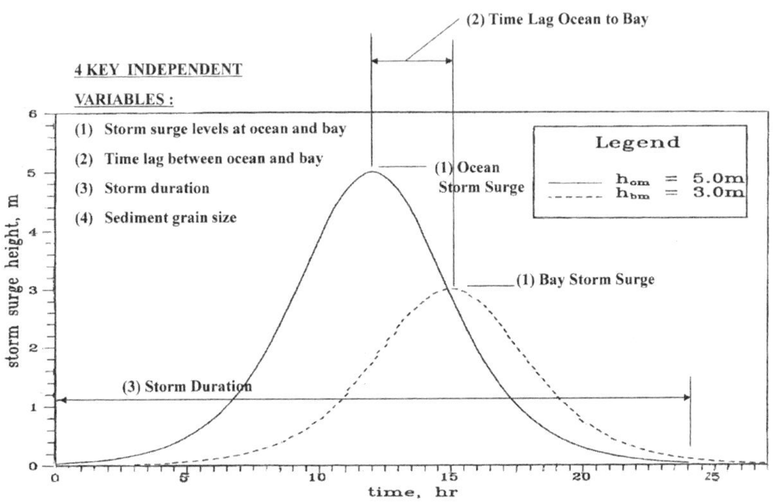

Figure 7.50. Ocean and bay storm surge hydrographs for one simulation of the 1D model (Basco and Shin, 1999).

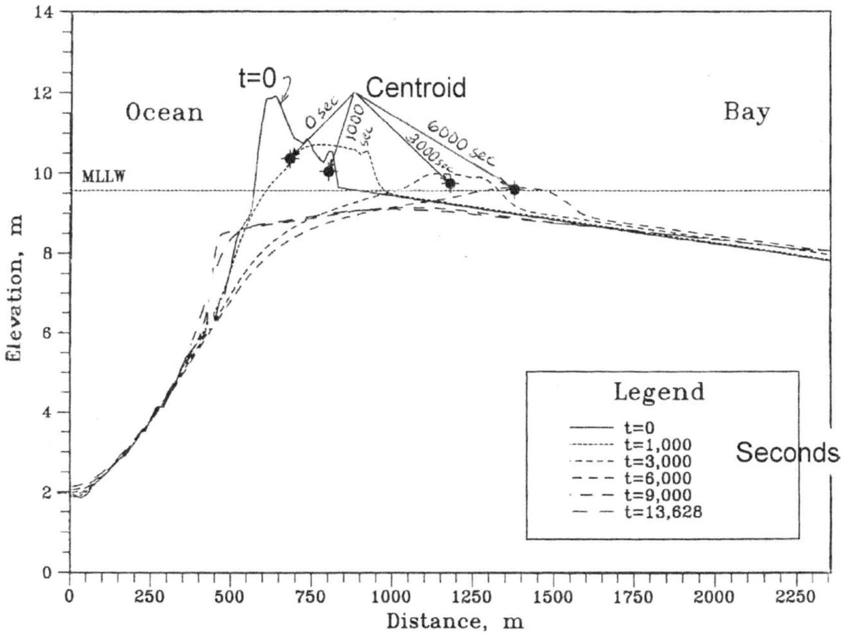

Figure 7.51. Profile change in time for the storm surge hydrographs shown in Figure 7.50 (Basco and Shin, 1999).

flatten and becomes lower beneath the MLLW elevation in stage (4) when the bay water returns to the ocean. The profile shape remains relatively the same from t = 9,000 s (2.5 hours) to t = 13,628 s (3.79 hours) of the dynamic simulation.

In short, it was learned that

- the dune erosion model for waves was not important,
- the peak storm surge elevation difference from ocean to bay is a key variable,
- sediments are transported in the landward direction to create a breach, and
- grain size greatly influences the results.

The peak storm surge difference Δh between ocean side h_{om} and bay side h_{bm} and the time lag t_{lag} between the peak storm surge levels were the key variables in producing a breaching event. Shorter duration storms produced strong ebbing flows and faster barrier retreat speeds. Median grain size of 0.4 mm produced maximum volume loss and retreat speed. In general, most sediment was transported landward by large peak storm surge differences with short time lags and longer duration storm events.

The main coastal processes that are modeled in Basco and Shin (1999) are essentially only the result of one-dimensional (1D), depth-averaged, open-channel flow and can be considered a 1D model. The profile elevation $z(y,t)$ varies in one-space dimension y and time t in this 1D model. In reality, the area elevation $z(x,y,t)$ varies to two horizontal space dimensions (x,y) in time.

7.4.3.2 *Delft 3D breaching model*

The original Delft 3D numerical model was developed by Delft Hydraulics (now Deltares) and is available as an open-source code (https://oss.deltares.nl). It is composed of four modules (WAVE, FLOW, SED, and MOR) for the wave dynamics, water flows, sediment transports, and morphological changes, respectively.

For barrier island breaching simulations in the early 2000s, Delft 3D employed an empirically derived breaching algorithm (van Kessel and Roelvink, 2002; Roelvink *et al.*, 2003) as illustrated in Figures 7.52(a) and 7.52(b). The breaching process moves sediment p_s in the main water-flow x-direction, but primarily causes slumping q_s in the lateral y-direction as the breach width $B(t)$ increases in time. Figure 7.52(a) presents the 3D geometry with an initial slice Δy where breaching begins. The empirical relationship between the volume of sediment eroded in the x-direction relative to the lateral volume removed (slumping) in the y-direction as the breach width, $B(t)$ grows over time step Δt is illustrated in Figure 7.52(b). Mass must be conserved so that

$$\text{Volume Dry}_{\text{removed}} = K\,(\text{Volume Wet}_{\text{eroded}}), \tag{7.68}$$

where K is the empirical coefficient $= 0.9$ and one wet/dry cell interface $= 0.45$ for two wet/dry cell interfaces.

The empirical coefficient K was found from field experiments as illustrated in Figure 7.53 reported by Visser (1998) with data normalized, labeled Zwin 94 (blue), and plotted in Figure 7.54. Using $K = 0.9$ in the Delft3D-MOR module produced the black curve for $B/B_{\max}$ versus $t/T_{\max}$ in Figure 7.54 and reasonably close to the Zwin 94 data. The field experiment results reported by Visser (1998) were conducted in the Zwin tidal estuary in the Netherlands in 1994. The maximum breach width $B_{\max}$

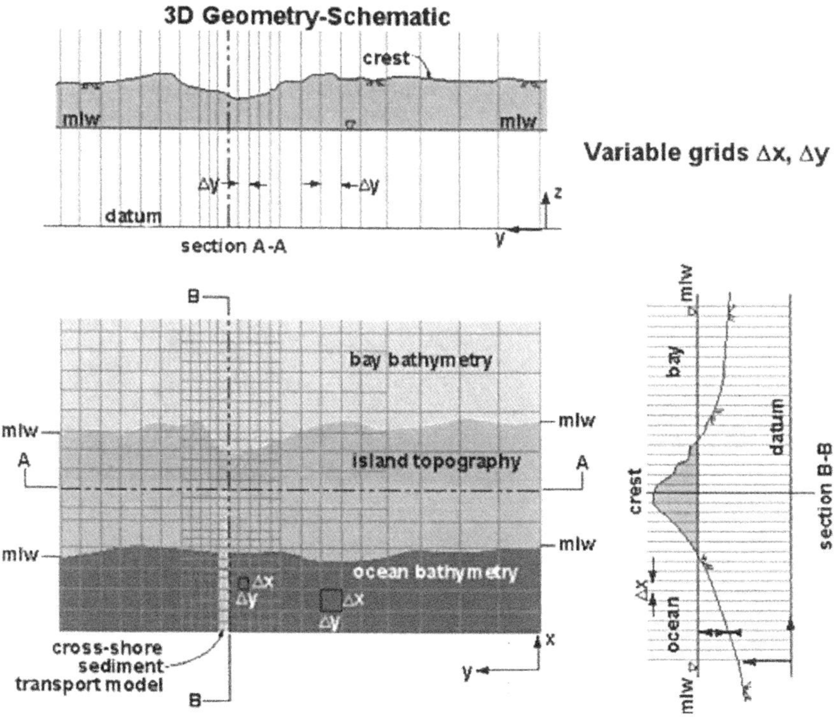

Figure 7.52(a). Three-dimensional schematic geometry of a barrier island with variable grid (Δx, Δy) (from Roelvink *et al.*, 2003).

was reached at time, $T_{\max}$. Other experiments gave similar results in the laboratory (Caan, 1996) and in the field (Rosov, 2003).

A theoretical explanation for the empirical relationship between the dry volume removed during the breach and wet volumes of sediment eroded in the breach was given by Basco (2004) as illustrated in Figure 7.55. The conservation form of the conservation of sediment momentum in the y-direction is shown. It is analogous to a similar expression for the conservation of water momentum and seldom discussed in the literature. It is for the unsteady transport rate of sediment $\partial q_s/\partial t$ on a level bed of sediment on the bottom and neglecting any bed shearing stresses. The spatial gradient of sediment momentum in the y-direction $\partial(q_s^2/h_s + gh_s^2/2)/\partial y$ is a moving sediment "jump" for the breaching process and analogous to a moving hydraulic jump. The speed of the moving sediment jump is the speed of the breach, $\partial B/\partial t$. If the sediment momentum in the sediment

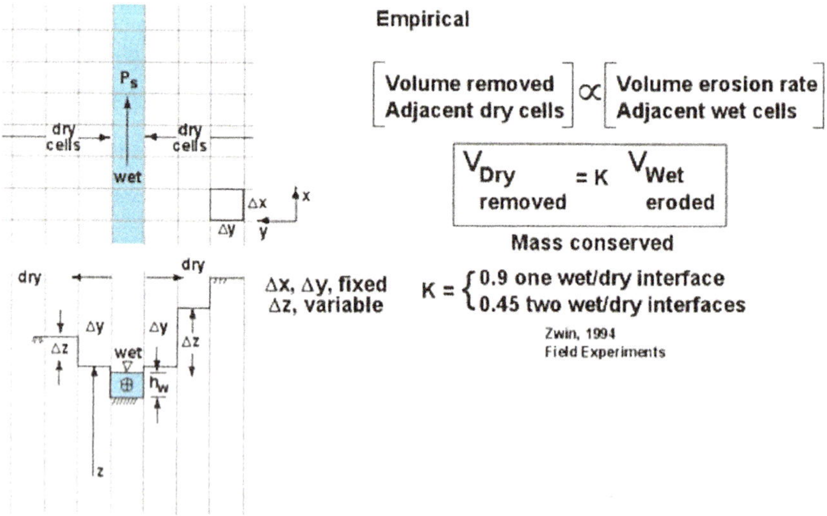

Figure 7.52(b). Empirical relationship between volume of sediment eroded in x-direction, p_s, relative to lateral volume slumping in y-direction, q_s, as breach width $B(t)$ grows over time step, Δt (from Roelvink *et al.*, 2003).

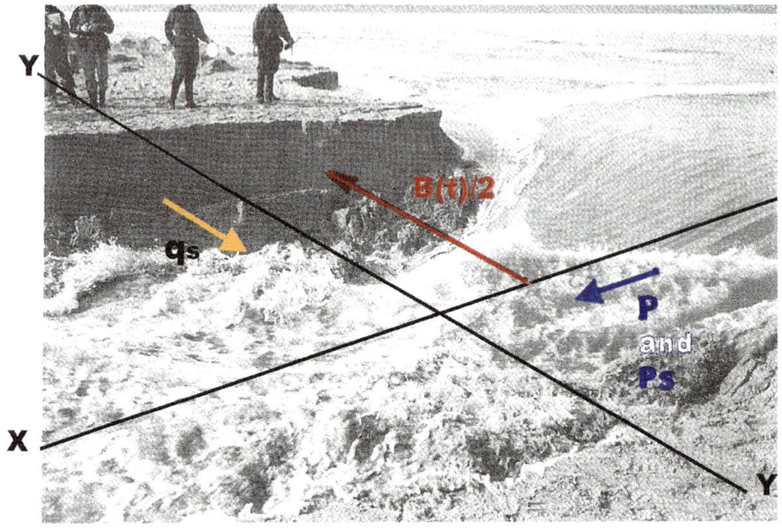

Figure 7.53. Photograph of breaching experiment showing water volume flux p and sediment flux p_s in the x-direction and sediment volume flux q_s in the y-direction coming from both sides of eroding channel width, $B(t)$ (from Visser, 1998).

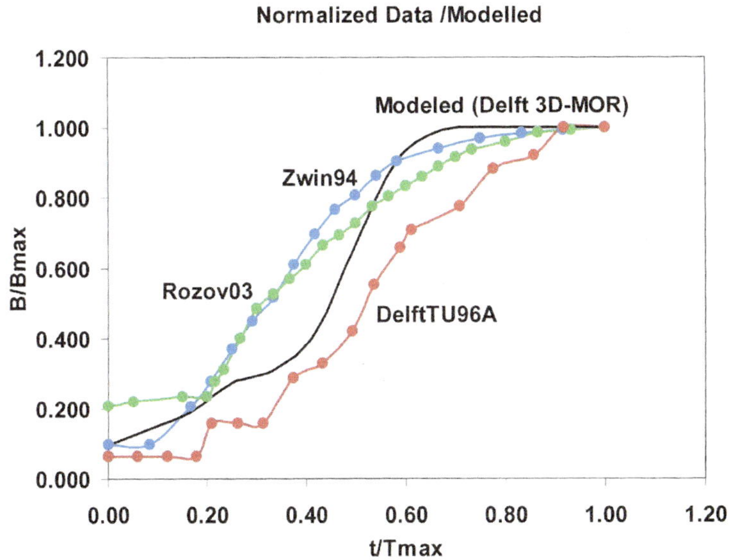

Figure 7.54. Normalized experimental data (points) and modeled curve (black) employed in Delft3D-MOR module using the empirical breaching relationship in Equation (7.68) with $K = 0.9$.

jump is conserved, then $\partial(q_s^2/h_s + gh_s^2/2)/\partial y = 0$ as in a moving hydraulic jump so that the terms contained within the box in Figure 7.55 remain. The depth-averaged sediment velocity $v_s = q_s/h_s$ in the y-direction is assumed constant in the x-direction so that the equation within the box in Figure 7.55 is the result. Over unit time interval Δt and unit width interval Δx, it is learned that the coefficient K in Equation (7.68) is a cross-momentum Courant number, i.e. the ratio of the depth-averaged sediment velocity v_s moving in the y-direction to the celerity of the numerical computation $\Delta x/\Delta t$ in the flow, x-direction. In this derivation, sediment mass and momentum are both conserved.

An example of the application of the Delft3D model using this breaching algorithm with $K = 0.9$ is shown in Figures 7.56 and 7.57 (from Canizares and Irish, 2008). The Corps of Engineers and their consultant Moffatt & Nichol, Inc., NYC, studied potential breach locations for Long Island, New York for the "Fire Island to Montauk Point (FIMP) Reformulation Study". The purpose was to determine the magnitude of storm surge increases along the mainland shoreline behind Great South Bay (GSB), Moriches Bay

CONSERVATION OF SEDIMENT MOMENTUM (Conservation Form)

Y-direction sediment momentum

$$\frac{\partial q_s}{\partial t} + \frac{\partial\left(q_s^2/h_s + g\, h_s^2/2 \right)}{\partial y} + \frac{\partial\left(p_s \cdot \frac{q_s}{h_s} \right)}{\partial x} = 0$$

or

$$\boxed{\frac{\partial q_s}{\partial t} + \bar{v}_s \frac{\partial p_s}{\partial x} = 0}$$

Unsteady flow
Neglect bed shear
Horizontal sediment bottom

Overtime interval, Δt and unit width, Δx

$$\Delta\forall_{DRY} = -\frac{\bar{v}_s}{\Delta x/\Delta t} \cdot \Delta\forall_{WET}$$

Let

$$K \equiv \frac{\Delta\forall_{DRY}}{\Delta\forall_{WET}} = \frac{|\bar{v}_s|}{\Delta x/\Delta t} =$$ Cross-Momentum Courant number for sediment

But $\bar{v}_s(t) =$ function of time

Hence $K(t)$ also variable with time

Figure 7.55. Equation for the conservation of sediment momentum in the y-direction (from Basco, 2004).

(MB), and Shinnecock Bay (SB) resulting from inundations and potential breaches of the barrier island.

Figure 7.56 shows the potential breach locations and Figure 7.57 displays the results of one "Realism" test for the 1938 hurricane at location MB1. The left side of Figure 7.57 is the before-breaching (top) and after-breaching (bottom) simulation results from the model and the right side are the prestorm (top) and post-storm (bottom) topography and bathymetry. These results while not perfect give confidence that the Delft 3D model with empirical breaching algorithm is producing results that are order-of-magnitude correct. The resulting storm surge hydrograph along the mainland shoreline directly behind the breach location (MB1) was 30 cm (1 ft) higher in peak magnitude as a result of the breach in the barrier island permitting more ocean water to penetrate into MB. The original beach nourishment and dune-rebuilding study of the Corps of Engineers was then reformulated to include the potential for barrier island breaching

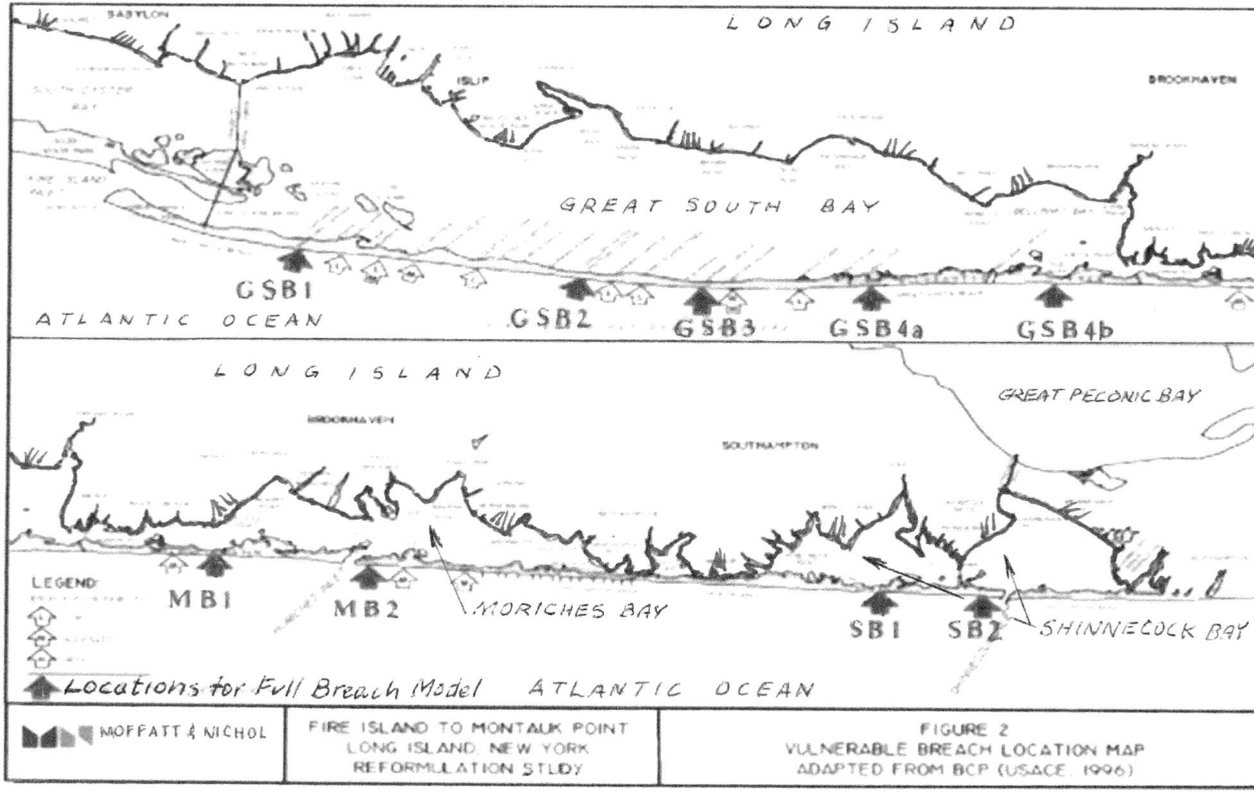

Figure 7.56. Potential breach locations on Long Island, NY (from Canizares *et al.*, 2004).

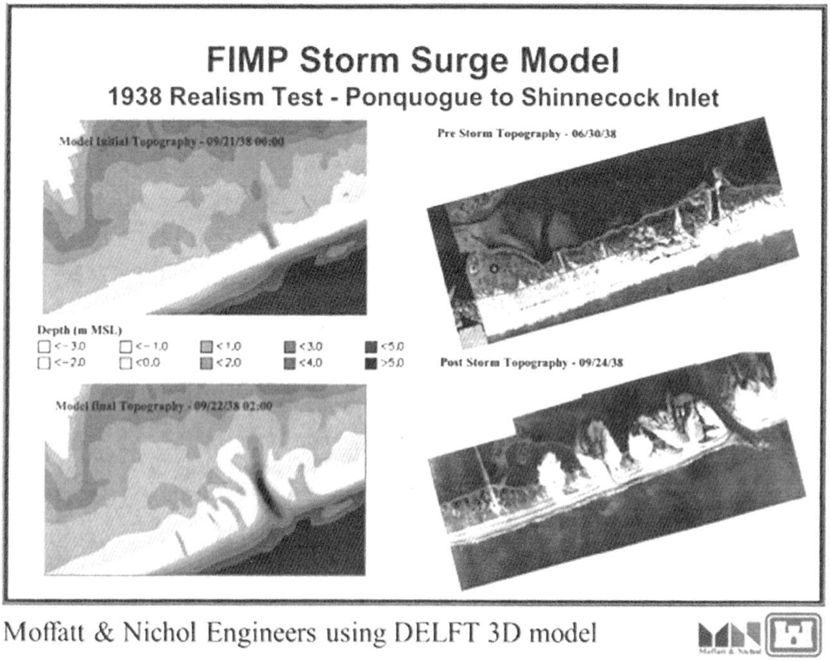

Figure 7.57. Results of 1938 hurricane simulation for DELFT 3D model with breaching (from Canizares *et al.*, 2004).

in the design. See USACE (1995, 2007) for details of the Fire Island to Montauk Point Reformulation Study.

In summary, major hurricanes in recent years along the Atlantic Ocean and Gulf of Mexico coastal barrier islands in the United States have produced numerous breaches of the relatively narrow landmass separating the ocean and the back bay. The frequency of these breaching events may increase with accelerated sea-level rise in the future. Roads along these barriers provide the only link for lifeline evacuation, emergency vehicles, daily transportation (schools, commerce, etc.). When breached, serious consequences result.

7.5 Beach Construction and Monitoring

7.5.1 *Introduction*

Beaches are constructed with sand transported by two methods: (1) truck hauls from inland sources and (2) by hydraulic dredging from deposits found beneath coastal seas. The most common method is by equipment (dredge plant) that excavates a water/sand (slurry) mixture for hydraulic transport through pipe lines with energy provided by centrifugal pumps and is called hydraulic dredging.

7.5.2 *Hydraulic dredging*

7.5.2.1 *Equipment (plant)*

When wave conditions permit, cutter-suction pipeline dredges are employed. Figure 7.58 schematically displays the floating, barge-like equipment that contains the dredge pump and motor for lifting the slurry mixture off the bottom and rotating the winches that swing the ladder supporting the suction pipe. On the pipe end is the cutterhead that is rotated on the bottom to cut-loose (dislodge) the dredged material. On the rear end are two vertical pipes (spuds) with one always embedded in the bottom. Swing winches pull the barge horizontally through the cut as it rotates about the embedded spud. The spuds are then alternated by lifting one and setting the other so that the dredging operation "walks" ahead to remove sand

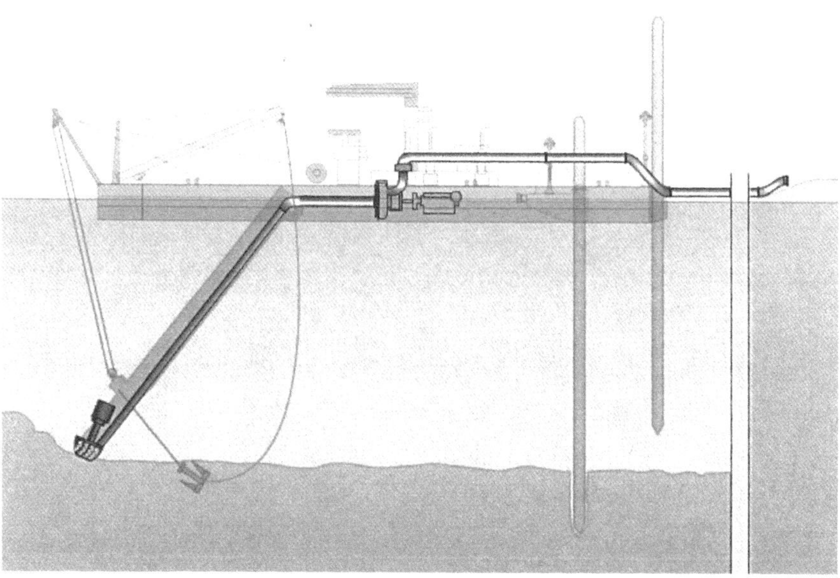

Figure 7.58. Schematic diagram of hydraulic, suction-cutter pipeline dredge (image from https://www.ghentdredging.be/en/cutter-suction-dredger).

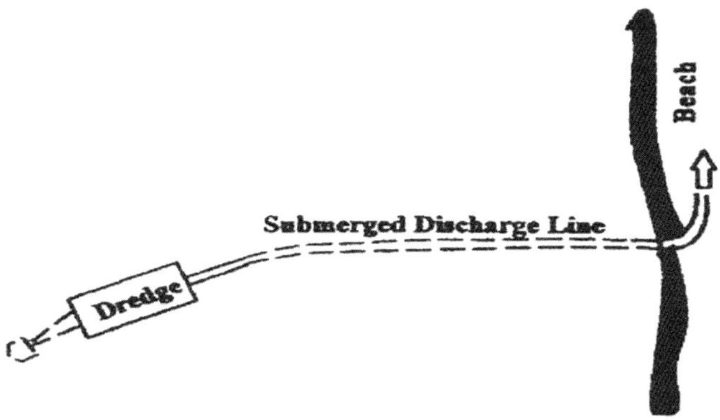

Figure 7.59. Hydraulic dredge with submerged discharge pipe for beach nourishment.

and send it to the beach via the discharge pipe as illustrated in Figure 7.59. Figure 7.60 are photos of hydraulic dredging for beach nourishment.

When nearshore wave conditions (heights and periods) make cutter-head, pipeline dredging impractical, then ocean-going, hopper dredges are employed. Figure 7.61 schematically displays the ship-like structure with large containers (hoppers) for storing the slurry mixture that has been lifted off the bottom through the draghead and flows through the dredge pipes (dragarms) to be deposited in the hoppers. Centrifugal pumps again provide the energy to transport the slurry mixture from the bottom to the hopper. The ship movement is by normal stern propellers during the dredging operation and to position the ship at the designated offshore borrow site location. Figure 7.62 is an artist's drawing of a hopper dredge courtesy of IHC Holland dredging company. Figures 7.63 and 7.64 show photos of hopper dredges transporting dredged material and working at sea with relatively low wave conditions.

The decision on when to use a cutter-suction, pipeline dredge or a hopper dredge depends on many factors and is best left up to the dredging contractor bidding for the beach nourishment project.

7.5.2.2 *Theory of hydraulic dredging*

The heart of a hydraulic dredging system (suction-cutter or hopper) is the centrifugal dredge pump. Designed to transport water–sediment (sand, mud, rocks, etc.) mixtures without clogging, the impeller has fewer vanes (3–4) with large clearances within the volute (casing) so that peak pump

(a) Hydraulic suction-cutter dredge "Jim Bean" off Pompano Beach, FL

(b) Sand-water (slurry) mixture at discharge pipe outlet on Pompano Beach, FL

Figure 7.60. Photos of hydraulic dredging for beach nourishment (Author's photos, 1971).

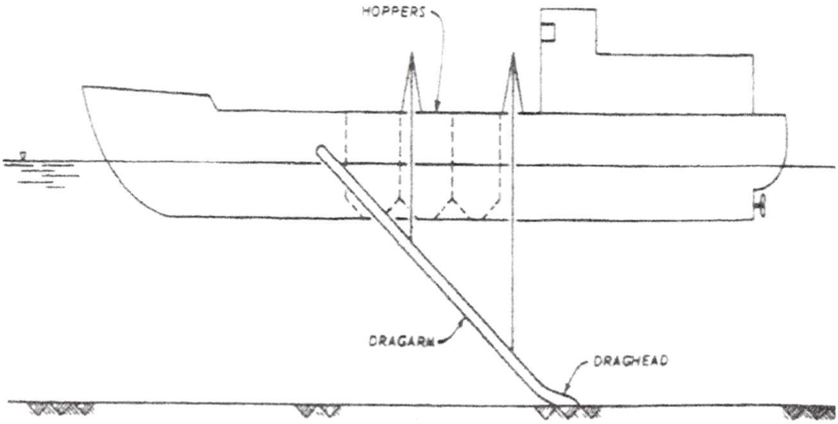

Figure 7.61. Schematic diagram of ocean-going, hopper dredge.

Figure 7.62. Artist's drawing of hopper dredge with dragarm and draghead digging along the bottom to dredge material and lift it into the hoppers (Courtesy, IHC Holland, the Netherlands).

Figure 7.63. A photo of the hopper dredge "Dodge Island" fully loaded and heading for the pump-out buoy for beach nourishment (Courtesy, Great Lakes Dredge and Dock, Chicago, Il).

efficiencies are low (≈70%). The Center for Dredging Studies at Texas A&M University pioneered dredge pump performance testing in the early 1970s (Herbich, 2000). For the first time, the dredge pump manufacturing industry obtained knowledge of the actual pump energy (head) versus volumetric flow rate (discharge) curves as a function of rotational speed (RPM) of the motor for various sizes and designs of dredge pump impellers. Hydraulic dredges are equipped with variable-speed drive motors to vary and control the output of the dredge pump.

The dredge pump performance curve at a given speed combined with the system curve for the elevation change and pipeline length determines the operating point, i.e. the pump head, flowrate, pump efficiency, and hence motor power requirements. The actual operating point of the dredge pump curve–system curve is not that found on the nameplate of the dredge pump. The pump manufacturer always puts the pump head, flowrate, and speed at the best efficiency point on the nameplate. Be careful here that you do not associate pump nameplate information with the actual operation of dredge pumps in the pipeline system.

Figure 7.64. Photo of the hopper dredge "Volvox Hollandria" dredging in the North Sea (Author's photo, 1971).

All the above-mentioned points are for clear water and become the starting point to quantify hydraulic dredging performance. Knowledgeable dredging contractors "calibrate" their equipment by determining the best pumping speeds for the line lengths required using clear water at the start. Then everything changes when pumping sediment through the pipeline system. The solids (sand, mud, rocks, etc.) replace water volume (typically, 15–20% sand for beach nourishment), require more power for transport, and generate more pipe friction in the dredge pipes. The line length is the limiting factor for the maximum pump power available on the dredge for solids production.

A second limiting factor is the digging depth. All centrifugal pumps have a cavitation limitation when the pressure on the suction-side entrance to the impeller approaches the vapor pressure of the water. Cavitation changes the water (liquid) to a gas (vapor) and reduces the energy head developed by the pump. The length of the suction pipe causes pipe-friction head losses that lower the suction-side pressure. Adding sediments to the water increases the friction losses but more importantly increases the mixture density to significantly reduce the energy head needed to overcome cavitation. The dredging industry learned this lesson the hard way when they found that solids' production dropped significantly when they used the same equipment in deeper water. Adding an extra, axial-flow pump on the ladder near the cutter or drag-head eliminated the cavitation problem and the digging depth limitation.

The third limiting factor for solids' production is the dislodgment of material from the bottom as it enters the cutter or drag-head. Maximum solids' output at the end of the pipeline is only as much at that entering the suction pipe at the entrance. And, related is the concern with too high solids' concentration that will cause settling and possibly plug the discharge pipe. The overall dredging production rate, solids' volume per day would be greatly reduced if the underwater discharge pipe became plugged and needed removal.

A summary review of all the above can be found in an analytical model of a hydraulic, cutter-head, pipeline dredging system giving all the details for various sand sizes by Basco (1975a, 1975b). See also Turner (1996) and Dean (2002, Appendix I). The *in situ* sand volume rate being transported and in place on the beach Q_{is} is related to the pumping volumetric rate, Q, the sand concentration percentage C_v (15–20%), and the sand porosity p (0.3–0.4) for normal sand packing density. The relationship is

$$Q_{is} = \frac{C_v Q}{1 - p}. \tag{7.69}$$

In all the previous sections, the beach volume, $\forall$ per unit length was involved and not the volumetric transport rate at which the *in situ* sediment volume is being delivered, i.e. Q_{is}. Also note that the upper practical limit for volume concentration C_v in a pipeline is 30% so that for sand of specific gravity 2.65 in seawater (s.g. = 1.03) the maximum slurry density is $\rho_{\text{slurry}} = 1.52$.

7.5.3 *Beach construction*

7.5.3.1 *Conventional methods*

In recent years, most borrow sites for beach nourishment have been found in the coastal waters offshore of sandy beaches where wave heights and periods dictate the use of hopper dredges. The conventional method is to locate a directional, pump-out "monobuoy" in water depths sufficient to moor and connect the dredge for pumping via an underwater pipeline to shore as illustrated in Figure 7.65.

The key element is the single-point mooring "monobuoy" as developed by the offshore oil industry for export and import of oil. It is a catenary anchor leg mooring (CALM) as illustrated in Figure 7.66(a). The hopper dredge is moored to the "monobuoy" fixed to the bottom by four anchor chains in 10–15 m (30–45 ft) water depths. Typically, the "monobuoy" pipe connection by floating hose attachment is located at the bow of the hopper dredge, not mid-ship as shown. Figure 7.66(b) displays details of the

Figure 7.65. Photograph of hopper dredge BE Lindholm (Weeks Marine, Inc.) connected to monobuoy for beach nourishment at Pine Knoll Shores, North Carolina (courtesy Weeks Marine, Inc.).

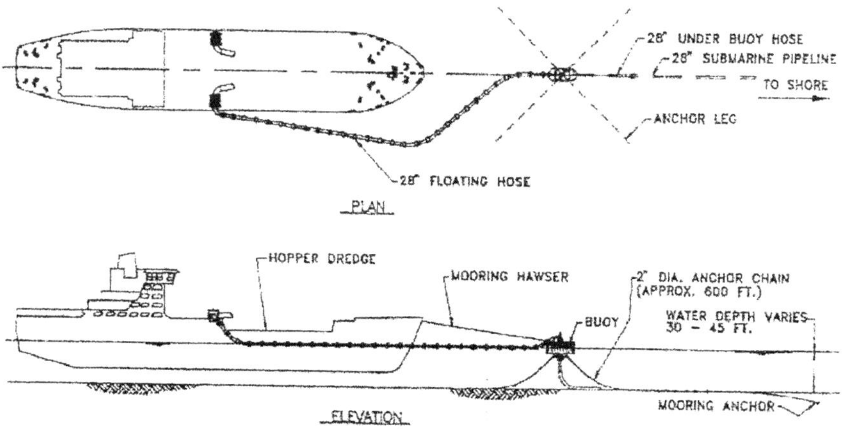

(a) Hopper dredge moored to "monobuoy"

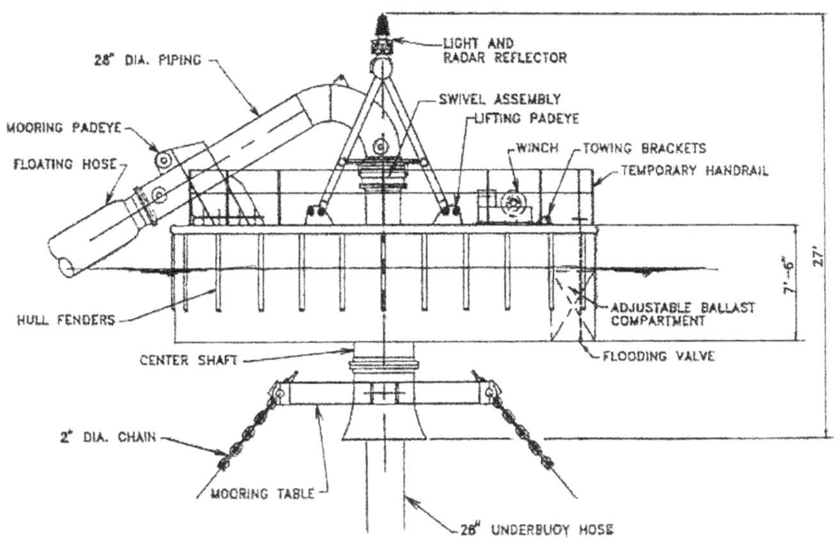

(b) Details of single-point "monobuoy" with swivel

Figure 7.66. Schematic diagram of single-point "monobuoy" for transfer of sand from hopper dredge to submerged pipeline to shore (courtesy Marine Insight).

single-point "monobuoy" with swivel to permit hopper dredge mooring facing wind direction and floating hose connection to hopper dredge at the center for vertical sand transport directly downward to underbuoy hose connection to submerged pipeline to shore.

As shown in Figure 7.67 (top-left), the hopper dredge "Dodge Island" (Figure 7.63) approaches the "monobuoy" for connection to the floating hose on the starboard side near the ship bow. Top-middle is the "monobuoy" located about 1,500 m (5,000 ft) from shore with Virginia Beach, VA hotels in the distance. At the borrow area, hoppers are continuously filled to overflowing until instrumentation indicates hoppers mainly filled with sand and not a sand–water mixture which would not be economic for return to the pump-out site. Then the sand in the hoppers must be refluidized by jets of clear water as illustrated in Figure 7.67 (top-right, bottom-left) to facilitate pumping the slurry through the submerged pipe that emerges onto the beach (bottom-middle). These photographs were taken in May 1989 when the Great Lakes Dredge and Dock Co. was renourishing the beach at Virginia Beach, VA. It was a good time for Old Dominion University graduate students in Coastal Engineering (John McCormick and Greg Williams, bottom-right) to visit a hopper dredge and see first-hand a beach construction project (Author's photos).

The pumped sand–water mixture exiting the pipe on the beach is 80–85% water and would quickly flow back out to sea so that temporary sand dikes are built to retain the slurry and help the sand remain on the beach. Figure 7.68 (top, middle) shows bulldozers building the sand dikes at the shoreline and sometimes also on the landward side (top-right). Pipe lengths are added as construction continues down the beach using front-end loaders to lift and move the pipe sections. The bottom three aerial photos in Figure 7.68 were taken in the Spring 2003 during operation "Big Beach" when Virginia Beach, Virginia was renourished to a design width of 100 m (300 ft) that require splitting the pipe in a Y-shape (middle). Note that construction continues during the tourist season with sand bridges crossing the dredge pipe for the many beach lovers to use to get close to the water. The final photo in Figure 7.68 (bottom right) is the finished beach looking north from Rudee Inlet to Cape Henry and the entrance to the Chesapeake Bay. We shall return to the Virginia beach nourishment project of 2003 in Section 7.7.

7.5.3.2 *Other construction methods*

The time involved to moor the hopper dredge to the "monobuoy" and then pump to shore is a significant amount of the cycle for the conventional method of beach nourishment. Use of bottom-dumping scows and hopper dredges eliminates the "monobuoy" and permits the sand to be placed in a

Figure 7.67. Photographs of hopper dredge "Dodge Island" (Great Lakes Dredge and Dock, Co, Figure 7.63) during beach nourishment for Virginia Beach, VA, May, 1989 (Author's photos).

Figure 7.68. Temporary sand dikes to retain sand from the discharge pipe (top) and the Virginia Beach, VA beach renourishment project aerial photos from May 2003 (bottom) (photo courtesy City of Virginia Beach).

nearshore "feeder" berm. Under the proper physical conditions, the dumped sand naturally moves landward to rebuild the beach.

A nearshore berm is defined as dredged sediment placed in either (1) deeper water as a stationary mound to attenuate waves as a "stable berm" or (2) as a shore-parallel bar in shallow water to attenuate waves and move shoreward as an active "feeder berm" (McLellan and Kraus, 1991). Efforts to develop a nearshore berm stability nomogram to distinguish the difference between active and stable berms was first presented by Hands and Allison (1991). Through field experiments, the goal is to develop methods to estimate berm response to the given wave climate, sediment characteristics, and initial conditions at a site. Field research has been possible by combining the need for maintenance dredging of nearby tidal inlet channels in the annual, Corps of Engineers, Maintenance Dredging Budgets with nearshore berm construction, and monitoring within the Coastal Inlets Research Program (CIRP) and Regional Sediment Management (RSM) programs. Active feeder berm research projects using sediments dredged from the San Francisco, CA navigation channel (Ocean Beach placement) and the Matanzas Pass, FL navigation channel (Ft. Myers Beach placement) are discussed in ERDC/CHL CHETN-XIV-10 (Beck *et al.*, 2012). The Mobile District dredged the Pensacola Pass, FL navigation channel and placed the sand as a nearshore (swash-zone) berm near Perdido Key, FL as shown in Figure 7.69.

The direct spraying of the sand–water mixture on the beach is also possible as shown in Figure 7.70. The floating hose connector at the bow of the ship has been fitted with a nozzle to jet-spray the slurry into an arc from where the hopper dredge sits in very shallow water.

The current state of knowledge of nearshore feeder berms is not sufficient to conclude that placement of a nearshore feeder berm is a reasonable and practicable alternative for beach nourishment.

7.5.3.3 *Borrow sites*

The onshore or offshore site (area) from which sand is removed to be placed on the beach is the "borrow area" for the beach-fill project. The mean grain size of "fill" material D_F must also be determined from sediment samples taken in the field at the borrow area. Four borrow area locations, land-based (e.g. riverine deposits), back bays (marshes, creeks, etc.), offshore (shoals), and navigation channels (tidal inlets) are the usual sites where beach nourishment projects have found sand that is compatible to put on

Figure 7.69. Example of a nearshore (swash-zone) feeder berm under construction at Perdido Key, Florida (photo courtesy of Mobile District, Corps of Engineers).

the native beach. Surface samples may suffice but sediment cores are needed to locate suitable sand deposits that may lie several meters beneath the surface materials composed of a mixture of fine sands, silts, and clay. Here, geophysical, seismic surveys are needed to help determine the locations for cores to be drilled. More than 95% of sediment placed in beach nourishment projects in the US comes from offshore sources (Dean, 2002).

Offshore (shoal) site investigation methods and analysis are demonstrated in Figure 7.71. Bathymetric survey maps (upper, left) reveal nearshore shoals that commonly contain beach-quality material since they may be relic formations from previous water levels (Cape Canaveral, FL from Field and Duane, 1974). The region is then surveyed in a zigzag (upper, middle) or grid (upper, right) pattern to obtain surface samples (Florida Coast, from Meisburger and Duane, 1971). At selected locations, seismic profile records are obtained (bottom, left) to determine the depths of sediments in various layers (from Anders and Hansen, 1990, for Ocean City, Maryland). Finally, an isopach map of sand thickness in the shoal

Figure 7.70. Example of beach nourishment by spraying the sand–water mixture directly on the beach (photo courtesy Great Lakes Dredge and Dock).

can be made (bottom, right) to estimate the total volume of beach quality material available for long-term use as a borrow site for the nearby beaches (from Anders and Hansen, 1990).

Equipment needed for the seismic operations and core-sample operations are beyond the scope of this book. Consult Coastal Engineering Manual (2006) for complete details. See also Dean (2002, Chapter 7) for general considerations including impacts of nearshore borrow sites on wave conditions and adjacent beaches.

7.5.4 *Monitoring of completed projects*

The purposes for monitoring of completed beach nourishment projects are to (1) ensure that the beach-fill template was filled to the design specifications for payment; (2) quantify the volume change in time over the entire project length for the management of future renourishments; and (3) quantify the volume of sand moving beyond the project boundaries and on to adjacent beaches, if present.

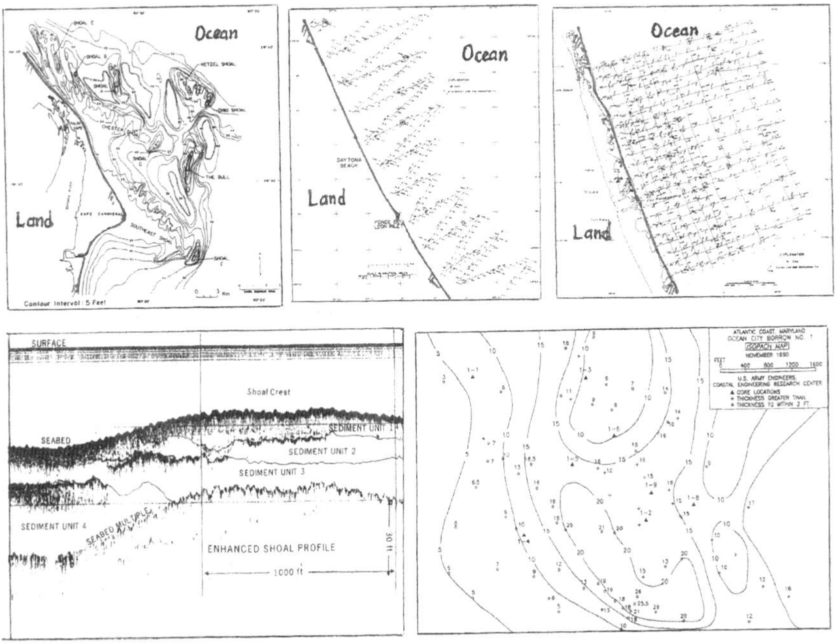

Figure 7.71. Offshore (shoal) site investigations for determination of quality and quantities of sand for potential use as beach nourishment (various references cited in text).

7.5.4.1 *Survey baseline and profile locations*

The same baseline and shore-normal beach profile system employed for design of the project should be used for monitoring after construction. Location of the initial baseline for design and monitoring should be carefully chosen and well back from the MHW shoreline to ensure future erosional trends and major storms do not cause the MHW shoreline to move landward of the baseline. This will result in negative numbers for location of the profile elevations and may create confusion in the resulting plotting of beach profiles. The goal is to have all beach profile elevations begin with 0.0 distance from the baseline.

Figure 7.72 is an example of the beach monitoring baseline, profile locations, and boundaries of the beach nourishment and dune construction employed for the US Navy, Fleet Combat Training Center, Atlantic (FCTCLANT) located at Dam Neck, VA in 1997 (Basco, 2000, Figure 3.1). The monitoring baseline is slightly curved to follow the shoreline over a distance

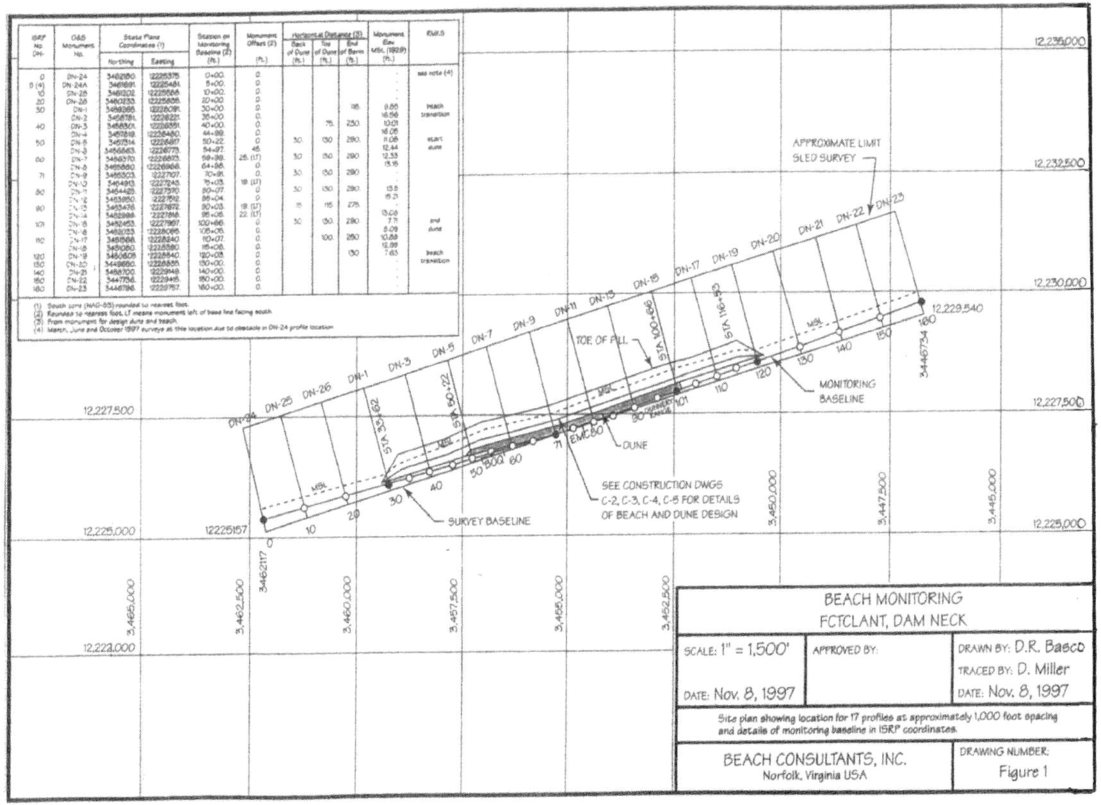

Figure 7.72. An example of the beach monitoring baseline, profile locations, and boundaries of the beach nourishment and dune construction employed for the US Navy, FCTCLANT located at Dam Neck, VA in 1997 (Basco and Shin (1996), Figure 3.1).

of 16,000 ft (4,877 m) and has 17 profiles at approximately 1,000-ft (305 m) spacing which convert to a station numbering system (STA 0+00, 10+00, 20+00, ... 160+00). This permits intermediate locations, e.g. STA 101+00 to be readily added e.g. STA 100+66 at the end of the rebuilt dune. The profile locations coincided with an existing, long-term beach monitoring program (DN-24, DN-24A, etc.) conducted by the City of Virginia Beach, VA. The location of the zero location on the baseline for each profile line is given in State Plane Coordinates in the accompanying table and graphically shown on the drawing. (*All US states have horizontal controls called "State Plane Coordinates" and they can be found from Lat, Lon coordinates.*) One end of the beach-fill project is STA 33+62 and the other end is STA 116+53 (8,300 ft) (2,530 m) before the 500-ft (150 m) tapered ends as sketched on the drawing. Therefore, 3–4 additional profiles on each end are routinely monitored following construction. The boundaries of the rebuilt dune are also shown and are from STA 50+22 to 100+66, a distance of about 5,000 ft (1,540 m). The outer limit of the profiles in the Atlantic Ocean are at about −22-ft (6.7 m) depth below MSL (1929 datum), the closure depth for this location and hence the limit for the sled survey. Further details are found in Section 7.6.

7.5.4.2 *Survey methods for beach profiles*

Section 7.2.4 discusses the closure depth and sediment budgets and Figure 7.15 schematically defines a beach profile out to the closure depth. Many beach profile survey methods have been developed over the years. The dynamic land–water interface at coasts with tides, waves, and readily movable bathymetry has made the development of hydrographic survey methods an active research area. Early problems with land-based profiles that do not match the water depth measurements from boats in the surf-zone area have been generally overcome, but diligence is still needed when these overlapping methods are employed.

A full discussion and summary of the many survey methods for making beach profiles out to closure depth is beyond the scope of this book. The interested reader should consult Cooper *et al.* (2007, online) and Klemas (2011) for summaries of many methods including LIDAR.

Figure 7.73 shows photos of the sea-sled method for the initial surveys for design and then the monitoring program after construction of Dam Neck beach renourishment and dune rebuilding in 1997. In the upper-left is the sea sled on the beach (Glenn and Sadler, consulting engineers) with the

Figure 7.73. Photos of the sea-sled method for the initial surveys for design and then the monitoring, program after construction of Dam Neck beach renourishment and dune rebuilding in 1997 (Author's photos).

Jet Ski driver getting ready to carry the haul line back out to the tow boat (top-middle) for the next profile. Figure 7.73 (top-right) displays the tractor with winch to slowly pull the sled shoreward from its far offshore location barely visible beyond the outer bar breaker line. The surveyor with Total Station Theodolite (TST) instrument on top of the dune (bottom left) is taking distance, angle, and elevation readings at specified locations marked on the haul line. Figure 7.73 (bottom-middle) shows the sled near the shoreline. The sled is towed on the beach to the next profile location, the TST surveyor moves to this location, and the tow line is again taken back out to the boat (bottom right) to repeat the beach profiling process. The sled survey method is equipment-and labor-intensive. It requires the survey sled, tow boat, tractor/winch, Jet Ski, and Total Station survey instrument. It also requires a minimum of six people for successful operation. The sea-sled method is one of the best methods for confidence in the beach profile elevations taken in the dynamic, surf-zone region where wave breaking and bar movements are rapidly occurring. The sea-sled must be strong and light weight for towing, have a mast with reflector prism tall enough to reach closure depth, and portable for truck haul to the beach. The author acknowledges Glenn and Sadler, Inc. (Norfolk, VA) and George Sadler for making possible the sea-sled survey method at Dam Neck.

7.6 Beach Management

Management of beaches requires many skills besides science and engineering. The goal is to keep the volume of natural materials (sand, pebbles, cobbles, etc.) sufficient for storm damage mitigation, the ecological benefits, and for recreation and tourism.

7.6.1 *Technical*

A brief introduction to beach engineering and the technical requirements is presented in the previous sections. However, other ways to manage sand volumes are employed that shift the existing volume to areas that have a sand deficit. Both natural and anthropogenic (man's activities) causes are reasons responsible for the deficit.

At many locations at property boundaries or near tidal inlets, terminal groins are constructed to mitigate the sand volume moving past the boundary or into the inlet and retain the sand on the up-drift beach. Large sand fillets develop over time with sand that has relocated from upland reaches which have a sand deficit as a result. One cost-effective method is to remove the excess sand in the fillet and transport it by truck back to the deficit reach. The sand management method has become known as "sand backpassing". Monitoring and a sand budget explain the success of this method.

An example of a very successful sand backpassing operation is Sea Island, Georgia (US). In 1990, over a million cubic yards of sand was pumped onto the central shores of Sea Island which is over 5,225 m (17,000 ft) long. Terminal groins were installed at both the north and south ends of Sea Island in conjunction with the beach nourishment project. Prior to renourishment, Sea Island beaches had very little sand in front of the concrete and rock revetments that lined most of the island. There was very little or no littoral drift and therefore no littoral drift to be trapped by the groins. The purpose of the groins was to retain this valuable resource placed by the renourishment in front of the revetments and to provide a catchment area for recycling back to central reaches of the island shores. Initially, about 100,000 cubic yards (75,000 m^3) of sand were recycled for the first four years. After the initial phase, about 150,000 cubic yards (112,000 m^3) were recycled every two to three years as required until 2007, and then again in 2013 and 2016. The recycling maintenance has been very successful as the volume of sand in the Project Areas is about 80% of the original volume (Oertel, 2016).

When navigation is of concern at tidal inlets, jetties may be built on both sides of the inlet to reduce wave energy and make ship passage safer through the inlet. The jetties may block the natural, net drift of sand passing the inlet to produce a positive sand fillet on the up-drift side and negative sand deficit on the down-drift side of the inlet. Dean (2002) attributes 95% of Florida's beach erosion problems as occurring on the near-down-drift sides of tidal inlets as produced by jetties blocking the natural littoral drift. For many years, the problem was compounded by the US Army Corps of Engineer dredging program that maintained navigable depths through the inlet. The "least-cost" site for sand dredged from the inlet was to dump the sand offshore and not place it on the down drift beach. Fortunately, this Corps policy has been changed so that Regional Sediment Management (RSM) is now a cornerstone of beach management of the US Federal government http://rsm.usace.army.mil.

Many other topics of technical interest can be found in Dean (2002).

7.6.2 *Environmental*

The environmental and ecological benefits of wide sandy beaches are many. They all begin with the microorganisms within the pore spaces as for example shown in Figure 7.7 which also shows shell and bits of coral scattered on an Indonesian beach. Then, up the ecological food chain with worms, crabs, birds, turtles, and fish. Compatible beach-quality sand from a physical standpoint means that the borrow-site material must be similar in median grain size and color to the native beach.

For turtles, these factors are important but so are temperature, compaction, sand moisture content, and the profile of the subaerial beach including the presence of scarps.

The loggerhead sea turtle shown in Figure 7.74 is the most common sea turtle species that makes its nests on sandy beaches of the United States. Florida beaches are one of the world's most abundant loggerhead turtle nesting sites. Beach nourishment increases the beach width and area that would otherwise be unavailable for turtle nesting, but it may also change parameters of the natural beach that affect nesting and reproductive success of turtles. Here again, the interested reader is referred to Dean (2002, Section 9.4) for a detailed summary of what the State of Florida

Figure 7.74. Loggerhead sea turtle (courtesy Pixabay.com, free sea turtle images).

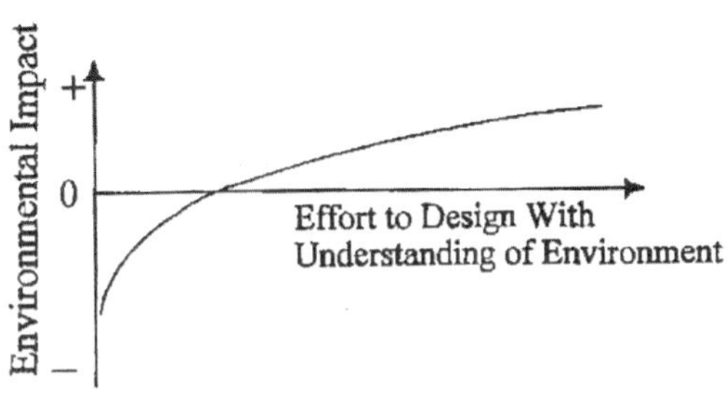

Figure 7.75. Research and knowledge of beach nourishment projects on the environment that produce a net positive impact (from Dean, 2002, Figure 9.1).

with over 2,000 km (1,200 m) of sandy beaches has learned over the years since monitoring of turtle nesting activity began in the 1970s. The State of Florida has Statutes that require all possible measures be taken to minimize the adverse impacts to turtles including restrictions on beach nourishment activities that coincide with turtle nesting seasons. As illustrated in Figure 7.75, a net positive environmental impact can be achieved when research and knowledge are applied in design with a factual understanding of the environment (from Dean, 2002, Figure 9.1).

7.6.3 *Social, political, and institutional*

Cultural, economic, recreational, and environmental benefits of beaches were introduced in Section 7.1.

Figure 7.76. The Blue Flag as the visual symbol of a program that meets the standards for water quality, environmental management, safety and services, and education at that beach (from "Number of English Beaches Gaining Blue Flag Awards Rises," *The Guardian*, May 15, 2012).

Figure 7.76 displays the "Blue Flag" which has become a worldwide, visual symbol (ecolabel) that the beach meets or exceeds standards for required criteria in the four areas that the general public and tourists have come to rely on when at the beach. It is trusted by millions of people around the world in 45 countries (Europe, Africa, Oceania, Asia, North America, and South America) and for 3,623 beach sites. The Blue Flag program is operated by the Foundation for Environmental Education (FEE) in Copenhagen, Denmark where it began in 1987. The four standards are for water quality, environmental (litter) management, safety (life guards) and services (food, rest rooms, changing rooms), and education (signs posted). The FEE is a non-profit, non-governmental organization that awards certificates on an annual basis by 65 organizations in the member countries. Spain has 597 Blue Flag beaches and has been a founding member since the beginning of the program. The United States of America presently does not have any organizations within the FEE and consequently there are no Blue Flag-designated beaches in the US.

Management of public beaches in the United States is a function of the city, county, state, and/or federal government political boundary wherein that beach is located. For example, the Public Works Department for the City of Virginia Beach, Virginia (US) has a Coastal Section with Beach Management Plan (Beaches and Waterways Commission, 2002) that is responsible for all four of the standards within the Blue Flag beach program. In addition, the City of Virginia Beach is also responsible for police, fire, traffic, parking, utilities, waste management, etc. issues surrounding the beach site. Most of the sandy beaches in the United States are public beaches that are free and open to the general public. Some private beaches exist that must provide limited public access, and one state (New Jersey) has a fee for beach use (known to the author at the time of this writing). Water quality concerns are the responsibility of the Federal Government, Environmental Protection Agency (EPA) for both the public and private beaches.

Beach management in the US is further complicated by the fact that sand movements along the coast do not recognize political boundaries; that beach nourishment designs originate from two separate institutions (Corps of Engineers and Civil/Coastal Consulting companies); and that public, private, and government-owned beaches all may have different economic standards for long-term management. Some examples are presented in Section 7.7 to illustrate the complexity of beach management in the US.

Beach management is a long-term responsibility of society with politics involved at some institutional level.

7.7 Design and Management Examples

Beach design and beach management are generally carried out by two separate institutions in the US. In other words, design and construction, and then the long-term management are never by the same organization. The design may be by either the US Army Corps of Engineers or a Civil/Coastal Engineering consulting firm. As a result, long-term beach management is affected by the institutional (Corps or consulting firms) memory regarding the physical environment through long-term storage, knowledge, and use of information regarding previous conditions of the beach.

The Corps of Engineers' beach nourishment projects must be approved by the US Congress as a Public Works Project in the Water Resources Development Act (WRDA) for that year. The project design life is generally 50 years with periodic maintenance a routine part of design. As a result, long-term records and knowledge are routinely maintained in the District Engineers' design office. Engineers move and retire but their replacements have access to data and what transpired in the past. The long-term memory regarding beach nourishment projects designed by the Corps of Engineers is excellent and often over-looked. Long-term beach management is by the non-federal, cost-sharing partner for the project except for the periodic renourishment decisions which remain a Corps of Engineers concern since funding is again within the WRDA authorization for that year. Funding is divided at 65% Federal government and 35% non-federal sponsor for all Corps of Engineers, beach nourishment projects.

The long-term memory regarding beach nourishment projects designed by consulting engineering firms is far less known and complicated by many factors. The design life is often left open so that routine maintenance is not discussed. The design company may change ownership or engineering direction or terminate business. Records maintained for the original design may be lost or not retained. The economic conditions of the ownership of the beach may change along with the original owner. The owner (city, private, etc.) may (or may not) retain any long-term records for the beach nourishment project including photographs.

On February 28, 2017, the ASBPA established a National Beach Nourishment Database (www.asbpa.org). The objective is to provide the general public with the following information regarding national (federal), state, and private beach nourishment projects for 28 US states. The database includes the following for each state:

- project name and number of nourishment events (separate calendar days),
- date of initial nourishment (year) and date of most recent event (year),
- total volume of sand placed for the project (cubic yards) and includes maintenance events,

- total known cost (US$); unfortunately, the year associated with this cost number is not specified,
- the known length (feet) for the beach nourishment project.

It is unfortunate that the volume of sand per unit length of beach (cy/ft, m^3/m) is not given in the National Database for each project.

Three different sources of sand and/or sources of funding for these projects were also identified and are listed in the database. They are as follows:

1. **Known Federal.** Typical projects designed and funded by the Corps of Engineers, FEMA, or Navy.
2. **Known RSM.** Sand dredged from navigation channels has come to be called beneficial use and part of the Corps' RSM program.
3. **Known Other.** Projects sponsored by private interests or local, or state governments with no federal government funding. Also, all unknown funding sources.

At the time of this writing, in the US, the total volume of sand placed as found in the ASBPA database is 1,514 million cubic yards (cy) (1,135 million m^3) for over 2,900 nourishment events covering about 790 mi (1,274 km). The total cost was $5.9 trillion but the year for this cost total is not revealed in the database. The earliest date given is for Coney Island, New York in 1923, but most projects began in the 1950s.

The following examples demonstrate the differences in design and management practices in the US.

7.7.1 *Southeast Virginia Coast (US)*

This coastline of Virginia was previously discussed to introduce shoreline change in Section 2.5.3 (Figure 2.38) and in Section 7.2 (Figure 7.20) for a sediment budget (Figure 7.21) from Sandbridge to Rudee Inlet.

Figure 7.77 again shows this coastline from the entrance to the Chesapeake Bay (top) down to the border with the state of North Carolina. Now with the city, state, and federal government property boundaries delineated, that span is 26.7 mi (43.1 km) of Atlantic Ocean beach. Table 7.2 defines each of the seven property segments and when beach nourishment began in that reach. The author is most familiar with the three beach nourishment projects along the southeast coast of Virginia as listed in Table 7.2.

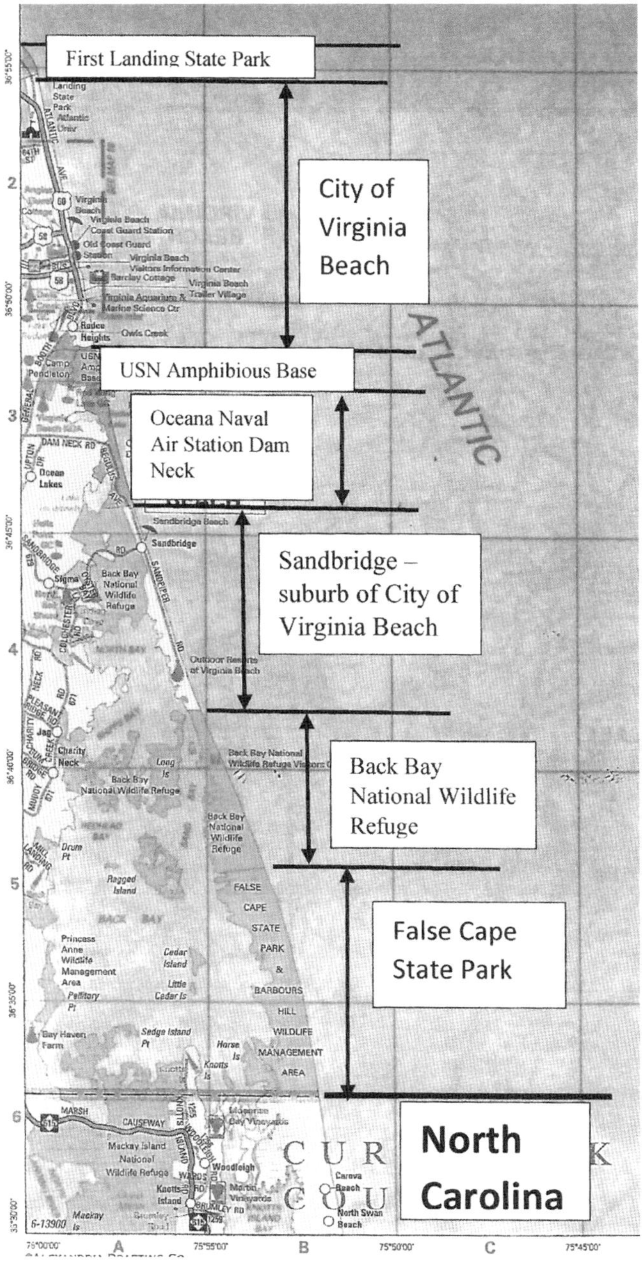

Figure 7.77. Southeast coastline of Virginia from entrance to the Chesapeake Bay to North Carolina border (courtesy Kappamapgroup for ADC Maps Inc.).

Table 7.2. Property boundary segments along Southeast Virginia coast.

Name	Government Level	Length, mi (km)	Beach Nourishment Start
First landing state park	State	0.83 (1.34)	—
City of Virginia Beach	City	7.06 (11.4)	1949 (2003)
USN amphibious base	Federal (navy)	0.83 (1.34)	—
Oceana Naval Air Station DAM NECK	Federal (navy)	3.09 (4.98)	1996
Sandbridge, suburb Virginia Beach	City (suburb)	5.10 (8.23)	1998
Back Bay national wildlife refuge	Federal (comm)	4.15 (6.70)	—
False Cape state park	State	5.64 (9.09)	—

7.7.1.1 *City of Virginia Beach*

Virginia Beach has been a popular tourist destination on the US East Coast for over 100 years. Three years before becoming incorporated (1952), it began the annual truck haul of about 100,000 cy (76,000 m^3) to attract tourists and protect hotels and businesses on the oceanfront. This small annual volume is analogous to filling potholes after the winter season to the many roads reaching the beach. Then starting in 1996, a major Corps of Engineers — City of Virginia Beach "Erosion Control and Hurricane Protection Project" began construction. Early design of a stepped, concrete revetment with curved seawall was rejected when the diminutive mayor of the City complained that she could not see the ocean when seated in the restaurants on the promenade (boardwalk). Her view was blocked by the crest elevation of the proposed seawall on the south end. The revised design (see Figure 3.15, Chapter 3) lowered the seawall elevation, modified the structural design, added interior storm water drainage to accommodate additional overtopping, and widened the nourished beach in front to mitigate storm energy. Operation "Big Beach" began that produced a beach about 300-ft wide (90 m) at a minimum berm elevation +8.5-ft (2.6 m) above MSL for a distance of about 4 mi (6.5 km) that required over 4 million cubic yards of sand ($D_{n50} = 0.29$ mm) from the navigation channels entering the Chesapeake Bay (Figure 7.68). The required fill volume density $\Delta y_o \approx 200$ cy/ft (500 m^3/m).

The combined new seawall/boardwalk, interior drainage, and wide beach nourishment project cost $125 million (dollar value in year 2000) with the federal government share at 65%. The City of Virginia Beach was the non-federal sponsor for the remaining 35%. The Corps incorporated advanced numerical models that increased the natural berm elevation and significantly widened the beach design width to reduce wave overtopping on the lowered seawall elevation. Annual renourishment level was projected at about 1 million cubic yards on a 3–4-year cycle for the 50-year design life. The benefit/cost ratio was of the order of 4:1.

Construction of the Big Beach ended in April 2003, and 6 months later Hurricane Isabel (September 18, 2003) struck Virginia Beach. The estimated dollar damages prevented to oceanfront property was \$82 million because of this project (Corps of Engineers, 2003). In other words, 65% of the initial construction costs were immediately realized as savings in damages prevented. The past 15 years have seen far many more storms and far many more dollar damages prevented by this project.

The first renourishment effort took place in 2013 (after Hurricane Sandy) that added 1.4 million cubic yds and restored the beach to its design dimensions. The dredging contractor was Weeks Marine (Camden, NJ) whose project manager (Paul Stewart) said

> *Coastal engineers need to come here and study the design and construction of this project. I've been working these beachfront re-nourishment projects up and down the Eastern Seaboard for decades, and most require sand replenishment about every three to four years. It's definitely a national model of how to build a smart beach erosion control and hurricane protection system that lasts.*

The beach nourishment project at Virginia Beach is a classic example of a Public Works project of the US Federal government as funded by Congress through the WRDA. Long-term data and design details are stored in the Norfolk District Office of the Corps of Engineers so that they are available for the design life of the project and beyond. The long-term memory of the beach nourishment design is available for future use.

7.7.1.2 *Dam Neck beach*

Further down the coast as seen in Figure 7.77 (Table 7.2) are the US Naval properties: Amphibious Base and the Oceana Naval Air Station — Dam Neck. The USN Amphibious Base is home to the famous Navy Seals training facility. And along the coast, the US Naval Fleet Combat Training Center is positioned within the Oceana Naval Air Station — Dam Neck. The coastal facility was established in 1942 to train gunners to fire 20-inch artillery. Over \$100 million (dollar value in 1995) of buildings and structures were threatened by severe winter storms in the early 1990s and the historic, long-term erosion rate of $-0.7\,\text{m/year}$ ($-2.3\,\text{ft/year}$).

The Naval Facilities' Command (NAVFAC) contracted with a private, civil/coastal engineering consulting firm, Glenn & Sadler, Inc. (Norfolk) to study the alternatives to mitigate future damage. Basco (2000) found that the "hard" revetment alternative was more expensive and far less environmentally acceptable to the Navy, than the "soft" beach nourishment and dune-rebuilding alternative for the 25-year design life. Native beach grain size $D_N = 0.29\,\text{mm}$ ($\sigma_N = 0.32\,\text{mm}$) was smaller than the Sandbridge Shoal borrow material, $D_F = 0.34\,\text{mm}$ ($\sigma_F = 0.69\,\text{mm}$) so that the coarse grain ($D_F > D_N$) intersecting profile Equation (7.39) was employed that resulted in the average section, beach-fill volume density, $\Delta v = 86\,\text{cy/ft}$ ($34\,\text{m}^3/\text{m}$). The design beach width $\Delta y_o = 125\,\text{ft}$ (38 m) required a construction width W = 155–160 ft (47–48 m) and 755,000 cy (566,000 m^3) of sand for the l = 8,800 ft average length of beach fill (including 500 ft tapers on each end). Construction ended in the fall 1996. See final report Basco (2000).

As discussed in Section 7.5, an extensive beach/dune monitoring program took place over the three-year period (November 1996–November 1999). Figure 7.78 shows Profile 60 just after construction (November 1996, dotted) and after three years (November 1999, solid) to illustrate how the profile shape adjusted in time. This volume change integrated across the profiles and summed for the nominal project length, l = 8,800 ft, gave the

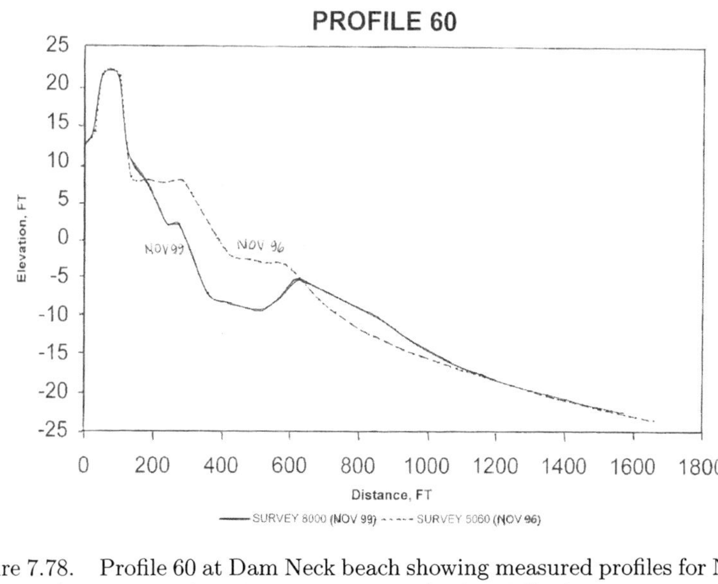

Figure 7.78. Profile 60 at Dam Neck beach showing measured profiles for November 1996 (start, dotted) and November 1999 (end, solid) of monitoring program.

Table 7.3. Results of Dam Neck beach surveys November 1996–November 1999 (includes background erosion rate).

Date	Volume, cy	$M(t)$	Volume/Unit Length	Loss Vol/ft	Remarks
Design	754,937	—	85.9 cy/ft	—	
Nov 1996	683,862	100.00	77.7	0	
Mar 1997	633,708	92.67	72.0	−5.7	
Jun 1997	622,180	90.98	70.7	−7.0	
Oct 1997	580,187	84.84	65.9	−11.8	
Mar 1998	437,726	64.01	49.7	−28.0	
Oct 1998	410,232	59.99	46.6	−31.1	
Apr 1999	312,506	45.70	35.5	−42.2	
Nov 1999	241,545	35.32	27.4	−50.3	Includes background erosion

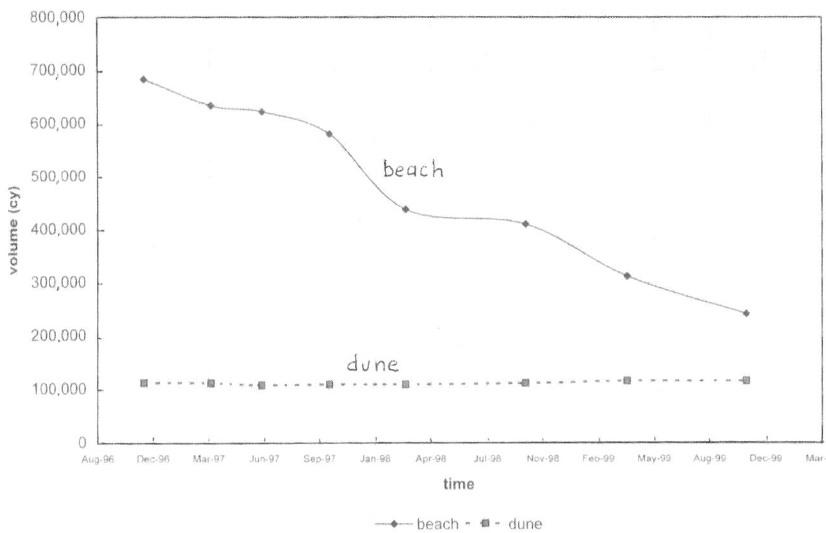

Figure 7.79. Volume (cy) within Dam Neck beach nourishment project limits (November 1996–November 1999).

results summarized in Table 7.3 for eight surveys with total volume remaining in column 2. The design survey volume was larger (row one) than that measured at the initial construction survey (November 1996, Row 2) for many reasons including the dredging contractor not filling the design template. Figure 7.79 is a plot of survey volumes measured over the three-year period.

Table 7.3 shows values for $M(t)$, the volume-change ratio (Column 3) that includes the background erosion rate of $-2.3\,\text{ft/year}$. Using 1 ft/year loss = 1 cy/ft loss gives 20,240 cy/year loss due to background erosion over the 8,800-ft project, over three years, that totals 60,720 cy. Adding this amount to the November 1999 total gives $241,545 + 60,720 = 302{,}265\,\text{cy}$. The corrected value for $M(t)$ after three years becomes $302,265/683,862 = 0.442$, or 44.2% remaining when background erosion is removed from $M(t)$ for November 1999 results. The theory as discussed herein assumes that background erosion has been removed.

In many respects, the wave year 1998 (October 1997–September 1998) was an extreme event. Twin extra-tropical storms (Northeasters) of late January and early February 1998 produced high storm surge elevations that permitted large waves to erode the subaerial beach. These extreme events dragged sand far offshore beyond the "closure depth" of $-21\,\text{ft}$ (NGVD 1929 datum). The $M(t)$ dropped from almost 66% in October 1997 to less than 50% the following March 1998 when over 140,000 cy moved offshore and beyond the lateral boundaries of the project.

Recall the diffusion Equation (7.47) for the spreading of sand for a beach nourishment project and repeated here as follows:

$$\frac{\partial y}{\partial t} = G\frac{\partial^2 y}{\partial x^2}, \qquad (7.46)$$

where G is the "longshore diffusivity" coefficient, and $y(x,t)$ is the location in time t and space x of the original design width $Y = \Delta y_o(t = 0)$ as schematically illustrated in Figure 7.29 by the dashed line for a rectangular beach nourishment project of length, l. This "spreading out" loss of sand is only a "loss" of sand within the project boundaries (Y, l). All the original sand mass is theoretically still present between infinite boundaries $\pm\infty$ along the $\pm x$-axis. Recall from Figure 7.71 that the Dam Neck beach monitoring program included seven additional profile locations outside the project boundaries. These profiles gained 212,196 cy of sand over the three-year period so that the total volume remaining (within the survey boundaries) was over 524,000 cy (77.7%) after correction for historic erosion losses.

The longhore diffusivity coefficient G has units of m^2/s (ft^2/s) and depends on the breaking wave height, H_b, the scale of the sand volume $(h_* + B)$ and all the same variables as found in the CERC formula for LST rate, Q in Equation (7.32). For a rectangular beach fill, the diffusion equation was solved to give $M(t)$, the percentage of beach fill remaining in time t.

$$M(t) = \frac{2\sqrt{Gt}}{l\sqrt{\pi}}(\exp^{-(l/2\sqrt{Gt})^2} - 1) + \text{erf}\left(\frac{l}{2\sqrt{Gt}}\right) \tag{7.51}$$

and $M(t)$ versus $\sqrt{Gt}/l$ is the solid curve plotted in Figure 7.31 (from Dean, 2002). The term "half-life" of a beach nourishment project is the time when 50% of the original sand volume remains, $M(t) = 0.5$. Inserting $M(t) = 0.50$ in Equation (7.52) and solving gives

$$\sqrt{Gt}/l = 0.481 \tag{7.52}$$

and solving for the half-life time t_{50} gives

$$t_{50} = (0.481)^2 \frac{l^2}{G} = 0.231 \frac{l^2}{G}. \tag{7.53}$$

Figure 7.80 shows the eight survey points (+) and three levels of the diffusivity coefficient G (low, design, and high) with resulting times for the half-life, t_{50}, of the project. Using linear regression for the data over the three-year monitoring period and adjusting for background erosion gave a best fit for an adjusted longshore diffusivity coefficient, $G^* = 0.1485\,\text{mi}^2/\text{year}$ (0.384

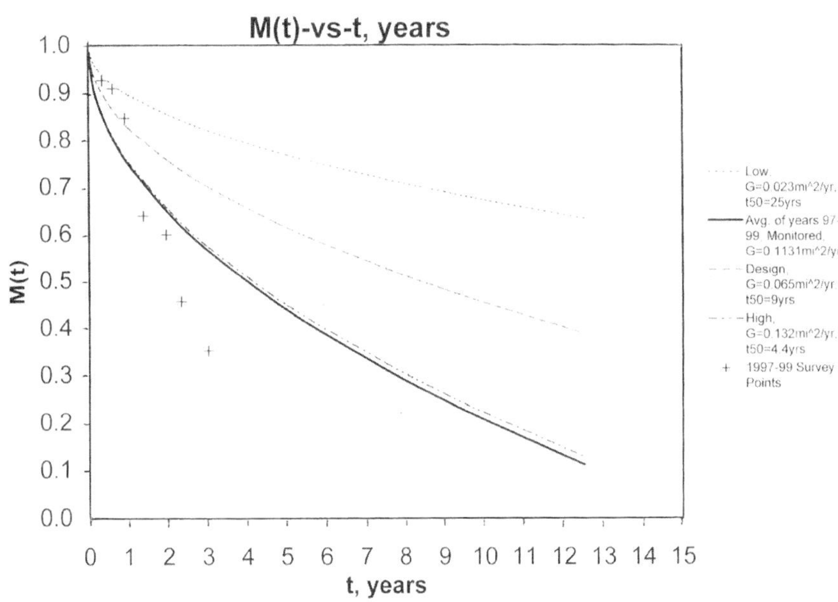

Figure 7.80. Theoretical $M(t)$ curves for three levels of the longshore diffusivity coefficient G and the resulting half-life time, t_{50}.

j = km^2/year). It is clear that the project design life will require more renourishment intervals than originally forecast. Use of any available information from similar beach nourishment projects in the area is far better than relying on estimates of average annual wave height values to estimate design values for G.

The enhanced dune design is discussed in detail in Section 7.4.2. A unique feature was the buried rock seawall/revetment structure within the dune (Figure 7.45). It was constructed well back inside the dune (Figure 3.16(a)) so that if a major storm caused severe dune erosion (like Hurricane Isabel, September 28, 2003), the sand volume would not be trapped behind the rock structure and would still be available for protection if a second storm occurred the same season. Construction cost (dollar value in 1996) was $6,882,250 for the beach, dune, and buried rock structure and was totally paid for by the US Navy. The first beach renourishment project (≈700,000 cy) occurred in 2004 (8-year interval) as financed by the Navy and a second renourishment project (700,000 cy) took place in 2015 (11-year interval) as financed by the federal government, Bureau of Ocean Energy Management (BOEM) within the US Department of the Interior.

At the time of this writing, the author has yet to identify the individual responsible for the NAVFAC — Atlantic Division records and future maintenance of Dam Neck beach. The consulting company that did the original design is no longer in business. The long-term memory of the beach nourishment design is unknown for future use.

Considering the relative amount of sand remaining three years after the initial construction (44.2%) (November 1999, Figure 7.79), the obvious question is how did Dam Neck beach survive by needing only two renourishments over almost 20 years? The answer is that the adjacent beach further south at Sandbridge also had beach nourishment projects that provided an additional source of sand that spread north to aid Dam Neck beach.

7.7.1.3 *Sandbridge beach*

The City of Virginia Beach instituted a Special Service Tax District for property owners living in its suburb Sandbridge (Ordinance No. 2297, November 1, 1994) where an extra $0.12 property tax per $100 assessed property valuation was collected and its use restricted to pay for beach nourishment costs. The first beach nourishment project occurred in 1998, cost $8 million, added 1 million cubic yards of sand, and 100% was financed by the City of Virginia Beach. Section 35.1.5 of this Ordinance includes the following:

> *to fund the creation and maintenance of a beach berm having an elevation approximately six (6) feet above mean sea level and a width of approximately fifty (50) feet from the nourishment line with a slope of approximately twenty to one (20:1) to the Atlantic Ocean. Implementation of the plan is expected to produce a designed dry beach area of approximately one hundred and twenty (120) feet, which in combination with the fifty-foot berm will yield a dry beach of approximately one hundred and seventy (170) feet at mean tide level. Actual or planned elevations, slopes and beach area may vary from time to time as determined by the district engineer of the Army Corp of Engineers, Norfolk District, and the city's director of public works*

In other words, recommendations for the actual beach design dimensions were specified in the Ordinance for the Special Service Tax District serving the 5.3 mi long (8 km) Sandbridge beach.

The borrow site for Sandbridge beach nourishment was the Sandbridge Shoal in nearshore waters about 3 mi (5 km) from the shoreline and the same as used for Dam Neck beach. Figure 7.81 displays the boundaries for the offshore borrow area employed (Nadal and Mazur, 2017). In the late 1980s, the Virginia Institute of Marine Science conducted nearshore bathymetric surveys, took surface grab samples, conducted seismic operations, and made core-boring and learned that over 75 million cubic yards of beach quality sand was available in the Sandbridge Shoal (Kimball and Dame, 1989). The original borrow area included all of regions A, B, and the cross-hatched, middle region where no dredging is now permitted. In 1996, during the first days of dredging this offshore borrow site for the initial Dam Neck project, the dredger was working in this middle region and cut some buried cables as employed by the US Navy for listening to underwater sounds of submarines in the region. The location of these cables was highly classified, and the US Navy was upset that they had been located and destroyed. The "No Dredging Within This Area" was then established as the cross-hatched region shown in Figure 7.81. The buried listening cable location is still classified, and it is not known if it remains in this no-dredging region. It is now called "buried communication cables" region in Corps of Engineers documents (Nadal and Mazur, 2017).

Renourishment of Sandbridge beach took place in 2003, 2007, 2013, and 2018. Sandbridge beach is now a fully authorized, 50-year project of the Corps of Engineers as established in (WRDA, 2000). The 35% local cost-sharing agreement with the Corps is paid by the City of Virginia Beach with

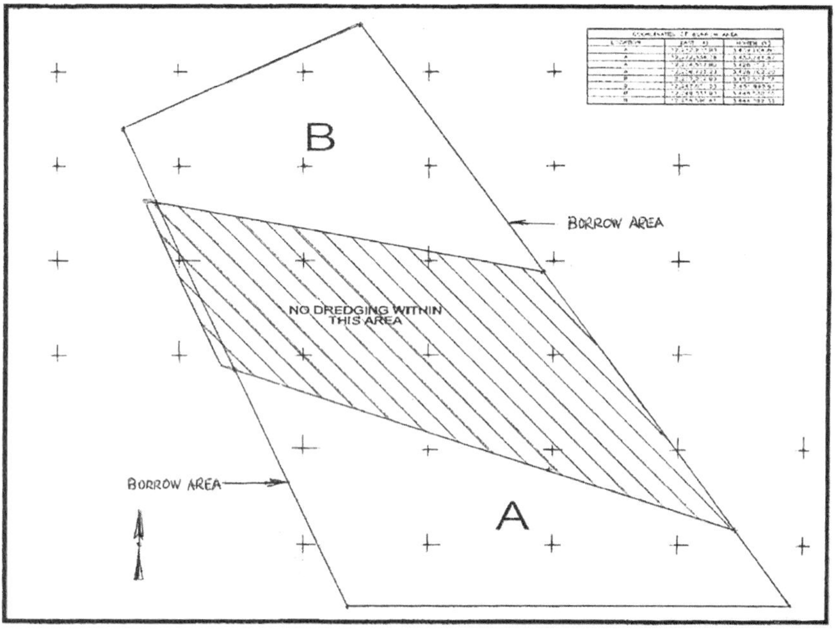

Figure 7.81. Borrow area in nearshore waters off Dam Neck and Sandbridge, VA employed for beach nourishment projects (from Nadal and Mazur, 2017).

full funding coming from the Sandbridge, Special Service Tax District. It is clear that the Dam Neck beach lying just north of Sandbridge beach will also benefit. Long-term data and design details are stored in the Norfolk District Office of the Corps of Engineers so that they are available for the design life of the project and beyond. The long-term memory of the beach nourishment design is available for future use.

Further discussion of Sandbridge beach is found in Chapter 10 when the interactions of the seawall/bulkheads with the beach are summarized as learned from nine years of monthly beach profile surveys.

7.7.2 *The "Outer Banks" of North Carolina — Nags Head Beach*

The "Outer Banks" of North Carolina is a chain of barrier islands along the Middle-Atlantic Ocean region on the East Coast of the United States. It begins below the border with Virginia (Figure 7.77) and extends south to Cape Hatteras, west to Ocracoke as shown in Figure 7.82, then extends

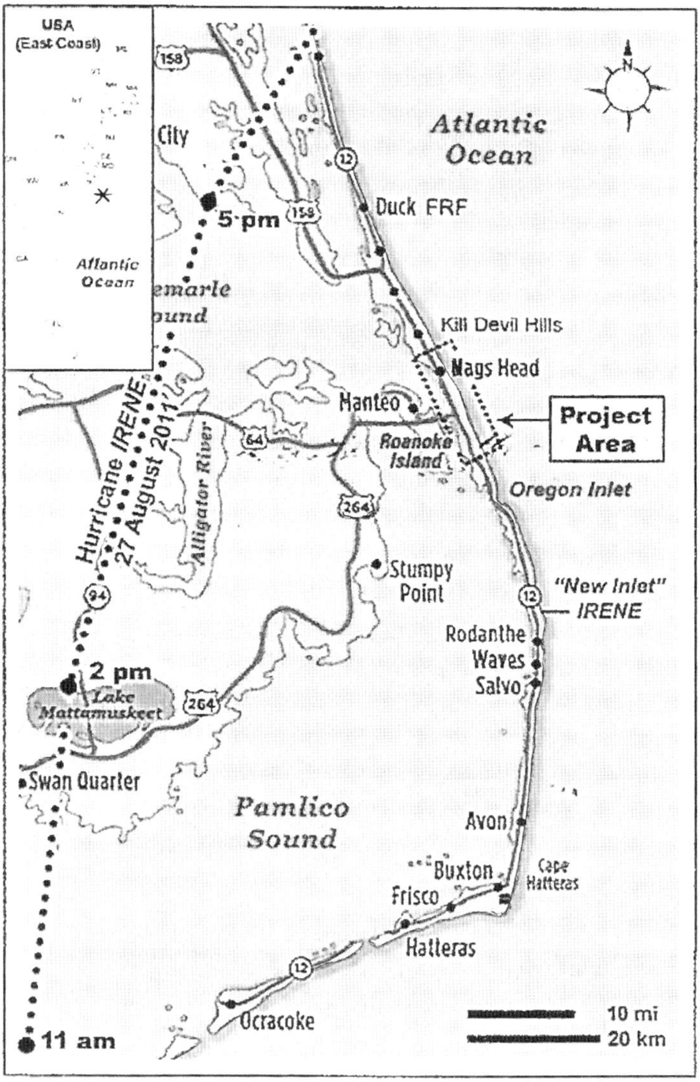

Figure 7.82. The "Outer Banks" of North Carolina and location of the town of Nags Head (from Kana and Kaczkowski, 2012).

further Southwest ending a Cape Lookout, NC. The famous, Corps of Engineers, FRF pier located at the Duck and Roanoke Island is the assumed home of the first English settlement (never located) in the late 1590s that preceded the Jamestown, Virginia settlement of 1607.

The Town of Nags Head (Dare County) fully financed the first, large, beach restoration project ever constructed on the "Outer Banks" (May 24–October 27, 2011) of North Carolina. The planning, design, construction, and performance monitoring was by Coastal Science & Engineering, Inc. (CS&E) Columbia, South Carolina, and this author is grateful to Dr. Tim Kana (president) for his help in summarizing this project. See also Kana and Kaczkowski (2012) for many more details.

7.7.2.1 *Design criteria and concerns*

The project was 16.1 km (10 mi) in length and coincided with the political boundaries of the town of Nags Head. Wave and water-level conditions are extremely well known since the Corps' FRF is only 32 km (20 mi) north, but they are not the five principal design concerns as determined by CS&E. They are listed in the following (italics) as directly found in Kana and Kaczkowski (2012a) and can be considered generically applicable to any beach nourishment project:

- *Determination of the average annual volumetric erosion rates, particularly the gradients of change alongshore.*
- *Calculation of the volume deficit with respect to an ideal or target profile volume.*
- *Identification and confirmation of suitable borrow area(s).*
- *Obtaining permits for hopper dredging outside the normal environmental protection windows.*
- *Formulation of a (beach) restoration plan meeting the goals, objectives, and budget of the local community.*

Each concern was addressed as summarized in the following.

7.7.2.2 *Annual volume erosion rates*

The North Carolina Department of Environment and Natural Resources (NCDENR, 2004) linear shoreline change rates obtained from aerial photographs were employed to determine the "official", 50-year, erosion rates as summarized in Figure 7.83 where values for each subsection are in meters per year (m/year). Over the project area, the erosion rate increases from −0.6 m/year (north end, right) to −2.0 m/year (south end, left) and even larger and then tapers off at Oregon Inlet. Figure 7.84 is an example of shoreline positions over time used to establish the erosion rates and previous locations of houses at the ocean front (online photos).

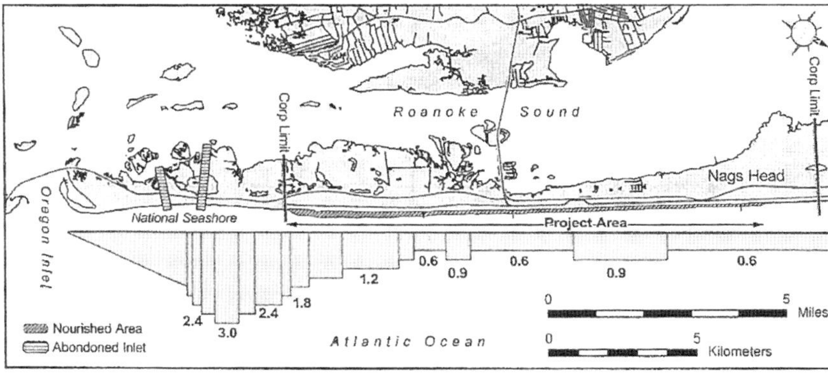

Figure 7.83. Erosion rates in meters/year (m/year) as determined from aerial photographs (from Kana and Kaczkowski, 2012).

Figure 7.84. Shoreline positions from maps and photos at Nags Head, NC (from Kana and Kaczkowski, 2012a).

To convert to annual, volume erosion rates requires beach profiles to correlate linear erosion distance to profile volume change. Along the US East Coast in the Middle-Atlantic region, Dean (2002) found that this relationship is approximately 1 m of beach erosion that equals $10\,m^3/m$ of volume change over the profile out to the closure depth (in Imperial units, 1 ft erosion = 1 cubic yard/ft volume change).

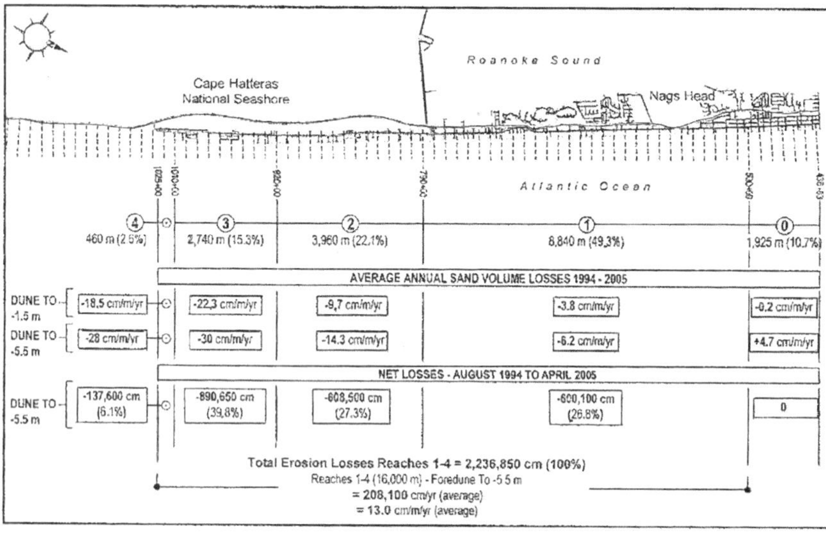

Figure 7.85. Volumetric erosion rates for five project reaches for the 1994–2005 period (from Kana and Kaczkowski, 2012). Note that cm = cubic meters.

The CS&E analysis employed this relationship to develop the results as shown in Figure 7.85 for the 1994–2005, 10-year period. They first reestablished a Corps of Engineers baseline (USACE, 2000) with stations (top, dashed vertical lines) at 1,000-ft (≈300-m) intervals and divided the project into five subsections based on different erosion rates. The lengths of each subsection are shown with Section 1 the longest (49.3%). Average annual sand volume loss rates are expressed as m^3/m/year (cm/m/year) with two values shown (left side of Figure 7.85) that depend on the location of the offshore profile cutoff depth. Using the North Atlantic Vertical Datum (NAVD), the first value is for the volume between the dune toe and −1.5 m (5.9 ft) NAVD which is considered the recreational beach. The second value is from the dune out to −5.5 m (19 ft) NAVD for this relatively short planning period of 10 years. The DOC for the long-term beach response is about 7.5 m (Birkemeier, 1985) for the FRF at Duck. For Section 1, the average annual sand volume loss rates are −3.8 m^3/m/year (cm/m/year) out to −1.8 m NAVD and −6.2 m^3/m/year (cm/m/year) out to the CS&E-adopted DOC of −5.5 m. Note that these values increase significantly in Sections 2 and 3 further south but over much shorter subsections of the project length.

The next row in Figure 7.85 integrates these average annual values over the 10-year period (1994–2005) and out to the CS&E-adopted DOC of

−5.5 m. We now learn that Section 3 which covers only 2,740 m (9,000 ft) of the project length (15.3%) accounts for 890,650 m^3 (cm) and almost 40% (39.8%) of the total erosion sand losses over 10 years. The total over the project length, Sections 1+2+3+4, is 2,236,850 m^3 for 10 years or about 208,000 m^3/year (274,000 cy/year) as the weighted averages. Per weighted average unit length, this is about 13.0 m^3/m/year (5.2 cy/ft/year) of volume loss. This Nags Head beach nourishment example demonstrates a design with a significant gradient in shoreline change rate.

7.7.2.3 *Beach nourishment volumes*

As discussed in this chapter, the normal practice is to restore the profile to achieve some subaerial beach width for storm damage mitigation and recreational use and to add some initial increase in volume in advance of expected annual losses over the design life or renourishment interval. This project was totally funded by the town of Nags Head and was considered an "interim" project in lieu of a future Corps of Engineers project receiving 65% funding from the federal government. Consequently, the beach nourishment volume employed by CS&E is based on a simple replacement volume plus a safety factor for the 10-year design life. This simple approach was supported by these three following reasons:

- The sand volume losses along the project length are relatively low compared with further south along Nags Head beach toward Oregon Inlet (see Figure 7.83).
- Funds were limited so that protective dunes and/or design widths based on storm damage mitigation to meet certain design criteria were not possible.
- Some dune growth and back beach buildup is expected.

The "interim" project volume was 3 million m^3 ±15% for the initial design or approximately 2.7–3.5 million m^3 (3.4–4.6 million cy). This included 10 years of erosion losses at 0.21 m^3/m plus a 50% safety factor. The estimated cost was approximately \$28–35 million (US) and the ±15% is Coastal Science and Engineering's standard practice for sand costs, volumes, and budgeted amounts may fluctuate ±15%. The anticipated unit costs were about \$10 m^3 (\$8 per cubic yard).

7.7.2.4 *Beach quality sand from borrow site*

Nags Head beach sand is predominately quartz and two native grain sizes were employed to characterize the (1) subaerial size $D1_N = 0.36$ mm and that (2) seaward of the MLW, $D2_N = 0.47$ mm. The CS&E experience with many projects has found that the subaerial size is more important because this area of the beach requires the most repair.

The Corps of Engineers (USACE, 2000) defined a large nearshore region (S1) shown in Figure 7.86 with over 75 million m^3 of beach-quality sand for potential use on Federal projects on the Outer Banks. CS&E conducted additional core borings and defined three subareas with areas 2 and 3 combined to yield over 6 million m^3 in the dredging depth range of 0.6–2.4 m (2–8 ft). The borrow material grain size $D_F = 0.42$–0.43 mm and was a close match with the native beach material in grain size, color, and concentration of gravel and shell. Content of mud was negligible. It was fortunate that this borrow area was located only about 2.8–3.8 km (1.7–2.4 mi) offshore in 12–18 m (30–60 ft) water depths suitable for hopper dredging. Subarea 1 in Figure 7.86 was not needed.

7.7.2.5 *Numerical model studies to refine initial design*

Coastal Science and Engineering also conducted numerical modeling studies to refine the design (Kaczkowski and Kana, 2012). The GENESIS model (Hanson, 1989) for shoreline change with waves simulated by the STWAVE model (see CEDAS, Veri-Tech, Inc) using the stations shown in Figure 7.85 and varying grid size dimensions. The optimum grid size was 30 m (100 ft) for both models. Wave climate analysis came for the FRF Duck data set and the Corps' WIS locations 221, 222, and 223. Model calibrations are discussed in detail in Kaczkowski and Kana (2012). The results were used to refine the preliminary nourishment plan as summarized above. CS&E beach profiles as shown in Figure 7.87 were employed to refine the relationship between shoreline change and volume required for beach fill per unit length. Using the March 2009 profile, the relationship employed was that, 1 ft of beach increase requires 1.11 cubic yards of sand.

7.7.2.6 *Final nourishment plan for recommended fill to meet budget*

State and Federal government permits for the project allowed up to 4.6 million cubic yards of material (4 million + 15%) to be placed along the 10.11-mi (53,400-ft) Nags Head town limits. The nourishment plan recommended is summarized in Table 7.4 and illustrates the non-uniform beach fill template employed for the four reaches with large differences in historic erosion rates (Figures 7.83 and 7.85). The average, modeled dry beach widths were found to be 68 ft after 1 year, 57 ft after 5 years, and 23 ft after

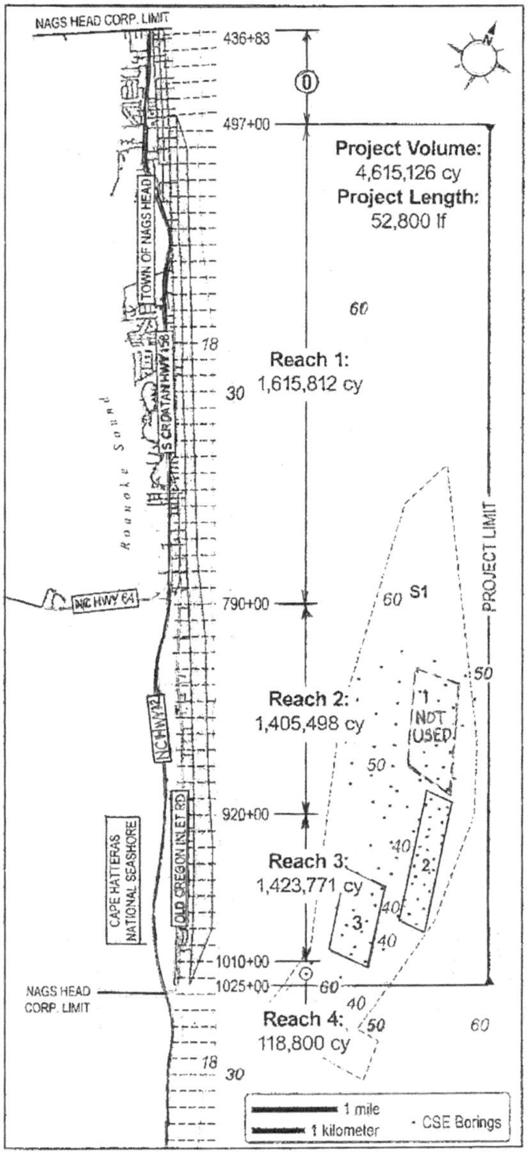

Figure 7.86. Location of borrow sites 1 and 2 employed for Nags Head beach nourishment (from Kana and Kaczkowski, 2012).

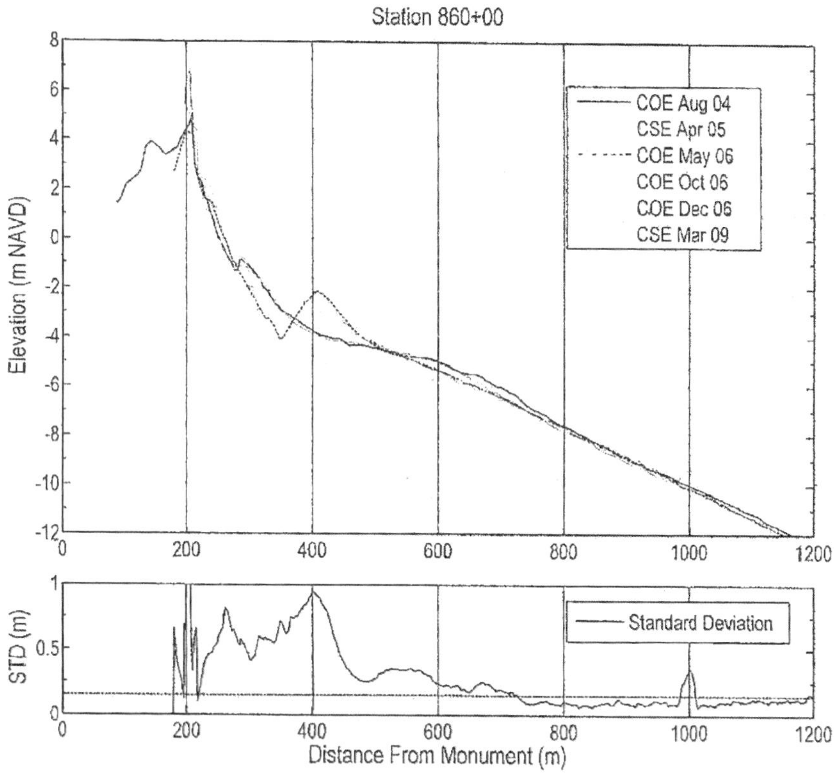

Figure 7.87. Beach profiles at Station 860+00 as typical for Nags Head Beach with closure depth about at −7 m NAVD (Kaczkowski and Kana, 2012).

Table 7.4. Nourishment plan for recommended beach fill.

Reach No.	Station Limits	Length, ft	Unit Volume, cy/ft	Total Reach Volume, cy
1–2	491+00 to 640+00	14,900	38	570,000
	640+00 to 740+00	10,000	50	500,000
2	740+00 to 860+00	12,000	65	780,000
	860+00 to 900+00	4,000	100	400,000
2–3	900+00 to 1000+00	10,000	150	1,500,000
3–4	1000+00 to 1025+00	2,500	100	250,00
Total	491+00 to 1025+00	53,400	75	4,000,000

Note: 1 ft ≈ 0.3 m, 1cy ≈ 0.76 m^3, and 1 cy/ft ≈ 2.5 m^3/m.

10 years of the design life. This expected retreat rate of ≈ -5 ft/year is similar to the historic, average erosion rate over the project area.

7.7.2.7 *Environmental permit and construction*

Sea turtle nesting was of environmental concern. Summer dredging was required, and nightly monitoring on the beach during construction and endangered species monitoring on board the dredge were required to obtain the environmental permit. Delays in environmental review for project approval meant it took over five years after the initial application to finally obtain the Environmental Permit.

Great Lakes Dredge & Dock Company submitted the low bid and below the Town of Nags Head's budget so that over 4.6 million cubic yards (3.52 million m^3) were placed on the beach. Three hopper dredges (Liberty, Padre, and Dodge Island) did the work with 85% of material placed over three summer months (May 24–August 27, 2011). Hurricane Irene (August 27, 2011) delayed completion of the beach nourishment at Nags Head (see Kana *et al.*, 2013 for details).

Project impacts of the borrow area on LST and of the beach nourishment project on Oregon Inlet shoaling (Figure 7.83) were studied with the numerical models and found negligible.

7.7.2.8 *Performance over the past six years*

As required, Coastal Science & Engineering has conducted annual (June) and mid-year, post-project surveys to monitor the volume of sand remaining on Nags Head Beach. The survey results can be found on the Town of Nags Head website (www.nagsheadnc.gov) and are an excellent example of how this technical data can be graphically displayed for the layperson.

Figure 7.88 (top, Graph A) presents three levels (lenses) of beach volume for analysis of profile volume change to graphically help the viewer understand the data displayed (bottom, Graph B) by bar charts for these three levels. All numbers are in cubic yards (cy). Year 1 (November 2012, purple) is after Hurricane Sandy, Year 5 (June 2016, brown is before), and then after (October 2016, gray) Hurricane Matthew. Note the large drop in total sand volume out to -19 ft NAVD resulting from the large storm surge and high waves in Hurricane Matthew (October 10, 2016). The project loss was 1,432,000 cy or about 30% of the nourishment volume. The last survey (July 2017) shows that about 200,000 cy of sand has recovered within the project since Hurricane Matthew. Relative to the initial placement volume,

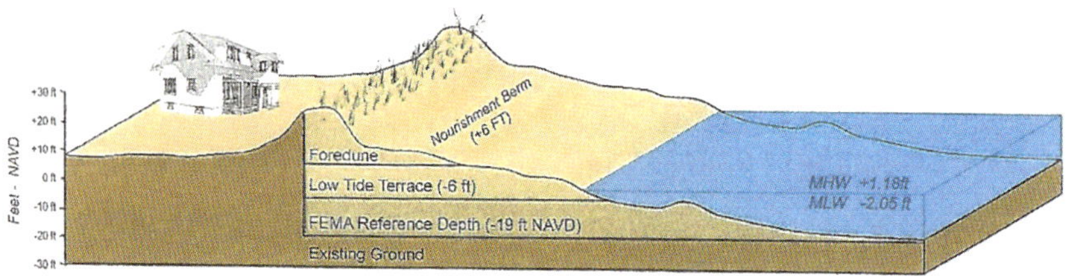

Graph A. Illustration of the three lenses used in the profile volume analysis for Nags Head.

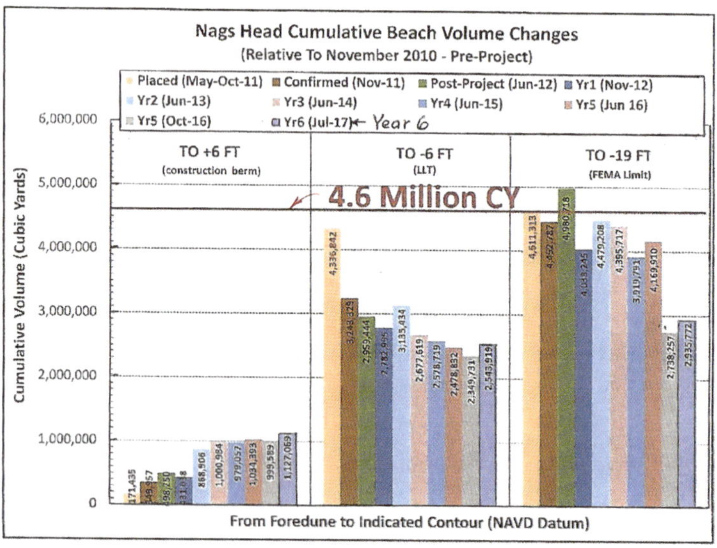

Graph B. Cumulative overall volume changes relative to November 2010 (pre-project) survey results between the foredune and indicated contours.

Figure 7.88. Graphical display of Graph A (top) showing three zones (lenses) of beach volume and Graph B (bottom) displaying the volumes of sand remaining within the three lenses and various survey dates (from Kaczkowski and Kana, 2012).

about 64% remains as of this last survey. This is about −5.5 cy/ft/year and slightly higher than the historic erosion rate (−5.2 cy/ft/year) employed in the project planning and design.

Figure 7.89 then shows these same survey results but divided into the project reaches as defined in Figure 7.85. All reaches lost sand from Hurricane Matthew.

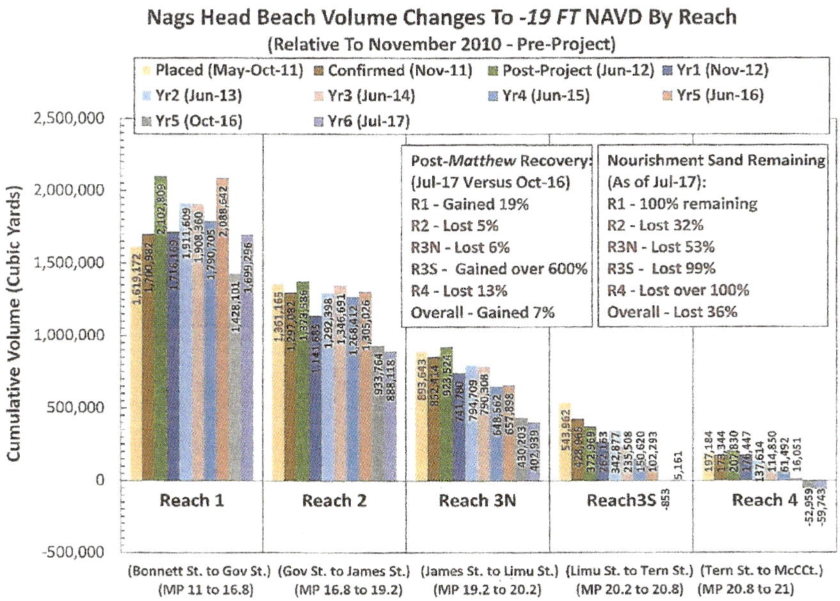

Figure 7.89. Total volume change to −19 ft depth in time for survey dates indicated and by Reach as designated in Figure 7.84 (from Kana and Kaczkowski, 2012).

Figure 7.90. Construction of Nags Head beach, North Carolina, Summer 2011 (Photo Credit: Nicholas Kusnetz/InsideClimate News).

7.7.2.9 *Future plans for Nags Head Beach*

At the time of this writing (Spring 2019), the Town of Nags Head was planning to have its first beach renourishment project in 2018 or 2019. The plan calls for a renourishment volume of 4 million cubic yards for the entire 10-mi project. However, as seen in Figure 7.89, Reach 3S (south end) and Reach 4 have experienced annual loss rates of over 20 cy/ft/year, which is significantly higher than the original design. This has been addressed in the planned renourishment project. The lowest bid (March 15, 2018) was by Great Lakes Dredge & Dock but was about 7% above the construction budget. Coastal Science & Engineering continues to serve as the Town of Nags Head's beach nourishment engineering firm. The overall goal is to continue to have a wide, healthy, sandy beach for all of Nags Head, NC as displayed in Figure 7.90 during initial construction in the summer of 2011.

Long-term data and design details are stored in the office of Coastal Science & Engineering (Columbia, SC), the Town of Nags Head files (website), and as published papers in coastal engineering journals. It is the firm belief of the author that the long-term memory of the beach nourishment design will be available for future use, in the form of such records. An overview of beach engineering in the United States is found in Basco (1999).

References

Abramowitz, M. and I. A. Stegun (1965) *Handbook of Mathematical Functions*, Dover Publications Inc., New York.

American Geological Institute (1966) The Beach: A River of Sand, Chicago, Ill.: Encyclopaedia Britannica Films in cooperation with American Geological Institute.

Bagnold, R. A. (1963) "Beach and Nearshore Processes Part I: Mechanics of Marine Sedimentation," In M. N. Hill ed., *The Sea: Ideas and Observations*, Vol. 3., New York, pp. 507–528.

Basco, D. R. (1975a) "Analytical Model of Hydraulic Pipeline Dredge," *Journal of the Waterways, Harbors, and Coastal Engineering Division*, Vol. 101, No. WW1, pp. 33–48.

Basco, D. R. (1975b) "Feedback from Field Studies of Hydraulic Dredges," *Journal of the Waterways, Harbors, and Coastal Engineering Division*, Vol. 101, No. WW3, pp. 277–289.

Basco, D. R. (1982) "Surf Zone Currents," US Army Corps of Engineers, Coastal Engineering Research Center, Misc. Rept. 82-7, Waterways Experiment Station, Vicksburg, MS, December.

Basco, D. R. (1983) "Surf Zone Currents," *Journal of Coastal Research*, Vol. 7, pp. 331–355.

Basco, D. R. (1998) "The Economic Analysis of 'Soft' Versus 'Hard' Solutions for Shore Protection," *Proceedings of the International Conference Coastal Engineering*, Vol. 2, ASCE, Copenhagen, Denmark.

Basco, D. R. (1999) "Overview of Beach Engineering in the United States of America," Final Report, Disaster Prevention Research Institute, Kyoto University, Uji, Japan, February.

Basco, D. R. (2000) "Beach Monitoring Results and Management Plan, FCT-CLANT, Dam Neck, Virginia Beach, VA," Final Report, Department of the Navy, NAVFAC and Glenn and Sadler, Norfolk, VA.

Basco, D. R. (2004) "Calibration of a Process Model for Barrier Island Breaching," *Proceedings of the 29th International Conference Coastal Engineering,* Vol. 2, ASCE, NY, pp. 2108–2119.

Basco, D. R., and B. E. Payne (2010) "Evaluation of FEMA's Guidelines for Coastal Dune Using a Numerical Model", Civil Engineering Department, Old Dominion University, Norfolk, VA.

Basco, D. R. and C. S. Shin (1996) "Dune Damage Curves and Their Use to Estimate Dune Maintenance Costs," in *Proceedings of the 25th International Conference on Coastal Engineering* (Orlando, FL), American Society of Civil Engineers.

Basco, D. R., and C.-S. Shin (1999) "A One-Dimensional Numerical Model for Storm-Breaching of Barrier Islands," *Journal of Coastal Research,* Vol. 15, No. 1, pp. 241–260.

Bayram, A., M. Larson, and H. Hanson (2007) "A New Formula for the Total Longshore Sediment Transport Rate," *Coastal Engineering*, Vol. 54, pp. 700–710

Beaches and Waterways Comm (2002) "Beach Management Plan," Public Works Department, City of Virginia Beach, Virginia.

Beck, T. M., J. D. Rosati, and Rosati, J. (2012) "An Update on Nearshore Berms in the Corps of Engineers: Recent Projects and Future Needs," Technical Note CHETN-XIV-10, Corps of Engineers, Vicksburg, MS.

Birkemeier, W. A. (1985) "Field Data on Seaward Limit of Profile Change," *Journal Waterway, Port, Coastal and Ocean Division*, Vol. III, No. 3, pp. 598–602.

Birkemeier, W. A., R. J. Savage, and M. W. Leffler (1988) "A Collection of Storm Erosion Field Data," Misc. Paper CERC 88-9, USACE, CERC, Fort Belvoir, VA.

Bodge, K. R. (1989) "A Literature Review of the Distribution of Longshore Sediment Transport Across the Surf Zone," *Journal of Coastal Research*, Vol. 5, No. 2.

Bowen, A. J. (1969) "The Generation of Longshore Currents on a Plane Beach," *Journal Geophysical Research,* Vol. 74, pp. 5479–5490.

Bowen, A. J., D. L. Inman, and V. P. Simmons (1968) "Wave Set-Down and Set-Up," *Journal Geophysical Research*, Vol. 73, pp. 2569–2577.

Brutsché, K. E., J. Rosati, C. E. Pollock, and B. C. McFall (2016) "Calculating Depth of Closure Using WIS Hindcast Data," ERDC/CHL CHETN-VI-45, Corps of Engineers, Vicksburg, MS.

Bruun, P. (1954) "Coast Erosion and the Development of Beach Profiles," Technical Memorandum No. 44, Corps of Engineers, Coastal Engineering Research Center, Fort Belvoir, VA, 82 p.

Bruun, P. (1962) "Sea Level Rise as a Cause of Shore Erosion," *Journal of Waterways and Harbors Division*, Vol. 88, pp. 117–130.

Caan, C. P. (1996) "Breach Growth, An Experimental Investigation of the Development of the Scour Hole," M.S. Thesis, Hydraulic and Geotechnical Divison, Delft University of Technology, The Netherlands.

Caldwell, J. M. (1956). "Wave Action and Sand Movement Near Anaheim Bay, California," U.S. Army, Beach Erosion Broad, Technical. Memorandum 68, 21 pp.

Campbell. T., B. De Sonneville, L. Benedet, and D. J. W. Walstra (2005) "Investigation of Morphosedimentary Processes on a Schematic Louisiana Barrier Island Using Process-Based Numerical Modeling," *Coastal Sediments '07*, ASCE, Washington, DC.

Cañizares R., S. Alfageme , and M. B. Gravens (2004) "Modeling of Storm Induced Barrier Island Overwash and Breaching in the Atlantic Coast of Long Island, New York," in *Proceedings of the 29th International Conference on Coastal Engineering* (Lisbon, Portugal), pp. 2146–2158.

Canizares, R., and J. L. Irish (2008) "Simulation of Storm-Induced Barrier Island Morphodynamics and Flooding," *Coastal Engineering*, Vol. 55, No. 12, pp. 1089–1101.

Carson, R. (1957) *Silent Spring*, Houghton Mifflin Publishing Co., New York.

CEM (2006) *Coastal Engineerng Manual*, US Army Corps of Engineers, Washington DC.

Cooper, N. J., D. J. Leggett, and J. P. Lowe (2007) "Beach-Profile Measurement, Theory and Analysis: Practical Guidance and Applied Case Studies," *Water and Environment Journal*, Vol. 14, No. 2, pp. 79–88.

Corps of Engineers (2003) "Hurricane Protection Project Saves $ Millions Damage from ISABEL," Newspaper Article, *Virginian Pilot, September*, Quoting Norfolk District, Corps of Engineers.

Dean, R. G. (1977) "Equilibrium Beach Profiles: US Atlantic and Gulf Coasts," Technical Report 12, Department of Civil Engineering, University of Delaware, Newark.

Dean, R. G. (1987) "Coastal Sediment Processes: Toward Engineering Solutions," *Proceedings of the Coastal Sediments*, ASCE, NY, pp. 1–24.

Dean, R. G. (1991) "Equilibrium Beach Profiles: Characteristics and Applications," *Journal of Coastal Research,* Vol. 7, pp. 53–84.

Dean, R. G. (2002) *Beach Nourishment: Theory and Practice*, World Scientific, Singapore.

Dean, R. G., and R. A. Dalrymple (2002) *Coastal Processes with Engineering Applications*, Cambridge University Press, Cambridge, UK.

Everts *et al.* (1983) "Shoreline Movements: Cape Henry Virginia to Cape Hatteras, NC, 1849–1980," Report No. 1, Technical. Reptort. CERC 83-1, Waterways Experiment Station, Vicksburg, MS, July, 123 p.

FEMA (1988) "Federal Register: New Rules and Definitions," 44CFR Sec 59.1 and 65.11, National Flood Insurance Program Regulation, October 1.

FEMA (2000) "Coastal Construction Manual," Federal Emergency Management Agency, Vol. I, Chapter 7, Washington, DC.

FEMA (2003) "Guidelines and Specifications for Flood Hazard Mapping Partners, Appendix D," Federal Emergency Management Administration, Flood Hazard Mapping Program, Washington DC.

FEMA (2017) "Guidelines and Standards for Flood Risk Analysis and Mapping," Appendix D, Federal Emergency Management Administration, Department of Homeland Security, Washington D.C. [Replaces FEMA's guidelines for mapping partners (2007)].

Field, M. E., and D. D. Duane (1974) "Geomorphology and Sediments of the Inner Continental Shelf, Cape Canaveral, Florida," Technical Memorandum 42, Coastal Engineering Research Center, Corps of Engineers, Waterways Experiment Station, Vicksburg, MS.

FSBPA (2017) http://www.coastalengineering.com/florida-shore-beach-preservation-association/.

Hallermeier, R. J. (1981) "A Profile Zonation for Seasonal Sand Beaches from Wave Climate," *Coastal Engineering*, Vol. 4, pp. 253–277.

Hallermeier, R. J. (1978). Uses for a Calculated Limit Depth to Beach Erosion," *Proceedings of the 16th Coastal Engineering Conference*, Vol. II, ASCE, NY, pp. 1493–1512.

Hallermeier, R. J., and P. E. Rhodes (1986) "Description and Assessment of Coastal Dune Erosion," Dewberry and Davis Consulting Engineers, Final Report, Fairfax, VA.

Hallermeier, R. J., and P. E. Rhodes (1988) "Generic Treatment of Dune Erosion for 100-Year Event," *Proceedings of the 21st International Conference Coastal Engineering*, Vol. II, ASCE, New York, pp. 1197–1211.

Hands, E.B., and M. C. Allison (1991) "Mound Migration in Deeper Water and Methods of Categorizing Active and Stable Depths," *Proceedings of the Coastal Sediments '91*, ASCE, pp. 1985–1999.

Hanson, H. (1989) "GENESIS — A Generalized Shoreline Change Numerical Model," *Journal Coastal Research,* Vol. 5, pp. 1–27.

Headland, J. R. (1991) "Erosion Study," Final Report, NAVFAC, Alexandria, VA.

Headland, J. R. (1992) "Design of Protection Dunes at Dam Neck, Virginia," *Proceedings of the Coastal Engineering Practice '92, Specialty Conference*, ASCE, Long Beach, CA.

Herbich, J. B. (ed.) (2000) *Handbook of Dredging Engineering*, 2nd edn., McGraw Hill, New York, NY.

Hillyer, T. M. (1996) "An Analysis of the US Army Corps of Engineers Shore Protection Program," Final Report, IWR 96-PS-1, Institute of Water Resources, Corps of Engineers, Fort Belvoir, VA.

Hillyer, T. M. (2003) "The Corps of Engineers and Shore Protection: History, Projects, Costs (2003)," Final Report, IWR 03-NSMS-1, Institute of Water Resources, Corps of Engineers, Fort Belvoir, VA.

Houston, J. R. (1995) "The Economic Value of Beaches," The CERCular, Coastal Engineering Research Center, CERC-95-4, Waterways Experiment Station, Vicksburg, MS.

Houston, J. R. (1996) "International Tourism & US Beaches," *Shore & Beach*, Vol. 63 (1), pp. 21–24.

Houston, J. R. (2002) "The Economic Value of Beaches — 2002 Update," *Shore & Beach*, Vol. 70, No. 1, pp. 9–12.

Houston, J. R. (2008) "The Economic Value of Beaches — 2008 Update," *Shore& Beach*, Vol. 76, No. 3, pp. 22–26.

Houston, J. R. (2013) "The Economic Value of Beaches — 2013 Update," *Shore & Beach*, Vol. 81, No. 1, pp. 1–10.

Inman, D. L., and R. A. Bagnold (1963) "Beach and Nearshore Processes Part II: Littoral Processes," In M. N. Hill (ed.), *The Sea: Ideas and Observations*, Vol. 3. New York, pp. 529–553.

Inman, D., and J. S. Shelton (1967) "The Beach: A River of Sand," Movie/Video, Encyclopedia Britannica Films, Inc. and American Geological Institute (WorldCat.org).

James, W. R. (1975) "Techniques in Evaluating Suitability of Borrow Material for Beach Nourishment," TM-60, Coastal Engineering. Research Center, Corps of Engineers, 81 pp.

Kaczkowski, H. L., and T. W. Kana (2012) "Final Design of the Nags Head Beach Nourishment Project Using Longshore and Cross-shore Numerical Models," *Proceedings of the 33rd International Conference on Coastal Engineering*, doi:1.10.9753/iece.v33.sedimont.64.

Kamphuis, J. W. (1991) "Alongshore Sediment Transport Rate Distributions," *Coastal Sediments '91*, ASCE, Washington DC, pp. 170–183.

Kamphuis, J. W. (2010) *Introduction to Coastal Engineering and Management* (2nd Edition), World Scientific, Singapore.

Kana, T. W. (1993) "The Profile Volume Approach to Beach Nourishment," In D. K. Stauble and N. C. Kraus (eds.), *Beach Nourishment Engineering and Management Considerations*, ASCE, NY, pp. 176–190.

Kana T. W., and H. L. Kaczkowski (2012a) "Planning, Preliminary Design, and Initial Performance of the Nags Head Beach Nourishment Project," *Proceedings of the 33rd International Conference Coastal Engineering*, Vol. 1, 33 pp.

Kana T. W., H. L. Kaczkowski, and S. B. Traynum (2017) "Building Blocks for a Sustainable Beach Nourishment Program," In Y. C. Kim (ed.), *Handbook of Coastal Engineering*, 2nd edn., World Scientific, pp. 1113–1138

Kana, T. W., S. B. Traynum, D. Gaudiano, H. L. Kaczkowski, and T. Hair (2013) "The Physical Condition of South Carolina Beaches 1980–2010," *Journal of Coastal Research, Special Issue*, Vol. 69, pp. 61–83.

Kimball, S. M., and J. K. Dame. (1989) "Geotechnical Evaluation of Sand Resources on the Inner Shelf of Southern Virginia," Vol. 1, Report and Appendices A-B, Final Report to City of Virginia Beach, VA, Virginia Institute of Marine Science, Gloucester, Pt. VA.

King, D. B. (2005) "Influence of Grain Size on Sediment Transport Rates with Emphasis on the Total Longshore Rate," CHETN-II-48, Engineering Research and Development Center, Corps of Engineers.

Klemas, V. (2011) "Remote Sensing of Wetlands: Case Studies Comparing Practical Techniques," *Journal of Coastal Research*, Vol. 27, No. 3, pp. 418–427.

Komar, P. D. (1971) "The Mechanics of Sand Transport on Beaches," *Journal Geophysical Research*, Vol. 76, No. 31, pp. 713–721.

Komar, P. D., and D. L. Inman (1970) "Longshore Sand Transport on Beaches," *Journal of Geophysical Research*, Vol. 75, No. 30, pp. 5514–5527.

Kraus, N. C., and T. O. Sasaki (1979) "Effect of Wave Angle and Lateral Mixing on the Longshore Current" *Marine Science Communications*, Vol. 5, pp. 91–126.

Kriebel, D. L., and R. G. Dean, (1993) "Convolution Method for Time-dependent Beach Profile Response," *Journal of Waterway, Port, Coastal and Ocean Engineering*, Vol. 119, No. 2, pp. 204–226.

Larson, M., H. Hanson, and N. C. Kraus (1987) "Analytical Solutions of the One-line Model of Shoreline Change," Technical Report, CERC-87-15, Corps of Engineers, Coastal Engineering Research Center.

Larson, M., H. Hanson, and N. C. Kraus (1997) "Analytical Solutions of One-line Model for Shoreline Change Near Coastal Structures," *Journal Waterway, Port, Coastal, Ocean Engineering* ASCE, Vol. 123, pp. 180–191.

Larson, M., and N. C. Kraus (1989) "SBEACH: Numerical Model for Simulating Storm Induced Beach Change," Technical Report CERC-89-9, US Army Corps of Engineers, Waterways Experiment Station, Vicksburg, MS.

Lenček, L., and G. Bosker (1998) *The Beach: the History of Paradise on Earth*, Penguin Publishing Inc., New York.

Longuet-Higgins, M. S. (1970) "Longshore Currents Generated by Obliquely Incident Sea Waves," *Journal Geophysical Research*, Vol. 75, pp. 6778–6801.

Longuet-Higgins, M. S., and R. W. Stuart (1964) "Radiation Stresses in Water Waves, A Physical Discussion With Applications," *Deep Sea Research*, Vol. 11, pp. 529–563.

McLellan, T. N., and N. C. Kraus (1991) "Design Guidance for Nearshore Berm Construction," *Proceedings of the Coastal Sediments '91*, ASCE, New York, pp. 2000–2011.

Meisberger, E. P., and D. B. Duane (1971) "Geomorphology and Sediments of the Inner Continental Shelf, Palm Beach to Cape Kennedy, Florida," Technical Memorandum No. 34, Coastal Engineering Research Center, U.S. Army Engineer Waterways Experiment Station, Vicksburg, MS.

Mil-Homens, J., R. Roshanka, J. S. M. van Thiel de Vries, and M. J. F. Stive (2013) Re-evaluation and Improvement of Three Commonly Used Bulk Longshore Sediment Transport Formulas," *Coastal Engineering*, Vol. 75, pp. 29–39.

Nadal, T., and K. Mazur (2017) "Sandbridge Beach Nourishment Project: Overview and Status," Norfolk District Office, Corps of Engineers, December.

NCDENR (2004) "Long-term Average Annual Shoreline Change Study and Setback Factors," North Carolina Department of Environment and Natural Resources, Raleigh.

NFIP (2006) "National Flood Insurance Program," administered by Federal Emergency Management Administration (FEMA).

Nicholls, R. J., M. Larson, M. Copobianco, and W. A. Birkemeier (1998). "Depth of Closure: Improving Understanding and Prediction," *Proceedings of the Coastal Engineering*, ASCE, New York, pp. 2888–2901.

Oertel, G. (2016) "Written Direct Testimony of George Oertel," Office of State Administrative Hearings, State of Georgia, Petitioners-versus-Sea Island Acquisitions, LLC and Shore Protection Committee, State of Georgia, April 25.

Pelnard-Considére, R. (1956) "Essai de Theorie de l'Evolution des Formes de Rivage en Plages de Sable et de Galets," *4th Journeess de l'Hydraulique, Les Energies de la Mer*, Question III, Rapport No.1.

Roehrs, P. (2003) Personal Communication, Beach Management Office, City of Virginia Beach, VA.

Roelvink, J. A., T. van Kessel, S. Alfageme, and R. Canizares (2003) "Modelling of Barrier Island Response to Storms," *Proceedings of the Coastal Sediments* 03, ASCE, New York.

Rozov, A. L. (2003) "Modeling of Washout of Dams," *Journal of Hydraulic Research*, Vol. 41, No. 6, IAHR, pp. 565–577.

Seabrook, J. (2013) "The Beach Builders: Can the Jersey Shore Be Saved?" *The New Yorker*, July 22, pp. 42–51.

Shin, C. S. (1999) "A One-Dimensional Model for Storm Breaching of Barrier Islands," PhD dissertation, Department of Civil and Environmental Engineering, Old Dominion University, Norfolk, VA.

SPM (1984) "The Shore Protection Manual," US Army Corp of Engineers, Washington DC.

Stronge W. B. (2005) "Economic Value of Beaches," in: Schwartz M.L. (ed.) Encyclopedia of Coastal Science, *Encyclopedia of Earth Science Series*, Springer, Dordrecht.

Svendsen, I. A. (2006) *Introduction to Nearshore Hydrodynamics*, World Scientific, Singapore.

Thomas, R. C., and A. E. Frey. (2013). "Shoreline Change Modeling Using One-line Models: General Model Comparison and Literature Review," ERDC/CHL CHETNII-55, US Army Engineer Research and Development Center, Vicksburg, MS.

Thornton, E. B. (1969) "Variation of Longshore Current Across the Surf Zone," *Proceedings of the 12th International Conference Coastal Engineering*, ASCE, Washington, DC, pp. 291–308.

Turner, T. M. (1996) *Fundamentals of Hydraulic Dredging* (2nd Edition), ASCE Press.

USACE (1995) "Breach Contingency Plan, Fire Island Inlet to Montauk Point, Long Island, New York, Reformulation Study," U.S. Army Corps of Engineers, New York District.

USACE (2000) "Final Feasibility Report and Environmental Impact Statement on Hurricane Protection and Beach Erosion Control: Dare County Beaches (Bodie Island Portion)," Dare County, North Carolina, Vol. I and II, US Army Corps of Engineers, Wilmington District, South Atlantic Division, 99 pp + appendices.

USACE (2007) "Baseline Conditions Storm Surge Modeling and Stage Frequency Generation: Fire Island to Montauk Point Reformulation Study," U.S. Army Corps of Engineers, New York District.

UNIBEST (2011) "UNIform BEach Sediment Transport- One-line Numerical Model," Deltares, Inc. Delft, The Netherlands (https://www.deltares.nl/).

Van de Graaff, J. (1989) "Probabilistic Design of Dunes; an Example from The Netherlands," *Coastal Engineering*, Vol. 9, pp. 479–500.

Van Kessel, T., and J. A. Roelvink (2002) "Modeling of Barrier Island Response to Storms," Special Contract Report, Task 2, Moffatt & Nichol Consulting Engineers, New York.

Verhagen, H. J. (1992) "Method for Artificial Beach Nourishment," *Proceedings of the 23rd Internatioinal Conference on Coastal Engineering*, Delft, ASCE, New York.

Visser, P. J. (1998) "Breach Growth in Sand-Dikes", Communications on Hydraulic and Geotechnical Engineering," Report No. 98.1, Delft University of Technology, The Netherlands, December, 172 pp.

Watts, G. M. (1953) "*A Study of Sand Movement at South Lake Worth Inlet, Florida*," US Army Corps of Engineers, Beach Erosion Board Technical Memo 42.

WRDA (2000) Water Resources Development Act, Public Works Projects, approved for Corps of Engineers, US Congress, Washington DC.

Wright *et al.* (1987) "Shoreline and Beach Dynamics of the Coastal Region from Cape Henry to False Cape Virginia," Technical Report, Virginia Institute of Marine Science, Gloucester Point, VA.

Chapter 8

Economics, Risk, and Resilience

8.1 Introduction

As introduced in Chapter 1 (Section 1.2.4) a key design constraint is economics. Each coastal damage mitigation and adaptation alternative has unique life-cycle costs, which are the sum of initial construction costs and maintenance costs over the design life of the project. It is important to have a clear definition of all terms when discussing costs. Nothing lasts forever. Advocates for permanent or "low" cost or retreat solutions for shore protection fail to understand the "cost" terms defined below and how they are applied in the professional practice of coastal engineering design economics. The estimation of initial and maintenance costs is the job and responsibility of professional engineers. Many individuals with formal training in the physical and biological sciences voice their opinions on the economics of coastal damage mitigation alternatives but with no formal education nor responsibilities to society in this regard. Their opinions should be viewed in this light.

8.1.1 *Definition of terms for "Costs"*

(a) **Costs.** The monetary value required for a project. Without additional qualifying words, the word alone is confusing and subject to misinterpretation.

(b) **Initial costs.** The total expense for all initial construction and design costs of a project. The year of construction should be noted so that inflationary cost aspects can be estimated in the future. Distinction should be made between estimates and actual contract costs.

(c) **Maintenance costs.** The estimated annual expenses required to maintain both the functional and structural integrity of the alternatives which are altered by storm damage and natural processes. Both deterministic estimates and risk-based calculation methods can be employed (see Section 8.4.2 for an example of the probabilistic method for rubble-mound breakwaters).

(d) **Alteration/removal costs.** The estimated expenses required to alter the design or completely remove the structure if there are significant, *down drift impacts.* Evidence from post-project monitoring and decision criteria mechanisms are required for implementation. This cost has been ignored for most projects and could be included in the category of maintenance costs.

(e) **Adaptation costs.** Because future water levels are unknown as a result of accelerating sea level rise in the world's oceans, a *new* "adaptation" cost category should be considered. Each alternative (e.g. rubble-mound, caisson, etc.) will have different future costs for adaptation to new, future sea levels and accompanying wave conditions. As a result, each alternative considered may have different life-cycle costs in the future at the end of the design life. Further discussion is found in Chapter 11.

(f) **Total, life-cycle cost.** The combined initial costs plus maintenance costs required over the design life of the project are typically the only life-cycle "costs" considered. Annual maintenance costs are converted to their present worth so that they can be directly combined with the initial construction costs. The present worth (value) is determined by multiplying the annual maintenance expense by the present worth factor, PWF. The PWF is a function of the design life and the interest (discount) rate (see further definitions below or any standard engineering economics text, e.g. Grant, 1950).

(g) **Design life, T.** An estimate of the number of years of useful life of the structure/alternative for the economic analysis. Usually 25–50 years is employed for well-designed projects. Design life selection includes structural life of materials (structural integrity), functional life (usefulness), technical life (technologically up-to-date), and aesthetics. Design life is employed for the economic analysis of the present worth of annual maintenance cost for the total, life-cycle cost comparison of all alternatives. It does not mean the length of time the project will last in the field. There are many examples of coastal structures that with proper maintenance continue to function well beyond the number of years employed in the economic analysis of life-cycle costs.

(h) **Present worth value of maintenance costs.** The sum of money set aside now to pay for all damages and functional loss over the design life (years) is the total present worth value of all future maintenance costs when applying principles of compound interest. It is simply the annual maintenance expense times the present worth factor (pwf).

(i) **Present worth factor (pwf).** The standard formula for the present worth factor (pwf) is:

$$\text{pwf} = \frac{[(1+i)^T - 1]}{i((1+i)^T)}, \tag{8.1}$$

where i = the interest (discount) rate per year expressed as a decimal (e.g. 3.5% = 0.035), T = the estimated (or required) design-life of the structure for economic analysis (e.g. 25, 50 years).

For example, when $i = 3.5\%$, then for $T = 25$ years, the pwf = 16.482.

(j) **Interest rate, *i*.** The second variable required to calculate the present worth of the annual maintenance costs. Often, the rate employed is related to the current, bank loan rate for construction projects. It can also be set by government policy.

(k) **Damage.** When energy levels in storms *exceed* the design levels, both structural damage and some loss in functional performance may occur. Repair is possible. Damage is an expected aspect of risk-based coastal design since there always exists some probability for storm energy greater than the design level (see Chapter 4).

(l) **Failure.** There are two definitions of "failure" as found in Coastal Engineering.

(1) **Design failure, f_D:** When storms with energy *below* the design level cause loss of structural integrity (damage) and/or functional performance, a "failure" in design (or construction) must be considered. A redesign may be needed before repair or reconstruction. Use of the word "failure" for loss of structural integrity or performance should be avoided until it can be proven that a design failure, f_D, took place under specified storm conditions that were less than that level used for the original design. Too often, coastal engineers misuse the word "failure" for a coastal structure following a storm event when damage occurs without knowledge of the design level employed in the original design.

(2) **Total system failure, f_{TS}:** Kamphuis (2010, 2nd Edition, Chapter 17) defines the physico-environmental subsystem (PES) as the combination of physical forces on the system and the environmental forces on the system. And that the probability of structural "failure", P_f, is the probability that the load exceeds the

resistance (see Chapter 4). The base of support for the "resistance" relative to the "load" as found in the PES subsystem design is then defined as the socio-economic subsystem (SES) that is made up of groups, governments, economics, and individuals so that the total system (TS) equals the sum of the PES and the SES, i.e. TS = PES + SES.

Kamphuis (2010) then defines a total system failure, f_{TS}, when the SES cannot bear the consequences (damage, death, environmental) of PES failure. There are many causes for failure of the PES part of the total system. One is simply that we cannot provide a low enough probability of failure, P_f, for the structure strength for the cost available for the structure. This will be further discussed in Section 8.6 when we deal with the consequences of total system failure and in Section 8.7 when the term Resilience is used in Contemporary Design. The bottom line is that "failure" must now be considered a fundamental part of the design process from the start, especially when facing the unknown future of accelerating sea level rise (Chapter 11).

(m) **Balanced design.** The most economical balance between the initial construction costs and maintenance (damage repair) costs so that the total, life-cycle cost is a minimum. Initial costs increase as the level of protection for more powerful (but rarer) storms increases, but maintenance costs decrease because damage is less frequent. The classical U-shaped, total cost curves result (see Section 8.4.2 for an example with rubble-mound breakwater design. See Section 8.2.3 for discussion of so-called "Low Cost" solutions for shore protection).

(n) **Optimized design.** The balanced design of all alternatives considering the benefit/cost analysis, environmental, and all other constraints. This includes calculating risk (see Section 8.6).

8.1.2 *Definitions of terms for "Benefits"*

Alternatives for coastal hazard mitigation are presented in Chapter 3. Reduction in flood damages and erosion losses along with wave attenuation are the primary benefits of these alternatives. Restoration, enhancement or creation of coastal ecosystems is another benefit. A full discussion of benefits is found in Sections 8.6 and 8.7 when Risk and Resilience, respectively, are treated in full. Then the costs and the benefits can be better understood as applied in the economic studies for coastal engineering design.

8.2 Initial Cost Estimates

Realistic estimates of the initial construction costs of all alternatives considered is the responsibility of the design engineer. It is not easy. Recent, similar constructions with actual cost data are the most accurate method. Location and year of construction differences can be adjusted using historical cost indexes, as for example using the quarterly, *RSMeans Construction Cost Indexes* (www.rsmeans.com) or the *Engineering News-Record* (www.enr.com). A wide variety of software is available to aid in cost analysis.

For all the alternatives considered, the volume of required material (rocks, concrete, steel, wood, sand, etc.) is a key variable in the cost. The second is the labor required to fabricate, transport, install/place, grade, etc. each alternative at the construction site. We herein only give some general information and observations regarding the making of initial cost estimates. Contacting local contractors is recommended.

8.2.1 *Rubble-mound structures*

Typically, costs from the quarry and/or rock supplier are given in terms of monetary units per weight of material. In the US, for example, the unit is \$ per ton (US), where 1 ton (US) = 2,000 pounds. In some cases, the rock supplier and labor costs for in-place construction are well established so that the in-place construction cost can be supplied by the contractor. For example, on the lower Chesapeake Bay, the in-place construction cost of rubble-mound groins, breakwaters, and revetments is about \$90 (US, 2014) per ton (Gunn, 2014). However, each type of rubble-mound structure may have a different construction unit weight (weight/unit volume) that must be specified before the structure cost can be determined. From experience, the following construction unit weights are employed for rubble-mound structures.

8.2.1.1 *Dense construction*

Use 1.8 tons (US) per cubic yard of rock volume.

As a check, this gives the unit weight $\gamma_{\text{dense}} = 133.3\,\text{lb/ft}^3$, so that for granite rock, with specific gravity of 2.65, the void ratio or porosity is about 19%, which appears reasonable. For dense construction, the stones can be placed manually for "best" fit or "chinked", using smaller rocks in the large void spaces.

8.2.1.2 *Loose construction*

Use 1.5 tons (US) per cubic yard of rock volume.

This is commonly employed for random or dumped rock placement. The unit weight $\gamma_{\text{loose}} = 111.1\,\text{lb/ft}^3$, so that again for granite rock, the porosity is about 33%, which looks correct.

When the construction density is not specified, then both values should be used, and a range of initial construction costs provided. In addition, the in-place cost will depend on how the rubble-mound material is transported to the site, which often depends on the source. Transport by land (train and truck haul) is often more expensive than transport by barge overwater. Again, contacting the local contractors is recommended.

8.2.2 *Concrete structures*

The main difference for coastal engineered structures is whether (or not) steel reinforcement is employed. Seawalls can be built with (or without) steel reinforcement (see Figure 1.8 for the Virginia Beach seawall with steel reinforcement and Figure 1.2 for the Galveston seawall without). Concrete armor units for groins, jetties, breakwaters, and revetments typically never include steel reinforcement. The concrete construction industry has vast experience in using concrete for roads, buildings, airports, bridges, etc. so that local cost data for these concrete structures will aid in determining costs for coastal concrete structures.

8.2.3 *"Low Cost" solutions for "shore protection"*

A "Shoreline Erosion Control Demonstration Act" in 1974 (Public Law 93-251) required the US Army Corps of Engineers to develop and demonstrate "low cost" methods of shore protection (USACE, 1981). From 1975 to 1980, 16 demonstration sites on all three coasts, in Alaska and on the Great Lakes, were studied for a wide variety of devices that resulted in the report "*Low Cost Shore Protection: A Guide for Engineers and Contractors*". In the introduction, it is stated:

> *The methods described in this report are usually appropriate for use only in sheltered waters. That is, they are generally not intended for open coast sites where they would be exposed to the undiminished attack of large oceanic waves. Use of most of these structures in such areas*

is definitely not recommended and entails a considerable risk of failure (USACE, 1981).

This general warning is important, but it is not the only cause for major alarm regards the shortcomings of this study and report. This study and report says **nothing about maintenance costs and life-cycle costs**. It is totally misleading in that it only deals with alternatives with relative *low initial construction costs* and does not address annual damage and resulting maintenance costs. In short, this study and the results are totally without merit and should be disregarded. The alternatives studied require substantial annual maintenance or replacement expenses that add up over time so that in most cases, the total, life-cycle costs far exceed more robust alternatives with higher initial costs but lower maintenance costs. In retrospect, it is extremely unfortunate that the US Army Corps of Engineers failed to take this into consideration when producing this study and writing the final report.

The study included some devices that were promoted by developers and marketed as new technology to control shoreline erosion. In many cases, they trapped or blocked sand moving laterally or cross-shore and produced negative sand volumes at another location. When questioned, their developers had no information regarding the strength of the device when exposed to high wave energy. Maintenance needs were unknown. The promoters of these devices are similar to "snake oil salesmen" of the last century selling products to cure all types of diseases without any concern for the initial sickness.

8.2.4 *Hidden costs*

Finally, each alternative considered has some impact on the surrounding environment. Chapter 10 focuses on the environmental impacts of alternatives. There is also the "do nothing" alternative and its cost. We return to this topic in Section 8.6 on Risk.

8.3 Estimates of Damage Curves

All alternatives require maintenance over the design life. And, all designs have some limit on structural stability for the level of probability of failure, P_f, that has been employed when all design constraints (scientific/technical, economic, environmental, etc.) have been considered. When the level of failure probability has been exceeded, the structure is damaged.

There are three damage/repair philosophies that can be considered as follows:

(1) **Immediate repair.** This is generally uneconomical for minor damage. The contractor is exposed to greater risk during the storm season, the work is harder and mobilization expense greater.
(2) **Annual repair.** All repairs are delayed until calm weather periods. This is the most common repair philosophy and recommended for estimates of annual maintenance costs.
(3) **No repair.** The risk for progressive increasing damage levels occurring once a given damage level is reached and ultimately structure failure makes this repair philosophy tenuous at best.

The repair philosophy considered will influence the estimates of maintenance costs for the design.

8.3.1 *Outline of probabilistic approach*

The probabilistic approach for estimating maintenance costs begins with the exceedance probability for the loading on the structure. This can be wave energy or water level or a combination of both. As the energy level increases, the probability of exceedance *decreases* for the wave energy as discussed in detail in Chapter 2, Section 2.3.8 using the POT method for extreme wave conditions. This is graphically introduced in Figure 8.1(a) with the exceedance probability, $P_>$, on the vertical axis for the frequency of load and the wave energy on the horizontal axis as the intensity of the load.

The middle curve in Figure 8.1(b) is the response of the structure as some measure of the percent of damage on the vertical axis which now *increases* as the intensity of the load, i.e. the wave energy, increases on the horizontal axis. This is called a "damage curve" for a coastal structure.

The product of the probability of exceedance, $P_>$, for the load (wave energy) and the percent of damage for load (wave energy) is a measure of the amount of damage for each incremental step change in wave energy as graphically illustrated in Figure 8.1(c). In this way, the lower wave energy levels occur more frequently but cause less damage, the middle levels of wave energy occur less often but cause more damage, and the high energy waves rarely occur and again cause less damage. The resultant probability levels of damage are a distribution of damage, that when summed together give the total amount of damage to the structure.

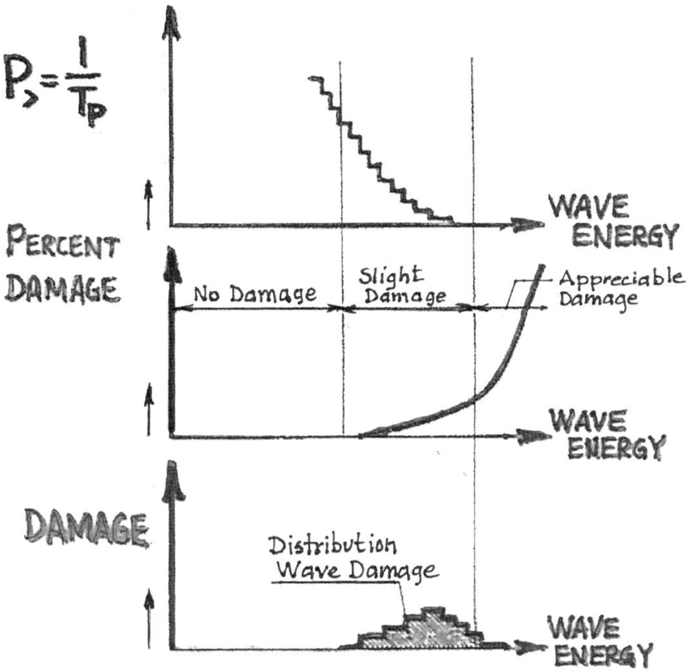

Figure 8.1. Schematic representation of the probability of exceedance of the load (a), the resultant damage response curve for the structure (b) and the distribution of structure damage (c) that all vary with the intensity of the load (wave energy).

The key variable is the damage curves for coastal structures. For property (homes, business, infrastructure, etc.) these are called "damage" curves, as discussed further in Section 8.6 and are a function of water level. The damage to the armor layer of rubble-mound structures from excessive wave energy has been studied the most since originally defined by Iribarren (1938) and then Hudson (1958, 1959).

8.3.2 *Damage curve definitions*

8.3.2.1 *Definition of armor layer damage for rock armor*

The outer layers of stones (armor layer) are most critical for stability. If the armor layer is eroded so that the smaller stones in the under layers are exposed, then the entire rubble-mound structure is vulnerable to structural failure. The relative level of the armor-layer damage can be defined several ways. See Table VI-5-19 in the CEM (2006) for five different definitions of the damage parameter. For rock armor, the eroded area, A_e, in Figure 8.2 relative to the square of the nominal, median rock size, D_{n50}, gives the relative eroded area, S (Equation (8.2)), as first defined by Broderick and Ahrens (1982) and employed by van der Meer (1987). This definition of the relative eroded area, S, for rock armor layers was introduced in Section 4.2.2 (Table 4.2) and then again in Section 6.2.3 (Table 6.2) and is repeated here as Table 8.1 for further discussion.

$$S = A_e/D_{n50}^2. \tag{8.2}$$

For natural rock, the various levels of damage, S, for a two-layer armor as determined by van der Meer (1988) are summarized in Table 8.1. Flatter slopes require a larger S value at the failure level.

The initial damage levels for S (2-3) are a practical lower limit recognizing that a few units may move and that taking $S = 0$ would cause Equations (6.5) and (6.6) to be zero. Hudson (1958) calls this the 0–5% damage level. The intermediate damage levels recognize that some rocks are displaced but without exposure of the under layer (or filter layer) by direct wave attack. The rubble-mound structural damage can be repaired, and the structure still performs its design function. The damage levels of S for failure (S = 8–17) means that the under layer (or filter layer) is exposed to wave attack and the structure may require redesign and may not perform its design functions as a breakwater, groin, jetty, or revetment. For the Hudson (1958) formula, this failure level was greater than 30% in terms of average eroded area from the profile relative to the original profile area. The definition for the relative eroded area, S (Equation (8.2)), is the preferred definition. For S values greater than 15–20, a rubble-mound structure with rock armor will form an S-shaped profile and should be analyzed as a dynamically stable structure, as discussed in Section 6.2.8 for berm breakwaters.

Table 8.1. Damage levels, S for two-layer, rock armor (van der Meer, 1988).

Unit	Slope, Cot α	Initial Damage, S	Intermediate Damage, S	Failure Damage Level, S
Rock	1.5	2	3–5	8
Rock	2	2	4–6	8
Rock	3	2	6–9	12
Rock	4–6	3	8–12	17

8.3.2.2 *Definition of armor layer damage for artificial, concrete units*

When artificial, concrete units are employed as the armor layer, then the number of displaced units within a unit width strip, D_n, widely defined as N_{od}, is the preferred relative damage level.

$$N_{od} = \frac{\text{number of units displaced out of armor layer width of tested section}}{D_n}. \tag{8.3}$$

For a two-layer concrete armor layer, van der Meer (1988) gave only the failure values for Cubes, Tetrapod and Accropode as $(N_{od})_{\text{failure}} = 2$, 1.5, and 0.5, respectively, when the structure slopes, cot α were 1.5, 1.5, and 1.33, respectively. The initial value of N_{od} was zero in all cases, but nothing was said about intermediate damage values for N_{od}. A relationship between S and N_{od} depends on structure slope, porosity, and concrete armor type with Table VI-5-20 in the CEM (2006) giving some values from model tests.

8.3.2.3 *Damage progression of rock armor layers*

Melby and Kobayashi (1998) and Melby (1999) developed a method for estimating the progression of damage to rock armor layers from a succession of storms with different wave conditions. See the CEM (2006) Part VI-5 and equations VI-5-62, 63, 64, and 65 for the results of model tests. Wave parameters near the structure toe and storm durations using probabilistic methods for long-term wave statistics are needed for reasonable sequences of storm conditions. These estimates are important if immediate repair is not practicable for the design and multiple storms may cause structural or functional failure of the structure.

8.3.3 *Damage curves for rubble-mound structures*

8.3.3.1 *Using van der Meer (1988) equations*

The equations for calculating the nominal diameter, D_n, for design of a rubble-mound structure with key coefficients given in Table 6.5 of Chapter 6 in shallow water for the van der Meer formulas are reproduced below (all mean values). Partial safety factors, $\gamma_z \gamma_H$, are found in Tables 6.3 and 6.4 and Equations ((6.18) and (6.19)) are used to find the reduction factor, f for moderate wave overtopping.

Plunging Waves ($\xi_m < \xi_{mc}$)

$$D_{n50} = \gamma_z \gamma_H \frac{H_s^T}{C_{pl}\, S^{0.2}\, P^{0.18} \Delta\, f(\xi_m^{-0.5})\, N_z^{-0.1}}. \tag{6.13}$$

Surging Waves ($\xi_m > \xi_{mc}$)

$$D_{n50} = \gamma_z \gamma_H \frac{H_s^T}{C_s\, S^{0.2}\, P^{-0.13} \Delta\, f(\xi_m^{P})\, N_z^{-0.1}\, (\cot\alpha)^{0.5}}. \tag{6.14}$$

Critical Surf Similarity Parameter

$$\xi_{mc} = \left(\frac{C_{pl}}{C_s}\, P^{0.31} (\tan\alpha)^{0.5}\right)^{1/(P+0.5)}. \tag{6.15}$$

Solving for the damage parameter, S gives

Plunging Waves ($\xi_m < \xi_{mc}$)

$$S^{0.2} = \gamma_z \gamma_H \frac{H_s^T}{C_{pl}\, D_{n50}\, P^{0.18} \Delta\, f(\xi_m^{-0.5})\, N_z^{-0.1}}$$

or

$$S = \left[\gamma_z \gamma_H \frac{H_s^T}{C_{pl}\, D_{n50}\, P^{0.18} \Delta\, f(\xi_m^{-0.5})\, N_z^{-0.1}}\right]^5. \tag{8.4}$$

Surging Waves ($\xi_m > \xi_{mc}$)

$$S^{0.2} = \gamma_z \gamma_H \frac{H_s^T}{C_s D_{n50}\, P^{-0.13} \Delta\, f(\xi_m^{P})\, N_z^{-0.1}\, (\cot\alpha)^{0.5}}$$

or

$$S = \left[\gamma_z \gamma_H \frac{H_s^T}{C_s D_{n50}\, P^{-0.13} \Delta\, f(\xi_m^{P})\, N_z^{-0.1}\, (\cot\alpha)^{0.5}}\right]^5 \tag{8.5}$$

with the same equation for the critical surf similarity parameter, ξ_{mc}, given by Equation (6.15).

8.3.3.2 *Damage level, D, and van der Meer level, S*

As shown in Figure 8.2, the eroded area, A_e, lies vertically below the armor layer surface. Another definition of damage is D, the percent of the original armor layer volume (Hudson, 1958).

$$D = \frac{\text{average eroded area from profile}}{\text{area of average original profile}} \times 100\%. \tag{8.6}$$

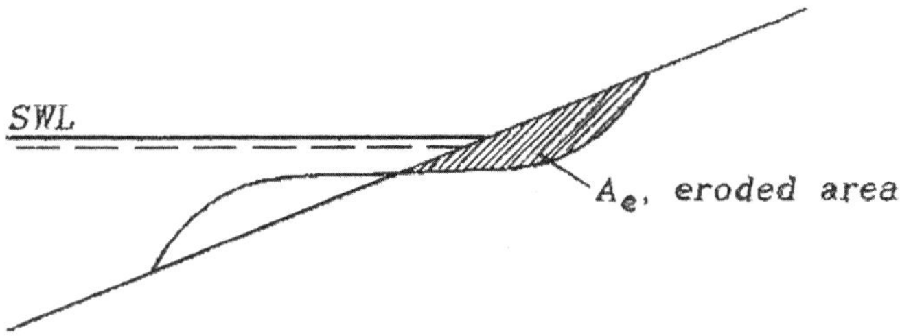

Figure 8.2. Definition of eroded area A_e in relative damage formulation, S.

Table 8.2. Damage levels, D, for two-layer, rock armor (Jackson, 1968).

Damage Level	Damage Range, D %
Initial	0–5
Intermediate	5–10
Failure	$\geq$20

Jackson (1968) suggested damage levels, D, for two-layer, rock armor (Jackson, 1968) on structure slopes cot $\alpha = 2$ and 3 shown in Table 8.2. The CEM (2006) states the values shown are percentage of eroded volume (Table VI-5-21).

We have developed the relationships between van der Meer's damage levels, S (area), in Table 8.1 and damage levels, D (volume), in Table 8.2 for initial, intermediate (range), and failure level conditions. Polynomial, curve-fit equations were developed for four structure slopes (cot $\alpha = 1.5$, 2.0, 3.0, and $\geq$4.0) that produced the following results summarized in Table 8.3. The equations are of the form with coefficients, a, b, and c so that once S is calculated from Equation (8.4) or (8.5), then the corresponding values of damage level, D, are determined. The last column in Table 8.3 is the regression coefficient, R^2, which was $\geq$0.998 for all four slope angles.

$$D = aS^2 + bS + c. \tag{8.7}$$

As an example, consider the same rubble-mound breakwater and non-overtopping conditions as found in Chapter 4 repeated here as

$$\cot\alpha = 2.0,$$

$$\rho_s = 2650\,\text{kg/m}^3,$$

$$\rho_w = 1025\,\text{kg/m}^3,$$

Table 8.3. Coefficients for use in polynomial Equation (8.7).

Cot α	a	b	c	R^2
1.5	0.1326	1.572	1.0795	0.9997
2.0	0.2231	0.0637	1.4002	0.9996
3.0	0.2027	1.1118	4.0071	0.9981
$\geq$4.0	0.0833	0.4167	3.000	1.0000

$$\Delta = 1.585,$$

$$P = 0.1,$$

$$S = 2(\text{dimensionless damage level}),$$

with storm and wave design conditions for deep water wave.

$$H_{sd} = 3.0\,\text{m at an annual probability of exceedance level of 50\%},$$

$$T_m = 6.0\,\text{s},$$

$$N_w = 2{,}000 \text{ waves}.$$

With partial safety factors, γ_H and $\gamma_Z = 1.0$, the rock diameter, $D_{n50} = 1.265\,\text{m}$, for a stable design. For these conditions, $\xi_m < \xi_{mc}$ and plunging waves apply so that using Equation (8.4) gives $S = 2.02$ and using Equation (8.7) gives $D = 2.44\%$. This is close to the initial damage level which was set at $D = 2.5\%$ and the average of 0–5% range shown in Table 8.2. Now taking H_s values greater than 3.0 m, we can calculate the increasing damage levels, S, using Equation (8.4) and resulting D levels using Equation (8.7). The results are plotted in Figure 8.3 for H_s/H_{sd} versus D (%) and noted as the S curve following van der Meer (1988). Note, the failure damage limit, $D = 20\%$, which means for $H_s/H_{sd} > 1.35$ structure failure is predicted. The generic damage curve from the Corps of Engineers model studies (Jackson, 1968) is also shown and gives the same general trend. Clearly, use of the van der Meer (1988) equations which include structure slope as an important variable and all the other variables (P, T_m, N_Z, plunging or surging, and partial safety factors) makes the van der Meer (1988) Equations (8.4) and (8.5) the correct way to estimate rubble-mound structure damage.

8.3.3.3 *Spreadsheets for damage calculations*

Development of spreadsheets with the ratio of the excessive significant wave height, H_s, to the design significant wave height, H_{sd}, i.e. (H_s/H_{sd}) the

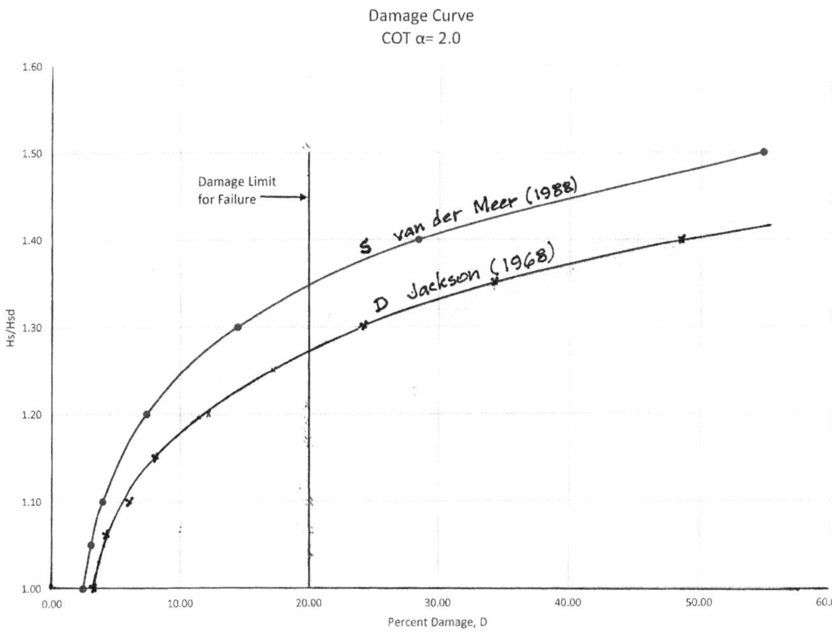

Figure 8.3. Damage curve from van der Meer (1988) for COT $\alpha = 2.0$ for example. Also shown is Corps of Engineers generic damage curve for any rubble-mound structure (Jackson, 1968).

key design variable greatly facilitates the application of these equations and tables for calculation of the damage level, S, and for estimate of D. They are found in the online supplementary materials accompanying this book. All employ the software product, Microsoft Excel and are located in the folder Structural Design/Rubble Mound/Armor Layer Damage. Figure 8.4(a) is an example for plunging waves under deep water conditions and Figure 8.4(b) is for surging waves. Separate spreadsheets are provided for both English and Metric unit systems with a Read Me page giving the formulas and additional information.

There is a considerable amount of information on these spreadsheets. But the pattern layout is the same for armor layer damage spreadsheets and follows the pattern (numbered) in Equation 8.4(a). Items 1, 2, and 3 are the same as in Figure 6.4 for calculation of the rock size and weight for various structure slopes. This information is then provided as "Design Data" [Item 4 in Figures 8.4(a) and 8.4(b)]. Item 5 is the calculated information and Item 6 is again a summary of the limitations of these formulas, e.g. $n = 2$ layers

1 **ARMOUR LAYER DAMAGE**
Rubble-Mound Structures and Non-Overtopping Slopes
(Metric System)

2 VAN DER MEER(1988) CEM(2000)
Table VI-6-5
Eqn VI-6-44 Plunging Waves
6 Plunging Coefficient, Cpl = 6.2
Quarry stone,rough angular
n=2 layers, random placement
Structure Trunk
COT alpha = 1.5 to 5.0
Damage conditions specified by S
Remarks: Eta must be less than Eta critical

3 **Enter Data**			Units
Design Wave Height,Hsd =		3.00	meters
Rho for rock	=	2650	Kn/m^3
Gravity	=	9.81	m/sec^2
Rho for water	=	1025	Kn/m^3
Delta = Sr-1	=	1.585	
Gamma Z	=	1.00	
Gamma H	=	1.00	

4 **Design Data**			Units
Nominal Diameter,Dn50=		**1.2650**	meters
Slope, COT α =		2.0	
Permeability,P	=	0.1	
No. of Waves,Nz	=	2000	
Wave Period,Tm	=	6.0	sec
Peak period,Tp	=	7.3	sec
5 **Calculated Information**			Units
Wave Length,Lom	=	56.64	meters
Steepness,Som	=	0.053	
Storm duration,t	=	3.35	hrs

7 SLOPE COT alpha	SURF PRA Eta	**DAMAGE S**	Rel Dam (Zero=2.0)	Eta critical	Remarks
2.0	2.1718612	2.019	**1.0**	3.574	**OK**

8 Hs/Hsd	Hs (storm)	Damage, S	COTα= 1.5 %Damage,D	COTα= 2.0 %Damage,D	COTα= 3.0 %Damage,D	COTα≥ 4.0 %Damage,D
1.00	3.00	2.02	2.63	2.44	2.59	2.50
1.05	3.15	2.58	3.85	3.05	2.49	2.48
1.1	3.30	3.25	5.43	3.97	2.54	2.53
1.2	3.60	5.02	10.17	7.35	3.54	3.01
1.3	3.90	7.50	18.16	14.42	7.06	4.56
1.4	4.20	10.86	31.63	28.40	15.84	8.30
1.5	4.50	15.33	54.19	54.82	34.61	16.19
1.7	5.10	28.67	152.96	186.57	138.72	59.51
2.0	6.00	64.61	654.00	936.81	778.31	323.80

9 **Partial Safety Factors= Plunging Waves**

Prob Failure	Sigma Waves = 0.5		Sigma Waves = 0.2	
0.01	1.6	1.04	1.9	1.00
0.05	1.4	1.02	1.5	1.06
0.10	1.3	1.00	1.3	1.10
0.20	1.2	1.00	1.2	1.06
0.40	1.0	1.08	1.0	1.10

10 **Coefficients for %Damage, D versus van der Meer, S values**

Slope, COT α	S^2 term	S term	Constant	R^2
1.5	0.1326	1.572	1.0795	0.9997
2.0	0.2231	0.0637	1.4002	0.9996
3.0	0.2027	1.1118	4.0071	0.9981
≥4.0	0.0833	0.4167	3.000	1.0000

(a)

ARMOUR LAYER DAMAGE
Rubble-Mound Structures and Non-Overtopping Slopes
(Metric System)

VAN DER MEER(1988) CEM(2000)
Table VI-6-6
Eqn VI-6-45 Surging Waves
Surging Coefficient, Cs= 1.0
Quarry stone,rough angular
n=2 layers, random placement
Structure Trunk
COT alpha = 1.5 to 5.0
Damage conditions specified by S
Remarks: Eta must be less than Eta critical

Enter Data			Units
Design Wave Height,Hsd =		3.00	meters
Rho for rock	=	2650	Kn/m^3
Gravity	=	9.81	m/sec^2
Rho for water	=	1025	Kn/m^3
Delta = Sr-1	=	1.585	
Gamma Z	=	1.00	
Gamma H	=	1.00	

Design Data			Units
Nominal Diameter,Dn50=		**1.2650**	meters
Slope, COT α =		2.0	
Permeability,P	=	0.1	
No. of Waves,Nz	=	2000	
Wave Period,Tm	=	6.0	sec
Peak period,Tp	=	7.3	sec
Calculated Information			Units
Wave Length,Lom	=	56.64	meters
Steepness,Som	=	0.053	
Storm duration,t	=	3.35	hrs

SLOPE COT alpha	SURF PRA Eta	**DAMAGE S**	Rel Dam (Zero=2.0)	Eta critical	Remarks
2.0	2.1718612	8.996	**4.5**	3.574	**PLUNGING Waves**

Hs/Hsd	Hs (storm)	Damage, S	COTα= 1.5 %Damage,D	COTα= 2.0 %Damage,D	COTα= 3.0 %Damage,D	COTα≥ 4.0 %Damage,D
1.00	3.00	9.0	23.79	20.03	10.41	5.99
1.05	3.15	11.5	34.45	31.54	17.96	9.20
1.1	3.30	14.5	49.53	49.15	30.45	14.45
1.2	3.60	22.4	100.55	114.61	80.68	35.41
1.3	3.90	33.4	199.35	252.41	193.00	82.01
1.4	4.20	48.4	385.34	526.68	424.66	177.81
1.5	4.50	68.3	725.05	1046.79	873.91	363.23
1.7	5.10	127.7	2362.90	3649.12	3168.78	1308.71
2.0	6.00	287.9	11439.04	18506.43	16480.26	6785.52

Partial Safety Factors= Surging Waves

Prob Failure	Sigma Waves = 0.05		Sigma Waves = 0.2	
0.01	1.70	1.00	1.90	1.02
0.05	1.30	1.10	1.60	1.00
0.10	1.30	1.02	1.40	1.04
0.20	1.10	1.10	1.20	1.08
0.40	1.00	1.08	1.10	1.00

Coefficients for %Damage, D versus van der Meer, S values

Slope, COT α	S^2 term	S term	Constant	R^2
1.5	0.1326	1.572	1.0795	0.9997
2.0	0.2231	0.0637	1.4002	0.9996
3.0	0.2027	1.1118	4.0071	0.9981
≥4.0	0.0833	0.4167	3.000	1.0000

(b)

Figure 8.4. (a) Spreadsheet for calculation of damage levels S and D for plunging waves on rubble-mound structures. Note, Remarks says "OK" for use of plunging waves; (b) Spreadsheet for calculation of damage levels S and D for surging waves on rubble-mound structures. Note, Remarks says use "PLUNGING Waves".

of armor stone. Item 7 now uses the design slope, cot α, to calculate the surf similarity parameter, ξ_m. In Figure 8.4(a), $\xi_m < \xi_{mc}$ so that plunging wave conditions are "OK". In Figure 8.4(b), these same conditions are not correct for using surging waves, so the Remarks say to use "PLUNGING Waves".

Item 8 is the key spreadsheet computation for a range of wave conditions with $H_s/H_{sd} \geq 1.0$ in column A. Then column B computes these wave heights, H_S, and then column C the corresponding damage level, S, using Equation (8.4). And then the damage level D using Equation (8.7) with coefficients in Table 8.3 is calculated. For this example spreadsheet, Item 10 coefficients are employed for cot $\alpha = 2.0$. Note that the failure level ($S = 8$, $D = 20$) lies between $1.3 \leq H_s/H_{sd} \leq 1.4$ and is about 1.35 as graphically displayed in Figure 8.3. Item 9 are the Partial Safety Factors and taken as 1.0 for this example.

A word of caution when using spreadsheets displayed in Figures 8.4(a) and 8.4(b). The original design data for D_{n50} and cot α should always be held constant when studying the calculated damage levels S (and D). The influence of the other variables (P, T_m, N_Z, plunging or surging, and partial safety factors) can then be adjusted and their influence determined. This may result in a redesign of the original D_{n50} and cot α design data.

8.3.4 *Damage curves for vertical-wall structures*

There are no simple damage curves for vertical-wall structures. In most cases, as wave energy increases above the design level ($H_s/H_{sd} \geq 1.0$), there is no visible "damage", i.e. a loss of structural material that forms the wall. However, at some extreme energy level, the wall may suddenly and completely "fail" in many different ways from many different mechanisms. Graphically, zero "damage" occurs until complete "failure" takes place.

Vertical-walled coastal structures are monolithic in cross-section consisting of one solid surface separating the water side from the interior. The water side is subject to hydrostatic water pressures plus dynamic wave pressures that integrated together over the water depth produce a horizontal force on the wall. Design of the structural integrity and stability of this vertical wall requires the ability to calculate this horizontal force on the wall. And, it is only one of other forces at work on the four main types of vertical-walled structures. The four types are mainly distinguished by their purpose (function) in coastal engineering design.

Seawalls mitigate water level and wave damage to what is on the interior side. Caissons are breakwaters that reduce wave heights on the leeward side for safe navigation and shoreline erosion mitigation. Bulkheads prevent loose soil materials in the interior side from eroding into the water on the sea side. And, flood walls mitigate water level damage to the interior side (like seawalls) but may have little or no wave forces to consider in their design. The design of all four types were reviewed in Chapter 6.

Proper seawall design requires the combined efforts of coastal engineers (wave and water level forces on the vertical wall; toe scour at the base of the wall to locate the lowest beach elevation; potential down drift impacts;

etc.); geotechnical engineers (soil properties behind the wall; active and passive soil pressures on the wall; active and passive soil forces and moments on the wall, etc.) and structural engineers (loads, stresses, strength of materials, etc.) for concrete, steel, and possible wood structures.. The design of seawalls is any extremely complicated process and we shall return to this subject in Chapter 11.

In contrast to seawalls, reports of structural damage and failure mechanisms of caissons are many, thanks to their extensive use for breakwater jetties in Japan for more than 75 years (Goda, 2010, 3rd Edition). Figure 8.5 illustrates 12 different failure mechanisms. The most dominant

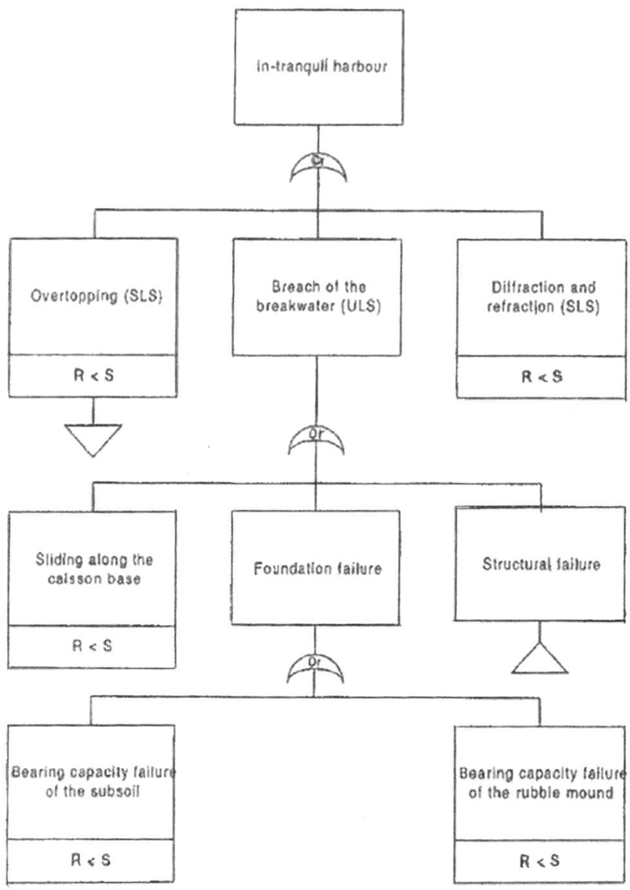

Figure 8.5. Failure mechanisms for caisson breakwaters (from Goda, 2010, 3rd Edition).

are: (3) translation (sliding), (5) rotation (overturning), and (9,11,12) foundation failures. A special issue on failure mechanisms for vertical breakwaters was published in *Coastal Engineering* (1994, Vol. 22).

Failure mechanisms for bulkheads and for flood walls were discussed in Sections 6.3.6 and 6.3.7, respectively.

8.3.5 *Damage curves for beaches and dunes*

At the opposite extreme are the fine-grained materials that form beaches and dunes. Fundamentals of the coastal hydrodynamics that produce sediment transport on beaches was covered in Section 7.2. Dune design is the topic of Section 7.4. In both cases, numerical models are employed to simulate cross-shore sediment movements that modify the original dune/beach profile to move sand in the offshore direction that could be considered "damage". Water levels are more important than wave conditions for dune erosion.

Figure 8.6 is an example of the use of the numerical model SBEACH to simulate dune cross-sectional area change, here called "percent

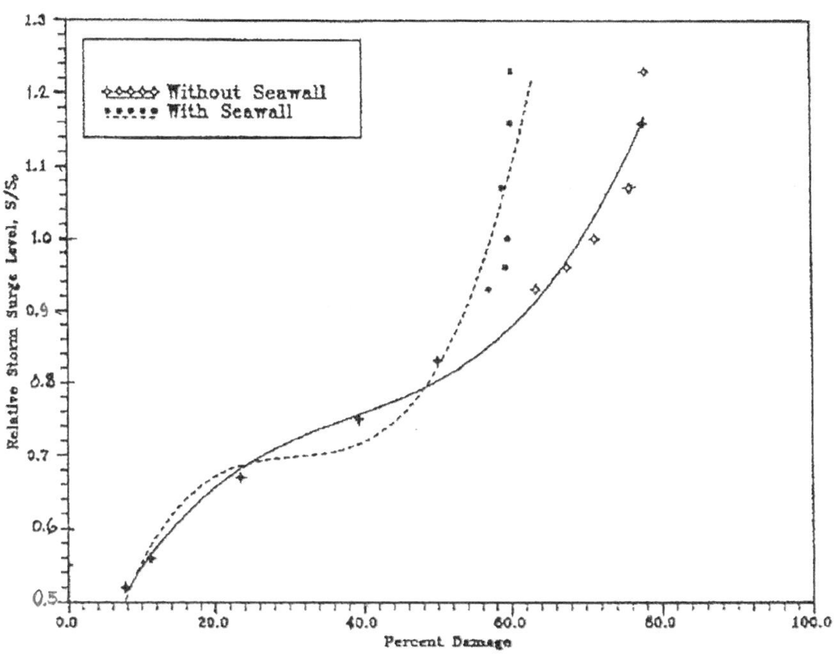

Figure 8.6. Dam Neck, VA dune design damage curve using SBEACH numerical model (Basco and Shin, 1996).

damage", for the Dam Neck beach/dune/buried seawall design discussed in Section 7.4.2.3. A key component in the cost analysis was the determination of dune damage curves to estimate dune maintenance costs (Basco and Shin, 1996). It was learned that a buried seawall within the dune section significantly prevented dune erosion for storm surge events, S, that were 80% and higher than the design storm surge level, S_D. Note that the dune began to erode (10% damage) when the relative storm surge S/S_D was about 50%. Dune failure is when the level of dune damage allows flooding of the interior behind the dune.

8.4 Maintenance Costs

8.4.1 *Introduction*

The probability that wave heights will be greater than the design wave height, i.e. the exceedance probability curve, $p_>$, versus H_s in any one year is the second variable needed to determine maintenance costs as graphically introduced in Figure 8.1(a). Chapter 2 presented the peak-over-threshold (POT) method of analysis for extreme (storm) wave data and the Wave Information System (WIS) method using numerical modeling of wind statistics to generate wave information in US coastal waters and develop exceedance probability curves. Chapter 4 included a discussion of the N-year probability, P_N, as the probability over a period of N-years that the given exceedance probability level $p_>$ used in design will be experienced. The N-year period is usually taken as the same as the design life.

The third variable needed is the cost, c, for repair of the annual structure damage, D. Since only the armor layer is damaged up to the failure level, the initial construction costs for the armor or cover layer of the structure, I_{cl}, should be considered. When repair takes place is also a factor. And, repair costs are generally double or more than the original construction cost, so a repair cost factor, z^*, should be considered. After failure, repair begins up to the level where the structure is destroyed, which requires special treatment to estimate structure maintenance costs by the probabilistic method.

8.4.1.1 *Maintenance costs to failure*

In mathematical form, the total annual damage repair cost, c, up to the wave height level where failure begins, H_{failure}, can be found from the expression

$$c = \int_{H_{sd}}^{H_{\text{failure}}} [(p_>)(D)(I_{cl} \times z^*)]d\,H_s, \tag{8.8}$$

where the three variables, $p_>$, D, and I_{cl} are all functions of H_s. Numerical integration is the simplest way to calculate c.

Dividing the probability of exceedance and damage curves into regions, we can reduce Equation (8.8) into the following formulation for spreadsheet solution.

$$c = \sum_{(H_s)_n}^{(H_s)_{n+1}} [(\Delta p_>)(\Delta \bar{D})(I_{cl} \times z^*)]\Delta H_s, \tag{8.9}$$

where for each region, $n = 1, 2, 3, \ldots$ to where H_{failure} begins.

$$\begin{aligned}
\bar{D} &= \text{ the average damage over the region, } \Delta\bar{D}\\
\Delta w &= (\Delta\bar{D})(I_{cl} \times z^*) = \text{the cost of repair for a region,}\\
\Delta p_> &= (p_>)_{n+1} - (p_>)_n = \text{the difference in the exceedance}\\
&\quad \text{probability over the region,}\\
\Delta c &= (\Delta p_>)(\Delta w) = \text{the annual repair cost for each}\\
&\quad \text{damage increment.}\\
\text{Then} \sum \Delta c &= \text{the total annual repair damage cost for all increments to}\\
&\quad \text{where failure begins.}
\end{aligned} \tag{8.10}$$

8.4.1.2 *Maintenance costs beyond failure for a destroyed structure*

Over the design life of the structure, N (years), the N-year probability to experience the design probability of exceedance, $p_>$, was discussed in Section 4.8.1 and given as Equation (4.43), repeated here

$$P_N = 1 - (1 - P_>)^N = 1 - \left(1 - \frac{1}{T_r}\right)^N \tag{4.43}$$

where $T_r = 1/p_>$, the recurrence interval.

For example, assume a design life N of 50 years and the design probability of exceedance $p_> = 0.2$. The N-year probability becomes

$$P_N = 1 - (1 - 0.2)^{50} = 1 - 0.0000143 \approx 1$$

which means that the design has almost a 100% chance of needing maintenance.

Now for the same design life, N of 50 years, assume the failure probability of exceedance $(p_>)_{\text{failure}} = 0.05$, so that the N-year probability becomes

$$P_N = 1 - (1 - 0.05)^{50} = 1 - 0.0769 = 0.923.$$

This means there is over a 92% chance that wave energy will be high enough over the design life to destroy the structure. For estimation of maintenance costs, the annual maintenance cost after failure to destruction, c^* is calculated as given by Equation (8.11). The damage level is taken at the start of failure; the initial construction cost (\$/m) for the entire structure, I, is employed for rebuilding; a rebuilding cost factor, z^{**}, tied to a construction cost index is used; and the exceedance probability at the level of total destruction is assumed very small and approximately zero.

$$c^* = D_{\text{failure}}(I \times z^{**})[(p_>)_{\text{failure}} - 0]. \tag{8.11}$$

Thus, c^* is the last region needed to estimate the total annual maintenance costs $(\sum \Delta c + c^*)$.

8.4.1.3 *Life-cycle maintenance costs*

The total present worth value, C_m, of all future maintenance costs when applying principles of compound interest is simply the annual maintenance costs $(\sum \Delta c + c^*)$ times the present worth factor (pwf) from Equation (8.2).

$$\begin{aligned} C_m &= \text{pwf} \times \left(\sum \Delta c + c^*\right) \\ &= \text{the present worth of the life-cycle maintenance costs.} \end{aligned} \tag{8.12}$$

This probabilistic method to estimate maintenance costs was first introduced to the author in 1975 at the Delft International Courses by Mr. Kees d'Angremond. It was first published by Kreeke and Paape (1964).

8.4.2 *Rubble-mound structure maintenance costs*

Application of Equations (8.9)–(8.12) is best illustrated by the same rubble-mound breakwater example as found in Chapter 4 and employed in Section 8.3.3 for calculation of the damage curve (Figure 8.3). The design wave height, $H_{sd} = 3.0\,\text{m}$ on a COT $\alpha = 2$ slope, required a rock diameter $D_{n50} = 1.265\,\text{m}$ for a stable design when both partial safety factors were taken as unity. Figure 8.7 now displays the percent damage, D, as a function of the wave height, H_s, greater than $H_{sd} = 3.0\,\text{m}$ and divides the damage curve into six regions for analysis. The number of regions taken is arbitrary. The wave heights at the boundaries of these six regions are then used to determine the exceedance probabilities $p_>$ as shown in Figure 8.8 for this example. A simple logarithmic, extreme wave height distribution is assumed with $H_{sd} = 3.0\,\text{m}$ at $p_> = 0.5$ and $H_s = 4.5\,\text{m}$ at the 100-year $(p_> = 0.01)$ level. See Figures 2.32 and 2.33 of Chapter 2 for real extreme wave height distributions that fit the simple logarithmic model.

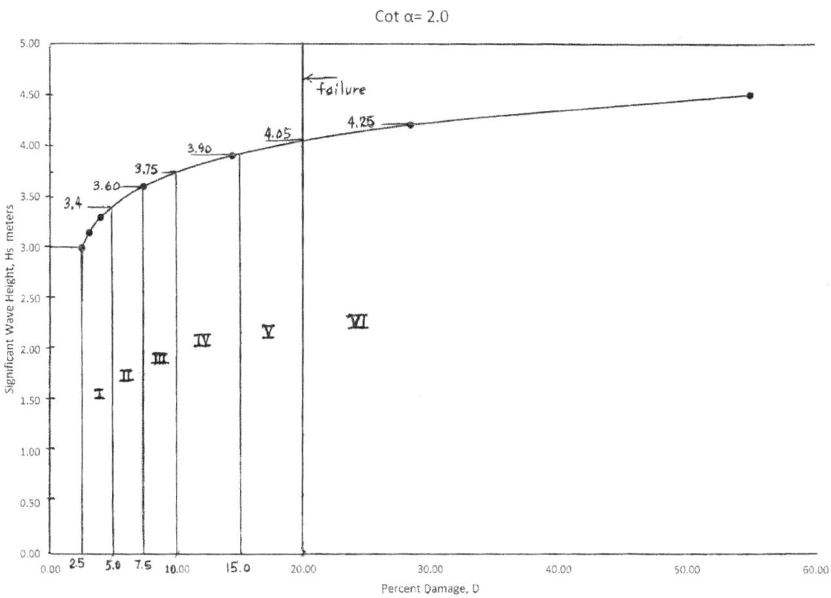

Figure 8.7. Damage level D versus wave height H_s, for example, showing six regions for calculation of maintenance costs.

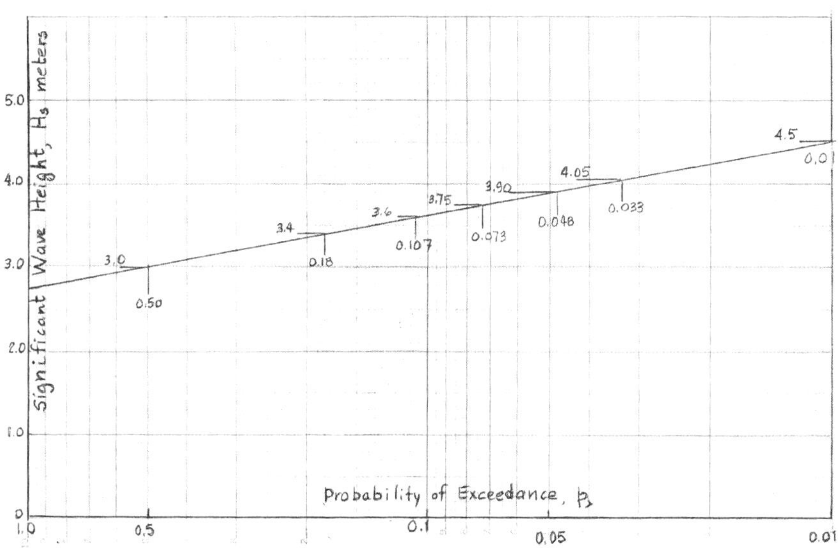

Figure 8.8. Probability of exceedance $p_>$ versus wave height, for example, in six regions for calculation of maintenance costs.

For this example, we assume the following additional cost information. It is also assumed that annual repair costs will be made.

$$
\begin{aligned}
I_{cl} &= \$10,000 = \text{cost of armor layer per unit meter},\\
z^* &= 2 = \text{repair cost factor for armor layer},\\
I &= \$25,000 = \text{initial cost of rubble-mound breakwater per unit meter}\\
&\quad \text{for } H_{sd} = 3.0\,\text{m design level},\\
z^{**} &= 1.5 = \text{rebuild cost factor for entire breakwater},\\
N &= 25 \text{ years design life},\\
i &= 3.5\% \text{ interest rate for present worth factor, pwf.}
\end{aligned}
$$

In region I:

$$
\begin{aligned}
\bar{D} &= \text{the average damage} = (0.025 + 0.05)/2 = 0.0375,\\
\Delta w &= (\Delta\bar{D})(I_{cl} \times z^*) = \text{the cost of repair} = 0.0375 \times 10,000 \times 2 = \$750,\\
\Delta p_{>} &= (p_{>})_{n+1} - (p_{>})_n = \text{the difference in the exceedance}\\
&\quad \text{probability} = 0.500 - 0.180 = 0.320,\\
\Delta c &= (\Delta p_{>})(\Delta w) = \text{the annual repair cost} = 0.320 \times \$750 = \$240.
\end{aligned}
$$

In region II:

$$
\begin{aligned}
\bar{D} &= \text{the average damage} = (0.05 + 0.075)/2 = 0.0625,\\
\Delta w &= (\Delta\bar{D})(I_{cl} \times z^*) = \text{the cost of repair} = 0.0625 \times 10,000 \times 2 = \$1,250,\\
\Delta p_{>} &= (p_{>})_{n+1} - (p_{>})_n = \text{the difference in the exceedance}\\
&\quad \text{probability} = 0.180 - 0.107 = 0.073,\\
\Delta c &= (\Delta p_{>})(\Delta w) = \text{the annual repair cost} = 0.073 \times \$1,250 = \$91.25.
\end{aligned}
$$

And similarly for regions III, IV, and V as shown in Table 8.4. Summing the Δc values gives

$$\sum \Delta c = \$505.75/\text{m}.$$

Region VI is the special region after failure begins ($D = 20\%$, $H_s = 4.05$) to destruction.

$$
\begin{aligned}
c^* &= D_{\text{failure}}(\text{I} \times z^{**})[(p_{>})_{\text{failure}} - 0]\\
&= 0.20 \times (25,000 \times 1.5) \times (0.033 - 0) = \$247.50/\text{m}.
\end{aligned}
$$

Table 8.4. Example calculation of C_m, the present-worth value of life-cycle maintenance costs.

Region	H, m	% Damage, D			Delta w	Exceedance Probability			Delta c
Design	3	Low	High	Average	\$	High	Low	Difference	\$
I	3.4	0.025	0.05	0.0375	750	0.5	0.18	0.32	240
II	3.6	0.05	0.075	0.0625	1,250	0.18	0.107	0.073	91.25
III	3.75	0.075	0.1	0.0875	1,750	0.107	0.073	0.034	59.5
IV	3.9	0.1	0.15	0.125	2,500	0.073	0.048	0.025	62.5
V	4.05	0.15	0.2	0.175	3,500	0.048	0.033	0.015	52.5
									$\sum$ delta $c = 505.75$
VI	4.25	0.2	0.2	0.2	7,500	0.033	0	0.033	247.5
									$\sum$ delta $c + c^* = 753.25$
									Maintenance Costs, $Cm = 12{,}414.701$
									Life-Cycle Costs, $C = \mathrm{I} + Cm = 37{,}414.701$

The total annual maintenance costs for all six regions is

$$\Sigma\Delta c + c^* = \$(505.75 + 247.50) = \$753.25/\text{m}.$$

The life-cycle maintenance costs over the 25-year design life then are

$$C_m = \text{pwf} \times (\Sigma\Delta c + c^*)$$

so that we must first compute the present worth factor, pwf (Equation (8.2))

$$\text{pwf} = \frac{[(1+i)^T - 1]}{i((1+i)^T)} = \frac{[(1\ +\ 0.035)\text{^}25 - 1]}{0.035(1+0.035)\text{^}25} = 16.482.$$

Finally, the life-cycle maintenance costs are

$$C_m = 16.482 \times 753.25 = \$12,415/\text{m}.$$

This example assumed that annual repairs were made to the breakwater. This produced higher maintenance costs than if annual repairs were not made. It is good engineering practice to include annual repairs in the maintenance costs for all alternatives to give conservative estimates.

8.4.3 *Dune maintenance costs*

As illustrated in Figure 8.6 for the example of percent dune damage at Dam Neck, VA, the storm surge water level, S, is now the key hydrodynamic variable and not the wave height. Figure 8.9 (left side) shows the S/S_D levels versus the percent dune cross-section damage, n, and eight regions for maintenance cost analysis. At Dam Neck, VA, the exceedance probability level $p_>$ is then also broken into eight Δp regions (Figure 8.9, middle).

The dune cross-sectional area was 30 cy/ft and sand unit costs were taken as \$15/cy so that the initial dune cost, IC, was \$450/linear ft of beach (Basco and Shin, 1996). A dune cost repair factor, z, was taken as unity. Using similar regional nomenclature as above for the rubble-mound structure:

$\Delta w = (\Delta\bar{n})(IC \times z) =$ the cost of repair for a region of average dune damage, $\Delta\bar{n}$,

$\Delta p_> = (p_>)_{n+1} - (p_>)_n =$ the difference in the exceedance probability over the same region,

$\Delta s = (\Delta p_>)(\Delta w) =$ the annual repair cost for each damage increment of the dune.

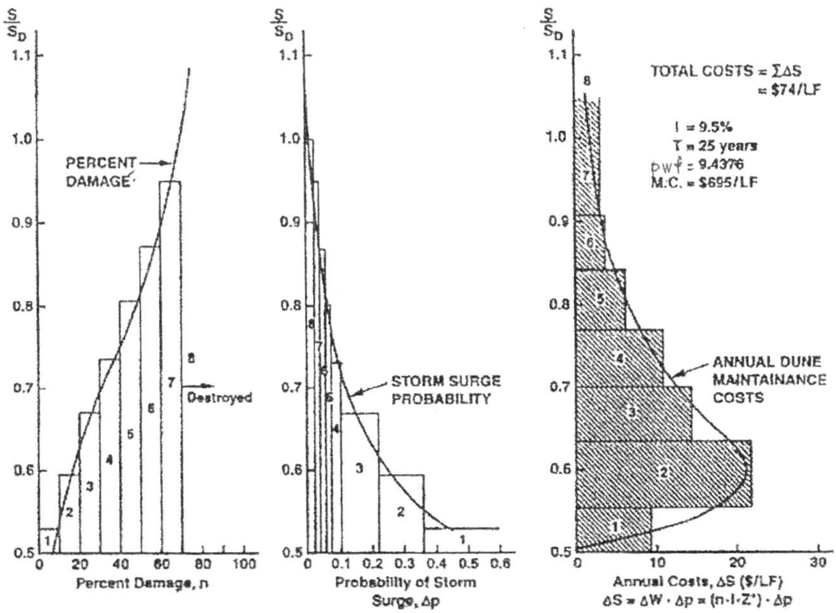

Figure 8.9. Probabilistic design method for dune maintenance costs for Dam Neck, VA (Basco and Shin, 1996).

Then $\sum \Delta s$ = the total annual repair damage cost for all increments. The letter "s" is now employed for dune damage repair costs to emphasize that it is due to storm surge water level, S, and not wave height. Table 8.5 presents a summary of the dune maintenance cost estimate for Dam Neck, VA (Basco and Shin, 1996), with Δs (\$/linear ft) versus S/S_D plotted in Figure 8.9 (right side). The total annual repair costs were \$74/linear ft.

In the early 1990s, interest rates in the US were very high. Taking $i = 9.5\%$ over a $T = 25$-year design life gave a pwf = 9.4376 so that the amortized maintenance costs MC = \$73.62 × 9.4376 = \$695/linear ft of beach. See Basco and Shin (1996) for complete details.

8.4.4 *Maintenance costs for vertical-walled structures*

Maintenance costs for vertical-walled coastal structures would be calculated in a similar way if the damage curves could be defined. But as discussed in Section 8.3.4, many failure mechanisms are possible (see Figure 8.5 for 12 different failure mechanisms of caissons). This makes the estimation of maintenance costs extremely difficult.

Table 8.5. Probabilistic calculation of dune maintenance costs for Dam Neck, VA (Basco and Shin, 1996).

Region	Storm Surge Elev. *, S feet	Percent Damage (Average) n	Storm Probability Difference Δp	Dune Repair Costs $\Delta W = n \cdot I \cdot Z\ S$	Annual Dune Maint. Costs, S $\Delta S = \Delta p \cdot \Delta W$
1	4.57	5	0.42	22.50	9.45
2	5.18	15	0.32	67.50	21.60
3	5.83	25	0.125	112.50	14.06
4	6.39	35	0.068	157.50	10.71
5	7.00	45	0.030	202.50	6.08
6	7.57	55	0.015	247.50	3.71
7	8.27	65	0.012	292.00	3.51
8	>8.7	>100 destroyed	0.010	450.00 rebuild	4.50
					$\sum \Delta S$ 73.62

*Zero datum (NGVD).
Z = repair cost factor (say 1).

8.4.5 *Influence of partial safety factors on maintenance costs*

For the example above, the damage curve results as shown in Figure 8.4(a) were obtained by taking both partial safety factors equal to 1.0. For partial safety factors ($\gamma_z = 1.06$ and $\gamma_H = 1.5$) at the $P_f = 0.05$ level and low-quality wave data an armor-layer rock diameter $D_{n50} = 2.01\,\text{m}$ is needed for stability using the plunging wave formulation of van der Meer (1988). Figure 8.10 displays the resulting damage values, S, which are the same as found in Figure 8.4(a) when using $D_{n50} = 1.265\,\text{m}$ and $\gamma_z = 1.00$ and $\gamma_H = 1.0$. Note, the design wave height, $H_{sd} = 3.0\,\text{m}$, is the same for both cases. So as a result, the use of partial safety factors does not change the maintenance costs but only the initial costs of the structure. However, the use of partial safety factors gives the designer knowledge of the probability of failure in any one year and the lifetime probability of failure as discussed below.

8.4.6 *Lifetime project probability of failure*

If only one failure mode was dominant for failure of a structure (for example the rock armor-layer of a rubble-mound structure) and only one load

VAN DER MEER(1988) CEM(2000)
Table VI-6-5
Eqn VI-6-44 Plunging Waves
Plunging Coefficient, Cpl = 6.2
Quarry stone,rough angular
n=2 layers, random placement
Structure Trunk
COT alpha = 1.5 to 5.0
Damage conditions specified by S
Remarks: Eta must be less than Eta critical

ARMOUR LAYER DAMAGE
Rubble-Mound Structures and Non-Overtopping Slopes
(Metric System)

Enter Data			Units
Design Wave Height,Hsd =		3.00	meters
Rho for rock	=	2650	Kn/m^3
Gravity	=	9.81	m/sec^2
Rho for water	=	1025	Kn/m^3
Delta = Sr-1	=	1.585	
Gamma Z	=	1.06	
Gamma H	=	1.50	

Design Data			Units
Nominal Diameter,Dn50=		**2.0100**	meters
Slope, COT α =		2.0	
Permeability,P	=	0.1	
No. of Waves,Nz	=	2000	
Wave Period,Tm	=	6.0	sec
Peak period,Tp	=	7.3	sec
Calculated Information			Units
Wave Length,Lom	=	56.64	meters
Steepness,Som	=	0.053	
Storm duration,t	=	3.35	hrs

SLOPE COT alpha	SURF PRA Xi	**DAMAGE S**	Rel Dam (Zero=2.0)	Xi critical	Remarks
2.0	2.1718612	2.0258	**1.0**	3.574	**OK**

			COTα= 1.5	COTα= 2.0	COTα= 3.0	COTα≥ 4.0
Hs/Hsd	Hs (storm)	Damage, S	%Damage,D	%Damage,D	%Damage,D	%Damage,D
1.00	3.00	2.03	2.65	2.44	2.59	2.50
1.05	3.15	2.59	3.87	3.06	2.49	2.48
1.1	3.30	3.26	5.46	3.98	2.54	2.53
1.2	3.60	5.04	10.21	7.39	3.55	3.02
1.3	3.90	7.52	18.25	14.50	7.11	4.58
1.4	4.20	10.90	31.79	28.58	15.96	8.35
1.5	4.50	15.38	54.48	55.18	34.87	16.30
1.7	5.10	28.76	153.84	187.81	139.73	59.93
2.0	6.00	64.83	658.07	943.09	783.77	326.05

Partial Safety Factors= Plunging Waves

Prob Failure	Sigma Waves = 0.5		Sigma Waves = 0.2	
0.01	1.6	1.04	1.9	1.00
0.05	1.4	1.02	1.5	1.06
0.10	1.3	1.00	1.3	1.10
0.20	1.2	1.00	1.2	1.06
0.40	1.0	1.08	1.0	1.10

Coefficients for %Damage, D versus van der Meer, S values

Slope, COT α	S^2 term	S term	Constant	R^2
1.5	0.1326	1.572	1.0795	0.9997
2.0	0.2231	0.0637	1.4002	0.9996
3.0	0.2027	1.1118	4.0071	0.9981
≥4.0	0.0833	0.4167	3.000	1.0000

Figure 8.10. Spreadsheet for calculation of damage levels S and D for plunging waves on rubble-mound structures using partial safety factors $\gamma_z = 1.06$ and $\gamma_H = 1.5$ for the nominal rock diameter $D_{n50} = 2.01$ m.

variable was dominant in the analysis of the Limit-State equation (for example, the significant wave height), then a simple project lifetime probability of failure, P_L, could be calculated as discussed in Section 4.8.2 for the "Lifetime Project Probability of Failure".

$$P_L = P_N \times P_f, \tag{4.44}$$

where P_N is the N-year probability for the significant wave height, $\hat{H}_s^T$, for the project design life, and P_f is the selected probability of failure, using the Level I, partial safety factors (e.g. in Table 4.1). Clearly, many combinations of P_N and P_f are possible to achieve a goal for P_L.

But as discussed in Section 4.7, coastal structures have many possible failure modes and two ways (series and parallel) to decompose their failure trees for determination of project failure of the system, $P_{f\,\text{system}}$. And, this analysis would only be for the *annual* probability of failure of the system. To apply Equation (4.43) for the N-year probability requires that each year's events are independent. For some failure modes of the "resistance", e.g. the concrete strength of the seawall, this requirement cannot be met. Therefore, use of the N-year probability equation together with the $P_{f\,\text{system}}$ will produce totally incorrect numbers for the lifetime project probability of failure, P_L, and is not recommended (see the CEM 2006, Part VI-6-4 and the two example problems of system probability of failure analysis for further discussion).

8.5 Total Life-Cycle Costs

Each alternative considered in coastal engineering design for flood damage and erosion mitigation has an initial construction cost, I, and a life-cycle maintenance cost, C_m, so that together the total, life-cycle cost of the alternative, C, is simply

$$C = I + C_m. \tag{8.13}$$

For rubble-mound structures and the example design in Section 8.4, $C =$ \$37,415 as shown in Table 8.4. For a design wave height, $H_{sd} = 3.0\,\text{m}$, the life-cycle maintenance costs are over 50% of the initial construction costs for the example stable design, $D_{n50} = 1.265\,\text{m}$. Increasing the design wave height will require a large rock size to increase the initial construction cost but possibly lower the life-cycle maintenance cost. The coastal engineering designer must therefore consider higher wave heights 3.5, 4.0, and 4.5 m with exceedance probabilities $p_>$ of 0.135, 0.036, and 0.01, respectively, to learn how the life-cycle maintenance costs will be reduced.

For another example, in general, as the design wave height increases, initial costs I will increase but C_m will decrease. Figure 8.11 schematically displays these trends. The resulting shape of the combined trend for the total life-cycle costs, C, is a U-shaped curve with some minimum value for some level of design wave height, H_{sd}. Note that the minimum, total, life-cycle cost is not necessarily at the rare recurrence interval, e.g. for $T_r = 100$ years ($p_> = 0.01$). This has come to be termed a "balanced" design. It is strictly based on the structural economics for the initial and maintenance costs of the alternative. Other "costs" such as functional performance and other design constraints (environmental, social, political, institutional) must also be considered.

8.6 Risk

8.6.1 *Definitions of risk*

Many definitions of "risk" are found in newspaper articles and the technical literature. We begin with some definitions of risk.

8.6.1.1 *Basic definition*

The dictionary definition of risk is simply "*Exposure (to a hazard) with the possibility of loss, injury, or other adverse or unwelcome circumstance*". The risk may be financial, require insurance, or due to natural, physical events. The hazard that causes the greatest exposure for coastal engineers is

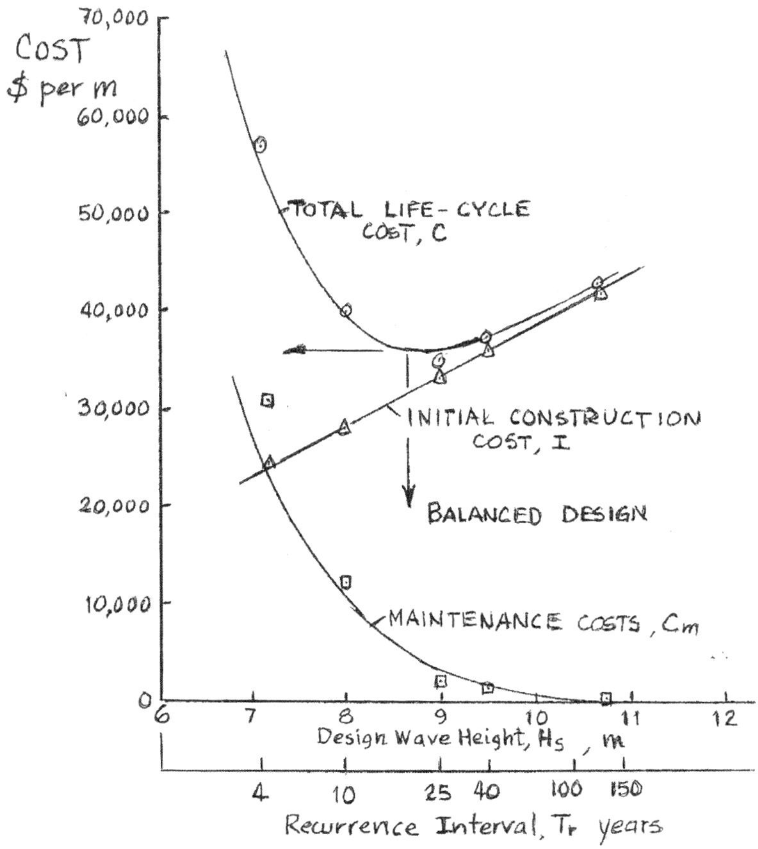

Figure 8.11. Example "Balanced Design" for economic costs of rubble-mound breakwater.

the coastal storm producing high water, high waves, strong winds, shoreline erosion, and ecological alteration of coastal waters.

8.6.1.2 *The Netherlands definition*

In Holland, the Dutch national water authority "Rijkswaterstaat" says simply: *Risk is a combination of probability and consequences* and includes a drawing reproduced here as Figure 8.12 to explain flood risk to the citizens of the Netherlands and Europe. FLOODsite (www.floodsite.net) was a project in the Global Change and Ecosystems program of the Sixth Framework of the European Commission (2004–2009).

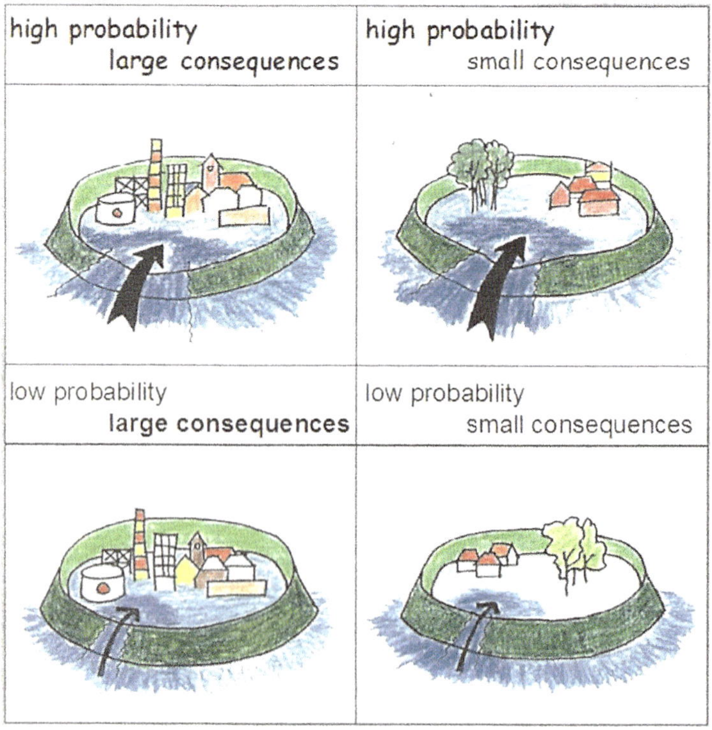

Figure 8.12. Drawing of four simple combinations of probability and consequences of the flooding hazard. (spillercapstone.blogspot.com May 9, 2013 "Sycamore Capstone, Andrew Spiller").

The concepts of "high" and "low" probability and "large" and "small" consequences are relative. In addition, the consequences depend on the vulnerability and value (economic, ecological, social) of what is flooded.

8.6.1.3 *IPET definition (Hurricane Katrina)*

An Interagency Performance Evaluation Task Force (IPET) was established by the US Federal government following Hurricane Katrina (August 2005) to study what went wrong with the Corps of Engineer's storm damage mitigation project for New Orleans, US. The existing Hurricane Protection System (HPS) that covered over 350 miles (560 km) was evaluated for system performance by modeling the risk to this existing (Pre-Hurricane Katrina) HPS under storm surge and waves.

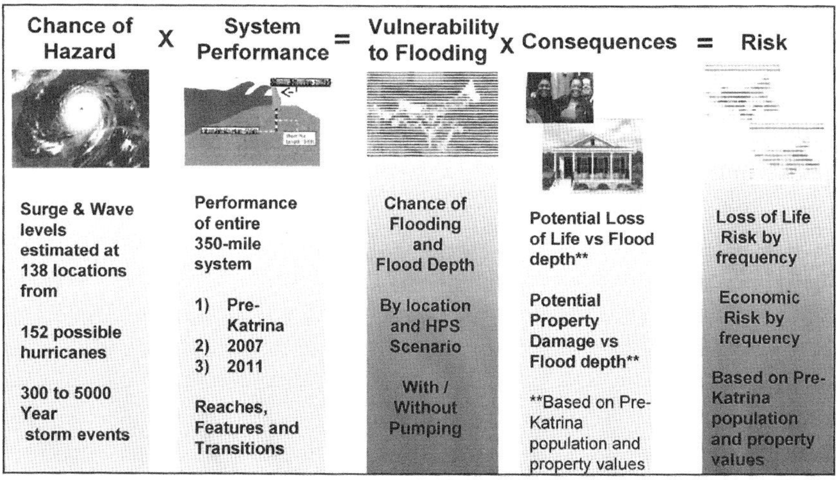

Figure 8.13. Risk assessment framework for New Orleans as employed in IPET (2009) (courtesy IPET, 2009 Vol VIII report).

Figure 8.13 graphically displays the risk assessment framework used in the IPET analysis for New Orleans. The first column "Chance of Hazard" is the storm surge and wave conditions; the next column "System Performance" is the geometry of the 350-mile HPS; the third column is the "Vulnerability to Flooding" and combines both the exceedance probability curves and the flood damage curves (explained below): the fourth column is the "Consequences" that relates flood damage in dollars ($$) to flood depths: and the last column is the "Risk" that includes both economic loss and loss of life. Both the pre-Katrina and post-Katrina risk (2007 conditions) were evaluated for how risk has changed with respect to the changed (damaged) conditions of the HPS. However, since population and property conditions are dramatically fewer after Hurricane Katrina in 2007, this risk reduction is not all due to the HPS. The formal risk assessment permitted a quantitative analysis of risk reduction when incorporating improvements to the original HPS for constant, pre-Katrina population and property data.

This effort produced a report *Performance Evaluation of the New Orleans and Southeast Louisiana Hurricane Protection System* (IPET, 2009). Volume VIII of this report was titled "Engineering and Operational Risk and Reliability Analysis" and included the following definition:

$$Risk = Event\ rate\ (or\ probability) \times Vulnerability \times Consequences\ of\ failure.$$

This report was an independent opinion of the IPET and was not the official position of the US Army Corps of Engineers. However, it became the starting point for the Corps of Engineers' efforts to define and quantify risk in evaluation of alternatives to reduce flooding risk in coastal areas.

8.6.1.4 *Risk index for flood hazard*

Efforts to place the relative risk of flood hazards on a scale has led to the development of flood risk indices. The UNESCO-IHP (International Hydrological Program) has developed the "Global Estimated Risk Index for Flood Hazard" as an interactive, web-based tool with five index levels (low, moderate, medium, high, and extreme) as color-coded graphics on world-wide screens. The user can zoom in to anywhere on the globe, but the pixel size is relatively coarse (http://ihp-wins.unesco.org). Other references can be found on the internet with a wide variety of index values ranging from 1 to 100 (See, e.g. Zonensein *et al.*, 2008).

We shall discuss the Corps of Engineers' "Exposure Index" as employed to estimate flood risk in considerable detail below.

8.6.1.5 *Residual risk*

Residual risk is the remaining risk after all risk-minimizing measures have been taken. It is the risk left over after efforts to eliminate them have been reduced by mitigation and adaptation measures. Civil/coastal engineering definitions of mitigation and adaptation of coastal hazards are:

(1) **Mitigation:** To *reduce* (mitigate) the damage caused by coastal storms (and increased sea level rise) using both structural and non-structural alternatives.

(2) **Adaptation:** To *change* (adapt) our present alternative methods (structural, non-structural) and decision-making methods as caused by predicted levels of coastal storms (and uncertainty in sea level rise).

Note that the non-structural mitigation alternative is not an adaptation strategy as often suggested by the science community and the media.

8.6.2 *Flood risk and the US government*

Many developed countries in the world have national plans for dealing with flooding. The author is most familiar with US Federal government programs as outlined in the following.

8.6.2.1 *FEMA Maps, Insurance, and Hazards*

(1) Flood Insurance Rate Maps (FIRMs)

In the US, the Federal Emergency Management Agency (FEMA) provides flood insurance rate maps for general use by the cities and general public. The exceedance probability, Q, is at the annual 1% level (100-year recurrence interval) within the flooding level boundaries displayed on the maps. The US Federal government provides flood insurance at actuarial rates to property owners within the 1% level. The flood insurance premiums (theoretically) pay for the flooding damages that occur each year. But the program currently must borrow money from the US Federal government to pay damages on premiums. The US Congress is working to correct this deficiency (Fall, 2019).

All new construction within the 1% chance flood zone must be above the 1% chance flood base level plus additional elevation for waves and wave set up. Figure 8.14 displays graphically the flood zones VE, Coastal AE, and AE, where both water levels (red dotted line) and wave heights (blue dotted line) are important in the "Flood Zone". Zone "X" is outside the mapped, 1% annual recurrence level. Section 2.2.4 in Chapter 2 also discusses FEMA, Flood Insurance Rate Maps (FIRMs).

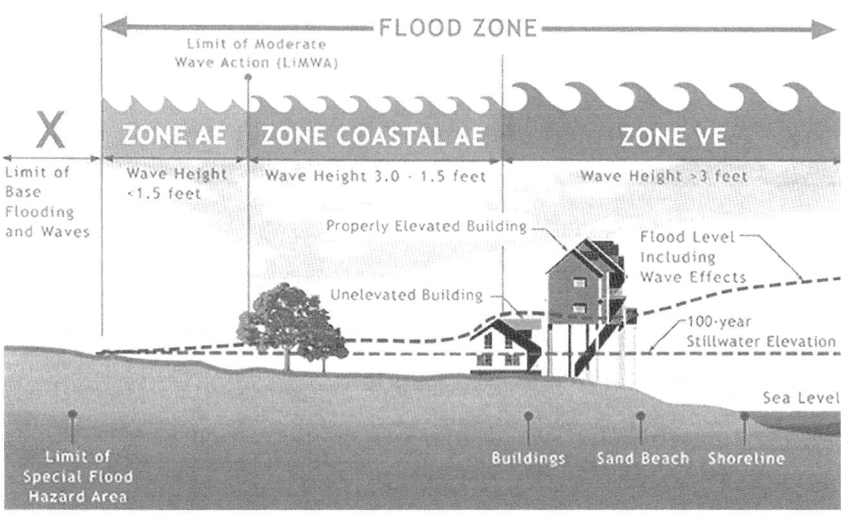

Figure 8.14. Schematic of FEMA defined "Flood Zone" boundaries for flood insurance rate maps (from FEMA "Coastal Analysis and Mapping" https://sites.google.com/site/region2coastal/coastal-mapping-basics).

The National Flood Insurance Program (NFIP) administered by FEMA is not a flood risk management program. The FIRMs show the probability of flooding at only the 1% annual chance level but not the *vulnerability* and *consequences* of flooding. This is clearly evident in Figure 8.14. Both houses are within the VE Zone and subject to the same water levels (100-year recurrence interval, still water level) and wave heights exceeding 3 ft (1 m) in height, i.e. they are exposed to the same *probability* of flooding. However, their vulnerability depends on the elevation of the first floor and the consequences depend on the values of the structural property and contents within. In addition, the unelevated building is vulnerable to flood damage at water levels below the 100-year, still water level and smaller wave heights at those water levels.

Unfortunately, many newspaper articles, companies advising property owners over the internet, and some science and engineering publications refer to FEMA's flood insurance rate maps when discussing flood risk and flood risk management. Clearly, the still water level at the 1% annual probability level is only one point on the probability curve and says nothing about the vulnerability and consequences to property at any location on the flood insurance rate map. The FEMA flood insurance rate map (FIRMs) is not an indicator of flood risk.

(2) Multi-hazard Loss Estimation (Hazus-MH)

In 1997, FEMA released a risk assessment software package termed HAZUS97. The current version is Hazus-MH 4.0 (MH = Multi-Hazard) and models multiple types of hazards: river flooding, coastal storm surge, tsunamis, and earthquakes. Hazus is built on geographic information system (GIS) technology (https://msc.fema.gov/portal/resources/hazus). But it requires ArcGIS with ArcView license for application. In addition, the ArcGIS Spatial Analyst extension is required for the flood model (https://www.arcgis.com). The risk is calculated using the same required three parts: probability, vulnerability, and consequences as defined in Equation (8.16) as explained in Section 8.6.3.

The Hazus-MH Technical Manual for Flooding is found online (https://www.fema.gov/hazus/), and provides an extensive Data Inventory for Direct Damage to buildings and facilities, transportation systems, lifeline utility systems, agricultural products (crops), vehicles, and others. In addition, it presents further guidelines for Indirect Economic Losses. This is extremely valuable information for application in risk models.

8.6.2.2 *The Corps of Engineers' NACCS Report*

The second major "watershed" event for the Corps of Engineers was the Superstorm Sandy (October 2012). (*Technically not a hurricane at landfall since wind speeds below 74* mph). The aftermath resulted in the "North Atlantic Coast Comprehensive Study" (NACCS, 2015). A full listing of all NACCS files [*Main Report, Engineering, Economic and Social Analysis, Planning Analyses, State and Wash DC Analyses* (January, 2015)] can be found online (http://www.nad.usace.army.mil/CompStudy/). These are extremely valuable documents. Elevated water levels occurred along the entire Atlantic Ocean east coast from Florida northward to Maine, but the study area was from the Virginia/North Carolina border to Maine that covers over 31,200 miles (50,300 km) of coastline.

The NACCS report develops a Framework of flood risk evaluation and management. Three levels of evaluation called tiers (1, 2, and 3) are defined by different regional size scales, different objectives to address flood risk, and different stakeholders for input and feedback. The NACCS report presents a large-scale application of the Framework in the evaluation of risk and exposure for the entire North Atlantic Coast study area (Tier 1). The large-scale datasets are likely not as refined as State or local datasets, which is why the steps are repeated as part of a Tier 2 (State and large watershed scale) and Tier 3 (local and small watershed scale) evaluation. A Tier 3 evaluation would likely include a benefit-to-cost ratio analysis leading to the selection of a plan. An example Tier 3 study for the City of Norfolk, Virginia is provided in the following. The Framework can also be used in anticipation of future storms and for climate change adaptation planning. Long-term flood risk and vulnerability should be considered when addressing flood risk and vulnerability solutions for today's water level and flooding conditions.

The NACCS report (and related studies) made five valuable contributions to the quantification of risk for the coastal flooding hazard and each one is further explained as follows.

(1) Determination of extreme water level probabilities

A statistical analysis employed historical tide gage observations together with Monte Carlo life-cycle simulation methodology. Non-exceedance probability distributions of extreme water levels for the central tendency estimate, a lower 10%, and upper 90% confidence intervals were determined for 23 locations along US North Atlantic coast (see Section 4.3.4 for review of confidence intervals associated with probability distributions).

The numerical models combined winds, waves, and water levels in Joint Probability methods within the Corps' Coastal Storm Modeling System (CSTORM-MS). Full details were published in Nadal-Caraballo and Melby, 2014 (ERDC/CHL TR-14-7) and in Nadal-Caraballo and Melby (2014) for water levels in Phase I and in Cialone *et al.* (2015) for Phase II water levels (ADCIRC) and waves (STWAVE).

Over 3 million computational points were employed in the numerical models. Then, over 18,000 locations have been made "Saved Points" along coastal boundaries. A Coastal Hazards System (CHS) website for storage and retrieval of storm surge, water level, wave height, wave period, wave direction, and current magnitude has been developed (https://chswebtool.erdc.dren.mil/). Save Point locations are LML files (KML is a file format used to display geographic data in an Earth browser such as Google Earth). Save Point data and locations are obtained from the CHS menu-driven Web Tool on the CHS website. The stored data can be readily accessed, downloaded, analyzed (mined), and plotted through the CHS web tool.

(2) Definition of areas flooded for risk analysis

The Saffir–Simpson Hurricane Wind Scale (SSHWS) was originally developed by Saffir (1973) and Simpson (1974) for the following sustained wind speed ranges within the eye of the hurricane for the five Category levels:

Category 1: 119–153 km/h (74–95 mph)
Category 2: 154–177 km/h (96–110 mph)
Category 3: 178–208 km/h (111–129 mph)
Category 4: 209–251 km/h (130–156 mph)
Category 5: 252 or higher (157 or higher)

Various levels of wind damage to property, infrastructure, trees, and flying debris were categorized including the possible risk of deaths to people, livestock, wild animals, and pets. In the spring of 2010, the National Hurricane Center (NHC) *removed* the estimates for storm surge as previously published for the five-category system to avoid confusion over storm surge and flooding predictions that failed to match what actually occurred as a hurricane made landfall. The storm surge levels no longer appear in NHC publications for the public and the media. However, the NHC continues to employ the hydrodynamic, numerical model (SLOSH) to calculate storm surge from hurricanes. A worst-case flood inundation level for a particular wind speed category is the maximum envelop of high-water level (MOM)

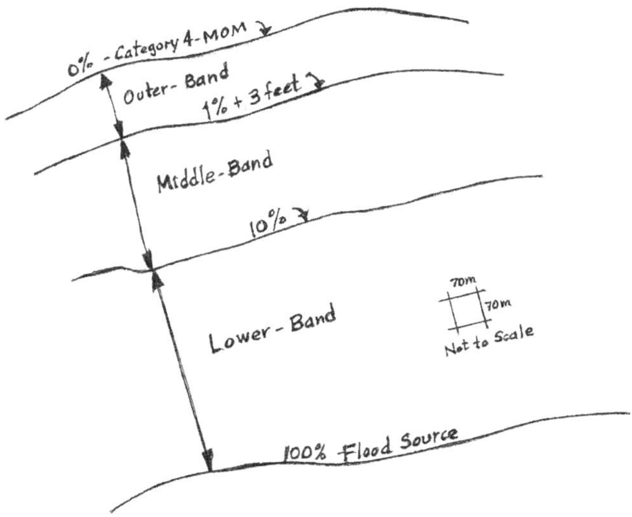

Figure 8.15. Schematic of three flood inundation areas and grid size employed in NACCS risk analysis.

considering high initial tide levels and critical forward speed and trajectory of the hurricane when applying the SLOSH model. The NHC is part of the US National Oceanic and Atmospheric Administration (NOAA) and coordinates flood risk management with the Corps of Engineers.

Coastal flood inundation mapped areas are employed to identify the specific areas at risk. Three bands of flooded areas were created. Figure 8.15 schematically illustrates the three bands of inundated area with boundary definitions. Also shown is the grid size (not to scale) which was 70 m (230 ft, 0.00833°) using ArcGIS software program (ESRI, 2012).

The Category 4-MOM from the SSHWS was taken as the upper limit of inundation for the NACCS area flooded for the first, outer-band. This upper limit attempted to avoid missing any residual risk if this upper limit of flood inundation was set too low. The probability of non-exceedance is essentially 0%.

The lower limit of this first, outer-band was the annual, 1% chance flood inundation level plus 3 ft (0.9 m). The 1% flood levels were obtained from the data stored in the Coastal Hazard System as "Share Points" discussed above. The extra 3 ft is related to possible sea level rise and will be further addressed in Chapter 11. It was recognized that the FEMA annual, 1% chance flood maps were not available for all regions throughout the study area. The FEMA is within the US Department of Homeland Security and also coordinates flood risk management with the Corps of Engineers.

This is also the upper-limit of the second, middle-band. The lower limit of this second, middle-band is simply inundation from a flood with annual, 10% probability of exceedance. And, this lower limit of the second band is also the upper-limit of the third and lowest band in the NACCS area for flood risk analysis. This lower-limit extends down to the flooding source along the coast, which has an annual probability of exceedance of 100%.

It is recognized that many other possibilities exist with more bands to sharpen the focus on the spatial extent of the risk analysis.

(3) Development of an NACCS-Exposure Index (EI) and NACCS-risk relationship

A framework for Coastal Storm Risk Management (CSRM) was developed that requires calculation of an Exposure Index (EI). Three component indices make up the Exposure Index.

(i) Population Density and Infrastructure Index (PDII)

Population density includes the number of people within a study area. Weighted values for population densities (number per acre) from the US Census (2011) are shown in Table 8.6.

Critical infrastructure needed to support that population in the area includes water, sewage, electricity, natural gas, trash, medical, safety, and many other considerations. The NACCS report listed 49 separate critical infrastructure categories as "point data" and their risk scores as illustrated in Table 8.7. Fire stations, hospitals, and water and wastewater facilities received the highest score (30). Bus stations and ferry terminals are given

Table 8.6. Weighted values for population density from Table VII-3 in the NACCS (2015) report. (One acre = 0.4 hectares = 4,047 m^2).

Table VII-3. Population Density

Population Density Per Acre (Census, 2011)	Weighted Value
0	0
0.1–12	40
13–72	50
73–193	60
194–687	70
688–3,072	80

Table 8.7. Weighted risk scores for critical infrastructure from Table VII-1 in NACCS (2015) report. They are listed as "point data" for the GIS layer.

Table VII-1. Infrastructure

Critical Infrastructure Point Data Layer (HSIP 2012)	Risk Score
Airport Boundaries	15
All Places of Worship	15
Amtrak Stations	15
Bus Stations	5
Cellular Towers	10
Colleges/Universities	15
Communication Centers	15
Dams	25
Electric Generating Units	20
Electric Power Generation Plants	25
EMS	25
Energy Distribution Control Facilities	20
Ferry	5
Fire Stations	30
Gas Stations	20
Hospitals	30
Intermodal Terminal Facilities	15
Law Enforcement Location	25
Local Emergency Operation Centers	20
National Shelter System	20
Natural Gas (LNG) Import Terminals	15
Natural Gas compressor Stations	15
Natural Gas Import/Export Points	5
Natural Gas Receipt and Delivery Points	10
Natural Gas Storage Facilities	15
Nuclear Power Plants	25
Nursing Homes	25
Oil and Natural Gas Interconnects	5
Oil Refineries	20
Pier/Wharf/Quay	15
Petroleum Pumping Stations	10
Pharmacies	15
POL Terminals/Storage Facilities/Tank Farms	15
Ports	15
Private Schools	10
Public Schools	15
Railroad Bridges	20
Railroad Stations	20
Railroad Tunnels	20
Railroad Yards	20
Receiving Hospitals	30
Road and Railroad Bridges	20
Service Providers	25
State Emergency Operation Centers	20
Substations	20
Urgent Care Facilities	20
Wastewater pump stations	30
Wastewater Treatment Plants	30
Water Treatment Facilities	30

Table 8.8. Weighted risk scores for critical infrastructure from Table VII-2 in NACCS (2015) report. They are listed as "linear data" for the GIS layer. The lines cross many sub-areas in the region of interest.

Table VII-2. Infrastructure

Critical Infrastructure Linear Data Layer (HSIP 2011)	Risk Score
Hurricane Evacuation Routes	20
Transmission Lines	20
Railroad	20
Road and Railroad Tunnels	20
Pipeline Distribution System	20
Canal	15
Channel	20
Oil and Natural Gas Pipelines	20
Ferry Route	10

low scores (5). They are included here to give the reader some help in defining critical infrastructure and in showing how the NACCS report was used for rating each category.

A second infrastructure category list for line (linear) data is reproduced here as Table 8.8. They almost all have the same score (20) and cross sub-areas in the study area.

These factors reflect the overall exposure of the built environment to the coastal flood hazard in the area. See the Planning Analysis document (Appendix C) for details of what is considered "critical infrastructure", obtained from the National Geospatial Intelligence Agency (HSIP, 2012). Weighting for population density ranged from 40 to 80 points in 10 point increments, and weighting for infrastructure data layers ranged from 5 to 30 points in 5 point increments in this first effort (Tier 1) of the NACCS report.

(ii) Social Vulnerability Characterization Index (SVCI)

A second component of indices making up the Exposure Index (EI) deals with how vulnerable people living in the area are when exposed to flooding. People aged 65 and over ($\%\text{Age}_{65+}$), aged 5 and under ($\%\text{Age}_{5-}$), with income below the poverty threshold ($\%\text{Income}_{\text{sub-poverty}}$), and those non-proficient in English (%non-proficient English) make up the Social Vulnerability Characterization Index (SVCI). It is simply the sum of these people, already expressed as percentage.

$$\begin{aligned}\text{SVCI} &= (\%\text{Age}_{65+}) + (\%\text{Age}_{5-}) + (\%\text{Income}_{\text{sub-poverty}})\\ &\quad + (\%\text{non-proficient English})\end{aligned}$$

The non-proficient English percentage was for people who speak a language other than English or not very well. The US Census Bureau, "Census Demographic Summary Profile Data" (US Census, 2011), was the source.

(iii) Environmental and Cultural Resources Index (ECRI)

Finally, the third component is a combination of the environment, habitat, and the cultural resources. Data provided by the US National Fish and Wildlife Service database, the National Wetlands Inventory, and the Nature Conservancy Ecoregional Assessments aided development of the Environmental and Cultural Resources Index (ECRI). Table 8.9 presents the categories used, the references cited, and the weight for each of the three criteria: Environmental (30%), Habitat (30%), and Cultural (40%) that add up to 100%. The column "Weight (0–25)" lists numbers that are relatively high, but they only count for 30–40% of the ECRI index. The least weight is 37 for rocky shore habitat and the highest is for estuarine emergent marsh grass (96).

These three independent indices are then simply summed together to develop the composite index for the overall NACCS Exposure Index (EI),

$$\text{EI (NACCS)} = 80\%\ \text{PDII} + 10\%\ \text{SCRI} + 10\%\ \text{ECRI} \tag{8.14}$$

Weightings employed in this NACCS version of an Exposure Index reflect the request of the US Congress, to ...*address flood risks of vulnerable coastal populations*, so that the emphasis was put at 80% for the PDII component. This can be modified as best meets the local needs for risk management at applicable sites.

The composite Exposure Index (EI) is a non-monitory unit of the product of vulnerability and consequences for the flood hazard. Risk as defined in this NACCS study, R_{NACCS}, is

$$R_{\text{NACCS}}(\text{flood}) = \text{EI (NACCS)} \times P(\text{flood}) \tag{8.15}$$

to recognize how the chance or probability of flooding, $P(\text{flood})$, enters a risk formulation.

As an example, consider EI (NACCS) $= 100$ for one grid cell, $70\,\text{m} \times 70\,\text{m} = 4{,}900\,\text{m}^2$ ($52{,}900\,\text{ft}^2$). The risk, R_{NACCS} will depend on which of the three flood probability of non-exceedance bands the grid cell is located

Table 8.9. Weighted risk scores for environmental, habitat, and cultural resources from Table VII-4 in NACCS (2015) report. The column "Weight (0–25)" lists numbers are then again weighted from 30% to 40% for the ECRI index.

Table VII-4. Environmental and Cultural Resources	
Environmental Criteria (30%)	Weight (0–25)
Priority Areas (USFWS 2014)	84
Coastal Barrier Islands under CBRA (USFWS 2014)	91
USFWS Refugees (USFWS 2014)	86
USFWS Protected Area (USFWS 2014)	65
USFWS Priority Areas (USFWS 2014)	91
Rare, Threatened, and Endangered Species (USFWS 2014)	89
Colonial Nesting Waterbirds (USFWS 2014)	94
The Nature Conservancy Conservation Areas (TNC 2011)	73
City, County, State, and Federal Parks > 100 acres (HSIP 2012)	44
Habitat (30%)	
Seagrass (TNC 2011; VIMS 2013a)	88
Estuarine Emergent Marsh (TNC 2011)	96
Forested Wetland (TNC 2011)	80
Scrub-Shrub Wetland (TNC 2011)	73
Freshwater Emergent Marsh (TNC 2011)	83
Freshwater Forested/Shrub Wetland (TNC 2011)	82
Riverine Wetlands (TNC 2011)	61
Rocky Shoreline (TNC 2011)	31
Unconsolidated Shore — mud, organic, flat (TNC 2011)	47
Unconsolidated Shore — sand, gravel, cobble (TNC 2011)	66
Cultural Resources (40%)	
Coastline Buffer (1,000 ft) (water layer including, lakes, ponds, river, streams, etc.) (TNC 2011)	50
National Monuments and National Historic Landmarks (HSIP 2012)	85
Historic Sites (HSIP 2012)	75

in as shown in Figure 8.14. Within the lower band, the probability of non-exceedance is 90% (100 – 10%) so that R_{NACCS} (lower-band) = 90. Within the middle band, the probability of non-exceedance is 9% (10 – 1%) which gives R_{NACCS} (middle-band) = 9. And finally, in the outer-band, with probability of non-exceedance of 1% (1 – 0%) then R_{NACCS} (outer-band) = 1. All three locations have the same Exposure Index, meaning they are equally exposed to flooding, but the risk is far different because the probability of getting flooded each year is far different. The flooding risk is far greater in the lower band because each year there is a 90% chance water levels will reach this area. The flooding risk is far less in the outer band because the annual chance of flooding is only 1%.

The final result of an evaluation of R_{NACCS} for all the grid cells within an NACCS planning reach (Blue box, NY_NJ1) is illustrated in Figure 8.16

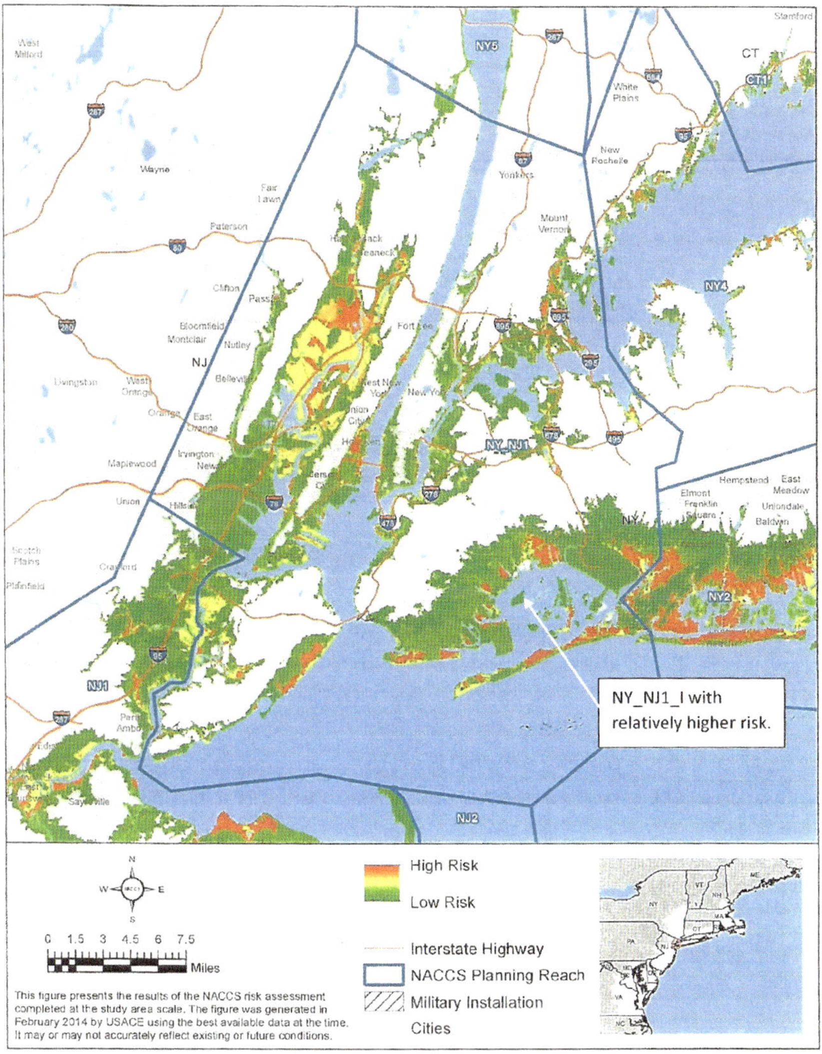

Figure 8.16. Color symbols of high (red) to low (green) NACCS flood risk calculations for region of New York City and New Jersey. The NY_NJ_I location is discussed as an example Tier 2 level evaluation (from NACCS, 2015 report).

for the New York–New Jersey region from the Atlantic Ocean, through the lower and upper New York Bay, and following the East River and Hudson River. High risk is colored red and low risk is green with the yellow color for medium risk. Actual risk numbers were not included in the NACCS (2015) report for this Tier 1 evaluation. Seventeen locations were then identified as having relatively higher risk and one was selected, "Southern Brooklyn and Queens-Jamaica Bay, Rockaway Peninsula (NY_NJ_I)" to illustrate a Tier 2 level evaluation as an example in the NACCS report (2015). Note that in this R_{NACCS} risk analysis method, most of Manhattan Island is at low risk.

The importance of the NACCS flood risk analysis method is that it includes population density and infrastructure, social vulnerability, and environmental and cultural resources in a quantifiable way that does not require a monetary estimate of value for these variables. At the highest level of risk evaluation (Tier 3), when design of coastal hazard mitigation and adaptation alternatives are being evaluated, then a Benefit to Cost (B/C) ratio is needed. The fourth contribution to the quantification of risk of the coastal flooding hazard are the damage functions needed to quantify the benefits of risk reduction by mitigation and adaptation alternatives.

(4) Development of damage functions for B/C ratio determination for alternatives

Corps of Engineers' CSRM projects are justified for approval by a cost, C, and benefit, B, analysis. Both the difference in the benefits minus the costs (B − C) and ratio of benefits/cost, B/C ratio (BCR), are employed. The Corps of Engineers now considers four types of benefits for flood hazard mitigation and adaptation alternatives. However, only benefits that increase value or productivity on a net *national* basis, called Net Economic Development (NED) benefits, are relatively easy to measure and attach economic ($$$) value. The other three are Regional Economic Development (RED), Environmental Quality (EQ), and Other Social Effects (OSE). As seen earlier, with the NACCS report and development of the composite Exposure Index, the Corps is working to include these other three benefits in risk analysis.

The NED benefits are the expected annual damages for existing site conditions (traditionally called the "without-project" conditions) minus the estimated annual damages for the mitigation and adaptation alternative considered (traditionally called the "with-project" conditions). These are

Table 8.10. Twelve structure types with damage functions found from Hurricane SANDY (from NACCS, 2015, Physical Depth–Damage Function, Summary Report).

Type	Number	Definition
Apartments	1A-1	One storey apartment, no basement
Apartments	1A-3	Three storey apartment, no basement
Commercial	2	Commercial, engineered
Commercial	3	Commercial, pre/non- engineered
High Rise	4A	Urban high rise
High Rise	4B	Beach high rise
Residence	5A	Single story, no basement
Residence	5B	Two story, no basement
Residence	6A	Single story, with basement
Residence	6B	Two story, with basement
Pile Foundation	7A	Building with open pile foundation
Pile Foundation	7B	Building with enclosed pile foundation

the reduction in damages between the without- and with-project conditions. The damages prevented are to structures (private homes, businesses, public buildings), structure contents, vehicles (cars, trucks, services), transportation delay costs, etc. Actual economic damages from Hurricane Sandy provided an opportunity to document damages from elevated water levels, water waves, and erosion that occurred to structures of various types listed in Table 8.10.

For example, 5B is a two-story residence with no basement and finished first floor about 3.0 ft above grade. Figure 8.17 is for inundation damage from water depths relative to the first-floor elevation (FFE), d_{FFE}, on the horizontal axis (0–10 ft) versus the percentage of damage to the structure value on the vertical axis (0–100%). A total of 1,542 data points (329 were slab on grade and 1,213 had a 3-ft crawl space beneath the first floor) were obtained using FEMA's Substantial Damage Estimator (SDE) tool (https://www.fema.gov/media-library/assests/documents/18692). The dashed black line is the linear line-best fit to the data with the equation

$$\text{Damage }\% = 9.571 \times d_{\text{FFE}} + 19.354 \text{ with } R^2 = 0.42$$

and relatively poor correlation with all the scatter in the data. This is a depth–damage function from the title of this summary report. Further analysis of this dataset gave the standard error of the regression estimate, $S_{\text{est}} = 14.8\%$. At the 95% confidence interval on the conservative side, +1.96

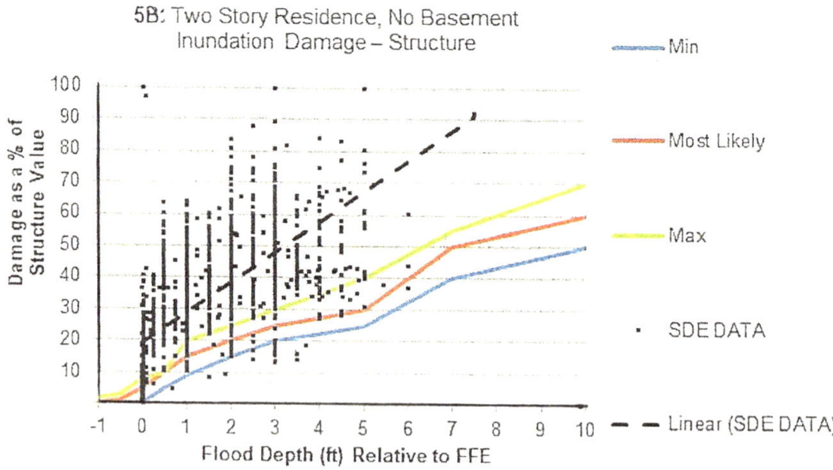

Figure 8.17. Structure inundation damage versus water depths above the first-floor elevation for 1,542 residential, two-story homes with no basement. (From NACCS, 2015, Physical Depth–Damage Function report).

$S_{\text{est}} = +29\%$. At the 50% damage level, the regression line gives $d_{\text{FFE}} = 3.2\,\text{ft}$, but at the 95% confidence interval, $d_{\text{FFE}} = 0.2\,\text{ft}$ or barely above the first-floor elevation. This is important for risk calculations.

But even more important are the three colored curves shown in Figure 8.17. These were determined by an expert-opinion elicitation team. Their results were used to develop generic damage functions for inundation, wave, and erosion damage applicable to homes and businesses along the North Atlantic coast of the US. As seen for this example in Figure 8.17, the elicitation team estimates for % Damage were very low over the complete range of flood depths. Figure 8.18(a) is a photograph of a typical two-story residence and Figure 8.18(b) displays the depth-vs-% damage table as reported by the elicitation team for minimum, most likely, and maximum levels of % damage (NACCS, 2015, Table 67 in Physical Damage Function Summary Report).

The damaged value of the contents of these same homes (Case 5B, two-story, no basement) as a result of flood inundation were also determined with the data plotted in Figure 8.19. In this case for contents, the linear regression line (black, dashed) is almost identical to the maximum (yellow) estimate of the elicitation team. Table 68 presents the elicitation team estimated result (not included herein). But the standard error of the

(a)

Table 67. Prototype 5B: Two-Story Residence, No Basement, Inundation Damage – Structure

Flood Depth	Min	Most Likely	Max
-2.0	0	0	0
-1.0	0	0	2
-0.5	0	1	3
0.0	0	5	8
0.5	5	10	10
1.0	9	15	20
2.0	15	20	25
3.0	20	25	30
5.0	25	30	40
7.0	40	50	55
10.0	50	60	70

(b)

Figure 8.18. (a) Photograph of typical two-story residence, (b) structure inundation damage table (from NACCS, 2015 report).

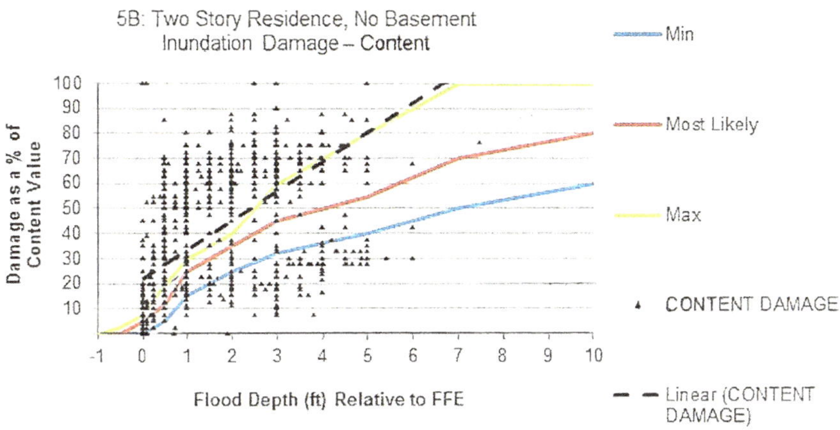

Figure 8.19. Content inundation damage versus water depths above the first-floor elevation for 1,542 residential, two-story homes with no basement. (From NACCS, 2015, Physical Depth–Damage Function report).

regression estimate, S_{est}, at the 95% confidence level will again be much higher and cause more content damage for the same flood depths.

For Case 5B, there were far less wave damage residences as illustrated in Figures 8.20 and 8.21. Of the 201 residences studied, 184 had 3 ft crawl spaces beneath the first flood elevation and the linear correlation line

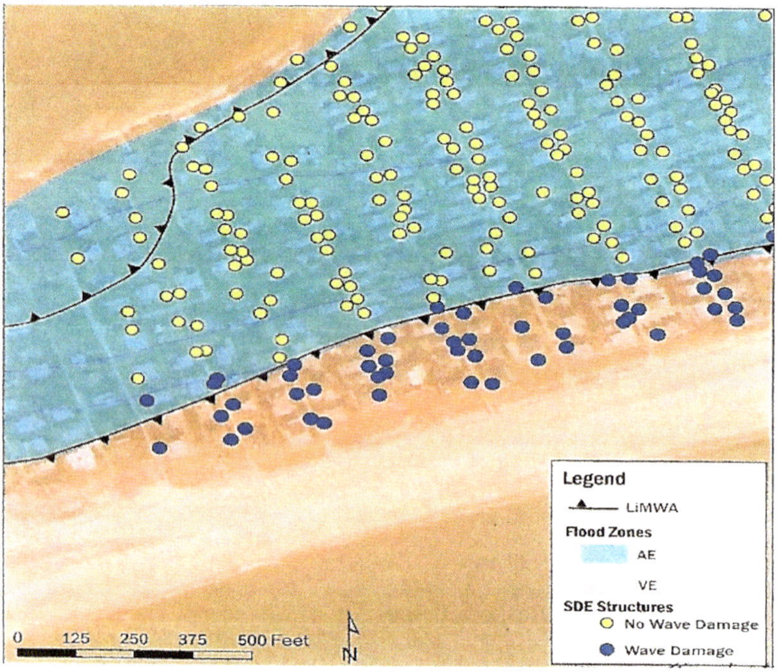

Figure 8.20. Example of structures assumed to have wave damage (NACCS, 2015).

(black, dashed) showed these had slightly less wave damage to the structure. The most-likely curve (Figure 8.21 red line) as postulated by the experts was now above the linear regression lines for the data. At the 50% structure damage level, both inundation and wave damage occur when the FFE was about 3 ft. It must be emphasized here, that the wave structure damage to those residences (blue circles) seaward of the FEMA VE Zone of moderate wave action (brown beach area) are *assumed* to have wave damage.

Erosion damage estimation is not possible with the FEMA SDE tool. Damage to foundations and beach profile change were investigated to possibly infer that structure damage was due to erosion but with inadequate level of certainty so that the erosion damage mechanism was not further considered.

In total, 74 damage functions for inundation and wave damage mechanisms for both structure and content damage in the form of damage tables

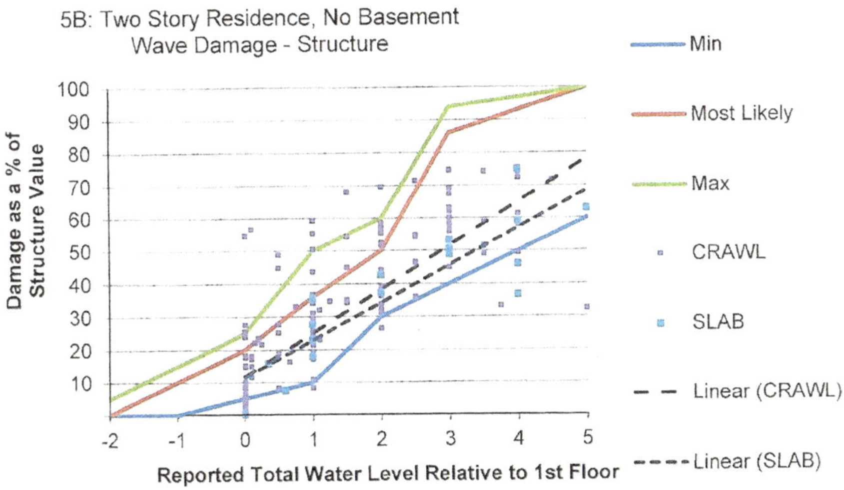

Figure 8.21. 5B: Two-story residence, no basement structure. Wave damage data compared with elicitation team (from NACCS, 2015 report).

(e.g. Figure 8.17(b)) were developed from measured field data in the aftermath of Hurricane Sandy. The knowledge gained for wave damage will be a valuable contribution since little previous information is available as in the literature for densely populated coastal areas. The author knows of no journal publication that has summarized these results.

The US Army Corps of Engineers' Institute for Water Resources (IWR) publishes Economic Guidance Memorandum (EGM) and has recommended generic flood depth–damage curves as discussed in Section 8.6.3.2.

(5) Other elements in risk benefit analyses

The NACCS also considered Loss of Life Projection, Emergency Costs, and Secondary and Tertiary Effects that could be determined as a result of Hurricane Sandy. See NACCS, 2015 (Appendix B — Economics and Social Analyses) and Department of Commerce (2013) for details.

8.6.3 *Coastal engineering quantification of risk*

8.6.3.1 *Coastal engineering definition of risk*

Risk, R, is the product of the *probability* of non-exceedance of the hazard, $p_<(hazard)$, times the *vulnerability* of what is exposed to the hazard, ν (*hazard*), and integrated over all the *consequences*, C_i, resulting from the hazard. In discrete mathematical terms:

$$\text{R (hazard)} = \text{Risk} = \sum_{j}^{jj} [(p_<(\text{hazard})) \times (\nu \ (\text{hazard})) \times C_i] \times \text{d (hazard)}, \tag{8.16}$$

where $j = 1, 2, 3, \ldots, jj$ is the range for discretization of the variable for the hazard, and $i = 1, 2, 3, \ldots, n$ is the number of consequences produced by the hazard.

The most important hazards are waves (wave height, H_s), water levels (storm surge, S), erosion (H_s, S, *duration*), and ecological change (ΔE). Three main components for risk calculation are

- Probability;
- Vulnerability (damage curves);
- Consequences.

We have already seen various forms of Equation (8.16) in this chapter. In Section 8.3, the vulnerability (damage curves) of rubble-mound structures was presented as introduction to maintenance cost estimates, c, by the probabilistic method in Section 8.4 and Equation (8.9), repeated here

$$c = \sum_{(H_s)_n}^{(H_s)_{n+1}} [(\Delta p_>)(\Delta \bar{D})(I_{cl} \times z^*)]\Delta H_s. \tag{8.9}$$

Clearly, the wave height (above the design wave height) H_s was the hazard, the exceedance probability (above the design wave height probability) $p_>$ produced the damage, the vulnerability was the rubble-mound damage curve, D, and the consequences, $(I_{cl} + z^*)$, were the costs to repair the damage. Maintenance costs are the economic risk of the chosen design alternative. A second consequence, C_2, could be the loss of time for loading and unloading ships if the rubble-mound structure was a breakwater that could not be utilized when damaged. This is a functional loss that can also be placed in monetary terms.

And again, the NACCS report defined risk in Section 8.6.2 (3) by Equation (8.15), repeated here:

$$R_{\text{NACCS}}(\text{flood}) = \text{EI (NACCS)} \times P(\text{flood}), \tag{8.15}$$

so that the Exposure Index (EI) was the combined vulnerability and consequences as produced by the flood hazard. Note that $R_{\text{NACCS}}(\text{flood})$ is not an estimate of economic risk since no monetary units are involved.

For a cost and benefit analysis, the life-cycle cost method applies Equation (8.9) to estimate maintenance costs in monetary units. The without- and with-project benefits are calculated using Equation (8.16) where the consequences, C_i, are the economic benefits in monetary units. Damage as a percentage of structure and content value is the vulnerability produced by the hazard and many studies have been conducted to estimate damage curves for use in Equation (8.16). Each of the three variables in Equation (8.16) influences the economic risk from flooding (probability, vulnerability, consequences) and each has a degree of uncertainty attached to it. The greatest uncertainty is due to the many variables that influence the damage curves when water depths are employed as the only variable in regression curves for flood depth–damage functions.

8.6.3.2 *Damage functions of Corps of Engineers*

Damage functions (aka curves, tables) have also been called "fragility curves" and defined as ...*functions that describe the probability of failure, conditioned on the load, over the full range of loads to which a system might be exposed* (Schultz *et al.*, 2010). Four approaches to developing damage curves are discussed: judgmental (expert-opinion solicitation), empirical (laboratory and field data), analytical, and hybrid. We've employed the analytical approach for the damage curves for rubble-mound structures (van der Meer, 1988) and seen some results for the judgmental and empirical approaches for flood damages from Hurricane Sandy (NACCS, 2015). Each approach has a high degree of uncertainty.

Uncertainty. There are many possible reasons for the large scatter in the data displayed in Figure 8.17 for structure damage from inundation. One is the fact that for the total number (1,542) of two-story, no basement residences shown by the data scatter, 1,213 had a 3 ft (1m) crawl space beneath the first-floor elevation, and the rest (329) were slab-on-grade. The linear regression lines and standard errors of the regression estimates are somewhat different for each. Damaging flood waters with currents existed beneath the residences with crawl space but were not present when the slab rested on the ground. This distinction was made for the damaged contents (Figure 8.18) but not for structure damage (Figure 8.17). There are many possible reasons for the data scatter that produce uncertainty.

(1) Inundation-damage curves

Table 8.11 presents data requirements for a structure inventory and we may never fully understand the uncertainty because of the great expense to gather the needed field data for the empirical approach.

Table 8.11. Structure inventory for data requirements to understand uncertainty in damage functions (from IWR, 2011, Table 7).

Table 7.
Structure Inventory Data Requirements

- Number of structures
- Land use (residential, commercial, industrial, public, etc.)
- Minimum, maximum, and most likely depreciated replacement values by structure, considering:
    - Type of structure: single family, multi-family, retail, etc.
    - Number of floors
    - Square feet
    - Foundation type: slab, pile size, etc.
    - Construction type: frame, brick, etc.
    - Basement present
    - GPS/GIS Coordinates
    - Age
- First floor elevation (min, max, most likely)
- Elevation of structure
- Ground elevation
- Minimum, maximum, and most likely content values
- Geographical location from the water or bluff (shore perpendicular distance, shore parallel distance)
- Parcel boundary

The Corps of Engineers' Institute for Water Resources (IWR) has been involved with risk management for rivers for many years and developing flood depth–damage functions for business (IWR, 1985), for residential (IWR, 1992), and more recently also for wave and erosion in coastal risk areas (IWR, 2011). For uniformity in district offices for standardized (nation-wide) flood–damage curves, a series of Economic Guidance Memoranda (EGM) for generic "Depth–Damage Relationships" have been produced:

- Residential Structures, Without Basements, EGM 01-03 (2000);
- Residential Structures, With Basements, EGM 04-01 (2003);
- Vehicles, EGM 09-04 (2009).

For example, Figure 8.22 displays the recommendation (EGM 01-03) for a two-story (or more) with no basement residential structure with depth (feet) versus mean damage (%) table and depth–damage curve. Also shown in the table for each depth is the standard deviation of the mean

Table 2 Two or More Stories-No Basement		
(feet) Depth	Mean of Damage	Standard Deviation of Damage
-2	0%	0.0%
-1	3.0%	4.1%
0	9.3%	3.4%
1	15.2%	3.0%
2	20.9%	2.8%
3	26.3%	2.9%
4	31.4%	3.2%
5	36.2%	3.4%
6	40.7%	3.7%
7	44.9%	3.9%
8	48.8%	4.0%
9	52.4%	4.1%
10	55.7%	4.2%
11	58.7%	4.2%
12	61.4%	4.2%
13	63.8%	4.2%
14	65.9%	4.3%
15	67.7%	4.6%
16	69.2%	5.0%

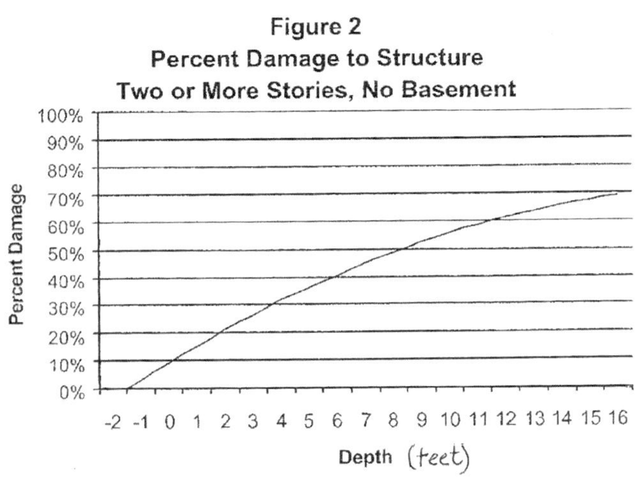

Figure 8.22. Recommendation for residential structure, depth–damage functions (from EGM, 2000).

damage in % units. When the water depth (assumed above the first-floor level) is 3 ft, the mean damage is 26.3% with standard development at 2.9%.

Comparison of expert-opinion elicitation curves shown in Figures 8.17 and 8.18 (b, Table) reveals that the results shown in Figure 8.22 (EGM, 2000) were probably consulted by the expert-opinion team to specify the min, mean, and max values in the NACCS (2015) report. For riverine, risk analysis since the 1990s, the Corps of Engineers has been using a triangular distribution as simple approximation to the probability density function (normal, bell-shaped curve) for the uncertainty surrounding these flood depth–damage curves. Figure 8.23 (IWR, 2011, corrected by author) is an example for single story, structure damage (no basement) at a depth of −1 ft where the values in EGM (2000) are mean damage, $\mu = 2.5\%$, and standard deviation of the mean damage, $\sigma = 2.7\%$. The peak of the simple triangle (red curve) is at the mean damage, $\mu = 2.5\%$, and the max value is shown at $3.33\sigma = 9\%$. This is not at the 95% confidence level, which

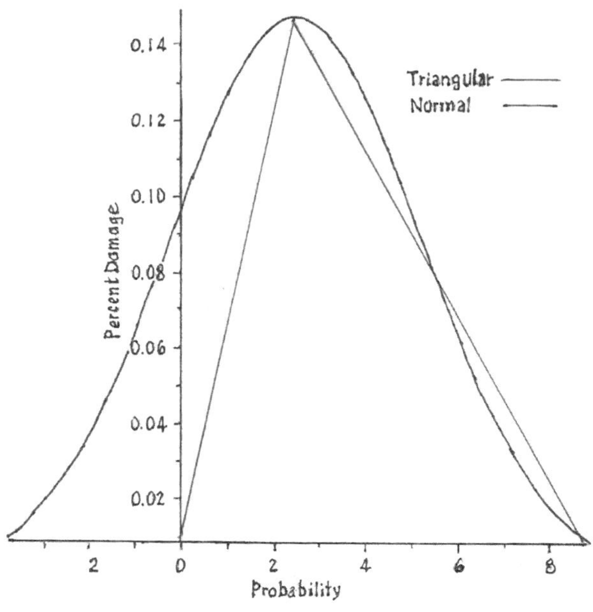

Figure 8.23. Triangular distribution for uncertainty in Corps of Engineers' generic, depth–damage curves. (From IWR, 2011, modified and corrected by author).

is $+1.96\ \sigma = +5.3\%$, but at a maximum estimated value by the expert-opinion team. The uncertainty in damage is not symmetrically distributed around the mean damage at lower values of mean damage. If symmetrical (dotted red line), this would produce negative damage, which is physically unrealistic. Therefore, the lower base of the triangle for this example is simply set at zero damage, i.e. min = 0. At higher damage levels, the min value forms a symmetrical, triangle shape distribution.

The data for the results for flood damage in the IWR Engineering Guidance Memos were collected from major river flooding events from 1996–2001 across the US. It is interesting to compare with coastal data from Hurricane Sandy. An analysis of the data for 1,542, two-story, no basement, residential structures (Figure 8.17) was made at a flood-depth of 3 ft (168 data points) and a histogram developed as shown in Figure 8.24 (data from Jones and Mulrain, 2017) for 10% increments in damage level. There were 48 structures with damage in the 45–54.99% range and the mean damage level of the data, μ, was 47.02% with standard deviation of the mean, $\sigma = 16.20\%$. These results were far different from that recommended by the IWR shown in Figure 8.22 for the same structure when the flood depth is 3 ft ($\mu = 26.3\%$, $\sigma = 2.9\%$) and employed by the expert-opinion elicitation team. Comparison of the normal distribution density functions calculated from both the IWR generic values ($\mu = 26.3\%$, $\sigma = 2.9\%$) and Hurricane Sandy data ($\mu = 47.0\%$, $\sigma = 16.2\%$) is shown in Figure 8.25. When the standard deviation of the mean is small (little uncertainty), it is relatively simple to devise a triangular distribution. Since an estimate (or data) for the standard deviation of the mean is given, use of the 95% confidence level

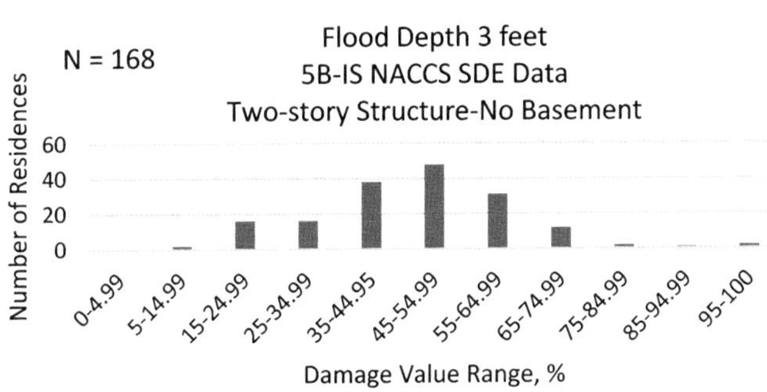

Figure 8.24. Histogram of flood structure damage data at depth of 3 ft for 168 two-story, no basement residences in Hurricane Sandy (data from Jones and Mulrain, 2017).

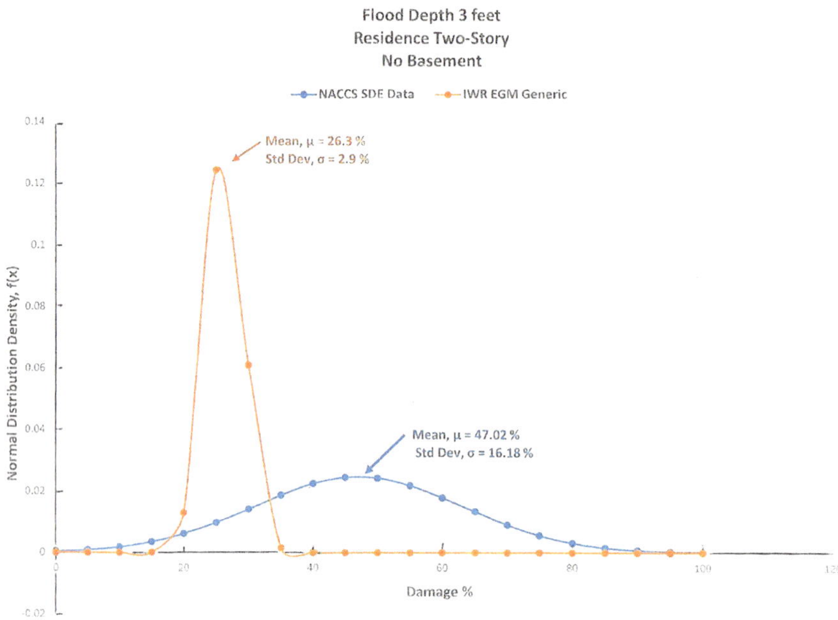

Figure 8.25. Normal distribution density functions calculated from IWR EGM generic tables (red) and from Hurricane Sandy, NACCS SDE data (blue).

in the uncertainty estimate ($\mu + 1.96\sigma$) would provide a more standard way to treat the uncertainty in the damage curves than continued use of a triangular distribution. An example economic risk calculation for Norfolk, VA, using both normal distribution curves in Figure 8.25 is presented below in Section 8.6.4.

(2) Wave-damage curves

Wave-damage curves are presented in the IWR (2011) report "Coastal Storm Risk Management" with no explanation on how they were obtained. Water depth above the first-floor elevation (FFE, feet) is again the forcing function with negative values down to −8 ft below FFE as illustrated in Figure 8.26. When the FFE is zero, wave damage to the structure increases rapidly above 20% and reaches 100% when the water depth above the FFE is only 1–2 ft for the Min, Most-likely, and Max curves. This report was published before the NACCS (2015) data results for wave damage (Figure 8.21). Practically, inundation and wave damage are combined so that double counting should not be done to the same structure. See the Beach-fx user manual for details.

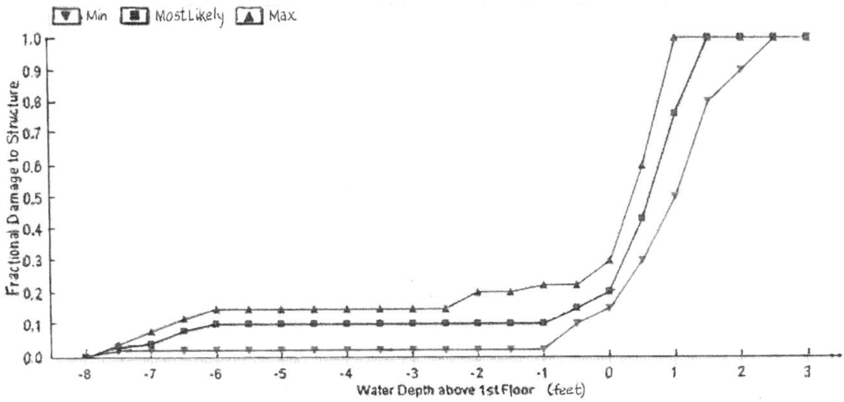

Figure 8.26. Generic, wave-damage curves (from IWR, 2011).

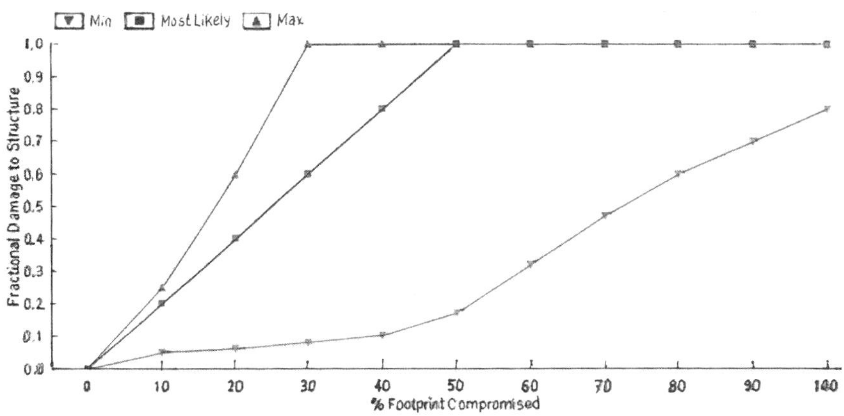

Figure 8.27. Generic, erosion–damage curves for a pile foundation (from IWR, 2011).

(3) Erosion footprint-damage curves

One erosion footprint-damage curve is presented in the IWR (2011) report "Coastal Storm Risk Management" and again with no explanation on how it was obtained. Figure 8.27 shows the results with the % Foot Compromised as the variable used to estimate the percent damage to the structure. The footprint is for a pile foundation. The erosion occurs around the piles and the more piles, the more the footprint that is compromised. This erosion-damage curve was developed for the Beach-fx software since many houses located at beaches are now constructed on pile foundations so that the first-floor elevation is above the 100-year flood plus waves, water–surface elevation.

Inundation- and wave-damage curves for pile foundations are also found in IWR (2011) as examples only and have not been reproduced herein.

(4) Combining inundation, wave, and erosion–damage curves.

To eliminate the potential for double counting when all three damage mechanisms are present (inundation factor, i; wave factor, w; and erosion factor, e), the IRW (2011) report gives the following equation for the value of a combined damage factor, d.

$$d = (i + w + e) - [(i \times w) + (i \times e) + (w \times e)] + (i \times w \times e). \quad (8.17)$$

The present-day market value of the structure is multiplied by d to compute the monetary value of the damages.

8.6.4 *Norfolk coastal storm risk management feasibility study*

"The North Atlantic Coast Comprehensive Study: Resilient Adaptation to Increasing Risk" (NACCS, Jan 2015), identified the city of Norfolk, Virginia, and the region (Appendix D-10 Commonwealth of Virginia) as a focus area for a finer, detailed evaluation (Tier 3). The NACCS recognized that many other cities in the Hampton Roads region are also at high risk of flooding. The city of Norfolk had funding and is an existing authorized Corps' project so became the nonfederal, local sponsor for the feasibility study. Federal funding was approved in 2016 Financial Year.

A Tier 3-level study includes determination of costs and benefits of all the "Risk Management Strategies" listed in Figure 8.28 (generic location). Color-coded labels for non-structural (orange), structural (black), and natural and nature-based (green) alternatives for coastal hazard mitigation for a generic location are shown. All are mentioned in this book and all were studied by the Norfolk District Office of the Corps of Engineers to develop a Tentatively Selected Plan (TSP) that is economically and environmentally feasible.

The coastal storm elevation for design was at the annual exceedance probability $p_{>} = 0.02$ level (50-year). This is a feasibility-level study and, if warranted, this level will be further evaluated and modified during the design stage for an authorized project. A +3 ft (1 m) sea level

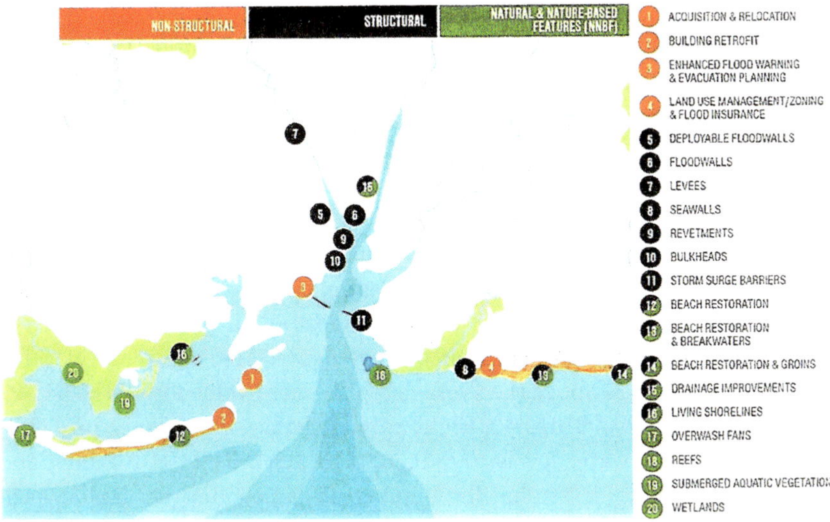

Figure 8.28. Twenty alternatives as risk management strategies for coastal communities as reviewed in the NACCS report (January, 2015).

rise was included for year 2076 considering a 50-year project starting in 2026 using the Corps of Engineers' "Intermediate Curve" for sea level rise (ER-1100-2-8162) and applying the Sea Level Rise Change Curve Calculator (www.corpsclimate.us/ccaces/curves.cfm). Chapter 11 has the details.

8.6.4.1 *Analysis of flood damages*

The economic evaluation tool was the Hydrological Engineering Center, Flood Damage Assessment (HEC-FDA), Corps-certified model. The process applies Monte Carlo Simulation (MCS) of the project life-cycle (typically 50 years) by determining the coastline and structure response to a set of storms (the events driving the process). Each simulation has a set of iterations. This simulation is repeated for many different project life-cycles represented by unique random sequences of storms, and the results are averaged or discussed as a range of possibilities. An advantage of MCS is the high level of detail that can be represented, the congruence of model entities with real-world entities, and the capability of including a variety of dynamic responses. MCS models also lend themselves to visualization. Disadvantages include the extensive time that may be needed to develop the simulation, potentially large data requirements, difficulty of verification and calibration, huge amounts of generated data, and the sometimes many iterations for convergence of results. The HEC-FDA automates much of this process. It was developed for riverine flood damage studies and does not include wave- or erosion-damage. However, for Norfolk, these limitations were not important since the Chesapeake Bay beaches of Norfolk were not included in this study (see Figure 8.29).

Damages to residences were calculated using the NACCS (2015) tables found in the Physical Depth Damage Function Summary Report. In the City of Norfolk, there are 54,591 residential structures of which 17,716 (32.5%) are two-story with no basement. Almost all have full crawl space of 3 ft (1m) above ground level. Their average depreciated replacement value was \$198,000. This is above the average replacement value for all the 54,591 residences, which was \$161,640. Eight annual chance, exceedance probability, $p_>$, values for water levels (0.5, 0.20, 0.10, 0.04, 0.02, 0.01, 0.005, and 0.002) were used to determine the number of damaged structures at each water level. At $p_> = 0.02$ (50-year event), the water levels damaged 7,466 residences. This means about 2,426 residences, assuming 32.5% are two-story with no basement. At 2017 price levels, the total damages to 7,466 residences was estimated at \$419,411,000 so that again assuming 32.5% (2,426) are two-story, no basement, the damage amount is over \$136,000,000 for the 50-year recurrence interval event.

In Figure 8.25, the normal distribution density curves were calculated for structure damage using water depths of 3 ft (1m) above the first-floor elevation. This is about 6 ft (3 ft crawl space +3 ft water depth) above ground level and assumed near the water levels in an annual exceedance probability, $p_> = 0.02$, event. Therefore, the Norfolk District employed the damage levels in Figure 8.18(b) at the 3 ft depth. Assume they took the % Damage, max value of 35% to obtain the \$136,000,000 damage amount. As shown in Figure 8.25, if they used the actual data from the NACCS report at the 95% confidence level, the % Damage $= (\mu + 1.96\sigma) = (47.0 + 1.96 \times 16.2) = 79\%$ and over twice the amount of damage as reported. The economic value for many other damages to structures (apartments, commercial, public, industrial, other), content, vehicles, and cost of debris removal were estimated and included. These with-out project damages are the benefits when damages from flooding are prevented. The total benefits of damages prevented by the TSP were estimated to be \$4.5 billion for the Norfolk Coastal Storm Risk Management feasibility study. This value would be higher if more realistic depth–damage curves were applied.

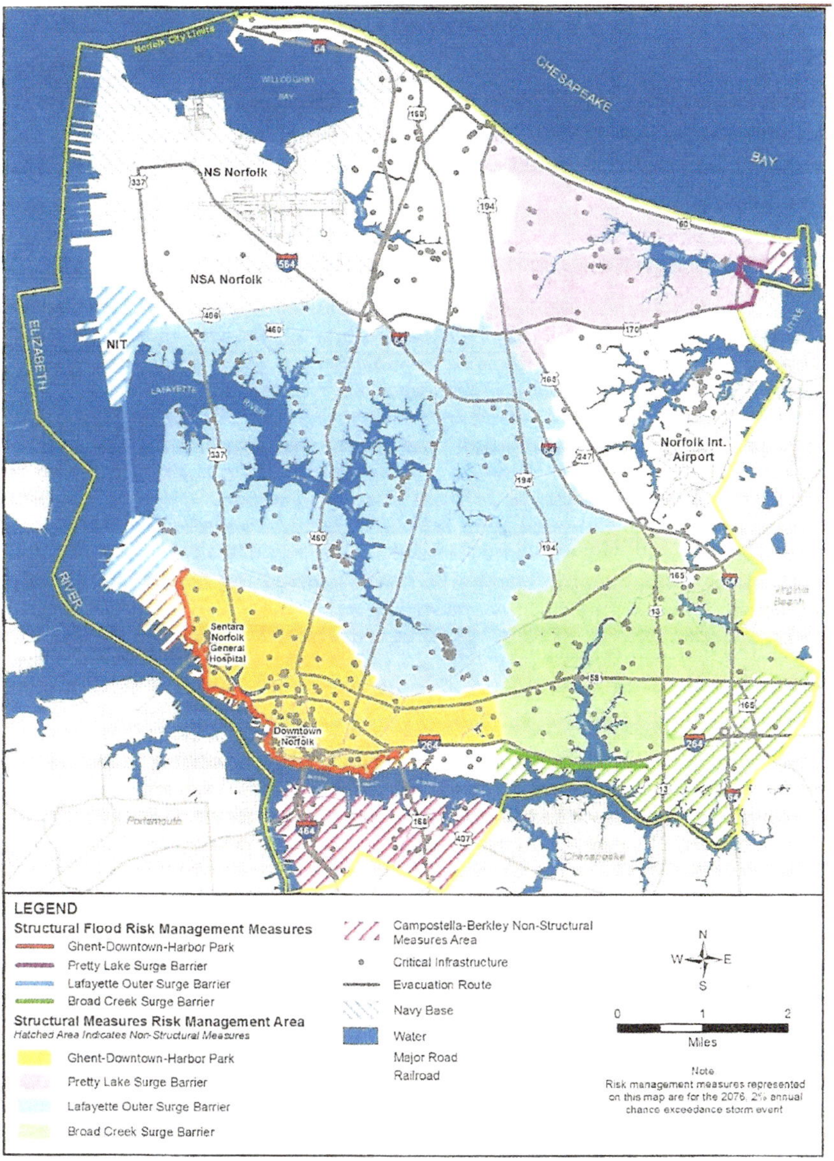

Figure 8.29. Map of Norfolk and description of Tentatively Selected Plan (TSP) of the feasibility study (from Norfolk Feasibility Study, 2017).

8.6.4.2 *Cost of alternatives to mitigate damages*

The City of Norfolk was divided into four main watershed areas that are tidally connected to the Chesapeake Bay. These are shown as colored areas in Figure 8.29. Two areas (yellow and green) are along the southern branch of the Elizabeth River tidal estuary; one area (blue) is the Lafayette River tidal estuary and one (pink) is the Pretty Lake tidal estuary at Little Creek Inlet to the Bay. The structural flood risk management alternative that proved most effective for *all four areas* was a storm surge barrier with earth dikes and floodwalls on the adjacent water/land boundaries. These are also shown in Figure 8.29 as red for the Ghent–Downtown–Harbor Park (yellow) area; dark green for the Broad Creek (green) area; blue for the Lafayette River (light blue) area; and magenta for the Pretty Lake (pink) area. Non-structural alternatives (including ring dikes) are planned for the colored, cross-hatched areas of the city. The Norfolk Naval Base is shown as the grey, cross-hatched region and was not included in this study.

The costs of all the many alternatives studied to mitigate these damages were estimated using standard spreadsheet tools of the Corps of Engineers. The alternatives studied resulted in determination of a Tentatively Selected Plan (TSP) which essentially is that described in Figure 8.29. A yellow line is shown around the city of Norfolk that represents the property boundary of the city. The Corps of Engineers was restricted to only consider flood damage mitigation alternatives for Norfolk.

Life-cycle cost for the Tentatively Selected Plan (TSP) was \$1.8 billion, giving a Benefit/Cost (B/C) ratio of $(4.5/1.8) = 2.5$. Annual Net Benefits were \$101 million, so that the TSP is economically feasible. The environmental impact study was conducted by the Virginia Institute of Marine Science (VIMS). A "Draft Integrated City of Norfolk Coastal Storm Risk Management Feasibility Study/Environmental Impact Statement" (DIFR-EIS, NCSRM) was released in October, 2017 (Norfolk Feasibility Study, 2017). Full results can be found at (http://www.nao.usace.army.mil/NCSRM/) including reports by VIMS regarding the environmental impacts.

8.6.5 *Generation II Coastal Risk Model (G2CRM)*

The need for a Corps of Engineers' risk-based, coastal hazard analysis model that uses GIS and readily available data was apparent after Hurricane Katrina (August 2005) and even more evident in the wake of Hurricane Sandy (October 2012). Damages that are expected with-out mitigation and with-mitigation alternatives need to be rapidly evaluated for selection of a possible "Tentatively Selected Plan" (TSP) for submittal to the US Congress for coastal hazard mitigation projects. This is especially important since the Corps has initiated the SMART planning process for feasibility studies ($3 \times 3 \times 3$) that requires 3 years, \$3 million, and has 3-levels of review before submitting to Congress.

As a result, an engineering-economic, planning-level tool labeled "Generation II Coastal Risk Model" (G2CRM) is being developed for certification by the Corps. It employs a probabilistic, life-cycle analysis using the event-driven, Monte Carlo Simulation (MCS) to overcome the problem of estimating repetitive damage during a life-cycle as employed in earlier (Generation I) modeling. It is similar to Beach-fx for beach/dune engineering-economic analysis and to HEC-FDA for riverine engineering-economic analysis. More details can be found in Moser *et al.* (2013).

8.6.6 *The literature surrounding coastal risk hazards*

There exists a wide variety and large list of literature regarding coastal risk analysis around the world.

Regarding flood depth–damage curves and uncertainty, see, e.g. Smith (1994, South Africa), Merz *et al.* (2004, Germany), Kang *et al.* (2005, Taipei), Freni *et al.* (2010, Italy), De Moel *et al.* (2011, the Netherlands), Pistrika *et al.* (2014, Greece), Hasanzadeh *et al.* (2015, Australia), and Prahl *et al.* (2016, Germany). For the European Commission, Huizinga *et al.* (2017) developed the methodology and database for "Global Flood Depth–Damage Functions".

Articles appear in journals of *Risk Analysis*, *Risk and Insurance*, *Environmental Planning and Management*, and more recently the *Journal of Flood Risk Management* founded by Professor David Balmforth in 2014 (Samuels, 2014). The United Nations Office for Disaster Risk Reduction and the online RISC-KIT established by the European Commission (www.riskit.eu/np4/home.html) should also be consulted with regards to policy for coastal resilience in Europe and the World.

8.6.7 *Summary*

This chapter began with definitions of risk and the quotation *Risk is a combination of probability and consequences* that comes from the Dutch national water authority, the "Rijkswaterstaat" in the Netherlands, which is comparable to the US Army Corps of Engineers in the US. Risk mitigation efforts of the Corps were discussed throughout. Residual risk is the risk left over after efforts to reduce risk from storm damage are enacted by levels of mitigation and adaptation measures that society is willing to finance and accept. How society then deals with this additional, residual risk is termed resilience.

8.7 Resilience

8.7.1 *Introduction*

Resilience is the ability of a system to maintain and/or recover its functional performance after absorbing an adverse event such as a coastal storm. As environmental and water level conditions change over time, the resilience of the system will also evolve to a different level. Schultz and Smith (2016) state:

> *Resilience is the ability to anticipate, prepare for, and adapt to changing conditions and withstand, respond to, and recover rapidly from disruptions. Methods and tools to quantify resilience are needed to provide actionable intelligence to plan, design, construct, and manage coastal systems.*

The key words are "methods and tools to quantify resilience". There are different types of coastal systems including property, infrastructure, social, ecological, and cultural, with each having its own level of resilience to be considered for "management". Coastal engineered alternatives for mitigation of coastal storm energy therefore increase the resilience of these systems for their functional performance.

Risk reduction from coastal storm damage is quantified economically by the Benefit/Cost analysis method for *structural* performance of coastal engineered alternatives. Resilience increase from coastal storm impacts can be quantified mathematically by the Bayesian Network analysis method for *functional* performance of coastal engineered alternatives. Together, they provide a quantitative and logical way to select the most appropriate mitigation alternative for both *structural* and *functional* performance of coastal engineered systems. In short, this section is how we have moved from risk management to disaster resilience and establishment of resilience metrics to quantify resilience on a systems scale.

8.7.2 *The role of "failure" in coastal engineering design*

The word "failure" is used in three different ways in coastal engineering design. When damage curves reach the critical level to cause "failure" of the structure is one definition (Section 8.1.1 (k)). When storms with energy *below* the design level cause loss of structural integrity, a design "failure" has occurred. Here we focus on the Kamphuis definition for total system "failure", f_{TS}, for it directly relates to the resilience of all types of coastal systems. Kamphuis (2010, 2nd edition) defined the Physico-Environmental Subsystem (PES) as the combination of physical forces on the system (water levels, waves, etc.) and the environmental forces on the system. The probability of "failure", P_f, is the probability that the load exceeds the resistance (see Chapter 4). The base of support for the "resistance" relative to the "load" is then defined as the Socio-Economic Subsystem (SES) that is made up of economics and governments, groups, and individuals. The Total System (TS) equals the sum of the PES and the SES, i.e. TS = PES + SES as schematically illustrated in Figure 8.30, with the SES base holding up the PES.

Kamphuis (2010) defined the product of the probability of non-exceedance of the hazard, $p_<$, times the vulnerability of what is exposed to the hazard, v, as the probability of PES failure, P_f.

$$P_f = (p_<(\text{hazard})) \times (v(\text{hazard}))$$

so that his definition of risk (Equation (8.18)) is equivalent to the full definition of Risk, Equation (8.16) in Section 8.6.3.

$$\text{Risk} = \sum_{j}^{jj} [P_f \times C_i] \times d(\text{hazard}), \tag{8.18}$$

where $j = 1, 2, 3, \ldots, jj$ is the range for discretization of the variable for the hazard, and $i = 1, 2, 3 \ldots, n$ is the number of consequences, C_i, produced by the hazard.

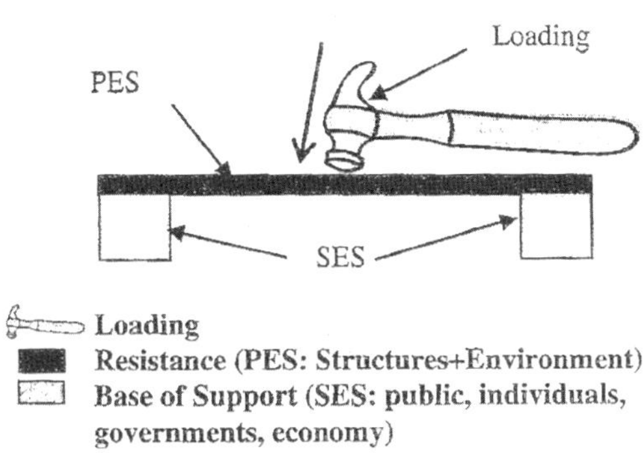

Figure 8.30. Schematic representation of load (PES) and resistance (SES) in total system (TS) after Kamphuis (2010).

There are many causes for P_f failure of the PES part of the total system. One is simply that we cannot provide a low enough probability of failure, P_f, for the structure strength from a study of alternatives for the cost available for design. Other causes of PES failure are an inadequate database to estimate $p_<$; concerns over age, initial design, and construction for the vulnerability, v (damage curves), and rampant real estate development in flood prone, coastal regions. Unknown secondary processes such as land subsidence and acceleration of sea level rise are also major factors as further discussed in Chapter 11.

The long standing, traditional engineering response to PES failure has been to build it stronger with design at a lower probability of failure. Kamphuis (2010) contends that coastal engineers must no longer continue this practice. The reason is simply that PES failures will continue in the future because the SES base of support when this happens has proven to be inadequate in the past. Contemporary Design is now to …*learn to live with PES failure.*

Contemporary System Design (CSD) in coastal engineering recognizes that risk is the product of the probability of PES failure, P_f, and the consequences, C_i, which are not by design. The consequences are SES failures due to historical evolution (development, population); individuals wanting safety at the lowest possible, P_f, and governments and clients wanting the lowest cost. Kamphuis (2010) then defines a total system failure, f_{TS}, when the SES cannot bear the consequences (damage, death, environmental) of PES failure.

Resilience is then defined as the ability to recover from this total system failure. There are three levels of resilient design as defined by Kamphuis (2010):

Level 1. Design resilient PES that does not completely collapse, can be repaired.

Level 2. Design resilient SES government interface that includes advanced warning, zoning, permitting, communications, transport, utilities, emergency services, etc.

Level 3. Design resilient SES base of support.

The bottom line is that "failure" must now be considered a fundamental part of the design process from the start, especially when facing the unknown future of accelerating sea level rise (Chapter 11). And that both the PES and the SES are parts of the total system failure, f_{TS}. Efforts to quantify resilience for the total system were not considered by Kamphuis (2010).

8.7.3 *Metrics to quantify resilience*

Many perspectives on resilience have appeared in the literature since the 1970s. Wang and Blackmore (2009) outlined three major concepts of resilience (Ecological, Engineering, Community) for water resource systems as summarized in Figure 8.31 (from Schultz *et al.*, 2010). Note, the emphasis is on functional performance of the initial state of the system for each concept.

These three concepts must be combined for coastal systems for the entire coastal watershed. Methods to quantify coastal system resilience for the US Army, Corps of Engineers are found in Rosati *et al.* (2015) as first defined by the Coastal Engineering Research Board (CERB, 2013) and depicted in Figure 8.32(a). The Corps' definition (color enhanced) is

> *Coastal resilience is defined as the ability of a system to prepare, resist, recover, and adapt to disturbances in order to achieve successful functioning through time.*

Figure 8.32(b) then illustrates how the functionality of the system quickly drops after the disturbance and then recovers over time to approach the pre-storm level. Efforts to increase resilience are then illustrated

Resilience Concept	Definition	Emphasis	Quantitative Measures	Estimation of Measures
Ecological resilience	The ability to resist being forced into an alternate state.	How the system functions are performed.	The force needed to push a system into an alternate steady state.	Theoretical. Not estimated in practice because of uncertainty in thresholds.
Engineering resilience	The ability to resist and recover from disturbance.	Functional performance.	Functions of the rate at which pre-disturbance performance levels are recovered.	Quantified in practice using simulation models or data on past performance.
Community resilience	The ability to preempt and avoid mishaps in organizations through learning and adaptation.	The ability to adapt, reorganize, or develop new functions specifically conditioned on the disturbance.	None known.	Conceptual. Not estimated quantitatively.

Figure 8.31. Three concepts of resilience from Schultz *et al.* (2010) as first suggested by Wang and Blackmore (2009) for water resource systems.

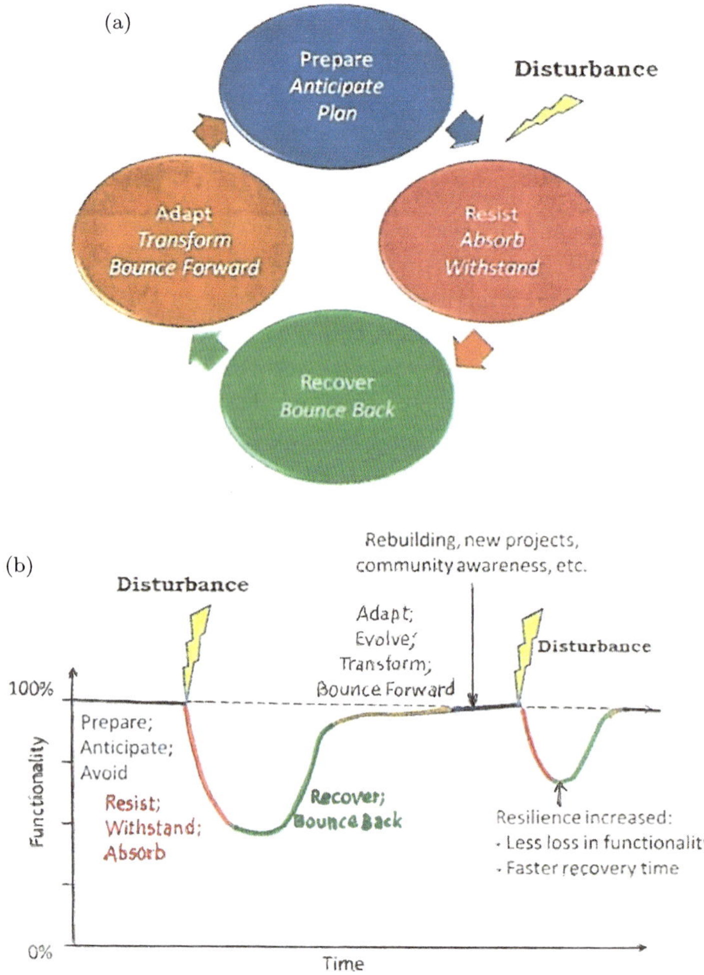

Figure 8.32. (a) Four steps to define resilience. (b) Time-dependent functionality of the system after a disturbance and after efforts to increase resilience to the next disturbance (from Rosati *et al.*, 2015).

following the next disturbance, in that, functionality is higher and recovery is quicker.

8.7.3.1 *Types of resilience assessment methods*

Assessment methods range from simple to complex and the type utilized depends on the availability of data, resources (time, money), and purpose. Initial use of the simple tools help frame the more detailed, complex methods. The Scorecard/index method for critical infrastructure and facilities was developed as a "Community Resilience Index" by NOAA and a Sea Grant Consortium (MS-AL), see the website: http://masgc.org/coastal-storms-program/resilience-index.

Two general matrix approaches are (1) expert elicitation (EE) and (2) data-driven (DD). See Rosati *et al.* (2015) for an excellent review of both types with examples from the US NOAA, EPA, and other methods from the general literature. For example, the Baseline Resilience Indicator for Communities (BIRC) tool (Cutter *et al.*, 2010; Cutter, 2014) is a DD approach with overall composite matrix for six sub-indices including social, economic, institutional, community, environmental, and infrastructural components.

The NACCS Exposure Index (EI) as discussed earlier and its relationship to risk could also be used as a resilience matrix.

8.7.3.2 *Probabilistic, Bayesian network approach*

This is the most complex, detailed approach. Site specific details (databases) for the dominant land use, land area, population, housing statistics, utility locations (water, wastewater, electrical power), designated evacuation zones, etc. are needed for community and infrastructure resilience studies. Numerical models for water levels, waves, and shoreline change are required. The terminology employed is perhaps new to the reader and what follows is found in Schultz and Smith (2016) for the probabilistic, Bayesian network approach.

(1) Definition of terms

First efforts to quantify key metrics illustrated in the levels of functionality and recovery time in Figure 8.32(b) are credited to Bruneau *et al.* (2003) and Chang and Shinozuka (2004). Functional performance, q, is normalized on a scale from 0 to 1, with $q_o = 1.0$ for the performance level before the disturbance and $q = 0$ for complete non-performance. The time when the disturbance begins, t_D, to the time when recovery is complete, t, permits the quantification of all intermediate and beyond metrics associated with a disturbance event as illustrated in Figure 8.33 (from Schultz and Smith, 2016). The ability of the system to maintain its functional performance is called "robustness" and the speed for recovery to pre-disturbance level is called "rapidity". Management objectives needed for both functional performance (robustness), q_s^*, and rapidity, r_s^*, as established before

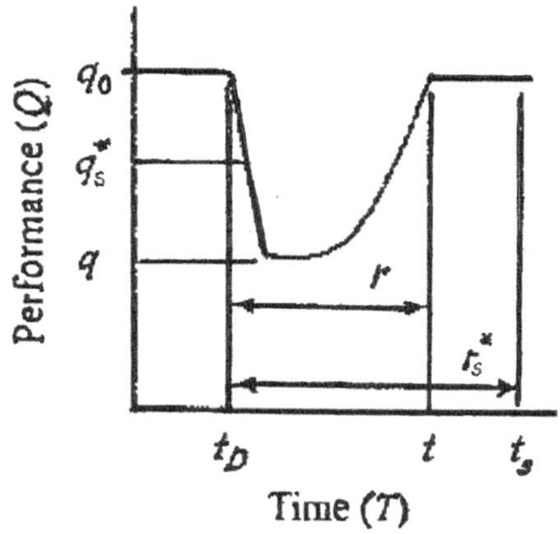

Figure 8.33. System functional performance over time after a disturbance at time, t_D, with robustness and rapidity objectives (q_S^*, r_s^*) (from Schultz and Smith, 2016).

the storm event are also shown in Figure 8.33. These are the limit state criteria for resilience management and vary with the strength of the storm event.

Figure 8.33 demonstrates an example when a coastal storm of magnitude s causes the functional performance level, q, so that the system function drops to level $d = q_o - q$ and then returns to full functional level, $r = t - t_D$. Performance (robustness) and recovery (rapidity) limit states (P_{LS}, R_{LS}) can then be defined as

$$P_{LS}, \text{ Performance limit state} = q - q_s^*, \tag{8.19a}$$

$$R_{LS}, \text{ Rapidity limit state} = r_s^* - r. \tag{8.19b}$$

For this example, the functional performance limit state was negative, meaning that the storm produced greater loss than management planned for and system robustness failed. However, the rapidity limit state was positive, meaning the system recovered more rapidly than management had planned. Four possibilities of Equation (8.19) are possible in the ± quadrants around 0.0

The probabilistic quantification of resilience acceptability, A, as defined by Chang and Shinozuka (2004) for a storm of magnitude s is then defined as

$$p(A \mid s) = p[(q \geq q_s^*) \cup (r \leq r_s^*)]. \tag{8.20}$$

A coastal system is functionality resilient, R, when $R = A$. The $\cup$ symbol means that the probability of union for both robustness and rapidity measures of resilience are achieved for the same storm of magnitude, s.

Subscript s can represent a full matrix of storm parameters for external forces acting on the coastal system. See Schultz *et al.* (2010) for more details with application of the Data Driven (DD) method. They also outlined the steps needed for a probabilistic approach using a Bayesian network.

(2) Bayesian networks

Bayes' theorem (Bayes and Price, 1763) is a formula that relates probabilities that an event A happens given the probability that a related but independent event B happens and is written $P(A\,|\,B)$. Both A and B have their own probabilities, $P(A)$ and $P(B)$, respectively. How often B happens given that A happens is written $P(B\,|\,A)$. The formula is derived from the equation of conditional probability defined as the probability of event A given event B and equals the probability of B and A happening together divided by the probability of B.

A graphical illustration in Figure 8.34 best illustrates conditional probability for two partially intersecting events A and B. Event A with probability, $P(A)$, is shaded white and event B with probability, $P(B)$ is shaded in red. Event B has happened and the part that influences the probability of A is that shaded in blue which is $A \cap B$. The probability of A given B in terms of colors is

$$P(A \cap B) = \frac{\text{(Blue Area)}}{\text{(Red Area)} + \text{(Blue Area)}}. \tag{8.21}$$

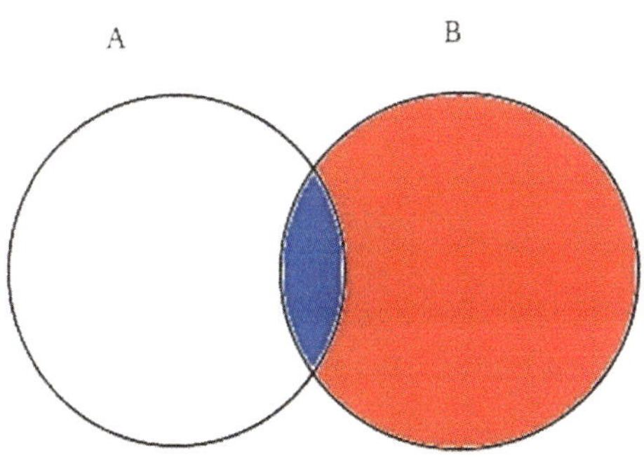

Figure 8.34. Graphical illustration example of conditional probability.

Therefore, we can write the formula for event B given A has already occurred by

$$P(B \mid A) = \frac{P(A \cap B)}{P(A)}, \tag{8.22}$$

or

$$P(A \mid B) = \frac{P(A \cap B)}{P(B)}. \tag{8.23}$$

Using Equation (8.22) in (8.23) gives

$$P(A \mid B) = \frac{P(B \mid A) \times P(A)}{P(B)}. \tag{8.24}$$

Equation (8.24) is the equation of conditional probability and Bayes' theorem when events A and B are independent but happen at the same time.

A Bayesian network is a probabilistic, directed, anti-cyclic (acyclic) graphical model that represents a set of random variables and their conditional dependencies via a directed acyclic graph (DAG). Figure 8.35 is a simple example for three nodes (sprinkler, rain, grass wet) with three directed edges (line with arrow) and three conditional probability tables (CPT). For a DAG, the directed edge can only be in one direction.

For the Rain node, the true (T) probability is 0.2 and the false (F) probability is 0.8, which must add up to 1.0. The probability that the Sprinkler node is running to wet the grass is conditional depending on

	SPRINKLER	
RAIN	T	F
F	0.4	0.6
T	0.01	0.99

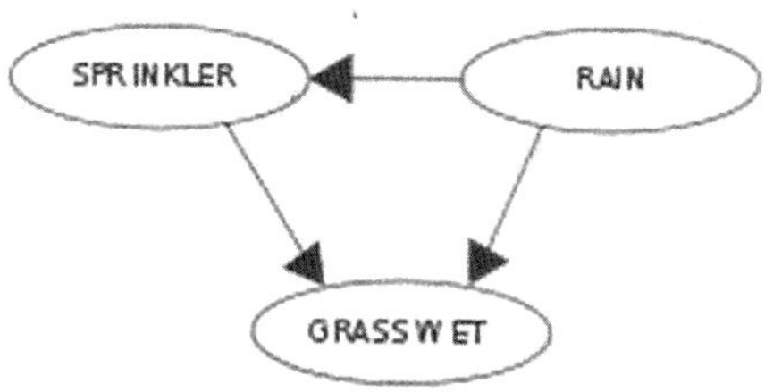

RAIN	
T	F
0.2	0.8

		GRASS WET	
SPRINKLER	RAIN	T	F
F	F	0.0	1.0
F	T	0.8	0.2
T	F	0.9	0.1
T	T	0.99	0.01

Figure 8.35. Example of directed acyclic graph (DAG) for Bayesian network.

whether (or not) it is raining. If it is raining (T), then the probability the sprinkler is not running (F) is 0.99. Note, the direction of the directed edge (line with black arrowhead) from the Rain node to the Sprinker node gives $P(S \mid R)$ as the CPT. Both the Rain and Sprinkler nodes have directed edges toward the Grass Wet node. Therefore, the Conditional Probability Table (CPT) for the Grass Wet node has four levels depending on the four possible combinations of the true (T) or false (F) conditions of each of the two nodes. Note that if it is both raining (T) and the sprinkler is on (T), then the joint probability of the grass being wet (T) is 0.99. Zeros are usually avoided in true/false tables so that (F) is 0.01.

Mathematically, a Bayesian network is... *a factorization of a joint probability distribution over a set of random variables, $X_1, X_2, X_3, \ldots, X_n$ with each describing an event within the system. Each variable can take two or more possible states that are mutually exclusive and collectively exhausted. For i nodes, the joint probability is calculated as*

$$p(X_1, X_2, X_3, \ldots, X_n) = \prod_{i=1}^{n} p(X_i \mid X_{pa(i)}), \tag{8.25}$$

where $\prod_{i=1}^{n}$ is the mathematical symbol for the product of all sub-script i terms from $i = 1$ to n; the subscript $pa(i)$ means X is a parent of node i; and the notation $p(X_i \mid X_{pa(i)})$ is the probability that X is in state i given the state of the parent nodes (Schultz and Smith, 2016).

For management of coastal systems, there are many nodes, directed edges, and Conditional Probability Tables that need to be defined and processed. And, they are best formulated for one or two of the three resilience areas [Engineering, Ecological, Community] listed in Figure 8.31. An example for community and engineering resilience best illustrates a Bayesian network.

(3) Pilot study for Jamaica Bay, NY, by the Corps of Engineers

Hurricane Sandy (October 2012) impacted the region known as Jamaica Bay shown in Figure 8.36 as part of New York City. The tidal inlet on the Atlantic Ocean is Rockaway Inlet and the bay is split by the Cross Bay bridge/causeway on Broad Channel island. John F. Kenney (JFK) Airport is on the eastern end. Active Corps of Engineers projects for coastal damage mitigation and wetlands restoration exist for the region so the Corps conducted a pilot study for assessing the infrastructure and community resilience using a Bayesian network. The results summarized below

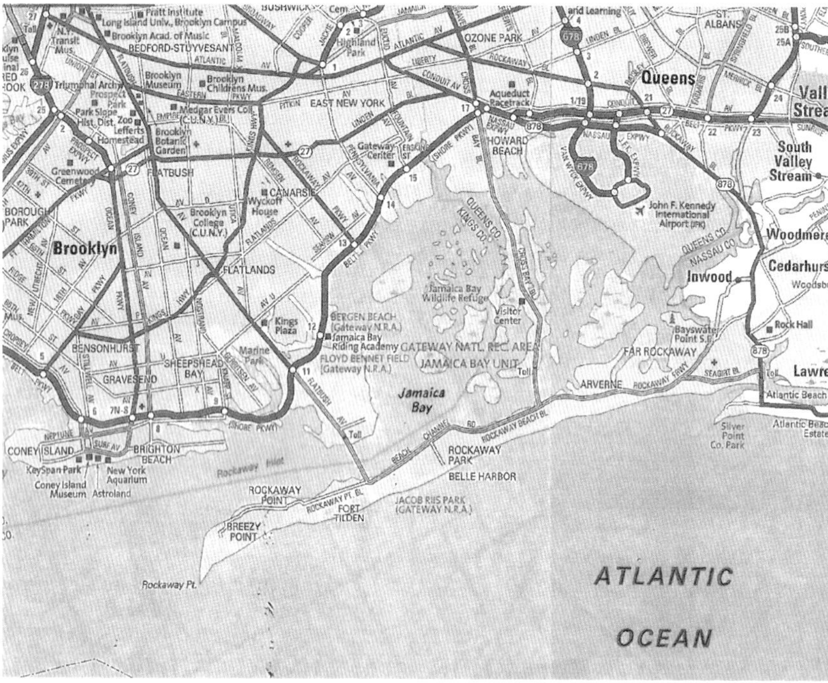

Figure 8.36. The region known as Jamaica Bay as part of Brooklyn and Queens of New York City on the Atlantic Ocean and extending from Brighton Beach, JFK Airport, and on the peninsula to Rockaway Pt (courtesy American Map Corporation, 2010 Road Atlas).

are found in the excellent paper with all the details by Schultz and Smith (2016).

The dominant land use is urban residential. Planning area studied was 87 km^2 (23% was JFK airport) with total shoreline length of 56.8 km (35.2 miles) and population of about 300,000. This study area was broken into 13 sub-areas with the goal to determine community functional resilience for (1) housing, (2) utilities (potable water, wastewater, electric power), (3) transportation (evacuation/return) and (4) life-safety. Table 8.12 shows the eight water surface elevations with annual return periods on the Atlantic side of Rockaway spit used in the ADCIRC numerical model (Luettich *et al.*, 1992) to calculate storm surge levels in Jamaica Bay. Resilience relative to these eight storm return periods (years) with respect to each of the four system functional areas and the entire Jamaica Bay area was determined.

Table 8.12. Water levels (NAVD88) for pilot study in Jamaica Bay (from Schultz and Smith, 2016).

Storm surge on the Atlantic side of the Rockaway split								
	Return Period (y)							
	1	10	25	50	100	250	500	1000
Water surface elevation (m)	0	2.13	3.60	4.21	4.57	5.18	5.70	6.19

System Function	Functional Performance Objective
Life safety	Minimize the fraction of residents exposed to floodwaters.
Housing	Maximize the fraction of residents who can return home and stay following the storm event.
Utilities	Maximize the fraction of residents with potable water, wastewater, and electric utility service.
Transportation	Minimize the increase in travel time between planning unit centroids and points on the planning area boundaries.

Figure 8.37. Performance objectives for functional resilience of four community and engineering (infrastructure) management areas (from Schultz and Smith, 2016).

To accomplish this, performance objectives were defined for each system function as listed in Figure 8.37. For example, for Housing, to maximize the fraction of residents who can return home and stay following the flood event, the limit-state criterion for *robustness* is the minimum acceptable fraction (Figure 8.38) which for the 100-year storm event was set at 0.90 as seen in Figure 8.39. For *rapidity* of recovery, the maximum length of time to repair damaged housing (in days) (Figure 8.38) following a 100-year storm event was set at seven days (Figure 8.39). Similarly, Figures 8.37–8.39 were employed in the analysis for the other system function areas and flooding recurrence values. Clearly, establishment of the limit-state criteria (Figure 8.39) is a key effort requiring input from all aspects of the Jamaica Bay community.

Figure 8.40 is the Directed Acyclic Graph (DAG) for the Bayesian network for Jamaica Bay. It consists of 37 nodes for the random variables.

Limit State Condition	System Function	Description of Limit State Criteria
Robustness	Life safety	Maximum acceptable fraction of the population located inside the maximum extent of the flood footprint
	Housing	Minimum acceptable fraction of the population that can return home and occupy their residences following the storm event
	Utilities	Minimum acceptable fraction of the population with water, wastewater, and electric utility service following the storm event
	Transportation	Maximum acceptable increase in travel time between planning unit centroids and planning unit boundaries (in minutes)
Rapidity	Life safety	Not applicable
	Housing	Maximum length of time to repair damaged housing (in days)
	Utilities	Maximum length of time to restore utility service (in days)
	Transportation	Maximum length of time to restore normal travel times (in days)

Figure 8.38. Descriptions of criteria for limit-state evaluation for robustness and recovery (rapidity) of resilience in four management areas (from Schultz and Smith, 2016).

Return Period (y)	System Function*							
	Life Safety		Housing		Utilities		Transportation	
	Robustness	Rapidity†	Robustness	Rapidity	Robustness	Rapidity	Robustness (min)	Rapidity
1	0.10	—	1.00	0	0.70	0	5	90
10	0.20	—	1.00	1	0.70	1	5	90
20	0.30	—	1.00	1	0.70	1	5	90
50	0.30	—	1.00	1	0.70	7	5	90
100	0.30	—	0.90	7	0.70	7	5	90
250	0.30	—	0.80	30	0.70	30	5	90
500	0.30	—	0.70	90	0.70	90	5	90
1000	0.30	—	0.60	365	0.70	365	5	90

Figure 8.39. Table of limit-state values employed for each resilience management objective listed in Figure 8.38 (from Schultz and Smith, 2016) and for each level of storm severity. Rapidity† means none for Life Safety.

The top one [PU(1)] represents the 13 sub-areas in Jamaica Bay with conditional probability table (CPT) described by the fraction of the population living in each sub-area. All the nodes have letters and numbers with descriptions, units, potential states, and parameterization details given in Table 6 of Schultz and Smith (2016) and are far too detailed to present here. For example, node [WWTP(2)] is the wastewater treatment plant inventory (Rockaway, Coney Island, Ward 26, and Jamaica), with conditional probabilities as the fraction of the population in each of the 13 sub-areas that is served by each wastewater treatment plant. Node [TIDE(6)] are the predicted tide elevations in meters for five storm surge intervals with conditional probabilities for the fraction of time the tidal elevation is within each interval.

Four area boxes (dotted) surround the nodes that only pertain to the four functional performance objectives (Life-Safety, Housing, Utilities, and Transportation). Near the bottom of each box are the four nodes for

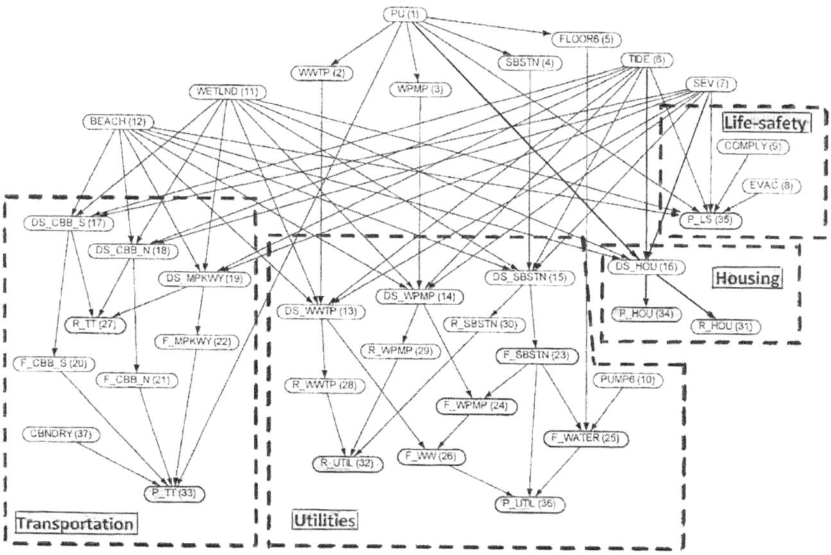

Figure 8.40. Graphical display of Bayesian network for Jamaica Bay (from Schultz and Smith, 2016). Note that all the direction lines only have arrows in one direction for a DAG.

variables [P_LS(35)], [P_HOU(34)], [P_UTL(36)], and [P_TT(33)] for the performance metrics (robustness) for these four objectives. For the length of time to recover to pre-storm levels, the variables shown are [R_HOU(31)], [R_UTL(32)], and [R_TT(27)] for housing, utilities, and transportation, respectively. There is no recovery time for Life-Safety.

As an example, CPT for Housing are shown in Figure 8.41. The CPT for performance robustness [P_HOU (34)] is Figure 8.41(a) for seven levels of state of damage and the probability levels associated with either being Habitable or Uninhabitable. As housing damage increases, the probability of being habitable decreases. For these same seven levels of housing damage, Figure 8.41(b) is the recovery time CPT for [R_HOU(31)] in six time periods (days). When over 50% of the residence is damaged, the recovery period can be from three months to a year, with a probability of 100% (1.00 in table).

The [DS_HOU(16)] node is the parent node with directed edges to both the robustness node [P_HOU] and recovery node [R_HOU]. Damage is discretized into seven possible states and is initially conditionally dependent upon three parent nodes with directed edges coming from node [PU(1)] which is a fraction of residential structures in the 13 sub-areas; node

Housing Damage State	Performance (P_HOU)	
	Habitable	Uninhabitable
0	1	0
$0 < x \leq 0.1$	0.9	0.1
$0.1 < x \leq 0.2$	0.75	0.25
$0.2 < x \leq 0.3$	0.5	0.5
$0.3 < x \leq 0.4$	0.25	0.75
$0.4 < x \leq 0.5$	0	1
$0.5 < x \leq 1.0$	0	1

(a)

Housing Damage State	Recovery Period (days)					
	0	0–1	1–7	7–30	30–90	90–365
0	1.00	—	—	—	—	—
$0 < x \leq 0.1$	—	0.50	0.50	—	—	—
$0.1 < x \leq 0.2$	—	—	0.50	0.50	—	—
$0.2 < x \leq 0.3$	—	—	0.25	0.50	0.25	—
$0.3 < x \leq 0.4$	—	—	—	0.10	0.80	0.10
$0.4 < x \leq 0.5$	—	—	—	—	—	1.00
$0.5 < x \leq 1.0$	—	—	—	—	—	1.00

(b)

Figure 8.41. (a) Conditional Probability Table (CPT) for housing robustness node [P_HOU] from Schultz and Smith, 2016. (b) Conditional Probability Table (CPT) for housing recovery node [R-HOU] from Schultz and Smith, 2016.

[TIDE(6)] is for tidal levels, and node [SEV(7)] is the severity of the potential storm event. Additional directed edges for node [BEACH] and node [WETLND] will be discussed below when these two alternatives for increasing resilience are discussed.

Damage is estimated using the same damage curves discussed in Section 8.6.3.2 for risk analysis. These damage functions estimate structural damage and the ratio of cleanup and repair costs to assessed tax value as a function of flood depth and housing characteristics (number of stories, basements, etc.).

The conditional probability relationships for Utilities and Transportation boxes are far more complex needing many more nodes. See Schultz and Smith (2016) for further details. Figure 8.40 displays the Bayesian network for the robustness and recovery (rapidity) metrics and must be further extended for management of resilience of the entire coastal system.

(4) Resilience management of the coastal system

Three more nodes are needed. One is for the limit-state values for each resilience management objective listed in the table shown in Figure 8.39 for each level of storm severity, s. The second is an outcome node to learn if management objectives are satisfied or not satisfied.

Let R_1 be the resilience outcome for robustness and R_2 be the resilience outcome for rapidity of recovery for a storm of severity s.

$$\text{For } R_1 \text{ (Housing and Utilities) } p(R_1 = 1 \,|\, s) = p(q \geq q_s^*), \tag{8.26a}$$

$$\text{and (Transportation and Life-Safety) } p(R_1 = 1 \,|\, s) = p(q \leq q_s^*), \tag{8.26b}$$

$$\text{For } R_2 \text{ (Housing, Utilities, Transportation) } p(R_2 = 1 \,|\, s) = p(T \leq t_s^*). \tag{8.27}$$

The values for the outcome node for $R_1$and R_2 take the value 1 only if the limit-state criteria are met and take the value of 0 otherwise.

The third additional node is for the resilience outcome when *both* robustness, R_1, and rapidity, R_2, management objectives are achieved or not. Functional resilience takes the value of 1 *when both* are 1 and the value of 0 when not. Resilience can be computed separately for each of the four functional objectives (Housing, Utilities, Transportation, Life-Safety) and for the combined, joint resilience. The node is parameterized using

$$R_i = \prod_k R_{ik}, \tag{8.28}$$

where i is the index of performance objectives and k is the index of resilience management objectives. When all functional performance objectives, k, satisfy both robustness and rapidity criteria, then

$$R = \prod R_i. \tag{8.29}$$

The value, R, of this third resilience node is the quantitative measure of resilience.

Table 8.13. Resilience, R, for Jamaica Bay pilot study.

Storm Period, years	1	10	25	50	100	250	500	1,000
Resilience, R	0.906	0.748	0.587	0.236	0.151	0.119	0.095	0.011

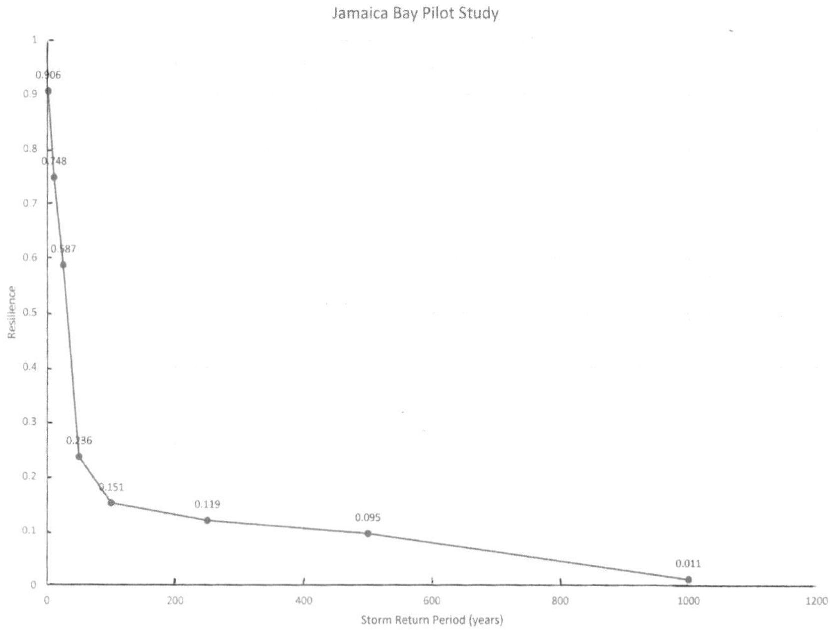

Figure 8.42. Resilience in Jamaica Bay for storm return periods from 1 to 1,000 years.

(5) Results for Jamaica Bay pilot study of resilience

When all the functional resilience management objectives are met for all levels of potential storm events (Figure 8.39) the overall resilience is 0.857. However, as coastal storm severity s increases as measured by the storm return periods, the resilience, R, decreases rapidly as summarized in Table 8.13 and plotted in Figure 8.42. This is the result of Rockaway Inlet conveying floodwaters into Jamaica Bay and the heavily populated residential region.

Resilience of the four system functions as varying with return period (years) is shown in Figure 8.43. (Graphic from Rosati *et al.*, 2018). The Utilities are the most vulnerable and even below unity for the water level with a one-year return period. Housing is the next most vulnerable with

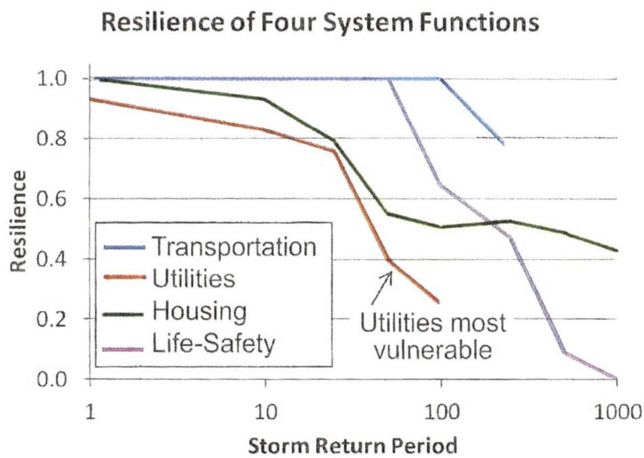

Figure 8.43. Resilience of four system functions (Schultz and Smith, 2016) as a function of storm return period (years). Graphic from Rosati *et al.* (2018).

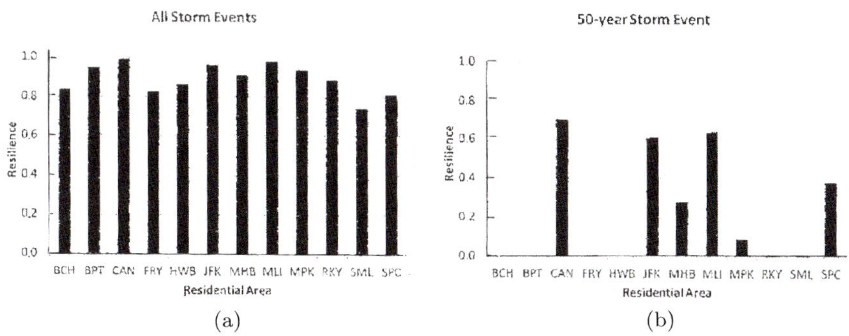

Figure 8.44. (a) Resilience of individual areas for all storm events and (b) for 50-year storm event (from Schultz and Smith, 2016).

transportation the least vulnerable. However, many major transportation system details were not considered (traffic lights, roadway flooding, roadway debris, etc.) that would increase travel times and should be considered in transportation system resilience.

Finally, resilience can also be separated and estimated for each of the 13 sub-areas that make up the entire Jamaica Bay pilot study. Figure 8.44(a) displays the results for all storm events when all functional management objectives are met (Figure 8.39) which produced the average value of 0.857 for overall resilience. The resilience of individual sub-areas is far different when displayed for specific storm events as illustrated in Figure 8.44(b) for the 50-year storm event. Six of the sub-areas have essentially no functional resilience regarding Housing, Utilities, Transportation, and Life-safety.

These estimates of resilience are based on existing conditions in and surrounding Jamaica Bay, NYC, for the Pilot Study example (status quo). They are meant to illustrate the advantages of the Bayesian network method to quantify resilience. The probabilistic approach estimates how likely (chance) it is that management objectives will be achieved. It quantitatively shows the change in resilience for flooding levels on the community and infrastructure as to which components are most vulnerable and which sub-areas of the community are most vulnerable. Decisions necessary to prioritize expenditure of limited funds in City budgets for capital improvements can be made on rational, factual information. Politics will always play a role, but the general public is better served when decisions are based on real data.

Coastal engineering design decisions for alternatives are typically based on risk-based (Benefit/Cost) methods. Quantification of functional resilience of alternatives using Bayesian networks offers an additional way to study different alternatives and complements the risk-based ($$) method.

(6) Alternatives to increase resilience for Jamaica Bay

The paper by Schultz and Smith (2016) included a discussion of hypothetical alternatives to illustrate how the Bayesian network can be employed to study coastal engineering alternatives to improve resilience. These were the two additional nodes [WETLND(12)], [BEACH(13)] in the Bayesian network graph, with many directed edges to the four functional areas, Figure 8.40. One alternative was extensive wetland restoration in Jamaica Bay and the other included increasing the dune height and beach width on the ocean side of Rockaway Spit. Together, they made very little improvement ($\approx$10%) in overall resilience and far less at increased water levels for high storm return periods. The purpose was to demonstrate the usefulness of Bayesian networks in the design process (not for Jamaica Bay design) and the results are not presented here. See Schultz and Smith (2016) for details. Many other alternatives are possible as discussed in this book.

(7) Limitations of Bayesian networks

There are some limitations for the Bayesian network approach. Recovery is measured as how fast it occurs (rapidity) but not how it will take place. Another limitation is that it combines all robustness and rapidity criteria simultaneously. The DAG constraint prohibits feedback in the modeled system, but that can be overcome. All variables must be discretized (tables). The development of the network is labor-intensive.

The time and work involved to set the limit-state criteria is also a challenge, but should not be viewed as a limitation but as a positive contribution to making realistic resilience management objectives. They can readily be adjusted to explore their sensitivity to the resilience results. Figure 8.45 illustrates the probabilistic nature of the target robustness in performance and the probabilistic nature of the target rapidity in recovery time (Kameshwar *et al.*, 2019). Input is needed from people living in flood-prone residential areas to establish practical guidelines for the limit-state criteria (target goals) for both robustness and rapidity to function normally.

(8) Tiered approach for quantification of resilience

To take advantage of the different ways to quantify resilience from simple (matrix) to complex (Bayesian network) and to match the range of studies conducted by the Corps of Engineers (scoping-level, feasibility-level, final-design, and construction level) a three-tiered approach has been developed. Tier-level 1 is a Resilience Matrix using Expert Elicitation (EE) from community, expert knowledge base to identify critical functions and vulnerabilities in the system. Tier-level 2 uses data, models, and community values for identification of critical needs. It creates a systems-wide scale of resilience assessment for feasibility studies. It is also a Resilience Matrix but

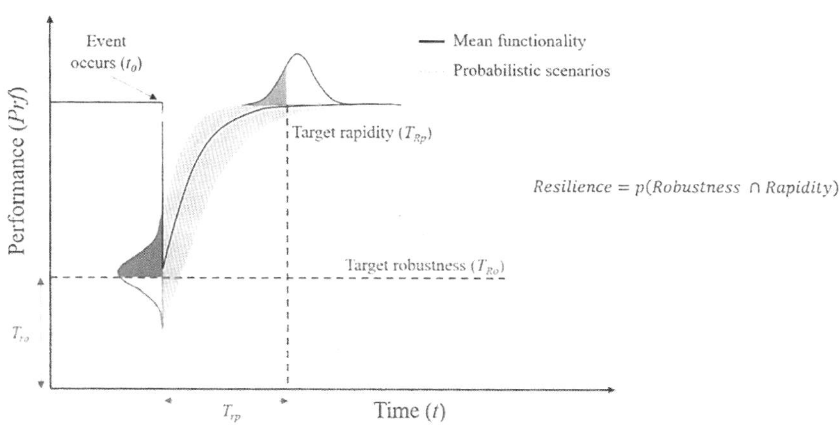

Figure 8.45. Schematic of probabilistic nature of limit-state (target) goals for robustness and rapidity (from Kameshwar *et al.*, 2019).

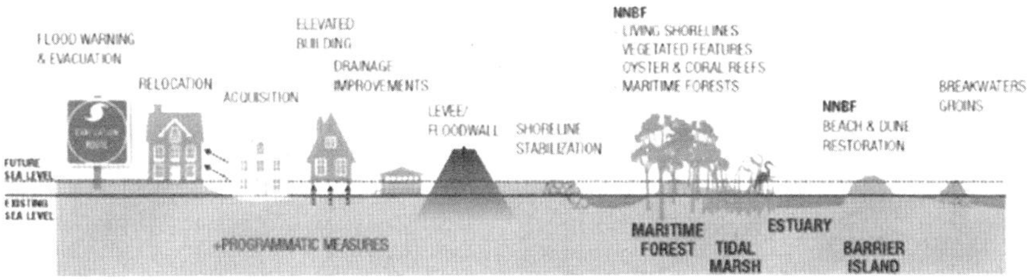

Figure 8.46. Coastal hazard mitigation and adaptation measures considered by the US Army Corps of Engineers (from NACCS, 2015). More than one alternative is redundancy to increase resilience and reduce risk.

is Data Driven (DD). Outcomes from Tier-level 2 are used to make decisions within feasibility studies for the Tentatively Selected Plan (TSP). The Bayesian network method is recommended for Tier-level 3 for final design and operations studies. See Rosati *et al.* (2015) for review of literature for resilience metrics and the tier-level approach.

It is expected that in the near future, as the Bayesian network approach is made more routine and less labor-intensive (e.g. cities will routinely include all the housing statistics needed in city-maintained websites), the Bayesian network methods will find greater use in feasibility studies (Tier-level 2).

(9) Redundancy

Figure 8.46 displays almost all the potential alternatives for coastal hazard mitigation and adaption to increase resilience and reduce the risk of damage from flooding, high waves, and erosion. When more than one alternative is considered at a site, the redundancy significantly increases the resilience and greatly reduces the risk, but at additional life-cycle cost. The acronym NNBF stands for Natural and Nature-Based Features and will be further discussed in Chapters 10 and 11.

8.8 Summary

There are three key formulations in this chapter.

8.8.1 *Economics*

For *economics*, each alternative considered in coastal engineering design for flood damage and erosion mitigation has an initial construction cost, I, and a life-cycle maintenance cost, C_m, so that together the total, life-cycle cost of the alternative, C, is simply as given in Equation (8.13).

$$C = I + C_m. \tag{8.13}$$

The total present worth value, C_m, of all future maintenance costs when applying principles of compound interest is simply the annual maintenance costs ($\sum \Delta c + c^*$) times the present worth factor (pwf) from Equation (8.2).

$$\begin{aligned} C_m &= \text{pwf} \times \left(\sum \Delta c + c^*\right) \\ &= \text{the present worth of the life-cycle maintenance costs.} \end{aligned} \tag{8.12}$$

The digital form for spread-sheet solution of annual damage repair cost, c, is found from

$$c = \sum_{(H_s)_n}^{(H_s)_{n+1}} [(\Delta p_>)(\Delta \bar{D})(I_{cl} \times z^*)]\Delta H_s. \tag{8.9}$$

Then $\sum \Delta c$ = the total annual repair damage cost for all increments to where failure begins Equation (8.10) and

$$c^* = D_{\text{failure}}(I \times z^{**})[(p_>)_{\text{failure}} - 0]. \tag{8.11}$$

Thus, c^* is the last region needed to estimate the total annual maintenance costs ($\sum \Delta c + c^*$).

Finally, the Benefit/Cost ratio (B/C) requires calculation of the hazard mitigation benefits which essentially is the calculation of the Risk reduction resulting from the design alternative considered.

8.8.2 *Risk*

Risk, R, is the product of the probability of non-exceedance of the hazard, $p_<(hazard)$, times the vulnerability of what is exposed to the hazard, ν $(hazard)$, and integrated over all the consequences, C_i, resulting from the hazard. In discrete mathematical terms:

$$R\ (\text{hazard}) = \text{Risk} = \sum_{j}^{jj}[(p_<(\text{hazard})) \times (\nu(\text{hazard})) \times C_i] \times d(\text{hazard}) \tag{8.16}$$

where $j = 1, 2, 3, \ldots, jj$ is the range for discretization of the variable for the hazard, and $i = 1, 2, 3 \ldots, n$ is the number of consequences produced by the hazard.

The most important hazards are waves (wave height, H_s), water levels (storm surge, S), erosion (H_s, S, *duration*), and ecological change (ΔE).

The design alternative(s) considered for risk reduction reduce both the vulnerability, v, of the hazard and the consequences, C_i. Hence, the value of risk reduction (i.e. Benefits) is quantified in terms of money saved so that a Benefit/Cost (B/C) ratio and the Net Benefits (B–C) can be calculated. However, many other benefits of the design alternative considered (community, life-saving, environmental, historical, etc.) cannot be put in terms of money saved. These are primarily the benefits of the functionality of the coastal system.

8.8.3 *Resilience*

Third and newest formulation in Coastal Engineering Design is the quantification of resilience through the development of Bayesian networks. A Bayesian network is a probabilistic, directed, anti-cyclic (acyclic) graphical model that represents a set of random variables and their conditional dependencies via a directed acyclic graph (DAG). Mathematically, a Bayesian network is

> *... a factorization of a joint probability distribution over a set of random variables, $X_1, X_2, X_3, \ldots, X_n$ with each describing an event within the system. Each variable can take two or more possible states that are mutually exclusive and collectively exhausted. For i nodes, the joint probability is calculated as*
>
> $$p(X_1, X_2, X_3, \ldots, X_n) = \prod_{i=1}^{n} p(X_i \,|\, X_{pa(i)}), \qquad (8.25)$$
>
> *where $\prod_{i=1}^{n}$ is the mathematical symbol for the product of all sub-script i terms from $i = 1$ to n; the subscript pa(i) means X is a parent of node i; and the notation $p(X_i \,|\, X_{pa(i)})$ is the probability that X is in state i given the state of the parent nodes.* (Schultz and Smith, 2016).

The random variables, X_n, can represent *both* those readily valued in monetary units and those that are not.

The ability of the system to maintain its functional performance is called "robustness" and the speed for recovery to pre-disturbance level is called "rapidity". Management objectives are needed for both functional performance (robustness), q_s^*, and rapidity, r_s^*, as established before the storm event. These are the limit state criteria for resilience management and vary with the strength of the storm event.

The probabilistic quantification of resilience acceptability, A, as defined by Chang and Shinozuka (2004) for a storm of magnitude s is then defined as

$$p(A \mid s) = p[(q \geq q_s^*) \cup (r \leq r_s^*)]. \tag{8.20}$$

A coastal system is functionality resilient, R, when $R = A$. The $\cup$ symbol means that the probability of union for both robustness and rapidity measures of resilience are achieved for the same storm of magnitude s. When *both* robustness, R_1, and rapidity, R_2, management objectives are achieved (or not), functional resilience takes the value of 1 *when both* are 1 and the value of 0 when not. Resilience can be computed separately for each of the functional objectives (e.g. Housing, Utilities, Transportation, Life-Safety, etc. as in the example) and for the combined, joint resilience. The node is parameterized using

$$R_i = \prod_k R_{ik} \tag{8.28}$$

where i is the index of performance objectives and k is the index of resilience management objectives. When all functional performance objectives, k, satisfy both robustness and rapidity criteria, then

$$R = \prod R_i. \tag{8.29}$$

The value, R, of this final resilience node is the quantitative measure of resilience.

8.8.4 *Alternatives for risk reduction and resilience increase*

The design of alternatives for coastal hazard risk reduction and increase of coastal resilience in the face of increasing hazards from rising seas is the focus of this book. See Section 11.3.8 for impacts of sea level rise in Chapter 8.

References

Basco, D. R., and C.-S. Shin (1996) "Dune Damage Curves and Their Use to Estimate Dune Maintenance Costs," in *Proceedings, 25th International Conference Coastal Engineering* (Orlando), ASCE. New York.

Bayes, T., and R. Price (1763). "An Essay Towards Solving a Problem in the Doctrine of Chance." By the late Rev. Mr. Bayes, communicated by Mr. Price, in a letter to John Canton, A. M. F. R. S. *Philosophical Transactions of the Royal Society of London.* Vol. 53, pp. 370–418.

Broderick, L., and J. P. Ahrens (1982). "Rip-rap Stability Scale Effects," Tech. Paper 82-3, *U.S. Army Engineer Waterways Experiment Station*, Coastal Engineering Research Center, Vicksburg, MS.

Bruneau *et al.* (2003) (10 co-authors) "A Framework to Quantitatively Assess and Enhance the Seismic Resilience of Communities," *Earthquake Spectra*, Vol. 19, No. 4.

CEM (2006) *Coastal Engineering Manual*, US Army Corps of Engineers, Engineering Research and Development Center, Vicksburg, MS.

Chang, S. E., and M. Shinozuka (2004) "Measuring Improvements in Disaster Resilience of Communities," *Earthquake Spectra*, Vol. 20, No. 3.

Cialone, M. A., N. Nadal-Caraballo, and C. Massey (2015) "Numerical Modeling and Statistics for NACCS," *Corps of Engineers*, Washington, DC.

Coastal Engineering (1994) "Special Issue Vertical Breakwaters", Vol. 22, Issues 1–2, Elsevier, pp. 1–183.

Cutter, S. L. (2014) "Building Disaster Resilience: Steps Toward Sustainability," *Challenges in Sustainability*, Vol. 1, No. 2, pp. 72–79.

Cutter, S. L., C. G. Burton, and C. T. Emrich (2010) "Disaster Resilience Indicators of Benchmarking Baseline Conditions," *Journal of Homeland Security and Emergency Management*, Vol. 7, No. 1.

De Moel, H., J. C. J. H. Aerts, and E. Koomen (2011) "Development of Flood Exposure in the Netherlands During the 20th and 21st Century," *Global Environment Change*, Vol. 21, pp. 620–627.

EGM (2000) "Generic Depth-Damage Relationships- 1, 2-Story Homes, No Basements", Economic Guidance Memorandum (EGM) No. 01-03, *Corps of Engineers, Institute for Water Resources*, Fort Belvoir VA.

ESRI (2012) "ArcGIS software," *Environment Rating Scales Institute*, Chapel Hill, NC.

Freni, G., G. la Loggia, and V. Notaro (2010) "Uncertainty in Urban Flood Damage Assessment Due to Urban Drainage Modeling and Depth-Damage Curve Estimation", *Water Science and Technology*, Vol. 61, No. 12, pp. 2979–2993.

Goda, Y. (2010) *Random Seas and Design of Maritime Structures*, 3rd edn. World Scientific Publishing Co. Singapore.

Gunn, J. (2014) *Coastal Design and Construction Co.* Gloucester, Virginia, personal communication.

Hasanzadeh, N. R., T. Ngo, and W. Lehman (2016) Calibration and Validation of FLFArs — A New Flood Loss Function for Australian Residential Structures," *Natural Hazards Earth System Science*, Vol. 16, pp. 15–27.

HSIP (2012) "Highway Safety Improvement Program: Fiscal Year 2012-2013," Traffic Engineering Division, Virginia Department of Transportation.

Hudson, R. (1958) "Design of Quarry-Stone Cover Layers for Rubble-Mound Breakwaters," Research Report No. 22, *Corps of Engineers, Waterways Experiment Station*, Vicksburg, MS.

Hudson, R. Y. (1959) "Laboratory Investigation of Rubble-mound Breakwaters," *Jour. of Waterways and Harbors Div., American Society of Civil Engineers*, Vol. 85, No. WW3, pp. 93–121.

Huizinga, J., H. de Moel, and W. Szewczyk (2017) "Global Flood Depth-Damage Functions: Methodology and Database with Guidelines", *JRC Technical Report, European Commission*, Brussels, (http://www.publications.jrc.ec.europa.eu/).

IPET (2009) "Performance Evaluation of the New Orleans and Southeast Louisiana Hurricane Protection System," *Interagency Performance Evaluation Task Force* (IPET), Corps of Engineers, Washington DC.

Iribarren, C. R. (1938) "A Formula for the Calculation of Rock-fill Dikes," Revista de Obras Publicas, translation found in *The Bulletin of the Beach Erosion Board*, Vol. 3, No. 1, p. 1949.

IWR (1985) "Business Depth-Damage Analysis Procedures," 85-R5, *Institute for Water Resources (IWR), Corps of Engineers*, Fort Belvoir, VA.

IWR (1992) "Catalog of Residential Depth-Damage Functions Used by the Army Corps of Engineers in Flood Damage Estimations," 92-R-3, *Institute for Water Resources (IWR), Corps of Engineers*, Fort Belvoir, VA. (By S. A. Davis and L. L. Skaggs)

IWR (2011) "Coastal Storm Risk Management National Economic Development Manual," 2011-R-09, *Institute for Water Resources (IWR), Corps of Engineers*, Fort Belvoir, VA.

Jackson, R. A. (1968) "Design of Cover Layers for Rubble-Mound Breakwaters Subjected to Non-breaking Waves," Research Report 2-11, *Corps of Engineers, Waterways Experiment Station*, Vicksburg, MS.

Jones, C. P., and S. Mulrain (2017) "Structural Damage Data Collected from NACCS Study," Excel Spreadsheet, *AECOM Consulting Engineers*, personal communication. (3/16/2017)

Kameshwar, S. D., T. Cox, A. R. Barbosa, K. Farokhnia, H. Park, M. S. Alam, and J. W. van de Lindt (2019) "Probabilistic Decision-support Framework for Community Resilience: Incorporating Multi-hazards, Infrastructure Interdependenies and Resilience Goals in a Bayesian Network," *Reliability and System Safety*, (in review).

Kamphuis, J. W. (2010) *Introduction to Coastal Engineering and Management*, 2nd edn. World Scientific, Singapore.

Kang, J. L., M. D. Su, and L. F. Chang (2005). "Loss Functions and Framework for Regional Flood Damage Estimation in Residential Area," *Journal of Marine Science and Technology*, Vol. 13, No. 3, pp. 193–199.

Kreeke, J. v. d., and A. Paape (1964) "Europoort Breakwater Structures; A Cost Comparison," *Ministry of Public Works, (Rijkswaterstaat), and Delft Hydraulics Laboratory*, W732 M748.

Luettich Jr., R. A., J. J. Westerink, and N. W. Scheffner, (1992) "ADCIRC: An Advanced Three-Dimensional Circulation Model for Shelves Coasts and Estuaries, Report 1: Theory and Methodology of ADCIRC-2DDI and ADCIRC-3DL." Dredging Research Program Technical Report DRP-92-6, *US Army Engineers Waterways Experiment Station*, Vicksburg, 137 p.

Melby, J. (1999) "Damage Progression on Rubble-mound Breakwaters," TR-CHL-99-17, Corps of Engineers. *Engineering Research and Development, Center, Coastal and Hydraulics Laboratory*, Vicksburg MS.

Melby, J. A., and N. Kobayashi (1998) "Progression and Variability of Damage on Rubble Mound Breakwaters," *Journal of Waterway, Port, Coastal and Ocean Engineering, ASCE*, Vol. 124, No. 6, pp. 286–294.

Merz, B., H. Kreibich, A. Thieken, and R. Schmidtke (2004) "Estimation of Uncertainty of Direct Monetary Flood Damage to Buildings," *Natural Hazards and Earth System Sciences*, Vol. 4, pp. 153–163.

Merz, B., H. Kreibich, R. Schwarze, and A. Thieken (2010) "Assessment of Economic Flood Damage," *Natural Hazards and Earth System Sciences*, Vol. 10, pp. 1697–1724.

Moser, D., M. Gravens, R. Patev, and J. Foster (2013) "Development of a Comprehensive Probabilistic Life Cycle Model for Planning Coastal Risk Management Projects," Proceedings, *World Environmental and Water Resources Congress*, pp. 2370–2381.

NACCS (2015) "North Atlantic Coast Comprehensive Study," *Corps of Engineers*, Washington DC.

Nadal-Caraballo, N. C., and J. A. Melby (2014) "Statistical Analysis and Storm Sampling Approach for the North Atlantic Coast Comprehensive Study: ERDC/CHL TR-14-7," *Corps of Engineers, Engineer Research and Development Center*, Vicksburg, MS.

Norfolk Feasibility Study (2017) "Draft Integrated City of Norfolk Coastal Storm Risk Management Feasibility Study/Environmental Impact Statement" (DIFR-EIS, NCSRM), *Norfolk District Office, Corps of Engineers.* October.

Pistrika, A., G. Tsakiris, and I. Nalbantis (2014) "Flood Depth-Damage Functions for Built Environment", *Environmental Process*, Vol. 1, pp. 553–572, online.

Prahl, B. F., M. Boettle, L. Costa, D. Rybski, and J. P. Kropp (2017) "Damage and Protection Cost Curves for Coastal Flooding at the 600 Largest Coastal Cities in Europe," PANGAEA, *Earth & Environmental Science*, online (https://doi.org/10.1594/PANGAEA.875254)

Rosati, J. D., K. F. Touzinsky, and W. J. Lillycrop (2015) "Quantifying Coastal Systems Resilience for US Army Corps of Engineers," *Environmental Systems and Decisions*, Vol. 35, No. 2, pp. 196–208.

Rosati, J., J. Smith, and M. Bryant (2018) "State of Science & Engineering for Natural and Nature-Based Features," *Short Course Presentation*, 36th ICCE, Baltimore, MD, July 29.

Saffir, H. S. (1973) "Hurricane Wind and Storm Surge," *The Military Engineer*, Vol. 65, No. 423, pp. 4–5.

Samuels, P. G. (2014) "Editorial: Journal of Flood Risk Management," *Journal of Flood Risk Management*, Vol. 7, No. 2, June (http://onlinelibrary.wiley.com/doi/10.1111/jfr3.12112/full).

Schultz, M. T., and E. R. Smith (2016) "Assessing the Resilience of Coastal Systems: A Probabilistic Approach," *Journal of Coastal Research*, Vol. 32, No. 5, pp. 1032–1050.

Schultz, M. T., B. P. Gouldby, J. D. Simm, and J. L. Wibowo (2010) "Beyond the Factor of Safety: Developing Fragility Curves to Characterize System

Reliability." ERDC SR10-1. *US Army Engineer Research and Development Center*, Vicksburg, MS.

Simpson, R. H. (1974) "The Hurricane Disaster Potential Scale," *Weatherwise Magazine*, Vol. 27, 169–186.

Smith, D. I. (1994) "Flood Damage Estimation — A Review of Urban Stage-Damage Curves and Loss Functions," *Water SA* Vol. 20, No. 3, July.

USACE (1981) "Low Cost Shore Protection: A Guide for Engineers and Contractors," *Corps of Engineers, Coastal Engineering Research Center*, Waterways Experiment Station, Vicksburg, MS.

US Census (2011) "Census Demographic Summary Profile Data," Department of Commerce, Census Bureau, Washington DC.

Van der Meer, J. (1987) "Rock Slopes and Gravel Beaches under Wave Attack," Ph. D. Thesis, *Civil Engineering Department, Delft Technical University*, Delft, The Netherlands.

Van der Meer, J. (1988) "Rock Slopes and Gravel Beaches Under Wave Attack," Delft Hydraulics Communication No. 396, *Delft Hydraulics Laboratory*, Delft, The Netherlands.

Wang, C., and J. M. Blackmore (2009) "Resilience Concepts for Water Resource Systems," *Journal of Water Resources Planning and Management*, Vol. 137, No. 6, pp. 528–536.

Wang, C.-H., J. Blackmore, X. Wang, K.-K. Yum, M. Zhou, C. Diaper, G. McGregor, and J. Anticev (2009) "Overview of Resilience Concepts, With Application to Water Resource Systems," *eWater Cooperative Research Centre Technical Report*, NSW, Australia.

Zonensein, J., M. G. Miguez, L. P. C. Magalhães, M. G. Valentin, and F. C. B. Mascarenhas (2008) "Flood Risk Index as an Urban Management Tool," in: *11th International Conference on Urban Drainage*, Edinburgh, UK: IAHR/IWA.

Chapter 9

Pile Supported Structures

9.1 Introduction

Coastal piers supported by slender piles in shallow water are designed by coastal engineers. In most cases, the piles are circular and driven vertically into the ground. Wave pressures on the pile surface and wave particle velocities moving past the pile cause fluid dynamic drag and momentum forces that vary from wave crest to the water bottom. The vertical wave force distribution produces wave overturning moments that must be resisted by geotechnical forces along the pile depth within the bottom materials. Breaking and non-breaking water wave hydrodynamics must both be considered for design. Pile sectional shape (circular, square, polygon, etc.), surface roughness, and driven angle are further factors to be considered.

Coastal piers have many functions that determine the pier surface elevation and the pier length (fishing, pipeline streets, boat and ship docks in marinas, research, tourism, etc.). Horizontal wave forces on the piles vary with pile location (water depths and distance from shore) which dictate the length of the pile. Vertical wave (uplift) forces also vary with location and the design of the under-surface at the water interface. Figure 9.1 illustrates a large breaking wave impacting square, concrete piles driven at an angle and hitting the under-surface of the pier to produce uplift forces on the pier.

In this chapter, we first review the fluid dynamics of drag and momentum forces in steady and unsteady flow. The majority of the chapter is about non-breaking waves on slender cylindrical piles, including the graphical solutions for nonlinear wave forces and moments. Breaking wave forces and example force and moment calculations are included.

Design of slender piles that support piers does not consider wave diffraction as customary for design of large, offshore piles for oil and gas platforms and for monopiles that support wind turbines.

Figure 9.1. Large breaking wave striking angled, slender piles and under-surface of pier (courtsey Flickr: Creative Commons).

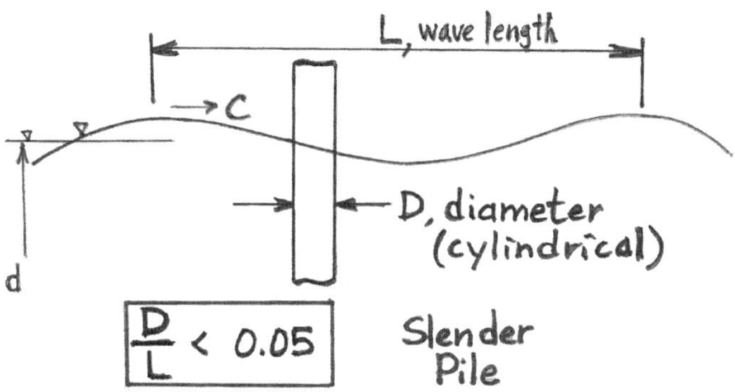

Figure 9.2. Definition of a slender pile when D/L is less than 0.05.

9.2 Fluid Dynamics Forces — Drag and Inertia

To begin, we first define what is meant by a "slender" pile. Consider a uniform wave of length L as defined by water depth, d, and wave period, T, with celerity, $C = L/T$ and a circular cylinder of diameter, D, as illustrated in Figure 9.2. A "slender" pile is when $D/L < 0.05$ so that the influence of the pile diameter on the lateral shape of the wave (diffraction) is very small and can be neglected.

$$\text{Slender Pile, } D/L < 0.05. \tag{9.1}$$

For example, for very short period waves, say $T = 3\,\text{s}$ and for $d = 5$–$10\,\text{m}$ (16–33 ft), then $L \approx 14\,\text{m}$ and the diameter, $D < 0.7\,\text{m}$ (2.3 ft). Many cylindrical piles are 30 cm (1 ft) in diameter and well within the definition of a slender pile. For longer period waves, L is larger, so D/L is far less than 0.05.

9.2.1 *Steady flow — Drag force*

The fluid dynamic drag force per unit length of pile, f_D, is the sum of the pressure drag and the skin-friction drag force. For bluff bodies (2D and 3D shapes), the pressure drag dominates. The steady flow drag force per unit length, f_D, is found from the standard drag equation of fluid dynamics.

$$f_D = \frac{1}{2}\rho C_D u^2 A_p, \tag{9.2}$$

where ρ is the fluid mass density, C_D is a drag coefficient, u is the upstream approach velocity, A_p is the projected area $= D * \Delta z$, D is the cylinder (or nominal) diameter, and Δz is the unit length of pile.

The drag coefficient in steady flow for smooth cylinders has been determined experimentally and depends on the Reynolds number, $\mathbf{R}$. For $10^3 < \mathbf{R} < 3 \times 10^5$, the drag coefficient is around 1.0 and then falls rapidly for higher Reynolds numbers when the wake collapses as the laminar boundary layer on the surface of the cylinder becomes turbulent (see any undergraduate text in fluid mechanics). In steady, uniform, turbulent, open channel flow, the mean velocity varies with depth z, $\overline{u}\,(z)$ and therefore so must the drag coefficient, $C_D(\overline{u})$. The total force on a cylinder, F_D, is the integral of f_D over the water depth, d

$$F_D = \frac{1}{2}\rho D \int_0^d [C_D(\overline{u})\overline{u}^2(z)]dz \tag{9.3}$$

and the total momentum about the bottom, M_D, is the integral of $(f_D \times z)$ over the water depth, d

$$M_D = \int_0^d [(f_D \times z)]dz. \tag{9.4}$$

9.2.2 *Unsteady flow — Inertia force*

To understand how the mass of fluid particles surrounding the cylinder from its radius, R, out to infinity in all radial directions are influenced, we first must define what has become known as the "inertia force", f_i, or sluggishness of this added mass of water. When the water motion becomes unsteady as in an oscillating wave, everything changes. The water velocity is now a function of depth, z, and time, t, $u(z, t)$ as we learned in Chapter 2 for a given location along the wave, x, of wave length, L. Therefore, in general, $u(x, y, z, t)$ since the fluid particles in the lateral, y-direction, are also moving.

9.2.2.1 *Fluid at rest — Cylinder starts to accelerate*

(1) Inertia force

To begin, consider a stationary body of water (at rest) of depth, d, and vertical cylinder with diameter, D, that begins to accelerate from rest at time, $t_o = 0$, as illustrated on the left side of Figure 9.3. A 2D slice of the pile, Δz, is initially also at rest, $\overline{V}(t) = 0$. All the fluid particles are also not moving so that $u(x, y, z, t_o) = 0$. When the cylinder starts to move ($t > t_o$), it must push on all the water particles that surround it, and accelerate this mass of water in a complicated pattern as illustrated on the right side of Figure 9.3.

From Newton's 2nd law of motion (force equals mass times acceleration), the force required to accelerate the water mass of fluid particles is simply,

$$F = m\frac{d\overline{V}(t)}{dt}, \tag{9.5}$$

where m is the mass of fluid $= \rho(\text{volume}) = \rho V$.

Therefore, over a 2D slice of fluid, this is the inertia force, f_i, per unit depth, dz.

$$f_i = \rho \int_{-\infty}^{\infty} \int_{-\infty}^{\infty} \frac{\partial u(x, y, t)}{\partial t} dx\, dy, \tag{9.6}$$

where, m is the $\rho\, dx\, dy\, dz$, $u(x, y, t)$ is the water particle velocity at any location (x, y) at time t.

Evaluation of the two integrals over infinite limits for some unknown velocity field surrounding the cylinder has been simplified by Lamb (1945).

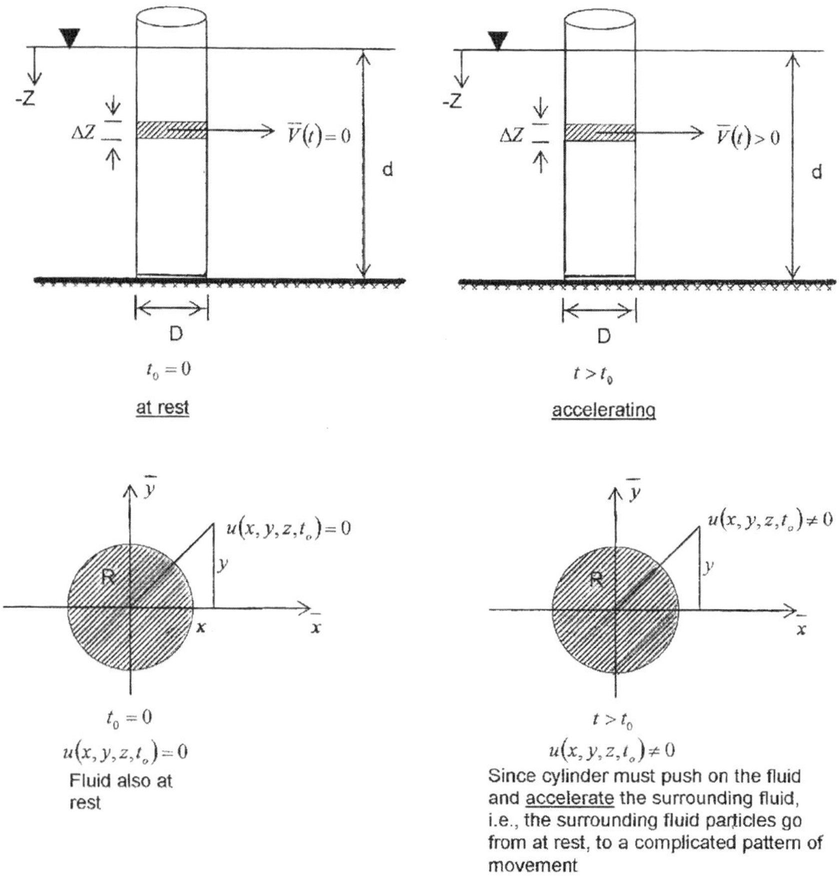

Figure 9.3. Fluid at rest (left side) and cylinder starts to accelerate (right side).

(2) Added mass — Lamb (1945)

Lamb's (1945) idea was to compute the energy in the flow pattern surrounding the cylinder at any time and set this equal to the energy of a fictitious and equivalent "added" mass of fluid particles moving with the same instantaneous velocity of the cylinder, $\overline{V}(t)$. The inertia force f_i for a 2D slice became after the energy balance

$$f_i = \rho[v_{\text{inf}} - v_{\text{cyl}}]\frac{d\overline{V}}{dt} = \rho[k_m v_{\text{cyl}}]\frac{d\overline{V}}{dt} \tag{9.7}$$

where $\rho k_m v_{\text{cyl}}$ is the "added" mass of all the accelerated fluid particles around the cylinder.

$$k_m = \text{the added mass coefficient.} \tag{9.8}$$

For a 2D slice of cylinder, the v_{cyl} is the area $(\pi/4)D^2$ giving

$$f_i = \rho k_m (\pi/4) D^2 \frac{d\overline{V}}{dt} \tag{9.9}$$

as the added inertia force to accelerate the cylinder in still water.

(3) Total force on cylinder

It is also necessary to include the force required to accelerate the cylinder in air, f_{air}, giving

$$f_{\text{air}} = \rho_{\text{cyl}} v_{\text{cyl}} \frac{d\overline{V}}{dt} \tag{9.10}$$

so that the total force per unit depth, f_{cyl}, on the cylinder is

$$f_{\text{cyl}} = f_i + f_{\text{air}} = (\rho_{\text{cyl}} + \rho k_m) v_{\text{cyl}} \frac{d\overline{V}}{dt}. \tag{9.11}$$

(4) Total force on pile of unit volume

In general, for any shaped pile of unit volume, v_{pile},

$$f_{\text{pile}} = (\rho_{\text{pile}} + \rho k_m^*) v_{\text{pile}} \frac{d\overline{V}}{dt}, \tag{9.12}$$

where k_m^* is an appropriate added mass coefficient for pile with unit volume, v_{pile}, and f_{pile} the total force on the pile as it starts to accelerate from rest in still water.

9.2.2.2 *Cylinder at rest — Water starts to accelerate*

Is the force on the pile, f_{pile}, as it starts to move the same as for a stationary pile with water moving past in a water wave? The answer is no. We must now also accelerate a volume of water equal (approximately) to the volume of water displaced by the pile, ρ_{pile}, in Equation (9.12) to give the total mass of water being accelerated for the inertia force per unit depth,

$$f_i = (\rho + \rho k_m) v_{\text{pile}} \frac{du}{dt} = (1 + k_m) \rho\, v_{\text{pile}} \frac{du}{dt} \tag{9.13}$$

and where the change in the velocity of water in the wave moving past the pile is simply $\overline{v} = u$.

9.2.2.3 *Inertia coefficient, C_m*

Now define a new inertia coefficient, C_m, as simply

$$C_m = 1 + k_m \tag{9.14}$$

giving for the inertia force

$$f_i = (C_m)\rho\, v_{\text{pile}}\frac{du}{dt}. \tag{9.15}$$

Lamb (1945) also used irrotational, inviscid, potential flow theory to find that the added mass coefficient, k_m for an elliptical cross-section as shown in Figure 9.4(a) is equal to

$$k_m = \frac{b}{a}, \tag{9.16}$$

where a is the one-half the length of the ellipse, and b is the one-half the width of the ellipse.

The inertia coefficient, C_m, is then shown in Figure 9.4(b) as a function of the a/b ratio giving $C_m = 2$ when $a = b$ for a circular cylinder.

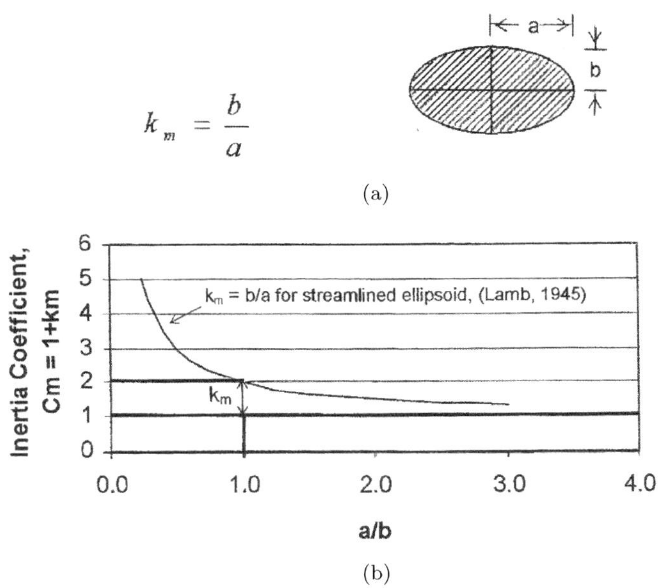

Figure 9.4. (a) Dimensions of an elliptical cross-section. (b) Inertia coefficient for an ellipsoid (Lamb, 1945).

9.3 Wave Forces on Slender Cylindrical Piles — Non-breaking Waves

9.3.1 *Morison equation*

It has been almost 70 years since a graduate student, J.R. Morison, conducted laboratory scale experiments at the University of California (Berkeley) under the supervision of professors M.P. O'Brien and Joe Johnson for wave forces on slender piles. Several publications and reports present the results, but the most often cited and available on the web is Morison *et al.* (1950).

This citation is preferred rather than Morison *et al.* (1950) to give full credit to those professors who directed the research work. In many references it is simply referred to as the "Morison Equation" and has stood the test of time. In water waves, both the steady-state drag force, f_D, and the unsteady inertia force, f_i, occur continuously in time so that the total wave force per unit length of pile, f, is

$$f = f_D + f_i, \tag{9.17}$$

$$f = \frac{1}{2}\rho C_D A_p u|u| + C_m \rho v_{\text{pile}} \frac{du}{dt} \quad \text{(Morison equation).} \tag{9.18}$$

The absolute value for $|u|$ insures that the drag force changes direction as the wave passes the pile. To solve Equation (9.18) for the force per unit length, f, at any depth beneath the wave surface, we need the following as illustrated in Figure 9.5: (a) velocity, $u(z,t)$, (b) acceleration, $du(z,t)/dt$, (c) drag coefficient, C_D, as a function of the Reynolds No. and, (d) inertia coefficient, C_m.

For cylindrical piles with diameter D per unit pile length, Δz,

$$f = \frac{1}{2}\rho C_D D u|u| + C_m \rho \frac{\pi}{4} D^2 \frac{du}{dt} \quad \text{(Morison equation).} \tag{9.19}$$

9.3.2 *Wave velocities and accelerations*

What wave theory and corresponding wave velocity theory should be employed in the Morison Equation (9.19)?

9.3.2.1 *Sinusoidal (linear) waves*

To begin, consider a sinusoidal wave passing the pile in Figure 9.5 with free surface defined by linear wave theory (Figure 9.6) with wave amplitude,

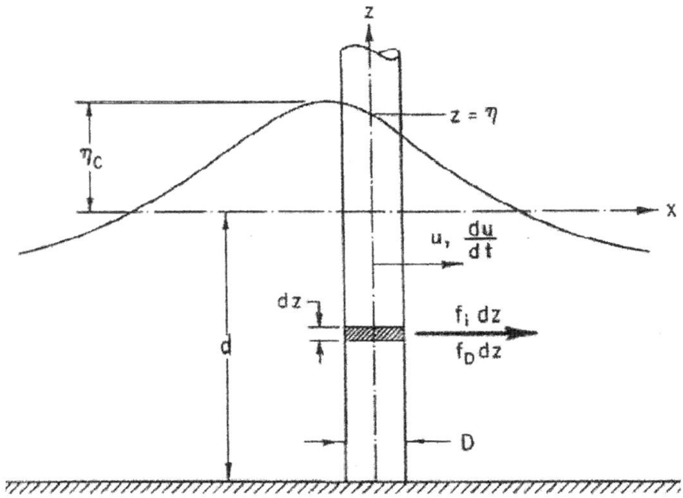

Figure 9.5. Schematic of cylindrical pile of diameter D, and unit pile length, dz, in water depth, d, and wave surface, $\eta(x,t)$ (from CEM, 2006).

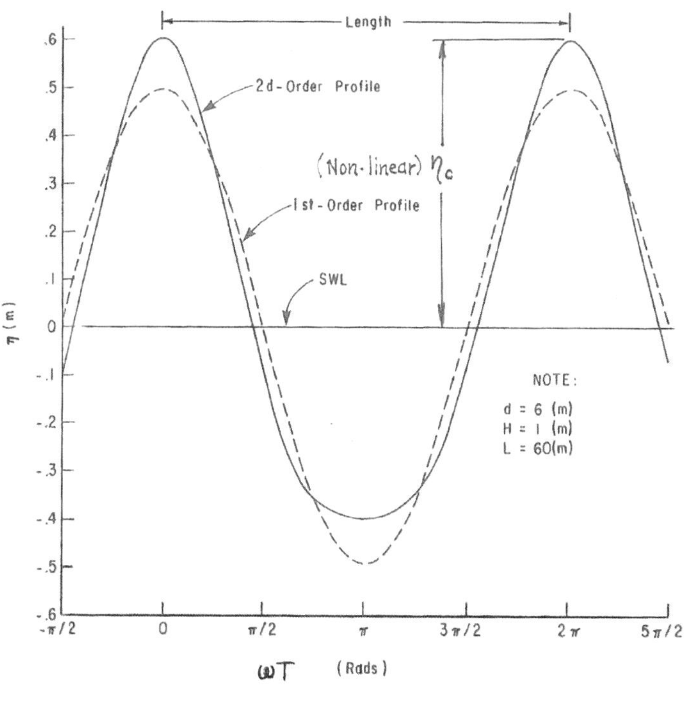

Figure 9.6. Surface profiles for linear (dotted) wave theory and nonlinear (solid) wave theory (from CEM, 2006).

$a = H/2$.

$$\eta(x,t) = a\cos\left(\frac{2\pi x}{L} - \frac{2\pi t}{T}\right) = a\cos(kx - \omega t). \tag{9.20}$$

As shown in Chapter 2 (Table 2.2), the corresponding horizontal wave particle velocity field, $u(x,z,t)$, is

$$u(x,z,t) = \frac{\pi a}{2T}\left[\frac{\cosh k(z+d)}{\sinh kd}\right]\cos(kx - \omega t) \tag{9.21}$$

and in phase with the free surface since both are a function of $\cos\omega t$. Taking $x = 0$ at the pile location and normalizing the horizontal velocity, u, we get

$$\frac{u}{a\,gT/L} = \cos\omega t$$

and

$$du/dt = \sin\omega t. \tag{9.22}$$

As illustrated in Figure 9.7, the nondimensional horizontal wave velocity varies as $\cos\omega t$ (top); the drag force varies as $\cos\omega t|\cos\omega t|$ (2nd from top); and the inertia force varies as sin ωt (3rd from top) over time t/T through the wave cycle. Adding the drag and inertia force variations in time produces the total wave force variation in time as shown in the bottom of Figure 9.7. The drag force and inertia forces are exactly 90° out-of-phase for regular (linear) wave theory, but they are not for nonlinear wave theory due to the nonlinearity in the definition of the total differential for the velocity component, $u(x,z,t)$,

$$\frac{D_u}{D_t} = \frac{\partial u}{\partial t} + u\frac{\partial u}{\partial x} + w\frac{\partial u}{\partial z}.$$

9.3.2.2 *Nonlinear waves*

The free surface variation of nonlinear waves in time is illustrated in Figures 9.6 (solid) and 9.8 showing higher, more peaked crests and flatter, less-lower troughs. Nonlinear water wave theories provide a far better fit to the free surface variation in time and space in shallow, coastal water depths. And, they are far more exact in the velocity, acceleration, and orbital motions that produce wave forces on slender piles.

Stokes wave theories (Stokes, 1847), Cnoidal wave theory (Korteweg and DeVries, 1895), and Solitary wave theory (Boussinesq, 1872) are nonlinear wave theories that are important but far beyond the needs for discussion in

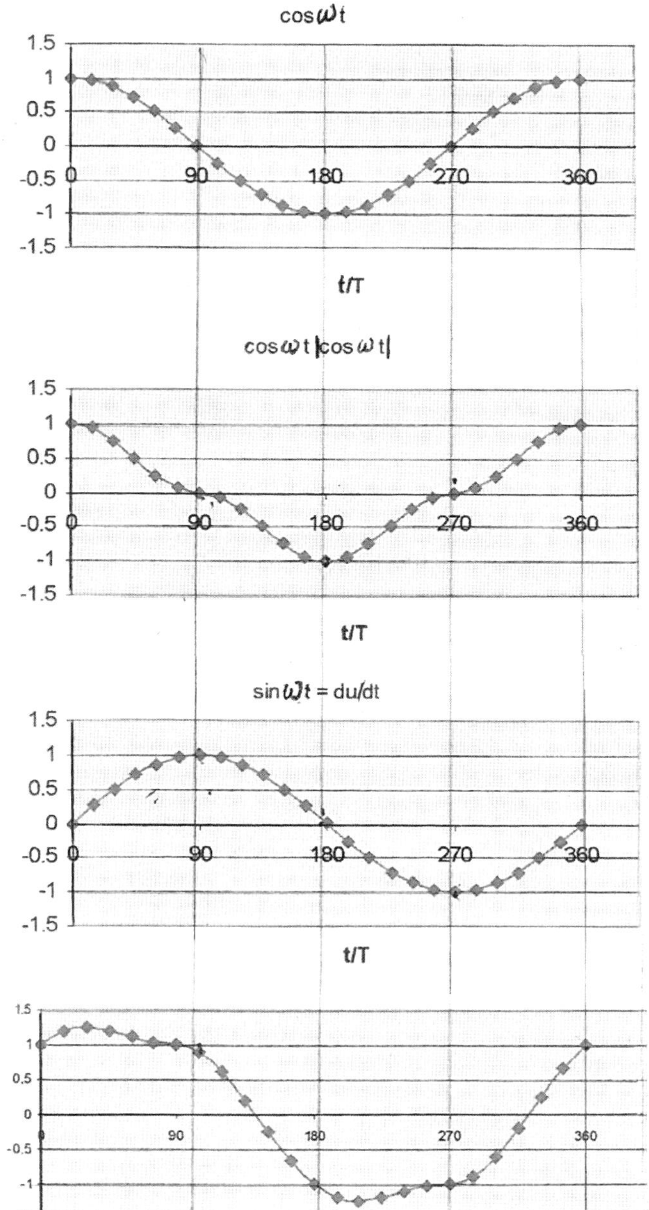

Figure 9.7. Normalized wave particle velocity, $u/(a\,g\,T/L)$, versus t/T for various phases of the wave (top), wave drag force (2nd), wave inertia force (3rd), and combined drag and inertia force (bottom).

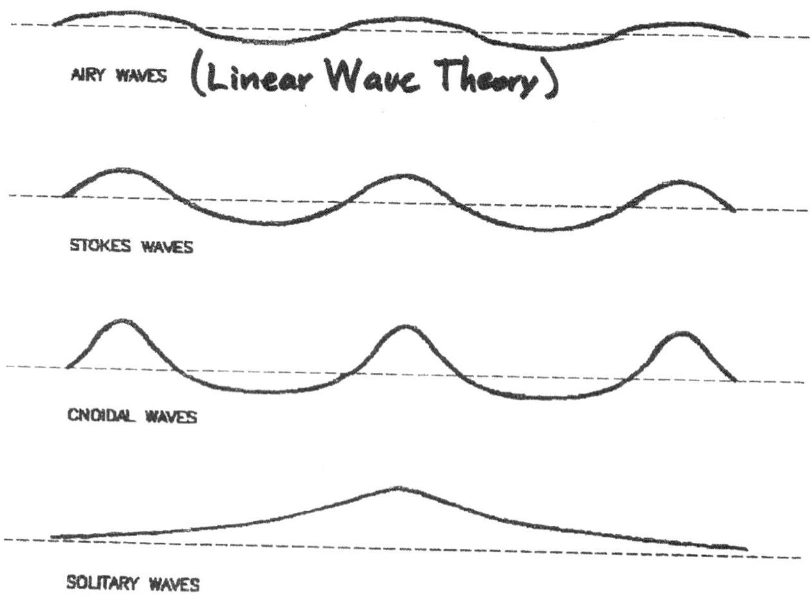

Figure 9.8. Schematic of water surface shape for linear (top) and three, nonlinear wave theories.

this book. Figure 9.9 presents the theoretical limits of these wave theories in terms of the ratio (d/gT^2) versus the ratio (H/gT^2) up to the theoretical breaking wave limit $H/d = 0.78$ in shallow and transitional water depths, and then in deep water where $H_o/L_o = 0.14$ is the breaking wave limit. Note that the deep water wave length $L_o = (g/2\pi)T^2$, so that the vertical axis is related to the wave steepness (H/L_0) and the horizontal axis is related to the relative water depth, (d/L_o). The boundary for Linear Wave Theory credited to Airy (1845) is also shown. See Dean and Dalrymple (1984) for complete details of these nonlinear wave theories.

For coastal engineering design of slender, cylindrical piles, the Stream Function Wave Theory (Dean, 1974) has proven to be most useful for it was used to produce nonlinear wave force theories and design curves that have stood the test of time.

9.3.2.3 *Stream function wave theory*

The Laplace equation is applicable to many disciplines in physics and engineering. For wave motions that are irrotational and non-divergent with

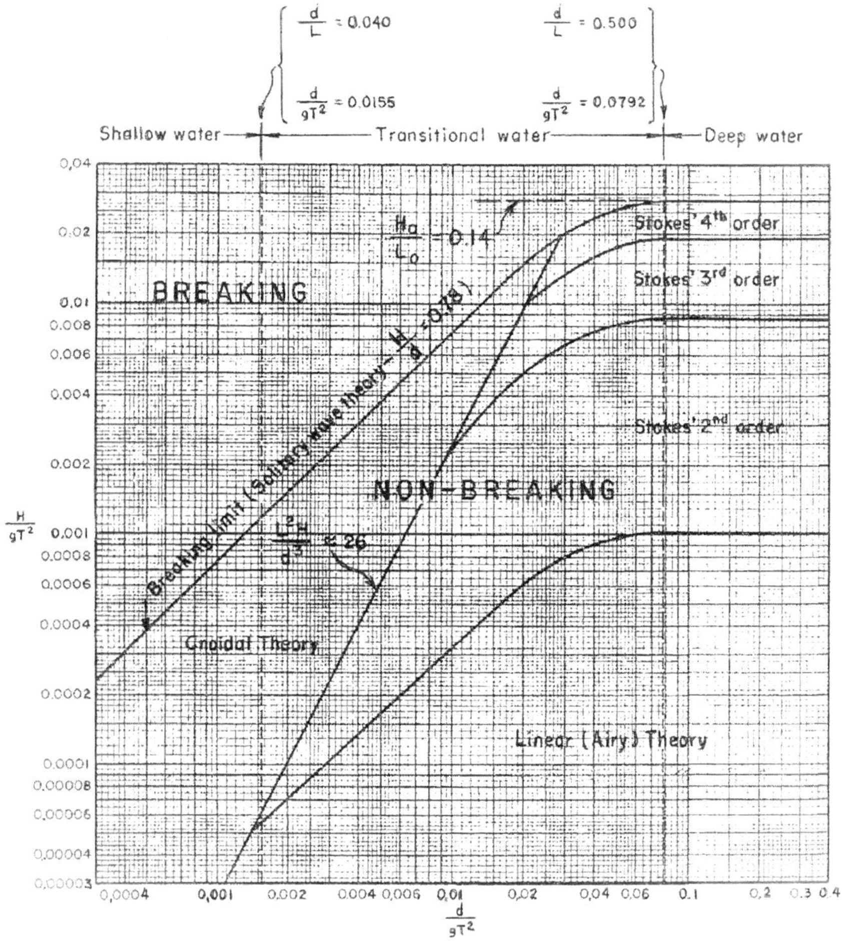

Figure 9.9. Regions of validity of various wave theories, and the breaking wave limit (from CEM, 2006).

incompressible water, the Laplace equation for the velocity potential function and stream function are the same. Solution of the Laplace equation depends on the dynamic, free surface (air–water interface) boundary condition, but the position of this upper boundary is not known *a priori* in solving for the wave theory. Dean (1974) used trial and error iterations on the computer to locate the water surface for 40 representative combinations of water depth, d, wave height, H, and wave period, T, with deepwater wave length, L_o, and tabulated the results in dimensionless form.

And, the corresponding water wave velocities, $u(x, z, t)$ and $w(x, z, t)$, were also found. See also Dean (1965) for an earlier stream function representation of nonlinear waves. The tables of velocity components were then used to calculate wave force and momentum design curves for slender piles as presented below.

9.3.3 *Graphical solution for maximum forces and moments*

Forty sets of dimensionless wave conditions were selected for tabulation and shown in Figure 9.10. Each of the 10 cases of relative depth ($0.002 < d/L_o < 2.0$) is characterized by four values of the wave breaking index (H/H_b) to cover the range from shallow to deep water. The parameter

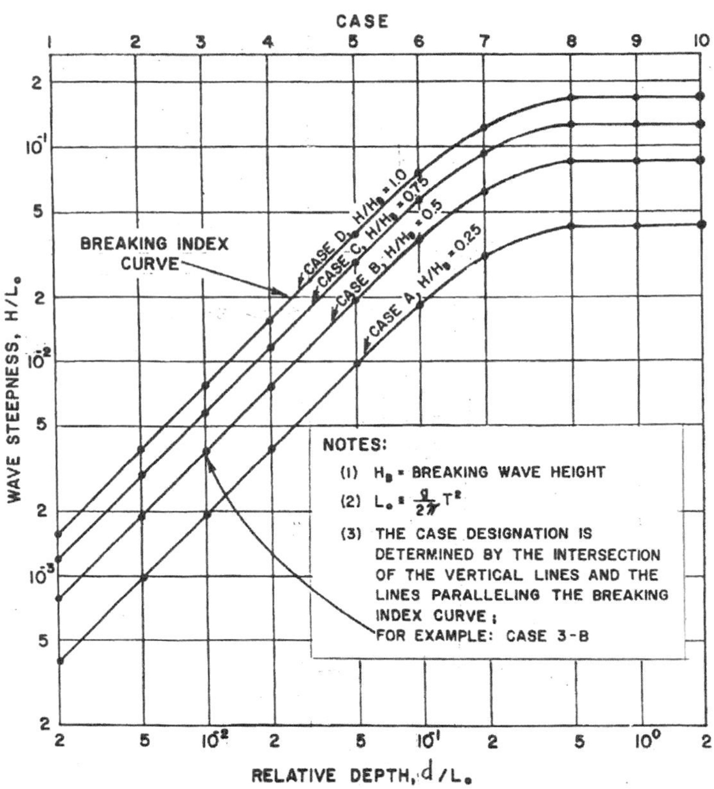

Figure 9.10. Forty wave characteristics selected for tabulation (from Dean, 1974).

Table 9.1. Relation between relative depth d/L_o and d/gT^2 for 10 cases (Dean, 1974).

Case No.	d/L_o	d/gT^2
1	0.002	0.000318
2	0.005	0.000796
3	0.01	0.0001592
4	0.02	0.003183
5	0.05	0.007958
6	0.1	0.015915
7	0.2	0.031831
8	0.5	0.079577
9	1.0	0.159155
10	2.0	0.31831

H/H_b included wave breaking indexes: Case $A = 0.25$, Case $B = 0.5$, Case $C = 0.75$, and Case $D = 1.0$. Wave breaking index cases are based on the degree of nonlinearity as originally defined by Dean (1974).

The coordinates in Figures 9.9 and 9.10 are simply related by 2 pi with Table 9.1 showing the actual values of d/gT^2 for the 10 Cases and d/L_o ratios employed by Dean (1974).

Variations of the drag force per unit length, f_D, and the inertia force per unit length, f_i, must be integrated over the water depth selected ($\eta > z > -d$) at some time (t/T) as the wave moves past the pile to find the total force, F, on the pile. This gives

$$F = \int_{-d}^{\eta} f_D \, dz + \int_{-d}^{\eta} f_i \, dz = F_D + F_i \tag{9.23}$$

and for the total momentum, M, about the mud line

$$M = \int_{-d}^{\eta} f_D (z + d) dz + \int_{-d}^{\eta} f_i (z + d) dz = M_D + M_i. \tag{9.24}$$

Dean (1974) performed these integrations using his stream function wave theory for the needed velocity distribution, $u(z, t)$, in the drag force equation and the needed acceleration, $du(z, t)/dt$, in the inertia force equation. Dean (1974) made the additional assumption that both the drag coefficient, C_D, and the inertia coefficient, C_m, are constant over the depth of integration. The results did not lend themselves to closed form presentation. Dean (1974) defined four dimensionless coefficients for plotting with the 40 cases of wave conditions so that the wave force drag and inertia components, F_D and F_i and M_D and M_i, could be independently computed.

For design, Dean (1974) went one-step further by recognizing that the maximum values for these four dimensionless coefficients were needed as the wave moved past the pile. As illustrated in the bottom of Figure 9.7, the maximums of the total force, F, and total momentum, M, occur at some relative time (t/T) in the wave that is not a simple fraction (1/4, 1/2 or 3/4) of the wave period, T. Therefore, Dean (1974) proposed the following formulas for the two components of F and the two components of M, namely,

$$F_{Dm} = C_D \frac{1}{2} \rho g D H^2 K_{Dm}, \tag{9.25}$$

$$F_{im} = C_m \rho g \frac{\pi}{4} D^2 H K_{im}, \tag{9.26}$$

$$M_{Dm} = F_{Dm} d S_{Dm}, \tag{9.27}$$

$$M_{im} = F_{im} d S_{im}, \tag{9.28}$$

where K_{Dm} is the maximum, dimensionless drag force coefficient integrated over pile, K_{im} is the maximum, dimensionless inertia force coefficient integrated over pile, S_{Dm} is the maximum, dimensionless drag momentum coefficient integrated over pile, and S_{im} is the maximum, dimensionless inertia momentum coefficient integrated over pile.

Figures 9.11–9.14 are the graphed results for K_{Dm}, K_{im}, S_{Dm}, and S_{im}, respectively, versus the d/gT^2 ratio and the four breaking height ratios, (1/4, 1/2, 3/4, and 1). $H = 0$ is linear, Airy theory.

As the breaking height ratio, H/H_b, approaches zero, the wave theory is Airy for $H = 0$. A simple example best illustrates the use of these figures which require using the correct H/H_b ratio.

Given: $H = 3.0m(9.8\,\text{ft})$

$T = 8\,\text{s}$

$d = 12.0\,\text{m}(39.4\,\text{ft})$

Find: H/H_b

Solution: $\dfrac{d}{gT^2} = \dfrac{12.0}{9.81\,(8)^2} = 0.0191$ which gives

from Figure 9.9 at the breaking limit curve, $H_b/d = 0.78$

$\dfrac{H_b}{gT^2} = 0.014$

Therefore, $H_b = 9.81(8)^2(0.014) = 8.79\,\text{m}(28.8\,ft)$

Giving, $\dfrac{H}{H_b} = \dfrac{3.0}{8.79} = 0.34.$

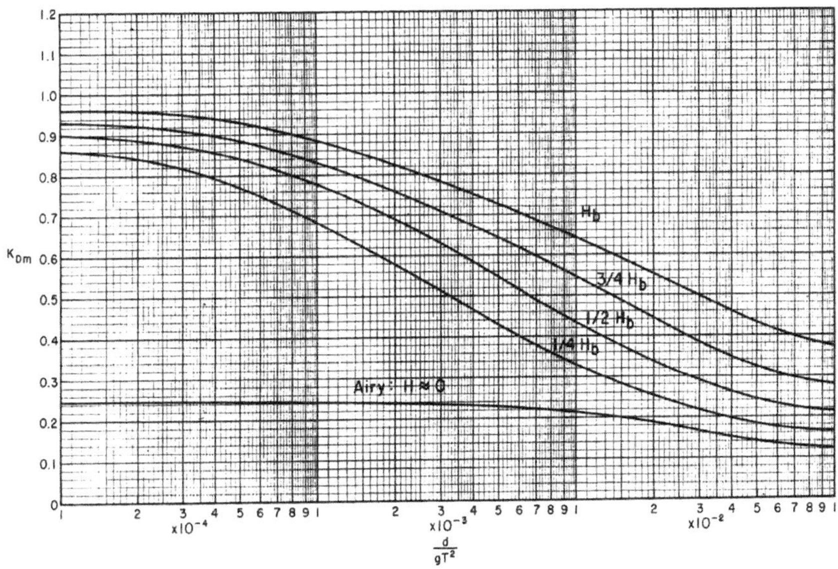

Figure 9.11. K_{Dm} versus the d/gT^2 ratio (from CEM, 2006).

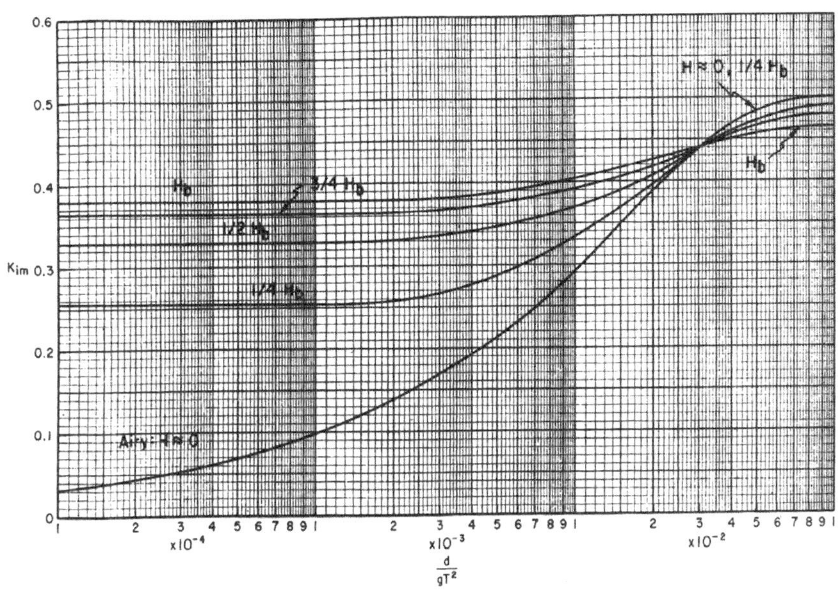

Figure 9.12. K_{im} versus the d/gT^2 ratio (from CEM, 2006).

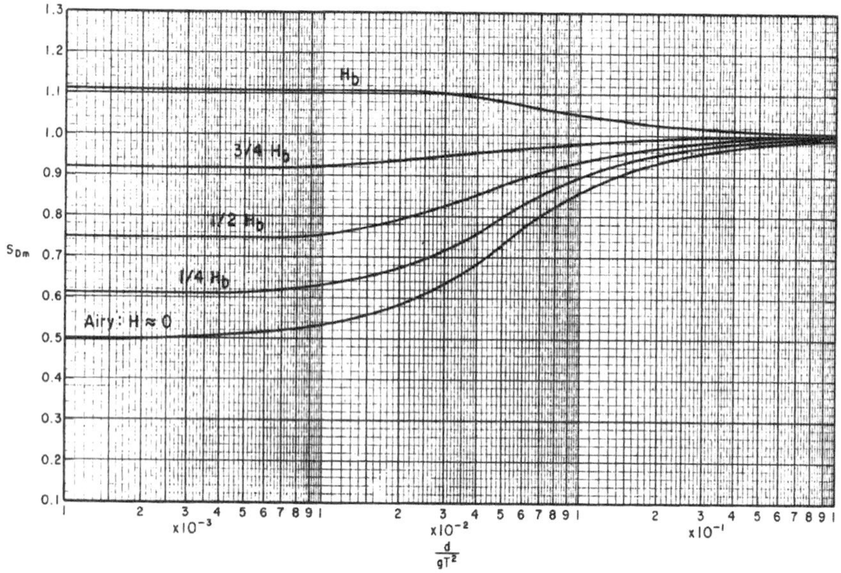

Figure 9.13. S_{Dm} versus the d/gT^2 ratio.

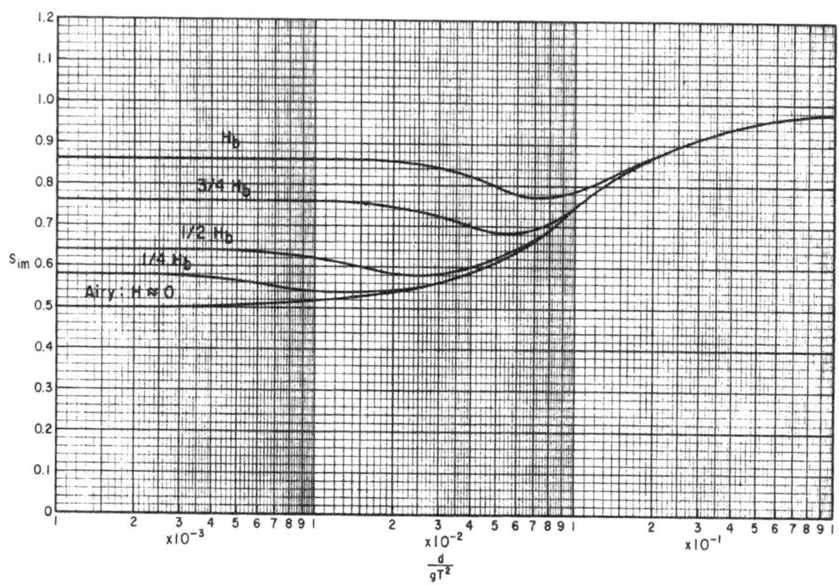

Figure 9.14. S_{im} versus the d/gT^2 ratio.

Use $H/H_b = 0.34$ when interpolating in Figures 9.11–9.14 to find the maximum force and momentum coefficients.

9.3.4 *Maximum total force and moments about the mud line*

The maximum force components (drag and inertia) found from Equations (9.25) and (9.26) must finally be combined to give the maximum total force, F_m, on the pile. Because of the nonlinearity in the velocity field, they cannot be directly added together. Therefore, Dean (1974) also presented additional curves to graphically determine a dimensionless total force coefficient, Φ_m, defined as

$$\Phi_m = \frac{F_m}{\rho g C_D \, H^2 D} \tag{9.29}$$

to calculate F_m.

However, because of the complexity of the graphical curves to estimate Φ_m, four separate graphs were needed with each one depending on a final coefficient, W, defined as

$$W = \frac{C_m D}{C_D H}. \tag{9.30}$$

Figures 9.15–9.18 are the results for $W = 0.05$, 0.10, 0.5, and 1.0, respectively. For intermediate values of W, interpolation is required to estimate an interpolated value for Φ_m.

The maximum momentum components (drag and inertia) found from Equations (9.27) and (9.28) must also be combined to give the maximum total momentum, M_m, on the mudline at the base of the pile. Because of the nonlinearity in the velocity field, they cannot be directly added together. Therefore, Dean (1974) also presented additional curves to graphically determine a dimensionless total momentum coefficient, α_m, defined as

$$\alpha_m = \frac{M_m}{\rho g C_D \, H^2 D \, d} \tag{9.31}$$

to calculate M_m.

However, because of the complexity of the graphical curves to estimate α_m, four additional separate graphs were needed with each one depending on the coefficient, W. Figures 9.19–9.22 are the results for $W = 0.05$, 0.10, 0.5, and 1.0, respectively. For intermediate values of W, interpolation is required to estimate an interpolated value for α_m.

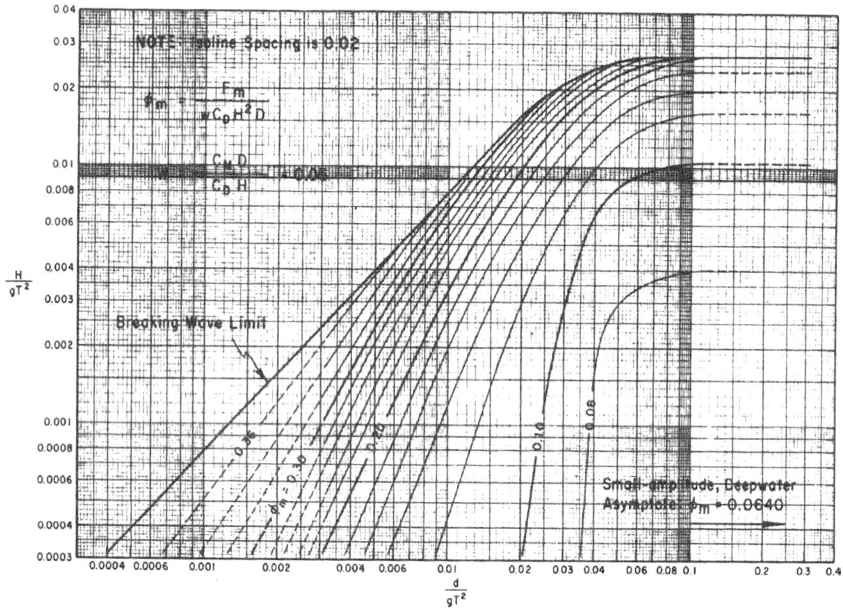

Figure 9.15. Isolines of Φ_m for $W = 0.05$ (from CEM, 2006).

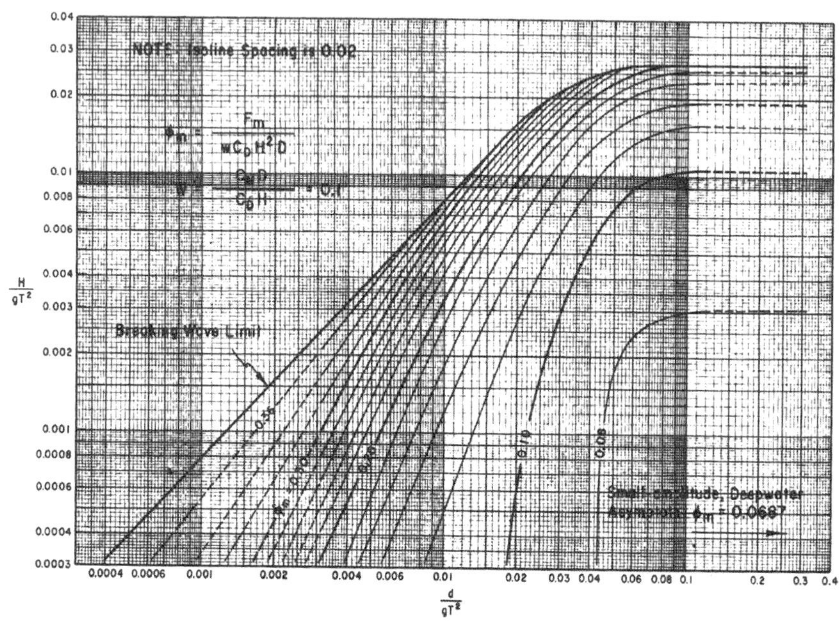

Figure 9.16. Isolines of Φ_m for $W = 0.10$ (from CEM, 2006).

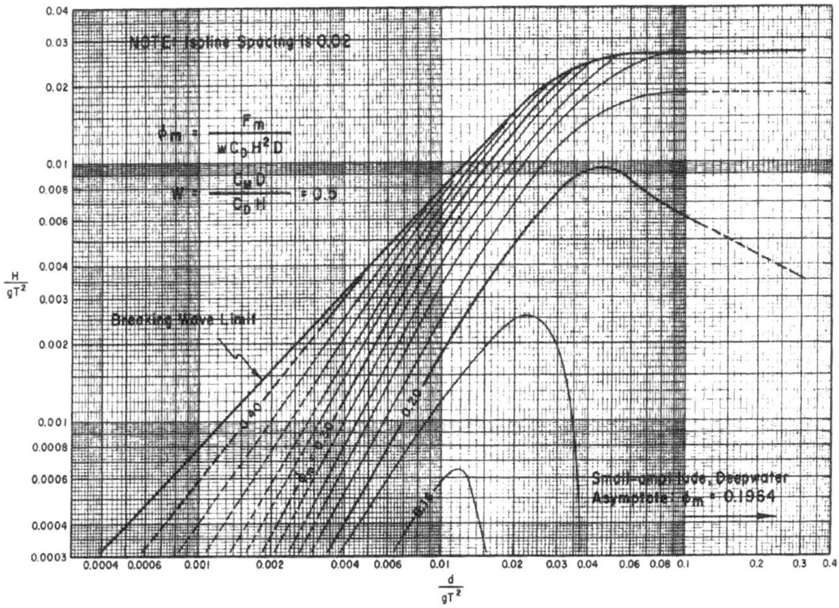

Figure 9.17. Isolines of Φ_m for $W = 0.50$ (from CEM, 2006).

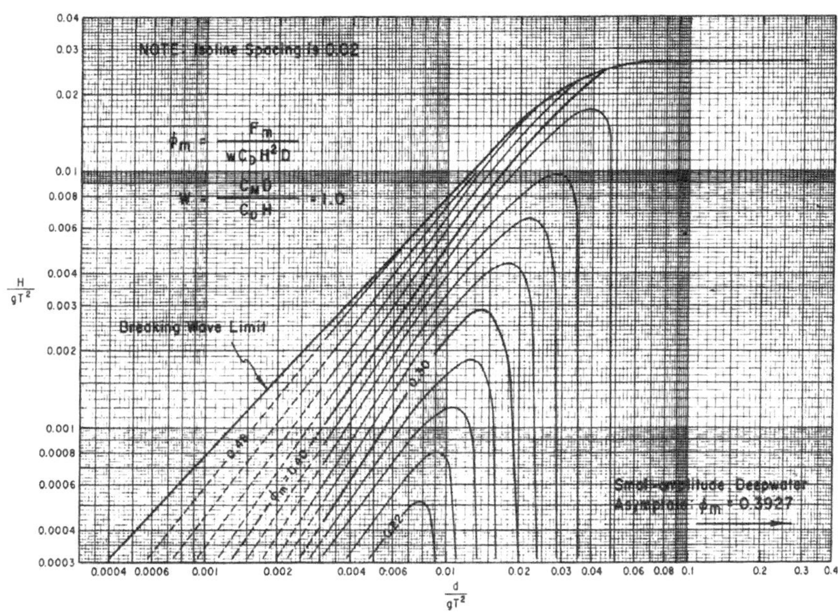

Figure 9.18. Isolines of Φ_m for $W = 1.0$ (from CEM, 2006).

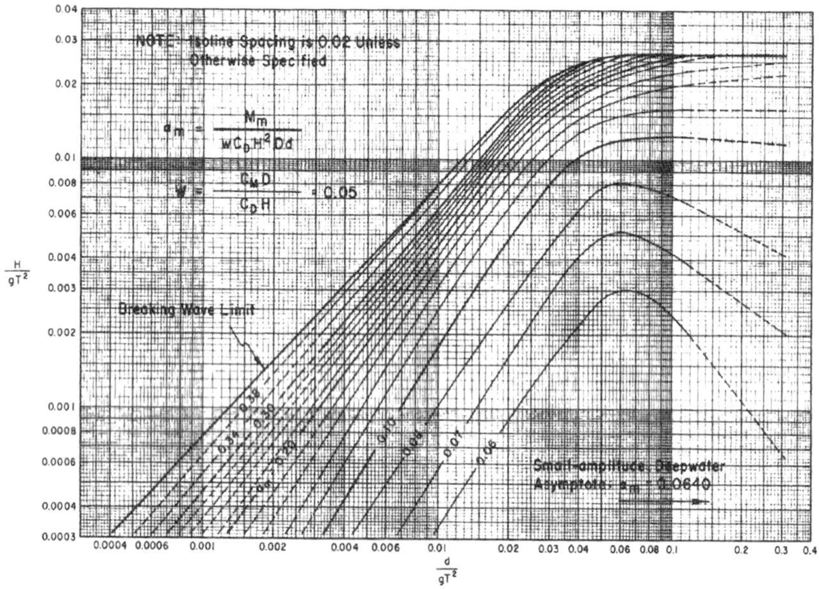

Figure 9.19. Isolines of α_m for $W = 0.05$ (from CEM, 2006).

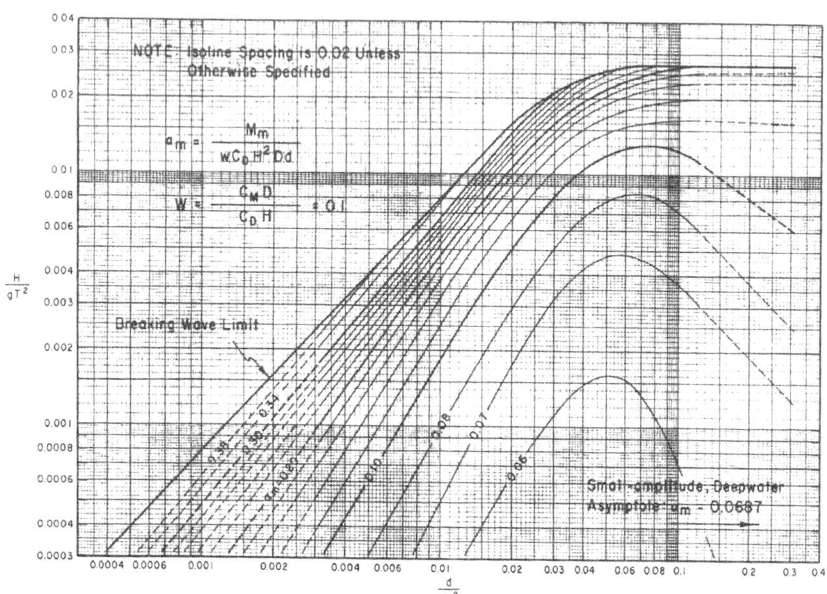

Figure 9.20. Isolines of α_m for $W = 0.10$ (from CEM, 2006).

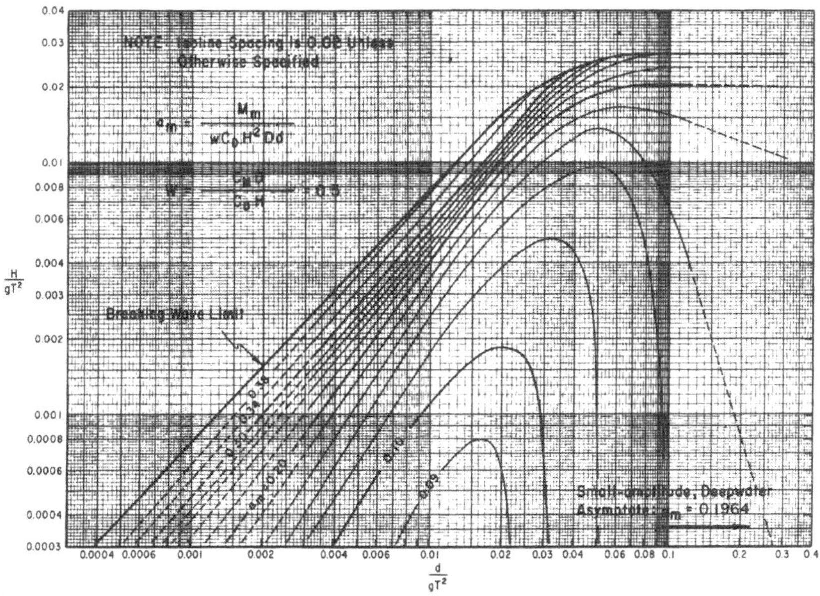

Figure 9.21. Isolines of α_m for $W = 0.50$ (from CEM, 2006).

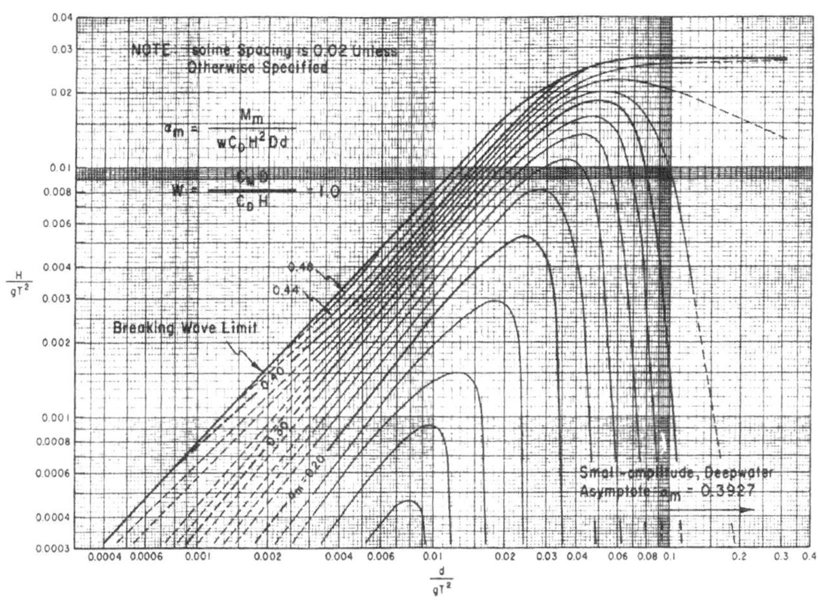

Figure 9.22. Isolines of α_m for $W = 1.0$ (from CEM, 2006).

Finally, the location for the moment arm, r_m, vertically above the mudline can be found from

$$r_m = \frac{M_m}{F_m}. \tag{9.32}$$

An example calculation for design of a slender pile in non-breaking waves is the best way to understand these many curves and will be presented in Section 9.3.6. But first, C_D and C_m coefficients are discussed.

9.3.5 *Typical values for C_D and C_m coefficients*

Equation (9.30) to calculate W requires knowledge of the drag coefficient, C_D, and the inertia coefficient, C_m. These two coefficients in the Morison equation have undergone many experimental studies by researchers all over the world. The measured forces and moments have been used to calculate the two coefficients using both measured velocities and various wave theories. To plot the results, several possibilities exist for making dimensionless combinations of the variables involved as the independent variable for the horizontal axis when plotting C_D and C_m on the vertical axis.

Dimensional analysis of the variables must include information about the wave (H, T), the fluid mass density, ρ, and fluid dynamic viscosity, ν, the pile diameter, D, and pile roughness coefficient, k. Piles in coastal waters accumulate marine growth that can add centimeters of thickness to the diameter. The key dimensionless numbers are the Reynolds number, R_e, the Keulegan–Carpenter number, KC, and the relative roughness, k/D, where u_m is the maximum horizontal wave velocity giving

$$R_e = \frac{u_m D}{\nu} \tag{9.33}$$

and

$$KC = \frac{u_m T}{D}. \tag{9.34}$$

Keulegan and Carpenter (1958) found some eddying to produce lateral forces (lift coefficient), but these are extremely small on rigid piles in shallow coastal water depths.

Clauss *et al.* (1992) suggested using the following Morison coefficients for smooth piles as shown in Table 9.2. Typically, the Reynolds number $R_e > 10^5$, so that $C_D = 0.6$ and $C_m = 1.5 - 2.0$, depending on the wave period for smooth cylinders. Barnacles growing on the pile in seawater can easily increase the roughness. Sarpkaya (1976) published curves for rough

Table 9.2. Coefficients for C_D and C_m in Morison equation for smooth cylinders (Clauss *et al.*, 1992).

	$R_e < 10^5$	$R_e < 10^5$	$R_e > 10^5$	$R_e > 10^5$
KC	C_D	C_m	C_D	C_m
<10	1.2	2.0	0.6	2.0
≥10	1.2	1.5	0.6	1.5

Table 9.3. Suggested design values for C_D and C_m in Morison equation for smooth and rough cylinders.

	Smooth	Smooth	Rough	Rough
	C_D	C_m	C_D	C_m
API	0.65	1.6	1.05	1.2
SNAME	0.65	2.0	1.0	1.8

cylinders from experiments with k/D ratios of 1/50, 1/180, 1/360, and 1/500. For Reynolds number $R_e > 10^5$, these curves gave the following logarithmic fits to the data for rough piles:

$$C_D = 0.227 \ln(k/D) + 2.64 \quad R^2 = 0.81 \tag{9.35}$$

so that as expected, roughness increased the drag coefficient from 1.1 to 1.7 for these experiments, and

$$C_m = -0.138 \ln(k/D) + 0.467 \quad R^2 = 0.80 \tag{9.36}$$

and the roughness decreased the inertia coefficient from 1.4 to 1.05 for these experiments. Several design codes have been published by the American Petroleum Institute (API) and the Society for Naval Architecture and Marine Engineering (SNAME) giving the simplest results in Table 9.3.

These suggested design values are consistent with Table 9.2 and Equations (9.35) and (9.36).

Safety factors for pile design are discussed in the Coastal Engineering Manual (CEM) (2006). Other factors regarding wave forces on slender piles such as pile groups and non-vertical piles should also be considered, including impact forces from boats and ice. Fendering systems for ports and harbors may be needed. The interested reader should consult the Corps of Engineers' CEM (2006, Part VI, Chapter 5) for details.

9.3.6 *An example calculation*

For an example, consider the following.

Given: $H_{\text{des}} = 3.0\,\text{m}$ (9.8 ft)

$T = 10\,\text{s}$

$D = 30\,\text{cm}$ ($\approx$1 ft) vertical pile

$d = 4.5\,\text{m}$ (14.8 ft) water depth

Assume: $C_D = 0.7$ (drag coefficient)

$C_m = 2.0$ (inertia coefficient)

$\rho = 1025\,\text{kg/m}^2$ (sea water)

Find: (1) Maximum total horizontal force, F_m

(2) Maximum total moment about mudline, M_m

(3) Location, r_m

Solution: 1. From Figure 9.9 for $\dfrac{H}{H_b}$

$\dfrac{d}{gT^2} = \dfrac{4.5}{9.81(10)^2} = 0.0046$, and giving at the breaking limit

$\dfrac{H_b}{gT^2} = 0.0034$, so that

$H_b = 0.0034gT^2 = 0.0034(9.81)(10)^2$

$H_b = 3.3\,\text{m}(10.8\,\text{ft})$ and therefore

$\dfrac{H}{H_b} = \dfrac{3.0}{3.3} = 0.91$ for use in the drag and inertia diagrams between Case C and D.

Check: $\left(\dfrac{H}{d}\right)_b = \dfrac{3.3}{4.5} = 0.73 < 0.78(\text{okay})$

2. From Figure 9.11 for K_{Dm}

$$K_{Dm} = 0.71 \text{ by interpolation}$$

3. From Figure 9.12 for K_{im}

$$K_{im} = 0.38 \text{ by interpolation}$$

4. From Equation (9.25) for F_{Dm}

$$F_{Dm} = C_D \frac{1}{2} \rho g D H^2 K_{Dm}$$
$$= 0.7(0.5)(1025 \times 9.81)(0.3)(3.0^2)(0.71)$$
$$F_{Dm} = 6740\,\text{N}$$

5. From Equation (9.26) for F_{im}

$$F_{im} = C_m \rho g \frac{\pi}{4} D^2 H K_{im}$$
$$= 2.0(1025 \times 9.81)(\pi/4)(0.3^2)(3.0)(0.38)$$
$$F_{im} = 1620\,\text{N}$$

6. From Equation (9.30) for W used for interpolation

$$W = \frac{C_m D}{C_D H} = \frac{2.0(0.3)}{0.7(3.0)} = 0.29$$

7. From Figure 9.16 for $W = 0.1$ for Φ_m

$$\Phi_m = 0.35$$

8. From Figure 9.17 for $W = 0.5$ for Φ_m

$$\Phi_m = 0.38$$

9. Linear interpolation when $W = 0.29$ gives

$$\Phi_m = 0.365$$

10. From Equation (9.29) for the maximum force, F_m

$$F_m = \Phi_m [\rho g C_D H^2 D]$$
$$= 0.365[(1025 \times 9.81)(0.7)(3.0^2)(0.3)]$$
$$F_m = 6930\,\text{N [Solution (1)]}$$

11. From Figure 9.20 for $W = 0.1$ for α_m

$$\alpha_m = 0.33$$

12. From Figure 9.21 for $W = 0.5$ for α_m

$$\alpha_m = 0.35$$

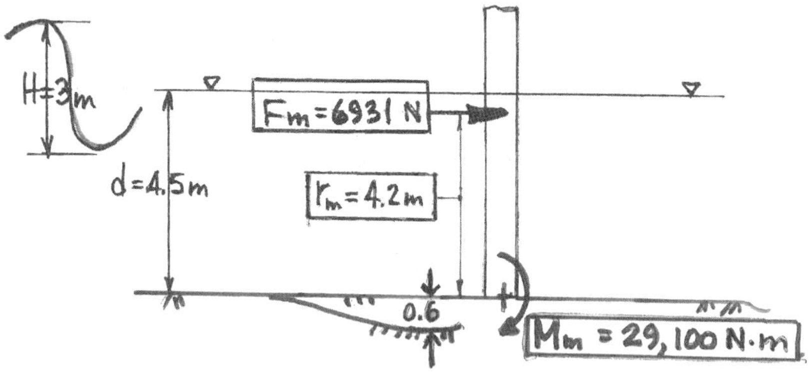

Figure 9.23. Graphic results of example problem for 3 m wave in 4.5 m water depth on a 30-cm pile.

13. Linear interpolation when $W = 0.29$ gives

$$\alpha_m = 0.34$$

14. From Equation (9.31) for the maximum moment, M_m

$$\begin{aligned} M_m &= \alpha_m[\rho g C_D H^2 D d] \\ &= 0.34[(1025 \times 9.81)(0.7)(3.0^2)(0.3)4.5] \\ M_m &= 29,100\,\text{Nm [Solution (2)]} \end{aligned}$$

15. From Equation (9.32) for the moment arm, r_m

$$r_m = \frac{M_m}{F_m} = \frac{29{,}100}{6{,}930} = 4.2\,\text{m [Solution (3)]}$$

The sketch in Figure 9.23 displays these results if there was no scour hole in front of the pile. Assuming a scour hole 0.6 m deep, and simply adding this to r_m would increase the overturning moment to 33,000 Nm.

As shown in Figure 9.1, breaking wave forces must also be considered.

9.4 Breaking Wave Forces on Slender Piles

9.4.1 *Introduction*

As introduced in Section 6.3.3 for vertical walls, breaking waves produce impact forces of short duration and are located near the crest. Figure 6.37 displayed the dimensionless, horizontal wave force, F_h^*, as a function of the time variation (t/T) for breaking wave impact loads where $F_{h,\max}^*$ was four times larger than for non-breaking waves. Large-scale, wave breaking experiments (GWK, Hannover) were employed and reported by Oumeraci *et al.* (2001) as produced for the PROVERBS (1994–1998) research project in the European Union.

9.4.2 *Large wave flume experiments*

9.4.2.1 *Relatively deep water breaking — Single focused wave*

This same large wave flume (GWK) of the Coastal Research Centre (FZK) in Hannover, Germany, has been employed to measure breaking wave impact forces on vertical (and inclined) slender piles (Wienke and Oumeraci, 2005). A steel cylinder of diameter 0.7 m (2.3 ft) and 7.3 m (24 ft) length was installed on a flat bottom. Bearings at top and bottom with strain gauges measured the horizontal load on the pile. Water depths varied between 4 and 4.25 m (13–14 ft) with focused, plunging-type breaker heights, H_b, of 2.2 to 2.8 m (7.2–9.2 ft) as illustrated in Figure 9.24. Wave crest heights, η_b, ranged from 1.7–2.0 m (5.6–6.6 ft) so that $\eta_b = (0.77\text{–}0.72)H_b$ above the still water level (SWL), or

$$\eta_{b,\text{ave}} \approx 0.75\ H_b. \tag{9.37}$$

These results are for single breaking waves generated by focusing wave packets at a fixed breaking point where the cylinder was located at 111 m (364 ft) from the wave paddle in the 309 m (1,014 ft) long flume. The natural, vibrational frequency of the cylinder was removed (filtered) from the recorded force signal, so that only the wave impact force variation with time was used to determine the key empirical coefficients required in the theory. The wave celerity, C, at breaking was on the order of 6 m/s (20 ft/s) and the

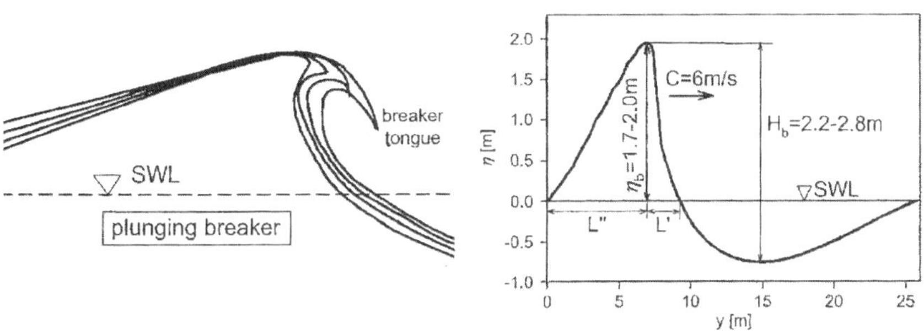

Figure 9.24. Plunging wave breaking dimensions in GWK experiments (from Wienke and Oumeraci, 2005).

wave period, T, was on the order of 4.2 s. Naturally breaking waves in deep water have a different asymmetrical shape that modifies Equation (9.37) and the wave celerity at breaking, but these differences were neglected. See Wienke and Oumeraci (2005) for all the details and formulas for design.

9.4.2.2 *Shallow water breaking — Progressive, regular waves*

Water depths ranged from 3.8 to 4.25 m in the flume and then decreased on a 10:1 slope to the front of a horizontal, steel berm (+2.3 m above the flume bottom) where the test cylinder was positioned. Regular waves with periods from 3 to 10 s and wave heights from 1.2 to 1.7 m shoaled on the slope to produce depth-limited wave breaking. The same steel cylinder with diameter 0.7 m was employed, but was only 5 m long in the vertical position. Bearings at top and bottom with strain gauges again measured the horizontal load on the pile. See Irschik *et al.* (2004) for complete details of the experimental setup.

9.4.3 *Theory of wave breaking impact force*

Wave breaking on the pile will produce an additional impact force, $F_I(t)$, of very short duration. The total force at impact is then simply the sum of the wave force from the Morison equation, F_{Morison}, and the impact force

$$F_{\text{total}} = F_{\text{Morison}} + F_I(t) \tag{9.38}$$

so that conservative design practice is to use Equation (9.29) to calculate the maximum force, $F_m = F_{\text{Morison}}$.

After a review of the literature (see both Wienke and Oumeraci, 2004; Irschik *et al.*, 2004) for the theory and verification of the theory by these large-scale experiments, the impact force can be calculated as

$$F_I(t) = \rho R (C_b)^2 C_s(t) \lambda\, \eta_b, \tag{9.39}$$

where ρ is the mass density of water, R is the radius of the pile, C_b is the wave celerity at breaking point, $C_s(t)$ is the slamming factor, λ is the curling factor, and η_b is the maximum water surface elevation at the breaking point with these definitions illustrated in Figure 9.25.

Wienke *et al.* (2001) found the highest forces when the wave breaks immediately in front of the pile giving the maximum curling factor, $\lambda = 0.46$. The slamming factor was highest at the beginning of impact ($t = 0$) and the experiments agreed with Wagner (1932) taking $C_s(t = 0) = 2\pi$. Figure 9.26 is one example measurement of the total measured response

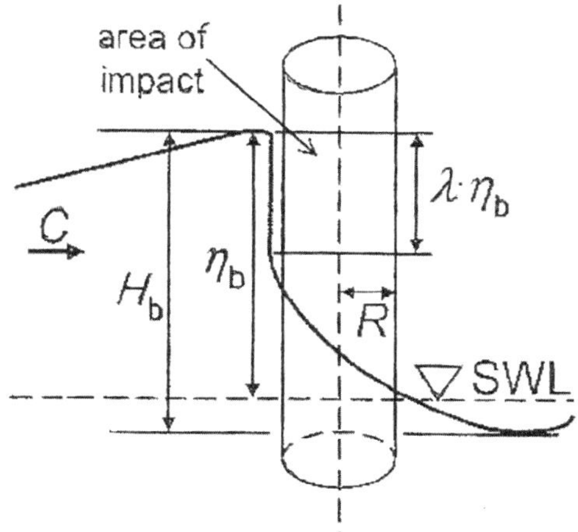

Figure 9.25. Schematic of wave breaking on cylindrical pile with definition sketch of variables.

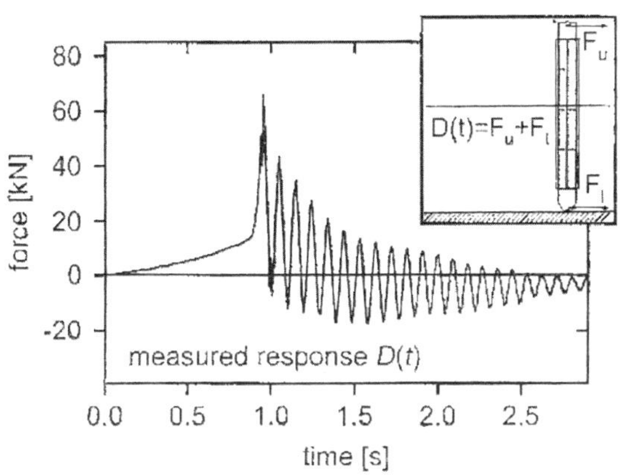

Figure 9.26. Example total measured force on the slender pile (from Wienke and Oumeraci, 2004).

force, F_{total} (kN), versus time (s) as the sum of the loads recorded on the upper and lower strain gauges on the bearing. The natural frequency of oscillation was removed from the signal and the wave force from the Morison equation, F_{Morison}, was filtered out leaving $F_I(t)$. Note that the peak force measured was about 70 kN and the maximum wave force just before the spike when the wave impacted the cylinder was about 10 kN, giving F_I equal to about 60 kN at $t = 0$ (from Wienke and Oumeraci, 2004).

The ratio of the impact force to the maximum wave force from Morison equation is then about $60/10 \approx 6$. The duration of this force spike when the breaking wave impacts the pile is in milliseconds. The force oscillations shown in Figure 9.26 are due to the natural frequencey of the pile and are not the duration of the maximum measured impact force. The location of this force spike on the pile is at the middle of the area of impact at $0.5(\lambda\eta_b)$, or a distance $0.77\eta_b$ above the SWL as illustrated in Figure 9.25.

Most slender piles are used beneath piers in shallow water. Using the following approximate relations for the variables in Equations (9.29) and (9.39):

$$\eta_{b,\text{ave}} \approx 0.75H_b, \tag{9.37}$$

$$C_b \approx \sqrt{g(d+\eta_b)}, \tag{9.40}$$

$$\gamma_b \approx \frac{H_b}{d_b}, \tag{9.41}$$

gives the ratio

$$\frac{F_I}{F_m} = \frac{1.084 + \left(\frac{1}{\gamma_b} + 0.75\right)}{\Phi_m C_D} \tag{9.42}$$

and taking $\gamma_b = 0.8$ for the breaking ratio in shallow water gives

$$\frac{F_I}{F_m} \approx \frac{2}{\Phi_m C_D}. \tag{9.43}$$

Using the same values as for the example above for Morison's equation

$$\Phi_m = 0.365 \text{ (dimensionless, total force coefficient)},$$
$$C_D = 0.7 \text{ (drag coefficient)},$$

gives $\frac{F_I}{F_m} \approx 8$ which is reasonable when compared to the results measured in Figure 9.26.

The author has conducted small scale laboratory measurements of breaking, regular wave forces on slender piles (Reddish, 1986; Basco and Niedzwecki, 1989) and compared them with large-scale, focused wave breaking results of Kjeldsen *et al.*, 1985. The nondimensional breaking wave force distribution over the vertical were also determined and found similar in both cases except for the vertical elevation above the SWL. A ratio of 7–8 was measured for the maximum impact force relative to the maximum wave force without breaking and found similar to that discussed above for the full-scale experiments in the GWK. See also Choi and Gudmestad (2013) for numerical model experimental results.

9.5 Traverse Forces and Non-vertical Piles

Lateral forces due to vortices in the turbulent wake of the pile in the transverse direction of wave propagation and similar to lift forces on airfoils may be present. An empirical lift coefficient, C_L, is then employed in an equation for the traverse force, F_L, proportional to the square of the horizontal wave velocity. Keulegan and Carpenter (1958) found that these traverse forces are extremely small on rigid piles in shallow coastal water depths.

All the above force and moment calculations for non-breaking and breaking waves are for vertical piles. When non-breaking waves move past non-vertical piles, the CEM (2006) recommends starting with the basic Morison equation (9.18), using an appropriate wave theory and selecting the needed coefficients, C_D and C_M. This would mean starting over. This author recommends simply using the design curves presented above in Figures 9.10–9.22 with the actual length of the non-vertical pile.

However, for the maximum breaking wave impact force, F_I, the yaw angle, α (degrees), can be readily incorporated in the calculations from the GWK experimental results. The curling factor, λ, increases for negative yaw angles as the pile tilts forward facing into the impacting wave front. Full details are found in Wienke and Oumeraci (2005).

9.6 Safety Factors in Pile Design

Safety factors should be considered for design of slender piles in shallow water. A safety factor of 2 is recommended and applied to the maximum wave force calculated using the Morison equation.

For pile design, the breaking wave impact force must be viewed with caution. It can be considered as one blow from a hammer striking the pile about 0.8 η_b above the SWL. This is for one direct blow by the breaking wave on the pile. Wave impact forces are less for waves breaking in front of the pile and for broken waves hitting the pile. Using the breaking wave impact force, which is 6–8 times larger than the maximum wave force calculated using the Morison equation, would essentially be using a safety factor 6–8 for the design.

More importantly for safety, is the corrosion and fouling by marine growth on the pile. Both the roughness and pile diameter are increased. See also the CEM (2006) Part VI-5-7 for information on pile groups.

9.7 Pier Uplift Wave Forces

The design of piers is beyond the scope of this book.

References

Airy, G. B. (1845) "Tides and Waves," *Encyclopaedia Metropolitana*, London, Vol. 5, pp. 241–396.

American Petroleum Institute (API) and the Society for Naval Architecture and Marine Engineering (SNAME).

Basco, D. R., and J. M. Niedzwecki (1989) "Breaking Wave Force Distributions and Design Criteria for Slender Piles," Proceedings No. 6009, Offshore Technology Conf., Houston, TX, May.

Boussinesq, J. (1872) "Theorie De Onde et Des Resous Qui Se Propagent Le Long d'un Canal Rectangulaire Horizontal, en Communiquant Au Liquide Contenu Dans Ce Canal des Vitesses Sensiblement Pareilles de la Surface au Fond," *Journal de Mathematics, Pure et Applied*, Deuxieme Serie, Vol. 17, pp. 55–108.

CEM (2006) *The Coastal Engineering Manual*, US Army Corps of Engineers, Washington, DC.

Choi, S.-J., and O. T. Gudmestad (2013) "Breaking Wave Forces on a Vertical Pile", Proceedings, Fluid Structure Interaction 2013 Conference, Vol. 129.

Clauss, G., E. Lehmann, and C. Östergaard (1992) *Offshore Structures: Conceptual Design and Hydromechanics*, Vol. 1, Springer Verlag.

Dean, R. G. (1965) "Stream Function Representation of Nonlinear Ocean Waves," *Journal Geophysical Research*, Vol. 70, No. 18.

Dean, R. G. (1974) "Evaluation and Development of Water Wave Theories for Engineering Application," Vols. 1 and 2, Special Report 1, *Corps of Engineers, Coastal Engineering Research Center*, Fort Belvoir, VA.

Dean, R. G., and R. A. Dalrymple (1984) *Water Wave Mechanics for Engineers and Scientists*, Prentice-Hall, New Jersey.

Irschik K., U. Sparboom, and H. Oumeraci (2004) "Breaking Wave Loads on a Slender Pile in Shallow Water," in *Proceeding, 29th International Conference Coastal Engineering* (Lisbon).

Keulegan, G. H., and L. H. Carpenter (1958) "Forces on Cylinders and Plates in an Oscillating Fluid," *Journal Research, National Bureau Standards*, Vol. 60, No. 5, May.

Kjeldsen, S. P. *et al.* (1985) "Wave Forces on a Pile in the Surf Zone from Wave Crest to the Wave Trough," *MARINTEK*, Report No. 1.7, Trondheim, Norway, April.

Korteweg, D. J., and G. DeVries (1895) "On the Change of Form of Long Waves Advancing in a Canal and On a New Type of Long Stationary Waves," *Philosophy Magazine*, Series 5, Vol. 39, pp. 422–443.

Lamb, H. (1945) *Hydrodynamics*, Dover Publications, New York.

Morison, J. R., M. P. O'Brien, J. W. Johnson, and S. A. Schaaf (1950) "The Force Exerted by Surface Waves on Piles," *Petroleum Transactions, American Institute of Mining Engineers*, Vol. 189, pp. 149–154.

Oumeraci, H., A. Kortenhaus, W. Allsop, M. deGroot, R. Crouch, H. Vrijling, and H. Voortman (2001) *Probabilistic Design Tools for Vertical Breakwaters (PROVERBS)*, Swets & Zeitlinger B. V. Lisse, Germany.

Reddish, H. J. (1986) "Laboratory Study of Plunging Breaker Wave Force Distribution on a Slender Pile," ME report, *Ocean Engineering Program, Texas A&M Univ.*, College Station, TX, May, 90p.

Sarpkaya, T. (1976) "Vortex Shedding and Resistance in Harmonic Flow About Smooth and Rough Circular Cylinders," *Proceedings, International Conference on Behavior of Offshore Structures*, BOSS '76, Vol. 1, pp. 220–235.

Stokes, G. G. (1847) "On the Theory of Oscillatory Waves," *Transactions, Cambridge Philosophical Society*, Vol. 8, pp. 441–473.

Wagner, H. (1932) "Über Stoß- und Gleitvorgänge an der Oberfläche von Flüssigkeiten," *Zeitschrift für angewandte Mathematik und Mechanik*, Vol. 12, No. 4, pp. 193–215 (in German).

Wienke, J., U. Sparboom, and H. Oumeraci (2001) "Breaking Wave Impact on a Slender Cylinder," Francis-Institute, University of Hannover, Germany.

Wienke J., and H. Oumeraci (2005) "Breaking Wave Impact Force on a Vertical and Inclined Slender Pile- Theoretical and Large-Scale Model Investigation," *Coastal Engineering*, Vol. 52, pp. 435–462.

Chapter 10

Environmental Impacts

10.1 Introduction

10.1.1 *Origins of the environmental constraint*

Where the land meets the sea, many individuals, organizations, and government institutions believe the best philosophy for our coasts is *let nature take its course*. Their rallying cry is *you can't fight mother nature*. All the many alternatives for coastal hazard mitigation as discussed in Chapter 3 are opposed from the outset. Therefore, the coastal engineer's responsibility for studying, proposing, designing, and constructing these alternatives for both private and public owners of the coastal lands is in direct opposition to this philosophy. In the US, a requirement for construction is a "Permit" that must be approved at local, state, and federal government (Corps of Engineers) levels. At all approval levels, individuals, organizations, and other government institutions have the right to question the environmental impacts of the proposed project. As a result of the "Permit" process, alterations may be made to the original permit application design (voluntary or required), the process is repeated, and the final result is either further delays, project approval, or project rejection (for negative environmental impacts).

Written documents, court testimony by experts, and/or permit process delays are often employed by opposition individuals and groups on a case-by-case basis. Some US states (e.g., North Carolina, South Carolina) prohibit seawalls and breakwaters on the open Atlantic Ocean coast. The US National Park Service prohibits all coastal structures on its coastal shorelines and is the author of the *let nature take its course* philosophy. Other federal government agencies in the US (Environmental Protection Agency, Fish, and Wildlife Service, etc.) must also review project permit applications submitted to the Corps of Engineers. It is a complicated and sometimes confusing business in the US. And each country of the world has its own laws and restrictions regarding the environmental impacts of coastal structures. Therefore, the best advice is to learn all about the site-specific rules, regulations, and restrictions for all potential coastal hazard mitigation alternatives at the *beginning* of the design process. Do not wait until the end during the permit process. In today's world, many hazard mitigation alternatives cannot be considered. A series of articles titled "The Law of the Coast in a Clamshell" have been published in *Shore and Beach* (American Shore and Beach Preservation Association (ASBPA)) by Peter Graber, Attorney at Law (Graber, 1985, Virginia) that should be consulted for US legislation with extensive references.

Since the 1960s, concerns regarding how the many alternatives for the reduction of coastal hazards (flooding, wave damage, and coastal erosion) have resulted in the "environmental" constraint for coastal engineering design. And, this has led to additional "social, political, and institutional" constraints by state and federal legislation.

10.1.2 *Scope and limitations of this chapter*

Concerns regarding the environment and ecology of the world's water resources and adjacent shorelines began circa 1950. This chapter is primarily a review of some of the published literature since the early 1980s; a summary of the author's research efforts regarding the impact of seawalls on adjacent beaches; the relatively recent emphasis on natural and nature-based features of "living shorelines"; and one example of environmental opposition to a permit application for a coastal hazard mitigation project.

Published information surrounding the potential, environmental, and ecological impacts of all the possible alternatives for coastal hazard mitigation is now found in many different journals. Extensive and thorough review of all the published literature has not been attempted and is a limitation of this chapter. An entire book (or two) could be written on this subject.

10.2 Literature Review

10.2.1 *Coastal geologists position paper (1982)*

Concerns of coastal geologists regarding the impacts of engineered structures (mainly seawalls, jetties, and groins) on beaches constructed along the Eastern Atlantic Ocean coast and the Gulf of Mexico coast were expressed in Pilkey *et al.* (1982). The title was "Saving the American

Beach: A Position Paper by Concerned Coastal Geologists" and a summary appeared in the publication *Shore and Beach* (Vol. 50, No. 2, pp. 3–4) of the ASBPA. The complete paper can be obtained from the Skidaway Institute of Oceanography, Savanah, Georgia (USA). These coasts contain many barrier islands and their argument is that these barrier islands are naturally "moving" landward with rising seas. In essence, they claimed that "*fixed coastal structures... almost always accelerate the natural rate of beach erosion... resulting in degradation of the beach... in the immediate vicinity of structures... or along adjacent shorelines miles away*". For the long term, this publication advocates a retreat philosophy for shoreline management. More discussion is found in Chapter 11 on Sea Level Rise.

The response of coastal engineers was immediately expressed by O'Brien (1982) in the same issue *of Shore and Beach* (Vol. 50, No. 2, pp. 6–8) and makes interesting reading today after over 35 years. Almost all of the coastal projects cited in the coastal geologist's "position paper" were downdrift of jettied inlets along the US Eastern seaboard. The simple reason was that US Federal government's first priority was for navigation through these tidal inlets, and impacts to adjacent shorelines were not addressed. This policy changed in 1968 along with placement of dredged material for the inlet as discussed in Chapter 8. Dr. O'Brien also discussed the future of rising sea levels and we defer further review until Chapter 11. [*Dr. M. P. O'Brien was Dean of the College of Engineering at the University of California, Berkeley, and President of the ASBPA at this time. In 1950, he helped organize the first International Conference on Coastal Engineering (ICCE) and coined the term Coastal Engineering. Because of his long career in the field, he is considered the founder of modern Coastal Engineering*].

10.2.2 *Summary of possible seawall effects on beaches (Dean, 1987)*

Additional allegations regarding the adverse effects of coastal armoring were presented in Walton and Sensabaugh (1979), Kana and Svetlichny (1982), Pilkey, Howard, and Kaufman (1985). The claims were encapsulated in the phrase "... *seawalls increase erosion and destroy the beach*" as made popular in the US media by geology professor Orrin H. Pilkey, Duke University, to promote his retreat philosophy for solving coastal problems (e.g. Pilkey and Wright, 1988). Dean (1987) critically examined all the commonly expressed concerns as summarized into nine possible effects in Figures 10.1(a) and 10.1(b) adapted from Basco (1987). Dean (1987) then made the following conclusions with numbers coinciding with those in Figure 10.1:

Armoring does NOT cause:

- Increase in long-term average erosion rate (3)
- Delay in beach recovery after storms (5)
- Profile steepening (6)
- Increase in longshore sediment transport (8)
- Sand transport far offshore (9)

Armoring CAN contribute to:

- Frontal effects of toe scour (1a)
- End-of-wall effects (1b)
- Diminished width of beach fronting armoring (2)
- Blockage of littoral drift when in surf zone (4)

A fundamental principle in this analysis is conservation of sand volume in the coastal system. Coastal armoring does not add nor remove sand from the system, but it may redistribute or prevent sand from entering the system. Let us consider these nine possible effects in numerical order.

(1a) Local Toe Scour. Wave orbital motions on the bottom in front of a seawall (or bulkhead) will cause the sediments to move seaward and create toe scour. The "rule-of-thumb" for design position of the lowest elevation of the bottom fronting the wall is that the depth of the scour hole below the storm profile is approximately equal to the wave height striking the wall as illustrated in Figure 10.2(b). The material removed is equal to the volume deposited seaward of the hole. In most cases, the wave height is depth-limited due to wave breaking so that claims of greater scour hole depth are not realistic. Kriebel *et al.* (1986) present measurements after Hurricane Elena (1985) that confirm this simple model for local toe scour. The author has stood behind the wall with a survey rod in the scour hole during the peak of a storm and found that the scour depth is less when this same measurement is taken after the storm is over. This is because sand has moved back into the hole when the wave periods are longer to create landward movement of sand along the bottom after the storm. See the CEM (2006, Part VI-5-6) for a detailed review of scour and scour protection as applied in the design of seawalls and bulkheads.

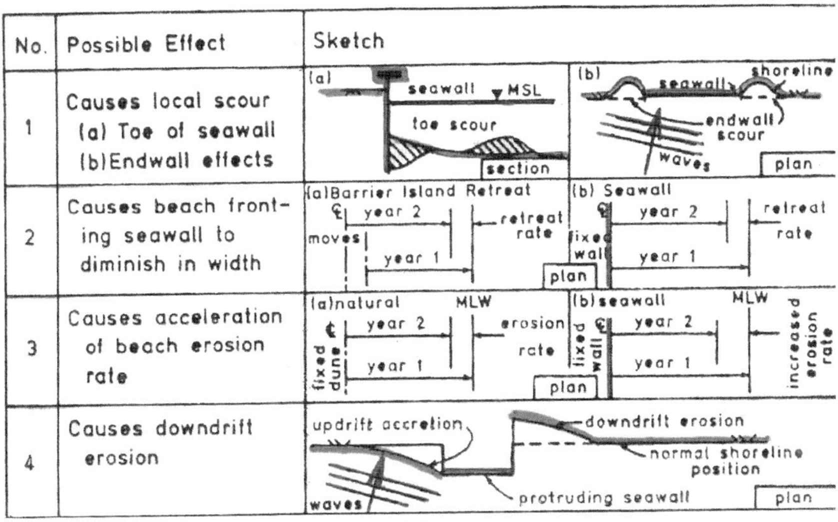

(a)

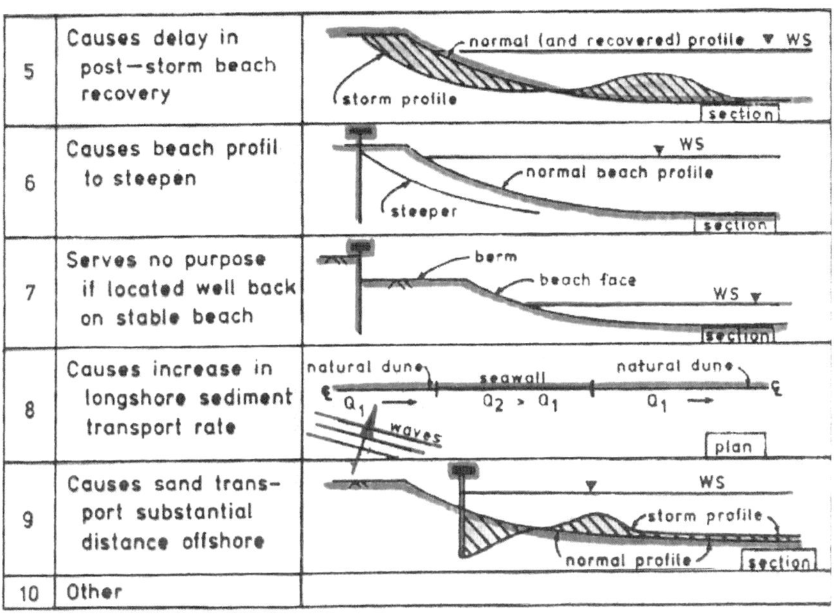

(b)

Figure 10.1. (a) Possible effects (1–4) of seawalls on adjacent shorelines and beaches (from Dean, 1987). (b) Possible effects (5–9) of seawalls on adjacent shorelines and beaches (from Dean, 1987).

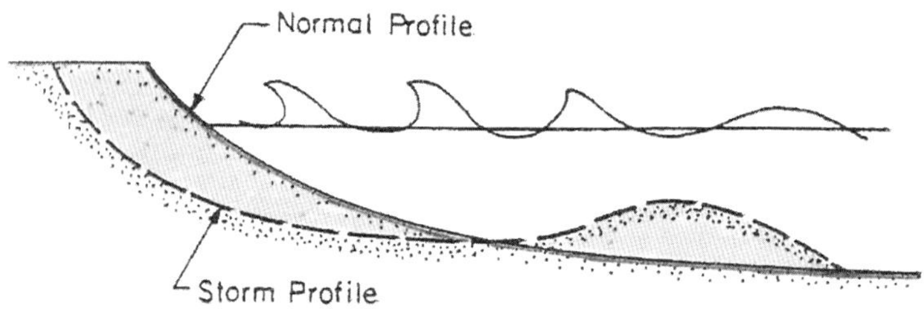

(a) Normal and Storm Profiles on a Natural Shoreline

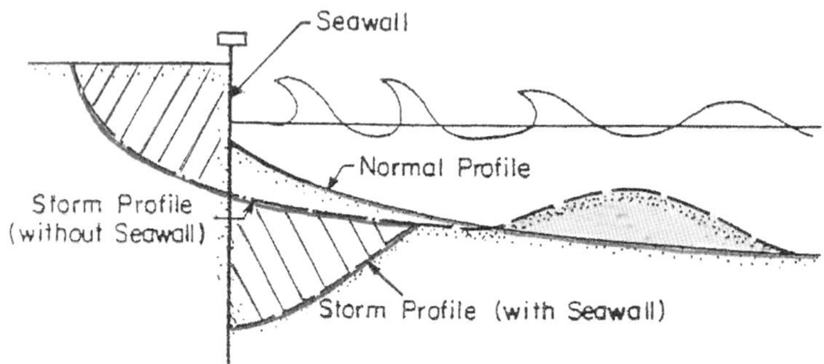

(b) Normal and Storm Profiles on a Seawalled Shoreline and Comparison with Profiles on a Natural Shoreline

Figure 10.2. Normal and storm beach profiles on (a) natural and (b) sea-walled shoreline illustrating local toe scour in front of wall (from Dean, 1987).

(1b) End Wall Effects. Shoreline changes at the end of seawalls are also called "flanking" effects. This is a key environmental concern for the potential impact is on the neighbor beach with no armoring.

Walton and Sensabaugh (1979) provide post-hurricane Eloise field observations of additional contour recession adjacent to seawalls at two sites in Florida. Figure 10.3 displays the data (14 points as excessive erosion, y (feet), versus structure length, L_s (feet)). For site 1 (Group I) data, y was between 11.5–43 ft for L_s from 100 to 230 ft. At site 2 (Group II), y ranged from 20 to 62 ft for L_s from 380 to 740 ft. Trends are difficult to determine. The excessive erosion was attributed to the adjacent seawalls. No other possible mechanisms were addressed.

McDougal *et al.* (1987) conducted small scale, "equilibrium" beach profile, laboratory experiments (nine data points) for regular waves with

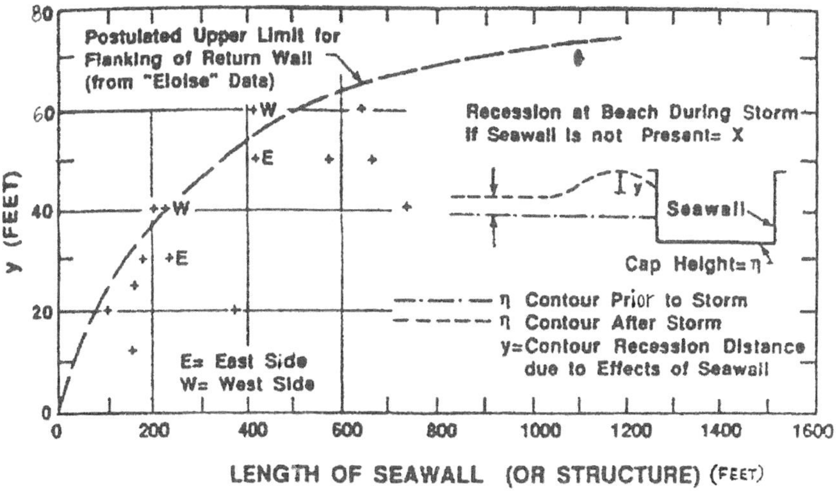

Figure 10.3. Excessive erosion adjacent to seawall relative to seawall length (from Walton and Sensabaugh, 1979).

7–14 cm wave heights at 1.1 s wave period and medium, grain-sized beach materials. The excess erosion, y, ranged from 3 to 10 cm for seawalls 65 to 105 cm long.

Unfortunately, these two datasets containing 23 data points have been erroneously combined in one plot (Komar and McDougal, 1988) as shown in Figure 10.4 in the metric system. The symbol r is now used for the excess erosion and requires logarithmic scales to capture the data. The trend line (red) connecting the two data groups has been included and this *incorrect* correlation suggests that "excess" flanking erosion, r, due to the presence of the seawall is about 10% of the seawall length, L_s. In fact, no such relationship can be "proven" by this data because of the well-known and accepted "scale" effects of the laboratory data. For over 100 years, physical model investigations with natural sediments have been conducted, but the results only qualitatively applied because of the great disparity between the Froude and Reynolds number scales for the sand particles. Simply put, if properly scaled, the sand particles would become silty, cohesive materials. However, Figure 10.4 with trend line continues to appear in the literature (e.g., Komar, 1998, 2nd Edition, Figure 12.22, p. 52) and continues to be accepted by state agencies for permitting coastal structures. This incorrect practice must be discontinued.

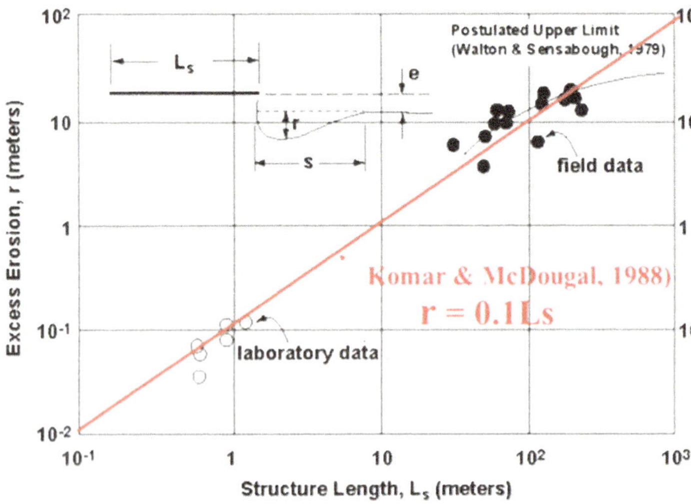

Figure 10.4. Incorrect combination of laboratory scale data (from McDougal *et al.* (1987) and field data (from Walton and Sensabaugh, 1979) on log-log scale. The $r = 0.1\ L_s$ relation is wrong!

Figure 10.5 correctly displays the laboratory data (near 0,0) location and the two sets of widely scattered field data with the incorrect trend line (red). One data point (red dot in Figure 10.3 at $L = 1{,}100$ ft) was ignored by Komar and McDougal (1988). There are many possible physical mechanisms (reasons) for the large scatter in the field data and these are explained as follows.

(i) Sand Trapping Theory. On historically erosional shorelines with no sediment source (river), Dean (1987) argued that the structure denies sand to the littoral system during storm events resulting in excess erosion of unprotected adjacent properties, as illustrated in Figure 10.6. The magnitude of the excess was linked to the seawall length. However, only the data in Figure 10.3 exists, and for very long seawalls, no correlation is physically plausible. The proposed linkage between excessive erosion, r, and seawall length, L_s, ignores the importance of the location of the seawall on the beach. Weggel (1988) defined six types of seawalls depending on their location on the beach and water depth at the base. Walls further seaward were said to have an increasing effect on coastal processes, e.g. as more sand is trapped. The proposed linkage between r and L_s also neglects the magnitude of the sand volume that is "active" in the cross-shore profile, and

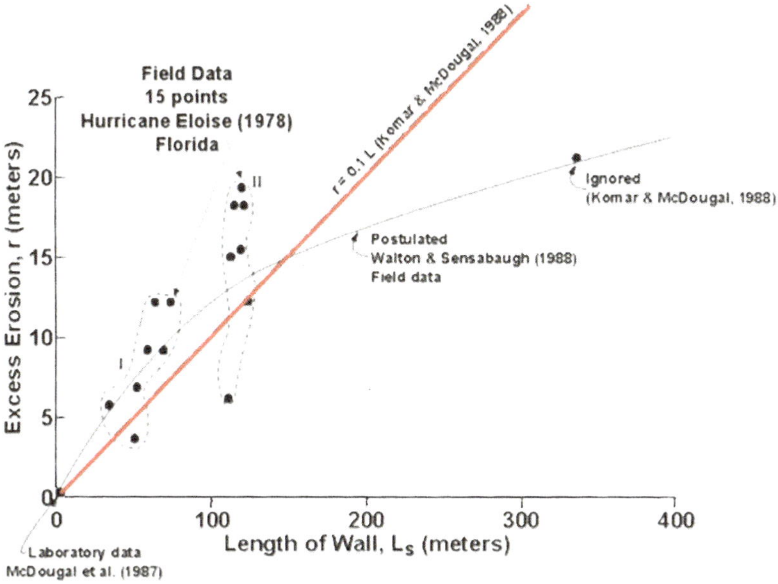

Figure 10.5. Field data of Walton and Sensabaugh (1979) from Hurricane Eloise.

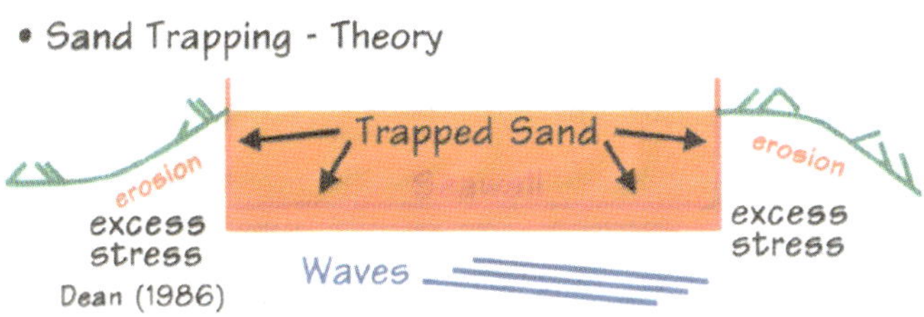

Figure 10.6. Sand trapping theory for excess erosion (flanking).

its distribution. Basco (2000) and Ozger (2002) discuss a wall trap ratio, WTR, as the ratio of actual volume removed to that active in the cross-shore volume during storm events. When the active volume is large, the tapped volume may only be a small percentage of the total, active sand volume in the coastal zone.

A few US states (Florida, California) have adopted mitigation policies and procedures to permit seawall construction and maintain a healthy beach. Formulas attempt to deterministically estimate the wall trap volume, but do not take into account the active sand volume so that the importance of the relative trapped volume is unknown. Property owners annually

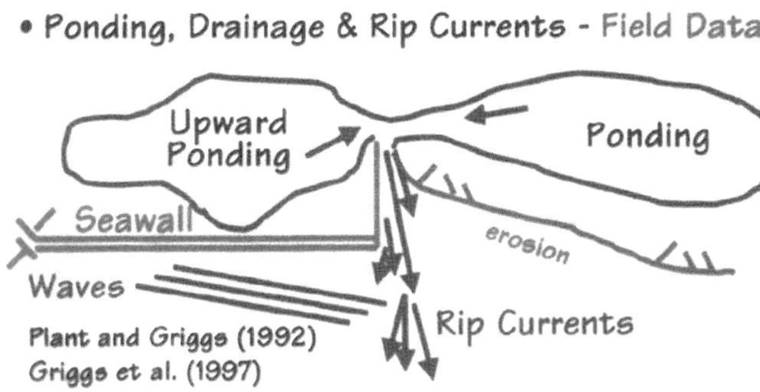

Figure 10.7. Ponding, drainage, and rip currents on end of seawall to produce end-of-wall impact.

replace the beach materials trapped behind the structure by placing new sand on the beach. Clearly, stochastic methods are needed to account for the statistical distribution of high water and wave energy events.

(ii) Rip Currents and Seaward Return Flows. McDougal *et al.* (1987) observed rip currents in their model tests and also field evidence to conclude that this mechanism may be more responsible for end-of-wall effects than trapping, as illustrated in Figure 10.7.

Similar observations of rip currents and return flows in the field were made by Plant (1990), Plant and Griggs (1992), and Griggs *et al.* (1994) at the interior and ends of armored sections. Figure 10.8(a) shows the location in Monterey Bay, CA (Aptos Seascape), and profile locations taken over eight years with profile 13 (red) and 18 (green, control) results averaged for both the summer and winter seasons in Figure 10.8(b). End-of-wall effects were attributed to wave overtopping, elevated beach water tables, and seaward, return flows. Griggs *et al.* (1994) concluded there were *no significant long-term effects....* These US West Coast field results are important, in that the waves, water levels, sand grain size is different from that found on East Coast and Gulf of Mexico beaches.

(iii) Blockage of Littoral Drift (Groin Effect). Griggs and Tait (1988) found significant, long-term flanking effects at one site on the California coast that lowered the beach profile over 150 m downcoast. It was proven that the up-coast end of this protruding wall produced sand impoundment, i.e. a groin effect as illustrated in Figure 10.9. Tove and Wang (1990)

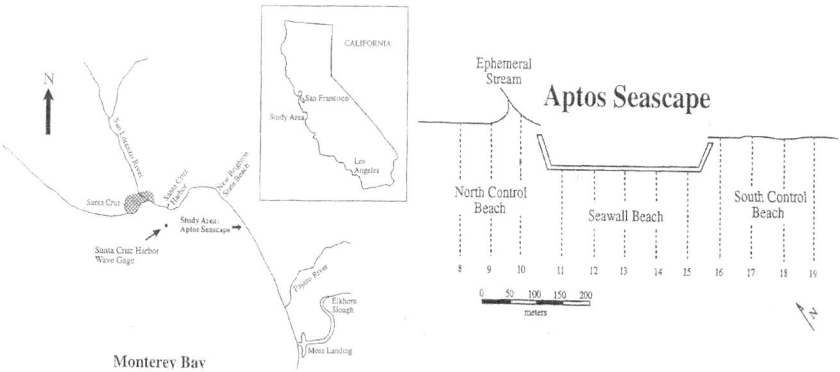

(a)

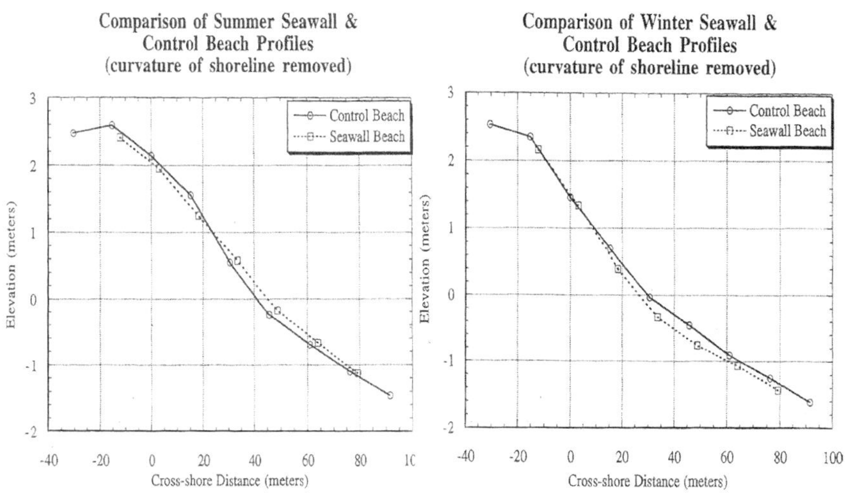

(b)

Figure 10.8. (a) Location of Aptos Seascape in Monterey Bay, CA, and beach profiles (Griggs *et al.*, 1997). (b) Seven-year average summer and winter beach profiles at location 13 in front of seawall (red) and at control location 18 (green) (from Griggs *et al.*, 1997).

conducted laboratory experiments with waves attacking walled and non-walled beaches at angles and concluded that downdrift impacts were a groin effect.

(iv) Headland, Parabolic Bay Beaches. Wave refraction, diffraction, shoaling, and breaking transformation processes create parabolic or

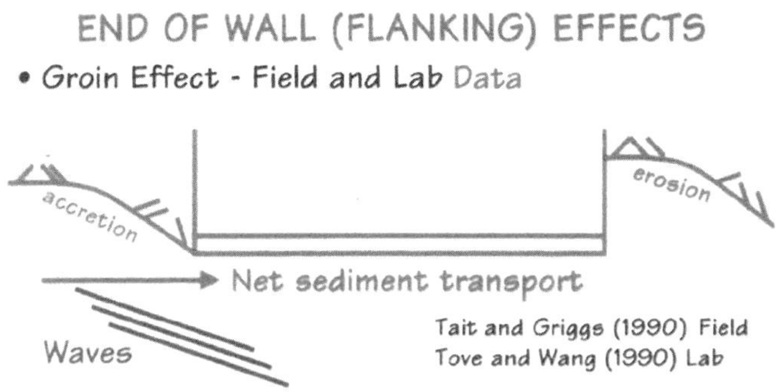

Figure 10.9. Blockage of littoral drift with seawall acting as a groin to produce excess erosion.

crenulated-shaped beach planforms adjacent to rocky headlands (Silvester and Hsu, 1993), as illustrated in Figure 5.32. For waves from a dominant direction, the planform evolves and reaches an equilibrium shape where all waves break normal to the beach, hence longshore transport ceases. Methods exist to predict the equilibrium planform shape, hence maximum indentation of the shoreline position for design (e.g. CEM, 2000). See Chapter 5 for details.

Seawalls and adjacent beaches could also be considered as headland, bay beach systems as photographs in Figure 10.10 illustrate. Note that this analogy means a maximum indentation, not connected with the "length" of the headland. And, for waves from many directions, the adjacent beach can fill with sand and the headland act as a groin. Not all seawalls are constructed at sites with eroding shorelines, but the excessive erosion misconception is unfortunately applied to many non-eroding situations.

(v) Natural Beach Recovery After Storms. Most beaches in the world naturally recover after erosional storm events. The wave characteristics (height, period) return to levels that cause the onshore direction of sand transport, and the wave direction may also change. Flanking effects, if any, due to the above cited mechanisms, or combinations, disappear as the sand returns from offshore or the long-shore transport direction changes toward the walled section. Walton and Sensabaugh (1978) did not return to the field sites in Florida (Figure 10.2) to determine the natural, post-hurricane Eloise recovery of the beach. McDougal *et al.* (1987) did not modify the wave conditions in the laboratory to determine the recovery of the beach in their laboratory investigations (Figure 10.3).

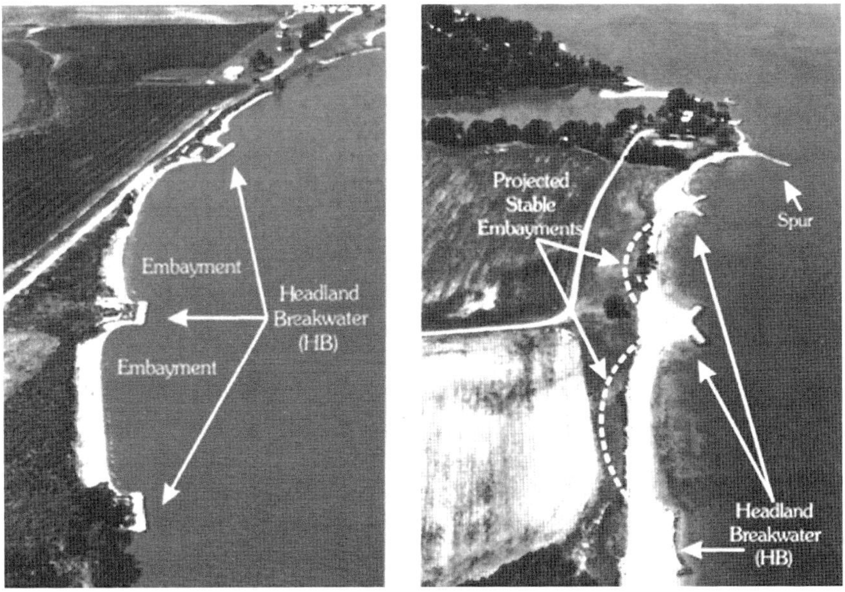

Figure 10.10. Photographs of headland, bay beach systems in Chesapeake Bay, VA (from Hardaway and Byrne, 1999).

Table 10.1. Possible mechanisms for flanking effect.

Possible Mechanism	References
Sand trapped behind wall	Dean (1987)
Rip currents and seaward return flows	McDougal *et al.* (1987), Griggs *et al.* (1997)
Blockage of littoral drift (groin effect)	Griggs and Tait (1988), Tove and Wang (1990)
Headland, parabolic bay system	—
Other	—

(vi) Summary. Possible mechanisms for the flanking effect as discussed above are summarized in Table 10.1. All these mechanisms are the reason for the wide scatter of the field data in Figure 10.4. The trapping mechanism of Dean (1987) is not the only one possible for causing end-of-wall effects.

(2) Beach Fronting Seawall Diminishes in Width. The second possible impact of hardening the shoreline (Figure 10.1(a)) is the beach width diminishing in time. This was the main concern voiced in the position paper by the coastal geologists (Pilkey *et al.*, 1982) and based on their experience with barrier islands. As shown in Figure 10.11 (from Dolan and Lins,

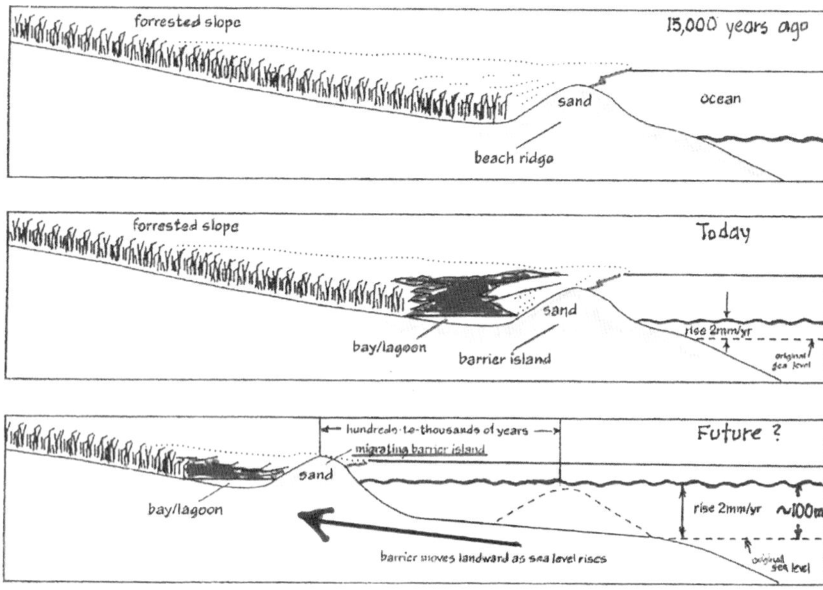

Figure 10.11. Schematic illustration of a migrating barrier island (from Dolan and Lins, 1987).

1987), as sea level rose over the past 15,000 years, sand ridges along the coast became sand barriers protecting bay/lagoon systems behind them and these "barrier islands" shifted landward (migrated) over hundreds to thousands of years of geologic time. The figure shows a sea level rise rate of 2 mm/year as illustration. Storm surges brought elevated water level events, high waves, and sand wash-overs that produced the transgression of the sand barrier so that it kept roughly the same sand volume in the cross-section. The sketch in Figure 10.1(a) for impact (2) illustrates this (a) barrier island "retreat rate" where the centerline (¢) moves laterally and then (b) where a fixed reference line, in this case a seawall, draws a reference location from which to quantify the retreat rate.

The fixed baseline or reference line used to measure the barrier island retreat rate could be the centerline (¢) of a road, as illustrated in Figure 10.12 (top). This retreat rate is then called coastal erosion and over the time history of measurements, t_0, t_1, t_2, it becomes the local coastal erosion rate, e. The road permitted construction of homes and then when barrier island retreat continued (now called "erosion") to possibly threaten the home, a seawall (or rock revetment) was constructed from which the

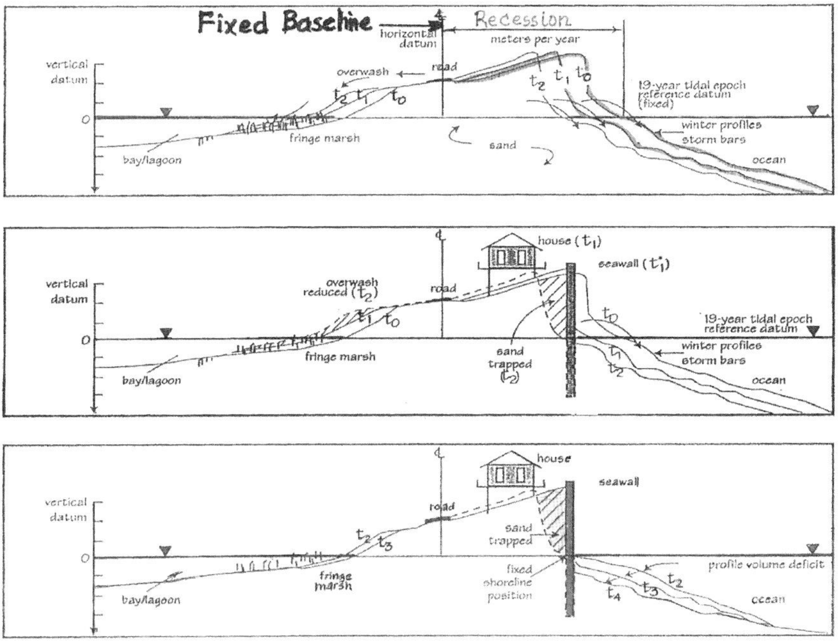

Figure 10.12. Schematic illustration of barrier island retreat with fixed reference baseline as the centerline of the road (top) and location of a hardened shoreline position (seawall, middle, bottom) from which coastal erosion can be quantified and the width of subaerial beach determined.

time history of the barrier island retreat could be measured (Figure 10.12, middle). Clearly, the visual, subaerial beach diminishes in width as the barrier island retreat continues (Figure 10.12, bottom) and this is the reason why Dean (1987) says armoring can contribute to a diminished width of beach fronting the hardened shoreline.

(3) Seawalls Cause Acceleration of Beach Erosion Rate. Opponents to hardening the shoreline claim that this will increase the local erosion rate. The only way to quantify this claim is through field measurements. A fixed reference (baseline) is needed from which to measure shoreline position (relative to a vertical datum) over time at locations with- and with-out the seawall. The locations should have similar water level and wave climates. As illustrated in Figure 10.1(a), for possible effect (3) if the erosion rate relative to a fixed location (road ¢, dune crest, property line, etc.) is e_n and the erosion rate relative to the fixed seawall location is e_s, and only if it

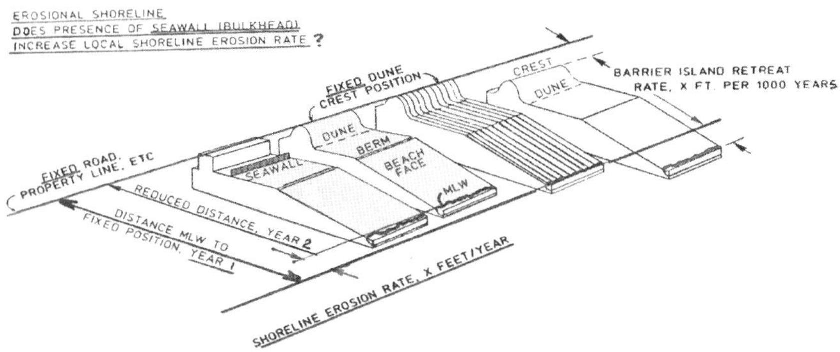

Figure 10.13. Public perceptions of seawall impacts on shoreline erosion rates.

can be shown through field data that $e_s > e_n$, then this result can be cited to support this claim.

Anecdotal evidence is a popular media entry into articles about beaches and seawalls. Beachfront home owners are often interviewed and quoted on how the beach width is diminishing in front of their oceanfront property. Previous owners constructed the rock revetment or seawall and that is now cited and blamed as *...increasing erosion and destroying the beach.* Figure 10.13 schematically displays this common public misconception. The seawall in front of their house is a common reference point to use when looking at the beach, especially for property owners who only spent a few weeks of summer vacation each year at this location. The road used to access the oceanfront property could also be used as this reference location as illustrated in Figure 10.14. Is the road *...increasing erosion and destroying the beach?* (see Section 10.3).

Dean (1987) states that armoring does NOT cause an increase in the long-term erosion rate. See the results of our monthly field measurements over 10 years at Sandbridge, VA (Section 10.3). We also concluded seawalls do NOT cause acceleration of beach erosion rates.

(4) Seawalls Cause Downdrift Erosion. This potential impact was discussed earlier as a possible reason for end-of-wall effects and will not be treated further. Yes, seawalls interacting with the surf zone and longshore transport can contribute to end-of-wall effects when acting as a groin. But they are not the only cause of downdrift changes in shoreline position.

(5) Seawalls Cause Delay in Post-storm Recovery of Beaches. No known data nor physical arguments support this concern (Dean, 1987). See Section 10.3 for data from Sandbridge, VA, that confirmed this conclusion.

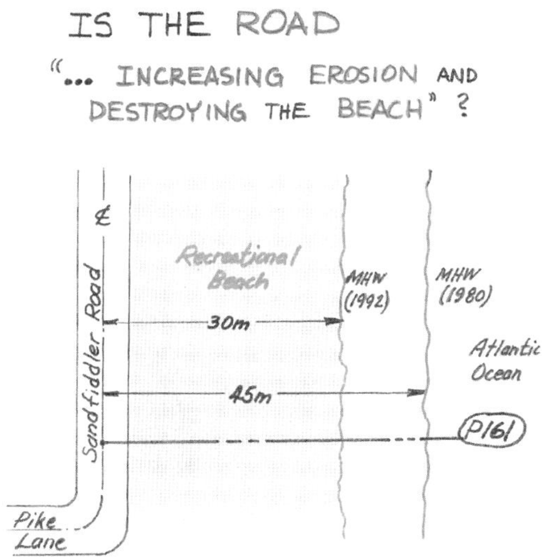

Figure 10.14. Is the road "...increasing erosion and destroying the beach"? (see Section 10.3).

(6) Seawalls Cause Beach Profile Steepening. As shown in Figure 10.12 (middle), coastal erosion will naturally create deeper water at a fixed distance from a reference line (i.e. seawall). This volume deficit was discussed in Chapter 8 for beach nourishment. No known data nor physical arguments support this concern (Dean, 1987).

(7) Seawalls Located Far Landward on Stable Beach Serve No Purpose. This statement was addressed by Dean (1987) who simply said that the seawall provides "insurance" for upland structures against severe storms. It has appeared in the scientific literature by well-meaning but ill-informed individuals who shall remain nameless herein.

(8) Seawalls Cause an Increase in the Longshore Sediment Transport Rate. As schematically illustrated in Figure 10.1(b), if the longshore transport rate upstream of the seawall is Q_1, and then if the seawall created a greater transport rate, $Q_2 > Q_1$, then the additional sediment volume would be deposited downstream of the seawall where the transport rate returns to Q_1. Proponents of this claim argue that the wave reflection from the seawall causes the increase in sediment transport. Dean (1987) provides momentum flux considerations that produce just the opposite result, in that wave reflection reduces the longshore transport in front of seawalls. Since waves come from both directions, the field data to support either view is difficult to locate in the literature.

(9) Seawalls Cause Sand Transport Far Offshore. No known data nor physical arguments support this concern (Dean, 1987).

Dean (1987) also offers the following recommendations for mitigation of the two possible impacts of coastal armoring. The mitigation would be the annual addition of sand to volumetrically compensate for that prevented from being part of the total sand volume in the local system. The key is identifying the annual sand placement volume that just offsets any adverse effects of the armoring. Formulas are given for the volume allegedly trapped by the armoring and the volume blocked by interruption of the possible longshore sediment transport. See Dean (1987) for this theory. See also California Coastal Commission (1996).

The 20th International Conference on Coastal Engineering (ICCE) was held in November 1986, in Taiwan with proceedings published in 1987 where Dean (1987) appears in the literature. That same year, the Coastal Sediments 1987 Conference was held in New Orleans. The "Effects of Seawalls on the Coast" was a subtheme of the conference. Two sessions included papers on the physical aspects of seawall and beach interactions and a panel session was convened to discuss the subject in a full plenary meeting. The great interest shown resulted in the material summarized in the following section. See also Dean *et al.* (1988) in a fine summary article that appeared in Geotimes (May).

10.2.3 *Journal of Coastal Research — Special Issue No. 4 (Autumn 1988)*

The Table of Contents of the *Journal of Coastal Research* (Special Issue No. 4, Autumn 1988) has been reproduced here as Figure 10.15. Co-editors were Nicholas Kraus, Coastal Engineering Research Center (CERC), Corps of Engineers and Orrin Pilkey, Department of Geology, Duke University. Contributors presented material for all four coastal regions of the US (Oregon, Komar and McDougal: California, Griggs and Tait: Texas, Morton: Great Lakes, Wood: and South Carolina, North Carolina and New Jersey, Pilkey and Wright). Missing was Florida and Dr. Dean as a contributor. Weggel suggested a classification system for coastal seawalls, Terchunian discussed permitting and Kraus presented a detailed literature

SPECIAL ISSUE NO. 4 (pages 1-146) AUTUMN 1988

CONTENTS

The Effects of Seawalls on the Beach: An Extended Literature Review	**Nicholas C. Kraus**	1
Seawalls: The Need for Research, Dimensional Considerations and a Suggested Classification	**Richard Weggel**	29
Seawalls Versus Beaches	**Orrin H. Pilkey and Howard L. Wright III**	41
Permitting Coastal Armoring Structures	**Aram V. Terchunian**	67
Coastal Erosion and Engineering Structures: The Oregon Experience	**Paul D. Komar and William G. McGougal**	79
The Effects of Coastal Protection Structures on Beaches Along Northern Monterey Bay, California	**Gary B. Griggs and James F. Tait**	95
Interactions of Storms, Seawalls, and Beaches of the Texas Coast	**Robert A. Morton**	115
Effects of Seawalls on Profile Adjustment Along Great Lakes Coastlines	**William L. Wood**	137

Published by the Coastal Education and Research Foundation, P.O. Box 8068, Charlottesville, Virginia

Figure 10.15. Table of Contents for Special Issue No. 4, *Journal of Coastal Engineering* (Autumn 1988).

review going back to Mathews (1913) and that included Dean (1987). This historically important reference is available online (www.cerf-jcr.org/index.php/) and is 146 pages in length.

Over 100 journal papers, conference proceedings, and technical reports were reviewed by Kraus (1988, co-editor). His conclusions were that

> *The majority of quantitative field studies indicates that seawalls neither accelerate nor enhance long-term erosion of beaches, and no systematic difference in post-storm recovery of beaches with and with-out seawalls was found if there was ample sediment or a wide surf-zone existed.*

He posed most of the same questions as cited in Figure 10.1 (Dean, 1987) and essentially made the same conclusions. He also addressed the longshore bar system in front of seawalls and found mixed results.

The paper title by the second co-editor, "Seawalls Versus Beaches" (Pilkey and Wright, 1988) immediately signals the confrontational content of the submittal. The second possible impact of hardening the shoreline (Figure 10.1(a)) is the beach width diminishing in time. This was the main concern voiced in the position paper by the coastal geologists based on their experience with moving barrier islands as discussed above and will not be repeated here. The term "passive" erosion was used for natural erosion (and barrier island retreat) and the term "active" erosion was used for man-made structural causes. However, Pilkey believes that arguments surrounding this distinction are irrelevant,

> *The real question and controversy is whether 'active' beach degradation occurs. But this argument, and interesting one to us, is of no consequence to the public. The question of concern to the public is whether beaches are degraded in front of seawalls. Whether... (natural) ... conditions are involved or not is beside the point and of no interest to them* (O.H. Pilkey, personal communication, 1990).

Pilkey and Wright (1988) state that there are only three alternatives for managing eroding shorelines (including retreating barrier islands), namely: (1) hard stabilization, (2) soft stabilization, and (3) retreat or relocation. They take the long-term perspective as coastal geologists and their views are based on one type (barrier islands) of the world's shorelines defined in Figure 10.16 (from Shepard, 1976). We shall return to the "retreat" alternative in Chapter 11 and the world's rising sea levels.

10.2.4 *Thirty years of environmental impacts research (1988–2018)*

Kraus and McDougal (1996) updated the literature review of Kraus (1988) as listed in Figure 10.15 with 40 additional references. The main conclusions stated by Dean (1987) remain unchanged. In addition, it was concluded that scour in front of seawalls may be more the result of longshore transport and wave overtopping water that returns beneath the wall than due to wave reflection. The author's own experience with this mechanism of scour depths in front of bulkheads at Sandbridge, VA, is seen in Figure 6.67 as discussed in Section 6.3.6(e) regarding bulkhead design guidance.

10.2.4.1 *Coastal scour prediction methods at seawalls*

The definition for the maximum scour depth, S_m, at a vertical seawall is illustrated in Figure 10.17 (from Kraus, 1988). Note that it is not necessarily at the base of the wall. Fowler (1992) proposed the following equation for estimating maximum scour depth of non-cohesive sediment due to irregular wave breaking on a mild approach slope to fit the data shown in Figure 10.18.

$$\frac{S_m}{(H_{mo})_o} = \left(22.7\frac{h}{(L_p)_o} + 0.25\right)^{0.5}, \tag{10.1}$$

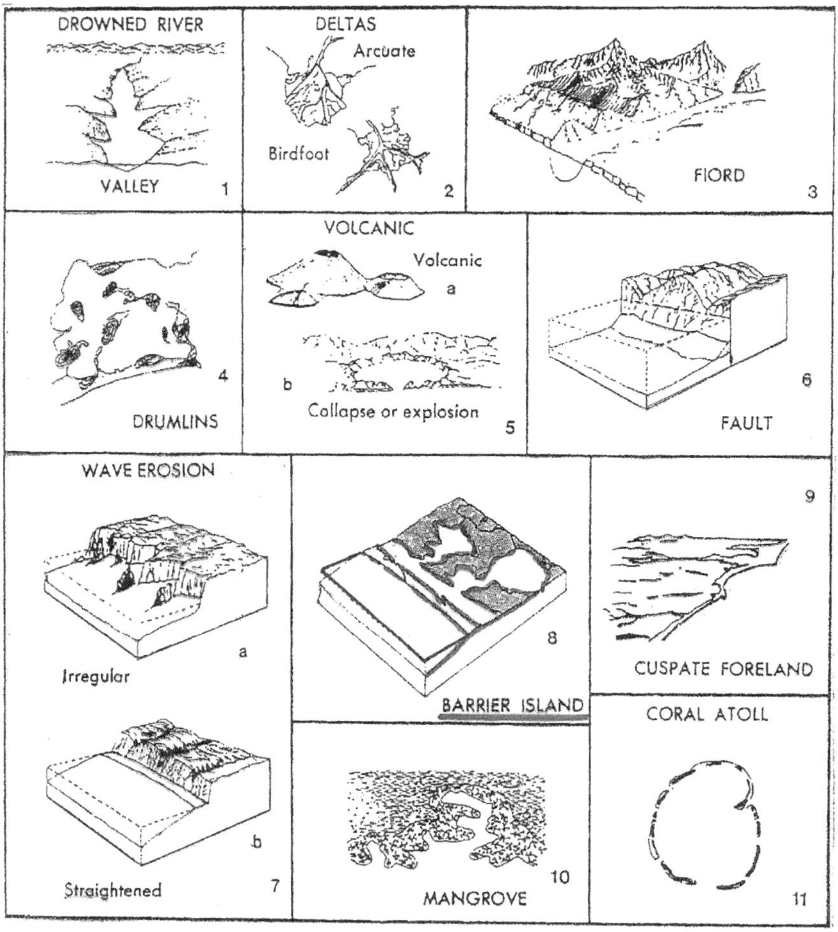

Figure 10.16. Eleven types of coastal shorelines (from Shepard, 1976).

where S_m is the maximum scour depth, $(H_{mo})_o$ is the zeroth-moment wave height in deep water, h is the pre-scour water depth at the vertical wall, $(L_p)_o$ is the deep water wave length depending on peak spectral wave period (T_p).

Note the range of relative water depths in Figure 10.18 which is the very shallow water and that the deep-water wave height is used to scale the scour depth. The rule-of-thumb using the maximum breaking wave height, $H_{\max}$, to estimate the maximum scour depth is still a good approximation

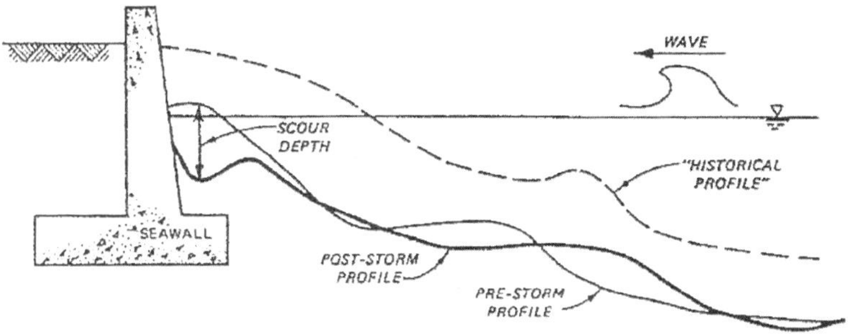

Figure 10.17. Definition of scour depth due to breaking waves at a vertical seawall (from Kraus, 1988).

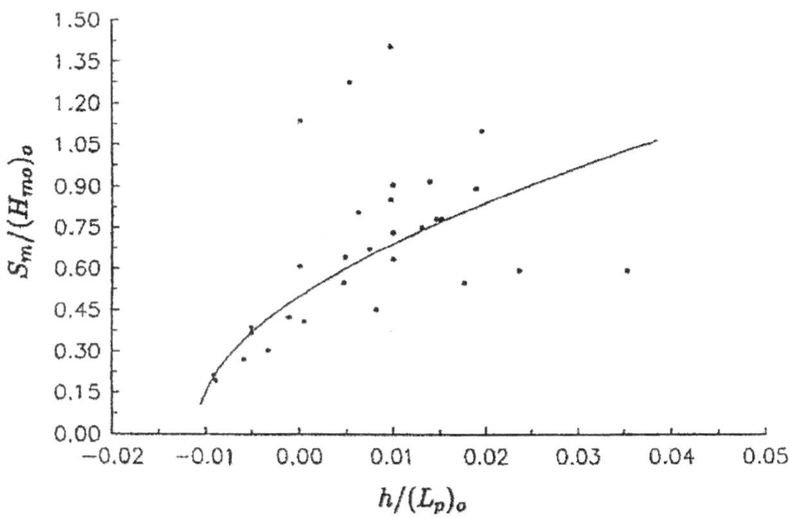

Figure 10.18. Relative maximum scour depth depending on the relative water depth (Fowler, 1992).

to begin.

$$S_m = H_{\max}. \tag{10.2}$$

As shown in Figure 10.17, the maximum scour hole generally is located near the plunging point of the breaking wave as tripped by the scour mound in front of the wall. Waves striking the wall at angles and on non-vertical walls produce less scour. The grain size of the sediment is not a factor. These results were validated in the SUPERTANK experiments as conducted in the large wave tank at Oregon State University. See McDougal *et al.* (1996) for details. They are also discussed in Fowler (1993) and in the CEM (2006, Part VI-5-6).

10.2.4.2 *Additional physical and geological impacts*

The Massachusetts Institute of Technology (MIT) conducted 3D laboratory experiments to study the effect of sea walls on beach erosion (Moody and Madsen, 1995). Scale affects the quantitative results (as illustrated in Figures 10.3 and 10.4) but the qualitative results are still valid. Figure 10.19(a) is the plan view and (b) is the section view with a divider wall separating the sloping-walled side (revetment) and plane beach side

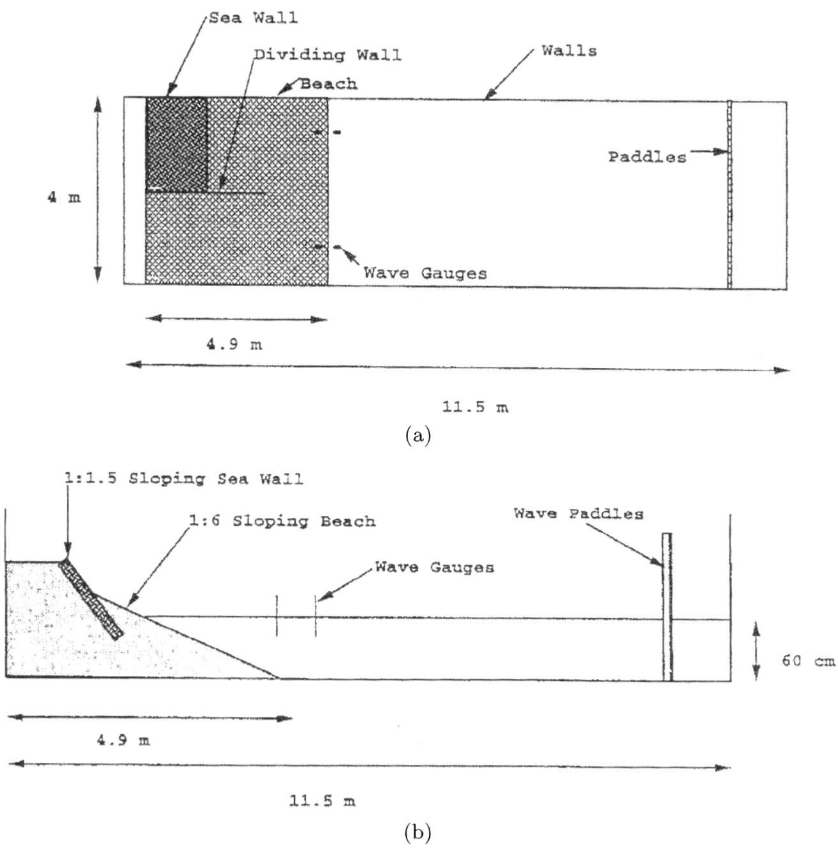

Figure 10.19. (a) Plan view of experimental section with dimensions (from Moody and Madsen, 1995). (b) Section view of experimental section with dimensions (from Moody and Madsen, 1995).

with sand dune the same shape as the walled side. Both are simultaneously impacted by normally directed, irregular waves (JONSWAP spectrum). The sand grain size was 0.2 mm. The conclusions were as follows regards the two beaches: one beach in front of the sloped revetment and one plane beach in front of a sloped sand-dune with no revetment.

- "*The two beaches were nearly identical seaward of the sea wall; there was no additional or excess erosion in the sea wall profile to match the erosion from the dune.*
- *The erosion and recovery rates for the two beaches were almost identical.*
- *During the sequences where waves were able to erode past the foreshore and start eroding the dune, there was some permanent loss from the dune to the offshore region.*
- *Erosion caused the nearshore sediment to move offshore and form a protective berm with nearly identical eroded profiles seaward of the base of the sea wall for both beaches.*
- *During recovery, the sand on both sides was brought back onshore with full recovery on the sea wall half of the beach and partial recovery on the dune half.*
- *Under storm wave conditions, the beach in front of the sea wall was not permanently lost, compared to the large permanent dune erosion of the unprotected beach.*

This small-scale laboratory investigation showed that beaches protected with sea walls behave as well as unprotected beaches under the limiting conditions of normally incident spectral waves. The cross-shore behavior of protected and unprotected beaches is very similar, except under extreme storm conditions where the unprotected beach can be severely and permanently eroded. Erecting a seawall on a beach does not condemn the immediate or adjacent beaches" (Moody and Madsen, 1995).

Griggs (2010) summarized the impacts of coastal armoring with emphasis on loss of sand supply from eroding bluffs and cliffs. Eroding cliffs (or bluffs) are an extensive coastal hazard with El Niño events (elevated water levels) and storm waves contributing to continuing erosion of the shoreline. About 10% (110 miles, 177 km) of the state of California coast has been armored on the Pacific Ocean. In California, 70–95% of beach sands come from rivers and streams (recall the "Beach A River of Sand", movie/video discussed in Chapter 2) with the remainder coming from eroding cliffs. The impact of coastal armoring on the sand budgets for two littoral, sediment-budget cells was studied by the California, State

Coastal Conservancy (2002). For the Santa Barbara littoral cell, cliff erosion under natural conditions contributed 0.4% of the total littoral material and for the Oceanside cell, cliff erosion supplied $\approx$ 10%. As noted above, in Figure 10.7, Griggs *et al.* (1997) found no additional "active" erosion due to the presence of armoring from his research efforts over eight years at Monterey Bay, CA. California has begun an active program to mine the sand volume located within water resources reservoirs on the rivers and transfer it to the shoreline to aid the beaches. This volume is much larger than that impacted by coastal armoring. See the report "Cumulative Losses of Sand to the California Coast by Dam Impoundment" (Slagel and Griggs, 2008) for details on this important study for the Pacific Ocean coast of the US.

Back on the Atlantic Ocean coast of the US, field observations in the aftermath of Superstorm Sandy (October 2012) revealed clear differences in storm impacts on two neighboring beaches on the New Jersey shore. A relic 1,260 m long seawall (circa 1880) buried beneath the dune along one shoreline (Bay Head, NJ) lessened the wave damage on oceanfront structures compared to the adjacent shoreline (Mantoloking, NJ) with no buried seawall. Over wash, sediment deposition, and dune breaching differences were also found between the buried seawall shoreline and that with only the dune (Irish *et al.*, 2013). A follow up study using the numerical model, XBeach (Roelvink *et al.*, 2009), driven by Superstorm Sandy water level and wave conditions found that the buried seawall protected Bay Head by dissipating wave energy during the peak of the storm. And, the buried seawall protected Bay Head from the rapidly increasing bay water levels that flooded the back-barrier region of Barnegat Bay. When the seawall was removed from the numerical simulation, the dune elevations were lowered permitting flooding of the bay that completely destroyed the remaining dune system. See Smallegan *et al.* (2016) for complete details. The buried seawall increased the resiliency of the Bay Head community by preserving the dune system. See Section 7.7.1(b) for a discussion of the Dam Neck, Virginia, design with a buried seawall beneath the dune.

Finally, the internet has been readily employed for both literature searches and for feedback from list-serve readers regarding the impacts of seawalls on adjacent beaches. In January, 2015, the Connecticut Department of Energy and Environmental Protection posted the following questions under the general title "Seawall Compensation" (Yue, 1/26/15, Univ of Delaware, Coastal_List):

(1) Examples/case studies of in-kind compensation for hardened shoreline structures;

(2) Examples/case studies of in-lieu fee-based compensation;
(3) Examples/case studies of incentive-based measures.

Responses were received from Australia, New Zealand, Italy, North Carolina, and California with the latter siting the California Coastal Commission study on seawall-shoreline impact assessment and associated mitigation fees (www.coastal.ca.gov). In general, there was little else presented to answer these questions. Section 10.3 should also be consulted in this regard.

10.2.4.3 *Ecological impacts — Estuarine bay areas*

Seawalls are the coastal flood damage mitigation alternative receiving the most attention. A review of the literature over the past 30 years (1988–2018) reveals a large increase in the number of references regarding ecological impacts of hardening shorelines in tidal estuaries. Figure 10.20 illustrates the results of a survey of scientific publications using three key words in the titles: (1) resilien*, (2) infrastructur*, and (3) flood* where the "*" is a wild character in the title (from Pearson *et al.*, 2018). The horizontal axis lists the names of 20 subject categories showing that "Engineering" and "Environmental Sciences & Ecology" publications lead the way, each

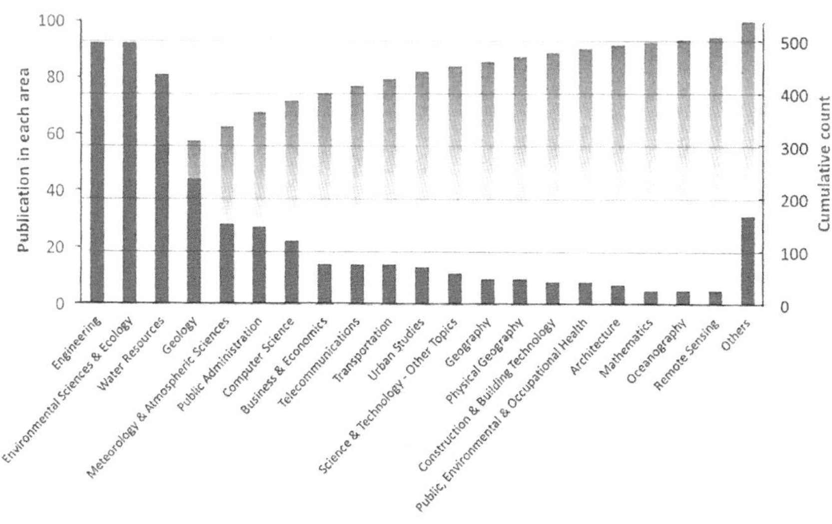

Figure 10.20. The number of scientific publications in 20 areas with resilience, infrastructure, and flooding in the title (from Pearson *et al.*, 2018).

with over 90 publications. This article was published in the *Environmental Systems and Decisions* journal (Springer, https://link.springer.com/journa l/10669) which began in March 1981.

Assessment and management of the ecology impacted by coastal-engineered, hazard mitigation alternatives depend on three main factors:

(i) the distribution of the sensitive habitats;
(ii) the tolerance of these habitats to changes in environmental conditions; and
(iii) the spatial and temporal distribution of the environmental change produced by the alternative.

Barrier beaches are *nature's* mechanism for protecting the bays, lagoons, and tidal estuaries that lie behind them. Together, barrier beach-bay systems make up about 35% of the US coastline. The reduced wave energy environment permits the retention of cohesive sediment and grasses to survive in the tidal marsh areas. The estuarine system consists of adjacent ocean water, tidal inlet, tidal estuary, bay, lagoon, and freshwater inflow. Salinity and current patterns over space and time dominate the life cycles of many marine animals. Simply stated, the fishes and crustaceans require various salinity and energy levels (waves and currents) during various stages in their development cycles. They also require the nutrients found in the waters and marshes of the estuary to sustain growth. The tidal estuary and their adjacent "wetlands" are nature's breeding grounds for much of marine life as we presently know it. They are also important nurturing areas for water fowl and breeding grounds for wading birds. The ecological productivity of the estuarine coastal zone is the highest of all land masses on the planet earth.

In 1979, the US Department of Interior (Fish and Wildlife Service) estimated that over a 20-year period (1950s to 1970s) roughly 460,000 acres (186,000 hectares) of wetlands were lost each year. The majority loss ($\approx$80%) was in Louisiana, Texas, and Florida due to conversion into open bays and sounds. In Louisiana, the losses were due to relative sea level rise, subsidence of the coastal plain, levee construction, channelization, and oil and gas extraction. Florida's coastal wetlands loss was primarily in urban areas due to dredging and filling operations for development.

Coastal marine biologists, ecologists, and wildlife experts led efforts to enact legislation against marsh filling that produced the estuarine/marsh mitigation act (US Congress, 1982) requiring the Corps of Engineers to review during the permits process. The basic rules for marsh replacement of lost wetlands with newly created areas of equal (or greater) value remains in place today. In 1988, it was extended to nontidal (upland) wetlands and was encapsulated in the phrase "no net loss of wetlands' during the presidential campaign that year.

10.2.4.4 *Estuarine/marsh loss due to barrier island migration*

Development on barrier islands also came into focus in the late 1970s within the Department of Interior. A study of over 40 US government programs that impacted development together with the coastal geomorphologists "barrier island migration" paradigm (see Figure 10.11) and position paper (Pilkey *et al.*, 1982) resulted in the Coastal Barrier Resources Act (CBRA) of 1982 that established a system of 186 coastal barriers along the Atlantic Ocean and Gulf of Mexico. This law prohibits most Federal expenditures that encourage development of barrier islands (e.g. for road, water, and sewer systems).

However, Leatherman (1988) has shown that two mechanisms produce barrier island migration, namely,

(1) roll over due to wash-over events during storms (Dolan and Lins, 1987; Pilkey *et al.*, 1982);
(2) the creation of new tidal inlets and sand trapping in the tidal deltas.

Leatherman (1988) concluded:

> *Over-wash is not the dominant process by which most barriers move landward since the amount of sediment transported by this means is too small. Inlet formation, when tidal currents cut a channel below sea level moves far greater quantities of sediment into a lagoon over the long term and is the major process for barrier migration.*

One example for this conclusion is shown in Figure 10.21 for the rotation of the northern end of Assateague Island, Maryland (MD), below Ocean City, MD, on the Atlantic Ocean (Leatherman, 1987). Pre-inlet, post-inlet (with jetties) and 1980 conditions are shown with shoreline positions in the 1840s and in 1980. Creation of the new inlet caused a three-fold increase in the average recession rate (3–10 m/year) that was also due to construction of the jetties and absence of sediment bypassing to mitigate jetty impacts. The US National Park Service are caretakers of the Assateague Island National Seashore and benefits of several alternatives are discussed to mitigate erosion in Leatherman (1987). But what about Sinepuxtent Bay that lies behind the Assateague Island as seen in Figure 10.21?

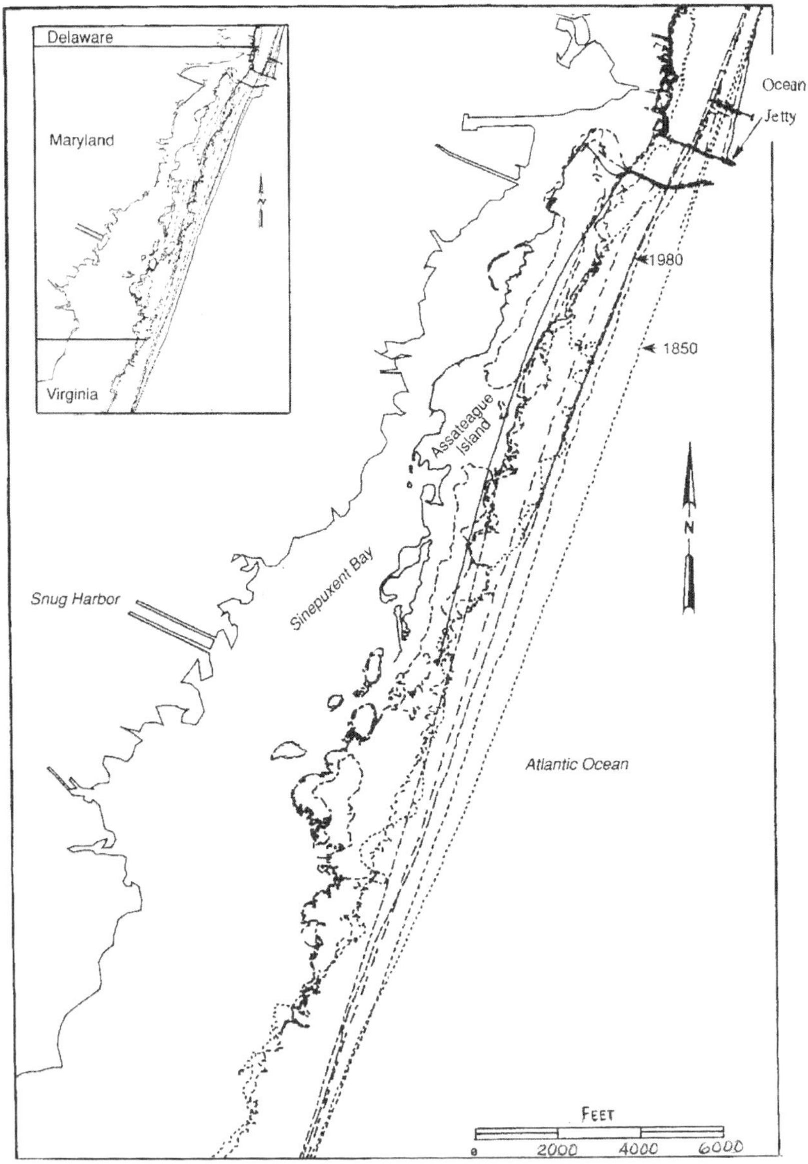

Figure 10.21. Rotation of the northern end of Assateague Island between 1850 and the 1980s with the resultant shrinkage of Sinepuxtent Bay (from Leatherman, 1987). The top and bottom edges of the figure are the borders with Delaware and Virginia, respectively.

Basco (1993) found that about 925 acres (1.45 sq. miles, 375 Hec) of surface bay area has been lost between 1840/1850 and 1980 due to island rotation. This is almost 25% of the original area of Sinepuxtent Bay in about 135 years. In this analysis, the barrier island width (400–700 ft, 122–215 m) was assumed constant with a fixed mainland bay-shore position as found by Leatherman (1987). Similar studies for Metompkin Bay behind Metompkin Island, Virginia (VA), found a net loss of bay surface area of 17 acres (7 hectares) in 135 years. Additional studies of Pleasant Bay behind Nauset Beach Spit, Chatham, Massachusetts (MA), and Back Bay behind Sandbridge Beach, VA, were discussed (Basco, 1993) where only coastal erosion and new inlet formation mitigation alternatives are considered. The loss of vegetated wetlands (coastal marshes and benthic communities) were not mentioned.

The "Barrier Island Migration" paradigm as embedded in the Coastal Barrier Resources Act (CBRA) preceeding Public Law 97–348 (1982) and the "No Net Wetlands Loss" paradigm as enacted by Congress in 1982 are in conflict. Basco (1993) argued that we need to base public policy on both the revelations of coastal geological and coastal biological/ecological science. We need to consider both:

(1) the economic and ecosystem value of nature's ecosystem *behind* barrier islands, and
(2) the economic value of human artifacts *on* barrier islands,

when making decisions regarding the stabilization of moving barrier islands. Benefits to humans regarding flood damage reduction and beach recreation and benefits to nature via saving of the most biologically productive marine areas on earth in the estuaries, bays, and lagoons behind barriers must be considered. More on this will be discussed in Chapter 11 when sea level rise impacts on barrier islands are reviewed.

10.2.4.5 *NSF — Long-Term Ecological Research (LTER)*

In 1980, the US National Science Foundation began funding long-term ecological research (LTER) sites to support ecological discovery on the influence of long-term and large-scale phenomenon. There are now 28 sites and almost 40 years of continuing observations made publicly available through the LTER Network (https://lternet.edu). The large-scale experiments provide starting conditions for process-level studies to test and parameterize models for future unknown conditions. Researchers working on short-term projects have trained hundreds of graduate students on collaborative and interdisciplinary teams.

In September 2018, a Special Issue of the publication *Estuaries and Coasts* (2018, Vol. 41 No. 1) with 15 articles focused on "Impacts of Coastal Land Use and Shoreline Armoring on Estuarine Ecosystems". Prosser *et al.* (2018) gives an introduction to this Special Issue and lists over 200 references, but not all regarding shoreline armoring and ecosystem impacts. Armoring is said to influence estuarine and nearshore marine ecosystem functions by degrading water quality, spreading invasive species, and destroying ecologically valuable habitat. Detrimental effects on the ecosystem function impact flora and fauna that are both ecologically and economically important, e.g. oysters, crabs, and shellfish. The majority of papers focus on the Chesapeake Bay region (US) that provides impacts of human activities on the local shoreline and the entire watershed ($\approx$167,000 km^2) scales. These papers focus on natural watershed land use versus armored shorelines on ecosystem properties and processes as well as on natural resources.

One LTER program at the University of California Santa Barbara studied the effects of seawalls on open-coast beaches and showed ecological effects on birds. The University of Georgia project studied small-scale armoring on salt marshes. And at the University of Virginia, Virginia Coast Reserve, the LTER concentrated on the use of constructed oyster reefs and living shorelines as coastal hazard mitigation alternatives. These three studies combined the different types of armoring and variety of sediment ecosystems to create and evaluate a new conceptual model. It examines six categories of ecological response for two armoring attributes: (1) the degree of exposure energy (water level, tides, waves, and currents) and (2) the degree of this hydrodynamic energy that is felt behind the armoring (if any). Over 60% of the 88 studies reviewed were from very low-energy environments (salt marshes, tidal creeks, mangroves), with 25% at middle-energy levels (harbors, tidal entrances, large estuaries) and the fewest, 15% at high-energy, open-coast, sandy beaches. The focus was mainly on changes in habitat and species distribution. Armoring impacts on nutrient cycling, connectivity, and productivity were not studied. The LTER director at the NSF stated (Garrison, 2018):

> *This is one of the first attempts to assess how engineering structures on beaches and other sedimentary environments affect the biota that inhabits these locations. With 40 percent of the nation's human population living in coastal counties, the study is certainly timely.*

The coastal engineering community should keep abreast of the latest results from the NSF-LTER.

10.2.4.6 *Summary — Literature review/environmental impacts*

The environmental constraint for design of coastal hazard mitigation and adaptation alternatives (Chapter 3) continues to evolve over time. Alleged physical impacts on sandy beaches now include ecological impacts on both open-coast and tidal estuarine shoreline sediments, biota, and marine life. Many believe that armored shorelines are inherently bad for the environment. However, there are many locations where both armoring (gray) and beach/vegetation (green) shorelines coexist. Figure 10.22 (top) shows a view from a backyard in the Larchmont neighborhood of Norfolk, VA, looking at the top of a rock revetment, the pier and waters of the Lafayette River, tidal estuary with a 1 m (3 ft) tidal range. Walking out on the pier and looking back (bottom) we see the house, backyard, the rock revetment, and the spartina alterniflora tidal marsh grasses in front of the revetment. The rock revetment and tidal marsh coexist at this location. In fact, the tidal marsh reduces the wave energy impacting the revetment that decreases maintenance costs.

Figure 10.23 (top) shows graduate students in coastal engineering at Old Dominion University hard at work to sort new rocks and then transport them for placement on the revetment (bottom). This hands-on approach reinforces the class lecture material. The author is standing on a vertical, stone seawall in Figure 10.23 (bottom) supervising rock placement. A marsh also exists in front of this wall.

Graduate students in Coastal Engineering at Old Dominion University also learned much about how seawalls interact with adjacent beaches from field work as described in Section 10.3.

10.3 Sandbridge Beach — Seawalls and Beaches Research

10.3.1 *Boundary conditions and long-term shoreline change rates*

Coastlines are either stable, accreting, or receding when viewed on a long-term basis relative to a fixed reference frame. The City of Virginia Beach, Virginia, is located on the Atlantic Ocean below the entrance to the Chesapeake Bay as shown in Figure 10.24. All three shoreline trends exist from the Virginia/North Carolina border to the entrance to the Bay. The relationship

Figure 10.22. Example of rock revetment and tidal marsh coexisting in low energy environment (Author's photos).

Figure 10.23. Old Dominion University graduate students in Coastal Engineering learning about maintenance of rock revetments (Author's photos, circa 2000).

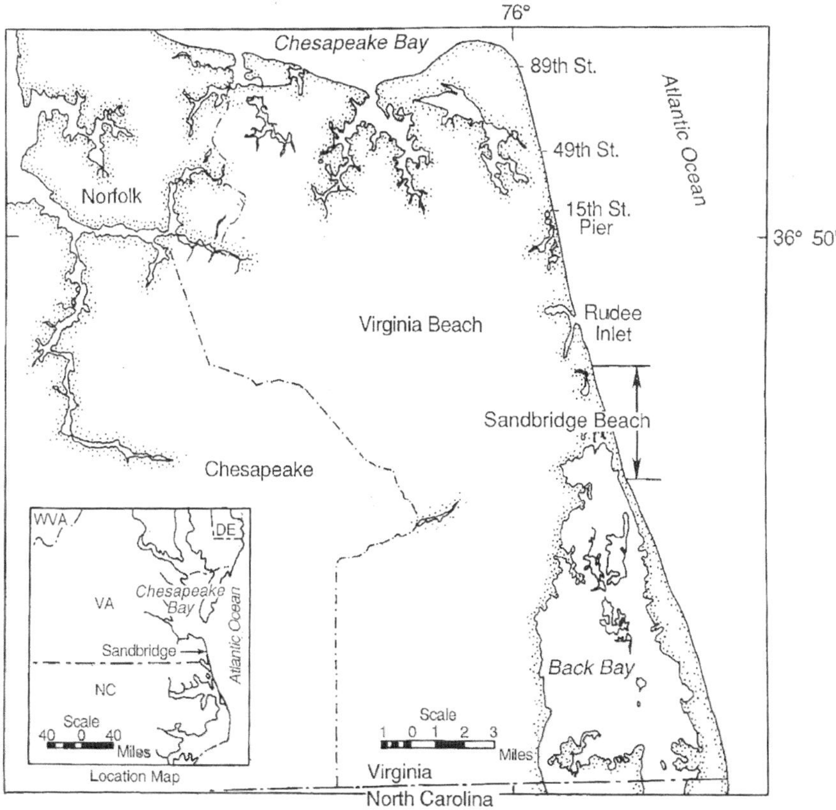

Figure 10.24. Location map for Virginia Beach, Virginia, and the subdivision, Sandbridge south of Rudee Inlet. A tourist beach north of Rudee Inlet has been in existence since the 1880s (from Basco, 1991).

between the offshore bathymetry, resulting wave climates and long-term shoreline change rates over 126 years is documented in Basco (1991). Offshore bathymetry (2-m isobath interval) is displayed in Figure 10.25 (middle) and Figure 10.26 shows the relatively flat profile at the tourist beach (Section 290 at 32nd street) relative to the steep profile at Section 220 in Sandbridge. As a result, breaking wave heights, $\overline{H}_b$, vary significantly along this shoreline.

Figure 10.25 (right) displays the average breaking wave heights for the year 1982 (from Wright *et al.*, 1987) using 55 different wave climate combinations (height, period, direction) and the computer model RCPWAVE

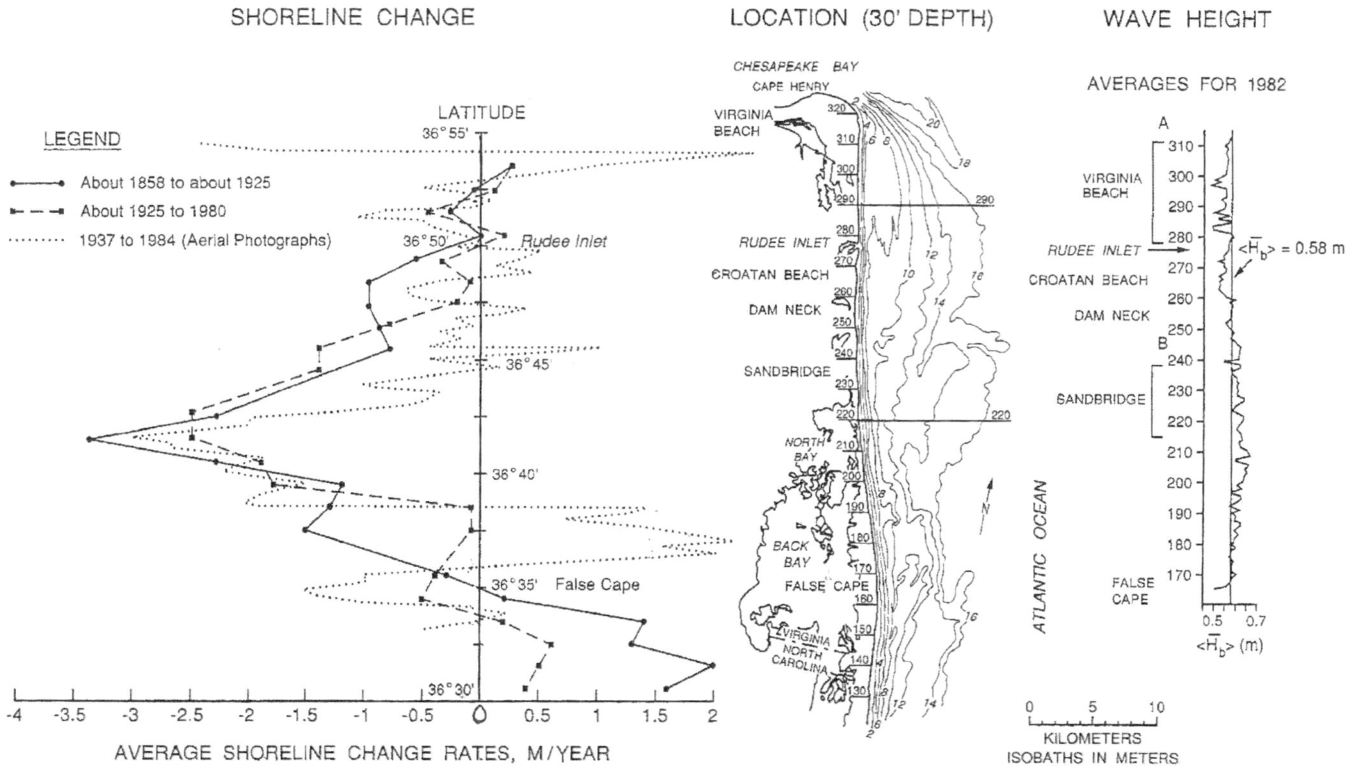

Figure 10.25. Composite of shoreline change rates (left), bathymetry (center), and breaking wave heights (right) and their variation along the 40 km (25 mi) shoreline of the city of Virginia Beach, Virginia (from Basco, 1991).

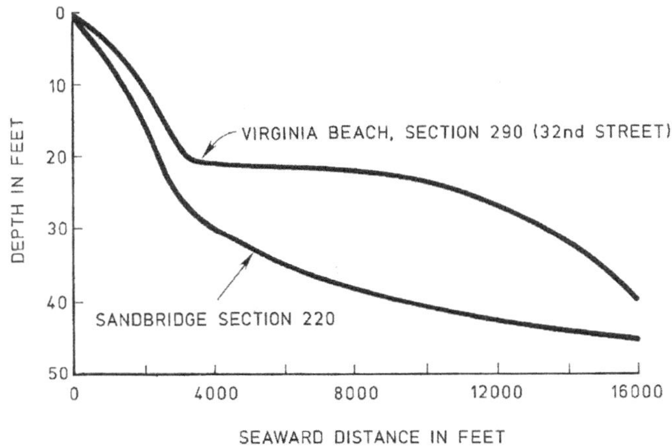

Figure 10.26. Nearshore profiles at Virginia Beach (Section 290) and Sandbridge (Section 220).

(Ebersole *et al.*, 1983). The mean for the entire study region was 0.58 m (1.9 ft) and Figure 10.25 (right) displays below-average conditions along the flat profile at Virginia Beach and the above-average breaking wave conditions at Sandbridge beach.

This combination of nearshore bathymetry and resulting breaking wave heights is responsible for the resulting long-term shoreline change rates as reproduced here in Figure 10.25 (left). The solid line is from Everts *et al.* (1983) from 1858 to 1925, the dashed line is from Everts *et al.* (1983) from 1925 to 1980 and the dotted line is from aerial photographs from Dolan (1985) for 1937 to 1984. Although variability exists, the pattern is clear. The tourist beach above Rudee Inlet has been roughly stable (some recession and accretion), while the 126-year data average (all sources) is about −3 m/year (≈−10 ft/year) recession at Sandbridge. This is attributed to the offshore boundary conditions producing a nodal point for sediment transport at this location. [*This same information has been employed in Chapter* 7 *as an example of how to calculate a sediment budget* (*see Figure* 7.20 *same as Figure 10.25 and Figure* 7.21 *for the sediment budget*).]

Bulkhead construction at Sandbridge began in 1978. The presence of this "seawall" is not responsible for a long-term average erosion rate of −3 m/year. If we neglect the offshore boundary conditions when making field studies of "armoring" sections of shorelines versus "natural" sections, we can reach completely misleading results. Thieler *et al.* (1992) wrote a discussion of Basco (1991) regarding offshore boundary conditions and shoreline change. Many of the same claims as found in Pilkey and Wright (1988) and mentioned earlier for the term "passive" erosion as used for natural erosion (and barrier island retreat) and the term "active" erosion as used for man-made structural causes were used in this discussion. But their main concern was stated as

> *...do seawalls degrade recreational beaches?* (Thieler *et al.*, 1992).

To answer this question, a 10-year research project was conducted at Old Dominion University for Sandbridge beach.

10.3.2 *Seawalls and beaches research at Sandbridge, Virginia*

10.3.2.1 *Geologic and hydrodynamic setting*

Sandbridge beach is located about 25 km (15 mi) south of the entrance to the Chesapeake Bay (Figure 10.24) and is literally a "sand bridge" about 7.7 km (4.8 mi) long and 250 m (825 ft) wide at its narrowest point. It lies at the northern end of a 120 km (75 mi) long barrier strip which teminates at Oregon Inlet, North Carolina. This barrier beach system protects the Albermarle Sound and Currituck Sound in North Carolina and Back Bay in Virginia. Figure 10.27 is a postcard view of Sandbridge Beach with Back Bay and marshes in the background.

The central section of Sandbridge is the terminus of an elevated topographic ridge dividing drainage towards the north and south elevations which are about 0.6 m (2 ft) lower than the middle section. Average, median grain size is 0.25 mm (Bellomo, 1993; Wright *et al.*, 1987). A clay layer lies about 3 m (10 ft) below mean sea level and clay outcroppings have been observed after winter storm (Bellomo, 1993). Barrier islands are not all uniform in elevation along their length and not completely sand volumes for analysis.

Mean tidal range is 1.04 m (3.4 ft). US Army Corps of Engineers' (1985) analysis of tidal gauge records yields a storm surge level of 2.65 m (8.7 ft) above the National Geodetic Vertical Datum (NGVD) datum for the 1% annual chance storm event (see Chapter 2, Figure 2.3, for exceedance frequency curve for Sewells Point tide gauge in Norfolk used for analysis of tidal data for Virginia Beach oceanfront).

Figure 10.27. Aerial view of Sandbridge Beach, City of Virginia Beach, VA (courtesy Zazzle, Inc.).

10.3.2.2 *Seawall construction history*

Sandbridge road to the Sandbridge subdivision on the oceanfront was first constructed in the early 1950s. In 1978, five homeowners banded together and built a timber, sheet-pile bulkhead, 137 m (450 ft) long at Station 148 as shown in Figure 10.28 (inset) and within the elevated, Middle Section in Figure 10.29 (*In what follows, we shall use the term "seawall" when in reality they are only "bulkheads"*). The stationing system extended 25,200 ft (7,680 m) to Station 252 + 00 (ft) at the South end near the Sandbridge pier. The survey Baseline was the centerline of Sandfiddler Road so that this first seawall was 89 ft (27 m) seaward of the Baseline. By the end of 1987, about 700 m (2,300 ft) of seawalls existed. Between 1988 and 1990, an additional 3,917 m (12,850 ft) of seawalls were built. Figure 10.30 presents the chronology of seawall construction at Sandbridge. By the end of 1995, there were 4,740 m (15,550 ft) of walled shoreline at 16 separate locations, which accounted for 61.6% of the total shoreline length of 7,680 m (25,200 ft).

Seawalls at Sandbridge are built to protect septic tanks, driveway concrete slabs, and property at ground level since most are elevated above the 100-year flood elevation as established by FEMA for flood insurance. They

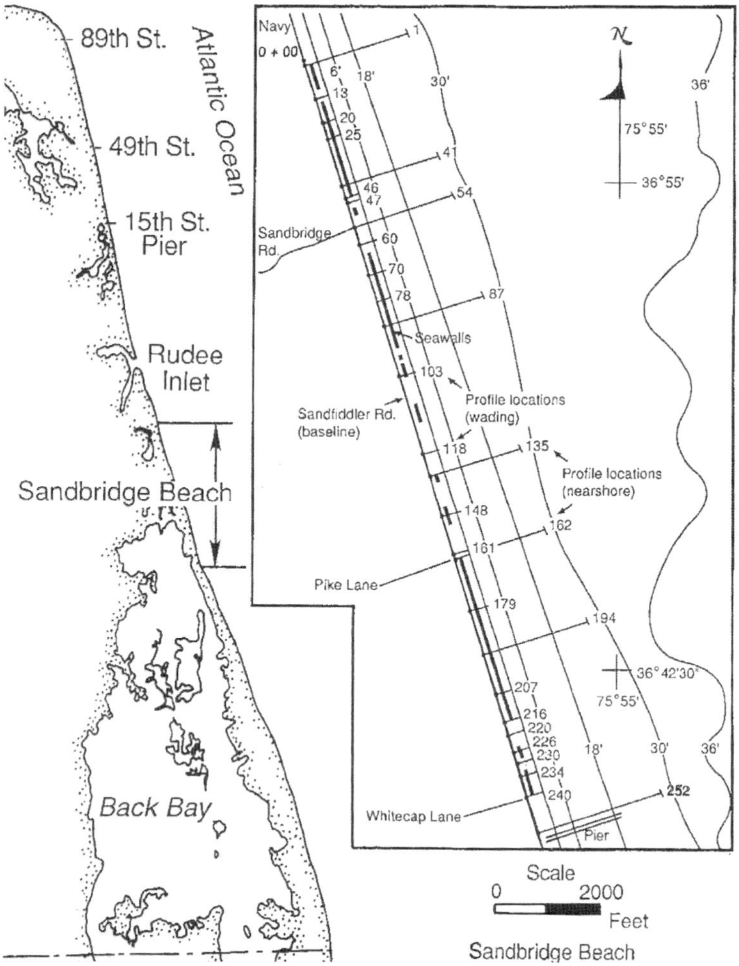

Figure 10.28. Seawall locations at Sandbridge from North (Station 0 + 00) to South (Station 252 + 00) end.

are built on private property using private funds and average seawall costs were US$ 40,000 (dollar value in 1995) per 27 m (90 ft) lot widths. The seawalls were built to mitigate damage to property behind the walls, not to protect the beach. Private property boundaries on deeds extend to the MLW elevation in the ocean but owners have never restricted general use of beaches in front of their seawalls. These seawalls are actually bulkheads constructed with timber, steel, and aluminum sheet-piles and have been

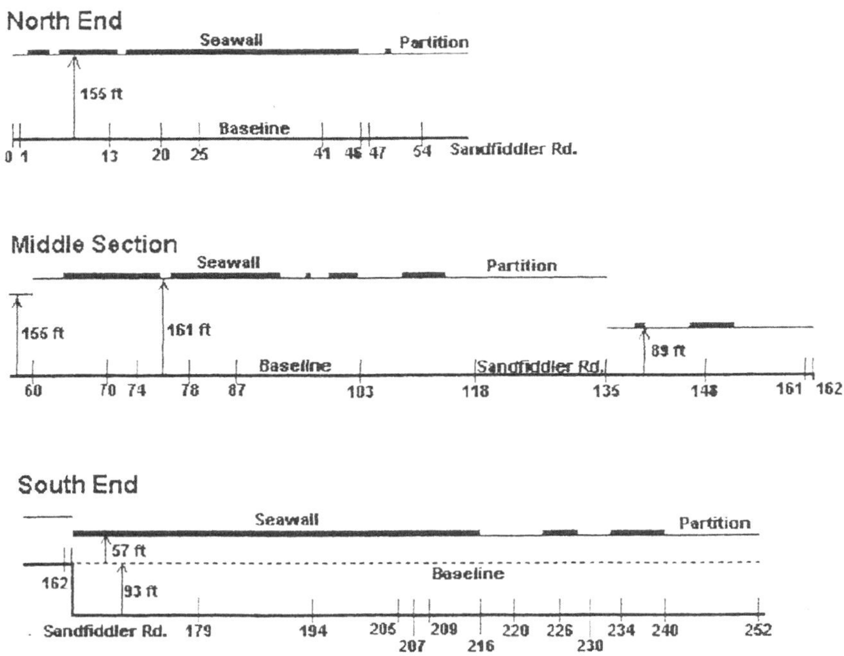

Figure 10.29. Seawall locations at Sandbridge along three sections with stations and distances from baseline (centerline of Sandfiddler Road).

damaged over the years from coastal storms. But that is another story. Figure 10.31 displays concrete, timber, steel, and steel with timber piles as typical seawalls constructed at Sandbridge Beach, which is a public beach in Virginia Beach.

10.3.2.3 *Datums, ODU surveys, and datasets*

The City of Virginia Beach's vertical datum (0.0) in the 1990s was mean sea level (MSL) in 1929 and equivalent to the NGVD with 1972 adjustment. Elevations reported below are relative to this datum. The city's current vertical datum and conversion is NAVD88 $= -0.80 +$ NGVD (72 Adj) in feet at Sandbridge (St. John, 2018, personal communication).

Old Dominon University began monthly measurements of nearshore beach profiles at 28 stations in August 1990 at locations shown in Figure 10.29. A limited number of profiles at stations 74, 107, and 131 were taken by the City of Virginia Beach. All 31 profile datasets are found summarized in Basco *et al.* (1997) ending in September 1995. Figure 10.32

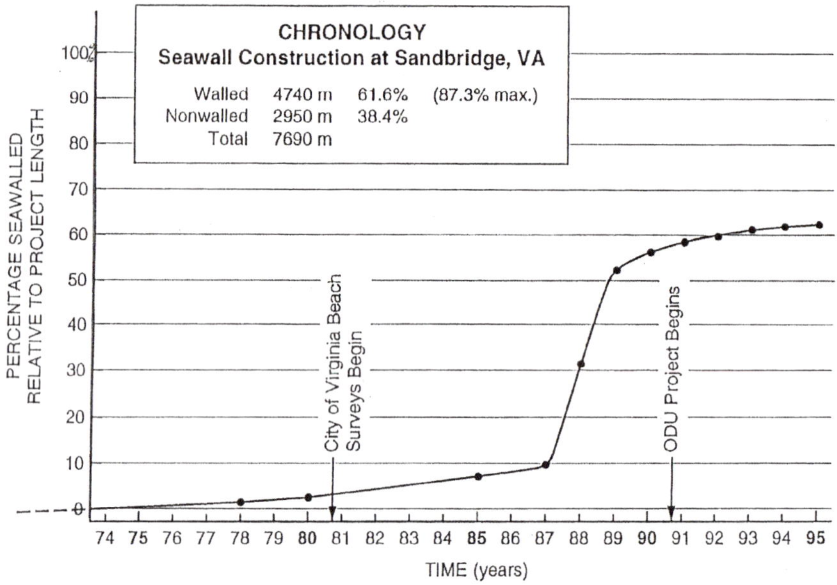

Figure 10.30. Chronology of seawall construction at Sandbridge, VA, up to 1995.

Figure 10.31. Typical type of bulkheads and materials used at Sandbridge Beach, a public beach in Virginia Beach (Author's photos).

Figure 10.32. The ODU survey crews using dumpy level, 25 ft telescoping rod, and cloth tape for measuring dune/beach profile elevation and distance pairs seaward from the survey Baseline to below MLW depth (Author's photos).

displays photographs of survey crew members using dumpy levels, a telescoping survey rod (25 ft) and cloth tape to measure and record distance-elevation pairs starting at the Baseline in the center of Sandfiddler Road. Each profile extended about 120 m (400 ft) seaward to elevations below MLW depths, which were −0.37 m (−1.2 ft) below vertical datum. In September 1995, there were 13 ODU profiles at non-walled locations, 11 profiles at walled locations, and four profiles at end-of-wall locations. Figure 10.32 (upper, middle) displays station 46, which was an end-of-wall location. Stations 13 and 25 had seawalls constructed during the five-year project.

Beach profile data were also obtained from regular surveys of the City of Virginia Beach since October 1980 and from Hardaway and Thomas (1990). Some stations have data from all three datasets. All survey data was archived in the Corps of Engineers' Interactive Survey Reduction Program (Birkemeier, 1984) and totaled more than 2,700 profiles for 53 locations over the full 15 years on record (October 1980–September 1995). Not all were employed in this analysis, because at some locations they were not surveyed regularly or for less than one year.

10.3.2.4 *Profile definitions and analysis methods*

Figure 10.33 displays 96 surveys at Profile (station) 25 with some before and some after wall construction in the Fall 1988. A similar "envelop" plot for Profile 252 at the southern end is presented in Figure 10.34 which is a dune/beach, non-walled location.

To quantify profile change in space and time, five profile parameters are defined as depicted in Figure 10.35. These five parameters are:

(1) V_S is the volume (per unit shoreline length) seaward of the seawall or partition location,
(2) V_L is the volume landward of the seawall or partition location,
(3) V_T is the $V_S + V_L$ = the total volume beneath each profile and bounded by the baseline and the MLW intercept.
(4) P is the horizontal distance from the Baseline to where the profile intesects the MHW elevation, and
(5) E_B is the "berm" elevation as the vertical distance from the 0.0 datum to the profile elevation at the seawall or partition.

The partition location is imaginary. Where no walls exist, the imaginary "partition" is the extension of a shore parallel line connecting adjacent

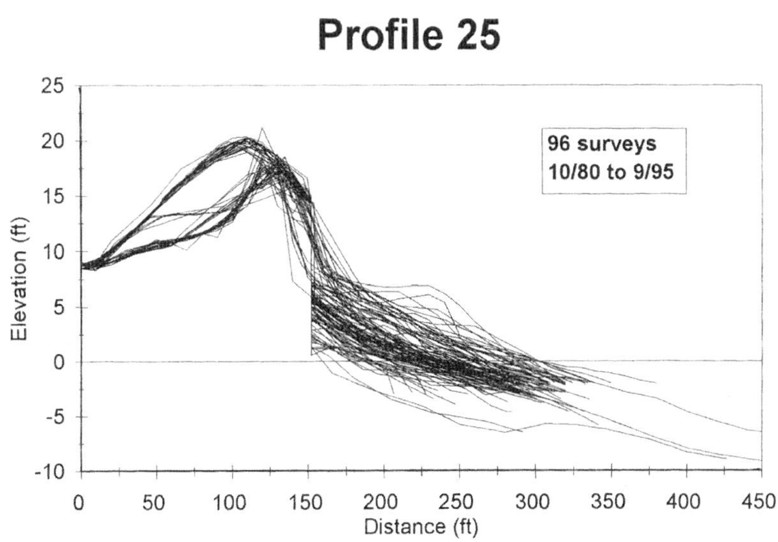

Figure 10.33. Subaerial beach profiles at location 25 between October 1980 and September 1995.

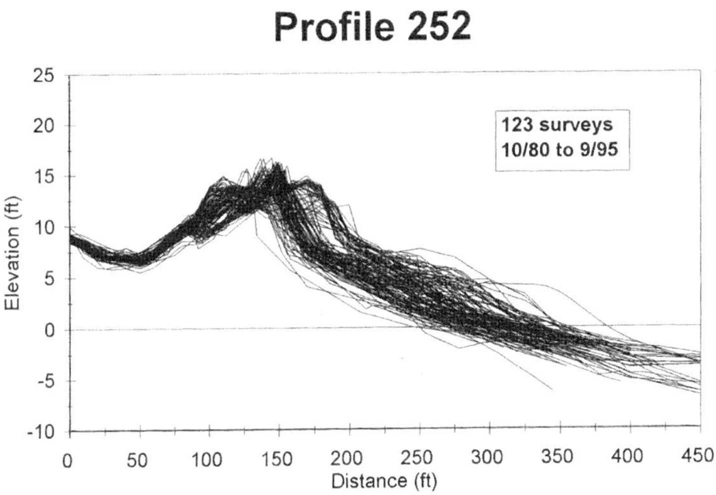

Figure 10.34. Subaerial beach profiles at location 252 between October 1980 and September 1995.

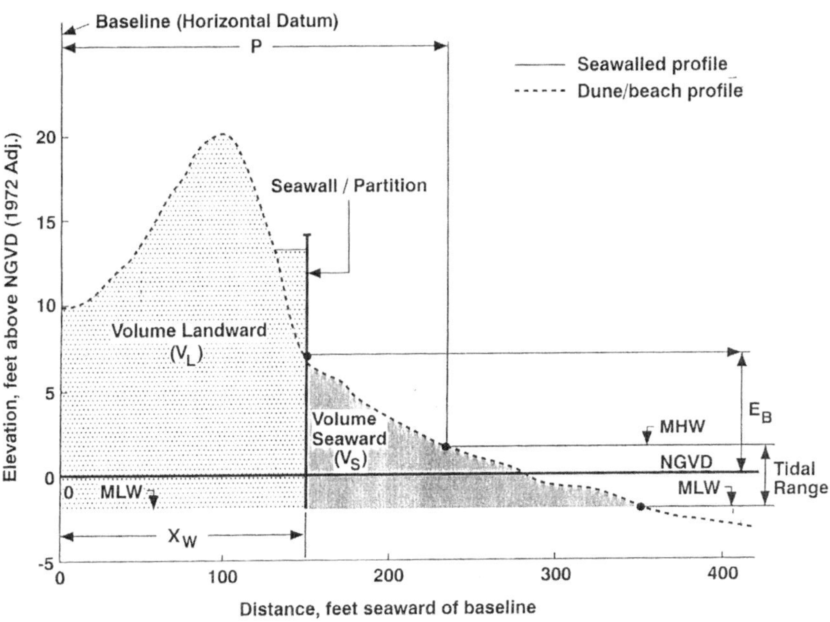

Figure 10.35. Definition of beach profile parameters, V_S, V_L, P, and E_B.

walls. The MHW elevation is +0.67 m (2.2 ft) above vertical datum. Volume change is the true measure of shoreline erosion (or accretion) at road or wall locations where the shoreline position P does not change.

Three basic analysis methods were employed to quantify changes in the profile parameters, namely,

(1) Weighted Average Method (WAM),
(2) Sectional Weighted Average Method (WAMSECT),
(3) Individual Profile Method (IPM),

with linear regression to determine the time rate-of-change for each method.

The WAM used only the 28 ODU profile locations for the entire study area and divided into either walled (15) or non-walled (13) locations. Each profile represented a length of shoreline for that type. For example, Figure 10.36(a) shows Profile 1 (P1) at the north end representing 54 m (177 ft) of non-walled shoreline. End-of-wall profiles were counted as non-walled and Figure 10.36(b) illustrates how a 157 m (515 ft) length of non-walled shoreline between walls was represented by P46 and P47 each 78.5 m (258 ft) in shoreline length. When a single profile occupied a section of walled or non-walled beach, that single profile was assigned the entire length for that section. See Basco *et al.* (1997) for section lengths for all profiles in WAM.

The total wall length was 4,738 m (15,545 ft) and the total non-walled (dune/beach) length was 2,953 m (9,689 ft). Each of the five profile parameters (V_S, V_L, V_T, P, and E_B) was determined for each profile and weighted by the representative shoreline length, and then divided by the total wall or non-walled length. The result is a single set of five weighted average

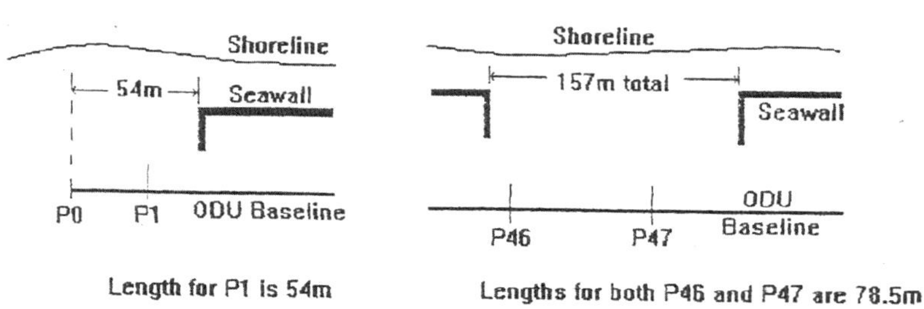

(a) (b)

Figure 10.36. Examples of representative shoreline lengths applied in the weighted average method.

parameters for each survey that represented how the profiles changed over time at Sandbridge.

The exact same procedure was employed for the Sectional, Weighted Average Method (WAMSECT). Figure 10.29 shows the northern section with Profiles 1–54 (8); the middle section with Profiles 60–162 (10) and the southern section with Profiles 179–252 (10).

For the Individual Profile Method (IPM), surveys dating back to October 1980 were employed (where available) and used to calculate the five profile parameters. This permitted the profile history to be analyzed beyond the five years of ODU data: allowed end-of-wall analysis; and rates of change before and after wall construction.

The three analysis methods are analogous to studying (1) a forest; (2) three subareas of the forest; and (3) individual trees in the forest relative to two types of trees and five parameters that characterize the health of each type of tree.

10.3.2.5 *Time scales and statistical analysis*

Simple, linear regression was employed as modeled by the equation

$$Y = b + m(t), \tag{10.3}$$

where Y is the parameter value (V_S, V_L, V_T, P, and E_B), b is the parameter initial value on October 1990 from the regression (constant), and m is the slope of the best fit, linear regression line for the rate of change of the parameter value.

Seasonal variations over the winter (October–March) and summer (April–September) were then modeled by the equation

$$Y_S = b + m(t) + a\sin(-2\pi(t - t_0)/365), \tag{10.4}$$

where a is the amplitude of a model sine wave, t is the time in number of days, and t_0 is the start date, October 1990.

After determining b and m from Equation (10.3) by linear regression, the amplitude, a, of the sine wave was found that produced the minimum total variance between the measured WAM (and WAMSECT) values and the predicted sine wave values, Y_S.

A null hypothesis test was then performed by calculating the Z-test statistic at the 95% confidence level (Bethea and Rhinehart, 1991) to determine if the parameter value change rates, m, were statistically different for the walled and non-walled results found in WAM and WAMSECT.

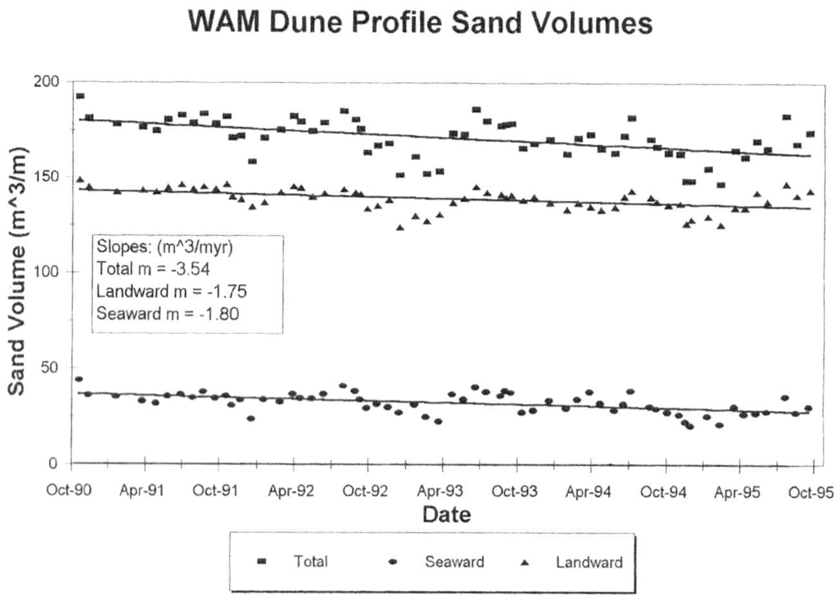

Figure 10.37. Sand volume per unit length versus time for non-walled (dune) sections using WAM.

10.3.2.6 *WAM results*

The sand volumes per unit length (m^3/m) for $V_S(\textit{seaward})$, $V_L(\textit{landward})$, and $V_T(\textit{total})$ for each survey data using WAM are displayed in Figure 10.37 for the non-walled (dune) sections and in Figure 10.38 for the walled sections. Total sand volumes are on the order of $150\,m^3/m$ for both types relative to the baseline location and profile down to MLW. All three volume change trends for the non-walled (labeled dune) are negative. Seasonal variations are evident. Similar plots for berm elevation (E_B) and shoreline position (P) display wide scatter and obscure the seasonal trend. Shoreline position is also greatly influenced by beach scraping and local short-term wave conditions. Volume change presents the best picture of subaerial beach change over time.

After five full years of monthly data (and after storms) characterized by two mild and two stormy winter seasons, and a summer season without normal recovery, the wall change rate seaward $\overline{V}_S$ (wall) $= -1.70$ (m^3/m) per year and the non-wall change rate seaward of the partition $\overline{V}_S$ (dune) $=$ $-1.80\,(m^3/m)$ per year. This difference in regression line slopes was NOT statistically significant. These results using profile data that was "weighted"

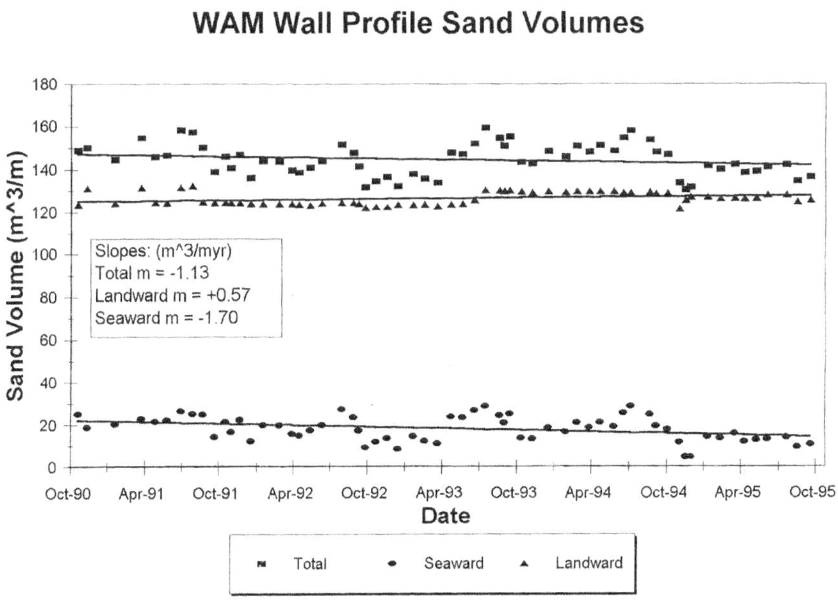

Figure 10.38. Sand volume per unit length versus time for walled sections using WAM.

and "averaged" over the entire study area are contrary to popular belief. The WAM results for each of the 62 surveys over five years used in the ODU dataset can be found in Basco *et al.* (1997), Appendix A, and include the change rates the volume landward, V_L, and total volume, V_T. More will be said about the volume landward in what follows.

Volume seaward trends at three time scales (five years, seasonal, and storms) are better demonstrated by plotting the volume *difference*, ΔV_S, where

$$\Delta V_S = V_S(t) - V_{So} \tag{10.5}$$

and V_{So} = the initial value taken as the regression line intercept for October 1, 1990, the starting point for wave year number one. The difference, ΔV_S, reflects the relative volume change with time over five years between winter and summer seasons and following storm events, as shown in Figure 10.39 for dune and Figure 10.40 for walled WAM results. Now the negative slopes and seasonal variations are clear. In general, sand is dragged off the sub-aerial beach by winter storms (October–March), deposited in nearshore bars, and then returns the next summer season (April–September) as rebuilt

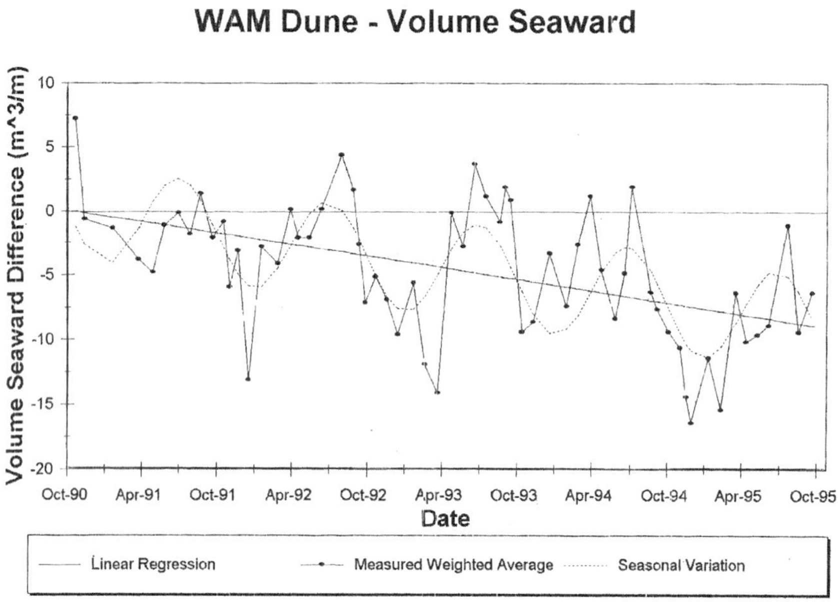

Figure 10.39. Volume seaward difference versus time for non-walled (dune) sections using WAM.

beach berms. The imbalance in the "dynamic equilibrium" is reflected in the negative slopes of the linear regression lines as given earlier. Winter storms in 1992–1993 and 1994–1995 were most severe with seaward volumes well below the regression lines. With the exception of the 1994–1995 wave year, the walled sections recovered to about $+7\,\text{m}^3/\text{m}$ (ΔV_S) each summer, which was the summer value prior to October 1, 1990. The dune sections continue to fall far below the sand volume present during the summer 1990 season.

Dotted lines in Figures 10.39 (dune) and 10.40 (wall) are the simple sinusoidal function in Equation (10.4) and gave a $= 3.8\,\text{m}^3/\text{m}$ (variance $=$ 13.9) for the dune and a $= 4.7\,\text{m}^3/\text{m}$ (variance $= 19.7$) I for the walled sections. The seawalls act as a barrier for landward sand transport during the summer season and produce greater seasonal variations in sand volumes on the subaerial beach.

Beach "recovery" is defined as the time when ΔV_S returns from *below* the regression line to *above* the regression line. The results are mixed since the walls recovered earlier in 1991 and later than the dune in 1992. Both recovered at the same time in 1993 and 1994. In 1995, the severely eroded beach never quite recovered for both. Little published research is

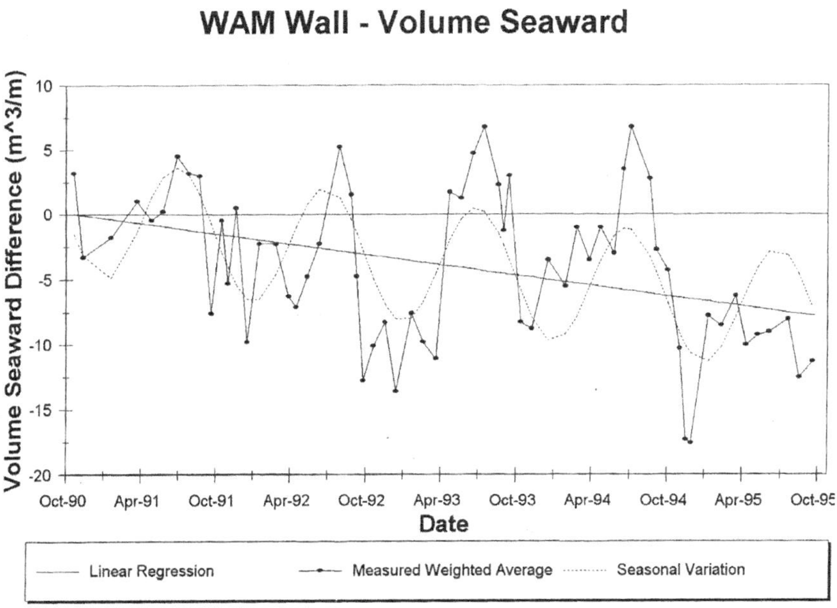

Figure 10.40. Volume seaward difference versus time for walled sections using WAM.

available for hydrodynamic conditions producing beach rebuilding. The results in Basco *et al.* (1997) for individual storms lend additional support for the walls producing greater seasonal variations in seaward sand volumes. Conversely, beach "erosion" from summer to winter seasons occurred at the same time for both types in 1990, 1992, 1993, and 1994. The dune sections eroded sooner in 1991 and 1995.

The volume landward *difference*, ΔV_L, for the dune sections is shown in Figure 10.41. Linear regression slope was negative, $-1.75\,\text{m}^3/\text{m}$ landward of the "partition" and nearly the same as the seaward volume change rate. Seasonal amplitude, $\text{a} = 4.93\,\text{m}^3/\text{m}$, and the storm volume change were generally larger (more negative) than the seaward volumes. Comparisons with the walled sections are meaningless for landward differences due to artificially manipulated sand volumes behind walls.

10.3.2.7 *WAMSECT results*

The exact same procedure was employed for the Sectional, Weighted Average Method (WAMSECT). Figure 10.29 shows the northern section with

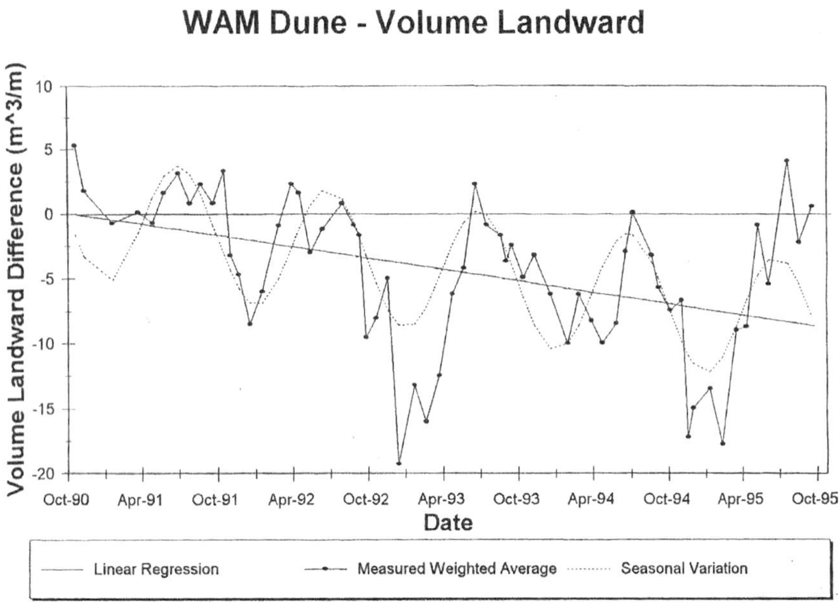

Figure 10.41. Volume landward difference versus time for non-walled (dune) sections using WAM.

Profiles 1–54 (8); the middle section with Profiles 60–162 (10) and the southern section with Profiles 179–252 (10). Appendix B in Basco *et al.* (1997) presents the WAMSECT plots for each sub-region for both profile types over the full five years of the study.

At the North end, the volume seaward rates of change for dunes and walls are negative and statistically equivalent. Amplitudes of the calculated seasonal variation are also equivalent. The volume landward trend for non-walled profiles possessed the greatest rate of change ($-5.0\,m^3/m$ per year) and seasonal amplitude ($9.1\,m^3/m$ per year) for the three subsections. However, owner removal of the dune at Profile 1 in 1994 influenced this negative trend. Over 78.8% are walls and only 21.2% are dunes.

Beach scraping and sand dumping along dune profiles in the Middle sub-section produced a slightly positive volume landward rate of change ($+0.46\,m^3/m$ per year). Volume seaward rates of change for dunes and walls were negative and statistically equivalent. The Middle sub-section is about 0.6 m (2 ft) higher than both North and South sections and needed fewer bulkheads to protect property landward of the partition location. The distribution is 43.3% walls and 56.7% dunes. Individual property owners are generally unaware that they live on higher ground and generally believe bulkheads are not needed at Sandbridge.

As in both the North and Middle sub-sections, the rates of change for the seaward volumes in the South end are statistically equivalent for both walled and non-walled profile types. The mix is 73.1% walls and 26.9% dunes in the South end similar to the North end.

In general, the results for each sub-section in the WAMSECT analysis are the same as for the WAM analysis.

10.3.2.8 *Individual profile method (IPM) results*

Only when individual profiles are considered can the full dataset dating back to October 1980 be utilized in the analysis. About 34 individual profiles were identified for analysis. Six locations had sufficient data to be included with the 28 ODU profile locations with all results for the five profile variables given in Table 10.2. Eleven locations provide the "before" and "after" data with time intervals for each period when the location or nearby location changed from a dune profile to a wall profile.

For example, Figure 10.42 shows Profile 1 with eight years "before" (10/80 to 7/88 and 24 surveys) and five years "after" (7/89 to 6/94 and 58 surveys) seawalls were constructed 27 m (89 ft) south on adjacent property in the Spring of 1989 (see Figure 10.36(a)). It appears that nearby seawall construction impacts both landward and seaward volumes to cause negative slopes. Figure 10.43 for Profile 161 displays the opposite trends for these same time periods and produced far less change in the rates of change for both landward and seaward volumes.

There is much information presented in Table 10.2 that may yield further insight on the impact of seawalls on adjacent beaches. We leave this for future graduate students in coastal science and engineering. Basco *et al.* (1997) employed the above analysis of the WAM, WAMSECT, and IPM to draw answers for three key questions, summarized as follows.

10.3.2.9 *Answers to three key questions*

Question No. 1. *Does the sand volume seaward of walls erode faster than the volume seaward of the "partition" for non-walled locations?*

The results at three time scales (five years, seasonal, storms) and from the three analysis methods all supported the same conclusion, namely: the volume erosion rates are *not higher* in front of seawalls. The elevated Middle

Table 10.2. IPM analysis summary of beach parameter change rates for 34 locations.

Profile Number	Type	Regression Dates	# Yrs	# Surveys	V_T (m^3/myr)	V_S (m^3/myr)	V_L (m^3/myr)	E_B(m/yr)	P (m/yr)
1 before	Dune	10/80 to 7/88	8	24	+0.3	−0.5	+0.8	−0.0	−0.1
1 after	Dune	7/89 to 6/94	5	58	−11.1	−2.1	−9.0	−0.1	−0.9
13 before	EOW	5/91 to 4/93	2	34	−14.4	−0.1	−14.3	−0.2	−0.7
13 after	Wall	5/93 to 4/94	1	29	−13.7	−12.4	−1.3	−0.4	−7.5
20	Wall	10/90 to 9/95	5	63	−1.8	−2.4	+0.6	−0.2	−1.6
25 before	Dune	10/80 to 9/88	8	24	+0.3	+0.7	−0.4	+0.1	+0.0
25 after	Wall	10/90 to 9/95	5	72	−7.0	−1.7	−5.3	−0.1	+0.0
41	Wall	10/90 to 9/95	5	69	−0.2	−0.7	+0.5	+0.0	−0.2
46	EOW	10/90 to 9/95	5	64	+1.7	+0.4	+1.2	−0.0	+0.7
47	Dune	10/90 to 9/95	5	63	−4.9	+0.5	−5.4	−0.0	+0.1
54	Dune	10/90 to 9/95	5	64	−5.9	−1.4	−4.5	−0.1	−1.6
60 before	Dune	10/80 to 7/85	4	17	−11.4	−6.3	−5.1	−0.5	−1.4
60 after	Dune	10/90 to 9/95	5	62	−6.5	−0.8	−5.7	−0.1	−1.4
70	Wall	10/90 to 9/95	5	66	−1.1	−0.0	−1.1	−0.0	−0.2
74 before	Dune	10/80 to 9/88	7	30	−11.4	−10.7	−0.8	−0.4	−2.0
74 after	Wall	7/89 to 7/93	4	9	+1.4	+1.4	−0.1	+0.1	+0.9
78	Wall	10/90 to 9/95	5	69	+0.9	−0.6	+1.5	−0.1	−0.6
87	Wall	10/90 to 9/95	5	63	−0.9	−0.4	−0.6	+0.1	−0.3
103	EOW	10/90 to 9/95	5	60	−2.6	−3.0	−0.5	−0.0	−1.0
107	Dune	10/80 to 7/85	4	18	+10.1	+5.3	+4.8	+0.1	+1.9
118	Dune	10/90 to 9/95	5	61	−5.9	−2.5	−3.4	−0.0	−1.2

(*Continued*)

Table 10.2. (*Continued*)

Profile Number	Type	Regression Dates	# Yrs	# Surveys	V_T (m^3/myr)	V_S (m^3/myr)	V_L (m^3/myr)	E_B(m/yr)	P (m/yr)
131	Dune	6/85 to 10/91	6	22	+13.3	+4.8	+8.6	+0.3	+1.6
135 before	Dune	10/80 to 9/88	8	24	+4.3	+1.4	+2.9	+0.0	+0.7
135 after	Dune	10/90 to 9/95	5	69	+5.4	+0.6	+4.8	+0.1	−0.2
148	Wall	10/90 to 9/95	5	62	−0.8	−0.7	−0.1	−0.3	−0.4
161 before	Dune	10/80 to 10/88	8	50	−4.2	−2.0	−2.2	−0.1	−1.0
161 after	Dune	10/90 to 9/95	5	70	−2.8	−2.0	−0.8	−0.1	−1.5
162	EOW	10/90 to 9/95	5	62	−4.7	−2.7	−2.0	−0.1	−2.4
179	Wall	10/90 to 9/95	5	59	−5.0	−4.0	−1.0	−0.1	−2.7
194 before	Dune	10/88 to 9/89	1	18	+3.9	+5.7	−1.9	+0.5	+4.3
194 after	Wall	10/90 to 9/95	5	64	−4.3	−3.1	−1.3	−0.4	−1.9
205	Dune	4/81 to 4/88	7	46	−6.0	−3.9	−2.1	−0.2	−1.7
207	Wall	10/90 to 9/95	5	60	−2.9	−2.2	−0.7	−0.3	−1.5
216 before	Dune	9/85 to 8/90	5	29	−7.6	−3.7	−3.9	−0.2	−2.4
216 after	EOW	10/90 to 9/95	5	65	−4.9	−1.4	−3.5	−0.2	−2.8
220 before	Dune	9/85 to 9/89	4	41	−16.5	−6.9	−9.6	−0.3	−3.9
220 after	Dune	10/90 to 9/95	5	65	−3.2	−0.4	−2.7	−0.1	−2.1
226	Wall	10/88 to 9/95	7	66	−4.0	−4.4	+0.4	−0.6	−2.6
230	Dune	10/88 to 9/95	7	80	−7.8	−4.0	−3.7	−0.2	−2.6
234	Wall	10/90 to 9/95	5	61	−5.9	−3.2	−2.7	−0.6	−1.8
240	EOW	10/90 to 9/95	5	62	−7.9	−5.0	−2.9	−0.2	−2.9
252 before	Dune	10/80 to 9/88	8	47	−2.1	−3.0	+0.9	+0.1	−1.5
252 after	Dune	10/90 to 9/95	5	71	−8.7	−7.2	−1.5	−0.4	−2.4

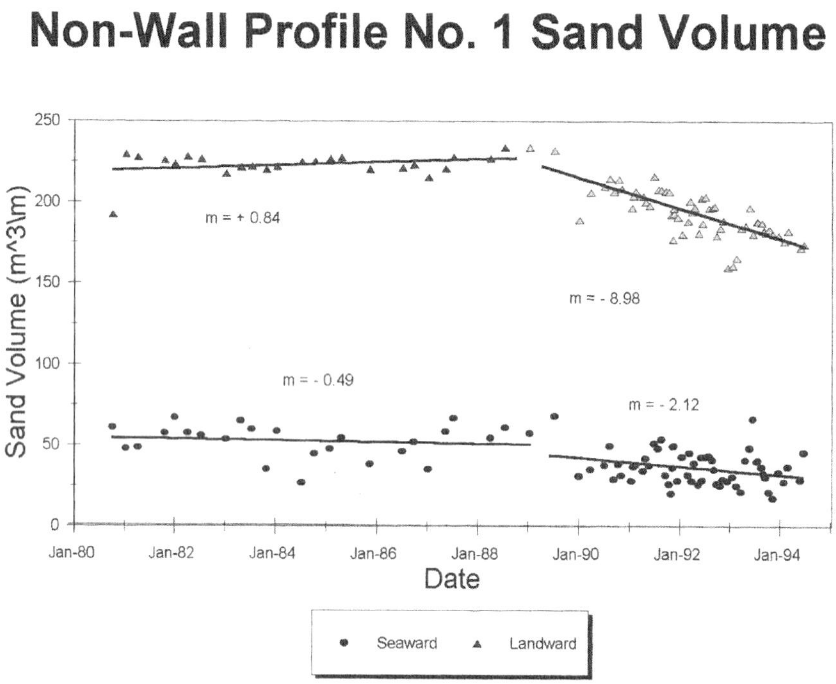

Figure 10.42. Seaward (circles) and landward (triangles) sand volumes at Profile 1. Seawalls constructed 27 m (89 ft) south early in 1990.

section shows evidence that non-walled (dune) locations eroded faster seaward of the partition in both the WAMSECT and IPM analyses, that is opposite of what was expected.

Seasonal variability of the sand volume in front of walls is generally greater than at non-walled locations. The walls create a narrow and higher energy zone for winter waves to drag more sand offshore. The walls also block the onshore transport of sand to return more sand during the summer season. Full wave years (winter and summer seasons) are needed. Too much emphasis in the literature is focused on storm waves removing more sand at walled locations. Little mention is made of summer swell waves returning to pile more sand up against the walls in beach rebuilding.

Question No. 2. *Do seawalls delay beach recovery*?

From the WAM and WAMSECT results, seawalled beaches recovered about the *same time* as non-walled locations. The abnormally high energy wave year (1994–1995) produced partial recovery of non-walled sections but the

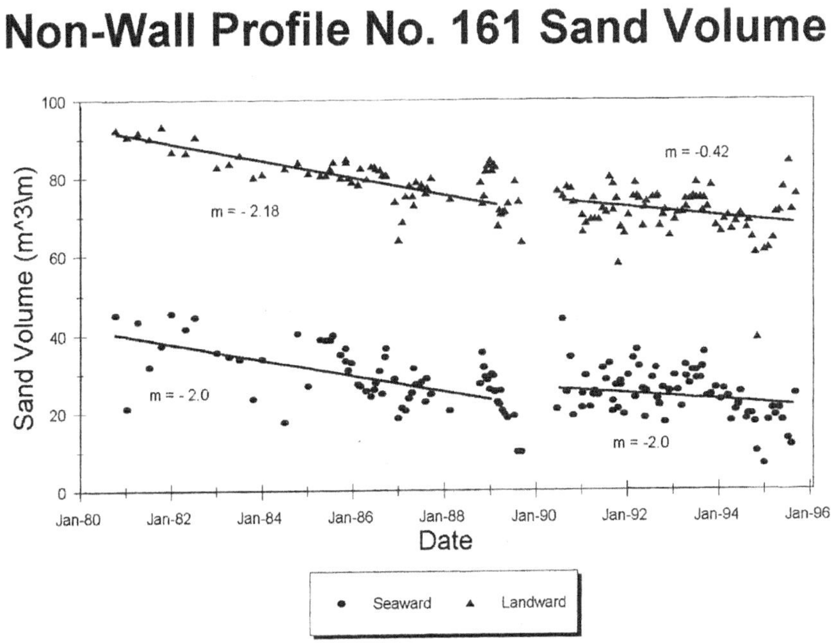

Figure 10.43. Seaward (circles) and landward (triangles) sand volumes at Profile 161. Seawalls constructed 30 m (100 ft) south early in 1990.

walled sections did not recover. The evidence also showed that beach transition from winter to summer volumes occurred simultaneously for walled and non-walled profiles. Over three of the five years, the walls recovered at the same time or earlier for the winter to summer transition.

Question No. 3. *Is the sand volume landward of the "partition" at non-walled locations eroding at a faster rate after construction of adjacent seawalls?*

The intermittent walled and non-walled reaches, lateral sand transport by waves in both the north and south directions, and the increased elevation of the Middle section all add considerable complexity to the Sandbridge site. Using the IPM results, some supporting evidence was found for the North end (P1, Figure 10.42). However, the opposite trend was observed in the Middle section and the evidence was inconclusive for the South end.

These results might be different if, e.g. the central two-thirds section was one continuous long wall, the barrier elevation was flat, and one dominant direction prevailed for littoral drift. There was no evidence of the walls acting as groins to trap sand on the "updrift" side.

After wall construction, sand trapped behind the wall is not available for transport to adjacent beaches during and after storm events. The authors (Basco *et al.*, 1997) believe variations in the seasonal signal mask this change and that more full wave years of profile data are needed to confirm (or negate) and quantify this effect. Sand is also bound beneath the road at non-walled locations.

10.3.2.10 *Nine full years of ODU research at Sandbridge*

Monthly profile data (and after storms) was collected at the 28 ODU locations for the next four years (October 1995 to August 1999) to complete nine full years for the ODU dataset. Sandbridge beach received its first ever beach nourishment project in the summer of 1998 with 1 million cubic yards of sand from the Dam Neck offshore shoal. See Chapter 7, Section 7.7.1 (c) for details. Figure 10.44 displays the WAM results for the volume seaward *difference* for the dunes and Figure 10.45 for the walled sections.

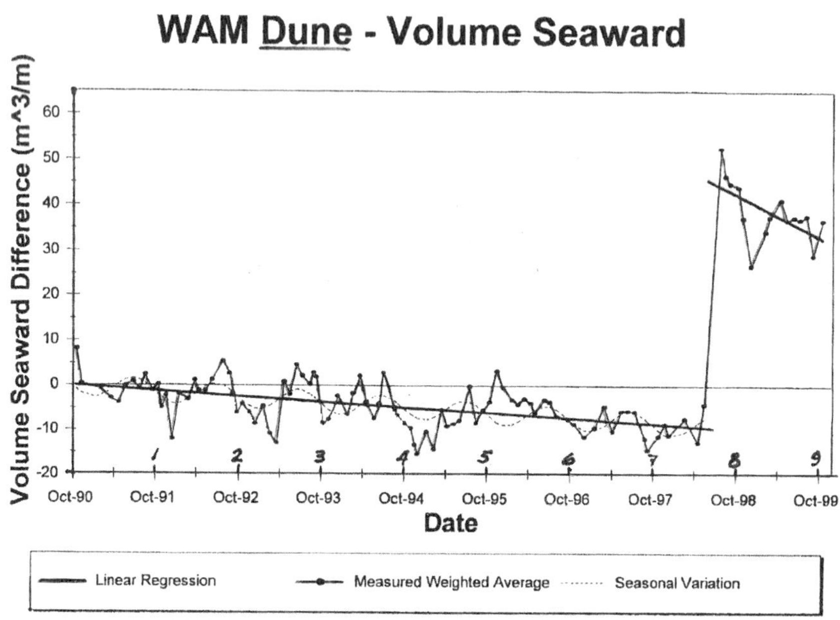

Figure 10.44. Volume seaward difference versus time for non-walled (dune) sections using WAM.

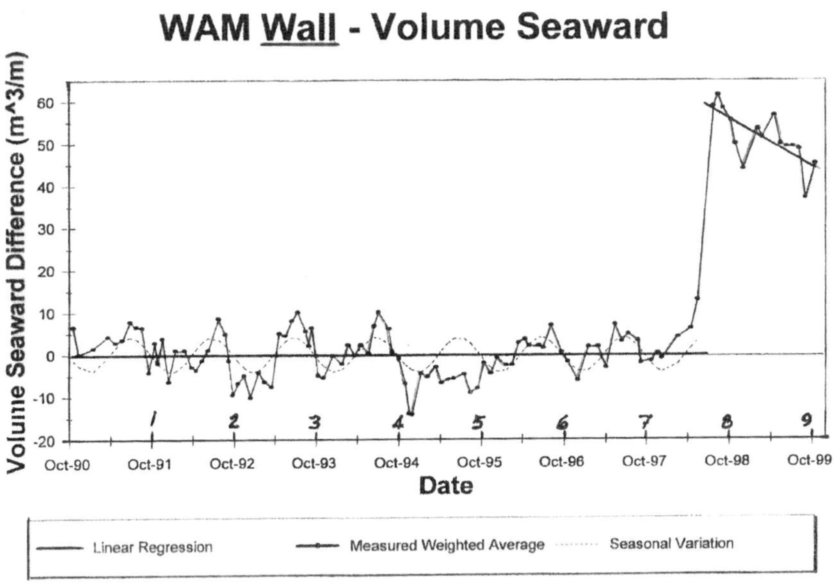

Figure 10.45. Volume seaward difference versus time for walled sections using WAM.

The volume increase seaward from beach nourishment is about 60 m^3/m (25 cy/ft) and clearly shown on these figures. The wall volume seaward has now recovered in years 6 and 7 so that the trend line for walled sections is almost flat, meaning the volume difference from the beginning (October 1990) is roughly zero. The trend line for the dune volume seaward difference continues on its downward slope, meaning the beach is eroding. Beach nourishment in 1998 modified the original beach volumes so that the ODU monitoring project was concluded in August 1999. However, the additional monitoring years did not change the conclusions for the three key questions above.

10.3.2.11 *Dune recession beyond South end seawall (Whitecap Lane)*

Figure 10.46 displays dune recession over 13–14 years for a 365 m (1,200 ft) section at the South end of Sandbridge. The data was taken from a series of aerial photographs by the Virginia Institute of Marine Science (Wright, 2003). The top curve (March 24, 1987) was before a wall was constructed in the winter of 1988/89 to the left (north) of zero. Beach access around the wall required sand fence removal and lowered the local elevations to permit

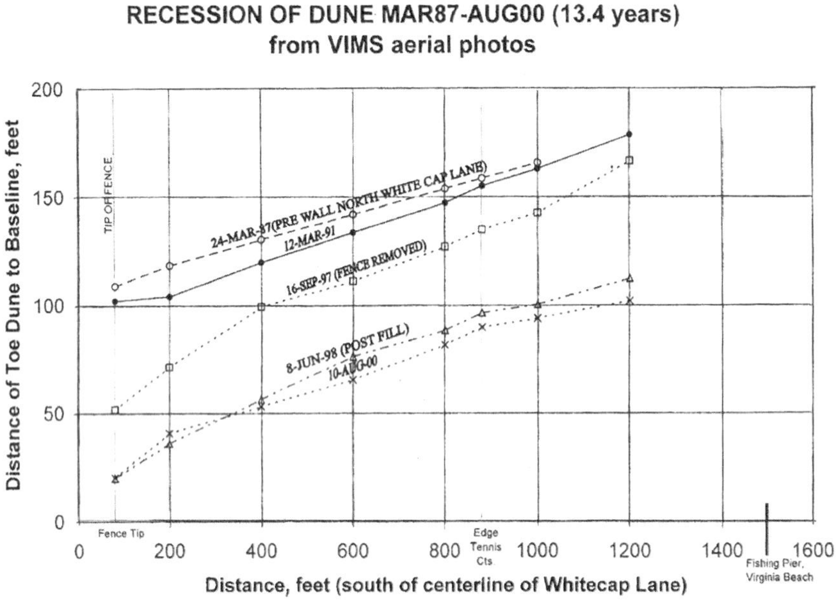

Figure 10.46. Toe of dune recession south of Profile 240 (Whitecap Lane) from VIMS aerial photographs over 13.4 years (March 1987–August 2000) (from Wright, 2003).

storm flooding and ebbing at the end of the wall. After 13.4 years, the entire dune toe has receded a similar distance landward, over the shoreline south of the wall end (zero). No discernible, end-of-wall effect is evident from the data. Figure 10.46 was employed to overturn the State of Virginia's denial of a permit application to extend the seawall approximately 100 m further south. Full details can be found in the article "Seawall Impacts on Adjacent Beaches: Separating Fact from Fiction" (Basco, 2003).

10.3.2.12 *Summary — Sandbridge beach research*

The influence of vertical bulkheads (seawalls) on the adjacent, subaerial beach volumes on a receding shoreline is the subject of nine years of research at Sandbridge, Virginia, on the Atlantic Ocean. The results vary over the project length (7.7 km, 25,200 ft) since the historic, long-term erosion rate varies from −1.1 m/year (−3.5 ft/year) on the north end (Navy fence) to −2.9 m/year (−9.5 ft/year) on the South end (City pier). And trends in the middle are influenced by the +0.6 m (+2 ft) increase in ground elevation. Over 60% of the length has bulkheads and since constructed, the long-term average erosion rate of about −1.8 m/year (−6 ft/year) has not increased. The devil is in the details as described earlier and in the references presented.

10.4 Natural and Nature-based Features (NNBF) — Living Shorelines

10.4.1 *NNBF*

The environmental impacts of alternatives for reducing coastal hazards are not all negative. In recent years, the categorization of alternatives as shown in Chapter 3, Figure 3.1 as (A) structural, (B) non-structural, and (C) combinations has been recast as follows:

- Structural,
- Non-structural,
- Natural, nature-based features (NNBF).

to support *the full array of measures employed to support coastal resilience and risk reduction* (USACE, 2013). The buzz words are now "Natural, Nature-Based Features" and identified by the acronym NNBF. Figure 10.47 displays the five commonly accepted components with "Benefits/Processes" and "Performance Factors" identified for each (from Miller *et al.*, 2014). We have devoted Chapter 7 in this book to Dunes and Beaches as one "Structural" (soft) alternative that is the only NNBF-type capable of being designed to mitigate flooding from elevated storm surge events. The other four NNBF-types only attenuate wave energy as their primary benefit.

The US Army Corps of Engineers, Environmental Laboratory (ERDC) began its Engineering With Nature (EWN) program in 2010 (www.engineeringwithnature.org). It is defined as the

> ... *intentional alignment of natural and engineering processes to efficiently and sustainably deliver economic, environmental and social benefits through collaborative processes.*

As a consequence of Hurricane Sandy, the Corps formed a team to develop a framework for identifying and evaluating opportunities for integrating NNBF (Bridges *et al.*, 2015). It was recognized from the start that nature-based features change under natural coastal processes (water levels, waves, sediment transports) and must be maintained by human intervention. The NNBF responds in many ways to storms and can provide support to coastal communities and ecosystems. It was also important to design nature-based

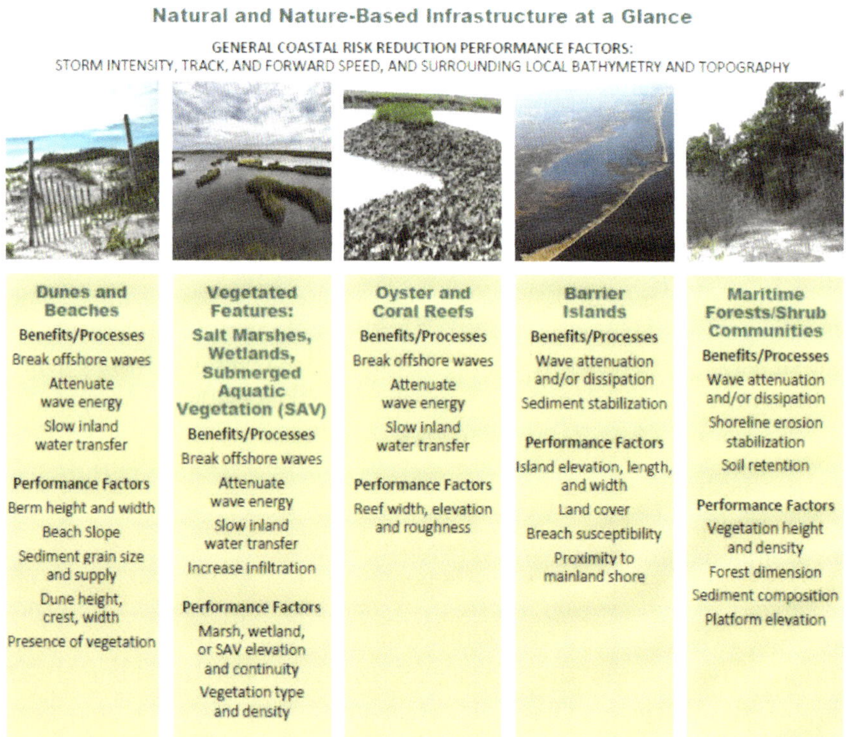

Figure 10.47. Five elements of NNBF as identified by the Corps of Engineers (Miller *et al.*, 2014).

features to establish and/or re-establish natural processes and become as self-sustaining as possible.

A strategy that combines NNBF with non-structural and structural measures represents an integrated approach to flood risk management that can deliver a broad array of ecosystem goods and services to local communities. For example, incorporating artificial reefs along a shoreline can attenuate waves — reducing coastal erosion, land loss, and flood damages — while also providing habitat for fish and wildlife. Engineered coastal landscapes that include wetlands, maritime forests, and structural measures (e.g. a levee), can reduce flood risks while providing fish and wildlife habitat and opportunities for recreation (e.g. fishing, bird watching, hiking).

The EWN initiative is working to expand opportunities for application of NNBF by supporting R&D, technology transfer, and stakeholder engagement. NNBF can be differentiated into two categories: natural features and nature-based features. Natural features (e.g. a reef) are comparatively "long-standing" in terms of age and created through the action of natural physical, biological, and chemical processes over time. Whereas nature-based features are created by human design, engineering, and construction to mimic natural features that provide similar, if not identical, services (Bridges *et al.*, 2015).

The *State of Science & Engineering for Natural and Nature-Based Features* as currently employed by the Corps of Engineers can be found from the presentation by Rosati *et al.* (2018) (https://ewn.el.erdc.dren.mil/workshops/2018_07-29-NNBF-short-course/ppt/1545-1615_Rosati-NNBF_RandD.pdf). See also Bridges (2018) for his presentation, also found on the EWN website.

10.4.2 *Living Shorelines*

The term "Living Shorelines" has an even earlier beginning around the year 2000. In 2005, the Ocean Studies Board of the National Research Council (the National Academies) formed a "Committee on Mitigating Shore Erosion Along Sheltered Coasts (NRC, 2007). The Chesapeake Bay has over 5,000 miles (8,000 km) of shorelines with limited fetch and Virginia's Institute of Marine Resources (VIMS) has been a leader for over 20 years in this effort. The Virginia legislative definition of a living shoreline states:

> *Living shoreline means a shoreline management practice that provides erosion control and water quality benefits; protects, restores or enhances natural shoreline habitat; and maintains coastal processes through the strategic placement of plants, stone, and sand fill, and other structural and organic materials.*

A *Living Shoreline Design Guidelines for Shore Protection in Virginia's Estuarine Environment* report was published (Hardaway and Milligan, 2010) that has stood the test of time and offers practical design considerations, case studies, and design examples. Figure 10.48 graphically illustrates by colors what has become the Green (Living Shorelines — Soft) versus Gray (Coastal Structures — Hard) approach. Most of the over 60 projects designed by the VIMS for the Chesapeake Bay use nearshore breakwaters and/or sills to reduce wave energy to the sandfills and tidal marshes constructed on the landward side. See Chapter 3, Figure 3.9 for a sill design as discussed in Hardaway and Byrne (1999).

The "Living Shoreline" concept has grown to many states throughout the US and in Canada. Consult the website (http://livingshorelines academy.org) for up-to-date details and especially the "Living Shorelines" Online Resources/Publications list.

Figure 10.48. Color graphic to illustrate the full range of alternatives for coastal erosion mitigation for low energy shorelines. The Green to Gray reference is now common shorthand for this array of possible designs (from Natural and Structural Measures for Shoreline Stabilization, https://coast.noaa.gov/data/digitalcoast/pdf/living-shoreline.pdf).

10.5 Downdrift Impacts of Groins

An overriding concern expressed by environmental groups is the potential for the impact of groins on the downdrift beach. Often cited is the classic argument of how groins trap longshore transport of sand on the updrift side to create a negative impact on the downdrift side of the groin. For this classic, "Beaches 101" classroom argument to be valid, the location must have a longshore sediment transport of sand. If it does not, then this classic argument must be considered relative to whether (or not) it is valid at this location.

At many locations with a long history of shoreline erosion of sandy beaches, individual home owners have constructed seawalls, bulkheads, and dikes with rock revetments to fix the seaward boundary of their property. Erosive forces continue to erode the sandy beach and in time longshore sand transport is greatly diminished because little (or no sand) remains in front of the hardened shoreline. At some locations such as Sandbridge, VA (see Figure 10.25), the hardened shoreline is a nodal point (region) with sand transport in both directions away from the nodal point. And, a beach nourishment project is then constructed in this region that becomes the source of sand for longshore sediment transport after the beach nourishment project is completed.

At some locations, terminal groins can then be constructed for the purpose of retaining the re-nourished beach sand on the updrift side in front of the hardened shoreline structures. Over time, the middle reaches in the nodal point (region) lose more sand since the original erosional processes

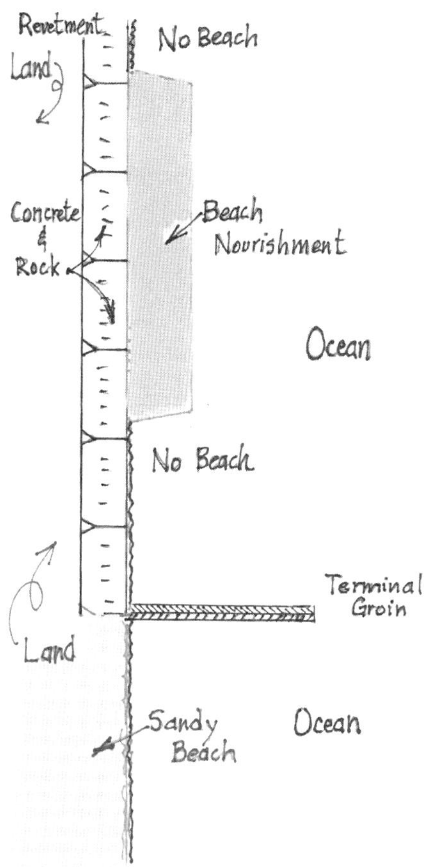

Figure 10.49. Author's schematic of conditions soon after South end groin construction and beach nourishment along middle reaches of revetment where no sediment transport exists prior to nourishment.

continue, but the terminal groins collect this sand in the up-drift fillet. The sand in the fillet is then recycled back to the middle regions by truck haul in a "back-pack" operation.

Figure 10.49 is the author's attempt to graphically depict this situation showing the revetments with no beach, the beach nourishment in the middle reaches, and the South end, terminal groin with no fillet before sand spreads south from the middle. The downdrift impact (offset) can now be properly addressed and is due to the following factors:

- The long-term erosion of the shoreline continues.
- Erosion continues on the beach south of the South end groin but not on the north side.
- The sand moving south from the beach nourishment forms a fillet on the north side of the groin.
- If there had been no beach nourishment project, there would have been no sand, no groin, and no fillet.

The offset distance is NOT solely due to the groin blocking the existing longshore sediment transport as found in the classic "Beaches 101" argument. Readers hopefully will remember this example and do some research to fully understand the situation when investigating the environmental impacts of groins. This is especially important if you are asked to be an "expert" consultant on either side of the argument.

10.6 Summary Chapter 10 — Environmental Impacts

The "Gray" *versus* "Green" mentality continues to grow via the conventional media and the "social" media, with many believing opinions and blog statements with no doubts regarding use of the word *versus* in the debate regarding the environmental impacts. In this chapter, we have tried to summarize the history of the debate going back over 30 years and then our own 10-year research effort at Sandbridge, VA, which separates fact from fiction. All is not negative, and the recent NNBF-Living Shorelines emphasis has many positive attributes for considering the complete spectrum from Gray to Green alternatives. Future sea level rise scenarios will require more hardened coastal structures (not less) for mitigation of flooding and erosion problems. Coastal engineers can expect to encounter more opposition from environment groups in the future.

See Section 11.3.10 for impacts of sea level rise on Chapter 10.

References

Basco, D. R. (1987) "The Effects of Seawalls on Coastlines and Beaches," Special Report, *the Coastal Engineering Institute*, Old Dominion University, Norfolk, VA.

Basco, D. R. (1993) "On the Loss of Estuarine Bay Areas Due to Barrier Island Migration," *Shore and Beach*, Vol. 61, No. 4, October, pp. 16–23.

Basco, D. R. (2000) "Shore Protection Planning," *The Coastal Engineering Manual*, Part V, Chapter 3, Wash, DC.

Basco, D. R. (2003) "Seawall Impacts on Adjacent Beaches: Separating Fact From Fiction," *Proceedings, 8th International Coastal Symposium*, Itajai, Brazil.

Basco, D. R., D. A. Bellomo, J. M. Hazelton, and B. N. Jones (1997) "Influence of Seawalls on Subaerial Beach Volumes with Receding Shorelines," *Contract Report, CHL-97-2, Waterways Experiment Station*, Corps of Engineers, Vicksburg, MS.

Basco, D. R. (1991) "The Effect of Seawalls on Long-Term Shoreline Change Rates for the Southern Virginia Ocean Coastline," *Proceedings, 22nd International Conference Coastal Engineering* (Delft), ASCE.

Bellomo, D. A. (1993) "Statistical Variations in Beach Parameter Change Rates for Walled and Non-walled Profiles at Sandbridge, Virginia," MS Thesis, Department of Civil and Environmental Engineering, *Old Dominion University*, Norfolk, VA.

Bethea, R. M., and R. R. Rhinehart (1991) *Applied Engineering Statistics*, Marcel Dekker, New York.

Birkemeier, W. (1984) "A User's Guide to ISRP: The Interactive Survey Reduction Program," Instruction Report, *CERC-84-1, Corps of Engineers*, Waterways Experiment Station, Vicksburg, MS.

Bridges *et al.* (15 co-authors) (2015) "Use of Natural and Nature-Based Features (NNBF) for Coastal Resilience," Corps of Engineers, ERDC-SR-15-1, *Engineering Research and Development Center*, Vicksburg, MS. Jan.

Bridges, T. S. (2018) "Natural Infrastructure: Past, Present and Future" slide presentation, available online (www.engineeringwithnature.org).

California Coastal Commission (1996) "Report on In-Lieu Fee Beach Sand Mitigation Program: San Diego County," by S. Sherilyn, L. Ewing, A. Roach, https://www.coastal.ca.gov/pgd/sand1.html.

California State Coastal Conservancy (2002) "California Beach Restoration Study," California Department of Boating and Waterways and State Coastal Conservancy, Sacramento, CA.

CBRA (1982) "Coastal Barrier Resources Act," Department of the Interior, Fish and Wildlife Service (CBRA, Public Law 97-348).

CEM (2000) *Coastal Engineering Manual*, US Army Corps of Engineers, Engineering Research and Development Center, Vicksburg, MS.

CEM (2006) *The Coastal Engineering Manual*, US Army Corps of Engineers, Washington, DC.

Dean, R. G. (1987) "Coastal Armoring: Effects, Principles, and Mitigation," *Proceedings, 20th International Conference Coastal Engineering*, ASCE, NY, Vol. 2, pp. 1843–1857.

Dean, R. G., O. H. Pilkey, and J. R. Houston (1988) "Eroding Shorelines Impose Costly Choices," *Geotimes*, Vol. 33, No. 5, pp. 9–14.
Dolan, R., and H. Lins (1987) "Beaches and Barrier Islands," *Scientific American*, July Vol. 255, No. 7, pp. 68–77.
Dolan, R. H. (1985) "Sandbridge Beach and Back Bay Virginia," Tech. Rept., *Coastal Research Assoc, Inc.*, Charlottesville, VA, 120p.
Ebersole, B. A., M. A. Cialone, and M. D. Prater (1983) "RCPWAVE- A Linear Wave Propagation Model for Engineering Use," Report 1, *Coastal Engineering Research Center, Corps of Engineers*, Vicksburg, MS.
Estuaries and Coasts (2018) "Impacts of Coastal Land Use and Shoreline Armoring on Estuarine Ecosystems," Vol. 41, No. 1 (special issue).
Everts *et al.* (1983). "Shoreline Movements: Cape Henry Virginia to Cape Hatteras, NC, 1849–1980". Report No. 1, Technical. Rept. *CERC 83-1, Waterways Experiment Station*, Vicksburg, MS, July, 123p.
Fowler, J. E. (1992) "Scour Problems and Methods for Prediction of Maximum Scour at Vertical Seawalls," *Technical Report CERC 92-16*, Corps of Engineers, Waterways Experiment Station, Vicksburg, MS.
Fowler, J. E. (1993) "Coastal Scour Problems and Methods for Prediction of Maximum Scour," *Technical Report, CERC-93-8*, Corps of Engineers, Waterways Experiment Station, Vicksburg, MS.
Garrison, D. (2018) "Long-Term Ecological Research (LTER)," website (https://www.nsf.gov/lter).
Graber, P. (1985, Virginia) "The Law of the Coast in a Clamshell," *Shore & Beach*, 26 articles (1975–1986).
Griggs, G. B., B. T. Benumof, and L. J. Moore, (1997) "The State of the Problem and the Problem of the State," *University of North Carolina-Chapel Hill.*
Griggs, G. B., J. F. Tait, and W. Corona (1994) "The Interaction of Seawalls and Beaches: Seven Years of Field Monitoring Monterey Bay, CA," *Shore and Beach*, Vol. 62, No. 3, pp. 21–28.
Griggs, G. B., and J. F. Tait (1988) "The Effects of Coastal Protection and Structures on Beaches Along Northern Monterey Bay, California," *Journal of Coastal Research*, Special Issue No. 4, pp. 93–111.
Griggs, G. B. (2010) "The Effects of Armoring Shorelines — The California Experience," in Shipman, H., Dethier, M. N., Gelfenbaum, G., Fresh, K. L., and Dinicola, R. S., eds., 2010, *Puget Sound Shorelines and the Impacts of Armoring — Proceedings of a State of the Science Workshop*, May 2009: U.S. Geological Survey Scientific Investigations Report 2010–5254, pp. 77–84.
Hardaway, C. S., and R. J. Byrne (1999) "Shoreline Management in Chesapeake Bay," Report, *Virginia Institute of Marine Science*, Gloucester Point, VA.
Hardaway, S., and D. Milligan (2010) "A Living Shoreline: Design Guidelines for Shore Protection in Virginia's Estuarine Environment," VIMS, Gloucester Point, VA.
Hardaway, S. C., and G. R. Thomas (1990) "Sandbridge Bulkhead Impact Study," Special Report No. 305, *Applied Science and Ocean Engineering, Virginia Institute of Marine Science*, Gloucester Point, VA.

Irish, J. L., P. Lynett, R. Weiss, S. Smallegan, and W. Cheng (2013) "Buried Relic Seawall Mitigates Hurricane Sandy's Impacts," *Coastal Engineering*, Vol. 80, pp. 79–82.

Journal of Coastal Research (1988) "The Effects of Seawalls on the Beach," Special Issue No. 4, Autumn 1988.

Kana, T. W., and M. Svetlichny (1982) "Artificial Manipulation of Beach Profiles," *Proceedings, 18th International Conference Coastal Engineering*, ASCE, NY, pp. 903–929.

Komar, P. D. (1998) *Beach Processes and Sedimentation* (2nd edn.), Prentice-Hall, New Jersey, 544p.

Komar, P. D., and W. G. McDougal (1988) "Coastal Erosion and Engineering Structures: The Oregon Experience," *Journal of Coastal Research*, Special Issue No. 4, pp. 77–92.

Kraus, N. C. (1988) "The Effects of Seawalls on the Beach: An Extended Literature Review," *Journal of Coastal Research*, Special Issue No. 4, pp. 1–28.

Kraus, N. C., and W. G. McDougal (1996) "The Effects of Seawalls on the Beach, Part I, and Updated Literature Review," *Journal of Coastal Research*, Vol. 12, No. 3, pp. 691–701.

Kriebel, D. L., W. R. Dally, and R. G. Dean (1986) "An Undistorted Froude Model for Surf Zone Sediment Transport," *Proceedings, 20th International Conference Coastal Engineering*, pp. 1296–1310.

Leatherman, S. P. (1988) *Barrier Island Handbook*, Laboratory for Coastal Research, University of Maryland, College Park, MD.

Leatherman, S. P. (1987) "Shoreline and Sediment Budget Analysis of North Assateague Island, Maryland." *Proceedings, Coastal Sediments*; 87, ASCE, NY, pp. 1460–1471.

Mathews, E. R. (1913) *Coastal Erosion and Protection*, Charles Griffin & Co., Ltd, London.

McDougal, W. G., M. A. Sturtevant, and P. D. Komar (1987) "Laboratory and Field Investigations of Shoreline Stabilization Structures on Adjacent Property," *Coastal Sediments '87*, pp. 961–973.

Miller, S. J., R. E. Henn, C. Fischenich, and T. S. Bridges (2014) "Nature-Based Features in a Systems Approach to Coastal Storm Risk Management," slide presentation, *Conference on Ecological and Ecosystem Restoration*, (https://conference.ifas.ufl.edu/CEER2014/Speaker), ERDC, Vicksburg, MS.

Moody, P. M., and O. Madsen (1995) "Laboratory Study of the Effect of Sea Walls on Beach Erosion," Final report, MITSG 95-03, *MIT Sea Grant*, Cambridge, MA.

O'Brien, M. P. (1982) "II-Memorandum Covering a Document Entitled 'Saving the American Beach: A Position Paper by Concerned Coastal Geologists', *Shore & Beach*, American Shore and Beach Preservation Association, April, pp. 6–8.

Ozger, S. (2002) "On the Active Sand Volume in the Cross-shore Beach Profile," MS Thesis, *Department of Civil Engineering*, Old Dominion University, Norfolk, Virginia.

Pearson, J., G. Punzo, M. Mayfield, G. Brighty, A. Parsons, P. Collins, S. Jeavons, and A. Tagg (2018) "Flood Resilience: Consolidating Knowledge Between and Within Critical Infrastructure Sectors," *Environmental Systems and Decisions*, Vol. 38, Issue 3, pp. 318–329.

Pilkey, O. H. (1990) Personal Communication.

Pilkey, O. H., and H. L. Wright (1988). "Seawalls Versus Beaches," *Journal of Coastal Research*, Special Issue No. 4, Autumn, pp. 41–46.

Pilkey, O. H., J. D. Howard, and W. Kaufman (1985) "National Strategy for Beach Preservation," *Proceedings, 2nd Skidaway Institute of Oceanography Conference*, Savannah, Georgia.

Pilkey, O. H., J. D. Howard, B. Brenninkmeyer, R. W. Frey, A. C. Hine, J. C. Kraft, R. Morton, D. Nummedal, and H. Wanless (1982) "I-Saving the American Beach: A Position Paper by Concerned Coastal Geologists," *Shore & Beach*, American Shore and Beach Preservation Association, April, pp. 3–4.

Plant, N. (1990) "The Effects of Seawalls on Beach Morphology and Dynamic Processes," MS Thesis, Institute of Marine Sciences, *University of California, Santa Cruz.*

Plant, N., and G. D. Griggs (1992) "Interaction Between Nearshore Processes and Beach Morphology Near a Seawall," *Journal of Coastal Research*, Vol. 8, No. 1, pp. 183–200.

Prosser *et al.* (2018) "Impacts of Coastal Land Use and Shoreline Armoring on Estuarine Ecosystems," Special Issue, *Estuaries and Coasts*, Vol. 41, No. 1, pp. 2–18.

Roelvink, J. A., A. Reniers, A. van Dongeren, J. van Thiel De Vries, R. McCall, and J. Lescinski (2009). "Modelling Storm Impacts on Beaches, Dunes and Barrier Islands," *Coastal Engineering*, Vol. 56, pp. 1133–1152.

Rosati, J., J. Smith, and M. Bryant (2018) "State of Science & Engineering For Natural and Nature-Based Features," slide presentation, *Short Course, International Conference Coastal Engineering*, Baltimore.

Shepard, F. P. (1976) "Coastal Classification and Changing Coastlines," *Geoscience and Man*, Vol. 14, pp. 53–65.

Silvester, R., and J. C. Hsu (1993) *Coastal Stabilization: Innovative Concepts*, Prentice-Hall, New Jersey.

Slagel, M. J., and G. B. Griggs (2008) "Cumulative Losses of Sand to the California Coast by Dam Impoundment," *Journal of Coastal Research*, Vol. 243, pp. 571–585.

Smallegan S. M., J. L. Irish, Ap R. Van Dongeren, and Joost P. Den Bieman (2016) "Morphological Response of a Sandy Barrier Island With a Buried Seawall During Hurricane Sandy," *Coastal Engineering*, Vol. 110, pp. 102–110.

Thieler, E. R., R. S. Young, and O. H. Pilkey (1992) "Discussion of: Boundary Conditions and Long-Term Shoreline Change Rates for the Southern Virginia Ocean Coastline," *Shore & Beach*, Vol. 60, No. 4, pp. 29–30.

Tove, T., and H. Wang (1990) "Three Dimensional Effects of Seawalls On the Adjacent Beach," *Proceedings of the 22nd International Conference on Coastal Engineering* (Delft) ASCE, pp. 2782–2795.

USACE (2013) "Natural and Nature-based Features (NNBF)-Living Shorelines," US Army Corps of Engineers, Environmental Laboratory, Engineering Research and Development Center, Vicksburg, MS.

Walton, T. L., and W. Sensabaugh (1979) "Seawall Design on the Open Coast," Final report, *University of Florida* Press.

Weggel, J. R. (1988) "Seawalls: The Need for Research, Dimensional Considerations and a Suggested Classification," *Journal of Coastal Research*, Special Issue No. 4, April.

Wright *et al.* (1987) "Shoreline and Beach Dynamics of the Coastal Region from Cape Henry to False Cape Virginia," Technical Report, *Virginia Institute of Marine Science*, Gloucester Point, VA.

Wright, D. (2003) "Aerial Photographs of Shoreline from Cape Henry to Virginia Border" March 1987 to August 2000, *Virginia Institute of Marine Science*, Gloucester Point, VA.

Chapter 11

Sea Level Rise

11.1 Background — Climate Change

The planet Earth is covered with over 71% water in its three forms (liquid, solid, and gas). Most of the liquid phase is saline water (96.5%) contained in the oceans. When the solid phase dominates (Ice Ages), the oceans are lower and cover less surface area. The sun and surface temperatures of both the land and the sea control the liquid and solid phases of the Earth's water.

The last maximum amount of ice in the Earth's glaciers was about 20,000 years ago. As a result, the global mean sea level (GMSL) was at its lowest level, depicted in Figure 11.1 as prepared by Robert A. Rohde for the Global Warming Art project (https://commons.wikimedia.org/wiki/File: Post-Glacial_Sea_Level-fr.svg) from published data (Fleming *et al.*, 1998; Milne *et al.*, 2009). One period of rapid glacier melting labeled "Meltwater Pulse 1A" occurred over 14,000 years ago due to a major meltdown in Antarctica. It is now generally agreed that this lowest level of the mean sea level was 130 ± 10 m (425 ft $\pm$ 33 ft) below present-day levels.

Geoscientists have labeled the period following the last Ice Age, the "Holocene," as the most recent geologic epoch over the past 11,700 years when the last major glacial retreat ended. As shown in Figure 11.2, global mean sea level rise over the past 7,000 years has been slow (3.5 m (11.5 ft)) due to a stable climate and stable ice sheets. The global mean sea level rise rate has been very flat over the past 2,000 years at about 1.9 mm/year from 1900 to 2000 (20th century).

Water, land, and air temperatures are rising (global warming) that are generally believed to have altered the fluid/solid dynamics of planet Earth. Figure 11.3 displays temperature anomaly (difference relative to 1,950°C) for land, sea surface, and ocean air temperature from 1850 to 2015 (top three graphs, left) showing the Earth's warning trend. The slope of the global mean sea level trend is increasing as the extent of the volume of arctic ice is decreasing (bottom two graphs, left).

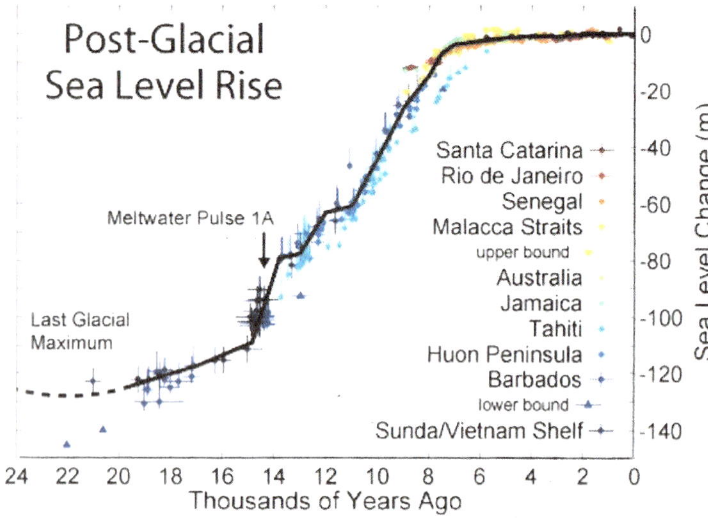

Figure 11.1. Post-glacial, global mean sea level (GMSL) rise since end of last Ice Age over 20,000 years ago (from Wiki commons, Global Warming Art project, credit Robert A. Rhode).

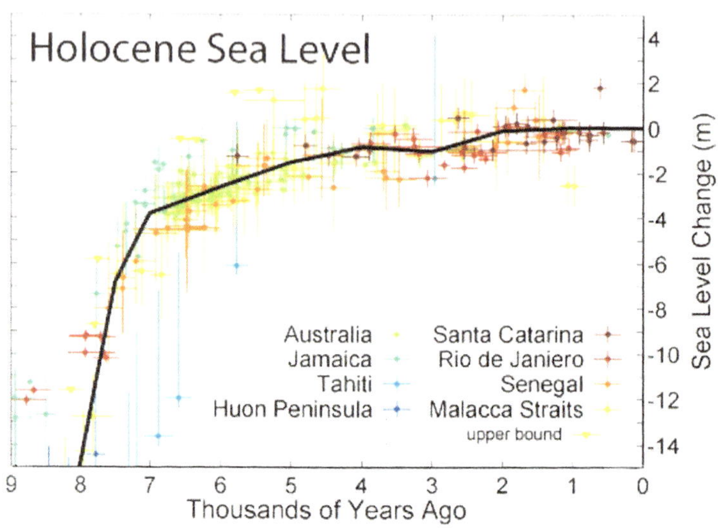

Figure 11.2. Global mean sea level change over the past 8,000 years in the "Holocene" geologic epoch (from Wiki commons, Global Warming Art project).

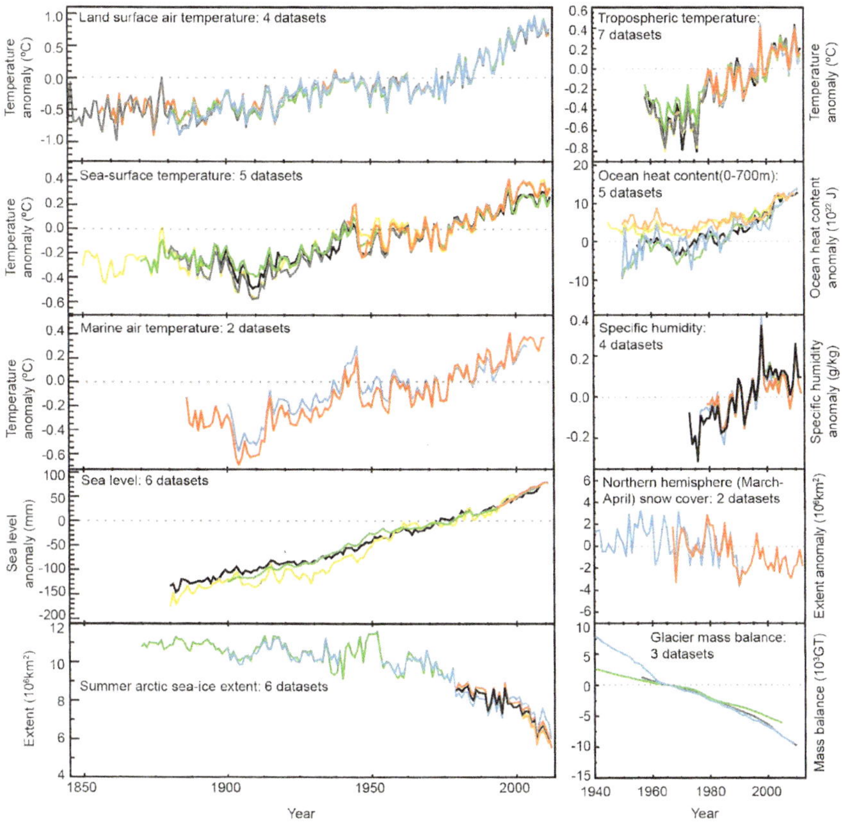

Figure 11.3. Earth's land-surface, sea-surface, and marine air temperature anomaly (relative to 1950) and global mean sea level increase and arctic ice area decrease over the past 150 years (IPCC (2013) Working Group 1, Assessment Report 5 (WG1AR5) Technical Summary).

Since 1880, the average global sea level has risen about 18 cm (7.1 in) from 1897 to 1997 (Douglas, 1997). More precise satellite-based data show about a 7.5 cm (3.0 in) accelerating rise in sea level from 1993 to 2017 (WCRP, 2018). This is due to global warming and accounted for by (1) thermal expansion of warmer sea water (42%) (2) melting temperate glaciers (21%) and ice caps (14%) and (3) ice loss from Greenland (15%) and Antarctica (8%) (Mengel *et al.*, 2016, WCRP, 2018).

In all of the above, the GMSL is presented. It is also known as the "eustatic" (global) water level. Elevations of the adjacent land may also change primarily due to adjustment of the Earth's crust due to pressure from the last ice sheet (rebound) and land subsidence. This is termed the "isostatic" land level. Therefore, at any land site on Earth the *local* "relative sea level" (RSL) change is the combination of the eustatic and isostatic components over a given time period. At some locations, the land may be rising so that the relative sea level change is negative. Positive, local relative sea level rise is the norm throughout the world.

11.2 Literature Review — Sea Level Rise Predictions to 2100

Our understanding of the Earth's climate system is imperfect. Climate research is improving through the use of computer models. One of the major outcomes of climate research are predictions of GMSLs for the year 2100 and beyond. It is instructive to review the published literature over the past 40 years since the concern was first articulated in the early 1980s when the issue of global warming was raised in the media. In general, all the predictions by the organizations cited in the following display a range of curves from lowest, low, medium, high, and highest levels in the year 2100. The ranges and the highest predictions for the year 2100 have *increased substantially* in time. We have purposely not included these earlier prediction graphs to avoid confusion. Anthropogenic impacts on climate change are debatable, but the potential for accelerated sea level rise is not.

11.2.1 *US Army Corps of Engineers (USACE)*

The US Army Corps of Engineers' Shore Protection Manual (SPM, 1984) was initially entitled "Shore Protection, Planning and Design" and originally published in 1954. Chapter 3, Part VIII (SPM, 1984) describes "Water Level Fluctuations," but says nothing about sea levels rising in the Earth's oceans. The latest version is called the Coastal Engineering Manual (2006) and Part IV-1-6 has a section labeled "Sea Level Changes". No curves are included for the 21st century. Figure IV-1-18 presents values of relative sea level rise rates (mm/yr.) based on tidal records from 1940 to 1980.

To provide design guidance, the Corps released Engineer Circular(s) (EC) in 2009 with update in 2011. In 2013 (Dec) the Corps released Regulation ER 1100-2-8162 "Incorporating Sea Level Change in Civil Works Programs" (USACE, 2014). Figure 11.4 displays the three scenarios (low, medium, and high) for GMSL relative to 1992 up to the year 2100 (*the start of 1992 was the median year of the tidal epoch from 1983–2001 used to establish tidal datums in the US by NOAA*). See Section 11.2.3 for the data

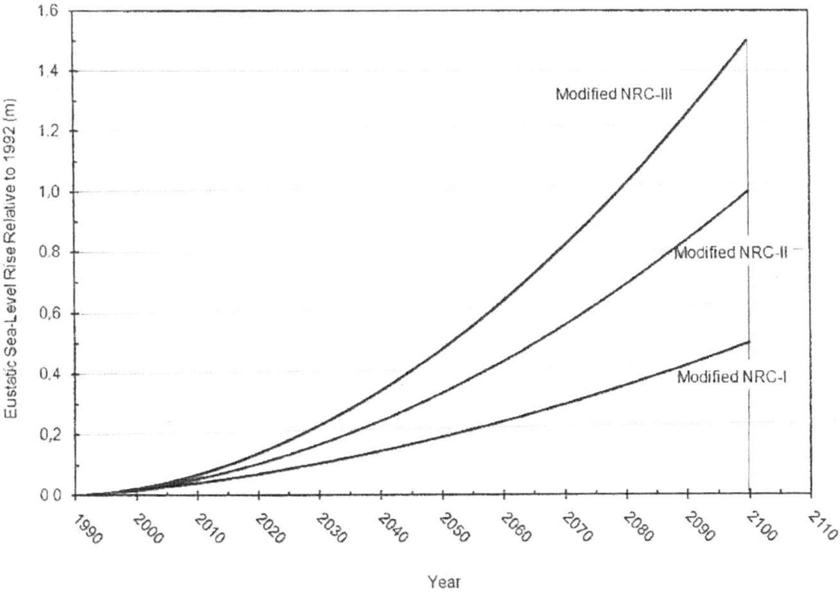

Figure 11.4. Sea level rise curves for three scenarios used by the US Army Corps of Engineers for design of Civil Works projects in the US (from Corps of Engineers, 2013).

and three scenarios originally proposed by the National Research Council (NRC, 1987) and modified by the Corps. For year 2100, the modified results give NRC-I of 0.5 m (1.6 ft), NRC-II of 1.0 m (3.3 ft), and NRC-III of 1.5 m (4.9 ft).

The equation that describes the relation between the global mean sea level rise, $E(t)$, for any time, t, for three curves in Figure 11.4 is parabolic and given as

$$E(t) = 0.0017t + bt^2, \tag{11.1}$$

where b is the 2.71E-5 for Modified NRC-I, b is the 7.00E-5 for Modified NRC-II, and b is the 1.13E-4 for Modified NRC-III.

The coefficient 0.0017 comes from the IPCC (2007) projections (see Section 11.2.5) as the most recent historic, GMSL change rate of 1.7 mm/yr in the published literature.

A "Sea-Level Change Curve Calculator" has also been developed and Version 2019.21 is available online at http://corpsmapu.usace.army.mil/rccinfo/slc/slcc_calc.html. It includes many modern features including the

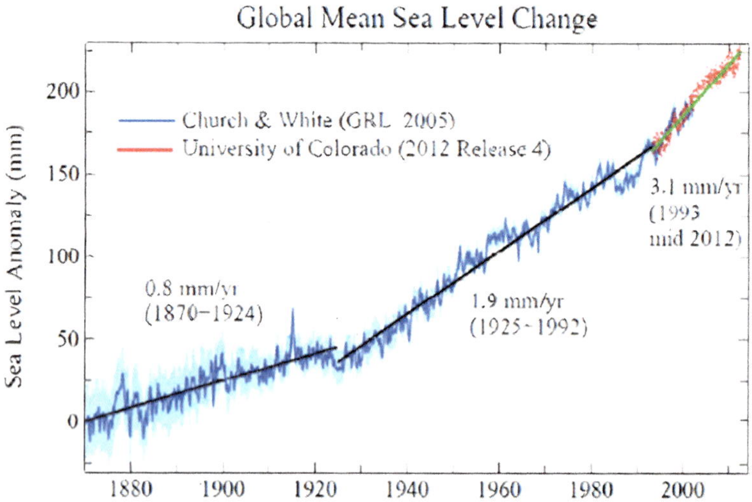

Figure 11.5. Global mean sea level rise rates for three time periods showing the increasing slope with time (http://www.cmar.csiro.au/sealevel/GMSL_SG_2011.html).

local rate of land change (uplifting or subsidence). The key coefficient is the GMSL change rate of 1.7 mm/yr. It can be readily changed in the Calculator. This will probably be revised upward for the next release of the online calculator. As shown in Figure 11.5, Church and White (2006 blue) believe 1.9 mm/yr. is appropriate up to 1992 and that it should be 3.1 mm/yr up to 2012 (https://www.cmar.csiro.au/sealevel/sl_data_cmar.html).

11.2.2 *US National Ocean and Atmospheric Administration*

The US Federal Government science agency "National Ocean and Atmospheric Administration" (NOAA) within the Department of Commerce maintains a Center for Operational Oceanographic Products and Services in their National Ocean Service (NOS). In January 2017, the NOAA Technical Report NOS CO-OPS 083 provided the latest future sea levels for six scenarios with probabilistic projections to the year 2100 as shown in Figure 11.6 (Sweet *et al.*, 2017). The levels shown are similar to those published in 2012 following Hurricane Sandy (NOS, 2012). The extreme level is projected to be 2.5m (8.2 ft) by year 2100.

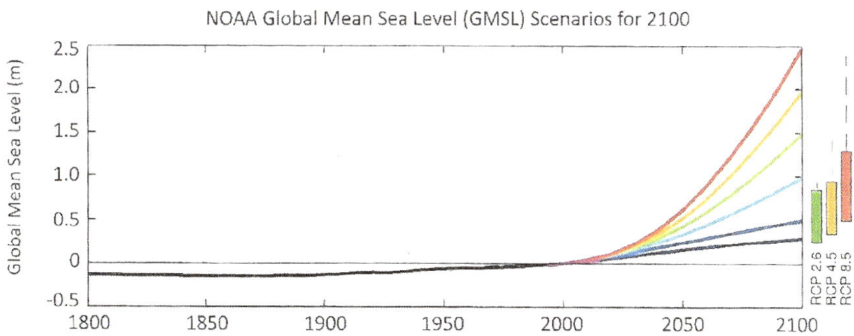

Figure 11.6. Global mean sea levels for six scenarios to year 2100 (from Sweet *et al.*, 2017).

11.2.2.1 *GMSL rise values*

The NOAA predictions (Sweet *et al.*, 2017) start in year 2000 surrounding the 1991 to 2009 tidal epoch. Figure 11.7(a) presents the GMSL rise values for the year 2100 for the six scenarios [Low, Intermediate-Low (0.5 m), Intermediate (1.0), Intermediate-High (1.5 m), High (2.0 m), and Extreme (2.5).]. Figure 11.7(c) gives mean GMSL values by decades to year 2100 and then to year 2200 and Figure 11.7(d) the sea-level rise rates (mm/yr) employed for the six scenarios. Note, for the Low estimate, a constant rise rate of 3 mm/yr was employed for the entire 100 years. These projection estimates of GMSL rise values are important and are based on the work of Kopp *et al.* (2016), Hay *et al.* (2015), Nerem *et al.* (2010), and other projections provided in the literature. All the GMSL rise values over time for the six scenarios are tied together with the year 2100 predictions.

11.2.2.2 *GMSL rise probabilities*

The fundamental premise is that global warming depends on green-house gas (carbon dioxide, methane, ozone, and fluorocarbon) emissions. Estimates of the conditional *probability* of GMSL change have been linked to emission pathways as "Representative Concentration Pathways" (RCPs) first defined by van Vuuren *et al.* (2011). Four RCPs have been evolved to implicitly represent scenarios in a future of uncertainty regarding sea-level rise. RCP 6.0 is similar to RCP 4.5 and omitted in the NOAA report (Sweet *et al.*, 2017) Three RCP levels are summarized in Table 11.1.

The key results presented in this NOAA report (Sweet *et al.*, 2017) are shown in Figure 11.7(c) with probabilities of exceedance in Figure 11.7(b). Consider the Low scenario with a 0.3-m sea-level rise in 100 years to 2100.

(a) GMSL rise scenarios, constraints (year 2100) and number of samples per scenario for year 2100

GMSL rise Scenario	Constraints	Number of samples
Low	2100 GMSL of 30 ± 2 cm; 2050 GMSL of 15 ± 2 cm; 2030 GMSL of 9 ± 1 cm	224
Intermediate-Low	2100 GMSL of 50 ± 2 cm	4407
Intermediate	2100 GMSL of 100 ± 2 cm	1040
Intermediate-High	2100 GMSL of 150 ± 5 cm	168
High	2100 GMSL of 200 ± 5 cm	23
Extreme	2100 GMSL of 250 ± 15 cm	21

(b) Probabilities of exceedance of the mean GMSL values in year 2100 (from Kopp *et al.* 2014)

GMSL rise Scenario	RCP2.6	RCP4.5	RCP8.5
Low (0.3 m)	94%	98%	100%
Intermediate-Low (0.5 m)	49%	73%	96%
Intermediate (1.0 m)	2%	3%	17%
Intermediate-High (1.5 m)	0.4%	0.5%	1.3%
High (2.0 m)	0.1%	0.1%	0.3%
Extreme (2.5 m)	0.05%	0.05%	0.1%

(c) Mean GMSL values to year 2200 for 19-yr tidal epoch centered on decades and starting in year 2000

GMSL Scenario (meters)	2010	2020	2030	2040	2050	2060	2070	2080	2090	2100	2120	2150	2200
Low	0.03	0.06	0.09	0.13	0.16	0.19	0.22	0.25	0.28	0.30	0.34	0.37	0.39
Intermediate-Low	0.04	0.08	0.13	0.18	0.24	0.29	0.35	0.4	0.45	0.50	0.60	0.73	0.95
Intermediate	0.04	0.10	0.16	0.25	0.34	0.45	0.57	0.71	0.85	1.0	1.3	1.8	2.8
Intermediate-High	0.05	0.10	0.19	0.30	0.44	0.60	0.79	1.0	1.2	1.5	2.0	3.1	5.1
High	0.05	0.11	0.21	0.36	0.54	0.77	1.0	1.3	1.7	2.0	2.8	4.3	7.5
Extreme	0.04	0.11	0.24	0.41	0.63	0.90	1.2	1.6	2.0	2.5	3.6	5.5	9.7

(d) Sea -level rise rates (mm/year for 19-yr average centered on decade) and associated with mean GMSL values in table (c)

GMSL Scenario Rates (mm/year)	2010	2020	2030	2040	2050	2060	2070	2080	2090
Low	3	3	3	3	3	3	3	3	3
Intermediate-Low	4	5	5	5	5	5	5	5	5
Intermediate	5	6	7	9	10	12	13	14	15
Intermediate-High	5	7	10	13	15	18	20	22	24
High	6	8	13	16	20	24	28	31	35
Extreme	6	10	15	20	25	30	35	40	44

Figure 11.7. (a) GMSL scenarios for year 2100, (b) exceedance probabilities in year 2100, (c) mean GMSL values by decades and beyond to year 2200, and (d) sea-level rise rates (mm/yr) by decades to year 2090 (from Sweet *et al.*, 2017).

Table 11.1. Definitions of RCP 2.6 (green), RCP 4.5 (yellow), and RCP 8.5 (red) shown in Figure 11.5 (from Sweet *et al.*, 2017). RCP 6.0 is similar to RCP 4.5 and omitted in the NOAA report (Sweet *et al.*, 2017).

RCP Level	Range for GMSL	Reduction of Emissions	Comments
RCP 2.6	Low-end	Strong mitigation	After 2080, negative emissions
RCP 4.5	Middle	Moderate mitigation	Stable emissions, declining after 2050
RCP 8.5	High-end	Fossil-fuel intensive	Business as usual

Notes: Four RCPs were used by the Intergovernmental Panel on Climate Change (IPCC) in Assessment Report 5 (AR5) in 2013 to provide four possible future greenhouse gas concentrations to the year 2300 (see Section 11.2.5 for further details). Each of the four RCP levels represents implicitly possible social, economic, and technological considerations to represent global mean temperatures as "likely" (>66% probability) for the reduction of emissions.

With strong emission regulations (RCP 2.6) the probability of exceedance is 94% and if little or nothing is done to reduce emissions (RCP 8.5) there is a 100% chance sea-levels will be greater than 0.3 m in year 2100. At the other end is the Extreme scenario with a predicted 2.5 m sea-level rise in 2100. If the world's governments can provide strong emission regulations (RCP 2.6) the probability of exceedance is low at 0.05% and 0.1% if business-as-usual is the norm (RCP 8.5). For the Intermediate (1.0 m) scenario with Moderate reduction of green-house gas emissions (RCP 4.5), the probability of exceedance is 3%, but if nothing is done to reduce emissions (RCP 8.5) the probability of exceedance is 17%. Plotting the exceedance probabilities relative to the GMSL values for the six scenarios for each of the RCP levels permits determination of the GMSL values for any probability. At the 1% exceedance probability level, RCP 2.6, RCP 4.5, and RCP 8.5 gave 1.2 m, 1.3 m, and 1.55 m, respectively. This means that even if strong mitigation was in place to significantly reduce emissions, (RCP 2.6) at the 1% annual exceedance level, the year 2100 GMSL would still be over 1 m (3.3 ft) above today's levels. See Section 11.2.6 for further discussion of uncertainty due to Greenland and Antarctic ice-sheet modeling effects on these conditional probability estimates.

11.2.2.3 *Regional climate-related relative sea-level predictions*

These GMSL rise scenarios are then employed to predict regional RSL levels on a 1° coastline grid for Alaska, Hawaii, the US Mainland, the Caribbean, and island territories in the Pacific Ocean. In addition, wherever NOAA maintains tidal stations. The GMSL values were adjusted to

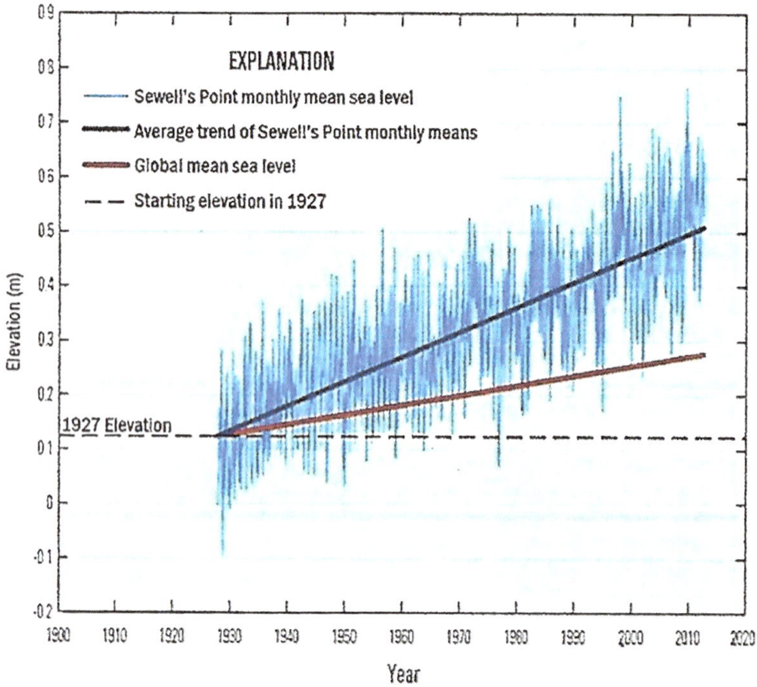

Figure 11.8. Tidal gauge rate of RSL rise for Norfolk, VA (blue), compared with GMSL rise (red) (from Eggleston and Pope, 2013). NOAA tide gauge No. 8638610 at Sewell's Point, Virginia. RSL rate = 4.2 mm/yr.

consider vertical land movements (subsidence or uplift), ocean circulation patterns, and the Earth's rotation and gravitational field. Findings are extremely detailed and beyond the scope of this book. Figure 11.8 is one example for Norfolk, Virginia (US), where subsidence is the main reason the RSL rise rate has been 4.2 mm/yr. Key findings as summarized in Sweet *et al.* (2017) were:

- *For almost all future GMSL rise scenarios, RSL rise is projected to be greater than the global average along the coasts of the US Northeast and the western Gulf of Mexico.*
- *Under the Intermediate and Low GMSL rise scenarios, RSL is projected to be less than the global average along much of the Pacific Northwest and Alaska coasts.*
- *Under the Intermediate-High, High, and Extreme GMSL rise scenarios, RSL is projected to be higher that the global average along almost all US coasts outside Alaska.*

Tidal driven nuisance flooding and NOAA's National Weather Service "flood warning" system are impacted by RSL rise. In general, when the water levels are about 0.75 m (2.5 ft) above the local, mean high, high water (MHHW) levels, the chance is 20% for annual "nuisance" flooding. It was learned that with only a 0.35 m (1.2 ft) increase, the annual frequency for flooding will be 25 times greater. See Section 11.3 for more discussion of nuisance flooding and sea level rise.

These results were a joint effort by NOAA, the US Geological Survey (USGS), the US Environmental Protection Agency (EPA), and Rutgers University. Other participating Federal Government agencies were the DoD, FEMA, NASA, and the USACE. They convene as the US Global Change Research Program (USGCRP) and the National Ocean Council (NOC).

11.2.3 *US national academies, congress, and executive office*

11.2.3.1 *National Academies Science Engineering Medicine, National Research Council*

In 1983, the US Environmental Protection Agency funded a study entitled "Projecting Future Sea Level Rise — Methodology, Estimates to the Year 2100 and Research Needs" (Hoffman *et al.*, 1983). This work was cited in an NSF/ONR (1984) Workshop and became the key finding of the National Acadamies of Science and Engineering's Marine Board report (NRC, 1987) "Responding to Changes in Sea Level: Engineering Implications". Figure 11.9 presents the results adopted in this report with the four data points (low, mid-low, mid-high, and high) from Hoffman *et al.* (1983). The three curves (I, II, and III) had GMSL values of 0.5, 1.0, and 1.5 m, respectively, for year 2100. A coefficient 0.0012 mm/yr was used in Equation (11.1) for the previous century with different b coefficients. This became the basis for the US Army Corps of Engineers modifications for the start year (1992) and the rise rate of 0.0017 mm/yr shown in Figure 11.4 and given by Equation (11.1). Curve III and curve Modified NRC-III give Year 2100 levels of 1.5 m for GMSL which is the same as the Intermediate-High value in Figure 11.7(c) from Sweet *et al.* (2017). The exceedance probability is 1.3% if limited emissions (RCP 8.5) are achieved.

11.2.3.2 *US Congress and Executive Office*

A Congressional Research Service report (CRS, 2016) cited the US National Climate Assessment with very high confidence (no probabilities) of global

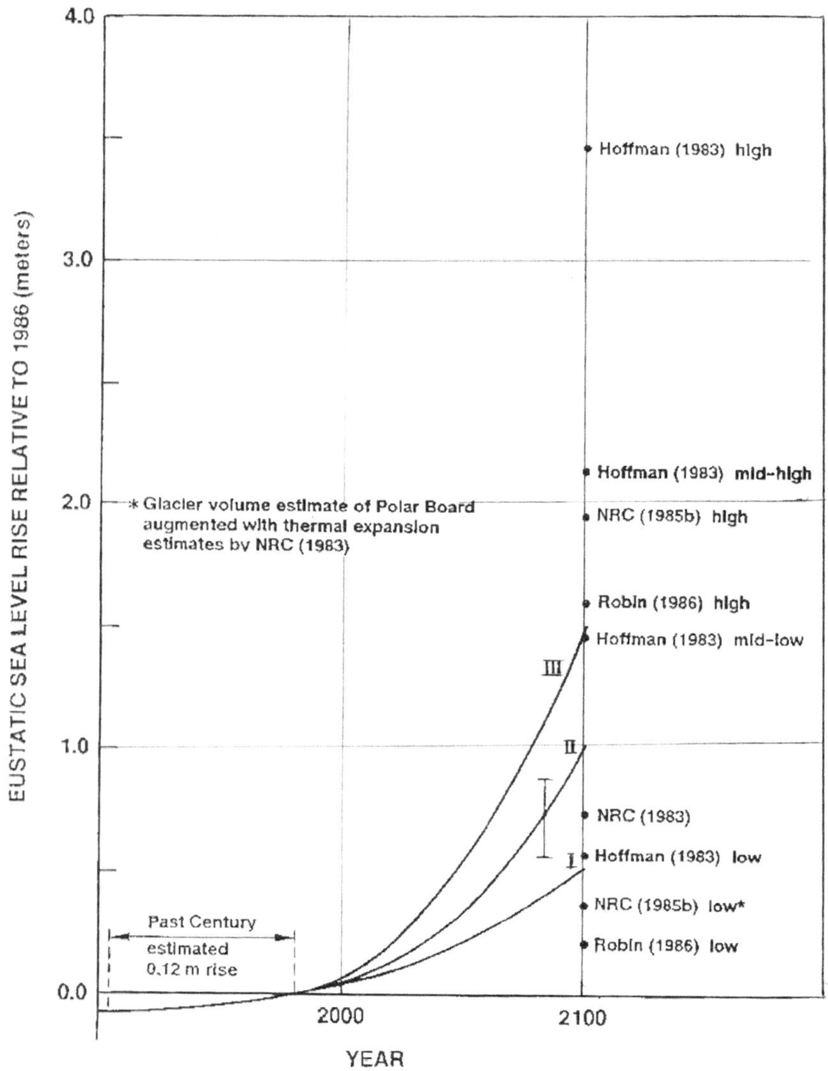

Figure 11.9. Three sea-level rise curves adopted in the National Academies report (NRC, 1987) relative to the year 1986.

sea level rise range from 0.2 m (0.67 ft) to 2.0 m (6.6 ft) by year 2100. This is an extremely wide range of 1.8 m (6 ft) and without associated probabilities, is not very useful for risk and resilience analysis for Coastal Engineering design. References are to IPCC (2013) reports as discussed in Section 11.2.5.

Similarly, the Executive Office of the President of the US has recently released a report (SOST, 2018) with five goals for the next decade of America's oceans. Goal V is "Develop Resilient Coastal Communities", but no specifics are given for sea levels.

11.2.4 *The Netherlands*

The Netherlands is a small, densly populated country in Western Europe. The surface area is 41,526 sq km (16,025 sq mi) about double the size of New Jersey in the US. It is the most densly populated in Europe (>500 people per sq km) with over 17 million people. Over 83% live in urban areas with 40% in Amsterdam, Rotterdam, The Hague, and Utrecht in the southwestern region sometimes called "Holland".

The coastline along the North Sea is 640 km (400 mi) long with over one-quarter (26%) of the land surface area below NAP (https://en.wikipedia.org/wiki/Amsterdam_Ordnance_Datum) which is about the same as GMSL. The purple, hatched region in Figure 11.10 shows this region below sea level along with that in the 0–1 m above sea level (tan) and the area 1 m and more above sea level (brown).

The world's coastal engineering community is interested in how rising sea levels in the North Sea threaten the Netherlands and the countries' projections for year 2100. Fortunately, the Dutch have centuries of history in dealing with rising seas both over the long term as seen in Figure 11.11 and during storm surge events as in 1953 that resulted in the Delta Plan (see Section 6.5 and Figure 6.93 for the Delta Works and Figure 6.94 for the Eastern Scheldt storm surge barrier with movable tidal flood gates). The dikes were raised and lengthened, dunes strengthened and barriers to the North Sea constructed that are also shown in Figure 11.10. Over 55% of the Dutch land surface is located in dike rings that are protected by these flood mitigation structures. Only 4% of the Netherlands surface area lies outside the dike rings and is not protected by dunes, dikes, and storm surge barriers across navigable inlets with movable tidal flood gates. Therefore, the Netherlands is well positioned to face the coming challenge of accelerated sea level rise. Figure 11.12 displays the area flooded if no flood hazard mitigation and adaptation alternatives existed today in the Netherlands. It is often said that "God made the world, but the Dutch made Holland".

In 2008, the Dutch Delta Commission's report said the country must plan for a rise in the North Sea up to 1.3 m (4.25 ft) by 2100 and plan for

Figure 11.10. A map of the Netherlands showing the region below sea level (purple, hatched) that is 26% of the surface area of the country (courtesy Maps, Inc.).

a 2–4 m (6.6–13 ft) rise by 2200. The IPCC (2007) reports were cited to arrive at these numbers, which will probably be revised using the latest projections of the Intergovernmental Panel on Climate Change (IPCC). At the time of this writing (2019), the Dutch Delta Commission has not released revised year 2100 projections using the IPCC (2013) results.

11.2.5 *The Intergovernmental Panel on Climate Change*

In 1988, the World Meteorological Organization (WMO) within the United Nations (UN) and aided by the UN Environmental Program (UNEP) established the Intergovernmental Panel on Climate Change (IPCC). The purpose is to provide scientifically verified and objective assessments of climate

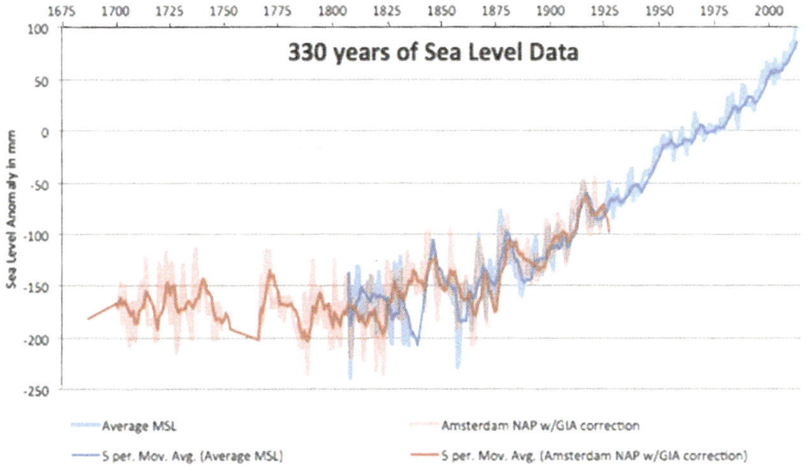

Figure 11.11. Sea level change for the Netherlands from 330 years of tide gauge data (from Caryl, online).

Figure 11.12. Blue area of the North Sea flooding the Netherlands in the "below sea level" regions if the Dutch did not construct dikes, dunes, and storm surge barriers (public domain from en.wikipedia.org/wiki/Flood_control_in_the_Netherlands).

change and its political and economic impacts to the worlds' community. The UN supports an international treaty on climate change called the "United Nations Framework Convention on Climate Change" with the objective to "*stabilize greenhouse gas concentrations in the atmosphere at a level that would prevent dangerous anthropogenic (human-induced) interference with the climate system*". The IPCC produces "Assessment Reports" (AR) that support this international treaty. As of 2018, there are 195 countries in the IPCC.

Four "working groups" (WG) each produce contributions on a five to six-year cycle. These are:

Working Group I (WGI) — The Physical Science Basis;
Working Group II (WGII) — Impacts, Adaptation, and Vulnerability;
Working Group III (WGIII) — Mitigation of Climate Change;
Synthesis Report (SR) — A Summary of the working group reports for the Assessment Report.

Figure 11.13 displays the covers of the Physical Science Basis (WGI) documents for the five IPCC Assessment Reports [AR1 (1990), AR2 (1995),

Figure 11.13. IPCC Assessment Reports, covers of the Physical Science Basis (WGI) contributions (IPCC, 2013 Media Report).

AR3 (2001), AR4 (2007), and AR5 (2013)] that are available on the IPCC website (https://www.ipcc.ch).

Thousands of scientists and experts voluntarily assess, review, and write the reports based on the *published* literature (peer reviewed and non-peer-reviewed sources) that must be approved by all governments. Climate monitoring and/or original research is not conducted by the IPCC. During intervening years between the Assessment Report cycles, Special Reports are also published. For example, *Global Warming of* 1.5°C was released on October 8, 2018, for the purpose of (IPCC, 2018):

> *Understanding the impacts of 1.5° C global warming above pre-industrial levels and related global emission pathways in the context of strengthening the response to the threat of climate change, sustainable development and efforts to eradicate poverty.*

Sea level rise results published in AR5 (IPCC, 2013) as determined by WGI for the range of years are shown in Table 11.2 relative to 1986–2005. Figure 11.14 is a graphical representation of these results in color showing the medians (50%) levels and the 95% confidence interval range. For example, scenario RCP 8.5 for the tidal epoch (2081–2100) has 0.82 m for exceedance probability of 5% and 0.45 m for exceedance probability of 95% for the 95% CI. For RCP 8.5 in the year 2100, the median is 0.74 m and the extreme rise is 0.98 m for exceedance probability of 5% (tidal epoch from 2091 to 2110). If greenhouse gas emissions are controlled near the end of the 21st century (RCP 2.6) the mean would be lowered to 0.44 m and the high end would be reduced to 0.61 m. Figure 11.14 is from the Summary for Policy Makers (SPM) in AR5.

Values shown in Table 11.2 for median and range are said to be "likely" results for year 2100. See the WG I report, Section 13.5.1 (p. 1179) for the quote:

> *...the 5 to 95% range of model results for each of the GMSL rise contributions that is projected... is interpreted as the likely range.*

Recall from Figure 4.4 of Chapter 4 for the normal distribution, the 90% confidence interval equals $\pm 1.64\,\sigma$, the 95% confidence interval equals $\pm 1.96\,\sigma$ and the 98% confidence interval $= \pm 2.33\,\sigma$ where the standard deviation of the mean, σ, is the normalizing variable from the data. The IPCC assumes all projections are normally distributed for their probability distributions.

Results presented in Table 11.2 and displayed in Figure 11.14 resulted from the efforts of over 80 authors from 15 countries all over the world.

Table 11.2. Projections of sea level rise relative to mid-range of two tidal epochs for GMSL (median) and range at the 90% confidence interval. (From Table 13.5, Chapter 13, Sea Level Change in AR5, WGI).

	GMSL, *m*			
	(2046–2065)		(2081–2100)	
Scenario	Median	Range	Median	Range
RCP 2.6	0.24	(0.17–0.32)	0.40	(0.26–0.55)
RCP 4.5	0.26	(0.19–0.33)	0.47	(0.32–0.63)
RCP 6.0	0.25	(0.18–0.32)	0.48	(0.33–0.63)
RCP 8.5	0.30	(0.22–0.38)	0.63	(0.45–0.82)

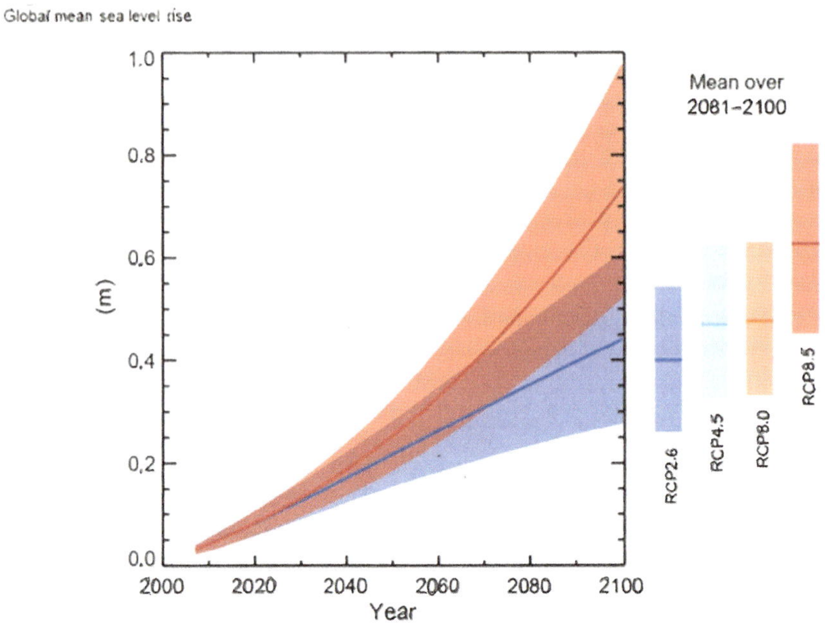

Figure 11.14. From IPCC (2013) AR5, Summary for Policy Makers (SPM), Figure SPM.9.

Chapter 13 "Sea Level Change" in AR5 (IPCC, 2013) produced by Working Group I included over 550 published references from the scientific literature.

From the beginning, the World Meteorological Organization (WMO) aided by the UN Environmental Program (UNEP), when establishing the IPCC, formulated "rules" for its Assessment Reports. The findings and conclusions therein must be based on the "published literature". For AR5 (IPCC, 2013), statements are made that the report does not rule out the possibility of higher sea levels, but there is insufficient evidence in the published literature. Models of the Earth's climate look at warming in the past and give higher projections for how much sea level will rise by the year 2100. These models predict over twice the levels stated in Table 11.2 and are discussed, but scientists do not consider them reliable enough to be included in AR5 (IPCC, 2013).

Figure 11.15 shows these results in Table 11.2 as a bar graph for the (2081–2100) tidal epoch and the five contributions to sea level rise, namely: thermal expansion (dark blue), glaciers (orange), Greenland (green), Antarctica (purple), and other (land/water storage, light blue) giving the totals for GMSL in 2100. (Note the total for RCP 8.5 should be 0.63 m). The values for the five contributions come from Table 13.5 (p. 1182) in Chapter 13, AR5/WGI (IPCC, 2013). The contribution from Antarctica (purple) is on the order of 5–10% for each RCP scenario, relatively small, but likely to increase substantially in the near future. The IPCC is presently in its Sixth Assessment cycle and AR6 is scheduled for 2022. Uncertainty

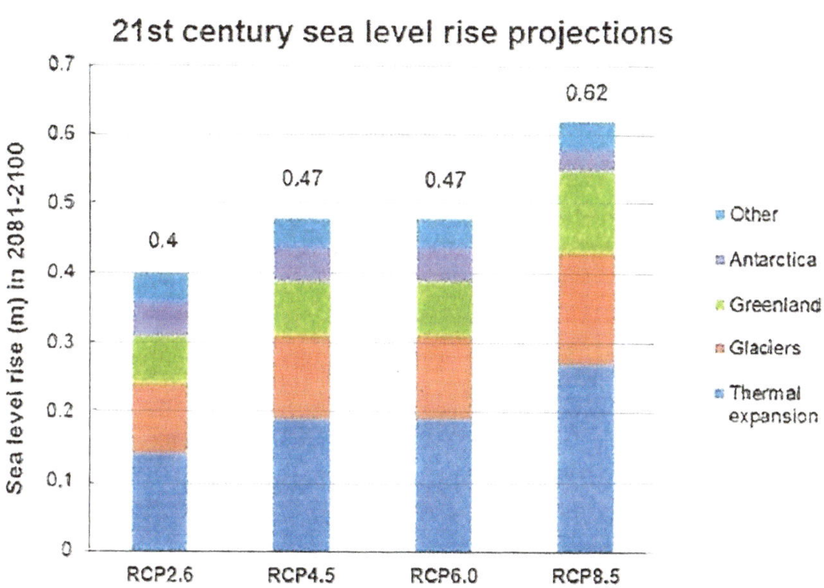

Figure 11.15. Contributions to project mean sea level rise in 2100 (from Table 13.5 in AR5/WG I, IPCC, 2013).

in sea level rise projections for year 2100 as found in the current literature are summarized in the following section and will be scientifically and objectively addressed in IPCC 2022.

11.2.6 *Uncertainty — The Anthropocene Era and the polar regions*

Uncertainty in sea level rise projections for the end of the 21st century is due to many factors. Two are fundamentally important. One is the magnitude of potential greenhouse gas concentrations that contribute to global warming and addressed by use of four "Representative Concentration Pathways" (RCPs) in the work of the IPCC scientist community and Working Group I. Each of the four RCP levels represents implicitly possible social, economic, and technological considerations to represent global mean temperatures as "likely" (>66% probability) for the reduction of emissions. Human influence on global warming has been dubbed the "Anthropocene". As seen in Figure 11.15, thermal expansion alone due to rising ocean water temperatures accounts for roughly 40% in all four RCPs. The second major factor in uncertainty is the two polar regions, Greenland and Antarctica.

11.2.6.1 *The Anthropocene Era*

The history of human activity on Earth has never dealt with the magnitude of climate change as now underway in what can be called the Anthropocene Era. One goal of the world's countries within the framework of the United Nations (started in 1945) is to "Promote Sustainable Development" (*development that promotes prosperity and economic opportunity, greater social wellbeing, and protection of the environment*) and offers the best path forward to improve the lives of people everywhere. The UN supports an international treaty on climate change called the "United Nations Framework Convention on Climate Change" (UNFCCC) with the objective to *stabilize greenhouse gas concentrations in the atmosphere at a level that would prevent dangerous anthropogenic (human-induced) interference with the climate system.* Climate change will impact the world's resources, lands and ecosystem, biodiversity, energy, food security, urban areas, infrastructure, industrial systems, tourism, etc.

Considering human nature and the world's complex political system, this writer believes the goals for greenhouse gas emissions, global temperatures, and hence climate change as represented by the four RCPs are laudable but will *not* be achieved by year 2100. Coastal engineering design should not expect the uncertainty to be reduced by a UN international treaty on climate change. (See for example, the Special Report "Global Warming of 1.5°C", IPCC 2018) (https://www.ipcc.ch/sr15). Coastal engineering design should therefore focus on the deep uncertainty associated with melting of the polar ice cap regions.

All five of the contributions to GMSL shown in Figure 11.15 have some statistical uncertainty, but the greatest is today's weak understanding of rate and volumes of ice-sheet melting to water and entering the oceans from Greenland and Antarctica.

11.2.6.2 *Ice-sheet volumes and sea level equivalents*

It is instructive to first understand the relationship between the volume of ice (km^3) and the global, sea level equivalent (SLE) as millimeters (mm) of sea level rise. Because of the large volumes of ice, the mass of ice is usually cited in metric gigatonne (Gt) units where 1 Gt $= 10^9$ metric tonnes (1 tonne $=$ 1,000 kg mass $=$ mass density of pure water). Therefore, because of some air in the ice mass voids, the mass density of glacier ice and ice sheets $= 916.7\,\text{kg/m}^3$ or $0.9167\,\text{Gt/km}^3$ and less than pure water. The mass of ice in Gt units will then simply be the mass density of ice (0.9167) times the ice volume measured in km^3 units.

In the IPCC 2013, AR5 report, the surface area of the oceans is $3.618 \times 10^8\,\text{km}^2$. If the global mean sea level (GMSL) increases 1 mm, it requires 10^{-3}m^3 of rise for each square meter of ocean surface or in kilometer units this is 1 mm in GMSL rise $= 10^{-6}$ km. Therefore, the volume (km^3) of water in the Earth's oceans required to raise GMSL by 1 mm is $3.618 \times 10^8\,\text{km}^2 \times 10^{-6}\,\text{km} = 361.8\,\text{km}^3$ of water.

Finally, since mass density of ice ($0.9167\,\text{Gt/km}^3$) is less than pure water, the equivalent volume of ice is $(361.8\,\text{km}^3/0.9167) = 394.67\,\text{km}^3$ of ice. In conclusion:

$$1\,\text{mm of GMSL rise} = 394.67\,\text{km}^3 \text{ of glacier ice and ice sheets.} \quad (11.1)$$

As an example, assume a calculated ice volume of $500\,\text{km}^3$. Then $500/394.67 = 1.27\,\text{mm}$ of GMSL rise. If the glacier ice volume is below sea level and not supported by the ground beneath the ice, then it should be subtracted from the ice volume conversion. Floating ice shelves, icebergs, and sea ice does not contribute to sea level rise after melting. See www.antarcticglaciers.org for more details.

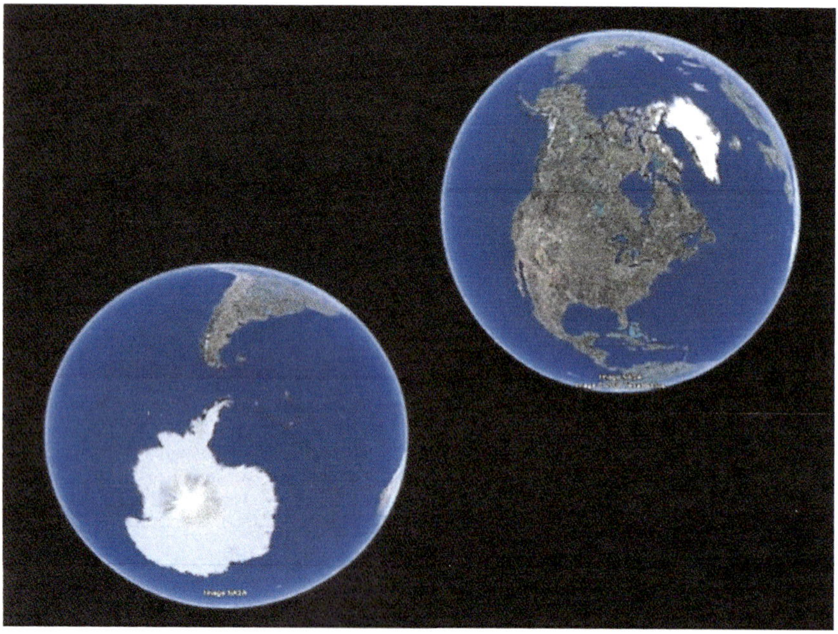

Figure 11.16. Greenland (upper right) and Antarctica (lower left) polar ice caps on Earth (Terra Metrics, NASA images, 2008 North America).

11.2.6.3 *Greenland and Antarctica Polar Regions*

Both Greenland and Antarctica are ice-sheets supported on solid earth at depths far below the sea surface. Therefore, when changing state from solid to liquid, the volume (mass) is added to the earth's ocean volume to increase the GMSL. (In contrast, ice floating in water that changes to liquid water does not increase the water volume (mass)). Figure 11.16 displays both polar ice regions on earth as white areas (from Englander, 2018, slide presentation). Finding out how fast these glaciers will collapse is one of the most important scientific questions in the world today.

(1) Greenland

The surface area of Greenland was 410,450 km^2 (158,500 mi^2) in 2016. Figure 11.17 displays Greenland (right side) in comparison to the size of the East coast of the US and Canada and the Atlantic Ocean (left side) and roughly the same size as Mexico. Melting at the surface of the roughly one-mile thick ice sheet has increased significantly in the 20th and early years of the 21st century. Over one-half of the ice-sheet water entering the

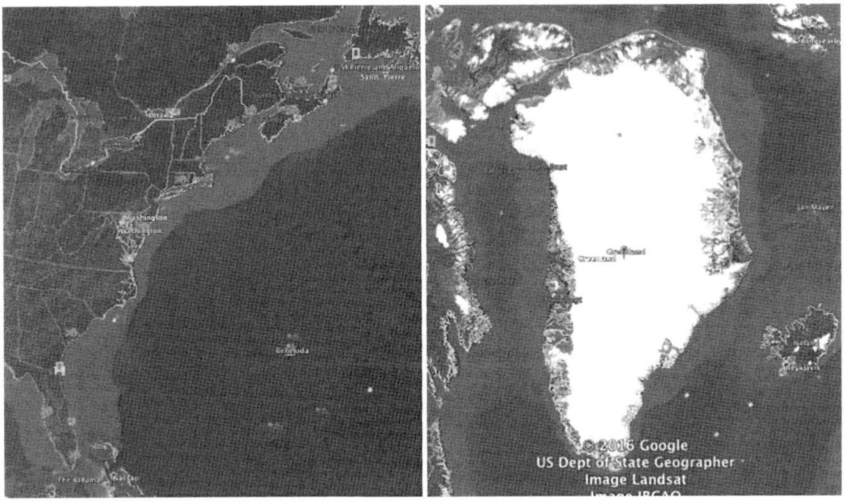

Figure 11.17. Approximate size comparison (Greenland) and North Atlantic Ocean/North America (Operation Ice Bridge, NASA George Hale).

ocean comes from surface runoff from glacial ice and melted snow on top of the ice sheet. The water volume in the surface runoff has been increasing at a nonlinear rate (Trusel *et al.*, 2018).

Recent ice core samples have permitted ice-sheet melt rates over the past 400 years to be determined and compared with today's rates. Scientists at Woods Hole Oceanographic Institution (WHOI) state that today's Greenland ice-melt rates are "off-the-chart" and attribute it to warmer air, land, and water temperatures in the summer. Small temperature changes cause exponential increase in melting and will only accelerate in the future (Trusel *et al.*, 2018).

If all the ice-sheet melted on Greenland, the GMSL would increase approximately 7.36 m (24.1 ft). As seen in Figure 11.15, the contribution under RCP 8.5 would be 0.13 m (21%) in 2100. Current research efforts to correlate increasing ice-melt rates to GMSL rise are being conducted. The contribution from Greenland will increase when the IPCC 2022 is released to reduce the uncertainty. However, the volume of ice is far less than that estimated to be melting from the Antarctic.

(2) Antarctica (the 5th Continent)

If all the ice-sheet that is the Antarctic melted, the estimated GMSL would rise an astounding 58.3 m (191.3 ft) (Vaughan *et al.*, 2013). This is

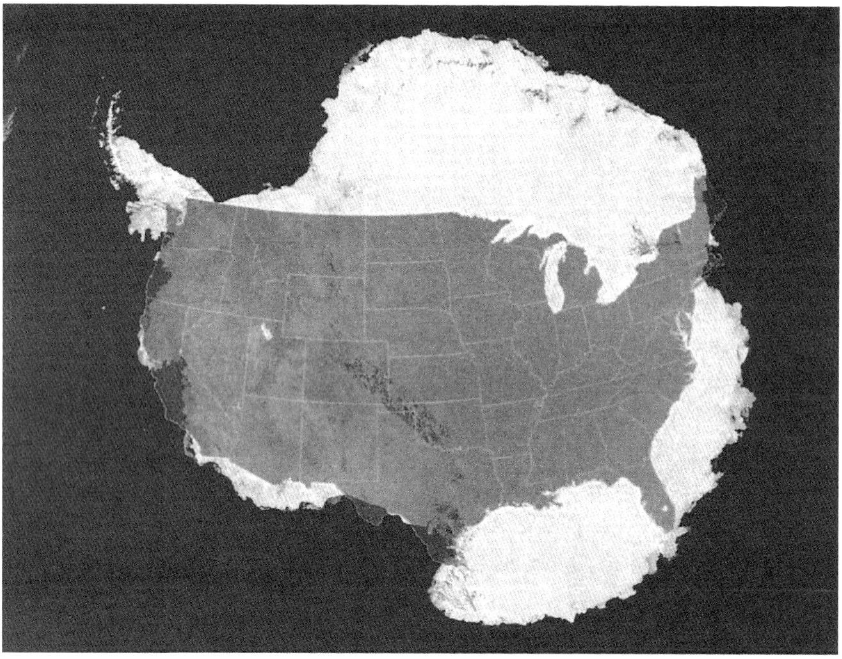

Figure 11.18. Approximate size comparison of the Antarctic (white) and US (Operation Ice Bridge, NASA George Hale).

because the Antarctic contains 90% of the ice on Earth within a surface area of 14.2 million km^2 and average ice thickness of 2,125 m (6,975 ft, 1.3 mi). Figure 11.18 shows the size compared against the US. Antarctica is the Earth's 5th continent and about twice the size of Australia.

Figure 11.19 divides the continent into East Antarctica, West Antarctica, and the Antarctica Peninsula regions that exhibit contrasting levels of ice-sheet physics. Forty countries maintain research stations staffed with approximately 1,000 staff in the winter and over 4,000 people during the summer seasons to conduct research for specific projects. The goal is to make probabilistic, GMSL projections in time [GMSL(t)] that are a function of the greenhouse gas emission scenarios (RCPs, SSPs, etc.) that include the joint probability of all five contributions to sea level rise, namely: thermal expansion, glaciers, land/water storage, Greenland, and Antarctica. Figure 11.20 displays the importance on West Antarctica as the focus of research since its contribution to the total from Antarctica has risen dramatically over the past 25 years.

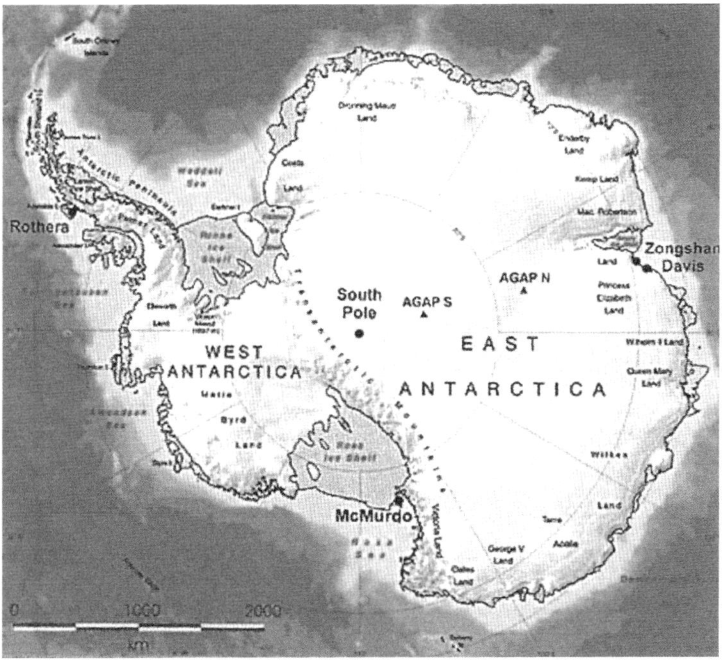

Figure 11.19. Continent of Antarctica and three major regions: Peninsula, West, and East Antarctica (NASA images, Antarctica).

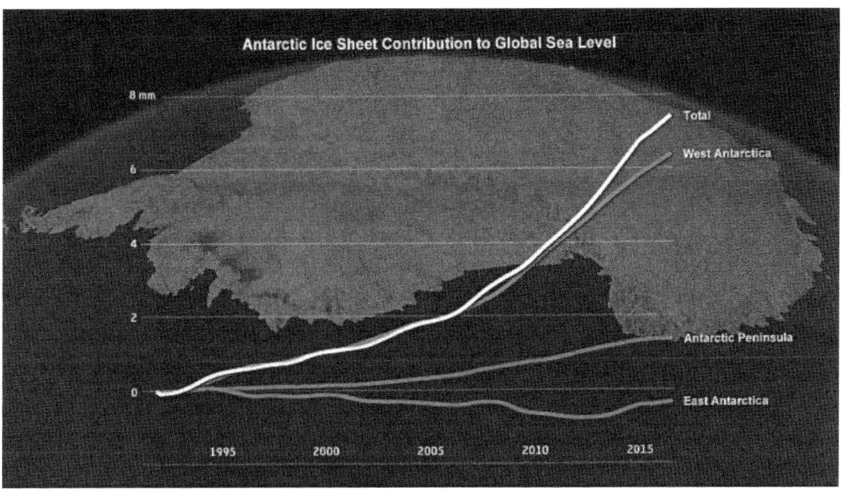

Figure 11.20. Antarctic ice sheet contribution to GMSL rise since 1990 (courtesy Georgia Tech News Center, July 8, 2019).

Since the IPCC (2013, AR5), the following publications have focused on Antarctica: Jackson and Jevrejeva (2016), Kopp *et al.* (2014, 2016), Le Bars *et al.* (2017), Mengel *et al.* (2016), Nauels *et al.* (2017), Slangen *et al.* (2014) and Wong *et al.* (2017). The publication of Le Bars *et al.* (2017) estimates "A high-end sea level probabilistic projection including rapid Antarctic ice sheet mass loss" in *Environmental Research Letters.* Wong *et al.* (2017) discuss "Impacts of Antarctic fast dynamics on sea level projections and coastal flood defense" in *Climate Change,* a publication of Springer publishing company. Figure 11.21 from Nauels *et al.* (2017) displays their findings for "Shared Socioeconomic Pathways" (SSP) as an alternative to the RCP. SSP5 is "business-as-usual" and similar to the RCP8.5 scenario dominated by ongoing fossil-fuel development and high energy demand. The High-end value (brown) predicted is over 1.8 m (5.9 ft) for 2100. Interest in the potential accelerated rate of ice sheet loss in West Antarctica was fueled by the new physical process model of DeConto and Pollard (2016) that included

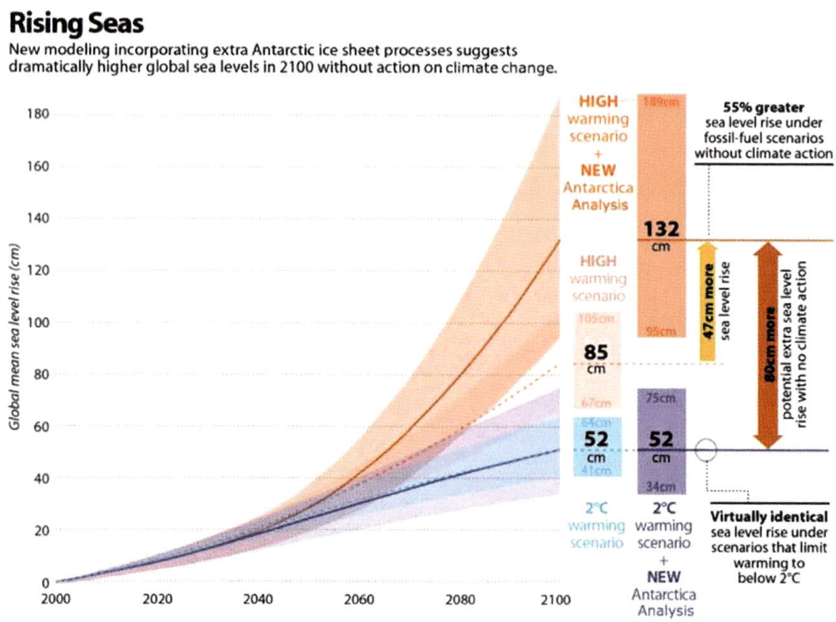

Figure 11.21. Enhanced results from Nauels *et al.* (2017) for Antarctic ice-sheet model results and Shared Socioeconomic Pathways (SSP) storylines (from Inside-Climate News).

ice-sheet hydrofracturing and structural collapse of tall, marine-terminating ice cliffs.

Publications of the American Geophysical Union (AGU) have longstanding relevance to Coastal Engineering design. Recently, the AGU publication *Earth's Future* has become important for keeping abreast of the science communities work in Antarctica. The research article "Evolving Understanding of Antarctic Ice-Sheet (AIS) Physics and Ambiguity in Probabilistic Sea-Level Projections" (Kopp *et al.*, 2017) is strongly recommended to understand the uncertainty in GMSL(t) predictions for year 2100 and beyond. Nine authors (Rutgers University, University of Massachusetts, Columbia University, Harvard University, Princeton University, Pennsylvania State University, and the Lamont Doherty Earth Observatory) contributed to the article and their work over the years has made significant contributions in the literature. The following is a brief summary of the publication by Kopp *et al.* (2017).

(i) Kopp *et al.* (2017)

The main focus of this work is to compare: (1) the conventionally constructed method using expert/structured assessment elicitations (AR5) for Antarctic ice-sheet (AIS) contributions (Bamber and Aspinall, 2013) as found in Kopp *et al.* (2014) labeled "K14" and (2) replacement of these AIS projections with the new, process models of the ice-shelf physics that included ice-sheet hydrofracturing and structural collapse of tall, ice cliffs found in DeConto and Pollard (2016) and labeled "DP16". Kopp *et al.* (2014) was employed in the NOAA report (Sweet *et al.*, 2017) discussed earlier.

Figure 11.22 helps to understand the physics of these new mechanisms in DeConto and Pollard (2016). The schematic shows a cantilevered "ice shelf" that is thinnest at the outer edge and grows in thickness until reaching the "Ridge" which is the grounding line for the cantilevered shelf. Ice shelf below water is supported by buoyancy so is "floating" to some degree, but mainly helps reduce the ice stresses in the cantilevered shelf. Warm ocean water melts the ice shelf from below and surface water runoff creates crevasses for penetration of meltwater called "hydrofracturing". Cliff collapse is triggered when the mass of loose ice "buttress" at the front end is lost causing the tall ice cliff stresses to exceed the yield strength of ice. Sections of the end of the ice shelf break off and the ice shelf retreats.

The grounding line position may also retreat and depends on both the soil material in the Continental shelf and the rate of subsurface ice melt near

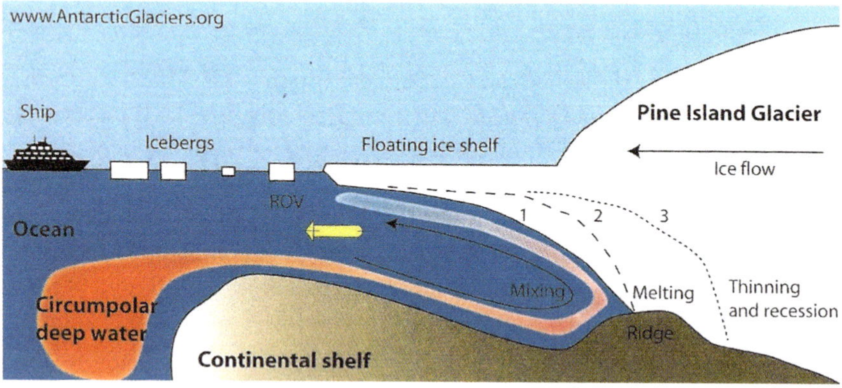

Figure 11.22. Schematic of ice shelf showing (1) warm water melting from below, (2) thinning of ice shelf, and (3) potential recession for grounding line that supports the cantilever stresses in the cantilevered shelf (from DeConto and Pollard, 2015).

this location. Clearly, its retreat to position 3 would significantly weaken the shearing strength of the cantilevered ice shelf to speed up the ice flow onto the shelf from the upland glacier. The schematic shown as Figure 11.22 says "Pine Island Glacier," but is the same for many glaciers in West Antarctica.

Figure 11.23 reproduces the main finding for projections of GMSL rise in centimeters (cm) as reproduced from Table 1 in Kopp *et al.* (2017). Four time periods are projected (2050, 2100, 2200, 2300) for three global warming temperature (forcing) scenarios for greenhouse emissions (RCP 8.5, RCP 4.5, and RCP 2.6) and five ranges of "simulation frequency distributions" that are similar to probability distributions. We are mainly interested in GMSL(t) projections to year 2100 to compare the results using the "K14" methodology and the "DP16" methods using the new physical processes model. The numbers in Figure 11.23 are shown as graphs in Figure 11.24(a) for K14 and in Figure 11.24(b) for DP16. Blue is for RCP 8.5 (business-as-usual), orange for RCP 4.5 (the middle ground to control emissions), and green for RCP 2.6 (strongest world response to control emissions).

Major conclusions from this research effort (Kopp *et al.*, 2017) are as follows:

(1) For year 2100, the central tendency (50%) GMSL rise under the RCP8.5 scenario shifts significantly upward for DP16 (146 cm) compared with

Projections of Global-Mean Sea-Level (GMSL) Rise (cm)

	50	17–83	5–95	1–99	99.9
K14 RCP 8.5					
2050	29	24–34	21–39	17–46	59
2100	79	62–101	51–123	40–159	232
2200	195	131–284	94–380	64–552	886
2300	318	175–516	98–737	37–1093	1929
RCP 4.5					
2050	26	21–31	18–35	15–41	55
2100	59	44–77	35–95	26–128	205
2200	126	70–197	36–278	8–433	780
2300	192	70–349	0–531	−55–900	1717
RCP 2.6					
2050	24	20–29	18–33	15–40	55
2100	49	36–66	28–84	20–120	203
2200	97	48–163	23–242	6–406	803
2300	142	32–288	−22–470	−57–847	1773
DP16 RCP 8.5					
2050	31	22–40	17–48	13–54	59
2100	146	109–209	93–243	80–267	297
2200	719	595–896	558–962	525–1049	1193
2300	1169	980–1409	913–1552	861–1751	2006
RCP 4.5					
2050	26	18–36	14–43	10–52	57
2100	91	66–125	50–158	39–180	197
2200	266	176–396	133–455	102–510	594
2300	421	275–595	211–696	163–806	995
RCP 2.6					
2050	23	16–33	12–41	9–50	54
2100	56	37–78	26–98	18–111	122
2200	110	70–161	47–206	30–250	314
2300	142	83–230	50–300	22–404	552

Figure 11.23. Projections of GMSL rise in time from year 2000 for K14 method and DP16 model for Australia (from Kopp *et al.*, 2017, Table 1).

K14 (79 cm). At the 95 % confidence interval level (5-95), the results were for DP 16 (93-243) and for K14 (51-123). The extreme levels approach 3 m for DP16 (297) and are 2.3 m for K14 (232).

(2) In the first half of the 21st century, both K14 and DP16 produced similar contributions for Antarctic projections to GMSL in year 2050.

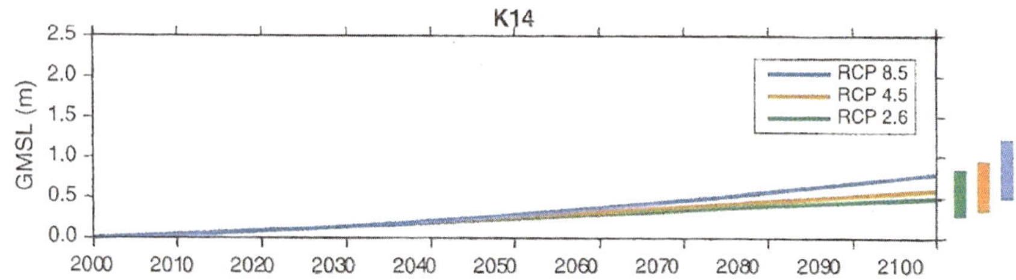

(a) Global mean sea level rise (meters) using K14 methodology to year 2100.

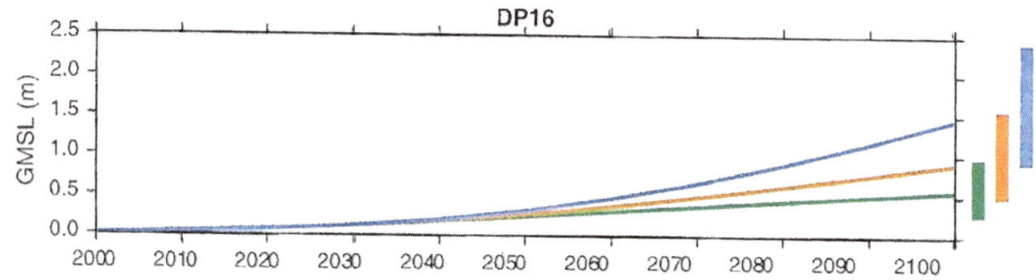

(b) Global mean sea level rise (meters) using DP16 process model to year 2100.

Figure 11.24. Graphical results from Figure 11.23 up to year 2100 (from Kopp *et al.*, 2017). Colored bars on right side are for three emission scenarios over (5–95) range of simulation frequency distribution.

And, both K14 and DP16 projections for GMSL in 2050 show minimal dependency on the RCP emissions scenarios.

(3) For the second half of the 21st century to year 2100 (and beyond), sea-level projections incorporating DP16 physics for Antarctica Ice Sheets (AIS) are significantly more dependent on the RCP emissions scenarios than the K14 projections.

These conclusions mean that if GMSL rise rates are moderate and current research improves the DP16 modeling effort, little change in year 2050 projections may be the result for both K14 and DP16 methods. However, "extreme" outcomes for year 2100 cannot be excluded. The consequences of these conclusions for coastal engineering design are discussed in Section 11.3. The DP16 model simulations increase our understanding of physical processes, but it only uses the single study efforts of DeConto and Pollard (2016) that may not include all the important processes such as geothermal heat spots at various locations beneath the glaciers (Schroeder *et al.*, 2014). Their modeling efforts and those of others will be enhanced by new research such as that recently started on the Thwaites Glacier.

(ii) Thwaites Glacier

The Thwaites Glacier has been under study since the 1980s. It is located on the west coast of West Antarctica as seen in Figure 11.25 (upper left side, red dot) and below the Pine Island Glacier on the Amundsen Sea (upper middle). This glacier is about the size of Britain or the state of Florida in the US. The amount of ice flowing from the 120 km (75 mi) wide mouth has doubled in the last 30 years. Warm water circulates beneath the shelf as illustrated in the schematic (Figure 11.25 upper right side) which also shows a grounding zone (not line) where subglacial lake and outflow may take place.

On April 30, 2018, the US National Science Foundation (NSF) together with the UK National Environmental Research Council (NERC) announced the start of a $50 million, five-year joint research project called the International Thwaites Glacier Collaboration (ITGC). Eight research projects illustrated in Figure 11.25 (lower right side) will be undertaken (acronym in parenthesis):

1. Geophysical habitat of subglacial sediment, hydrology, and bedrock underlying the glacier (GHOST).
2. Melting processes at ice-ocean interface of glacier (MELT).
3. Measuring the ocean circulation and glacier thinning beneath the floating shelf (TARSAN).
4. Geological history of constraints on grounding zone retreat (GHC).
5. Understanding glacier margins and what controls width and speed of retreat (TIME).
6. Computer modeling of calving and associated processes to cause collapse and rapid retreat (DOMINOS).
7. Modeling the history and evolution of processes, drivers, and predictions (PROPHET).

The project goal is to collect field data needed to understand whether the glacier could begin to collapse in the next few decades or centuries. See "The International Thwaites Glacier Collaboration" (http://thwaitesglacier.org) for more details. Some published results should be available for IPCC 2022, AR6 for use in making the next round of GMSL rise predictions for 2100.

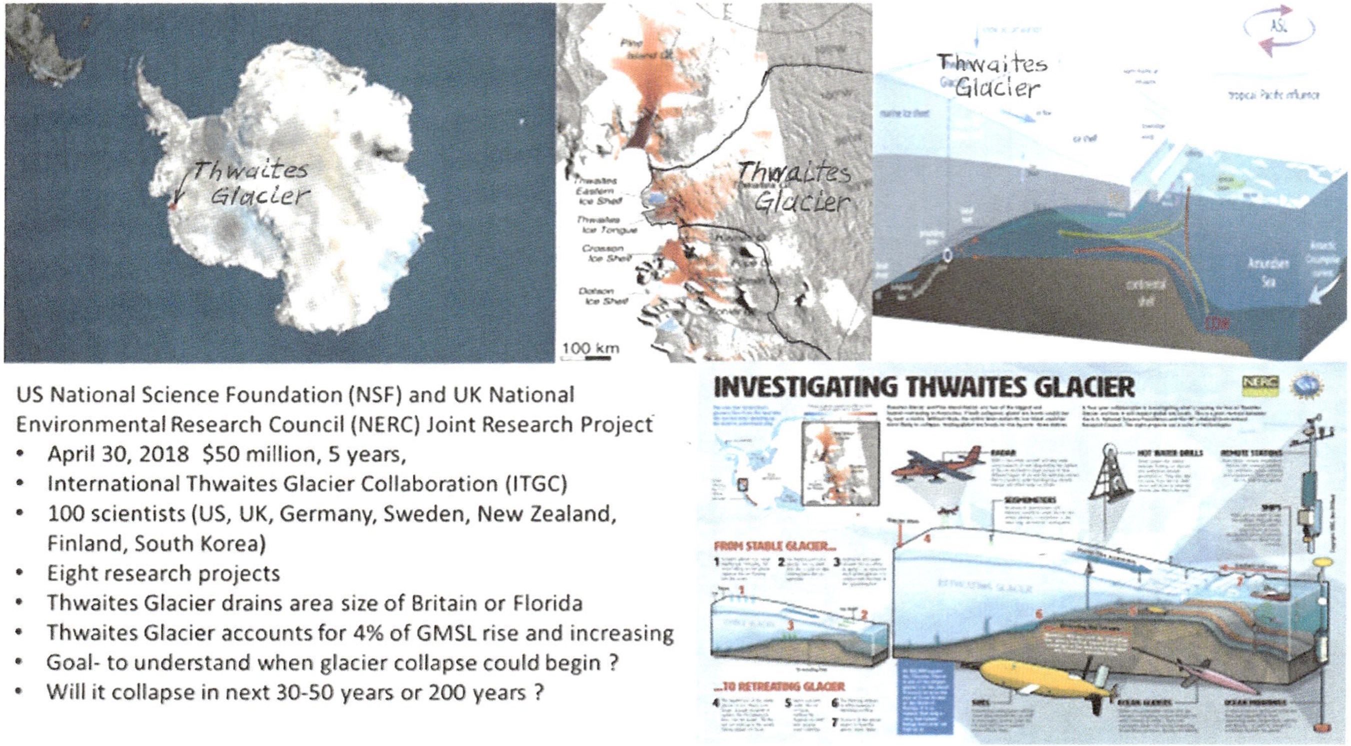

Figure 11.25. Thwaites Glacier location, schematic and overview of new, joint US and UK research project (Authiors composite from references listed for Thwaites Glacier).

11.2.7 *Summary*

The High-end GMSL (meters) projections for the year 2100 from six sources reviewed earlier are summarized by chronological year of publication in Table 11.3. Included are estimates of exceedance probabilities at the 5% level (95% confidence interval) if given in the references cited. These six sources represent US Federal Government organizations (USACE, NOAA, National Academies), the Netherlands, the United Nations' Intergovernmental Panel on Climate Change (IPCC, 2013) and one recent scientific publication (Kopp *et al.*, 2017) that explores the "deep uncertainty" of year 2100 predictions surrounding the West Antarctic ice-shelves and glaciers. The values are for RCP 8.5, the business-as-usual scenario.

Since 2013, predictions for year 2100 with 5% exceedance probability have been increasing. As more is learned about Antarctica's and the Arctic's contributions from ongoing research, this trend in GMSL predictions will probably increase in time. Similar findings are found in Garner *et al.* (2018).

Predictions of GMSL(t) for intermediate years can be made using Equation (11.1) (and see, Equation (11.2)). Predictions for regional, climate-related, relative sea-level rise (RSL) begin with the world's global mean (GMSL) and depend on many local factors such as land subsidence, land rising, etc. and are not addressed herein. In the US, the NOAA reference (Sweet *et al.*, 2017) and the USACE sea level rise "Calculator" can be employed to estimate RSL at your location. See Section 11.3 for how to combine local, historically determined annual, exceedance probability distributions of water levels (e.g. Figure 2.3) with predictions of future levels GMSL(t).

Table 11.3. Chronologic summary of High-end GMSL projections for year 2100 and RCP 8.5.

Source	Year Publication	High-end GMSL, m	Year Start	Probability Exceedance, $p_>$	Reference
National Academies	1987	1.5	1986	unknown	NRC (1987)
Netherlands	2008	1.3	unknown	unknown	Dutch Delta Comm
USACE	2013	1.5	1992	unknown	NRC-III modified
IPCC	2013 (Oct)	0.98	2000	5%	AR5, WG I
NOAA	2017 (Jan)	1.3	2000	5%	Sweet *et al.* (2017)
Kopp *et al.* (RCP8.5)	2017 (Dec)	2.43	2000	5%	DeConto and Pollard (2016)

Selection of a value for GMSL (m) in the year 2100 for use in the *Design of Coastal Hazard Mitigation Alternatives for Rising Seas* (this book) is addressed in the next section.

11.3 Sea Level Rise — Impact on this Book

Sea level rise will continue for many centuries in the future. The major concern is the accelerating rate of rise due to many factors (natural and anthropogenic) and as a result we cannot predict what the future GMSL(t) will be, based on past history. What the future holds for the world's sea level is uncertain. This fact impacts all the previous 10 chapters in this book. Therefore, in this section, we examine what this means for each chapter because today's designs must be undertaken within these circumstances. Each previous chapter refers to Chapter 11 and Section 11.3 so that readers and users of this book may have already been introduced to the material in the following. Here, we will only discuss those parts in each chapter that are impacted by a future of unknown rate of sea level rise. Fortunately, the vast majority of the book remains unchanged. But a word of warning. We are entering a new Anthropocene Era when non-stationarity is the new norm. Old methods that rely on time series of "stationary" coastal processes (water levels, waves, sediment transport, etc.) will no-longer work (see Section 11.3.2 below) unless they are "corrected" for predicted GMSL in the future. This will be an exciting time for Civil/Coastal engineering.

11.3.1 *Chapter 1 — Introduction*

Four time periods are introduced for "coastal engineering" in Section 1.1.1. They are as follows:

(1) "naturalistic" antique age (circa 3000BCE–20BCE),
(2) "experience" technology age (20BCE–1950CE),
(3) "engineering/science" computer age (1950CE–1990CE),
(4) "internet" digital information age (circa 1990CE–present).

The fourth (4) roughly coincides with the beginning of the Anthropocene Era (circa 1980s–future).

Defining the problem is the first step (Table 1.1) in a Feasibility Study (Section 1.2). Coastal engineers must become involved in the community planning process, networks, etc. and work with resilience officials because residents of coastal urban communities will be most affected by coastal flooding. What is "coastal engineering"? should not continue to be the first question asked by the people and the media when beginning a feasibility study to mitigate and adapt to flooding problems.

Design constraints are discussed in Section 1.2.4. The first one is (a) "Scientific and engineering understanding of nature". As summarized above in Section 11.2.6, the "deep uncertainty" surrounding the future melting of the world's polar ice regions should become standard knowledge for all coastal engineers. It is the number one scientific and engineering design constraint. GMSL predictions are evolving upward in time so that keeping abreast of the literature is important.

Major urban coastal cities (New York, Boston, Houston) have undertaken feasibility studies for dealing with future ocean water levels and flooding. The alternatives considered to mitigate and adapt in these studies should be reviewed for selection of the "Tentatively Selected Plan" (TSP), its cost, and the benefit cost ratio (see also Sections 11.3.3 and 11.3.8). It is important to keep up with what other coastal cities are doing in this regard.

Government agencies and consulting engineering firms responsible for design should appoint at least one individual to keep abreast and active in each of the three areas (community involvement, scientific literature, plans of other cities) mentioned earlier.

11.3.2 *Chapter 2 — Design criteria*

Water depth is arguably the most important variable for design of alternatives for hazard reduction and new adaptation strategies. Kamphuis (2010) stated *Coastal design is primarily a function of water levels.* Four methods to determine annual, exceedance probability curves for coastal water elevations resulting from tides, storm surges, and other factors (e.g. El Nino events) are presented in Section 2.2.4. An example annual, exceedance probability curve is shown in Figure 2.3 for Norfolk, Virginia. Historical data for 51 years (1927–1978) for the Sewells Pt. tide gauge (station 8638610) was employed in the analysis and includes local subsidence as shown in Figure 11.7. The annual, exceedance probability distribution curve depends on the length of record when historical tidal is employed and local, relative sea level rise. Climate change is increasing GMSL and also expected to increase the number and strength of coastal storms so that the impacts of future storm surges and coastal precipitation events will change the annual, exceedance probability levels. Methods have been developed to combine today's local exceedance probability distributions with future predictions of sea level rise.

The common thought is to simply add a constant level of predicted sea level rise to all levels in today's local exceedance probability curve. For extreme water levels, this is incorrect. The correct method is shown below as Equation (11.9) and depends on the annual exceedance probability level of interest and the N-year probability level selected for the design life, N (years), of the project.

As noted in Table 11.3, some predictions for GMSL in year 2100 have annual, exceedance probabilities attached, and some do not. The correct method as developed by Houston (2013a,b) requires exceedance probabilities for the predicted sea level rise levels (IPCC, 2013; NOAA, 2017) as summarized below.

11.3.2.1 *Water level — Extremes*

The method of Houston (2013b) employs the basic IPCC assumption that sea level rise (and temperature) projections are normally distributed. Normal probability distributions permit the use of the mean value, μ, standard deviations of the mean, σ, and confidence intervals (C.I.) for defining limits of the GMSL rise prediction. As shown in Figure 4.4 of Chapter 4, for the normal distribution, the 90% confidence interval (C.I.) level $= \pm 1.64\,\sigma$. Using the probability distribution equations, the 95% C.I. $= \pm 1.96\,\sigma$ and the 98% C.I. $= \pm 2.33\,\sigma$. Formulas for the normal (standard) probability distribution function, $P(z)$, cumulative distribution $[(1 - P(z)]$, and the probability, $Q(z)$, that $P(z)$ will be equal or exceeded can be found in Benjamin and Cornell (1970) or any standard text on statistics and probability. The percentage value of the confidence interval is the area beneath the normal distribution curve.

The IPCC employs sea level rise projections in time, $SR_{p\%}$ (2100), at the 95% and 50% confidence interval levels for their year 2100 projections. This means $\mu = SR_{50\%}$ (2100) and $SR_{95\%}$ (2100) is the upper bound of the 95% confidence interval. As a result,

$$SR_{95\%}(2100) = \mu + 1.96\,\sigma$$
$$= SR_{50\%}(2100) + 1.96\,\sigma,$$

so that

$$SR_{95\%}(2100) - SR_{50\%}(2100) = 1.96\,\sigma \tag{11.2}$$

similarly,

$$SR_{98\%}(2100) - SR_{50\%}(2100) = 2.33\,\sigma. \tag{11.3}$$

For the 98% confidence interval, 1% falls above and 1% falls below the interval. Therefore, if the annual exceedance probability, $P_>$, is at the 1% level (100-yr recurrence interval) for the design flood, taking the ratio of Equations (11.2) and (11.3) gives

$$SR_{98\%}(2100) - SR_{50\%}(2100) = [SR_{95\%}(2100) - SR_{50\%}(2100)] \times (2.33/1.96). \quad (11.4)$$

Sea level rise projections in time, $SR_{p\%}(t)$, are then employed that must have a confidence interval level attached and be for a certain time in the future.

Houston (2013a) used the IPCC quadratic curve for sea level rise projections in time given by

$$SR_{p\%}(t) = a_o + [a_1 \times (t - 1990)] + \left\{\frac{1}{2}a_2 \times (t - 1990)^2\right\} \quad (11.5)$$

because the first IPCC report (IPCC, AR1) was published in 1990. Values and formulas for the three coefficients a_o(m), a_1(m/yr), and a_2 (m^2/yr) are found in Houston (2013a) as taken from Bindoff *et al.*, (2007) for IPCC (2007), AR4 and as summarized herein.

If the projection start time is 1990 as in Equation (11.5), then $a_o = 0$. Over the 10-year period (1992–2012), tidal data and satellite altimeter data produce a GMSL rise rate of 3 mm/yr. Therefore, set $a_1 = 0.003$ m/yr. For local RSL rise, use the local rate of sea level rise, as for example shown in Figure 11.7 for Norfolk where $a_1 = 0.0042$ m/yr due to subsidence. Note, the start dates for the last three entries in Table 11.3 are year 2000.

Given $SR_{98\%}$ (2100) and $SR_{50\%}$ (2100), Equation (11.5) is then employed in reverse to solve for coefficient $a_{2(p\%)}$.

$$\begin{aligned} a_{2(98\%)} &= (2/110^2) \times [SR_{98\%}(2100) - (0.003 \times 110)] \\ &= 0.000165 \times [SR_{98\%}(2100) - 0.33] \end{aligned} \quad (11.6)$$

and

$$a_{2(50\%)} = 0.000165 \times [SR_{50\%}(2100) - 0.33]. \quad (11.7)$$

Using these values in Equation (11.5) gives equations for $SR_{98\%}(t)$ and $SR_{50\%}(t)$, for each year, t from 1990 to 2100. These two equations form the basis of the method and what is required are predictions for GMSL for year 2100 ($SR_{50\%}$) and $SR_{95\%}$ as shown in Table 11.3 using Equation (11.3) to find $SR_{98\%}$ (2100).

The second step is to employ the current annual, exceedance probability curve as determined by any of the four methods (Section 2.2.4) for the location of interest. If the annual, probability of exceedance is 1% for the design level, then at the 98% confidence interval, $F_{98\%}$ is the elevation at the exceedance probability level, $P_{>} = 1\%$. The median (central tendency) probability level elevation is $F_{50\%}$. Both $F_{50\%}$ and $F_{98\%}$ are assumed known and not time-dependent with nomenclature following Houston (2013b). Figure 2.3 in Chapter 2 is an example annual exceedance probability distribution curve for the Sewells Pt. tide gauge in Norfolk, VA. When possible, extreme value theory (see Section 2.3.8) should be considered to fit a mathematical equation to the distribution.

The final step is to combine the sea level rise predictions, $SR_{p\%}(t)$, with the annual, exceedance probability distribution curve for each year to give the total elevation, $T_{p\%}(t)$. The mean $F_{50\%}$ is assumed independent of time so that adding means for the total, $T_{50\%}(t)$, gives

$$T_{50\%}(t) = F_{50\%} + SR_{50\%}(t). \tag{11.8}$$

At the 1% design level and 98% confidence level, for each year of total sea level rise, $T_{98\%}(t)$, the error intervals must be added together by quadrature (Castrop, 2004).

$$T_{98\%}(t) = T_{50\%}(t) + \text{SQRT}[\{F_{98\%} - F_{50\%}\}^2 + \{SR_{98\%}(t) - SR_{50\%}(t)\}^2]. \tag{11.9}$$

This is the total, combined coastal flood and sea level rise *for each year* at the *annual* probability of exceedance level of 0.01 (1%). This is the fundamental reason for the IPCC using the normal probability distribution for the predicted probability distribution of GMSL(t). And, Equation (11.9) is the main reason why simply adding a constant level for GMSL to the local, annual exceedance probability distribution curve is incorrect. If $\{SR_{98\%}(t) - SR_{50\%}(t)\} \approx 0$, then this simplified method would be approximately correct.

At other levels of annual probability desired for the total combined flood level, other values of F, SR, and T can be employed for the desired value of $p_{\%}$ keeping in mind that new values for σ will be needed in Equation (11.3).

Finally, in Section 4.8.1, we defined the N-year probability, P_N, as the probability over a period of N years for the annual, probability of exceedance, $P_{>}$, of the variable of interest. The recurrence interval $T_r = 1/P_{>}$. We used the symbol T_r here with subscript for the recurrence interval to distinguish from wave period T.

The N-year probability P_N is then

$$P_N = 1 - (1 - P_>)^N = 1 - \left(1 - \frac{1}{T_r}\right)^N \qquad (4.43) \text{ and } (11.10)$$

For example, for a $N = 50$-year period, the $P_> = 1\%$ annual chance combined coastal flood and sea level rise will have an N-year probability of 39.5% $[(1-(1-0.01)^{50} = 0.39499]$. The so-called 100-yr recurrence interval event (1% chance in any one year) will have a 39% chance of occurrence in 50 years. Equation (4.43) and here (11.10) are related to the binomial distribution and each annual probability must be independent each year. The number of encounter years, N, could be related to the design life of the project. The number of years N has no relationship to the recurrence interval, T_r.

For flood insurance rates in the US, FEMA issues flood maps showing the 1% annual chance flood boundaries. Residences within the flood zone with a 30-yr mortgage have a 26% chance of being flooded by the 100-yr flood event $[(1 - (1 - 0.01)^{30} = 0.2603]$ over 30 years. The mortgage lender and FEMA by setting the flood insurance rates accept this amount of risk.

For coastal flooding damage mitigation design, some level of N-year probability may be specified for the predicted SR(t) at the 1% chance, annual exceedance probability level. Houston (2013b) suggest making an *initial*, arbitrary estimate of total elevation for T_{initial} for $t = N$, the last year in the sea level rise projection needed for the design.

But as before with Equations (11.2) and (11.3), we can take

$$T_{\text{initial}} - T_{50\%}(t) = z(t)\sigma, \qquad (11.11)$$

$$T_{98\%}(t) - T_{50\%}(t) = 2.33\,\sigma \qquad (11.12)$$

and taking the ratio of (11.11) and (11.12) and solving for $z(t)$ for each year gives

$$z(t) = 2.33 \times [T_{\text{initial}} - T_{50\%}(t)]/[T_{98\%}(t) - T_{50\%}(t)], \qquad (11.13)$$

where $z(t)$ = the unknown, time varying, standard deviation of the mean multiplier in the normal distribution. Then solving Equation (11.13) for $z(t)$, and using it in the equation for the exceedance probability $P_>(z)$,

$$P_>(z) = P_>[z(t)] = \frac{1}{2}[1 - \text{erf}\{z(t)/\sqrt{2}\}] \qquad (11.14)$$

and then using $P_>$ in Equation (11.10) to determine the N-year probability, P_N, for design. Using an *initial*, arbitrary estimate, this elevation will exceed elevations with 0.01 exceedance probability in the initial years for the N-yr period, and then be less than elevations with 0.01 exceedance probability in the final years of the N-yr period. Therefore, if the design requires a given life-cycle, N-yr probability, the process becomes one of continued applications of Equations (11.11), (11.14), and (11.9) until the final result for $T_{98\%}(t)$ gives the desired N-yr probability.

Note that for any other desired exceedance probability, $T_{p\%}(t)$, the same procedure applies using appropriate $z(t)$ values associated with that probability level.

Houston (2013b) presents an example for The Battery, New York City, in the state of New York using local probability of exceedance distribution from Zervas (2005) and IPCC (2007) values for scenario A1F1 for the year 2100.

$$\begin{aligned} \text{Given:}\quad F_{50\%} &= 0.75\,\text{m, and} \\ F_{98\%} &= 1.58\,\text{m} \\ SR_{50\%}(2100) &= 0.60\,\text{m, and} \\ SR_{95\%}(2100) &= 0.82\,\text{m.} \end{aligned}$$

Find: The period of interest is from 2012 to 2062 (50 years) and the goal is to determine the 1% annual flood elevation being equaled or exceeded for an N-yr probability of 39% over the 50 year "design life".

Solution:

(1) Using Equation (11.4)

$$\begin{aligned} [SR_{98\%}(2100) - SR_{50\%}(2100)] &= [SR_{95\%}(2100) - SR_{50\%}(2100)] \\ &\quad * (2.33/1.96) \\ &= (0.82 - 0.60) * 1.18878 \\ &= 0.26 \\ \text{so that}\quad SR_{98\%}(2100) &= 0.26 + 0.60 \\ &= 0.86\,\text{m.} \end{aligned}$$

(2) Equation (11.5) is then used to find $SR_{98\%}(t)$ and $SR_{50\%}$ values from 1990 to 2100.

$$SR_{p\%}(t) = a_o + a_1 \times (t - 1990) + \frac{1}{2}(a_2)^2 \times (t - 1990)^2 \qquad (11.5)$$

As first published in Bindoff *et al.* (2007), for scenario A1F1 for the year 2100, $a_o = 0$ (1990) and $a_1 = 0.003\,\text{m/year}$ using satellite altimeter data to establish GMSL.

For the 98% values at $t = 2100$, Equation (11.6) gives

$$\begin{aligned} a_{2(98\%)} &= 0.000165 \times [SR_{98\%}(2100) - 0.33] \\ &= 0.000165 \times [0.86 - 0.33] \\ &= 0.00008745. \end{aligned}$$

Equation (11.7) gives

$$\begin{aligned} a_{2(50\%)} &= 0.000165 \times [SR_{50\%}(2100) - 0.33] \\ &= 0.000165 \times [0.60 - 0.33] \\ &= 0.00004455 \end{aligned}$$

for all years from 1990 to 2100.

(3) Equations (11.11) and (11.12) are then used to determine initial estimates for $T_{50\%}(t)$ and $T_{98\%}(t)$ values for all the years from 1990 to 2100.

(4) Equation (11.13) is then used for $z(t)$ values, Equation (11.14) for $P_>[z(t)]$ values, and finally, Equation (11.10) to learn if the N-yr probability equaled 39%. Houston's initial guess of $T_{\text{initial}} = 1.73\,\text{m}$ gave $P_N = 44\%$ for the period 2012 to 2062. Taking higher values in steps reduces the N-yr probabilities and produced the desired goal of $P_N = 39\%$ so that $T_{98\%}(2012\text{–}2062) = 1.8\,\text{m}$. The combination of the predicted sea level rise in year 2062 at the annual, probability of exceedance level of 1% is 1.8 m for $SR_{50\%}(2100) = \text{GMSL}~(2100) = 0.60\,\text{m}$.

For other N-yr exceedance probabilities, Houston (2013b) presented the results shown in Figure 11.26 for the 50-yr period (2012–2062) for The Battery, New York City, in the state of New York. For example, if the N-yr/P_N level is reduced to 15%, the needed level for design increases to about 1.95 m. It must be noted that this curve is **NOT** the original $F_{50\%} = 0.75\,\text{m}$, and $F_{98\%} = 1.58\,\text{m}$ annual, exceedance probability curve with added SLR predicted to the end year, 2062. Sea levels are rising each and every year, modifying today's probability distribution curves which are based on historic data.

Houston (2013b) also presented an additional figure for this example as reproduced in Figure 11.27. It displays four similar curves if the GMSL values, i.e. $SR_{50\%}$ (2100), increased to 1.5 m, 2.0 m, 2.5, or 3.0 m. At the

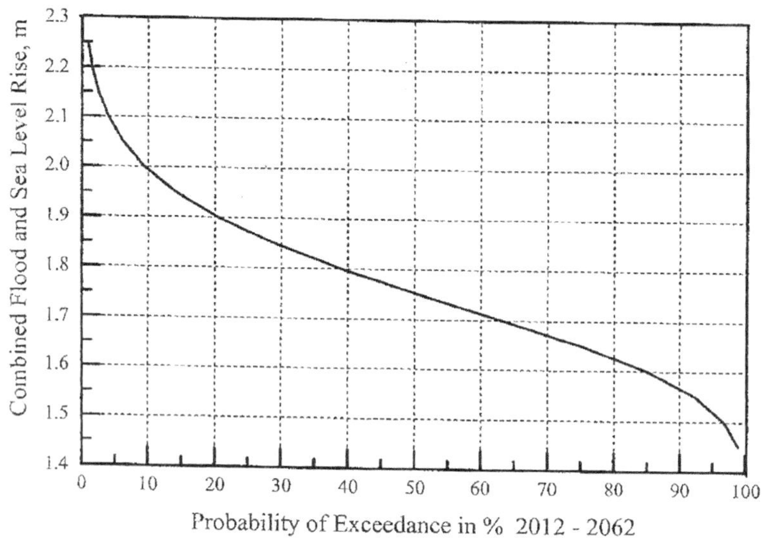

Figure 11.26. N-yr probability of exceedance curve for period (2012–2016) for The Battery, New York City, in the state of New York, when annual exceedance level is 1% (example from Houston, 2013b).

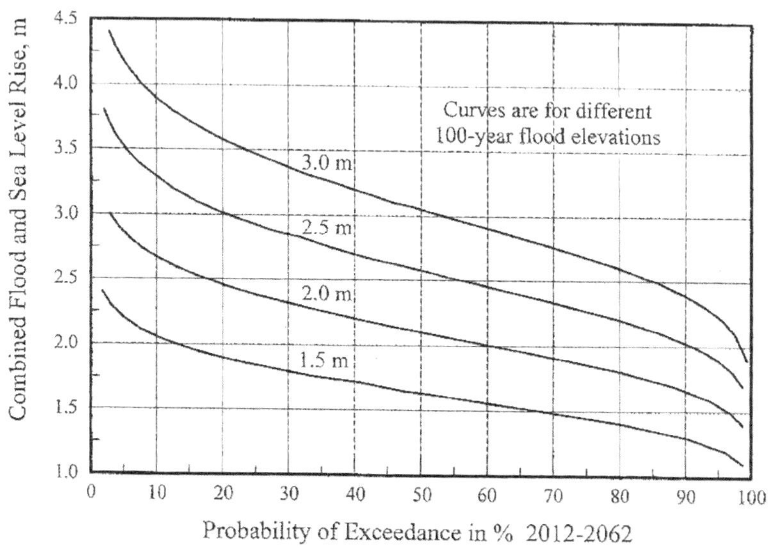

Figure 11.27. N-yr probability of exceedance curves for period (2012–2016) for The Battery, New York City, in the state of New York, when annual exceedance level is 1% (example from Houston, 2013b). Four curves shown for different 100-yr flood elevations with four different SLR elevations.

N-year, $P_N = 39\%$ for the 50-yr design (2012–2062), the combined flood and sea level rise values for the 1% (100-yr) annual flood level would be 1.75 m, 2.25 m, 2.75 m, and 3.25 m, respectively.

Houston (2013b) did this work when the global warming studies used different scenarios, then presently applied in IPCC (2013, AR5). Scenario A1F1 emphasizes fossil-fuels and is the business-as-usual scenario similar to RCP 8.5. This work focused on the low frequency end (1%) of the annual, exceedance probability distribution curve.

At the other end of the annual, exceedance probability distribution are high frequency flooding events that have come to be called "nuisance flooding".

11.3.2.2 *Water level — Nuisance flooding elevations*

Relatively low levels of flooding depths that inconvenience the public's ability to function normally such as frequent road closures; backing up of storm drainage system; and disruption of functional services that infrastructure routinely provides (water, sewerage, power, communications, internet, etc.) has come to be called "nuisance flooding". Flooding levels are not sufficient to cause structural damage to roadways, bridges, storm water pipes, and infrastructure facilities, but high enough that the National Weather Service issues "flood warnings" via the broadcast media. The term "King tide" for routine, spring (>MHHW) tidal conditions is a colloquial term that has no scientific context, but has also found usage in the print and social media regarding nuisance flooding.

The extra high elevation tidal events are caused by seasonal and decadal changes in mean sea level and minor wind events causing minor storm surges along the coast. To begin, all coastal communities should determine the flood elevation above some vertical datum, herein called f_{nuisance} that correlates with the general public's interpretation of when "nuisance flooding" begins. As global mean sea level (GMSL) increases, the annual number of flooding events above f_{nuisance} will also increase.

(i) Kriebel *et al.* (2015)

Kriebel *et al.* (2015) employed a time series of the monthly, extra-high, water measurements (above NAVD88) for the NOAA tide gauge at Annapolis, Maryland (US) for 1928–2012. Extreme water levels from major storm events were removed from the record. They computed the cumulative probability distribution (CDF) after detrending the time series for RSL rise over

85 years and fitted a generalized extreme value distribution (Pareto distribution) to capture the highest 10% of the observed monthly high-water events. Random resampling this empirically fit distribution 10,000 times permitted determination of the mean, $\mu = 0.54\,\text{m}$, with standard deviation $\sigma = \pm 0.16\,\text{m}$. The elevation of the critical nuisance flooding, f_{nuisance}, was set at 1.02 m (3.3 ft) and estimated to be equal to the detrended mean of the monthly, extra high-water measurements plus three standard deviations of the mean (Kreibel and Geiman, 2013). This elevation correlated well with roadway and minor building flooding in Annapolis.

The three US Army Corps of Engineers sea level rise curves (scenarios) discussed in Section 11.2.1, shown in Figure 11.4, and with equations given by Equation (11.1), were then used as projections of RSL(t), except the coefficient employed was 0.0035 m/yr. as determined from the Annapolis tidal record. A fourth curve (NRC-0) was taken as a simple linear extension of this existing slope from 2012 to 2100. No probabilities are attached to these curves in contrast to that employed by Houston (2013a) as discussed above.

Combining the RSL(t) projections with the monthly maxima as randomly sampled for the empirical Pareto probability distribution was by simple addition. This is equivalent to repeatedly using Equation (11.8) and taking Equation (11.9) with $\{SR_{98\%}(t) - SR_{50\%}(t)\} = 0$. Figure 11.28 presents the results for the four Corps' sea level rise curves showing the time (years) versus the probability the monthly exceedance flooding will occur in Annapolis.

These results may be interpreted in several ways. The probability for monthly flooding for scenario NRC-0 in year 2100 is 16% (business-as-usual) but occurs at this same monthly probability level much sooner for NRC-I (yr-2062), NRC-II (yr-2044), and NRC-III (yr-2035), which is only 20 years from the starting year (2012). In 50 years (2062, red line), the probabilities for monthly flooding increase dramatically from 4% (RCP-0) to 16% (RCP-I) to 73% (RCP-II) and reach 100% for RCP-III.

(ii) Kriebel (2018)

Kriebel (2018) presents a simpler, approximate way to make nuisance flooding projections with future sea level rise. The following steps are needed in the analysis.

(1) Use measured, hourly water elevations (above vertical datum) from NOAA tide records (online).
(2) Remove mean sea level trend from the record.

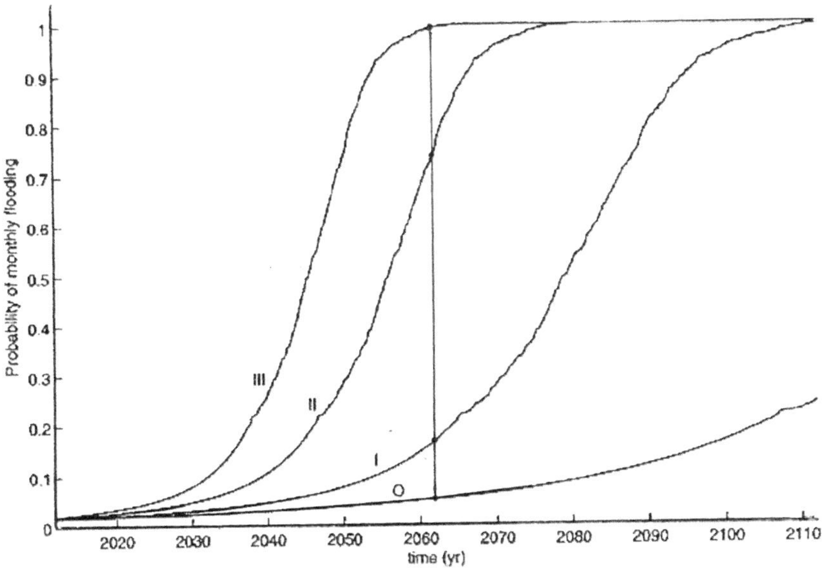

Figure 11.28. Probability of monthly exceedance of flooding elevations above f_{nuisance} versus time for Annapolis, MD (from Kriebel *et al.*, 2015).

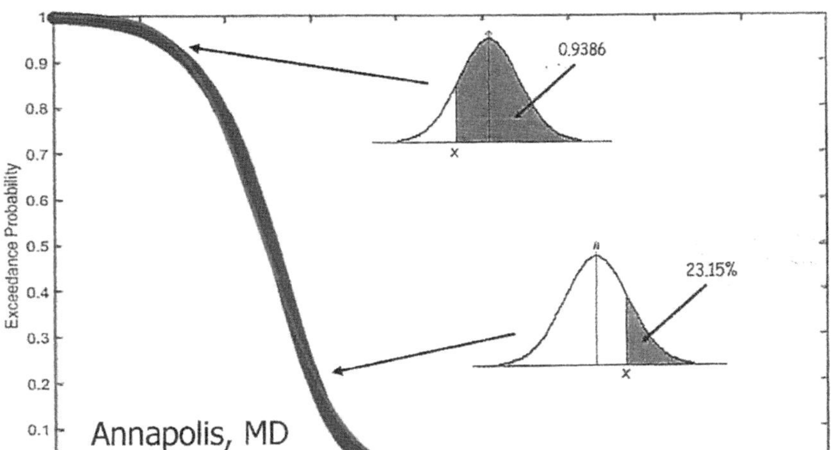

Figure 11.29. Annual exceedance probability distribution for elevation of high tide peaks above MSL datum for Annapolis, MD, tide gauge (from Kriebel, 2018).

(3) Identify high tide peaks over length of record. Filter extreme high levels produced by major storms (hurricane, Northeasters, etc.) since only interested in nuisance flooding.
(4) Develop exceedance probability distribution of high tide amplitudes. With extreme events removed, data give a normal cumulative probability curve. See Figure 11.29 as an example (Annapolis, MD) for amplitude of the high tide peaks (above MSL vertical datum) versus the cumulative exceedance probability.
(5) Convert from annual exceedance probability to annual number of high tide events for existing curve (Figure 11.29, change vertical datum) and simply add small increments of relative sea level rise. Figure 11.30 shows 0.5 ft (15 cm), 1.0 ft (30 cm), 1.5 ft (46 cm), and 2.0 ft (61 cm) additions for Annapolis, MD.
(6) Determine the "flooding stage" definitions and elevations for minor, moderate, and major stages of flooding in your area. In the US, the NOAA, National Weather Service defines the following stages: *Minor stage* — issues public flooding advisories/warning; *Moderate stage* — experiences some road and residence/structure inundation; and *Major stage* — extensive inundation, flood damage, evacuation needed.

    Specifically, the elevation for the mean $\bar{f}_{\text{nuisance}}$ and a range of standard deviation of the mean σ should be determined. Invert elevations for low-lying storm drainage pipes that connect to tidal waters is a good place to start for quantification of $\bar{f}_{\text{nuisance}}$ and σ. These elevations are then superimposed as also shown in Figure 11.30 for Annapolis. One f_{nuisance} elevation (1.98 ft, NAVD88) is given for McNair Road (dashed line) on the US Naval Academy (USNA) property.
(7) The numbers of annual future nuisance flooding events can then be found as the intersection of these vertical elevation lines for individual f_{nuisance} locations; or using the mean $\bar{f}_{\text{nuisance}}$ elevation; and/or the NOAA, National Weather Service definitions for *minor* and *moderate* flooding events versus sea level rise levels relative to today's level. Figure 11.31 shows one result for McNair Road with $f_{\text{nuisance}} = 1.98$ ft (60 cm) (NAVD88 datum). A 0.5 ft (6 inches, 15 cm) increase in RSL resulted in a $90/9 = 10$-fold increase in the number of annual, nuisance flooding events. A 1.0 ft (12 inches, 30 cm) increase is a $275/9 = 30$-fold increase.[1]

[1]This process was presented by Dr. Kriebel at the 36th ICCE Short Course (Baltimore MD) on July 29, 2018. He called it "A Game of Inches". Thanks Dave, for permission to include in this book.

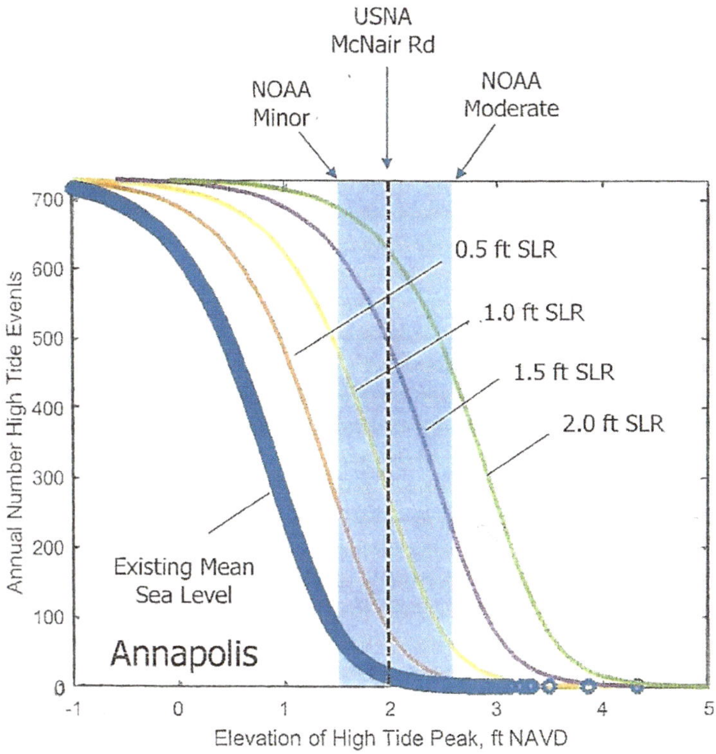

Figure 11.30. Number of annual high tide peak events (from Figure 11.29) versus peak elevations for existing MSL conditions and sea level rise of 0.5 ft, 1.0 ft, 1.5 ft, and 2.0 ft (from Kriebel, 2018).

(8) Interpret these results relative to your location's functional resilience to nuisance flooding.

(9) Consider adaptation and mitigation alternatives (Chapter 3) to increase the elevation of f_{nuisance}. For example, in Figure 11.31, increasing f_{nuisance} by 1.0 ft to 2.98 ft (90 cm) decreases the number of annual high tide events from 275 to about 20 for a future RSL rise of 1.0 ft (30 cm).

(10) Develop a curve as shown in Figure 11.32 and discuss these results with your area's Public Works engineer and the flooding resilience office when presenting alternatives to mitigate nuisance flooding.

11.3.2.3 *Water level — Duration of extreme and nuisance flooding*

The length of time that water levels exceed f_{nuisance} elevations is also a factor for both extreme events that produce residence and structural damage

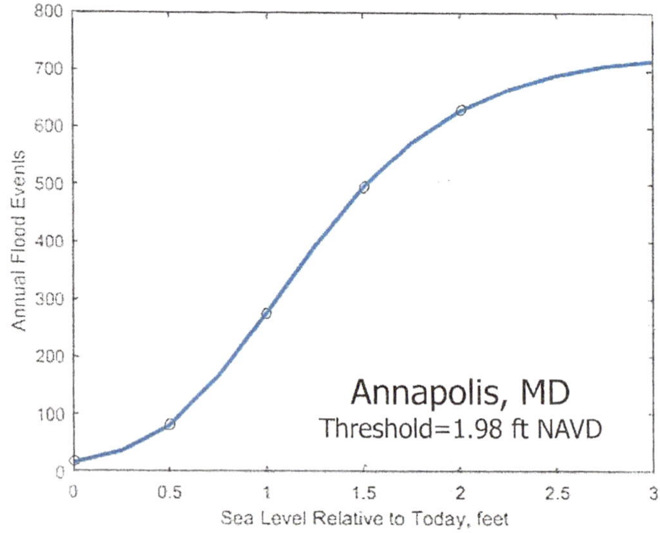

Figure 11.31. Number of annual flood events relative to sea level rise for $f_{\text{nuisance}} = 1.98\,\text{ft}$ (60 cm) (from Kriebel, 2018).

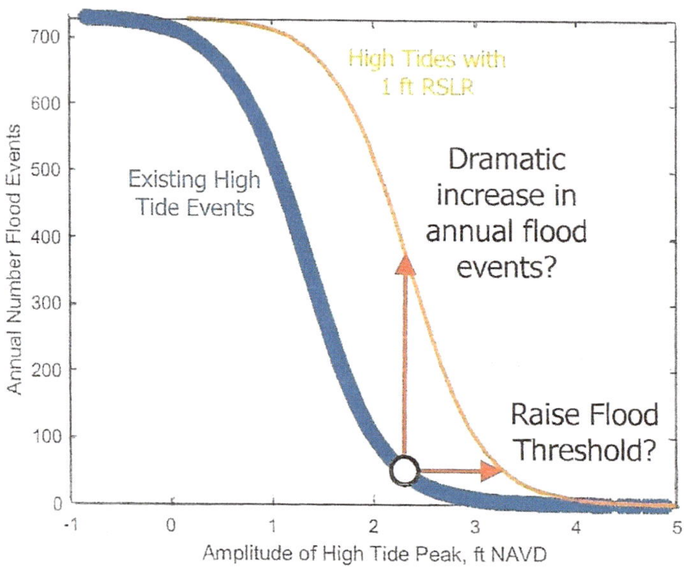

Figure 11.32. Schematic of decision graphic regarding mitigation and adaptation alternatives to increase f_{nuisance} elevation and reduce annual number of nuisance flooding events (from Kriebel, 2018).

and for nuisance flooding events that disrupt normal functions of urban communities. Figure 11.33(a) illustrates the measured (red) extreme water elevations during Hurricane Isabel (September 18, 2003) at the Sewells Pt. tide gauge in Norfolk, VA. This was the 2% (50-yr) annual exceedance probability, storm surge flooding event in Norfolk and above elevation 3.5 ft (MLLW datum) for about 15 hours. As illustrated in Figure 11.33(b), a 1 ft (30 cm) rise of sea level and this same storm surge event would increase the flooding duration to over 67 hours. A nominal, 24-inch (60 cm) diameter circle illustrates vehicle problems on this roadway.

A source for evaluating durations of flooding events in the US is the NOAA "Inundation and Analysis Tool" available at https://tidesandcurrents.noaa.gov/inundation/. Users provide a value for the f_{nuisance} elevation and the program gives histograms for *both* elevations above and durations (hours) for measured tidal elevations at NOAA tide gauge locations in the US over specified time intervals. See the above link for an "inundation analysis user's guide". Elevation-versus-Duration scatter plots are also provided to help filter extreme flooding elevations for nuisance elevation and duration studies. However, the impact of future sea level rise predictions is not included in this NOAA product.

Pezza and White (2020, in review) have studied the impact of sea level rise on the duration of coastal nuisance flooding in Norfolk, VA. Using the same Sewells Pt. tidal data as in Figure 11.33, they developed the duration, exceedance probability distributions for hourly gauge readings from July 1, 1927 to May 31, 2018 (8,760 hours per normal year, >90 years) for flood elevations above $f_{\text{nuisance}} = 1.25$ m (4.10 ft, NAVD88). This location is the sill elevation of a tidal flood gate located at Brooke Avenue as shown in Figure 6.72 for the Norfolk flood wall. Sea level rise elevations of 0.5 ft (15 cm), 1.0 ft (30 cm), 1.5 ft (45 cm), 2.0 ft (60 cm), and 2.5 ft (75 cm) were studied for both Intermediate (NRC-II) and High (NRC-III) scenarios for the Corps of Engineers curves shown in Figure 11.4 and starting in 2018. The results are presented in Table 11.4 for the number of flood duration hours for six annual exceedance probability levels (99%, 50%, 10%, 4%, 1%, and 0.2%) for both the NRC-II and NRC-III scenarios. The years when this SLR level is reached are also given in Table 11.4.

Figure 11.34 displays the durations for a range of annual exceedance probability levels when the NRC-II "Intermediate" SLR scenario elevations are employed.

In 2018, there were 14 h with a chance of 1% each year for the flood elevations to be above the sill elevation at the Brooke Avenue tidal gate in

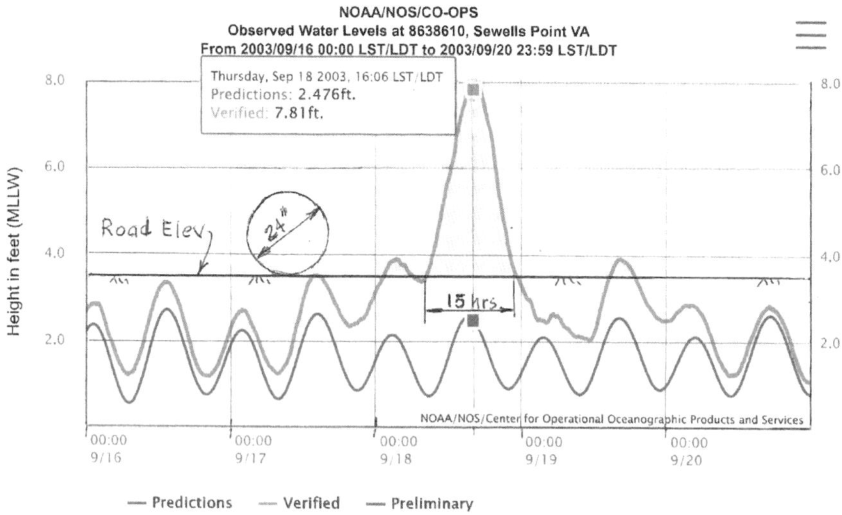

(a)

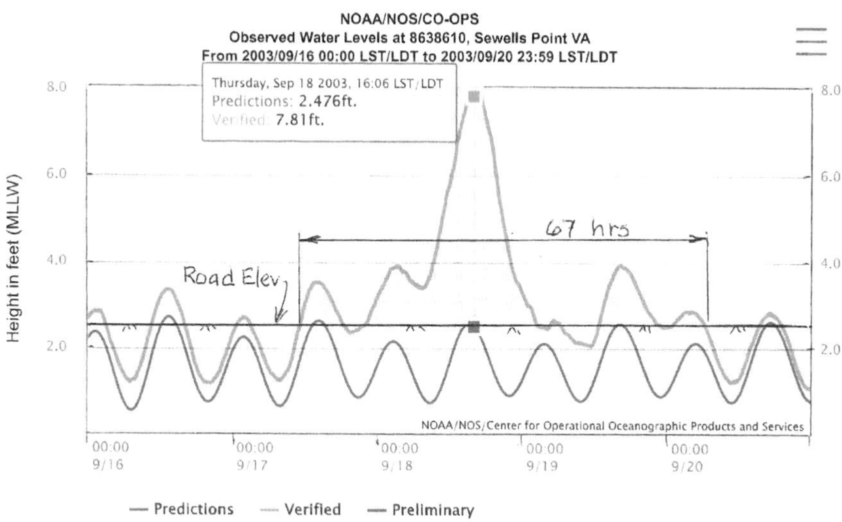

(b)

Figure 11.33. (a) Illustration of duration of road blockage during 2% (50-yr) flooding event under today's sea level. Duration of road blockage is 15 h. (b) Illustration of duration of road blockage during 2% (50-yr) flooding event with 1 ft (30 cm) of sea level rise. Duration of road blockage increase to over 67 h.

Table 11.4. Number of flood duration hours at specified, annual exceedance probability levels for five future sea level rise elevations under both the years reached for both the Intermediate (NRC-II) and High (NRC-III) climate change scenarios (from Pezza and White, 2020, in review).

SLR cm (ft)	99% Hours	50% Hours	10% Hours	4% Hours	1% Hours	0.2% Hours	Intermediate NRC-II, year	High NRC-III, year
0.0 (0.0)	0	2	6	9	14	21	2018	2018
15 (0.5)	2	4	9	14	24	40	2042	2031
30 (1.0)	5	7	13	19	29	45	2062	2041
45 (1.5)	7	10	20	28	43	69	2080	2051
60 (2.0)	11	15	28	38	58	91	2096	2059
75 (2.5)	17	22	39	51	76	116	>2100	2067

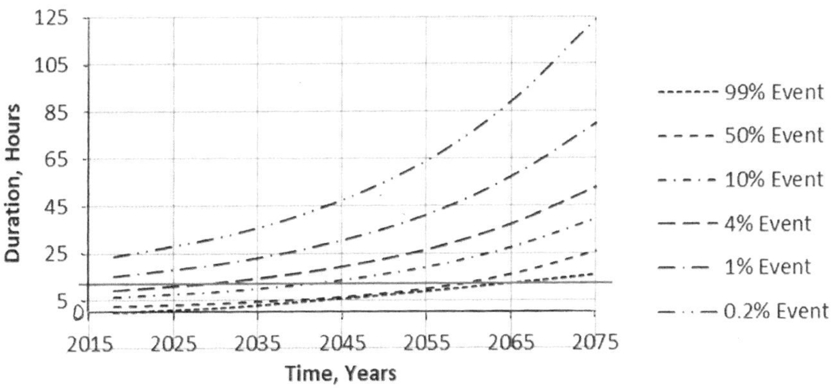

Figure 11.34. Flood duration (hours) for elevations above $f_{\text{nuisance}} = 1.25\,\text{m}$ (4.10 ft, NAVD88) at the Brooke Ave. tidal flood gate in the Norfolk flood wall (from Pezza and White, 2020, in review).

the Norfolk flood wall. As sea level rises, the chance increases to 4% each year with a 15-cm (0.5 ft) rise; the chance increases to 10% each year with a 30-cm (1.0 ft) rise, etc. There are other ways of looking at these trends, but the results are clear. As sea levels rise, the number of hours that water levels will be above given thresholds each year will also increase.

The NOAA should include RSL in their "Inundation and Analysis Tool" as described above. And, the rigorous methodology analogous to that employed by Houston (2013b) for extreme water level elevations should be employed for durations of these elevated events.

11.3.2.4 *Water wave climate — Ocean waves*

After water level, the second most important variable is water waves as specified by their significant wave height, H_s, the peak wave period, T_p, their direction, Θ, and the water depth, d, at which these wave conditions are given for the local wave climate. Predicted wave climates relative to climate change scenarios have evolved over the past 15 years. Wang *et al.* (2004) predicted increases in significant wave heights during fall and winter seasons in the northeast Atlantic Ocean in the 21st century. Sea surface temperatures are the main driving force for wave climate changes (Shimura *et al.*, 2016). Recent publications by the Ocean Atmosphere Climate Science Center in Australia (Hemer *et al.*, 2018) suggest just the opposite trends. He predicts that most of the world's oceans will likely see lower wave heights and calmer sea conditions. Only the Southern Ocean will have an increase in wave energy, and especially Australia. His organization is a part of the IPCC "Coordinated Ocean Wave Climate Project" (COWCLIP) at https://cowclip.org. Figure 11.35 is a summary of annual difference in H_s meters from ocean wave, numerical models in five organizations (Australia, Japan, European Union, US, and Canada). Color coded differences (Future-Present) are clearly seen. Lower wave climates (blue) cover most of the world's oceans. Visit the COWCLIP for details in your area and coming conferences and workshops that will highlight the impacts of climate change (Hemer *et al.*, 2012).

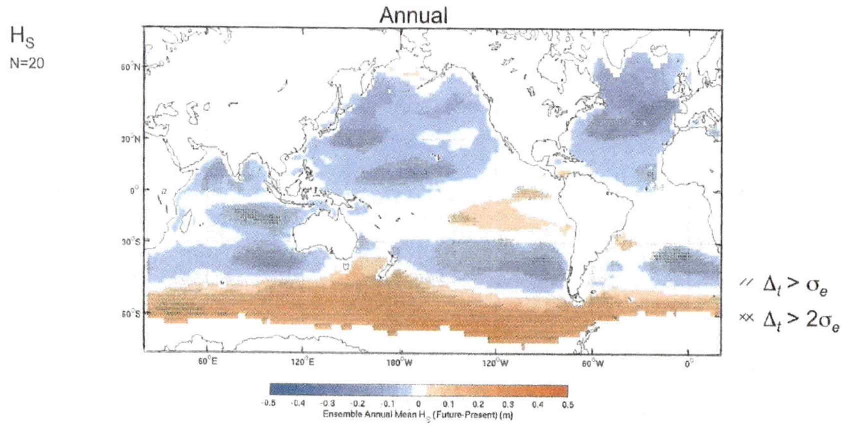

Figure 11.35. Combined predictions for annual, significant wave heights in the world's oceans from five contributions of numerical models (from Hemer *et al.*, 2018).

Here is a clear case of the departure from "stationary" physical processes governing the long-term, wave climate statistics (see Section 2.3.7 Wave Hindcasts from Wind Statistics and Section 2.3.8 Extreme Waves) and the new, non-stationary norm that will alter the future wave climate.

11.3.2.5 *Breaking waves, wave transformations, and time-averaged waves impacts in coastal waters*

Future rising seas will alter the wave lengths for given wave periods that change wave breaking processes and wave shoaling, refraction, and reflection as presented in Sections 2.3.4 and 2.3.5. Wave set-down, set-up, and wave-induced currents will also be changed (Section 2.3.6). Form and functional designs of coastal structures (nearshore breakwaters, groins) will be altered (Chapter 5). In shallow waters with depth-limited wave-breaking physics (Equation (2.25)), deeper water means bigger wave heights will be needed for structural designs discussed in Chapter 6.

11.3.2.6 *Sediments and coastal geomorphology*

Finally, Chapter 2 on Design Criteria presents soil (sediment) characteristics, sediment budgets, and introduction to shoreline change. The impacts of sea level rise for these factors will be further discussed below in Chapter 7 on Beach Engineering (Section 11.3.7).

11.3.3 *Chapter 3 — Alternatives for hazard mitigation*

Both structural and non-structural alternatives for hazard mitigation were introduced in Chapter 3. Figure 11.36 displays these alternatives relative to existing sea level (solid line) and future sea level (dotted line) as "programmatic measures" of the US Army Corps of Engineers (USACE online graphic).

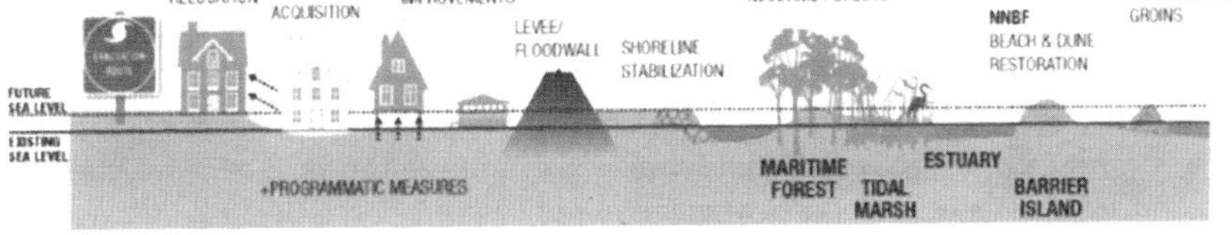

Figure 11.36. Structural and non-structural alternatives to mitigate flooding, waves, and erosion damage from future sea level rise (from USACE, 2013).

It is important for civil/coastal engineers to have a clear definition and understanding of terms.

MITIGATION — To **REDUCE** (mitigate) the damage caused by coastal storms and increased sea level rise using **both** structural and non-structural alternatives.

ADAPTATION — To **CHANGE** (adapt) our present alternative methods (structural and non-structural) and decision-making methods as caused by the *predicted levels and uncertainty* in sea level rise.

The non-structural mitigation alternative currently utilized is NOT adaptation. The non-structural mitigation alternative has been employed by coastal engineers for over 70 years since the beginning of the profession in 1950. As discussed in Section 10.4, the Natural, Nature-Based Features (NNBF) or "Living Shorelines" language has been added to the definition of "mitigation" in an effort to explain to the public and the media, *the full array of measures employed to support coastal resilience and risk reduction* (USACE, 2013). In general, mitigation means reducing the threat of damage from the hazard and preventing something from happening (flood damage, wave damage, beach and shoreline erosion, etc.) and the alternatives are the ways this can be accomplished. Adaptation is more focused on changes in how the alternative is designed and used (functions) and in changes in societies' laws and behavior to future hazards.

Unfortunately, the terms mitigation and adaptation have different meanings to scientists, societal communities, and the media and no general consensus has emerged. The IPCC defines "climate change mitigation" as *the process of mitigating (reducing) climate change by reducing the atmospheric concentration of Greenhouse Gases.* Social communities define "climate change adaptation" as *the process of enabling the built environment, social communities, and the ecosystems to function and exist under the changing climate conditions that could not be mitigated.*

In the US, the FEMA definition of "Mitigation" is

> *Mitigation is the effort to reduce loss of life and property by lessening the impact of disasters. It is action taken now — before the next disaster — to reduce human and financial consequences later. Effective mitigation requires that we all understand local risks, address the hard choices, and invest in long-term community wellbeing. Without mitigation actions, we jeopardize our safety, financial security and self-reliance* (http://www.fema.gov/what-mitigation).

This definition is consistent with the insurance industry, where the term adaptation is not widely used.

In the United Kingdom (UK), a dialogue has evolved regarding the growing population of coastal cities, climate change, environmental risks, and increasing dependencies on infrastructure systems. The pressing challenge of the 21st century is how to create, manage, and maintain sustainable cities. The debate has been condensed to the simple phrase (sound bite for the media)

"ADAPTATION WITHOUT MITIGATION IS IMMORAL"

As seen in Figure 6.97, for the vertical rotating gates in the Thames estuary near London and discussed in Sections 3.3.1, storm surge barriers will find increasing application in the future of sea level rise. However, the cost is large, and critics of the costs will argue for "adaptation" without "mitigation". The moral obligation of society to include both mitigation and adaptation measures to reduce risk and increase resilience to flooding events in the future will become debatable all over the world. See Section 11.3.8 for further discussion.

The American Society of Civil Engineers (ASCE) contributes to the confusion on the use of the terms mitigation and adaptation. The recent ASCE Manual *Climate-Resilient Infrastructure: Adaptive Design and Risk Management* (ASCE Manual 140, 2018) includes a Section 7.9 "Coastal Adaptive Design and Adaptation" and gives examples for what it labels: Protect, Accommodate, and Retreat. The word mitigation is missing from this ASCE Manual No. 140. See Section 11.3.6 for further discussion.

11.3.4 *Chapter 4 — Probabilistic design*

We no longer live in a stationary, stochastic process world where unconditional, joint probability distributions and parameters such as the mean and variance do not change over time. And, we can no longer easily transform a non-stationary process by removing the underlying trend (detrending).

Long-term sea state uncertainty is discussed in Section 4.5.2. The Peak-Over-Threshold (POT) method for statistical analysis of long-term wave heights (e.g. 34.7 years, Lake Huron, Kamphuis, 2010, Section 2.3.8) will change. Time series of water wave data is now a non-stationary stochastic process that must be analyzed differently. We must now look at decades of data to first discern if (when) the data has been impacted by climate change. The POT method is still valid, but now the results will be for POT(t).

Fortunately, the main focus of Chapter 4 is on Partial Safety Factors for structural design and this will not change. Application will be affected as discussed in Section 11.3.6.

11.3.5 *Chapter 5 — Functional and planform design*

11.3.5.1 *Part I — Functional design*

Part I of Chapter 5 is on the functional design of coastal structures and primarily concerning wave run-up, wave overtopping, and wave transmission. The wave climate will impact run-up, but the greatest impact of sea level rise will be the increase in still water depth, h_s (SWL), and resulting decrease in freeboard, R_c, for a given structure crest elevation as shown in Figure 5.12. This combination will increase the time-averaged, wave overtopping rate, q, as calculated from the formulas presented in Section 5.3. Continued developments of the Wave EurOtop Manual (Die Küste, 2007) as discussed in Section 5.3.4 should be followed for the latest developments and especially refinements using a parapet (bullnose) on vertical seawalls at the crest to reflect run-up flows in the seaward direction and reduce the overtopping rate as illustrated in Figure 5.17. Tolerable mean discharge rates and maximum volumes from the EurOtop Manual as presented in Table 5.8 may also change under rising sea conditions.

Wave transmission heights past coastal structures will also be impacted by sea level rise. Many existing rubble-mound breakwaters will become Low-Crested Structures (LCS) as illustrated in Figure 5.28 so that the transmission coefficient, K_t, will increase as given by Equation (5.47). It is anticipated that more research using large-scale wave tanks will be conducted to improve knowledge regarding wave transmission heights and wave period changes past LCS.

11.3.5.2 *Part II — Planform design*

Part II of Chapter 5 focuses on the planform, layout dimensions for parabolic shoreline forms and shoreline stabilization structures (nearshore breakwaters and groin systems).

Wave refraction in coastal waters is significantly impacted by water depth and consequently modifies the basic wave crest angle, β, relative to the control line, R_o that are the two independent variables in the parabolic shoreline model given by Equation (5.53) as shown in Figure 5.34. The resultant dependent variable pair (R, Θ) will change giving a different shoreline position. Coefficients C_o, C_1, C_2 also depend on the wave crest angle β in the parabolic model.

Whether salients or tombolos form behind nearshore breakwaters is independent of water depth. But the structure distance, Y, offshore and resulting water depth at this location may be dictated by structural design loads (Chapter 6) which will change with water depth. Therefore, indirectly, whether salients or tombolos form depends on the length of structure, L_s, to Y ratios shown in Figures 5.38 and 5.39.

Ten basic "rules" for functional design of groins are presented in Section 5.9.3. The key variable is the minimum, dry beach width, $Y_{\min}$, for upland protection during storm events. As RSL rises, this volume of sand will decrease so that future, end-of-design life conditions must be considered.

11.3.6 *Chapter 6 — Structural design*

The four basic principles of coastal structure design are as follows:

(1) Determine the position (location) of the structure on the coast;
(2) Determine the crest elevation of the structure;
(3) Determine the dimensions and materials of construction to make a cost estimate;
(4) Determine how the structure interacts with the adjacent coastal environment.

All four concerns are impacted by rising seas.

11.3.6.1 *Rubble-mound structures (breakwaters, groins, jetties, and revetments)*

The key variable in the van der Meer formulas for armor layer stability is $\widehat{H}_S^T$, the probabilistic, mean value of the significant wave height at some design value for the recurrence interval, T_r. For non-overtopping designs, there are two criteria that govern which formulas to apply. Both depend on the water depth. One is the critical, surf similarity parameter, ξ_{mc}, that separates plunging and surging wave formulas. It depends on the wave steepness, s_m, which is dependent on the mean wave period, T_m. Equation ((6.10), plunging) and Equation ((6.11), surging) depend on the water depth for these "deep water" equations.

The second is the water depth at the toe of the structure, h_{toe}. When "shallow water" wave conditions prevail, $h_{\mathrm{toe}} < 3\widehat{H}_S^T$ (Rock Manual, 2007),

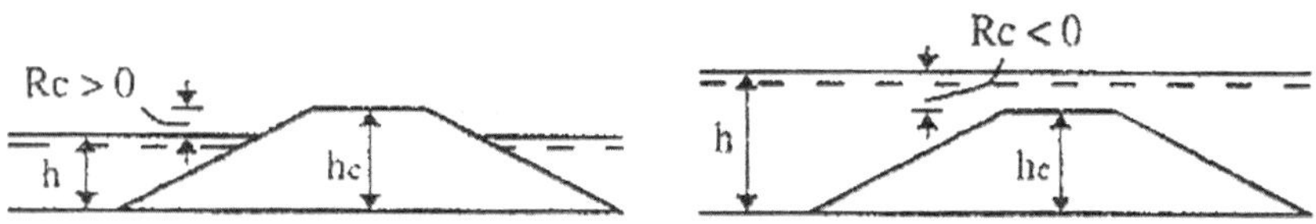

Figure 11.37. Definition of freeboard, $R_c = h_c - h$. (from Figure 6.9).

key coefficients change (Table 6.5) and the spectral wave period, $T_{m-1.0}$ is employed for Equation ((6.13), plunging) and Equation ((6.14), surging).

When wave overtopping and submerged conditions prevail as illustrated in Figure 6.9 (repeated here as Figure 11.37), the full impact of sea level rise is felt for the structural stability of rubble mound structures. Section 6.2.5 presents all the details. When their function to prevent wave overtopping is impaired, the crest elevation may be increased, and the structure "adapted" for rising seas as discussed in Section 11.4.

11.3.6.2 *Vertical-walled structures (seawalls, caissons, bulkheads, and flood walls)*

In Section 6.3, the water depth is now defined as d and is a key variable in the Goda formula, Equation (6.48) for the horizontal wave force, F_H, on a vertical wall. For the four types of wave impact loads illustrated in Figure 6.37 and four structure types identified and studied in PROVERBS, the still water depth at the structure toe is very important (Section 6.3.4). For caisson design (Section 6.3.5), the Goda formula together with wave impact loads in PROVERBS are key forces causing overturning moments that must be resisted for stability. Similarly, bulkhead and flood wall designs both vary with water depths in front of these vertical sections.

11.3.6.3 *Dikes (levee) and revetments*

In these coastal structures, the ground water elevations behind the dike or revetment will also vary depending on long-term sea surface elevations.

11.3.6.4 *Storm surge barriers*

Section 6.5 presents a brief overview of storm surge barriers with movable tidal flood gates with six types shown in Figure 6.96. All have limits on the fixed, maximum crest elevations that will be approached as sea levels rise. The one possible exception is the horizontal rotating gate which has been designed with a floating structure at Maeslant on the Rotterdam waterway

(Figure 6.93) in the Netherlands and at St. Petersburg (Russia). Limits are still present for the maximum vertical distance the structure can float.

11.3.6.5 *Special structures*

Finally, floating breakwaters, artificial reef breakwaters, and breakwaters as wave energy converters are discussed in Section 6.6. All are designed to function as "breakwaters" to reduce wave energy on the leeward side of the structure. All have the ability to naturally or manually "adapt" to rising sea level conditions as discussed in Section 11.4.

11.3.7 *Chapter 7 — Beach engineering*

Sandy beaches are continually moving in time and in all three dimensions due to wave orbital motions, turbulence in breaking waves, wave-induced currents, tidal currents, and ocean circulations. Seasonal variations in Mean Water Level (MWL) are generally much greater than GMSL rise and a stable beach in dynamic equilibrium responds seasonally with greater volume in the summer than in the winter months. GMSL rise rates will impact sediment transport rates in space and time. It's about time.

11.3.7.1 *Bruun rule*

Over long time periods, as the mean of the ocean's sea level rises, the local shoreline position moves landward (retreats). The seasonal beach profiles may be in dynamic equilibrium so that "erosion" is not occuring but shoreline retreat is still occuring. The horizontal recession, R, of the beach profile (of any shape) responses to sea level rise, S, and the relationship between R and S as put forward by Bruun (1962) has come to be called the "Bruun Rule" as illustrated in Figure 11.38. The width, L_*, of the beach profile with active sand movement is usually taken as the horizontal distance to the closure depth, h_*, for the equilibrium beach profile with berm height, B.

For many locations in the world, the average slope of the active beach profile is small and on the order of $\tan\theta = 0.01$–0.02 so that the retreat rates are on the order of 50–100 times the sea level rise rate as estimated by the Bruun Rule. Local sea level changes may be relative to local land movements (subsidence or lifting) and not the global average. Beaches adjust to baseline sea level conditions, but local, site-specific factors control long-term erosion (or accretion) rates that have generally been much greater than long-term sea level rise rates.

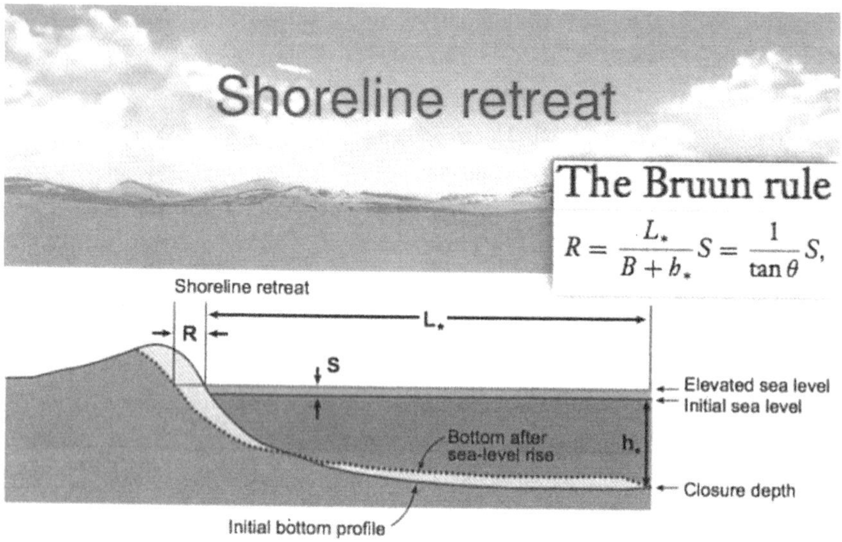

Figure 11.38. The Bruun rule to estimate shoreline retreat, R, for relative increase in sea level rise, S (from Cooper and Pilkey, 2004).

Both existing beach nourishment projects and new projects will be impacted by accelerating sea level rise rates.

11.3.7.2 *Existing beach nourishment projects*

Coastal Science and Engineering Inc. (CSE) in Columbia, SC (US), and president/coastal geologist, Dr. Tim Kana have a long history of successful beach nourishment projects from New York to Georgia along the Atlantic Ocean. Since the early 1980s, they have designed and managed over 47 projects totaling 35 million cubic yards of sand volume (26.8 million cubic meters). Many journal publications by Dr. Kana and colleagues have been published in the literature over the years. See Kana *et al.* (2015) for their experience under sea level rise conditions these past 40 years.

For example, an important, often overlooked factor is the depth of closure (DOC), h_*, as discussed in Section 7.2.4. It is primarily the limiting depth between material in the nearshore region and offshore material. Beach profile surveys are the best way to establish the local DOC at the site. Repetitive surveys can then be used to determine the standard deviation of the mean (SD) for each distance in the profile. A threshold, near-zero SD can then be used to quantitatively establish the DOC (Kana *et al.*, 2015).

Using profiles from healthy beaches which have ...*demonstrably weathered storms without significant adverse impacts to the foredune or landward development* the unit profile volume to DOC is calculated that becomes a "benchmark" volume to quantify eroding (deficit) sections with far different DOC. Rating curves for beach volumes to overcome these volume deficits and their related beach widths can then also be related to the DOC. The DOC also changes (increases) after beach nourishment to reduce the retreat rate for a given sea level rise (Bruun Rule).

This was followed by a more recent publication *Building Blocks for a Sustainable Beach Nourishment Program* (Kana *et al.*, 2017) where five "building blocks" were identified, namely,

- Shoreline inventory,
- Erosion database,
- Conceptual geomorphic models,
- Target beach condition,
- Identifying potential borrow area.

Although this last paper deals only with beach nourishment projects in one state, South Carolina (US), it provides some valuable insights for project implementation and sustainability under future sea level rise conditions. It was learned that in South Carolina, the local community alone can finance the annual cost of beach re-nourishments if less than 1% of the value of oceanfront property per year. This is without any financial subsidy from the state or federal (Corps of Engineers) governments. Review of the five "building blocks" is recommended to supplement the material presented in Chapter 7.

11.3.7.3 *New beach nourishment projects*

Beach nourishment costs over the design-life will continue to increase due to inflation and sea level rise. Maintenance intervals will decrease adding to the life-cycle costs. The volume of sand needed in the maintenance cycle to accommodate RSL rise over the past 40 years has been a small fraction of the deficit volume resulting from the local erosion rate. A rule-of-thumb is that when the local, RSL rise is greater than 60 cm (2 ft), it becomes an important factor in the maintenance cycle and maintenance costs.

Dr Kana (2018) believes that beach restoration at most sites will continue to be sustainable with future sea level rise rate as long as the annual average sand replacement cost is less than 1% of the oceanfront property value per year. However, a planning time frame of decades or less is recommended. Continued beach performance monitoring and the value of the properties at risk is needed to make informed decisions on ...*whether to continue nourishing the beach or to abandon the property* (Kana *et al.*, 2017; Kana, 2018).

Keeler *et al.*, 2018, using beachfront communities on the Outer Bank's of North Carolina's (US) beaches as the example, argue that beach nourishment as risk reduction projects provide positive "feedbacks" that encourage higher-value real estate investments, which inhibits long-term adaptation to sea level rise. They suggest ways to weaken the positive feedback cycle and regional/national planning for managed retreat.

11.3.7.4 *Dunes*

Dunes are the primary structural element that prevent inundation and flooding of the Netherlands. The design criteria for an erosion event is 1:10,000-year recurrence interval of the water level elevation. A sensitivity analysis of the impacts of RSL rise on dune erosion volume was made at two sites along the Dutch coast (de Winter and Ruessink, 2017). Using the numerical model XBeach, it was found that there was a linear relation between dune erosion volume and sea level rise. The increased sea level allowed waves to attack the dune at higher levels. They also learned that offshore wave direction angles produced similar magnitude of dune erosion volume as found for sea level rise. And finally, the volume and location of beach nourishment sand fronting the dune as a mitigation measure had a strong effect on reduction of the dune erosion volume. Details are site specific, but the general findings are important for all locations where the combination of dunes and beach nourishment is employed for flood damage mitigation.

11.3.8 *Chapter 8 — Economics, risk, and resilience*

11.3.8.1 *Risk and resilience metrics*

The life-cycle costs for each coastal hazard mitigation alternative over a specified, economic-design-life is the focus of this chapter. The ratio of the benefits in hazard mitigation (risk reduction) relative to the life-cycle costs of each alternative (B/C ratio) is then used to rate each alternative for feasibility of design. The alternative with the highest B/C ratio that also satisfies the environmental constraint is then generally selected for detailed design. However, many functional benefits to society cannot be readily quantified in monetary units (e.g. Section 8.6.2 (b), Corps of Engineers, NACCS — Exposure Index for population density and infrastructure index; social vulnerability index; and environmental and cultural resources index). As a result, the quantification of a resilience metric (Section 8.7.3) is now an active research area because the functional benefits to society of these areas can be quantified as an increase in resilience for each alternative. The example for Jamaica Bay using Bayesian Networks is a step in this direction (Schultz and Smith, 2016).

At present, because only the damage reduction benefits are quantifiable in monetary units, socially and economically disadvantaged citizens are not accounted for in the B/C analysis method. For example, consider a storm surge elevation that produces flood damage at the 20% level for residential property in an urban community. Figure 8.22 presents depth-damage table and curve for a two-or-more story residence with no basement. Two elderly citizens reside in a property worth $1 million in the affluent section of town and the allotted benefit is 20% of $1,000,000 or $200,000 toward the cost of alternatives to prevent this level of damage. In an adjacent neighborhood that also receives the same storm surge elevation, eight people (parents, kids, grandparents) reside in a $100,000 valued home. The allotted benefit is 20% of $100,000 or $20,000 which is 10 times *less* and perhaps does not justify any flood damage mitigation alternative to be implemented for that neighborhood. The Risk reduction metric as measured by the B/C ratio favors the wealthy residences and neglects the poorer residences and the people living there. The situation will only get worse as sea levels rise and the next time the same storm intensity will produce flood damage greater than at the 20% level. The old adage, *the squeaky wheel gets the grease* could be modified in this case to *the big wheel gets the shock absorber* and *the little wheel gets rusty.*

Social justice will increase by quantifying the Resilience metric. Regarding residential housing, assume at the 20% flooding level, citizens can return home immediately because the functional level has not been impaired. For this one example, eight people have a place to reside (they may not afford to go anywhere else) whereas only two people can return home (if they choose to do so or can stay away because they evacuated elsewhere). Assuming Utilities (water, wastewater, power, communications) and Transportation systems were also functioning normally, the Resilience metric would again assign four times more functional value- to-society (people) for alternatives that preserved the residences, utilities, and transportation functions.

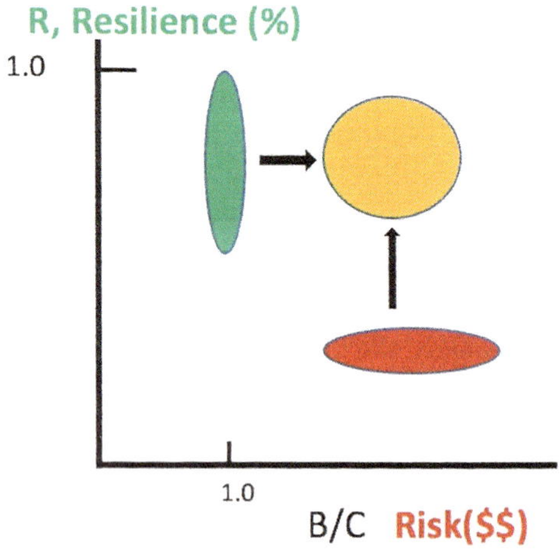

Figure 11.39. Schematic of combined Risk and Resilience metrics for selection of coastal hazard mitigation alternatives (from Basco, 2020).

Risk reduction is about damage mitigation quantified in monetary units. Resilience increase is about functional mitigation quantified in units of all levels of society (people) aided by the hazard mitigation alternative.

Graphically, Figure 11.39 displays the Risk metric in terms of the Benefit/Cost (B/C) ratio on the X-axis as the red region with B/C values increasing above 1.0 for damage reduction (mitigation) alternatives. The Y-axis is the Resilience metric in terms of the percentage increase in robustness and rapidity of recovery as the green region with values increasing for functional resilience alternatives toward 1.0. Community (social), Utility/Infrastructure and Ecological resilience can be combined to demonstrate total resilience (%). Together, the selected alternative (yellow region) may not be the one with the greatest B/C ratio for risk reduction, but the one with the greatest increase in resilience (Basco, 2019, in review).

11.3.8.2 *Total, life-cycle costs in Anthropocene Era*

Total life-cycle costs depend on the sum of initial construction cost, I, and the present worth of maintenance costs, C_m, over the design-life of the project (Equations (8.12) and (8.13). A new "Adaptation Cost", C_{Adapt}, was introduced (Section 8.1.1 (e)) because of the future, unknown sea level rise. Some suggested adaptations for existing alternatives for functional and structural design are presented in Section 11.4.

In the Anthropocene Era, the total life-cycle cost, C_{AE}, for each hazard mitigation alternative should be

$$C_{\mathrm{AE}} = I + C_m + C_{\mathrm{Adapt}}. \tag{11.15}$$

Each adaptation and its cost, C_{Adapt}, should be considered at the initial stages of design so that all alternatives can be evaluated equally. For example, in relatively "deep" water for breakwater design, the caisson type structure may be more economic when future adaptation costs are not considered. However, incorporation of future adaptation costs during the initial design stage to achieve the same level of wave height reduction in the future may significantly increase the total, life-cycle cost, C_{AE}, so that the rubble-mound design will be less expensive. The initial design life should be 30 years to year 2050 as discussed in Section 11.5.

11.3.8.3 *Adaptations to future levels of the oceans*

Adaptation to climate change means different things to different people. The reader is forewarned to be careful when seeing the word "adaptation" in the article title. Neumann *et al.* (2011) discuss coastal armoring and beach nourishment in their article "The Economics of Adaptation Along Developed Coastlines" when they should have used the word "mitigation" in the title. We have only used the word "adaptation" for some possible adaptations for existing mitigation alternatives for functional and structural design as presented in Section 11.4.

11.3.9 *Chapter 9 — Pile supported structures*

The Morison Equation (9.18) and Dean's Stream Function Theory for nonlinear, nonbreaking wave velocities were combined (Dean, 1974) to produce curves on graphs of wave force and wave momentum *coefficients* to apply in the basic Morison Equation for slender piles. The graphs for individual maximum force and momentum components (drag and inertia) employ the dimensionless ratio d/gT^2 on the X-axis to estimate the *coefficients* (Y-axis) as a function of fractions (1/4, 1/2, 3/4, and 1.0) of the breaking wave height, H_b. Airy wave theory ($H = 0$) is also shown for completeness. The graphs for combined total force and total momentum use H/gT^2 as the Y-axis coordinate. The *coefficients* are found on these graphs as isolines. Sea level rise will change the design levels employed for water depth, d, wave height, H, and breaking wave height, H_b. Since most slender piles support piers constructed in shallow coastal waters, predicted future sea levels and the end of the design-life should be employed.

Breaking wave forces must also be considered. As discussed in Section 9.4.3, this means adding an additional wave impact force, $F_I(t)$, to the wave force calculated using the Morison Equation, F_m. Sea level increase impacts the variables in the resulting equations, but the final result that $F_I \approx 8\ F_m$ remains unchanged.

11.3.10 *Chapter 10 — Environmental impacts*

How will sea level rise alter the environmental impact of coastal structures as summarized in Chapter 10?

The major impact will be on armored shorelines. The horizontal distance from the RSL shoreline to the structure (seawall, bulkhead, dike/revetment) will decrease and the vertical profile elevations fronting the structure will deepen in the two-dimensional view neglecting longshore sediment transport processes. Time rates for these changes to occur will increase depending on the sea level rise rate. The same is true for any fixed position near the shoreline (roadway, railway, airport runway, pipelines, powerlines, etc.). In general, the conclusions provided by Dean (1987) as shown in Figure 10.1 and for the past 30 years as summarized in Section 10.2.4 will not change.

However, one major effect of sea levels rising rapidly will be the ecological impacts in estuarine-bay areas as presented in Sections 10.2.4.3 and 10.2.4.5. The natural upward growth of marsh elevation over time requires deposition of silty/sandy sediments to keep pace with rising mean sea level. If the sea level rise rate is too fast, the volume of sediments coming into estuarine-bay areas may not be large enough and the marsh grasses will be drowned. Silty/sandy sediments from upland sources could then be found in "borrow sites" and transported for spreading over the marsh area. This "marsh nourishment" process would be analogous to beach nourishment and require research to understand the economics of "marsh engineering". The NSF- Long-Term Ecological Research (LTER) program is in place to conduct this research effort.

Section 10.4 reviews Natural and Nature-Based Features (NNBF) and "living shorelines". As illustrated in Figure 10.48, the "Green to Gray" reference is now common social media shorthand for the full range of alternatives for low energy shorelines. Nearshore breakwaters, sills, and "edges" together with marsh grass vegetation define "Living Shorelines" for many people.

In the Anthropocene Era of rapidly rising GMSL, coastlines and people living along the shoreline have been identified by the US National

Science Foundation for special research attention. A "Coastlines and People" (CoPe) research workshop was held by the NSF (CoPe, 2018) to identify research needs, broaden participation, and scope out research hubs. On high energy, open coast shorelines such as along the Eastern Seaboard of the Atlantic Ocean, it may be time to consider building nearshore breakwaters together with beach nourishment projects. People would benefit by local recreation, beach tourism, and reduced costs for the life-cycle of beach nourishment. A possible tagline would be "CoPe Shorelines" to signal that people are part of the ecology of the region and that if blamed for global warming, people should also benefit from global warming.

Finally, the environmental impacts of the relocation/retreat alternative could be significant but little research has been conducted in this regard.

11.4 Adaptation of Coastal Hazard Mitigation Alternatives

In this section, we focus on the unknown future for today's new designs and ways to incorporate adaptation strategies and ways to change (adapt) our present-day hazard mitigation alternative design methods to estimate the "adaptation cost", C_{Adapt}. In short, the functional goals for mitigation of the hazard (flooding, wave heights, erosion, ecological restoration) remains the same. The adaptation method is implemented at some unknown time in the future to maintain the original functional goals of the original design.

Existing coastal structures can also be modified at some future time for uncertain levels of the sea. Some of the adaptation methods discussed below can be applied to existing projects. However, the expense will be much greater than if the adaptation method is incorporated in today's new design.

11.4.1 *Increase crest elevation*

11.4.1.1 *Rubble-mound structures*

Rubble-mound breakwater crest elevations can be readily increased by adding more armor stones or concrete armor units. Figure 11.40 illustrates this method as applied to the damaged breakwater at Lajes Air Field breakwater (Figure 6.35) in Chapter 6. Wave heights for design will increase so that additional armor layer(s) may require larger stone size or artificial armor units. The cross-sectional foot print will be larger after adaptation requiring both geotechnical information to ensure stability and environmental permit approval.

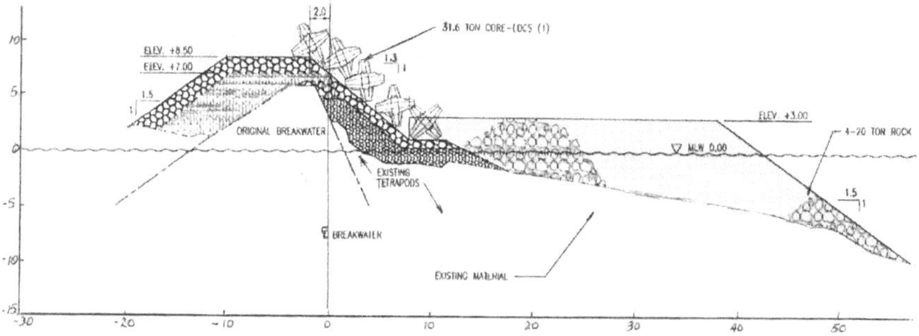

Figure 11.40. Repair of Lajes Air Field breakwater, Terceria Island, Azores, Portugal (Figure 6.35).

Nearshore breakwaters with primary function to increase the fill-life of the adjacent beach regularly experience a wide range of water levels including overtopping and submergence. Clearly, the functional purpose of rubble-mound structures (breakwaters, groins, jetties, sills, and revetments) varies widely so that the adaptation method and strategy may also be different for each type.

Sergent *et al.* (2015) employed analytical and statistical methods and physical models to show that for rubble-mound breakwaters, the increase in crest height, ΔD, is relative to the increase in mean sea level, Δs, i.e. the ratio $\Delta D/\Delta s$ was in the range 1.4–1.7 for a constant overtopping discharge rate, q. Figure 11.41 is included here as good example of how the presentation of results can be displayed and not for the actual numbers for this one breakwater site in Deauville, France. Sea levels are shown along the X-axis and offshore, significant wave heights on the Y-axis that together produce the joint-probability, family of exceedance probability curves with corresponding recurrence intervals (see e.g. Hawkes *et al.* (2002) for how to calculate the joint probability of waves and water levels in coastal engineering design). Then an analytical formula for wave height transformation (shoaling) across the nearshore profile (e.g. Equation (2.31) for irregular wave shoaling, Goda, 2010) and depth-limited breaking (e.g. Equation (2.23)) are applied to calculate wave overtopping rate, q, for a given crest elevation. The red crosses are the number of events giving overtopping discharge rates greater than the intersection of the two curves when $q = 0.05\,\mathrm{m}^3/\mathrm{s}$ per linear meter (mL) of breakwater from existing data at the site. Increasing the crest elevation will reduce the wave overtopping rates from the family of curves shown for this one existing breakwater (from Sergent *et al.*, 2015).

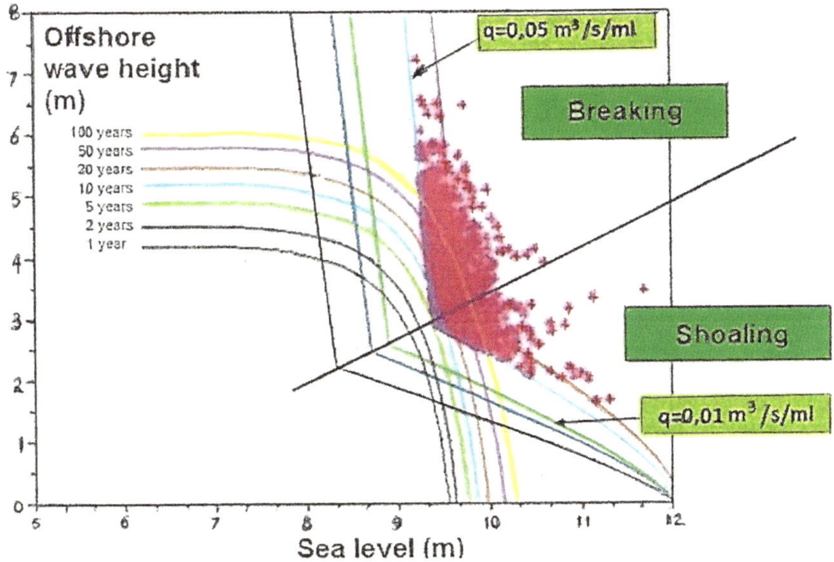

Figure 11.41. Curves of joint-probability (sea level and wave height) of recurrence intervals (years) and curves of overtopping discharge rates, q, for one breakwater site (from Sergent *et al.*, 2015).

11.4.1.2 *Impervious, vertical-walled structures*

Increasing the crest elevation for vertical, concrete structures is more difficult and expensive. Figure 11.42 is one proposed design offered by a company for application along the Danube river in Grein, Austria. Clearly, the rock revetment, foundation, substructure, and internal mechanism for storing then raising the vertical wall/gate will require a large expense per meter length of flooding protection for the road/walkway.

At the low-cost end of the cost spectrum is the concept of a movable cap design that can be explained by noting the actual designs of existing seawalls in Figure 11.43. Here the total concrete cap with steel reinforcing is about 2.5 m (8 ft) in length. See Figures 1.7 and 1.9 in Chapter 1. The initially constructed cap would still be reinforced concrete but shorter in length (1.5 m, 5 ft) with reinforcing bars sticking out the top for future use when needed. The movable cap would be steel (1.0 m, 3 ft) and fitted over the exposed reinforcing bars so that the vertical wall was intact up to the desired crest elevation and the required 2.5 m (8 ft) cap length. This steel cap at the top of the vertical wall could also hold any required hand railing system. It would be manufactured in lengths compatible for lifting and

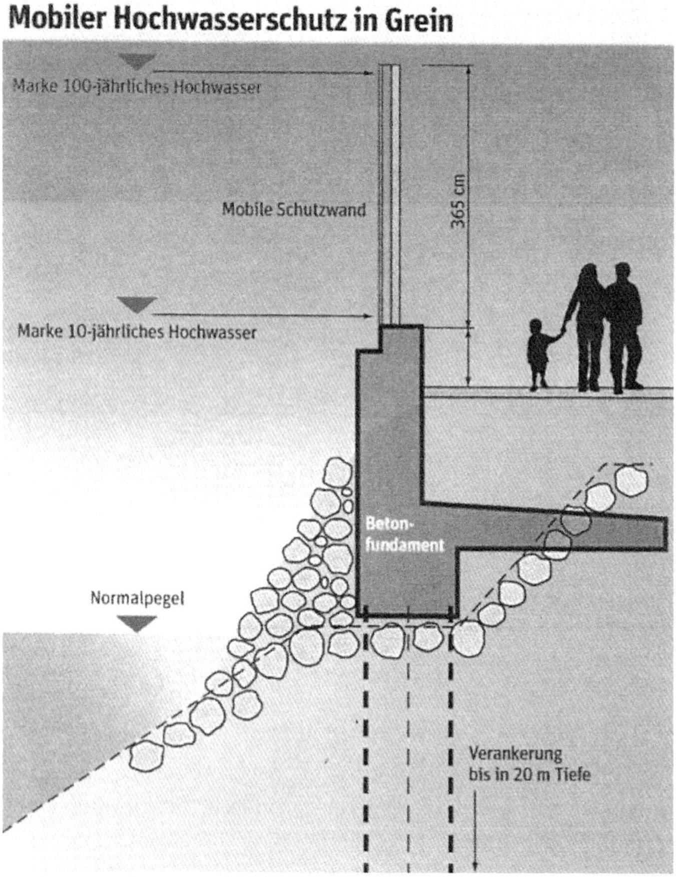

Figure 11.42. Impervious, vertical structure with mechanized system to increase the crest elevation (courtesy The Cellar Image of the Day, cellar.org).

transport. When an increase in the crest elevation was needed, the steel cap sections would be removed, and additional reinforced concrete added on top of the original design concrete cap. The exposed reinforcing bars would now be embedded in the new, added concrete section and exposed reinforcing bars would again be added at the top of the new concrete section. When complete, the movable steel cap would again be placed on top. If the added new concrete section was 1.0 meters (3 ft) in length, the total cap length would now be 3.5 m (11.5 ft). Future sea wall crest elevations could be increased in the same way. Adaptation cost, C_{Adapt}, would be the difference in initial costs for this movable cap design and the original design

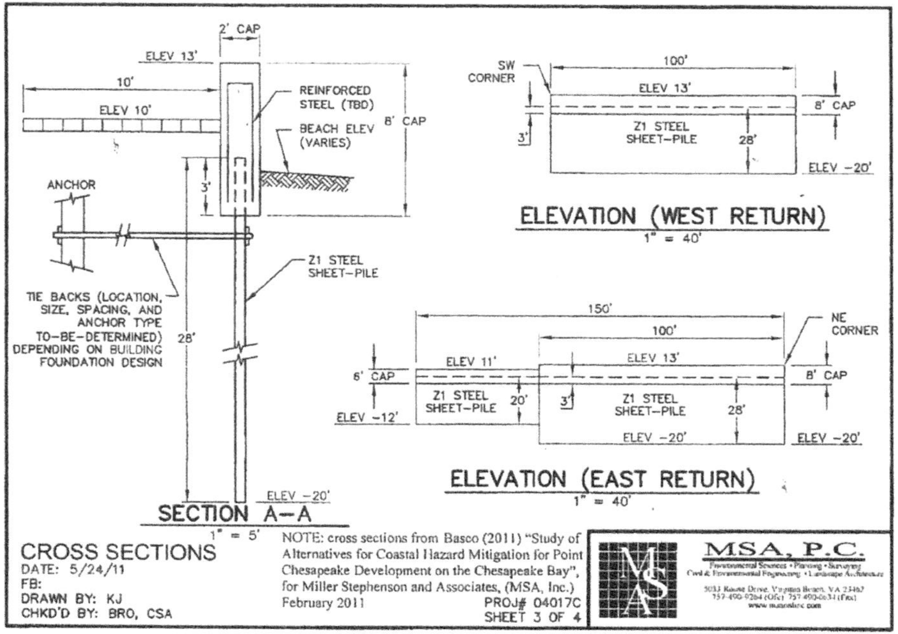

Figure 11.43. Vertical seawall design for Chesapeake Point, VA (left, Figure 1.7) and for Virginia Beach, VA (right, Figure 1.9) as described in Chapter 1. The reinforced concrete cap is 2.5 m (8 ft) in length.

for the sea wall alternative. Leakage would be handled by increased interior drainage.

11.4.1.3 *Parapet, wave return crests (bullnose)*

Section 5.3.4 presents a discussion of overhanging parapets (aka, wave return, bullnose, recurved crest) as shown in Figure 5.17. The parapet coefficient, k, is the ratio of the wave overtopping discharge rates with and without the parapet, Equation (5.32). Figure 5.16 displayed data of k-vs-relative freeboard (R_c/H_s) as determined by Pearson *et al.* (2004). A "decision chart" is needed for three regions of relative freeboard to estimate the k coefficient.

Increasing sea levels decrease the actual freeboard, R_c, and potentially increase the wave height, H_s, at the structure toe so that the relative freeboard (R_c/H_s) decreases at a faster rate. The k coefficient varies from about 0.05 to 1.0 in Region 2. Decreasing the relative freeboard increases the parapet coefficient, k, and increases the overtopping rate. The height of the vertical wall below the parapet is defined as P_c. Three variables define the parapet geometry, namely,

h_r = height of parapet (design variable) $= R_c - P_c$,

B_r = horizontal extension or parapet in front of vertical wall (design variable),

α = angle of parapet, $\tan\alpha = h_r/B_r$.

If the goal is to design an adaptative parapet, its geometry must be modified to recover the original design parapet coefficient, k, after sea level rise. This would be comparable to increasing the crest elevation of the vertical wall to keep the original wave overtopping rate. Formulas provided in Figure 5.18 for Region 2 permit the calculation of geometry modification to recover the original parapet coefficient, k, for rising seas. A movable steel cap design could incorporate the parapet design and its modification more economically constructed than a concrete cap with parapet.

Other parapet shapes are possible than studied in Pearson *et al.* (2004). Research is needed.

11.4.2 *Decrease frontal wave heights*

Wave overtopping rates could also be mitigated under rising sea conditions by decreasing the wave heights at the toe of the structure. One way is to construct a berm in front of the sloping rubble-mound or vertical impervious structure as illustrated in Figure 11.40 for the repair at Lajes Air Field. Shallow water depths would cause depth-limited, wave breaking heights that produce less wave runup and wave overtopping. The frontal berm could be the reshaping type. When shallow water depths already are present seaward of the structure, a fixed bar to trip the waves is also possible. The cross-sectional foot print will be larger after adaptation requiring both geotechnical information to ensure stability and environmental permit approval during initial design stages for the project.

11.4.3 *Other potential adaptation strategies*

For future rising sea levels and fixed crest elevations for flood mitigation, early consideration for interior drainage to handle increased wave overtopping rates is required. Costs for replacement of old, urban drainage pipe systems with robust pipe systems to handle potential, future interior drainage (including rain-fall runoff events) will become more economical if incorporated today and not in some unknown future time.

Floating breakwaters will become more economical even with the added expense for mooring systems. Floating storm surge barriers across tidal inlets will be given more attention because of the potential for large flood damage prevention benefits to offset the large initial costs. Figure 11.44 is the floating storm surge barrier at Maastricht, the Netherlands.

Combinations of wave energy reduction structures (breakwaters) with wave energy conversion features will be considered at more sites. Wave power production income in the future will offset the added costs for the initial construction plus the future adaptation costs for keeping the breakwater function under rising sea levels.

Combinations of beach nourishments and nearshore breakwater systems will become more acceptable on open ocean shorelines. The time will come when "CoPe shorelines" will be just as important as "Living shorelines".

A number of new companies are advertising "Adaptation Technologies". See Appendix D in ASCE Manual 140 for products, names, and websites in this regard (ASCE Manual 140, 2018).

11.4.4 *Mitigation of marine ice shelf instability*

The time has also come to "think big". Slowing the rate of the fastest ice melt and flows into the oceans would buy time (centuries) for facing the reality of climate change and reduce sea level rise rates to mitigate flooding damage to coasts. Focus is on the polar glacier sources of marine ice sheets

Figure 11.44. Floating storm surge barrier at Maastricht, the Netherlands (Dreamstime stock photos https://www.dreamstime.com/photos-images/surge-barrier.html) (courtesy Flickr: Creative Commons).

in Greenland and the Antarctic. Three potential ways to delay the loss of ice sheets were recently presented by an international team of scientists at Princeton University (Moore *et al.*, 2018).

11.4.4.1 *Blocking warm ocean water*

Figure 11.45 illustrates one concept to block warm water in the ocean from reaching the underside of the ice shelf, and reduce its melt-rate. Initial conditions (a) show the grounding line for the massive, ice-sheet glacier (grey–blue) and its attachment to the bottom (dark brown) at the tipping point and thinning ice shelf (purple) melting from below. Ice flux from the glacier is turning into sea level rise. Instability of the ice sheet, Figure 11.45(b) due to warm ocean water near the bottom (red) has caused the grounding line to retreat and the ice shelf (purple) to reduce in size and strength and allow an increase in the ice flux from the glacier into the ocean. This is labeled marine ice sheet instability.

Figure 11.45(c) graphically shows an artificially constructed "sill" (light brown) located at the original, elevated tipping point to block the warm ocean water from reaching the underside of the ice shelf. Melt-rate for the

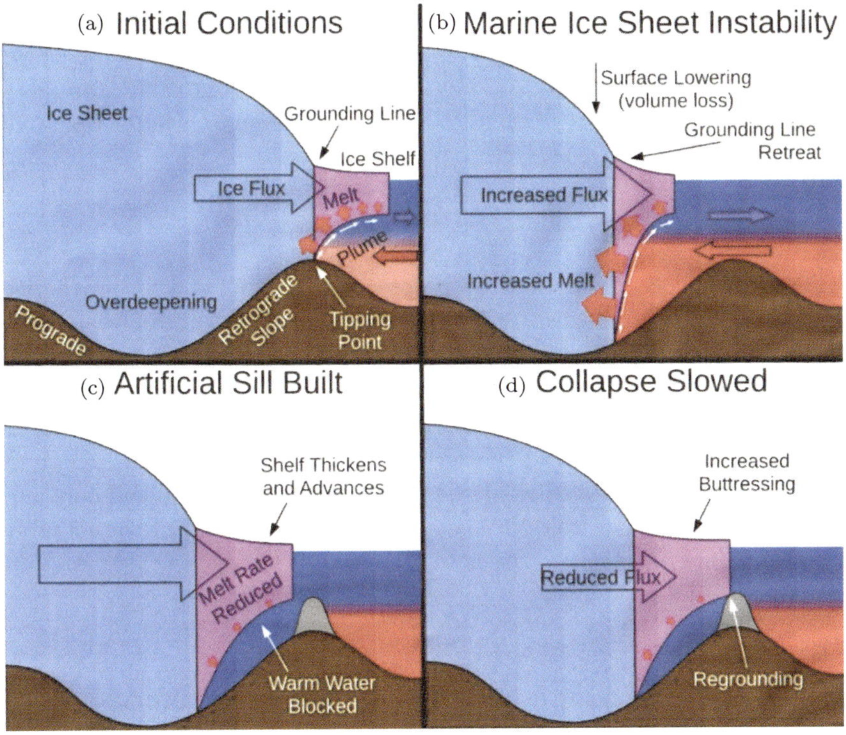

Figure 11.45. Schematic of concept of artificial dike to block warm water from melting glacier ice shelf (from Moore *et al.*, 2018).

ice shelf is reduced. Natural ice formation processes in the polar regions cause the ice shelf to grow in length and thickness and slow glacier collapse to reduce the flux of ice from the glacier and lower the rate of sea level rise as illustrated in (d) where the ice shelf would again be grounded on top of the dike.

The Jakobshavn glacier in Greenland in a relatively narrow (5 km, 3 mi) fjord was proposed as a possible location for further feasibility study of the engineering, economic, environmental, and political constraints. A 100 m (330 ft) high dike (sloping sides) would require about 1×10^8 m^3 (1.3×10^8 cy) of sand and gravel dredged from the continental shelf. This large volume is 10 times *less* than excavated for the Suez Canal and the Hong Kong airport required three times more volume as landfill. The authors used the term "glacial geoengineering" in describing this effort. No cost estimate was made.

However, as noted in Section 11.2.6, Greenland's contribution to GMSL is relatively minor (10–20 cm) for year 2100. Antarctica is the largest contributor and would require far larger and more diverse projects because of the vast lengths of coastlines with ice shelves holding back glaciers from the sea.

11.4.4.2 *Support for ice shelves*

The second concept proposed by Moore *et al.* (2018) was to pin the ice shelves in front of the two main glaciers (Pine Island and Thwaites) in West Antarctica. To do this, both artificial islands and under-water berms were suggested. At Pine Island, with an existing Jenkins Ridge, the berm on top of the ridge would again be over 100 m tall but require 60 times more volume than for Jakobshavn glacier in Greenland because of its length of 300 km (180 mi). An island would be over 300 m (1,000 ft) tall and only jam the ice shelf at one location. No cost estimate was provided but considerably more than in Greenland.

11.4.4.3 *Reducing the below-ice shelf water flow*

Figures 11.45(a) and (b) also shows some small arrows (white) directed seaward along the subsurface of the ice shelf. The third suggestion is to drill holes from the surface for pumping stations to either extract or freeze the water at the glacier base to slow the sliding ice-water streams entering the sea.

The paper by Moore *et al.* (2018) contains many more interesting details including countries that may fund feasibility studies of these "geoengineering" schemes. They will not mitigate global warming from greenhouse gases. The future of the polar ice sheets will continue to rest on emission reductions. The long-term goal is to possibly manage the collapse of the ice sheets to reduce the sea-level rise rate, to buy time to facilitate mitigation and adaptation.

11.5 Philosophy for Design

11.5.1 *Coastal engineering design to year 2050*

Coastal engineering design for the Anthropocene Era under uncertainty in GMSL(t) predictions for the next 30 years (to year 2050) in the polar ice-cap regions has led this author to the following conclusions:

(1) Human nature and the world's complex political system will mean that greenhouse gas emission reductions will not be achieved.
(2) The IPCC, "Representative Concentration Pathway" link of emission pathways to future sea level rise conditions [(RCP 8.5) Business-as-usual] is the conservative choice for design.
(3) Understanding the Antarctic Ice-Sheet (AIS) physical contribution to probabilistic sea-level rise projections will determine the acceleration rates of GMSL(t) predictions.
(4) The research results of Kopp *et al.* (2017) as shown in Figures 11.23 and 11.24 and their major conclusions provide valuable insights for coastal engineering design.

    (a) For the next 30 years up to year 2050, both empirical-based models and process-based models of AIS physics show little difference in probabilistic projections of GMSL.
    (b) At the median (50%) level, predictions for GMSL rise are (29–31) cm for both models.
    (c) At the (5–95%) range, predictions for GMSL (2050) are [(17–21)-(39–48)] cm for both models.

(5) Coastal hazard mitigation design should be based on a design life of 30 years (to year 2050–2060). **Use a GMSL rise of 0.45 m (1.5 ft) for year 2050** at the probability of exceedance level of 5% (90% Confidence Interval) for major projects. Relate the GMSL rise to the RSL rise at your site.
(6) Selection of the hazard mitigation alternative should consider the feasibility and economics of adaptation of the alternative to meet GMSL rise projections to year 2100. Adaptations for both functional (Chapter 5) and structural, strength (Chapter 6) performance objectives to reduce risk and increase resilience must be considered. See Section 11.4 for further details regarding adaptation of both structural and non-structural mitigation alternatives for risk reduction.
(7) Include the flooding resilience metric to quantify the increase in resilience for all mitigation alternatives. Conditional probability theory (Bayesian networks) should be employed.
(8) Both risk reduction in terms of the Benefit/Cost ratio in monetary terms and the increase in resilience in terms of the probability for improved robustness and rapidity of recovery must be considered when

making a final selection of the hazard mitigation alternative. People from all socio-economic levels living in coastal regions must be considered and accounted for in the final design.

11.5.2 *Long-term coastal engineering design (2050–2100)*

Coastal engineering design for the Anthropocene Era under the "deep" uncertainty in GMSL(t) predictions for the following 50 years up to year 2100 must recognize that physics-based model predictions are far higher and show wide-spread results depending on the RCP emissions scenarios. And, projections of GMSL rise to year 2100 will be vastly improved in 2050 so that a 50-year economic design-life may again be possible. New studies of coastal hazard mitigation alternatives should be undertaken in 2050.

11.6 Summary — Sea Level Rise

In this chapter we have reviewed climate change and the literature for GMSL rise predictions from today to year 2100. The uncertainty generated by the rapidly changing ice-sheet physics in the polar regions of Greenland and Antarctica is of greatest concern. Impacts of GMSL rise for design are then summarized for the first 10 chapters and some suggestions provided for adapting (changing) our alternatives for coastal hazard mitigation in today's design to the unknown future levels of the sea. This is all summarized in our "Philosophy for Design" in Section 11.5.

Coastal engineers will play an ever-expanding role in making future decisions regarding mitigation, accommodation, and relocation/retreat from the world's coasts. The timing of these decisions will depend on the following three areas of concern. Timing is everything.

11.6.1 *Tipping points*

Complex systems (political units, organizations, societies, ecosystems, and global civilization) are largely self-regulated by internal feedback loops which act to maintain them in a stable state. When stressed beyond the stable state, the complex system reaches a "tipping point" which causes it to be altered into a new state. Coastal urban cities are complex systems of systems. The property tax system brings revenue to pay for the public works system (utilities, transportation), school system, police and fire protection systems, etc. However, stable conditions in coastal urban cities tell us little about when a "tipping point" has been reached because we have never experienced them in the past.

Waterfront property market value is perhaps the first indicator of approaching a "tipping point" in the real estate industry. As the world's oceans have been rising, the value of homes "on the water" has been sinking. A study by the non-profit, First Street Foundation (https://firststreet.org) found that eight US Atlantic Ocean and Gulf coastal states (CN, NJ, NY) in the north and (VA, NC, SC, GA, and FL) in the south, lost $14.1 billion in property values from 2005 to 2017 because of increasing flooding risks of vulnerable homes (First Street, 2018). For cities in the south, Miami Beach, FL, lost the most (over $337 million) and Norfolk, VA, was number 20 in cities with total property value lost of over $55 million averaging about $4.5 million/year.

Norfolk, VA, is an interesting example of the relative importance of the residential property assessments which make up 65% of the cities taxable assessed value. Residential property assessments include condominiums, but exclude residential vacant land and apartments. However, Norfolk is "built out", meaning little vacant land exists for new residential development. The 2019 Budget for Norfolk is $880.8 million of which $243 million (27.6%) comes from real estate taxes. Real estate assessments are the largest local generated source of General Fund revenue for the city. For the 2019 Budget, the city has increased the real estate tax rate by 10 cents, from $1.15 to $1.25 per $100 of assessed value. The city tax rate has been as high as $1.35 to $1.40 per $100 assessed value from 1990–2005. What is the "tipping point" for the maximum value of the city tax rate before the General Fund revenue is no longer sufficient to pay for all the city services that people expect from the City? When will the stable state become unstable and the domino effect begin where lost property value lowers the tax base and cripples routine, normal functions of the city? Development of community resilience guidance for "best practices" for risk mitigation and long-term adaptive management is just beginning. Decision-support guidance is needed to understand "tipping points" in the decision-making process.

For many people who would like to live "on the water", they have already decided against the prospect of buying waterfront property. It no longer is increasing in value each year and they may not be able to sell when they are ready. They have already reached their own "tipping point" in making this decision.

Waterfront property *rentals* may be a "solution" for the city and for these people. Private enterprise buys at fair market value and then rents property on n-year lease to families desiring to "live on the water". The business plan includes all expenses [mortgage and insurance (fire, flood), property taxes, maintenance, improvements, and fair profit] and strategy to retreat at some time in the future. City property tax assessment division, mortgage lending industry, real estate industry, and media all have to adapt present-day thinking regarding residential property that is increasingly vulnerable to flooding risk. The bottom-line is that the city continues to obtain property tax income to function and home owners are able to sell their waterfront property at fair market value. People, businesses, and cities all benefit from added time to select mitigation alternatives and make long-term adaptive management decisions.

Design of alternatives to mitigate current and future flooding resulting from rising seas requires financial planning.

11.6.2 *Financial plan*

The world's coastal countries, states, and cities now face a financial crisis for survival. Planning, construction, maintenance, and adaptation costs for flood damage mitigation require large expenditures for funding the selected alternative.

The world looks to the Netherlands for advice on how it solves flooding problems and should also consider how the Dutch finance these solutions. The 17 million citizens, businesses, and industry all understand that their country has a long history of survival against the sea because they apply the age-old adage *who benefits equals who pays.* The Dutch "Delta Plan" and its current improvements for future sea level conditions is *funded by everyone living and working in Holland.* Every country in the world has its own economy and is now considering how to finance mitigation and adaptation plans for the future.

In the United States, the Federal Government's Administration prepares an annual budget that includes Public Works Projects designed by the Army Corps of Engineers. Then the US Congress (Senate and House of Representatives) approves the annual Budget that may (or may not) include all the projects recommended by the Corps of Engineers. The longstanding policy is that the Federal Government pays 65% of the initial construction cost and the local government (state, county, city) pays 35%. Who pays life-cycle maintenance costs is more complicated. In a future of uncertain sea level rise, this institutional policy needs to be changed for a fairer distribution of costs.

One example is for a proposed storm surge barrier across the entrance to the Hampton Roads region of Virginia on the lower Chesapeake Bay. It would mitigate current high-water flood events and future sea level rise,

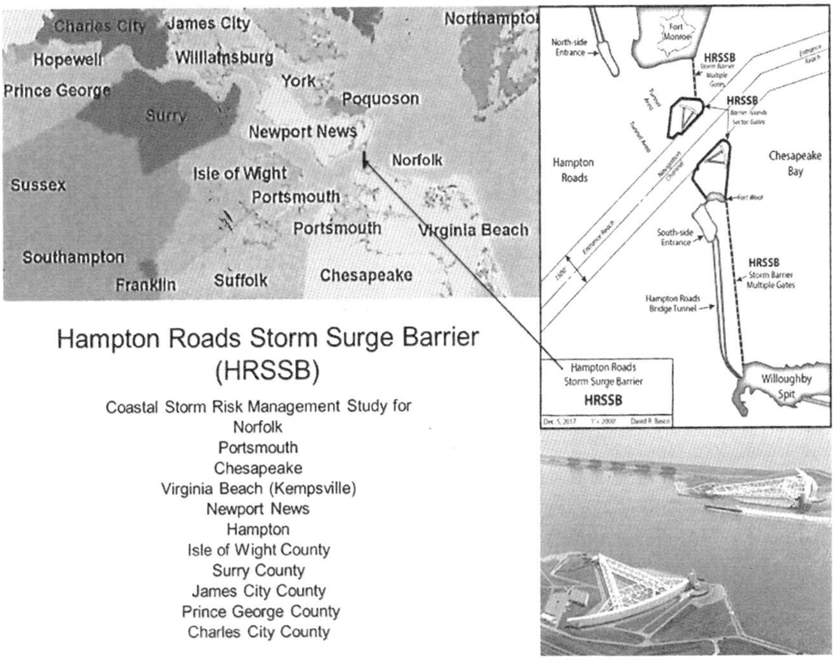

Figure 11.46. Location map and drawing of proposed Hampton Roads Storm Surge Barrier with photo of Maeslant barrier on the Rotterdam waterway (Basco, 2017).

protecting all the low-lying cities on the Elizabeth River and James River as illustrated in Figure 11.46 (Basco, 2017). The flood reduction benefits would reach billions of dollars and far exceed the costs. The Hampton Roads Storm Surge Barrier (HRSSB) would include movable, horizontally rotating gates similar to that constructed 20 years ago in Holland (photo, lower right side, see Section 6.5) so that all navigation requirements through the Hampton Roads channel (drawing, upper right side) remain the same.

In the Hampton Roads region of Virginia, more than one-third of the economy relies on military spending with the bulk for the Norfolk Naval Station and for the Huntington Ingalls ship building company located in Newport News, VA. The Norfolk Naval Station is the largest navy base in the world and Huntington Ingalls ship building company is the largest in the United States with outlines of their locations shown in Figure 11.47. Both would benefit by construction of an HRSSB which would reduce high water events on their properties. They both could construct flood walls and dikes to surround their properties but that would not protect the over

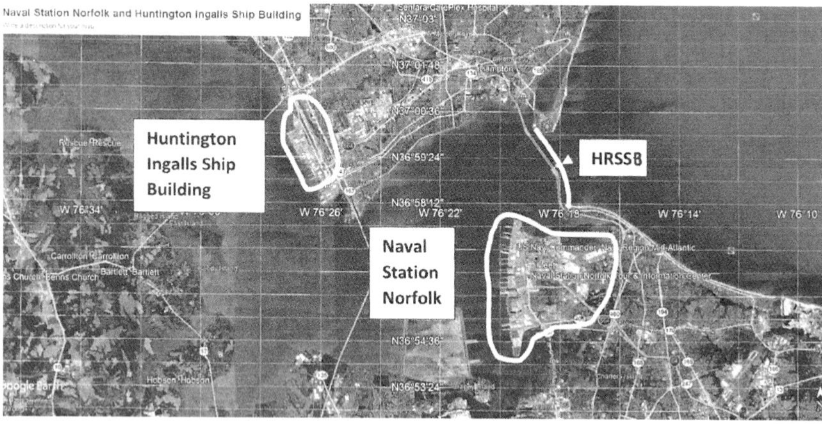

Figure 11.47. Google Earth map of Hampton Roads region showing the HRSSB across the entrance to the Elizabeth River and James River and outlines of the Norfolk Naval Station and Huntington Ingalls Ship Building company (courtesy Map data: Google, Maxar Technologies).

15,000 civilians and military personnel living in the region that work at the Naval Base or the over 25,000 employees who work at the ship yard and live in the cities of the region. They travel to work from these cities.

The unknown future of sea level rise should be the catalyst for serious analysis and debate on how best to fund projects such as the HRSSB. The age-old adage *who benefits equals who pays* for a fairer distribution of costs should be the goal. Today's outdated policy is the social, political, and institutional constraint facing coastal engineers. Over 330 million people live in the US and all benefit by having a Norfolk Naval Station for docking, servicing, and deploying the many ships at this location. These ships are built at the Huntington Ingalls ship yard. The Naval Base and ship yard must be located at coastal harbor locations where low wave energy and deep draft water depths are present. The US Department of Defense (DOD) accounts for 22% ($0.904 trillion) of the Federal Government budget ($4.11 trillion) for 2019. Some percentage of the 65% Federal Government share for the HRSSB could come from the DOD budget. A strong, modern US Navy is a benefit to all citizens of the United States.

Adaptation of the social, political, and institutional policies of the coastal countries, states, and cities of the world must take place to deal effectively with the uncharted water elevations of the future. These policy changes will take time for decisions to be made.

Design of alternatives for *mitigation* (reduction) of current and future flooding and *adaptation* (change) of current policies resulting from rising seas requires both strategies. For implementation and to become effective, people must have a clear definition of each term. The coastal engineering profession should make a strong effort to educate the public, politicians, and the media in this regard.

11.6.3 *Adaptation without mitigation is immoral*

Talk is cheap. The democracies of the world, in general, elect officials (mayors, representatives, senators, etc.) promising "no new taxes" to serve and make decisions for the relatively short terms (few years) of the positions they hold in office. *Adaptation* to sea level rise is easy to talk about and has become the "low cost" solution. Alternatives to *mitigate* damage from today's storm surges and future sea levels are expensive. Decisions are needed for the long-term that today's elected officials are unwilling to face because they will require tax increases.

Consider Norfolk, VA, as an example. Figure 11.48 graphically depicts by color, the four types of long-term strategy the Planning Department has published for their NORFOLKVISION 2100 plan (City of Norfolk, 2016). Norfolk is one of the 100 Resilient Cities selected by the Rockefeller Foundation.

The areas colored yellow are designated "Adapting to Rising Waters". This planning document states:

> *These locations... in the yellow areas all need additional protection from rising waters. However, the sheer cost and environmental challenges inherent in walling off such large, complex waterfront areas from rising waters likely make the heavily-engineered solutions ...infeasible here* (p. 32).

These yellow areas are single-family, developed neighborhoods and property owners are encouraged to invest in *smaller-scale techniques* such as living shorelines, rain gardens, and green roofs. The City of Norfolk is to develop *mitigation steps that are recommended for residents to pursue.* The bottom line is that although residents in these yellow areas with "waterfront property" pay higher property taxes (assessed at higher value), they are asked to go it alone when it comes to citywide mitigation planning for flooding and sea level rise in the future. The red areas are "economic engines" and include downtown Norfolk (bottom, left side) and the Norfolk Naval Station (top, left side), worthy of flood damage mitigation.

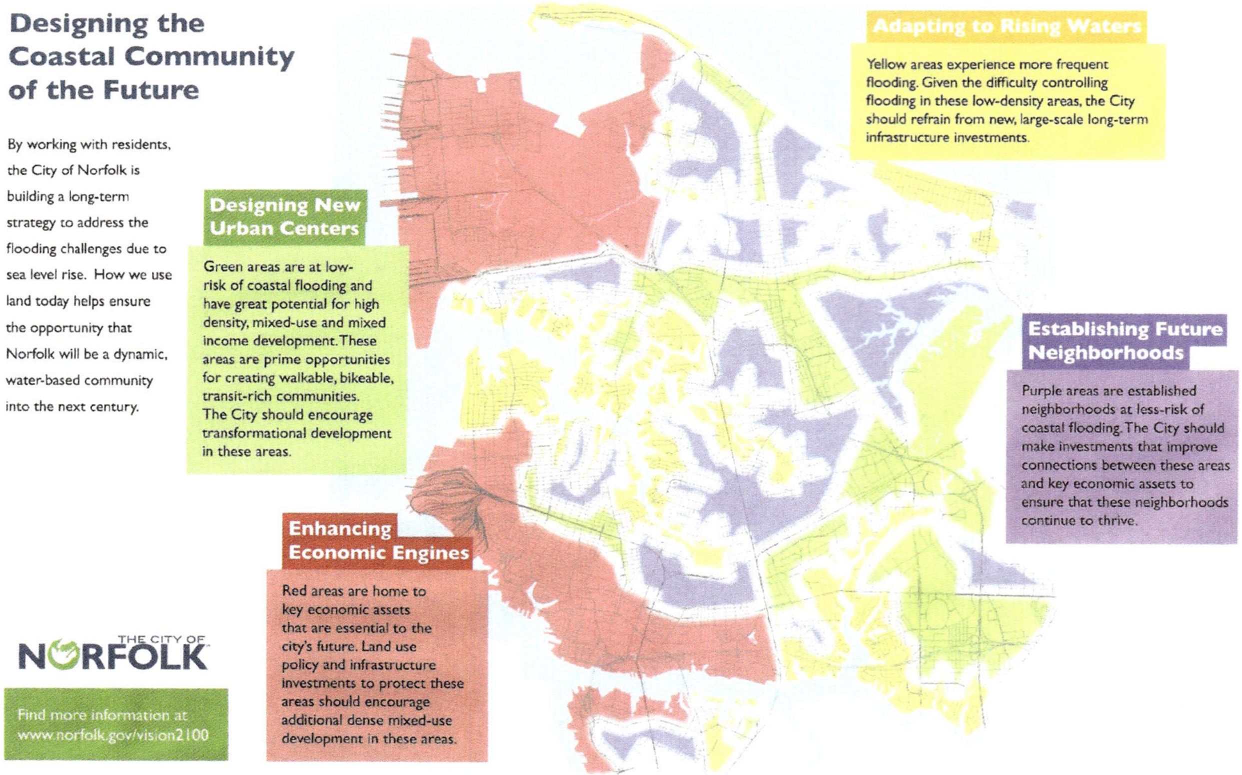

Figure 11.48. Color representation regions and four types of long-range plans adopted by the City of Norfolk, VA, for the NORFOLKVISION 2100 plan. Yellow is "Adapting to Rising Waters".

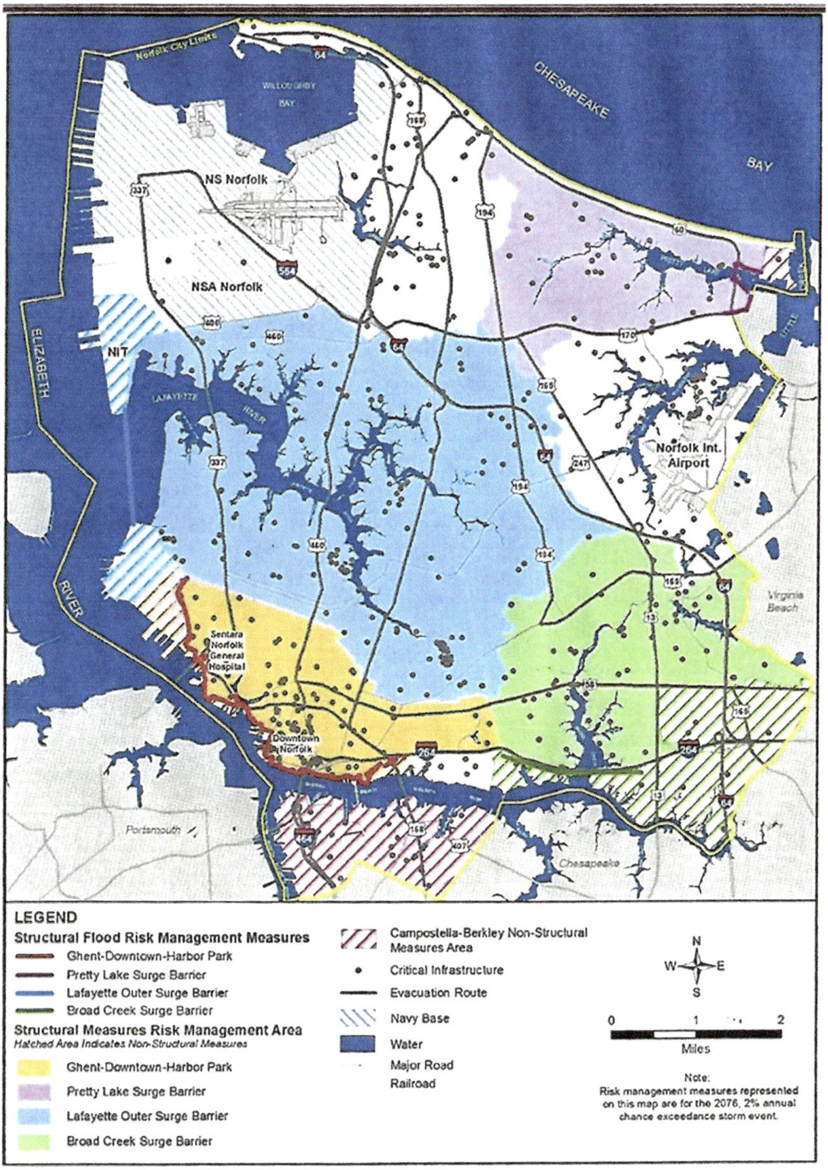

Figure 11.49. Graphic illustration of the TSP for flood damage mitigation (from Draft, Integrated, Coastal Flood Risk Management Feasibility Report/ Environmental Impact Statement, DIFS-EIS, 2017).

In direct contrast is the Corps of Engineers recently completed feasibility study to determine the Tentatively Selected Plan (TSP) to mitigate flooding in Norfolk (Norfolk D.O., 2017). Figure 11.49 is the Corps' graphic of the TSP with different colors. Light blue here is the Lafayette River watershed and essentially the main "Adapting to Rising Waters" yellow area as published by the City of Norfolk in Figure 11.48. Many alternatives were studied in this effort. Structural flood risk mitigation measures selected are listed in the Legend. A storm surge barrier is illustrated as a blue line across the entrance to the Lafayette River. Flood damages would be significantly reduced for all the waterfront property (and adjacent low-lying property) that is adjacent to the Lafayette River tidal estuary. Similar storm surge barriers are recommended across the entrances to Broad Creek (green area) and Pretty Lake (purple area). Flood walls, dikes, and a storm surge barrier across the Hague (red line) was the structural measure for the Ghent–Downtown–Harbor Park (yellow) area.

This Corps of Engineers study was released in October 2017, with headline in the local paper:

> *Federal study calls for $1.8 billion in upgrades to reduce Norfolk flooding, but funding is far away.*

The news article failed to mention that the flood damage reduction benefits totaled $4.5 billion giving a B/C ratio of 2.5 and that the net annual benefits (B-C) were $101 million.

As of this writing, the City of Norfolk Vision 2100 plan approved in November 2016 has not considered the Corps of Engineers' feasibility study results released in October 2017. As a result, the citizens of Norfolk are left wondering what "adaptation" and "mitigation" mean for them when facing a future of sea level rise. Families living in the socio-economically disadvantaged section of the city labeled "Campostella–Berkeley Non-structural Measures Area" are especially vulnerable to future flooding. (Cross-hatched in red near the bottom of Figure 11.49). This is a good example of shortcomings of the risk reduction metric as quantified by the B/C ratio. Damage reduction benefits for the relatively wealthy property owners living around the Lafayette River are large because the assessed value of the property is large. The B/C ratio is sufficient to justify a storm surge barrier across the entrance to the Lafayette River. At the other end of the economic spectrum, low property assessment value in the Campostella-Berkeley region cannot provide sufficient damage reduction benefits to justify structural measures (dikes) to reduce flooding damage. In this case, a new-age adage should be "*the big-wheel gets the shock absorber and the little-wheel gets rusty*".

Adding a second metric by quantifying flooding resilience that considers the increase in functional performance (housing, utilities, transportation, etc.) for all the people may produce a different TSP by the Corps of Engineers (see Figure 11.39).

Design of alternatives for *mitigation* (reduction) of current and future flooding and *adaptation* (change) of current policies resulting from rising seas requires *both* strategies. Many scientists, managers, media articles confuse the two and use adaptation when mitigation is correct. For implementation and to become effective, people must have a clear definition of each term. The coastal engineering profession should make a strong effort to educate the public, politicians, and the media in this regard.

In summary, all coastal cities face tipping points that trigger instability in vital functions as a consequence of GMSL rise. Governments at all levels have a moral obligation to do the right thing for all the people. At the national (federal) government level, both the risk reduction metric (B/C ratio) and the functional resilience increase metric should be employed so that people at all socio-economic levels benefit. Who benefits equals who pays should be the goal to finance flood damage mitigation projects. Flooding does not recognize political nor institutional boundaries, so that military and industry must pay their fair share. At the regional (state) government level, many adjacent cities benefit from the same flood mitigation project (e.g. the HRSSB) so that the state government should be the "local" sponsor of national government studies, not just one city (e.g. Norfolk). At the local (city) government level, individual home owners with shoreline property have been assessed with higher property taxes. City governments have a financial need to maintain this revenue stream. And, they have a moral obligation to help these "waterfront property" owners in a future of sinking property values. They must embrace *both* financial plan options for flood damage mitigation, and long-term adaptation strategies.

Adaptation without mitigation is immoral at all government levels when facing the unknown future of sea level rise. The design of coastal hazard mitigation alternatives for rising seas is the title and focus of this book.

References

ASCE Manual 140 (2018) "Climate-Resilient Infrastructure," Manual of Practice No. 140, *American Society of Civil Engineers*, Wash DC. 308 p.

Bamber, J. L. and W. P. Aspinall (2013) "An Expert Judgement Assessment of Future Sea Level Rise from the Ice Sheets", *Nature Climate Change*, Vol. 3, pp. 424–427.

Bambera, J. L., M. Oppenheimer, R. E. Kopp, W. P. Aspinall, and R. M. Cook (2019) "Ice Sheet Contributions to Future Sea-level Rise from Structured Expert Judgment," *Proceedings of the National Academy of Sciences of the USA (PNAS) June 4 2019*, Vol. 116, No. 23, pp. 11195–11200.

Basco, D. R. (2017) "Storm Surge Barrier for Hampton Roads," Opinion Page, *Virginian-Pilot* newspaper, Dec 30.

Basco, D. R. (2019) "Quantifying Coastal Flooding Resilience for Rising Seas," *Earth's Future* (in review)

Benjamin, J. R. and C. A. Cornell (1970) *Probability, Statistics, and Decision for Civil Engineers*, McGraw-Hill, New York.

Bindoff *et al.* (2007) "The Physical Science Basis," *IPCC Climate Change 2007*, S. Solomon *et al.*, editor, Cambridge University Press, Cambidge, U.K., pp. 385–432.

Bruun, P. (1962) "Sea Level Rise as a Cause of Shore Erosion," *Journal of Waterways and Harbors Division, American Society of Civil Engineers*, Vol. 88, pp. 117–130.

Castrop, H. (2004) "Estimating and Combining Uncertainties," in *8th Annual International Test and Evaluation Association Instrumentation Workshop*, online.

Church, J. A., and N. J. White (2006) "A 20th Century Acceleration in Global Sea-level Rise," *Geophysical Research Letters*, Vol. 33, p. 160.

Church, J. A., and N. J. White (2011) "Sea-level Rise from the Late 19th to the Early 21st Century," *Surveys in Geophysics*, Vol. 32 No. 4–5, 585–602.

City of Norfolk (2016). "Opportunity, Collaboration, Vision," NORFOLKVISION2100, City of Norfolk Virginia, website (https://www.norfolk.gov/Vision2100).

CoPe (2018) "Coastlines and People," Scoping Workshop (Chicago), US National Science Foundation.

Corps of Engineers (2013) *Incorporating Sea Level Change in Civil Works Programs*, ER 1100-2-8162. Washington, D.C.

CRS (2016) US Congressional Research Service report (online, Dec 2018).

Dean, R. G. (1974) "Evaluation and Development of Water Wave Theories for Engineering Application", Vols. 1 and 2, Special Report 1, *Corps of Engineers, Coastal Engineering Research Center*, Fort Belvoir, VA.

Dean, R. G. (1987) "Coastal Armoring: Effects, Principles, and Mitigation", *Proceedings of the 20th International Conference Coastal Engineering*, American Society of Civil Engineers, New York, Vol. 2, pp. 1843–1857.

DeConto, R. M., and D. Pollard (2016) "Contribution of Antarctica to Past and Future Sea-level Rise," *Nature*, Vol. 531, pp. 591–597.

De Winter, R. C., and B. G. Ruessink (2017) "Sensitivity Analysis of Climate Change Impacts on Dune Erosion: A Case Study for the Dutch Holland Coast," *Climate Change*, Vol. 141, pp. 685–701.

Die Küste (2007) "EurOtop: Wave Overtopping of Sea Defences and Related Structures Assessment Manual", Heft 73, Boyens Medien GmbH & Co. August (in English).

Douglas, Bruce C. (1997) *Global Sea Rise: A Redetermination*, Surveys in Geophysics, Vol. 18.

Eggleston, J., and J. Pope (2013) "Land Subsidence and Relative Sea-level Rise in the Southern Chesapeake Bay region," *U.S. Geological Survey Circular 1392*, 30 p.

Englander, John (2018) "Rising Sea Level and the Coming Coastal Crisis," Introduction (slide presentation) at *36th ICCE Short Course, Design of Coastal Hazard Mitigation and Adaptation Alternatives for Rising Seas*, Baltimore, MD

Fleming, Kevin, P., D. Johnston, D. Zwartz, Y. Yokoyama, K. Lambeck, and J. Chappell (1998) "Refining the Eustatic Sea-level Curve Since the Last Glacial Maximum Using Far- and Intermediate-field Sites," *Earth and Planetary Science Letters*, Vol. 163, pp. 327–342.

Garner, A. J., J. L. Weiss, A. Parris, R. E. Kopp, R. M. Horton, J. T. Overpeck, and B. P. Horton (2018) "Evolution of 21st Century Sea Level Rise Projections," *Earth's Future, AGU100* online, Nov 25.

Hawkes, P. J., B. Gouldby, J. A. Tawn, and M. W. Owen (2002) "The Joint Probability of Waves and Water Levels in Coastal Engineering Design," *Journal of Hydraulic Research*, Vol. 40, No. 3, pp. 241–251.

Hay, C. C., E. Morrow, R. E. Kopp, and J. X. Mitrovica (2015) "Probabilistic Reanalysis of Twentieth Century Sea-level Rise," *Nature*, Vol. 517, No. 7535, pp. 481–484.

Hemer, M. A., X. L. Wang, R. Weisse, and V. R. Seail (2012) "Advancing Wind-Waves Climate Science: The COWCLIP Project," *American Meterorological Society*, online (https://doi.org/10.1175/BAMS-D-11-00184.1).

Hemer, M., R. Manasseh, I. Penesis, K. McInnes, T. Pitman (2018). "Perspectives on a Wave Forward for Ocean Renewable Energy in Australia," *Renewable Energy*, Vol. 127, pp. 733–745.

Hoffman, J. S., D. Keyes, and J. G. Titus (1983) "Projecting Future Sea Level Rise; Methodology, Estimates to Year 2100 and Research Needs," *US Environmental Protection Agency*, Washington D.C. 121p.

Houston, J. R. (2013a) "Global Sea Level Projections to 2100 Using Methodology of the Intergovernmental Panel on Climate Change," *Journal Waterway, Port, Coastal, and Ocean Engineering, ASCE*, Vol. 139, No. 2, pp. 82–87.

Houston, J. R. (2013b) "Methodology for Combining Coastal Design-Flood Levels and Sea-Level Rise Projections," *Journal Waterway, Port, Coastal and Ocean Engineering, ASCE*, Vol. 139, No. 5, pp. 341–345.

IMBIE Team (2018) "Mass Balance of the Antarctic Ice Sheet from 1927–2017," *Nature*, Vol. 558, No. 7709, pp. 219–222.

IPCC (2007) *The Physical Science Basis. Contribution of Working Group I, Fourth Assessment Report of the Intergovernmental Panel on Climate Change* [Solomon, S., D. Qin, M. Manning, Z. Chen, M. Marquis, K.B. Averyt, M. Tignor, and H.L. Miller (eds.)], Cambridge University Press, Cambridge, United Kingdom and New York, NY, USA.

IPCC (2013) *The Physical Science Basis. Contribution of Working Group I Fifth Assessment Report of the Intergovernmental Panel on Climate Change* [Stocker, T.F., D. Qin, G.-K. Plattner, M. Tignor, S.K. Allen, J. Boschung, A. Nauels, Y. Xia, V. Bex, and P.M. Midgley (eds.)], Cambridge University Press, Cambridge, United Kingdom and New York, NY, USA, 1535 pp.

IPCC (2018) "Global Warming of 1.5°," Special Report, *Intergovernmental Panel on Climate Change*, Cambridge University Press, Cambridge, United Kingdom and New York, NY, USA.

Jackson, L. P., and S. Jevrejeva (2016). "A Probabilistic Approach to 21st Century Regional Sea-level Projections Using RCP and High-end Scenarios," *Global and Planetary Change*, Vol. 146, pp. 179–189.

Kamphuis, J. W. (2010) *Introduction to Coastal Engineering and Management*, Advanced Series on Ocean Engineering, Vol. 30, World Scientific Publ. Co., Singapore, London, NJ.

Kana T. W., H. L. Kaczkowski, and S. B. Traynum (2015) "An Empirical Approach to Beach Nourishment Formulation," ed. Y. C. Kim, *Design of Coastal Structures and Sea Defenses*, Vol. 2, Series on Coastal and Engineering Practice, World Scientific, Chapter 4, pp. 105–144.

Kana T. W., H. L. Kaczkowski, and S. B. Traynum (2017) "Building Blocks for a Sustainable Beach Nourishment Program" ed. Y. C. Kim, *Handbook of Coastal Engineering — 2nd Edition*, World Scientific, Chapter 38, pp. 1113–1138.

Kana, T. W. (2018) "Lessons Learned from Sustainable Beach Nourishment Programs (CSE's Empirical Approach to SLR in Beach Nourishment Programs," 5th Lecture (slide presentation) at *36th ICCE Short Course, Design of Coastal Hazard Mitigation and Adaptation Alternatives for Rising Seas*, Baltimore, MD, July 29.

Keeler, A. G., D. E. McNamara, and J. L. Irish (2018) "Responding to Sea Level Rise: Does Short-Term Risk Reduction Inhibit Successful Long-Term Adaptation," *Earth's Future*, Vol. 6, pp. 618–621.

Kopp, R. E., A. C. Kemp, K. Bittermann, B. P. Horton, J. P. Donnelly, W. R. Gehrels, and S. Rahmstorf (2016) "Temperature-driven Global Sea-level Variability in the Common Era," *Proceedings of the US National Academy of Sciences*, Vol. 113, pp. 1434–1441.

Kopp, R. E., R. M. DeConto, D. A. Bader, C. C. Hay, R. M. Horton, S. Kulp, M. Oppenheimer, D. Pollard, and B. H. Strauss (2017) "Evolving Understanding of Antarctic Ice-Sheet Physics and Ambiguity in Probabilistic Sea-Level Projections," *Earth's Future, AGU*, online, Dec 14.

Kopp, R. E., R. M. Horton, C. M. Little, J. X. Mitrovica, M. Oppenheimer, D. J. Rasmussen, B. Strauss, and C. Tebaldi (2014) "Probabilistic 21st and 22nd Century Sea-level Projections at a Global Network of Tide-gauge Sites," *Earth's Future*, Vol. 2, No. 8, pp. 383–406.

Kriebel, D. (2018) "Nuisance Flooding: A Game of Inches," 2nd Lecture (slide presentation) at *36th ICCE Short Course, Design of Coastal Hazard Mitigation and Adaptation Alternatives for Rising Seas*, Baltimore, MD, July 29.

Kriebel, D. L., and J. D. Geiman (2013) "A Coastal Flood Stage to Define Existing and Future Sea-level Hazards," *Journal Coastal Research*, Vol. 30, No. 5, pp. 1017–1024.

Kriebel, D. L., J. D. Geiman, and G. R. Henderson (2015) "Future Flood Frequency Under Sea-level Rise Scenarios," *Journal Coastal Research*, Vol. 31, pp. 1078–1083.

LeBars, D., S. Drijfhout, and H. DeVries (2017) "A High-end Sea Sea-level Rise Probabilistic Projection Including Rapid Antarctic Ice sheet Mass-loss," *Environmental Research Letters*, Vol. 12, No. 4.

Mengel, Matthias, A. Levermann, K. Frieler, A. Robinson, B. Marzeion, and R. R. Winkelmann (2016) "Future Sea Level Rise Constrained by Observations and Long-term Commitment," *Proceedings of the National Academy of Sciences*, Vol. 113, No. 10, pp. 2597–2602.

Milne, G. A., W. R. Gehrels, C. W. Hughes, and M. E. Tamisiea (2009) "Identifying the Causes of Sea-level Change," *Nature Geosciences*, Vol. 2, pp. 471–478.

Moore, J. C., R. Gladstone, T. Zwinger, and M. Wolovick (2018) "Geoengineering Polar Glaciers to Slow Sea-level Rise," *Nature, International Journal of Science*, Vol. 555.

Nauels, A. J. Rogelj, C.-F. Schleussner, M. Meinshausen, and M. Mengel (2017) "Linking Sea Level Rise and Socio-economic Indicators Under the Shared Socio-economic Pathways," *Environmental Research Letters*, Vol. 12, No. 11.

Nerem, R. S., D. Chambers, C. Choe, and G. T. Mitchum (2010) "Estimating Mean Sea Level Change from the TOPEX and Jason Altimeter Missions," *Marine Geodynamics*, Vol. 33, pp. 435–446.

Neumann, J., D. Hudgens, J. Herter, and J. Martinich (2011) "The Economics of Adaptation Along Developed Coastline," *Climate Change*, Vol. 2, No. 1, pp. 89–99.

Norfolk, D. O. (2017) "Draft, Integrated, Coastal Flood Risk Management Feasibility Report/Environmental Impact Statement," for City of Norfolk, VA, US Army *Corps of Engineers, Norfolk District Office, Norfolk, VA.*

NOAA (2017) "Global and Regional Sea Level Rise Scenarios for the United States," Technical Report, NOS CO-OPS 083, National Ocean and Atmospheric Agency, Washington, D.C., US.

NOS (2012) "NOAA Responding to Hurricane Sandy, 2012," National Ocean Service (www.oceanservice.noaa.gov).

NRC (1987) "Sea Level Rise to Year 2100," *National Research Council*, National Academy of Sciences, Washington, D.C.

Pearson, J. M., A. Kortenhaus, T. Bruce, and W. Allsop (2004) "Influence of Parapets and Recurves on Wave Overtopping and Wave Loading on Complex Vertical Walls," Proceedings, *Coastal Structures (2003), ASCE.*

Pezza, D. and J. M. White (2020 in review) "Impact of the Duration of Coastal Flooding on Infrastructure," *Public Works Management & Policy*, submitted for review.

Schroeder, D. M., D. D. Blankenship, D. A. Young, and E. Quartini (2014) "Evidence for Elevated and Spatially Variable Geothermal Flux Beneath the West Antarctic Ice Sheet," *Proceedings of the National Academy of Sciences.*

Schultz, M. T., and E. R. Smith (2016) "Assessing the Resilience of Coastal Systems: A Probabilistic Approach," *Journal of Coastal Research*, Vol. 32, No. 5, pp. 1032–1050.

Sergent, P., G. Prevot, G. Mattarolo, and P. Mallet (2015) "Adaptation of Coastal Structures to Mean Sea Level Rise," *La Houille Blanche*, Vol. 6, No. 2014, pp. 54–61.

Shimura, T., N. Mori, and M. A. Hemer (2016) "Variability and Future Decreases in Winter Wave Heights in the Western North Pacific," *Geophysical Research Letters*, AGU 100, Vol. 43, No. 6.

Slangen, A. B. A., M. Carson, C. A. Katsman, R. S. W. van de Wal, A. Köhl, L. L. A. Vermeersen, and D. Stammer (2014) "Projecting Twenty-first Century Regional Sea-level Changes," *Climatic Change*, Vol. 124, Nos. 1–2, pp. 317–332.

SOST (2018) "Science and Technology for America's Oceans: A Decadal Vision, 2018," National Science Foundation, Subcommittee on Ocean Science and Technology, Washington, DC.

SPM (1984) "*Shore Protection Manual*," US Army Corps of Engineers, Washington, D.C.

Sweet, W. V., R. E. Kopp, C. P. Weaver, J. Obeysekera, R. M. Horton, E. R. Thieler, and C. Zervas (2017) "Global and Regional Sea Level Rise Scenarios for the United States," *NOAA Technical Report NOS CO-OPS 083*, January.

Trusel, L. D., S. B. Das, M. B. Osman, M. J. Evans, B. E. Smith, X. Fettweis, J. R. McConnell, B. P. Y. Noel, and M. R. van den Broeke (2018) "Nonlinear Rise in Greenland Runoff in Response to Post-Industrial Arctic Warming," *Letter 104 in Nature*, Vol. 564, Dec.

USACE (2013) online graphic.

USACE (2014) "Procedures to Evaluate Sea-level Change: Impacts, Responses, and Adaptation," U.S. *Army Corps of Engineers*, Technical Letter No. 1100-2-1, 254 p.

Vaughan, D. G., J. C. Comiso, I. Allison, J. Carrasco, G. Kaser, R. Kwok, P. Mote, T. Murray, F. Paul, J. Ren, E. Rignot, O. Solomina, K. Steffen, and T. Zhang (2013) "Observations: Cryosphere," In: *Climate Change 2013: The Physical Science Basis. Contribution of Working Group I to the Fifth Assessment Report of the Intergovernmental Panel on Climate Change* [Stocker, T. F., D. Qin, G.-K. Plattner, M. Tignor, S. K. Allen, J. Boschung, A. Nauels, Y. Xia, V. Bex and P. M. Midgley (eds.)]. Cambridge University Press, Cambridge, United Kingdom and New York, NY, USA.

Wang, X. L., F. W. Zwiers, and V. R. Swail (2004) "North Atlantic Ocean Wave Climate Change Scenarios for the Twenty-First Century," *Journal of Climate*, Vol. 17, pp. 2368–2383.

WCRP (2018) "Global Sea-level Budget 1993–present," Global Sea Level Budget Group, *Earth System Science Data.*

Wong, T. E., A. M. R. Bakker, and K. Keller (2017) "Impacts of Antarctica Fast Dynamics on Sea-level Projections and Coastal Flooding Defense," *Climate Change*, Vol. 144, pp. 347–364.

Zervas, C. E. (2005) "Response of Extreme Storm Tide Levels to Long-term Sea Level Change." *Proc. Marine Technology Society/Institute of Electrical and Electronic Engineers, Oceans Conf.*, Vol. 3, pp. 2501–2506.

Index